DIFFERENTIAL GEOMETRY AND LIE GROUPS FOR PHYSICISTS

Differential geometry plays an increasingly important role in modern theoretical physics and applied mathematics. This textbook gives an introduction to geometrical topics useful in theoretical physics and applied mathematics, including manifolds, tensor fields, differential forms, connections, symplectic geometry, actions of Lie groups, bundles and spinors.

Having written it in an informal style, the author gives a strong emphasis on developing the understanding of the general theory through more than 1000 simple exercises, with complete solutions or detailed hints. The book will prepare readers for studying modern treatments of Lagrangian and Hamiltonian mechanics, electromagnetism, gauge fields, relativity and gravitation.

Differential Geometry and Lie Groups for Physicists is well suited for courses in physics, mathematics and engineering for advanced undergraduate or graduate students, and can also be used for active self-study. The required mathematical background knowledge does not go beyond the level of standard introductory undergraduate mathematics courses.

Marián Fecko, currently at the Department of Theoretical Physics, Comenius University, Bratislava, Slovakia and at the Institute of Physics, Slovak Academy of Sciences, Bratislava, works on applications of differential geometry in physics. He has over 15 years' experience lecturing on mathematical physics and teaching courses on differential geometry and Lie groups.

[illegible]FERENTIAL [illegible] A[illegible] [illegible]ROUPS
[illegible]

Differential geometry [illegible] role in modern [illegible]
and applied mathematics. [illegible] for [illegible] and [illegible]
theoretical physics [illegible]
[illegible]
[illegible] the [illegible]
[illegible] through [illegible]
[illegible] readers [illegible]
[illegible]
[illegible].

[illegible] Geometry [illegible] is well suited for courses in physics, mathematics and engineering for advanced undergraduate or graduate students, and can also be used for active self-study. The required [illegible] knowledge does not go beyond the level of standard [illegible] undergraduate [illegible] courses.

[illegible] is currently at the [illegible] of [illegible] Physics [illegible] Comenius [illegible] of Sciences [illegible] over 15 years [illegible] physics [illegible]

DIFFERENTIAL GEOMETRY AND LIE GROUPS FOR PHYSICISTS

MARIÁN FECKO
Comenius University, Bratislava, Slovakia
and
Slovak Academy of Sciences, Bratislava, Slovakia

CAMBRIDGE UNIVERSITY PRESS
Cambridge, New York, Melbourne, Madrid, Cape Town,
Singapore, São Paulo, Delhi, Tokyo, Mexico City

Cambridge University Press
The Edinburgh Building, Cambridge CB2 8RU, UK

Published in the United States of America by Cambridge University Press, New York

www.cambridge.org
Information on this title: www.cambridge.org/9780521187961

First published 2006
First paperback edition 2011

A catalogue record for this publication is available from the British Library

ISBN 978-0-521-84507-6 Hardback
ISBN 978-0-521-18796-1 Paperback

Contents

Preface

This is an introductory text dealing with a part of mathematics: modern differential geometry and the theory of Lie groups. It is written from the perspective of and mainly for the needs of physicists. The orientation on physics makes itself felt in the choice of material, in the way it is presented (e.g. with no use of a definition–theorem–proof scheme), as well as in the content of exercises (often they are closely related to physics).

Its potential readership does not, however, consist of physicists alone. Since the book is about mathematics, and since physics has served for a fairly long time as a rich source of inspiration for mathematics, it might be useful for the mathematical community as well. More generally, it is suitable for anybody who has some (rather modest) preliminary background knowledge (to be specified in a while) and who desires to become familiar in a comprehensible way with this interesting, important and living subject, which penetrates increasingly into various branches of modern theoretical physics, "pure" mathematics itself, as well as into its numerous applications.

So, what is the minimal background knowledge necessary for a meaningful study of this book? As mentioned above, the demands are fairly modest. Indeed, the required mathematical background knowledge does not go beyond what should be familiar from standard introductory undergraduate mathematics courses taken by physics or even engineering majors. This, in particular, includes some calculus as well as linear algebra (the reader should be familiar with things like partial derivatives, several variables Taylor expansion, multiple Riemann integral, linear maps versus matrices, bases and subspaces of a linear space and so on). Some experience in writing and solving simple systems of ordinary differential equations, as well as a clear understanding of what is actually behind this activity, is highly desirable. Necessary basics in algebra in the form used in the main text are concisely summarized in Appendix A at the end of the book, enabling the reader to fill particular gaps "on the run," too.

The book is intentionally written in a form which makes it possible to be fully grasped also by a self-taught person – anybody who is attracted by tensor and spinor fields or by fiber bundles, who would like to learn how differential forms are differentiated and integrated, who wants to see how symmetries are related to Lie groups and algebras as well as to their representations, what is curvature and torsion, why symplectic geometry is useful in Lagrangian and Hamiltonian mechanics, in what sense connections and gauge fields realize the same idea, how Noetherian currents emerge and how they are related to conservation laws, etc.

Clearly, it is highly advantageous, as the scope of the book indicates, to be familiar (at least superficially) with the relevant parts of physics on which the applications of various techniques are illustrated. However, one may derive profit from the book (in terms of geometry alone) even with no background from physics. If we have never seen, say, Maxwell's equations and we are not aware at all of their role in physics, then although we will not be able to understand *why* such attention is paid to them, nevertheless we will understand perfectly *what* we do with these equations here from the technical point of view. We will see how these partial differential equations may be reformulated in terms of differential forms, what the action integral looks like in this particular case, how conservation laws may be derived from it by means of the energy–momentum tensor and so on. And if we find it interesting, we may hopefully also learn some "traditional" material on electrodynamics later.

If we, in like manner, know nothing about general relativity, then although we will not understand from where the concept of a "curved" space-time endowed with a metric tensor emerged, still we will learn the basics of what space-time is from a geometrical point of view and what is generally done there. We will not penetrate into the physical heart of the Einstein equations for the gravitational field, we will see, however, their formal structure and we will learn some simple, though at the same time powerful, techniques for routine manipulations with these equations. Mastering this machinery then greatly facilitates grasping the physical side of the theory, if later we were to read something written about general relativity from the physical perspective.

The key qualification asked of the future reader is a real interest in learning the subject treated in the book not only in a Platonic way (say, for the sake of an intellectual conversation at a party) but rather at a working level. Needless to say, one then has to accept a natural consequence: it is not possible to achieve this objective by a passive reading of a "noble science" alone. On the contrary, a fairly large amount of "dirty" self-activity is needed (an ideal potential reader should be *pleased* by reading this fact), inevitably combined with due investment of time. The formal organization of the book strongly promotes this method of study.

A specific feature of the book is its strong emphasis on developing the general theory through a large number of simple exercises (more than a thousand of them), in which the reader analyzes "in a hands-on fashion" various details of a "theory" as well as plenty of concrete examples (the proof of the pudding is in the eating). This style is highly appreciated, according to my teaching experience, by many students.

The beginning of an exercise is indicated by a box containing its number (as an example, [14.4.3] denotes the third exercise in Section 4, Chapter 14), the end of the exercise is marked by a square □. The majority of exercises (around 900) are endowed with a hint (often quite detailed) and some of them, around 50, with a worked solution. The symbol • marks the beginning of "text," which is not an exercise (a "theory" or a comment to exercises). Starred sections (like 12.6*) as well as starred exercises may be omitted at the first reading (they may be regarded as a complement to the "hard core" of the book; actually they need not be harder but more specific material is often treated there).

This book contains a fairly large amount of material, so that a few words might be useful on how to read it efficiently. There are several ways to proceed, depending on what we actually need and how much time and effort we are willing to devote to the study.

The basic way, which we recommend the most, consists in reading systematically from cover to cover and solving (nearly) all the problems step by step. This is the way in which we may make full use of the text. The subject may be understood in sufficient broadness, with a lot of interrelations and applications. This needs, however, enough motivation and patience.

If we lack either, we may proceed differently. Namely, we will solve in detail only those problems which we, for some reason, regard as particularly interesting or from which we crucially need the result. Proceeding in this way, it may happen here and there that we will not be able to solve some problem; we are lacking some vital link (knowledge or possibly a skill) treated in the material being omitted. If we are able to locate the missing link (the numbers of useful previous exercises, mentioned in hints, might help in doing so), we simply fill this gap at the relevant point.

Yet more quickly will proceed a reader who decides to restrict their study to a particular direction of interest and who is interested in the rest of the book only to the extent that it is important for his or her preferred direction. As an aid to such a reader we present here a scheme showing the logical dependence of the chapters:

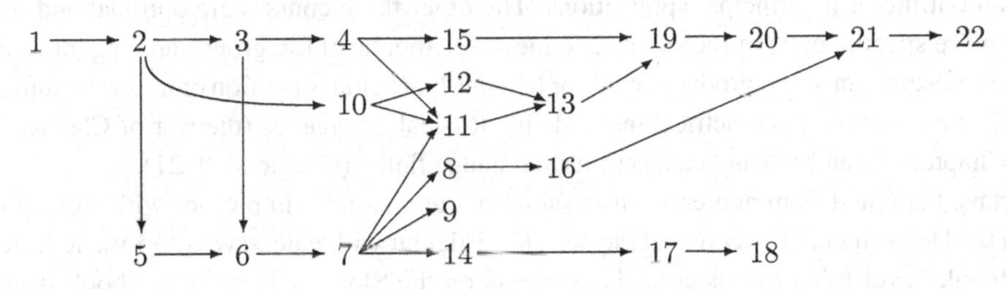

(The scheme does not represent the dependence completely; several sections, short parts or even individual exercises would require the drawing of additional arrows, making the scheme then, however, virtually worthless.)

To be more explicit, one could mention the following possible particular directions of interest.

1. The geometry needed for the fundamentals of **general relativity** (**covariant derivatives**, **curvature tensor**, **geodesics**, etc.).

   One should follow the line $1 \to 2 \to 3 \to 4 \to 15$ (similar material goes well with advanced **continuum mechanics**). If we want to master working with forms, too (to grasp, as an example, Section 15.6, dealing with the computation of the Riemann tensor in terms of Cartan's structure equations, or Section 16.5 on Einstein's equations and their derivation from an action integral), we have to add Chapters 5–7.

2. **Elementary** theory of **Lie groups** and their **representations** ("(differential) geometry-free mini-course").

   The route might contain the chapters (or only the explicitly mentioned sections of some of them) $1 \to 2.4 \to 10 \to 11.7 \to 12 \to 13.1\text{–}13.3$.

3. **Hamiltonian** mechanics and **symplectic manifolds.**
   The minimal itinerary contains Chapters 1 → 2 → 3 → beginning of 4 → 5 → 6 → 7 → 14. Its extension (the formulation of Lagrangian and Hamiltonian mechanics on the fiber bundles TM and T^*M respectively) takes place in Chapters 17 and 18. If we have the ambition to follow the more advanced sections on symmetries (Sections 14.5–14.7 and 18.4), we need to understand the geometry on Lie groups and the actions of Lie groups on manifolds (Chapters 11–13).
4. Basics of working with **differential forms.**
   The route could be 1 → 2 → 3 → beginning of 4 → 5 → 6 → 7 → 8 → 9, or perhaps adding the beginning of Chapter 16.

This book stems from (and in turn covers) several courses I started to give roughly 15 years ago for theoretical physics students at the Faculty of Mathematics and Physics in Bratislava. It has been, however, extended (for the convenience of those smart students who are interested in a broader scope on the subject) as well as polished a bit (although its presentation often still resembles more the style of informal lectures than that of a dry "noble-science monograph"). In order to mention an example of **how the book may be used by a teacher**, let me briefly note what **four** particular formal **courses** are covered by the book. The first, fairly broad one, is compulsory and it corresponds roughly to (parts of) Chapters 1–9 and 14–16. Thus it is devoted to the essentials of general differential geometry and an outline of its principal applications. The other three courses are optional and they treat more specific parts of the subject. Namely, (elementary) Lie groups and algebras and their representations (it reproduces more or less the "particular direction of interest" number 2, mentioned above), geometrical methods in classical mechanics (the rest of Chapter 14 and Chapters 17 and 18) and connections and gauge fields (Chapters 19–21).

I have benefited from numerous discussions about geometry in physics with colleagues from the Department of Theoretical Physics, in particular with Pal'o Ševera and Vlado Balek.

I thank Pavel Bóna for his critical comments on the Slovak edition of the book, Vlado Bužek and Vlado Černý for constant encouragement during the course of the work and the former also for the idea to publish it abroad.

Thanks are due to E. Bartoš, J. Buša, V. Černý, J. Hitzinger, J. Chlebíková, E. Masár, E. Saller, S. Slisz and A. Šurda for helping me navigate the troubled waters of computer typesetting (in particular through the subtleties of TEX) and to my sons, Stanko and Mirko, for drawing the figures (in TEX).

I would like to thank the helpful and patient people of Cambridge University Press, particularly Tamsin van Essen, Vincent Higgs, Emma Pearce and Simon Capelin. I would also like to thank all the (anonymous) referees of Cambridge University Press for valuable comments and suggestions (e.g. for the idea to complement the summaries of the individual chapters by a list of the most relevant formulas).

I am indebted to Arthur Greenspoon for careful reading of the manuscript. He helped to smooth out various pieces of the text which had hardly been continuous before.

Finally, I wish to thank my wife, L'ubka, and my children, Stanko, Mirko and Danka, for the considerable amount of patience displayed during the years it took me to write this book.

I tried hard to make *Differential Geometry and Lie Groups for Physicists* error-free, but spotting mistakes in one's own writing can be difficult in a book-length work. If you notice any errors in the book or have suggestions for improvements, please let me know (fecko@fmph.uniba.sk). Errors reported to me (or found by myself) will be listed at my web page

http://sophia.dtp.fmph.uniba.sk/~fecko

Bratislava Marián Fecko

I tried hard to make [illegible], but spotting mistakes in one's own writing can be difficult in a book-length work. [illegible]

[illegible]

Introduction

In physics every now and then one needs something to differentiate or integrate. This is the reason why a novice in the field is simultaneously initiated into the secrets of differential and integral calculus.

One starts with functions of a single variable, then several variables occur. Multiple integrals and partial derivatives arrive on the scene, and one calculates plenty of them on the drilling ground in order to survive in the battlefield.

However, if we scan carefully the structure of expressions containing partial derivatives in real physics formulas, we observe that some combinations are found fairly often, but other ones practically never occur. If, for example, the frequency of the expressions

$$\frac{\partial^2 f}{\partial x^2}+\frac{\partial^2 f}{\partial y^2}+\frac{\partial^2 f}{\partial z^2} \quad \text{and} \quad \frac{\partial^3 f}{\partial x^3}+\frac{\partial^2 f}{\partial y \partial z}+4\frac{\partial f}{\partial z}$$

is compared, we come to the result that the first one (Laplace operator applied to a function f) is met very often, while the second one may be found only in problem books on calculus (where it occurs for didactic reasons alone). Combinations which do enter real physics books, result, as a rule, from a computation which realizes some *visual local geometrical* conception corresponding to the problem under consideration (like a phenomenological description of diffusion of matter in a homogeneous medium). These very conceptions constitute the subject of a systematic study of *local differential geometry*. In accordance with physical experience it is observed there that there is a fairly small number of truly interesting (and, consequently, frequently met) operations to be studied in detail (which is good news – they can be mastered in a reasonably short time).

We know from our experience in general physics that the same situation may be treated using *various kinds of coordinates* (Cartesian, spherical polar, cylindrical, etc.) and it is clear from the context that the *result* certainly *does not depend* on the choice of coordinates (which is, however, far from being true concerning the *sweat involved* in the computation – the very reason a careful choice of coordinates is a part of wise strategy in solving problems). Thus, both objects and operations on them are independent of the choice of coordinates used to describe them. It should be not surprising, then, that in a properly built formalism a great deal of the work may be performed using *no coordinates* whatsoever (just what part of the computation it is depends both on the problem and on the level of mastery of a particular user). There are several advantages which should be mentioned in favor of these "abstract" (coordinate-free) computations. They tend to be considerably shorter and more transparent, making repeated checking, as an example, much easier, individual steps may be better understood visually and so on. Consider, in order to illustrate this fact, the following equations:

$$\begin{aligned} \mathcal{L}_\xi g = 0 \quad &\leftrightarrow \quad \xi^k g_{ij,k} + \xi^k{}_{,i} g_{kj} + \xi^k{}_{,j} g_{ik} = 0 \\ \nabla_{\dot\gamma} \dot\gamma = 0 \quad &\leftrightarrow \quad \ddot x^i + \Gamma^i_{jk} \dot x^j \dot x^k = 0 \\ \nabla g = 0 \quad &\leftrightarrow \quad g_{ij,k} - \Gamma_{ijk} - \Gamma_{jik} = 0 \end{aligned}$$

We will learn step by step in this book that the pairs of equations standing on the left and on the right side of the same line always tell us *just the same*: the expression on the right may be regarded as being obtained from that on the left by expressing it in (arbitrary) coordinates. (The first line represents *Killing equations*; they tell us that the Lie derivative of g along ξ vanishes, i.e. that the metric tensor g has a *symmetry* given by a vector field ξ. The second one defines particular curves called geodesics, representing uniform motion in a straight line (= its acceleration vanishes). The third one encodes the fact that a linear connection is metric; it says that a scalar product of vectors remains unchanged under parallel translation.)

In spite of the highly efficient way of writing of the coordinate versions of the equations (partial derivatives via commas and the summation convention – we sum on indices repeating twice (dummy indices) omitting the $\sum$ sign), it is clear that they can hardly compete with the left side's brevity. Thus if we will be able to *reliably manipulate* the objects occurring on the left, we gain an ability to manipulate (indirectly) fairly complicated expressions containing partial derivatives, always keeping under control what we *actually* do.

At the introductory level calculus used to be developed in Cartesian space $\mathbb{R}^n$ or in open domains in $\mathbb{R}^n$. In numerous cases, however, we apply the calculus in spaces which *are not* open domains in $\mathbb{R}^n$, although they are "very close" to them.

In analytical mechanics, as an example, we study the motion of pendulums by solving (differential) Lagrange equations for coordinates introduced in the pendulum's configuration spaces, regarded as functions of time. These configuration spaces are not, however, open domains in $\mathbb{R}^n$. Take a simple pendulum swinging in a plane. Its configuration space is clearly a *circle* S^1. Although this is a one-dimensional space, it is intuitively clear (and one may prove) that it is essentially *different* from (an open set in) $\mathbb{R}^1$. Similarly the configuration space of a spherical pendulum happens to be the two-dimensional sphere S^2, which differs from (an open set in) $\mathbb{R}^2$.

Notice, however, that a sufficiently *small neighborhood* of an arbitrary point on S^1 or S^2 is practically indistinguishable from a sufficiently small neighborhood of an arbitrary point in $\mathbb{R}^1$ or $\mathbb{R}^2$ respectively; they are in a sense "locally equal," the difference being "only global." Various applications of mathematical analysis (including those in physics) thus strongly motivate its extension to more general spaces than those which are simple open domains in $\mathbb{R}^n$.

Such more general spaces are provided by *smooth manifolds*. Loosely speaking they are spaces which a *short-sighted observer* regards as $\mathbb{R}^n$ (for suitable n), but globally ("topologically," when a pair of spectacles are found at last) their structure may differ profoundly from $\mathbb{R}^n$.

We can regard as an enjoyable bonus the fact that the formalism, which will be developed in order to perform coordinate-free computations, happens to be at the same time (free of charge) well suited to treating *global* geometrical problems, too, i.e. we may study the objects and operations on them, being well defined *on the manifold* as a whole. Therefore, we speak sometimes about *global analysis*, or the analysis on manifolds. All the above-mentioned equations $\mathcal{L}_\xi g = 0$, $\nabla_{\dot\gamma}\dot\gamma = 0$ and $\nabla g = 0$ represent, to give an example, equations on manifolds and their solutions may be defined as objects living on manifolds, too.

The key concept of a manifold itself will be introduced in Chapter 1. The exposition is mainly at the intuitive level. A good deal of material treated in detail in mathematical texts on differential *topology* will only be mentioned in a fairly informative way or will even be omitted completely. The aim of this introductory chapter is to provide the reader with a minimal amount of material which is necessary to grasp (fully, already at the working level) the main topic of the book, which is differential *geometry* on manifolds.

1
The concept of a manifold

• The purpose of this chapter is to introduce the concept of a smooth manifold, including the ABCs of the technical side of its description. The main idea is to regard a manifold as being "glued-up" from *several* pieces, all of them being very simple (open domains in $\mathbb{R}^n$). The notions of a *chart* (local coordinates) and an *atlas* serve as essential formal tools in achieving this objective.

In the introductory section we also briefly touch upon the concept of a topological space, but for the level of knowledge of manifold theory we need in this book it will not be used later in any non-trivial way.

(From the didactic point of view our exposition leans heavily on recent scientific knowledge, for the most part on ethnological studies of Amazon Basin Indians. The studies proved convincingly that even those prodigious virtuosos of the art of survival within wild jungle conditions make do with only intuitive knowledge of smooth manifolds and the medicine-men were the only members within the tribe who were (here and there) able to declaim some formal definitions. The fact, to give an example, that the topological space underlying the smooth manifold should be *Hausdorff* was observed to be told to a tribe member just before death and as eyewitnesses reported, when the medicine-man embarked on analyzing examples of *non-Hausdorff* spaces, the horrified individual preferred to leave his or her soul to God's hands as soon as possible.)

1.1 Topology and continuous maps

• *Topology* is a useful structure a set may be endowed with (and at the same time the branch of mathematics dealing with these things). It enables one to speak about continuous maps. Namely, in order to introduce a topology on a set X, one has to choose a system $\{\tau\}$ of subsets τ of the set X, such that

1. $\emptyset \in \{\tau\}, \ X \in \{\tau\}$;
2. the union (of an arbitrary number) of elements from $\{\tau\}$ is again in $\{\tau\}$;
3. the intersection of a finite number of elements from $\{\tau\}$ is again in $\{\tau\}$.

(So that the system necessarily contains the empty set as well as the set X itself, and is closed with respect to arbitrary unions and finite intersections.) The elements of $\{\tau\}$ are called *open sets* and the pair $(X, \{\tau\})$ is a *topological space*. Given two topological spaces $(X, \{\tau\})$ and $(Y, \{\sigma\})$, a map

$$f : X \to Y$$

is said to be *continuous* if $f^{-1}(A) \in \{\tau\}$ for any $A \in \{\sigma\}$, that is to say if the inverse image[1] of any open set is again an open set.[2] Moreover, if the map f happens to be bijective and f^{-1} is continuous as well, f is called a *homeomorphism* (topological map); X and Y are then said to be homeomorphic.

1.1.1 Verify that the "weakest" (coarsest) possible topology on a set X is given by the *trivial topology*, where $\emptyset$ and X represent the only open sets available, whereas the "strongest" (finest) topology is the *discrete topology*, where *every* subset is open (in particular, this is also true for every point $x \in X$); all other topologies reside "somewhere between" these two extreme possibilities. □

1.1.2 Let $\{\tau\}_0, \{\tau\}_1$ be the trivial and the discrete topology respectively (1.1.1). Describe all *continuous* maps

$$f : (X, \{\tau\}_a) \to (Y, \{\tau\}_b) \qquad a, b \in \{0, 1\}$$

realizing thus that continuity of a map depends, in general, on the choice of topologies both on X and Y. (For $a = 1$ (b arbitrary) and for $a = 0 = b$ *all* maps are continuous; for $a = 0, b = 1$ the only continuous maps are *constant* maps ($x \mapsto y_0$, the same for all x).) □

1.1.3 Let

$$X \xrightarrow{f} Y \xrightarrow{g} Z,$$

f, g being continuous. Show that the composition map

$$g \circ f : X \to Z$$

is continuous, too. □

1.1.4 Check that the notion of homeomorphism introduces an equivalence relation among topological spaces (reflexivity, symmetry and transitivity are to be verified). □

• The reader may find it helpful to visualize homeomorphic spaces as being made of rubber; Y can then be obtained from X by means of a deformation alone (neither cutting nor gluing are allowed). Example: a circle, a square and a triangle are all homeomorphic, the figure-of-eight symbol is not homeomorphic to the circle (provided that the intersection in the middle of it is regarded as a *single* point).[3]

[1] Recall that f^{-1} does not mean the inverse map here (this may not exist at all); $f^{-1}(A)$ denotes the collection of all elements in X which f sends into A, i.e. the *inverse image* of the set A.

[2] In elementary calculus continuity used to be defined in terms of *distance*; this turns out to be a particular case of the above definition (the distance induces a topology, to be mentioned later).

[3] *Differential Topology* by A. H. Wallace can be recommended as a nice introductory text about topology (see the Bibliography for details).

One usually restricts oneself (for purely technical reasons, in order not to allow for manifolds of some fairly complicated objects that we do not want to be concerned with) to *Hausdorff* topological spaces. In these spaces (by definition), given any two points x, y, there exist non-intersecting neighborhoods of them (open sets A, B, such that $x \in A, y \in B, A \cap B = \emptyset$); one can thus *separate* any two points by means of open sets. From now on Hausdorff spaces will be understood automatically when speaking about topological spaces.

The fact that the *Cartesian space* $\mathbb{R}^n$ (ordered n-tuples of real numbers) represents a topological space (where open sets coincide with those used in the elementary calculus of n real variables) will be important in what follows.

1.1.5 Let $d(x, y)$ be the standard Euclidean distance between two points $x, y \in \mathbb{R}^n$, i.e.

$$d^2(x, y) := (x_1 - y_1)^2 + \cdots + (x_n - y_n)^2$$

and let

$$D(a, r) := \{x \in \mathbb{R}^n, d(x, a) < r\}$$

(open ball $\equiv$ disk, centered at a, the radius being r). A set $A \in \mathbb{R}^n$ is open if for any point $x \in A$ there exists an open ball centered at x which lies entirely in A. Check that this definition of an open set meets the axioms of a topological space. This topology is called the *standard topology in* $\mathbb{R}^n$. □

1.2 Classes of smoothness of maps of Cartesian spaces

- Let A be an open set in $\mathbb{R}^n[x^1, \dots, x^n]$ and

$$f : A \to \mathbb{R}^m[y^1, \dots, y^m]$$

This means that we are given m functions of n variables

$$y^a = y^a(x^1, \dots, x^n) \qquad a = 1, \dots, m$$

If all partial derivatives up to order k exist and are continuous, then f is called a map of *class* C^k. In particular, it is called *continuous* ($k = 0$), *differentiable* ($k = 1$), *smooth* ($k = \infty$) and (real) *analytic* (if for all $x \in A$ the Taylor series of $y^a(x)$ converges to the function $y^a(x)$ itself: $k = \omega$). In general, there clearly holds

$$C^0(A, \mathbb{R}^m) \supset C^1(A, \mathbb{R}^m) \supset \cdots \supset C^\infty(A, \mathbb{R}^m) \supset C^\omega(A, \mathbb{R}^m).$$

Far less trivial is the fact that not a single inclusion is in fact equality.

1.2.1 Consider the function $f : \mathbb{R} \to \mathbb{R}$, given by

$$f(x) = e^{-\frac{1}{x}} \qquad x > 0$$
$$f(x) = 0 \qquad x \leq 0$$

Use this function to prove that in general $C^\omega(A, \mathbb{R}^m) \neq C^\infty(A, \mathbb{R}^m)$.

Hint: show, that $f^{(n)}(0) = 0$ for $n = 0, 1, 2, \ldots$ (so that the Taylor series in the neighborhood of $x = 0$ gives a function which vanishes *for positive* x, too). □

1.3 Smooth structure, smooth manifold

- A tourist map may be regarded as a true *map* (in the mathematical sense of the word)

$$\varphi : \text{TD} \to \text{SP}$$

where TD is a tourist district and SP is a sheet of paper. If the sheet of paper happens to be in fact in a square paper exercise book, we have another map

$$\chi : \text{SP} \to \mathbb{R}^2[x^1, x^2]$$

and their composition results finally in

$$\psi : \text{TD} \to \mathbb{R}^2 \qquad \psi \equiv \chi \circ \varphi$$

For a good map ψ should be a bijection and this makes it possible to assign a pair of real numbers – its coordinates – to any point in TD.

In an effort to map a bigger part of a country, an atlas[4] (a collection of maps) has proved to be helpful. A good atlas should be consistent at all overlaps: if some part of the land happens to be on two (or more) maps (close to the margins, as a rule), information obtained from them must not be mutually contradictory.

If we enlarge the region to be mapped (district $\mapsto$ country $\mapsto$ continent, etc.), we first observe annoying *metric* properties of the maps – the continents become (in comparison with their shape on the globe) somewhat deformed and the intuitive estimation of the *distances* becomes unreliable. This is a manifestation of the fact that ψ fails to be an *isometry* (see Section 4.6); as a matter of fact such an isometry (of a part of the sphere to a part of a sheet of paper) *does not exist* at all.[5] Topologically, however, everything is still all right – even if TD = all of America, ψ still remains a homeomorphism (the latter *need not* preserve distances). But even this ceases to be the case abruptly at the moment we try to display *all* the globe on a *single* map. It turns out, once again, that such a map (a bijective and continuous map of a sphere onto a plane) *does not exist*; that is to say, *more than* one single map – an atlas – is *inevitable*. An optimistic element in these contemplations lies in the fact that *in spite of* the topological complexity of the sphere S^2 (as compared with the plane), its mapping is fairly simple when an atlas containing *several* maps is used. In a similar way one can construct (highly practical) atlases of some other two-dimensional surfaces, like T^2 = the surface of a tire (repairmen in a tire service will then be happy to mark the exact position of a puncture into this atlas) or the exotic (1.5.9) Klein bottle K^2 (appreciated by orienteering fans, mainly in sci-fi).

[4] Atlas, the brother of Prometheus, hero of Greek mythology, keeps (as he used to do) the cope of heaven on his shoulders on the title page of a series of detailed maps of various parts of Europe. They were published in 1595, one year after the death of the author, Gerhard Kramer (Gerardus Mercator in Latin). Since then, *every* series of maps has been called an "atlas."

[5] There are several characteristics preserved by isometries and the sphere and the sheet of paper *differ* in some of them (see, e.g., the result of the computation of the Lie algebras of Killing fields in (4.6.10) and (4.6.13) or of the scalar curvature in (15.6.11)).

The aim, now, will be to formalize the idea of an atlas. This will result in the definition of the crucial concept of a smooth manifold.

Let $(X, \{\tau\})$ be a topological space and $\mathcal{O} \subset X$ an open set. A homeomorphism

$$\varphi : \mathcal{O} \to \mathbb{R}^n[x^1, \dots, x^n]$$

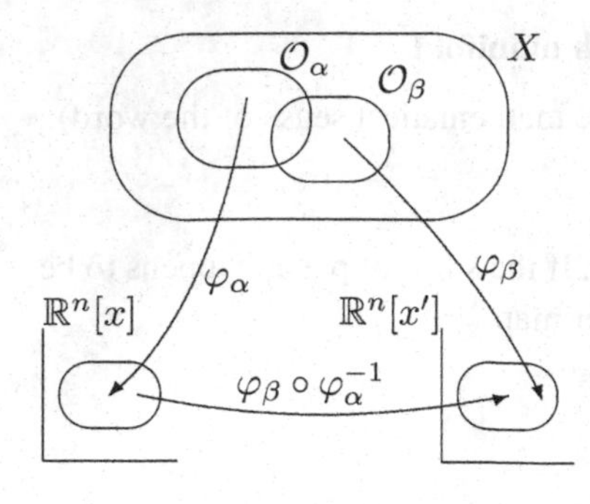

is called a *chart*, or alternatively *local coordinates*. Each point $x \in \mathcal{O} \subset X$ is then uniquely associated with an n-tuple of real numbers – its coordinates. The set $\mathcal{O}$ is known as a *coordinate patch* in this context. So far we have introduced coordinates in a single coordinate patch – in $\mathcal{O}$. If we want to assign coordinates to all points from X, we need an *open covering* $\{\mathcal{O}_\alpha\}$ of the space X (i.e. $\bigcup_\alpha \mathcal{O}_\alpha = X$) and local coordinates for each domain $\mathcal{O}_\alpha$

$$\varphi_\alpha : \mathcal{O}_\alpha \to \mathbb{R}^n$$

(n being the same for all α). A collection of charts $\mathcal{A} \equiv \{\mathcal{O}_\alpha, \varphi_\alpha\}$ is called an *atlas* on X. If the intersection $\mathcal{O}_\alpha \cap \mathcal{O}_\beta$ is non-empty, a map

$$\varphi_\beta \circ \varphi_\alpha^{-1} : A \to \mathbb{R}^n, \qquad A \equiv \varphi_\alpha(\mathcal{O}_\alpha \cap \mathcal{O}_\beta) \subset \mathbb{R}^n$$

called a *change of coordinates* is induced. Since it is a map of Cartesian spaces (see Section 1.2), it makes sense to talk about its class of smoothness. Automatically (check (1.1.3)) its class is C^0, but it might be higher. If, given an atlas, *all* maps of this type happen to be C^k or higher, it is called a C^k*-atlas* $\mathcal{A}$.

An atlas may be supplemented by additional maps, provided that the consistency with the maps already present is assured. A map

$$\mu : \mathcal{O} \to \mathbb{R}^n$$

is said to be C^k*-related* (and it may be added to $\mathcal{A}$), if it is consistent with all maps $(\mathcal{O}_\alpha, \varphi_\alpha)$ on the intersections $\mathcal{O} \cap \mathcal{O}_\alpha$, i.e. if the class of the map $\varphi_\alpha \circ \mu^{-1}$ is C^k or higher. If a C^k-atlas $\mathcal{A}$ is supplemented consecutively with *all* maps, we are left with a unique *maximal* C^k*-atlas* $\hat{\mathcal{A}}$. This in turn endows X with a C^k*-structure*. A pair $(X, \hat{\mathcal{A}})$ is called an (n-dimensional) C^k*-manifold* (in particular, topological, differentiable, ..., smooth, analytic). In this book we will be concerned exclusively with[6] *smooth manifolds*, or here and there (when Taylor series are used) even *analytic manifolds*. The essential structure to be used implicitly throughout the book and assumed to be available in all discussions and constructions is the *smooth structure* on a manifold X.

Since an atlas $\mathcal{A}$ leads to the unique maximal atlas $\hat{\mathcal{A}}$, for the practical construction of a manifold it suffices to give the atlas $\mathcal{A}$. In spite of this fact the definition of a manifold

[6] This highly convenient option is offered by the result of the Whitney ("embedding") theorem, to be mentioned later, see Section 1.4.

refers to the maximal atlas. This emphasizes the formal *equality of rights* of all charts (local coordinates). The constitution (= definition) unambiguously states that the initial charts from $\mathcal{A}$ are by no means privileged in $\hat{\mathcal{A}}$ with respect to those coming later (so that there is no fear of them usurping a privileged position at any later moment). This does not at all mean that privileged coordinates are of no importance in differential geometry. If the *smooth* structure is the *only* structure available, all charts are to be treated equally. In applications, on the other hand, there are typically additional structures on manifolds. Then, of course, particular coordinates tailored to these structures (*adapted coordinates*) would play a privileged role from the *practical* point of view.

The simplest n-dimensional manifold is clearly $\mathbb{R}^n$ itself. A possible atlas is comprised of a *single* chart, given by the *identity* map

$$\varphi \equiv \mathrm{id} : \mathbb{R}^n[x^1, \dots, x^n] \to \mathbb{R}^n[x^1, \dots, x^n]$$

This atlas is trivially smooth (or analytic as well; there are no intersections to spoil it) and the maximal atlas generated by this atlas defines the *standard smooth structure* in $\mathbb{R}^n$. Any other chart from this atlas corresponds to *curvilinear coordinates* in $\mathbb{R}^n$ (like the polar coordinates in a part of the plane $\mathbb{R}^2$).

The next two exercises deal with the construction of smooth atlases on spheres and projective spaces.

1.3.1 On a circle S^1 of radius R we introduce local coordinates x, x' as shown on the figure (this is called the *stereographic projection*). On higher-dimensional spheres $S^2, \dots, S^n$ a natural generalization of this idea results in coordinates $\mathbf{r}, \mathbf{r}'$. Verify that:

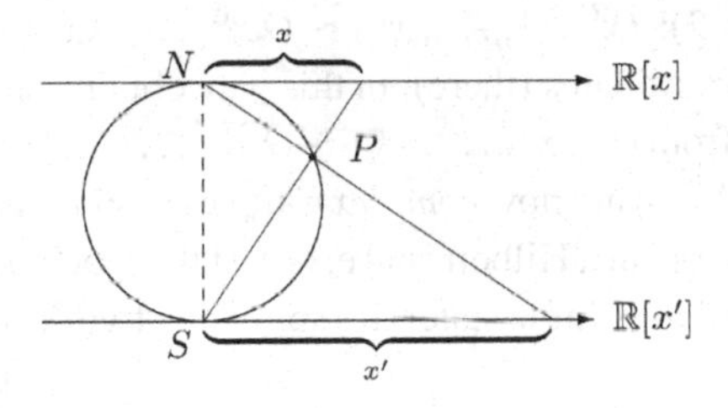

(i) on the intersection of the patches, where the primed and unprimed coordinates are in operation, we find for S^1 and S^n respectively the following explicit transition relations:

$$x' = \frac{(2R)^2}{x} \qquad \mathbf{r}' = \frac{(2R)^2}{r}\frac{\mathbf{r}}{r}$$

(ii) in this way an analytic atlas composed of two charts has been constructed on S^n – the sphere S^n is thus an n-dimensional analytic manifold;

(iii) if the complex coordinates z and z' are introduced on S^2

$$\mathbf{r} \leftrightarrow (x, y) \leftrightarrow z \equiv x + iy \qquad \mathbf{r}' \leftrightarrow (x', y') \leftrightarrow z' \equiv x' + iy'$$

then the transition relations are

$$z' = (2R)^2/\bar{z} \qquad \bar{z} \equiv x - iy$$

Hint: on S^n a projection is to be performed onto n-dimensional mutually parallel *planes*, touching the north and south poles respectively (in these planes $\mathbf{r} \equiv (x^1, \dots, x^n)$ represent common Cartesian coordinates centered at the poles). Then $\mathbf{r}' = \lambda\mathbf{r}$ and one easily finds λ

from the observation that in the (two-dimensional) plane given by the poles and the point P the situation reduces to S^1. □

1.3.2 The real *projective space* $\mathbb{R}P^n$ is the set of all lines in $\mathbb{R}^{n+1}$ passing through the origin. The complex projective space $\mathbb{C}P^n$ is introduced similarly – one should replace $\mathbb{R} \mapsto \mathbb{C}$ in the preceding definition. (Here, a complex line consists of all *complex* multiples of a fixed (non-vanishing) complex vector (point of $\mathbb{C}^{n+1}$) z, so that it is a *two-dimensional* object from a real point of view.)

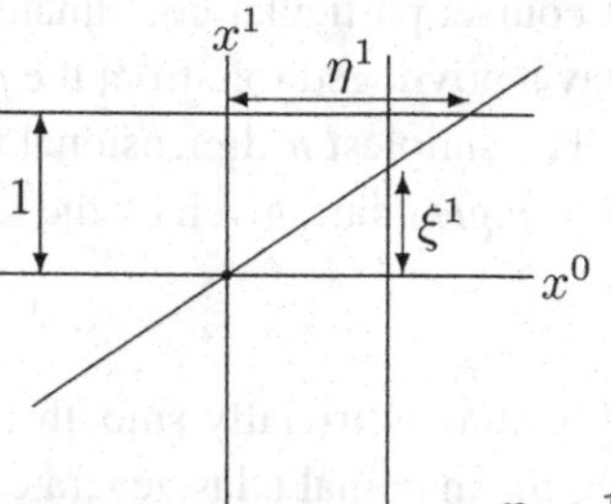

(i) Introduce the structure of an n-dimensional smooth manifold (= local coordinates) on $\mathbb{R}P^n$.
(ii) The same for $\mathbb{C}P^n$ (it is $2n$-dimensional).
(iii) Show that the states of an n-level system in quantum mechanics are in one-to-one correspondence with the points of $\mathbb{C}P^{n-1}$.
(iv) Show that $\mathbb{C}P^1 = S^2$ (in the sense of (1.4.7)) $\Rightarrow$ states with spin $\frac{1}{2}$ correspond to unit vectors $\mathbf{n}$ in $\mathbb{R}^3$.

Hint: (i) one line (a point from $\mathbb{R}P^n$) consists of those points of $\mathbb{R}^{n+1}$ which may be obtained from a fixed $(x^0, x^1, \ldots, x^n)$ using the freedom $(x^0, x^1, \ldots, x^n) \sim (\lambda x^0, \ldots, \lambda x^n)$; in the part of $\mathbb{R}^{n+1}$ where $x^0 \neq 0$ the freedom enables one to make 1 from the first entry of the array (visually this means that the point of intersection of the line with the plane $x^0 = 1$ has been used as a representative of the line); the other n numbers are to be used as local coordinates on $\mathbb{R}P^n$ (they are the coordinates in the plane $x^0 = 1$ mentioned above; see the figure for $n = 1$, try to draw the case $n = 2$): $(x^0, x^1, \ldots, x^n) \sim (\lambda x^0, \ldots, \lambda x^n) \sim (1, \xi^1, \ldots, \xi^n)$ for $x^0 \neq 0$, $\Rightarrow (\xi^1, \ldots, \xi^n)$ are coordinates (there); in this way obtain step-by-step $(n+1)$ charts,[7] with the last one coming from $(x^0, x^1, \ldots, x^n) \sim (\lambda x^0, \ldots, \lambda x^n) \sim (\eta^1, \ldots, \eta^n, 1)$ for $x^n \neq 0$; (ii) in full analogy, $\xi, \ldots, \eta$ are now *complex* n-tuples, giving rise to $2n$ real coordinates; (iii) two non-vanishing vectors in a Hilbert space, one of them being a complex constant multiple of the other, correspond to a single state; (iv) spin $\frac{1}{2}$ is a two-level system. □

• From two given manifolds $(X, \hat{A})$ and $(Y, \hat{B})$, we can form a new manifold called the *Cartesian product*. This new manifold is denoted by the symbol $X \times Y$. As a set, it is the Cartesian product $X \times Y$ (points being ordered pairs), an atlas is constructed in the exercise.

1.3.3 Let $(X, \hat{A})$ and $(Y, \hat{B})$ be smooth manifolds and let

$$\varphi_\alpha : \mathcal{O}_\alpha \to \mathbb{R}^n \qquad \psi_a : \mathcal{S}_a \to \mathbb{R}^m$$

represent two charts on X and Y respectively. Show that

$$\varphi_\alpha \times \psi_a : \mathcal{O}_\alpha \times \mathcal{S}_a \to \mathbb{R}^{m+n}$$
$$(x, y) \mapsto (\varphi_\alpha(x), \psi_a(y)) \in \mathbb{R}^{n+m} \qquad x \in \mathcal{O}_\alpha, \quad y \in \mathcal{S}_a$$

[7] In this context the coordinates $(x^0, x^1, \ldots, x^n)$ in $\mathbb{R}^{n+1}$ are said to be the *homogeneous coordinates* (of the points in $\mathbb{R}P^n$). Note that they are *not* local coordinates on $\mathbb{R}P^n$ in the sense of the definition of a manifold, since they are not in *one-to-one* correspondence with the points (they *are* official coordinates only in $\mathbb{R}^{n+1}$).

introduces a smooth atlas on $X \times Y$, so that we have a smooth manifold of dimension $n + m$. This means, in plain English, that given $(x^1, \dots, x^n)$ and $(y^1, \dots, y^m)$ local coordinates on X and Y respectively $(x^1, \dots, x^n, y^1, \dots, y^m)$, *may be used* as local coordinates on $X \times Y$. □

1.3.4 Show that the following manifolds represent the configuration spaces of simple mechanical systems mentioned below: a plane double pendulum $S^1 \times S^1$, a free wheel on a road $\mathbb{R}^2 \times SO(3)$ and a wheel of a car $\mathbb{R}^2 \times S^1 \times S^1$.

Hint: a free wheel: $\mathbb{R}^2$ for the centre and $SO(3)$ (see (10.1.8)) for the orientation in space; a car: the wheel is to be perpendicular to the road (= vertical). □

• We have seen in (1.3.1) in the example of a sphere S^2 how two real coordinates (x, y) can be encoded compactly into a single complex coordinate z. This can be clearly generalized trivially to any *even*-dimensional manifold, so that charts may be regarded then as the maps into $\mathbb{C}^n$ (rather than $\mathbb{R}^{2n}$ in the real language).

However, it is far from always that the *additional* requirement can be fulfilled; namely, to make all the transition relations of the complex coordinates be given by *holomorphic* functions. A manifold is called a *complex manifold* (of complex dimension n, real dimension $2n$) if this is possible. Complex manifolds may thus be regarded at the same time as ordinary (necessarily even-dimensional) "real" manifolds, but the converse may not be true (there are even-dimensional manifolds where it is *not possible* to introduce the above-mentioned "holomorphic" atlas). The theory of complex manifolds is rich and interesting; however, in this introductory text we shall not take this subject further.

1.3.5 Show that the two-dimensional sphere S^2 *is a complex* manifold (of complex dimension 1)

Hint: consider an atlas with charts w, w', where (using notation from (1.3.1)) $w \equiv \bar{z}$, $w' = z'$; then $w'(w) = 4R^2/w$, and this *is* a holomorphic relation (*where it is needed*). □

1.4 Smooth maps of manifolds

• When manifolds appear in some context, they nearly always go hand in hand with various *mappings* of them. This may happen in a direct and overt way sometimes, or, in contrast, in a modest and inconspicuous way other times (and this by no means indicates that the mappings are less important in those cases). Reminding the reader that the virtue a manifold is especially proud of is its *smooth* structure, those mappings respecting (in some sense) the smooth structure will surely present particular interest. Such mappings are called *smooth*. A closer look at

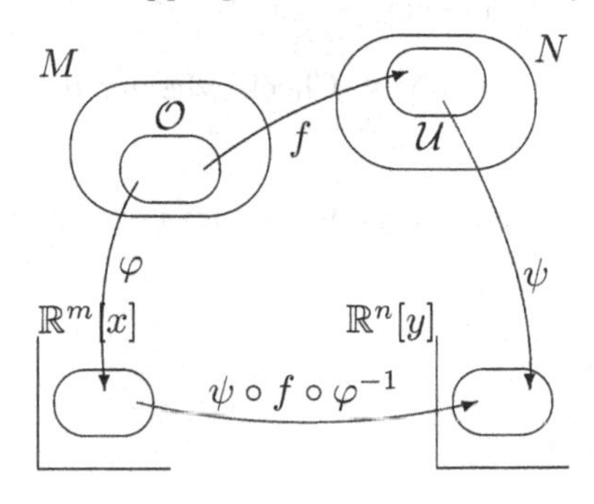

what is meant by this exactly is necessary. Given $(M, \hat{\mathcal{A}})$ and $(N, \hat{\mathcal{B}})$, two smooth manifolds of dimensions m and n, respectively, and a mapping

$$f : M \to N$$

let

$$\varphi : \mathcal{O} \to \mathbb{R}^m[x^1, \dots, x^m] \qquad \mathcal{O} \subset M$$
$$\psi : \mathcal{U} \to \mathbb{R}^n[y^1, \dots, y^n] \qquad \mathcal{U} \subset N$$

be local coordinates such that $f(\mathcal{O}) \subset \mathcal{U}$. Then the composition map

$$\hat{f} \equiv \psi \circ f \circ \varphi^{-1} \; : \; \mathbb{R}^m \to \mathbb{R}^n$$

is induced, which is called the *coordinate presentation* of the mapping f. In technical terms one has a collection of n functions of m variables

$$y^a = y^a(x^1, \dots, x^m) \qquad a = 1, \dots, n$$

1.4.1 Reconsider the domains of all the mappings involved and refine, if necessary, the figure in this respect. □

• Since $\hat{f}$ is a map of Cartesian spaces (see Section 1.2), it makes sense to be interested in its class of smoothness. By definition, f *itself* is said to be smooth (or C^k more generally), if its coordinate presentations with respect to *any* pair of charts (from $\hat{\mathcal{A}}$ and $\hat{\mathcal{B}}$) happen to be smooth (C^k).

1.4.2 Let $\mathcal{A}$ and $\mathcal{B}$ be *finite* atlases on M and N respectively, which generate maximal atlases $\hat{\mathcal{A}}$ and $\hat{\mathcal{B}}$. Show that if the coordinate presentations are smooth with respect to $\mathcal{A}, \mathcal{B}$, then they are smooth with respect to $\hat{\mathcal{A}}, \hat{\mathcal{B}}$ as well (so that f *is smooth*).

Hint: one has to check that if $y^a(x^i)$ is smooth and $x'^i(x^j)$, $y'^a(y^b)$ represent changes of coordinates on M and N respectively, then $y'^a(x'^i)$ is smooth, too. □

• Since a manifold needs in general an atlas consisting of several charts, several coordinate presentations correspond to a single mapping $f : M \to N$.

1.4.3 Let

$$f : \mathbb{R}^2 \times \mathbb{R}^2 \to \mathbb{R}^2 \qquad f(z, w) = zw$$

be the map induced by the multiplication of complex numbers. Check whether it is a C^∞-map. □

1.4.4 Let $M = \mathbb{R}^2 \setminus (0, 0)$ and consider the map defined in terms of complex coordinates as follows

$$f : M \to M \qquad f(z) = z^{-1}$$

Is this a C^∞-map? □

1.4.5 Write down in coordinates the map (*canonical projection*)

$$\pi : \mathbb{R}^3 \setminus (0) \to \mathbb{R}P^2$$

which assigns to a point from $\mathbb{R}^3$ the line (1.3.2) on which the point is situated. Show that the preimages of any two points of $\mathbb{R}P^2$ are diffeomorphic (diffeomorphism is defined a few lines later). ($\xi^1(x, y, z) = y/x$, $\xi^2(x, y, z) = z/x$; for the remaining two charts on $\mathbb{R}P^2$, similarly.) □

1.4.6 Show that $(z, w) \mapsto z/w$ (where $z, w \in \mathbb{C}; |z|^2 + |w|^2 = 1$) can be interpreted as a map

$$f : S^3 \to S^2$$

(The *Hopf mapping*, see also (20.1.7)–(20.1.10).)

Hint: $(z, w) \mapsto (z/w, 1) \leftrightarrow z/w$ is a coordinate presentation of the map $\mathbb{C}^2 \to \mathbb{C}P^1 \sim S^2$ ((1.3.2), cf. a similar situation in (1.4.4)); the *extended* complex plane (the result of the quotient being *there*) = (Riemannian) sphere S^2. □

• If $\dim M = \dim N \equiv n$, f is a bijection and if both f and f^{-1} happen to be smooth, then the mapping f is called a *diffeomorphism* and M and N are said to be *diffeomorphic manifolds.*

1.4.7 Check that

(i) the concept of a diffeomorphism defines an equivalence relation
(ii) all diffeomorphisms $M \to M$ form a group (it is denoted by Diff (M)). □

1.4.8 Let T^2 be a two-dimensional *torus* ≡ surface of a tire, ○ a circle ≡ S^1 and □ a square. Show that (≈ meaning a diffeomorphism)

$$S^1 \times \mathbb{R} \approx \text{surface of a cylinder} \qquad \mathbb{R}^n \times \mathbb{R}^m \approx \mathbb{R}^{m+n}$$
$$S^1 \times S^1 \approx T^2 \qquad \square \approx \bigcirc$$

(Caution: □ at the end of a hint is a sign of the end of the exercise and there is nothing to be proved for it.)

Hint: consider coordinates on a square obtained by means of radial rays from the inscribed (or circumscribed) circle. □

• Consider a smooth map $f : M \to N$, with $m \equiv \dim M \leq n \equiv \dim N$. Let $x \in M$ be mapped to $y \in N$. This map (locally $y^a = y^a(x^1, \dots, x^m)$, $a = 1, \dots, n$) is said to be an *immersion* if in some neighborhood $\mathcal{U}$ of each point $y \in f(M) \subset N$ there exist local coordinates $y^1, \dots, y^n$ such that for a sufficiently small neighborhood $\mathcal{O}$ of x ($f(\mathcal{O}) \subset \mathcal{U}$) it holds there that the subset $f(\mathcal{O})$ is given by the system of equations

$$y^{m+1} = \cdots = y^n = 0$$

that is to say, if the *image* of this immersion may be *locally* expressed in terms of the *vanishing* of some *part of the coordinates* on N. It is possible to show that this requirement on the subset $f(M) \subset N$ prohibits the existence of "corners" (edges) and cusps (this is the interesting point about immersions). In this form the definition is not always suitable for a practical test (in particular, for a proof that a given f *is not* an immersion; if this *is* the case and coordinates with required properties are easily found, it is suitable). One can show that the following statement holds:[8]

$$f : M \to N \text{ is an immersion} \qquad \Leftrightarrow \qquad \text{rank of } J^a_i \equiv \frac{\partial y^a}{\partial x^i} \text{ is maximal } (\equiv m) \text{ on } f(M)$$

Moreover, if f is injective, it is called an *embedding* (then $f(M)$ has no self-intersections).

1.4.9 Zero and eight (or infinity), when drawn on a sheet of paper, may be regarded as $f(M)$ for $M = S^1$, $N = \mathbb{R}^2$. Decide whether the mapping f in these two cases is an immersion, or perhaps even an embedding. □

• If $f : M \to N$ happens to be an embedding, then the subset $f(M) \subset N$ is naturally endowed with the structure of a manifold ($y^1, \dots, y^m$, i.e. those y^a which *do not vanish* on $f(M)$ serve as local coordinates) and it is called a *submanifold* of a manifold N.[9]

1.4.10 Given a smooth map $\hat{f} : M \to N$ we define the map

$$f : M \to M \times N \qquad m \mapsto (m, \hat{f}(m))$$

Show that the *graph* of the map $\hat{f}$, i.e. $f(M) \subset M \times N$, is a smooth submanifold in $M \times N$. Draw a picture for $M = N = \mathbb{R}$ and $M = \mathbb{R}^2$, $N = \mathbb{R}$.

Hint: check that f is an immersion. (For $x^i \mapsto (x^i, y^a(x))$, there is an $m \times m$ unit block in the Jacobian matrix.) □

1.4.11 Let L be an arbitrary n-dimensional linear space over $\mathbb{R}$. Show that it is an n-dimensional manifold which is diffeomorphic to $\mathbb{R}^n$.

Hint: if e_a is any basis in L, then $L \ni x = x^a e_a$ defines an atlas consisting of a single global chart; $x \mapsto x^a$ is a diffeomorphism $L \to \mathbb{R}^n$. □

1.4.12 Let $\pi_M : M \times N \to M$, $(m, n) \mapsto m$ be the projection on the factor M and let a smooth map $f : M \to M \times N$ satisfy

$$\pi_M \circ f = \text{id}_M$$

[8] The geometrical meaning of the requirement on the rank of the *Jacobian matrix* J^a_i is related to the mapping of *tangent vectors*, cf. (3.1.2); technically, the *implicit function theorem* from several variables differential calculus is behind this.

[9] Not all subsets $X \subset N$, although being themselves manifolds, are thus submanifolds in N. They need to be "very nicely" placed in N – with no edges, cusps or self-intersections – in order to satisfy the strict criteria of the "submanifold club" membership. All this is guaranteed by the existence of an embedding f. Both a circle and a square, for example, represent (diffeomorphic) smooth one-dimensional manifolds. The circle is a submanifold in $\mathbb{R}^2$, too, but the square fails to meet the submanifold requirements (since it has corners).

Show that

(i) f induces a map $\hat{f}: M \to N$ (and vice versa it is fully given by this map)
(ii) f is an embedding
(iii) the submanifold $f(M)$ is diffeomorphic to M.

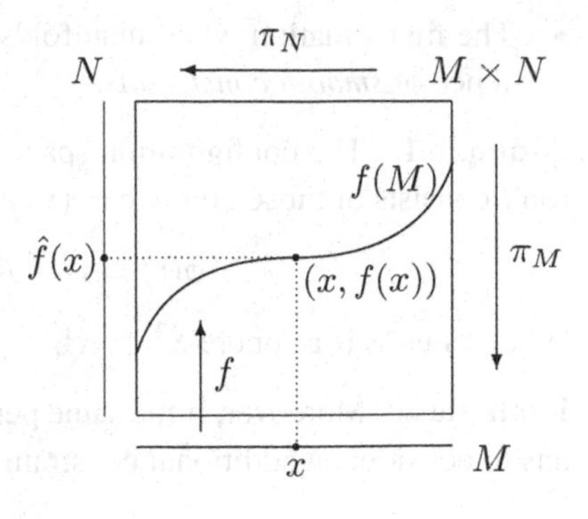

Note: this situation is the simplest realization of an important geometrical structure, a *fiber bundle* (namely the *product* fiber bundle here). We will encounter this concept in more detail later (in Chapter 17 and beyond). The map f is a *section* of the bundle, π_M serves as the *canonical projection*, M is its *base* and $M \times N$ its *total space*.

Hint: (i) $\hat{f} = \pi_N \circ f$; (ii) (1.4.10); (iii) $f: M \to f(M)$ is a diffeomorphism. □

1.4.13 Let f be a diffeomorphism of M to itself and let $\mathcal{A} \equiv \{\mathcal{O}_\alpha, \phi_\alpha\}$ be an atlas on M. Show that

(i) f induces a new atlas $\hat{\mathcal{A}} \equiv \{\hat{\mathcal{O}}_\alpha, \hat{\phi}_\alpha\}$ on M, which may be regarded as the "shifted" version (by the diffeomorphism f) of the original atlas $\mathcal{A}$

$$\hat{\mathcal{O}}_\alpha = f(\mathcal{O}_\alpha) \qquad \hat{\phi}_\alpha = \phi_\alpha \circ f^{-1}$$

(ii) in new coordinates, the f-image of any object (e.g. the equation of a circle) has the same form that the initial object had in the initial coordinates.

Hint: draw a picture to see what is going on. □

• The official definition of a manifold (a topological space, plus ...) is fairly abstract and one cannot be sure, a priori, whether "nice surfaces," like spheres, projective spaces, etc., which ultimately motivated this definition, are the only objects which satisfy all the properties required. Maybe some ugly creatures, which nobody needed, are compatible with the definition as well. Fortunately, there is a useful *"embedding" theorem*, which fully protects our slumber. We therefore mention its content (without proof) here.

Theorem (*Whitney*) Each C^1 n-dimensional manifold is diffeomorphic to some C^ω submanifold of $\mathbb{R}^{2n+1}$.

Lesson: each manifold M may be realized (up to a diffeomorphism) as a *nicely* (no edges, cusps or self-intersections) located "surface" in *Cartesian* space, one can estimate its dimension *from above* by $2n+1$ (if dim $M = n$) and if there is an atlas of class C^1, it can be *improved* up to class C^ω (i.e. if a C^1 atlas $\mathcal{A}$ generates a maximal C^1 atlas $\hat{\mathcal{A}}$, one can choose from $\hat{\mathcal{A}}$ a "better" atlas $\tilde{\mathcal{A}}$ of class C^ω). This enables us to restrict our interest directly to C^ω (or smooth, C^∞) manifolds and, in addition, to regard all manifolds as nice smooth surfaces in Cartesian spaces from the very beginning. This is the reason we now embark on the study of the two basic ways of treating such smooth surfaces.

1.5 A technical description of smooth surfaces in $\mathbb{R}^n$

• The first situation when manifolds frequently enter the scene as surfaces in $\mathbb{R}^n$ is the existence of *smooth constraints.*

Example 1 The configuration space of a mathematical pendulum with the length of the rod l consists of those points $\mathbf{r} = (x, y, z) \in \mathbb{R}^3$ which obey

$$\phi(\mathbf{r}) = \mathbf{r}^2 - l^2 \equiv x^2 + y^2 + z^2 - l^2 = 0$$

What remains is a sphere $S^2 \subset \mathbb{R}^3$.

Example 2 Moreover, if the same pendulum can swing only in a fixed vertical plane, one has to consider an additional constraint, so that together

$$\phi^1(\mathbf{r}) \equiv x^2 + y^2 + z^2 - l^2 = 0$$
$$\phi^2(\mathbf{r}) \equiv y = 0$$

What remains is only a circle $S^1 \subset S^2 \subset \mathbb{R}^3$, now.

Example 3 A state of an ideal gas is given by a triple of real numbers $(p, V, T) \subset \mathbb{R}^3$, constrained through the equation of state

$$\phi(p, V, T) = pV - RT = 0$$

(plus some further restrictions on the realistic intervals for the numbers p, V, T are to be added). What remains is a particular two-dimensional surface in $\mathbb{R}^3$.

In general, several functions $\phi^1, \ldots, \phi^m$ (constraints) in $\mathbb{R}^n[x^1, \ldots, x^n]$ are often available and the interesting subset consists of those points in $x \in \mathbb{R}^n$ where all of the functions vanish ($\equiv$ satisfy the constraints):

$$\phi^1(x) = \cdots = \phi^m(x) = 0$$

If certain conditions are satisfied, one really obtains a smooth manifold (and even submanifold in $\mathbb{R}^n$) in this way.

Theorem (on a submanifold defined implicitly) Let $A \subset \mathbb{R}^n[x^1, \ldots, x^n]$ be an open set,

$$g : A \to \mathbb{R}^m[\phi^1, \ldots, \phi^m] \qquad m \leq n$$

a smooth map such that the rank of the Jacobian matrix

$$J_i^a(x) \equiv \frac{\partial \phi^a(x)}{\partial x^i}$$

is constant[10] (and equals k, $0 \neq k \leq m$) on the set $M \equiv g^{-1}(0)$. Then M is an $(n-k)$-dimensional manifold,[11] which is a submanifold of $\mathbb{R}^n$ (an atlas may be constructed by means of the implicit function theorem).

[10] If $k = m$ (*maximal* rank), the constraints are said to be *independent* (at a given point).

[11] More generally one can consider the inverse image of *any* point in the target space (not necessarily *zero*). Such an inverse image is called a *level surface*. As an example *contours* (level curves) on a map represent the level surfaces for the "height function," regarded as a mapping $\mathbb{R}^2[x, y] \to \mathbb{R}^1[z]$.

For $m = 1$, the manifold M is said to be a *hypersurface* (codimension 1 submanifold). The Jacobian matrix then reduces to a single row ($1 \times n$ matrix) and the requirement on the rank ($k \equiv 1$) means that the row does not vanish (i.e. at least one element is non-zero; just which one saves the good reputation of the Jacobian matrix may depend on the point $x \in M$).

1.5.1 Show that the sphere $S_r^n \subset \mathbb{R}^{n+1}$ of radius r

$$S_r^n := \{x \in \mathbb{R}^{n+1} \mid (x^1)^2 + \cdots + (x^{n+1})^2 = r^2\}$$

is a hypersurface in $\mathbb{R}^{n+1}$.

Hint: $\phi(x^1, \ldots, x^{n+1}) = (x^1)^2 + \cdots + (x^{n+1})^2 - r^2$. □

1.5.2 The same as in (1.5.1) for the ellipsoid with half-axes $a^1, \ldots, a^{n+1}$. □

1.5.3 Let $SL(n, \mathbb{R})$ denote the set of real $n \times n$ matrices A with unit determinant (see also (10.1.7)). Show that it can be regarded as a hypersurface in $\mathbb{R}^{n^2}$.

Hint: define $g : \mathbb{R}^{n^2} \to \mathbb{R}^1$ as

$$g(A_{11}, A_{12}, \ldots, A_{nn}) := \det A - 1$$

and check that the relevant row is indeed non-zero (row expansion of a determinant says that the row in question is given by *minors* and it is impossible for *all* minors to vanish, since the determinant is *non-zero* on M; or use the result of (5.6.7). □

1.5.4 Let $M = \{(x, y) \in \mathbb{R}^2 \mid y = f(x)\}$, where $f : \mathbb{R} \to \mathbb{R}$. Show that M is a submanifold in $\mathbb{R}^2$ and introduce local coordinates on M (see also (1.4.10)). Draw for $f(x) = \tanh x$.

Hint: $\phi(x, y) = y - f(x)$; coordinate x. □

• It turns out that it is not possible to treat all manifolds by means of constraints (implicitly). One can show (see (6.3.4)) that a manifold constructed by this method is necessarily *orientable*. There are, however, non-orientable manifolds, too. A more general approach is offered by a *parametric expression* of the latter. Within this scheme a manifold appears as the *image* of a smooth mapping

$$f : A \to \mathbb{R}^n[x^1, \ldots, x^n], \qquad A \text{ is an open domain in } \mathbb{R}^m[u^1, \ldots, u^m],\ m \leq n$$

with the maximum ($\equiv m$) rank of the Jacobian matrix

$$J_a^i(x) \equiv \frac{\partial x^i(u)}{\partial u^a}$$

That is to say

$$M := \operatorname{Im} f \equiv f(A) \subset \mathbb{R}^n$$

($\mathbb{R}^m$ is the parameter space, the coordinates $u^1, \ldots, u^m$ are parameters).[12]

[12] A comparison of the implicit and parametric ways of defining a manifold: in both cases mappings of Cartesian spaces $\mathbb{R}^m \to \mathbb{R}^n$ play an essential role. In the implicit way $m \geq n$ holds and the resulting manifold appears as the subset *on the left* (as the *inverse image* of (say) zero), in the parametric case $m \leq n$ and the manifold appears as the subset *on the right* (as the *image* of the map).

1.5.5 Consider a map

$$f : \mathbb{R}^1[u] \to \mathbb{R}^2[x, y] \qquad u \mapsto (\cos u, \sin u) \equiv (x(u), y(u))$$

Verify that

(i)

$$\mathrm{Im} f \equiv M = \{(x, y) \in \mathbb{R}^2 \mid x^2 + y^2 = 1\} \equiv S^1 \subset \mathbb{R}^2$$

i.e. the manifold M is a sphere S^1 (circle)

(ii) the fact that the *sphere* S^1 appears *on the right* could be recognized (in advance) in the parameter space (*on the left*) as well.

Hint: (ii) introduce the equivalence in $\mathbb{R}^1[u]$ as follows:

$$u \sim u' \quad \Leftrightarrow \quad f(u) = f(u')$$

One can see easily that they are equivalence *classes* on the left, which are in one-to-one relation with the *points* of S^1 on the right (1.5.6). What do they look like? We have $u \sim u + 2\pi k$ ($k \in \mathbb{Z}$), so that (a) it is enough to restrict oneself to the interval $\langle 0, 2\pi \rangle$ and (b) $0 \sim 2\pi$, so that the ends of the interval $\langle 0, 2\pi \rangle \subset \mathbb{R}^1[u]$ are to be *identified* ($\equiv$ glued together); a figure homeomorphic to a circle is obtained. □

• Exercises (1.5.7)–(1.5.11) treat some *two-dimensional* surfaces in a similar way. Instead of the interval $\langle 0, 2\pi \rangle$ one has a basic *square* in the parameter plane $\mathbb{R}^2[u, v]$, now, and the formulas of the mapping induce equivalence relations on the boundary of the square, i.e. the rules of gluing the boundary in order to obtain the resulting two-dimensional surface. Standard conventions are used for that: "like" gluing of the opposite sides by $\uparrow\uparrow$, "reverse" gluing (first turn, then glue) by $\uparrow\downarrow$; see the figure in exercise (1.5.11).

1.5.6 Let $f : M \to N$ be a mapping of sets. Define a relation $\sim$ on M as follows:

$$m \sim m' \quad \Leftrightarrow \quad f(m) = f(m')$$

Show that it is an equivalence and that $f(M) = M/\sim$ (where $=$ means bijection and $M/\sim$ denotes the factor-set of M with respect to $\sim$, that is to say, the elements of $M/\sim$ are the equivalence classes in M). □

1.5.7 Let $f : \mathbb{R}^2 \to \mathbb{R}^3$,

$$(u, v) \mapsto ((a + b \sin v) \cos u, (a + b \sin v) \sin u, b \cos v) \qquad 0 < b < a$$

Show that

(i) in the sense of $\sim$ from (1.5.6) it holds that

$$(u, v) \sim (u + 2\pi n_1, v + 2\pi n_2)$$

for any $n_1, n_2 \in \mathbb{Z}$

(ii) the image of the parametric plane $\mathbb{R}^2[u, v]$ in the target space $\mathbb{R}^3$ is the two-dimensional *torus*, $f(\mathbb{R}^2) \approx T^2$. What is the visual meaning of the constants a, b and the parameters u, v?

Hint: see (1.5.6), (3.2.2), (1.5.11) and (1.4.8). □

1.5.8 Consider the mapping $f : \mathbb{R}^2 \to \mathbb{R}^4$, given by

$$(u, v) \mapsto (\cos u, \sin u, \cos v, \sin v)$$

Show that

(i) in the sense of $\sim$ from (1.5.6) it holds that

$$(u, v) \sim (u + 2\pi n_1, v + 2\pi n_2)$$

for any $n_1, n_2 \in \mathbb{Z}$

(ii) the image of the parametric plane $\mathbb{R}^2[u, v]$ in the target space $\mathbb{R}^4$ is the two-dimensional *torus* (once again), $f(\mathbb{R}^2) \approx T^2$.

Hint: see (1.5.6), (3.2.3) and (1.5.11). □

1.5.9 Given $0 < b < a$ define a map $f : \mathbb{R}^2 \to \mathbb{R}^4$ as follows:

$$(u, v) \mapsto \left((a + b\cos v)\cos u, (a + b\cos v)\sin u, b\sin v\cos\frac{u}{2}, b\sin v\sin\frac{u}{2}\right)$$

Show that the equivalence in the sense of (1.5.6) in $\mathbb{R}^2$ corresponds to the *Klein bottle* K^2, i.e. to the third figure in exercise (1.5.11), so that the image of the parametric plane $\mathbb{R}^2[u, v]$ is the Klein bottle embedded into $\mathbb{R}^4$. □

1.5.10 Try to visualize the Klein bottle from (1.5.11), when realized in $\mathbb{R}^3$, and see that there is no way to avoid a self-intersection. The mapping from (1.5.9) realizes the bottle in $\mathbb{R}^4$ *without* self-intersection.

Hint: according to the figure in (1.5.11) the first gluing (identification of the top and bottom lines) gives a (surface of a) cylinder, by the second one (right and left lines) one is to join two circles; however, taking into account their orientations, one has to approach one by another "from the inside" (unlike T^2); this needs self-intersection. □

1.5.11 Think out (visually) that there are three possibilities altogether to identify the opposite sides of a square in a "like/reverse" way (see above), namely

$$= S^1 \times S^1 = T^2 \qquad = \mathbb{R}P^2 \qquad = K^2$$

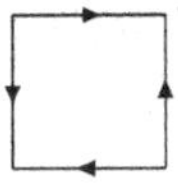

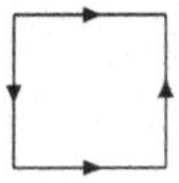

(the last picture is to be understood as a *definition* of $K^2 \equiv$ Klein bottle).

Hint: $\mathbb{R}P^2$: first deform the square onto a *disk* (opposite points of the boundary are to be identified); then blow bottom-up to the disk in order to deform it onto a hemisphere in $\mathbb{R}^3$, whose opposite points of the boundary (circle) are identified; finally realize that each point of the resulting entity corresponds uniquely to a line passing through the origin (1.3.2). □

Summary of Chapter 1

The *smooth manifold* is *the* basic playing field in differential geometry. It is a generalization of the Cartesian space $\mathbb{R}^n$ (or an open domain in the latter) to a more elaborate object, which (only) *locally* looks like $\mathbb{R}^n$, but its global structure can be much more complicated. It is, however, always possible to contemplate it as a whole in which *several* pieces homeomorphic to $\mathbb{R}^n$ are glued together; the number n, which is the same for all pieces, is called the dimension of the manifold. The technical realization of these ideas is achieved by the concepts of a *chart* (local coordinates) and an *atlas* (consisting of several charts). The Cartesian product $M \times N$ of two manifolds is a new manifold, constructed from two given ones M and N. Any manifold admits a realization as a surface, which is nicely embedded in a *Cartesian* space of sufficiently large dimension.

$(x_1 - y_1)^2 + \cdots + (x_n - y_n)^2$	Euclidean distance between two points $x, y \in \mathbb{R}^n$	(1.1.5)
$\varphi : \mathcal{O} \to \mathbb{R}^n[x^1, \dots, x^n]$	Chart (local coordinates) in a patch $\mathcal{O} \subset (X, \{\tau\})$	Sec. 1.3
$\varphi_\beta \circ \varphi_\alpha^{-1}$	Change of coordinates in a patch $\mathcal{O}_\alpha \cap \mathcal{O}_\beta$	Sec. 1.3
$(x, y) \mapsto (\varphi_\alpha(x), \psi_a(y)) \in \mathbb{R}^{n+m}$	Atlas for the Cartesian product $X \times Y$	(1.3.3)
$\hat{f} \equiv \psi \circ f \circ \varphi^{-1} : \mathbb{R}^m \to \mathbb{R}^n$	Coordinate presentation of $f : M \to N$	Sec. 1.4
$y^{m+1} = \cdots = y^n = 0$	Immersion (some coordinates on N vanish)	Sec. 1.4
$f(M) \subset N$	$f(M)$ is a submanifold of N (f = embedding)	Sec. 1.4
$\phi^1(x) = \cdots = \phi^m(x) = 0$	Smooth constraints (manifold as a surface in $\mathbb{R}^n$)	Sec. 1.5
$x^i(u^1, \dots, u^m),\ i = 1, \dots, n \geq m$	Parametric expression of a manifold	Sec. 1.5

2
Vector and tensor fields

• From elementary physics we know vectors as being arrows, exhibiting direction and length. This means that they have both a head *as well as a tail*, the latter being drawn as a point *of the same space* in which the physics is enacted. A vector, then, is equivalent to an ordered *pair of points* in the space. Such a conception works perfectly on the common plane as well as in three-dimensional (Euclidean) space.

However, in general this idea presents difficulties. One can already perceive them clearly on "curved" two-dimensional surfaces (consider, as an example, such a "vector" on a sphere S^2 in the case when its length equals the length of the equator). Recall, however, the various contexts in which vectors enter the physics. One comes to the conclusion that the "tail" point of the vector has no "invariant" meaning; only the *head* point of the vector makes sense as a point of the space. Take as a model case the concept of the (instantaneous) *velocity* vector $\mathbf{v}$ of a point mass at some definite instant of time t. Its meaning is as follows: if the point is at position $\mathbf{r}$ at time t, then it will be at position $\mathbf{r} + \epsilon\mathbf{v}$ at time $t + \epsilon$. However long the vector $\mathbf{v}$ is, the point mass will be *only infinitesimally* remote from its original position. The (instantaneous) velocity vector $\mathbf{v}$ thus evidently carries only "local" information and it is related in no reasonable way to any "tail" point at *finite* distance from its head.

And the transition from (say) a plane to a sphere (or any other curved surface) changes practically nothing in this reasoning: although we may visualize the velocity as an arrow *touching* the surface at a given place, it makes no sense to take seriously its tail as a second point on the surface (within a finite distance from the first one), since all the velocity vector informs us about is the behavior of the trajectory within the nearest (infinitesimal) time interval and over such a short time interval all that we manage to do is to move to a point infinitesimally near to the first one. Consequently, the *second point* (the tail of the vector) *plays no invariant role* in this business. The velocity vector is thus to be treated as a concept which is strictly *confined to a point*. A similar analysis of other vectors in physics (acceleration, force, etc.) leads to the same result. Vectors are objects which are to be treated as being "point-like" entities, i.e. as existing at *a single point*.

That means, however, that our approach to vectors on a manifold has to take into account this essential piece of information. Fortunately, such an approach does exist; in fact, there are even several equivalent ways of reaching this goal, as described in Section 2.2.

Before doing this, we undertake a short digression on the concepts of a curve and a function on a manifold, since they play (in addition to being important enough in themselves) essential roles in the construction of a vector. The simple machinery of multilinear algebra (see Section 2.4) then makes it possible to take a (long) step forward, introducing objects of great importance in physics as well as in mathematics – tensor fields on a manifold.

2.1 Curves and functions on M

- A *curve* on a manifold M is a (smooth) map

$$\gamma : \mathbb{R}[t] \to M \qquad t \mapsto \gamma(t) \in M$$

or, more generally,

$$\gamma : I \to M$$

$I \equiv (a, b)$ being an open interval on $\mathbb{R}[t]$. Note that a definite *parametrization* of points from $\operatorname{Im} \gamma \subset M$ is inherent in the definition of a curve, and two curves which differ by the parametrization alone are to be treated as being different (in spite of the fact that their image sets $\operatorname{Im} \gamma$ on the manifold M coincide). If

$$\varphi : \mathcal{O} \to \mathbb{R}^n[x^1, \dots, x^n]$$

is a chart (i.e. x^i are local coordinates on $\mathcal{O} \subset M$), one obtains a *coordinate presentation* of a curve γ,

$$\hat{\gamma} \equiv \varphi \circ \gamma : \mathbb{R}[t] \to \mathbb{R}^n[x^1, \dots, x^n]$$

i.e. a curve on $\mathbb{R}^n$

$$t \mapsto (x^1(t), \dots, x^n(t)) \equiv (x^1(\gamma(t)), \dots, x^n(\gamma(t)))$$

In general, a curve may convey several coordinate patches, so that several coordinate presentations are sometimes needed for a single curve.

A *function* on a manifold M is a (smooth) map

$$f : M \to \mathbb{R} \qquad x \mapsto f(x) \in \mathbb{R}$$

If

$$\varphi : \mathcal{O} \to \mathbb{R}^n[x^1, \dots, x^n]$$

is a chart, one obtains a *coordinate presentation* of a function f

$$\hat{f} \equiv f \circ \varphi^{-1} \; : \mathbb{R}^n \to \mathbb{R}$$

i.e. a function on (a part of) $\mathbb{R}^n$

$$(x^1, \dots, x^n) \mapsto \hat{f}(x^1, \dots, x^n) \in \mathbb{R}$$

so that $\hat{f}$ is a common "function of n variables." We will frequently identify the function with its coordinate presentation in what follows. What will be "really" meant should be clear from the context (the same holds for curves).

[2.1.1] Show that the prescription

$$A \mapsto \det A \equiv f(A)$$

defines a smooth function on the manifold of all real $n \times n$ matrices ($\sim \mathbb{R}^{n^2}$).

Hint: The determinant is a polynomial in the matrix elements. □

2.2 Tangent space, vectors and vector fields

• The concept of a vector in a point $x \in M$ is undoubtedly one of the most fundamental notions in differential geometry, serving as the basis from which the whole machinery of tensor fields (in particular, differential forms) on a manifold is developed with the aid of the standard methods of multilinear algebra (to be explained in Section 2.4).

A word of caution is in order. Although the *actual computations* with vectors (as well as vector and tensor fields) are very simple and indeed "user friendly," the *definition* of a vector is, in contrast, a fairly subtle and tricky matter for the beginner and it might need some time to grasp the ideas involved in full detail. Our recommendation is not to be in a hurry and reserve due time to digest all the details of the exposition. A clear understanding of what a vector *is* in differential geometry saves time later, when vectors are used in more advanced applications.

There are several (equivalent) ways in which the concept of a vector at a point $x \in M$ may be introduced. In what follows we mention four of them. In different contexts different definitions turn out to be the most natural. That is why it is worth being familiar with *all* of them.

Each approach reveals the key fact that one can naturally associate an n-dimensional vector space with each point P on an n-dimensional manifold M. The elements of this vector space (the tangent space at P) are then treated as vectors at the point $P \in M$.

The first approach generalizes naturally the concept of the instantaneous velocity $\mathbf{v}(t) = \dot{\mathbf{r}}(t)$ of a point mass moving along a trajectory $\mathbf{r}(t)$, mentioned at the beginning of the chapter. The essential idea is that of tangent curves.

Definition Given two curves γ_1, γ_2 on M, we say that γ_1 is *tangent* to γ_2 at the point $P \in M$ if

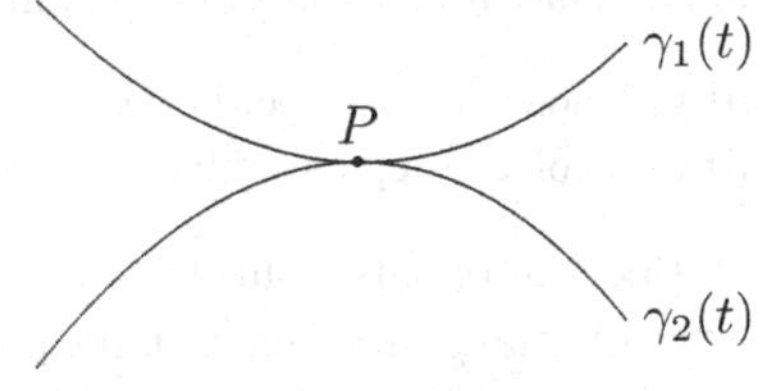

1. $\gamma_1(0) = \gamma_2(0) = P$
2. $\frac{d}{dt}\big|_0 x^i(\gamma_1(t)) = \frac{d}{dt}\big|_0 x^i(\gamma_2(t))$

(x^i being arbitrary local coordinates in the neighborhood of P). When expressed in the terminology of analytical mechanics, the definition says that at the moment $t = 0$ the positions of two fictitious points in a configuration space M, moving along trajectories $\gamma_1(t)$ and $\gamma_2(t)$ respectively, happen to coincide (they are both in P) and, in addition, the values of their generalized velocities are the same. The curves (trajectories), which are tangent at $t = 0$, thus have (at $t = 0$) the same values of *both* generalized *coordinates and velocities*. It is clear, then, that the motions along these trajectories are *up to the first order* in time (within the interval from 0 to ϵ) *equal*. (Note that the particular choice $t = 0$ actually plays no distinguished role in this concept; the curves may be tangent at any other "time" as well.)

2.2.1 Show that

(i) the definition does not depend on the choice of local coordinates in a neighborhood of P
(ii) the relation "to be tangent in P" is an equivalence on the set of curves on M obeying $\gamma(0) = P$
(iii) the Taylor expansion (the class of smoothness C^ω is assumed here) of equivalent curves in a neighborhood of $t = 0$ is as follows:

$$x^i(\gamma(t)) = x^i(P) + ta^i + o(t)$$

where $x^i(P), a^i \in \mathbb{R}$ are common for the whole equivalence class.

Hint: (i) $\frac{d}{dt}\big|_0 x'^i(\gamma(t)) = \frac{\partial x'^i}{\partial x^j}(P) \left.\frac{dx^j(\gamma(t))}{dt}\right|_0$, i.e. $a'^i = J^i_j(P)a^j$. □

• It turns out that the *equivalence classes* $\dot{\gamma} := [\gamma]$ of curves γ are endowed with a natural linear structure, which may be introduced by means of representatives.

2.2.2 Given $T_P M$ the set of equivalence classes in the sense of (2.2.1), let $v, w \in T_P M$ and γ, σ be two representatives of these classes ($v = \dot{\gamma} \equiv [\gamma]$, $w = \dot{\sigma} \equiv [\sigma]$), such that

$$x^i(\gamma(t)) = x^i(P) + ta^i + o(t)$$
$$x^i(\sigma(t)) = x^i(P) + tb^i + o(t)$$

Show that the prescription

$$v + \lambda w \equiv [\gamma] + \lambda[\sigma] := [\gamma + \lambda\sigma]$$

where

$$x^i((\gamma + \lambda\sigma)(t)) := x^i(P) + t(a^i + \lambda b^i) + o(t)$$

introduces by means of representatives into $T_P M$ the well-defined structure of an n-dimensional linear space, i.e. that the definition does not depend on

(i) the choice of local coordinates
(ii) the choice of representatives γ, σ of the classes v, w. □

• Because of this result we may for good reasons (and justly indeed) call the elements $v \in T_P M$ (tangent) *vectors* at the point $P \in M$; the space $T_P M$ itself is called the *tangent space* at the point P. From the definition of linear combination in (2.2.2) one can see that *all* vectors at the point P share the same values of $x^i(P)$ and the property by which they can be

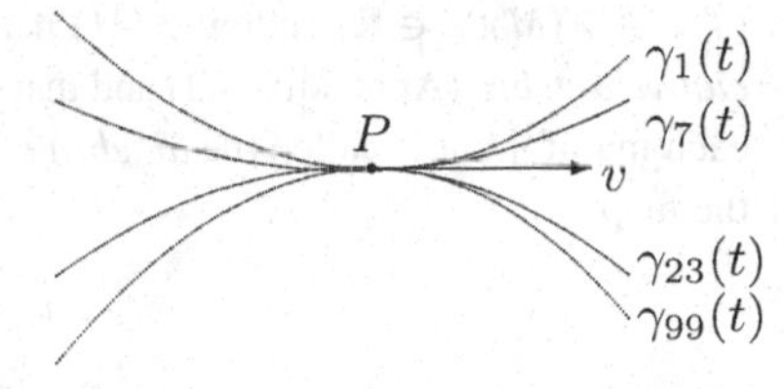

distinguished from one another is by the values of the coefficients $a^i \equiv \dot{x}^i(0)$. Note that a vector "uses" only the first two terms of the Taylor expansion of its coordinate presentation (the zeroth and the first derivatives), the higher terms being completely arbitrary. This means that a single vector corresponds to an *infinite number* of curves which represent this particular vector (which should be clear in advance from the intuitive vision of *all* the curves being tangent to one another), so that there are an infinite number of representatives of each equivalence class. If we would like to visualize the concept of a vector in the sense of an equivalence class of tangent curves, we should assign something like a "bunch" or a "sheaf" of curves, all of them firmly bound together at the point P. And a good old arrow, which cannot be thought of apart from the vector, could be put at P in the direction of this bunch, too (so that it does not feel sick at heart that it had been forgotten because of some dubious novelties).

2.2.3 Verify that

(i) if $\dim M = n$, then $T_P M$ is an n-dimensional space
(ii) equivalence classes of *coordinate curves* $\gamma_j(t)$, i.e. the curves obeying $x^i(\gamma_j(t)) = x^i(P) + \delta^i_j t$ (the value of the jth coordinate is the only one that varies (namely linearly) with t) constitute a basis of $T_P M$.

Hint: (i) $v \leftrightarrow a^i$ is an isomorphism $T_P M \leftrightarrow \mathbb{R}^n$; (ii) check that $v \equiv [\gamma] = a^i[\gamma_i]$. □

• The definition of a vector in terms of curves is intuitively easy to grasp. From the point of view of practical manipulations with vectors (and tensors) another one proves to be convenient, too. It is based on the idea of the *directional derivative* of a function and leans heavily on *algebraic* properties of functions and their directional derivatives.

2.2.4 Let $\mathcal{F}(M) := \{f : M \to \mathbb{R}\}$ denote the set of (smooth) functions on M, $f \in \mathcal{F}(M)$, $v \in T_P M$. Define the map (derivative of f in the direction of v)

$$\hat{v} : \mathcal{F}(M) \to \mathbb{R} \qquad f \mapsto \hat{v}(f) := \left.\frac{d}{dt}\right|_0 f(\gamma(t)) \qquad v = [\gamma]$$

Prove that $\hat{v}$ does not depend on the representative γ in the class $[\gamma] = v$ (i.e. correctness of the definition). □

• It turns out that this map has interesting algebraic properties, enabling one to give an alternative definition of the concept of a vector at the point $P \in M$.

2.2.5 Check that

(i) the prescriptions

$$(f + \lambda g)(x) := f(x) + \lambda g(x)$$
$$(fg)(x) := f(x)g(x)$$

($f, g \in \mathcal{F}(M), \lambda \in \mathbb{R}$) endow $\mathcal{F}(M)$ naturally with the structure of an (∞-dimensional) *associative algebra* (Appendix A.2) and that this algebra turns out to be *commutative* ($fg = gf$) for each manifold; it is called the *algebra of functions on a manifold M*

(ii) the map

$$\hat{v} : \mathcal{F}(M) \to \mathbb{R}$$

from exercise (2.2.4) is a *linear functional* on $\mathcal{F}(M)$, i.e. it behaves on linear combination according to the rule

$$\hat{v}(f + \lambda g) = \hat{v}(f) + \lambda \hat{v}(g)$$

(iii) in addition this functional has the property (behavior on a product)

$$\hat{v}(fg) = \hat{v}(f)g(P) + f(P)\hat{v}(g) \qquad \text{(Leibniz's rule)}$$

(iv) such linear functionals (obeying Leibniz's rule associated with the point P) constitute a linear space (we denote it as $\hat{T}_P M$, here), if one defines

$$(\hat{v} + \lambda \hat{w})(f) := \hat{v}(f) + \lambda \hat{w}(f)$$

(v) the map

$$\psi : T_P M \to \hat{T}_P M \qquad v \mapsto \hat{v}$$

is linear and bijective (i.e. it is an *isomorphism*).

Hint: (v) surjectivity: if $\hat{v}x^i =: a^i$, the inverse image is $v = a^i[\gamma_i]$. □

• Because of the existence of the (canonical) isomorphism $T_P M \leftrightarrow \hat{T}_P M$, these spaces are completely equivalent, so that one may alternatively *define* a vector at the point $P \in M$ as a linear functional on $\mathcal{F}(M)$, behaving according to Leibniz's rule on the product, too.

2.2.6 Define the elements $e_i \in \hat{T}_P M$, $i = 1, \ldots, n$ as follows:

$$e_i(f) := \left.\frac{\partial f}{\partial x^i}\right|_P \equiv \partial_i|_P f$$

or symbolically

$$e_i := \partial_i|_P$$

Check that

(i) the e_i belong to $\hat{T}_P M$, indeed
(ii) the e_i happen to be just the images of vectors $[\gamma_i] \in T_P M$ (which constitute a basis of $T_P M$) with respect to the map ψ from exercise (2.2.5)
(iii) any vector $\hat{v} \in \hat{T}_P M$ may be uniquely written in the form

$$\hat{v} = a^i e_i \qquad \text{where} \quad a^i = \hat{v}x^i$$

(iv) under the change of coordinates $x^i \mapsto x'^i(x)$, the quantities a^i and e_i transform as follows:

$$a^i \mapsto a'^i = J^i_j a^j \qquad e_i \mapsto e'_i = (J^{-1})^j_i e_j$$

where

$$J^i_j = \frac{\partial x'^i}{\partial x^j}(P) \equiv J^i_j(P) = \text{Jacobian matrix of the change of coordinates}$$

(v) the "whole" $\hat{v} \equiv a^i e_i$ is not altered under the change of coordinates (it is invariant)

$$a^i e_i = a'^i e'_i = \hat{v}$$

(vi) the transformation rules for a^i as well as e_i meet the consistency condition on the intersection of *three* charts (coordinate patches): the composition $x \mapsto x' \mapsto x''$ is to give the same result as the direct way $x \mapsto x''$. □

• These results enable one to introduce immediately *another two* definitions of a vector at the point P (and the tangent space as well). The first possibility is to declare as a vector a first-order differential operator with constant coefficients, i.e. an expression $a^i \; \partial_i|_P$, with linear combinations being given by

$$a^i \; \partial_i|_P + \lambda b^i \; \partial_i|_P := (a^i + \lambda b^i) \; \partial_i|_P$$

The second possibility is the definition adopted by classical differential geometry: a vector at a point $P \in M$ is an n-tuple of real numbers $a^i, i = 1, \ldots, n$, associated with the coordinates x^i in a neighborhood of P; under change of coordinates the n-tuple should transform (by definition) according to the rule

$$x^i \mapsto x'^i(x) \quad \Rightarrow \quad a^i \mapsto J^i_j(P) a^j$$

Altogether we gave *four equivalent* definitions (one can even add more) of a vector: a vector as being

1. an equivalence class of curves (with respect to the equivalence relation "being tangent at the point P")
2. a linear functional on $\mathcal{F}(M)$, which behaves on a product according to Leibniz's rule
3. a first-order differential operator (together with the evaluation of the result at the point P)
4. an n-tuple of real numbers a^i, which transform in a specific way under the change of coordinates.

2.2.7 Check in detail their equivalence: given a vector in any of these four ways, associate with it corresponding vectors in the other three senses. In particular, make explicit the correspondence between the *basis* vectors in all four languages. □

• Taking into account the equivalence of the four definitions mentioned above, we may regard a vector as being given in *any* of the possible realizations, from now on. The corresponding tangent space will be denoted by a *common* symbol $T_P M$, as well. The basis $e_i \equiv \partial_i|_P \leftrightarrow [\gamma_i] \leftrightarrow \ldots$ is said to be the *coordinate basis* in $T_P M$ and the numbers a^i constitute the *components* of a vector v with respect to the basis e_i.

(Note that the linear combination has only been defined for vectors sitting *at the same point* of a manifold (i.e. in a single tangent space $T_P M$). The spaces $T_P M$ and $T_{P'} M$ for $P \neq P'$ are to be regarded as *different* vector spaces. It is true that they are isomorphic (both being n-dimensional), but there is no *canonical* isomorphism (there exist *infinitely many* isomorphisms, but none is distinguished, in general) so that there is no *natural* (preferred) correspondence between vectors sitting at different points. The fact that vectors *are routinely* linearly combined, in spite of sitting at different points, in physics (the momenta of a collection of particles are *added* in order to obtain the total momentum vector of the system, to give an example) is justified by *particular additional* structure inherent in the *Euclidean* space – so-called complete parallelism (to be discussed in Chapter 15).)

We say that a *vector field* on M has been defined if a rule is given which enables one to choose exactly one vector residing at each point of a manifold M. Only the fields which "do not differ too much" at two "neighboring" points will be of interest for us in what follows (what we need is smoothness of the field). It turns out that this property is most easily formulated after one learns how vector fields act on (the algebra of) functions, i.e. by looking at the matter from an algebraic perspective.

One can apply a vector *field* V to a function f so that at each point $P \in M$ the *vector* $V_P \in T_P M$ (regarded as a linear functional on $\mathcal{F}(M)$, here) is applied to f. In this way we get a *number* $V_P(f)$ residing at each point P of a manifold M, i.e. a *new function* altogether. A vector field thus may be regarded *as a map* (operator)

$$V : \mathcal{F}(M) \to \mathcal{F}(M) \qquad f \mapsto Vf \qquad (Vf)(P) := V_P(f)$$

V is said to be a *smooth vector field* ($\equiv C^\infty$-field) if the image of the map V above is indeed in $\mathcal{F}(M)$, that is to say, if a smooth function results whenever acted on a smooth function by V. The set of (smooth) vector fields on M will be denoted by $\mathfrak{X}(M) \equiv \mathcal{T}^1_0(M)$ (the reason for the second notation will be elucidated in Section 2.5).

2.2.8 Show that the map

$$V : \mathcal{F}(M) \to \mathcal{F}(M) \qquad f \mapsto Vf$$

obeys

$$V(f + \lambda g) = Vf + \lambda Vg$$
$$V(fg) = (Vf)g + f(Vg)$$

($f, g \in \mathcal{F}(M)$, $\lambda \in \mathbb{R}$). The first property alone says that V is a *linear operator* on $\mathcal{F}(M)$; when taken both together they say that V is a *derivation of the algebra of functions* $\mathcal{F}(M)$ (in the sense of Appendix A.2).[13] □

• As is the case for vectors, components may be assigned to vector fields, too. In a given coordinate patch $\mathcal{O}$ with coordinates x^i, a vector field V may be written, according to (2.2.6), in the form

$$V = V^i(x)\partial_i \equiv V^i(x)\frac{\partial}{\partial x^i}$$

[13] The converse is true, too: given *any* derivation D of the algebra of functions $\mathcal{F}(M)$, there exists a vector field V such that $D = V$. This makes it possible to *identify* vector fields on M with derivations of the algebra of functions $\mathcal{F}(M)$.

since the coefficients of a decomposition of a vector with respect to the coordinate basis may be, in general, different at different points (a^i, denoted here as V^i, depend on x). The functions $V^i(x)$ are called the *components* of the field V. The vector fields (!) ∂_i are called the *coordinate basis* of vector fields.

We came to the conclusion, then, that a first-order differential operator with *non-constant* coefficients corresponds to a vector *field* and the action of V on f in coordinates may be expressed simply as

$$f(x) \mapsto (Vf)(x) = V^i(x)(\partial_i f)(x) \equiv V^i(x)\frac{\partial f(x)}{\partial x^i}$$

2.2.9 Prove that V is smooth if and only if its components $V^i(x)$ are smooth functions and that this criterion does not depend on the choice of local coordinates.

Hint: smooth functions are closed with respect to linear combinations and product (= operations in $\mathcal{F}(\mathcal{O})$) elements of $J^i_j(x)$ are smooth. □

2.2.10 Show that under the change of coordinates $x \mapsto x'(x)$ the components of a vector field transform as follows:

$$V'^i(x') = J^i_j(x)V^j(x)$$

Hint: see (2.2.6); $V'^i(x')\partial'_i = V^i(x)\partial_i$. □

2.2.11 Write down the vector field $V = \partial_\varphi$ (in polar coordinates in the plane $\mathbb{R}^2$) in Cartesian coordinates and try to visualize at various points the direction of the vectors given by this field.

Hint: see (2.2.10); ($V = \partial_\varphi = x\partial_y - y\partial_x$). □

• One should understand clearly the difference between the algebraic properties of a vector and a vector field: a vector is a linear *functional* on $\mathcal{F}(M)$ (a map into $\mathbb{R}$), a vector field is a linear *operator* on $\mathcal{F}(M)$ (a map into $\mathcal{F}(M)$). We have learned in exercise (2.2.5) that the linear functionals on $\mathcal{F}(M)$ comprise a vector space over $\mathbb{R}$, i.e. linear combinations with coefficients from $\mathbb{R}$ are permitted. This kind of combination is permitted for vector fields as well (so that they comprise a real (albeit ∞-dimensional) vector space, too). It turns out, however, that the life of vector fields is *considerably richer*; in particular, one can form linear combinations with coefficients from the *algebra* $\mathcal{F}(M)$. This means (Appendix A.4) that vector fields actually comprise a *module* over the algebra of functions $\mathcal{F}(M)$.

2.2.12 Given $V, W \in \mathfrak{X}(M)$ and $f \in \mathcal{F}(M)$, check that a linear combination $V + fW$ is a vector field, too, if one defines it in terms of a *pointwise combination* of the constituent vectors

$$(V + fW)_P := V_P + f(P)W_P$$

or equivalently (in terms of the action on functions) as

$$(V + fW)g := Vg + f(Wg)$$ □

• If we say, then, that the fields ∂_i constitute a *basis* for vector fields, what we have in mind is that this is a basis *in the sense of a module* (as opposed to a linear space over $\mathbb{R}$). This means that any vector field in a coordinate patch $\mathcal{O} \leftrightarrow x^i$ (this may not hold for the manifold as a whole) may be uniquely decomposed with respect to ∂_i as $V = V^i \partial_i$, the coefficients of the decomposition (components) V^i being, however, from the *algebra* $\mathcal{F}(\mathcal{O})$ ($\mathbb{R}$ is not enough, in general). Thus $\mathfrak{X}(\mathcal{O})$ is an ∞-dimensional linear space over $\mathbb{R}$, but it is, at the same time, *finitely generated* as a module over $\mathcal{F}(\mathcal{O})$. Namely, it has n *generators* (∂_i, for example), from which it may be generated completely by means of the algebra $\mathcal{F}(\mathcal{O})$ in full analogy with an n-dimensional linear space, which may be generated from an arbitrary basis $e_1, \ldots, e_n$ with the help of the *field*[14] of real numbers $\mathbb{R}$.

2.2.13* Let L be an n-dimensional linear space over $\mathbb{R}$. Show that

(i) there exists the canonical (independent of the choice of basis in L) isomorphism of L itself and a tangent space $T_x L$ (x being an arbitrary point in L), so that a linear space L may be canonically identified with the tangent space at an arbitrary point

(ii) if a fixed vector $v \in L$ is successively mapped into *all* tangent spaces in this way, the vector *field* V is obtained on L; explicitly (in coordinates introduced in (1.4.11), $v = v^a e_a$) it reads $V = v^a \partial_a$.

Hint: (i) $L \ni v \mapsto (d/dt)_0 (x + tv)$; a picture might be helpful in order to visualize what is going on. □

2.2.14 Let $M \times N$ be a manifold, which is the Cartesian product of two other manifolds M and N. Show that

(i) there is a canonical decomposition of tangent spaces at any point (m, n) into the sum of two subspaces, each of them being isomorphic to the tangent spaces at points m and n respectively of the initial manifolds

$$T_{(m,n)}(M \times N) = T_m M \oplus T_n N$$

(ii) any vector field V on $M \times N$ may be uniquely decomposed into the sum of two vector fields $V = V_M + V_N$, where V_M "is tangent to" M and V_N "is tangent to" N.

Hint: (i) consider the curves $t \mapsto (m(t), n)$ and $t \mapsto (m, n(t))$; in coordinates from (1.3.3) the subspaces span ∂_i and ∂_a; (ii) pointwise realization of (i); $V = A^i(x, y)\partial_i + B^a(x, y)\partial_a \equiv V_M + V_N$. □

2.3 Integral curves of a vector field

• *Lines of force field* provide an aid for visualizing the field; they are essentially a map of the field. A momentary glance at the pattern of lines provides rich information concerning the field itself, since if $\mathbf{F}(\mathbf{r})$ is the field in question, we know that (by definition) the vector $\mathbf{F}$

[14] The field $\mathbb{R}$ is hidden in the algebra $\mathcal{F}(\mathcal{O})$ in terms of *constant functions*, so that the algebra $\mathcal{F}(\mathcal{O})$ is a much richer object than $\mathbb{R}$ is – this is the reason why far fewer generators are needed to reach the same goal.

at a point $\mathbf{r}$ is *tangent* to the line of force at $\mathbf{r}$. The concept of an integral curve adds a definite *parametrization* to this idea (it is a *curve* rather than a *line*), the latter being irrelevant in the case of force lines: its orientation is all they need.

A vector field V on M determines a vector $V_P \in T_P M$ at each point $P \in M$. On the other hand, we know from Section 2.1 that a vector V_P may be regarded as an equivalence class of curves, each representative of the class "being the same" in the immediate vicinity of the point P (up to order ϵ). An *integral curve* of a vector field V is then the curve γ on M, such that at each point of its image, the equivalence class $[\gamma]$ given by the curve, coincides with the class V_P, given by the value of the field V in P. Put another way, from each point it reaches, it moves away exactly in the direction (as well as with the speed) dictated by[15] the vector V_P. All this may be written as a succinct geometrical equation

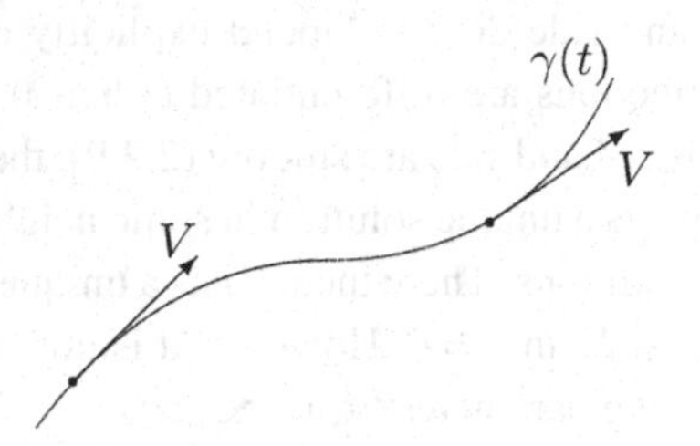

$$\dot{\gamma} = V \qquad \text{i.e.} \quad \dot{\gamma}(P) = V_P$$

(this is the equation for finding an integral curve γ of a vector field V in a "coordinate-free" form), where the symbol $\dot{\gamma}(P)$ denotes the *tangent vector* to the curve γ at the point P (i.e. the equivalence class $[\gamma]$, given by the curve γ at the point P). If the vectors on both sides of this equation are decomposed with respect to a coordinate basis, a system of *differential equations* for the functions $x^i(t) \equiv x^i(\gamma(t))$ (for the coordinate presentation of the curve to be found) is obtained.

2.3.1 Show that the differential equations for finding an integral curve γ of a vector field V have the form

$$\dot{x}^i = V^i(x) \qquad i = 1, \dots, n$$

i.e. in more detail

$$\dot{x}^1(t) = V^1(x^1, \dots, x^n)$$
$$\dots$$
$$\dot{x}^n(t) = V^n(x^1, \dots, x^n)$$

Hint: $\dot{\gamma}(\gamma(t)) = \dot{x}^i(t)\, \partial_i|_{\gamma(t)}$, $V_{\gamma(t)} = V^i(x(t))\, \partial_i|_{\gamma(t)}$. □

2.3.2 Write down and solve the equations for integral curves of the field V from exercise (2.2.11), both in polar and in Cartesian coordinates. Draw the solutions ($\dot{r} = 0, \dot{\varphi} = 1$; $\dot{x} = -y, \dot{y} = x$). □

[15] Like a well-disciplined hiker, always walking in the direction of arrows on destination signs and obediently following the instructions concerning time indications given there (how many minutes he or she would need to reach the next arrow).

2.3.3 Find integral curves of the field $V = \partial_x + 2\partial_\varphi$ on $\mathbb{R}[x] \times S^1[\varphi]$ (the surface of a cylinder). Draw the results. □

• We may see that, in general, one has to do with a system of n first-order ordinary differential equations for n unknown functions $x^i(t)$. Moreover, the system is *quasi-linear* (linear in the highest (= the first, here) derivatives), *autonomous* (functions on the right-hand side do not depend explicitly on the variables with respect to which the unknown functions are differentiated (t here)) and, in general, coupled. Since the functions on the right-hand side are smooth (2.2.9), the theory of equations of this type guarantees that there exists a unique solution in some neighborhood of the point, which corresponds to the initial conditions. There then exists a unique integral curve of a field V, which starts at (any given) $P \in M$ in $t = 0$. However, it is not, in general, possible to extend this curve for all values of the parameter $t \in (-\infty, \infty)$.

2.3.4 A vector field V on M is said to be *complete* if for any point $P \in M$ the integral curve $\gamma(t)$, which starts from P, may be extended to all values of the parameter t. Show that the vector fields $V = \partial_x$ on $M = (-1, 1)$ and $W = x^2\partial_x$ on $N = \mathbb{R}$ are *not complete* (and learn a lesson from these two examples, what some problems with such an extension might look like). □

2.3.5 Given $\gamma(t)$, an integral curve of a vector field V on M, let $\hat{\gamma}(t) := \gamma(\sigma(t))$ be a *reparametrized* curve. Find the most general dependence $\sigma(t)$, so that $\hat{\gamma}$ will be an integral curve of the vector field V, too.
Hint: $(d/dt)\, f(\gamma(\sigma(t))) = \sigma'(t)(d/dt)\, f(\gamma(t))$, so that $\dot{\hat{\gamma}} = \sigma'\dot{\gamma}$; $[\sigma(t) = t + \text{constant}]$. □

• This result is easy to understand. Consider $\gamma(t)$ as being a trajectory. Then $\hat{\gamma}$ is another trajectory, such that we traverse the same set of points on M at different moments of time. Put another way, the *path* remains unchanged, but the (instantaneous) *speed* of traversing the path may be different.[16] Just how much different depends on the point and the result of the exercise shows that the new speed is $\sigma'(t)$ times the old one at any point $\gamma(t)$. (As an example, for $\sigma(t) = 2t$, the new speed is *twice* the old one at each point.) Since the velocity vector of an integral curve *may not* be changed (it is given by V uniquely), $\sigma'(t) = 1$ results. This means that the only possibility to change the trajectory is to traverse the same path either *sooner* or *later*. This freedom ($t \mapsto t +$ constant) enables one to set an arbitrary value of the parameter t (time) at the starting point P.

2.3.6 Let γ be an integral curve of a vector field V on M, which starts from $P \equiv \gamma(0) \in M$. Show that the integral curve (of the same field V) $\hat{\gamma}$, which starts from $Q \equiv \gamma(a)$, is $\hat{\gamma}(t) := \gamma(t + a)$.

Hint: see (2.3.5). □

[16] In fact, we have not enough structure, yet, to speak of the "speed" (a metric tensor, to be introduced later, is needed for it). In spite of this, we *can* speak of the *ratio* of two speeds, since our velocity vectors are *proportional*.

• The result of (2.3.1) admits a different interpretation, too. It shows that *each* system of equations of the type (2.3.1) may be regarded as a system for finding integral curves of the particular vector field (we read out its components from the right-hand side of the equations). This is the important observation, since it provides a key to the investigation of properties of solutions of such equations by powerful geometrical and topological methods – corresponding vector fields (or other objects associated with them) are studied instead of the equations themselves. We will see this, for example, in Chapter 14, where Hamiltonian systems will be discussed.

2.3.7 Find a vector field V on $\mathbb{R}^{2n}[q^1, \dots, q^n, p_1, \dots, p_n]$, which corresponds to the *Hamilton equations*

$$\dot{q}^a = \frac{\partial H}{\partial p_a} \qquad \dot{p}_a = -\frac{\partial H}{\partial q^a} \qquad a = 1, \dots, n$$

$(V = (\partial H/\partial p_a)\partial/\partial q^a - (\partial H/\partial q^a)\partial/\partial p_a)$. □

• A vector field V on a manifold M gives rise to a new and interesting structure, a *congruence* of integral curves on M: the manifold M is "densely" filled by a system of (infinitely many) curves, which never intersect and the "speed" of motion along them is completely determined by the field V. This situation may be conveniently visualized as the *flow of a river*. This flow is *stationary* (the velocity vector in a given point being always the same; in particular, the river does not flow at the points where the field vanishes) and for particular types of fields (e.g. for Hamiltonian fields) the fluid is in addition *incompressible* (14.3.6). Integral curves correspond to the streamlines of the flow. If one fine (and hot) afternoon we do not resist the temptation and let ourselves waft downstream, we get from $P \equiv \gamma(0) \in M$ to the point $Q \equiv \gamma(t) \in M$; naturally a one-parameter class of mappings

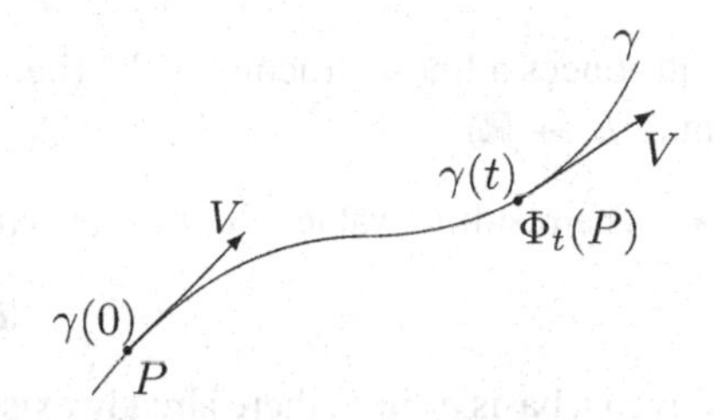

$$\Phi_t : M \to M \qquad P \equiv \gamma(0) \mapsto \gamma(t)$$

arises, called a (local) *flow* generated by the vector field V. We will return to this important concept in more detail later, in Chapter 4 and beyond.

2.3.8 Justify the statement mentioned above, that integral curves never intersect (nor are tangent to one another).

Hint: from a point P one has to make a move in the direction of V_P (uniquely). □

2.3.9 Express the results of exercises (2.3.2) and (2.3.3) in the form of a flow Φ_t : $x^i \mapsto x^i(t) \equiv \Phi_t(x^i)$ $((r, \varphi) \mapsto (r, \varphi + t)$ or $(x, y) \mapsto (x \cos t - y \sin t, x \sin t + y \cos t)$; $(x, \varphi) \mapsto (x + t, \varphi + 2t))$. □

2.4 Linear algebra of tensors (multilinear algebra)

• It turns out that each linear space L automatically gives rise to a fairly rich algebraic structure "above" L – a whole infinite cascade of further and further linear spaces, the spaces of *tensors in* L and an ∞-dimensional associative graded algebra, the *tensor algebra* $T(L)$, associated with them. In this section we will become familiar with tensors at the level of linear algebra, and in the next section we shift to manifolds and introduce the concept of a tensor field.

Within this section we consider arbitrary n-dimensional linear space L over the field of real numbers $\mathbb{R}$.

First, we observe that *linear forms* on L, i.e. linear maps such that

$$\alpha : L \to \mathbb{R} \qquad \alpha(v + \lambda w) = \alpha(v) + \lambda \alpha(w) \quad v, w \in L, \lambda \in \mathbb{R}$$

form a linear space in its own right, the *dual space* L^*. Its elements are called *covectors* in L.

2.4.1 Check that the prescription

$$(\alpha + \lambda \beta)(v) := \alpha(v) + \lambda \beta(v)$$

introduces a linear structure in L^* (i.e. check that the linear combination is indeed a linear map $L \to \mathbb{R}$). □

• The resulting value of $\alpha(v) \in \mathbb{R}$ will be denoted, as a rule, in the form

$$\langle \alpha, v \rangle := \alpha(v)$$

Given a basis e_a in L, there already exists the distinguished basis in L^* (tailored to the basis e_a in L).

2.4.2 Let e_a be a basis in L and let $v = \sum_{b=1}^{n} v^b e_b \equiv v^b e_b$. Verify that

(i) the maps

$$e^a : L \to \mathbb{R} \qquad e^a(v^b e_b) := v^a \quad a = 1, \ldots, n$$

are covectors and, in addition, they constitute a basis in L^* (called the *dual basis* with respect to e_a):

$$\alpha = \alpha_a e^a \qquad \alpha_a := \langle \alpha, e_a \rangle$$

(ii) an equivalent definition of the dual basis is

$$\langle e^a, e_b \rangle = \delta^a_b$$

(iii) a change of the basis in L given by a matrix A results in the change of the dual basis given by the *inverse* matrix A^{-1}

$$e_a \mapsto e'_a = A^b_a e_b \quad \Rightarrow \quad e^a \mapsto e'^a = (A^{-1})^a_b e^b$$

(iv) the dimension of a dual space equals the dimension of the original space: $\dim L^* = \dim L \ (= n)$.

Hint: (i) check that α and $\langle \alpha, e_a \rangle e^a$ are equal linear maps; (iv) consider the number of elements of the dual basis. □

• Since L^* is an n-dimensional vector space in its own right, the whole story may be repeated again and one can construct the dual space $(L^*)^*$. It turns out, however, that this space is (for finite-dimensional L) in a sense redundant. The reason is that it is *canonically* isomorphic to the original space L. What do we mean by this and how can one profit from it?

In general, any two n-dimensional linear spaces are isomorphic, but there are an infinite number of equally good isomorphisms available ($e_a \mapsto E_a$, for *arbitrary* choice of basis E_a), so that there is no reasonable (independent of arbitrary choices) way to choose a preferred one. This is true, in particular, for the relation $L \leftrightarrow L^*$. (Try, for example, to describe your favorite isomorphism to a remote extraterrestrial, who is well educated in linear algebra and understands all the steps you dictate.) Exercise (2.4.3) shows, however, that for $L \to (L^*)^*$ the situation is essentially different. In this case, there is a *distinguished* isomorphism f, which *can* be described to our remote extraterrestrial friend and he or she or it *will know* what maps into what. This isomorphism suggests using a standard mathematical trick – *identification* of the spaces L and $(L^*)^*$, and, by analogy then, the nth with the $(n-2)$th dual spaces. Only the first two members, L and L^*, thus survive from the threatening looking, potentially infinite chain of still higher and higher dual spaces. (This, in a moment, will result in the fact that we will make do with only *two kinds* of indices, "lower" and "upper," on general tensors.)[17] If a non-degenerate bilinear form were *added* to L, the situation would change significantly, since it *would* be possible already to identify L with L^* in a *canonical* way (via the "raising and lowering of indices" procedure, see (2.4.13).)

2.4.3 Prove that the space $(L^*)^*$ is canonically isomorphic to the space L.

Hint: the canonical isomorphism $f : L \to (L^*)^*$ is $\langle f(v), \alpha \rangle := \langle \alpha, v \rangle$. □

2.4.4 Imagine we have defined a "canonical" isomorphism $L \leftrightarrow L^*$ with the help of *dual* bases by

$$f(e_a) := e^a$$

(i.e. $v \leftrightarrow \alpha$, if they have equal coefficients of decomposition with respect to e_a and e^a respectively). Check that if we change the basis as $e_a \mapsto A^b_a e_b$, the isomorphism above will be *changed* (and since in general L all bases are equally good, no distinguished f is given in this way). □

[17] This step saves the huge number of higher dual spaces as well as various kinds of indices for future generations, so it can be regarded as highly satisfactory far-sighted behavior from an ecological point of view; one should not lavishly waste any non-renewable resources, including mathematical structures.

• Let us have a look at one aspect, common for linear spaces L, L^* and $\mathbb{R}$. One may, in all three cases, regard their elements as *linear maps into* $\mathbb{R}$, namely

1. $\alpha \in L^*$ maps $v \mapsto \langle \alpha, v \rangle \in \mathbb{R}$ $(v \in L)$
2. $v \in L$ maps $\alpha \mapsto \langle \alpha, v \rangle \in \mathbb{R}$ $(\alpha \in L^*)$
3. $a \in \mathbb{R}$ maps $\emptyset \mapsto a \in \mathbb{R}$ (*no* input and a real number as output).

Although item 3 might look fairly far-fetched, it proves convenient to incorporate it as a gear-wheel into a device, which in general operates as follows: *several vectors* as well as *covectors* are inserted and (after a crank is turned, of course) a real number drops out. Moreover, if this number depends *linearly* on *each* argument (which holds for all three cases, albeit trivially for the third case), we get a *tensor*.

Definition Let L be an n-dimensional linear space and L^* its dual space. A *tensor of type* $\binom{p}{q}$ in L is a *multilinear* ($\equiv$ *polylinear* := linear in each argument) map

$$t : \underbrace{L \times \cdots \times L}_{q} \times \underbrace{L^* \times \cdots \times L^*}_{p} \to \mathbb{R}$$

$$(\underbrace{v, \ldots, w}_{q}; \underbrace{\alpha, \ldots, \beta}_{p}) \mapsto t(v, \ldots, w; \alpha, \ldots, \beta) \in \mathbb{R}$$

$$t(\ldots, v + \lambda w, \ldots) = t(\ldots, v, \ldots) + \lambda t(\ldots, w, \ldots)$$

(and similarly for an arbitrary covector argument). A collection of tensors of type $\binom{p}{q}$ in L will be denoted by $T^p_q(L)$, and for $p = q = 0$ we set $T^0_0(L) := \mathbb{R}$.

2.4.5 Check that

(i) for $t, \tau \in T^p_q(L)$, $\lambda \in \mathbb{R}$, the rule

$$(t + \lambda\tau)(v, \ldots; \alpha, \ldots) := t(v, \ldots; \alpha, \ldots) + \lambda\tau(v, \ldots; \alpha, \ldots)$$

introduces a linear structure into $T^p_q(L)$ (i.e. the linear combination displayed above indeed happens to be a multilinear map)

(ii) some special instances are given by

$$T^0_0(L) = \mathbb{R} \qquad T^0_1(L) = L^* \qquad T^1_0(L) \approx L$$

$$T^1_1(L) \approx \mathrm{Hom}\,(L, L) \approx \mathrm{Hom}\,(L^*, L^*) \qquad T^0_2(L) = \mathcal{B}_2(L)$$

where $\mathrm{Hom}\,(L_1, L_2)$ denotes all linear maps from L_1 into L_2, $\mathcal{B}_2(L)$ are bilinear forms on L and $\approx$ denotes canonical isomorphism.

Hint: $\binom{0}{0}$, $\binom{0}{1}$ and $\binom{0}{2}$ definitions, $\binom{1}{0}$ (2.4.3); $\binom{1}{1}$: the isomorphisms $\mathrm{Hom}(L, L) \to T^1_1(L)$ and $\mathrm{Hom}\,(L^*, L^*) \to T^1_1(L)$ read

$$t(v; \alpha) := \langle \alpha, A(v) \rangle \qquad \text{and} \qquad t(v; \alpha) := \langle B(\alpha), v \rangle$$

or, equivalently (in the opposite direction),

$$A(v) := t(v; \,\cdot\,) \qquad B(\alpha) := t(\,\cdot\,; \alpha)$$

□

• Taking into account (multi)linearity, a tensor $t \in T^p_q(L)$ is known completely if we know its values on all possible combinations of basis vectors e_a and covectors e^a. This collection of numbers

$$t^{c...d}_{a...b} := t(e_a, \dots, e_b; e^c, \dots, e^d)$$

is said to form the *components* of the tensor t with respect to e_a. The mnemonic rule of the notation $\binom{p}{q}$ should finally be clear: a tensor t is in the space $T^p_q(L)$ if its components have p upper indices and q lower indices.

[2.4.6] Check that

(i) in components, the rule for performing linear combinations from (2.4.5) reduces to

$$(t + \lambda\tau)^{c...d}_{a...b} = t^{c...d}_{a...b} + \lambda\tau^{c...d}_{a...b}$$

(ii) $\dim T^p_q(L) = n^{p+q} \equiv (\dim L)^{p+q}$ (the number $(p+q)$ is known as the *rank* of a tensor)
(iii) under the change of basis in L, components of a tensor transform as follows:

$$e_a \mapsto A^b_a e_b \equiv e'_a \quad \Rightarrow \quad t'^{c...d}_{a...b} = (A^{-1})^c_k \dots (A^{-1})^d_l A^r_a \dots A^s_b t^{k...l}_{r...s}$$

(iv) if $v = v^a e_a$, $\alpha = \alpha_a e^a \dots$ represent the decompositions of arguments, then

$$t(v, \dots, w; \alpha, \dots, \beta) = t^{c...d}_{a...b} v^a \dots w^b \alpha_c \dots \beta_d$$

(v) three different applications of a $\binom{1}{1}$-type tensor t from (2.4.5) in components look like

$$(v^b, \alpha_a) \mapsto t^a_b v^b \alpha_a \qquad v^a \mapsto t^a_b v^b \qquad \alpha_a \mapsto t^b_a \alpha_b$$

Hint: (ii) $t \mapsto t^{c...d}_{a...b}$ is the isomorphism $T^p_q(L) \to \mathbb{R}^{n^{p+q}}$ (each of $(p+q)$ indices takes n values); (iii) $t'^{c...d}_{a...b} := t(e'_a, \dots, e'^d) +$ linearity in each argument. □

• Thus we have learned that L induces an infinite number of further linear spaces – for each pair (p, q) of non-negative integers there is the n^{p+q}-dimensional space $T^p_q(L)$. (This means that if we envisage tensor spaces as a "tower," the tower dilates in the upward direction, like a pyramid does on a photograph snapped in Giza by a distrait yogi, forgetting he has just performed a headstand.)

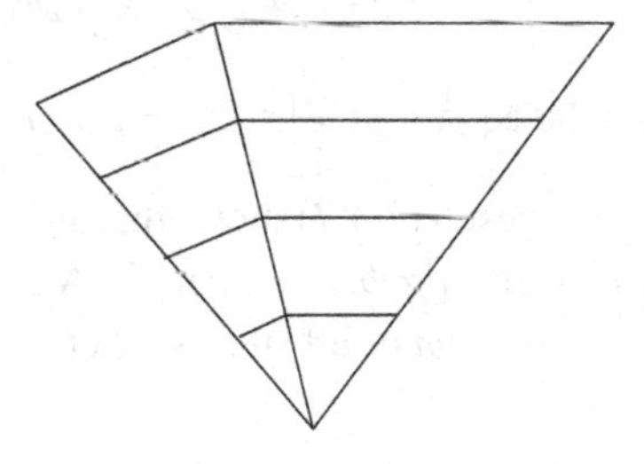

If we combine components with a suitable basis, we get "complete" tensors. It turns out that a suitable basis may be constructed out of the basis for vectors and covectors, if an additional operation on tensors is introduced, the *tensor product*. It may be regarded as a map

$$\otimes : T^p_q(L) \times T^{p'}_{q'}(L) \to T^{p+p'}_{q+q'}(L)$$

i.e. two tensors of *arbitrary* types $\binom{p}{q}$ and $\binom{p'}{q'}$ are multiplied – contrary to linear combination, where both types have to be equal – and the resulting tensor is of type $\binom{p+p'}{q+q'}$. The definition is as follows:

$$(t \otimes \sigma)(v_1, \dots, v_q, w_1, \dots, w_{q'}; \alpha_1, \dots, \alpha_p, \beta_1, \dots, \beta_{p'})$$
$$:= t(v_1, \dots v_q; \alpha_1, \dots, \alpha_p)\sigma(w_1, \dots, w_{q'}; \beta_1, \dots, \beta_{p'})$$

(here the indices label *complete* vectors and covectors, rather than their components!). Stated in words, we first insert the arguments of both types into the first (left) tensor, until it is filled completely; the rest we put into the second (right) one. The resulting two *numbers* are then simply multiplied.

2.4.7 Verify that

(i) the result of the multiplication $t \otimes \sigma$ is a tensor, indeed (i.e. check multilinearity)
(ii) at the level of components the multiplication $\otimes$ gives

$$(t \otimes \sigma)^{c\dots dr\dots s}_{a\dots bk\dots l} = t^{c\dots d}_{a\dots b}\sigma^{r\dots s}_{k\dots l}$$

(iii) the multiplication $\otimes$ is associative (we need not bother about brackets in *multiple* products), bilinear and *non*-commutative
(iv) tensors of type (p, q)

$$e^a \otimes \cdots \otimes e^b \otimes e_c \otimes \cdots \otimes e_d \in T^p_q(L)$$

constitute the basis of $T^p_q(L)$ with respect to which components have been defined above, i.e. an arbitrary tensor $t \in T^p_q(L)$ may be decomposed as

$$t = t^{c\dots d}_{a\dots b} e^a \otimes \cdots \otimes e^b \otimes e_c \otimes \cdots \otimes e_d \qquad t^{c\dots d}_{a\dots b} := t(e_a, \dots, e_b; e^c, \dots, e^d)$$

Hint: (iv) one has to check that the "original" tensor and its decomposition represent *the same map*; since they are (multi)linear, it is enough to check it for the basis; as an example

$$\left(t^c_d e^d \otimes e_c\right)(e_a; e^b) = t^c_d \langle e^d, e_a\rangle\langle e^b, e_c\rangle = t^b_a = t(e_a; e^b)$$

thus the equality $t^c_d e^d \otimes e_c = t$ *of maps* (= tensors) has been proved. □

• The result (2.4.7) shows that all tensors constitute an (∞-dimensional non-commutative) *associative algebra* (Appendix A.2), called the *tensor algebra* $T(L)$. As a linear space, it is a *direct sum* of all spaces $T^p_q(L)$

$$T(L) := \bigoplus_{r,s=0}^{\infty} T^r_s(L)$$
$$\equiv T^0_0(L) \oplus T^1_0(L) \oplus T^0_1(L) \oplus T^2_0(L) \oplus T^1_1(L) \oplus T^0_2(L) \oplus \cdots$$

(up to infinity), i.e. an element from $T(L)$ may be regarded as a linear combination of tensors of all types $\binom{p}{q}$. Multiplication $\otimes$ is defined as a linear extension of the definition of $\otimes$ on *homogeneous terms* (terms with fixed $\binom{p}{q}$), i.e. according to the rule "everybody with everybody":[18]

$$(k+v+\alpha+\cdots)\otimes(q+w+\beta+\cdots):=k\otimes q+k\otimes w+k\otimes\beta+\cdots$$
$$+v\otimes q+v\otimes w+\cdots$$

Furthermore, this algebra is ($\mathbb{Z}\times\mathbb{Z}$)-*graded* (Appendix A.5): its "homogeneous" subspaces $T^p_q(L)$ are labelled by a *pair* of integers (p,q), i.e. (we define $T^p_q(L):=0$ for negative p,q) by an element of group $\mathbb{Z}\times\mathbb{Z}$, and multiplication in algebra $T(L)$ is compatible with the grading: the product of any two elements from the subspaces $\leftrightarrow (p,q)$ and $(p',q')\in\mathbb{Z}\times\mathbb{Z}$ is homogeneous, too, belonging to the subspace which corresponds to a *product in the sense* of $\mathbb{Z}\times\mathbb{Z}$, i.e. $(p+p',q+q')$.

Operations producing tensors from tensors, are said to be *tensor operations*. So far we have met linear combination and tensor product. One further important tensor operation is provided by *contraction*. It is defined (for $p,q\geq 1$) as follows:

$$C: T^p_q(L)\to T^{p-1}_{q-1}(L) \qquad t\mapsto Ct:=t(\ldots,e_a,\ldots;\ldots,e^a,\ldots)$$

where the exact position of arguments e_a and e^a is to be specified – it forms a part of the definition (there are several (pq) various possible contractions, in general, and one has to state *which one* is to be performed).

2.4.8 Check that

(i) the result is indeed a tensor (multilinearity)
(ii) C does not depend on the choice of the basis e_a (when e_a has been fixed, however, e^a is to be the dual)
(iii) in components the rule for C looks like[19]

$$t^{\cdots\cdots}_{\cdots\cdots}\mapsto t^{\cdots a\cdot}_{\cdot a\cdots} \qquad \text{i.e.} \quad \text{as a } \textit{summation} \text{ with respect to a pair of upper and lower indices}$$

(iv) independence of a choice of basis results from the component formula, too.

Hint: (ii) see (2.4.2); (iv) see (2.4.6). □

2.4.9 Show that

(i) the prescription

$$\hat{1}(V;\alpha):=\langle\alpha,V\rangle$$

defines a $\binom{1}{1}$-type tensor, the *unit* tensor
(ii) its components with respect to *any* basis e_a (e^a being dual, as usual) are given by

$$\hat{1}^a_b=\delta^a_b \qquad \text{so that} \quad \hat{1}=e^a\otimes e_a$$

[18] The maximum promiscuity rule.
[19] Each contraction thus unloads a tensor by two indices. It breathes with fewer difficulties immediately (fewer indices = fewer worries), it feels like after a rejuvenation cure. This human aspect of the matter is reflected sensitively in German terminology, where the word *Verjüngung* (rejuvenescence) is used.

(iii) it realizes the *unit* operator ($v \mapsto v, \alpha \mapsto \alpha$) if it is interpreted as a map

$$\hat{1} : L \to L \qquad \text{and} \qquad \hat{1} : L^* \to L^*$$

respectively

(iv) its contraction (2.4.8) gives

$$C\hat{1} = n \equiv \dim L$$

Hint: (iii) see (2.4.5). □

2.4.10 Show that the evaluation of a tensor on arguments may be regarded as a composition of tensor product and contractions; as an example, for a $\binom{1}{1}$-type tensor it is

$$t(v, \alpha) = CC(t \otimes v \otimes \alpha) = (t \otimes v \otimes \alpha)^{ab}_{ba} \equiv (t \otimes v \otimes \alpha)(e_b, e_a; e^a, e^b)$$

In particular, (see exercise 2.4.8),

$$\hat{1}(v, \alpha) \equiv \langle \alpha, v \rangle = C(\alpha \otimes v)$$

□

• A *metric tensor* in L is a symmetric non-degenerate tensor of type $\binom{0}{2}$, i.e. $g \in T^0_2(L)$ such that

$$g(v, w) = g(w, v) \qquad \text{symmetric}$$
$$g(v, w) = 0 \text{ for all } w \Rightarrow \; v = 0 \qquad \text{non-degenerate}$$

2.4.11 Check that

(i)

$$g_{ab} = g_{ba} \qquad \det g_{ab} \neq 0$$

(ii) conditions in (i) do not depend on the choice of basis e_a. □

• Sometimes one demands that g meets stronger requirements, namely to be *positive definite*,[20] so that

$$g(v, v) \geq 0 \qquad \text{(and equality holds } only \text{ for } v = 0)$$

and (metric) tensors, which *are not* positive definite, are said to be *pseudo-metric* tensors. We will use, in what follows, the nomenclature *metric* tensor also for g, which is *not* positive definite,[21] and if some statement relies heavily on the positive definiteness of the latter (i.e. "true" metric tensor), it will be specially emphasized.

As is well known from linear algebra, one can bring a matrix of a general symmetric bilinear form by a suitable (non-unique) choice of basis e_a to the canonical form

$$b_{ab} = \text{diag}(\underbrace{1, \dots, 1}_{r}, \underbrace{-1, \dots, -1}_{s}, \underbrace{0, \dots, 0}_{l})$$

[20] Then $(v, w) := g(v, w)$ has the properties of a *scalar product* in L, see (2.4.13).

[21] This is the case both in special and in general relativity, where one speaks of a "metric" in situations where in finer terminology *pseudo*-metric tensor (or even tensor *field*) should be used.

where the numbers (r, s, l) are inherent properties of the form (*Sylvester's theorem*). Non-degeneracy adds $l = 0$ (why?), so that the canonical form of a *metric tensor* reads as

$$g_{ab} = \eta_{ab} \equiv \text{diag}\,(\underbrace{1, \ldots, 1}_{r}, \underbrace{-1, \ldots, -1}_{s})$$

or, in other words,

$$\begin{aligned} g &= g_{ab} e^a \otimes e^b \\ &= e^1 \otimes e^1 + \cdots + e^r \otimes e^r - e^{r+1} \otimes e^{r+1} - \cdots - e^{r+s} \otimes e^{r+s} \end{aligned}$$

In this case we will speak about a metric tensor with *signature* (r, s).[22] Thus, the positive definite case corresponds to $s = 0$ (terms with a minus sign are not present in the canonical form). Any basis $e_a \leftrightarrow e^a$ in which this canonical form of g is obtained is called an *orthonormal basis*.

2.4.12 Given e_a an arbitrary basis and $g_{ab} = g(e_a, e_b)$, define g^{ab} as elements of the *inverse matrix* to g_{ab}, i.e.

$$g^{ac} g_{cb} := \delta^a_b$$

Prove that

(i) g^{ab} constitute the components of a (symmetric) $\binom{2}{0}$-type tensor (so that they indeed deserve two upper indices)

$$g \equiv g_{ab} e^a \otimes e^b \in T^0_2(L) \quad \Rightarrow \quad g^{-1} := g^{ab} e_a \otimes e_b \in T^2_0(L)$$

(ii) matrix g^{ab} is non-singular.

Hint: (i) check the transformation law of g^{ab} under a change of basis. □

2.4.13 Consider the maps $\flat_g$ and $\sharp_g$ given by

$$\begin{aligned} \flat_g &: L \to L^* \qquad v \mapsto \flat_g v := g(v, \cdot) \\ \sharp_g &: L^* \to L \qquad \alpha \mapsto \sharp_g \alpha := g^{-1}(\alpha, \cdot) \end{aligned}$$

Check that

(i) they are linear (and canonical) isomorphisms
(ii) when expressed in bases and in components, they look like

$$\begin{array}{llll} \flat_g : & e_a \mapsto g_{ab} e^b & v^a \mapsto v_a := g_{ab} v^b & v^a e_a \mapsto v_a e^a \\ \sharp_g : & e^a \mapsto g^{ab} e_b & \alpha_a \mapsto \alpha^a := g^{ab} \alpha_b & \alpha_a e^a \mapsto \alpha^a e_a \end{array}$$

(iii) they are inverse to each other:

$$\flat_g \circ \sharp_g = \text{id}_{L^*} \qquad \sharp_g \circ \flat_g = \text{id}_L$$

[22] Sometimes, the number $r - s$ is called the signature, too.

(iv) if *scalar products* in L and L^* are introduced[23] by

$$(v, w) := g(v, w) \equiv g_{ab} v^a w^b \qquad (\alpha, \beta) := g^{-1}(\alpha, \beta) \equiv g^{ab} \alpha_a \beta_b$$

then both $\flat_g$ and $\sharp_g$ are *isometries*, i.e. $(\flat_g v, \flat_g w) = (v, w)$, $(\sharp_g \alpha, \sharp_g \beta) = (\alpha, \beta)$. □

• The maps $\flat_g$ and $\sharp_g$ are known as *lowering* and *raising of indices* (with the help of g), respectively. The quantities v_a, v^a are often called covariant and contravariant *components* of (the same) vector v. We will not adopt this nomenclature, however. We will always strictly discriminate between a *vector* $v = v^a e_a$ and a *covector* $v_a e^a$ (as being elements of L and L^*) and interpret the operations of raising and lowering of indices as maps between two *different* spaces $L \leftrightarrow L^*$. Note that the graphical expressions used for these maps originate from well-known musical symbols.[24]

The metric tensor makes it possible to change the position of indices on higher rank tensors, too, for example

$$t^a_{bc} \mapsto t_{abc} := g_{ad}\, t^d_{bc} \qquad R^{ab}_{cd} \mapsto R_{abcd} := g_{ae}\, g_{bf}\, R^{ef}_{cd}$$

This belongs to basic exercises of *index gymnastics*.[25]

2.4.14 Prove the validity of the exercise

$$t^{\dots a.}_{..a\dots} = t^{..a\dots}_{\dots a.}$$

Hint: do you intend to base your proof upon the fact that the total potential energy remains unchanged? (Red herring.) □

• There are several possibilities of how to raise or lower indices on second or higher rank tensors, differing in the order of the indices on the resulting tensor. As an example, there are four places below where one can lower the index on the fourth rank tensor R^a_{bcd}

$$R_{abcd} := g_{aj}\, R^j_{bcd} \qquad R_{abcd} := g_{bj}\, R^j_{acd} \qquad \dots$$

The indices are sometimes written so as to have *only one* index on each vertical line, being *either* upper *or* lower, e.g. $R_a{}^b{}_{cd}$. Within this particular convention, it is always clear where exactly any upper index should be lowered.

It is useful to realize that symmetry of the metric tensor g is of no importance for raising and lowering of indices, the *only* property that matters being its *non-degeneracy*. These operations might as well be defined by virtue of an *anti*symmetric tensor $\omega_{ab} = -\omega_{ba}$, provided that it happens to be non-degenerate ($\det \omega_{ab} \neq 0$). We will see in what follows that this possibility is indeed exploited, the most prominent applications being in symplectic geometry (to be discussed in Chapter 14 and beyond) and in the theory of two-component spinors (12.5.3).

[23] They are positive definite for *Euclidean* g only!

[24] Namely "flat" and "sharp." Thoughtful graduates of schools of music might recall that *no g was present* on sharps and flats they had read in sheets of music – this is simply because the validity of *Euclidean* geometry is normally assumed in concert halls, so that musical flats and sharps are conventionally associated with *this Euclidean g* (and are not indicated explicitly).

[25] It should be performed, as is the case for arbitrary gymnastics, at an open window, never directly after a substantial meal.

Finally, let us contemplate whether the lowering and raising of indices does change the numerical values of components. The formula $v^a \mapsto v_a \equiv g_{ab}v^b$ shows that the numbers v^a and v_a are the same only in the case where g is given, in a given basis, by the *identity* matrix, $g_{ab} = \delta_{ab}$. This (only) happens to be true in the *positive definite* case in the *orthonormal* basis; in the indefinite case, this happens *in no basis*. Therefore, when working with vectors in Euclidean spaces E^2 or E^3, one may safely ignore the detailed position (upper/lower) of indices with respect to an *orthonormal basis*.[26] On the other hand, one should pay due attention to this issue in all cases when non-orthonormal bases or indefinite metrics are used. In Minkowski space, for example, the lowering and raising of indices *always* changes numerical values of (some) components; in an orthonormal basis this change reduces to the change of a *sign* (of some of them), but it may be more complicated in general.

2.4.15 Check that raising and lowering of indices

(i) are tensor operations
(ii) may be regarded as compositions of a tensor product (with the tensor g) and contractions.

Hint: e.g. $\flat_g v \equiv g(v, \cdot) = C(g \otimes v)$. □

• The last tensor operations to be mentioned are symmetrizations and antisymmetrizations in various subgroups of indices. Let us illustrate this on just two indices.

2.4.16 Given $t \in T_2^0(L)$, define

$$t^{\mathrm{S}} := \frac{1}{2}(t_{ab} + t_{ba})e^a \otimes e^b \equiv t_{(ab)}e^a \otimes e^b$$
$$t^{\mathrm{A}} := \frac{1}{2}(t_{ab} - t_{ba})e^a \otimes e^b \equiv t_{[ab]}e^a \otimes e^b$$

(symmetric and antisymmetric part of the tensor t respectively). Check that

(i)

$$t \mapsto t^{\mathrm{S}} \equiv \pi^{\mathrm{S}} t \qquad t \mapsto t^{\mathrm{A}} \equiv \pi^{\mathrm{A}} t$$

are tensor operations, independent of the choice of e_a

(ii) tensors, for which $t = t^{\mathrm{S}}$ or $t = t^{\mathrm{A}}$ is true, constitute subspaces in $T_2^0(L)$

(iii) π^{S} and π^{A} satisfy

$$\pi^{\mathrm{S}} \circ \pi^{\mathrm{S}} = \pi^{\mathrm{S}} \qquad \pi^{\mathrm{S}} \circ \pi^{\mathrm{A}} = \pi^{\mathrm{A}} \circ \pi^{\mathrm{S}} = 0$$
$$\pi^{\mathrm{A}} \circ \pi^{\mathrm{A}} = \pi^{\mathrm{A}} \qquad \pi^{\mathrm{S}} + \pi^{\mathrm{A}} = \hat{1}$$

so that they serve as projection operators on the subspaces of the symmetric and antisymmetric tensors mentioned above, the whole space $T_2^0(L)$ being the direct sum of these two subspaces (only). □

• Finally, two more useful concepts will be introduced at the end of this section on multilinear algebra, namely those of a dual map and an induced metric tensor.

[26] That is, at the level of components one is allowed to make no difference between a vector and the associated covector, like the gradient as a covector and a gradient as a vector, see the end of Section 2.6.

2.4.17 Let $A : L_1 \to L_2$ be a linear map, e_i a basis of L_1 and e_a a basis of L_2. The *rank* of the map A is defined as a dimension of the image of the space L_1 in L_2, i.e. rank $A := \dim \operatorname{Im} A$. Show that

(i) by the prescription

$$\langle A^*(\alpha_2), v_1\rangle := \langle \alpha_2, A(v_1)\rangle \qquad \alpha_2 \in L_2^*, v_1 \in L_1$$

a linear map

$$A^* : L_2^* \to L_1^*$$

is defined (*dual map*)

(ii) on the basis it gives

$$e_i \overset{A}{\mapsto} A_i^a e_a \quad \Rightarrow \quad e^a \overset{A^*}{\mapsto} A_i^a e^i$$

i.e. matrices of the maps A, A^* are *transposes* of each other

(iii)

$$\text{rank } A = \text{rank of the } \textit{matrix} \text{ of a map } A$$

$$\text{rank } A^* = \text{rank of the } \textit{matrix} \text{ of a map } A^*$$

(iv) rank A = rank A^* ($\Rightarrow$ that the *row* and *column* ranks of a matrix happen to coincide).

Hint: (iv) use *adapted* bases: a part of e_i is a basis of the *kernel* Ker A of the map (those v for which $v \mapsto 0 \in L_2$), the rest are chosen arbitrarily to complete a basis; in L_2 take images of the remaining part (they span Im A) + complete a basis. □

2.4.18 Given $A : L_1 \to (L_2, h)$, $\dim L_1 \leq \dim L_2$ a maximum rank linear map (2.4.17) (h being a metric tensor in L_2), show that

(i) by the rule

$$g := A^* h \qquad (A^* h)(v, w) := h(Av, Aw)$$

a metric tensor g in L_1 is defined (*induced metric tensor*)

(ii) if $e_i \in L_1$ and $e_a \in L_2$ are bases, then

$$g_{ij} = A_i^a h_{ab} A_j^b \qquad Ae_i =: A_i^a e_a \qquad \text{(in matrix notation } g = A^{\mathrm{T}} h A)$$

Hint: (i) (among others) one has to check the maximum rank (2.4.13) of the map

$$\flat_g : L_1 \to L_1^* \qquad v \mapsto g(v, \cdot\,) \equiv \tilde{v} \equiv \flat_g v$$

($\equiv$ non-degeneracy of g). This map is a composition of

$$\flat_g = A^* \circ \flat_h \circ A \qquad L_1 \overset{A}{\to} L_2 \overset{\flat_h}{\to} L_2^* \overset{A^*}{\mapsto} L_1^* \qquad (A^* \text{ in the sense of (2.4.17))}$$

(since $e_i \mapsto g_{ij} e^j = A_i^a h_{ab} A_j^b e^j$), all factors in the composition do have maximum rank and $\dim L_1 \leq \dim L_2 \Rightarrow \flat_g$ is a maximum rank ($= \dim L_1$) map, too. □

2.4.19 Let V be a linear space with a *distinguished subspace* $W \subset V$. Show that in the dual space V^* the associated distinguished subspace $\hat{W} \subset V^*$ of dimension dim V minus dim W is given canonically; it is said to be an *annihilator* of the subspace W.

Hint: consider covectors $\sigma \in V^*$ annihilated by vectors from W, i.e. such that $\langle \sigma, w \rangle = 0$ for all $w \in W$ (see also (10.1.13)). □

2.5 Tensor fields on M

• In Section 2.2 we showed that there is a vector space associated with each point P of a manifold M, the tangent space $T_P M$. In Section 2.4 we learned how to construct tensors of type $\binom{p}{q}$, starting from an *arbitrary* finite-dimensional vector space L. If we now take L to be the space $T_P M$, we immediately get (with practically no labor – it simply suffices to harvest the crop sown earlier in Section 2.4) tensors at the point $P \in M$. In particular, the dual space to $T_P M$, the space of *covectors* in $P \in M$, is called the *cotangent space* in P and it is denoted by $T_P^* M$.

Equally naturally the concept of a *tensor field* of type $\binom{p}{q}$ on M appears. In full analogy with the special case of a vector field, one has to choose exactly one tensor of type $\binom{p}{q}$ residing at each point of a manifold M. Once again, we restrict to fields which vary smoothly from point to point. In order to formulate this succinctly, an algebraic perspective is useful. In particular, one should realize what kind of *maps* tensor fields actually are.

An individual tensor of type $\binom{p}{q}$ in $P \in M$ takes as its arguments vectors and covectors in P, and the result is a number which depends linearly on each of the arguments. At the level of fields, this happens in each point $P \in M$. It is convenient to regard it as if we inserted vector and covector *fields* as arguments of a tensor *field*, obtaining a number at each point, i.e. a *function*. Since at each point linearity *over* $\mathbb{R}$ is required, one has to demand linearity *over* $\mathcal{F}(M)$ for fields. Let us clarify this subtle point in more detail. Consider a covector field α. At each point P we have α_P, and the value V_P of a vector field V is inserted in it as an argument. In this way we obtain a function

$$\langle \alpha, V \rangle \in \mathcal{F}(M) \qquad \langle \alpha, V \rangle(P) := \langle \alpha_P, V_P \rangle \in \mathbb{R}$$

Since α_P is a covector, for any $\lambda \in \mathbb{R}$ it holds that

$$\langle \alpha_P, V_P + \lambda W_P \rangle = \langle \alpha_P, V_P \rangle + \lambda \langle \alpha_P, W_P \rangle$$

At a different point $Q \neq P$ we have

$$\langle \alpha_Q, V_Q + \lambda W_Q \rangle = \langle \alpha_Q, V_Q \rangle + \lambda \langle \alpha_Q, W_Q \rangle$$

Both results should be valid, however, for *arbitrary* λ, so that λ present in the formula corresponding to the point P may be completely *different* from λ in the formula corresponding to the point Q – a "constant" λ may depend on a point, and therefore for any *function*

$f \in \mathcal{F}(M)$ we must have

$$\langle \alpha, V + fW \rangle = \langle \alpha, V \rangle + f \langle \alpha, W \rangle$$

This is said to be the *$\mathcal{F}(M)$-linearity* of the map α, which should be contrasted with the weaker requirement of *$\mathbb{R}$-linearity*. At the same time, we see the important fact that the property of being $\mathcal{F}(M)$-linear ultimately springs from the *pointwise* character of the construction (the expression $\langle \alpha, V \rangle$ is in fact $\langle \alpha_P, V_P \rangle$ performed in each point P). The $\mathcal{F}(M)$-linearity means that the arguments (vector fields in the case of a covector field) constitute a module over the algebra $\mathcal{F}(M)$ and the map

$$\alpha : \mathcal{T}_0^1(M) \to \mathcal{F}(M)$$

is linear *in the sense of modules*.

In terms of these maps the smoothness of a covector field is easily stated: α is said to be *smooth* (of class C^∞) if the function $\langle \alpha, V \rangle$ is smooth for any smooth vector field V. Smooth covector fields on M will be denoted by $\mathcal{T}_1^0(M)$.

2.5.1 Given $\alpha, \beta \in \mathcal{T}_1^0(M)$, $f \in \mathcal{F}(M)$, check that also $\alpha + f\beta \in \mathcal{T}_1^0(M)$, if the linear combination is defined as

$$\langle \alpha + f\beta, V \rangle := \langle \alpha, V \rangle + f \langle \beta, V \rangle.$$

□

• This means that not only vector fields, but also covector fields constitute an $\mathcal{F}(M)$-module. Now, it is clear from this perspective that a tensor field of type $\binom{p}{q}$ may be regarded *as a map*

$$t : \underbrace{\mathcal{T}_0^1(M) \times \cdots \times \mathcal{T}_0^1(M)}_{q} \times \underbrace{\mathcal{T}_1^0(M) \times \cdots \times \mathcal{T}_1^0(M)}_{p} \to \mathcal{F}(M)$$

which is $\mathcal{F}(M)$-linear in each argument. If the resulting function happens to be smooth for arbitrary smooth arguments, the field t is said to be smooth. Smooth tensor fields of type $\binom{p}{q}$ on M will be denoted by $\mathcal{T}_q^p(M)$, the case of $\mathcal{T}_0^0(M)$ being identified with $\mathcal{F}(M)$. (This makes the notation $\mathcal{T}_0^1(M)$ comprehensible for vector fields, too.)

2.5.2 Check that each $\mathcal{T}_q^p(M)$ is naturally endowed with the structure of an $\mathcal{F}(M)$-module.

□

• If we make a comparison between tensors in L and tensor fields on M, we can say that virtually everything goes the same way, if we substitute $T_q^p(L)$ by $\mathcal{T}_q^p(M)$, linear spaces by $\mathcal{F}(M)$-modules and $\mathbb{R}$-linearity by $\mathcal{F}(M)$-linearity.

In particular, let us look more closely at the properties of tensor algebra. This concept may be readily transferred to a manifold, after performing the substitutions mentioned above: one takes the direct sum of all *modules* $\mathcal{T}_q^p(M)$

$$\mathcal{T}(M) := \bigoplus_{p,q=0}^{\infty} \mathcal{T}_q^p(M)$$

(it is an $\mathcal{F}(M)$-module, too) and defines there a pointwise product $\otimes$, just like in Section 2.4. This algebra, the *algebra of tensor fields on* M, is ∞-dimensional (which looks much the same as for $T(L)$), but here already each homogeneous part $\mathcal{T}^p_q(M)$ is ∞-dimensional (over $\mathbb{R}$; the most salient difference occurs for the lowest degree $\binom{0}{0}$: $\mathbb{R} \leftrightarrow \mathcal{F}(M)$). On higher degrees, the situation is repeated in the form we met already in Section 2.2: although the spaces $\mathcal{T}^p_q(\mathcal{O})$ are ∞-dimensional even on "sufficiently small" domains $\mathcal{O} \subset M$ (e.g. in coordinate patches $\mathcal{O} \leftrightarrow x^i$), when regarded as linear spaces, they are finitely generated, when regarded as modules. And what do the basis tensor fields actually look like, with respect to which decomposition is to be performed?

We have seen in Section 2.4 that the most natural basis in L^*, with respect to a given basis e_a in L, is the dual basis e^a. At the same time, for vector fields we know a *coordinate* basis ∂_i. What does a basis for covector fields look like which is dual (in each point) to this particular basis?

2.5.3 Let $f \in \mathcal{F}(M)$, and let x^i be local coordinates in $\mathcal{O} \subset M$. Check that

(i) by the prescription

$$\langle df, V\rangle := Vf$$

a *covector field* df on M is defined. This field is called the *gradient* of the function f

(ii) gradients of coordinates (= functions!) $dx^i \in \mathcal{T}^0_1(\mathcal{O})$ constitute a basis for covector fields on $\mathcal{O}$, i.e. any $\alpha \in \mathcal{T}^0_1(\mathcal{O})$ may be decomposed in the form

$$\alpha = \alpha_i(x)\, dx^i \qquad \alpha_i(x) := \langle \alpha, \partial_i\rangle \ (\equiv \textit{components} \text{ with respect to the basis } dx^i)$$

and, in particular, for a gradient we have

$$df = f_{,i}\, dx^i \equiv \frac{\partial f}{\partial x^i}\, dx^i$$

(iii) covectors $dx^i|_P$ constitute a basis for covectors in P, which is dual to the coordinate basis $\partial_i|_P$ for vectors in P (the basis dx^i is said to be a *coordinate basis*, too)

(iv)

$$\langle \alpha, V\rangle = \alpha_i(x) V^i(x)$$

(v) under the change of coordinates one has (J being, as usual, the Jacobian matrix)

$$x^i \mapsto x'^i(x) \ \Rightarrow\ dx^i \mapsto dx'^i = J^i_j(x)\, dx^j \qquad \text{and} \qquad \alpha_i(x) \mapsto \alpha'_i(x') = (J^{-1})^j_i(x)\alpha_j(x)$$

Hint: (i) see (2.2.12); (v) set $f = x'^i$ in (ii). □

• Since we already have the dual basis dx^i to ∂_i, we may write down component decompositions of arbitrary tensor fields.

2.5.4 Check that if $t \in \mathcal{T}^p_q(M)$, then

(i) locally (in $\mathcal{O} \leftrightarrow x^i$) it holds that

$$t = t^{i\dots j}_{k\dots l}(x)\, dx^k \otimes \cdots \otimes dx^l \otimes \partial_i \otimes \cdots \partial_j$$

(ii) under the change of coordinates $x \mapsto x'$ components transform according to the formula

$$x^i \mapsto x'^i(x) \quad \Rightarrow \quad t^{i\dots j}_{k\dots l}(x) \mapsto t'^{i\dots j}_{k\dots l}(x')$$
$$\equiv J^i_r(x)\dots J^j_s(x)(J^{-1})^u_k(x)\dots(J^{-1})^v_l(x)t^{r\dots s}_{u\dots v}(x)$$

□

2.5.5 Prove that the module $\mathcal{T}^p_q(\mathcal{O})$ has n^{p+q} generators.
Hint: see (2.5.4) and (2.4.6); $t^{i\dots j}_{k\dots l} \in \mathcal{F}(\mathcal{O})$. □

• The result given in (2.5.4) might serve as a basis for an independent *definition* of a tensor field on M (definition of classical differential geometry; refer to definition no. 4 of a vector in Section 2.2): the tensor field of type $\binom{p}{q}$ on M is a collection of *functions* $t^{i\dots j}_{k\dots l}(x)$ associated with coordinates x^i defined in patches $\mathcal{O} \leftrightarrow x^i$, transforming under the changes of coordinates according to the rule given in (2.5.4). Note that a *global* object on M is defined here in terms of its pieces (components $t^{i\dots j}_{k\dots l}(x)$ on $\mathcal{O} \subset M$) as well as a rule of how to *globalize* them, i.e. how to glue these pieces together consistently so as to obtain a desired whole. In order to make this method work, one has to ensure that the rule for transition from one piece to another satisfies a consistency condition on *triple overlap* of charts (see (2.2.6)): two steps $x \mapsto x' \mapsto x''$ are to lead to the same result as a single one $x \mapsto x''$. This may be regarded actually as a requirement, namely that the rule should have particular *group properties* – coordinate changes on triple overlaps are naturally endowed with the structure of a group (multiplication being realized as a composition of the two transformations involved) and the transformation rules are to have the properties of "action" of the group (in particular, its *representation* in linear spaces, as is the case here; see Section 12.1). Some of these rules may be fairly complicated (e.g. the rule for Christoffel symbols of a linear connection, see (15.2.3)), but the property of group action is necessary for a globally defined object (and sufficient as well).

2.5.6 Check that the rule given in (2.5.4) for transformation of components of a tensor field *meets* the requirement of consistency on triple overlaps of charts.

Hint: consider the behavior of Jacobian matrices for the transitions $x \mapsto x' \mapsto x''$. □

2.5.7 Prove that a tensor field is smooth if and only if its components happen to be smooth (and this does not depend on the choice of coordinates). □

2.6 Metric tensor on a manifold

• On a manifold M, tensor fields of arbitrary type $\binom{p}{q}$ may be introduced. The only *canonical* (existing automatically) tensor field on a general manifold is the *unit* tensor field $\hat{1}$ of type $\binom{1}{1}$ (its other names being the *contraction tensor* or *canonical pairing*; note that the tensor product of several copies of this tensor as well as all possible symmetrizations and antisymmetrizations of such products are canonical, too)

$$\hat{1}(V,\alpha) := \langle \alpha, V\rangle \qquad \text{i.e.} \quad \hat{1}(V,\cdot\,) = V, \quad \hat{1}(\,\cdot\,,\alpha) = \alpha$$

2.6.1 Check that

(i) in coordinates

$$\hat{1} = dx^i \otimes \partial_i \qquad \text{i.e.} \quad \hat{1}^i_{\ j} = \delta^i_j$$

(ii) the expression in (i) does not depend on the choice of coordinates (see (2.4.9)). □

• All other tensor fields on a manifold have to be specially defined and they provide *additional* structure on M. What particular manifold we choose and what tensor fields it is endowed with depend ultimately on the physical context in which the tools of differential geometry are intended to be used (they represent input data, which characterize the problem in geometric language). In the majority of physically interesting applications of geometry (although not in all of them) a metric tensor on a manifold enters the scene, i.e. a field $g \in \mathcal{T}^0_2(M)$ such that for each point P it is a metric tensor in $T_P M$ in the sense of (2.4.11). It is a fairly "strong" structure, indeed, which enables one to perform various operations directly (such as lowering and raising of indices, association of lengths and angles with vectors, etc.), but it also *induces* various additional structures (linear connection, volume form, etc.) as well. A manifold endowed with a metric tensor, i.e. a pair (M, g), is said to be the *Riemannian manifold* and the branch of geometry which treats such manifolds is *Riemannian geometry*. If g is not positive definite (see the text just after (2.4.11)), one sometimes speaks about the *pseudo-Riemannian* manifold and geometry and, in particular, about the *Lorentzian* manifold and geometry for signature $(+,-,\cdots-)$ or $(-,+,\cdots+)$.

2.6.2 Check that in the coordinate basis it holds that

$$\flat_g(V^i\partial_i) = V_i\,dx^i \qquad \sharp_g(\alpha_i\,dx^i) = \alpha^i\partial_i$$

where

$$V_i := g_{ij}V^j \qquad \alpha^i := g^{ij}\alpha_j$$

Hint: see (2.4.13). □

The simplest n-dimensional manifold is given by Cartesian space $\mathbb{R}^n$. Here the *standard (flat) metric tensor* of signature (r,s) $(r+s=n)$ is introduced; by definition, in *Cartesian* coordinates we put

$$g_{ij} = \eta_{ij} \equiv \text{diag}(\underbrace{1,\ldots,1}_{r},\underbrace{-1,\ldots,-1}_{s})$$

i.e.

$$g = \eta_{ij}\,dx^i\otimes dx^j = dx^1\otimes dx^1 + \cdots + dx^r\otimes dx^r - dx^{r+1}\otimes dx^{r+1} - \cdots - dx^n\otimes dx^n$$

This manifold will be denoted by $(\mathbb{R}^n, \eta_{ij}) \equiv E^{r,s}$ from now on (and called the *pseudo-Euclidean space*), and, in particular, in the positive definite case $(\mathbb{R}^n, \delta_{ij}) \equiv E^n$ (the *Euclidean space*).

Let us have a closer look at the motivation for this definition in the most mundane spaces E^2 and E^3. In a common plane E^2 it says, for example, that the length of the two vectors ∂_x and ∂_y is (at each point) 1 and that these vectors are orthogonal to each other. For $|\partial_x|^2 = g(\partial_x, \partial_x) = (dx \otimes dx + dy \otimes dy)(\partial_x, \partial_x) = 1$, the rest similarly. This shows that the definition nicely matches our intuitive conception of metric conditions in the usual plane.

2.6.3 Write down the metric tensor in the common plane E^2 in *polar* coordinates. ($g = \underbrace{dx \otimes dx + dy \otimes dy}_{\text{Cartesian}} = \underbrace{dr \otimes dr + r^2 d\varphi \otimes d\varphi}_{\text{polar}}$.) □

2.6.4 Write down the metric tensor g in the common three-dimensional space E^3 in Cartesian, cylindrical and spherical polar coordinates.
Result:

$$\begin{aligned} g &= dx \otimes dx + dy \otimes dy + dz \otimes dz && \text{Cartesian coordinates} \\ &= dr \otimes dr + r^2 d\varphi \otimes d\varphi + dz \otimes dz && \text{cylindrical coordinates} \\ &= dr \otimes dr + r^2\, d\vartheta \otimes d\vartheta + r^2 \sin^2 \vartheta\, d\varphi \otimes d\varphi && \text{spherical polar coordinates} \end{aligned}$$

□

• This kind of computation can be done either making use of transformational properties of tensor components (i.e. reading components from its expression in Cartesian coordinates, using (2.5.4) or (2.4.18) and "gluing together" a new coordinate basis with new components), or computing new "differentials" (= gradients of coordinates), first, according to (2.5.3), e.g. in (2.6.3) $dx = x_{,r}\, dr + x_{,\varphi}\, d\varphi = \cos\varphi\, dr - r \sin\varphi\, d\varphi$, and then exploiting bilinearity of the tensor product. As a rule, this alternative method is quicker for simple metric tensors. In elementary situations (like that mentioned above) one can see, after a bit of practice, the result directly from the visual conception of what the geometry is about on a particular manifold, see (3.2.11) and (3.2.12).

2.6.5 Check that the non-Cartesian coordinate bases in (2.6.3) and (2.6.4) are ortho*gonal*, but they are not ortho*normal*.

Hint: see the text prior to (2.4.12). □

• If some local coordinates on (M, g) induce at each point the orthogonal coordinate basis of the tangent space, they are said to be *orthogonal coordinates*. We have learned above that, besides Cartesian coordinates, also polar coordinates in E^2 and spherical polar as well as cylindrical coordinates in E^3 (and various others, too; e.g. see (3.2.2)–(3.2.7)) deserve to be titled by this prestigious nomenclature.

A manifold $(\mathbb{R}^4, \eta_{ij}) \equiv E^{1,3}$ with signature $(1, 3)$ is called *Minkowski space* and it plays a featured role in the special theory of relativity (being the *space-time* there; see more in

Chapter 16). Cartesian coordinates are usually labelled in this particular case as (x^0, x^i), $i = 1, 2, 3$, $x^0 = t$ being time and x^i corresponding to Cartesian coordinates in our good old $\mathbb{R}^3$ (the choice of units with $c = 1$ is adopted).

2.6.6 Write down the Minkowski metric η in spherical polar and cylindrical coordinates (i.e. $(t, r, \vartheta, \varphi)$ and (t, r, φ, z) respectively instead of (t, x, y, z)). ($\eta = dt \otimes dt - h$, h from (2.6.4).) □

• An important metric tensor is unobtrusively hidden in the expression for the kinetic energy of a system of particles.

2.6.7 Given $(\mathbf{r}_1(t), \ldots, \mathbf{r}_N(t))$ a trajectory of a system of N point masses in mechanics, we may regard it as a *curve* $\Gamma(t)$ on a manifold $M \equiv \mathbb{R}^3 \times \cdots \times \mathbb{R}^3 = \mathbb{R}^{3N}$. Check that the *kinetic energy* of this system induces the particular metric tensor $h \in \mathcal{T}_2^0(M)$ on $\mathbb{R}^{3N}$ (being different from the standard one, in general) by

$$\text{kinetic energy} \equiv T = \frac{1}{2} h(\dot{\Gamma}, \dot{\Gamma})$$

Hint: if (x_k, y_k, z_k) are Cartesian coordinates of the kth point, then $h = m_1 h_1 + \cdots + m_N h_N$, where $h_k := dx_k \otimes dx_k + dy_k \otimes dy_k + dz_k \otimes dz_k$. □

2.6.8 Write down the kinetic energy of *a single* point mass in Cartesian, cylindrical and spherical polar coordinates.

Hint: see (2.6.7) and (2.6.4); for a single point mass, h is *only a multiple* of the standard metric tensor; one obtains

$$\begin{aligned} T &= \frac{1}{2} m(\dot{x}^2 + \dot{y}^2 + \dot{z}^2) && \text{Cartesian coordinates} \\ &= \frac{1}{2} m(\dot{r}^2 + r^2\dot{\varphi}^2 + \dot{z}^2) && \text{cylindrical coordinates} \\ &= \frac{1}{2} m(\dot{r}^2 + r^2\dot{\vartheta}^2 + r^2 \sin^2\vartheta\, \dot{\varphi}^2) && \text{spherical polar coordinates} \end{aligned}$$

□

• The metric tensor turns out to be the essential element for introducing the concept of the *length of a curve* on (M, g), too. Let us begin in E^3. If a point moves along a trajectory $\mathbf{r}(t)$ in our usual space E^3, it traverses (to first order in ϵ) the distance $ds = |\mathbf{v}|\epsilon = \epsilon\sqrt{\dot{x}^2 + \dot{y}^2 + \dot{z}^2}$ within the time interval between t and $t + \epsilon$ (according to the theorem of Pythagoras; this is the place, of course, where the metric tensor in E^3 is hidden). Note, however, that one can write this as $\epsilon\sqrt{g(\dot{\gamma}, \dot{\gamma})}$ for $\gamma \leftrightarrow (x(t), y(t), z(t))$. The length of a finite segment between $P = \gamma(t_1)$ and $Q = \gamma(t_2)$ is given by $\int dt\, \sqrt{g(\dot{\gamma}, \dot{\gamma})}$. The most interesting feature of this expression consists in the fact that one cannot see from it that $(M, g) = E^3$ and Cartesian coordinates are used. It is then natural to use this very expression for the definition of the length of a curve *in general*. One should understand that even in this general case its meaning remains just the same – *for small pieces*, the relation "distance = speed × time interval" is used, and the result is summed over all small pieces (i.e. integrated).

It is a suitable time now to contemplate the visual meaning of the concept of the length of a vector V itself. The following is meant by this notion: if we proceed a parametric distance ϵ along the vector V, we travel (in the positive definite case) a distance (in the sense of the length of the curve)[27] $\epsilon|V| \equiv \epsilon\sqrt{g(V,V)}$. Keeping this in mind one is often able to derive explicit forms of metric tensors on two-dimensional surfaces in E^3 simply by a "rule of thumb" (see (3.2.11); the same is true for curves in E^3 as well, being fairly useful, for example, in computing line integrals of the first kind, see (7.7.4)).

There is an alternative way of displaying the metric tensor, which is frequently used in general relativity, and may be ultimately traced back to the connection between the length of a curve and a metric tensor. In this convention one writes directly the "square of the distance" dl^2 between two points which are infinitesimally close to one another (i.e. points with values of coordinates being x^i and $x^i + dx^i$ respectively), where dx^i denote infinitesimal *increments* of the values of coordinates (so that *they are not* our base covector fields (!)). For metric tensors from exercise (2.6.4), as an example, we have

$$\begin{aligned} dl^2 &= dx^2 + dy^2 + dz^2 \\ &= dr^2 + r^2\,d\varphi^2 + dz^2 \\ &= dr^2 + r^2\,d\vartheta^2 + r^2\sin^2\vartheta\,d\varphi^2 \end{aligned}$$

Although we will not, as a rule, use this convention in the course of the book, it is fairly common in texts on relativity and one should understand clearly its precise meaning.

2.6.9 Let $t \mapsto t(\sigma)$ be a *reparametrization* of a curve γ, i.e. $\hat\gamma(\sigma) := \gamma(t(\sigma))$. Check that the functional of the *length of a curve* (refer to (4.6.1), (7.7.5) and (15.4.8))

$$\text{length of a curve } \gamma \;\equiv l[\gamma] := \int_{t_1}^{t_2} dt\,\sqrt{g(\dot\gamma,\dot\gamma)}$$

is *reparametrization invariant*, $l[\gamma] = l[\hat\gamma]$, i.e. this expression depends on the image set of a curve (i.e. on the *path*; recall that the curve is *a map*) rather than on a particular parametrization of this set (on a *curve*).

Hint: according to (2.3.5) $\hat\gamma' = (dt/d\sigma)\dot\gamma$, therefore $d\sigma\sqrt{g(\hat\gamma',\hat\gamma')} = dt\sqrt{g(\dot\gamma,\dot\gamma)}$. □

• Finally, we mention the possibility of introducing the *gradient* as a *vector* field. The gradient df as a *covector* field has been defined in (2.5.3). If a metric tensor is available, we can find a *vector* field, simply by raising the index on the covector df. The resulting vector field is called the gradient (of a function f), too, and will be denoted by $\operatorname{grad} f$ or ∇f

$$\operatorname{grad} f \equiv \nabla f := \sharp_g df \equiv g^{-1}(df,\cdot\,) \qquad \text{i.e.} \quad (\nabla f)^i := g^{ij}(df)_j \equiv g^{ij} f_{,j}$$

A well-known example is provided by the *potential force field* in mechanics. It is the gradient of the (by definition negative) *potential energy* of a system. Here, indices are raised by means

[27] Remember that the vector V officially resides as a whole at a single point x and its length is $g_x(V,V)$. This length (in the sense of a scalar product in T_xM) now becomes related with a formally *different* length, namely the length of a small piece of a *curve* $\gamma(t)$ defined by the vector, the representative of a class specified by the vector V. Both computations need g and the definitions are intentionally designed *so as to* make the results coincide.

of the *standard* metric tensor on $M \equiv \mathbb{R}^3 \times \cdots \times \mathbb{R}^3 = \mathbb{R}^{3N}$, that is by $h_1 + \cdots + h_N$ (as opposed to (2.6.7), where masses are present, too).

2.6.10 Find the lines of electric field of a point charge and of an elementary dipole.

Hint: first, write down equations for *integral curves* of the electric field $\mathbf{E} = -\nabla\Phi$, i.e.

$$\dot{x}^i = -g^{ij}\Phi_{,j}$$

for

$$\Phi(r, \vartheta, \varphi) = \frac{\alpha}{r} \qquad \text{resp.} \quad \Phi(r, \vartheta, \varphi) = \alpha\frac{\mathbf{p}\cdot\mathbf{r}}{r^3} \equiv (\alpha p)\frac{\cos\vartheta}{r^2}$$

($\alpha \in \mathbb{R}$) and then disregard parametrization (eliminate dt in the separation of variables procedure; see also (8.5.13)). □

Summary of Chapter 2

For each point x of an n-dimensional manifold M there is the canonically defined n-dimensional linear space T_xM, the tangent space at the point x. Its elements are called vectors at x. There are several mutually equivalent definitions of this concept, useful in different contexts. A vector *field* on a manifold M is a smooth assignment of a vector to each point $x \in M$. The integral curve of a vector field is the curve whose motion at each point is just that dictated by the vector of the field at this point. Standard constructions of multilinear algebra (construction of tensors of type $\binom{p}{q}$ for a given vector space L) lead to the notion of a *tensor field* of type $\binom{p}{q}$ on a manifold. In particular, one has functions (type $\binom{0}{0}$), vector and covector fields (type $\binom{1}{0}$ and $\binom{0}{1}$), fields of bilinear form (type $\binom{0}{2}$, in the symmetric non-degenerate case the metric tensor) and linear operators (type $\binom{1}{1}$).

$\gamma : \mathbb{R} \to M$	A curve γ on a manifold M	Sec. 2.1
$f : M \to \mathbb{R}$	A function f on a manifold M	Sec. 2.1
$e_i := \partial_i\vert_P$	Coordinate basis of T_PM	(2.2.6)
$a^i \mapsto a'^i = J^i_j(P)a^j$	Transformation of components of a vector in P	(2.2.6)
$V(fg) = (Vf)g + f(Vg)$	Leibniz rule for action of vector fields	(2.2.8)
$\dot{x}^i = V^i(x) \ \ (\dot{\gamma} = V)$	Equations for finding integral curves of V	(2.3.1)
$v = \sum_{b=1}^{n} v^b e_b \equiv v^b e_b$	Summation convention	(2.4.2)
$\langle e^a, e_b\rangle = \delta^a_b$	The base e^a is dual with respect to e_a	(2.4.2)
$t^{c\ldots d}_{a\ldots b} := t(e_a, \ldots, e_b; e^c, \ldots, e^d)$	Components of tensor $t \in T^p_q(L)$	(2.4.6)
$v_a := g_{ab}v^b \ , \ \alpha^a := g^{ab}\alpha_b$	Lowering and raising of indices by means of g	(2.4.13)
$\langle df, V\rangle := Vf$	Gradient of a function f as a covector field	(2.5.3)
$T = \frac{1}{2}h(\dot{\Gamma}, \dot{\Gamma})$	Kinetic energy of a system of N point masses	(2.6.7)
$l[\gamma] := \int_{t_1}^{t_2} dt\sqrt{g(\dot{\gamma}, \dot{\gamma})}$	Functional of the length of a curve γ	(2.6.9)
$(\nabla f)^i := g^{ij} f_{,j} \ \ (\nabla f := \sharp_g\, df)$	Gradient of a function f as a vector field	Sec. 2.6

3

Mappings of tensors induced by mappings of manifolds

• Nearly all situations in geometry, as we will see in this text over and over again, are closely related to maps of manifolds $f : M \to N$; in particular, often $M = N$. It turns out that each mapping of *points* of manifolds automatically leads to a mapping of *tensors* at these points and (provided some restrictions are satisfied) also of tensor *fields* on M or N. Some of them move in the same direction as the points under the action of f, that is from M to N, while others *reverse the arrow* and move against the direction of f. A clear understanding of this transport of tensors serves as a ticket into a number of following chapters.[28]

3.1 Mappings of tensors and tensor fields

• We begin with the simplest case, a function. Let us assume that we have functions on both M and N denoted by χ and ψ respectively, so that altogether three maps are involved. This situation may be visualized as

$$\mathbb{R} \overset{\chi}{\leftarrow} M \overset{f}{\to} N \overset{\psi}{\to} \mathbb{R}$$

We want to find out whether χ induces some function on N or, alternatively, ψ induces some function on M. Put another way, whether there is some combination of the three maps under consideration which is a map $M \to \mathbb{R}$ *other* than χ, or a map $N \to \mathbb{R}$ *other* than ψ. A short inspection shows that the answer is *yes* to the first question and *no* to the second question. The composition map

$$M \overset{f}{\to} N \overset{\psi}{\to} \mathbb{R}$$

is an effective arrow from M to $\mathbb{R}$, i.e. a new function on M

$$\psi \circ f : M \to \mathbb{R}$$

but one cannot compose the maps χ and f since the *reverse* of the arrow f is needed for that (f^{-1} should exist), but we *do not assume* this: f is a general smooth map for which

[28] Fortunately, the price/value relation of this ticket is very favorable and since the penalty for fare dodgers is high, there is no point in trying to travel without paying.

the inverse map *may not* exist. We thus conclude that one can naturally transport functions (tensor fields of type $\binom{0}{0}$) *against* the direction of the arrow; functions are "pulled back."

3.1.1 Let $f : M \to N$ be a smooth map and $\psi : N \to \mathbb{R}$ a function on N. Show that

(i) the prescription

$$f^*\psi := \psi \circ f$$

induces a function on M (*pull-back* of the function ψ)

(ii) in local coordinates

$$f : x^i \mapsto y^a(x) \quad \Rightarrow \quad (f^*\psi)(x) = \psi(y(x))$$

(iii) for the composition of maps of manifolds $M \xrightarrow{f} N \xrightarrow{g} S$ we get the simple rule

$$(g \circ f)^* = f^* \circ g^*$$

(iv) if $\mathcal{F}(M)$, $\mathcal{F}(N)$ are the algebras of functions on M, N, then

$$f^* : \mathcal{F}(N) \to \mathcal{F}(M)$$

is a morphism of algebras, i.e.

$$f^*(\psi_1 + \lambda\psi_2) = f^*\psi_1 + \lambda f^*\psi_2 \qquad f^*(\psi_1\psi_2) = (f^*\psi_1)(f^*\psi_2)$$ □

• Let us proceed to vectors and vector fields. We will show that vectors are transported naturally in the direction of the arrow f. There are two equivalent ways to describe this and the choice depends on how one represents the vector itself, either via curves or via the algebra of functions (Definitions 1 and 2 in Section 2.2). The idea of these two ways (to be developed in detail in the exercise) is that a transport of *points* results automatically in a transport of *curves* (in the direction of the arrow) and consequently of vectors; and a transport of functions backwards enables one to introduce the action of the transported vectors on functions *on* N as the action of the original vectors on the function *pulled back* to M.

3.1.2 Let $f : M \to N$ be a smooth map of manifolds, x^i local coordinates on M, y^a local coordinates on N, $J^a_i \equiv \partial y^a(x)/\partial x^i$. We define a map (usually called a *differential of the map* f, *push-forward* or *tangent map*)

$$T_x f \equiv f_* : T_x M \to T_{f(x)} N$$

by the relation

$$f_*[\gamma] := [f \circ \gamma]$$

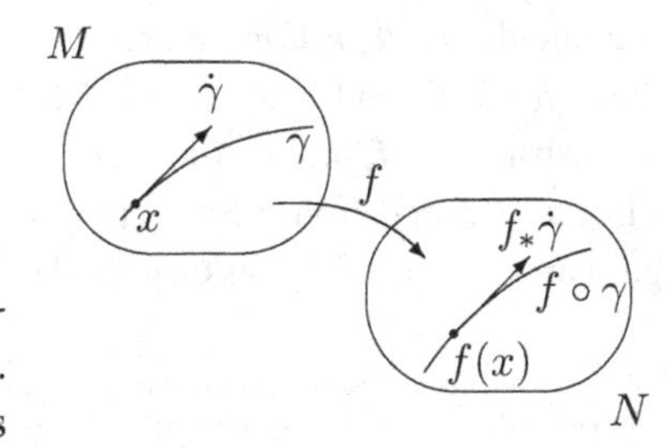

In plain English, the resulting vector is simply identified with the tangent vector to the mapped curve. This formula may also be expressed equivalently as follows:

$$f_* \left(\frac{d}{dt}\bigg|_{t=0} \gamma(t) \right) := \frac{d}{dt}\bigg|_{t=0} (f \circ \gamma)(t) \qquad \text{or briefly} \quad f_*\dot{\gamma} = (f \circ \gamma)^{\dot{}}$$

Show that

(i) the map is well defined (it does not depend on the choice of the representative γ of the class $[\gamma]$ of curves tangent to each other)
(ii) it is linear
(iii) on the coordinate basis it gives

$$f_*\partial_i = J_i^a \partial_a \quad \text{so that} \quad f_*(V^i\partial_i) = \left(J_i^a V^i\right)\partial_a$$

(iv) the rank of the Jacobian matrix determines whether the mapped basis vectors are linearly independent or not; namely for $\dim M \leq \dim N$ there holds

$$J_i^a(x) \text{ has maximum rank} \quad \Leftrightarrow \quad f_* \text{ is injective} \quad \Leftrightarrow \quad \text{the vectors } f_*\partial_i \text{ are linearly independent}$$

(v) for compositions of maps of manifolds $M \xrightarrow{f} N \xrightarrow{g} S$ we get the simple rule

$$(g \circ f)_* = g_* \circ f_*$$

(vi) if $\psi : N \to \mathbb{R}$ is a function then

$$(f_*\dot{\gamma})\psi = \dot{\gamma}(f^*\psi)$$

This means that an independent (and equivalent) way *to define* f_* is given by the formula[29]

$$(f_*V)\psi := V(f^*\psi)$$

Hint: (ii) using (vi); (iv) see (2.4.17). □

• Let us now see whether this transport can be extended to vector *fields*, too. We find immediately that we encounter problems if it is not possible to invert f (if f is not a diffeomorphism). If f is not surjective, there are no vectors outside the image $\operatorname{Im} f$ of the manifold M and so there is no transported field on the *entire* manifold N. A much more serious problem arises, however, if f is not injective. If x_1, x_2 are any two preimages of the point $y = f(x_1) = f(x_2) \in N$, then there are *two* vectors sitting in y, transported from x_1 and x_2, and there is no reason for them to coincide in general. In the case of non-injective maps, e.g. projections, only the transport of very special vector fields makes sense, namely of *projectable fields*, for which the results of the transport of vectors from all preimages do coincide *by definition*. When a vector field W on N is an f_*-image of a field V on M ($W = f_*V$), W and V are said to be *f-related*. For a general f it is, however, *not possible* to construct the f_*-image of a general vector field V.

Let us proceed from vectors to covectors. They move *against* the direction of the arrow again and the idea follows exactly the lines of the idea of transport of vectors with the help

[29] Whether you put the shoe on your foot or you put your foot in the shoe, the result is the same – your foot is in the shoe. It is thus possible to define the procedure both ways.

of functions: if we know how to transport vectors *forward*, we are able to transport covectors *backward*, too. A closer look reveals that the same trick (the shoes, see (3.1.2)) enables one to transport even *arbitrary* strictly *covariant* tensors (:= with *lower* indices only, i.e. of the type $\binom{0}{p}$).

3.1.3 Let $f : M \to N$ be a smooth map of manifolds, x^i local coordinates on M and y^a local coordinates on N. Define a map

$$f^* : T^*_{f(x)}N \to T^*_x M$$

by the relation

$$\langle f^*\alpha, V\rangle := \langle \alpha, f_* V\rangle$$

(f_* is from (3.1.2)). Show that

(i) it is linear
(ii) on the coordinate basis

$$f^* dy^a = J^a_i \, dx^i \quad \text{so that} \quad f^*(\alpha_a \, dy^a) = \left(\alpha_a \, J^a_i\right) dx^i$$

(iii) for a composition of maps of manifolds $M \xrightarrow{f} N \xrightarrow{g} S$ we get the simple rule

$$(g \circ f)^* = f^* \circ g^*$$

□

3.1.4 Let $f : M \to N$ be a smooth map of manifolds, x^i local coordinates on M and y^a local coordinates on N. Define a map (*pull-back* of the covariant tensor field)

$$f^* : \mathcal{T}^0_p(N) \to \mathcal{T}^0_p(M)$$

by the relation

$$(f^* t)(U, \ldots, V) := t(f_* U, \ldots, f_* V)$$

Show that

(i) on the linear combination and tensor product there holds

$$f^*(t + \lambda s) = f^* t + \lambda f^* s \qquad t, s \in \mathcal{T}^0_p(N), \quad \lambda \in \mathbb{R}$$
$$f^*(t \otimes \tau) = (f^* t) \otimes (f^* \tau)$$

(ii) on the coordinate basis

$$f^*(dy^a \otimes \cdots \otimes dy^b) = J^a_i \ldots J^b_j \, dx^i \otimes \cdots \otimes dx^j = (f^* dy^a) \otimes \cdots \otimes (f^* dy^b)$$

and consequently

$$t = t_{a\ldots b}(y) \, dy^a \otimes \cdots \otimes dy^b \quad \mapsto \quad f^* t = t_{a\ldots b}(y(x)) J^a_i(x) \ldots J^b_j(x) \, dx^i \otimes \cdots \otimes dx^j$$

(iii) for a composition of maps of manifolds $M \xrightarrow{f} N \xrightarrow{g} S$ we get

$$(g \circ f)^* = f^* \circ g^*$$

(iv) exercise (3.1.3) is a special case for $p = 1$
(v) exercise (3.1.1) is a special case for $p = 0$. □

• Note that, as a matter of fact, in (3.1.4) we have introduced directly the (pointwise) transport of *fields*, and no surjectivity or injectivity of f was needed for that (see the text after (3.1.2)).

3.1.5 Be sure that you understand that in the case of pull-back of strictly covariant tensor fields *no problems* occur for any smooth map f (i.e. f need be neither surjective nor injective in order to be well defined). □

• If one needs to transport tensor *fields* with *upper* indices, too, f *has to be* the diffeomorphism (we encountered this fact already when dealing with vector fields). Then the strategy can be based on the fact (established empirically by centuries of diverse human activities) that to move forward via f is the same thing as to move backward via f^{-1} (and vice versa).

3.1.6 Let $f : M \to N$ be a *diffeomorphism*. Define a map (*pull-back* of a general tensor field)

$$f^* : \mathcal{T}^r_s(N) \to \mathcal{T}^r_s(M)$$

by the relation

$$(f^*t)(U, \dots, V; \alpha, \dots, \beta) := t(f_*U, \dots, f_*V; (f^{-1})^*\alpha, \dots, (f^{-1})^*\beta)$$

(the above-mentioned trick was applied to covector arguments). Show that

(i) on linear combination and tensor product there holds

$$f^*(t + \lambda s) = f^*t + \lambda f^*s \qquad t, s \in \mathcal{T}^r_s(N), \quad \lambda \in \mathbb{R}$$
$$f^*(t \otimes \tau) \;= (f^*t) \otimes (f^*\tau)$$

(ii) on the coordinate basis

$$f^*(dy^a \otimes \cdots \otimes \partial_b) = J^a_i \dots (J^{-1})^j_b \, dx^i \otimes \cdots \otimes \partial_j$$

and consequently

$$t = t^{a\dots}_{\dots b}(y)\, dy^b \otimes \cdots \otimes \partial_a \quad \mapsto \quad f^*t = t^{a\dots}_{\dots b}(y(x)) J^b_i(x) \dots (J^{-1})^j_a(x)\, dx^i \otimes \cdots \otimes \partial_j$$

(iii) for a composition of maps of manifolds $M \xrightarrow{f} N \xrightarrow{g} S$ we get

$$(g \circ f)^* = f^* \circ g^*$$

(iv) if the *push-forward* of a general tensor field is *defined* by

$$f_* := (f^{-1})^* : \mathcal{T}^r_s(M) \to \mathcal{T}^r_s(N)$$

then this f_* coincides on $\mathfrak{X}(M)$ with the pointwise extension of the construction from (3.1.2)

(v)
$$(g \circ f)_* = g_* \circ f_*$$

(vi) the definition mentioned at the beginning can also be written as

$$(f^*t)(V, \ldots, ; \alpha, \ldots) = t(f_* V, \ldots, f_* \alpha, \ldots)$$

(vii)

$$f^* : \mathcal{T}(N) \to \mathcal{T}(M)$$

is an (iso)*morphism* of tensor algebras (it reduces to (3.1.1) on the component $\mathcal{T}_0^0(N) \equiv \mathcal{F}(N)$), i.e. it is a bijective map which respects the degree and commutes with linear combination and (tensor) product. □

3.1.7 Let $f : M \to N$ be a diffeomorphism. Show that the pull-back of a tensor field commutes with (any) contraction, i.e.

$$f^* \circ C = C \circ f^*$$

Hint: for $t \in \mathcal{T}_s^r(N)$

$$(C \circ f^*t)(V, \ldots; \alpha, \ldots) = t(f_* V, \ldots, e_i, \ldots; f_* \alpha, \ldots, e^i, \ldots)$$
$$(f^* \circ Ct)(V, \ldots; \alpha, \ldots) = t(f_* V, \ldots, f_* e_i, \ldots; f_* \alpha, \ldots, f_* e^i, \ldots)$$

For a diffeomorphism $f_* e_i$ is a basis, too ($f_* e^i$ being dual to $f_* e_i$); see (2.4.8). □

3.1.8 Prove that for a diffeomorphism f

$$f^*(t(V, \ldots; \alpha, \ldots)) = (f^*t)(f^*V, \ldots; f^*\alpha, \ldots)$$

and, in particular,

$$f^*\hat{1} = \hat{1}$$
$$f^*\langle \alpha, V \rangle = \langle f^*\alpha, f^*V \rangle$$

Hint: see (2.4.10) and (3.1.7). □

3.1.9 Let

$$d : \mathcal{T}_0^0(M) \to \mathcal{T}_1^0(M) \qquad \psi \mapsto d\psi$$

be the operation of the *gradient* of a function (2.5.3). Verify that d commutes with arbitrary pull-back, i.e. that if $f : M \to N$ then the diagram commutes

$$\begin{array}{ccc} \mathcal{T}_0^0(N) & \xrightarrow{f^*} & \mathcal{T}_0^0(M) \\ d \downarrow & & \downarrow d \\ \mathcal{T}_1^0(N) & \xrightarrow[f^*]{} & \mathcal{T}_1^0(M) \end{array} \qquad \text{i.e.} \quad df^* = f^*d$$

Hint: the question is whether for every function ψ and vector field V there holds $\langle df^*\psi, V \rangle = \langle f^*d\psi, V \rangle$ or equivalently $V(f^*\psi) = (f_*V)\psi$; this is, however, exactly what the ("independent") definition of f_* in (3.1.2) says. □

3.1.10 Let $\psi : M \to \mathbb{R}$ be a function on M. Show that its *gradient* $(d\psi)_x$ at the point $x \in M$ can be *canonically* identified with the *differential* (= push-forward) ψ_{*x} of a map ψ at the point x (that is why the notions of the gradient and the differential of a function are often freely interchanged).

Hint: the gradient gives a *number* $(\partial_i \psi)(x)$ on a vector ∂_i and the differential a *vector* $(\partial_i \psi)(x)\partial_\psi$ (the function is treated as a map $M[x^i] \to \mathbb{R}[\psi], x^i \mapsto \psi(x^i)$). The vector field ∂_ψ is, however, canonical on $\mathbb{R}[\psi]$. □

3.1.11 Show that f_* can be used to characterize $f : M \to N$ ($\dim M \leq \dim N$) as an *immersion* or *embedding* in the following way:

$$f : M \to N \text{ is an immersion} \quad \Leftrightarrow \quad f_* \text{ is (for each } x \in M\text{) injective}$$
$$f : M \to N \text{ is an embedding} \quad \Leftrightarrow \quad f_* \text{ as well as } f \text{ are injective}$$

Hint: see Section 1.4 and (3.1.2). □

3.2 Induced metric tensor

• The construction of an *induced metric tensor* provides an important example of maps of covariant tensor fields. It enables one to endow an "empty" manifold M with a metric tensor, using an embedding of M into a manifold N where a metric tensor already exists.

3.2.1 Let

$$f : M \to (N, h)$$

be an embedding of M into a (pseudo-)Riemannian manifold (N, h). Show that

(i)
$$g := f^* h$$
is a (pseudo-)metric tensor on M, i.e. (M, g) is a (pseudo-)Riemannian manifold

(ii) in coordinates

$$g_{ij}(x) = J_i^a(x) h_{ab}(y(x)) J_j^b(x) \equiv y^a{}_{,i}(x) h_{ab}(y(x)) y^b{}_{,j}(x)$$

Hint: see (3.1.4) and consider non-degeneracy (see 2.4.18). □

• Let us have a closer look at how the induced metric tensor actually works. By the definition in (3.1.4), the scalar product of two vectors V, W in the sense of g on M is

$$g(V, W) \equiv (f^* h)(V, W) := h(f_* V, f_* W)$$

We can see from this formula that if we use the induced metric tensor the result is the same as if we first transported the vectors V, W onto N and then performed the computation of the scalar product *in the sense of* h there.[30]

[30] The following analogy with computer networks could be helpful: M and N are computers "here" and "there", h is a useful piece of software there (we are sitting here). We have to make a choice: either to run the program there (which might be inconvenient, if the work is to be done at the time when the network is overloaded), or first to download the software onto our disk (f^* serves as, say, *ftp*), obtaining $(M, f^* h)$ ($\leftrightarrow$ our computer endowed with useful downloaded software), and then run the program (for performing scalar products and computing expressions containing them) conveniently at any time *here*.

3.2.2 Induce a metric tensor on a torus T^2 from its embedding (1.5.7) into E^3

$$x = (a + b\sin\psi)\cos\varphi \qquad y = (a + b\sin\psi)\sin\varphi \qquad z = b\cos\psi$$

$(g = (a + b\sin\psi)^2\, d\varphi \otimes d\varphi + b^2\, d\psi \otimes d\psi)$. □

3.2.3 Induce a metric tensor on a torus T^2 from its embedding (1.5.8) into E^4 (*flat torus*)

$$x^1 = \cos\alpha \qquad x^2 = \sin\alpha \qquad x^3 = \cos\beta \qquad x^4 = \sin\beta$$

$(g = d\alpha \otimes d\alpha + d\beta \otimes d\beta)$. □

3.2.4 Induce a metric tensor on a sphere S^2 from its embedding into E^3

$$x = R\sin\vartheta\cos\varphi \qquad y = R\sin\vartheta\sin\varphi \qquad z = R\cos\vartheta$$

$(g = R^2(d\vartheta \otimes d\vartheta + \sin^2\vartheta\, d\varphi \otimes d\varphi))$. □

3.2.5 Induce a metric tensor on a sphere S^3 from its embedding into E^4

$$x = R\sin\vartheta\cos\varphi \qquad y = R\sin\vartheta\sin\varphi \qquad z = R\cos\vartheta\cos\psi \qquad w = R\cos\vartheta\sin\psi$$

Show that the coordinates $(\vartheta, \varphi, \psi)$ (they are called *biharmonic coordinates*) are orthogonal $(g = R^2(d\vartheta \otimes d\vartheta + \sin^2\vartheta\, d\varphi \otimes d\varphi + \cos^2\vartheta\, d\psi \otimes d\psi))$. □

3.2.6 Let r, z, φ be cylindrical coordinates in E^3 and consider a rotational surface $\mathcal{S}$ given by both expressions $r(z)$ and $z(r)$. Induce a metric tensor (in coordinates z, φ and r, φ respectively) on $\mathcal{S}$. Specify for the surface of a cylinder and a cone as well as for both kinds of rotational hyperboloids and rotational paraboloids $(g = (1 + (r'(z))^2)\, dz \otimes dz + r^2(z)\, d\varphi \otimes d\varphi = (1 + (z'(r))^2)\, dr \otimes dr + r^2\, d\varphi \otimes d\varphi)$. □

3.2.7 Let $E^{1,2}$ be $1 + 2$ Minkowski space (signature $+ - -$).

(i) Induce a metric tensor g on the *pseudosphere* (2-sheeted hyperboloid endowed with the metric from $E^{1,2}$), i.e. the set of points satisfying

$$\eta_{\mu\nu}x^\mu x^\nu \equiv (x^0)^2 - (x^1)^2 - (x^2)^2 = R^2.$$

(ii) Verify that the pseudosphere is a *space-like hypersurface*, i.e. that its metric is *negative* definite.

Hint: use coordinates α, φ such that

$$x^0 = \pm R\cosh\alpha \qquad x^1 = R\sinh\alpha\cos\varphi \qquad x^2 = R\sinh\alpha\sin\varphi$$

$(g = -R^2(d\alpha \otimes d\alpha + \sinh^2\alpha\, d\varphi \otimes d\varphi))$. □

3.2.8 Find a coordinate expression of the *loxodrome*, i.e. a path on the ocean surface (= a part of (S^2, g)) traced by a ship keeping the course (*azimuth* α = angle with a local north-oriented meridian) *fixed*. Compute the length of the loxodrome with given $\vartheta_{\text{initial}}$ and ϑ_{final}.

Hint: the curve to be found has $\dot\gamma = \dot\vartheta\partial_\vartheta + \dot\varphi\partial_\varphi$, a north-oriented meridian has $\dot\sigma = -\partial_\vartheta$; write down the condition

$$\cos\alpha = \frac{g(\dot\gamma,\dot\sigma)}{\sqrt{g(\dot\gamma,\dot\gamma)}\sqrt{g(\dot\sigma,\dot\sigma)}} \qquad g \text{ is from (3.2.4)}$$

and solve the resulting differential equation $(\tan(\vartheta/2) = \tan(\vartheta_{\text{initial}}/2)\exp\{\pm(\varphi - \varphi_{\text{initial}})/\tan\alpha\}$, where $\pm = -\cos\alpha/|\cos\alpha|$; $dl = -R\,d\vartheta/\cos\alpha \;\Rightarrow\; l = -R(\vartheta_{\text{final}} - \vartheta_{\text{initial}})/\cos\alpha\ (\geq 0))$. □

• The induced metric tensor also occurs in theoretical mechanics, namely in the definition of the kinetic energy of constrained systems. In (2.6.7) we have seen that the kinetic energy of a system of N point masses can be written in the form

$$T = \frac{1}{2}h(\dot\Gamma,\dot\Gamma)$$

If one imposes smooth *constraints* on the possible positions of the point masses (see Section 1.5), the motion becomes restricted to a *configuration space* $M \subset \mathbb{R}^{3N}$. This space may be thought of as an *abstract* manifold M and we can forget about its origin from $\mathbb{R}^{3N}$ if we remember, however (just before we start the forgetting procedure), the most important chunk of information concerning this "big" ambient space, which is precisely the metric tensor needed for the expression of the kinetic energy.[31] Put another way, we need to map (via pull-back) a metric tensor h onto a manifold M. In technical language, an embedding of M into $\mathbb{R}^{3N}$ is given by a parametrization of position "vectors" of all particles in terms of *generalized coordinates*

$$f : M \to \mathbb{R}^{3N} \qquad (q^1,\dots,q^n) \overset{f}{\mapsto} (\mathbf{r}_1(q^1,\dots,q^n),\dots,\mathbf{r}_N(q^1,\dots,q^n))$$

(being *arbitrary* local coordinates on M). If we represent the motion in configuration space M as a curve $\gamma(t) \leftrightarrow q^a(t), a = 1,\dots,n$ on M, its image with respect to the embedding f into $\mathbb{R}^{3N}$ is $\Gamma(t) = (f\circ\gamma)(t) \;\leftrightarrow \mathbf{r}_k(q(t))$, and kinetic energy may be written in *two equivalent* ways

$$T = \frac{1}{2}h(\dot\Gamma,\dot\Gamma) = \frac{1}{2}g(\dot\gamma,\dot\gamma) \qquad g := f^*h$$

This is *the same* kinetic energy (the same number of joules on both sides of the equality sign), but the second expression is written entirely in "intrinsic" terms of the configuration space M, namely in terms of the curve γ and the metric tensor $g \equiv f^*h$ on it.

3.2.9 Verify that

(i) the standard expression of analytical mechanics for kinetic energy on a configuration space

$$T = \frac{1}{2}T_{ab}(q)\dot q^a\dot q^b \qquad T_{ab}(q) = \sum_{k=1}^{N} m_k \frac{\partial\mathbf{r}_k(q)}{\partial q^a}\cdot\frac{\partial\mathbf{r}_k(q)}{\partial q^b}$$

is nothing but the pull-back of (2.6.7) onto M

[31] The potential energy should be remembered, too; see (3.2.9).

(ii) the standard expression for *potential* energy in terms of generalized coordinates

$$U(q^1,\dots,q^n) := \mathcal{U}(\mathbf{r}_1(q^1,\dots,q^n),\dots,\mathbf{r}_N(q^1,\dots,q^n))$$

is the *pull-back* of a function U from $\mathbb{R}^{3N}$ on M, too

(iii) one can summarize the situation by saying that

$$\text{on } \mathbb{R}^{3N}: \quad \mathcal{L}(\mathbf{r}_k,\dot{\mathbf{r}}_k) = \frac{1}{2}h(\dot{\Gamma},\dot{\Gamma}) - \mathcal{U}$$

$$\text{on } M: \quad L(q^a,\dot{q}^a) = \frac{1}{2}g(\dot{\gamma},\dot{\gamma}) - U$$

where

$$g = f^*h \qquad U = f^*\mathcal{U} \qquad \Gamma = f\circ\gamma$$

□

3.2.10 Write down the metric tensor on a torus, entering the kinetic energy of the double (plane) mathematical pendulum (in coordinates where φ^1,φ^2 are the angles of displacement of two material points comprising the pendulum with respect to the vertical direction) ($g = \frac{1}{2}(m_1+m_2)l_1^2\, d\varphi^1\otimes d\varphi^1 + \frac{1}{2}m_2l_2^2\, d\varphi^2\otimes d\varphi^2 + m_2l_1l_2\cos(\varphi^1 - \varphi^2)(d\varphi^1\otimes d\varphi^2 + d\varphi^2\otimes d\varphi^1)$). □

3.2.11 Derive by a "rule of thumb" (use your intuitive understanding of geometry in E^3) metric tensors on the two-dimensional surfaces treated in exercises (3.2.2), (3.2.4) and (3.2.6). The solution for (3.2.2): according to the definition of coordinates φ, ψ we proceed along the lines given in the text before (2.6.9) as follows: a step of (parametric) length ϵ in a coordinate ψ (i.e. in the direction of ∂_ψ) induces in E^3 a step of (true) length $b\epsilon$ (the arc of a circle of radius b) $\Rightarrow |\partial_\psi| = b \Rightarrow g_{\psi\psi} = b^2$. The same in coordinate φ leads to a step of length $(a + b\sin\psi)\epsilon$ (the arc of a circle again) $\Rightarrow |\partial_\varphi| = (a+b\sin\psi) \Rightarrow g_{\varphi\varphi} = (a+b\sin\psi)^2$. These two steps are always mutually orthogonal $\Rightarrow \partial_\psi\cdot\partial_\varphi \equiv g_{\psi\varphi} = 0$.

□

3.2.12 Derive by a "rule of thumb" (use your intuitive understanding of the geometry of E^2 and E^3) expressions for the standard metric tensors in E^2 and E^3 in polar, spherical polar and cylindrical coordinates.

Hint: see (3.2.11), (2.6.3) and (2.6.4). □

Summary of Chapter 3

Each (smooth) mapping of the points of manifolds $f: M\to N$ induces a mapping of tensors living on them. It is denoted by f_* if it pushes tensors forward (in the same direction as f, from M to N) and f^* if it pulls tensors back (in the opposite direction, from N to M). For diffeomorphisms it is possible to define both f_* and f^* for tensor fields of arbitrary type; if f is not the diffeomorphism, several kinds of problems may occur. There always exists a pull-back map f^* for tensor fields of type $\binom{0}{p}$. In particular, one can induce (via pull-back) a metric tensor on M from a Riemannian manifold (N, h), giving rise to a Riemannian manifold (M, g), $g = f^*h$. The most common instance of this procedure is that one induces a metric tensor onto a submanifold M of the Euclidean space $N = E^n$ (or more generally $E^{r,s}$), starting from the canonical metric tensor $h = \eta$ on N.

$f^*\psi := \psi \circ f$	Pull-back of a function ψ	(3.1.1)
$f_*[\gamma] := [f \circ \gamma]$	Push-forward of a vector $[\gamma]$	(3.1.2)
$(f_*V)\psi := V(f^*\psi)$	Push-forward of a vector V	(3.1.2)
$(f^*t)(U, \alpha) := t(f_*U, (f^{-1})^*\alpha)$	Pull-back of a tensor field	(3.1.6)
$(g \circ f)^* = f^* \circ g^*$	Pull-back for the composition of maps	(3.1.6)
$(g \circ f)_* = g_* \circ f_*$	Push-forward for the composition of maps	(3.1.6)
$f^* \circ C = C \circ f^*$	Pull-back commutes with contractions	(3.1.7)
$df^* = f^*d$	Pull-back commutes with gradient	(3.1.9)
$g := f^*h$	Induced metric tensor ($f\colon M \to (N, h)$)	(3.2.1)
$g_{ij} = J^a_i h_{ab} J^b_j \equiv y^a{}_{,i} h_{ab} y^b{}_{,j}$	Induced metric tensor (components)	(3.2.1)
$T = \frac{1}{2}g(\dot{\gamma}, \dot{\gamma})$	Kinetic energy on a configuration space	(3.2.9)

4
Lie derivative

• Various equations in physics contain partial derivatives of components of tensors. A possible combination of such derivatives corresponds to an important geometrical object known as the *Lie derivative* of a tensor field.

If we speak about a derivative of a tensor field, we should compare (subtract) its values at infinitesimally close points. However, two tensors at different points (no matter how close they are to one another) represent elements of *completely different* linear spaces and therefore it is *not possible* to perform their subtraction (linear combination) *straight from the definition* (if no tricks are used). A general way to validate the required comparison should consist in some kind of *transport* of the tensor from one point to another. Making use of the concept of transport, comparison may be defined as follows: given two tensors sitting at two nearby points, one of them is to be transported to the point where the other resides. In this way two tensors are now available at *the same point*. If the two tensors happen to coincide, we may infer that their values at the original points "are equal" (in the sense of the particular rule of transport) and, consequently, that the derivative of the tensor (field) in the direction given by the two points *vanishes*. If the two tensors do not coincide, we get a non-zero derivative.

In this chapter we thrash out the question of how to carry out this simple idea in the case where *Lie* transport is used in the above-mentioned scheme. A highly important and useful way of differentiating tensor fields emerges from these considerations, namely the *Lie derivative*. Later (in Chapter 15), we will return to this idea once again. Another way of transporting tensors will be introduced there, so-called *parallel* transport. Consequently, another kind of derivative will enter the scene, which is known as the *covariant* derivative.

4.1 Local flow of a vector field

• At the end of Section 2.3 we encountered an important concept associated with a vector field, namely that of a local flow. Recall briefly the main idea of this notion.

A vector field "tears up" a manifold into a system of integral curves. If *each* point $P \in M$ moves a *parametric* distance t along "its own" integral curve, we get a map

$$\Phi_t : M \to M \qquad P \equiv \gamma(t_0) \mapsto \gamma(t_0 + t)$$

which is called the *local flow* generated by the field V.

(Here, the term *local* indicates a slightly tricky point in these ideas; namely it means that Φ_t need not be defined for arbitrarily large t, but rather in general only in some neighborhood of $t = 0$ and this neighborhood may, in turn, depend on $P \in M$; the reader should contemplate the flow given by the first field in (2.3.4) in order to feel the issue clearly. If a flow Φ_t exists for $t \in (-\infty, \infty)$, so that the field V is complete, one speaks about a global flow, or simply a flow. A local flow is enough for the definition of the central concept of this chapter, the Lie derivative. Therefore, in what follows we will often omit the word "local" and speak about a "flow" in spite of being only local.)

4.1.1 Check that this map does not depend on the value of the parameter t_0 which we assign to the point P, so that the definition is correct in this sense.

Hint: see (2.3.5). □

• The essential feature of a flow Φ_t is its "composition" property with respect to the parameter t.

4.1.2 Show that the flow Φ_t of a vector field V satisfies

$$\Phi_{t+s} = \Phi_t \circ \Phi_s$$

Hint: let γ be the integral curve of the field V from $P \equiv \gamma(t_0) \in M$, with Γ being the integral curve of V starting from $Q \equiv \gamma(t_0 + t)$. By means of (2.3.6) show that $\Phi_s(Q) = \Phi_{t+s}(P)$. □

• Sometimes a flow is expressed in terms of a map

$$\Phi : M \times \mathbb{R} \to M \qquad (x, t) \mapsto \Phi_t(x)$$

4.1.3 Rewrite the "composition" property (4.1.2) using the language of the map Φ. □

• There is a one-to-one correspondence between vector fields and their flows: with each vector field we may associate a flow Φ_t (in the way described above) and vice versa, any flow Φ_t uniquely determines a field V, by which it is in turn generated ($\Phi_t(P)$ is the integral curve of the field V and by means of the derivative V is itself then recovered).

4.1.4 Check that if

$$\Phi_t : x^i \mapsto x^i(t; x)$$

is the coordinate expression of a flow, then

$$V = \dot{x}^i(0; x)\partial_i$$

Test the method on the result of exercise (2.3.9). □

4.1.5 Check that the prescription

$$\mathbf{r} \mapsto e^{\lambda t}\mathbf{r} \equiv \Phi_t(\mathbf{r}) \qquad \lambda \in \mathbb{R}$$

describes a (global) flow on $\mathbb{R}^3$; find the vector field V which generates this flow and draw some of its integral curves.

Hint: see (4.1.4) ($V = \lambda(x\partial_x + y\partial_y + z\partial_z) \equiv \lambda\mathbf{r} \cdot \nabla$). □

4.1.6 Show that *fixed points* of the flow Φ_t (i.e. those points on a manifold which do not move under the maps Φ_t for *all* values of t) coincide with *zero points* of the generating vector field V (i.e. the points $P \in M$ such that $V_P = 0$ holds). Check this interrelation on concrete flows and generators that we have encountered (or will encounter in the near future). □

4.1.7 Check that the prescription

$$\Phi_t : (x, y, z) \mapsto (x(t), y(t), z(t)) := (x \cos t - y \sin t, x \sin t + y \cos t, z)$$

gives a (global) flow on $\mathbb{R}^3$, find the vector field V which generates this flow and draw some of its integral curves. Write down both Φ_t and V in cylindrical as well as in spherical polar coordinates ($V = -y\partial_x + x\partial_y = \partial_\varphi = \partial_\varphi$ in Cartesian, cylindrical and spherical polar coordinates). □

• In the following account it is essential to realize that Φ_t is a *diffeomorphism* $M \to M$ (at least for some neighborhood of $t = 0$; often this turns out to be the case, however, for a fairly large interval, or even for the whole $\mathbb{R}[t]$). The inverse map for Φ_t is (according to (4.1.2)) clearly given by Φ_{-t} and the statement about the smoothness of Φ_t is a (non-trivial) theorem from the theory of ordinary differential equations of the type studied here (a smooth dependence of solutions on the initial conditions).

The map Φ_t is also known as a *one-parameter group of transformations* of a manifold M; the following exercise elucidates the reason for this terminology.

4.1.8 Check that

$$\hat{\phi} : (\mathbb{R}, +) \to \mathrm{Diff}\,(M) \qquad t \mapsto \Phi_t$$

is a homomorphism of groups (or, more precisely, only of *local* groups, i.e. a homomorphism of some neighborhood of the unit element on the left (small enough t) into some neighborhood of the unit element on the right (those diffeomorphisms which are close to the identity on M)). □

• If a diffeomorphism $f : M \to N$ is available, a flow may be easily shifted from M to N.

4.1.9 Let $f : M \to N$ be a diffeomorphism and Φ_t a flow on M. Show that

(i) $\psi_t := f \circ \Phi_t \circ f^{-1}$ is a flow (on N, however), too
(ii) if the flow Φ_t is generated by the field V, then the flow ψ_t is generated by the field $f_* V$ (i.e. the generators happen to be f-*related*). □

4.1.10 Let $f : M \to M$ be a diffeomorphism and let $\gamma(t)$ be the integral curve of a field V which starts in $x \in M$. Show that the curve $f(\gamma(t))$ is then the integral curve of the field $f_* V$ which starts in $f(x)$ (so that we obtain the f-image of the initial situation).

Hint: an assumption is $\dot{\gamma} = V$, the aim is $f_* \dot{\gamma} = f_* V$. □

• The flow of a vector field induces specific local coordinates on a manifold, which correspond to an observer who "drifts" with the flow.

4.1.11 Let Φ_t be a flow on M and let $\mathcal{A} \equiv \{\mathcal{O}_\alpha, \phi_\alpha\}$ be an atlas (local coordinates x^i) on M. Check that

(i) the flow Φ_t induces (for each t) a new atlas $\mathcal{A}^t \equiv \{\mathcal{O}^t_\alpha, \phi^t_\alpha\}$ on M, which results from the "shift" by the flow Φ_t of the initial atlas

$$\mathcal{O}^t_\alpha = \Phi_t(\mathcal{O}_\alpha) \qquad \phi^t_\alpha = \phi_\alpha \circ \Phi_t^{-1}$$

(it uses coordinates $x^i_t := \Phi_t^* x^i \equiv x^i \circ \Phi_t$)

(ii) in new coordinates the Φ_t-image of any object has the same coordinate expression as the original object had in the original coordinates

(iii) these coordinates correspond to (are used by) an *observer drifting in the flow* Φ_t; in (4.1.7), say; this is an observer who *rotates uniformly* around the z-axis.

Hint: see (1.4.13); see also (4.6.26). □

• The concept of a flow promotes the clarification of the local structure of a vector field. Given an n-dimensional manifold M, let V be a vector field which is *non-vanishing at the point* P. It is then non-vanishing on some *neighborhood* of the point P as well (the property of smooth components $V^i(x)$). Fix any $(n-1)$-dimensional submanifold $\mathcal{S}$ on this neighborhood which is "transversal" with respect to V, i.e. such that the ($(n-1)$-dimensional) tangent space to this submanifold is at each point complementary to the one-dimensional subspace given by the vector V. If we let the points of this submanifold drift away by means of the infinitesimal flow Φ_t of the field V ($-\epsilon < t < \epsilon$), we find a neighborhood of P which happens to be an n-dimensional manifold equipped with coordinates $(x^1 \equiv t, x^2, \dots, x^n)$, where $(x^2, \dots, x^n)$ are the coordinates on $\mathcal{S}$ and $x^1 \equiv t$ informs us how "far" we have moved by means of the flow of the field V.

4.1.12 Consider as a manifold Cartesian space $\mathbb{R}^3[x, y, z]$ and as a vector field $V = -y\partial_x + x\partial_y$. Check that

(i) this field is non-vanishing everywhere outside the z-axis

(ii) for any point P apart from the z-axis we may take as a submanifold $\mathcal{S}$ a small piece of a plane (around P) given by the z-axis and the point P

(iii) an infinitesimal flow Φ_t generates a three-dimensional manifold $\mathcal{U}$ which has the shape of a "cylinder over $\mathcal{S}$" ($\mathcal{S}$ expanded in the perpendicular direction). □

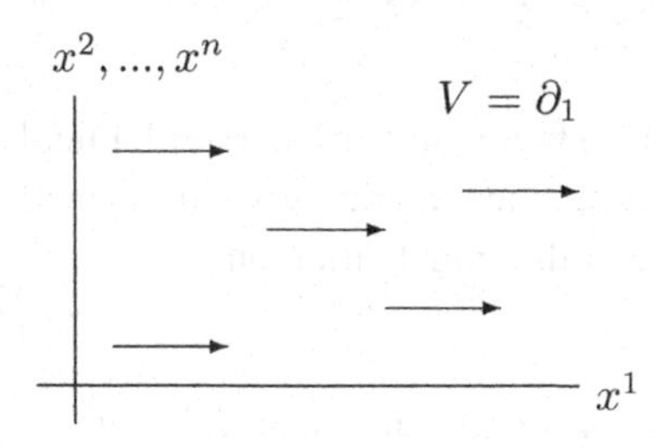

• Now, it is clear from the general construction described above what is the essential property of the coordinates introduced in just this way: the vector field V looks like

$$V = \partial_1 \qquad \text{i.e.} \qquad V^1(x^1, \dots, x^n) = 1,$$
$$\text{remaining} \quad V^j(x^1, \dots, x^n) = 0.$$

In these coordinates the field V "straightens out" (in the small patch under consideration), its integral curves being *straight lines* (the first coordinate curves)

$$x^1(t) = x^1(0) + t, \qquad \text{remaining} \quad x^j(t) = x^j(0)$$

This statement is known as the *straightening out lemma*.[32] It ensures that the "local structure" of a vector field in a (small enough) neighborhood of a point in which it *does not vanish* is always the same, and moreover it is very simple: each field is "locally straight."

4.1.13 Consider the situation treated in (4.1.12). Check that the field V in $\mathcal{S}$ straightens out in *cylindrical* coordinates,

$$V = \partial_1 \equiv \partial_\varphi \qquad x^1 = \varphi, \quad x^2 = r, \quad x^3 = z$$

□

• And what about the local structure of a vector field in a neighborhood of a point, in which it *does vanish*? The situation is less boring, here, allowing for more possibilities.

4.1.14* Given a vector field V which vanishes at a point P (being non-zero, however, in a neighborhood of the point), consider any coordinates centered in this point ($x^i(P) = 0$; this may always be arranged by means of a shift). Check that if the situation is *linearized* in a small neighborhood of P, then

(i) locally (in a small neighborhood of P) the field is characterized (with respect to coordinates x^i) completely[33] by a matrix k^i_j with *numerical* entries

(ii) a change of coordinates (all of them always being centered at P) is encoded in a *non-singular numerical* matrix A^i_j and the matrix k^i_j then transforms as a type (1, 1) tensor

$$k^i_j \mapsto k'^i_j \equiv A^i_s k^s_r (A^{-1})^r_j$$

so that by means of an appropriate change of coordinates one can get a canonical form of the matrix k^i_j; possible types of behavior of the field V in a neighborhood of P are thus classified by the possible canonical forms of the tensor k^i_j of type (1, 1).

Hint: (i) $V^i(x) = k^i_j x^j + \cdots$; (ii) $x'^i(x) = A^i_j x^j + \cdots$, so that $J^i_j(P) = A^i_j$ and the rule $k'^i_j x'^j \equiv V'^i(x') = J^i_j(x) V^j(x)$ (2.2.10) gives $k'^i_j = A^i_s k^s_r (A^{-1})^r_j$. □

4.1.15* Classify all the possible types of behavior of a vector field in the neighborhood of its zero point on a *two-dimensional* manifold (with non-singular matrix k^i_j). For each possibility write down explicitly the corresponding field as well as its integral curves.

Hint: using appropriate A the symmetric part of k^i_j may be diagonalized and the antisymmetric part gets multiplied by the determinant of A; thus the resulting k^i_j is given by the sum of the diagonal part diag (a, b) and an antisymmetric part $c\epsilon_{ij}$, so that locally the vector field looks like $V = ax\partial_x + by\partial_y + c(y\partial_x - x\partial_y)$; if, for example, $a = -b = 1, c = 0$, the field is $V = x\partial_x - y\partial_y$, the integral curves being $x(t) = x(0)e^t$, $y(t) = y(0)e^{-t}$; for

[32] The previous lines are not to be understood as a *proof* of this important lemma, but rather as a visual explanation of its content.

[33] This holds when the matrix k^i_j described in the hint happens to be *non-singular*. Otherwise the *higher-order* terms matter and the situation gets fairly complicated.

$a = b = 0, c = 1$ we obtain $V = y\partial_x - x\partial_y$ and the integral curves rotate around the origin (try to draw corresponding pictures and learn how they differ for various combinations of a, b, c = positive/negative/vanishing). □

4.2 Lie transport and Lie derivative

• Let V be a vector field on M and let

$$\Phi_t : M \to M$$

be the corresponding flow. Since Φ_t is a diffeomorphism, it induces, according to (3.1.6), the mapping (pull-back) of *tensor fields* of arbitrary type on M:

$$\Phi_t^* : \mathcal{T}_q^p(M) \to \mathcal{T}_q^p(M)$$

This mapping is known as *Lie transport* (or, sometimes, *Lie dragging*) of tensor fields. Note that the fields are transported a parametric distance t along the integral curves of the field V *against* the direction of the flow Φ_t (if a transport *in the direction* of the flow is needed, one clearly has to use Φ_{-t}^*).

4.2.1 Check that Φ_t^* is (for each t) a *linear operator* on $\mathcal{T}_q^p(M)$. □

• Let us have a look, to start with the simplest example, at how this map works visually on functions (*scalar* fields, $p = q = 0$).

4.2.2 Consider a function ψ on M. Imagine it is drawn in the form of a *graph*, i.e. as a hypersurface $(x, \psi(x)) \subset M \times \mathbb{R}$.

(i) For $M = \mathbb{R}$, $V = \partial_x$, $\psi(x) = e^{-x^2}$, draw the graph of the Lie transported function $\Phi_t^*\psi$

(ii) do the same for $M = \mathbb{R}^2$, $V = -y\partial_x + x\partial_y$, $\psi(x, y) = e^{-[(x-2)^2+(y-3)^2]}$

(iii) take a lesson from these particular examples and realize that in general the graph of a function $\Phi_t^*\psi$ may be obtained from the graph of ψ simply by a shift of the former by a parameter t *against* the integral curves of the field V. □

• A simple trick – the use of the *field lines* (or, more precisely, the integral curves of the field) – enables one to visualize the Lie transported *vector* fields as well.

4.2.3 Given $\Phi_t \leftrightarrow V$, let $\gamma(\tau)$ be the integral curves of a field W. Justify the idea that the integral curves $\Gamma(\tau)$ of the Lie transported vector field $\Phi_t^* W$ are given simply as the Φ_{-t}-images of the initial curves $\gamma(\tau)$.
Hint: differentiate $\Phi_{-t} \circ \gamma(\tau) = \Gamma(\tau)$ with respect to τ, see (3.1.6). □

4.2.4 Consider two electrostatic fields, the homogeneous field $\mathbf{E}_{(1)} = E\partial_x$ directed along x and the Coulomb field of a point charge $\mathbf{E}_{(2)} = (k/r^2)\,\partial_r$. Consider, in addition, three vector fields, generating (three different) flows in E^3, namely $V = \partial_x$, $U = \partial_y$ and $W = y\partial_x - x\partial_y$. Sketch (performing no calculations at all) the field lines of

(i) the initial electrostatic fields $\mathbf{E}_{(1)}$ and $\mathbf{E}_{(2)}$

(ii) the electrostatic fields, obtained by the Lie transport along the three vector fields V, U, W (by some fixed values of the parameter t, e.g. $t = \pi/2$; altogether $2 \times 3 = 6$ cases are to be discussed).

Hint: (ii) see (4.2.3); the drawings resulting from (i) are to be *shifted* by $\pi/2$ in the direction of x and y respectively (for V, U) and *rotated* by $\pi/2$ around the z-axis (for W). □

- It may happen, for particular tensor fields, that one gets[34]

$$\Phi_t^* A = A \qquad t \in \mathbb{R}$$

Such specific fields are said to be *invariant* with respect to the flow Φ_t (or vector field V), or alternatively, *Lie dragged*. This condition means that the value of A in $x \in M$ coincides with the value of A being transported into x from an arbitrary point lying on the integral curve passing through x.

(So we obtained in exercise (4.2.4) that the field $\mathbf{E}_{(1)}$ is invariant (Lie dragged) with respect to *translations* along both the x and y axes and $\mathbf{E}_{(2)}$ is in turn invariant with respect to *rotations* around the z-axis; one easily verifies that $\mathbf{E}_{(1)}$ is, in fact, invariant with respect to translations in *any* direction and $\mathbf{E}_{(2)}$ is invariant with respect to rotations around *any* axis passing through the origin.)

There is *no reason* for a *general* tensor field A, however, to be constant along the integral curves of a field V: the tensor $(\Phi_t^* A)(x)$, which has been transported into x from the point $\Phi_t(x)$, in general depends on t. A convenient measure of this dependence (i.e. of Lie *non*-constancy $=$ *non*-invariance with respect to V) is given by the object

$$\mathcal{L}_V A := \left.\frac{d}{dt}\right|_0 \Phi_t^* A$$

which is called the *Lie derivative* of a tensor field[35] A along a vector field V. This derivative "palpates" the changes of tensor fields induced by a tiny Lie transport along V: first, the value of the field A at the "slightly drained away" point $\Phi_\epsilon(x)$ is transported back into x and then it is compared with the initial value of A in x. The comparison := their difference (which *makes sense already*, since both tensors, the one transported back as well as the original one, sit at *a single* point, i.e. they represent elements of a *single* linear space), divided by the increment of the parameter ϵ, resulting in a quantity measuring just the "change of the tensor field per unit value of the parameter t" (or the "rate of change of the field" A along V).

Right from the definition it follows that

$$\mathcal{L}_V : \mathcal{T}_q^p(M) \to \mathcal{T}_q^p(M)$$

(it preserves the degree of a tensor field) and

$$\mathcal{L}_V A = 0 \quad \Leftrightarrow \quad A \text{ is invariant (Lie dragged) with respect to } V$$

[34] This occurred in three out of six cases in exercise (4.2.4), namely when $\mathbf{E}_{(1)}$ was transported along both V and U and when $\mathbf{E}_{(2)}$ was transported along W.

[35] In Arnold's monograph the Lie derivative is also mentioned under the well-turned name the *fisherman's derivative*: a fisherman stands in a river and differentiates tensor fields, floating around him. Unfortunately, the present-day status of the human environment makes this juicy bon mot barely intelligible to the younger generation. The lamentable quality of water causes tensor fields of higher ranks to simply not be able to survive in the overwhelming majority of rivers and the exciting stories narrated by our grandfathers on how they (when small boys) used "to guddle fifth-rank completely antisymmetric tensors in a spruit behind a village" may seem to be typical *fish stories*, today.

In the next section the general properties of $\mathcal{L}_V$ (i.e. how it behaves in some standard situations) will be studied in detail. This will result, in particular, in explicit formulas for the component calculation of the Lie derivative of an arbitrary tensor. Moreover, the appropriate use of these properties alone provides an efficient way to compute a number of useful expressions with *no reference* to components.

4.3 Properties of the Lie derivative

• As we will see in a while (4.3.4), the component expression of the Lie derivative of a general tensor field is a sum of several pieces, each one carrying a number of indices. The overall structure is given by a system of clear rules; the resulting expression looks, however, fairly complicated at first glance. All the properties of $\mathcal{L}_V$ may, in principle, be derived[36] from its component expression, but the use of simple *algebraic* properties of the Lie derivative (which may be ultimately traced back to the simple properties of the pull-back Φ_t^*) turns out to be both more efficient and more instructive.

Recall that pull-back with respect to a diffeomorphism $M \to M$ is an *isomorphism* of the tensor algebra $\mathcal{T}(M)$, which commutes with contractions, see (3.1.6) and (3.1.7). A simple (and very useful) consequence of this is the fact that the Lie derivative is a *derivation* of the tensor algebra, which commutes with contractions.

4.3.1 Check that

(i) for $|\epsilon| \ll 1$

$$\Phi_\epsilon^* A = A + \epsilon \mathcal{L}_V A + o(\epsilon^2)$$

(ii) $\mathcal{L}_V$ preserves the degree and satisfies

$$\mathcal{L}_V(A + \lambda B) = \mathcal{L}_V A + \lambda \mathcal{L}_V B$$
$$\mathcal{L}_V(A \otimes B) = (\mathcal{L}_V A) \otimes B + A \otimes (\mathcal{L}_V B)$$

i.e. (see Appendix A.2) that $\mathcal{L}_V$ happens to be a *derivation of the tensor algebra* $\mathcal{T}(M)$

(iii) if C is an arbitrary contraction and $\hat{1}$ denotes the unit tensor (field), then

$$\mathcal{L}_V \circ C = C \circ \mathcal{L}_V \qquad \mathcal{L}_V \hat{1} = 0$$

i.e. $\mathcal{L}_V$ commutes with contractions

(iv)

$$\mathcal{L}_V(A(W, \dots; \alpha, \dots)) = (\mathcal{L}_V A)(W, \dots; \alpha, \dots) + A(\mathcal{L}_V W, \dots; \alpha, \dots) + \cdots$$
$$+ A(W, \dots; \mathcal{L}_V \alpha, \dots) + \cdots$$

and, in particular,

(v)

$$\mathcal{L}_V \langle \alpha, W \rangle = \langle \mathcal{L}_V \alpha, W \rangle + \langle \alpha, \mathcal{L}_V W \rangle$$

[36] It used to be done in this way in older textbooks. As an example, the walls of Altamira and Lascaux caverns have been reported to be densely covered by such fairly long component expressions. Let us remark, as a nice illustration of the inventiveness of the primeval hunters in masterful use of terrain irregularities, that in caves of calcite, limestone and dolomite they used stalactites for the location of upper indices, stalagmites for lower indices and stalagnates as the most convenient places for the contraction of a pair of indices.

Hint: (i) right from the definition; (ii) see (3.1.6) and (2.4.7) and (i) here; (iii) see (3.1.7); (iv) see (3.1.8); (v) see Exercise (2.4.9), put $A = \hat{1}$ in (iv). □

• The result of (iv) reveals that a function, a vector field and a covector field are all one really needs to be able to compute $\mathcal{L}_V A$ of an arbitrary tensor field. According to (v), a function and a vector *or* a covector field is enough.

4.3.2 Consider the *arbitrary* derivative D of the tensor algebra $\mathcal{T}(M)$, which commutes with contractions. Show that it is completely specified once its action on degree $\binom{0}{0}$ and *either* $\binom{1}{0}$ *or* $\binom{0}{1}$ is given.

Hint: see (2.4.10), apply D on a tensor with all slots being filled by arguments. □

• So we now embark on the derivation of explicit expressions for the action of $\mathcal{L}_V$ on a function and a covector field.

4.3.3 Check that the Lie derivative $\mathcal{L}_V$ acts as follows:

(i) on functions

$$\mathcal{L}_V \psi = V\psi \qquad (\equiv V^i(x)\psi_{,i}(x))$$

(ii) on covector fields, which happen to be gradients of functions

$$\mathcal{L}_V(d\psi) = d(\mathcal{L}_V \psi) \equiv d(V\psi)$$

(iii) on general covector fields $\alpha = \alpha_i(x)\,dx^i$

$$\mathcal{L}_V \alpha = \left(V^j \alpha_{i,j} + V^j_{\ ,i}\alpha_j\right) dx^i$$

Hint: (i) see (3.1.1); (ii) see (3.1.9); (iii) $\alpha = \alpha_i\, dx^i \equiv \alpha_i \otimes dx^i$. □

• Since we learned how to cope with functions and covector fields, we are in a position, according to (4.3.2) and (4.3.1), to derive a component expression of the Lie derivative of an arbitrary rank tensor field.

4.3.4 Check that

(i) the Lie derivative of the coordinate basis fields reads

$$\mathcal{L}_V\, dx^i = V^i_{\ ,j}\, dx^j \qquad \mathcal{L}_V \partial_i = -V^j_{\ ,i}\partial_j$$

(ii) this results in the following component expression of the Lie derivative of an arbitrary rank tensor field:

$$(\mathcal{L}_V A)^{i\dots j}_{k\dots l} = V^m A^{i\dots j}_{k\dots l,m} + V^m_{\ ,k} A^{i\dots j}_{m\dots l} + \cdots - V^j_{\ ,m} A^{i\dots m}_{k\dots l}$$

i.e. there is the first term (flat amount), plus there is one term to be added for each index of the tensor (with a + sign for a lower index and a − sign for an upper one). These rules may be concisely summarized in the form of a table – a recipe for cooking the house speciality $(\mathcal{L}_V A)^{i\dots j}_{k\dots l}$;

compare with (15.2.7):

	for preparation of $\mathcal{L}_W A$
first put on the bottom of a pan	$W A^{\cdots}_{\cdots} \equiv W^m A^{\cdots}_{\dots,m}$
plus for each $A^{\dots i \dots}$ add	$-W^i_{\ ,m} A^{\dots m \dots}$
plus for each $A_{\dots i \dots}$ add	$+W^m_{\ ,i} A_{\dots m \dots}$

Hint: (i) see (4.3.3) for $\psi = x^i$ and (4.3.1) for $\alpha = dx^j$ and $W = \partial_i$; (ii)

$$\begin{aligned}(\mathcal{L}_W A)^{i \dots j}_{k \dots l}\, dx^k \otimes \dots \otimes \partial_j &:= \mathcal{L}_W \big(A^{i \dots j}_{k \dots l}\, dx^k \otimes \dots \otimes \partial_j\big) \\ &= \big(\mathcal{L}_W A^{i \dots j}_{k \dots l}\big)\, dx^k \otimes \dots \otimes \partial_j \\ &\quad + A^{i \dots j}_{k \dots l} (\mathcal{L}_W\, dx^k) \otimes \dots \otimes \partial_j \ + \cdots \\ &\quad + A^{i \dots j}_{k \dots l}\, dx^k \otimes \dots \otimes (\mathcal{L}_W \partial_j) = \cdots\end{aligned}$$ □

4.3.5 Write down explicit component expressions of the Lie derivative of tensors of rank 0, 1 and 2 (six formulas altogether). □

4.3.6 Show that

(i) the Lie derivative of a vector field turns out to be

$$\mathcal{L}_V W = [V, W]$$

where the vector field on the right-hand side is called the *commutator* (or *Lie bracket*) of the vector fields V and W and it is defined as follows:

$$[V, W]\psi := V(W\psi) - W(V\psi)$$

(ii) if V, W are two vector fields, then their "product" VW ($(VW)\psi := V(W\psi)$) *is not* a vector field, whereas their *antisymmetrized* product (= commutator) *is* a vector field

(iii) a *product* of first-order differential operators is a second-order differential operator, in general, but their commutator happens to be only the first-order operator

(iv) "the same thing" has been said in (ii) and (iii)

(v) the collection of all vector fields endowed with the bracket operation $[\,\cdot\,,\cdot\,]$ constitutes a (∞-dimensional) *Lie algebra*, i.e. the following hold (cf. Appendix A.3):

$$\begin{aligned}[V, W] &= -[W, V] && \text{antisymmetry} \\ [V_1 + \lambda V_2, W] &= [V_1, W] + \lambda [V_2, W] && \text{(bi)linearity} \\ 0 &= [[V, W], U] + [[U, V], W] + [[W, U], V] && \text{Jacobi identity}\end{aligned}$$

Hint: (i) compute $[V, W]$ in components and compare with (4.3.4); (ii) apply on a product of two functions. □

• The properties of $\mathcal{L}_V$ mentioned up to now were related to its behavior with respect to particular *arguments*. They may be summarized concisely as a statement that the Lie derivative $\mathcal{L}_V$ is a derivation of the tensor algebra, which commutes with contractions (4.3.1) as well as with the operator of gradient d (4.3.3).

Some additional (very useful) properties of the Lie derivative $\mathcal{L}_V$ are related to its behavior with respect to its "*parameter*," the vector field V (along which it is computed). Here, an algebraic approach turns out to be the most effective way of reasoning as well. We start with the following property of arbitrary derivations of the tensor algebra (useful in its own right, too).

4.3.7 Let D_1, D_2 be two derivations of the tensor algebra. Check that

(i) their linear combination as well as the commutator

$$D := D_1 + \lambda D_2 \qquad \text{resp.} \quad D := [D_1, D_2] \equiv D_1 D_2 - D_2 D_1$$

happen to be derivations of the tensor algebra, too

(ii) if D_1, D_2 commute with contractions, then this is true for linear combination and the commutator, too. □

• This may be rephrased as the statement that the collection of all derivations of the tensor algebra is naturally endowed with the structure of the *Lie algebra* (it is denoted by Der $\mathcal{T}(M)$) and, furthermore, the derivations which commute with contractions constitute a subalgebra. This elementary observation provides a simple proof of the following useful proposition.

4.3.8 Prove that

(i)

$$\begin{aligned}\mathcal{L}_{V+\lambda W} &= \mathcal{L}_V + \lambda \mathcal{L}_W \\ \mathcal{L}_{[V,W]} &= [\mathcal{L}_V, \mathcal{L}_W] \equiv \mathcal{L}_V \mathcal{L}_W - \mathcal{L}_W \mathcal{L}_V\end{aligned}$$

(ii) the mapping

$$\mathcal{L} : \mathfrak{X}(M) \to \text{Der}\, \mathcal{T}(M) \qquad V \mapsto \mathcal{L}_V$$

is a homomorphism of Lie algebras.

Hint: (i) according to (4.3.7) we are to prove the equality of two derivations of the tensor algebra which commute with contractions, or equivalently (after reshuffling of all terms to one side of the equation), that a certain derivation of this type *vanishes*. By (4.3.2) it is enough to verify this on functions and vector fields, which is easy (4.3.6); (ii) just this is asserted in (i). □

4.4 Exponent of the Lie derivative

• The Lie derivative $\mathcal{L}_V$ has been defined in Section 4.2 in terms of the pull-back of a flow Φ_t^*. It turns out that the pull-back Φ_t^* may in turn often be expressed in terms of the Lie derivative in a useful form of the exponent $\Phi_t^* = e^{t\mathcal{L}_V}$. Let us have a look, first, at what this formula actually says in the simplest case. Then we prove its validity in a more general setting. (One should realize that if the map Φ_t itself is problematic globally, which is the case for (only) *local* flows (see the note before (4.1.1)), the formula is problematic as well. The reader is once again referred to contemplate the particular example (2.3.4), which is mentioned in the note.)

4.4.1 Let $M = \mathbb{R}[x]$, $V = \partial_x$. Check that the Taylor expansion of a function

$$\psi(x+t) = \psi(x) + t\psi'(x) + \frac{t^2}{2!}\psi''(x) + \cdots$$

may be expressed in the form

$$\Phi_t^*\psi = e^{t\mathcal{L}_V}\psi \qquad e^{t\mathcal{L}_V} := \hat{1} + t\mathcal{L}_V + \frac{t^2}{2!}\mathcal{L}_V\mathcal{L}_V + \cdots$$

i.e. that in this particular case there holds

$$\Phi_t^* = e^{t\mathcal{L}_V}$$

Hint: see (3.1.1) and (4.3.3). □

4.4.2 Let Φ_t be the flow generated by a vector field V. Starting from the definition

$$\mathcal{L}_V := \left.\frac{d}{dt}\right|_0 \Phi_t^*$$

prove that

(i)

$$\frac{d}{dt}\Phi_t^* = \Phi_t^*\mathcal{L}_V$$

(ii) for C^ω tensor fields there holds

$$\Phi_t^* = e^{t\mathcal{L}_V} \equiv 1 + t\mathcal{L}_V + \frac{t^2}{2!}\mathcal{L}_V\mathcal{L}_V + \cdots$$

Hint: (i) $\frac{d}{dt}\Phi_t^* = \frac{d}{ds}\big|_{s=0}\Phi_{t+s}^*$, (4.1.2); (ii) $\left(\frac{d}{dt}\right)^n\Phi_t^* = \Phi_t^*(\mathcal{L}_V)^n$. □

• This formula proves very useful in providing the tool for a systematic expansion of the *infinitesimal* flow Φ_ε^* in powers of the parameter ε. If, for example, we need second-order accuracy, we may write $\Phi_\varepsilon^* = e^{\varepsilon\mathcal{L}_V} \equiv 1 + \varepsilon\mathcal{L}_V + \frac{\varepsilon^2}{2!}\mathcal{L}_V\mathcal{L}_V$. We will make use of this particular result in the next section in order to grasp the visual meaning of the commutator of two vector fields.

Pull-back of a flow Φ_t^* enables one to write down explicitly solutions of linear first-order partial differential equations in terms of initial conditions, too.

4.4.3 Let $V = V^i(x)\partial_i$ be a complete vector field on M and let Φ_t be the corresponding flow. Consider a first-order linear partial differential equation on $M \times \mathbb{R}[t]$ of the form

$$(\partial_t - V^i(x)\partial_i)f(x,t) = 0$$

together with an initial condition

$$f(x,0) = h(x)$$

Check that the solution may be written as

$$f(x,t) = h(\Phi_t(x))$$
$$\equiv (\Phi_t^* h)(x)$$

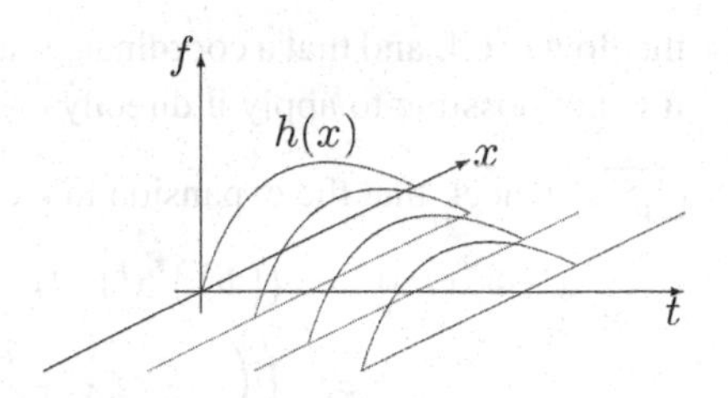

Hint: see (4.4.2); this means (in visual terms) that the graph which corresponds to the initial conditions $h(x)$ is moved against the direction of integral curves of the vector field V on M. For $(M, V) = (\mathbb{R}, \partial_x)$, this looks (for particular $h(x)$) something like in the figure displayed here, since in this case we explicitly obtain $f(x,t) = h(\Phi_t(x) = h(x+t))$. □

4.5 Geometrical interpretation of the commutator $[V, W]$, non-holonomic frames

• We encountered the concept of the commutator $[V, W]$ of two vector fields V and W when the Lie derivative of a vector field was computed. Here, we would like to examine consequences of the fact that the commutator does not vanish for particular vector fields, i.e. to grasp the geometrical meaning of the commutator.

Suppose we undertake two infinitesimal journeys, both of them starting at a point $P \in M$. The first one consists of motion by ϵ along V and subsequently by ϵ along W, the second one performs the same steps in reversed order. Now, the question arises as to whether or not we reach the same point. It turns out that the answer is positive within first-order accuracy in ϵ, but it already happens to be negative within order ϵ^2, and the necessary *correction* (i.e. a step to be *added* to the first journey in order to arrive at the end of the second one) in this order consists in a motion by ϵ^2 along the *commutator* $[W, V]$. There is an equivalent formulation of the same problem, which is represented by the schematic drawing here. The question is whether we return to the same point or not if we, after reaching the end of the first journey, keep on traveling further, by ϵ along $(-V)$ and subsequently by ϵ along $(-W)$. If the flows corresponding to the fields V and W are denoted by Φ_t^V and Φ_t^W respectively, the question about the closure of such a "circular tour" may be written as follows:

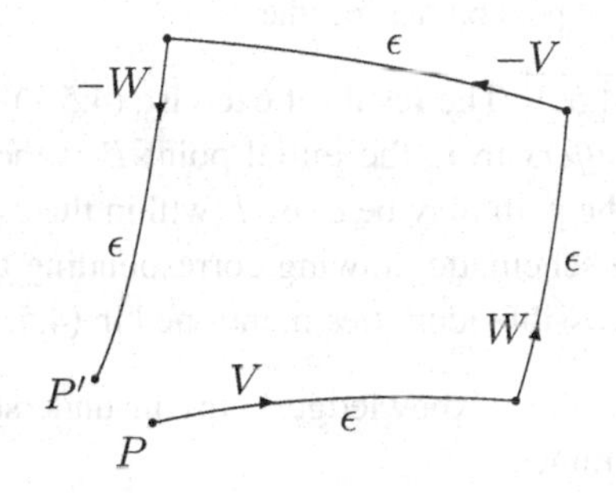

$$P' \equiv \left(\Phi_{-\epsilon}^W \circ \Phi_{-\epsilon}^V \circ \Phi_{\epsilon}^W \circ \Phi_{\epsilon}^V\right)(P) \stackrel{?}{=} P$$

Once again, a commutator correction is needed within ϵ^2 accuracy. Thus the geometrical role of a commutator is "to close a parallelogram, which fails to close (a bit), yet." These statements are easily proved. In order to compute the left-hand side within the desired accuracy we may exploit the result (4.4.2), provided that *coordinates* of the expression on the left are computed. (Remember that (4.4.2) refers to the *pull-back* of a flow, rather than the flow itself, and that a coordinate is a *function* $\Rightarrow$ pull-back may be applied to it, whereas it is not possible to apply it directly on a point.)

4.5.1 Check that the expansion in ϵ explicitly gives

$$\begin{aligned} x^i(\Phi^V_\epsilon(P)) &= ((\Phi^V_\epsilon)^* x^i)(P) \\ &= \left(\left(1+\epsilon\mathcal{L}_V + \frac{\epsilon^2}{2!}\mathcal{L}_V\mathcal{L}_V + \cdots\right)x^i\right)(P) \\ x^i(\Phi^V_\epsilon \circ \Phi^W_\epsilon(P)) &= ((\Phi^W_\epsilon)^*(\Phi^V_\epsilon)^* x^i)(P) \\ &= \left(\left(1+\epsilon\mathcal{L}_W + \frac{\epsilon^2}{2!}\mathcal{L}_W\mathcal{L}_W + \cdots\right)\left(1+\epsilon\mathcal{L}_V + \frac{\epsilon^2}{2!}\mathcal{L}_V\mathcal{L}_V + \cdots\right)x^i\right)(P) \end{aligned}$$

etc.

Hint: see (3.1.1) and (4.4.2). □

4.5.2 Check that

(i) to within ϵ^2 accuracy the following identity holds:

$$\Phi^W_{-\epsilon} \circ \Phi^V_{-\epsilon} \circ \Phi^{[V,W]}_{-\epsilon^2} \circ \Phi^W_\epsilon \circ \Phi^V_\epsilon = \hat{1} + o(\epsilon^2)$$

(ii) the term containing the commutator may be reshuffled into any other place on the left (five possibilities together). □

4.5.3 The result of exercise (4.5.2) shows that already in second order in ε, the point P' *differs* from the initial point P if the commutator $[V, W]$ does not vanish in P and that the path may be *closed* (within the same accuracy) by an appropriate small piece. Outline a schematic drawing corresponding to the closed path composed of *five* pieces to all five possible identities mentioned in (4.5.2). □

• This knowledge helps in understanding the situation with so-called non-holonomic frames.

In Chapter 2 the concept of a *coordinate basis* ∂_i for vector fields (as well as the dual basis dx^i for covector fields) was introduced. We know from linear algebra, however, that each basis is equally good in a general linear space and, moreover, an arbitrary basis may be obtained from any other one by "mixing" (making linear combinations) the elements of the first basis with the help of a non-singular matrix:

$$e_a \mapsto e'_a = A^b_a e_b$$

This means that it is not necessary to use coordinate frames for the decomposition of tensor fields. Instead we are free to use any non-singular linear combinations (depending on x)

$$e_a(x) = e^i_a(x)\partial_i \qquad e^a(x) = e^a_i(x)\,dx^i$$

the only two requirements concerning the new frame fields e_a, e^a being smoothness (resulting in the smoothness of the matrices $e^i_a(x)$, $e^a_i(x)$) and linear independence (leading to non-singularity of the matrices).

4.5.4 Check that the requirements regarding duality and completeness of new frames $e_a \leftrightarrow e^a$ result in

$$e^a_i(x)e^i_b(x) = \delta^a_b \qquad e^i_a(x)e^a_j(x) = \delta^i_j$$

i.e. the matrices $e^i_a(x)$, $e^a_i(x)$ are to be *inverse* to each other at each point x.
Hint: see (2.4.2) and (2.4.9); $\delta^a_b = \langle e^a, e_b\rangle = \cdots$, $\hat{1} = dx^i \otimes \partial_i \overset{!}{=} e^a \otimes e_a = \cdots$. □

4.5.5 Check that

$$e^a_i(x) = \langle e^a, \partial_i\rangle = \hat{1}(\partial_i, e^a) \equiv \hat{1}^a_i$$
$$e^i_a(x) = \langle dx^i, e_a\rangle = \hat{1}(e_a, dx^i) \equiv \hat{1}^i_a$$

$\Rightarrow$

$$\hat{1} = e^a_i(x)\, dx^i \otimes e_a = e^i_a(x) e^a \otimes \partial_i$$

This means that the functions $e^i_a(x)$ and $e^a_i(x)$ may also be regarded as components of the *unit* tensor with respect to *non*-dual frames. □

• As we will see later, the appropriate choice of a *frame field* $e_a(x)$ and a *coframe field* $e^a(x)$ may strongly simplify both reasoning and computation in various situations. Important examples are provided by *orthonormal* frame fields on Riemannian manifolds (see, for example, Section 15.6) or *left-invariant* fields on Lie groups (see Section 11.1). In the general theory of relativity a frame field (appropriately chosen, most often orthonormal) is usually called a *tetrad field*[37] and a formalism working with components of tensors with respect to this kind of frame field is known as the *tetrad formalism* (see, for example, (15.6.20) and Sections 16.5 and 22.5).

4.5.6 Find the coefficients $e^i_a(x)$, $e^a_i(x)$, if x^i = Cartesian coordinates in $\mathbb{R}^3$ and $e_a = \partial_a$ = coordinate basis with respect to the *spherical polar* coordinates in $\mathbb{R}^3$. □

• Imagine we were given (only) the *result* of the last exercise, not being told, however, that the new frame field e_a (mixing well the old Cartesian frame field ∂_i) is, in fact, the coordinate one, too (with respect to other coordinates, of course; here spherical polar). Is it possible to reveal this fact from the structure of e_a? And, more fundamentally, is it possible to construct a frame field, for which *no coordinates* y^a exist at all, such that $e_a = \partial_a$?

It turns out that the correct answer to both questions is *yes*. Thus there are frame fields which are not generated by coordinates (i.e. such that $e_a \neq \partial_a$), and if there *are* some

[37] Since a space-time (M, g) is a *four-dimensional* manifold; in general, the nomenclature *vielbein field* is widely used, i.e. a "manypod" or "manyvet field;" a frame in *three* dimensions resembles (with a bit of fantasy, no doubt a fairly useful instrument in the realm of mathematics as such) a *dreibein* ≡ a tripod or a trivet, so that a tetrad is the same thing as a *vierbein*.

coordinates hidden behind a frame field, it is an easy job to recognize this fact (one can even compute these coordinates explicitly).

Consider first the case, when $e_a \equiv e^i_a(x)\partial_i = \partial_a$, i.e. when the frame *is* a coordinate one (with respect to y^a). Then

$$[e_a, e_b] = [\partial_a, \partial_b] = 0$$

since the order of partial derivatives is (on the class of functions we are working with) not relevant. This means that just one non-vanishing commutator reveals that the frame under consideration is *non-holonomic* = *non-coordinate*, i.e. there are no coordinates y^a such that $e_a = \partial_a$ ($\Rightarrow e^a = dy^a$).

4.5.7 Check that

(i) the coordinate frame fields corresponding to polar coordinates in $\mathbb{R}^2$ and both spherical polar and cylindrical coordinates in $\mathbb{R}^3$ happen to be ortho*gonal*, but they fail to be ortho*normal*
(ii) if their lengths are "corrected" so as to *be* ortho*normal*, the resulting frame fields turn out to already be non-holonomic.

Hint: for polar coordinates $|\partial_r| \equiv \sqrt{g(\partial_r, \partial_r)} \equiv g_{rr} = 1$, but $|\partial_\varphi| = \dots = r \ (\neq 1) \Rightarrow$ "orthonormal polar" frame consists of $e_r = \partial_r$, $e_\varphi = r^{-1}\partial_\varphi$ and $[e_r, e_\varphi] \neq 0$. □

• The vanishing of all commutators $[e_a, e_b]$ is thus a *necessary* condition for a frame field to be holonomic = coordinate. The question whether this condition is at the same time *sufficient* remains, however, open. This problem may be tackled in the language of vector fields and the answer is *yes*. There is, though, a simpler way to demonstrate the same fact, using differential forms, namely the so-called Poincaré lemma (9.2.11); we will return to this issue later.

The simpler question alone, whether it is possible to mix a coordinate frame ∂_i so as to generate a non-coordinate one, may be resolved by "counting of degrees of freedom," too. A change of coordinates $x^i \mapsto y^a(x)$ provides "n degrees of freedom," namely the choice of new functions $y^a(x)$. Then the transformation of the (co)frame field is already fixed: $dx^i \mapsto dy^a = J^a_i(x)dx^i \equiv e^a_i(x)dx^i$. Notice, however, that the matrix $e^a_i(x)$ has a very specific structure here – it is the *Jacobi* matrix, clearly carrying *less* freedom (n functions only) in comparison with a general non-singular matrix $e^a_i(x)$ (encoding n^2 functions). In these terms the question is whether it is possible to choose a matrix $e^a_i(x)$ so as *not to be* a Jacobian matrix for any choice of new coordinates $y^a(x)$. Since $n^2 > n$ (for $n \geq 2$), the answer reads *yes*, non-holonomic frames do exist.

We started this section with the problem of whether we return to the point of departure after taking a (particular) circular tour, i.e. whether two flows generated by vector fields V and W commute. The lesson from the analysis is that this issue may be reduced to the much simpler problem of investigating the commuting of *generators* of the flows, the vector fields V and W.

4.5.8 Let Φ_t^V and Φ_s^W be two flows generated by vector fields V, W. We say that the flows commute, if *for each* t, s there holds

$$\Phi_t^V \circ \Phi_s^W = \Phi_s^W \circ \Phi_t^V$$

Prove that the flows Φ_t^V and Φ_s^W commute if and only if their generators V, W commute (in the sense of vector fields).

Hint: see (4.4.2) and (4.3.8); $f^* = g^* \Leftrightarrow f = g$. □

4.6 Isometries and conformal transformations, Killing equations

• Imagine several geometrical figures (triangles, rectangles, circles, etc.) drawn in a plane $\mathbb{R}^2$. We search for bijective maps of the plane to itself (transformations) such that all the figures look "the same" (both in shape and in size), after the transformation.

Think about how this requirement might be reformulated in terms of the concepts we have met before.

Geometrical figures are composed of lines (possibly curved), which have a *length* and some of them intersect under some *angles*. There is, however, a *metric tensor* beyond both the lengths and angles, see (2.6.9) and (3.2.8). Let us examine in detail how, for example, the length of a curve γ changes under a general transformation f.

4.6.1 Let γ be a curve on M, f a transformation of M (= diffeomorphism $f : M \to M$) and $\hat{\gamma} := f \circ \gamma$ the curve transformed by f. Denote by $l[\gamma, g]$ the functional of the *length of a curve* (2.6.9) on a manifold (M, g), i.e.

$$l[\gamma, g] := \int_{t_1}^{t_2} dt \sqrt{g(\dot{\gamma}, \dot{\gamma})}$$

Check that one obtains for the length of the transformed curve the following simple expression[38]

$$l[f \circ \gamma, g] = l[\gamma, f^* g]$$

Hint: $\gamma \mapsto f \circ \gamma \Rightarrow \dot{\gamma} \mapsto f_* \dot{\gamma} \Rightarrow \sqrt{g(\dot{\gamma}, \dot{\gamma})} \mapsto \sqrt{g(f_* \dot{\gamma}, f_* \dot{\gamma})} = \sqrt{(f^* g)(\dot{\gamma}, \dot{\gamma})}$. □

• If we require that the length of *any* curve γ should *not* change, we have to restrict the class of the maps under consideration to

$$f : M \to M \qquad \text{such that} \quad f^* g = g$$

These transformations of M are called *isometries*[39] of a (Riemannian) manifold (M, g).

[38] The length of a transformed curve differs, in general, from the length of the initial one since the new curve (= the image of the initial one) is situated in a domain characterized by quite different "metric conditions." The same effect is achieved, however (the trick with a shoe, see (3.1.2)), if we pull back the metric conditions from the domain where the new curve is situated. Put another way, instead of traveling there we simulate "here" the metric conditions which are valid "there."

[39] More generally, given *two* Riemannian manifolds (M, g) and (N, h), a map $f : M \to N$ is called an *isometry* if $f^* h = g$ (then the length of the f-image of any curve on M happens to be the same as the length of the curve itself). In particular, for $M = N$, $g = h$ we find that the isometries are transformations of *a single* Riemannian manifold.

4.6.2 Check that isometries automatically preserve the *angles* under which *arbitrary* curves intersect, too.

Hint: let two curves intersect in $x \in M$ at an angle α and let the vectors v, w be tangent (of any lengths) to the curves in x. Then there holds $x \mapsto f(x)$, $v \mapsto f_* v$, $w \mapsto f_* w$ and $\cos\alpha \mapsto \cos\alpha' = \cos\alpha$, since according to the definition of an angle

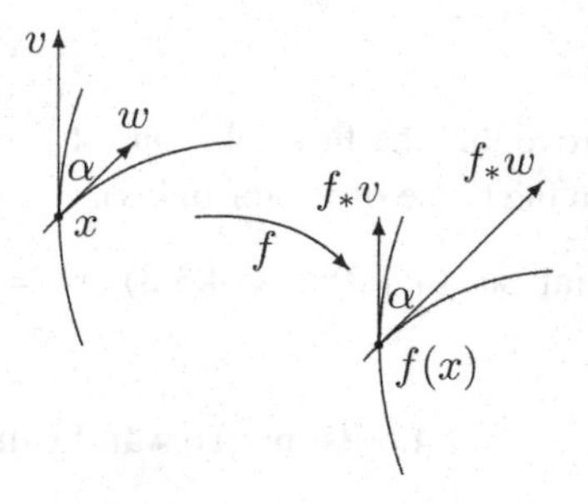

$$\cos\alpha := \frac{g(v, w)}{\sqrt{g(v, v)}\sqrt{g(w, w)}}$$

we find for the dashed angle

$$\cos\alpha' := \frac{g(f_* v, f_* w)}{\sqrt{g(f_* v, f_* v)}\sqrt{g(f_* w, f_* w)}} = \frac{(f^* g)(v, w)}{\sqrt{(f^* g)(v, v)}\sqrt{(f^* g)(w, w)}} = \cos\alpha$$

using $f^* g = g$. Intuitively, it looks fairly reasonable that all angles are preserved if the lengths of *all* lines remain unchanged. Simply imagine the angle being realized in some (infinitesimal, in order that our Euclidean intuition works) triangle and realize that it is impossible to affect its angles if the lengths of all of its sides are preserved. □

• From the expression for the change of $\cos\alpha$ we can see, however, that a *weaker* requirement is enough for the preservation of (all) the *angles alone* (i.e. if we do not insist at the same time on preserving the lengths of all lines), namely

$$f^* g = \sigma g \qquad 0 < \sigma : M \to \mathbb{R}, \text{arbitrary}$$

Such transformations are called *conformal transformations* of a manifold (M, g) (in particular, for σ = constant we have *homotheties* and for $\sigma = 1$ they reduce to *isometries*).

4.6.3 Check that conformal transformations constitute a group, homotheties form a subgroup and isometries are a subgroup of the group of homotheties. □

4.6.4 Count up the "degrees of freedom" and check that a *general* manifold (M, g) has no *non-trivial* isometries (= differing from $f = \mathrm{id}_M$).

Hint: there are n levers (coordinates of new points as functions of the initial ones) at our disposal to meet $n(n+1)/2$ conditions (in components $f^* g = g$ turns into the equality of two *symmetric* matrices), i.e. $f^* g = g$ results in an *overdetermined system* of equations (see (4.6.6)). □

• A highly effective tool for finding a relevant part of all isometries (namely those isometries which may be obtained by a smooth deformation of the trivial isometry = identity) provides the infinitesimal approach. In the first step, the strategy consists in finding all *infinitesimal* isometries $\Phi_\epsilon : M \to M$ (differing only slightly from the identity; in coordinates $x^i \mapsto x^i + \epsilon \xi^i(x)$) and then, in the second step, one obtains the finite ("large") maps by iteration of the infinitesimal ones. (A rotation by an angle α may be, as an example, regarded as the N-fold repetition of rotation by an angle α/N.) In this way we get a whole *one-parameter group* = *flow* of isometries $\Phi_t : M \to M$ with the generator of the flow being the vector field $\xi = \xi^i(x)\partial_i$. We now embark on the derivation of the equations which specify the key vector field.

4.6.5 Let $\Phi_t : M \to M$ be a one-parameter group (flow) of isometries, generated by a vector field ξ. Show that ξ satisfies the *Killing equations*

$$\mathcal{L}_\xi g = 0$$

Hint: differentiate the defining equation $\Phi_t^* g = g$ with respect to t in $t = 0$. □

4.6.6 Check that

(i) a component expression of the Killing equations reads

$$\xi^k g_{ij,k} + \xi^k{}_{,i} g_{kj} + \xi^k{}_{,j} g_{ik} = 0$$

(ii) it is an overdetermined system of equations for unknown functions $\xi^1(x), \dots, \xi^n(x) \Rightarrow$ no (non-vanishing) solution is guaranteed, in general

(iii) given ξ and η two solutions of the Killing equations, then both

$$\xi + \lambda\eta \qquad \text{and} \qquad [\xi, \eta]$$

represent solutions, as well

(iv) *Killing vectors* (i.e. solutions of the Killing equations) constitute a subalgebra of the Lie algebra of all vector fields on M.

Hint: (i) see (4.3.5); (ii) $n(n+1)/2$ equations for n unknowns, see also (4.6.4); (iv) see (4.3.8). □

• The Lie algebra of Killing vectors is, unlike the Lie algebra of all vector fields, always *finite-dimensional* and one can show that its dimension may be (for an n-dimensional manifold) at most $n(n+1)/2$ (this maximal value is achieved, as an example, on (pseudo-) Euclidean spaces $E^{r,s}$ as well as on spheres, see (4.6.10) and (4.6.11)).

The fact that the space of solutions of the partial differential equations under consideration (Killing equations) is endowed with the structure of the Lie algebra[40] may be used, sometimes, for finding additional solutions, when only some solutions are known: we simply form all the possible commutators of the solutions which are known so far and, if we are lucky enough, a new solution drops out in this way.[41]

4.6.7 Find Killing vectors and the corresponding flows for the ordinary Euclidean plane.

Hint: denote $\xi^1(x, y) \equiv A(x, y)$, $\xi^2(x, y) \equiv B(x, y)$. Then the Killing equations read

$$\begin{aligned} A_{,x} = 0 = B_{,y} \quad &\Rightarrow \quad A(y), B(x) \\ A_{,y} = -B_{,x} \quad &\Rightarrow \quad A'(y) = -B'(x) = \text{constant} \end{aligned}$$

[40] Linear combinations are trivial, since the equations are linear; however, the commutator is non-trivial.

[41] If we are a bit less lucky, we only find linear combinations of the solutions we already know, in particular the zero field solution.

so that the general solution is

$$\xi \equiv A\partial_x + B\partial_y = k_1 e_1 + k_2 e_2 + k_3 e_3,$$

e_1, e_2 and e_3 being three linearly independent solutions

$$e_1 = \partial_x \qquad e_2 = \partial_y \qquad e_3 = -y\partial_x + x\partial_y$$

(they are linearly independent *over* $\mathbb{R}$; this is a basis of the Lie algebra of Killing fields, *not* to be confused with a basis (in the sense of a frame field) of vector fields in $\mathbb{R}^2[x, y]$!). Their flows are translations along the x and y directions and rotations around the origin $(0, 0)$ respectively. □

4.6.8 Let $x' = x - x_0$, $y' = y - y_0$ be the coordinates in $\mathbb{R}^2$ with respect to the origin, which is translated into (x_0, y_0).

(i) Check that a general Killing vector, expressed in the initial coordinates (x, y) as well as the new coordinates (x', y'), reads

$$\begin{aligned}\xi &= k_1\partial_x + k_2\partial_y + k_3(-y\partial_x + x\partial_y)\\ &= (k_1 - k_3 y_0)\partial_{x'} + (k_2 + k_3 x_0)\partial_{y'} + k_3(-y'\partial_{x'} + x'\partial_{y'})\end{aligned}$$

(ii) give an interpretation of this computation

Hint: (ii) unless the isometry (which may be obtained by the deformation of the identity) is a pure translation (i.e. $k_3 \neq 0$), it may be regarded as a *pure rotation* around the appropriate point (x_0, y_0) (this point is obtained by equating the coefficients of the generators of translations $\partial_{x'}$, $\partial_{y'}$ to zero, or using (4.1.6)). □

4.6.9 Guess (and then test your intuition by plugging the guess into Killing equations) a Killing vector for a general rotational surface discussed in (3.2.6).

Hint: the surface is symmetric with respect to rotations around the z-axis; see (4.1.7). □

4.6.10 Find all Killing vectors for the *(pseudo-)Euclidean space*, i.e. for $E^{p,q} \equiv (\mathbb{R}^n, \eta)$, where η is the Minkowskian metric with the signature (p, q), $p + q = n$. Show that there are three types of flows: translations, rotations and hyperbolic rotations (for $p = 1, q = 3$ they are known as *Poincaré transformations*, for $q = 0$ *Euclidean transformations*, see also (10.1.15) and (12.4.8)).

Hint: in Cartesian coordinates the Killing equations read

$$\xi_{i,j} + \xi_{j,i} = 0 \qquad \xi_i \equiv \eta_{ij}\xi^j$$

Differentiation with respect to x^k gives

$$\xi_{i,jk} + \xi_{j,ik} = 0$$

In full analogy we get

$$\xi_{i,kj} + \xi_{k,ij} = 0 \qquad \xi_{j,ik} + \xi_{k,ji} = 0$$

Then,

$$\xi_{i,jk} = -\xi_{j,ik} = \xi_{k,ij} = -\xi_{i,kj} \quad \Rightarrow \quad \xi_{i,jk} = 0$$

$\Rightarrow$

$$\xi^i = A^i_{\ j} x^j + a^i \qquad A, a = \text{constant}$$

and plugging into the initial equations leads to the restriction for the matrix A

$$(\eta A) + (\eta A)^{\mathrm{T}} = 0 \quad \Rightarrow \quad A \in so(p,q)$$

(see (11.7.6)), i.e.

$$\xi = \left(A^i_{\ j} x^j + a^i\right)\partial_i \equiv \xi^{(A,a)} \qquad (A,a) \in so(p,q) \ltimes \mathbb{R}^n$$

One can check that

$$\xi^{(A,a)} \leftrightarrow -\begin{pmatrix} A & a \\ 0 & 0 \end{pmatrix}$$

is an isomorphism of the Killing algebra with the semidirect sum $so(p,q) \ltimes \mathbb{R}^{p+q}$ (see (12.4.9)). We can verify as well that the field $\xi^{(A,a)}$ may be written in the form

$$\xi^{(A,a)} = \frac{1}{2}(A\eta)^{ij} M_{ji} + a^i P_i$$

($(A\eta)^{ij} = -(A\eta)^{ji}$ being a consequence of $(\eta A) + (\eta A)^{\mathrm{T}} = 0$) where the vector fields

$$M_{ij} \equiv -M_{ji} \equiv x_i \partial_j - x_j \partial_i \qquad P_i \equiv \partial_i \quad x_i \equiv \eta_{ij} x^j$$

constitute a basis of the Killing algebra. Flows: solve the equations for the flow of M_{ij} and P_j respectively. The fields P_j correspond to translations, M_{ij} yield rotations and hyperbolic rotations in the plane (ij), depending on the sign of the product $\eta_{ii}\eta_{jj}$ (not to be summed; $+1$ rotations, -1 hyperbolic rotations (*boosts*)). □

4.6.11 Find the Killing vectors on the standard sphere (S^2, g) from (3.2.4).

Hint: the first possibility is to solve the Killing equations directly in coordinates ϑ, φ. Another (instructive) way: it is clear intuitively that isometries of a sphere are given by all the possible rotations around its centre. The only problem is how to *write down* their generators in coordinates ϑ, φ. This may be achieved in the following way: the generators of the flows corresponding to rotations in $\mathbb{R}^3$ in Cartesian coordinates are known from (4.6.10) (they are M_{ij}). The only thing to do is to express them in *spherical polar* coordinates in $\mathbb{R}^3$ (convince yourself that they are tangent to the spheres centered in the origin) and set[42] $r = 1$ in these formulas. We obtain three vector fields on (S^2, g) and it is now a simple matter to check that they indeed provide solutions of the Killing equations (see also (13.4.6)). □

[42] In principle; in practice there are no rs present.

4.6.12 Find the Killing vectors on a torus (T^2, g) treated in (3.2.2) and (3.2.3).

Hint: on a "curved" torus in $\mathbb{R}^3$ show (by solving the Killing equations) that the only solution is given by the generator of rotations around the z-axis (this is intuitively clear in advance, see (4.6.9)). On a "flat" torus the Killing equations *coincide* with the equations in the Euclidean plane (4.6.7), but the tricky point is that the counterpart of the field e_3, i.e. $-\beta\partial_\alpha + \alpha\partial_\beta$ is *not acceptable*, here, since this field is not (even) continuous on T^2 (its components are not periodic). This means that the (global) topology of the torus selects only part of the solutions offered by (local) Killing equations. □

• A lesson we learned from the last example is that the global characteristics of a manifold may sometimes force us to abandon some Killing vectors we have obtained by local analysis (solving differential Killing equations), so that finally we are left with only *part* of the solutions. In the language of Lie algebras this means that the initial Lie algebra of all solutions of Killing equations reduces to some *subalgebra*, in general.[43] The initial (bigger) algebra carries (*invariant*, coordinate-independent) information about the *local metric* situation on a manifold – the resulting (smaller) one already encodes the *global metric* conditions. It is clear that if two manifolds *differ* in their local Killing algebras, they cannot be *locally isometric* (isometric within sufficiently small domains; if they were, one could choose local coordinates such that both the metric tensors looked *identical*, so that the solutions of Killing equations were the same).

4.6.13 Show that both the (surface of the) cylinder and a cone (3.2.6) happen to be locally isometric with an "ordinary" plane (consequently they may be, after being *slit* – which alters its *global* properties only – painlessly unfolded into the plane). Does this hold for the sphere S^2 (3.2.4), too?

Hint: find a change of coordinates, making metrics of the cylinder and the cone look the same as the metric of the plane (in Cartesian or polar coordinates). For a sphere try to do the same in a reasonable time and then give it up, recalling that the sphere's (both local and global) Killing algebra turned out to be $so(3)$ (see (4.6.11) and (11.7.6)), which *differs* from the Euclidean plane's $e(2)$ (see an argument based on different scalar curvatures in (15.5.7), too). □

• So far we have interpreted Killing equations $\mathcal{L}_\xi g = 0$ in the following way: g is a *given* metric tensor and ξ is an *unknown* generator (*to be determined*) of the symmetry of g (i.e. the isometry of (M, g)). The same equations may be used, however, for just the *opposite* task: for finding the most general form of g, possessing a *prescribed* set of isometries (say, a rotationally invariant metric tensor g). In this case, the same equations are to be solved, just the role of unknowns and known objects has to be interchanged.[44]

[43] On a flat torus, as an example, one starts with a three-dimensional algebra of the plane $e(2)$, but it is to be reduced to its two-dimensional "translation" subalgebra (the whole algebra $e(2)$ happens to be a semidirect sum of a two-dimensional translational and a one-dimensional rotational part, see (12.4.9)).

[44] Note that Killing equations (4.6.6) contain the first derivatives of both the components ξ^i and g_{ij}, so that if we treat any of these objects as being given, we get a system of first-order partial differential equations for the other one.

More generally, *any* equation of the form $\mathcal{L}_\xi A = 0$, where A is a tensor field (not necessarily a metric tensor) may be interpreted in this dual way. We either look for the symmetries of a given tensor field (given A, unknown ξ), or for the most general tensor field A possessing prescribed symmetries (given ξ, unknown A). The latter point of view is especially important for finding exact solutions of complicated partial differential equations; here, one often looks for solutions with the particular type of symmetry. This involves, first, finding the *most general* expression exhibiting this particular type of symmetry and then using this expression in the role of an *ansatz* (a tentative solution containing some freedom, which is then fixed by plugging into the equations). This procedure is often implicitly assumed when one says "*let us search for the solution in the form....*"

4.6.14 Find the most general $\binom{0}{2}$-type tensor in E^n, which is both translation and rotation invariant (*homogeneous* and *isotropic*). What is exceptional about the case $n = 2$? In particular, the metric tensor.

Hint: let h be a tensor to be determined, then $\mathcal{L}_\xi h = 0$ is needed for $\xi = P_i$ and M_{ij} from (4.6.10). P_i lead to constancy of components (with respect to the Cartesian coordinate frame), M_{ij} then results in $h_{ij} = \lambda \delta_{ij}$. For $n = 2$ M_{ij} yields a more general expression $h_{ij} = \lambda_1 \delta_{ij} + \lambda_2 \varepsilon_{ij}$, where ε_{ij} is the two-dimensional Levi-Civita symbol (5.6.1). □

4.6.15 Find the most general rotationally invariant vector field in E^3 ($W = f(r)\partial_r$ in spherical polar coordinates (as expected intuitively)). □

• And what about an analog of the Killing equations for the case of *conformal* transformations?

4.6.16 Let $\Phi_t : M \to M$ be a one-parameter group (flow) of conformal transformations, generated by a vector field ξ. Show that

(i) ξ satisfies the *conformal Killing equations*

$$\mathcal{L}_\xi g = \chi g \qquad \chi \text{ "arbitrary" (unknown) function}$$

(ii) $\chi =$ constant corresponds to homotheties.

Hint: differentiate the defining equation $\Phi_t^* g = \sigma(x, t) g$ with respect to t in $t = 0$; $\chi \equiv \partial_t|_0 \sigma(x, t)$. □

4.6.17 Check that

(i) the component expression of conformal Killing equations reads

$$\xi^k g_{ij,k} + \xi^k{}_{,i} g_{kj} + \xi^k{}_{,j} g_{ik} = \chi g_{ij}$$

(ii) it is an overdetermined system of equations for the unknown functions $\xi^1(x), \dots, \xi^n(x) \Rightarrow$ no (non-vanishing) solution is guaranteed, in general

(iii) given ξ and η two solutions of the Killing equations, then both

$$\xi + \lambda \eta \qquad \text{and} \qquad [\xi, \eta]$$

represent solutions as well

(iv) *conformal Killing vectors* (= solutions of conformal Killing equations) constitute a subalgebra of the Lie algebra of all vector fields on M and Killing vectors, in turn, constitute a subalgebra of the Lie algebra of conformal Killing vectors.

Hint: see (4.6.6). □

• We mentioned already (see the text after (4.6.6)) the Lie algebra of Killing vectors is always finite-dimensional. As a rule, the Lie algebra of conformal Killing vectors is finite-dimensional, too; however, there exist important exceptions.

4.6.18 Check that the conformal Killing algebra of an ordinary Euclidean plane happens to be *infinite*-dimensional.

Hint (cf. (4.6.7)): denote $\xi^1(x, y) \equiv A(x, y), \xi^2(x, y) \equiv B(x, y)$. Then the conformal Killing equations read

$$A_{,x} = B_{,y} \ (= \chi/2) \qquad A_{,y} = -B_{,x}$$

These are, however, just *Cauchy–Riemann relations* for a complex function

$$\psi(z) := A(x, y) + iB(x, y) \quad z = x + iy$$

This means that *any holomorphic* function $f(z) = u + iv$ yields a conformal Killing vector $\xi = u(x, y)\partial_x + v(x, y)\partial_y$. In particular, the powers $z^n, n = 0, 1, \dots$ generate an infinite number of solutions (with the first degree polynomial $P_1(z) = ik_3 z + k_1 + ik_2, k_1, k_2, k_3 \in \mathbb{R}$ corresponding to *isometries* from (4.6.7)). □

• A connection between conformal transformations of the Euclidean plane and holomorphic functions may be understood in an alternative way, too.

The complex plane $\mathbb{C}$ may be regarded as a (two-dimensional real) manifold $\mathbb{R}^2$. We use either Cartesian coordinates (x, y) or complex[45] coordinates $(z, \bar{z})$ on it, the latter being defined standardly as

$$z = x + iy, \quad \bar{z} = x - iy \qquad \text{or} \quad x = \frac{1}{2}(z + \bar{z}), \quad y = \frac{1}{2i}(z - \bar{z})$$

Consider (smooth) functions on $\mathbb{C} \equiv \mathbb{R}^2$ with values *in* $\mathbb{C}$, *too*, $f : \mathbb{C} \to \mathbb{C}$. There are two "ordinary" (real-valued) functions $f(z, \bar{z}) = u(x, y) + iv(x, y)$ "hidden" in it. A key restriction is given by the introduction of *holomorphic* functions.

4.6.19 We say that f is a *holomorphic function*, if it "does not depend on $\bar{z}$," i.e. if it satisfies

$$\partial_{\bar{z}} f = 0 \qquad \text{so that} \quad f = f(z)$$

Check that

[45] Strictly speaking, this already needs an extension of the formalism to "V-valued" tensors (V being a vector space; here $V = \mathbb{C}$), to be discussed in more detail in Sections 6.4 and 8.6.

(i) in coordinates (x, y) this yields

$$(\partial_x + i\partial_y)f = 0$$

(ii) in terms of the functions u, v this results in the *Cauchy–Riemann relations*

$$\partial_{\bar{z}} f = 0 \quad \Leftrightarrow \quad \partial_x u = \partial_y v \quad \partial_x v = -\partial_y u$$

□

4.6.20 We may use either coordinate frame fields dx, dy or $dz \equiv dx + i\, dy, d\bar{z} \equiv dx - i\, dy$ for decomposition of a general covector field on $\mathbb{C}$. Check that

(i) the coordinate expression of the standard metric tensor on the Euclidean plane E^2 in coordinates $z, \bar{z}$ reads

$$g = \frac{1}{2}(dz \otimes d\bar{z} + d\bar{z} \otimes dz)$$

(ii) if we consider a map $F : E^2 \to E^2$ of the (appropriate part of the) plane given by a *holomorphic* function $w(z)$ (i.e. a map $z \mapsto w(z)$, obeying $\partial w(z)/\partial z \neq 0$), the metric tensor transforms as follows:

$$g \mapsto F^* g = \sigma g \qquad \left|\frac{\partial w(z)}{\partial z}\right|^2 \equiv \sigma > 0$$

so that *each* such map of (an appropriate part of) the plane is a *conformal* map (it preserves all angles of mutually intersecting lines)

(iii) if we take, as an example, the (holomorphic) function $w(z) = z^2$, the corresponding conformal map (of the first quadrant onto the upper half-plane) reads (when expressed in polar coordinates) as $(r, \varphi) \mapsto (r^2, 2\varphi)$ and we get $\sigma = 4|z|^2 \equiv 4r^2$.

Hint: $F^* dz = dw(z) = (\partial w(z)/\partial z)\, dz$. □

• Conformal Killing algebras corresponding to higher-dimensional Euclidean spaces turn out to be already finite-dimensional.

4.6.21* Find the conformal Killing vectors for E^n, $n \neq 2$. Show that the corresponding Lie algebra is isomorphic to $so(1, n+1)$ (11.7.6).

Hint: modify the procedure used in (4.6.10). Manipulating the equations one has to show that $\xi_{i,jkl} = 0$ here, so that $\xi^i = A^i_{jk} x^j x^k + A^i_j x^j + A^i, \ldots$. □

4.6.22 Check that the vector field V encountered in (4.1.5) is a conformal Killing vector in E^3, which corresponds to a *homothety*. □

4.6.23 Let M be the surface of a cone in E^3, which makes an angle of 45° with the z-axis, endowed with a metric induced from E^3. Check that there exists a conformal Killing vector on M of the form $V = f(z)\partial_z$ (the remaining coordinate being φ). Find the flow of the field V (the corresponding finite conformal transformations of M).

Hint: see (3.2.6), $z(r) = r$. □

4.6.24 In the mechanics of elastic continua one introduces the *strain tensor* in the following way: when the points in the continuum are (infinitesimally) displaced according to $\mathbf{r} \mapsto \mathbf{r} + \mathbf{u}(\mathbf{r})$ (a vector field[46] $\mathbf{u}(\mathbf{r})$ is called the *displacement* (field)), the corresponding deformation is encoded in a second-rank tensor (field) with components (in Cartesian coordinates)

$$\varepsilon_{ij} := \frac{1}{2}(\partial_i u_j + \partial_j u_i)$$

Check that the coordinate-free expression of this tensor reads

$$\varepsilon = \frac{1}{2}\mathcal{L}_{\mathbf{u}} g$$

where g is the (standard) metric tensor in E^3 and that it follows from the definition of the Lie derivative as well as from the context that a *deformation* of the medium (a shift of points, which *alters* distances between them) is measured by the Lie derivative of a metric tensor ($\varepsilon = 0 \Leftrightarrow$ a deformation did not take place $\Leftrightarrow$ it is an *isometry*).

Hint: see (4.6.5). □

4.6.25 In the hydrodynamics of viscous fluids we encounter a tensor (field), which resembles the strain tensor, with *velocity field* $\mathbf{v}$ of the fluid's flow replacing, however, the displacement $\mathbf{u}$. It is called the *rate of deformation tensor* or the *strain-rate tensor*. Namely, the η-multiple (η = coefficient of viscosity) of this tensor stands for an inner friction (viscosity) part of the *stress tensor*. The full stress tensor of the viscous fluid then reads (in Cartesian coordinates)

$$\sigma_{ij} = -p\delta_{ij} + \eta(\partial_i v_j + \partial_j v_i)$$

Check that

(i) the coordinate-free expression of this tensor is

$$\sigma = -pg + \eta\mathcal{L}_{\mathbf{v}} g$$

g being the (standard) metric tensor in E^3

(ii) the Lie derivative of g just corresponds intuitively to Newton's idea of a phenomenological description of viscosity: a term responsible for viscosity is to be proportional to the "relative velocity of the nearby points" (the force is due to the friction between adjacent layers of the fluid; if they are not moving with respect to each other, there is no reason for the frictional force to arise; a quicker motion results in a larger transfer of momentum, so that the resulting force increases).

Hint: $\mathcal{L}_{\mathbf{v}} g$ measures the rate of deformation, i.e. the rate of change of relative distances of points in the fluid. □

4.6.26 Given Φ_t a flow of isometries on M ($\Phi_t^* g = g$) let $\mathcal{A}'$ be an atlas (local coordinates) on M, which arises from $\mathcal{A}$ through a displacement by the flow (it uses coordinates $x_t^i :=$

[46] A shift $\mathbf{r} \mapsto \mathbf{r} + \mathbf{u}(\mathbf{r})$ is interpreted as an infinitesimal *flow* generated by a vector field $\mathbf{u}(\mathbf{r})$.

$\Phi_t^* x^i \equiv x^i \circ \Phi_t$ of the *co-moving observer* (drifted by the flow Φ_t), see (4.1.11)). Check that

(i) it holds then that

$$g = g_{ij}(x)\, dx^i \otimes dx^j = g_{ij}(x_t)\, dx_t^i \otimes dx_t^j$$

so that the components of the metric tensor are the *same* both in initial and in transformed coordinates[47]

(ii) if the components have been *constant* in the initial coordinates, they still remain constant (being *the same* constants)

(iii) in particular, for the Euclidean metric in E^n, the matrix of components turns out to be the identity matrix in Cartesian coordinates used by the arbitrarily oriented observer.

Hint: (4.1.11), (1.4.13); (i) $g \equiv g_{ij}(x)\, dx^i \otimes dx^j = \Phi_t^* g \equiv g_{ij}(x_t)\, dx_t^i \otimes dx_t^j$. □

• This result is used sometimes (mainly in strictly coordinate sources) for a derivation of the Killing equations: one looks for a collection of functions $\xi^i(x)$ such that the functional form of the components $g_{ij}(x)$ remains (up to first order) unchanged under the infinitesimal transformations of coordinates $x^i \mapsto x_\epsilon^i(x) \equiv x^i + \epsilon \xi^i(x)$. What we get defines the (infinitesimal) flow $\Phi_\epsilon : x^i \mapsto x^i + \epsilon \xi^i(x)$ (and eventually the "finite" one $\Phi_t \leftrightarrow \xi$).

Summary of Chapter 4

Each vector field V on M naturally induces a map $\Phi_t : M \to M$, which translates a point x along the integral curve starting in x by the parametric distance t. It is called the flow generated by V or, taking into account its composition property $\Phi_{t+s} = \Phi_t \circ \Phi_s$, a one-parameter group of transformations. According to the results of Chapter 3 the map Φ_t of a manifold M onto itself induces a mapping of tensor fields Φ_t^*, which is called the *Lie transport* of tensors (along the integral curves of the field V). The natural measure of sensitivity of a tensor field A to Lie transport is the *Lie derivative*. One can assign to any two vector fields V, W a third one, their commutator $[V, W]$ (which happens to coincide with $\mathcal{L}_V W$). Two fields commute if and only if their flows do; non-commuting of vector fields thus results in anholonomy phenomena (dependence on the path). A Killing vector is a vector field with respect to which the metric tensor is Lie constant. The flow of a Killing vector is the isometry of a Riemannian manifold (M, g), i.e. a map of M onto itself which preserves all lengths and angles. If the angles alone are preserved, we speak of conformal transformations and the corresponding generators are called conformal Killing vectors.

$\Phi_{t+s} = \Phi_t \circ \Phi_s$	"Composition" property of a flow	(4.1.2)
$\Phi_t^* A = A$	A is Lie invariant (dragged)	Sec. 4.2
$\mathcal{L}_V A := (d/dt)_0 \Phi_t^* A$	Lie derivative of A along $V \leftrightarrow \Phi_t$	Sec. 4.2

[47] There are two Jacobian matrices there, in general; now, the situation is fairly specific and they drop out (in matrix notation $G \mapsto J^T G J = G$).

$\mathcal{L}_V(A+\lambda B)=\mathcal{L}_V A+\lambda\mathcal{L}_V B$	Lie derivative of a linear combination	(4.3.1)
$\mathcal{L}_V(A\otimes B)=\mathcal{L}_V A\otimes B+A\otimes\mathcal{L}_V B$	Lie derivative of a tensor product	(4.3.1)
$\mathcal{L}_V\circ C=C\circ\mathcal{L}_V$	Lie derivative commutes with contractions	(4.3.1)
$\mathcal{L}_V W=[V,W]$	Lie derivative of W along V	(4.3.6)
$\mathcal{L}_{V+\lambda W}=\mathcal{L}_V+\lambda\mathcal{L}_W$	Lie derivative along a linear combination	(4.3.8)
$\mathcal{L}_{[V,W]}=[\mathcal{L}_V,\mathcal{L}_W]$	Lie derivative along a commutator	(4.3.8)
$\Phi_t^*=e^{t\mathcal{L}_V}\equiv 1+t\mathcal{L}_V+\cdots$	Exponent of the Lie derivative	(4.4.2)
$\Phi_{-\epsilon}^W\circ\Phi_{-\epsilon}^V\circ\Phi_{-\epsilon^2}^{[V,W]}\circ\Phi_\epsilon^W\circ\Phi_\epsilon^V=\hat{1}+\cdots$	Interpretation of the commutator $[V,W]$	(4.5.2)
$l[f\circ\gamma,g]=l[\gamma,f^*g]$	Behavior of the length functional	(4.6.1)
$f^*g=g$	f is an isometry of (M,g)	(4.6.2)
$f^*g=\sigma g$	f is a conformal transformation of (M,g)	(4.6.3)
$\mathcal{L}_\xi g=0$	Killing equations (ξ generates isometries)	(4.6.5)
$f^*\eta=\eta$	f is the Poincaré transformation	(4.6.10)
$\mathcal{L}_\xi g=\chi g$	Conformal Killing equations	(4.6.16)
$\varepsilon=\frac{1}{2}\mathcal{L}_{\mathbf{u}}g$	Strain tensor (elastic continuum)	(4.6.24)
$\frac{1}{2}\mathcal{L}_{\mathbf{v}}g$	Strain-rate tensor (viscous fluids)	(4.6.25)

5
Exterior algebra

• In Chapter 2 we met tensor fields on a manifold. It turns out that a prominent role is played in geometry by a specific class of tensor fields, namely the totally antisymmetric (reversing sign under interchange of any pair of arguments), fully covariant (with lower indices only) tensor fields. They are known as *differential forms*, or simply *forms*. The power and beauty of forms ultimately springs from a simple observation in linear algebra (see Section 5.1) that just these objects (their linear space prototypes) provide the ideal tool in order to introduce the concept of the *volume* of a parallelepiped in a linear space. The volume of the (infinitesimal) parallelepiped is a key element within the context of integration[48] and the integral calculus is closely related to differential calculus. These are the reasons why differential forms occur naturally as objects of the highest importance in differential as well as in integral calculus on manifolds.

As we will see, there are several algebraic and differential operations which are specific for forms and, in a sense, forms represent the *only* objects one can integrate at all (i.e. *each* integral may be regarded as an integral of a differential form; in this sense differential forms may be understood simply as the quantities under the integral sign, too).

In this chapter a *linear algebra* of forms will be discussed, which is just a part of the theory of tensors. Thus, it may be regarded as a continuation of Section 2.4, which dealt with the algebra of *general* tensors of type $\binom{p}{q}$. Here we restrict ourselves to the features that are specific for totally antisymmetric tensors of type $\binom{0}{p}$.

5.1 Motivation: volumes of parallelepipeds

• A place where the introduction of forms is most natural is the computation of the volume of the parallelepiped. Let us have a look at how this takes place.

5.1.1 Given three vectors **a**, **b** and **c** in $L \equiv \mathbb{R}^3$, imagine we want to compute the volume of a parallelepiped spanned by them. After a short browse through a suitable reference book entitled "Mathematical formulas" (or on Google; experts browse in their memory, true experts derive the formula quickly from scratch) we come to the result $V = |(\mathbf{a} \times \mathbf{b}) \cdot \mathbf{c}|$.

[48] An integral of a function equals a limit of sums of numbers, each of them being the product of the volume of an infinitesimal parallelepiped and the value of the function somewhere inside this parallelepiped.

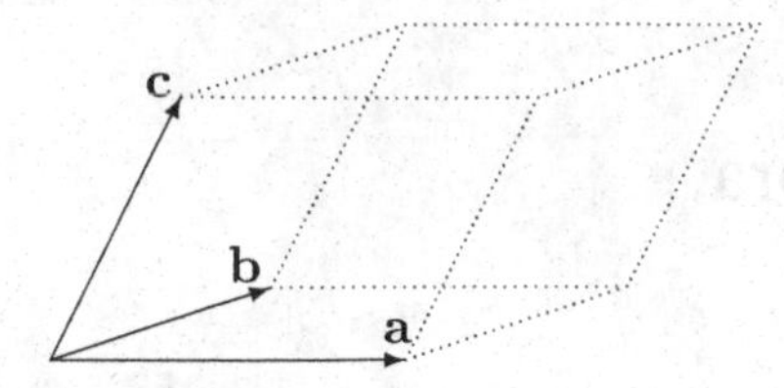

This expression has a fairly remarkable structure, especially if we concentrate on the part *inside* the absolute value (the latter only guarantees non-negativity of the volume). Check that

(i) the map

$$\tilde{V} : L \times L \times L \to \mathbb{R} \qquad \tilde{V}(\mathbf{a}, \mathbf{b}, \mathbf{c}) := (\mathbf{a} \times \mathbf{b}) \cdot \mathbf{c}$$

is a tensor of type $\binom{0}{3}$

(ii) the tensor is completely antisymmetric, i.e. the interchange of any two arguments results in the change of sign of the resulting number

$$\tilde{V}(\mathbf{a}, \mathbf{b}, \mathbf{c}) = -\tilde{V}(\mathbf{b}, \mathbf{a}, \mathbf{c}) = -\tilde{V}(\mathbf{c}, \mathbf{b}, \mathbf{a}) = -\tilde{V}(\mathbf{a}, \mathbf{c}, \mathbf{b})$$

(iii) if $\mathbf{e}_1, \mathbf{e}_2, \mathbf{e}_3$ is any right-handed (to be defined more precisely later, see Section 5.5) orthonormal basis in L, then the value $\tilde{V}$ on this particular triplet is $\tilde{V}(\mathbf{e}_1, \mathbf{e}_2, \mathbf{e}_3) = 1$

(iv) interpret (iii).

Hint: (iv) the volume of a unit cube is equal to 1. □

5.1.2 Repeat the analysis from (5.1.1) for $L = \mathbb{R}^2$, i.e. for the area $P(\mathbf{a}, \mathbf{b})$ of a parallelogram spanned by two vectors $\mathbf{a}, \mathbf{b} \in L = \mathbb{R}^2$.

Hint: an explicit formula may be obtained from $P(\mathbf{a}, \mathbf{b}) = \tilde{V}(\mathbf{a}, \mathbf{b}, \mathbf{e}_3)$. □

• These results enable one to *define* the volume of a parallelepiped spanned by an n-tuple (be wise, generalize) of vectors $v, \dots, w \in L$ ($n = \dim L$) in a natural way as

$$P(v, \dots, w) := |\tilde{P}(v, \dots, w)|$$

with $\tilde{P}$ being a *tensor* of type $\binom{0}{n}$, which is completely antisymmetric

$$\tilde{P}(\dots, v, \dots, w, \dots) = -\tilde{P}(\dots, w, \dots, v, \dots)$$

The expression $\tilde{P}(v, \dots, w)$ itself is known as the *oriented volume* of the parallelepiped spanned by the vectors $v, \dots, w$. The oriented volume may be both positive and negative, its sign depends on the order of the arguments and the "usual" volume is given by its absolute value. Note that from the perspective of linear algebra the concept of the oriented volume is, in fact, *simpler* than the "usual" one (the absolute value *spoils* both multilinearity and antisymmetry).

5.1.3 A parallelepiped is said to be *degenerate* if it is spanned by a system of linearly dependent vectors.

(i) Find the visual meaning of this for $n = 2, 3$

(ii) check that the volume of such a parallelepiped vanishes

(iii) check that, conversely, the requirement of vanishing of the volume of *any degenerate* parallelepiped plus the linearity yields antisymmetry (which is a natural motivation to pay attention to just such tensors).

Hint: (ii) antisymmetry of $\tilde{V}$; (iii) first show that $\tilde{P}(\dots, v, \dots, v, \dots) = 0$, then put $v = u + w$. □

5.2 p-forms and exterior product

• This section includes material to familiarize the reader with the algebra of completely antisymmetric tensors with lower indices.

Given an n-dimensional (real) linear space L, consider $T^0_p(L)$, the space of tensors of type $\binom{0}{p}$ over L (cf. Section 2.4). A tensor $\alpha \in T^0_p(L)$ will be called a *p-form* in L if it is *completely antisymmetric*, i.e. if

$$\alpha(\dots, v, \dots, w, \dots) = -\alpha(\dots, w, \dots, v, \dots)$$

The collection of p-forms in L will be denoted by $\Lambda^p L^*$ (the origin of the star becomes clear in Section 5.3). The definition makes sense only for $p \geq 2$ (when there is something to be interchanged). The structure to be obtained extends, however, to $p = 0, 1$, too, if one *defines*

$$\Lambda^0 L^* := T^0_0(L) \equiv \mathbb{R} \qquad \Lambda^1 L^* := T^0_1(L) \equiv L^*$$

0-forms thus being simply real numbers and 1-forms coinciding with covectors.

5.2.1 Thus, in general, p-forms are those tensors of type $\binom{0}{p}$ which happen to be completely antisymmetric *whenever it makes sense*. Check that

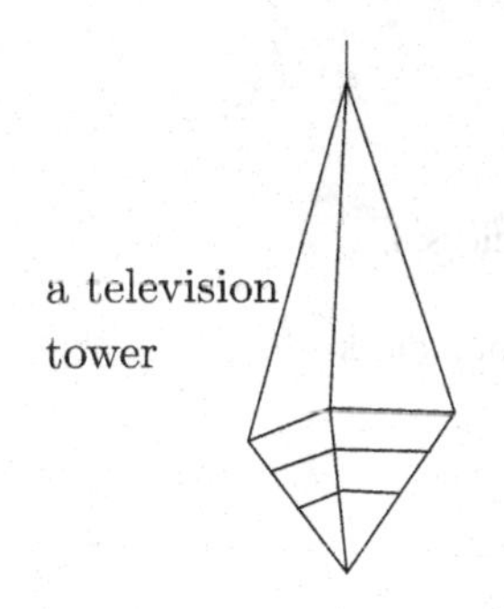

(i) in components

$$\alpha \in \Lambda^p L^* \quad \Leftrightarrow \quad \alpha_{\dots a \dots b \dots} = -\,\alpha_{\dots b \dots a \dots}$$

(ii) $\Lambda^p L^*$ is a *subspace* of $T^0_p(L)$

(iii) the dimension of the space of p-forms in n-dimensional space is

$$\dim \Lambda^p L^* = \binom{n}{p} \equiv \frac{n!}{(n-p)!p!}$$

(iv) $\dim \Lambda^p L^* = \dim \Lambda^{n-p} L^*$

(v) non-zero p-forms can only exist for $p = 0, 1, \dots, n$.

Forms thus resemble a television tower rather than (an infinite modification of) an inverted pyramid from Giza (cf. the note before (2.4.7)). The tower widens only up to half its height and then starts to narrow down; item (v) then guarantees its finite height.

Hint: (iii) from (i) it follows that a component $\alpha_{a\dots b}$ may be non-vanishing only if there are *no repeating* values of indices in it (all the indices take *different* values). The number of mutually independent components of a general form α ($\equiv \dim \Lambda^p L^*$) thus coincides with the number of ways in which one can choose p numbers out of n numbers, i.e. $\binom{n}{p}$; (iv) a property of the combinatorial numbers $\binom{n}{p}$; (v) for $p > n$ at least two indices necessarily have the same value. □

• Thus we have learned an important, albeit fairly simple fact that, unlike the infinite "tower" of general tensors, non-trivial (non-vanishing) p-forms *terminate* at $p = n = \dim L$.

5.2.2 Given a tensor $t \in T^0_p(L)$ we define the tensor $\pi^A t \in T^0_p(L)$ as follows:

$$(\pi^A t)(v, \ldots, w) := \frac{1}{p!} \sum_\sigma (\operatorname{sgn} \sigma) t(\sigma(v, \ldots, w))$$

with σ being a permutation and $\operatorname{sgn} \sigma$ being $+1$ for even permutations and -1 for odd ones. Check that

(i) π^A is a tensor operation (i.e. $\pi^A t$ is indeed a tensor)
(ii) π^A is a *projector* onto p-forms, i.e.

$$\pi^A : T^0_p(L) \to \Lambda^p L^* \subset T^0_p(L) \qquad \pi^A \circ \pi^A = \pi^A$$

The tensor $\pi^A t$ is said to represent the *antisymmetric part* (= skew-symmetric part) of the tensor t (2.4.16)
(iii) for $p = 1, 2, 3$ we have explicitly

$$(\pi^A t)(v) = t(v)$$
$$(\pi^A t)(v, w) = \frac{1}{2!}(t(v, w) - t(w, v))$$
$$(\pi^A t)(u, v, w) = \frac{1}{3!}(t(u, v, w) + t(w, u, v) + t(v, w, u) - t(v, u, w) - t(w, v, u) - t(u, w, v))$$

(iv) in components π^A reads

$$t_{a\ldots b} \mapsto (\pi^A t)_{a\ldots b} = t_{[a\ldots b]}$$

where the square brackets denote *complete antisymmetrization* in indices, i.e.

$$t_{[a\ldots b]} := \frac{1}{p!}(t_{a\ldots b} \pm \text{all the remaining permutations of indices})$$

(+ for even, − for odd permutations)
(v) for $p = 2, 3$ explicitly

$$t_{[ab]} = \frac{1}{2!}(t_{ab} - t_{ba})$$
$$t_{[abc]} = \frac{1}{3!}(t_{abc} + t_{cab} + t_{bca} - t_{bac} - t_{cba} - t_{acb})$$

Hint: (ii) permutations constitute a group S_p with respect to composition, $\sigma \mapsto \operatorname{sgn} \sigma$ is a homomorphism $S_p \to \{1, -1\}$, an interchange of two arguments is realized by a *transposition* $\hat\sigma$, which has $\operatorname{sgn} \hat\sigma = -1$, $\sum_\sigma 1 = p!$ □

• The fact that p-forms constitute a subspace of the space of all tensors of type $\binom{0}{p}$ means that they are closed with respect to linear combinations. But tensors may be multiplied (tensorially) with one another too, so that a natural question arises as to whether the forms are closed with respect to this kind of multiplication. One easily verifies that *they are not*.

5.2.3 Check that the tensor product of a p-form and a q-form is no longer a form in general (it is "only" *a tensor* of type $\binom{0}{p+q}$), i.e.

$$\otimes : \Lambda^p L^* \times \Lambda^q L^* \to T^0_{p+q}(L)$$

Hint: a tensor $\alpha_{a\ldots b}\beta_{c\ldots d}$ is antisymmetric within both subgroups $(a\ldots b)$ and $(c\ldots d)$, but it is not within the whole group $(a\ldots bc\ldots d) \Rightarrow \alpha\otimes\beta \notin \Lambda^{p+q}L^*$. □

• However, since the resulting object happens to be a tensor with lower indices only, we may project out its antisymmetric part with the help of π^A, which already *yields* a form! In this way one arrives at the definition of a new product, which is specific for forms. It is called the *exterior product*, denoted by $\wedge$ and[49] it is defined by

$$\wedge : \Lambda^p L^* \times \Lambda^q L^* \to \Lambda^{p+q} L^* \qquad (\alpha,\beta) \mapsto \alpha\wedge\beta$$

$$\wedge := \frac{(p+q)!}{p!q!}\,\pi^A \circ \otimes \qquad \text{i.e.} \quad \alpha\wedge\beta := \frac{(p+q)!}{p!q!}\,\pi^A(\alpha\otimes\beta)$$

□

(Awkward factors, containing factorials, occur inevitably in the theory of forms. They originate in combinatorics and (unfortunately) one cannot avoid them, indeed. There are two main conventions, differing in *where exactly* these factors do appear (one of them is used in this book). If we had not used, for example, the above-mentioned factor in the definition of the exterior product (which is possible), several factors would emerge elsewhere. This should be borne in mind, in particular, when using various sources dealing with forms: in different conventions "the same" formulas may contain different factors.)

5.2.4 Verify that the exterior product has the following properties:

(i) bilinearity

$$\alpha\wedge(\beta+\lambda\tau) = \alpha\wedge\beta + \lambda\alpha\wedge\tau \qquad \lambda\in\mathbb{R}$$
$$(\beta+\lambda\tau)\wedge\alpha = \beta\wedge\alpha + \lambda\tau\wedge\alpha$$

(ii) associativity

$$(\alpha\wedge\beta)\wedge\gamma = \alpha\wedge(\beta\wedge\gamma)$$

(iii) *$\mathbb{Z}$-graded commutativity* (see Appendix A.5)

$$\alpha\wedge\beta = (-1)^{pq}\beta\wedge\alpha \qquad \alpha\in\Lambda^p L^*,\ \beta\in\Lambda^q L^*$$

Hint: (ii) in addition to prefactors one should check that $\alpha_{[[a\ldots b}\beta_{c\ldots d]}\gamma_{e\ldots f]} = \alpha_{[a\ldots b}\beta_{[c\ldots d}\gamma_{e\ldots f]]}$, for which (5.2.6) may come in handy; (iii) there holds

$$(\alpha\wedge\beta)_{a\ldots bc\ldots d} \sim \alpha_{[a\ldots b}\beta_{c\ldots d]} = \beta_{[c\ldots d}\alpha_{a\ldots b]} \sim (\beta\wedge\alpha)_{c\ldots da\ldots b}$$

($\sim$ means that factorials are not written explicitly, but we take the sign seriously; think over the validity of $=$). Now one has to interchange the group $(c\ldots d)$ with $(a\ldots b)$ (in order to

[49] Sometimes, this is read as *wedge*.

get the $(a \dots bc \dots d)$-component of the result), from where a factor $(-1)^{pq}$ emerges (each interchange of a pair of indices results in (-1)). □

• Both associativity and bilinearity of $\wedge$ are properties inherited from the tensor product $\otimes$ (2.4.7). Graded commutativity, to the contrary, is specific for the exterior product and it has no counterpart within tensor multiplication (the tensors $\alpha \otimes \beta$ and $\beta \otimes \alpha$ are not related, in general). This means, from the practical point of view, that two forms *commute* as a rule, with the exception of both degrees being odd (both p and q odd) when they *anticommute*.

5.2.5 Check that

(i) in components we have

$$(\alpha \wedge \beta)_{a\dots bc\dots d} = \frac{(p+q)!}{p!q!} \alpha_{[a\dots b}\beta_{c\dots d]}$$

(ii) multiplication by a scalar may be regarded as an *exterior* product:

$$\lambda\alpha = \lambda \wedge \alpha = \alpha \wedge \lambda \qquad \lambda \in \mathbb{R} \equiv \Lambda^0 L^*$$

□

• The following exercises in *index gymnastics*[50] will prove to be useful in what follows.

5.2.6 Justify the legitimacy of the following steps (α, β, A, t being arbitrary indexed objects commuting one with another, such as the components of tensors)

(i)

$$\alpha_{[a\dots b]}\beta^{a\dots b} = \alpha_{[a\dots b]}\beta^{[a\dots b]} = \alpha_{a\dots b}\beta^{[a\dots b]}$$
$$\alpha_{(a\dots b)}\beta^{a\dots b} = \alpha_{(a\dots b)}\beta^{(a\dots b)} = \alpha_{a\dots b}\beta^{(a\dots b)}$$

(ii)

$$A^a_{[c} \dots A^b_{d]} = A^{[a}_{[c} \dots A^{b]}_{d]} = A^{[a}_{c} \dots A^{b]}_{d}$$
$$A^a_{(c} \dots A^b_{d)} = A^{(a}_{(c} \dots A^{b)}_{d)} = A^{(a}_{c} \dots A^{b)}_{d}$$

(iii)

$$t^{\dots}_{[\dots a\dots[\dots b\dots c\dots]\dots d\dots]} = t^{\dots}_{[\dots a\dots b\dots c\dots d\dots]}$$
$$t^{\dots}_{(\dots a\dots(\dots b\dots c\dots)\dots d\dots)} = t^{\dots}_{(\dots a\dots b\dots c\dots d\dots)}$$

(iv)

$$t^{\dots}_{[\dots a\dots(\dots b\dots c\dots)\dots d\dots]} = t^{\dots}_{(\dots a\dots[\dots b\dots c\dots]\dots d\dots)} = 0$$

where the round brackets represent complete *symmetrization* (all the terms on the right are to be summed *with a plus sign* in the definition from (5.2.2)). The idea of (i)–(iii) is to recognize typical situations, in which some (anti)symmetrizations may be omitted (or conversely added formally), since they are ensured automatically by means of other (anti)symmetrizations;

[50] They used to be fairly popular in those fitness centers in which both square and round brackets are installed.

(iv) says that a symmetrization, when performed *inside* an antisymmetrization (and vice versa), gives zero. □

• We know from (2.4.7) that the tensors $\overbrace{e^a \otimes \cdots \otimes e^b}^{p \text{ pieces}}$ may be used as a basis of $T^0_p(L)$. Their number equals n^p, just matching the dimension of the space $T^0_p(L)$. Since the dimension of the subspace of p-forms is (for $p \geq 2$) *lower* than this ($\dim \Lambda^p L^* = \binom{n}{p} < n^p = \dim T^0_p(L)$), it is clear that although p-forms may be decomposed with respect to the basis of "general" tensors of type $\binom{0}{p}$, we can make do very well with a more economical basis, containing just $\binom{n}{p}$ members.

5.2.7 Check that

(i) the *antisymmetrized* tensor basis

$$e^{[a} \otimes \cdots \otimes e^{b]} \equiv \pi^{\mathrm{A}}(e^a \otimes \cdots \otimes e^b)$$

is enough to decompose any p-form

(ii) those tensors $e^{[a} \otimes \cdots \otimes e^{b]}$ which obey $a < \cdots < b$ form a basis of $\Lambda^p L^*$.

Hint: (i) $\alpha = \alpha_{a \cdots b} e^a \otimes \cdots \otimes e^b = \alpha_{[a \cdots b]} e^a \otimes \cdots \otimes e^b \overset{(5.2.6)}{=} \alpha_{a \cdots b} e^{[a} \otimes \cdots \otimes e^{b]}$; (ii) the rest may differ at most by a sign. □

• It turns out that these objects may be expressed in a simple and very useful form, containing just *exterior* products of the basis 1-forms (covectors) e^a.

5.2.8 Given two *1-forms* α and β, check that

(i)

$$\alpha \wedge \beta = \alpha \otimes \beta - \beta \otimes \alpha$$

(ii) in particular, for basis 1-forms we have

$$e^a \wedge e^b = 2!\, e^{[a} \otimes e^{b]} \equiv 2!\, \pi^{\mathrm{A}}(e^a \otimes e^b)$$

Hint: (i) evaluate both sides on general arguments u, v, making use of (5.2.5) and (2.4.7). □

5.2.9 Check that

(i)

$$\underbrace{e^a \wedge \cdots \wedge e^b}_{p \text{ entries}} = p!\, e^{[a} \otimes \cdots \otimes e^{b]} \equiv p!\, \pi^{\mathrm{A}}(e^a \otimes \cdots \otimes e^b)$$

(ii) any p-form may be written in the following standard way:

$$\alpha = \frac{1}{p!} \alpha_{a \ldots b}\, e^a \wedge \cdots \wedge e^b$$

(iii) there holds

$$\underbrace{e^a \wedge \dots \wedge e^b}_{p \text{ entries}}(v, \dots w) = p!\, v^{[a} \dots w^{b]}$$

Hint: (i) induction with respect to p, (5.2.5) and (5.2.6); (ii) (5.2.7) and (i). □

• Expressing forms in terms of *exterior* products of basis 1-forms turns out to be highly convenient, indeed, making practical manipulations with forms so simple as to border on the trivial. To make this clearer, let us look at an algorithm for computation of an exterior product $\alpha \wedge \beta$ of two forms. The properties of $\wedge$ (discussed in problems (5.2.4) and (5.2.5)) result in the following (amazingly simple) instructions for use:

1. juxtapose the forms (both expressed in terms of exterior products alone)
2. multiply out all terms
3. reshuffle all constants to the left
4. *delete* those of the resulting terms which contain some basis covector e^a *more than once* (such terms *vanish* because of the *anti*commutation of the basis covectors: $e^a \wedge e^b = -e^b \wedge e^a \Rightarrow e^1 \wedge e^1 = e^2 \wedge e^2 = \dots = 0$).

As an illustration, consider $\dim L = 3$, a basis of L^* being e^1, e^2, e^3,

$$\alpha = 2e^1 + e^3 \qquad \beta = -3e^1 \wedge e^3 + 4e^2 \wedge e^3$$

Then,

$$\begin{aligned}\alpha \wedge \beta &= (2e^1 + e^3) \wedge (-3e^1 \wedge e^3 + 4e^2 \wedge e^3) \\ &= -6\underbrace{e^1 \wedge e^1}_{0} \wedge e^3 + 8e^1 \wedge e^2 \wedge e^3 - 3\underbrace{e^3 \wedge e^1}_{-e^1 \wedge e^3} \wedge e^3 + 4\underbrace{e^3 \wedge e^2}_{-e^2 \wedge e^3} \wedge e^3 \\ &= 8e^1 \wedge e^2 \wedge e^3 + 3e^1 \wedge \underbrace{e^3 \wedge e^3}_{0} - 4e^2 \wedge \underbrace{e^3 \wedge e^3}_{0} \\ &= 8e^1 \wedge e^2 \wedge e^3\end{aligned}$$

We see that three out of four terms drop out (vanish). After some practice, such unlucky terms are immediately recognized and one displays directly the non-vanishing part of the result alone. Note that it is the highly effective (and merciless) mechanism no. 4 which bears full responsibility for the fact that so many (innocent and agreeable) terms are not allowed to survive.[51]

This method of computation of the exterior product is very convenient; indeed, it is much quicker than working with components (i.e. applying the result of (5.2.5)).

[51] "Heterogeneity" turns out to be a strong evolutionary advantage within the population of exterior forms: $e^1 \wedge e^2 \wedge e^3$ survives, $e^1 \wedge e^1 \wedge e^3$ is not fit enough (its mortal sin being "repeating e^1"). Remarkably, five years on the Beagle (1831–1836) seemed to be not enough for young Charles Darwin to notice this simple example of how natural selection works (although, in those times, there was a flourishing colony of exterior forms living in the Galapagos, their multiplication being routine activity, well known to native people; nor did Alfred Russel Wallace use it in his independent speculations). It was observed only by a teacher of "Gymnasium" in Stettin (today's Szczecin in Poland), Hermann Grassmann, in 1844. Because of the lukewarm response to his work, however, he was so frustrated as to leave this battlefield and set his brain to the understanding Sanskrit (where he was fairly successful, at last). The ideas of Grassmann were fully appreciated and then developed by Elie Cartan.

5.2.10 Repeat the computation of the product $\alpha \wedge \beta$ (treated above) *in components* and convince yourself how cumbersome the component method is in comparison with the way presented above.

Hint: starting with the standard expressions (5.2.9) $\alpha = \alpha_a e^a$ and $\beta = \frac{1}{2}\beta_{ab} e^a \wedge e^b$, identify first the components α_a, β_{ab}, then plug them into (5.2.5), thus computing $(\alpha \wedge \beta)_{abc}$ and finally reconstruct the whole form $\frac{1}{3!}(\alpha \wedge \beta)_{abc} e^a \wedge e^b \wedge e^c$; in the course of the computation, do your best to avoid (in spite of the temptation being increasingly hard to resist) shouting highly substandard words (all the more accurate, however), unworthy of a true lady (or gentleman). □

5.2.11 Reproduce the component result from (5.2.5) by direct multiplication of forms, both of them being represented according to (5.2.9).

Solution:

$$\begin{aligned}
\alpha \wedge \beta &= \left(\frac{1}{p!}\alpha_{a\dots b} e^a \wedge \dots \wedge e^b\right) \wedge \left(\frac{1}{q!}\beta_{c\dots d} e^c \wedge \dots \wedge e^d\right) \\
&= \frac{1}{p!q!}\alpha_{a\dots b}\beta_{c\dots d} e^a \wedge \dots \wedge e^b \wedge e^c \wedge \dots \wedge e^d \\
&= \frac{1}{p!q!}\alpha_{[a\dots b}\beta_{c\dots d]} e^a \wedge \dots \wedge e^b \wedge e^c \wedge \dots \wedge e^d \\
&\overset{!}{=} \frac{1}{(p+q)!}(\alpha \wedge \beta)_{a\dots bc\dots d} e^a \wedge \dots \wedge e^b \wedge e^c \wedge \dots \wedge e^d
\end{aligned}$$

so that we indeed get (5.2.5). □

5.2.12 Check that

$$(\overbrace{e^a \wedge \dots \wedge e^b}^{p \text{ entries}})_{c\dots d} \equiv (e^a \wedge \dots \wedge e^b)(e_c, \dots e_d) = p!\, \delta^a_{[c} \dots \delta^b_{d]} \equiv p!\, \delta^{a\dots b}_{c\dots d}$$

Hint: (5.2.9), (5.2.6); $\delta^{a\dots b}_{c\dots d}$ is defined in (5.6.2). □

5.2.13 Check that

$$\alpha \equiv \frac{1}{p!}\alpha_{a\dots b}\, e^a \wedge \dots \wedge e^b = \sum_{a<\dots<b} \alpha_{a\dots b}\, e^a \wedge \dots \wedge e^b$$

Hint: in the sum on the left, we always have $p!$ terms, differing only in the *order* of indices, thus being equal. As an example, for $p = n = 2$ one has

$$\alpha = \frac{1}{2!}\,(\alpha_{12} e^1 \wedge e^2 + \alpha_{21} e^2 \wedge e^1) = \alpha_{12} e^1 \wedge e^2$$

A basis of $\Lambda^p L^*$ is given by the terms $e^a \wedge \dots \wedge e^b$ with $a < \dots < b$ *alone*, but since *all* terms without *any restrictions whatsoever* are present in the decomposition (5.2.9), the factor $1/p!$ appears. □

• Let us close this section with a very simple, but nevertheless fairly useful, criterion of the linear independence of a set of (co)vectors.

5.2.14 Let $\alpha, \beta, \dots, \sigma$ be any covectors (1-forms). Verify the validity of the following criterion to test their linear independence:

$$\alpha, \beta, \dots, \sigma \text{ are linearly independent} \quad \Leftrightarrow \quad \alpha \wedge \beta \wedge \cdots \wedge \sigma \neq 0$$

Hint: $\Rightarrow$: denote $\alpha \equiv e^1, \beta \equiv e^2, \dots, \sigma \equiv e^k$ and complete arbitrarily to a basis e^a, $a = 1, \dots, n$. According to (5.2.9), the k-form $e^1 \wedge \cdots \wedge e^k$ represents an element of the basis of the space of all k-forms (thus being necessarily non-vanishing: e.g. if e_a is the dual basis, then $(e^1 \wedge \cdots \wedge e^k)(e_1, \dots, e_k) = 1 \neq 0$); the opposite direction by a contradiction: being linearly dependent, some of them may be expressed in terms of the others $\Rightarrow$ after plugging this expression back, each term necessarily contains two identical 1-forms. □

5.3 Exterior algebra ΛL^*

• In Section 2.4 we introduced the concept of the tensor algebra $T(L)$. If we restrict ourselves to p-forms and at the same time replace the tensor product by the exterior product, we get in like manner another interesting object, the *exterior algebra* ΛL^* of a space L^*. Regarded as a linear space, it is the direct sum

$$\Lambda L^* := \bigoplus_{p=0}^{n} \Lambda^p L^* \equiv \Lambda^0 L^* \oplus \Lambda^1 L^* \oplus \cdots \oplus \Lambda^n L^*$$

so that its general element is a linear combination of forms with various values of $p = 0, 1, \dots, n$. Such a general element is called an *inhomogeneous form* and the elements of a subspace with fixed value of p (p-forms) are then said to be *homogeneous*. The number p is called the *degree* (of the form) in this context and it is usually denoted as $\deg \alpha = p$ (for $\alpha \in \Lambda^p L^*$).

The definition of $\Lambda^p L^*$ may be readily extended to *all integers* p by defining $\Lambda^p L^* := 0$ for $p < 0$, $p > n$ (we have encountered a similar extension in dealing with $T(L)$). What we obtain in this way is a $\mathbb{Z}$-*graded algebra* (Appendix A.5); that is to say ΛL^* is a $\mathbb{Z}$-graded linear space (it is a direct sum of *homogeneous* subspaces labeled by elements of the group $(\mathbb{Z}, +)$) endowed with a multiplication rule compatible with the grading (the product of two homogeneous elements with degrees p and q respectively is homogeneous as well, its degree being $p + q$, which just equals the result of the *product* of integers in the group $(\mathbb{Z}, +)$), i.e.

$$\Lambda^p L^* \wedge \Lambda^q L^* \subset \Lambda^{p+q} L^*$$

5.3.1 Find the dimension of the linear space ΛL^* ($\dim \Lambda L^* = \sum_{p=0}^{n} \dim \Lambda^p L^* \overset{(5.2.1)}{=} 2^n$). □

• The concept of an exterior algebra is canonically induced by any (finite-dimensional) linear space. It may also be introduced in the following way. Let $e_1, \dots, e_n$ be (any) basis in L. We define a formal multiplication, satisfying the relations

$$e_a e_b = -e_b e_a$$

The elements of the algebra are then given as the results of arbitrary products and linear combinations[52] of the elements e_a and 1 (we say that e_a constitute a set of *generators* of the algebra; there are $n = \dim L$ of them). This algebra is denoted by ΛL.

5.3.2 Check that

(i) this construction of exterior algebra does not depend on the choice of the basis e_a in L
(ii) the algebra resulting from this construction is isomorphic to the algebra ΛL^* from the beginning of the section.

Hint: a basis of ΛL is given by the elements 1 for $\Lambda^0 L \equiv \mathbb{R}$, e_a for $\Lambda^1 L$, $e_a e_b$ for $\Lambda^2 L, \dots$ (the indices in $e_a \dots e_b \in \Lambda^p L$ fulfilling $a < \dots < b$), an isomorphism is $e_a \dots e_b \leftrightarrow e^a \wedge \dots \wedge e^b$. □

• It is seen now that the above-mentioned algebra "is actually" just the exterior algebra of the space L^* (its generators being e^a, the particular realization of a formal product being given by the antisymmetrized tensor product $\wedge$, which just fulfills the relations needed by the definition). This is the reason why a star occurs in denoting ΛL^*.

5.3.3 Given an exterior algebra ΛL^* one defines a linear operator

$$\hat{\eta} : \Lambda L^* \to \Lambda L^* \qquad \hat{\eta}\alpha = (-1)^p \alpha \equiv (-1)^{\deg \alpha}\, \alpha \qquad \text{for } \alpha \in \Lambda^p L^*$$

Prove that $\hat{\eta}$ is an automorphism of the exterior algebra ΛL^* (called its *main automorphism*), i.e. a bijection of ΛL^* onto itself, such that

(i) it respects the $\mathbb{Z}$-grading

$$\hat{\eta}(\Lambda^p L^*) \subset \Lambda^p L^*$$

(ii) it respects the structure of an algebra (both the linear structure and the product)

$$\hat{\eta}(\alpha + \lambda\beta) = \hat{\eta}\alpha + \lambda\hat{\eta}\beta \qquad \hat{\eta}(\alpha \wedge \gamma) = \hat{\eta}\alpha \wedge \hat{\eta}\gamma$$

(iii) In addition, it obeys

$$\hat{\eta}^2 = \hat{1}$$

□

• One further useful point of view consists in regarding the exterior algebra as the result of appropriate factorization of the tensor algebra. Consider a linear space L and its "purely covariant" tensor algebra, i.e. the linear space

$$T_{(\cdot)}(L) := \bigoplus_{r=0}^{\infty} T_r^0(L) \equiv T_0^0(L) \oplus T_1^0(L) \oplus T_2^0(L) \oplus \cdots$$
$$\equiv \mathbb{R} \oplus L^* \oplus T_2^0(L) \oplus \cdots \text{ (up to infinity)}$$

endowed with the product induced by the tensor product of the homogeneous terms (just like in Section 2.4). In this ∞-dimensional algebra consider, now, the *two-sided ideal* I

[52] With possible simplifications of multiple products by means of the definition relations, e.g. $e_1 e_2 e_1 = -e_1 e_1 e_2 = 0$.

(see Appendix A.2) generated by elements of the form $\alpha \otimes \alpha$, where $\alpha \in L^*$. The ideal I thus consists (by definition) of all sums of terms of the form

$$t_1 \otimes (\alpha \otimes \alpha) \otimes t_2 \qquad t_1, t_2 \in T_{(\cdot)}(L), \quad \alpha \in L^*$$

Both the algebra $T_{(\cdot)}(L)$ and the ideal I under consideration are ∞-dimensional; however, the *factor-algebra* $T_{(\cdot)}(L)/I$ turns out to be finite-dimensional.

5.3.4* Consider the algebra $T_{(\cdot)}(L)$ and the set I discussed above. Convince yourself that

(i) the set I is indeed a two-sided ideal of the algebra $T_{(\cdot)}(L)$
(ii) the same ideal is also generated by elements of the form

$$\alpha \otimes \beta + \beta \otimes \alpha \qquad \alpha, \beta \in L^*$$

(iii) the factorization under consideration effectively annihilates *symmetric* parts of tensors and leaves unchanged just their *completely antisymmetric* part; this is the reason why the factor-algebra $T_{(\cdot)}(L)/I$ is isomorphic to the exterior algebra ΛL^* (as defined before)[53]

$$T_{(\cdot)}(L)/I = \Lambda L^*$$

(iv) the ideal I is generated by *homogeneous* elements of the initial $\mathbb{Z}$-graded algebra $T_{(\cdot)}(L)$; this gives rise to the fact that the factor-algebra *inherits* the *$\mathbb{Z}$-grading*.

Hint: (i) $t \otimes (\cdots + t_1 \otimes (\alpha \otimes \alpha) \otimes t_2 + \cdots) = \cdots + (t \otimes t_1) \otimes (\alpha \otimes \alpha) \otimes t_2 + \cdots \equiv \cdots t_1' \otimes (\alpha \otimes \alpha) \otimes t_2 \cdots \in I$, so that I is a *left* ideal; the right one in full analogy; (ii) $I \ni (\alpha + \beta) \otimes (\alpha + \beta) = \alpha \otimes \alpha + \beta \otimes \beta + (\alpha \otimes \beta + \beta \otimes \alpha)$; (iii) according to the meaning of a factorization process the elements of an ideal become "vanishing"; here $e^a \otimes e^b + e^b \otimes e^a \in I$, so the rules of computation "by means of representatives" lead to $[\alpha] + \lambda[\beta] := [\alpha + \lambda\beta]$ and $[\alpha][\beta] := [\alpha \otimes \beta]$, so that finally

$$0 = [e^a \otimes e^b + e^b \otimes e^a] = [e^a][e^b] + [e^b][e^a]$$

We obtained the key rule characterizing the *exterior* product. From tensors themselves we get *forms*, since (for a homogeneous term)

$$[\alpha] = [\alpha_{a\ldots b} e^a \otimes \cdots \otimes e^b] = \alpha_{a\ldots b}[e^a] \cdots [e^b] = \alpha_{[a\ldots b]}[e^a] \cdots [e^b]$$

so that only the completely antisymmetric part of a tensor survives. This is the origin of the fact that the resulting factor-algebra is *finite-dimensional*: completely antisymmetric tensors with sufficiently high rank necessarily vanish. If square brackets denoting classes are not explicitly displayed and the resulting multiplication is denoted as $\wedge$, we just get the algebra known as ΛL^*. (iv) Since each generator $z := \alpha \otimes \alpha$ is *homogeneous* (namely, it has degree 2), all elements of the ideal $i = \cdots + t_1 \otimes z \otimes t_2 + \cdots$ *may be* written as a sum of *homogeneous* terms $i = \sum i_k$, *each* i_k being from the ideal, so that the ideal itself (not only the algebra $T_{(\cdot)}(L)$ as a whole) is $\mathbb{Z}$-graded, too; next see Appendix A.5. □

[53] This algebra, as a matter of fact, uses a slightly different convention for the numerical prefactors present in the multiplication rule of forms (see the note after (5.2.3); as an example, here we get for the product of classes of basis 1-forms $[e^a][e^b] := [e^a \otimes e^b] = [e^{[a} \otimes e^{b]}] + [e^{(a} \otimes e^{b)}] = [e^{[a} \otimes e^{b]}] = \frac{1}{2}\,[e^a \wedge e^b]$ so that the product used here differs by a factor of $\frac{1}{2}$ from the "wedge-product" $\wedge$ introduced previously.

• It turns out that in addition to the $\mathbb{Z}$-grading discussed until now, the exterior algebra ΛL^* is naturally endowed with a "coarser" (less distinguishing) grading, too. Namely, it may be regarded as a $\mathbb{Z}_2$-graded algebra ($\mathbb{Z}_2$ being the group with two elements, realized as the set of integers, with multiplication given by addition *modulo 2*; this means there are two elements there, [0] and [1] (the square bracket denotes an equivalence class), and the rules of an addition read $[0]+[0]=[0], [0]+[1]=[1], [1]+[0]=[1], [1]+[1]=[0]$, see (13.2.11)). The new grading arises by dividing the forms (only) into those with even and odd degrees first and observing then that the exterior product is compatible *with this* grading, *too*. Now, the grading is given by the direct sum

$$\Lambda L^* = \Lambda^{[0]}L^* \oplus \Lambda^{[1]}L^*$$

where

$$\Lambda^{[0]}L^* := \Lambda^0 L^* \oplus \Lambda^2 L^* \oplus \cdots \qquad \Lambda^{[1]}L^* := \Lambda^1 L^* \oplus \Lambda^3 L^* \oplus \cdots$$

and there holds

$$\Lambda^{[i]}L^* \wedge \Lambda^{[j]}L^* \subset \Lambda^{[i+j]}L^*$$

Such a $\mathbb{Z}_2$-grading plays an essential role in *supermathematics* (*super*spaces, *super*-algebras, *super*manifolds, *super*symmetries, etc.) and *super*symmetric field theories (the latter being the main source of inspiration for establishing supermathematics itself). The exterior algebra ΛL^*, regarded as $\mathbb{Z}_2$-graded space, provides a simple example of a (associative) superalgebra.

(In supermathematics, the subspaces of degrees 0 and 1 are said to be the subspaces of *even* and *odd* elements, respectively; supersymmetries in physical theories relate bosons (represented by even variables) to fermions (represented by odd variables); namely, supersymmetric models are invariant with respect to particular transformations, *mixing* bosons with fermions.)

5.4 Interior product i_v

• The interior product is a simple, albeit fairly useful and important, operation on forms. For a given vector $v \in L$ this is a map $\alpha \mapsto i_v\alpha$, which consists in inserting v as the first argument into a p-form α, i.e.

$$\begin{aligned} (i_v\alpha)(u,\ldots,w) &:= \alpha(v,u,\ldots,w) \qquad && \alpha \in \Lambda^p L^*,\ p\geq 1 \\ i_v\alpha &:= 0 && p=0 \end{aligned}$$

5.4.1 Check that

(i)

$$i_v i_w = -i_w i_v \quad (\Rightarrow (i_v)^2 = 0) \qquad i_{v+\lambda w} = i_v + \lambda i_w \qquad (i_v\alpha)_{a\ldots b} = v^c \alpha_{ca\ldots b}$$

(ii)

$$i_v = v^a i_a \qquad i_a := i_{e_a}$$

(iii) the computation of $i_a(e^b \wedge \cdots \wedge e^c)$ may be performed in the following formal way: if a is not among $(b, \dots c)$, it is zero; otherwise the corresponding (co)vector e^a is to be *reshuffled to the leftmost position* (there is some $\pm$ because of this) and then *deleted*. It thus resembles a "partial differentiation" of a homogeneous "polynomial" $e^b \dots e^c$ with respect to e^a, modulo a slight sign complication caused by the reshuffling (see the comment at the end of Section 5.6 about "anticommuting variables" θ^a, too). □

5.4.2 Check that i_v is a derivation of ΛL^* of degree (-1) (see Appendix A.5), i.e. that there holds[54]

(i) $i_v \; : \; \Lambda^p L^* \to \Lambda^{p-1} L^*$	it lowers the degree by 1
(ii) $i_v(\alpha + \lambda\beta) = i_v\alpha + \lambda\, i_v\beta$	it is linear
(iii) $i_v(\alpha \wedge \beta) = (i_v\alpha) \wedge \beta + (\hat{\eta}\alpha) \wedge (i_v\beta)$	it obeys a graded Leibniz rule

Hint: (iii) because of the linearity and the possibility of renaming the basis elements it is sufficient to verify ($i_1 \equiv i_{e_1}$)

$$i_1\{(\underbrace{e^a \wedge \cdots \wedge e^b}_{p \text{ entries}}) \wedge (e^c \wedge \cdots \wedge e^d)\} = \{i_1(e^a \wedge \cdots \wedge e^b)\} \wedge (e^c \wedge \cdots \wedge e^d)$$
$$+(-1)^p(e^a \wedge \cdots \wedge e^b) \wedge \{i_1(e^c \wedge \cdots \wedge e^d)\}$$

Four cases are to be analyzed (in the spirit of the mnemonics learned in (5.4.1)); namely, when e^1 is/is not present in the first/second factor. □

5.4.3* Prove that *all* derivations of ΛL^* of degree (-1) are of the form i_v (for some vector v).

Hint: given any such derivation, it is uniquely defined by its action on the subspaces with degrees 0 and 1; on degree 0 it is necessarily zero ($D1 = 0$, no (-1)-forms available), on degree 1 we have by definition $De^a =: v^a \in \mathbb{R}$; the same result is clearly obtained, however, when acted on by i_v with $v := v^a e_a$. □

- The operation of the interior product has a useful interpretation in the integral calculus of (differential) forms; we will return to this point in the problem (7.6.11).

5.5 Orientation in L

- It turns out that $E(L)$, the set of all bases in L, may be naturally divided into two "equally large" halves. To introduce an *orientation in L* means declaring one of these two halves of $E(L)$ to represent *right-handed* bases. One should remember, however, that the two halves are completely equivalent, so that there is *no preferred choice* in fact. In practice, to declare

[54] The interior product $i_v\alpha$ is often denoted by $v \lrcorner \alpha$; the graded Leibniz rule then reads $v \lrcorner (\alpha \wedge \beta) = (v \lrcorner \alpha) \wedge \beta + (\hat{\eta}\alpha) \wedge (v \lrcorner \beta)$.

a *single basis* as being right-handed is clearly enough to make the choice mentioned above (the half containing the right-handed basis is then regarded as being right-handed).[55] A linear space endowed with an orientation will be denoted by (L, o).

5.5.1 Let $E(L)$ be the set of all bases of a vector space L and let $f \in E(L)$. Then any basis e may be uniquely expressed as

$$e_a = f_b A^b_a \qquad \text{i.e.} \quad e = fA \qquad A \in GL(n, \mathbb{R})$$

($GL(n, \mathbb{R})$ being the set of all non-singular $n \times n$ real matrices, see (10.1.3) and beyond). Show that

(i) each basis falls either into $E(L)_+$ or into $E(L)_-$, i.e.

$$E(L) = E(L)_+ \cup E(L)_- \qquad E(L)_+ \cap E(L)_- = \emptyset$$
$$E(L)_\pm = \{e \in E(L); \det A \gtrless 0\}$$

(ii) $E(L)_+$ and $E(L)_-$ are "equally large," i.e. there exists a bijection of $E(L)_+$ onto $E(L)_-$

(iii) dividing of $E(L)$ into $E(L)_+$ and $E(L)_-$ does not depend on the choice of $f \in E(L)$, i.e. if e and $\tilde{e}$ share the same half with respect to f, they share the same half with respect to any other reference basis $\hat{f} \in E(L)$.

Hint: (ii) $(e_1, e_2, \dots, e_n) \leftrightarrow (-e_1, e_2, \dots, e_n)$; (iii) $\det(AB) = \det A \det B$. □

• In the Cartesian linear space $\mathbb{R}^n$ (ordered n-tuples of numbers) a *standard* orientation is introduced by declaring the "canonical basis" $e_1 = (1, 0, \dots, 0)$, $e_2 = (0, 1, 0, \dots, 0)$, ... to be right-handed.

5.6 Determinant and generalized Kronecker symbols

• Generalized Kronecker symbols $\delta^{a\dots b}_{c\dots d}$ (p-delta symbols) play a similar role in the machinery of forms as does the ordinary Kronecker delta symbol δ^a_b for vectors or covectors. In this section several useful identities involving p-deltas are derived.[56] Furthermore, we will learn how they are related to some other useful objects, like the Levi-Civita symbol and the determinant.

5.6.1 The *Levi-Civita symbol* $\epsilon_{a\dots b}$ (carrying n indices, each running from 1 to n) is uniquely defined by just two properties:

$$\epsilon_{a\dots b} = \epsilon^{a\dots b} = \epsilon_{[a\dots b]} = \epsilon^{[a\dots b]} \qquad \epsilon_{12\dots n} := 1$$

(i.e. it is completely antisymmetric and one of its components is explicitly given). Check that

(i) these data indeed fully determine its value for any other n-tuple of indices

[55] Orientation may also be introduced by means of a *volume form*; see (5.7.5).

[56] A reader who suffers from index sickness might use a half tablet of an anti-indexicum or, preferably, skip this section completely.

(ii)

$$\epsilon_{a...b}\,\epsilon^{a...b} = n!$$

(iii) if ω is an n-form in n-dimensional space L, then its components may be written in the form

$$\omega_{a...b} = \lambda\,\epsilon_{a...b} \qquad \lambda \in \mathbb{R}$$

Hint: (iii) $\lambda = \omega_{1...n}$. □

(For the Levi-Civita *symbol* there holds (by definition) $\epsilon_{a...b} = \epsilon^{a...b}$, i.e. the position of indices (upper or lower) *does not matter* – it is a symbol, not a tensor. A word of caution is in order, however. The same symbol often denotes something else in the literature, namely components of so-called volume form, which *is* a tensor, to be denoted by ω (see (5.7.3)) in this book. The *only* reason to use a particular position of indices (upper or lower) on the Levi-Civita symbol is to make the typographical layout of a formula more transparent; for example, in (ii) the indices are displayed both in upper and lower position in order to indicate that a summation convention is understood there.)

5.6.2 Let us define the *p-delta* symbol (the generalized Kronecker symbol $\delta(p)$) as a completely antisymmetric (with respect to upper and lower indices separately) tensor of type $\binom{p}{p}$, which serves as the unit operator on p-forms, i.e. fulfilling

$$\delta^{a...b}_{c...d} = \delta^{[a...b]}_{c...d} = \delta^{a...b}_{[c...d]} = \delta^{[a...b]}_{[c...d]}$$

$$\delta^{c...d}_{a...b}\alpha_{c...d} = \alpha_{a...b} \qquad \text{or in brief} \qquad \delta(p)\alpha = \alpha \qquad \alpha \in \Lambda^p L^*$$

Show that

(i) it may be composed of ordinary delta symbols:

$$\delta^{a...b}_{c...d} = \delta^a_{[c} \ldots \delta^b_{d]} \equiv \delta^{[a}_c \ldots \delta^{b]}_d \equiv \delta^{[a}_{[c} \ldots \delta^{b]}_{d]}$$

(ii) for the Levi-Civita symbol there holds

$$\varepsilon_{a...b}\varepsilon^{c...d} = n!\,\delta^{c...d}_{a...b} \qquad \text{or in brief} \quad \varepsilon\varepsilon = n!\,\delta(n)$$

(iii)

$$\delta^{a...b}_{r...s}\,\delta^{r...s}_{c...d} = \delta^{a...b}_{c...d} \qquad \text{or in brief} \quad \delta(p)\delta(p) = \delta(p)$$

(iv) the projection π^{A} from (5.2.2) (antisymmetrization) may be written as

$$t_{a...b} \mapsto \delta^{c...d}_{a...b}t_{c...d} \qquad \text{or in brief} \quad t \mapsto \delta(p)t$$

(v)

$$(\underbrace{e^a \wedge \cdots \wedge e^b}_{p \text{ entries}})_{c...d} = p!\,\delta^{a...b}_{c...d}$$

Hint: (i) see (5.2.6); (ii) according to (5.6.1) the components of arbitrary n-form are given by $\alpha_{a\dots b} = \lambda\varepsilon_{a\dots b}$. Then,

$$\varepsilon_{a\dots b}\varepsilon^{c\dots d}\alpha_{c\dots d} = \lambda\varepsilon_{a\dots b}\underbrace{(\varepsilon^{c\dots d}\varepsilon_{c\dots d})}_{n!} \equiv n!\,\alpha_{a\dots b}$$

(v) (5.2.12). □

• Sometimes one encounters *contractions* of Levi-Civita or p-delta symbols in manipulations with forms. We will derive several useful results for such contractions.

5.6.3* Prove the identity

$$\underbrace{\overbrace{\delta^{ra\dots b}_{rc\dots d}}^{p}}_{p} = \frac{n-p+1}{p}\underbrace{\overbrace{\delta^{a\dots b}_{c\dots d}}^{p-1}}_{p-1}$$

or in the notation $\delta(p) \leftrightarrow p$-delta, $\delta_k(p) \leftrightarrow k$-times contracted p-delta, the identity

$$\delta_1(p) = \frac{n-p+1}{p}\delta(p-1)$$

Hint: according to (5.6.2)

$$\delta^{ra\dots b}_{rc\dots d} = \delta^r_{[r}\delta^a_c \dots \delta^b_{d]}$$

Write down explicitly the sum indicated by [ ] on lower indices; perform the sum separately for the terms which begin with r $(= \frac{n}{p}\delta^{a\dots b}_{c\dots d})$, with c $(= -\frac{1}{p}\delta^{a\dots b}_{c\dots d})$, ... and finally with d $(= -\frac{1}{p}\delta^{a\dots b}_{c\dots d})$. □

5.6.4* Use the notation from (5.6.3) and prove that

(i)

$$\delta_k(n) = \frac{1}{\binom{n}{k}}\delta(n-k) \qquad \text{or alternatively} \quad \delta(p) = \binom{n}{p}\delta_{n-p}(n)$$

(ii)

$$(\varepsilon\varepsilon)_k = k!\,(n-k)!\,\delta(n-k)$$

i.e. in detail

$$\underbrace{\varepsilon_{\;a\dots b}}_{k}{}_{i\dots j}\varepsilon^{a\dots br\dots s} = k!\,(n-k)!\,\delta^{r\dots s}_{i\dots j}$$

(iii)

$$\delta_k(p) = \frac{\binom{n}{p}}{\binom{n}{n-p+k}}\delta(p-k)$$

Hint: (i) $\delta_1(n) = \frac{1}{n}\delta(n-1)$, $\delta_2(n) = \frac{1}{n}\delta_1(n-1) = \frac{1\cdot 2}{n(n-1)}\delta(n-2), \dots$; (ii) see (5.6.2). □

5.6.5 Define the *determinant* of a square $n \times n$ matrix[57] with entries A^a_b by the formula

$$\det A := \varepsilon_{a\dots b} A^a_1 \dots A^b_n$$

Check that

(i) this definition leads to the same result as your favorite definition yields (in case it differs from the one given here)

(ii) one has also

$$\det A = \frac{1}{n!} \varepsilon_{a\dots b} \varepsilon^{c\dots d} \, A^a_c \dots A^b_d$$

from which it is clear that the determinant of the *transposed* matrix equals the determinant of the original one

(iii)

$$\det A = \delta^{c\dots d}_{a\dots b} \underbrace{A^a_c \dots A^b_d}_{n \text{ matrices}} \qquad \text{or in brief} \quad \det A = \delta(n) \underbrace{A \dots A}_{n \text{ matrices}}$$

(iv) for matrices close to the unit matrix we get a useful first term of the expansion

$$\det(\mathbb{I} + \varepsilon C) = 1 + \varepsilon \operatorname{Tr} C + \cdots$$

(v)

$$\varepsilon_{a\dots b} A^a_c \dots A^b_d = \det A \; \varepsilon_{c\dots d}$$

(vi)

$$\underbrace{A^{[a}_{[c} \dots A^{b]}_{d]}}_{n \text{ matrices}} = \det A \; \delta^{a\dots b}_{c\dots d}$$

Hint: (iv) set $A^a_b = \delta^a_b + \varepsilon C^a_b$ and use (5.6.4); (v) the expression on the left is completely antisymmetric ⇒ we may use (5.6.1). □

5.6.6* Prove that

(i) the elements of the *inverse* matrix may be written explicitly as

$$(n-1)! \; \det A \; (A^{-1})^a_b = \varepsilon^{ak\dots l} \varepsilon_{br\dots s} A^r_k \dots A^s_l \;\; =: \;\; (n-1)! \; \Delta^a_b$$

⇒ for $\det A \neq 0$

$$(A^{-1})^a_b = \frac{1}{\det A} \; \Delta^a_b$$

(ii) Δ^a_b is the (a, b)th *minor*, i.e. the determinant of a matrix which we get from A when the ath column and the bth row are deleted

(iii) it holds that

$$A^a_b \Delta^b_c = \delta^a_c \; \det A$$

[57] This formal definition is the output of a computation, based on an *intuitively comprehensible* interpretation of the determinant as a coefficient, by which all *volumes* are multiplied under the linear transformation A, see (5.7.6).

which gives for $a = c$ the expansion of a determinant with respect to the ath row and

$$\Delta^a_b A^b_c = \delta^a_c \det A$$

which gives for $a = c$ the expansion of a determinant with respect to the ath column.

Hint: (i) see (5.6.5) and (5.6.4); (ii) $\varepsilon^{ak\dots l}$ with fixed a effectively behaves as $\varepsilon^{k\dots l}$ in $(n-1)$-dimensional space. □

5.6.7* Prove the formula for the partial derivative of a determinant with respect to a matrix element

$$\frac{\partial(\det A)}{\partial A^a_b} = (\det A)(A^{-1})^b_a \equiv \Delta^b_a \qquad \text{so that} \qquad d(\det A) = (\det A)\,\mathrm{Tr}\,(A^{-1}dA)$$

Hint: see (5.6.5) and (5.6.6). □

5.6.8* Let L be an even-dimensional linear space (its dimension being $n = 2m$ $(m \geq 1)$), e_a a basis in L, e^a the dual basis in L^*, $A_{ab} = -A_{ba}$, $B^a_b \in GL(n,\mathbb{R})$, $(B^{\mathrm{T}}AB)_{ab} \equiv B^c_a A_{cd} B^d_b$, $(eB)_a \equiv e_b B^b_a$, and finally

$$\alpha^e_A := \frac{1}{2} A_{ab}\, e^a \wedge e^b \in \Lambda^2 L^*$$

One defines then the *Pfaffian* Pf(A) of the matrix A by

$$\alpha^e_A \wedge \dots \wedge \alpha^e_A =: m!\,\mathrm{Pf}(A)\, e^1 \wedge \dots \wedge e^{2m}$$

Prove that

(i) the explicit expression for the Pfaffian is

$$\mathrm{Pf}(A) = \frac{1}{2^m m!}\, \varepsilon^{\overbrace{ab\dots cd}^{2m}} \underbrace{A_{ab} \dots A_{cd}}_{m \text{ matrices}}$$

(ii)

$$\alpha^e_{B^{\mathrm{T}}AB} = \alpha^{eB}_A \qquad \mathrm{Pf}\,(B^{\mathrm{T}}AB) = (\det B)\mathrm{Pf}\,(A)$$

(iii) the numerical value of the square of the Pfaffian is equal to the determinant

$$(\mathrm{Pf}\,(A))^2 = \det A$$

Hint: (i) see (5.6.1); (ii) see (5.7.2); (iii) make use of the existence of the *canonical form* of a skew-symmetric matrix: $\exists C \in GL(n,\mathbb{R})$ such that

$$A \mapsto C^{\mathrm{T}}AC \equiv \mathcal{A} \equiv \begin{pmatrix} 0 & 1 & 0 \\ -1 & 0 & 0 \\ 0 & 0 & 0 \end{pmatrix}$$

(a block expression of the matrix $\mathcal{A}$ is used), and directly check that $\mathrm{Pf}\,(\mathcal{A}) = \pm 1$ (or 0 if $\mathcal{A}$ happens to be singular, i.e. if *indeed* there is a zero block displayed in the rightmost down position). □

5.6.9 * Introduce the exponent of an arbitrary form β by making use of the formal expansion

$$e^{\beta} := 1 + \beta + \frac{1}{2!}\beta \wedge \beta + \cdots$$

The resulting object is an *inhomogeneous form*, i.e. an element of the *exterior algebra* ΛL^*. Check that

(i) the series necessarily *terminates*, so that it is, in fact, a *finite* expression rather than an infinite series (unless the degree of the form equals 0, of course)
(ii) for forms of *odd* degree it contains only the *first two* terms
(iii) if we substitute the 2-form α^e_A from problem (5.6.8) as β, the series looks like

$$e^{\alpha^e_A} := 1 + \alpha^e_A + \frac{1}{2!}\alpha^e_A \wedge \alpha^e_A + \cdots + \frac{1}{m!}\alpha^e_A \wedge \cdots \wedge \alpha^e_A$$

so that the *Pfaffian* enters the $2m$-form

$$\{e^{\alpha^e_A}\}^{\text{top}} = \text{Pf}\,(A)\, e^1 \wedge \cdots \wedge e^{2m} \equiv \sqrt{\det(A)}\, e^1 \wedge \cdots \wedge e^{2m}$$

which is a *top degree form* of the series. □

• This simple result has an important application in the quantum theory of fermion fields, where it is found under the name "integral (of an exponent of a quadratic form) over *anticommuting variables.*" The correspondence is as follows: the "anticommuting variables" θ^a from there match our e^a, our symbol $\wedge$ of the exterior product is omitted and as the "integral" $\int d\theta^1 \dots d\theta^n(\cdots)$ of a general "function of anticommuting variables" (an element of the exterior algebra here) one *defines* the coefficient of $\theta^1 \dots \theta^n$ in this function (the only component of the top degree form). The result from (5.6.9) may be then written in the form[58]

$$\int d\theta^1 \dots d\theta^n e^{\frac{1}{2}A_{ab}\theta^a\theta^b} = \sqrt{\det A}$$

In this notation the interior product from (5.4.1) may be written as

$$i_a(e^b \wedge \cdots \wedge e^c) \quad \leftrightarrow \quad \frac{\partial}{\partial\theta^a}(\theta^b \dots \theta^c)$$

5.7 The metric volume form

• At the beginning of the chapter we came to the conclusion that a completely antisymmetric tensor of type $\binom{0}{n}$, i.e. an n-form in L, is needed to enable one to compute the volume of a parallelepiped spanned by n vectors in an n-dimensional space L. This is the reason why such a (non-vanishing) n-form used to be called a *volume form* in L. We know from

[58] A similar integral for *ordinary commuting* variables (over the whole $\mathbb{R}^n$) is the well-known *Gaussian integral* (the matrix A_{ab} then being *symmetric* and negative definite), and it gives the result which very much resembles that obtained here, the square root of the determinant being, however, in the *denominator* rather than in the numerator, as is the case here.

(5.2.1) that the space of n-forms $\Lambda^n L^*$ is, in fact, *one-dimensional*. This means that any two volume forms may differ at most by a (non-vanishing) constant factor.

5.7.1 Let (e^a) be an arbitrary basis in L^*. Check that

(i)

$$\underbrace{e^a \wedge \cdots \wedge e^b}_{n \text{ pieces}} = \epsilon^{a\ldots b}\, e^1 \wedge \cdots \wedge e^n$$

(ii) the most general n-form ω may be expressed as

$$\omega = \lambda e^1 \wedge \cdots \wedge e^n \qquad \lambda \in \mathbb{R}$$

(iii) if $e_a \mapsto \hat{e}_a \equiv e_b A^b_a$, then

$$\omega \equiv \lambda e^1 \wedge \cdots \wedge e^n = \hat{\lambda}\hat{e}^1 \wedge \cdots \wedge \hat{e}^n$$

where

$$\hat{\lambda} = (\det A)\, \lambda$$

A quantity which transforms in this way under a change of basis is called a *scalar density* (of weight -1; see the text after (6.3.7) and problem (21.7.10)).

Hint: (ii) (5.2.9), (5.6.1), $\lambda = \omega_{1\ldots n}$; (iii) (2.4.2), (5.6.5). □

• So there is a freedom in a *single parameter* λ in the formula for computation of the volume of a parallelepiped in L. This parameter may be fixed by ascribing a definite value of the volume to any *one particular* (non-degenerate) parallelepiped. In a "general" linear space (endowed with no additional structure, like a metric tensor), however, *all* (non-degenerate) parallelepipeds are *completely equivalent* (a parallelepiped is given by an n-tuple of vectors and all vectors are equivalent) and there is no reason for preferring some of them for the purpose of fixing the constant λ. Put another way, there is no *natural scale* of volumes. *All* the volume forms and, consequently, all the formulas for computation of volumes (i.e. with any choice of λ) based on them are equivalent. We can speak of a *ratio* of two volumes rather than of "the" volume itself.[59]

The state of affairs changes substantially, however, in (L, g, o), i.e. if L is endowed with a *metric tensor* and *orientation*, too, since the additional structures (g, o) single out

[59] Intense and merciless advertisements, hammering us day after day, try to make us think that an individual has not the remotest chance of surviving without a credit card, wireless phone and a metric tensor. Some of us, however, never shared this opinion. John Lennon, as an example, expressed his visionary dreams about a life in a linear space with no metric tensor (a situation one nowadays can hardly imagine, indeed) in his famous composition *Imagine*. In the original version we might hear the courageous verse

Imagine there's no metric	Imagine there's no countries
It isn't hard to do	It isn't hard to do
No way to measure angles	Nothing to kill or die for
No lengths of vectors too	And no religion too

The time was, however, not ripe and people not mature enough to be able to accept such a far-reaching idea in those times; censorship (closely intertwined to the tensor lobby, of course) forced him (under pressure) to revise substantially the first strophe and the result is well known today: in the new innocent first strophe, which occurred at the shop counters and which we like to sing up to the present day, no reference to the metric tensor has remained at all.

a distinguished class of bases; namely, they enable one to speak about *orthonormal right-handed* bases. A parallelepiped spanned by an arbitrary such basis is a good old (oriented) *unit cube* and a natural choice is to assign the unit volume just to this figure (as is customarily done from time immemorial for $n = 1, 2, 3$). The unique volume form resulting from this way of fixing the freedom mentioned above is called the *metric volume form* (or, more precisely, the volume form compatible with the metric *and orientation*) and will be denoted by $\omega_{g,o}$, or more frequently only by ω_g.

5.7.2 Let $e \equiv (e_a)$, $a = 1, \dots, n$ be a basis in a vector space L, $E(L)$ the collection of all bases in L and (e^a) the dual basis in L^*. Define the maps

$$\omega : E(L) \to \Lambda^n L^* \qquad \omega(e) := e^1 \wedge \dots \wedge e^n$$
$$R_A : E(L) \to E(L) \qquad (R_A e)_a := e_b A^b_a \equiv (eA)_a \qquad A \in GL(n, \mathbb{R})$$

Show that

(i) R_A is a *right action* of $GL(n, \mathbb{R})$ on $E(L)$ (see more in Section 13.1 and in problem (13.2.7)), i.e. that there holds

$$R_{AB} = R_B \circ R_A$$

(ii) ω responds to the right action R_A in the following way:[60]

$$\omega(eA) = (\det A)^{-1}\omega(e) \qquad \text{i.e.} \quad \omega \circ R_A = (\det A)^{-1}\omega$$

(iii) a straightforward consequence of this behavior of ω and R_A is (a well-known fact)

$$\det\,(AB) = (\det A)(\det B)$$

(iv) in particular, for the matrices which belong to the *special* orthogonal group (10.1.8) we have

$$A \in SO(r, s) \quad \Rightarrow \quad \omega(eA) = \omega(e)$$

(v) $(\omega(e))(e_1, \dots, e_n) = 1$.

Hint: (ii) see (2.4.2) and (5.6.5); (v) see (5.2.12). □

5.7.3 Let (L, g, o) be an n-dimensional vector space endowed with a metric tensor g and an orientation o, $e \equiv (e_a)$ and $\hat{e} \equiv (\hat{e}_a)$, two right-handed orthonormal bases respectively, $f \equiv (f_a)$ an arbitrary basis and $\omega(f) := f^1 \wedge \dots \wedge f^n$ (5.7.2). Prove that

(i)

$$\omega(e) = \omega(\hat{e})$$

i.e. $\omega_g := \omega(e)$ does not depend on the choice of right-handed orthonormal basis

(ii) its expression in terms of the *arbitrary* basis f reads

$$\omega_g \equiv \omega(e) = o(f)\sqrt{|g|}\,\omega(f)$$

where $o(f)$ is $+1$ or -1 depending on whether f is right-handed or left-handed and $|g| \equiv |\det g(f_a, f_b)|$

[60] Using the terminology introduced in Section 13.5, this is a 0-form on $E(L)$ of type $\hat{\rho} = \det$, i.e. $R_A^*\omega = (\det A)^{-1}\omega$.

(iii)

$$\omega_g(f_1, \ldots, f_n) = o(f)\sqrt{|g|} \qquad \text{so that} \quad \omega_{a\ldots b} = o(f)\sqrt{|g|}\,\varepsilon_{a\ldots b}$$

(iv)

$$\omega^{a\ldots b} = o(f)\,\text{sgn}\,g \frac{1}{\sqrt{|g|}}\,\varepsilon^{a\ldots b}$$

where sgn g is the sign of the determinant of the matrix $g_{ab} \equiv g(f_a, f_b)$ (for a metric tensor with signature (r, s) (see the text before (2.4.12)) this is $(-1)^s$)

(v) a change of orientation in L results in the change

$$\omega_g \mapsto -\omega_g \qquad \text{i.e.} \quad \omega_{g,-o} = -\omega_{g,o}$$

if $(-o)$ is the orientation which is opposite with respect to o.

Hint: (ii) let f_a be the arbitrary basis, $f_a = e_b B^b_a$ (i.e. $f = eB$). Then

$$g(f_a, f_b) \equiv g_{ab} = B^c_a \eta_{cd} B^d_b \equiv (B^{\mathrm{T}}\eta B)_{ab} \Rightarrow \det g = (\det B)^2 \det \eta$$
$$\Rightarrow \det B = \pm\sqrt{|\det g|}$$

The sign is given (since e is right-handed) by the orientation of the basis f, so that $\det B = o(f)\sqrt{|\det g|}$. According to (5.7.2) then

$$\omega(e) = \omega(fB^{-1}) = \det B\,\omega(f) = o(f)\sqrt{|\det g|}\,\omega(f)$$

(iv) $\omega^{a\ldots b} \equiv g^{ac}\ldots g^{bd}\omega_{c\ldots d} = \ldots$ (5.6.5); (v) the only change is $o(f) \mapsto -o(f)$. □

- The form

$$\omega_g \equiv \omega(e) = e^1 \wedge \cdots \wedge e^n = o(f)\sqrt{|g|}\,f^1 \wedge \cdots \wedge f^n$$

is the *metric volume form* in (L, g, o) mentioned above. We see that its explicit expression is especially simple in terms of an (arbitrary) orthonormal right-handed basis, being merely a product of basis 1-forms. In a general basis, there is a "correction factor" in front of the product of the basis 1-forms, which is the square root of the (absolute value of the) determinant of the component matrix of the metric tensor with respect *to this basis* (and possibly the minus sign, if the basis is left-handed).

Note that the letter g in these formulas denotes the *determinant of the matrix* corresponding to the metric tensor (with lower indices), rather than the metric tensor itself. The actual meaning of the letter g in any particular formula should always be clear from the context.

5.7.4 Let ω_g be a metric volume form and f an arbitrary basis. Check that the (oriented) volume of the parallelepiped $\mathcal{P}$ spanned by the vectors $(v, \ldots, w)$ may be written as

$$\text{(oriented) volume of the parallelepiped } \mathcal{P} = o(f)\sqrt{|g|}\,\det A$$

where A denotes the matrix with the columns (or, equivalently, rows) given by the components of the vectors $v, \ldots, w$ with respect to the basis f. (In particular, for a right-handed orthonormal basis the term $\det A$ alone stands on the right.)

Hint: the volume $:= \omega(v, \dots, w) = \omega_{a \dots b}\, v^a \dots w^b = \dots$, (5.7.3), (5.6.5). □

5.7.5 Prove that any volume form may be used to define an *orientation* on L, moreover in such a way that the form $\omega_{g,o} \equiv \omega_g$ from (5.7.3) will give just the orientation with which it is compatible.

Hint: for $0 \neq \alpha \in \Lambda^n L^*$ a basis f will be right-handed (see Section 5.5), if $\alpha(f_1, \dots, f_n) > 0$. □

• Let us devote some time, in closing the section, to the interpretation of the concept of the determinant of a matrix. We present two useful (and closely related) ways in which it may be understood.

The first one says that the determinant is a factor by which the volume (in the sense of an arbitrary volume form) of a parallelepiped is multiplied when all of its "constituent" vectors undergo a linear map A. The second one introduces the determinant in terms of a lift of the map A to the (one-dimensional linear) space of forms of top degree.

Note that in both approaches an invariant notion of the determinant of a linear *map* is introduced as a primary concept and it is then a matter of computation to show that they actually coincide with a common expression for the determinant of the *matrix* of the map (with respect to an *arbitrary* basis).

5.7.6 Let $A : L \to L$ be a linear map. Consider a non-degenerate parallelepiped, spanned by the vectors $(u, \dots, v)$. Its volume (in the sense of an arbitrary but fixed volume form ω) is

$$\text{volume}\,(u, \dots, v) \equiv \omega(u, \dots, v)$$

Define the *determinant of the map* A as the factor by which the volume of the initial parallelepiped is to be multiplied in order to get the volume of the parallelepiped spanned by the vectors $(Au, \dots, Av)$

$$\text{volume}\,(Au, \dots, Av) =: (\det A)\ \text{volume}\,(u, \dots, v)$$

Show that

(i) the number $\det A$ does not depend on the choice of parallelepiped (so that it informs us about the factor by which *each* volume in L gets multiplied under the map A)
(ii) the determinant of a map A may be computed as the "ordinary" determinant of the *matrix* A^a_b of the map (with respect to an arbitrary basis)
(iii) the determinant of the product (composition) of maps is the product of determinants

$$\det(AB) = \det A\ \det B$$

Hint: (ii) see (5.6.5); if $\omega = \lambda e^1 \wedge \dots \wedge e^n$, then

$$\begin{aligned}\omega(Au, \dots, Av) &= \lambda\epsilon_{a \dots b}(Au)^a \dots (Av)^b = \dots = (\widetilde{\det} A)\lambda\epsilon_{c \dots d}u^c \dots v^d \\ &\equiv (\widetilde{\det} A)\omega(u, \dots, v)\end{aligned}$$

where $\widetilde{\det} A$ already denotes the determinant of the *matrix*; (iii) under a map B each volume increases q times ($q \equiv \det B$), with the additional map A causing each volume to increase k times ($k \equiv \det A$), so that the volume, which is already q times bigger, increases altogether kq times. □

5.7.7 Let $A : L \to L$ be a linear map and let A^a_b be its matrix with respect to a basis e_a (associated by the standard relation $Ae_a := A^b_a e_b$). Consider the dual map $A^* : L^* \to L^*$; on the dual basis then $A^* e^a := A^a_b e^b$ (2.4.17). We introduce the *lift* (prolongation, induced map) of A to the space of arbitrary p-forms, $\hat{A} : \Lambda^p L^* \to \Lambda^p L^*$, by

$$(\hat{A}\alpha)(u, \dots v) := \alpha(Au, \dots Av)$$

(for $p = 1$ there holds $\hat{A} = A^*$), so that in components

$$A : w^a \mapsto A^a_b w^b \quad \Rightarrow \quad \hat{A} : \alpha_{a\dots b} \mapsto A^c_a \dots A^d_b \alpha_{c\dots d}$$

In particular, on n-forms ($n = \dim L^*$) this is a (linear) map on a *one-dimensional* linear space; it is then given by *a single number*. This number is defined to be the *determinant of a map A*

$$\det A \equiv \hat{A} : \Lambda^n L^* \to \Lambda^n L^*$$

Show that

(i) the determinant of the map A may be computed as an "ordinary" determinant of the *matrix* A^a_b of the map (with respect to an arbitrary basis)
(ii) the lift of a map copies (in the reversed order) the composition[61] of the initial maps, i.e.

$$\widehat{AB} = \hat{B}\,\hat{A}$$

(iii) the determinant of the product (composition) of maps is the product of determinants

$$\det(AB) = \det A \; \det B$$

(iv) the definition of the determinant of a map A presented here is *equivalent* to the definition given in (5.7.6) (and thus to that from (5.6.5) as well).

Hint: (ii) for a general n-form $\alpha_{a\dots b} = \lambda \epsilon_{a\dots b}$, so that

$$(\det A)\alpha_{a\dots b} \equiv (\hat{A}\alpha)_{a\dots b} = \alpha(Ae_a, \dots, Ae_b) = A^c_a \dots A^d_b \alpha_{c\dots d} = \cdots$$

(5.6.5); (iii) the special case of (ii) for $p = n$; (iv)

$$\begin{aligned}(\det A)_{(5.7.6)} \text{volume}\,(u, \dots, v) &= \text{volume}\,(Au, \dots, Av) = \omega(Au, \dots, Av)\\ &= (\hat{A}\omega)(u, \dots, v) = (\det A)_{(5.7.7)}\omega(u, \dots, v)\\ &= (\det A)_{(5.7.7)} \text{volume}\,(u, \dots, v)\end{aligned}$$

□

[61] For non-singular maps the prescription $A \mapsto \hat{A}$ thus provides an (anti)*representation* of the group $GL(L)$ in the space of p-forms $\Lambda^p L^*$, see Section 12.1.

5.8 Hodge (duality) operator $*$

• From the result of problem (5.2.1) it follows that the dimensions of the spaces of p-forms and $(n-p)$-forms in an n-dimensional space L happen to coincide, being in both cases $\binom{n}{p}$. This means that the spaces are isomorphic, albeit not canonically, in general. It turns out, however, that if L is endowed with a *metric tensor and orientation*, a *canonical* isomorphism is available, which is of great importance in various applications of forms.

5.8.1 Let (L, g, o) be an n-dimensional linear space with a metric tensor and orientation and let $\omega \equiv \omega_{g,o}$ be the metric volume form from (5.7.3). Define a map $* \equiv *_{g,o}$ (called the *Hodge operator* or the *duality operator*)

$$\alpha \mapsto *\alpha \qquad \alpha \in \Lambda^p L^*$$

$$(*\alpha)_{a\dots b} := \frac{1}{p!}\,\alpha^{c\dots d}\,\omega_{c\dots da\dots b} \qquad \alpha^{c\dots d} \equiv g^{cr}\dots g^{ds}\,\alpha_{r\dots s}$$

Check that

(i)

$$*: \Lambda^p L^* \to \Lambda^{n-p} L^*$$

(ii) $*$ is a linear map

$$*(\alpha + \lambda\beta) = *\alpha + \lambda * \beta$$

(iii) on a (general) basis it gives

$$*(\underbrace{e^a \wedge \dots \wedge e^b}_{p \text{ entries}}) = \frac{1}{(n-p)!}\,\omega^{a\dots b}{}_{c\dots d}\, e^c \wedge \dots \wedge e^d$$

$$\text{where} \quad \omega^{a\dots b}{}_{c\dots d} \equiv g^{ar}\dots g^{bs}\,\omega_{r\dots sc\dots d}$$

(iv) on a right-handed orthonormal basis it gives

$$*(\underbrace{e^a \wedge \dots \wedge e^b}_{p \text{ entries}}) = \frac{1}{(n-p)!}\,\eta^{ac}\dots\eta^{bd}\,\varepsilon_{c\dots dr\dots s}\, e^r \wedge \dots \wedge e^s$$

(v) in particular, for $p = 0, n$ we have

$$*_g 1 = \omega_g \qquad * \,\omega_g = \operatorname{sgn} g$$

(vi) a change of orientation in L results in

$$*_g \mapsto - *_g \qquad \text{i.e.} \quad *_{g,-o} = - *_{g,o}$$

if $(-o)$ denotes the orientation, which is opposite to o.

Hint: (iii) see (5.6.2); (v) $*\omega = *(e^1 \wedge \dots \wedge e^n) = \omega^{1\dots n} \overset{(5.7.3)}{=} \operatorname{sgn} g$ holds in a right-handed orthonormal basis. □

5.8.2 Check that

(i) repeated application of the duality operator

$$\Lambda^p L^* \xrightarrow{*} \Lambda^{n-p} L^* \xrightarrow{*} \Lambda^p L^*$$

happens to coincide (modulo a sign) with the identity map in $\Lambda^p L^*$, i.e. that there holds (in detail)

$$*_g *_g = \operatorname{sgn} g\ (-1)^{p(n-p)} \qquad \text{on } \Lambda^p L^*$$

(ii)

$$(-1)^{p(n-p)} = (-1)^{p(n+1)}$$

and so[62] one can also write this as

$$*_g *_g = \operatorname{sgn} g\ \hat{\eta}^{n+1} \qquad \text{so that} \quad *_g^{-1} = \operatorname{sgn} g\ *_g\ \hat{\eta}^{n+1} \qquad (\text{it fulfills } *_g *_g^{-1} = \mathbf{1})$$

(iii) $*$ is a canonical *isomorphism*

Hint: (i) a direct computation in components gives

$$(**\alpha)_{a\ldots b} = \cdots = \frac{(-1)^{p(n-p)}}{p!(n-p)!}\ \omega^{c\ldots dr\ldots s}\ \omega_{c\ldots da\ldots b}\ \alpha_{r\ldots s}$$

Complete the calculation with the help of (5.7.3), (5.6.4) and (5.3.3) (most easily in a right-handed orthonormal basis); (iii) the non-trivial kernel contradicts $** \sim \hat{1}$. □

5.8.3 A *conformal rescaling* of a metric is a replacement $g \mapsto \kappa g$ ($0 \neq \kappa \in \mathbb{R}$; the angles between vectors remain unchanged,[63] but their lengths do change under this transformation). Prove that for the Hodge operator with respect to a conformally rescaled metric one has a simple formula

$$*_{\lambda^2 g} = \lambda^{n-2p} *_g$$

Hint: in a fixed basis $g_{ab} \equiv g(f_a, f_b) \mapsto \lambda^2 g_{ab} \Rightarrow \sqrt{|g|} \mapsto \lambda^n \sqrt{|g|} \Rightarrow \omega_{\lambda^2 g} = \lambda^n \omega_g$, $g^{ab} \mapsto \lambda^{-2} g^{ab} \overset{(5.8.1)}{\Rightarrow}$ just what is needed. □

• The Hodge operator enables one to express a scalar product in the space of p-forms (as well as in the exterior algebra ΛL^*, then) in a compact and component-free way. This scalar product plays an essential role in many applications (like in the action integrals in field theory). The way one can easily arrive at this concept might look like this: given α and β as two p-forms, $\alpha \wedge *\beta$ is always an n-form (for all $p = 0, 1, \ldots, n$), depending linearly on both α and β. The space of n-forms is one-dimensional, however, so that any n-form may be regarded as a multiple of some fixed (reference) n-form. If one chooses the metric volume form ω_g to serve this purpose (which is clearly the most natural choice in (L, g, o)), we get $\alpha \wedge *\beta = (\alpha, \beta)\omega_g$, where the coefficient $(\alpha, \beta) \in \mathbb{R}$ depends linearly on both α and β.

[62] From this expression one can see that in odd-dimensional space the resulting sign does not depend on the degree of a form (in particular, in ordinary E^3 we always have $** = \hat{1}$). In $(1+3)$-Minkowski space $E^{1,3}$ we have $** = -\hat{\eta}$.

[63] If a conformal *transformation of a manifold* (4.6.3) is given, it results in a conformal rescaling of $g(x)$ at each point $x \in M$.

5.8.4 Given α and β two p-forms in an n-dimensional space (L, g, o), define $(\alpha, \beta)_g \in \mathbb{R}$ by

$$\alpha \wedge *_g \beta =: (\alpha, \beta)_g \omega_g$$

Check that

(i) in components this gives

$$(\alpha, \beta)_g = \frac{1}{p!} \alpha_{a...b} \beta^{a...b} \equiv \frac{1}{p!} g^{ac} \dots g^{bd} \alpha_{a...b} \beta_{c...d}$$

(ii) (α, β) is a symmetric,[64] non-degenerate bilinear form in the space $\Lambda^p L^*$, i.e. that

$$(\alpha, \beta) = (\beta, \alpha) \qquad (\alpha + \lambda\gamma, \beta) = (\alpha, \beta) + \lambda(\gamma, \beta)$$

$$(\alpha, \beta) = 0 \text{ for all } \beta \quad \Rightarrow \quad \alpha = 0$$

(iii) for *Euclidean* space (L, g) (positive definite g) (α, β) is positive definite, i.e.

$$(\alpha, \alpha) \geq 0 \qquad \text{and the equality occurs only for } \alpha = 0$$

$\Rightarrow (\alpha, \beta)$ is the *scalar product* in $\Lambda^p L^*$

(iv) for $p = 0$ and 1 we get the ordinary product in $\mathbb{R}$ and the product from (2.4.13) respectively.

Hint: (i) compute in an orthonormal right-handed basis, make use of the results of (5.7.1), (5.7.3) and (5.6.4); (iii) (α, α) turns out to be the sum of squares in the orthonormal basis. □

5.8.5* Prove that

(i) for $\alpha, \beta \in \Lambda^p L^*$

$$(*\alpha, *\beta) = \operatorname{sgn} g \; (\alpha, \beta)$$

(ii) in the Euclidean case $*$ is an *isometry*

(iii) for $\alpha \in \Lambda^p L^*, \gamma \in \Lambda^{n-p} L^*$

$$(*\alpha, \gamma) = (\alpha, *\hat{\eta}^{n+1} \gamma)$$

Hint: $(*\alpha, *\beta)\omega = (*\alpha) \wedge * * \beta = \cdots$ (5.8.2), (5.2.4), (5.8.4); (ii) this is just what (i) says for this case; (iii) $(*\alpha, \gamma) = \operatorname{sgn} g \; (* * \alpha, *\gamma) = \cdots$. □

5.8.6 Check that

$$(i_v \alpha, \beta) = (\alpha, j_v \beta)$$

where

$$j_v \beta := \tilde{v} \wedge \beta \qquad \tilde{v} \equiv \flat_g v \equiv g(v, \cdot)$$

[64] Note that its symmetry with respect to $\alpha \leftrightarrow \beta$ is *not evident* at all from the definition relation alone. It *is* clear, however, from the component expression (i).

Hint: according to (5.4.1), (5.8.4) and (5.2.6)

$$(i_v\alpha, \beta) = \frac{1}{(p-1)!} v^a \alpha_{ab\ldots c}\, \beta^{b\ldots c} = \frac{1}{(p-1)!} \alpha^{ab\ldots c}\, v_{[a}\beta_{b\ldots c]} = (\alpha, \tilde{v} \wedge \beta)$$

□

5.8.7* Given α and β two p-forms, check that

(i) under the conformal rescaling of a metric $g \mapsto \lambda^2 g$ the n-form $\alpha \wedge *\beta$ transforms as follows:

$$\alpha \wedge *_{\lambda^2 g}\beta = \lambda^{n-2p}\, \alpha \wedge *_g \beta$$

(ii) in particular, the forms of "middle" degree in an even-dimensional space ($n = 2p$; e.g. 2-forms in four-dimensional space) this n-form turns out to be *conformally invariant*

(iii) if one defines appropriately a scaling of *forms*, $\sigma \mapsto \lambda^{f(p)}\sigma$ for $\sigma \in \Lambda^p L^*$, the n-form $\alpha \wedge *_g \beta$ remains *unchanged*; explicitly

$$\begin{aligned} g &\mapsto \lambda^2 g \equiv \hat{g} \\ \alpha &\mapsto \lambda^{(p-n/2)}\alpha \equiv \hat{\alpha} \qquad \Rightarrow \qquad \hat{\alpha} \wedge *_{\hat{g}} \hat{\beta} = \alpha \wedge *_g \beta \\ \beta &\mapsto \lambda^{(p-n/2)}\beta \equiv \hat{\beta} \end{aligned}$$

(iv) the situation in (ii) is a special case of (iii).

Hint: see (5.8.3). □

• The bilinear form (α, β), as well as the linear operators defined before (like $i_v, *, \ldots$), may be naturally regarded as being defined on the *whole* exterior algebra ΛL^*.

5.8.8* Consider inhomogeneous forms $\alpha, \beta \in \Lambda L^*$, $\alpha = \alpha_{(0)} + \alpha_{(1)} + \cdots + \alpha_{(n)}$, $\alpha_{(p)} \in \Lambda^p L^*$ and similarly β. Define

$$(\alpha, \beta) := (\alpha_{(0)}, \beta_{(0)}) + (\alpha_{(1)}, \beta_{(1)}) + \cdots + (\alpha_{(n)}, \beta_{(n)})$$

where

$$(\alpha_{(p)}, \beta_{(p)}) = \text{according to (5.8.4)}$$

Check that

(i) in components this gives

$$(\alpha, \beta) = \alpha_{(0)}\beta_{(0)} + (\alpha_{(1)})_a(\beta_{(1)})^a + \frac{1}{2!}(\alpha_{(2)})_{ab}(\beta_{(2)})^{ab} + \cdots + \frac{1}{n!}(\alpha_{(n)})_{a\ldots b}(\beta_{(n)})^{a\ldots b}$$

(ii) (α, β) is a symmetric, non-degenerate bilinear form in the space ΛL^*

(iii) for Euclidean space (L, g) it is positive definite (a scalar product in ΛL^*).

Hint: see (5.8.4). □

5.8.9* Consider a linear extension to the whole of ΛL^* of the operators discussed up to now; for example,

$$*\alpha \equiv *(\alpha_{(0)} + \alpha_{(1)} + \cdots + \alpha_{(n)}) := *\alpha_{(0)} + *\alpha_{(1)} + \cdots + *\alpha_{(n)}$$

$$i_v\alpha \equiv i_v(\alpha_{(0)} + \alpha_{(1)} + \cdots + \alpha_{(n)}) := i_v\alpha_{(0)} + i_v\alpha_{(1)} + \cdots + i_v\alpha_{(n)}$$

etc. Check the following identities:

(i)

$$\hat{\eta}* = (-1)^n * \hat{\eta}$$

(ii)

$$*i_v = -j_v * \hat{\eta} \qquad i_v * = *j_v \hat{\eta}$$

(iii)

$$i_v \hat{\eta} = -\hat{\eta} i_v \qquad j_v \hat{\eta} = -\hat{\eta} j_v$$

(iv)

$$i_v j_w + j_w i_v = g(v, w)\hat{1}$$

(v)

$$(i_v \alpha, \beta) = (\alpha, j_v \beta) \qquad (*\alpha, \beta) = (\alpha, *\hat{\eta}^{n+1}\beta) \qquad (*\alpha, *\beta) = \operatorname{sgn} g\ (\alpha, \beta)$$

Hint: (ii) on homogeneous terms: according to (5.8.4) and (5.8.6) there holds $(i_v\alpha) \wedge *\beta = \alpha \wedge *j_v\beta$. Making use of (5.4.2) on the left and realizing that $\alpha \wedge *\beta = 0$ (since it is an $(n+1)$-form), we get $\alpha \wedge i_v * \hat{\eta}\beta = \alpha \wedge *j_v\beta$, from which ($\alpha, \beta$ arbitrary) $i_v * \hat{\eta} = *j_v$. Similarly for the second relation. Extend to inhomogeneous terms by linearity. (iv) Using (5.4.2) and (5.8.6) we have $i_v j_w \alpha = i_v(\tilde{w} \wedge \alpha) = (i_v \tilde{w}) \wedge \alpha - \tilde{w} \wedge (i_v \alpha) = \cdots$.

□

5.8.10* Let e_a be an orthonormal basis in $L = E^{r,s}$, $\eta_{ab} = g(e_a, e_b)$, e^a the dual basis in L^* and $\omega_g \equiv *_g 1$ the metric volume form. Define the operators (see (5.8.6) and (5.4.1))

$$\begin{aligned} i_a &:= i_{e_a} & i^a &:= \eta^{ab} i_b \equiv i_{\sharp_g e^a} \\ j_a &:= j_{e_a} & j^a &:= \eta^{ab} j_b \equiv j_{\sharp_g e^a} \\ \gamma_a &:= i_a + j_a & \gamma^a &:= i^a + j^a \end{aligned}$$

Show that

(i)

$$i_a j^b + j^b i_a = \delta_a^b \hat{1} \qquad i_a j_b + j_b i_a = \eta_{ab} \hat{1} \qquad i^a j^b + j^b i^a = \eta^{ab} \hat{1}$$

(ii)[65]

$$\gamma_a \gamma_b + \gamma_b \gamma_a = 2\eta_{ab} \hat{1} \qquad \gamma^a \gamma^b + \gamma^b \gamma^a = 2\eta^{ab} \hat{1}$$

(iii)

$$e^a \wedge \cdots \wedge e^b = j^a \dots j^b 1$$

(iv)

$$*(e^a \wedge \cdots \wedge e^b) = i^b \dots i^a * 1 \equiv i^b \dots i^a \omega$$

[65] The operators γ_a realize a (real, reducible) representation of the *Clifford algebra* $C(r, s)$ (see Section 22.1) in the exterior algebra $\wedge L^*$.

(v)

$$i_a\omega = d\Sigma_a \qquad i_a i_b\omega = d\Sigma_{ab}$$

etc. (the forms $d\Sigma_a, d\Sigma_{ab}, \dots$ are defined in (6.3.11) and they play an important role then in integral calculus, see (8.2.8)).

Hint: (i) see (5.8.6); (iv) see (5.8.9); (v) $\omega = *1$. □

5.8.11* Prove that for $\alpha \in \Lambda^p L^*$

$$j^a i_a\alpha = p\alpha \qquad i_a j^a\alpha = (n-p)\alpha$$

Hint: computation on a monomial $\underbrace{e^c \wedge \cdots \wedge e^d}_{p \text{ entries}}$ + linearity, (5.8.10). □

5.8.12* Let i_a, j^a be the operators defined in (5.8.10). Show

(i) that

$$j^a i_b : \Lambda L^* \to \Lambda L^*$$

is a derivation of ΛL^* of degree 0 (see (5.4.2) and Appendix A.5)

(ii) that if A^a_b are the components of a tensor of type $\binom{1}{1}$, then

$$\hat{A} \equiv A^a_b j^b i_a : \Lambda L^* \to \Lambda L^*$$

is a derivation of ΛL^* of degree 0, which already does not depend on the choice of a frame field (it may not be orthonormal, as is needed in (5.8.10))

(iii) how $\hat{A}$ acts on $\Lambda^0 L^*$, $\Lambda^1 L^*$ and $\Lambda^2 L^*$.

□

5.8.13* Let B be a linear operator in ΛL^* and $(\cdot, \cdot)$ the bilinear form on ΛL^* discussed in (5.8.8). Define the *adjoint operator* B^+ by the standard formula

$$(B^+\alpha, \beta) := (\alpha, B\beta)$$

Check that

$$i_v^+ = j_v \qquad j_v^+ = i_v \qquad \hat{\eta}^+ = \hat{\eta} \qquad *^+ = *\hat{\eta}^{n+1}$$

Hint: see (5.8.6). □

5.8.14* Consider a linear space L with the *Euclidean* scalar product ($\eta_{ab} = \delta_{ab}$). Check that ΛL^* realizes the Hilbert space of *n fermions*. Namely, check that

(i) the operators i_a and $j_a \equiv (i_a)^+ \equiv j^a$ act as the annihilation and the creation operators of the ath fermion respectively
(ii) the subspace of p-forms corresponds to the p-particle states
(iii) $\hat{N} \equiv j^a i_a \equiv i_a^+ i_a$ acts as the operator of the total number of particles.

Hint: (i) (5.8.10) and (5.8.13) say that $i_a(i_b)^+ + (i_b)^+ i_a = \delta_{ab}\hat{1}$; (iii) see (5.8.11). □

5.8.15* Let $\alpha \in \Lambda^p L^*$. Define a linear map

$$\hat{\alpha} : L \to \Lambda^{p-1} L^* \qquad v \mapsto i_v \alpha$$

The rank of this map (2.4.17) is called the *rank of the form* α. A p-form α is said to be *decomposable* if it can be written as a product of p 1-forms (a general p-form can only be written, according to (5.2.9), as a *sum of several* such products). Check that

(i) the minimal rank of a (non-vanishing) p-form is p
(ii) a form is decomposable if and only if it has minimal rank.

Hint: Let $(e_A) \equiv (e_i, e_a)$ be a basis of L, which is adapted to the *kernel* of the map $\hat{\alpha}$ ($e_i \in \operatorname{Ker}\hat{\alpha}$). Then α decomposes with respect to the e^as alone (since $p!\alpha = (i_B \dots i_A \alpha) e^A \wedge \dots \wedge e^B$). By definition the rank of α equals the number of entries of e^a. If we are to compose a p-form from them, their number should be at least p. If there are exactly p of them, we have $\alpha = (ke^1) \wedge \dots \wedge e^p$, so that it is decomposable. □

5.8.16* The *characteristic subspace* $L^{(\alpha)}$ of a form α is the kernel of the map $\hat{\alpha}$ introduced in (5.8.15), i.e.

$$L^{(\alpha)} := \{v \in L \mid i_v \alpha = 0\} \equiv \operatorname{Ker}\hat{\alpha}$$

Check that

(i) the characteristic subspaces of the forms α and $*\alpha$ are *orthogonal* to one another

$$L^{(\alpha)} \perp_g L^{(*_g \alpha)}$$

i.e. the characteristic subspace of the form $*\alpha$ is a part (subspace) of the *orthogonal complement* to the characteristic subspace of the initial form α ($\perp_g$ denotes orthogonality in the sense of g)

(ii) for a *decomposable* form α

$$L^{(*\alpha)} = (L^{(\alpha)})^{\perp_g}$$

i.e. in this case $*\alpha$ corresponds to the *entire* orthogonal complement ("geometrical meaning" of the operator $*_g$).

Hint: (i) let $i_v \alpha = 0,\ i_w * \alpha = 0$. Then,

$$0 = i_w * \alpha \overset{(5.8.9)}{=} *j_w \hat{\eta} \alpha \ \Rightarrow\ j_w \alpha = 0 \ \Rightarrow\ g(v, w)\alpha \overset{(5.8.9)}{=} i_v j_w \alpha + j_w i_v \alpha = 0$$
$$\Rightarrow g(v, w) = 0$$

(ii) make use of a *right-handed orthonormal* basis (e_i, e_a) (adapted to the subspace $L^{(\alpha)}$); if α happens to be decomposable, it is of the form $\alpha = ke^1 \wedge \dots \wedge e^p$ due to (5.8.15); according to (5.8.1) $*\alpha = \hat{k} e^{p+1} \wedge \dots \wedge e^n$ so that L is a direct sum of (the orthogonal subspaces) $L^{(\alpha)}$ and $L^{(*_g \alpha)}$. □

Summary of Chapter 5

The computation of volumes of parallelepipeds (and consequently the integration procedure, where the values of functions are multiplied by the volumes of *infinitesimal* parallelepipeds) singles out completely antisymmetric fully covariant tensors, usually called forms. This chapter makes the reader acquainted with forms at the level of linear algebra. Forms enjoy several important unique properties (not shared with general tensors). They are naturally $\mathbb{Z}$-graded, one can multiply them one with another via the (graded commutative) exterior (wedge) product $\wedge$ (giving rise to a graded *exterior* = Grassmann algebra) and with vectors via the interior product i_v (which turns out to be a derivation of degree -1 of the exterior algebra). If a vector space is endowed with a metric tensor and orientation, there are also the canonical volume form and Hodge star operator $*$ on forms available. The determinant is naturally related to these concepts.

Formula	Description	Equation
$\wedge := \dfrac{(p+q)!}{p!q!}\pi^{\mathrm{A}} \circ \otimes$	Exterior (wedge) product of forms	(5.2.4)
$(\beta + \lambda\tau) \wedge \alpha = \beta \wedge \alpha + \lambda\tau \wedge \alpha$		
$\alpha \wedge (\beta + \lambda\tau) = \alpha \wedge \beta + \lambda\alpha \wedge \tau$	Bilinearity of $\wedge$	(5.2.4)
$(\alpha \wedge \beta) \wedge \gamma = \alpha \wedge (\beta \wedge \gamma)$	Associativity of $\wedge$	(5.2.4)
$\alpha \wedge \beta = (-1)^{pq} \beta \wedge \alpha$	$\mathbb{Z}$-graded commutativity of $\wedge$	(5.2.4)
$\alpha = (1/p!)\, \alpha_{a\dots b}\, e^a \wedge \dots \wedge e^b$	Expression of a p-form in terms of e^a	(5.2.9)
$\hat{\eta}\alpha := (-1)^{\deg \alpha} \alpha$	Main automorphism of $\wedge L^*$	(5.3.3)
$(i_v \alpha)(u, \dots, w) := \alpha(v, u, \dots, w)$	Interior product (of v and α)	(5.4.1)
$(i_v \alpha)_{a\dots b} = v^c \alpha_{ca\dots b}$	Component expression of i_v	(5.4.1)
$i_v(\alpha \wedge \beta) = (i_v \alpha) \wedge \beta + (\hat{\eta}\alpha) \wedge (i_v \beta)$	Graded Leibniz rule for i_v	(5.4.2)
$\delta^{a\dots b}_{c\dots d} = \delta^a_{[c} \dots \delta^b_{d]} \equiv \delta^{[a}_c \dots \delta^{b]}_d \equiv \delta^{[a}_{[c} \dots \delta^{b]}_{d]}$	p-delta (generalized Kronecker) symbol	(5.6.2)
$n!\ \det A = \varepsilon_{a\dots b} \varepsilon^{c\dots d}\ A^a_c \dots A^b_d$	Determinant and Levi-Civita symbol	(5.6.2)
$\omega_g = o(f)\sqrt{\lvert g \rvert}\ f^1 \wedge \dots \wedge f^n$	Metric volume form	(5.7.3)
$\mathrm{vol}\,(Au, \dots, Av) =: (\det A)\ \mathrm{vol}\,(u, \dots, v)$	Determinant of a linear map A	(5.7.6)
$p!(*\alpha)_{a\dots b} := \alpha^{c\dots d}\ \omega_{c\dots da\dots b}$	Hodge star (duality) operator	(5.8.1)
$*_g *_g =\ \mathrm{sgn}\, g\ (-1)^{p(n+1)}$	Star squared is $\pm$ the unity	(5.8.2)
$\alpha \wedge *_g \beta =: (\alpha, \beta)_g \omega_g$	Scalar product $(\alpha, \beta)_g$ of forms	(5.8.4)
$p!(\alpha, \beta)_g = \alpha_{a\dots b}\ \beta^{a\dots b}$	Component expression of $(\alpha, \beta)_g$	(5.8.4)

6

Differential calculus of forms

6.1 Forms on a manifold

• In Section 2.5 we described how one may progress from the linear algebra of tensors (Section 2.4) to tensor *fields* on a manifold M. Since p-forms in L are nothing but special tensors in L, the same construction brings us to tensor fields on M again, namely to completely antisymmetric tensor fields of the type $\binom{0}{p}$ in this particular case. Such objects[66] are called (differential) *p-forms on M* and the space of p-forms on M will be denoted by $\Omega^p(M)$.

The straightforward pointwise approach of Section 2.5 thus enables us to carry all the objects and operations, introduced at the level of linear algebra in Chapter 5, to the manifold. In particular, we get a *Cartan algebra* of differential forms[67] on M

$$\Omega(M) := \bigoplus_{p=-\infty}^{\infty} \Omega^p(M) \equiv \Omega^0(M) \oplus \Omega^1(M) \oplus \cdots \oplus \Omega^n(M)$$

and $\mathcal{F}(M)$-linear operators i_V, j_V, $*_g$ and $\hat{\eta}$ on it (V is a vector *field* on M and g is the *field* of a metric tensor[68] on M).

6.1.1 Check that an arbitrary p-form α on M may be written locally (in a coordinate patch $\mathcal{O} \leftrightarrow x^i$) as

$$\alpha = \frac{1}{p!}\alpha_{i\ldots j}(x)\underbrace{dx^i \wedge \cdots \wedge dx^j}_{p \text{ entries}}$$

Hint: see (5.2.9) and (2.5.4). □

6.1.2 Write down the most general forms on $M = \mathbb{R}^2[x,y]$ ($\Omega^0(\mathbb{R}^2) \ni f(x,y)$, $\Omega^1(\mathbb{R}^2) \ni \alpha = \alpha_1(x,y)\,dx + \alpha_2(x,y)\,dy$, $\Omega^2(\mathbb{R}^2) \ni \beta = \hat{\beta}(x,y)\,dx \wedge dy$). □

6.1.3 Let $M = \mathbb{R}^3[x,y,z]$, $\alpha = x\,dy - y\,dz$, $\beta = z^2\,dx \wedge dz - dy \wedge dx$, $V = (xy)^2\partial_x + \partial_y$. Compute

$$\alpha \wedge \beta \qquad i_V\alpha \qquad i_V\beta \qquad i_V(\alpha \wedge \beta)$$

[66] As in Section 5.2 we put $\Omega^0(M) := \mathcal{T}_0^0(M) \equiv \mathcal{F}(M)$, $\Omega^1(M) := \mathcal{T}_1^0(M)$.

[67] Its origin being in the exterior algebra ΛT_P^*M (cf. $\mathcal{T}(M) \leftrightarrow T(T_PM)$ in Section 2.5); the subspaces of degrees $p < 0$ and $p > n$ are trivial and so they need not be displayed explicitly.

[68] The introduction of $*$ needs, in fact, an *oriented* manifold, see Section 6.3.

Hint: the calculation before (5.2.10) and (5.4.2); $(-(xz^2+y)\,dx\wedge dy\wedge dz,\ x,\ -dx+(xy)^2dy+(xyz)^2,\ (xz^2+y)\,dx\wedge dz-(xz^2+y)(xy)^2\,dy\wedge dz)$. □

• Taking into account the result of (3.1.5) we see that differential forms (being special strictly covariant tensors) behave nicely under (smooth) maps of manifolds (one can always pull them back). Since the exterior product is based on the tensor product, pull-back behaves very simply on the former, too.

6.1.4 Check that a map of manifolds $f: M\to N$ induces a *morphism of their Cartan algebras*, i.e. the map $f^*:\Omega(N)\to\Omega(M)$ (pull-back of differential forms), which respects grading, linear structure and product, so that there holds

$$f^*:\Omega^p(N)\to\Omega^p(M)\qquad f^*(\alpha+\lambda\beta)=f^*\alpha+\lambda f^*\beta$$
$$f^*(\alpha\wedge\sigma)=f^*\alpha\wedge f^*\sigma$$

Hint: see (3.1.4), the definition of $\wedge$. □

6.1.5 Check that if a map $f:M\to N$ is given in coordinates as $x^i\mapsto y^a(x)$ and if $\alpha\in\Omega^p(N)$, then

$$f^*\alpha\equiv f^*\left\{\frac{1}{p!}\,\alpha_{a\dots b}(y)\underbrace{dy^a\wedge\dots\wedge dy^b}_{p\text{ entries}}\right\}=\frac{1}{p!}\,\alpha_{a\dots b}(y(x))\,dy^a(x)\wedge\dots\wedge dy^b(x)$$
$$=\frac{1}{p!}\,\alpha_{a\dots b}(y(x))\,J^a_i(x)\dots J^b_j(x)\underbrace{dx^i\wedge\dots\wedge dx^j}_{p\text{ entries}}$$

Hint: see (3.1.4). □

6.1.6 Let $f:S^2\to\mathbb{R}^3$ be the standard realization of the two-dimensional sphere in the three-dimensional space (3.2.4). Compute $f^*\beta$ for the 2-form β discussed in (6.1.3) for $R=1$ (the unit sphere).

Hint:

$$f^*(z^2\,dx\wedge dz-dy\wedge dx)=z^2(\vartheta,\varphi)\,dx(\vartheta,\varphi)\wedge dz(\vartheta,\varphi)-dy(\vartheta,\varphi)\wedge dx(\vartheta,\varphi)=\cdots$$
$$=\sin\vartheta\cos\vartheta(1-\sin\vartheta\cos\vartheta\sin\varphi)\,d\vartheta\wedge d\varphi$$

□

• We come now to more technical results concerning properties of derivations of the Cartan algebra. This algebra, as already mentioned above, is $\mathbb{Z}$-graded and it is also *graded commutative* (5.2.4). Let us state two fairly simple, albeit very useful, results in the abstract language of such algebras.

6.1.7 Let A be a $\mathbb{Z}$-graded and graded commutative algebra and let D_k and D_l be its derivations of degree k and l respectively; so there holds

$$A=\bigoplus_{p=-\infty}^{\infty}A_i\qquad a_ia_j=(-1)^{ij}a_ja_i\quad a_i\in A_i\quad a_j\in A_j$$
$$D_k:A_i\to A_{i+k}\qquad D_k(a_ib)=(D_ka_i)b\ +\ (-1)^{ik}a_i(D_kb)\quad a_i\in A_i,\ \ b\in A$$

Show that their *graded commutator*

$$[D_k, D_l] := D_k D_l \; - \; (-1)^{kl} D_l D_k$$

(being actually a commutator, unless both derivations are of *odd* degree, when it becomes the *anti*commutator[69]) is a derivation of the algebra A (of degree $k + l$), too.

Hint: brute force (apply $[D_k, D_l]$ on the product $a_i b$ and make use of the definitions). □

6.1.8 Let A be a $\mathbb{Z}$-graded and graded commutative algebra, D_k a derivation of degree k and let a_r be an element of degree r (i.e. $a_r \in A_r$). Show that

$$D \equiv a_r D_k : A_i \to A_{i+r+k} \qquad b \mapsto a_r(D_k(b))$$

is a derivation of the algebra A (of degree $k + r$), too.

Hint: as in (6.1.7). □

The result (6.1.7) will be used as early as in the next section (see, for example, (6.2.8)), (6.1.8) will be used in Chapter 15, which deals with a linear connection (15.6.17).

We will close this section by mentioning two concepts, which will not be used directly in what follows. They are, however, fairly common in modern mathematical physics.

The objects treated in problem (6.1.7) provide an example of a *graded Lie algebra* (Appendix A.5). In the case under consideration, its underlying linear space is given as a direct sum of linear spaces of *derivations* of degree k and a *graded commutator* is then introduced into this space according to (6.1.7) (extended to non-homogeneous elements by linearity). If a coarser $\mathbb{Z}_2$ grading were to be considered (i.e. if forms and their derivations were divided only into *even and odd*), it would result in a *Lie superalgebra*. The corresponding $\mathbb{Z}_2$-graded commutator is usually called the *supercommutator*, being (like in (6.1.7)) actually a commutator, unless both elements are *odd*, when it becomes the *anti*commutator). It obeys the super-Jacobi identity, details of which are left to the ambitious reader (less ambitious readers may find it in Appendix A.5).

The problem (6.1.8) provides a basis of another useful trick, frequently met in supermathematics, namely the use of *odd parameters*. We see that if an expression $a_r D_k$ is combined with r and k such that their sum is *even*, the resulting derivation turns out to be *even*. In the $\mathbb{Z}_2$-case this opens up the possibility of getting rid of considering odd derivations (one always combines them with auxiliary odd "parameters," i.e. with odd elements of an auxiliary algebra A) as well as anticommutators (for even derivatives *ordinary* commutators suffice).

6.2 Exterior derivative

• In addition to algebraic operations on forms on a manifold ($*, i_V, j_V, \hat{\eta}$) being merely pointwise extensions of corresponding operations at each point $P \in M$, a *differential* operation of highest importance appears on $\Omega(M)$, namely the exterior derivative. As we will see

[69] Although it is written as an *ordinary* commutator, in graded algebra this means automatically the *graded* commutator (since the latter is much more important than the former).

later in Section 8.5, this derivative represents a "common base" (core) hidden behind all the basic differential operations known from vector analysis in E^3 (gradient, divergence, curl and Laplace operator), but also of various far-reaching generalizations. In the next chapter, which deals with the integral calculus of forms, we will derive the (very simple, general and useful) Stokes' theorem, which relates "volume integrals" to particular "surface integrals" (over the boundaries of the volumes) and it turns out that the exterior derivative happens to play a prominent role in this theorem, too.[70]

The exterior derivative is often introduced in an axiomatic way (see items 1–5 in (6.2.5)), but we will try, as is done frequently in this book, to arrive at its definition "bottom up," through a rough motivation first, and possibly a further improvement of the raw result, afterwards.

Consider the following problem: given a tensor field of type $\binom{0}{p}$, examine whether the operation of partial differentiation of its components

$$t_{i\dots j} \mapsto t_{i\dots j,k} \equiv \partial_k t_{i\dots j}$$

represents a tensor operation. We see that *an additional lower index* has appeared, so that it might represent a map $\mathcal{T}^0_p(M) \to \mathcal{T}^0_{p+1}(M)$. A short calculation, however, cures us quickly of these expectations.

6.2.1 Given a tensor field $t_{i\dots j}(x)$ of type $\binom{0}{p}$, check that

$$s_{i\dots jk}(x) := t_{i\dots j,k} \equiv \partial_k t_{i\dots j}$$

fails to be a tensor field (of type $\binom{0}{p+1}$) in general.

Hint: this object transforms under a change of coordinates $x^i \mapsto x^{i'}$ as follows ($t_{i'\dots j',k'} \equiv \partial_{k'} t_{i'\dots j'}$):

$$t_{i'\dots j',k'} = \left(\frac{\partial x^k}{\partial x^{k'}}\partial_k\right)\left(\frac{\partial x^i}{\partial x^{i'}}\cdots\frac{\partial x^j}{\partial x^{j'}}t_{i\dots j}\right)$$

$$= \underbrace{\frac{\partial x^k}{\partial x^{k'}}\frac{\partial x^i}{\partial x^{i'}}\cdots\frac{\partial x^j}{\partial x^{j'}}t_{i\dots j,k}}_{\text{tensorial (good) terms}} + \underbrace{\left(\frac{\partial^2 x^i}{\partial x^{k'}\partial x^{i'}}\cdots\frac{\partial x^j}{\partial x^{j'}} + \cdots + \frac{\partial x^i}{\partial x^{i'}}\cdots\frac{\partial^2 x^j}{\partial x^{k'}\partial x^{j'}}\right)t_{i\dots j}}_{\text{non-tensorial (bad) terms}}$$

□

6.2.2 Check that non-tensorial terms actually *do not appear* if:

(i) $p = 0$ (so that $f \mapsto f_{,j}$ *is* a genuine tensor operation; namely, the good old gradient $f \mapsto df$)
(ii) one restricts to *affine* changes of coordinates $x^{i'} = A^i_j x^j + a^i$ (in particular, *linear* changes for $a^i = 0$).

Hint: see (6.2.1) and (2.5.3). □

[70] The exterior derivative may also be *defined* in terms of this theorem (and this way of acquainting oneself with it is fairly instructive, too; see the reputable monographs by Arnold and by Misner, Thorne and Wheeler for more details. In this approach the exposition of integral calculus of forms *precedes* that of the differential calculus; the idea generalizes well-known procedures leading to divergence in terms of (the limit of) a flux of a vector field for the boundary of the region and the curl in terms of (the limit of) a circulation of a vector field for the boundary of a surface, given standardly in the courses of hydrodynamics and electrodynamics (or in textbooks devoted to vector analysis, such as that by Marder).

• So, in general, mere partial differentiation of components of tensor fields *does not lead* in turn to components of tensor fields. The structure of "bad" (non-tensorial) terms present in the expression above, however, strongly suggests a simple (but non-trivial) way to get rid of them.[71] Let us consider the case when $t_{i\dots j}$ is a completely *antisymmetric* tensor (i.e. a *p-form*) and let us *arrange* the antisymmetry of the result as well:

$$t_{i\dots j} \equiv t_{[i\dots j]} \mapsto t_{[i\dots j,k]}$$

6.2.3 Check that if $\alpha \in \Omega^p(M)$, then the component rule

$$d_0 : \alpha_{i\dots j} \mapsto \alpha_{[i\dots j,k]}$$

provides a map

$$d_0 : \Omega^p(M) \to \Omega^{p+1}(M)$$

(so that d_0 actually *is* a tensor operation).

Hint: antisymmetrization cancels out "bad" terms in (6.2.1), since they are *symmetric* (each one in a pair of indices), (5.2.6). □

6.2.4 Check that the map d_0 from (6.2.3) enjoys the following properties:

(i) $d_0(\alpha + \lambda\beta) = d_0\alpha + \lambda d_0\beta \qquad \lambda \in \mathbb{R}$
(ii) $d_0 f = df \qquad f \in \Omega^0(M)$, d is the gradient from (2.5.3)
(iii) $d_0 d_0 = 0$
(iv) if $\alpha \in \Omega^p(M),\ \beta \in \Omega^q(M)$, then on their product we have

$$d_0(\alpha \wedge \beta) = A(p,q)(d_0\alpha) \wedge \beta + B(p,q)\alpha \wedge d_0\beta$$

where

$$A(p,q) = \frac{p+1}{p+q+1}(-1)^q \qquad B(p,q) = \frac{q+1}{p+q+1}$$

Hint: (iii) $\alpha_{[[i\dots j,k]l]} = \alpha_{[i\dots j,(kl)]} \overset{(5.2.6)}{=} 0$; (iv) a direct computation of components right from the definition. □

• The first three properties are very simple. The fourth one would be simple, too (it resembles Leibniz's rule), if it did not contain the awkward factors A and B. We may get rid of these factors easily, however (without losing the nice properties of d_0), if a new operation d, the *exterior derivative*, is defined as being just an appropriate *multiple* of d_0 (depending on the degree of a form, i.e. $d = C(p)d_0$ on $\Omega^p(M)$).

6.2.5 Check that the component rule (with respect to the coordinate basis)

$$(d\alpha)_{i\dots jk} := (-1)^p\,(p+1)\,\alpha_{[i\dots j,k]} \qquad \alpha \in \Omega^p(M)$$

[71] Another, technically more involved, possibility is to introduce a *covariant derivative* (Chapter 15). Here, further terms occur in addition to partial derivatives, resulting in exact cancellation with the non-tensorial terms from (6.2.1). There is an essential difference between these two lines of reasoning: while the covariant derivative needs an *additional structure* on a manifold (namely a linear connection), the exterior derivative makes do with a "bare" manifold (endowed with a smooth structure alone).

(i.e. the choice[72] $C(p) = (-1)^p\,(p+1)$) defines a map on forms, enjoying the properties

1.	$d : \Omega^p(M) \to \Omega^{p+1}(M)$	a map of degree +1
2.	$d(\alpha + \lambda\beta) = d\alpha + \lambda d\beta$	it is $\mathbb{R}$-linear ($\lambda \in \mathbb{R}$)
3.	$df = df$	$f \in \Omega^0(M)$, d on the right being the gradient
4.	$dd = 0$	it is nilpotent
5.	$d(\alpha \wedge \beta) = (d\alpha) \wedge \beta + (\hat{\eta}\alpha) \wedge d\beta$	graded Leibniz's rule

i.e. put all together, d is a *derivation of the Cartan algebra of degree* +1 (items 1, 2 and 5; see Appendix A.5 or (6.1.7)), which moreover happens to be nilpotent (item 4) and which coincides with the gradient of a function on degree 0.

Hint: item 5: see (6.2.6). □

6.2.6 Check that properties 1–5 fully characterize the operator d and that, in particular, they already result in the component formula displayed in problem (6.2.5).

Hint: since (6.2.5) yields $ddx^i = 0$ (!), we find

$$\begin{aligned} d\alpha &= d\left(\frac{1}{p!}\alpha_{i\ldots j}\,dx^i \wedge \cdots \wedge dx^j\right) = \frac{1}{p!}d\alpha_{i\ldots j} \wedge dx^i \wedge \cdots \wedge dx^j \\ &= \frac{1}{p!}\alpha_{i\ldots j,k}\,dx^k \wedge dx^i \wedge \cdots \wedge dx^j = \frac{(-1)^p}{p!}\,\alpha_{[i\ldots j,k]}\,dx^i \wedge \cdots \wedge dx^j \wedge dx^k \end{aligned}$$

□

6.2.7 Compute $d\alpha$, $d\beta$ for α, β given in (6.1.3).

Hint: making direct use of the *properties* mentioned in (6.2.5) turns out to be a much quicker method of computation than the *component formula* from (6.2.5). Here, for example, we may write

$$d(x\,dy - y\,dz) = dx \wedge dy + x \wedge ddy - dy \wedge dz - y \wedge ddz = dx \wedge dy - dy \wedge dz$$

since $ddy = 0 = ddz$ (compare with the similar situation when the exterior product was discussed before (5.2.10)). □

6.2.8 Show that the Lie derivative of *differential forms* may be expressed in the following (very useful) form:[73]

$$\mathcal{L}_V = i_V\,d + d\,i_V \qquad \textit{Cartan's identity}$$

Hint: according to (6.1.7) this is an equality of two derivations (of degree 0) of the algebra $\Omega(M) \Rightarrow$ it suffices to verify it in degrees 0 and 1, where it is easy (e.g. in components). □

6.2.9 Prove the validity of the (fairly useful) identity

$$[\mathcal{L}_V, i_W] \equiv \mathcal{L}_V\,i_W - i_W\,\mathcal{L}_V = i_{[V,W]}$$

Hint: just like in (6.2.8). □

[72] Making use of *another* choice of $C(p)$ an "opposite" convention may be arranged, by which $\hat{d}(\alpha \wedge \beta) = \alpha \wedge (\hat{d}\beta) + (\hat{d}\alpha) \wedge \hat{\eta}\beta$ ($\hat{d} = d\hat{\eta}$ is enough for that). This convention is often adopted in the context of *super*manifolds.

[73] The operators which enter this formula may be given a visual meaning in the *integral calculus* of forms and this identity itself may be interpreted in terms of Stokes' theorem, see (7.8.2).

6.2.10 Prove that the exterior derivative commutes with the Lie derivative (along an arbitrary vector field)

$$[d, \mathcal{L}_V] \equiv d\,\mathcal{L}_V - \mathcal{L}_V\, d = 0$$

Hint: just like in (6.2.8), or use the *result* of (6.2.8). □

6.2.11 Prove that the exterior derivative commutes with pull-back (with respect to an arbitrary smooth map $f : M \to N$); that is to say, the following commutative diagram holds:

$$\begin{array}{ccc} \Omega^p(N) & \xrightarrow{f^*} & \Omega^p(M) \\ \downarrow d & & \downarrow d \\ \Omega^{p+1}(N) & \xrightarrow[f^*]{} & \Omega^{p+1}(M) \end{array} \qquad \text{i.e.} \quad [d, f^*] \equiv d\, f^* - f^* d = 0$$

Hint: denote $\hat{A} := [d, f^*]$ and check that it is linear and on a product it gives (see (6.1.4) and (6.2.5))

$$\hat{A}(\alpha \wedge \beta) = (\hat{A}\alpha) \wedge f^*\beta + (\hat{\eta} f^*\alpha) \wedge \hat{A}\beta$$

$\Rightarrow$ it suffices to verify $\hat{A} = 0$ in degrees 0 and 1; degree 0 is treated in (3.1.9), from degree 1 just $d\psi$ is enough for an arbitrary function ψ, which results immediately from $dd = 0$ and (once again) (3.1.9); or everything in components. □

6.2.12 Derive (6.2.10) from (6.2.11). Generalize to the following statement: *each* operation (not only d) which is invariant with respect to diffeomorphisms commutes with the Lie derivative.

Hint: for d: differentiate (in $t = 0$) $[d, \Phi_t^*] = 0$ for $\Phi_t^* \leftrightarrow V$; in general: invariant with respect to diffeomorphisms means that it *commutes* with the latter; the Lie derivative is a generator (of the pull-back) of diffeomorphisms. □

• Often one needs to evaluate the exterior derivative $d\alpha$ of a form α on *general* arguments (for a 2-form, as an example, to evaluate $d\alpha(U, V, W)$ on arbitrary vector fields U, V, W). Although this certainly can be done in components (= values on *special* arguments, a coordinate basis), in many cases of interest "Cartan formulas" (to be discussed in the following problem) prove to be much more efficient.

6.2.13 Prove

(i) the validity of the *Cartan formulas* (from a practical point of view one makes do with the particular cases $p = 0, 1, 2$ mostly) for the evaluation of the exterior derivative of a form on arbitrary arguments (vector fields), i.e. not only on a coordinate basis

$$\begin{aligned} d\alpha(X_1, \dots, X_{p+1}) &= \sum_{j=1}^{p+1} (-1)^{j+1} X_j \alpha(X_1, \dots, \hat{X}_j, \dots, X_{p+1}) \\ &\quad + \sum_{i<j} (-1)^{i+j} \alpha([X_i, X_j], \dots, \hat{X}_i, \dots, \hat{X}_j, \dots, X_{p+1}) \end{aligned}$$

where $\alpha \in \Omega^p(M)$ and the hat indicates that corresponding arguments are to be omitted

(ii) that on the coordinate basis they yield just (6.2.5)
(iii) that for $p = 0, 1, 2$ they explicitly read as

$$\begin{aligned} df(U) &= Uf \\ d\alpha(U, V) &= U(\alpha(V)) - V(\alpha(U)) - \alpha([U, V]) \\ d\beta(U, V, W) &= U(\beta(V, W)) - V(\beta(U, W)) + W(\beta(U, V)) \\ &\quad - \beta([U, V], W) + \beta([U, W], V) - \beta([V, W], U) \end{aligned}$$

(iv) that in the case where $p = 2$ it may also be written as

$$d\beta(U, V, W) = \{U(\beta(V, W)) - \beta([U, V], W)\} + \text{cycl.}$$

Hint: first show that the formulas to be proved are equivalent to[74] the identities

$$\begin{aligned} i_U d &= \mathcal{L}_U && \text{on } \Omega^0(M) \\ i_V i_U d &= \mathcal{L}_U i_V - \mathcal{L}_V i_U - i_{[U,V]} && \text{on } \Omega^1(M) \\ i_W i_V i_U d &= (\mathcal{L}_U i_W i_V + \text{cycl.}) - (i_W i_{[U,V]} + \text{cycl.}) && \text{on } \Omega^2(M) \\ & && \text{etc.} \end{aligned}$$

In order to prove these identities (as well as further ones, i.e. $i_{X_{p+1}} \dots i_{X_1} d = \cdots$) one has to commute in successive steps d through the interior products, making use of (6.2.8) and (6.2.9); the term with d at the leftmost position vanishes (why?). □

6.2.14 * Check that for the Lie derivative of *forms* along a field σV, where σ is a function, there holds

$$\mathcal{L}_{(\sigma V)}\alpha = \sigma \mathcal{L}_V \alpha + d\sigma \wedge i_V \alpha$$

Hint: see (6.2.8) and (5.4.1). □

6.3 Orientability, Hodge operator and volume form on M

• In Chapter 5 we encountered the concepts of orientation (Section 5.5), the Hodge operator (Section 5.7) and the volume form (Section 5.8) at the level of linear algebra. Now we would like to carry these objects onto a manifold.

Each tangent space $T_P M$, $P \in M$, is a linear n-dimensional space and one may introduce an orientation there. As explained in detail in Section 5.5, given a space L there exist *just two* possible orientations in it (one particular basis is declared to be either right-handed or left-handed), i.e. the orientation in L turns out to be a *discrete* quantity. If one intends to set an orientation on each tangent space on a (smooth) manifold, an additional requirement arises. Namely, it is natural to restrict to *smooth* choices of orientations (so that two "nearby" points, roughly speaking, do not have opposite, i.e. "the farthest possible," orientations of their tangent spaces). This may be stated simply within a single coordinate patch as follows: coordinates $x^i \leftrightarrow \mathcal{O}$ induce the coordinate basis in each tangent space in $\mathcal{O}$. To

[74] A possible point of view is that Cartan formulas express the *exterior* derivative in terms of the *Lie* derivative.

make a smooth choice of orientation in $\mathcal{O} \subset M$ means to declare *all* the coordinate bases to be, say, right-handed. If the coordinate basis $\{\partial_i\}$ is right, the *coordinates themselves* are said to be right-handed.

On a manifold $\mathbb{R}^n$ the *standard orientation* is introduced by declaring *Cartesian* coordinates to be right-handed.[75]

6.3.1 Let two coordinates x^i and $x^{i'}$ be available in a patch $\mathcal{O}$. Check that

(i) it is the sign of the Jacobian matrix $J(x) := \det J^i_j \equiv \det(\partial x^{i'}/\partial x^j)$ which determines the *relative* orientations of the coordinates x^i with respect to $x^{i'}$
(ii) the interchange of any pair of coordinates (like $x^1 \leftrightarrow x^2$) changes the orientation.

Hint: see (2.2.6) and Section 5.5 and use the properties of the determinant. □

• If we try, however, to introduce in this way an orientation onto the *manifold as a whole*, insurmountable problems may arise. Imagine there are only two charts on the manifold and let their intersection $\mathcal{O} \cap \mathcal{O}'$ be connected. Then if one chooses an orientation in $\mathcal{O}$, the orientation $\mathcal{O}'$ is induced automatically (making use of the consistency on the overlap) and so on the whole manifold $M = \mathcal{O} \cup \mathcal{O}'$ as well.

Consider a case with $M = \mathcal{O} \cup \mathcal{O}'$ still, where the intersection $\mathcal{O} \cap \mathcal{O}'$ is, however, *no longer* connected. Now, the orientation from $\mathcal{O}$ gets to $\mathcal{O}'$ via *two* (or more) channels and it might happen that the results which stem from these two sources will *contradict* one another. A simple[76] example illustrating that this threat is real is provided by the *Möbius band.*

6.3.2 Take two bands P, P' cut out of a square paper exercise book (so that they are both endowed with Cartesian coordinates, x, y and x', y' respectively). Denote by $A, B \leftrightarrow P$ and $A', B' \leftrightarrow P'$ respectively their (two-dimensional) marginal regions. Now put B and B' over each other and glue together; a (longer) band with margins A, A' results. On the regions B, B' (when glued together) a natural change of coordinates arises, $x'(x, y) = x + c$, $y'(x, y) = y$, with *positive* Jacobian matrix (= 1). Now put A and A' over each other and glue together again. Check (by experiment) that

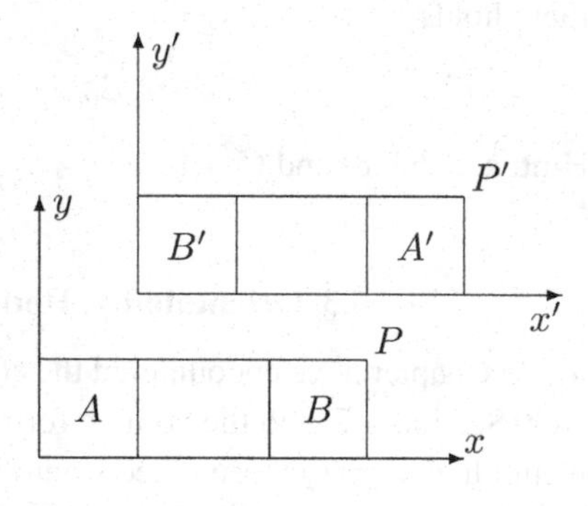

(i) if A and A' are glued together "as it is proper" (no flip over ⇒ a ring results), the Jacobian of the change of coordinates in this region is positive, too
(ii) if we *flip* the band *over* at A' by the angle π around the longitudinal axis of the band, first, and only then glue together A and A' (⇒ the Möbius band results), the change of coordinates in this region is $x'(x, y) = x + c_1$, $y'(x, y) = -y + c_2$ with *negative* Jacobian $(= -1)$
(iii) in general, if we flip the band over by an angle $2k\pi$, the Jacobian is positive, for $(2k + 1)\pi$ it is negative. □

[75] Since a single chart is enough on $\mathbb{R}^n$, the orientation is fixed by this on the whole manifold.
[76] It is also popular, see Escher's art works Band van Möbius I, II.

• We see that the orientations of the band P', induced from P via the channels $A = A'$ and $B = B'$ respectively, contradict one another. This actually means that we obtain *no consistent* global orientation on the *union* (on the whole Möbius band). One can prove that this is really an *inherent* problem with the Möbius band itself.[77]

(Here we encountered the tiny tip of a huge iceberg on our voyage, the volume of its underwater part being, as is well known, much bigger than that of its visible part. Unfortunately the majority of the iceberg will remain under the surface until the end of the book. What we are speaking about is the close relation between the differential geometry and the topology of manifolds. We see that global topological properties of manifolds may, as an example, *obstruct* the introduction of some particular geometrical structures (here the orientation or, equivalently (6.3.5), a volume form). Similar "topological conditions" are imposed by several other celebrities of the geometrical heaven, such as spinor fields or a metric tensor with Lorentzian signature (the latter cannot be globally defined on the ordinary sphere S^2). They might be more modest and follow the example of such a reputable and useful quantity as the "ordinary" (positive definite) metric tensor: without any idle talk it gladly allows itself to be defined, when nicely asked, on an *arbitrary* manifold.)

Manifolds like the Möbius band are said to be *non-orientable manifolds*.[78] In contrast, a manifold is called *orientable* if it can be endowed with an *oriented atlas*, which is an atlas in which the Jacobian of the change of coordinates happens to be positive on every non-empty overlap of coordinate patches. Thus, if we are given an atlas on a manifold, we may try to improve it (as regards orientability) by interchanging, if necessary, the order of coordinates (6.3.1) and, if the manifold is orientable, we end up with the oriented atlas.

6.3.3 Check that the spheres S^n are orientable manifolds.

Hint: inspect the structure of overlapping region(s?) of charts in the atlas consisting of stereographic coordinates (1.3.1); or use the result of (6.3.4). □

6.3.4* Check that each manifold which can be defined implicitly (in terms of constraints; (see Section 1.5)) is necessarily orientable.

Hint: in a *neighborhood* of the manifold M (given by m independent constraints in $\mathbb{R}^{n+m}$) we may use as local coordinates the constraints $\phi^1, \dots, \phi^m$ plus some n additional coordinates $z^1, \dots, z^n$ (the latter then provide an atlas for M itself). Order these further n coordinates so that $(\phi^1, \dots, \phi^m, z^1, \dots z^n)$ is right-handed (in the sense of the *standard* orientation in $\mathbb{R}^{n+m}$). Check that the atlas on M constructed in this manner is oriented. □

• The test of the orientability of a manifold by direct construction of the oriented atlas may turn out to be far from simple. There is, fortunately, another practical criterion[79] based

[77] It is *unavoidable* by any trick like, say, some ingenious choice of coordinate patches and the structure of their overlaps; see (6.3.6).

[78] There are long lasting heated disputes among scientists as to whether non-orientability of a manifold is congenital, unalterable by upbringing at all, or results from an emotionless approach within babyhood (some claim even during the prenatal period, when being glued from trivial pieces).

[79] This criterion enables one to prove in an elegant way the orientability of each *Lie group* (11.1.6) as well as of each *phase space* in mechanics (14.3.6).

on a volume form. A *volume form* on an n-dimensional manifold M is an *arbitrary* (globally defined, smooth and) everywhere non-vanishing n-form on M (i.e. $\omega \in \Omega^n(M)$, $\omega(x) \neq 0, x \in M$).[80]

6.3.5 Prove the following statement:

$$M \text{ is orientable} \quad \Leftrightarrow \quad \exists \text{ a volume form } \omega \text{ on } M$$

Hint: $\rightarrow$ let $x \equiv x^i$, $y \equiv y^a$, $z \equiv z^\alpha$ be charts of an oriented atlas. In the coordinate patch of z define

$$\omega := dz \equiv dz^1 \wedge \cdots \wedge dz^n$$

Then on $z \cap y$ it is

$$\omega = \underbrace{J(y \mapsto z)}_{>0} \, dy \neq 0$$

$\Rightarrow$ it may be extended (being non-vanishing) to the region y, etc. Consistency on a triple overlap $x \cap y \cap z$ follows from

$$J(x \mapsto z) = J(x \mapsto y) J(y \mapsto z)$$

$\leftarrow$ any atlas may become oriented as follows: in the patches of x and y we accomplish (by the interchange $x^1 \leftrightarrow x^2$ or $y^1 \leftrightarrow y^2$, if necessary)

$$\omega\left(\frac{\partial}{\partial x^1}, \ldots, \frac{\partial}{\partial x^n}\right) > 0 \qquad \omega\left(\frac{\partial}{\partial y^1}, \ldots, \frac{\partial}{\partial y^n}\right) > 0$$

(5.7.5); since

$$J(x \mapsto y) \underbrace{\omega\left(\frac{\partial}{\partial y^1}, \ldots, \frac{\partial}{\partial y^n}\right)}_{>0} = \underbrace{\omega\left(\frac{\partial}{\partial x^1}, \ldots, \frac{\partial}{\partial x^n}\right)}_{>0}, \qquad \text{we have} \quad J(x \mapsto y) > 0$$

□

• One more criterion of orientability stems from the idea of the continuous transport of a *frame* along a curve. Consider a point $x \in M$ and let $(\mathcal{O}, x^i)$ be its coordinate neighborhood. Let γ be a curve passing through $x = \gamma(0)$. If there is a frame $e(x)$ at the point x, it may be decomposed with respect to the coordinate frame field, $e_a = A^i_a \partial_i$. When speaking about the continuous transport of a frame along a curve (its part in $\mathcal{O}$), we will understand such a frame field on $\gamma(t)$ which, when decomposed with respect to the local coordinate basis, leads to a *continuous* (matrix valued) function $A^i_a(t)$ (where $A^i_a(0) = A^i_a$). If a frame is transported in this way along *a loop* which lies entirely in $\mathcal{O}$, it is clear that the frame carried back has the same orientation that it had at the beginning (since the determinant of A does not vanish anywhere it cannot change sign along the way). If, however, the loop traverses several coordinate patches, the matter gets more complicated.

[80] Thereby defining a volume form in the sense of Section 5.7 in each *tangent* space.

6.3.6 Think over the fact that

(i) a manifold is orientable if and only if a frame, after being *continuously* transported along an *arbitrary* loop, preserves its orientation
(ii) the Möbius band is non-orientable.[81]

Hint: (i) given two coordinate patches with non-empty overlap $\mathcal{O}_\alpha \cap \mathcal{O}_\beta$, an orientation may be carried from one to the other by constructing a loop which lies in both of the patches and transporting an arbitrary frame along it; (ii) consider a frame which is transported along the central line of the Möbius band, e_1 being directed along the motion and e_2 always in the same half with respect to the line; after finishing the circuit e_2 lies on the opposite side. □

• On an oriented Riemannian manifold (M, g, o) we may introduce (by pointwise construction) the metric volume form (see Section 5.7) as well as the Hodge (duality) operator (see Section 5.8). Things are now easy, the essential part of the work having been done in the linear algebra.

6.3.7 Check that

(i) *any* volume form on M may be expressed in local coordinates as

$$\omega = f(x)\, dx^1 \wedge \cdots \wedge dx^n \qquad f \neq 0 \text{ on } \mathcal{O} \leftrightarrow x^i$$

(ii) in particular, the *metric* volume form on M reads as follows

$$\omega_g = o(x)\sqrt{|g(x)|}\, dx^1 \wedge \cdots \wedge dx^n$$

where $o(x)$ (being ± 1) is given by the orientation of coordinates x^i and $g(x) := \det g_{ij}(x)$.

Hint: see (5.7.1) and (5.7.3). □

• The "function" $f(x)$, which arises as a (single independent) component of the volume form, *does not* stand for a genuine *function* on M. Namely, it transforms under the change of coordinates according to the rule

$$x \mapsto x' \quad \Rightarrow \quad f(x) \mapsto f'(x') = J^{-1}(x'(x)) f(x)$$

whereas, as we know, the Jacobian is absent for a true function (= "scalar field"). Such a quantity is called a *scalar density* (of weight -1) on M (see (5.7.1) and (21.7.10)).

6.3.8 Write down the metric volume form in E^2 in Cartesian and polar coordinates and in E^3 in Cartesian, spherical polar and cylindrical coordinates.
[E^2: $\omega_g = dx \wedge dy = r\, dr \wedge d\varphi$;
E^3: $\omega_g = \underbrace{dx \wedge dy \wedge dz}_{\text{Cartesian}} = \underbrace{r\, dr \wedge d\varphi \wedge dz}_{\text{cylindrical}} = \underbrace{r^2 \sin\vartheta\, dr \wedge d\vartheta \wedge d\varphi}_{\text{spherical polar}}$] □

[81] Note that the criterion in terms of transporting frames along loops is particularly well suited to proving *non-orientability* of a manifold (*a single* suitable loop is enough for that), whereas the criterion in terms of a volume form serves well for the proof of *orientability* (*a single* volume form guarantees its orientability).

6.3.9 Write down the metric volume form in Minkowski space $E^{1,3}$ in Cartesian, spherical polar and cylindrical coordinates ($\omega_g = dt \wedge \hat{\omega}$, where $\hat{\omega}$ are the expressions valid for E^3 from (6.3.8)). □

6.3.10 Check that the metric volume forms ω_g on manifolds from problems (3.2.2)–(3.2.7) read

$$
\begin{aligned}
(a + b\sin\psi)b\, d\varphi \wedge d\psi & \qquad \text{for } T^2 \subset E^3 \\
d\alpha \wedge d\beta & \qquad \text{for } T^2 \subset E^4 \\
R^2 \sin\vartheta\, d\vartheta \wedge d\varphi & \qquad \text{for } S^2 \subset E^3 \\
R^3\, |\sin\vartheta\cos\vartheta|\, d\vartheta \wedge d\varphi \wedge d\psi & \qquad \text{for } S^3 \subset E^4 \\
r(z)\sqrt{(1 + (r'(z))^2)}\, dz \wedge d\varphi & \qquad \text{for a rotational surface} \\
R^2\, |\sinh\alpha|\, d\alpha \wedge d\varphi & \qquad \text{for the pseudosphere}
\end{aligned}
$$

□

6.3.11 Let (M, g, o) be an n-dimensional (pseudo-)Riemannian oriented manifold and let $\omega \equiv \omega_g$ be the metric volume form on M. Show that

(i) the following results are true in a (right-handed) *coordinate* basis[82]

$$
\begin{aligned}
*dx^i &= g^{ij} d\Sigma_j \\
*(dx^i \wedge dx^j) &= g^{ik} g^{jl} d\Sigma_{kl}
\end{aligned}
$$

etc.

where the $(n-1)$, $(n-2)$, ...-forms on the right are defined as follows:

$$
\begin{aligned}
d\Sigma_i &:= \frac{1}{(n-1)!}\omega_{ij\dots k}\, dx^j \wedge \dots \wedge dx^k \equiv \frac{1}{(n-1)!}\sqrt{|g(x)|}\, \varepsilon_{ij\dots k}\, dx^j \wedge \dots \wedge dx^k \\
d\Sigma_{ij} &:= \frac{1}{(n-2)!}\omega_{ijk\dots l}\, dx^k \wedge \dots \wedge dx^l \equiv \frac{1}{(n-2)!}\sqrt{|g(x)|}\, \varepsilon_{ijk\dots l}\, dx^k \wedge \dots \wedge dx^l
\end{aligned}
$$

etc.

(ii) in a (right-handed) *orthonormal* basis we have

$$*e^a = \eta^{ab}\, d\Sigma_b$$

etc.

where

$$d\Sigma_a := \frac{1}{(n-1)!}\epsilon_{ab\dots c} e^b \wedge \dots \wedge e^c$$

etc.

Hint: see (5.8.1). □

[82] Caution: the letter d in $d\Sigma_i$ (just like in $d\Sigma_{ij}, \dots$) is conventional and *does not* denote (!) the exterior derivative, here (see (6.3.12)).

6.3.12 Check that the forms $d\Sigma_i$ and $d\Sigma_{ij}$ come out in E^3 as follows:

(i) in Cartesian coordinates

$$d\Sigma_x = dy \wedge dz \qquad d\Sigma_y = dz \wedge dx \qquad d\Sigma_z = dx \wedge dy$$
$$d\Sigma_{xy} = dz \qquad d\Sigma_{yz} = dx \qquad d\Sigma_{zx} = dy$$

(ii) in spherical polar coordinates

$$d\Sigma_r = r^2 \sin\vartheta \, d\vartheta \wedge d\varphi \qquad d\Sigma_\vartheta = r^2 \sin\vartheta \, d\varphi \wedge dr \qquad d\Sigma_\varphi = r^2 \sin\vartheta \, dr \wedge d\vartheta$$
$$d\Sigma_{r\vartheta} = r^2 \sin\vartheta \, d\varphi \qquad d\Sigma_{\vartheta\varphi} = r^2 \sin\vartheta \, dr \qquad d\Sigma_{\varphi r} = r^2 \sin\vartheta \, d\vartheta$$

(iii) verify that "d" in their labeling does not stand for the exterior derivative (find such $d\Sigma_i$ (or $d\Sigma_{ij}$), for which $d(d\Sigma_i) \neq 0 \Rightarrow d\Sigma_i \neq d(\ldots)$ in the sense of the exterior derivative).

Hint: (6.3.11); (iii) $d(d\Sigma_r) = 2r dr \wedge \sin\vartheta d\vartheta \wedge d\varphi \neq 0$. □

6.3.13 Show that a volume form on E^3 which is translationally invariant is automatically rotationally invariant (i.e. homogeneity yields isotropy). Is the opposite true? Compare with the metric tensor (4.6.14).

Hint: translations are generated (in Cartesian coordinates) by the basis vector fields ∂_i; then if $\omega = f(x, y, z) dx \wedge dy \wedge dz$, translational invariance gives f = constant. (The opposite does not hold. The form $\Phi(r)\hat{\omega}$, $\Phi(r)$ being any non-vanishing function of the radial spherical coordinate and $\hat{\omega}$ being the standard (metric) volume form, is rotationally invariant albeit it is not (unless Φ = constant) translationally invariant.) □

6.4 V-valued forms

• In a number of applications of forms (Lie groups, fiber bundles, connections, etc.) one actually encounters slightly more general objects than we have discussed up to now, namely forms with values in a (finite-dimensional) vector space V. They represent a simple generalization of "ordinary" ($\mathbb{R}$-valued) forms, which are, however, a fairly convenient tool, economizing both the conceptual and computational sides of the matter. We will treat V-valued forms first at the level of linear algebra and then carry them onto a manifold.

Recall that p-forms in L (the elements of $\Lambda^p L^*$, see Section 5.2) were introduced as the multilinear, completely antisymmetric maps

$$\sigma : \underbrace{L \times \cdots \times L}_{p} \to \mathbb{R} \qquad \underbrace{(v, \ldots, w)}_{p} \mapsto \sigma(v, \ldots, w) \in \mathbb{R}$$

Multilinearity itself is perfectly meaningful, however, in a more general case, when the target space is an arbitrary vector space V rather than (the simplest one) $\mathbb{R}$. Consider accordingly multilinear, completely antisymmetric[83] maps

$$\alpha : \underbrace{L \times \cdots \times L}_{p} \to V \qquad \underbrace{(v, \ldots, w)}_{p} \mapsto \alpha(v, \ldots, w) \in V$$

[83] Complete antisymmetry is not required, as usual, for $p = 0, 1$ (when it makes no sense).

Such maps will be called *V-valued p-forms in L* and the space of such forms[84] will be denoted by $\Lambda^p(L^*, V)$.

6.4.1 Let $\alpha \in \Lambda^p(L^*, V)$, e_i be a basis in L, E_A be a basis in V and e^i, E^A be corresponding dual bases. Show that

(i) by the rule

$$\alpha^A(v, \dots, w) := \langle E^A, \alpha(v, \dots, w)\rangle$$

N pieces of "ordinary" ($\mathbb{R}$-valued) p-forms in L are defined, where $N \equiv \dim V$; they are called *component forms* and together they carry the same information as α does

(ii) the component forms depend on the choice of a basis in V (but not in L) and

$$E_A \mapsto \hat{E}_A \equiv A^B_A E_B \;\Rightarrow\; \alpha^A \mapsto \hat{\alpha}^A \equiv (A^{-1})^A_B \alpha^B$$

(iii) if the form α is decomposed as

$$\alpha = \alpha^A E_A \quad \text{i.e.} \quad \alpha(v, \dots, w) = \alpha^A(v, \dots, w)E_A \qquad \text{then} \qquad \alpha^A E_A = \hat{\alpha}^A \hat{E}_A$$

from which we see that the *whole* does not depend on the choice of a basis (as should be the case)

(iv) a complete decomposition (including a basis in L) of the form α reads

$$\alpha \equiv \alpha^A E_A = \left(\frac{1}{p!}\alpha^A_{i\dots j} e^i \wedge \dots \wedge e^j\right) E_A$$

so that the components $\alpha^A_{i\dots j}$ (already being real numbers) of these forms carry one upper index of "type A" in addition to the usual lower indices of "type i" (with respect to which they are completely antisymmetric).[85] □

• The vector space V, in which forms take their values, is often endowed with some supplementary structure (besides the linear one); this enables one to perform various supplementary operations (besides linear combinations). If, to give an example, the target space V turns out to be an *algebra* (either associative or Lie), one can multiply its elements and this is reflected in the possibility of introducing the *exterior product* of such forms. The simplest way to define this product is in terms of the decompositions from (6.4.1).

6.4.2 Let $\alpha \in \Lambda^p(L^*, \mathcal{A})$ and $\beta \in \Lambda^q(L^*, \mathcal{A})$ be $\mathcal{A}$-valued forms ($\mathcal{A}$ being an *algebra*), and let the product in $\mathcal{A}$ be given by the relations $E_A E_B = c^C_{AB} E_C$ with respect to a basis E_A (see Appendix A.2). Show that

(i) the rule (*exterior product* of such forms)

$$\wedge : \Lambda^p(L^*, \mathcal{A}) \times \Lambda^q(L^*, \mathcal{A}) \to \Lambda^{p+q}(L^*, \mathcal{A})$$

$$\alpha \wedge \beta \equiv (\alpha^A E_A) \wedge (\beta^B E_B) := (\alpha^A \wedge \beta^B) E_A E_B \equiv \left(c^C_{AB} \alpha^A \wedge \beta^B\right) E_C$$

is well defined (= does not depend on the choice of E_A; the exterior product of the component forms is assumed to be known already)

[84] For $p = 0, 1$, we set by definition $\Lambda^0(L^*, V) := V$, $\Lambda^1(L^*, V) := \mathrm{Hom}(L, V)$ (linear maps from L to V).

[85] If we return to the special case where $V = \mathbb{R}$, i.e. when we consider "ordinary" forms, the index A takes just a single value, becoming a futile luxury, and it is therefore *omitted*; again only the lower indices remain.

(ii) if the multiplication in the algebra $\mathcal{A}$ happens to be *symmetric* ($E_A E_B = E_B E_A$), then there holds

$$\alpha \wedge \beta = (-1)^{pq} \beta \wedge \alpha$$

(iii) if the multiplication in the algebra $\mathcal{A}$ is *antisymmetric* ($E_A E_B = -E_B E_A$), then[86] an extra minus sign arises

$$\alpha \wedge \beta = -(-1)^{pq} \beta \wedge \alpha$$

Hint: (i) see (6.4.1); (ii) and (iii) see (5.2.4). □

• We pass to the manifold from linear algebra in just the same way as we did in Section 6.1, when treating ordinary forms (for the fields $L \mapsto T_x M$ at each point of a manifold).

6.4.3 Check that

(i) the V-valued forms on a manifold M may also be decomposed in terms of a basis in V and the component forms

$$\alpha = \alpha^A E_A \qquad \alpha^A(U, \dots, W) := \langle E^A, \alpha(U, \dots, W) \rangle$$

with α^A already being *ordinary* forms on a manifold and $U, \dots, W$ being vector fields

(ii) a "complete" decomposition of the forms of this type (with respect to a frame field on a manifold, too) reads

$$\alpha \equiv \alpha^A E_A = \left(\frac{1}{p!} \alpha^A_{i \dots j} e^i \wedge \dots \wedge e^j \right) E_A$$

or, in particular, for a coordinate frame field

$$\alpha \equiv \alpha^A E_A = \left(\frac{1}{p!} \alpha^A_{i \dots j} \, dx^i \wedge \dots \wedge dx^j \right) E_A$$

so that the components $\alpha^A_{i \dots j}(x)$ ("functions" already) of such forms carry one upper index of "type A" in addition to the usual lower indices of "type i" (with respect to which they are completely antisymmetric). □

• The space of V-valued p-forms on M will be denoted by $\Omega^p(M, V)$; in particular, ordinary forms are $\Omega^p(M, \mathbb{R}) \equiv \Omega^p(M)$. Then the decomposition reads

$$\Omega^p(M, V) \ni \alpha = \alpha^A E_A \qquad \alpha^A \in \Omega^p(M)$$

Numerous operations, which we know already from the context of ordinary forms, may be performed with V-valued forms as well, if they are applied on the component forms alone.

6.4.4 Show that the following operations are well defined (they do not depend on the choice of a basis in E_A):

[86] The algebra $\mathcal{A} = \mathbb{R}$ (corresponding to ordinary forms) provides an example for the symmetric product, Lie algebras in turn provide an example for the antisymmetric "product" (it is usually denoted as a commutator $E_A E_B \equiv [E_A, E_B]$ there). In a general algebra it is neither symmetric nor antisymmetric (like in the complete matrix algebra), which means that $\alpha \wedge \beta$ and $\beta \wedge \alpha$ *are not related* at all.

(i) exterior derivative[87]

$$d : \Omega^p(M, V) \to \Omega^{p+1}(M, V) \qquad d\alpha \equiv d(\alpha^A E_A) := (d\alpha^A) E_A$$

and $dd = 0$ still holds

(ii) interior product (U being a vector field on M)

$$i_U : \Omega^p(M, V) \to \Omega^{p-1}(M, V) \qquad i_U\alpha \equiv i_U(\alpha^A E_A) := (i_U\alpha^A) E_A$$

(iii) pull-back (for any map $f : M \to N$)

$$f^* : \Omega^p(N, V) \to \Omega^p(M, V) \qquad f^*\alpha \equiv f^*(\alpha^A E_A) := (f^*\alpha^A) E_A$$

(iv) if $V = \mathcal{A}$ is an algebra, then the exterior product is just like in (6.4.2) and the multiplication of a form by an element of the algebra ($a^A \in \mathbb{R}$)

$$a\alpha \equiv (a^A E_A)(\alpha^B E_B) := (a^A \alpha^B) E_A E_B \equiv \left(c^C_{AB} a^A \alpha^B\right) E_C$$

(v) on a Riemannian oriented manifold the Hodge operator $*$

$$* : \Omega^p(M, V) \to \Omega^{n-p}(M, V) \qquad *\alpha \equiv *(\alpha^A E_A) := (*\alpha^A) E_A$$

(vi) if there is a scalar product $h = h_{AB} E^A \otimes E^B$ in V and $\alpha, \beta \in \Omega^p(M, V)$, then

$$h_{AB}\alpha^A \wedge *\beta^B \equiv h_{AB}(\alpha^A, \beta^B)\omega_g$$

is an *ordinary* n-form on M. □

• An especially important particular case of the forms treated in this section is provided by forms with values in the algebra of *complex numbers*. There we have a basis $E_1 = 1$, $E_2 = i$, so that each form may be decomposed as

$$\sigma = \sigma^A E_A \equiv \alpha + i\beta \qquad \sigma \in \Omega^p(M, \mathbb{C}), \quad \alpha, \beta \in \Omega^p(M)$$

6.4.5 Check that the operations mentioned above reduce to

$$\begin{aligned} d(\alpha + i\beta) &= d\alpha + i d\beta \\ i_U(\alpha + i\beta) &= i_U\alpha + i(i_U\beta) \\ f^*(\alpha + i\beta) &= f^*\alpha + i f^*\beta \\ (\alpha + i\beta) \wedge (\hat\alpha + i\hat\beta) &= (\alpha \wedge \hat\alpha - \beta \wedge \hat\beta) + i(\alpha \wedge \hat\beta + \beta \wedge \hat\alpha) \\ *(\alpha + i\beta) &= *\alpha + i * \beta \\ (k + iq)(\alpha + i\beta) &= (k\alpha - q\beta) + i(q\alpha + k\beta) \end{aligned}$$

Hint: see (6.4.4). □

6.4.6 * Consider two forms on M with values in *two* (possibly different) vector spaces, $\alpha \in \Omega^p(M, V_1)$ and $\beta \in \Omega^q(M, V_2)$. Let the bases in these spaces be $E_a \in V_1$ and $E_A \in V_2$. Show that

[87] Another way to express the same idea: notice that $\Omega(M, V) = \Omega(M) \otimes V$ and $d = d \otimes \mathrm{id}$, where d on the left is the new one and d on the right is the good old one acting on ordinary forms. A similar approach may be used for some other operators.

(i) the rule

$$\alpha \overset{\otimes}{\wedge} \beta \equiv (\alpha^a E_a) \overset{\otimes}{\wedge} (\beta^A E_A) := (\alpha^a \wedge \beta^A)(E_a \otimes E_A)$$

correctly introduces their product, being a form with values in the *tensor product* $V_1 \otimes V_2$ (Appendix A.1)

(ii) for the exterior derivative and interior product of such forms we have

$$d(\alpha \overset{\otimes}{\wedge} \beta) = d\alpha \overset{\otimes}{\wedge} \beta + \hat{\eta}\alpha \overset{\otimes}{\wedge} d\beta \qquad i_V(\alpha \overset{\otimes}{\wedge} \beta) = i_V\alpha \overset{\otimes}{\wedge} \beta + \hat{\eta}\alpha \overset{\otimes}{\wedge} i_V\beta$$

By analogy various further operations mentioned above may be expressed on such a product of forms. □

Summary of Chapter 6

Forms are treated as fields on a manifold (*differential* forms). All the algebraic constructions known from Chapter 5 still work, but a new differential operation of crucial importance enters the scene, the *exterior derivative*. It turns out to be a nilpotent ($dd = 0$) derivation of degree $+1$ of the Cartan algebra $\Omega(M)$ of forms on a manifold. A simple (but useful) generalization of ordinary forms is provided by arbitrary vector space valued forms (the ordinary ones being $\mathbb{R}$-valued).

$\alpha = (1/p!)\,\alpha_{i\ldots j}(x)\,dx^i \wedge \cdots \wedge dx^j$	Coordinate expression of a form	(6.1.1)
$D_k(a_i b) = (D_k a_i)b + (-1)^{ik} a_i (D_k b)$	D_k is a derivation of degree k	(6.1.7)
$(d\alpha)_{i\ldots jk} := (-1)^p\,(p+1)\,\alpha_{[i\ldots j,k]}$	Exterior derivative in coordinates	(6.2.5)
$dd = 0$	Exterior derivative is nilpotent	(6.2.5)
$d(\alpha \wedge \beta) = (d\alpha) \wedge \beta + (\hat{\eta}\alpha) \wedge d\beta$	Graded Leibniz's rule for d	(6.2.5)
$\mathcal{L}_V = i_V\, d + d\, i_V$	Cartan's identity	(6.2.8)
$[d, \mathcal{L}_V] \equiv d\,\mathcal{L}_V - \mathcal{L}_V\, d = 0$	Exterior and Lie derivatives commute	(6.2.10)
$[d, f^*] \equiv d\, f^* - f^*\, d = 0$	Exterior derivative commutes with pull-back	(6.2.11)
$d\alpha(U, V) = \cdots$	Cartan formula (for $p = 1$)	(6.2.13)
$d\beta(U, V, W) = \cdots$	Cartan formula (for $p = 2$)	(6.2.13)
$\alpha = \alpha^A E_A$	V-valued form on M	(6.4.1)

7
Integral calculus of forms

• In this chapter we adopt a highly useful point of view on integral calculus (in particular, on line, surface, volume, etc. integrals), in which the quantities under the integral sign are regarded as *differential forms*. The reader is expected to already have elementary experience with the concept of the multiple (Riemann) integral and our intention will be to build a bridge over troubled waters that flow between this standard background knowledge and the machinery of differential forms, which we have learned in the previous chapter.

$$\int_{\text{over something}} \text{of something}$$

The graphical presentation of the integral reflects the fact that the integral combines two independent and completely different objects: a quantity under the integral sign (*what* is to be integrated) and a domain of integration (*over what* is the first object to be integrated). The quantities under the integral sign are discussed in Section 7.1. The main observation will be the fact that it is very natural to regard these quantities as differential forms (already familiar to us from the last chapter). The structure of the domains of integration is studied in Sections 7.2. and 7.3. We restrict ourselves to domains which can be *triangulated*, i.e. decomposed into simple parts, known as *simplices*.[88] We will see that the simplices (and their formal linear combinations: chains) are endowed with an interesting algebraic structure, resembling very much the structure which we encountered when dealing with differential forms (a degree is defined and a nilpotent operator, altering the degree by one unit (thus resembling d on forms), operates there): they form a (chain) *complex* in the language of Section 9.3. In the rest of the chapter the key concept of the *integral of a form over a chain* is introduced, its essential properties are examined and a central theorem of the integral calculus of forms on manifolds (Stokes' theorem) is proved.

7.1 Quantities under the integral sign regarded as differential forms

• Let us begin with a simple example. Consider a function $f(x, y)$ in $\mathbb{R}^2[x, y]$. Recall that if we compute its integral over the disc $\bigcirc \equiv \{(x, y)|x^2 + y^2 \leq R^2\}$, we save some labor, as a rule, by passing to *polar* coordinates r, φ (the integration domain is simply stated in

[88] It turns out that all "reasonable" domains occurring in standard theoretical physics belong to this class.

these coordinates – it becomes a *rectangle* $\square \equiv [0, R] \times [0, 2\pi]$). It reads

$$\int_{\bigcirc} f(x, y)\, dx\, dy = \int_{\square} \tilde{f}(r, \varphi)\, r\, dr\, d\varphi \qquad \tilde{f}(r, \varphi) \equiv f(x(r, \varphi), y(r, \varphi))$$

One usually says that the "volume element" is to be re-expressed in new coordinates

$$dx\, dy \mapsto r\, dr\, d\varphi$$

Note that in new coordinates the "product of differentials" $dr\, d\varphi$ should be supplemented by the factor r (in general, the *Jacobian* of a change of coordinates).

In geometry we used expressions like $dx, dy, \ldots$ to denote covector fields (*1-forms*). Here, under the integral sign, these objects are "multiplied" somehow and it might be interesting to investigate whether the manipulations with the differentials *regarded as 1-forms* are related in some way or not with the changes of the "volume elements" under the integral sign. Two of the products already mentioned are eligible: the tensor product $dx \otimes dy$ and the exterior product $dx \wedge dy$.

7.1.1 Check that the tensor product does not result in anything interesting here, whereas the *exterior* product of differentials yields just the needed expression.[89]

Hint: for $dx = \cos\varphi\, dr - r \sin\varphi\, d\varphi$ and $dy = \sin\varphi\, dr + r \cos\varphi\, d\varphi$ we have $dx \wedge dy = r\, dr \wedge d\varphi$. □

- The lesson so far is that if we regard $dx\, dy$ under the integral sign as the *exterior* product $dx \wedge dy$, a standard conversion of this 2-form to polar coordinates leads just to the needed expression $r\, dr\, d\varphi \leftrightarrow r\, dr \wedge d\varphi$. A simple computation reveals even more.

7.1.2 Let $\Phi : (r, \varphi) \mapsto (x, y)$ be the coordinate change under consideration (polar $\mapsto$ Cartesian). Check that

(i)

$$\Phi(\square) = \bigcirc$$

(ii)

$$\Phi^* f = \tilde{f}$$

(iii)

$$\Phi^*(f(x, y)\, dx \wedge dy) = \tilde{f}(r, \varphi) r\, dr \wedge d\varphi$$

so that if $\alpha := f(x, y)\, dx \wedge dy \in \Omega^2(\mathbb{R}^2[x, y])$, we may write down the above-mentioned equality of integrals in an amazingly simple and transparent way:

$$\int_{\Phi(\square)} \alpha = \int_{\square} \Phi^* \alpha$$

□

[89] The relation between forms and the concept of a volume, discussed at the beginning of Chapter 5 and in Section 5.1, is behind this result; we recommend that the reader contemplate this *visually* in more detail.

• What is remarkable in this formula is that the result which is obtained in integral calculus with *no reference* to the machinery of forms acquires a particularly clear expression in the language of differential forms – conversion of the quantity under the integral sign is nothing but a standard operation on forms – namely its *pull-back* with respect to the diffeomorphism corresponding to the change of coordinates. One easily verifies that this fact is not limited to the particular case under consideration, but rather holds in general.

7.1.3 Let $\Phi : (y^1, \dots, y^n) \mapsto (x^1, \dots, x^n)$ be a change of coordinates $y \mapsto x(y)$ and let $\alpha := f(x^1, \dots, x^n)\, dx^1 \wedge \dots \wedge dx^n$ be an arbitrary n-form. Check that

(i)

$$\Phi^*\alpha \equiv \Phi^*(f(x)\, dx^1 \wedge \dots \wedge dx^n) = f(x(y))J\, dy^1 \wedge \dots \wedge dy^n$$

where $J \equiv J(x(y))$ denotes the Jacobian of the change of coordinates, so that

(ii) a general expression of the well-known result concerning the change of coordinates in the multiple integral may be written in the language of forms as follows:

$$\int_{\Phi(D)} \alpha = \int_D \Phi^*\alpha$$

(D being an integration domain).

Hint:

$$\Phi^*(dx^1 \wedge \dots \wedge dx^n) = x^1_{,i} \dots x^n_{,j} \underbrace{dy^i \wedge \dots \wedge dy^j}_{\varepsilon^{i \dots j}\, dy^1 \wedge \dots \wedge dy^n} \overset{(5.6.5)}{=} J\, dy^1 \wedge \dots \wedge dy^n$$

□

• Thus, in general, the quantities under the integral sign transform under the change of coordinates (i.e. when "the same" integral is expressed in two sets of coordinates) just like differential forms do,[90] and therefore it is quite natural to *identify* these objects and regard *all* integrals as being the *integrals of differential forms*. The fact that this is not a red herring will be confirmed later, when we learn that other natural operations on forms (exterior derivative, interior product, etc.) play an important part in integral calculus as well.

7.2 Euclidean simplices and chains

• In this section we begin with a systematic study of the objects which in the theory of integration of forms will play the role of the domains of integration. The building blocks from which general integration domains will be composed are provided by simplices on manifolds. Before this, however, we say what the simplices in $\mathbb{R}^n$ are.

[90] Under the correspondence $f\, dx^1 \dots dx^n \leftrightarrow f\, dx^1 \wedge \dots \wedge dx^n$.

In $\mathbb{R}^n$, consider points $P_0, \ldots, P_p$ $(p \leq n)$, for which the "relative vectors" $\overrightarrow{P_0 P_a}$ are linearly independent $(a = 1, \ldots, p)$. Denote by $s_p \equiv (P_0, \ldots, P_p)$ a set of points of the form[91]

$$s_p \equiv (P_0, \ldots, P_p) := \{P \in \mathbb{R}^n \mid P = t_\mu P_\mu \equiv t_0 P_0 + \cdots + t_p P_p\}$$
$$t_\mu \geq 0 \quad (\mu = 0, \ldots, p), \quad \sum_{\mu=0}^{p} t_\mu = 1$$

This set will be called the *Euclidean p-simplex*. An *orientation* is ascribed to this object, too. A simplex which has the opposite orientation with respect to $(P_0, \ldots, P_p)$ is denoted by $-(P_0, \ldots, P_p)$; by definition, the interchange of an arbitrary pair of points alters the orientation, for example $(P_0, P_1) = -(P_1, P_0)$.

7.2.1 Check that the 0-simplex is an oriented point, the 1-simplex is an oriented segment, the 2-simplex is an oriented triangle (including its interior), the 3-simplex is an oriented tetrahedron (including its interior), etc. Draw suitable figures. □

- The formal[92] linear combinations

$$c = c_i s_p^i \qquad c_i \in \mathbb{R}, \ s_p^i = i\text{th } p\text{-simplex}$$

of p-simplices are called *p-chains*[93] and the corresponding (∞-dimensional) linear space is denoted by C_p.

7.2.2 Define a linear map (the *boundary operator*)

$$\partial : C_p \to C_{p-1}$$

by the prescription

$$\partial \left(c_i s_p^i\right) := c_i \left(\partial s_p^i\right) \qquad \partial P_0 := 0 \qquad \partial(P_0, \ldots, P_p) := \sum_{\mu=0}^{p} (-1)^\mu \, (P_0, \ldots, \hat{P}_\mu, \ldots, P_p)$$

(where the hatted points are to be *omitted*). Check that this operator happens to be *nilpotent* or, as it is standardly rephrased, that the *boundary has no boundary*

$$\partial\partial = 0$$

Hint: if $\partial\partial$ is applied on a simplex $(P_0, \ldots, P_p)$, one obtains a sum of terms of the form

$$(P_0, \ldots, \hat{P}_\mu, \ldots, \hat{P}_\nu, \ldots, P_p);$$

each term appears twice there, with mutually opposite signs. □

[91] Here, the points P_μ are regarded as the tips of *vectors*, so that their linear combinations are permitted.

[92] A linear space may be specified by enumerating the basis elements. If $\mathcal{A}$ is an apple and $\mathcal{P}$ is a pear, we may introduce the two-dimensional linear space of elements of the form $v = v^1 \mathcal{A} + v^2 \mathcal{P}$ (the apple and the pear constitute its basis). In the case under consideration, the basis consists of simplices.

[93] The sum of simplices $s_p^1 + s_p^2$ may be visualized by drawing both of them in the plane *at the same time* and the simplex $3s_p$ by three simplices s_p. Less visual is the direct interpretation of the expressions with coefficients which are not natural numbers, but (as we will see in Section 7.4) in the context of integration (a simplex being regarded as a domain of integration) this "problem" does not manifest itself at all.

7.2.3 Check that

(i) explicitly we have

$$\begin{aligned}\partial(P_0, P_1) &= P_1 - P_0\\ \partial(P_0, P_1, P_2) &= (P_1, P_2) + (P_2, P_0) + (P_0, P_1)\\ \partial(P_0, P_1, P_2, P_3) &= (P_1, P_2, P_3) + (P_0, P_3, P_2) + (P_0, P_1, P_3) + (P_0, P_1, P_2)\end{aligned}$$

(ii) these expressions say that ∂ deserves its name, i.e. it *indeed* yields the boundary of a simplex (including the intuitively perceived orientation).

Hint: (ii) draw the pictures. □

7.2.4 Given two n-simplices $(P_0, \dots, P_n)$ and $(Q_0, \dots, Q_n)$ in $\mathbb{R}^n$, prove that

(i) there exists the unique *affine* transformation $\chi : \mathbb{R}^n \to \mathbb{R}^n$, such that it just matches the vertices of these simplices, i.e.

$$\chi : P_\mu \mapsto Q_\mu \qquad \text{for} \quad \mu = 0, 1, \dots, n$$

(ii) in general (including the interior points) then

$$\chi(t_\mu P_\mu) = t_\mu Q_\mu$$

Hint: an arbitrary affine transformation $x \mapsto Ax + a$ has the freedom in $n^2 + n$ parameters (a matrix A and a column a), the condition $P_i \mapsto Q_i$ gives $n(n+1)$ linear equations for the parameters. □

- This enables one to regard an arbitrary n-simplex in $\mathbb{R}^n$ as the image of some preferred "canonical" n-simplex with respect to an appropriate affine map. The role of this preferred simplex is usually played by the *standard n-simplex in* $\mathbb{R}^n$ with vertices

$$P_0 = (0, \dots, 0) \qquad P_1 = (1, \dots, 0) \qquad \dots \qquad P_n = (0, \dots, 1)$$

which will be denoted by $\overline{s}_n$.

7.2.5 Write down explicitly the affine transformation which maps the standard 2-simplex $\overline{s}_2 \equiv (P_0, P_1, P_2)$ onto the 2-simplex (Q_0, Q_1, Q_2), where $Q_0 = (1, 1)$, $Q_1 = P_2$ and $Q_2 = P_1$ ($(x, y) \mapsto (-x + 1, -y + 1)$). □

7.2.6 Check that

(i) the standard 2-simplex may also be characterized as a set by

$$\begin{aligned}\overline{s}_2 &= \{(x, y) \mid x \geq 0, y \geq 0;\ x + y \leq 1\}\\ &= \{(x, y) \mid x \geq 0, y \geq 0;\ x \leq 1;\ y \leq (1 - x)\}\end{aligned}$$

(ii) the standard n-simplex may also be characterized as a set by

$$\bar{s}_n = \left\{ (x^1, \dots, x^n) \mid x^i \geq 0, i = 1, \dots, n; \sum_{i=1}^{n} x^i \leq 1 \right\}$$
$$= \left\{ (x^1, \dots, x^n) \mid x^i \geq 0, i = 1, \dots, n; \sum_{j=1}^{n-1} x^j \leq 1; \; x^n \leq \left(1 - \sum_{j=1}^{n-1} x^j\right) \right\}$$

□

7.3 Simplices and chains on a manifold

• Now we take a step forward and introduce the concept of a p-simplex (and a p-chain) on the n-dimensional manifold M. Provisionally, this is a pair (s_p, Φ), where s_p is the *Euclidean* p-simplex and

$$\Phi : s_p \to M$$

is a smooth map.[94] Within the context of integration, it is convenient to regard a simplex on a manifold as the *image* $\sigma_p := \Phi(s_p)$ of a simplex in $\mathbb{R}^n$, together with the map Φ, which provides a parametrization on this image.[95] Now, if there is another "provisional" simplex (s'_p, Φ'), such that the points, matching in the sense of the affine map χ from (7.2.4), share the same image,

$$\Phi(t_i P_i) = \Phi'(t_i P'_i) \equiv \Phi' \circ \chi(t_i P_i)$$

we declare it to be equivalent[96] to the simplex (s_p, Φ)

$$(s_p, \Phi) \sim (\chi(s_p), \Phi') \; \Leftrightarrow \; \Phi = \Phi' \circ \chi$$

and we define the true simplex σ_p on a manifold to be given as the whole *equivalence class* $[(s_p, \Phi)]$ of such provisional simplices.

A formal linear combination of p-simplices on a manifold M is a *p-chain on M* and the (∞-dimensional) space of all chains is denoted by $C_p(M)$. In what follows such chains will serve as domains of integration.

Making use of the map Φ, the *boundary operator* ∂ is carried onto a manifold, too. Natural definitions read

$$\partial c \equiv \partial \left(c_i \sigma_p^i\right) := c_i \partial \sigma_p^i \qquad \partial \sigma_p \equiv \partial(\Phi(s_p)) := \Phi(\partial s_p)$$

7.3.1 Check that this boundary operator is

[94] More precisely, it is even a *triple* $(s_p, \Phi, \mathcal{U})$, where $\mathcal{U}$ is some *neighborhood* of s_p and $\Phi : \mathcal{U} \to M$ (i.e. the map Φ is to be defined on a slightly larger domain than s_p itself). In this way one avoids possible problems with the differentiability at the boundary.

[95] This parametrization (in contrast to local coordinates) may fail to be one-to-one (Φ may not be injective). The domain σ_p thus may also be "covered" more than once. For example, $\Phi : t \mapsto e^{it}$, $t \in s_1 = \langle 0, 4\pi \rangle$ defines a 1-simplex on the manifold $S^1 \subset \mathbb{C}$, being a *twofold* covering of the image $\sigma_1 \equiv \Phi(s_1) = S^1$.

[96] Compare with the equivalence from problem (1.5.6).

(i) a linear map

$$C_p(M) \to C_{(p-1)}(M)$$

(ii) which is *nilpotent* (the boundary has no boundary)

$$\partial\partial = 0$$

(iii) and which indeed yields the boundary of a chain in the intuitive sense.

Hint: see (7.2.2) and (7.2.3). □

7.4 Integral of a form over a chain on a manifold

• Now we have learned all we need about the objects of both types (quantities under the integral sign as well as domains of integration) and we put them together in this section.

From the elementary experience with performing integrals in $\mathbb{R}^2$ and $\mathbb{R}^3$ we know that

1. if the domain of integration is one-dimensional (a line integral), the quantity under the integral sign contains differentials of coordinates linearly
2. if the domain of integration is two-dimensional (a surface integral), the quantity under the integral sign contains a "product" of two differentials of coordinates
3. if the domain of integration is three-dimensional (a volume integral), the quantity under the integral sign contains a "product" of three differentials of coordinates, etc.

From the perspective of our new interpretation of the quantities under the integral sign and the domains of integration this means that if the integral is performed over a p-chain, the quantity under the integral sign should be a p-form. Clearly, the next thing to do then is to introduce officially the most important concept of this chapter: the *integral of a p-form on M over a p-chain on M*

$$\int_c \alpha \qquad c \in C_p(M), \quad \alpha \in \Omega^p(M)$$

The definition consists of a sequence of three steps. In the first step[97] one reduces the definition by the prescription

$$\int_c \alpha \equiv \int_{c_i\sigma_p^i} \alpha := c_i \int_{\sigma_p^i} \alpha$$

from a p-chain to a p-simplex, in the second step[98]

$$\int_\sigma \alpha \equiv \int_{\Phi(\bar{s}_p)} \alpha := \int_{\bar{s}_p} \Phi^*\alpha$$

[97] This part of the definition is related to the *additivity* of an integral with respect to the domain of integration.
[98] Here the integral is carried from an abstract object, as a manifold is, into a parameter space – the good old Cartesian space $\mathbb{R}^p$; the rule is consistent with the special case of a *change of coordinates* (7.1.3).

from a p-simplex on a manifold to the standard p-simplex in $\mathbb{R}^p$. Now, a general p-form in $\mathbb{R}^p$ reads (6.3.7)

$$\alpha = f(x^1, \dots, x^p)\, dx^1 \wedge \dots \wedge dx^p$$

and if we keep the order of the differentials from 1 to p, the (only independent) component $f(x) \equiv f(x^1, \dots, x^p)$ is unique. In the third step, finally, the integral of the form reduces by the rule

$$\int_{\bar{s}_p} \alpha \equiv \int_{\bar{s}_p} f(x^1, \dots, x^p)\, dx^1 \wedge \dots \wedge dx^p := \int_{\bar{s}_p} f(x^1, \dots, x^p)\, dx^1 \dots dx^p$$

to a standard p-fold Riemann integral of the *function* $f(x^1, \dots, x^p)$ over the domain $\bar{s}_p$ (reducing then through Fubini's theorem to p consecutive ordinary integrals). This ends the definition of the integral of an arbitrary p-form on a manifold over an arbitrary p-chain on the manifold.

7.4.1* Check that

(i) the integral defined above may be regarded as a map

$$I \equiv \int : C_p(M) \times \Omega^p(M) \to \mathbb{R} \qquad (c, \alpha) \mapsto \int_c \alpha \equiv I(c, \alpha)$$

(ii) this map actually gives a *bilinear pairing* (Appendix A.1) of the linear spaces involved and it turns out to be non-degenerate with respect to the chain slot – if the integral of a form vanishes for *all* chains, the form is necessarily zero (this is *not* true for the form slot; just consider a chain consisting of going there and back along the same path). □

7.5 Stokes' theorem

• Stokes' theorem fully deserves to be appreciated as a culmination point of this chapter; it is the most important theorem of the integral calculus on manifolds, with numerous important applications, many of them in physics. In fact, all theorems which equate two integrals, the domain of integration of one of them being the boundary of the domain of integration of the second one, may be ultimately traced back to be merely particular cases of this theorem (see Chapter 8 for more details). In general, the theorem states the following.

Theorem Given $c \in C_{p+1}$ and $\alpha \in \Omega^p(M)$, there holds

$$\int_c d\alpha = \int_{\partial c} \alpha$$

or, making use of notation from (7.4.1),

$$I(c, d\alpha) = I(\partial c, \alpha)$$

(so that the operators d and ∂ are "adjoint" to each other in the sense of the pairing I; the reason for the quotation marks is that this pairing is actually degenerate (as we mentioned in problem 7.4.1)).

7.5.1 Check that the validity of Stokes' theorem in the general case follows easily from its validity in a very special case, namely from the statement

$$\int_{\overline{s}_{p+1}} d\eta = \int_{\partial \overline{s}_{p+1}} \eta$$

where

$$\eta := f(x^1, \dots, x^{p+1})\, dx^1 \wedge \dots \wedge dx^p \ \in \Omega^p(\mathbb{R}^{p+1})$$

and $\overline{s}_{p+1}$ is the *standard* $(p+1)$-simplex in $\mathbb{R}^{p+1}$.

Hint: by definitions from Section 7.4 we have

$$\int_{c=c_i\sigma^i_{p+1}} d\alpha \overset{?}{=} \int_{\partial(c_i\sigma^i_{p+1})} \alpha \quad \Leftrightarrow \quad \int_\sigma d\alpha \overset{?}{=} \int_{\partial\sigma} \alpha$$

(reduction of the proof from a chain to a simplex),

$$\int_{\sigma=\Phi(\overline{s}_{p+1})} d\alpha \overset{?}{=} \int_{\partial\sigma} \alpha \quad \Leftrightarrow \quad \int_{\overline{s}_{p+1}} d\beta \overset{?}{=} \int_{\partial\overline{s}_{p+1}} \beta$$

(reduction from a general simplex on M to the standard simplex in $\mathbb{R}^{p+1}$; $\beta \in \Omega^p(\mathbb{R}^{p+1})$); finally, for $\beta \mapsto \eta$ it suffices to realize that β is a *sum* of terms of type η (by renaming of coordinates, if necessary, one can make the omitted differential just dx^{p+1}). □

7.5.2 Prove the "very special case" from (7.5.1) for $p = 1$.

Solution: we live in $\mathbb{R}^2[x, y]$, now; then $\eta = f(x, y)\, dx$, $d\eta = -\frac{\partial f}{\partial y} dx \wedge dy$, so that

$$\int_{\overline{s}_2} d\eta = -\int_{\overline{s}_2} \frac{\partial f}{\partial y} dx \wedge dy = -\int_0^1 dx \int_0^{1-x} \frac{\partial f}{\partial y}\, dy = \int_0^1 dx\, (f(x,0) - f(x, 1-x))$$

The boundary of $\overline{s}_2$ consists of three 1-simplices in $\mathbb{R}^2$, which may be written as the images of the *standard* 1-simplex $\overline{s}_1 = \langle 0, 1 \rangle$ in $\mathbb{R}^1[t]$ with respect to the maps

$$\Phi_1 : t \mapsto (t, 0) \qquad \Phi_2 : t \mapsto (1-t, t) \qquad \Phi_3 : t \mapsto (0, 1-t)$$

(drawing of a simple picture is highly recommended). Since

$$\Phi_1^*\eta = f(t, 0)\, dt \qquad \Phi_2^*\eta = -\, f(1-t, t)\, dt \qquad \Phi_3^*\eta = 0$$

we obtain

$$\int_{\partial \bar{s}_2} \eta = \int_{\bar{s}_1} (\Phi_1^* \eta + \Phi_2^* \eta + \Phi_3^* \eta) = \int_0^1 dt\,(f(t,0) - f(1-t,t))$$
$$= \int_0^1 dt\,(f(t,0) - f(t,1-t))$$

□

[7.5.3] Prove the "very special case" from (7.5.1) for arbitrary p.

Hint: just like in (7.5.2) we have

$$\eta = f(x^1, \dots, x^{p+1})\, dx^1 \wedge \dots \wedge dx^p$$
$$d\eta = (-1)^p \frac{\partial f}{\partial x^{p+1}}\, dx^1 \wedge \dots \wedge dx^p \wedge dx^{p+1}$$

⇒

$$\int_{\bar{s}_{p+1}} d\eta = (-1)^p \int_{\bar{s}_{p+1}} \frac{\partial f}{\partial x^{p+1}}\, dx^1 \dots dx^p dx^{p+1} \overset{(7.2.6)}{=} \dots$$
$$= (-1)^p \int_{\substack{\sum_{i=1}^p x^i \leq 1 \\ x^i \geq 0}} dx^1 \dots dx^p \left(f\left(x^1, \dots, x^p, 1 - \sum_{i=1}^p x^i \right) - f(x^1, \dots, x^p, 0) \right)$$

For the computation of

$$\int_{\partial \bar{s}_{p+1}} \eta$$

one should realize that

$$\partial \bar{s}_{p+1} \equiv \partial(P_0, P_1, \dots, P_{p+1}) = (P_1, \dots, P_{p+1}) + (-1)^{p+1}(P_0, P_1, \dots, P_p)$$
$$+ \text{other faces}$$

Making use of the maps Φ_j, $j = 1, \dots, p+2$ (cf. (7.5.2)) show that the simplices referred to as "other faces" yield zero (which justifies the use of this impersonal, even dishonorable, name), whereas the explicitly mentioned two faces lead just to the required two terms. □

7.6 Integral over a domain on an orientable manifold

• In real applications, one seldom encounters *directly* a formal chain as a domain of integration in $\int_D \alpha$; rather some "well-behaved" subset $D \subset M$ (a cube, a ball, a sphere, etc.) on an *orientable* manifold M occurs there. Such subsets may, however, be naturally made to behave like chains: it is enough to divide them, if necessary, into several parts such that each of them may be regarded as a simplex on M. In order that the integral introduced along this line really produces what is intuitively expected from it, one should be careful about the *orientation* of the simplices.

Let us explain first what is meant in general by saying that a map of manifolds preserves orientation. Consider a map $f : M \to N$ of manifolds, both being oriented and of equal dimension. Then we say that f *preserves orientation* if for each $x \in M$ the induced map f_* always sends a right-handed basis in x into a right-handed basis in $f(x) \in N$ (i.e. if the orientation is preserved at the level of *tangent spaces*).

7.6.1 Let (M, o_M) and (N, o_N) be oriented manifolds of equal dimension endowed with volume forms ω_M and ω_N compatible with orientations o_M and o_N respectively (cf. (5.7.5) and (6.3.5)). Define a function $\varphi : M \to \mathbb{R}$ by

$$f^*\omega_N =: \varphi\, \omega_M$$

Check that

$$f \text{ preserves orientation} \quad \Leftrightarrow \quad \varphi > 0$$

□

• Now, consider an *n-dimensional* domain D on an n-dimensional oriented manifold (M, o_M). The maps $\Phi : \mathbb{R}^n \to D \subset M$ which occur in simplices (realizing D as a chain) then happen to be just of the type under consideration (the *standard* orientation being understood in "parameter" space $\mathbb{R}^n$). We adopt the convention that D should always be represented by a chain in which *all* the maps Φ *preserve orientation* (then we say that the simplices themselves preserve orientation).

□ = (P, Q, S triangle) + (S, R, Q triangle)

□′ = (P, Q, S triangle) + (S, R, Q triangle)

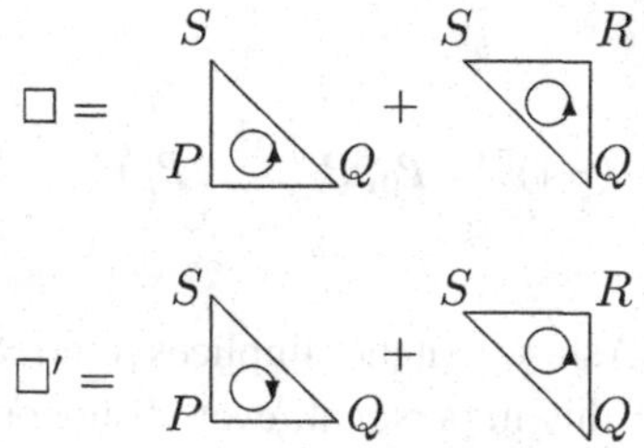

What is the meaning of this condition? Contemplate an example, which is so simple as to border on the trivial, the computation of the area of a square $\square \equiv PQRS$ in $\mathbb{R}^2[x, y]$, where $P = (0, 0)$, $Q = (1, 0)$, $R = (1, 1)$ and $S = (0, 1)$. The area is given, as is well known, by the integral $\int_\square dx\, dy$. Since the square is a two-dimensional figure, we know that an integral of a *2-form* in $\mathbb{R}^2[x, y]$ is to be evaluated

$$\int_\square dx\, dy \leftrightarrow \int_\square \omega \qquad \omega \equiv dx \wedge dy$$

The square on the right is to be realized as a *2-chain* in $\mathbb{R}^2[x, y]$; choose, for example, the following two possibilities

$$\square = (P, Q, S) + (Q, R, S) \equiv \sigma_2^1 + \sigma_2^2$$
$$\square' = (Q, P, S) + (Q, R, S) \equiv \sigma'^1_2 + \sigma_2^2$$

[7.6.2] Check that if we evaluate the integrals of the form $dx \wedge dy$ over the two chains under consideration (*both* of them match the square perfectly from the "naive area point of view"), we get the following results:

(i)

$$\int_{\square} \omega \equiv \int_{\sigma_2^1} \omega + \int_{\sigma_2^2} \omega = \frac{1}{2} + \frac{1}{2} = 1 \quad = \text{just the expected area of the square}$$

$$\int_{\square'} \omega \equiv \int_{\sigma'^1_2} \omega + \int_{\sigma_2^2} \omega = -\frac{1}{2} + \frac{1}{2} = 0 = \text{something different from what we need}$$

(ii) the simplices σ_2^1 and σ_2^2 preserve orientation, whereas σ'^1_2 fails to preserve it (so that the chain $\square'$ does not meet the above-mentioned convention).

Hint: (P, Q, S) is the standard simplex (i.e. $\Phi : (x, y) \mapsto (x, y)$), for (Q, R, S) one may use $\Phi : (x, y) \mapsto (1 - y, x + y)$ and for (Q, P, S) similarly $\Phi : (x, y) \mapsto (y, x)$. □

• In general, one defines the *volume of a domain D* on (M, ω), an oriented manifold endowed with a volume form, as the number

$$\text{vol}\,(D) \equiv \text{volume of } D := \int_D \omega \qquad \omega = \text{volume form on } M$$

If each piece (simplex) of D is to contribute to the total volume with the same sign (positive, usually), all the maps Φ have to preserve the orientation.

[7.6.3] Let $P_0 = (a, c)$, $P_1 = (b, c)$, $P_2 = (b, d)$, $P_3 = (a, d)$ be the points in $\mathbb{R}^2$. Then the 2-chain $\sigma = (P_0, P_1, P_3) + (P_1, P_2, P_3)$ represents a rectangle. Show by direct computation that for a 2-form $\alpha = f(x, y)\, dx \wedge dy$ there holds

$$\int_\sigma \alpha = \int_a^b dx \int_c^d f(x, y)\, dy$$

Lesson: a domain on a manifold which turns out to be a *rectangle* in coordinates (a *cuboid* in general dimension) need not be explicitly divided into simplices (!), but rather we perform the integration over it in just the same way as we have done until now. □

• There is yet another motivation to meet the orientation convention mentioned above: it greatly simplifies the computation of the boundary ∂D of a domain D. Consider again a simple example.

[7.6.4] Check that in the case of the chains $\square$ and $\square'$ from problem (7.6.2) there holds

(i) for $\square$

$$\partial \square = (P, Q) + (Q, R) + (R, S) + (S, P)$$

i.e. the *inner* part of the boundary (the diagonal), which occurs (twice) in the formal computation of $\partial\square$, *cancels out* and only the "true boundary" (= its *outer* part) contributes

(ii) for $\square'$ the inner part of the boundary fails to cancel out (it even doubles instead). □

• The canceling out of all "inner parts of the boundary" occurs in all those cases when one satisfies the convention concerning orientations of simplices which add up to D. This is especially important in the context of Stokes' theorem – in order to evaluate the integral over the boundary $\int_{\partial D} \alpha$, it is enough to restrict to the "true" (outer) boundary of the domain D.

Since ∂D is the boundary of a *chain*, the orientation $o_{\partial D}$ of all relevant (outer) simplices, which add up to ∂D, is unambiguously given (by the formula which defines the action of the boundary operator). This orientation may also be described independently, namely in terms of an *outer normal*. The latter is defined as an arbitrary vector ν on the boundary ∂D, which "sticks out" of the domain[99] D, i.e. it is given by an equivalence class of curves for which there holds $\gamma(0) = x \in \partial D, \gamma(\varepsilon) \notin D$ for small enough $\varepsilon > 0$. Given an orientation o_M on an n-dimensional manifold M, the orientation $o_{\partial D}$ of the boundary ∂D of the domain[100] D is introduced in terms of the outer normal as follows: a basis $(e_2, \dots, e_n)$ in the tangent space $T_x(\partial D)$ is declared to be right-handed if the basis $(\nu, e_2, \dots, e_n)$ in the whole tangent space $T_x M$ happens to be right-handed.

7.6.5 Draw a picture of the standard 2-simplex in $\mathbb{R}^2$ and check that the orientation of its boundary, which is given by the formula from (7.2.2), coincides with the orientation in the sense of an outer normal. Do the same for the standard 3-simplex in $\mathbb{R}^3$. □

7.6.6 Let D be a (two-dimensional) domain in $\mathbb{C} \equiv \mathbb{R}^2$ endowed with the standard orientation ($\leftrightarrow \ \omega = dx \wedge dy$) and let $\Gamma \equiv \partial D$ be its boundary (*contour*). Check that the orientation of Γ may also be characterized as being such that whenever one moves along Γ in the "positive" (right) direction, the domain D is situated *on the left*. □

7.6.7 Contemplate this until it is clear: Stokes' theorem (see Section 7.5) may be specified for the case of integration over an n-dimensional domain D on an n-dimensional oriented manifold (M, o_M) in the form

$$\int_D d\alpha = \int_{\partial D} \alpha \qquad \alpha \in \Omega^{(n-1)}(M)$$

∂D being the "true" boundary alone, oriented according to the rule of the outer normal. □

• A few words will be said about the situation where a p-form is integrated over a p-dimensional submanifold S of an n-dimensional manifold M, p being *less than n* (as an example, if a 2-form is integrated over a two-dimensional surface in $\mathbb{R}^3$). Here, one should realize first that no preferred orientation is induced from M on such a submanifold S, in general.[101] Consequently, we are to *choose* the orientation (arbitrarily) first (if this is not

[99] This vector *need not* be *perpendicular* (i.e. "normal") to the boundary; since no reference was made to a metric tensor, it makes no sense to speak about orthogonality; this will change in the context of Gauss' theorem (8.2.8), where a "true" (perpendicular) normal occurs.

[100] Recall that here we are concerned with an n-dimensional domain D and its $(n-1)$-dimensional boundary ∂D.

[101] An exception is provided, as discussed above, by the case of an $(n-1)$-dimensional submanifold S which happens to be the *boundary* of an n-dimensional domain D, $S = \partial D$ (like a sphere S^2, which is the boundary of a ball D^3 in $\mathbb{R}^3$); such *closed surfaces* may be oriented in terms of the outer normal.

specified, the integral is only given up to the sign). Secondly, one should *restrict* the form α (to be integrated) from M to S. This results in a p-form on S, reducing the situation to the standard case – there is a p-form on a p-dimensional oriented domain.

7.6.8 Let M be an n-dimensional manifold and let α be a p-form on M. What is then meant when we say to *restrict the form* α to a p-dimensional submanifold S ($\alpha \mapsto \alpha|_S$)? It is an irrevocable vow that from this time forth only vectors tangent to S (i.e. vectors from the subspace $T_xS \subset T_xM,\ x \in S \subset M$) will be offered (as arguments) to α. (So, the first restriction concerns *points*, $x \in S \subset M$, and the second one *vectors* at allowed points, $v \in T_xS \subset T_xM$.) Check that if

$$j : S \to M \qquad x \mapsto x$$

is the *canonical embedding* of S into M, then

$$\alpha|_S = j^*\alpha$$

i.e. if in coordinates we have $j : q^a \mapsto x^i(q^a),\ a = 1, \dots, p,\ i = 1, \dots, n$, then $\alpha \mapsto \alpha|_S$ reads as follows:

$$\frac{1}{p!}\alpha_{i\dots j}(x)\,dx^i \wedge \dots \wedge dx^j \mapsto \frac{1}{p!}\alpha_{i\dots j}(x(q))J^i_a(q)\dots J^j_b(q)\,dq^a \wedge \dots \wedge dq^b$$

Hint: the only allowed arguments are of the form j_*v, where $v \in T_xS$; see (6.1.5). □

- The restriction is also *implicit* in the integral on the right-hand side in *Stokes' theorem*: in full it should read

$$\int_D d\alpha = \int_{\partial D} \alpha|_{\partial D}$$

7.6.9 Check the validity of Stokes' theorem for $M = \mathbb{R}^3$, $D \equiv S^2_+ =$ the upper unit half-sphere and the form $\alpha = x\,dy$.

Hint: the orientation on S^2_+ is to be (arbitrarily) chosen, the orientation of the boundary (the equator S^1) is then already given by the outer normal; if we choose $(\partial_\vartheta, \partial_\varphi)$ to be right-handed on D, then ∂_φ is right-handed on $\partial D \equiv S^1 \subset S^2$. A straightforward computation (using parametrization from (3.2.4) with $R = 1$) yields explicit expressions for $\alpha|_{S^1} = \cos^2\varphi\,d\varphi$ and $(d\alpha)|_{S^2} = \sin\vartheta\cos\vartheta\,d\vartheta \wedge d\varphi$ and eventually one should check whether

$$\int_{S^2_+} (d\alpha)|_{S^2} \equiv \int_0^{\pi/2} d\vartheta \int_0^{2\pi} d\varphi \sin\vartheta\cos\vartheta \overset{?}{=} \int_0^{2\pi} \cos^2\varphi\,d\varphi \equiv \int_{S^1} \alpha|_{S^1}$$

□

7.6.10 Consider L-valued differential forms (L being a vector space); check that the integral of a p-form over a p-dimensional domain

$$\int_D : \Omega^p(M, L) \to L \qquad \int_D \alpha \equiv \int_D (\alpha^A E_A) := \left(\int_D \alpha^A\right) E_A$$

(the result being the *element of* L) is well defined (does not depend on the choice of a basis E_A in L) and that Stokes' theorem holds for this integral.

Hint: see (6.4.4). □

• We will close this section by a useful visual interpretation of the operation of the interior product i_V within the *integral* calculus of forms.

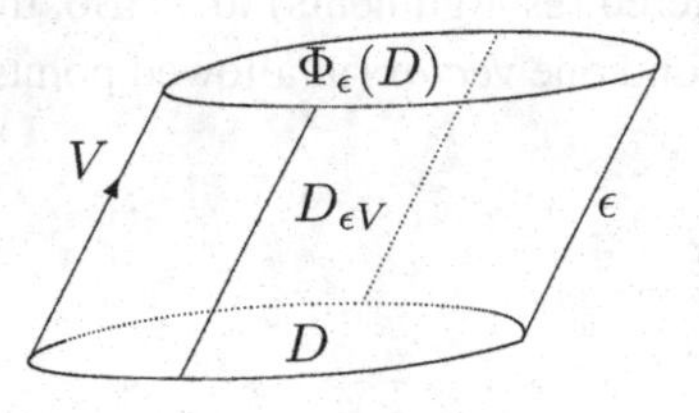

7.6.11 Let α be a $(p+1)$-form on M, V a vector field and let $i_V\alpha$ denote their interior product (which is a p-form on M). Check the validity of the following[102] "coin interpretation" of the form $i_V\alpha$: the integral of $i_V\alpha$ over a p-dimensional domain D is related to the integral of α itself over the $(p+1)$-dimensional domain $D_{\epsilon V}$, the "coin over D of thickness ϵ." Namely (up to first-order accuracy in ϵ), there holds

$$\epsilon \int_D i_V\alpha = \int_{D_{\epsilon V}} \alpha$$

where $D_{\epsilon V}$ may be obtained from D by "expansion by ϵ" along the field V; thus it is bounded from below by D, from above by $\Phi_\epsilon(D)$ (the image of D with respect to the flow $\Phi_\epsilon \leftrightarrow V$) and from the side by the (infinitesimal) integral curves of the field V, which start at the boundary ∂D of the domain D.[103]

Hint: instead of D (regarded as a chain) consider rather a single simplex $\sigma = \Phi(\overline{s}_p)$ and let x^i be any right-handed coordinates on σ. Then the "coin" $\sigma_{\epsilon V}$ which corresponds to this simplex admits the parametrization by coordinates (t, x^i), $t \in \langle 0, \epsilon\rangle$ (these are just the coordinates discussed at the end of Section 4.2, t corresponds to the flow); the orienting volume form in $\sigma_{\epsilon V}$ is $-dt \wedge dx^1 \wedge \cdots \wedge dx^p$ (since the outer normal on D is $-\partial_t$). The form α reads $\alpha = -dt \wedge a(t,x)\,dx^1 \wedge \cdots \wedge dx^p$, so that $i_V\alpha = -a(t,x)\,dx^1 \wedge \cdots \wedge dx^p$ (recall that the field V itself looks like $V = \partial_t$). Then

$$\begin{aligned}\int_{\sigma_{\epsilon V}} \alpha &= -\int_{\sigma_{\epsilon V}} dt \wedge a(t,x)\,dx^1 \wedge \cdots \wedge dx^p \\ &= -\int_0^\epsilon dt \int_\sigma a(t,x)\,dx^1 \wedge \cdots \wedge dx^p = +\epsilon \int_\sigma i_V\alpha\end{aligned}$$

□

[102] Warning: this is not official terminology.

[103] The orientation in $D_{\epsilon V}$ is such that D is its "bottom" face (the coin is *over* D), i.e. it enters the boundary with a *negative* sign: $\partial(D_{\epsilon V}) = -D + \cdots$ (the outer normal is oriented *downwards* on D); note that then the *whole* boundary $= \Phi_\epsilon(D) - D+$ the "side face," the latter also being *the coin*, however, over ∂D. The analysis of the orientation shows that the side face enters the boundary as the coin with *negative* sign (the rules of the outer normal and of the coin lead to opposite results); that is to say that $\partial(D_{\epsilon V}) = \Phi_\epsilon(D) - D - (\partial D)_{\epsilon V}$.

7.7 Integral over a domain on an orientable Riemannian manifold

• Now consider the integration on an orientable *Riemannian* manifold. A new element – a distinguished volume form – enters into play in this situation, namely the *metric* volume form ω_g (i.e. compatible with the metric and the orientation, see (6.3.7)). This then results in the distinguished definition of the *volume of a domain* D ($\dim D = \dim M$)

$$\operatorname{vol} D \equiv \text{the volume of } D := \int_D \omega_g$$

7.7.1 Compute the area of the sphere S_R^2 of radius R in E^3 and of (two-dimensional) tori in E^3 and E^4; all of them by the direct integration of the metric volume forms.

Hint: first induce the metric tensors onto the corresponding submanifolds from E^3 and E^4 respectively (see (3.2.2)–(3.2.4)), then associate the metric volume forms with them and finally integrate the forms over the submanifolds; compare the results with those obtained by elementary mathematics. □

7.7.2 On the sphere S_R^2 of radius R (endowed with the metric induced from E^3) draw a disk of radius r using the *gardener's technique*; namely, fasten two pins at the ends of a rope of length r, stick one of them in the ground and draw the boundary circle of the disk by means of the remaining pin (keeping the rope tight). Compute (first by direct integration of the volume form and then check visually by common sense)

(i) the area $A(r)$ of the disk
(ii) the perimeter $P(r)$ of the circle (the boundary of the disk).

Compare with the corresponding results in the Euclidean plane E^2. What do we obtain in the limit $R \to \infty$? Why?

Hint: like in (7.7.1); recall that both the concepts of *area* and *length* are just particular cases of the "*volume*" discussed here. □

7.7.3 Show that the area of a rotational figure given by the function $r(z)$, $z \in \langle a, b\rangle$ (in cylindrical coordinates in E^3) reads

$$A = 2\pi \int_a^b r(z)\sqrt{1 + [r'(z)]^2}\, dz$$

Apply this to the surface of a cone and a cylinder.

Hint: see (3.2.6) and (6.3.10). □

• The volume form ω_g enables one also to define the *integral of a function* f over a domain D using the prescription[104]

$$\int_D f := \int_D f\omega_g \equiv \int_D *_g f$$

Such integrals already occur frequently in elementary physics, namely in the situations where one speaks about line, surface and volume *densities* of physical quantities. Consider, as an example, the electric charge continuously distributed on a curve, a surface or in a volume. The total amount of charge is then computed using the integrals

$$\int_l \rho_1\, dl \qquad \int_S \rho_2\, dS \qquad \text{or} \qquad \int_V \rho_3\, dV$$

where dl, dS and dV denote the line, surface and volume "elements" respectively and ρ_1, ρ_2 and ρ_3 represent the corresponding charge densities. These "elements" correspond in the language of forms just to the (metric) volume forms ω_g on the curve,[105] the surface or the volume. In more detail we mean on the curve and surface the 1-form or the 2-form $\omega_{\hat{g}}$ respectively, where $\hat{g}$ is the *restriction* (compare with (7.6.8)) of the metric tensor g to the curve l or the surface S, i.e.

$$j : l(\text{or } S) \to (M, g) \qquad \Rightarrow \qquad \hat{g} := j^* g$$

The reader may recall such integrals from introductory calculus, too; they are called (line or surface) *integrals of the first kind*, whereas *integrals of the second kind* have a "general" form as the expression under the integral sign, e.g. (in $\mathbb{R}^3$)

$$\int_l A\, dx + B\, dy + C\, dz \qquad \text{or} \qquad \int_S a\, dx \wedge dy + b\, dx \wedge dz + c\, dy \wedge dz$$

(Note that we *do not need* a metric tensor to perform integrals of the second kind, but it *is needed* for integrals of the first kind.)

7.7.4 Let $M = E^3$ and assume that we are given (arbitrary) parametrizations using the variables u, (u, v) or (u, v, w) of a curve l, a surface S or of a volume V respectively. That is to say we are given the maps

$$\text{for } l : u \mapsto \mathbf{r}(u) \qquad \text{for } S : (u, v) \mapsto \mathbf{r}(u, v) \qquad \text{for } V : (u, v, w) \mapsto \mathbf{r}(u, v, w)$$

(i) Check the validity of the following standard formulas (expressed in terms of the "vector product" in E^3, cf. (8.5.8)):

$$dl \equiv \omega_{\hat{g}} = \left| \frac{\partial \mathbf{r}}{\partial u} \right| du \qquad dS \equiv \omega_{\hat{g}} = \left| \frac{\partial \mathbf{r}}{\partial u} \times \frac{\partial \mathbf{r}}{\partial v} \right| du \wedge dv$$

$$dV \equiv \omega_g = J du \wedge dv \wedge dw = \left| \left(\frac{\partial \mathbf{r}}{\partial u} \times \frac{\partial \mathbf{r}}{\partial v} \right) \cdot \frac{\partial \mathbf{r}}{\partial w} \right| du \wedge dv \wedge dw$$

(J being the Jacobian of the map $\mathbf{r}(u, v, w)$)

[104] In Section 7.4 the integral of a p-form over a p-chain was defined. One sometimes *extends* formally the definition of the integral to the case of a q-form and a p-chain to be *zero* for $p \neq q$. Here we introduce the integral of a function (= 0-form) over an n-chain as a *non-zero* number in such a way that the *n-form* $f\omega_g \equiv *_g f$ is actually integrated rather than the 0-form f (we say we do something but *in fact* we do something else, as often happens).

[105] The word "curve" is used here to denote only a (unparametrized) line (1-chain).

(ii) re-derive in this way the "surface element" on the sphere, $dS = R^2 \sin\vartheta \, d\vartheta \wedge d\varphi$.

Hint: (i) for dl we have $\hat{g}_{uu} = \cdots = J^i_u \delta_{ij} J^j_u = \frac{\partial \mathbf{r}}{\partial u} \cdot \frac{\partial \mathbf{r}}{\partial u} \Rightarrow dl \equiv \omega_{\hat{g}} \equiv \sqrt{|\hat{g}_{uu}|}\, du = \ldots$; (ii) $\mathbf{r}(\vartheta, \varphi) =$ from (3.2.4), compare with the result of (6.3.10). □

- Consider an n-dimensional manifold (M, g) together with a (lower-dimensional) submanifold ("surface") $\mathcal{S}$; we may then induce (by restriction) the metric tensor $\hat{g} \equiv j^* g$ on $\mathcal{S}$. The integral of the first kind with the *unit* function $f = 1$ yields the *volume of the submanifold* $\mathcal{S}$ in the sense of the (restriction of the) metric tensor g on M

$$\text{the volume of } \mathcal{S} := \int_{\mathcal{S}} \omega_{\hat{g}}$$

When the dimension of $\mathcal{S}$ is $1, 2, \ldots$ this "volume" reduces to the length of a curve, the area of a two-dimensional surface, etc.

7.7.5 Check that for a curve γ and a two-dimensional surface S, parametrized as

$$\gamma : t \mapsto x^\mu(t) \qquad S : (u^1, u^2) \mapsto x^\mu(u^1, u^2)$$

we get[106]

$$\text{the length of } \gamma = \int \sqrt{g_{\mu\nu} \dot{x}^\mu \dot{x}^\nu}\, dt \equiv \int \sqrt{g(\dot{\gamma}, \dot{\gamma})}\, dt$$

$$\text{the area of } S = \int \sqrt{\det(g_{\mu\nu} x^\mu_{\ ,a} x^\nu_{\ ,b})}\, du^1 \wedge du^2 \qquad x^\mu_{\ ,a} \equiv \frac{\partial x^\mu}{\partial u^a}$$

Hint: $(\gamma^* g)_{tt} = g_{\mu\nu} \dot{x}^\mu \dot{x}^\nu$, $(S^* g)_{ab} = g_{\mu\nu} x^\mu_{\ ,a} x^\nu_{\ ,b}$. □

- The volume integral (of the first kind) of a function (of the density ρ of some quantity) enables one also to introduce the concept of the *mean value* of the (scalar) quantity over a domain D as[107]

$$\langle \rho \rangle_D := \frac{\int_D \rho \omega_g}{\int_D \omega_g} \equiv \frac{\int_D \rho \omega_g}{\text{vol } D} \qquad \text{i.e.} \qquad \int_D \rho \omega_g = \langle \rho \rangle_D \text{ vol } D$$

7.8 Integral and maps of manifolds

- The language of differential forms proves to be extremely convenient for analyzing the behavior of the integral with respect to maps of manifolds.

7.8.1 Let $f : M \to N$ be a map and let $\Phi : \bar{s}_p \to M$ be a p-simplex σ_p on M. By composition with the map f we then get the p-simplex $\hat{\sigma}_p := f(\sigma_p)$ on N (given by $f \circ \Phi : \bar{s}_p \to N$) and the linear extension to p-chains

$$f(c) \equiv f\left(c_i \sigma^i_p\right) := c_i f\left(\sigma^i_p\right)$$

[106] The formula for the length of a curve thus coincides with the expression (2.6.9) given earlier.

[107] This makes sense for a domain with *finite* volume; for infinite domains appropriate limits are needed.

induces the map

$$\hat{f} : C_p(M) \to C_p(N) \qquad c \mapsto f(c)$$

(we will not distinguish explicitly between f and $\hat{f}$ in what follows). Check that

(i) if α is an arbitrary p-form on N, there holds

$$\int_{f(c)} \alpha = \int_c f^*\alpha$$

(ii) in terms of the pairing introduced in (7.4.1) this result may also be written as

$$I(f(c), \alpha) = I(c, f^*\alpha)$$

(iii) the special case is provided by the formula (7.1.3) for the change of coordinates in the integral.

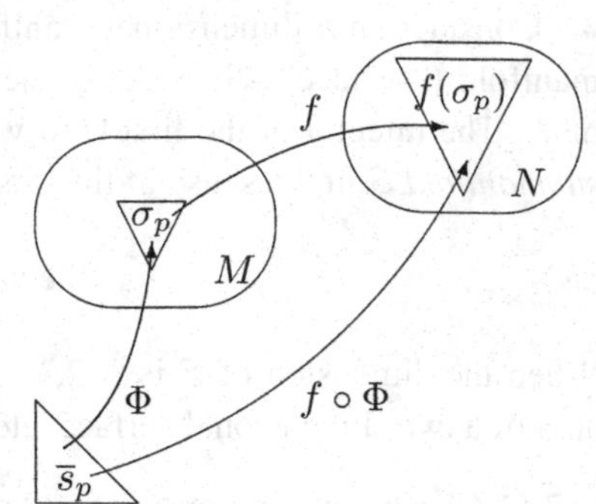

Hint: (i) according to the definition of the integral and (6.1.4) we obtain for the simplex σ_p

$$\int_{f(\sigma_p)} \alpha \equiv \int_{f\circ\Phi(\bar{s}_p)} \alpha = \int_{\bar{s}_p} (f\circ\Phi)^*\alpha = \int_{\bar{s}_p} \Phi^* f^*\alpha = \int_{\sigma_p} f^*\alpha$$

for the chain by linearity; (iii) see (7.1.3). □

• As a small application (many further applications will be encountered later) we show how this result, when combined with the "coin" interpretation of the interior product i_V (7.6.11), makes it possible to gain insight into the visual meaning of the good old identity $\mathcal{L}_V = di_V + i_V d$ from (6.2.8) and to understand its close relation to Stokes' theorem.[108]

7.8.2* Show that Stokes' theorem is hidden behind the (differential) identity $\mathcal{L}_V = di_V + i_V d$.

Hint: let $D_{\epsilon V}$ be the "coin" based on D (7.6.11). If $\Phi_t \leftrightarrow V$, then using $\Phi_\epsilon^* = \hat{1} + \epsilon\mathcal{L}_V + \cdots$ and $\partial(D_{\epsilon V}) = \Phi_\epsilon(D) - D - (\partial D)_{\epsilon V}$ there holds

$$\begin{aligned}\int_{D_{\epsilon V}} d\alpha &\overset{1}{=} \epsilon \int_D i_V\, d\alpha \\ &\overset{2}{=} \int_{\partial D_{\epsilon V}} \alpha = \int_{\Phi_\epsilon(D)} \alpha - \int_D \alpha - \int_{(\partial D)_{\epsilon V}} \alpha \\ &= \int_D \Phi_\epsilon^*\alpha - \int_D \alpha - \epsilon \int_{\partial D} i_V\alpha = \epsilon \int_D (\mathcal{L}_V - di_V)\alpha\end{aligned}$$

so that $\int_D (di_V + i_V d - \mathcal{L}_V)\alpha = 0$, from which (both D and α being arbitrary) we finally get $di_V + i_V d = \mathcal{L}_V$. □

[108] I learned this from Pal'o Ševera, which I warmly appreciate.

Summary of Chapter 7

Inspection of several simple examples and facts from elementary integral calculus leads to the conclusion that the objects under the integral sign should be treated as differential forms from Chapter 6. The crucial concept of the integral of a form over a chain is introduced, assuming the standard background knowledge on basics of the Riemann multiple integral. Stokes' theorem for differential forms is presented. It relates the integral of a form over the *boundary* of a chain to the integral of the *exterior derivative* of the form over the chain itself. Reinterpretation of the integral over a domain on an oriented manifold in terms of the integral over the chain (including Stokes' theorem) is given and particular features of integration over a *Riemannian* manifold are mentioned. The remarkably simple behavior of the integral with respect to maps between manifolds is revealed.

$c = c_i s^i_p$	Euclidean p-chain	Sec. 7.2
$\partial(P_0, \dots, P_p) = \cdots$	Action of the boundary operator on a simplex	(7.2.2)
$\partial\partial = 0$	Boundary has no boundary	(7.2.2)
$\int_c d\alpha = \int_{\partial c} \alpha$	Stokes' theorem	Sec. 7.5
$\text{vol}\,(D) := \int_D \omega$	Volume of a domain D on (M, ω)	Sec. 7.6
$\epsilon \int_D i_V\alpha = \int_{D_{\epsilon V}} \alpha$	A "coin interpretation" of the form $i_V\alpha$	(7.6.11)
$\int_D f := \int_D f\omega_g$	Integral of the first kind on (M, g, o)	Sec. 7.7
$\int \sqrt{\det(g_{\mu\nu}x^\mu{}_{,a}x^\nu{}_{,b})}\, du^1 \wedge du^2$	Area of a two-dimensional surface	(7.7.5)
$\langle\rho\rangle_D := \dfrac{\int_D \rho\omega_g}{\int_D \omega_g}$	Mean value of the (scalar) quantity ρ over D	Sec. 7.7
$\int_{f(c)} \alpha = \int_c f^*\alpha$	Integral and maps of manifolds	(7.8.1)

8

Particular cases and applications of Stokes' theorem

• As already mentioned at the beginning of Section 7.5, Stokes' theorem

$$\int_c d\alpha = \int_{\partial c} \alpha$$

is actually hidden behind all theorems which equate two integrals, the domain of integration of one of them being the boundary of the domain of integration of the other. In this chapter we discuss explicitly the particular cases one encounters most frequently.

8.1 Elementary situations

• By far the simplest version of Stokes' theorem is provided by the *Newton–Leibniz formula*.[109]

8.1.1 The most natural definition of the integral of a 0-form (function) f over a 0-simplex (point) is given by[110]

$$\int_P f := f(P)$$

Check that the *Newton–Leibniz formula*

$$\int_a^b f'(x)dx = f(b) - f(a)$$

may then be regarded as a particular case[111] of Stokes' theorem.

Hint: $M = \mathbb{R}^1$, $c = [a, b]$. □

• The next unostentatious application of Stokes' theorem is hidden within the context of the calculation of the area under a given curve. Recall that this particular calculation

[109] Stokes' theorem may be regarded as a far-reaching generalization of just this fundamental formula.

[110] The intuitive meaning of the integral, as is well known, is the sum of the values of the function in infinitesimal domains (resulting from the division of the total domain of integration) multiplied by the volumes of these domains. If the total domain reduces to *a single point P*, there is nothing to be divided and it suffices to take the value of the function right at this point. Note that in doing this the volume of the point P is effectively regarded to be 1 ($\int_P 1 = 1(P) = 1$).

[111] Although in a sense a *tautological* one – the definition of the integral of a 0-form has been *extended* in such a way as to make the theorem hold.

is usually presented as the very first problem which motivates the integral calculus itself. Then, after becoming familiar with multiple integrals, one learns that the same area may alternatively be computed as an appropriate *double* integral and that the same "duplication" is also true for the calculation of the volume under a given surface. What is the origin of this "duplication?" It turns out that this is nothing but a simple manifestation of Stokes' theorem.

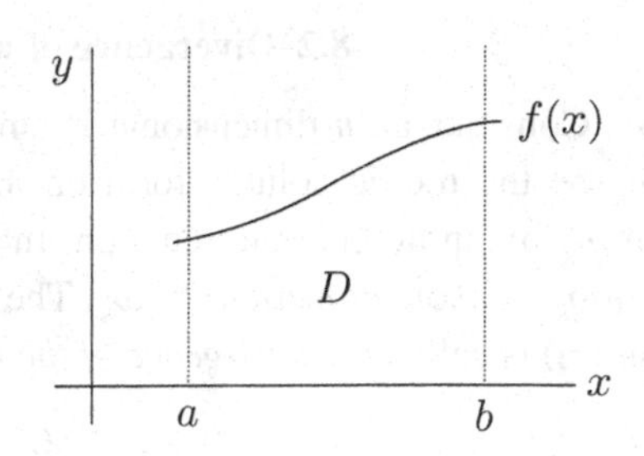

8.1.2 Let D be the domain in the xy plane which is bounded by the straight lines $x = a$, $x = b$, $y = 0$ and by the curve $y = f(x)$ from above. In full analogy let V be the domain in the space xyz which lies above the domain S in the plane xy and which is bounded from above by the surface $z = f(x, y)$. Then we know that the area of the domain D may be computed in two (completely different) ways, namely either as

$$\int_a^b f(x)\,dx \qquad \text{or as} \qquad \int_D dx\,dy$$

Similarly, the volume of the domain V may be computed in two (completely different) ways, namely as $\int_S f(x, y)\,dx\,dy$ or as $\int_V dx\,dy\,dz$. Show that both cases may be regarded as a manifestation of Stokes' theorem.
Hint: $dx \wedge dy = d(-y\,dx)$, $dx \wedge dy \wedge dz = d(z\,dx \wedge dy)$. □

- The well-known trick of integration *by parts* (*per partes*) turns out to be another simple consequence of Stokes' theorem and the behavior of the exterior derivative *on the* (exterior) *product* of forms.

8.1.3 Let α, β be two forms on an n-dimensional manifold M, with their degrees being such that $\deg\alpha + \deg\beta + 1 = n$ and let D be an n-dimensional domain. Check that

(i) the following identity holds:

$$\int_D d\alpha \wedge \beta = -\int_D \hat{\eta}\alpha \wedge d\beta + \int_{\partial D} \alpha \wedge \beta$$

(ii) the formula representing the "by parts" method of integration

$$\int_a^b f'(x)g(x)\,dx = -\int_a^b f(x)g'(x)\,dx + [fg]_a^b$$

is but a simple special case of this identity.

Hint: (i) see (6.2.5) and (7.6.7); (ii) $M = \mathbb{R}$, $D = [a, b]$, $\alpha = f$, $\beta = g$. □

- Finally we mention the fairly popular *Green's theorem* in the plane $\mathbb{R}^2$.

8.1.4 Given two functions $f(x, y), g(x, y) \in \mathcal{F}(\mathbb{R}^2)$ and a (two-dimensional) domain D, let $C \equiv \partial D$ be its (oriented) boundary (closed curve, the *contour*). Show that there holds

$$\oint_C f\,dx + g\,dy = \int_D (\partial_x g - \partial_y f)\,dx \wedge dy$$

Hint: set $\alpha = f\,dx + g\,dy$ in (7.6.7). □

8.2 Divergence of a vector field and Gauss' theorem

• Consider an n-dimensional Riemannian manifold with orientation (M, g, o) and let ω_g be the metric volume form on M. Any other n-form is then necessarily a multiple of ω_g by an appropriate function. In particular, given any vector field V on M, we have $\mathcal{L}_V \omega_g =$ (some function) $\times\ \omega_g$. The function specified by this equation (depending on V and g) is called the *divergence of the vector field V*

$$\mathcal{L}_V \omega_g =: (\text{div } V)\,\omega_g$$

8.2.1 Show that

(i)

$$\mathcal{L}_V \omega_g = d(i_V \omega_g) = d(*_g \tilde{V}) \qquad \tilde{V} \equiv g(V, \cdot) \equiv \flat_g V$$

(ii)

$$\text{div } V = \overset{-1}{*_g} d *_g \tilde{V} = \text{ sgn } g *_g d *_g \tilde{V}$$

(iii) in (right-handed) local coordinates the following explicit formula may be used:

$$\text{div } V = \frac{1}{\sqrt{|\,g\,|}} (\sqrt{|\,g\,|}\, V^k)_{,k}$$

Hint: (i) see (6.2.8), (5.4.1) and (5.8.1); (iii) see (6.3.7) and (6.3.11) or (14.3.7). □

• What is the visual meaning of the function div V? In this section we will mention two basic interpretations of this concept. The first relates the divergence to the change of local volumes due to the flow of the field V (this aspect is treated in a more general setting in problem (14.3.7)).

8.2.2 Given a flow $\Phi_t \leftrightarrow V$ on M let $\Phi_t(D) \equiv D(t)$ denote the domain D drifted by the flow Φ_t a (parametric) distance t. Check that

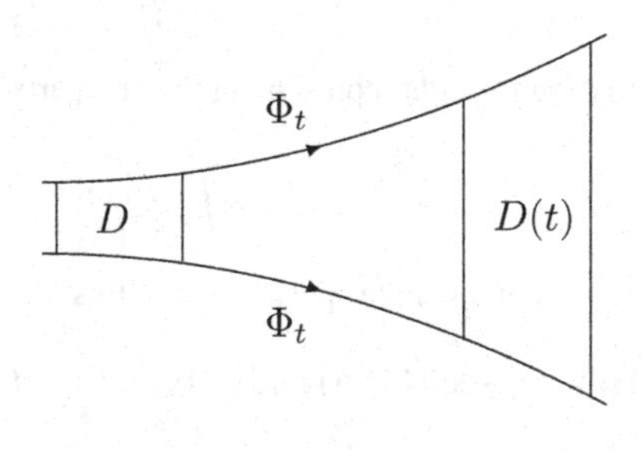

(i) the volume of the drifted domain may be expressed in terms of the infinite series

$$\text{vol } D(t) \equiv \int_{D(t)} \omega_g = \int_D \Phi_t^* \omega_g$$

$$= \int_D \left(1 + t\mathcal{L}_V + \frac{t^2}{2!}\mathcal{L}_V^2 + \cdots\right)\omega_g$$

(ii) so that for $\varepsilon \ll 1$ we get

$$\text{vol}\, D(\varepsilon) = \text{vol}\, D + \varepsilon \int_D (\text{div}\, V)\, \omega_g + \cdots \equiv \text{vol}\, D + \varepsilon \langle \text{div}\, V \rangle_D \text{vol}\, D + \cdots$$

(iii) the relative change of the volume within the time interval ϵ turns out to be

$$\frac{\text{vol}\, D(\varepsilon) - \text{vol}\, D}{\text{vol}\, D} = \varepsilon \langle \text{div}\, V \rangle_D + o(\varepsilon^2) \qquad \text{or} \qquad \langle \text{div}\, V \rangle_D = \frac{d}{dt}\bigg|_{t=0} \frac{\text{vol}\, D(t)}{\text{vol}\, D}$$

so that the mean value of the divergence of the field V over the domain D measures the *rate of relative change of the volume* of the domain D (at "time" $t = 0$) due to the flow Φ_t

(iv) the divergence of the field V at the point x measures the rate of relative change of the volume (at "time" $t = 0$) *in the neighborhood* of the point x due to the flow Φ_t.

Hint: (i) see Section 7.7 and (7.8.1), (4.4.2); (iv) take the infinitesimal neighborhood of the point x as the domain D. □

8.2.3 Check that

$$\text{vol}\, D(t) = \text{vol}\, D \qquad \Leftrightarrow \qquad \text{div}\, V = 0$$

The vector fields with *vanishing divergence* thus generate the flows Φ_t which *preserve the volumes*.

Hint: see (8.2.2); see also (14.3.7). □

• Consider the "hydrodynamical" interpretation of the flow (so that the vector field V is interpreted as the *velocity field* of the flow of a fluid (4.6.25)). The general vector field V on M then corresponds to a situation which is a rather particular one and which is called the *stationary flow* of a fluid in physics. (This is, by definition, the flow which always remains the same; namely, the velocity field V does not depend on time.)[112] Under these fairly general conditions the fluid may not preserve its volume (it may happen to be compressible). From the considerations of this section it should now be clear (see (8.2.3)) that *vanishing of the divergence* of the velocity field V may also be used as the *condition of incompressibility* of the fluid:

$$\text{incompressible fluid} \qquad \Rightarrow \qquad \text{div}\, V = 0$$

(Recall that the standard form of the *continuity equation* in fluid dynamics reads $\partial_t \rho + \text{div}\, \mathbf{j} = 0$, where $\rho(\mathbf{r}, t)$ is the *mass density* of the fluid and $\mathbf{j}(\mathbf{r}, t) = \rho \mathbf{v}$, with $\mathbf{v}(\mathbf{r}, t)$ being the velocity field (see also (16.2.4)). For the incompressible fluid $\rho(\mathbf{r}, t) =$ constant and we find $\text{div}\, \mathbf{v} = 0$.)

8.2.4 Let D be the unit ball centered at the origin of E^3 and consider the vector field V from problem (4.1.5).

(i) Compute explicitly vol $D(t)$.

[112] Since it is a field *on* M, it is given locally as $V = V^i(x)\partial_i$. For a *non*-stationary flow $V = V^i(x, t)\partial_i$ is needed, which may be regarded either as a one-parameter family of vector fields on M, or (from a different point of view) as a field *on* $M \times \mathbb{R}[t]$.

(ii) Compute the divergence of V both from the definition and using the component formula derived in (8.2.1).

(iii) Check the interpretation of the divergence mentioned in (8.2.2) (iii) in this particular case. □

8.2.5 Let $M = E^3 \setminus (0)$, $V = f(r)\partial_r$ (in the spherical polar coordinates r, ϑ, φ). Choose the function f so that $\operatorname{div} V = 0$ holds. Write down the resulting flow $\Phi_t \leftrightarrow V$.
Hint: according to (6.3.8) and (8.2.1), in spherical polar coordinates one obtains

$$(\operatorname{div} V)\omega_g \equiv d(i_V \omega_g) = d(f(r)r^2 \sin\vartheta \, d\vartheta \wedge d\varphi) = \cdots = (f'r^2 + 2rf)\, \omega_g$$

$(\Phi_t : (r, \vartheta, \varphi) \mapsto ((r^3 + kt)^{1/3}, \vartheta, \varphi), k \equiv 3f(1).)$ □

8.2.6 Let V be the vector field from (8.2.5), $D_{a,b}$ the spherical layer between $r = a$ and $r = b$ and $\Phi_t(D_{a,b})$ its image with respect to the flow of the field V. Compute the volume $\Phi_t(D_{a,b})$ and explain why it does not depend on t.

Hint: see (8.2.3). □

• We now embark on the derivation of Gauss' theorem. This theorem relates the volume integral of the divergence of a vector field to a certain ("surface") integral over the boundary of the volume. There are two forms in which the integral over the boundary may be written down; one of them may be used "always" and the other one encounters (insurmountable) problems, sometimes. We first mention the version which is "generally valid."

8.2.7 Let D be an m-dimensional domain on an m-dimensional (possibly pseudo-) Riemannian manifold (M, g), ∂D its boundary oriented in the sense of the outer normal and V a vector field on M. From Stokes' theorem $\int_D d\alpha = \int_{\partial D} \alpha$ derive as a particular case *Gauss' theorem*:

$$\int_D (\operatorname{div} V)\, \omega_g = \int_{\partial D} V^i \, d\Sigma_i|_{\partial D}$$

Hint: according to (5.4.1), (5.8.10) and (8.2.1) there holds $(\operatorname{div} V)\omega_g = \mathcal{L}_V \omega_g = d(i_V \omega_g) = d(V^i d\Sigma_i)$, from where Stokes' theorem already yields immediately the result being sought. □

• The alternative way to write down the integral over the boundary may be used in those cases in which the *unit* outer normal field n exists ($n \perp \partial D$, $g(n, n) = 1$) over all the boundary ∂D (except for at most a set of "measure zero") and in addition the (non-degenerate) metric tensor $\hat{g}$ is induced on the (whole) boundary ∂D; then the metric volume form $\omega_{\hat{g}}$ on the boundary is also available. This requirement is not fulfilled, for example, when a part of the boundary in the case of mixed signature contains "isotropic directions" (vectors such that their lengths vanish).[113] In the positive definite case these problems do not occur.

[113] Consider, for example, a triangle in the two-dimensional Minkowski plane $\mathbb{R}^2[t, x]$ with the vertices $A = (0, 0)$, $B = (1, 0)$ and $C = (1, 1)$; then on the part CA of the boundary the induced metric $\hat{g}$ *vanishes* and we lack the unit normal there as well.

8.2.8 Let n be the field of the outer normal on ∂D, which is in addition perpendicular to the boundary and normalized to unity and let $\omega_{\hat{g}}$ be the volume form which corresponds to the induced metric $\hat{g}$. Show that Gauss' theorem may then also be written in the form

$$\int_D (\text{div}\, V)\, \omega_g = \int_{\partial D} (V \cdot n)\, \omega_{\hat{g}}$$

Hint: in a neighborhood of a point $x \in \partial D$ choose an orthonormal right-handed frame field e_a such that $e_1 = n$ (so that $e_2, \dots e_n || \partial D$ then); let e^a be the dual coframe field. Check that $e^1 = \tilde{n} := g(n, \cdot\,) \equiv \flat_g n$ and that if ω_g and $\omega_{\hat{g}}$ represent the metric volume forms on M and ∂D respectively, they may be expressed as follows:

$$\omega_g = \tilde{n} \wedge e^2 \wedge \dots \wedge e^n \qquad \omega_{\hat{g}} = (e^2 \wedge \dots \wedge e^n)|_{\partial D}$$

Then

$$i_V \omega_g = (n \cdot V) e^2 \wedge \dots \wedge e^n - \tilde{n} \wedge i_V (e^2 \wedge \dots \wedge e^n) \qquad n \cdot V := g(n, V) \equiv V_\perp \equiv \langle \tilde{n}, V \rangle$$

so that the restriction (7.6.8) to ∂D gives

$$(i_V \omega_g)|_{\partial D} = (n \cdot V)\, \omega_{\hat{g}}$$

(upon restriction the arguments e_a with $a = 2, \dots, n$ get into the 1-form $\tilde{n}$). By comparison we can also see that $d\Sigma_i|_{\partial D} = n_i\, \omega_{\hat{g}}$. □

• Let us also mention how the expressions mentioned above may be transcribed from the noble hieroglyphic writing into the demotic writing used by common people. Common people use the notation

$$\omega_g \leftrightarrow d\Omega \equiv \sqrt{|g|}\, d^n x \qquad \omega_{\hat{g}} \leftrightarrow dS \qquad d\Sigma_i \leftrightarrow dS_i \leftrightarrow d\mathbf{S}$$

in which Gauss' theorem looks like

$$\int_D (\text{div}\, V)\, d\Omega \equiv \int_D (\text{div}\, V) \sqrt{|g|}\, d^n x = \oint_{\partial D} (\mathbf{V} \cdot \mathbf{n})\, dS \equiv \oint_{\partial D} \mathbf{V} \cdot d\mathbf{S} \equiv \oint_{\partial D} V^i\, dS_i$$

with the small circle around the integral sign indicating that the integral is performed over a *closed* "surface" (the boundary[114] of the domain D). Again it is true (see the note in (6.3.11)) that in general *neither* $d\Omega$ *nor* dS are exterior derivatives of anything else; this is nothing but the conventional way to write down such objects (here "d" is related to the conception of being "infinitesimal").

The divergence often enters the scene in situations where some volume *flows out* from a given domain across its boundary.

8.2.9 We adopt the "hydrodynamical" point of view once again, i.e. we regard V as the *velocity field* of a (stationary) *flow* of a fluid.[115] Check that

[114] So that *it is not*, as some books mistakenly claim, the trendy jewelry known as (body) *piercing*.

[115] An abstract "fluid" on an n-dimensional Riemannian manifold M in general; in particular, also a real fluid.

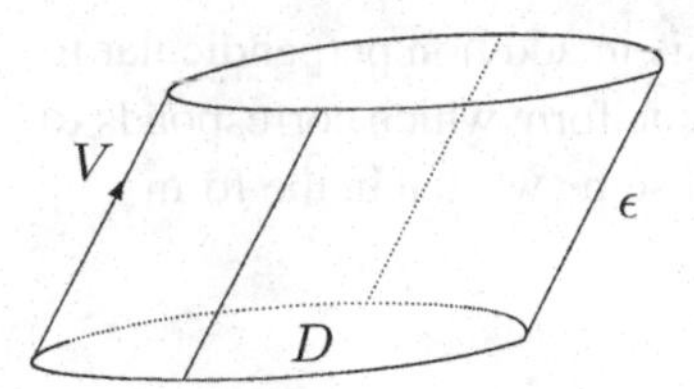

(i) the volume of the fluid which flows out from a *fixed* domain D within an infinitesimal time ϵ is given by the integral

$$\epsilon \int_{\partial D} i_V \omega_g \equiv \epsilon \oint_{\partial D} V^i \, dS_i \equiv \epsilon \oint_{\partial D} (\mathbf{V} \cdot \mathbf{n}) \, dS$$

so that the expression $V^i \, dS_i \equiv (\mathbf{V} \cdot \mathbf{n}) \, dS$ corresponds to the volume of the fluid which flows out per unit time across the surface element dS (the volume which flows inwards is to be counted with a negative sign); the expression $\int_S W^i \, dS_i$ is often called the *flux of a vector field* W for the surface S and, in general, the field W may not be related to any "flow" whatsoever (in electrodynamics one computes, say, the flux of the *electric* field for some surface)

(ii) Gauss' theorem then relates the volume which flows out to the volume integral of the divergence of the velocity field

$$\begin{aligned} \epsilon \langle \operatorname{div} V \rangle_D &= \frac{\text{the volume which flows out from } D \text{ within time } \epsilon}{\text{the volume of the domain } D} \\ &= \epsilon \frac{\text{the flux of the field } V \text{ for the boundary } D}{\text{the volume of the domain } D} \end{aligned}$$

so that the mean value of the divergence of the field W over the domain D is (also) the ratio of the flux of the field for its boundary to the volume of the domain[116]

$$\begin{aligned} \langle \operatorname{div} W \rangle_D &= \frac{\oint_{\partial D} W^i \, dS_i}{\int_D \omega_g} \\ &\equiv \frac{\text{the flux of } W \text{ for } \partial D}{\text{the volume of } D} \end{aligned}$$

(iii) this fact may also be obtained from the first interpretation of the divergence (as the rate of relative change of the volume due to the flow corresponding to the field V).

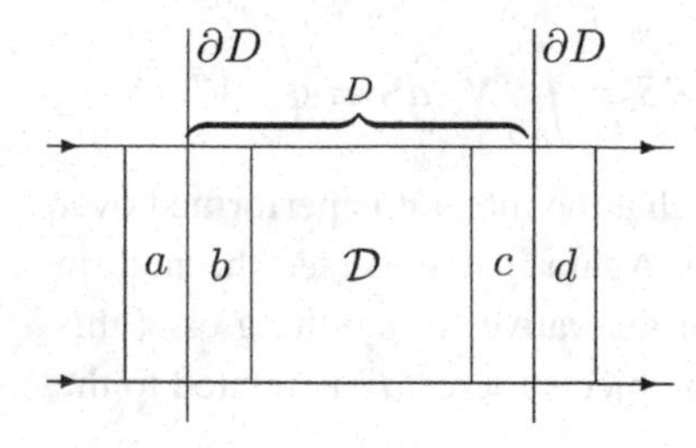

Hint: (i) if a small surface dS spans[117] the vectors $e_2, \ldots, e_n$, then the volume which flows out across dS within the time interval ϵ is (by definitions of both the volume form and the velocity field) given as $\omega_g(\epsilon V, e_2, \ldots, e_n) \equiv \epsilon(i_V \omega_g)(e_2, \ldots, e_n)$; alternatively use the "coin" interpretation of i_V (7.6.11); (iii) observe the motion of the fluid during the time interval ϵ across the domain D (we follow the "tube" of the fluid on the picture, which moves across D to the right). Let a be the region which enters (the part under consideration of) D, thus substituting the volume of the former $b \equiv a(\epsilon) \subset D$ and let $c \subset D$

[116] In the limit where the volume of D approaches zero, we get another well-known interpretation of the divergence *in the point* x.

[117] The small surface does not actually depend on the individual vectors $e_2, \ldots, e_n$, but rather on their exterior product alone or, put another way, on the $(p-1)$-*vector* given by them.

be the region which flows out, thus becoming $d \equiv c(\epsilon)$. The volumes of these regions are of order ϵ and *to this order* there holds (according to (8.2.2)) the volume of a = the volume of b, the volume of c = the volume of d. Then $D = b + \mathcal{D} + c$, $D(\epsilon) = \mathcal{D} + c + d$ (with $\mathcal{D}$ being the interior of D except b, c). According to (8.2.2) the volume of $D(\epsilon)$ = {the volume of D}$\{1 + \epsilon \langle \text{div } V \rangle_D\}$, so that

$$\epsilon \langle \text{div } V \rangle_D = \frac{\text{vol } D(\epsilon) - \text{vol } D}{\text{vol } D} = \cdots = \frac{\text{the volume of } d - \text{the volume of } a}{\text{the volume of } D}$$

$$= \frac{\text{the net volume flowing out from } D}{\text{the volume of } D}$$

□

• The value of the divergence of a vector field W at the point x thus informs us about two closely related characteristics of the field: about the rate of relative change of the volume due to the flow Φ_t generated by the field W as well as about the limit of the ratio of the net flux $\oint_{\partial D} W^i dS_i$ for the boundary of an infinitesimal domain around x to the volume of the domain.

8.2.10* Prove that for the divergence of the *commutator* of two fields there holds

$$\text{div } [V, W] = V(\text{div } W) - W(\text{div } V)$$

Hint: see (4.3.8) and (8.2.1). □

8.2.11* Let f be a function which has a finite integral over a manifold M ($\int_M f\omega_g < \infty$) and let $\Phi_t \leftrightarrow V$ be an (at least local) flow on M. Prove that the following integral vanishes:

$$\int_M (Vf + f\text{div } V)\omega_g = 0$$

and, in particular, that given f an integrable (smooth) function in $\mathbb{R}^n$, the integral

$$\int_{\mathbb{R}^n} (x^i \partial_i f + nf) d^n x = 0$$

vanishes.

Hint: since $\Phi_\epsilon(M) = M$, we have $\int_M f\omega_g = \int_{\Phi_\epsilon(M)} f\omega_g = \int_M \Phi_\epsilon^*(f\omega_g) = \int_M f\omega_g + \epsilon \int_M \mathcal{L}_V(f\omega_g) + \cdots = \int_M f\omega_g + \epsilon \int_M (Vf + f\text{div } V)\omega_g + \cdots$. □

8.3 Codifferential and Laplace–deRham operator

• We learned in (5.8.4) that if a linear space L is endowed with a metric tensor g and an orientation o, a natural *scalar product* in the space of p-forms[118]

$$(\cdot\,,\cdot) : \Lambda^p(L^*) \times \Lambda^p(L^*) \to \mathbb{R}$$

[118] If the metric tensor is not positive definite, it is "only" a non-degenerate bilinear form on $\Lambda^p(L^*)$.

may be defined by the prescription

$$\alpha \wedge *_g \beta =: (\alpha, \beta)_g \omega_g \;=\; \frac{1}{p!}\, \alpha_{a\ldots b}\, \beta^{a\ldots b}\, \omega_g$$

By integration of this expression over the domain $D \in M$ we get a scalar product (or at least a non-degenerate bilinear form) $\langle \cdot\,, \cdot \rangle$ on $\Omega^p(M)$. This opens up the possibility of defining in a standard way the *adjoint* operator A^+ to each linear operator A acting on forms by the rule

$$\langle A^+\alpha, \beta\rangle := \langle \alpha, A\beta\rangle.$$

The case $A = d$ turns out to be especially important in applications. The adjoint operator to the "differential" d (= exterior derivative) is called the *codifferential* $\delta = d^+$ and the (self-adjoint) combination $-(dd^+ + d^+d)$ is known as the *Laplace–deRham operator* ("Laplacian on forms").

8.3.1 Let D be a domain on an n-dimensional Riemannian oriented manifold (M, g, o). For all $p = 0, 1, \ldots, n$ we define the map

$$\langle \cdot\,, \cdot \rangle : \Omega^p(M) \times \Omega^p(M) \to \mathbb{R}$$

by the prescription[119]

$$\begin{aligned}\langle \alpha, \beta\rangle &:= \int_D \alpha \wedge *_g \beta \\ &= \int_D (\alpha, \beta)_g \omega_g = \int_D \frac{1}{p!}\, \alpha_{a\ldots b}\, \beta^{a\ldots b}\, \omega_g \qquad \alpha, \beta \in \Omega^p(M)\end{aligned}$$

Check that

(i) it is a symmetric, non-degenerate bilinear form on $\Omega^p(M)$ (regarded as an ∞-dimensional linear space over $\mathbb{R}$)
(ii) moreover, for positive definite g it happens to be positive definite ($\Rightarrow$ the *scalar product* on $\Omega^p(M)$)
(iii) for $p = 0$ it reduces to a good old[120] scalar product on functions

$$\langle \psi, \chi\rangle := \int_D \psi\chi\, \omega_g \equiv \int_D \psi\chi\, \sqrt{|g|}\, d^n x$$

(iv) an analogous role for V-valued forms is played by the expression

$$\langle \alpha, \beta\rangle := \int_D h_{AB}\alpha^A \wedge *\beta^B$$

Hint: (iv) see (6.4.4) and (7.6.10). □

[119] This map is not to be confused with the canonical *pairing* $\langle \cdot\,, \cdot \rangle$, which is denoted in the same way (see Section 2.6). The pairing assigns a function to a *vector* and a *covector* field, the bilinear form introduced here assigns a number to a *pair of p-forms*.

[120] This is well known, for example, from the theory of Fourier series or special functions; for the wave functions in quantum mechanics the *complex conjugation* of the function ψ is to be added since $\mathbb{C}$-valued functions, the linear spaces *over* $\mathbb{C}$ and the ($\mathbb{C}$-valued) *hermitian* scalar product, are used there.

8.3.2 Check that for $\alpha \in \Omega^{p-1}(M)$, $\beta \in \Omega^p(M)$ there holds

(i)

$$d\alpha \wedge *\beta = \alpha \wedge *\delta\beta + d(\alpha \wedge *\beta)$$

where the *codifferential* $\delta \equiv \delta_g$ is defined as

$$\delta_g : \Omega^p(M,g) \to \Omega^{p-1}(M,g) \qquad \delta_g := *_g^{-1} d *_g \hat{\eta}$$

(ii)

$$\langle d\alpha, \beta \rangle = \langle \alpha, \delta\beta \rangle + \int_{\partial D} \alpha \wedge *\beta$$

Hint: (i) $d(\alpha \wedge *\beta) = d\alpha \wedge *\beta + (\hat{\eta}\alpha) \wedge d * \beta$. But $(\hat{\eta}\alpha) \wedge d * \beta = (\hat{\eta}\alpha) \wedge * *^{-1} d * \beta = -\alpha \wedge *(*^{-1} d * \hat{\eta}\beta) \equiv -\alpha \wedge *\delta\beta$; (ii) use Stokes' theorem. □

• From this result we can see that if for some reason the integral over the boundary of the domain D *vanishes*, the codifferential δ would be the *adjoint*[121] operator d^+ to the operator d in the sense of $\langle \cdot , \cdot \rangle$. There are actually several reasons why the integral often vanishes:

1. we consider the space of forms which vanish on ∂D (this is the case in applications in the *calculus of variations*, see (16.3.2) or (21.5.2))
2. $D = M =$ a *compact manifold* (like the sphere S^n, torus T^n, groups $U(n)$ and $O(n)$, ...); then $\partial D = \partial M = \emptyset$
3. $D = M$ is not compact, but we consider the space of forms with compact support or the forms which decrease "sufficiently quickly at infinity" (for example, for $r \to \infty$ in $\mathbb{R}^n$ the expression under the integral sign has to decrease rapidly enough in comparison with the increasing of the area of the sphere S_r^{n-1}).

In all of these cases the codifferential δ is the adjoint operator to the differential d ($\delta = d^+$) and we will automatically assume in what follows that this *is* the case.

8.3.3 Check that the *Laplace–deRham operator*

$$\Delta_g := -(\delta_g d + d\delta_g) \equiv -(d^+ d + dd^+)$$

is *self-adjoint* in the sense of $\langle \cdot , \cdot \rangle$, i.e. there holds

$$\langle \Delta_g \alpha, \beta \rangle = \langle \alpha, \Delta_g \beta \rangle$$

Hint: see (8.3.2) or $(d^+ d + dd^+)^+ = \cdots$. □

• The codifferential is also known as the "generalized divergence." The reason should be clear from the next two exercises. First we will notice that on 1-forms this "indeed *is*" the divergence, then we will see that the component expression for δ in the general case

[121] Both d and δ change the degree of forms so that the fact that they happen to be adjoint to each other in more detail looks like $\langle d\alpha, \beta \rangle_p = \langle \alpha, \delta\beta \rangle_{p-1}$, where $\langle \cdot , \cdot \rangle_p$ is the scalar product in the space of *p-forms*. Since the official definition of the adjoint operator $\langle A^+\alpha, \beta \rangle := \langle \alpha, A\beta \rangle$ assumes a *single* scalar product in a *single* linear space one should actually regard the Cartan algebra as this single space, i.e. the *direct sum* $\Omega(M) := \oplus_{p=0}^n \Omega^p(M)$ endowed with the scalar product which is a natural extension to inhomogeneous forms (i.e. after the fashion of (5.8.8)).

(on p-forms) may be regarded as a generalization of the corresponding expression for the divergence.

8.3.4 Check that if V is a vector field and $\tilde{V} = g(V, \cdot\,) \equiv \flat_g V$ is the corresponding covector field, then

$$\delta \tilde{V} = -\text{div}\, V$$

so that in coordinates

$$\delta\alpha \equiv \delta(\alpha_i\, dx^i) = -\frac{1}{\sqrt{|\,g\,|}}(\sqrt{|\,g\,|}g^{kj}\alpha_j)_{,k} \qquad \alpha \in \Omega^1(M)$$

Hint: see (8.2.1) and (8.3.2). □

8.3.5 Check that

(i) the coordinate expression of the codifferential reads[122]

$$(\delta_g\alpha)^{i\ldots j} = -\frac{1}{\sqrt{|\,g\,|}}(\sqrt{|\,g\,|}\alpha^{ki\ldots j})_{,k} \qquad (\delta_g\alpha)^{i\ldots j} := g^{ir}\ldots g^{js}(\delta_g\alpha)_{r\ldots s}$$

(ii) the coordinate expression of the Laplace–deRham operator *on functions* (0-forms) is[123]

$$\Delta f = \frac{1}{\sqrt{|\,g\,|}}(\sqrt{|\,g\,|}g^{kj}f_{,j})_{,k} \qquad f \in \Omega^0(M) \equiv \mathcal{F}(M)$$

Hint: (i) see (6.2.5), (5.8.1), (5.8.2), (6.3.7) and (5.6.4); (ii) here $\Delta = -\delta d$ (only). □

• Recall (6.2.11) that the differential d commutes with the pull-back of *arbitrary* smooth maps, including diffeomorphisms $f : M \to M$. It turns out that a fairly simple behavior with respect to diffeomorphisms is also true for the codifferential as well as for the Laplace–deRham operator. However, since both of them depend (unlike d!) on the metric tensor, it should not be surprising that they nevertheless prefer (behave more simply with respect to) those diffeomorphisms which *respect* the metric tensor, namely the isometries (or at least conformal transformations). We will now investigate this behavior explicitly.

If an operator $A = A_B$ is available which depends on a tensor B and which in addition behaves with respect to a diffeomorphism $f : M \to M$ according to the rule[124]

$$f^* A_B = A_{f^*B} f^*$$

we say that the operator A is *natural with respect to diffeomorphisms*. This (fairly simple) behavior occurs for several operators we have already encountered or we will encounter in the future.

8.3.6 Check that the operators i_V and j_V are

[122] The codifferential δ (as well as the differential d itself) may also be conveniently expressed in terms of the *covariant* derivatives, see (15.6.17).

[123] This operator is also known as the *Laplace–Beltrami operator*.

[124] That is, the commuting of f^* with A leaves certain persistent effects on A; namely, the change $B \mapsto f^*B$ occurs.

(i) natural with respect to diffeomorphisms, i.e. that there holds

$$f^* i_V = i_{f^*V} f^* \qquad f^* j_V = j_{f^*V} f^*$$

(ii) adjoint to each other in the sense of $\langle \cdot , \cdot \rangle$

$$i_V = (j_V)^+, \qquad j_V = (i_V)^+ \qquad \text{i.e.} \quad \langle j_V \alpha, \beta \rangle = \langle \alpha, i_V \beta \rangle$$

Hint: (i) apply to a p-form α, then "fill by" arguments and use definitions; (ii) see (5.8.6).

□

8.3.7 Check that the operator of the Lie derivative $\mathcal{L}_V$ (on arbitrary tensor fields, in particular on forms) is natural with respect to diffeomorphisms, i.e. that there holds

$$f^* \mathcal{L}_V = \mathcal{L}_{f^*V} f^*$$

Hint: in general making use of (4.1.9):

$$f^* \circ \mathcal{L}_V := \frac{d}{dt}\bigg|_0 f^* \circ \Phi_t^* = \frac{d}{dt}\bigg|_0 (f^{-1} \circ \Phi_t \circ f)^* \circ f^* = \mathcal{L}_{f^*V} \circ f^*$$

on forms alternatively using (6.2.8) and (8.3.6). □

8.3.8 Check that the following operators on forms (or vector fields) on (M, g, o) happen to be natural with respect to diffeomorphisms:

(i) the raising and lowering index operators $\sharp_g$ and $\flat_g$ and the Hodge star operator $*_{g,o}$, i.e.

$$f^* \sharp_g = \sharp_{f^*g} f^* \qquad f^* \flat_g = \flat_{f^*g} f^* \qquad f^* *_{g,o} = *_{f^*g, f(o)} f^*$$

($f(o) = \pm o$ if f preserves (does not preserve) an orientation o on M)

(ii) the codifferential δ_g and the Laplace–deRham operator Δ_g, i.e.

$$f^* \delta_g = \delta_{f^*g} f^* \qquad f^* \Delta_g = \Delta_{f^*g} f^*$$

Hint: (i) $\sharp_g \alpha = C(g^{-1} \otimes \alpha)$, $\flat_g V = C(g \otimes V)$, (3.1.7); $*_{g,o}$: compute according to (5.8.1) in a right-handed orthonormal basis e^a in the sense of (g, o) and use the fact that $E^a := f^* e^a$ is right-handed orthonormal in the sense of $(f^*g, f(o))$; (ii) the definitions and (6.2.11).

□

8.3.9 Check that the codifferential as well as the Laplace–deRham operator (unlike the Hodge operator) *do not depend* on the choice of the orientation on (M, g, o) (i.e. they actually feel (M, g) alone)

$$\delta_{g,o} = \delta_{g,-o} \equiv \delta_g \qquad \Delta_{g,o} = \Delta_{g,-o} \equiv \Delta_g$$

Hint: according to (5.8.1) $\delta_{g,-o} := \overset{-1}{*}_{g,-o} d *_{g,-o} \hat{\eta} = (-\overset{-1}{*}_{g,o}) d (-*_{g,o}) \hat{\eta} = + \ \overset{-1}{*}_{g,o} d *_{g,o} \hat{\eta} \equiv \delta_{g,o}$. □

8.3.10 Check that the volume form $\omega_{g,o}$, which is compatible with the metric and orientation, is natural with respect to diffeomorphisms, i.e. there holds

$$f^*\omega_{g,o} = \omega_{f^*g, f(o)}$$

Hint: according to (8.3.8) $f^*\omega_{g,o} = f^* *_{g,o} 1 = *_{f^*g,f(o)} f^*1 = \omega_{f^*g,f(o)}$. □

8.3.11 Let ξ be a conformal Killing vector, so that $\mathcal{L}_\xi g = \chi g$. Show that

(i) the function χ is then a constant multiple of the *divergence* of the field ξ, namely $\chi = (2/n)\operatorname{div}\xi$ (where $n = \dim M$), so that the conformal Killing equations may be written in the form

$$\mathcal{L}_\xi g = \left(\frac{2}{n}\operatorname{div}\xi\right) g$$

(ii) any Killing vector is necessarily *divergence-less.*

Hint: (i) for the infinitesimal flow Φ_ϵ which corresponds to ξ we have

$$\begin{aligned}\Phi_\epsilon^*\omega_g &\overset{1}{=} (1+\epsilon\mathcal{L}_\xi)\omega_g = (1+\epsilon\operatorname{div}\xi)\omega_g\\ &\overset{2}{=} \omega_{\Phi_\epsilon^* g} = \omega_{g+\epsilon\chi g} = *_{(1+\epsilon\chi)g}1 = *_{\lambda^2 g}1 = \lambda^n *_g 1 = \lambda^n\omega_g\end{aligned}$$

where $\lambda = 1+\chi\epsilon/2$ and we have also used (5.8.3). Since $\lambda^n = 1+\chi n\epsilon/2$, by comparison we get what is needed; (ii) $\chi = 0$; or with no computation: Killing vectors generate isometries, preserving (even *more* than) the (metric!) volume so that by (8.2.2) its divergence necessarily vanishes. □

8.3.12 Check that the codifferential is (just like the "differential" d) a *nilpotent* operator, i.e.

$$\delta\delta = 0$$

Hint: $*^{-1}d*\hat\eta *^{-1} d*\hat\eta = \pm *^{-1} d**^{-1}d* = \pm *^{-1} dd*$. □

8.3.13 Let $f: M\to M$ be a conformal transformation of a *two-dimensional* (pseudo-)Riemannian oriented manifold (M,g,o) and let $f^*g = \sigma^2 g$ $(\sigma^2 > 0)$. Show that

(i) the Laplace operator on *functions* is "conformally covariant" in the sense that

$$f^*\Delta_g = \sigma^{-2}\Delta_g f^*$$

(ii)

$$u = \text{harmonic} \quad\Rightarrow\quad f^*u = \text{harmonic}$$

(u = a *harmonic function* means that it satisfies the *Laplace equation* $\Delta_g u = 0$)

(iii) if u is a solution of the *Dirichlet problem* for the Laplace equation

$$\begin{aligned}\Delta_g u &= 0 &&\text{in the domain } \mathcal{U}\\ u|_S &= 0 &&\text{on } S\equiv\partial\mathcal{U}\end{aligned}$$

then f^*u is the solution of the Dirichlet problem with $\mathcal{U}\mapsto f^{-1}(\mathcal{U})$ and $S\mapsto f^{-1}(S)$

(iv) a similar statement (which one?) is true for the *Poisson equation* $\Delta_g u = v$.

Hint: (i) check that (on functions, i.e. for $\Delta_g = -\delta_g d$) there holds $f^*(\Delta_g u) = \sigma^{-2}\Delta_g(f^*u) + (n-2)(d\ln\sigma, df^*u)_g$ so that for $n=2$ we are finished; (iv) $\Delta_g(f^*u) = \sigma^2 f^*v$. □

8.3.14 Consider a Killing vector ξ on (M, g). Check that the Lie derivative $\mathcal{L}_\xi$ commutes with the operators $\flat_g, \sharp_g, *_g, \delta_g$ and Δ_g

$$[\mathcal{L}_\xi, \hat{A}_g] = 0 \qquad \hat{A}_g = \flat_g, \sharp_g, *_g, \delta_g, \Delta_g$$

Hint: ξ generates isometries $\Rightarrow \Phi_t^* g = g$; all the operators are natural with respect to diffeomorphisms (8.3.8), so that $[\Phi_t^*, \hat{A}_g] = 0$; the differentiation with respect to t at $t=0$ gives just $[\mathcal{L}_\xi, \hat{A}_g] = 0$ (which is nothing but an infinitesimal version of $[\Phi_t^*, \hat{A}_g] = 0$).

8.4 Green identities

• The next useful direct consequence of Stokes' theorem is provided by the Green identities.

8.4.1 Let u, v be two functions on an m-dimensional oriented Riemannian manifold (M, g, o), D an m-dimensional domain on M and ∂D its boundary oriented in the sense of the outer normal. Prove the *Green identities*:

(i)

$$\langle du, dv\rangle + \langle u, \Delta v\rangle = \int_{\partial D} u * dv \qquad \text{"ordinary" Green identity}$$

$$\langle u, \Delta v\rangle - \langle v, \Delta u\rangle = \int_{\partial D} (u * dv - v * du) \qquad \text{"symmetric" Green identity}$$

(ii) if the field n of the "unit perpendicular" outer normal on ∂D exists ($n \perp \partial D$, $g(n,n) = 1$, cf. the text before (8.2.8)), then there holds

$$\int_{\partial D} u * dv = \int_{\partial D} u \frac{\partial v}{\partial n} dS$$

with $\partial f/\partial n := nf$ being the *normal derivative* of f (the derivative of the function f along the unit outer normal) so that the Green identities may also be expressed in the form

$$\int_D (\nabla u \cdot \nabla v + u\Delta v)\, d\Omega = \int_{\partial D} u\frac{\partial v}{\partial n} dS$$

$$\int_D (u\Delta v - v\Delta u)\, d\Omega = \int_{\partial D} \left(u\frac{\partial v}{\partial n} - v\frac{\partial u}{\partial n}\right) dS$$

Hint: (i) set $\alpha = u$, $\beta = dv$ in (8.3.2) or $V = u\nabla v$ in (8.2.8), then subtract the equations with $u \leftrightarrow v$; (ii):

$$*dv|_{\partial D} = v_{,i}\,(*dx^i)\big|_{\partial D} \overset{(6.3.11)}{=} v_{,i} g^{ij}\, d\Sigma_j\big|_{\partial D} \overset{(8.2.7)}{=} n_i g^{ij} v_{,j} dS \equiv (nv)\, dS =: \frac{\partial v}{\partial n} dS$$

and (8.2.8) ($d\Omega \equiv \sqrt{|g|}\, d^n x$ denotes the "volume element," i.e. it coincides with ω_g). □

8.4.2 Check the validity of the (first) Green identity for $M = E^3$ and

(i)

$$D = \{(x, y, z)|x^2 + y^2 + z^2 \leq 1, y \geq 0\}, \quad u = z^2, \quad v = x^2 + y^2 - z^2$$

(ii)

$$D = \{(x, y, z)|x^2 + y^2 \leq 1, 0 \leq z \leq 1\}, \quad u = 2x^2, \quad v = x^2 + z^2$$

Hint: (i) spherical polar coordinates; (ii) cylindrical coordinates. □

8.4.3 Making use of the Green identity on (M, g) prove the uniqueness of the solution of the *Poisson equation* in a domain D

$$\Delta u = f \qquad f \text{ a given function}$$

with the *Dirichlet* or the *Neumann* boundary condition

$$u|_{\partial D} = g \quad \text{or} \quad \left.\frac{\partial u}{\partial n}\right|_{\partial D} = h$$

(with g and h being given functions on ∂D).

Hint: given two solutions u, v, $w := u - v$ turns out to be a solution of the problem

$$\Delta w = 0 \qquad w|_{\partial D} = 0 \quad \text{or} \quad \left.\frac{\partial w}{\partial n}\right|_{\partial D} = 0$$

The ("ordinary") Green identity then yields (in both cases)

$$\langle dw, dw \rangle \equiv \int_D (\nabla w)^2 \, d\Omega = 0$$

$\Rightarrow$ (in the positive definite case) $\nabla w = 0 \Rightarrow w =$ constant. For the Dirichlet case the constant $= 0$, in the Neumann case the constant is arbitrary, being the freedom which is clear from the outset. □

8.5 Vector analysis in E^3

• Vector analysis in the "usual" three-dimensional space E^3 serves as an indispensable mathematical tool in physics. Let us start our contemplation on this topic with the inspection of two key formulas of vector analysis, namely the integral identities

$$\oint_{\partial S} \mathbf{A} \cdot d\mathbf{r} = \int_S (\text{curl } \mathbf{A}) \cdot d\mathbf{S} \qquad \oint_{\partial D} \mathbf{A} \cdot d\mathbf{S} = \int_D (\text{div } \mathbf{A}) \, dV$$

They both relate a pair of integrals over domains of neighboring dimensions (volume $\leftrightarrow$ surface $\leftrightarrow$ line), raising thus a suspicion that the *Stokes' theorem* (for differential forms) might be behind them.[125] Notice also that we see explicitly *scalar and vector* fields under the integral sign, but we know that, in fact, *differential forms* are there (e.g. a 2-form in the surface integral). It seems a possibility exists to *parametrize* differential forms in terms of scalar and vector fields; the differential operations applied on scalar and vector fields, gradient, curl and divergence could then be the "effective" operators resulting from this parametrization.

In this section we learn that this is really the case – after the above-mentioned parametrization the whole machinery of vector analysis "drops out" as a simple special case of the *standard* operations on forms (d, δ_g, $*_g$, $\wedge$) and Stokes' theorem (for forms).

Let us begin with the possibility of encoding scalar and vector fields into forms and vice versa. If we have an n-dimensional manifold endowed with a metric tensor and orientation, the canonical isomorphisms $\sharp \equiv \sharp_g, \flat \equiv \flat_g$ (the raising and lowering of indices) and $* \equiv *_{g,o}$ (the Hodge operator) are available. One can then identify the spaces of vector fields, 1-forms and $(n-1)$-forms, as well as the spaces of 0-forms and n-forms. This means that it is easy to encode scalar and vector fields into forms, but we are not able to express *forms of all degrees* in terms of scalar and vector fields (it is possible for the "marginal pairs" 0, 1, $(n-1)$, n, but it is not for forms of "inner" degrees $2, 3, \ldots, (n-2)$). There exists an important exception, however, namely *three-dimensional* manifolds (the most interesting from the practical point of view being undoubtedly the simplest one, the good old Euclidean space E^3), where the "inner" degrees are simply missing.[126]

Thus, on a three-dimensional manifold endowed with a metric tensor and orientation one can parametrize *all forms* in terms of the scalar and vector fields (0- and 3-forms via the scalar fields and 1- and 2-forms via the vector fields).[127]

8.5.1 Be sure to understand that on a *three*-dimensional manifold (M, g, o) the following canonical identifications are possible:[128]

$$\mathfrak{X}(E^3) \underset{\sharp}{\overset{\flat}{\rightleftarrows}} \Omega^1(E^3) \underset{*^{-1}}{\overset{*}{\rightleftarrows}} \Omega^2(E^3) \qquad \Omega^0(E^3) \underset{*^{-1}}{\overset{*}{\rightleftarrows}} \Omega^3(E^3)$$

Hint: see (2.6.2) and (5.8.1). □

8.5.2 Show that if we define in E^3 the forms (x^i being arbitrary coordinates)

$$dS_i := \frac{1}{2}\omega_{ijk}\, dx^j \wedge dx^k \qquad \text{(see (6.3.11))}$$

$$dV := \frac{1}{3!}\omega_{ijk}\, dx^i \wedge dx^j \wedge dx^k \equiv \omega \equiv \omega_{g,o} \qquad \text{(the metric volume form)}$$

[125] This suspicion probably occurred in the reader's mind independently as far as he or she did not fail to observe that this section is a part of the chapter entitled "Special cases and applications of *Stokes' theorem*."

[126] And they are also missing of course on one- and two-dimensional manifolds; on these manifolds there thus exists a (simplified version of) "vector analysis," too (it may also be regarded as the vector analysis "diluted up to homeopathic concentrations"). After reading this section the interested reader can work up the details of the corresponding theory as a simple exercise by him(her)self.

[127] To be more precise, one should add *pseudo*scalar and *pseudo*vector fields, too, since the operator $*$ changes the sign as a result of a change of orientation of a manifold, (5.8.1).

[128] According to (5.8.2) $** = 1$ on E^3 holds, i.e. $*^{-1} = *$.

then the most general 0-, 1-, 2- and 3-forms read

$$f \qquad \mathbf{A}\cdot d\mathbf{r} \equiv A_i\,dx^i = A^j g_{ji}\,dx^i \qquad \mathbf{B}\cdot d\mathbf{S} \equiv B^i\,dS_i \qquad h\,dV$$

i.e. that one can take $1, dx^i, dS_i$ and dV as the basis for differential forms of degrees 0, 1, 2 and 3 on E^3.

Hint: dS_i happen to be the images under isomorphism of the basis dx^i, thus they are linearly independent. □

• From these expressions we see explicitly how arbitrary differential forms in E^3 may be parametrized in terms of the scalar (0- and 3-forms) or the vector (1- and 2-forms) fields.

8.5.3 The bases defined in (8.5.2) turn out to be very convenient in that they are adapted to those of the identifications (isomorphisms) of (8.5.1) which concern forms (half of the forms being, in fact, the $*$-images of the other one). Check that

$$*f = f\,dV \qquad *(h\,dV) = h$$

$$*(\mathbf{B}\cdot d\mathbf{S}) = \mathbf{B}\cdot d\mathbf{r} \qquad *(\mathbf{A}\cdot d\mathbf{r}) = \mathbf{A}\cdot d\mathbf{S}$$

so that the identifications read

$$A^i\partial_i \underset{\sharp}{\overset{\flat}{\rightleftarrows}} A^i g_{ij}\,dx^j \equiv \mathbf{A}\cdot d\mathbf{r} \underset{*}{\overset{*}{\rightleftarrows}} A^i\,dS_i \equiv \mathbf{A}\cdot d\mathbf{S} \qquad f \underset{*}{\overset{*}{\rightleftarrows}} f\,dV$$

or at the level of bases

$$1 \underset{*}{\overset{*}{\rightleftarrows}} dV \qquad dx^i \underset{*}{\overset{*}{\rightleftarrows}} g^{ij}\,dS_j$$

□

• The fact that we may parametrize forms of *all (non-trivial)* degrees is of vital importance: it means that also an arbitrary *operation* on forms may be translated into the language of the operations on scalar and vector fields.[129] The most important one turns out to be in this respect the translation of the operator d (exterior derivative) and Stokes' theorem closely related to it.

As we know, the operator d raises the degree of a form by one unit. On a three-dimensional manifold it thus acts as follows:

$$\Omega^0(M) \xrightarrow{d} \Omega^1(M) \xrightarrow{d} \Omega^2(M) \xrightarrow{d} \Omega^3(M)$$

If one applies d, say, on a 1-form, a 2-form results. This means, however (since the spaces of both 1- and 2-forms are identified with the space of vector fields), that the operator d induces some (differential) operation which assigns a vector field to a vector field. A similar reasoning for the remaining two objects (0- and 2-forms) reveals that altogether *three* differential operators are induced, namely of the type scalar $\mapsto$ vector, vector $\mapsto$

[129] If for some degree p the identification with vectors or scalars did not exist, one would not be able to translate the operations on forms producing a p-form.

vector and vector $\mapsto$ scalar. We will see that they are nothing but the three well-known operations of vector analysis: the *gradient*, the *curl* and the *divergence*.

8.5.4 On an arbitrary *three-dimensional* manifold (M, g, o) let us define the operations grad, curl and div via the commutative diagram

$$\begin{array}{ccccccc}
\Omega^0(M) & \xrightarrow{d} & \Omega^1(M) & \xrightarrow{d} & \Omega^2(M) & \xrightarrow{d} & \Omega^3(M) \\
\downarrow \text{id} & & \downarrow \sharp & & \downarrow \sharp * & & \downarrow * \\
\mathcal{F}(M) & \xrightarrow[\text{grad}]{} & \mathfrak{X}(M) & \xrightarrow[\text{curl}]{} & \mathfrak{X}(M) & \xrightarrow[\text{div}]{} & \mathcal{F}(M)
\end{array}$$

Show that

(i) their "abstract" expressions read

$$\text{grad} = \sharp\, d \qquad \text{curl} = \sharp * d\, \flat \qquad \text{div} = * d * \flat \equiv -\delta\, \flat$$

(ii) in local coordinates

$$\begin{aligned}
\text{grad} f &\equiv \nabla f = (g^{ij} f_{,j})\partial_i \\
\text{curl}\,(A^i \partial_i) &= (\omega^{ijk}(g_{kl}A^l)_{,j})\partial_i \equiv (\text{curl}\,\mathbf{A})^i \partial_i \\
\text{div}\,(A^i \partial_i) &= \frac{1}{\sqrt{|g|}}(\sqrt{|g|}A^i)_{,i}
\end{aligned}$$

(iii) on E^3 they *indeed do* represent the operations gradient, curl and divergence known from vector analysis

(iv) the following identities hold:

$$\text{curl grad} = 0 \qquad \text{div curl} = 0$$

(v) there holds

$$\begin{aligned}
df &= (\text{grad} f) \cdot d\mathbf{r} \equiv (\nabla f) \cdot d\mathbf{r} & d(\mathbf{A} \cdot d\mathbf{r}) &= (\text{curl}\,\mathbf{A}) \cdot d\mathbf{S} \\
d(\mathbf{B} \cdot d\mathbf{S}) &= (\text{div}\,\mathbf{B})\, dV & d(h\, dV) &= 0
\end{aligned}$$

Hint: (i) compose the arrows according to the diagram; (iii) write down in *Cartesian* coordinates and compare with the output of your memory (see also their presence in the integral identities (8.5.6)); (iv) use (i) or just have a quick look at the defining diagram (since there holds $dd = 0$ in the upper line, the composition of the corresponding operators in the bottom line should also vanish); (v) cf. the hint to (i). □

8.5.5 Show that

(i) the commutative diagram with differential from (8.5.4) results in a similar diagram with codifferential

$$\begin{array}{ccccccc}
\Omega^0(M) & \xleftarrow{\delta} & \Omega^1(M) & \xleftarrow{\delta} & \Omega^2(M) & \xleftarrow{\delta} & \Omega^3(M) \\
\downarrow \text{id} & & \downarrow \sharp & & \downarrow \sharp * & & \downarrow * \\
\mathcal{F}(M) & \xleftarrow[-\text{div}]{} & \mathfrak{X}(M) & \xleftarrow[\text{curl}]{} & \mathfrak{X}(M) & \xleftarrow[-\text{grad}]{} & \mathcal{F}(M)
\end{array}$$

(ii) there holds

$$\delta(\mathbf{A}\cdot d\mathbf{r}) = -\mathrm{div}\,\mathbf{A}$$
$$\delta(\mathbf{B}\cdot d\mathbf{S}) = (\mathrm{curl}\,\mathbf{B})\cdot d\mathbf{r}$$
$$\delta(h\,dV) = -(\mathrm{grad}\,h)\cdot d\mathbf{S} \equiv -(\nabla h)\cdot d\mathbf{S}$$

□

• The combination of the results of exercise (8.5.4) with the general Stokes' theorem for differential forms gives us immediately the three fundamental integral theorems of vector analysis.

8.5.6 Prove the following integral theorems of the vector analysis on E^3:

(i) let C be a curve (1-chain) from the point A to B and f a function. Then

$$\int_C (\nabla f)\cdot d\mathbf{r} = f(B) - f(A)$$

(ii) let S be a two-dimensional surface (2-chain) in E^3, ∂S its (one-dimensional) boundary and $\mathbf{A}$ a vector field. Then *Stokes' theorem* holds

$$\oint_{\partial S} \mathbf{A}\cdot d\mathbf{r} = \int_S (\mathrm{curl}\,\mathbf{A})\cdot d\mathbf{S}$$

(iii) let D be a three-dimensional domain (3-chain) in E^3, ∂D its (two-dimensional) boundary and $\mathbf{A}$ a vector field. Then *Gauss' theorem* holds[130]

$$\oint_{\partial D} \mathbf{A}\cdot d\mathbf{S} = \int_D (\mathrm{div}\,\mathbf{A})\,dV$$

Hint: see (7.6.7) and (8.5.4). □

8.5.7 Check that the Laplace–deRham operator $\Delta = -(\delta d + d\delta)$ acts here as follows:

$$\Delta f = \mathrm{div}\,\mathrm{grad}\,f$$
$$\Delta(\mathbf{A}\cdot d\mathbf{r}) = (-\,\mathrm{curl}\,\mathrm{curl}\,\mathbf{A} + \mathrm{grad}\,\mathrm{div}\,\mathbf{A})\cdot d\mathbf{r}$$
$$\Delta(\mathbf{B}\cdot d\mathbf{S}) = (-\,\mathrm{curl}\,\mathrm{curl}\,\mathbf{B} + \mathrm{grad}\,\mathrm{div}\,\mathbf{B})\cdot d\mathbf{S}$$
$$\Delta(h dV) = (\mathrm{div}\,\mathrm{grad}\,h)\,dV \equiv (\Delta h)\,dV$$

Hint: see (8.3.3), (8.5.4) and (8.5.5). □

• Also the *algebraic* operations performed on forms induce some algebraic operations on functions and vector fields. The most interesting novelty consists in the *vector product* of two vectors resulting from the *exterior* product of the forms.

8.5.8 Check that

[130] Clearly this version of Gauss' theorem is only a particular case of (8.2.8) for $m = 3$ and the divergence of a vector field introduced in (8.5.4) is a particular case of (8.2.1). All of these theorems are, as we see, special cases of the general Stokes' theorem for forms (7.6.7) or see Section 7.5 (including Stokes' theorem presented in this section).

(i) the exterior products of forms on E^3 give

$$(\mathbf{A}\cdot d\mathbf{r})\wedge(\mathbf{B}\cdot d\mathbf{r}) = (\mathbf{A}\times\mathbf{B})\cdot d\mathbf{S}$$
$$(\mathbf{A}\cdot d\mathbf{r})\wedge(\mathbf{B}\cdot d\mathbf{S}) = (\mathbf{A}\cdot\mathbf{B})\,dV$$

where $\mathbf{A}\times\mathbf{B}$ and $\mathbf{A}\cdot\mathbf{B}$ are the common vector and scalar products of two vectors respectively, i.e.

$$(\mathbf{A}\times\mathbf{B})^i := g^{ij}\omega_{jkl}A^kB^l \qquad \mathbf{A}\cdot\mathbf{B} := g_{ij}A^iB^j$$

and that from this (and from (8.5.3)) it follows that the objects under the integral signs in the scalar product of type (8.3.1) are

$$f\wedge *h = (f\,dV)\wedge *(h\,dV) = (fh)\,dV$$
$$(\mathbf{A}\cdot d\mathbf{r})\wedge *(\mathbf{B}\cdot d\mathbf{r}) = (\mathbf{A}\cdot d\mathbf{S})\wedge *(\mathbf{B}\cdot d\mathbf{S}) = (\mathbf{A}\cdot\mathbf{B})\,dV$$

(ii) the interior products lead to

$$i_{\mathbf{A}}(\mathbf{B}\cdot d\mathbf{r}) = \mathbf{A}\cdot\mathbf{B}$$
$$i_{\mathbf{A}}(\mathbf{B}\cdot d\mathbf{S}) = (\mathbf{B}\times\mathbf{A})\cdot d\mathbf{r}$$
$$i_{\mathbf{A}}(h\,dV) = h\mathbf{A}\cdot d\mathbf{S}$$

(iii) the behavior of $i_{\mathbf{A}}$ on the exterior product of forms gives the vector identity

$$(\mathbf{A}\times\mathbf{B})\times\mathbf{C} = (\mathbf{A}\cdot\mathbf{C})\mathbf{B} - (\mathbf{B}\cdot\mathbf{C})\mathbf{A}$$

Hint: (iii) see (5.4.2). □

• All of the component expressions for the operations mentioned in this section use the *coordinate* components of vector fields. In vector analysis the *orthonormal* (non-coordinate, in general) components are used, however, quite frequently. As a rule one uses a "corrected coordinate frame field" (see, for example, (4.5.7)). This means that one starts with *orthogonal* coordinates (see Section 2.6), in which the metric tensor reads

$$g = h_1^2\,dx^1\otimes dx^1 + h_2^2\,dx^2\otimes dx^2 + h_3^2\,dx^3\otimes dx^3 \equiv e^1\otimes e^1 + e^2\otimes e^2 + e^3\otimes e^3$$

for

$$e^1 := h_1\,dx^1 \qquad e^2 := h_2\,dx^2 \qquad e^3 := h_3\,dx^3$$

The functions h_1, h_2 and h_3 are known as the *Lamé coefficients* corresponding to given (orthogonal) coordinates.

8.5.9 Determine the Lamé coefficients for Cartesian, spherical polar and cylindrical coordinates.

Hint: see (2.6.4) (Cartesian 1, 1, 1; spherical polar 1, r, $r\sin\vartheta$; cylindrical 1, r, 1). □

8.5.10 Let us denote the components with respect to an *orthonormal* basis by a hat

$$\mathbf{A} = A^i\partial_i = A^{\hat{i}}e_{\hat{i}} \qquad e_{\hat{1}} = \frac{1}{h_1}\partial_1 \qquad \text{etc.}$$

Check that if one plugs the orthonormal components of vectors into the component expressions from (8.5.4), the following formulas result:[131]

$$\begin{aligned}
\text{grad } f &= \frac{1}{h_1}(\partial_1 f)e_{\hat{1}} + \frac{1}{h_2}(\partial_2 f)e_{\hat{2}} + \frac{1}{h_3}(\partial_3 f)e_{\hat{3}} \equiv \nabla f \\
\text{curl } (A^{\hat{i}} e_{\hat{i}}) &= \frac{1}{h_2 h_3}[\partial_2(h_3 A^{\hat{3}}) - \partial_3(h_2 A^{\hat{2}})]e_{\hat{1}} + \frac{1}{h_1 h_3}[\partial_3(h_1 A^{\hat{1}}) - \partial_1(h_3 A^{\hat{3}})]e_{\hat{2}} \\
&\quad + \frac{1}{h_1 h_2}[\partial_1(h_2 A^{\hat{2}}) - \partial_2(h_1 A^{\hat{1}})]e_{\hat{3}} \\
\text{div } (A^{\hat{i}} e_{\hat{i}}) &= \frac{1}{h_1 h_2 h_3}[\partial_1(h_2 h_3 A^{\hat{1}}) + \partial_2(h_1 h_3 A^{\hat{2}}) + \partial_3(h_1 h_2 A^{\hat{3}})] \\
\Delta f &= \frac{1}{h_1 h_2 h_3}\left[\partial_1\left(\frac{h_2 h_3}{h_1}(\partial_1 f)\right) + \partial_2\left(\frac{h_3 h_1}{h_2}(\partial_2 f)\right) + \partial_3\left(\frac{h_1 h_2}{h_3}(\partial_3 f)\right)\right]
\end{aligned}$$

Hint: $g_{ij} = h_i^2 \delta_{ij}$ (no summation), $|g| = (h_1 h_2 h_3)^2$. □

8.5.11 Write down these formulas explicitly for Cartesian, spherical polar and cylindrical coordinates.

Hint: see (8.5.9). □

8.5.12 Check that the Laplace operator (on functions) in spherical polar coordinates has the following structure:

$$\Delta = \Delta_r + \frac{1}{r^2}\Delta_{\vartheta,\varphi}$$

where

$$\Delta_r := \frac{1}{r^2}\partial_r r^2 \partial_r \qquad \Delta_{\vartheta,\varphi} := \frac{1}{\sin\vartheta}\partial_\vartheta \sin\vartheta\, \partial_\vartheta + \frac{1}{\sin^2\vartheta}\partial_\varphi^2$$

Hint: see (8.5.11). □

8.5.13 Let us return for a while to the problem of finding the lines of force for a given electrical field. Show that

(i) one can write the differential equations for the lines of the field **E** as

$$\mathbf{E} \times d\mathbf{r} = \mathbf{0} \qquad \text{i.e.} \quad \nabla\Phi \times d\mathbf{r} = \mathbf{0}$$

(ii) in (coordinate) components this is equivalent to[132]

$$\epsilon_{ijk} g^{jl} \Phi_{,l}\, dx^k = 0$$

[131] They are frequently found in the appendices of various books on theoretical physics. Usually the authors do not use hats, however, since they work with the components of vectors with respect to an orthonormal basis *alone* (here we also display the formulas using *coordinate* components of vectors (8.5.4) and we need to distinguish somehow between these two kinds of components).

[132] The expressions dx^i here are not to be regarded as differential forms but rather as small increments of coordinates *along the curve*; the advantage of this form of equations consists in avoiding the (redundant) parameter of the integral curves.

(iii) in particular, in spherical polar coordinates this results in the system of equations

$$\sin^2\vartheta\,\Phi_{,\vartheta}\,d\varphi = \Phi_{,\varphi}\,d\vartheta \qquad r^2\Phi_{,r}\,d\vartheta = \Phi_{,\vartheta}\,dr \qquad r^2\sin^2\vartheta\,\Phi_{,r}\,d\varphi = \Phi_{,\varphi}\,dr$$

(iv) for the dipole field from (2.6.10) this reduces to

$$2r\cos\vartheta\,d\vartheta = \sin\vartheta\,dr \qquad d\varphi = 0$$

the solution being

$$r(\vartheta) = r_{\max}\sin^2\vartheta \qquad \varphi(\vartheta) = \text{constant}$$

Hint: (i) according to (2.6.10) we have $\dot{\mathbf{r}} = \mathbf{E}$, i.e. $d\mathbf{r} = \mathbf{E}\,dt \Rightarrow \mathbf{E}\times d\mathbf{r} = \mathbf{0}$; (ii) (8.5.8).

□

8.6 Functions of complex variables

• In this section we will try to indicate briefly how the differential forms (and, in particular, Stokes' theorem) are hidden in the machinery of elementary complex analysis (the theory of functions of a complex variable).

In this theory one deals with complex valued functions of a complex variable; that is to say one studies maps $\mathbb{C} \to \mathbb{C}$. The complex plane $\mathbb{C}$, where such functions are defined, may be regarded as the (two-dimensional real) manifold $\mathbb{R}^2$, the ordinary Euclidean plane. In this plane now consider $\mathbb{C}$-valued functions, which we will treat as a particular case of the "V-valued forms" discussed in Section 6.4. To summarize, we contemplate 0-forms on $\mathbb{R}^2$ with values in the *algebra* $\mathbb{C}$. Since the vectors $1, i$ constitute the (most natural) basis in this algebra, we may express an arbitrary function (0-form) in terms of this basis and corresponding *component* functions (0-forms)

$$f = u(x, y) + iv(x, y) \qquad f : \mathbb{C} \to \mathbb{C}$$

Particular (and important) examples are provided by the "coordinate functions" themselves

$$z = x + iy \qquad \bar{z} = x - iy$$

Consider 1-forms now. Each $\mathbb{C}$-valued 1-form may be written as

$$\sigma = \alpha + i\beta$$

with the component forms α and β being already "ordinary" ($\mathbb{R}$-valued) 1-forms.

In general, V-valued 1-forms may be expressed (for more details see Section 6.4) as

$$\sigma = \sigma^A E_A \equiv \sigma^A_\mu\,dx^\mu\,E_A$$

so that the V-valued 1-forms $dx^\mu\,E_A$ may serve as a "basis" in the space of such forms (in the sense of generators of a *module* over the algebra of functions, see the text after (2.2.12)). In the particular case treated here the ($\mathbb{C}$-valued) 1-forms

$$dx, dy, i\,dx, i\,dy$$

may be used as the generators (over the algebra of *real* functions). Instead of these four generators, *just two* other (properly chosen) generators would be enough, however, if we extend the algebra of ($\mathbb{R}$-valued) functions to the algebra of $\mathbb{C}$-valued functions (if the number of generators is reduced, the set of coefficients is to be extended in order to reach the same goal).

8.6.1 Introduce two $\mathbb{C}$-valued 1-forms on $\mathbb{R}^2$ as the exterior derivatives (in the sense of (6.4.4)) of the functions $z, \bar{z}$, i.e.

$$dz = dx + i\,dy \qquad d\bar{z} = dx - i\,dy$$

Check that an *arbitrary* $\mathbb{C}$-valued 1-form on $\mathbb{R}^2$ may be decomposed with respect to this basis (set of generators), if $\mathbb{C}$-*valued* functions are allowed as coefficients

$$\sigma = f dz + \hat{f} d\bar{z} \qquad f, \hat{f} : \mathbb{C} \to \mathbb{C}$$

Hint: $\sigma = A(x,y)\,dx + B(x,y)\,dy + C(x,y)\,i\,dx + D(x,y)\,i\,dy \overset{?}{=} (u(x,y) + iv(x,y))\,dz + (\hat{u}(x,y) + i\hat{v}(x,y))\,d\bar{z}$. □

- Since for a general 1-form the functions $f, \hat{f}$ are *arbitrary* (smooth) functions in the xy plane, *both* coordinates z *and* $\bar{z}$ are needed in order to rewrite the functions in terms of the variables $z, \bar{z}$ rather than x, y (for example, $x^2 + y^2 = z\bar{z}$, so that neither z nor $\bar{z}$ alone is enough). Put another way, the expression of a general (smooth) $\mathbb{C}$-valued 1-form in the xy plane reads in more detail as

$$\sigma = f(z,\bar{z})\,dz + \hat{f}(z,\bar{z})\,d\bar{z}$$

so that it may be parametrized by *two* independent complex functions of *both* complex coordinates $z, \bar{z}$,

$$f(z,\bar{z}) = u(x,y) + iv(x,y) \qquad \hat{f}(z,\bar{z}) = \hat{u}(x,y) + i\hat{v}(x,y)$$

Such 1-forms may be integrated (as all 1-forms may) over 1-chains; here, in particular, over the curves in the plane.[133]

Now recall that usually *simpler* expressions are found under the integral sign in complex analysis, however; namely, the expressions $f(z)\,dz$. A question then arises as to what exactly is special about just these forms, i.e. about the 1-forms σ such that the part $\hat{f}(z,\bar{z})\,d\bar{z}$ is absent at all and, moreover, the coordinate $\bar{z}$ is missing in the only function which remains, so that f happens to be a *holomorphic function*.[134]

It turns out that the first specific property mentioned above is related to the concept of "(anti-)self-duality" (to be explained in the next exercise; here it manifests itself in the fact that the two real 1-forms α, β cease to be independent) and the second one adds simply the *closedness* of both α and β (this concept is essential for Chapter 9; a form is said to be *closed* if its exterior derivative vanishes).

[133] The integral is understood in the sense of (7.6.10), its result being a *complex* number.

[134] Recall (4.6.19) that for a holomorphic function there holds $(\partial_x + i\partial_y)f = 0$, which may also be written as $\partial_{\bar{z}} f = 0$, and that this condition is equivalent to the validity of the *Cauchy–Riemann relations* $\partial_x u = \partial_y v,\ \partial_x v = -\partial_y u$ for the function u, v.

8.6.2 A p-form α on an n-dimensional manifold (M, g, o) is called (anti-)*self-dual* if it is an eigenvector of the Hodge operator $*$, i.e. if

$$*\alpha = \lambda\alpha \qquad \lambda = \text{ the eigenvalue (a number)}$$

Check that

(i) α is necessarily a form of the "middle degree" on an *even*-dimensional manifold ($n = 2p$)
(ii) the eigenvalue λ may be only ± 1 or $\pm i$; namely, for the metric tensor with signature (r, s) it is

$$\lambda = \pm 1 \qquad \text{if } s + p = \text{ even}$$
$$\lambda = \pm i \qquad \text{if } s + p = \text{ odd}$$

($\Rightarrow$ the *real* self-dual (non-vanishing) forms may occur *only* in the case where $s + p =$ even; for odd $s + p$ we necessarily need $\mathbb{C}$-valued forms)
(iii) an arbitrary p-form on a $2p$-dimensional manifold may be uniquely decomposed into the sum of the *self-dual* ($\lambda = +1$ or $+i$) and the *anti-self-dual* ($\lambda = -1$ or $-i$) parts
(iv) if an (anti-)self-dual form happens to be closed, it is then automatically also *coclosed* and *harmonic*

$$d\alpha = 0 \quad \Rightarrow \quad \delta\alpha = 0 \text{ and } \Delta\alpha \equiv -(\delta d + d\delta)\alpha = 0$$

Hint: (i) $\deg \alpha \equiv p = \deg *\alpha \equiv n - p$; (ii) $** = \cdots$ (5.8.2); (iii) there holds $\alpha = (P_+ + P_-)\alpha = \alpha_+ + \alpha_-$, where the projection operators to the self-dual and anti-self-dual parts are given by

$$P_+ := \frac{1}{2}(1 + *) \qquad P_- := \frac{1}{2}(1 - *) \qquad \text{for } s + p = \text{ even}$$
$$P_+ := \frac{1}{2}(1 - i*) \qquad P_- := \frac{1}{2}(1 + i*) \qquad \text{for } s + p = \text{ odd}$$

□

8.6.3 Consider the standard metric tensor and orientation in $\mathbb{C} \equiv \mathbb{R}^2$ (we are thus indeed in the Euclidean plane) and the corresponding Hodge operator $*$. Check that

(i) for the basis $dz, d\bar{z}$ there holds

$$*dz = -i\, dz \qquad *\, d\bar{z} = i\, d\bar{z}$$

so that the form dz happens to be anti-self-dual and $d\bar{z}$ is self-dual
(ii) the presentation (8.6.1) of the most general 1-form just corresponds to the sum of its anti-self-dual and self-dual parts

$$\sigma = f\, dz + \hat{f}\, d\bar{z} \qquad \Rightarrow \qquad f\, dz \text{ is the anti-self-dual part}$$
$$\Rightarrow \qquad \hat{f}\, d\bar{z} \text{ is the self-dual part}$$

(iii) if we restrict to the anti-self-dual 1-form $f\, dz$ and associate the component 1-forms α and β (the "real and imaginary parts") with it

$$f(z, \bar{z})\, dz = \alpha + i\beta$$

then α and β happen to be closely related, namely

$$\beta = *\alpha \qquad \text{i.e.} \quad f(z)\,dz = \alpha + i * \alpha$$

Hint: (i) $*dx = dy,\ *dy = -dx$; (ii) $*f\,dz = f * dz$; (iii) combine (i) and $f\,dz = \alpha + i\beta$. □

• In Section 4.6 we learned that the *holomorphic functions* play an important role in treating the *conformal* mappings of the plane. Now we encounter another important situation where these functions are essential: for holomorphic $f(z)$ the 1-form $f(z)\,dz$ happens to be *closed* (with all subsequent consequences).

8.6.4 Consider a $\mathbb{C}$-valued 1-form of the form $\sigma = f(z)\,dz$ in the plane $\mathbb{C}$, so that in comparison with the general case its self-dual part vanishes and, moreover, f is holomorphic. Check that

(i) this form is *closed*

$$d(f(z)\,dz) = 0 \qquad \text{i.e. for the component forms} \quad d\alpha = 0 = d\beta$$

(ii) the form α is, in addition, *coclosed* and *harmonic*

$$\delta\alpha = 0 \qquad \Delta\alpha = 0 \quad (\Delta = -(\delta d + d\delta))$$

(iii) the functions u and v themselves turn out to be harmonic, too

$$\Delta u = 0 = \Delta v$$

Hint: (i) $d(f\,dz) = d\{(u + iv)(dx + i\,dy)\} = \cdots = 0$ due to the Cauchy–Riemann relations (or alternatively $d(f(z)\,dz) = f'(z)\,dz \wedge dz = 0$); (ii) the combination with the anti-self-duality: $0 = d\beta = d * \alpha \Rightarrow \delta\alpha = 0$; (iii) Cartesian coordinates. □

• If we now apply Stokes' theorem to the closed 1-form $\sigma = f(z)\,dz$, we immediately arrive at the important *Cauchy theorem*.

8.6.5 Let f be a holomorphic function in the domain $D \subset \mathbb{C}$. Show that then[135]

$$\oint_{\partial D} f(z)\,dz \equiv \oint_{\partial D} \alpha + i \oint_{\partial D} \beta = 0$$

Hint: $\oint_{\partial D} \alpha = \int_D d\alpha = 0$ due to (8.6.4). □

Summary of Chapter 8

One often encounters the general Stokes' theorem for differential forms from Chapter 7 as hidden behind one of its numerous classical versions. Here we demonstrate this, in

[135] If there is a *pole* in the domain D, the integral over the "outer" boundary may be expressed in terms of the *residue* at the pole. This procedure is *also* based on the application of Stokes' theorem, namely to a modified domain $\tilde{D} = D$ minus the infinitesimal disk centered at the pole. For this modified domain the theorem again holds (remember that the form failed to be smooth if f had a pole) and it enables one to replace the integral over the outer boundary by the integral over the boundary of the small disk (∂D consists of the outer boundary plus the boundary of the disk, the total integral over ∂D being zero).

particular, for the divergence (Gauss') theorem, Green's identities, the "common" Stokes' theorem known from vector analysis or some well-known facts from elementary complex analysis. The codifferential δ is introduced (as the operator adjoint to the differential $d =$ exterior derivative) and the self-adjoint combination $\Delta = -(d\delta + \delta d)$, the Laplace–deRham operator (a generalization of the Laplace operator on functions to forms of arbitrary degree). In the section devoted to standard vector analysis we learn that the essence of the well-known operations of gradient, curl and divergence is simply the exterior derivative acting on forms of all non-trivial degrees in three-dimensional space.

$\mathcal{L}_V \omega_g =: (\text{div } V)\, \omega_g$	Definition of the divergence of V	(8.2.1)
$\text{div } V = \frac{1}{\sqrt{\mid g \mid}} (\sqrt{\mid g \mid}\, V^k)_{,k}$	Coordinate expression of div V	(8.2.1)
$\langle \text{div } V \rangle_D = \left.\frac{d}{dt}\right\rvert_{t=0} \frac{\text{vol } D(t)}{\text{vol } D}$	Interpretation of div V	(8.2.2)
$\langle \text{div } V \rangle_D = \frac{\text{the flux of } V \text{ for } \partial D}{\text{the volume of } D}$	Another interpretation of div V	(8.2.9)
$\int_D (\text{div } V)\, \omega_g = \int_{\partial D} V^i d\Sigma_i \rvert_{\partial D}$	Gauss' theorem	(8.2.7)
$\langle \alpha, \beta \rangle := \int_D \alpha \wedge *\beta$	Scalar product of forms on (M, g)	(8.3.1)
$\delta := *^{-1} d * \hat{\eta}$	Definition of the codifferential δ	(8.3.2)
$\langle d\alpha, \beta \rangle = \langle \alpha, \delta\beta \rangle + \int_{\partial D} \alpha \wedge *\beta$	Basic property of the codifferential δ	(8.3.2)
$\Delta := -(\delta d + d\delta) \equiv -(d^+ d + d d^+)$	Laplace–deRham operator	(8.3.3)
$\Delta f = -\delta d f \equiv \frac{1}{\sqrt{\mid g \mid}} (\sqrt{\mid g \mid} g^{kj} f_{,j})_{,k}$	Laplace–Beltrami operator	(8.3.5)
$\langle du, dv \rangle + \langle u, \Delta v \rangle = \int_{\partial D} u * dv$	"Ordinary" Green identity	(8.4.1)
$\langle u, \Delta v \rangle - \langle v, \Delta u \rangle = \int_{\partial D} (u * dv - v * du)$	"Symmetric" Green identity	(8.4.1)
$f,\ \mathbf{A} \cdot d\mathbf{r},\ \mathbf{B} \cdot d\mathbf{S},\ h\, dV$	Differential forms on E^3	(8.5.2)
$d(\mathbf{A} \cdot d\mathbf{r}) = (\text{curl } \mathbf{A}) \cdot d\mathbf{S}$	A definition of curl $\mathbf{A}$	(8.5.4)
$(\mathbf{A} \cdot d\mathbf{r}) \wedge (\mathbf{B} \cdot d\mathbf{r}) = (\mathbf{A} \times \mathbf{B}) \cdot d\mathbf{S}$	How the vector (cross) product appears	(8.5.8)
$g = h_1^2 dx^1 \otimes dx^1 + \cdots$	Lamé coefficients	(8.5.9)
$d(f(z)\, dz) = 0$	Why the Cauchy theorem holds	(8.6.5)

9

Poincaré lemma and cohomologies

• One of the most remarkable properties of the exterior derivative operator

$$d : \Omega^p(M) \to \Omega^{p+1}(M)$$

is its *nilpotence* (6.2.5), the property $dd = 0$. So if some form α is the differential of another form β, then the differential of the form α necessarily vanishes

$$\alpha = d\beta \quad \Rightarrow \quad d\alpha = 0$$

However, the issue of the possible validity of the *converse* of this statement (the opposite implication) is an independent problem, with an a priori unclear solution (in no way does it result directly from $dd = 0$)

$$d\alpha = 0 \quad \overset{?}{\Rightarrow} \quad \alpha = d\beta$$

In this chapter we adopt the standard terminology which is widely used in this context. Namely, we introduce the concepts of a *closed form* (such α that $d\alpha = 0$) and an *exact form* (α such that $\alpha = d\beta$ for some form β; the form β is then called the *potential* of the form α). The formulation of the statements in terms of these concepts then reads

$$\begin{array}{lcll} \alpha \text{ is exact} & \Rightarrow & \alpha \text{ is closed} & (dd = 0) \\ \alpha \text{ is closed} & \overset{?}{\Rightarrow} & \alpha \text{ is exact} & (\text{the converse of } dd = 0) \end{array}$$

and the key question is: does a potential always exist for a closed form? If not, under what conditions is its existence guaranteed?

In physics there are numerous situations in which it is tacitly assumed that the *converse* of $dd = 0$ also holds.

In mechanics, as an example, one infers the existence of the potential energy from the vanishing of the work done by a force on an object along an arbitrary closed path C, so the deduction is

$$\oint_C \mathbf{F} \cdot d\mathbf{r} = 0 \quad \Rightarrow \quad \mathbf{F} = -\nabla U$$

In thermodynamics, in a similar way, one often adopts the following reasoning: since the integral of a 1-form along an arbitrary cycle C vanishes, the *state quantity* necessarily exists, i.e. such a *function* (0-form) whose differential just equals the 1-form under consideration. In symbols

$$\oint_C (\dots) = 0 \quad \Rightarrow \quad (\dots) = d(\text{state quantity})$$

It turns out that the converse of $dd = 0$ is behind both of these situations. For if D is an arbitrary two-dimensional domain, we may write for its boundary ∂D

$$0 = \oint_{\partial D} (\dots) = \int_D d(\dots) \quad \overset{(7.4.1)}{\Rightarrow} \quad d(\dots) = 0 \quad \overset{?}{\Rightarrow} \quad (\dots) = d(\dots)'$$

Finally, the standard reasoning in electrodynamics (see Section 16.3 for more details) contains the deductions of the existence of both vector and scalar potentials from the structure of a part of the equations of motion (Maxwell equations):

$$\operatorname{div} \mathbf{B} = 0 \quad \Rightarrow \quad \exists \mathbf{A} \text{ such that } \mathbf{B} = \operatorname{curl} \mathbf{A}$$

$$\operatorname{curl} (\mathbf{E} + \partial_t \mathbf{A}) = \mathbf{0} \quad \Rightarrow \exists \Phi \text{ such that } \mathbf{E} = -\nabla \Phi - \partial_t \mathbf{A}$$

In light of (8.5.4) one easily sees that here also the converse of $dd = 0$ is the key element.

In this chapter we first convince ourselves using simple examples that the converse of $dd = 0$ *cannot* be true *in general*. This is a rather frustrating piece of knowledge, but fortunately we will then learn that under some circumstances (on topologically sufficiently simple manifolds) the statement actually *is* true (which restores our equanimity again). In this important case (a manifold contractible to a point)[136] the converse of $dd = 0$ is known as the *Poincaré lemma* and we will derive the explicit formula for computation of the potential for a given closed form. After this we adopt the algebraic perspective and learn what role cohomology theory plays in this context.

9.1 Simple examples of closed non-exact forms

• A whole family of examples of forms, which *contradict* the converse of $dd = 0$, is provided by the following general scheme.

9.1.1 Let (M, ω) be a *compact* n-dimensional manifold ($\Rightarrow \partial M = \emptyset$) endowed with a volume form ω. Show that

(i) the integral of an arbitrary *exact* n-form over the whole manifold M necessarily vanishes
(ii) the form ω is *closed* and its integral over M is non-zero, so that the form ω may serve as a counterexample to the general validity of the converse of $dd = 0$.

Hint: if the n-form σ is exact, $\sigma = d\beta$, then

$$\int_M \sigma = \int_M d\beta = \int_{\partial M} \beta = \int_\emptyset \beta = 0.$$

[136] One also often speaks about the *local* validity of the converse of $dd = 0$: it is true in some *neighborhood* of any point (however complicated the manifold is).

The form ω is *trivially* closed (it is of maximum degree for non-vanishing forms). The volume of the manifold is by definition (see Section 7.7) $\int_M \omega$ and it is clearly non-zero by the meaning of the concept of volume. □

• Another simple concrete example is given by an appropriate 1-form on the circle $S^1 \subset \mathbb{R}^2$, or more generally on the spheres $S^n \subset \mathbb{R}^{n+1}$.

9.1.2 Consider as a manifold the *circle* $M = S^1 = \partial D^2 \subset \mathbb{R}^2$, where $D^2 = \{(x, y) \in \mathbb{R}^2 \mid x^2 + y^2 \leq 1\}$ is the unit *disk*. Show that

(i) the 1-form $\alpha := (x\,dy)|_{S^1}$ has a non-vanishing integral over $S^1 \Rightarrow$ it is a closed albeit non-exact form on S^1

(ii) the same is true for the form $\hat{\alpha} := (x\,dy - y\,dx)|_{S^1}$, being now the special case of (9.1.1)

(iii) in terms of the standard local coordinate φ we find

$$\alpha = \cos^2\varphi\, d\varphi \qquad \hat{\alpha} = d\varphi$$

(iv) the form $\hat{\alpha}$ *is not* exact in spite of the way it is written.

Hint: (i) $0 \neq \pi = \int_{D^2} d(x\,dy) = \int_{S^1} \alpha$, similarly for $\hat{\alpha}$; (ii) $\hat{\alpha} = \omega_g$ for $g = d\varphi \otimes d\varphi$; (iv) $\varphi \notin \mathcal{F}(S^1)$ (it is not continuous at the point $\varphi = 0 \sim 2\pi$). □

9.1.3 Generalize item (i) from problem (9.1.2) to $S^2 = \partial D^3 \subset \mathbb{R}^3, \ldots, \; S^n = \partial D^{n+1} \subset \mathbb{R}^{n+1}$.

Hint: $\alpha := (x\,dy \wedge dz)|_{S^2}$ and $\alpha := (x^1\,dx^2 \wedge \cdots \wedge dx^{n+1})\big|_{S^n}$. □

• Thus the converse of $dd = 0$ is indeed not always valid and it fails to hold even on such simple manifolds, like spheres or tori. In Section 9.2 we will learn that a class of manifolds where in contrast the converse of $dd = 0$ *does hold* is given by the manifolds which are *contractible to a point*,[137] like $\mathbb{R}^n$ or all the manifolds diffeomorphic to $\mathbb{R}^n$, the most important representative being a *neighborhood* (e.g. a coordinate patch) of an arbitrary point on a (general) manifold.

9.2 Construction of a potential on contractible manifolds

• In this section we derive the explicit formula for a potential of an arbitrary closed form on a particular class of manifolds, namely on manifolds contractible to a point. A manifold M is said to be (smoothly) *contractible* to a point, if there exists a vector field ξ on the manifold such that the flow $\Phi_t \leftrightarrow \xi$ gradually shrinks the manifold to the point $x_0 \in M$, i.e. it fulfills

$$\Phi_0 = \mathrm{id}_M \qquad \lim_{t\to\infty} \Phi_t(x) = x_0$$

where the point x is arbitrary and x_0 is *fixed* (it does not depend on x). Put another way, the flow Φ_t "begins" (at $t = 0$) with the identity, it is then, however, deformed more and more

[137] Also indirectly by this that the spheres S^n and the tori T^n *are not* contractible to a point.

and finally it "ends" (at $t = \infty$) with the *trivial* map, which sends the whole manifold to one of its points x_0.[138] Each point x gradually moves to x_0 along a smooth curve $\gamma(t) = \Phi_t(x)$ (the integral curve of the field ξ), which lies completely in M.

9.2.1 Show that the manifold $\mathbb{R}^n$ is indeed contractible to a point.

Hint: the generator of the flow is, for example, $\xi = -x^i \partial_i$ (in Cartesian coordinates), the flow itself (the shrinking homotopy) is $\Phi_t : x^i \mapsto e^{-t} x^i$ and it shrinks the whole $\mathbb{R}^n$ into the origin. By a translation (i.e. by the field $\xi = -(x^i - c^i)\partial_i$) one can achieve the shrinking into any other point c^i. □

9.2.2 Show that an arbitrary manifold which is *diffeomorphic* to $\mathbb{R}^n$ (in particular, an *open neighborhood* of an arbitrary point x on an arbitrary, possibly non-contractible, manifold) is also contractible to a point.

Hint: if $f : \mathbb{R}^n \to M$ is a diffeomorphism, then $\hat{\Phi}_t := f \circ \Phi_t \circ f^{-1}$ shrinks M into $f(0)$; its generator is according to (4.1.9) the field $\xi = f_*(-x^i \partial_i)$. □

- In the next two problems we (constructively) show that on a contractible manifold each closed form has a potential, i.e. that here the converse of $dd = 0$ *does hold*.

9.2.3 Let M be contractible to a point, the shrinking homotopy (the flow Φ_t) being generated by the vector field ξ. Define the *homotopy operator* by[139]

$$\hat{h} : \Omega^p(M) \to \Omega^{(p-1)}(M) \qquad \hat{h}\alpha := -\int_0^\infty dt\, \Phi_t^* i_\xi \alpha$$

Check that the following identity holds:

$$d \circ \hat{h} + \hat{h} \circ d = \Phi_0^* - \Phi_\infty^* = \hat{1} \equiv id_{\Omega^p(M)}$$

Hint: there holds $d\hat{h}\alpha = -\int_0^\infty dt\, \Phi_t^* d i_\xi \alpha$ and so making use of (6.2.8) and (4.4.2) we get

$$\begin{aligned}(d\hat{h} + \hat{h}d)\alpha &= -\int_0^\infty dt\, \Phi_t^*(di_\xi + i_\xi d)\alpha = -\int_0^\infty dt\, \Phi_t^* \mathcal{L}_\xi \alpha \\ &= -\int_0^\infty dt \frac{d}{dt}(\Phi_t^* \alpha) = -(\Phi_\infty^* \alpha - \Phi_0^* \alpha)\end{aligned}$$

Now $\Phi_0^* = \mathrm{id}_M^* = id_{\Omega^p(M)}$ and one easily verifies (in coordinates, say; alternatively: the pull-back is actually from a *point*!) that $\Phi_\infty^* \alpha = 0$. □

[138] If some one-parameter family of maps $f_t : M \to N$ smoothly joins two maps f_{t_1} and f_{t_2} (both of them being from M to N), it is called (smooth) *homotopy* and f_{t_1} and f_{t_2} themselves are called (smoothly) *homotopic maps* (they may be smoothly deformed into one another). On a contractible manifold the identity map is homotopic to the trivial one (the homotopy is provided by the flow Φ_t).

[139] The convergence of this integral is not evident. This "problem" does not occur if one adopts a more general approach (a clear (and warmly recommended) exposition may be found, for example, in the book by Flanders), where the contractibility itself is introduced in a slightly different way (it may not be realized by a flow and it lasts only a *finite time*, so that the issue of the convergence of the integral becomes trivial). The computation is a bit lengthier, however, and that was the reason we preferred the approach given here.

[9.2.4] Prove that on a contractible manifold the *Poincaré lemma*[140] holds: each closed form is necessarily exact.

Hint: if $d\alpha = 0$, then due to (9.2.3) $\alpha = d(\hat{h}\alpha) \equiv d\beta$. □

• The form of the miraculous homotopy operator $\hat{h} := -\int_0^\infty dt\, \Phi_t^* i_\xi$, which indeed did a very good job, appeared all at once with no attempt at motivation or derivation (we merely plucked the operator out of thin air and it is not clear at all why it looks as it does and does what it does). At the end of the section we sketch a simple geometrical picture which is hidden behind the formula and which leads by a straightforward computation to its structure presented here (9.2.13).

[9.2.5] Let M be an arbitrary manifold (not necessarily contractible). Show that

(i) a potential β of an exact form α ($\alpha = d\beta$) is not given uniquely, but rather one has the freedom of performing a "gauge transformation"

$$\beta \mapsto \beta' = \beta + \rho \qquad \rho \text{ closed}$$

(ii) in particular, on a contractible manifold the freedom is

$$\beta \mapsto \beta' = \beta + d\sigma \qquad \sigma \text{ arbitrary}$$

(iii) $\beta \sim \beta'$ is the equivalence on the set of all potentials of a given (exact) form α.

Hint: $dd = 0$, see (9.2.4). □

• A condition which is imposed on a potential and which *is not* invariant with respect to the transformations discussed in (9.2.5) (so that *not all* potentials equivalent to each other meet this condition), is called a *gauge condition*.[141]

[9.2.6] Check that

(i) the potential β given by the formula $\beta = \hat{h}\alpha$ from the problem (9.2.4) satisfies the gauge condition (not very often used in physics)

$$i_\xi \beta = 0$$

(ii) the conditions

$$d\beta = \alpha \qquad i_\xi \beta = 0$$

still do not fix the potential β uniquely.

Hint: the freedom $\beta \mapsto \beta' = \beta + d\sigma$ remains, but only with $d\sigma$ such that $i_\xi\, d\sigma = 0$ holds. □

[140] The "reversed" terminology may also be occasionally found in the literature: the statement $dd = 0$ is called the Poincaré lemma and what we call the Poincaré lemma is then said to be "the converse" of the Poincaré lemma.

[141] It is also known as "fixing of the gauge"; the fixation may not be complete, in fact, still leaving some, although restricted, freedom.

• Let us now compute the explicit coordinate expression of the potential given by the "abstract" formula (9.2.4) for the case of shrinking of $\mathbb{R}^n$ as discussed in (9.2.1).

9.2.7 Given the field $\xi = -x^i \partial_i$ on $\mathbb{R}^n$ check that

(i) if the components with respect to Cartesian coordinates of a (closed) p-form α in $\mathbb{R}^n$ are $\alpha_{ki\dots j}(x)$, then (for $p \geq 1$)

$$(\Phi_t^* i_\xi \alpha)_{i\dots j}(x) = -(e^{-t})^p x^k \alpha_{ki\dots j}(e^{-t}x)$$

(ii) the component formula for the potential β then reads

$$\beta_{i\dots j}(x) \equiv (\hat{h}\alpha)_{i\dots j}(x) = x^k \int_0^\infty dt\, (e^{-t})^p \alpha_{ki\dots j}(e^{-t}x)$$
$$= x^k \int_0^1 d\lambda\, \lambda^{p-1} \alpha_{ki\dots j}(\lambda x)$$

(iii) here the gauge condition from problem (9.2.6) reduces to

$$x^i \beta_{i\dots j}(x) = 0$$

Hint: use the field ξ from (9.2.1). □

• Let us see now what the general formulas yield in some simple situations mentioned at the beginning of the chapter. Namely we compute, making use of these formulas, the potential energy $U(\mathbf{r})$ for a given (conservative) force field $\mathbf{F}(\mathbf{r})$ as well as the scalar and vector potentials $(\Phi, \mathbf{A})$ for given electric and magnetic fields $(\mathbf{E}, \mathbf{B})$.

9.2.8 Let $\mathbf{F}(\mathbf{r})$ be a conservative force (vector) field in $\mathbb{R}^3$ and let $\mathbf{F} \cdot d\mathbf{r}$ be the corresponding covector field (in the sense of (8.5.1)). According to the introduction to this chapter there thus holds $d(\mathbf{F} \cdot d\mathbf{r}) = 0$. Check that

(i) in "vector notation" this is the condition $\operatorname{curl} \mathbf{F} = \mathbf{0}$

(ii) here the formula for the potential from (9.2.7) leads to the well-known result

$$\mathbf{F} = -\nabla U \qquad U(\mathbf{r}) = -\mathbf{r} \cdot \int_0^1 d\lambda\, \mathbf{F}(\lambda \mathbf{r}) \equiv -\int_0^{\mathbf{r}} \mathbf{F}' \cdot d\mathbf{r}' \quad \mathbf{F}' \equiv \mathbf{F}(\mathbf{r}')$$

where the integral on the right is to be performed along the straight line[142] from the origin to the point $\mathbf{r}$ (and it has the physical meaning of minus the *work done by the force* $\mathbf{F}$ on an object along the path from $\mathbf{0}$ to $\mathbf{r}$)

(iii) here the freedom from problem (9.2.5) yields

$$U'(\mathbf{r}) = U(\mathbf{r}) + \text{a constant function}$$

Hint: (i) (8.5.4); (ii) $\mathbf{r}(\lambda) = \mathbf{r}\lambda$. □

9.2.9 Check that

[142] However, just the condition $d(\mathbf{F} \cdot d\mathbf{r}) = 0$ guarantees that the same result would be obtained if the integral were computed along any other curve sharing its ends with the straight line discussed above.

(i) the electrostatic equation curl $\mathbf{E} = \mathbf{0}$ leads to the scalar potential $\Phi(\mathbf{r})$ and that (9.2.7) gives

$$\mathbf{E} = -\nabla\Phi \qquad \Phi(\mathbf{r}) = -\int_0^{\mathbf{r}} \mathbf{E}' \cdot d\mathbf{r}'$$

(ii) the magnetostatic equation div $\mathbf{B} = 0$ (it is valid in electro*dynamics* as well) leads to the vector potential $\mathbf{A}$ and that (9.2.7) gives

$$\mathbf{B} = \operatorname{curl}\mathbf{A} \qquad \mathbf{A}(\mathbf{r}) = -\mathbf{r} \times \int_0^1 \lambda\, d\lambda\, \mathbf{B}(\lambda\mathbf{r}) \equiv -\frac{1}{r}\int_0^{\mathbf{r}} d\mathbf{r}' \times r'\mathbf{B}' \qquad \mathbf{r}\cdot\mathbf{A} = 0$$

Hint: (9.2.8), (8.5.4); $A_i(\mathbf{r}) = x^k \int_0^1 \lambda\, d\lambda\, B_{ki}(\lambda\mathbf{r}),\ B_{ki} = \varepsilon_{kij}B^j$. □

• Recall that we have to square up to the reader concerning a debt from Chapter 4. We spoke about non-holonomic frames there (see, for example, (4.5.7)) and it turned out that the *necessary* condition that the frame e_a is holonomic (coordinate) consists in the vanishing of all the mutual commutators. Now we have mastered the formalism which enables us to cope with the problem easily and to prove that this condition is actually both necessary and *sufficient*.

9.2.10 The *coefficients of anholonomy* $c^c_{ab}(x)$ of the frame field e_a are defined by the relations

$$[e_a, e_b] = c^c_{ab}(x)e_c$$

Show that

(i) in spite of their suggestive graphical form they *do not form* the components of a tensor

(ii) for the dual coframe field the following equations hold:

$$de^a + \frac{1}{2}c^a_{bc}(x)e^b \wedge e^c = 0$$

Hint: (i) the vanishing or non-vanishing of the components of a tensor does not depend on the choice of the frame and for the *coordinate* frame field the coefficients vanish; (ii) $de^a(e_b, e_c) = \cdots$ by means of Cartan formulas (6.2.13). □

9.2.11 Prove the criterion for a frame field e_a to be holonomic (coordinate):

$$e_a = \partial_a \quad \Leftrightarrow \quad [e_a, e_b] = 0$$

Hint: to the right: once upon a time (see before (4.5.7), trivial); to the left: $c^c_{ab}(x) = 0 \Rightarrow$ due to (9.2.10) $de^a = 0 \Rightarrow$ according to the Poincaré lemma there are functions (0-forms) x^a such that $e^a = dx^a$; since e^a is a (co)frame field, we have

$$e^1 \wedge \cdots \wedge e^n \equiv dx^1 \wedge \cdots \wedge dx^n \neq 0$$

$\Rightarrow$ the functions x^a happen to be independent and we may use them as coordinates. The dual basis is then $e_a = \partial_a$. □

• It turns out to be useful to realize the following simple fact concerning the *co*differential.

9.2.12 Prove that (also) for the codifferential (8.3.2) the analog of the Poincaré lemma holds: from the *coclosedness* ($\delta_g \alpha = 0$) there locally follows the *coexactness* ($\alpha = \delta_g \beta$).

Hint: $*^{-1} d * \hat{\eta} \alpha = 0 \Rightarrow d * \alpha = 0 \Rightarrow *\alpha = d\sigma \equiv d * \hat{\eta} \beta \Rightarrow \alpha = *^{-1} d * \hat{\eta} \beta$. □

• We will return now to the homotopy operator introduced in (9.2.3) and explain (as promised there) its geometrical origin. Let us contemplate a slightly more general situation where the flow Φ_t corresponding to the vector field ξ does not shrink the manifold M necessarily to a point, but possibly only to some submanifold $N \subset M$. (The point is a special case. For example, the field

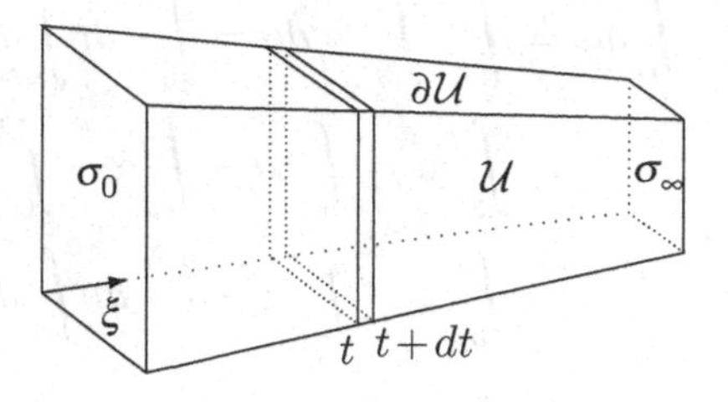

$$\xi = (1-r)\partial_r$$

shrinks E^3 without the origin only to the unit sphere $S_1^2 \subset E^3$.) If there is a p-simplex $\sigma \equiv \sigma_0$ on M, it will move on M due to the flow Φ_t and it will end as $\sigma_\infty \equiv \Phi_\infty(\sigma)$. This creates a "full tube" $\mathcal{U}$. It is bounded by σ from the left and by $\Phi_\infty(\sigma)$ from the right, and the side faces are formed from the integral curves of the field ξ, emanating from the boundary $\partial\sigma$. Our aim now is to compare the integral of an arbitrary p-form α over the end image of the simplex with the integral over the initial simplex itself, i.e. to compare the integrals $\int_{\Phi_\infty(\sigma)} \alpha$ and $\int_\sigma \alpha$.[143] Since both $\Phi_\infty(\sigma)$ and σ are parts of the *boundary* of $\mathcal{U}$, both the integrals occur in *Stokes' theorem* written for the form α and the domain $\mathcal{U}$. In addition to the longed-for two integrals, Stokes' theorem appends two more terms, one of them being a "volume" integral over $\mathcal{U}$ and also a "surface" integral over the "side faces" of the boundary $\partial\mathcal{U}$. So we are expected to be able to compute these two integrals. The key idea lies in the observation that the tube (as well as the side faces of its boundary) may be *put together from infinitesimal slices*[144] of thickness dt (put together as $\int_0^\infty dt \ldots$). Stokes' theorem thus yields the equation in which there are two integrals, which we need, plus two additional integrals which contain the procedure of putting together the slices $\int_0^\infty dt \ldots$. The last crucial technical point is to realize that the slices may be actually regarded as the "coins" from the problem (7.6.11), so that we may profit from the "coin interpretation" of the interior product i_V.

9.2.13* Carry out the steps indicated above and convince yourself that they indeed result in

$$\int_{\sigma_\infty} \alpha = \int_{\sigma_0} \alpha - \int_{\sigma_0} (d\hat{h} + \hat{h}d)\alpha$$

or, in terms of the forms themselves (with no integrals involved),

$$\Phi_\infty^* \alpha = \alpha - (d\hat{h} + \hat{h}d)\alpha$$

143 Or, what results simply from that, the pulled-back form $\Phi_\infty^* \alpha$ and the form $\alpha \equiv \Phi_0^* \alpha$ itself.
144 Just like a piece of ham (or rather a carrot for us vegetarians) may be cut into thin slices.

where the *homotopy operator* did not fall from heaven, now (neither did we pluck it out of thin air), but rather it *comes out* just in the form

$$\hat{h} = -\int_0^\infty \Phi_t^* i_\xi \, dt$$

Hint:

$$\int_{\mathcal{U}} d\alpha \overset{1}{=} \int_0^\infty \int_{(\sigma_t)_{dt\xi}} d\alpha = \int_0^\infty dt \int_{\sigma_t} i_\xi \, d\alpha = \int_0^\infty dt \int_{\Phi_t(\sigma_0)} i_\xi \, d\alpha = \int_{\sigma_0} \int_0^\infty dt \, \Phi_t^* i_\xi \, d\alpha$$

$$\overset{2}{=} \int_{\partial\mathcal{U}} \alpha = \int_{\sigma_\infty} \alpha - \int_{\sigma_0} \alpha - \int_0^\infty \int_{(\partial\sigma_t)_{dt\xi}} \alpha = \int_{\sigma_\infty} \alpha - \int_{\sigma_0} \alpha - \int_0^\infty dt \int_{\partial\sigma_t} i_\xi \alpha$$

$$= \int_{\sigma_\infty} \alpha - \int_{\sigma_0} \alpha - \int_0^\infty dt \int_{\sigma_0} d\Phi_t^* i_\xi \alpha = \int_{\sigma_\infty} \alpha - \int_{\sigma_0} \alpha - \int_{\sigma_0} d \int_0^\infty dt \, \Phi_t^* i_\xi \alpha$$

so that we have the required equality of integrals $-\int_{\sigma_0} \hat{h} \, d\alpha = \int_{\sigma_\infty} \alpha - \int_{\sigma_0} \alpha + \int_{\sigma_0} d\hat{h} \, \alpha$. If this is written in the form $\int_{\sigma_0} (\Phi_\infty^* \alpha - \alpha + d\hat{h} \, \alpha + \hat{h} \, d\alpha) = 0$ we may infer (since the simplex σ_0 is arbitrary) the vanishing of the form under the integral sign, whence finally $\Phi_\infty^* \alpha - \alpha + d\hat{h} \, \alpha + \hat{h} \, d\alpha = 0$. □

• Now if the flow Φ_t shrinks the manifold to a *point*, the pull-back $\Phi_\infty^* \alpha$ vanishes (unless α happens to be a 0-form; the forms are then pulled back from a zero-dimensional manifold), so that only

$$\alpha = d\hat{h} \, \alpha + \hat{h} \, d\alpha$$

remains. And if at last the form α is *closed*, we finish with the result which provides the formula for the potential,

$$\alpha = d(\hat{h}\alpha) \equiv d\beta$$

9.3* Cohomologies and deRham complex

• Recall that this chapter started with the contemplation of the *relation between closed and exact forms* on a manifold. The easy part of the issue was the observation that whenever a form happens to be exact, it is necessarily also closed ($dd = 0$). The hard part concerns the converse of this statement.

There is a specialized and mature branch of mathematics called *cohomology theory* (which is a part of algebraic topology) which studies things like this as a full-time job. It thus provides a convenient language as well as sophisticated and elaborate computational machinery tailored to the very relation between the objects generalizing "closed and exact forms." In this section we say a few words about the concepts needed for understanding how cohomology theory is related to the issue of the converse of $dd = 0$.

The basic object in the theory is called a *complex* (C, d). It is a $\mathbb{Z}$-graded vector space

$$C := \bigoplus_{r=-\infty}^{\infty} C^r \equiv \cdots \oplus C^{-1} \oplus C^0 \oplus C^1 \oplus C^2 \oplus \cdots \oplus C^n \oplus \cdots$$

(often only a finite number of subspaces C^r are non-trivial), in which a family of linear maps (operators) between subspaces with "adjacent" degrees is defined

$$\dots \xrightarrow{d_{-1}} C^0 \xrightarrow{d_0} C^1 \xrightarrow{d_1} C^2 \xrightarrow{d_2} \dots \xrightarrow{d_{n-1}} C^n \xrightarrow{d_n} \dots$$

obeying moreover the condition $d_{k+1} \circ d_k = 0$. The whole family of these operators is called the *differential* d of the complex (C, d) and the condition mentioned above is often written concisely (with no indices) in the form $dd = 0$, so that the differential is by definition a *nilpotent* operator.[145]

The differential singles out two types of distinguished subspaces in C.

9.3.1 Let (C, d) be a complex. Define the subspaces

$$Z^p := \operatorname{Ker} d_p \equiv \{c \in C^p \mid dc = 0\}$$
$$B^p := \operatorname{Im} d_{p-1} \equiv \{c \in C^p \mid c = d\hat{c}; \hat{c} \in C^{p-1}\}$$

The elements of Z^p are called *p-cocycles* and those of B^p are called *p-coboundaries*. Sometimes we will use the terminology (the reason for that will be made clear in a while) *closed* and *exact* elements.[146] Check that

(i)
$$B^p \subset Z^p \subset C^p$$
(so that each coboundary is necessarily a cocycle)

(ii) $c \sim c' := c + d\hat{c}$ defines an equivalence in the space Z^p. We thus regard as being equivalent those p-cocycles which differ from each other by (at most) a p-coboundary. The elements which are equivalent in this sense are said to be *cohomologous*

(iii) the p-coboundaries are cohomologous to *zero*.

□

• Although the reader must already understand why the concept of a complex is *the* structure which is relevant in the context of the Poincaré lemma, we will formulate it explicitly.

9.3.2 Contemplate the following items until they are clear:

(i) $(\Omega(M), d)$, i.e. the space of differential forms on a manifold endowed with the exterior derivative operator constitutes a (cochain) complex

$$\Omega^0(M) \xrightarrow{d} \Omega^1(M) \xrightarrow{d} \Omega^2(M) \xrightarrow{d} \dots \xrightarrow{d} \Omega^n(M)$$

(with only degrees 0 to $n \equiv \dim M$ being represented non-trivially); this complex is called the *deRham complex* of a manifold M

[145] The differential may also act in the opposite direction (to lower the degree – see, for example, (9.3.2) and (9.3.4)); then its nilpotency looks in detail like $d_k \circ d_{k+1} = 0$ and the homogeneous subspaces are denoted by C_r.

[146] The standard convention is that in the case when the differential *raises* the degree by one unit we speak of the *cochain complex*, *cocycles* Z^p, *coboundaries* B^p and *cohomologies* H^p, whereas in the case when the differential *lowers* the degree by one unit we use the lower indices and the concepts of a *chain complex*, *cycles* Z_p, *boundaries* B_p and *homologies* H_p.

(ii) in this complex the role of p-cocycles is played by the closed p-forms and the role of p-coboundaries is played by the exact p-forms (in the sense of the beginning of the chapter)
(iii) the validity of the Poincaré lemma on M is equivalent to the statement that for each p there holds $B^p(M) = Z^p(M)$, so that all cocycles are *cohomologous to zero*
(iv) another (chain) complex related to the manifold M is the complex $(C(M), \partial)$ of its *singular chains*

$$C_0(M) \overset{\partial}{\leftarrow} C_1(M) \overset{\partial}{\leftarrow} C_2(M) \overset{\partial}{\leftarrow} \cdots \overset{\partial}{\leftarrow} C_n(M)$$

(again with only degrees 0 to n being represented non-trivially); here the subspace of degree p is the space $C_p(M)$ from Section 7.3, the space of p-chains on a manifold M and as a differential (which *lowers* the degree here) we take the *boundary operator* ∂ from problem (7.3.1). □

• According to (9.3.1) the spaces B^p and Z^p may not be identical, in general; the only thing we know a priori is that B^p is a *subspace* of Z^p. The existence of closed (but) non-exact forms, which we illustrated explicitly in problems (9.1.1)–(9.1.3) shows that (at least for the *deRham* complex) one encounters cases when B^p is a *non-trivial* subspace of Z^p ($B^p \neq Z^p$). It turns out that highly useful information is carried by the structure of the *equivalence classes* in the sense of the equivalence from (9.3.1), i.e. so-called *cohomology classes*.

9.3.3 Check that

(i) if all p-cocycles happen to be p-coboundaries ($Z^p = B^p$), there will be (for given p) only *a single* class (the class $[0]$).
(ii) if there is a *non-trivial* p-cocycle z (such that it is not a p-coboundary), then all of its (non-zero) multiples λz are also non-trivial; moreover, the multiples by *different* numbers are *inequivalent*.

Hint: (i) if $z \in Z^p = B^p$, then $z = dw$ for some $w \in C^{p-1}$. Then $z = dw = 0 + dw \Rightarrow [z] = [0]$, and this is true[147] for *each* z; (ii) $z \neq d(\dots) \;\Rightarrow\; \lambda z \neq d(\dots)'$; if $\lambda_1 z = \lambda_2 z + d(\dots)$, then $(\lambda_1 - \lambda_2)z = d(\dots) \Rightarrow z = d(\dots)'$, which is a contradiction. □

• So if there exists a non-trivial p-cocycle z, automatically we are given an infinite number of additional p-cocycles, non-equivalent to one another (non-zero multiples of z) which span altogether a one-dimensional subspace in Z^p. From the perspective of finding a useful measure of "all non-equivalent p-cocycles," however, it is not too interesting to distinguish between vectors from the same subspace, since the whole subspace (ray) may be completely reconstructed from a single representative; what really matters is the subspace alone. There may exist, of course, other subspaces of non-trivial p-cocycles (linearly independent of the first one). The relevant quantitative information which succinctly encodes the relation between the cocycles and the coboundaries (closed and exact elements) of a complex is thus given in a certain *linear space*. This space may be formally defined as the *factor*

[147] Herein the author would like to thank the Indians for inventing the concept of zero (as well as all nations, individuals and firms that have merit in putting it on the market). In this (as well as in numerous other) proof(s) it came in handy.

space[148] of the space of p-cocycles Z^p with respect to the subspace of p-coboundaries B^p

$$H^p := Z^p / B^p$$

and it is called the pth *cohomology group.*[149] The $\mathbb{Z}$-graded vector space

$$H := \bigoplus_{r=-\infty}^{\infty} H^r \equiv \cdots \oplus H^0 \oplus H^1 \oplus H^2 \oplus \cdots \oplus H^n \oplus \cdots$$

is called the *cohomology* of a complex (C, d). A complex is said to be *coexact at* C^p if $Z^p = B^p$ and *coexact* if this happens for all p. Thus the cohomology of a cochain complex is a measure of the lack of coexactness of the cochain complex. The dimension of the space H^p is known as the (pth) *Betti number* $b^p := \dim H^p$. This number simply says how many linearly independent non-trivial p-cocycles there are in Z^p (or, in particular, for the case of the *deRham* complex, how many linearly independent non-exact (albeit closed) differential p-forms there are on a manifold).

9.3.4 Two simple examples of complexes are provided by $(\Lambda L^*, i_v)$ and $(\Lambda L^*, j_v)$, i.e. the exterior algebra of a linear space, where the differential is given either by the interior product i_v (5.4.1) or by the (adjoint) operator j_v introduced in (5.8.6)

$$\Lambda^0 L^* \overset{i_v}{\leftarrow} \Lambda^1 L^* \overset{i_v}{\leftarrow} \Lambda^2 L^* \overset{i_v}{\leftarrow} \cdots \overset{i_v}{\leftarrow} \Lambda^n L^*$$

$$\Lambda^0 L^* \overset{j_v}{\rightarrow} \Lambda^1 L^* \overset{j_v}{\rightarrow} \Lambda^2 L^* \overset{j_v}{\rightarrow} \cdots \overset{j_v}{\rightarrow} \Lambda^n L^*$$

(recall that $(i_v)^2 = 0 = (j_v)^2$). Check that

(i)

$\hat{h} := |v|^{-2} j_v$ plays the role of a homotopy operator for the complex $(\Lambda L^*, i_v)$
$\hat{h} := |v|^{-2} i_v$ plays the role of a homotopy operator for the complex $(\Lambda L^*, j_v)$

(ii) both of the complexes have *trivial* cohomologies ($b^p = 0$ for all p).

Hint: (i) $i_v j_v + j_v i_v = |v|^2 \hat{1}$ due to (5.8.9), then (9.2.3), (9.2.4): for example if $i_v \alpha = 0$, then $\alpha = i_v \beta \equiv i_v(|v|^{-2} j_v \alpha)$ (so that $\beta \equiv |v|^{-2} j_v \alpha$ is a "potential" for α). □

• A slightly more involved complex, which is related to the representations of Lie algebras, will be discussed in Section 12.5. Let us return to manifolds, however, i.e. to the deRham complex of differential forms on a manifold.

9.3.5 Justify the statement that in the case of the *deRham* complex $(\Omega(M), d)$ the *zeroth* cohomology $H^0(M)$ carries (only) the information about the number of *connected*

[148] See Appendix A.1; we simply need to "subtract" those dimensions in Z^p which are "irrelevant" within the context under discussion; just this is done by the formal procedure of factorization.

[149] H^p is a linear space, which *is* (a bit more than) an *Abelian group* (= a commutative group; we can multiply the elements by scalars from $\mathbb{R}$ in addition to the structure of a group). The term cohomology *group* stems from a more general definition of a complex, where C^p are (only) Abelian groups and d_p are homomorphisms of groups. In this situation we get as H^p (indeed only) Abelian *groups*.

components of the manifold M (see (11.7.15)), namely

$$b^0 \equiv \dim H^0(M) = \text{the number of connected components of the manifold } M$$

Hint: there are no exact 0-forms, the closed forms coincide with the *locally* constant functions on M; if M has several connected components, the value of the constants *may differ* on them, so that a particular closed 0-form on M is given by a point of $\mathbb{R}^k$ (a *k-tuple of numbers*, the values of the function on k connected components). □

9.3.6 Consider again the situation from before problem (9.2.13), where a manifold M was shrunk by the flow $\Phi_t \leftrightarrow \xi$ to some submanifold $N \subset M$. Moreover, let the flow Φ_t be such that the points of the submanifold N itself remain fixed in the course of the shrinking (such shrinking is called the *deformation retraction* and the resulting manifold $N \subset M$ is the *deformation retract* of the manifold M). Check that

(i) the cohomology class $[\alpha]$ of an arbitrary (closed) form α on M is completely specified by the restriction of the form α to N, i.e. the information about the class is actually encoded in a far *simpler* form, the restriction of α to the deformation retract[150]

(ii) the unit n-dimensional sphere is the deformation retract of the space $\mathbb{R}^{n+1}$ minus the origin, so that the cohomological classes of the forms in $\mathbb{R}^{n+1}$ minus the origin are given by their restriction on the sphere S^n

(iii) for $n = 1$ complete information about the class of a (general closed) 1-form $\alpha = a(r, \varphi)\, dr + b(r, \varphi)\, d\varphi$ in the plane minus the origin is in the 1-form *on the circle*

$$\hat{\alpha} = b(1, \varphi)\, d\varphi$$

(iv) there are no (cohomologically) non-trivial 2-forms in the plane minus the origin.

Hint: (i) if we denote by $\pi \equiv \Phi_\infty$ the projection of the whole manifold M onto the deformation retract $N \subset M$ (indeed $\pi \circ \pi = \pi$ holds), then due to (9.2.13) we have $\pi^*\alpha = \alpha - d(\hat{h}\alpha)$, so that $[\alpha] = [\pi^*\alpha] = [\pi^*(\alpha|_N)]$; (ii) $\pi : \mathbf{r} \mapsto \mathbf{r}/r$; (iii) in polar coordinates $\pi : (r, \varphi) \mapsto (1, \varphi)$; (iv) there are no non-zero 2-forms on the *circle*. □

• The knowledge of the Betti numbers of a given manifold (i.e. the Betti numbers for the deRham complex of the manifold M) for $p \geq 1$ yields a (more than) exact quantitative answer to the question about the validity of the converse of $dd = 0$: the converse holds for p-forms if and only if $b^p(M) = 0$. On a manifold which is contractible to a point the converse holds for all $p \geq 1 \Rightarrow$ *all* the Betti numbers (for $p \geq 1$) vanish:

$$M \text{ is contractible} \quad \Rightarrow \quad b^1 = \cdots = b^n = 0 \quad (\text{and } b^0 = 1)$$

The contractibility is, however, an unreasonably strong request in order that the potential exist for a closed form of some fixed degree p: evidently the vanishing of a *single* Betti number, $b^p = 0$ alone, is sufficient. For example, one can prove that for the sphere S^2 the Betti numbers read $b^0 = 1 = b^2$, $b^1 = 0$. The fact that $b^1 = 0$ means that each closed 1-form has a potential, but non-vanishing b^2 indicates that for closed 2-forms this is no

[150] In particular if the deformation retract of M happens to be a *point*, the classes are specified by the restrictions of the forms on the *point*, so that they necessarily *vanish* (unless they are 0-forms).

longer true (in accordance with the result of problem (9.1.3); $b^2 = 1$ moreover says that there are *no other* non-trivial classes except the one we revealed in the problem). For the plane the Betti numbers read $b^0 = 1$ (it is connected) and all remaining ones vanish (it is contractible), but for the plane minus the origin there already holds $b^1 \neq 0$, since this is true *on the circle* (9.3.6), (9.1.3).

From the pragmatic point of view the formalism of the deRham complex would say nothing about the issue of the converse of $dd = 0$ if an *independent* method of computation of the cohomological groups of a manifold did not exist. Yet this method does exist. It turns out that one may also associate some *completely different* complexes with a manifold, which may be identified with the deRham complex, however, *at the level of cohomologies* (i.e. they result in the same cohomologies in spite of being completely dissimilar as complexes). And what is essential, the cohomologies of some of them may be computed fairly easily for numerous cases of considerable interest. Such statements about coincidence of cohomologies of two different complexes associated with a manifold are non-trivial and very useful mathematical theorems. Let us mention at least the complex of "singular chains" (M, ∂) or the complex given by "cell decomposition of a manifold"; see the books by Nash and Sen or Schwartz.

Summary of Chapter 9

A form is closed if its exterior derivative vanishes, and exact if it is itself the exterior derivative of some other form (its *potential*). Since the operator d is nilpotent (i.e. $dd = 0$), each exact form is necessarily closed. Simple counterexamples show that the converse of this statement, freely used in elementary physics, does not hold *in general*. It does hold, however, on *contractible* manifolds. In particular it holds *locally*, i.e. within a sufficiently small neighborhood of any point on any manifold; this statement is known as the Poincaré lemma. An explicit formula for the potential is then given. A more subtle treatment of the issue is provided by cohomology theory, namely by cohomologies of the *deRham* complex.

$\hat{h} = -\int_0^\infty dt\, \Phi_t^* i_\xi$	Homotopy operator	(9.2.3)
$d \circ \hat{h} + \hat{h} \circ d = \hat{1}$	Essential property of $\hat{h}$	(9.2.3)
$\alpha = d(\hat{h}\alpha) \equiv d\beta$	$\beta \equiv \hat{h}\alpha$ is a potential of α	(9.2.4)
$x^k \int_0^1 d\lambda\, \lambda^{p-1} \alpha_{ki\dots j}(\lambda x)$	Coordinate expression of $(\hat{h}\alpha)_{i\dots j}(x)$	(9.2.7)
$[e_a, e_b] = c^c_{ab}(x) e_c$	Coefficients of anholonomy of e_a	(9.2.10)
$e_a = \partial_a \quad \Leftrightarrow \quad [e_a, e_b] = 0$	When a frame field is holonomic (coordinate)	(9.2.11)
$e^a = dx^a \quad \Leftrightarrow \quad de^a = 0$	When a coframe field is holonomic	(9.2.11)
$Z^p := \operatorname{Ker} d_p$	p-cocycles	(9.3.1)
$B^p := \operatorname{Im} d_{p-1}$	p-coboundaries	(9.3.1)
$H^p := Z^p / B^p$	pth cohomology group	Sec. 9.3
$b^p := \dim H^p$	pth Betti number	Sec. 9.3
$\Omega^0(M) \xrightarrow{d} \Omega^1(M) \xrightarrow{d} \cdots \xrightarrow{d} \Omega^n(M)$	deRham complex of a manifold M	(9.3.2)

10
Lie groups: basic facts

• Groups may be ranked among the most important objects in mathematics and indirectly in theoretical physics as well. This reputation is by no means accidental, but it is related to the fact that they serve as a formal tool for utilizing *symmetries*; the importance of groups is then simply a reflection of the importance of symmetries.

From the perspective of differential geometry a special class of groups turns out to be of particular interest, namely the *Lie* groups. They represent objects in which their two distinct aspects peacefully coexist in a happy symbiosis, shoulder to shoulder: algebraic (they are groups) and geometrical or differential-topological (they are smooth manifolds). These two aspects restrict one another,[151] but (as the world goes in a good partnership) they also immensely enrich one another – the richness of the *geometry* on Lie groups ultimately springs from the existence of the *algebraic* structure of groups.

10.1 Automorphisms of various structures and groups

• Groups occur in all applications by means of their *actions*, as *groups of transformation* of something. The transformations of an arbitrary set X (bijective maps $g : X \to X$; for finite sets the permutations) are endowed naturally with the structure of a group (with respect to the composition of maps). A simple, albeit highly important, observation is that if we add some structure on X (being for now only a set), $X \mapsto (X, \mathrm{s})$, then the transformations which *preserve* (respect) the structure,[152] the *automorphisms* of the structure (X, s), also constitute a group, which is clearly a subgroup of the group of all transformations of the "bare" set X; we will denote this group by G.

10.1.1 Consider X to be the four-element set $X = \{a, b, c, d\}$. The group of its transformations is S_4 = the permutation group of four elements. Introduce the structure (X, s) as

[151] We will see, for example (11.1.6), that a manifold, which yearns to become a Lie group, has to first vow that for all its life it will be *parallelizable* (a *global frame field* should exist on it, i.e. $n = \dim M$ nowhere vanishing vector fields, being moreover linearly independent at each point) and not all manifolds are disposed to bind themselves by oath to this. It turns out that on the common sphere S^2, to give an example, there is *not a single* nowhere vanishing vector field, so that it is not possible to introduce the structure of a Lie group on S^2.

[152] A structure may be defined as a rule in terms of some *maps*; then to preserve the structure means that the transformation *commutes* with these maps. The analysis of several particular situations is the best way to gain a feeling for this point.

follows: we regard the elements of X as the vertices of a square. Find the transformation group $G \subset S_4$ which preserves this structure (i.e. from a square makes a square again). Hint: such $g \in S_4$ which do not produce "non-neighbors" from neighbors. □

10.1.2 Let (X, s) be a smooth manifold M, i.e. "s" is the *smooth* structure on a manifold. Be sure you understand that in this case the group is $G = \text{Aut}\, M$, the group of *diffeomorphisms* of the manifold.[153] □

10.1.3 Let (X, s) be a (finite-dimensional real) *linear space* V. The group G of automorphisms of this structure is denoted by $\text{Aut}\, V \equiv GL(V)$ (GL = general linear). Show that

(i) it is the group of *invertible* linear operators on V
(ii) it is (*non*-canonically) isomorphic to the group $GL(n, \mathbb{R}) \equiv GL(\mathbb{R}^n)$ of non-singular real $n \times n$ matrices.

Hint: (i) here preserving "s" means commuting with the operation of making linear combinations, $g(v + \lambda w) = g(v) + \lambda g(w)$, i.e. it means the *linearity* of the transformations, being bijective, yields the invertibility; (ii) a basis $e_a \in V$ (an arbitrary choice) in a standard way identifies $V \leftrightarrow \mathbb{R}^n$ and the operators on V with the square matrices (if $Ae_a = A^b_a e_b$, then $A \leftrightarrow A^b_a$). □

10.1.4 Let (X, s) be a linear space V endowed with a *bilinear form* $h : V \times V \to \mathbb{R}$ and let $(X, s) := (V, h)$. The group G of automorphisms of this structure consists of those invertible linear operators A on V (10.1.3), which moreover satisfy the condition

$$h(Av, Aw) = h(v, w)$$

Show that

(i) such operators are indeed closed with respect to group operations
(ii) at the level of their *matrices* we get

$$A^a_c A^b_d h_{ab} = h_{cd} \qquad \leftrightarrow \qquad A^{\mathrm{T}} h A = h$$

Hint: (ii) insert a basis ($h_{ab} := h(e_a, e_b)$). □

10.1.5 If the bilinear form h from (10.1.4) is *symmetric* and *non-degenerate* (i.e. a metric tensor in V) with signature (r, s), the corresponding group is denoted by $O(r, s)$ and it is called the *pseudo-orthogonal group*. Show that

(i) at the level of matrices it may be characterized by the condition

$$A^{\mathrm{T}} \eta A = \eta \qquad \eta = \text{diag}\,(\underbrace{1, \dots, 1}_{r}, \underbrace{-1, \dots, -1}_{s})$$

[153] The group Diff(M) is *not* a (finite-dimensional) Lie group. However, some of its subgroups which will be of great importance for us *are* Lie groups.

The matrices A which satisfy this condition are called *pseudo-orthogonal matrices*. The best known example is the *Lorentz group* $\equiv O(1,3)$.

(ii) in particular for $s = 0$ (or equivalently $r = 0$) we get the *orthogonal matrices* defined by the condition

$$A^{\mathrm{T}}A = 1$$

(and the *orthogonal group* $O(n;\mathbb{R}) \equiv O(n) := O(n,0)$).[154]

Hint: (i) choose a basis in which h has the canonical form ($= \eta_{ab}$). □

10.1.6 If the bilinear form h from (10.1.4) happens to be *antisymmetric* and *non-degenerate* (i.e. a non-degenerate 2-*form* in V), we get the *symplectic group* $\mathrm{Sp}(m,\mathbb{R})$ (for $\dim V = 2m$). Show that

(i) the dimension of the space V must indeed be *even* ($\dim V = 2m$)
(ii) at the level of matrices we have $2m \times 2m$ matrices A characterized by the condition

$$A^{\mathrm{T}}\omega A = \omega \qquad \omega := \begin{pmatrix} 0 & -\mathbb{I} \\ \mathbb{I} & 0 \end{pmatrix}$$

(the elements of ω are $m \times m$ blocks; it is the *canonical form* of h under consideration).

Hint: (i) $\omega^{\mathrm{T}} = -\omega \Rightarrow \det \omega^{\mathrm{T}} = \cdots$; (ii) (5.6.8). □

10.1.7 Let (X, s) be a linear space V endowed with a *volume form* ω (see Section 5.7), $(X, \mathrm{s}) := (V, \omega)$. The group G of automorphisms of this structure consists of the invertible linear operators A on V (10.1.3), such that they in addition satisfy the condition

$$\omega(Av, \dots, Aw) = \omega(v, \dots, w)$$

i.e. the parallelepiped spanned by $(v, \dots, w)$ is transformed to the parallelepiped spanned by $(Av, \dots, Aw)$ with *the same* (oriented) *volume*. The corresponding group is denoted by $SL(V)$ (*special* linear). Show that

(i) such operators are indeed closed with respect to group operations
(ii) at the level of *matrices* we have

$$A^a_c \dots A^b_d \varepsilon_{a\dots b} = \varepsilon_{c\dots d} \quad \leftrightarrow \quad \det A = 1$$

This group of matrices is standardly denoted as $SL(n,\mathbb{R})$.

Hint: (ii) insert a basis, (5.7.1) and (5.7.6) □

10.1.8 Show that

(i) for the determinant of the (pseudo-)orthogonal matrices we get

$$A \in O(r,s) \quad \Rightarrow \quad \det A = \pm 1$$

[154] The (pseudo-)orthogonal groups exist in both versions $\mathbb{R}$ and $\mathbb{C}$. In the first case the matrices are real, in the second one they are complex.

(ii) there is a bijection between the parts of the group characterized by $\det A = +1$ and $\det A = -1$
(iii) the part with $\det A = +1$ constitutes a subgroup (denoted by $SO(r, s)$: the *special* (pseudo-) orthogonal matrices)

$$A^{\mathrm{T}}\eta A = \eta, \ \det\ A = +1 \quad \leftrightarrow \quad SO(r, s)$$
$$A^{\mathrm{T}}A = 1, \ \det\ A = +1 \quad \leftrightarrow \quad SO(n)$$

(iv)

$$SO(r, s) = O(r, s) \cap SL(r + s, \mathbb{R})$$

Hint: (i) take the determinant of both sides of the defining equation; (ii) for example, $A \mapsto \sigma A, \sigma := \operatorname{diag}(-1, 1, \ldots, 1)$. □

10.1.9 Let (X, s) be a linear space V with an *orientation* (see Section 5.5), $(X, s) := (V, o)$. The group $GL_+(V)$ of the automorphisms of this structure consists of the invertible linear operators A on V (10.1.3) which satisfy the condition

$$e_a \text{ is a right-handed basis} \quad \Rightarrow \quad e'_a := Ae_a \text{ is also a right-handed basis}$$

Show that

(i) such operators are indeed closed with respect to group operations
(ii) at the level of matrices we get

$$\det A > 0$$

(this group is standardly denoted as $GL_+(n, \mathbb{R})$).

Hint: (ii) see (5.5.1). □

10.1.10 Let (X, s) be an n-dimensional *complex linear space* V (a linear space over $\mathbb{C}$). Show that

(i) the group G consists in this case of the invertible $\mathbb{C}$-linear operators in V
(ii) G is (*non*-canonically) isomorphic to the group $GL(n, \mathbb{C}) \equiv GL(\mathbb{C}^n)$ of non-singular complex $n \times n$ matrices.

Hint: like (10.1.3) but with $\lambda \in \mathbb{C}$. □

• It turns out that the situation from the last problem may also be expressed in "real" language.

10.1.11 A point $z \in \mathbb{C}^n$ (a column of n complex numbers) may be naturally identified with the point $(x, y) \in \mathbb{R}^{2n}$ (the column of $2n$ real numbers; $z^n = x^n + iy^n$). If the matrices acting in $\mathbb{C}^n$ are written in the same spirit as $A = B + iC$, a $\mathbb{C}$-linear map in $\mathbb{C}^n$, $z \mapsto Az$, then induces a certain $\mathbb{R}$-linear map in $\mathbb{R}^{2n}$. Show that

(i) explicitly we get

$$z \mapsto Az \quad \leftrightarrow \quad \begin{pmatrix} x \\ y \end{pmatrix} \mapsto \begin{pmatrix} B & -C \\ C & B \end{pmatrix} \begin{pmatrix} x \\ y \end{pmatrix}$$

(ii) the map

$$\rho : GL(n, \mathbb{C}) \to GL_+(2n, \mathbb{R}) \qquad A \equiv B + iC \mapsto \begin{pmatrix} B & -C \\ C & B \end{pmatrix}$$

is induced by this, which is an injective homomorphism of groups; its image $\mathrm{Im}\,\rho \subset GL(2n, \mathbb{R})$ is then an isomorphic copy of $GL(n, \mathbb{C})$ in $GL(2n, \mathbb{R})$

(iii) for $n = 1$ we obtain a useful realization of the complex *numbers* themselves in the form of 2×2 matrices

$$z \equiv x + iy \mapsto \begin{pmatrix} x & -y \\ y & x \end{pmatrix} \equiv x \begin{pmatrix} 1 & 0 \\ 0 & 1 \end{pmatrix} + y \begin{pmatrix} 0 & -1 \\ 1 & 0 \end{pmatrix}$$

(iv) the matrices from $\mathrm{Im}\,\rho$ ("complex" matrices realized as double size real matrices) may also be characterized uniquely as such matrices $D \in GL(2n, \mathbb{R})$ which commute[155] with the matrix J

$$[D, J] := DJ - JD = 0 \qquad J := \rho\,(0 + i1) \equiv \begin{pmatrix} 0 & -1 \\ 1 & 0 \end{pmatrix}$$

Hint: (i) $(B + iC)(x + iy) = (\dots) + i(\dots)$; (ii) we have $\det \rho(A) = |\det A|^2 > 0$, since

$$\det \begin{pmatrix} B & -C \\ C & B \end{pmatrix} = \det \begin{pmatrix} B - iC & -C \\ C + iB & B \end{pmatrix} = \det \begin{pmatrix} B - iC & \dots \\ 0 & B + iC \end{pmatrix} = |\det (B + iC)|^2$$

□

10.1.12 Let (X, s) be a complex linear space $\mathbb{C}^n$ endowed with the *Hermitian scalar product* $(n = p + q)$

$$h(z, w) := \eta_{ab}\bar{z}^a w^b \equiv \bar{z}^1 w^1 + \dots + \bar{z}^p w^p - \bar{z}^{p+1} w^{p+1} - \dots - \bar{z}^{p+q} w^{p+q}$$

i.e. $(X, s) := (\mathbb{C}^n, h)$. Show that

(i) the group $G \equiv U(p, q)$ of automorphisms of this structure consists of those non-singular complex matrices (10.1.10) which satisfy the condition

$$A^+ \eta A = \eta$$

(they are known as *pseudo-unitary matrices*)

(ii) in particular, for $q = 0$ we get the group $U(n) := U(n, 0)$ of *unitary* matrices

$$A^+ A = 1 \qquad \leftrightarrow \qquad A \in U(n)$$

(iii) the determinant of $A \in U(p, q)$ obeys

$$|\det A| = 1 \quad \text{so that} \quad \det A = e^{i\alpha}, \ \alpha \in \mathbb{R}$$

[155] *Not all* matrices from $GL(2n, \mathbb{R})$ thus correspond to some complex linear operators; it is easy to check that those which *anti*commute with J correspond to *anti*linear operators and that each matrix may be uniquely written as a sum of two terms with this simple behavior with respect to J.

(iv) the condition $\det A = 1$ singles out a subgroup in $U(p,q)$ (it is denoted by $SU(p,q)$); for $q = 0$ it is the group (fairly ubiquitous in physics) of *special unitary* matrices $SU(n)$

$$A^+A = 1, \ \det A = 1 \quad \leftrightarrow \quad A \in SU(n)$$

Hint: (i) $h(Az, Aw) = h(z, w)$ gives straightforwardly $A^+\eta A = \eta$. □

10.1.13 Let (X, s) be a (finite-dimensional real) linear space V in which a *subspace* $W \subset V$ is singled out. The group G which preserves this structure consists of those $A \in \text{Aut}V \equiv GL(V)$ for which $A(W) \equiv \text{Im}A \subset W$ (they take vectors from W into W again; the subspace W is said to be *invariant* with respect to the action of such As).

(i) Check that at the level of matrices corresponding to the basis $e_a = (e_i, e_\alpha)$ which is *adapted* to the subspace W (i.e. $e_i \in W$, $i = 1, \dots, p \equiv \dim W$, then to be completed arbitrarily by vectors e_α to a basis of V), it is the group of non-singular matrices of structure (in block notation)

$$A = \begin{pmatrix} B & C \\ 0 & D \end{pmatrix}$$

i.e. explicitly

$$(e_i, e_\alpha) \mapsto (\hat{e}_i, \hat{e}_\alpha) \equiv (Ae_i, Ae_\alpha) = \left(B_i^j e_j, C_\alpha^i e_i + D_\alpha^\beta e_\beta\right)$$

(ii) determine the dimensions of the blocks involved and justify the non-singularity of B and D
(iii) the dual basis (e^i, e^α) is transformed in the following way:

$$(e^i, e^\alpha) \mapsto (\hat{e}^i, \hat{e}^\alpha) = \left((B^{-1})^i_j e^j - (B^{-1}CD^{-1})^i_\alpha e^\alpha, (D^{-1})^\alpha_\beta e^\beta\right)$$

i.e. by the inverse matrix

$$A^{-1} = \begin{pmatrix} B^{-1} & -B^{-1}CD^{-1} \\ 0 & D^{-1} \end{pmatrix}$$

(iv) in the dual space V^* a subspace $\hat{W} \subset V^*$ of dimension $\dim V - \dim W$ is singled out canonically, which may be characterized as the *annihilator* of the subspace W, i.e. it contains those covectors $\sigma \in V^*$ which *annihilate* vectors from W, $\langle \sigma, w \rangle = 0$ for all $w \in W$
(v) the covectors e^α constitute a basis in $\hat{W}$.

Hint: (ii) $\det A = \det B \det D$; (iii) see (2.4.2); (iv) see (2.4.19). □

10.1.14 Let (X, s) be a (finite-dimensional real) linear space V, which is a *direct sum* (Appendix A.1) of two subspaces, $V = V_1 \oplus V_2$. This means that any vector from V may be *uniquely* decomposed into the sum $v = v_1 + v_2$ such that $v_1 \in V_1$ and $v_2 \in V_2$. The group G which preserves this structure consists of $A \in \text{Aut}\,V \equiv GL(V)$ such that $A(V_i) \subset V_i$, $i = 1, 2$, so that *both* the subspaces V_1 and V_2 are invariant with respect to A. Check that at the level of matrices with respect to a basis e_a which is *adapted* to the direct sum structure $(e_1, \dots, e_p \in V_1$, $p \equiv \dim V_1$, the rest $\in V_2)$, this is the group of non-singular matrices of the structure (in block notation)

$$A = \begin{pmatrix} B & 0 \\ 0 & D \end{pmatrix}$$

□

• All the groups of automorphisms mentioned in problems (10.1.3)–(10.1.14) were subgroups of $GL(k, \mathbb{R})$ for appropriate k,[156] i.e. we treated the *real matrix* groups. In this book we will be concerned virtually always with groups which are of this type either directly, or they are isomorphic to groups of this type.[157] Let us mention an example (important enough in its own right) of a group which is *not* a matrix group by its very definition, which may be, however, easily replaced by an isomorphic copy which *is* already a matrix group.

10.1.15 Consider the *affine transformations* of $\mathbb{R}^n$, i.e. the transformations

$$\mathbb{R}^n \ni x \mapsto Ax + a \equiv (A, a)x \qquad A \in GL(n, \mathbb{R}), \quad a \in \mathbb{R}^n$$

Check that

(i) they constitute a group (we will denote it by $GA(n, \mathbb{R})$, the general *affine group*), the linear group being a subgroup, $GL(n, \mathbb{R}) \subset GA(n, \mathbb{R})$
(ii) the multiplication in $GA(n, \mathbb{R})$ turns out to be[158] (see also (12.4.8))

$$(A, a) \circ (B, b) = (AB, Ab + a)$$

(iii) the map

$$\rho : GA(n, \mathbb{R}) \to GL(n + 1, \mathbb{R}) \qquad (A, a) \mapsto \begin{pmatrix} A & a \\ 0 & 1 \end{pmatrix}$$

is an injective homomorphism of groups ⇒ its image $\operatorname{Im} \rho$ is an isomorphic copy of $GA(n, \mathbb{R})$ in $GL(n + 1, \mathbb{R})$ and it may therefore be used as a fully fledged *substitute* for the "original" $GA(n, \mathbb{R})$, being, moreover, already a *matrix* group. □

10.2 Lie groups: basic concepts

• The majority of the groups (all but the first two) which we encountered in Section 10.1 were endowed with the structure of a smooth manifold in addition to the structure of a group. The first sign of this is that their group elements did not constitute a discrete set (like in (10.1.1)), but rather a "continuum" in which *coordinates* may be introduced in order to label the points. We already know, however, that they are exactly the coordinates which are at the heart of the concept of a manifold. Let us inspect some simple examples.

[156] According to (10.1.11) this also applies to "complex" groups, if we interpret the words "were subgroups . . . ," as is often the case, in the sense of "were, up to an isomorphism, subgroups . . . "
[157] The geometrical machinery to be developed in the near future will, however, *not* be dependent on this at all. Sometimes one can still profit significantly from the fact that the group elements happen *to be* matrices, see Section 11.7.
[158] The expression on the left of the equals sign is to be understood in such a way (as it is common when treating maps) that the transformation on the *right* of the multiplication sign is performed *first*.

10.2.1 Consider the group $SO(2)$, i.e. (10.1.8) the 2×2 real matrices, which satisfy the conditions

$$A^T A = 1 \qquad \det A = 1$$

Show that

(i) each such matrix may be written in the form

$$A = A(\alpha) \equiv \begin{pmatrix} \cos\alpha & \sin\alpha \\ -\sin\alpha & \cos\alpha \end{pmatrix} \qquad \alpha \in \mathbb{R}$$

(ii) $SO(2)$ is a smooth manifold which is diffeomorphic to the circle (sphere S^1).

Hint: see (1.5.5). □

10.2.2 Consider the group $U(1)$, i.e. (10.1.12) the 1×1 complex matrices (i.e. numbers), which satisfy the condition

$$A^+ A = 1 \qquad \text{i.e. actually} \quad |A| = 1, \; A \in \mathbb{C}$$

Show that

(i) each such "matrix" may be written in the form

$$A = A(\alpha) \equiv e^{i\alpha} = \cos\alpha + i \sin\alpha \qquad \alpha \in \mathbb{R}$$

(ii) $U(1)$ is a smooth manifold which is diffeomorphic to the circle (sphere S^1).

Hint: (i) $\cos\alpha + i\sin\alpha \leftrightarrow (\cos\alpha, \sin\alpha) \in \mathbb{R}^2$; (ii) see (1.5.5). □

10.2.3 Consider the group $GL(n, \mathbb{R})$ itself. Show that it is an n^2-dimensional manifold.

Hint: the matrix elements A^a_b may be used as coordinates for $A \in GL(n, \mathbb{R})$ (a single chart is enough). □

10.2.4 Consider the group $GL(n, \mathbb{C})$. Show that it is a $2n^2$-dimensional manifold.[159]

Hint: write $A \equiv B + iC \in GL(n, \mathbb{C})$; then the (real) matrix elements B^a_b, C^a_b (a single chart) may be used as coordinates. □

• The examples we have analyzed here were very simple. The same result could be obtained (with the devotion of more time and labor), however, for all the remaining groups we treated in the previous section. All of them happen to be smooth manifolds, moreover the sub*groups* of $GL(n, \mathbb{R})$ mentioned there actually also turn out to be sub*manifolds* of $GL(n, \mathbb{R})$. This is ample motivation for introducing a separate concept, which combines

[159] Notice that this group is even endowed with the (stronger) structure of a *complex* manifold of dimension n in the sense of the text at the end of Section 1.3. One can use (complex) matrix elements A^a_b as the coordinates for $A \in GL(n, \mathbb{C})$ in this case (1-chart).

the structure of a group with the structure of a smooth manifold. This is exactly what a *Lie group* is.

If two different structures are expected to live together peacefully in a common household, they have to agree on the terms of this coexistence; they have to be *compatible*. One of the parties involved, the *smooth* structure on a manifold, insists on meeting at any moment only maps which are smooth. Its imperative towards the group (the other party involved) thus consists in always bringing home only smooth maps (if any).[160] The group cannot imagine its life without *three* key maps (so that it will certainly bring them home); thus, first and foremost, the request of smoothness concerns these three maps.

10.2.5 Let G be a group which is at the same time a (smooth) manifold. Show that

(i) the definition of a group may be rephrased in terms of three maps[161]

$$m : G \times G \to G \qquad i : G \to G \qquad j : G \to G$$
$$m(g,h) := gh \qquad i(g) := g^{-1} \qquad j(g) := e \equiv \text{the unit element of the group}$$

which satisfy some identities (try to write down all of them!). The map m is known as the *composition law* on a group, i defines the inverse element and j singles out (as the image $\text{Im}\, j$) the most distinguished element in the group, the unit element

(ii) the map j is automatically smooth (so that m and i are the only maps to bother about)

(iii) the demand on the smoothness of all three maps m, i, j may be equivalently reformulated as a demand on the smoothness of *a single* (combined) map

$$f : G \times G \to G \qquad f(g,h) := gh^{-1} \equiv m(g, i(h))$$

Hint: (i) what (for example) does $m(g, i(g)) = j(g)$ say? (ii) It is "constant." □

10.2.6 Find the explicit coordinate presentation of the composition law m on the groups

(i) $G = (\mathbb{R}, +)$, $GL(1, \mathbb{R})$, $GL(n, \mathbb{R})$
(ii) $GA(1, \mathbb{R})$, $GA(n, \mathbb{R})$
(iii) the group of upper-triangular matrices, i.e. matrices of the form

$$A(x,y,z) := \begin{pmatrix} 1 & x & z \\ 0 & 1 & y \\ 0 & 0 & 1 \end{pmatrix}$$

Hint: (i) $m(x,y) = x + y$, $m(x,y) = xy$, $(m(x,y))^i_j = x^i_k y^k_j$; (ii) under the correspondence $(x,y) \leftrightarrow \begin{pmatrix} x & y \\ 0 & 1 \end{pmatrix}$ (for $n = 1$, in general similarly) it is $m((a,b),(x,y)) = (ax, ay + b)$, $m((a,b),(x,y)) = (a^i_k x^k_j, a^i_k y^k + b^i)$; (iii) $m((a,b,c),(x,y,z)) = (a + x, b + y, c + z + ay)$. □

[160] Some requirements of the group structure towards the manifold were mentioned in Section 10.1.

[161] Instead of j it is more common to consider $\hat{j} : \{a\} \to G$, $a \mapsto e \in G$ (where $\{a\}$ is a zero-dimensional manifold, which contains just a single point).

• In this way we eventually come to the official *definition* of a *Lie group* as an object which is a group and at the same time a manifold, the operations of multiplication and the transition to the inverse element being smooth maps (which is sometimes formulated in terms of smoothness of the map $(g, h) \mapsto gh^{-1}$).

10.2.7 Check the compatibility of the smooth structure with the structure of the group for $G = GL(n, \mathbb{R})$ and $G = GL(n, \mathbb{C})$, i.e. prove that both these groups are indeed Lie groups.

Hint: $G = GL(n, \mathbb{R})$: any matrix element of the result of AB^{-1} is a quotient of two polynomials in matrix elements of the initial matrices, the denominator being all right (since the determinant of B is non-vanishing); $G = GL(n, \mathbb{C})$ in full analogy (this group is even a *complex Lie group*, since the corresponding coordinate presentation of the map $(g, h) \mapsto gh^{-1}$ is given by *holomorphic* functions). □

• Note that there is a subtlety in the notion of a *Lie subgroup*. This object is, as one would expect, a subgroup from the algebraic point of view. However, the requirement of being a *closed* subset is added lest we avoid certain pathological constructions that are waiting for their chance (a textbook example is given by the so-called irrational wrapping of the torus T^2). This concerns, in particular, the concept of a *matrix group*: it is a subgroup G of the group $GL(n, \mathbb{R})$, which is at the same time a *closed subset* in $GL(n, \mathbb{R})$. All of the cases we will encounter are all right from this point of view.

Summary of Chapter 10

Groups enter into play both in physics and in mathematics as *symmetry* groups, i.e. (in mathematical terms) as groups of *automorphisms* of various structures. Several structures leading to the common "classical groups" (general linear, orthogonal, symplectic, unitary, etc.) are discussed from this point of view. The *Lie group* combines in a single object the algebraic concept of a group with the differential-topological notion of a manifold. All of the above-mentioned groups (as well as some others) are examples of Lie groups.

$G = \text{Aut}\,(X, s)$	Group of automorphisms of a structured set	Sec. 10.1
$h(Av, Aw) = h(v, w)$	A preserves the bilinear form h	(10.1.4)
$A^a_c h_{ab} A^b_d = h_{cd}$	Component expression of the same fact	(10.1.4)
$A^{\mathrm{T}} h A = h$	Matrix expression of the same fact	(10.1.4)
$\omega(Av, \dots, Aw) = \omega(v, \dots, w)$	A preserves the volume form ω	(10.1.7)
$A^a_c \dots A^b_d \varepsilon_{a \dots b} = \varepsilon_{c \dots d}$	Component expression of the same fact	(10.1.7)
$\det A = 1$	Matrix expression of the same fact	(10.1.7)
$m(g, h) := gh$	Composition law in a group	(10.2.5)
"Classical" matrix groups introduced	They are summarized in problem	(11.7.6)

11
Differential geometry on Lie groups

• Lie groups should be mentioned as role models in all handbooks on the "art of living" (*savoir vivre*) – from the point of view of differential geometry they indeed live life to the full. There are several canonical geometrical objects living on them and some specific procedures may be performed only on them. This richness of the Lie group as a manifold is due to the group properties of the Lie group, i.e. it can ultimately be traced back to the symbiosis of its algebraic and differential-topological structures. Numerous constructions to be discussed in what follows are based on the fundamental concept of an *(left-) invariant field*. We will learn about this in the next section.

11.1 Left-invariant tensor fields on a Lie group

• On any (not only Lie) group an important role is played by special maps $G \to G$, which are called *left* (and right) *translations*.

11.1.1 Given any element $g \in G$ define the maps

$$\begin{aligned} L_g : G \to G \qquad & h \mapsto L_g h := gh \qquad \text{left translation} \\ R_g : G \to G \qquad & h \mapsto R_g h := hg \qquad \text{right translation} \end{aligned}$$

i.e.

$$L_g := m(g, \cdot) \qquad R_g := m(\cdot, g)$$

Show that

(i) both maps are bijective
(ii) for Lie groups both maps are smooth, so that (together with (i)) they are *diffeomorphisms* of G to itself
(iii) the following relations hold:

$$\begin{aligned} L_{gh} &= L_g \circ L_h \qquad & R_{gh} &= R_h \circ R_g \\ L_g^{-1} &= L_{g^{-1}} \qquad & R_g^{-1} &= R_{g^{-1}} \end{aligned}$$

(iv) the right and left translations *commute* with one another (for arbitrary g, h)

$$L_g \circ R_h = R_h \circ L_g$$

(v) a diffeomorphism $f : G \to G$ which commutes *with all* left translations is necessarily the right translation $f = R_k$, namely by the element $k = f(e)$.

Hint: (ii) $L_g h = m(g, h)$, with m being smooth (10.2.5); (v) the requirement $f \circ L_g = L_g \circ f$ when applied on e yields $f(g) = R_{f(e)}(g)$. □

11.1.2 Write down the explicit coordinate presentation of the map L_g for the groups

(i) $G = (\mathbb{R}, +), GL(1, \mathbb{R}), GL(n, \mathbb{R})$
(ii) $GA(1, \mathbb{R}), GA(n, \mathbb{R})$.

Hint: see (10.2.6); (i) $L_a x = a + x$, $L_a x = ax$, $(L_a x)^i_j = a^i_k x^k_j$; (ii) $L_{(a,b)}(x, y) = (ax, ay + b)$ or $L_{(a,b)}(x, y) = (a^i_k x^k_j, a^i_k y^k + b^i)$. □

• Let us digress from the main path towards *Lie* groups for a while in order to see a simple application to *finite* groups, namely the classification of groups containing a (very) small number of elements. Such a group may be displayed lucidly by its *multiplication table*. In this table (a square matrix) the ijth entry is defined to be the result of the product of the ith and jth group elements.

11.1.3 Given G a finite group, show that

(i) in its multiplication table no element occurs more than once in any column or row (so that an arbitrary column or row actually represents some permutation of the group elements)
(ii) this enables one to classify easily the groups of (very) low orders (the *order of a (finite) group* := the number of its elements); in particular, check that there is just one group (up to isomorphism) of orders 1, 2, 3 and two non-isomorphic groups of order 4.

Hint: (i) the kth row is the image of G with respect to L_{g_k}, with L_{g_k} being a bijection (permutation); for columns use the same idea with R_g; (ii) for $n = 1, 2, 3$ they are $\{e\}$, $\mathbb{Z}_2$, $\mathbb{Z}_3$, for $n = 4$ either $\mathbb{Z}_4$ or $\mathbb{Z}_2 \times \mathbb{Z}_2$ (for the explanation of $\times$ see Section 12.4). □

• Let us return to the main object of our interest, namely to Lie groups. Since L_g is (for each $g \in G$) a diffeomorphism $G \to G$, its pull-back L_g^* may be applied to an arbitrary tensor field on a group G, the result being again a tensor field on G. In particular, we may raise the question of whether there exist tensor fields which do not change under such a pull-back map thus being L_g-invariant. A closer analysis shows that there even exist fields which are invariant in this sense *with respect to all $g \in G$ at once*.

11.1.4 A tensor field T of type $\binom{p}{q}$ on G is said to be *left-invariant* if it satisfies[162]

$$L_g^* T = T \qquad \text{for all } g \in G$$

Show that the left-invariant fields are

[162] In what follows we will develop the whole machinery for the *left* translation L_g, but everything may always be repeated almost verbatim for the *right* translation R_g as well. One obtains then, for example, the right-invariant fields, which satisfy $R_g^* T = T$ and so on. If a situation occurs where the difference is important, we specify it explicitly.

(i) uniquely specified by its value at the unit element $e \in G$ of the group (or at any other point $h \in G$ as well)
(ii) smooth.

Hint: (i) since $L^*_{g^{-1}} = L_{g*}$, the definition in detail says $T(gh) = L_{g*}T(h)$, which for $h = e$ gives $T(g) = L_{g*}T(e)$, so that the tensor T in the point g is to be the L_{g*}-image of its value at the point e. Check that if in turn we *define* the tensor field T by the prescription $T(g) := L_{g*}T(e)$, it is left-invariant; (ii) a consequence of the smoothness of multiplication. □

• The fact that the whole left-invariant field may be reconstructed from its value at a single point $e \in G$ means that its information content is necessarily much poorer than that of a general field. There should then be a very "small number" of left-invariant fields in comparison with general fields where the values at various points are completely independent. Recall that the tensor fields of arbitrary (fixed) type $\binom{p}{q}$ constitute an ∞-dimensional linear space (see Section 2.5). It turns out that the left-invariant fields always constitute only a *finite-dimensional* subspace.

11.1.5 Let $T \in \mathcal{T}^p_q(G)$ denote the left-invariant field on G which is generated by its value $T(e)$ in e. Show that

(i) the prescription $T(e) \mapsto T$ defines an isomorphism of the space $(T_e)^p_q$ of *tensors* of type $\binom{p}{q}$ in $e \in G$ and the space of *left-invariant* tensor *fields* on G (regarded as linear spaces over $\mathbb{R}$)
(ii) the dimension of the linear space of left-invariant fields of type $\binom{p}{q}$ is thus $n^{(p+q)}$, where $n = \dim G$
(iii) the left-invariant *functions* (the tensor fields of type $\binom{0}{0}$) are just all *constant* functions on G (the space of such functions being evidently one-dimensional, as needed).

Hint: (i) L_g is a diffeomorphism $\Rightarrow L_{g*}$ is a linear isomorphism of the spaces of tensors in e and g respectively; (ii) see (2.4.6); (iii) $(L^*_g f)(e) \equiv f(g) \overset{!}{=} f(e)$. □

• We learned that the spaces of left-invariant tensor fields are just the isomorphic images of the (finite-dimensional) spaces of tensors at the point e. Let us study the structure and convenient expression of such fields in more detail. The fields of type $\binom{0}{0}$ (functions) were already discussed and we came to the conclusion that they are just constant functions. Let us take a step further, to the tensors of type $\binom{1}{0}$, i.e. to the left-invariant *vector* fields. According to the general result (11.1.5) they constitute an n-dimensional linear space, which copies exactly the tangent space in the unit element of the group.

11.1.6 Let E_a be a basis of the tangent space in the unit element and let e_a denote the left-invariant fields on G generated by E_a; there thus holds

$$e_a(g) = L_{g*}E_a \qquad E_a = e_a(e)$$

Show that

(i) the fields e_a constitute a global *frame field* on G
(ii) any Lie group is a *parallelizable* as well as *orientable* manifold

(iii) the vector field $V = V^a e_a$ is left-invariant if and only if it has constant components V^a with respect to the left-invariant frame field e_a
(iv) if $\hat{e}_a = A^b_a e_b$ is any other left-invariant frame field, then the transition matrix A^b_a is necessarily *constant*.

Hint: (i) a linear dependence of the vectors $e_a(g)$ at a point g would need the linear dependence of E_a (contradiction); (ii) the definition of being parallelizable; a volume form is, for example, given by $e^1 \wedge \cdots \wedge e^n$, (6.3.5); (iii) $L^*_g(V^a e_a) = (L^*_g V^a) e_a \overset{!}{=} V^a e_a$, so that the components V^a are to be left-invariant, i.e. according to (11.1.5) constant; (iv) consider the consequence of item (iv). □

• It turns out that the explicit expressions of the left-invariant vector fields for concrete groups may be found most easily in a slightly roundabout way, making use of the left-invariant 1-*forms* (see (11.1.11)). So let us take another step further, to the tensor fields of type $\binom{0}{1}$, i.e. to left-invariant 1-forms on G.

11.1.7 Let E^a be the basis of the cotangent space in the unit element which is dual to the basis E_a and let e^a denote the left-invariant 1-form on G generated by E^a; there thus holds

$$\langle E^a, E_b \rangle = \delta^a_b \qquad e^a(g) = L_{g*} E^a \qquad E^a = e^a(e)$$

Show that

(i) the fields e^a constitute a global *coframe field* on G
(ii) e^a and e_a from (11.1.6) happen to be the dual coframe and frame fields to one another, i.e.

$$\langle e^a(g), e_b(g) \rangle = \delta^a_b \qquad \text{at each point } g \in G$$

(iii) the 1-form $\alpha = \alpha_a e^a$ is left-invariant if and only if it has constant components α_a with respect to the left-invariant coframe field e^a
(iv) a vector field V is left-invariant if and only if

$$\langle \alpha, V \rangle = \text{constant for } \textit{any left-invariant} \text{ form } \alpha$$

(v) a 1-form α is left-invariant if and only if

$$\langle \alpha, V \rangle = \text{constant for } \textit{any left-invariant} \text{ field } V$$

Hint: see (11.1.6). □

• Now it is clear how things look in general for tensor fields of type $\binom{p}{q}$.

11.1.8 Let E_a and E^a be mutually dual bases in the tangent and cotangent space in the unit element of a group and let e_a, e^a be the left-invariant frame and coframe fields respectively generated by E_a and E^a. Show that

(i) an arbitrary tensor field T (in particular, an arbitrary p-form α) on G may be written as

$$T = T^{a\dots b}_{c\dots d} e^c \otimes \cdots \otimes e^d \otimes e_a \otimes \cdots \otimes e_b \qquad \alpha = \frac{1}{p!} \alpha_{a\dots b} e^a \wedge \cdots \wedge e^b$$

(ii) these fields are left-invariant if and only if they have constant components (with respect to this basis)

$$T^{a\dots b}_{c\dots d} = \text{constant} \qquad \text{or} \qquad \alpha_{a\dots b} = \text{constant}$$

(iii) left-invariance of tensor fields is preserved under the tensor product $\otimes$, linear combinations (over $\mathbb{R}$) and the Lie derivative $\mathcal{L}_V$ with respect to the *left-invariant* field V

(iv) left-invariance of forms is preserved under the exterior product $\wedge$, the interior product i_V with a *left-invariant* field V and the exterior derivative d

(v) a tensor field T is left-invariant if and only if

$$T(V,\dots;\alpha,\dots) = \text{constant}$$

for any *left-invariant* arguments $V,\dots,\alpha\dots$. □

• Now, let us find explicitly the left-invariant 1-forms and vector fields on the group $GL(n,\mathbb{R})$. As we will see later, this easy computation is the key element of a very convenient method of computation of these objects on all matrix groups.

11.1.9 Find all left-invariant 1-forms on the group $G = GL(n,\mathbb{R})$. In order to do this show in successive steps that

(i) in coordinates x^i_j (the matrix elements) for $x \in GL(n,\mathbb{R})$ the left translation by $A \in GL(n,\mathbb{R})$ reads

$$x^i_j \mapsto (L_A x)^i_j = (Ax)^i_j = A^i_k x^k_j$$

(ii) the condition of left-invariance $L^*_A\alpha = \alpha$ for a general 1-form[163]

$$\alpha = a^i_j(x)\,dx^j_i \equiv \operatorname{Tr}\{a(x)\,dx\}$$

leads to the requirement

$$a^i_k(Ax)A^k_j = a^i_j(x) \qquad \text{or in matrix notation} \qquad a(Ax)A = a(x)$$

(iii) the most general solution is

$$a^i_j(x) = (Cx^{-1})^i_j \equiv C^i_k(x^{-1})^k_j \qquad \text{or in matrix notation} \qquad a(x) = Cx^{-1}$$

where $C \equiv a(I_n)$ is an *arbitrary constant* $n \times n$ matrix

(iv) the most general left-invariant 1-form is parametrized by a matrix C and reads

$$\alpha \equiv \alpha_C = C^i_k(x^{-1})^k_l\,dx^l_i \equiv \operatorname{Tr}(Cx^{-1}\,dx)$$

(v) if C is decomposed with respect to the *Weyl basis*[164]

$$(E^i_j)^k_l = \delta^i_l\delta^k_j \qquad C = C^i_j E^j_i$$

[163] It is convenient to label the coordinates by a *pair* of indices here (since the composition law is very simple then), which naturally *doubles* the number of all indices on the components of tensors in comparison with a common situation. In particular, the components of 1-forms have (as many as) *two* indices.

[164] E^i_j is the matrix where the jith entry equals unity and all remaining elements vanish. It is clear that such matrices indeed form a basis in the n^2-dimensional linear space of all $n \times n$ matrices.

then

$$\alpha_{E^i_j} =: \hat{\alpha}^i_j = (x^{-1})^i_k \, dx^k_j \equiv (x^{-1}\, dx)^i_j$$

($\alpha^i_j(x)$, being functions (*components* of the forms), should not be confused with $\hat{\alpha}^i_j$, which are the 1-*forms* themselves)

(vi) the 1-forms $\hat{\alpha}^i_j$ may be used as a basis of the n^2-dimensional space of left-invariant 1-forms on $GL(n, \mathbb{R})$ and

$$\alpha_C = C^i_j \hat{\alpha}^j_i \equiv \mathrm{Tr}\,(C\hat{\alpha})$$

Hint: (i) see (11.1.2); (iii) set $x = I_n \equiv$ the identity matrix; (vi) the forms dx^i_j are linearly independent (being the coordinate basis), $\hat{\alpha}^i_j$ constitute their *non-singular* linear combinations (although different at different points, but everywhere non-singular). □

11.1.10 Find all left-invariant vector fields on the group $G = GL(n, \mathbb{R})$. In order to do this show in successive steps that

(i) the most general left-invariant vector field is parametrized by a matrix C and reads

$$V = V_C = x^i_k C^k_j \partial^j_i \equiv \mathrm{Tr}\,(xC\partial)$$

(ii) if C is decomposed with respect to the Weyl basis $(E^i_j)^k_l = \delta^i_l \delta^k_j$ as $C = C^i_j E^j_i$, then

$$V_{E^i_j} =: \hat{V}^i_j = x^i_k \partial^k_j \equiv (x\partial)^i_j$$

(iii) the fields $\hat{V}^i_j$ may be used as a basis of the n^2-dimensional space of left-invariant vector fields on $GL(n, \mathbb{R})$ and

$$V_C = C^i_j \hat{V}^j_i \equiv \mathrm{Tr}\,(C\hat{V})$$

($V^i_j(x)$, being functions (*components* of the fields), should not be confused with $\hat{V}^i_j$, which are the vector *fields* themselves)

(iv) if α_D is the most general left-invariant 1-form, then

$$\langle \alpha_D, V_C \rangle = C^i_j D^j_i \equiv \mathrm{Tr}\,(CD) \qquad \text{and, in particular,} \qquad \langle \alpha^i_j, V^k_l \rangle = \delta^i_l \delta^k_j$$

so that $\hat{V}^i_j$ and $\hat{\alpha}^k_l$ are global (dual to one another) frame and coframe fields on $GL(n, \mathbb{R})$ respectively (they were mentioned in a general setting in (11.1.7)).

Hint: (i) set $V = V^i_j(x)\partial^j_i$ and analyze the expression $\langle \alpha_D, V \rangle$ (11.1.7); for the remaining items see the hint to (11.1.9). □

• Concerning the group $GL(n, \mathbb{R})$ the matter is settled. And what about the remaining matrix groups – the *sub*groups G of $GL(n, \mathbb{R})$? It turns out that this matter is virtually settled as well. The convenient way for finding left-invariant fields is contained in the following simple observation.

11.1.11 Let $f : G \to H$ be a homomorphism of Lie groups and let α be a left-invariant form on H. Then prove that $f^*\alpha$ is a left-invariant form on G. In particular, check that

(i) if $\hat{L}_h$ and L_g are the left translations on H and G respectively, then they are related by the identity

$$f \circ L_g = \hat{L}_{f(g)} \circ f$$

(ii) from this it follows that $(\forall h \in H, \forall g \in G)$

$$\hat{L}_h^* \alpha = \alpha \quad \Rightarrow \quad L_g^*(f^* \alpha) = f^* \alpha$$

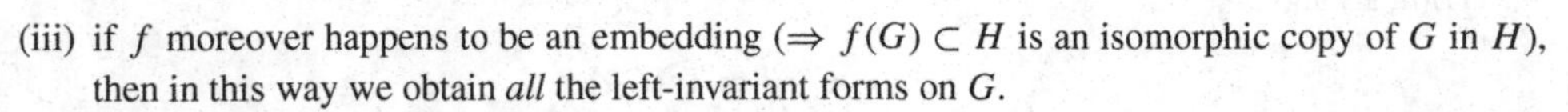

(iii) if f moreover happens to be an embedding ($\Rightarrow f(G) \subset H$ is an isomorphic copy of G in H), then in this way we obtain *all* the left-invariant forms on G.

Hint: (i) $f(g\tilde{g}) = f(g)f(\tilde{g})$; (iii) f_* realizes an isomorphism of the tangent space in the unit element on G onto its image in the tangent space in the unit element on H, which is enough according to (11.1.5). □

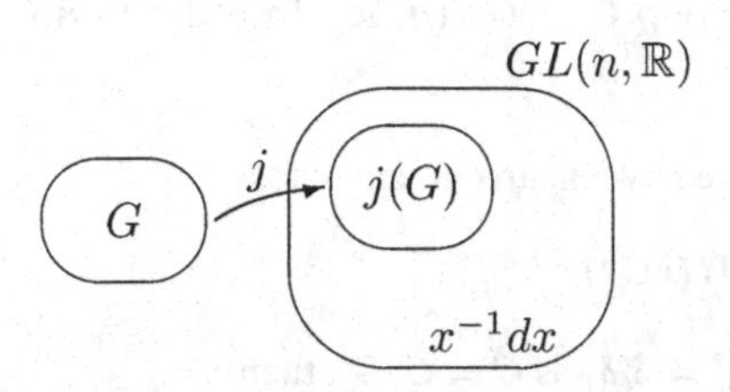

• Now if we take the group $GL(n, \mathbb{R})$ (where we already know all the left-invariant 1-forms) as H and some subgroup (i.e. an arbitrary matrix group) as G, the pull-back with respect to the canonical embedding of G as a subgroup of $GL(n, \mathbb{R})$ then gives us all the left-invariant 1-forms (and according to (11.1.8) trivially also the higher degree forms or tensors) on G. A general scheme looks as follows. Given a matrix group $G \subset GL(n, \mathbb{R})$ endowed with local coordinates z^μ, we may write down explicitly the embedding

$$j : G \to GL(n, \mathbb{R}) \qquad z^\mu \mapsto x^i_j(z^\mu)$$

Since we know from (11.1.9) that a basis of the space of the left-invariant 1-forms on $GL(n, \mathbb{R})$ is given by the forms $(x^{-1})^i_k \, dx^k_j$, all of them being stored in the matrix $x^{-1} \, dx$, it suffices to make the pull-back

$$j^*(x^{-1} \, dx) = x^{-1}(z) \, dx(z)$$

and take then all the linearly independent 1-forms from the resulting matrix (a slightly different point of view on the same material will be presented in Section 11.7). Let us illustrate how it works with some simple examples.

11.1.12 Find explicitly a basis of the left-invariant 1-forms as well as vector fields on $SO(2)$.

Solution: $SO(2) \subset GL(2, \mathbb{R})$, the parametrization (embedding) being presented in problem (10.2.1). We then have

$$x(\varphi) = \begin{pmatrix} \cos\varphi & \sin\varphi \\ -\sin\varphi & \cos\varphi \end{pmatrix} \Rightarrow x^{-1}(\varphi) = \begin{pmatrix} \cos\varphi & -\sin\varphi \\ \sin\varphi & \cos\varphi \end{pmatrix}$$

$\Rightarrow$

$$x^{-1}(\varphi) \, dx(\varphi) = \begin{pmatrix} \cos\varphi & -\sin\varphi \\ -\sin\varphi & \cos\varphi \end{pmatrix} d \begin{pmatrix} \cos\varphi & \sin\varphi \\ -\sin\varphi & \cos\varphi \end{pmatrix} = \begin{pmatrix} 0 & d\varphi \\ -d\varphi & 0 \end{pmatrix}$$

so that the bases are: the left-invariant basis 1-form on $SO(2)$ is $e^1 = d\varphi$ and the duality condition then yields that the left-invariant basis vector field is $e_1 = \partial_\varphi$. □

11.1.13 Find on $GA(1, \mathbb{R})$ explicitly a basis of the left-invariant 1-forms, vector fields and 2-forms (i.e. here the volume forms)

Solution: according to (10.1.15) or (11.1.2)

$$A(x, y) = \begin{pmatrix} x & y \\ 0 & 1 \end{pmatrix} \Rightarrow A^{-1}\,dA = \cdots = \begin{pmatrix} x^{-1}\,dx & x^{-1}\,dy \\ 0 & 0 \end{pmatrix}$$

so that we may take as the corresponding bases, for example,

for 1-forms	$e^1 = x^{-1}\,dx$	$e^2 = x^{-1}\,dy$
for vector fields	$e_1 = x\partial_x$	$e_2 = x\partial_y$
for 2-forms	$e^1 \wedge e^2 = x^{-2}\,dx \wedge dy$	

(since the points with $x = 0$ do not belong in the group, the denominators are all right). □

• We mentioned already that for the *right*-invariant objects everything works similarly. For the sake of completeness let us examine what changes occur for the group $GL(n, \mathbb{R})$ and then for our well-studied simple example $GA(1, \mathbb{R})$.

11.1.14 Show that

(i) a basis of the *right*-invariant 1-forms on $GL(n, \mathbb{R})$ is provided by the elements of the matrix $(dx)x^{-1}$

(ii) the most general right-invariant 1-form reads

$$\alpha_C = \mathrm{Tr}\,(x^{-1}C\,dx)$$

(iii) the most general right-invariant vector field reads

$$V_C = \mathrm{Tr}\,(Cx\partial)$$

(iv) there holds

$$\langle \alpha_C, V_D \rangle = \mathrm{Tr}\,(CD)$$

Hint: see (11.1.9) and (11.1.10). □

11.1.15 Find on $GA(1, \mathbb{R})$ explicitly a basis of the right-invariant 1-forms, vector fields and 2-forms.

Hint: making use of (11.1.13) and (11.1.14) you should end up with

for 1-forms	$f^1 = x^{-1}\,dx$	$f^2 = dy - yx^{-1}\,dx$
for vector fields	$f_1 = x\partial_x + y\partial_y$	$f_2 = \partial_y$
for 2-forms	$f^1 \wedge f^2 = x^{-1}\,dx \wedge dy$	

so that they may be expressed in terms of the *left*-invariant objects as

$$\begin{aligned} f^1 &= e^1 & f^2 &= -ye^1 + xe^2 \\ f_1 &= e_1 + yx^{-1}e_2 & f_2 &= x^{-1}e_2 \\ f^1 \wedge f^2 &= xe^1 \wedge e^2 \end{aligned}$$

□

11.2 Lie algebra $\mathcal{G}$ of a group G

• In problem (11.1.8) we already learned that some standard operations performed on tensors (in particular, on forms), like $\otimes, \wedge, d, \ldots$, do not spoil the left-invariance of the objects on which they act so that they may be restricted to the corresponding finite-dimensional subspaces given by the left-invariant fields. Here we first convince ourselves that the *commutator* of vector fields also has this remarkable property. This turns out to be very important. It means that the left-invariant vector fields on a Lie group constitute a *finite-dimensional Lie algebra.* A Lie group thus always induces canonically a simple algebraic object: *its* Lie algebra.[165] More detailed analysis shows that this algebra is not only a rich source of information about the group itself, it also serves as an efficient tool to exploit the strength of the group in applications.

11.2.1 Check that

(i) if $f : M \to M$ is a diffeomorphism and V, W arbitrary vector fields, then

$$f^*[V, W] = [f^*V, f^*W]$$

(ii) if V, W happen to be *left-invariant* vector fields on G, then both their linear combinations (over $\mathbb{R}$) and the commutator are, in turn, left-invariant vector fields; thus the space $\mathfrak{X}_L(G)$ of left-invariant vector fields on a group G is a (finite-dimensional) Lie algebra

(iii) if e_a is a left-invariant frame field on G, then

$$[e_a, e_b] = c^c_{ab} e_c$$

the coefficients of anholonomy of the frame field e_a now being *constants*

(iv) explain why the object

$$c := c^c_{ab} e^a \otimes e^b \otimes e_c \equiv \left(\frac{1}{2} c^c_{ab} e^a \wedge e^b\right) \otimes e_c$$

(e_a being a left-invariant frame field) *is* now a tensor field,[166] whilst in the general case (9.2.10) it was not.

[165] We know (see (4.3.6)), that *all* vector fields on the Lie group (as is the case on an arbitrary smooth manifold) constitute the Lie algebra $\mathfrak{X}(G)$. It is, however, ∞-dimensional, i.e. much more complicated. The Lie algebra $\mathfrak{X}_L(G) \subset \mathfrak{X}(G)$ which we are speaking about here is, in contrast, *finite* dimensional. (This resembles the Lie algebra of *Killing* vectors (4.6.6) on a Riemannian manifold; in both cases some specific property asked of vector fields (and preserved by the commutator) reduces the Lie algebra of specific fields to be finite-dimensional.)

[166] Notice that it is a *canonical* tensor field of type $\binom{1}{2}$ on G.

Hint: (i) $f^*[V, W] = f^*(\mathcal{L}_V W)$, (8.3.7); (ii) $f = L_g$; (iii) (9.2.10), the expression on the right represents a decomposition with respect to the left-invariant basis $\Rightarrow$ (11.1.6); (iv) restriction of the definition on the *left-invariant* fields; although we may change the basis fields at will, of course, the components will no longer coincide with the coefficients of anholonomy of the new basis in a general frame. □

• The Lie algebra $\mathfrak{X}_L(G)$ of left-invariant vector fields on G is sometimes already called the Lie algebra of a group G. More often, however the relation between the left-invariant fields and the tangent space in the unit element is used and the definition is adopted in which *the* Lie algebra $\mathcal{G}$ of a group G is identified with the tangent space (in the unit element) itself, $T_e G =: \mathcal{G}$. In this space there is clearly a linear structure, but we still lack a commutator there; for two vectors in a single tangent space on a *general* manifold there is indeed no commutator (there is nothing to be used for the required definition). However, it turns out that on a *Lie group* this *may* be arranged.

11.2.2 Let $X, Y \in \mathcal{G}$ be two vectors from the tangent space $T_e G =: \mathcal{G}$ in the unit element of the Lie group G and let $L_X, L_Y \in \mathfrak{X}_L(G)$ be the left-invariant vector *fields* induced by them, so that $X = L_X(e)$, $Y = L_Y(e)$. Show that

(i) the definition[167]

$$[X, Y] := [L_X, L_Y](e) \quad \text{or equivalently} \quad L_{[X,Y]} := [L_X, L_Y]$$

introduces a commutator (with all the needed properties) into $\mathcal{G}$, so that $\mathcal{G}$ becomes a Lie algebra. This algebra is called *the Lie algebra $\mathcal{G}$ of the group G*

(ii) if $E_a \leftrightarrow e_a$, then

$$[e_a, e_b] = c^c_{ab} e_c \quad \Leftrightarrow \quad [E_a, E_b] = c^c_{ab} E_c$$

so that the *same* coefficients c^c_{ab} = the *structure constants* of the Lie algebra $\mathcal{G}$ occur in both formulas

(iii) the structure constants

$$c^c_{ab} \equiv \langle E^c, [E_a, E_b] \rangle$$

constitute the components of a tensor (in the Lie algebra)

$$\hat{c} := c^c_{ab} E^a \otimes E^b \otimes E_c \in T^1_2(\mathcal{G})$$

the tensor *field* c from (11.2.1) being the left-invariant tensor field on G generated just by $\hat{c}$.

Hint: (i) all the necessary properties are inherited from the commutator of the *fields*; (iii) the map $\mathcal{G} \times \mathcal{G} \to \mathcal{G}$, $(X, Y) \mapsto [X, Y]$ is bilinear. □

• We know from (11.2.1) that the structure constants c^c_{ab} actually coincide with the coefficients of anholonomy of the left-invariant frame field e_a so that they carry information not only about the objects at the point e, but at least in some neighborhood of this point. Recall

[167] One associates first "their" left-invariant vector fields to the vectors X, Y, *these fields* are commuted *as vector fields* then and finally the result is in turn read off in the unit element, so that we end up again in $\mathcal{G}$.

that a left-invariant field is uniquely "extended" from its prescribed value $E_a = e_a(e)$ at e to points apart from e by a left translation $L_g = m(g, \cdot)$, which in turn depends on the composition law $m : G \times G \to G$ on a group G (i.e. we get different left-invariant fields and consequently different structure constants for the same vectors E_a ($\Rightarrow$ different Lie algebra) if we modify the composition law). We see then that "genetic" information of vital importance about the *composition law on a group* (being "the heart" of the group) is encoded in a concentrated form in the structure constants (or, as a matter of fact, in the Lie algebra $\mathcal{G}$ itself)[168] even though they are formally expressed in terms of objects living at a single point (the unit element e) alone.

The following formula reveals another "habitat" of the structure constants c^a_{bc} in nature (as well as another useful method of how to compute them explicitly).

11.2.3 Show that the structure constants of the Lie algebra $\mathcal{G}$ of a group G also enter the following relations and are valid for the dual left-invariant *co*frame field

$$de^a + \frac{1}{2}c^a_{bc}e^b \wedge e^c = 0 \qquad \text{the } \textit{Maurer–Cartan formula}$$

Hint: see (9.2.10) and (11.2.1). □

11.2.4 Find the structure constants of the Lie algebra $ga(1, \mathbb{R})$ of the group $GA(1, \mathbb{R})$ both in terms of the left-invariant vector fields and by the Maurer–Cartan formula.

Hint: see (11.1.13) and (11.2.3) ($c^2_{12} = -c^2_{21} = 1$, all the remaining ones being 0). □

11.2.5 Show that $dd = 0$ leads (via the Maurer–Cartan formula) to the identity

$$c^f_{d[a}c^d_{bc]} = 0$$

for the structure constants (the same identity also being a consequence of the Jacobi identity for the commutator).

Hint: $0 = dde^a = \dots$ (11.2.3); $0 = [E_a, [E_b, E_c]] + \text{cycl.}$, (11.2.2). □

• We now introduce another canonical object on a Lie group, the Lie algebra valued 1-form θ. If the notation from Section 6.4 is adopted, then $\theta \in \Omega^1(G, \mathcal{G})$. This form will play an important role in connection theory, where it occurs in the formula describing the change of gauge potentials under gauge transformations (see (21.2.3)).

11.2.6 Given a Lie group G, define the $\mathcal{G}$-valued *canonical 1-form*[169] $\theta \in \Omega^1(G, \mathcal{G})$ by the prescription

$$\langle \theta(g), v \rangle := L_{g^{-1}*}v \qquad v \in T_gG$$

[168] Some analogy may be found with the relation between some foodstuffs (a Lie group) and their "*powdered*" versions (its Lie algebra); the powdered form represents a "simplified" ("compressed", "zipped") version of the original one, preserving essential (total according to advertisements) parts of the properties of the original.

[169] It is also known as the *Maurer–Cartan 1-form*.

Put another way, we simply map the vector v by appropriate (unique) left translation from g to e (so that $\theta(g) : T_g G \to T_e G \equiv \mathcal{G}$). Show that

(i) an equivalent definition is (L_X is the left-invariant field generated by $X \in \mathcal{G}$)

$$\langle \theta, L_X \rangle := X$$

(both sides being $\mathcal{G}$-valued functions on G)

(ii) this form is left-invariant[170]

$$L_g^* \theta = \theta$$

(iii) if E_a is a basis in $\mathcal{G}$ and if e^a is the left-invariant coframe field generated by E^a, then

$$\theta = e^a E_a$$

so that the *component* 1-forms (see Section 6.4) of the form θ coincide with the left-invariant 1-forms e^a

(iv) if for $\mathcal{G}$-valued forms we introduce the operation (see (6.4.2) and the footnote therein)

$$[\alpha \wedge \beta] := (\alpha^a \wedge \beta^b)[E_a, E_b] \equiv \left(c^c_{ab}\alpha^a \wedge \beta^b\right) E_c \qquad \alpha = \alpha^a E_a, \beta = \beta^b E_b$$

then for θ there holds

$$d\theta + \frac{1}{2}[\theta \wedge \theta] = 0$$

and this formula is equivalent to the Maurer–Cartan formula from (11.2.3).

Hint: (i) by (11.1.4) there is unique X such that $v = L_X(g) \equiv L_{g*}X$; (ii) $\langle (L_h^*\theta)(g), v\rangle = \cdots$; (iii) each $\alpha \in \Omega^1(G, \mathcal{G})$ may be written in the form $\alpha = (\alpha^a_b e^b)E_a$; the definition of θ then gives $\alpha^a_b = \delta^a_b$; (iv) by (6.4.4) and (11.2.3) $d\theta = (de^c)E_c = (-\frac{1}{2}e^a \wedge e^b)c^c_{ab}E_c = -\frac{1}{2}(e^a \wedge e^b)[E_a, E_b] \equiv -\frac{1}{2}[\theta \wedge \theta]$. □

• For practical computation of the canonical form θ it is convenient to exploit the virtue of the matrix formalism, which we will learn about in Section 11.7.

11.3 One-parameter subgroups

• On each Lie group there exist distinguished curves, which are called *one-parameter subgroups*. These curves $\gamma(t)$ may be uniquely characterized by the properties

$$\gamma(t+s) = \gamma(t)\gamma(s) \qquad \gamma(0) = e \quad t, s \in \mathbb{R}$$

Let us formulate the same thing differently and see what consequences follow from this.

11.3.1 Check that any one-parameter subgroup may be regarded (or defined as well) as the image of a homomorphism $\gamma : (\mathbb{R}, +) \to G$, i.e. as a homomorphic image of the additive group of real numbers in the group G. □

[170] It also behaves nicely under *right* translations (we need the concept of the representation of a group, however, to express it explicitly, see (12.3.4)).

11.3.2 Let us divide the interval $[0, t]$ into a large number N of equal pieces of length $\epsilon = t/N$. Check that

$$\gamma(t) = \underbrace{\gamma(\epsilon)\dots\gamma(\epsilon)}_{N\text{ times}}$$

Hint: $\gamma(t/2)\gamma(t/2) = \gamma(t)$. □

• This means that we can reach the point $\gamma(t)$ so that we will multiply N times the starting point e (from the right or from the left, it does not matter) repeatedly by the same group element "close to the unit element" $\gamma(\epsilon)$. The whole curve thus depends in a sense (which becomes clearer in (11.3.3)) on this "small" element alone.[171] The method of construction mentioned above is evidently far from being convenient. Fortunately it turns out that there is a much easier way based upon the use of the left-invariant vector fields on G.

11.3.3 Let L_X be the left-invariant vector field on G which is generated by a vector $X \in \mathcal{G}$. Show that

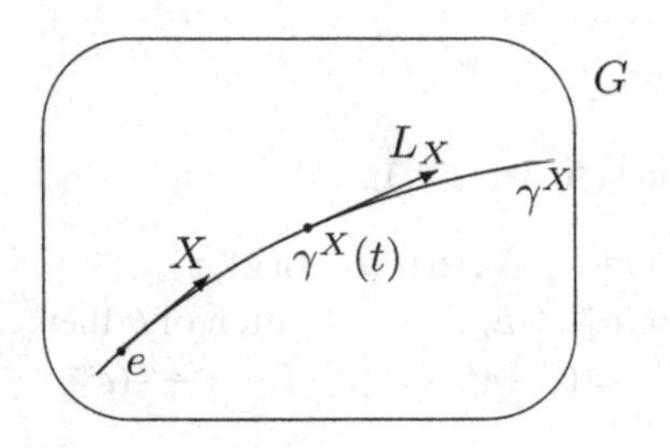

(i) its integral curve $\gamma^X(t)$ starting from e is a one-parameter subgroup

$$\gamma^X(t+s) = \gamma^X(t)\gamma^X(s) \qquad \gamma^X(0) = e$$

(ii) if, in turn, $\gamma(t)$ is an arbitrary one-parameter subgroup, then it is necessarily the integral curve of the left-invariant[172] field L_X with $X \equiv L_X(e) = \dot{\gamma}(0)$; the *complete* trajectory $\gamma(t)$ then turns out to be given by its *initial velocity* (in which direction and how fast does it rush forth), i.e. by the tangent vector $\dot{\gamma}(0) = X$ at the starting point e.

Solution: (i) the curve $\Gamma(t) := \gamma^X(t+s)$ is by (2.3.6) the integral curve of L_X starting at $\gamma^X(s)$. Since $L_{g*}L_X = L_X$, the curve $\Gamma(t)$ is also the integral curve of the field $L_{\gamma(s)*}L_X$. By (4.1.10) this is true for the curve $L_{\gamma(s)}\gamma(t) \equiv \gamma(s)\gamma(t)$. Put together $\gamma^X(t+s) = \gamma^X(t)\gamma^X(s)$; (ii) we have

$$\begin{aligned}\dot{\gamma}(t) &= \left.\frac{d}{ds}\right|_{s=0} \gamma(t+s) = \left.\frac{d}{ds}\right|_0 \gamma(t)\gamma(s) = \left.\frac{d}{ds}\right|_0 L_{\gamma(t)}\gamma(s) = L_{\gamma(t)*}\left.\frac{d}{ds}\right|_0 \gamma(s) = L_{\gamma(t)*}X \\ &= L_X(\gamma(t))\end{aligned}$$

□

[171] Since $\gamma(0) = e$, we expect (on the grounds of continuity) that $\gamma(\epsilon)$ would not be far from e even though we do not specify this distance more precisely by introducing some metric tensor.

[172] It came to our knowledge (as top-secret information; don't spread, please!) that the same curve leaves nothing to its fate ("a curve never can tell") and that it systematically prepares for the time when in all textbooks throughout the country authors will give attention mainly to *right-invariant* objects. In secrecy, completely unwitnessed even now it is *at the same time* the integral curve of the *right-invariant field R_X as well*, which any investigative journalist may easily convince himself/herself of by a computation. Individuals of a less inquisitive nature are recommended to consult (11.4.7).

11.3.4 Be sure to understand that there is a one-to-one correspondence

elements of the Lie algebra of the group	$\leftrightarrow$	one-parameter subgroups	$\leftrightarrow$	left-invariant vector fields

so that each element $X \in \mathcal{G}$ may be associated uniquely with *its* one-parameter subgroup as well as with *its* left-invariant vector field.

Hint: $X \leftrightarrow \gamma^X(t) \leftrightarrow L_X$. □

11.3.5 Find the explicit form of the one-parameter subgroups on the group $GL(n, \mathbb{R})$. In order to do this check that

(i) the equations for integral curves of a general left-invariant field V_C read

$$\dot{x}^a_b(t) = x^a_c(t)C^c_b \quad \text{or in matrix form a single equation} \quad \dot{x}(t) = x(t)C, \ x(0) = \mathbb{I}_n$$

(ii) the solution of this (matrix) equation is[173]

$$x(t) = \exp tC \equiv e^{tC} := \mathbb{I}_n + tC + \frac{t^2}{2!}C^2 + \cdots$$

Hint: (i) (11.1.10). □

11.3.6 Find the explicit form of the one-parameter subgroups on $GA(1, \mathbb{R})$. Namely, check that

(i) the equations for integral curves of a general left-invariant field $L_X = k^1 e_1 + k^2 e_2$ (for $X = k^1 E_1 + k^2 E_2$) read

$$\dot{x} = k^1 x \quad \dot{y} = k^2 x \quad x(0) = 1 \quad y(0) = 0$$

(ii) the solution for $k^1 \neq 0$ is

$$x(t) = e^{k^1 t} \quad y(t) = \frac{k^2}{k^1}\left(e^{k^1 t} - 1\right)$$

(for $k^1 = 0$ it is $x(t) = 1$, $y(t) = k^2 t$, being just the limit of the expression given above)

(iii) in the plane (x, y) (with the y-axis excluded) they represent the radial *lines* emanating from the unit element of the group, the point $(1, 0) \leftrightarrow e \in GA(1, \mathbb{R})$.

Hint: (i) (11.1.13); (iii) $y(t) = \text{constant}\,(x(t) - 1)$. □

11.4 Exponential map

• Consider the one-parameter subgroup $\gamma^X(t)$ generated by the element X of the Lie algebra $\mathcal{G}$. Thus, it satisfies

$$\gamma^X(0) = e \qquad \gamma^X(t+s) = \gamma^X(t)\gamma^X(s) \qquad \dot{\gamma}^X(0) = X$$

[173] The diligent reader is invited to prove the convergence of the series (i.e. the convergence of all the numerical series $(e^X)_{ij}$) for any finite matrix X.

Define the *exponential map* by

$$\exp : \mathcal{G} \to G \qquad X \mapsto \exp X \equiv e^X := \gamma^X(1)$$

The resulting point (the image of the element X) thus lies on the one-parameter subgroup $\gamma^X(t)$ at a parametric distance of 1 from the unit of the group. This map enjoys several remarkable properties.

11.4.1 Show that

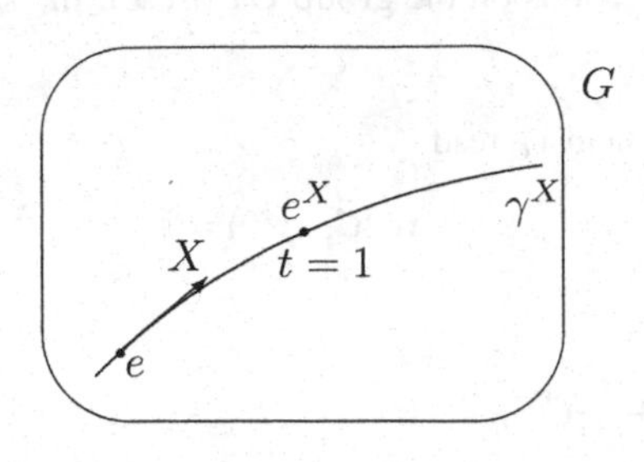

(i) the one-parameter subgroup satisfies

$$\gamma^X(kt) = \gamma^{kX}(t) \qquad k \in \mathbb{R}$$

(ii) this enables one to express the one-parameter subgroup in terms of an exponential map in the form

$$\gamma^X(t) = \exp tX \equiv e^{tX}$$

(iii) this expression says that the motion along a one-parameter subgroup is the image with respect to exp of the "uniform straight-line motion" in the Lie algebra.

Hint: (i) both sides are one-parameter subgroups with the same initial conditions, (11.3.3); (ii) $t \mapsto 1, k \mapsto t$; (iii) $t \mapsto tX$ is clearly the motion in $\mathcal{G}$ which *is* both uniform and straight-line. □

11.4.2 Show that

$$\exp 0 = e \qquad \exp(-X) = (\exp X)^{-1} \qquad \exp(t+s)X = \exp tX \exp sX$$

□

• The origin (point 0) of a Lie algebra is thus mapped to the unit element of the group. It turns out (15.4.10), that one can say even more; namely, that some *neighborhood* of the origin is *diffeomorphically* mapped onto some *neighborhood* of the unit element of the group.[174] Since the Lie algebra is a linear space $\sim \mathbb{R}^n$, we have all that is needed for the introduction of local coordinates on some neighborhood of the unit element of the group.

11.4.3 Let E_a be a basis of $\mathcal{G}$, $X = X^a E_a$. Define a chart in a neighborhood of $e \in G$ as follows: assign to the point $g = \exp X$ the n-tuple $x^a := X^a$. The coordinates x^a constructed in this way are called the *normal coordinates*[175] on (a part of) G. Show that the coordinate presentation of the one-parameter subgroups is extremely simple in normal coordinates, namely

$$\gamma^X(t) \quad \leftrightarrow \quad x^a(t) = X^a t$$

so that it actually coincides with the coordinate presentation of the *straight lines* in $\mathcal{G}$, being the exp-preimages of $\gamma^X(t)$ (see (11.4.2)).

[174] The map $\exp_{*0} : T_0\mathcal{G} \to T_eG$ is an isomorphism (its kernel vanishes).

[175] It is a special case of the (Riemann) normal coordinates constructed in terms of geodesics (see (15.4.11)); we will see that one can introduce a connection on a Lie group such that one-parameter subgroups turn out to be exactly the geodesics.

Hint: $\gamma^X(t) \equiv \exp tX \Rightarrow$ we assign the coefficients of the decomposition tX^a. □

11.4.4 Write down explicitly the exponential map for $GL(1, \mathbb{R})$. Show that we get the *ordinary* exponential function.

Solution: according to (11.1.10) the left-invariant field is $V = kx\partial_x \Rightarrow$ the one-parameter subgroup reads $x(t) = e^{kt} \Rightarrow$ the exp map is $k \mapsto e^k$. □

11.4.5 Write down explicitly the exponential map for $GA(1, \mathbb{R})$ $(X = k^1 E_1 + k^2 E_2 \mapsto (e^{k^1}, k^2(k^1)^{-1}(e^{k^1} - 1))$, (11.3.6)).

• Now we will discuss the flows generated by the left-invariant and right-invariant vector fields. □

11.4.6 Denote by $\Phi_t^{L_X} : G \to G$ the flow generated by the left-invariant field L_X and $\Phi_t^{R_X} : G \to G$ the flow generated by the right-invariant field R_X. Check that

$$\Phi_t^{L_X} = R_{\exp tX} \qquad \Phi_t^{R_X} = L_{\exp tX}$$

or

$$\Phi_t^{L_X}(g) = g(\exp tX) \equiv g e^{tX} \qquad \Phi_t^{R_X}(g) = (\exp tX)g \equiv e^{tX} g$$

so if we go from g along the left-invariant field L_X by t, we arrive at ge^{tX}; if we go along the right-invariant field R_X, we end up at $e^{tX}g$. Thus, the map $\Phi_t^{L_X}$ is the *right* translation by the element $\exp tX$, whereas $\Phi_t^{R_X}$ is the *left* translation by (the same) element $\exp tX$. We say that the left-invariant fields generate right translations, whereas the right-invariant fields generate left translations.

Hint: $L_X \leftrightarrow \Phi_t^{L_X} \Rightarrow$ by (4.1.9) $L_{g*}L_X \leftrightarrow L_g \circ \Phi_t^{L_X} \circ L_g^{-1}$; but $L_{g*}L_X = L_X \Rightarrow L_g \circ \Phi_t^{L_X} \circ L_g^{-1} = \Phi_t^{L_X}$, i.e. $\Phi_t^{L_X}$ commutes with all left translations $\Rightarrow$ by (11.1.1) it is the right translation R_k by the element $k = \Phi_t^{L_X}(e) = \gamma^X(t) = \exp tX$; the flow $\Phi_t^{R_X}$ in full analogy. □

11.4.7 Show that it follows from (11.4.6) that the one-parameter subgroup $\gamma^X(t)$ happens to be *at the same time* the integral curve of both the left-invariant as well as the right-invariant field generated by the element $X \in \mathcal{G}$ (i.e. of both L_X and R_X).

Hint: $\Phi_t^{L_X}(e) = \Phi_t^{R_X}(e) = \gamma^X(t)$. □

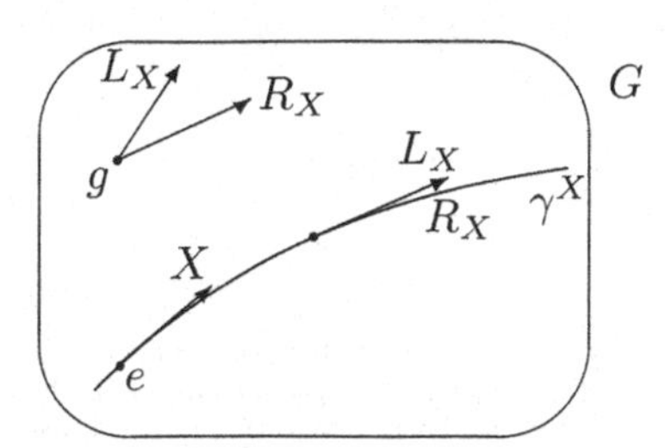

11.4.8 The result of (11.4.7) implies that although the right-invariant and the left-invariant fields differ *in general*, they necessarily coincide on the one-parameter subgroups, i.e. that both the fields L_X and R_X have the same values *at the curve* $\gamma^X(t)$ (X being *the same* in all three cases). Check this explicitly on the group $GA(1, \mathbb{R})$.

Hint: see (11.1.14) and (11.3.6). □

11.5 Derived homomorphism of Lie algebras

• Let $f : G \to H$ be a homomorphism of Lie groups. It turns out then that a homomorphism of the corresponding Lie algebras is automatically induced. It is known as the derived homomorphism. Let us start with a simple observation.[176]

11.5.1 Let $f : G \to H$ be a homomorphism of Lie groups and let

$$g(t) = \gamma^X(t) \equiv e^{tX}$$

be the one-parameter subgroup on G generated by the element $X \in \mathcal{G}$. Check that its f-image on H is also a one-parameter subgroup.

Hint: if $h(t) := f(g(t))$, then $h(t+s) = \cdots = h(t)h(s)$ and $h(0) = e_H$. □

• However in (11.4.1) we learned that *each* one-parameter subgroup is necessarily of the form e^{tY}, Y being an element of a Lie algebra (in the present case $\mathcal{H}$). In this way we obtain the assignment $X \mapsto Y$, which is a map $\mathcal{G} \to \mathcal{H}$.

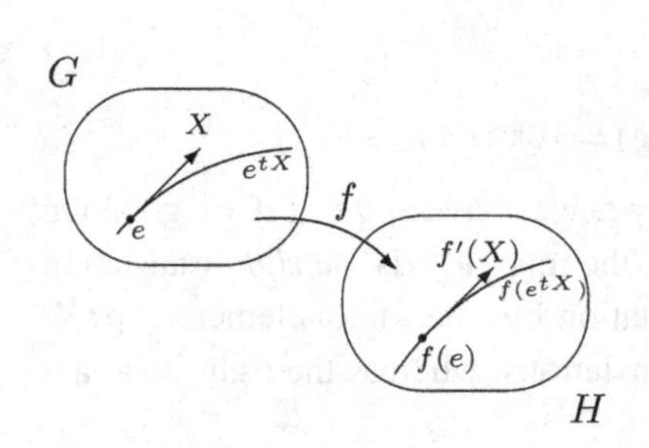

11.5.2 Let $f : G \to H$ be a homomorphism of Lie groups. Show that

(i) the one-parameter subgroup $f(g(t)) \equiv f(e^{tX}) \equiv e^{tY}$ generates the element

$$Y \equiv f'(X) := f_*X \in \mathcal{H} \qquad X \in \mathcal{G}$$

(ii) the left-invariant field on H generated by this element is f_*L_X (or possibly the extension of this field to the whole group H, if f fails to be surjective)

(iii) the field L_X is *projectable* with respect to f, so that the f-related field f_*L_X on $\operatorname{Im} f \subset H$ *is* well defined (f need not be injective so that in principle this question may be addressed).

Hint: (i) for $g(t) = e^{tX}$ we have $X = \frac{d}{dt}\big|_0 e^{tX}$ so that for $f(g(t) =: e^{tf'(X)}$ it is $f'(X) = \frac{d}{dt}\big|_0 f(g(t)) \equiv f_{*e}X$; (iii) the text after (3.1.2); here one should check that if $f(g) = f(\tilde{g})$ (i.e. if $(f \circ L_g)(e) = (f \circ L_{\tilde{g}})(e)$), then $f_*L_X(g) = f_*L_X(\tilde{g})$. □

11.5.3 Let $f : G \to H$ be a homomorphism of Lie groups. Show that

(i) the map $f' : \mathcal{G} \to \mathcal{H}$ from (11.5.2) is a homomorphism of Lie algebras (*derived homomorphism*)
(ii) we have the commutative diagram

$$\begin{array}{ccc} G & \xrightarrow{f} & H \\ \uparrow{\scriptstyle \exp} & & \uparrow{\scriptstyle \exp} \\ \mathcal{G} & \xrightarrow[f']{} & \mathcal{H} \end{array} \qquad \text{i.e.} \qquad f \circ \exp = \exp \circ f' \qquad f(e^X) = e^{f'(X)}$$

[176] This observation is always to be performed under a soft red light in a thoroughly blacked-out room. In the literature one can also find the recommendations to prepare in advance some hemp yarn, a magnifying (or reducing) glass, a diode and a bit of lip-salve. However, the rich and long-lasting personal experience of the author of this book shows that one can always somehow get along without the facilities listed above and they might be used equally well in the course of some other amazing observation.

(iii) for the composition of two homomorphisms we have a simple rule

$$(f_1 \circ f_2)' = f_1' \circ f_2'$$

Hint: (i) preservation of the operations: f' consists of two steps: first f_* of left-invariant fields (resulting in their homomorphic images – also left-invariant fields on H), then the evaluation of the result in $e \in H$, i.e. $(f_* L_X)(e) \in T_e H \equiv \mathcal{H}$; the first step preserves both linear combinations and the commutator; (ii) on the argument tX it is the definition of f'; (iii) (3.1.2). □

11.6 Invariant integral on G

• An important special case of the left-invariant forms on a Lie group is provided by a left-invariant form of top degree, a left-invariant *volume form* on G.

11.6.1 Show that the left-invariant volume form on G always exists and that it is moreover given uniquely up to a constant (non-zero) factor as

$$\omega_L = \lambda e^1 \wedge \cdots \wedge e^n \qquad 0 \neq \lambda \in \mathbb{R}$$

with e^a being an arbitrary left-invariant coframe field on G.

Hint: see (11.1.5) and (11.1.6); the space of n-forms in the unit element is one-dimensional. □

• Since we have the distinguished volume form on G, we may also integrate functions over G. The definition of the integral reads (see the text after (7.7.3))

$$\int_G f := \int_G f\omega_L \qquad f \in \mathcal{F}(G)$$

The integral introduced in this way is specific in that it is *left-invariant* in the following sense.

11.6.2 Let $f \in \mathcal{F}(G)$ be an arbitrary function on G and assume that the integral $\int_G f = \int_G f\omega_L$ exists. Show that then

$$\int_G f \circ L_g = \int_G f$$

i.e.

$$\int_G (f \circ L_g)\omega_L \equiv \int_G (L_g^* f)\omega_L = \int_G f\omega_L$$

where L_g is an arbitrary left translation on G.

Hint: (7.8.1), $L_g(G) = G$ (L_g is a diffeomorphism); therefore

$$\int_G (L_g^* f)\omega_L = \int_G L_g^*(f\omega_L) = \int_{L_g(G)} f\omega_L = \int_G f\omega_L$$

□

11.6.3 Check that the general statement (11.6.2) leads in particular cases to the following elementary results:

(i) for $G = U(1)$

$$\int_0^{2\pi} f(\alpha + \beta)\, d\alpha = \int_0^{2\pi} f(\alpha)\, d\alpha \qquad f(\alpha + 2\pi) = f(\alpha)$$

(ii) for $G = GL(1, \mathbb{R})$

$$\int_{-\infty}^{\infty} f(ax)\frac{dx}{x} = \int_{-\infty}^{\infty} f(x)\frac{dx}{x}$$

(iii) for $G = GA(1, \mathbb{R})$

$$\int_{-\infty}^{\infty}\int_{-\infty}^{\infty} f(ax, ay + b)\frac{dx\, dy}{x^2} = \int_{-\infty}^{\infty}\int_{-\infty}^{\infty} f(x, y)\frac{dx\, dy}{x^2}$$

Hint: (ii) see (11.1.2) and (11.1.9); (iii) see (11.1.2) and (11.1.13). □

• In full analogy one obtains the *right*-invariant volume form ω_R. This form is *different in general* from the left-invariant form ω_L (see for example (12.3.16)), and consequently also the result of the right-invariant integral (defined as $\int_G f\omega_R$) is different. It is a fairly important and useful fact that on each *compact* Lie group G (see problem (12.3.17)) the two volume forms happen to *coincide*: $\omega_L = \omega_R$. On these groups there thus exists a *bi-invariant integral.*

11.7 Matrix Lie groups: enjoy simplifications

• Numerous facts about *matrix* Lie groups and their Lie algebras may also be obtained with no use at all of the geometrical ideas like left-invariant fields, their integral curves, etc. Here we try to "derive" this standard simplified machinery from the formalism adopted up to now and to work up convenient and quick algorithms,[177] which we then apply to analyzing some particular common Lie groups mentioned in Section 10.1.

First, we look at the matrix Lie algebras as such, with no relation whatsoever to the Lie algebras of groups.

11.7.1 Consider End V, the set of all *endomorphisms* of a linear space V, i.e. the set Hom (V, V) of *all linear* maps from V to V. Check that

(i) it is naturally endowed with the structure of an *associative* algebra[178]
(ii) for $V = \mathbb{R}^n$ we may identify this (associative) algebra with the algebra $M_n(\mathbb{R}) \equiv \mathbb{R}(n)$ of real $n \times n$ matrices

[177] A natural and valid objection might be raised as to why we actually started with a "complicated" geometrical exposition, when there is a "simple" matrix formalism available on the market. We can allege *ad defendendum* before the Court of Conscience that (i) after ten chapters we have got chummy with differential geometry in so far as we can understand its reasoning with equal ease as we understand matrix multiplication and (ii) some issues look more complicated from the perspective of matrix fundamentalism or they need fairly non-ecological (highly paper-consuming) computations.

[178] Note that the subset given by the *auto*morphisms of V (being *bijective* ⇒ *invertible* linear maps) is not closed with respect to linear combinations and so it does not constitute a subalgebra; its closure with respect to the product gives rise naturally to a *group*, as we already know from (10.1.3).

(iii) the prescription (see Appendix A.3)

$$[A, B] := AB - BA$$

makes from End V (and, in particular, also from $M_n(\mathbb{R})$) a *Lie* algebra as well

(iv) some subsets, like the traceless matrices or the antisymmetric matrices, happen to be closed with respect to the operations of Lie algebra, and thus constitute Lie algebras (subalgebras of $M_n(\mathbb{R})$) in their own right.

Hint: (i) product = a composition of the maps. □

• Now, it turns out that the matrix Lie algebras happen to be closely related to the Lie algebras of groups, which we discussed. In order to make this relation clear recall the results of problem (11.1.10) concerning the structure of the left-invariant vector fields on $GL(n, \mathbb{R})$.

11.7.2 According to (11.1.10) a general left-invariant vector field on the group $GL(n, \mathbb{R})$ in coordinates x^i_j reads $V \equiv V_C = x^i_k C^k_j \partial^j_i \equiv \operatorname{Tr}(xC\partial)$ so that it is parametrized by the $n \times n$ real matrix C. Show that

(i) for a linear combination and the commutator of two left-invariant vector fields on $GL(n, \mathbb{R})$ we get

$$V_C + \lambda V_D = V_{C+\lambda D} \qquad [V_C, V_D] = V_{[C,D]}$$

(ii) the maps

$$\operatorname{End}\mathbb{R}^n \to \mathfrak{X}_L(GL(n, \mathbb{R})) \qquad C \mapsto V_C$$
$$\operatorname{End}\mathbb{R}^n \to gl(n, \mathbb{R}) \equiv T_e GL(n, \mathbb{R}) \qquad C \mapsto V_C(e)$$

are Lie algebra isomorphisms.

Hint: (i) for example, making use of the coordinate expression from (11.1.10). □

• This means, however, that the official and starchy Lie algebra $gl(n, \mathbb{R})$, defined abstractly as the tangent space in the unit element endowed with a sophisticated commutator, may be *replaced* by its humanized *isomorphic copy* $M_n(\mathbb{R}) \equiv \operatorname{End}\mathbb{R}^n$, which is the mundane $n \times n$ matrices with a simple commutator $CD - DC$.

11.7.3 Show that the elements of the matrix C may be regarded as the components of the vector $V_C(e)$ (i.e. the most general element of the Lie algebra $gl(n, \mathbb{R})$ of the group $GL(n, \mathbb{R})$) with respect to the coordinate basis.

Solution: for the unit element e we have $x^i_j = \delta^i_j \Rightarrow V_C(e) = \delta^i_k C^k_j \partial^j_i\big|_e = C^i_j \partial^j_i\big|_e$. □

• Now, given an arbitrary matrix group $G \subset GL(n, \mathbb{R})$, its Lie algebra is officially realized as a subspace of the *tangent space* of the group $GL(n, \mathbb{R})$ in the unit element. With regard to the isomorphism $C \leftrightarrow V_C(e)$ from (11.7.1) this means, however, that we may also consider its isomorphic copy instead, the subspace of the *space of $n \times n$ matrices*. To summarize, there are already *three* realizations of the Lie algebra of a Lie group G available:

1. $\mathcal{G}$ as the set of left-invariant vector fields on G,
2. $\mathcal{G}$ as the tangent space in the unit element (the most official definition),
3. $\mathcal{G}$ as a subalgebra of the algebra of $n \times n$ matrices $\operatorname{End}\mathbb{R}^n$.

The isomorphisms $V_C \leftrightarrow V_C(e) \leftrightarrow C$ found above enable one, as usual, *not to discriminate* between "the real" Lie algebra and its "isomorphic copies," but rather to work with them as with truly equivalent objects. The concrete choice of the realization will depend, as was the case for example with the definition of the tangent vector in Section 2.2, on the concrete context. The natural consequence then is to use the same nomenclature for *any* of the triple of the mutually isomorphic algebras so that, for example, the labeling $gl(n, \mathbb{R})$ will also, in what follows, be used for the Lie algebra of all $n \times n$ real matrices $\operatorname{End}\mathbb{R}^n$ (whereas $M_n(\mathbb{R}) \equiv \mathbb{R}(n)$ will be reserved for the corresponding *associative* algebra of all $n \times n$ real matrices).

Let us illustrate this with the example of the orthogonal group $G = O(n, \mathbb{R}) \subset GL(n, \mathbb{R})$ how (easily) the corresponding subspace of $n \times n$ matrices may be found.

11.7.4 Let $G = O(n, \mathbb{R}) \subset GL(n, \mathbb{R})$. Show that

(i) if we identify the tangent space in the unit element of the group $GL(n, \mathbb{R})$ (i.e. the Lie algebra $gl(n, \mathbb{R})$) with the $n \times n$ matrices, then the subspace corresponding to the subgroup $G = O(n, \mathbb{R}) \subset GL(n, \mathbb{R})$ (i.e. the Lie algebra $o(n, \mathbb{R})$) corresponds to *antisymmetric* $n \times n$ matrices
(ii) all antisymmetric matrices happen to be closed with respect to both the linear combinations and the commutator, so that they constitute a Lie algebra in their own right.

Hint: (i) let $x(t)$ be the coordinate presentation of an arbitrary curve, which starts at the unit element and which is situated (for any t) in $O(n, \mathbb{R})$ (i.e. not only in $GL(n, \mathbb{R})$); then $x(0) = \mathbb{I}_n = x^{\mathrm{T}}(0)$, $x^{\mathrm{T}}(t)x(t) = \mathbb{I}_n$; differentiation with respect to t in $t = 0$ for $\dot{x}(0) =: C$, $\dot{x}^{\mathrm{T}}(0) = C^{\mathrm{T}}$ yields the restriction on the matrices C in the form $C^{\mathrm{T}} + C = 0$. □

• We see that finding the (isomorphic copies of the) Lie algebras of matrix groups by this method is indeed simple. In general, we may summarize the algorithm for finding the matrix versions of the Lie algebras as follows: the matrices C from the *algebra* are obtained by *differentiation* (with respect to t in $t = 0$) of the defining relations valid for the matrices $x(t)$ from the *group*, in particular

1. we consider the curves in $G \subset GL(n, \mathbb{R})$ starting from the unit element; in coordinates $x(t) \in G \subset GL(n, \mathbb{R})$, $x(0) = \mathbb{I}_n$
2. the matrix version of the Lie algebra is constituted exactly by all the matrices $C := \dot{x}(0)$.

11.7.5 Check that this algorithm may also be reformulated in the following way: one writes an element of the group in the form $x = \mathbb{I}_n + \varepsilon C$, plugs the ansatz into the defining relations of the group and leaves only the terms within first-order accuracy in ε.

Hint: if $x(t) = x(0) + tC + \cdots$ then $\dot{x}(0) = C$. □

• Let us now see what this algorithm gives for the most important matrix Lie groups from Section 10.2.

11.7.6 The matrix groups from Section 10.1 may be lucidly summarized as follows:

$GL(n,\mathbb{R})$	$\det A \neq 0$	non-singular matrices
$O(r,s)$	$A^{\mathrm{T}}\eta A = \eta$	pseudo-orthogonal matrices
$SO(r,s)$	$A^{\mathrm{T}}\eta A = \eta, \ \det A = 1$	pseudo-orthogonal unimodular matrices
$O(n)$	$A^{\mathrm{T}}A = \mathbb{I}$	orthogonal matrices
$SO(n)$	$A^{\mathrm{T}}A = \mathbb{I}, \ \det A = 1$	orthogonal unimodular matrices
$SL(n,\mathbb{R})$	$\det A = 1$	unimodular matrices
$U(n)$	$A^{+}A = \mathbb{I}$	unitary matrices
$SU(n)$	$A^{+}A = \mathbb{I}, \ \det A = 1$	unitary unimodular matrices

Show that the corresponding matrix Lie algebras are given in terms of the following restrictions:

$gl(n,\mathbb{R})$	nothing	all $n \times n$ matrices
$o(r,s)$	$(\eta C)^{\mathrm{T}} = -\eta C$	pseudo-antisymmetric matrices
$so(r,s)$	just as $o(r,s)$	just as $o(r,s)$
$o(n)$	$C^{\mathrm{T}} = -C$	antisymmetric matrices
$so(n)$	just as $o(n)$	just as $o(n)$
$sl(n,\mathbb{R})$	$\operatorname{Tr} C = 0$	traceless matrices
$u(n)$	$C^{+} = -C$	antihermitean matrices
$su(n)$	$C^{+} = -C, \operatorname{Tr} C = 0$	antihermitean traceless matrices

Hint: see (10.1.5), (10.1.7) and (10.1.12); the condition $\operatorname{Tr} C = 0$ is already automatically fulfilled for $so(r,s)$ and $so(n)$; by (5.6.5) we have $\det(\mathbb{I}_n + \varepsilon C) = 1 + \varepsilon \operatorname{Tr} C$. □

11.7.7 Show that the dimensions of the Lie algebras are (for $r + s \equiv n$)

$$\dim o(r,s) = \dim so(r,s) = \dim o(n) = \dim so(n) = \frac{n(n-1)}{2}$$

$$\dim sl(n,\mathbb{R}) = n^2 - 1 \qquad \dim u(n) = n^2 \qquad \dim su(n) = n^2 - 1$$

Hint: $n(n-1)/2$ is the number of elements above the diagonal in an $n \times n$ matrix; there are imaginary numbers on the diagonal of antihermitean matrices, above the diagonal we have $2n(n-1)/2$ *real* independent numbers $\Rightarrow$ for $u(n)$ this gives $2n(n-1)/2 + n \equiv n^2$ real independent numbers altogether. □

11.7.8 Explain why the dimension of the Lie algebra $o(r,s) \equiv so(r,s)$ does not depend on the detailed splitting of n into $r + s$, but rather only on the sum n.

Hint: $C \mapsto \hat{C} := \eta C$ is a linear (which is enough here) isomorphism $o(n) \leftrightarrow o(r,s)$. □

11.7.9 Check explicitly in all cases in (11.7.6) that the sets of matrices which satisfy the restrictions listed in the table are indeed closed with respect to linear combinations and commutators, therefore being Lie subalgebras of the algebra $gl(n, \mathbb{R}) \equiv \operatorname{End} \mathbb{R}^n$. □

11.7.10 Use the algorithm from before (11.7.5) to find the Lie algebra $ga(1, \mathbb{R})$ of the group $GA(1, \mathbb{R})$ and compare the result with (11.2.4).

Solution: by (11.1.13) the curve in $GA(1, \mathbb{R})$ needed for determining of C reads

$$A(t) = \begin{pmatrix} x(t) & y(t) \\ 0 & 1 \end{pmatrix} \qquad x(0) = 1, \quad y(0) = 0$$

Then the corresponding element of the Lie algebra $ga(1, \mathbb{R})$ reads

$$C := \dot{A}(0) = \begin{pmatrix} \dot{x}(0) & \dot{y}(0) \\ 0 & 0 \end{pmatrix} \equiv \begin{pmatrix} k^1 & k^2 \\ 0 & 0 \end{pmatrix} = k^1 \begin{pmatrix} 1 & 0 \\ 0 & 0 \end{pmatrix} + k^2 \begin{pmatrix} 1 & 0 \\ 0 & 0 \end{pmatrix} \equiv k^1 E_1 + k^2 E_2$$

The Lie algebra $ga(1, \mathbb{R})$ thus consists of the matrices of the form $\left(\begin{smallmatrix} k^1 & k^2 \\ 0 & 0 \end{smallmatrix}\right)$ and a basis E_1, E_2 may be chosen. One easily verifies that $[E_a, E_b] = c^c_{ab} E_c$, the structure constants being identical with those from (11.2.4). □

• The explicit construction of one-parameter subgroups and the exponential map reduces in the matrix formalism to the computation of *exponentials of matrices*, i.e. the infinite series $e^X := \mathbb{I}_n + X + (1/2!)X^2 + \cdots$, where X is an $n \times n$ matrix (an element of the Lie algebra corresponding to the subgroup). We already encountered this in problem (11.3.5) for the case of $GL(n, \mathbb{R})$, but it is clear from the solution of the problem that it holds for any subgroup $G \subset GL(n, \mathbb{R})$ (i.e. matrix group G) as well, if X belongs to the subalgebra $\mathcal{G} \subset gl(n, \mathbb{R})$.

11.7.11 Find the one-parameter subgroups and the exponential map for $GA(1, \mathbb{R}) \subset GL(2, \mathbb{R})$ by explicit summation of the series e^{tC}, where C is the matrix from the problem (11.7.10). Show that in this way one obtains formulas which coincide with those from (11.3.6) and (11.4.5), i.e.

$$\exp\left\{t \begin{pmatrix} k^1 & k^2 \\ 0 & 0 \end{pmatrix}\right\} = \begin{pmatrix} x(t) & y(t) \\ 0 & 1 \end{pmatrix} \qquad x(t) = e^{k^1 t}, \quad y(t) = \frac{k^2}{k^1}(e^{k^1 t} - 1)$$

so that

$$\exp : (k^1, k^2) \mapsto \left(e^{k^1}, (k^2/k^1)(e^{k^1} - 1)\right)$$

Hint: explore several powers and gain insight into the structure of the series (what the nth power looks like). □

• The series of the form e^{tC} may also be fairly easily summed explicitly in some other cases. In the problems discussed below this will be carried out for $SU(2)$ and $SO(3)$ (the results may come in handy one day) as well as for the group $SL(2,\mathbb{R})$, from which one can learn that the exponential map need not always be surjective (there may exist elements in G which cannot be written as e^X for $X \in \mathcal{G}$).

11.7.12 Let $\mathcal{G} = su(2)$. Show that

(i) given σ_j, $j = 1, 2, 3$ the *Pauli matrices*

$$\sigma_1 = \begin{pmatrix} 0 & 1 \\ 1 & 0 \end{pmatrix} \qquad \sigma_2 = \begin{pmatrix} 0 & -i \\ i & 0 \end{pmatrix} \qquad \sigma_3 = \begin{pmatrix} 1 & 0 \\ 0 & -1 \end{pmatrix}$$

we may use the matrices $e_j := -\frac{1}{2} i\sigma_j$ as a basis in $su(2)$

(ii) the structure constants with respect to this basis read $c^k_{ij} = \varepsilon_{ijk}$ (compare with (11.7.13))

(iii) a general element $X \in su(2)$ may be written in the form

$$X = -\frac{i}{2}\alpha\mathbf{n}\cdot\boldsymbol{\sigma} \equiv -\frac{i}{2}\alpha n_j\sigma_j \qquad \alpha \in \mathbb{R}, \quad |\mathbf{n}| = 1$$

(iv) the one-parameter groups in $SU(2)$ are

$$A(t) = e^{-\frac{1}{2}it\alpha\mathbf{n}\cdot\boldsymbol{\sigma}} \qquad \alpha \in \mathbb{R}, \quad |\mathbf{n}| = 1$$

or, when written in an alternative form

$$A(t) = \mathbb{I}_2 \cos\frac{t\alpha}{2} - i(\mathbf{n}\cdot\boldsymbol{\sigma})\sin\frac{t\alpha}{2}$$

(v) the most general matrix $A \in SU(2)$ reads

$$A = \begin{pmatrix} z & -\bar{w} \\ w & \bar{z} \end{pmatrix} \qquad |z|^2 + |w|^2 = 1 \quad (z, w \in \mathbb{C})$$

or, equivalently

$$A = \mathbb{I}_2 a_0 + i(\mathbf{a}\cdot\boldsymbol{\sigma}) \equiv \begin{pmatrix} a_0 + ia_3 & a_2 + ia_1 \\ -a_2 + ia_1 & a_0 - ia_3 \end{pmatrix} \qquad a_0^2 + \mathbf{a}\cdot\mathbf{a} \equiv a_0^2 + a_1^2 + a_2^2 + a_3^2 = 1$$

from which we can see immediately that topologically $SU(2) = S^3$

(vi) the general $A \in SU(2)$ may always be written in the form e^X, $X \in su(2)$, i.e. the exponential map "covers" the *whole* group $SU(2)$ (it is surjective).

Hint: (iv) since (check) $(\mathbf{n}\cdot\boldsymbol{\sigma})^2 \equiv (n_i\sigma_i)(n_j\sigma_j) = \mathbb{I}_2$, the identity $e^{i\alpha\mathbf{n}\cdot\boldsymbol{\sigma}} = \mathbb{I}_2\cos\alpha + i(\mathbf{n}\cdot\boldsymbol{\sigma})\sin\alpha$ holds; (vi) the form of the one-parameter groups of item (iv) exactly matches item (v) for $a_0 = \cos(t\alpha/2)$, $\mathbf{a} = -\mathbf{n}\sin(t\alpha/2)$; the most general solution of the requirement $a_0^2 + \mathbf{a}\cdot\mathbf{a} = 1$ is, however, just $a_0 = \cos\beta$, $\mathbf{a} = \mathbf{n}\sin\beta$ ($|\mathbf{n}| = 1$) so that each matrix from $SU(2)$ lies in some one-parameter subgroup. □

11.7.13 Let $\mathcal{G} = so(3)$. Show that

(i) we may use the 3×3 matrices l_i with the elements

$$(l_i)_{jk} := -\varepsilon_{ijk}$$

i.e. explicitly

$$l_1 = \begin{pmatrix} 0 & 0 & 0 \\ 0 & 0 & -1 \\ 0 & 1 & 0 \end{pmatrix} \qquad l_2 = \begin{pmatrix} 0 & 0 & 1 \\ 0 & 0 & 0 \\ -1 & 0 & 0 \end{pmatrix} \qquad l_3 = \begin{pmatrix} 0 & -1 & 0 \\ 1 & 0 & 0 \\ 0 & 0 & 0 \end{pmatrix}$$

as a basis in $so(3)$

(ii) the structure constants with respect to this basis read $c_{ij}^k = \varepsilon_{ijk}$

(iii) a general element $X \in so(3)$ may be expressed as

$$X = \alpha \mathbf{n} \cdot \mathbf{l} \equiv \alpha n_i l_i \qquad \alpha \in \mathbb{R}, \quad |\mathbf{n}| = 1$$

(iv) the one-parameter subgroups in $SO(3)$ are $A(t) = R(t\alpha, \mathbf{n})$, where

$$R(\alpha, \mathbf{n}) := e^{\alpha \mathbf{n} \cdot \mathbf{l}}$$

(v) the explicit form of the 3×3 matrices $R \equiv R(\alpha, \mathbf{n})$ is

$$R_{ij} \equiv (R(\alpha, \mathbf{n}))_{ij} = \delta_{ij} \cos\alpha + (1 - \cos\alpha) n_i n_j - \varepsilon_{ijk} n_k \sin\alpha$$

so that if we multiply a column $\mathbf{x} \in \mathbb{R}^3$ from the left by such a matrix,

$$x_i \mapsto x'_i := R_{ij} x_j$$

the result may be written as

$$\mathbf{x}' = \mathbf{x} \cos\alpha + (1 - \cos\alpha)(\mathbf{n} \cdot \mathbf{x})\mathbf{n} + (\mathbf{n} \times \mathbf{x}) \sin\alpha$$

(vi) this formula describes the result of the *rotation* of the vector $\mathbf{x}$ by an angle α around the axis given by the vector $\mathbf{n}$, so that the matrix $R(\alpha, \mathbf{n})$ corresponds to the *most general rotation* in the Euclidean space E^3 $\Rightarrow$ the most general matrix from $SO(3)$ $\Rightarrow$ the exponential map "covers" the *whole* group $SO(3)$ (it is surjective, each matrix $A \in SO(3)$ is of the form e^X for some $X \in so(3)$)

(vii) motion along the one-parameter subgroup $A(t) = R(t\alpha, \mathbf{n})$ corresponds to a uniform rotation with angular velocity $\omega = \alpha \mathbf{n}$.

Hint: (v) compute several powers and gain insight into the structure of the series (what the nth power looks like); for example, $X_{ij} = \alpha n_k (l_k)_{ij} = -\alpha \varepsilon_{ijk} n_k$, $X_{ij}^2 = X_{ik} X_{kj} = \cdots$, (5.6.4); (vii) in time t we turn by $t\alpha$ around (always the same) $\mathbf{n}$. □

11.7.14 Let $\mathcal{G} = sl(2, \mathbb{R})$. Show that

(i) a general element of the Lie algebra has the form[179]

$$X = \begin{pmatrix} a & b \\ c & -a \end{pmatrix} \qquad a, b, c \in \mathbb{R}$$

(ii) for the square of the matrix X we get

$$X^2 = \varkappa \mathbb{I}_2 \qquad \varkappa \equiv a^2 + bc \equiv -\det X \in \mathbb{R}$$

[179] A word of caution for beginner relativists is in order: c in this expression has *nothing to do* with the velocity of light in vacuum.

(iii) the one-parameter subgroups $A(t) \equiv e^{tX}$ read as

$$A(t) = \begin{pmatrix} \cosh|k|t + \frac{a}{|k|}\sinh|k|t & \frac{b}{|k|}\sinh|k|t \\ \frac{c}{|k|}\sinh|k|t & \cosh|k|t - \frac{a}{|k|}\sinh|k|t \end{pmatrix} \qquad \text{for } \varkappa \equiv k^2 > 0$$

$$A(t) = \begin{pmatrix} 1+ta & tb \\ tc & 1-ta \end{pmatrix} \qquad \text{for } \varkappa = 0$$

$$A(t) = \begin{pmatrix} \cos|k|t + \frac{a}{|k|}\sin|k|t & \frac{b}{|k|}\sin|k|t \\ \frac{c}{|k|}\sin|k|t & \cos|k|t - \frac{a}{|k|}\sin|k|t \end{pmatrix} \qquad \text{for } \varkappa \equiv -k^2 < 0$$

(iv) the traces of the matrices come out as follows in these three cases:

$$\begin{aligned} &\mathrm{Tr}\,A(t) = 2\cosh|k|t \geq 2 && \varkappa > 0 \\ &\mathrm{Tr}\,A(t) = 2 && \varkappa = 0 \\ &\mathrm{Tr}\,A(t) = 2\cos|k|t \in \langle -2, 2\rangle && \varkappa < 0 \end{aligned}$$

so that we always get $\mathrm{Tr}\,A(t) \geq -2$

(v) the matrix

$$B = \begin{pmatrix} -2 & 0 \\ 0 & -\frac{1}{2} \end{pmatrix}$$

is from $SL(2,\mathbb{R})$ and $\mathrm{Tr}\,B = -\frac{5}{2} < -2$. Therefore the image $\exp(sl(2,\mathbb{R}))$ is not the whole $SL(2,\mathbb{R})$, i.e. the exponential map "does not cover" the whole group $SL(2,\mathbb{R})$. □

• There may sometimes be a simple *topological* reason why the exponential map fails to "cover" the whole group – the non-connectedness of the group. A topological space M is said to be (arcwise) *connected* if for any two points $a, b \in M$ there is a *continuous* map of the interval $\langle 0, 1\rangle$ to M such that the image of 0 is the point "a" and the image of 1 is the point "b." Visually this means that any two points may be linked together by a (continuous) curve, all of its points being in M.

11.7.15 Be sure to understand that

(i) the relation "may be linked together..." is an equivalence on M; one may then factorize with respect to this equivalence and get the factorspace (classes of this particular equivalence are called the *connected components*)
(ii) the space M is connected $\Leftrightarrow$ there is only a single equivalence class (the whole space is thus a single connected component)
(iii) the connected components are already connected.

Hint: (i) if $a \sim b$, $b \sim c$, a curve from a to c is given by the "linkage" of the curves *plus a reparametrization* (in order that 1 maps to c). □

11.7.16 Check that the groups $GL(n,\mathbb{R})$ and $O(n,\mathbb{R})$ *are not* connected, so that they have *at least* two connected components (actually both have *just two*).

Hint: the map $GL(n, \mathbb{R}) \to \mathbb{R}$, $A \mapsto \det A$ is continuous, so that one cannot link two points (non-singular matrices) which differ in the sign of the determinant (the determinant of a joining curve $A(t)$ is a continuous function of t, which has different signs at $t = 0$ and $t = 1$ $\Rightarrow$ it should vanish somewhere inside the interval, which corresponds to a *singular* matrix; so the curve *is not wholly* in $GL(n, \mathbb{R})$). For $A \in O(n, \mathbb{R})$ we have $\det A = \pm 1$. □

• A general Lie group thus may have more than one connected component and the identity element of the group lies in one of them. It is called the *connected component of the identity*. In the context we are interested in now it is important to realize that the image of the exponential map cannot contain more than this component of the group.

11.7.17 Show that the image of the exponential map always lies in the connected component of the identity, i.e. the group elements which are not lucky enough to live in this component cannot be written as e^X for any $X \in \mathcal{G}$.

Hint: the one-parameter subgroup e^{tX} links together the identity with the element $g = e^X$. □

11.7.18 For the group $GA(1, \mathbb{R})$ show explicitly (and also draw a picture) that

(i) it has two connected components
(ii) no one-parameter subgroup reaches "the other" component (not belonging to the identity)
(iii) the exponential map "covers" the whole component of the identity.

Hint: (i) non-singularity of $\begin{pmatrix} x & y \\ 0 & 1 \end{pmatrix}$ excludes the y-axis in the plane xy; (ii), (iii) see (11.3.6). □

• Now we should appreciate much more the non-triviality of the result (11.7.14). We learned there that the exponential map on $SL(2, \mathbb{R})$ does not "cover" the whole group. We mention without proof that this group happens to be *connected*. Non-surjectivity of the exponential map is thus not related to what was discussed above and it provides the lesson that even on connected Lie groups it may not be possible to write down each element as e^X.

In connection with all of this the term *exponential group* is sometimes introduced (if $G = \exp \mathcal{G}$). We learned that $SU(2)$ and $SO(3)$ *are* exponential groups, whereas $SL(2, \mathbb{R})$ as well as any non-connected group *are not* exponential. Notice that if a group happens to be exponential, it is necessarily also connected (we get from e^X to e^Y along the path consisting of two parts, first by e^{tX} to $e \in G$ and then by e^{tY} to e^Y).

Now we would like to make good the election pledges we gave to our electorate at the end of Section 11.2 (and as an indication in a note before (11.1.12)). This concerned the technique of finding the *canonical 1-form* θ on G as well as the left-invariant 1-forms. The first step is to realize that the matrix $x^{-1}dx$ (with 1-form elements), which we encountered in (11.1.9), may be actually regarded as *the* canonical 1-form *on* $GL(n, \mathbb{R})$.

11.7.19 Be sure to understand that the canonical 1-form $\hat{\theta}$ on $GL(n, \mathbb{R})$ is indeed given by the matrix (-valued 1-form)

$$\hat{\theta} = x^{-1}dx = (x^{-1}dx)^i_j E^j_i \equiv (x^{-1})^i_k dx^k_j E^j_i$$

and check explicitly its left-invariance.

Solution: it is a matrix-valued (i.e. a $gl(n, \mathbb{R})$-valued) 1-form on $GL(n, \mathbb{R})$. It satisfies the definition relation (11.2.6): $\langle \hat{\theta}, V_C \rangle = x^{-1}\langle dx, V_C \rangle = x^{-1}V_C x = C$. Finally, $L^*_A(x^{-1}\,dx) = (Ax)^{-1}\,d(Ax) = x^{-1}\,dx$. □

• Now we will take a step forward and try to elucidate the relation between the canonical forms on G and H, provided that $j : G \to H$ is an injective homomorphism.

11.7.20 Let $j : G \to H$ be a homomorphic embedding of Lie groups (so that $j(G) \subset H$ is an isomorphic copy of G in H) and let $\hat{\theta} = \hat{\theta}^a E_a$ be the canonical 1-form on H (E_a being a basis in $\mathcal{H}$). Show that its pull-back $j^*\hat{\theta} \equiv (j^*\hat{\theta}^a)E_a$ is "almost" the canonical 1-form on G; the difference lies *only* in the fact that it is not ("true") $\mathcal{G}$-valued, but its values are in the isomorphic copy $j_{*e}\mathcal{G} \subset \mathcal{H}$.

Hint: making use of (11.1.11) and (11.2.6) we get $(j^*\hat{\theta})(g) = \hat{\theta}(j(g)) \circ j_* = \hat{L}_{j(g)^{-1}*} \circ j_* = (L_{j(g)^{-1}} \circ j)_* = j_{*e} \circ L_{g^{-1}*} \equiv j_*\theta(g)$. If $\mathcal{E}_i$ is a basis in $\mathcal{G}$, $j_{*e}\mathcal{E}_i$ is its image in $\mathcal{H}$ (its linear envelope being the copy of $\mathcal{G}$ in $\mathcal{H}$), so that in detail $(j^*\hat{\theta}^a)E_a = \theta^i(j_{*e}\mathcal{E}_i)$. This means that, in general, the form $j^*\hat{\theta}$ does not need the *whole* basis E_a of the algebra $\mathcal{H}$ for its decomposition, but rather only $j_{*e}\mathcal{E}_i$, which is just the basis of the *subalgebra* $j_{*e}\mathcal{G} \subset \mathcal{H}$. □

• This already gives a straightforward conclusion concerning the situation with arbitrary matrix groups. We simply take $GL(n, \mathbb{R})$ as H and some of its subgroups as G.

11.7.21 Be sure to understand that

(i) if we introduce local coordinates z^μ on a matrix group, i.e. if

$$j : G \to GL(n, \mathbb{R}) \qquad z^\mu \mapsto x^i_j(z^\mu)$$

is an injective homomorphism, then the 1-form on G

$$\theta := j^*(x^{-1}\,dx)$$

may be actually regarded as the canonical 1-form on G

(ii) its decomposition into component forms

$$\theta = \theta^a E_a \equiv \left(\theta^a_\mu\,dz^\mu\right) E_a$$

contains only the matrices E_a, which constitute a basis of the subalgebra $\mathcal{G} \subset gl(n, \mathbb{R})$ (rather than the whole Weyl basis E^i_j of the algebra $gl(n, \mathbb{R})$).

Hint: see (11.7.19) and (11.7.20). □

• Let us review some facts we have already encountered earlier from this new point of view.

[11.7.22] Express the results of the problems (11.1.12) and (11.1.13) in the language of problem (11.7.21).

Solution: for $SO(2)$

$$\theta = \begin{pmatrix} 0 & d\varphi \\ -d\varphi & 0 \end{pmatrix} = d\varphi \begin{pmatrix} 0 & 1 \\ -1 & 0 \end{pmatrix} \equiv \theta^1 E_1$$

and for $GA(1,\mathbb{R})$

$$\theta = \begin{pmatrix} x^{-1}dx & x^{-1}dy \\ 0 & 0 \end{pmatrix} = e^1 \begin{pmatrix} 1 & 0 \\ 0 & 0 \end{pmatrix} + e^2 \begin{pmatrix} 0 & 1 \\ 0 & 0 \end{pmatrix} \equiv \theta^1 E_1 + \theta^2 E_2$$

where $\theta^a = e^a$ from (11.1.13) and E_a is the basis from (11.7.10), which matches the decomposition in (11.2.6). □

The technique of finding explicit expression of the canonical 1-form θ (and consequently the left-invariant 1-forms e^a) on a matrix group G may be thus summarized in general as follows:

(i) we parametrize the group elements $x^i_j(z^\mu)$ in terms of arbitrary local coordinates z^μ
(ii) we form the expression (a matrix, whose entries are 1-forms) $\theta := x^{-1}(z)\, dx(z)$
(iii) this object may be decomposed uniquely as $\theta = \theta^a E_a$ with respect to the matrices E_a, which constitute a basis of $\mathcal{G}$; the "coefficients of the decomposition" turn out to be just the left-invariant[180] 1-forms e^a, moreover just corresponding to this very basis ($e^a \leftrightarrow E^a \leftrightarrow E_a$).

• Now, let us apply this technique to a new situation, the computation of the canonical 1-form and left-invariant fields on the group $SU(2)$.

[11.7.23] Find the canonical 1-form θ on the group $SU(2)$ and the left-invariant 1-forms corresponding to the basis $E_a = -\frac{1}{2}i\sigma_a$ in $su(2)$. In order to do this show in successive steps that

(i) the constraint $|z|^2 + |w|^2 = 1$ from (11.7.12) may be solved by parametrization in terms of the *Euler angles*

$$z = \cos\frac{\vartheta}{2} e^{-\frac{1}{2}i(\psi+\varphi)} \qquad w = \sin\frac{\vartheta}{2} e^{-\frac{1}{2}i(\psi-\varphi)}$$
$$\vartheta \in \langle 0,\pi\rangle, \quad \varphi \in \langle 0, 2\pi\rangle, \quad \psi \in \langle 0, 4\pi\rangle$$

(ii) on $SU(2)$ we have $A^{-1} = A^+$ and so the canonical 1-form may also be written as

$$\theta \equiv A^{-1}\, dA = A^+\, dA \qquad A = A(\vartheta,\varphi,\psi)$$

(iii) the left-invariant 1-forms e^a, given by the decomposition

$$A^{-1}\, dA = A^+\, dA \overset{!}{=} e^a E_a \equiv -\frac{i}{2} \begin{pmatrix} e^3 & e^1 - ie^2 \\ e^1 + ie^2 & -e^3 \end{pmatrix}$$

[180] To find the *right*-invariant forms f^a we start with the matrix $(dx)x^{-1}$ and make a decomposition $(dx)x^{-1} = f^a E_a$.

and the dual left-invariant vector fields e_a turn out to be

$$\begin{aligned} e^1 &= \sin\psi\, d\vartheta - \sin\vartheta\cos\psi\, d\varphi & e_1 &= \sin\psi\,\partial_\vartheta + \cot\vartheta\cos\psi\,\partial_\psi - \frac{\cos\psi}{\sin\vartheta}\partial_\varphi \\ e^2 &= \cos\psi\, d\vartheta + \sin\vartheta\sin\psi\, d\varphi & e_2 &= \cos\psi\,\partial_\vartheta - \cot\vartheta\sin\psi\,\partial_\psi + \frac{\sin\psi}{\sin\vartheta}\partial_\varphi \\ e^3 &= d\psi + \cos\vartheta\, d\varphi & e_3 &= \partial_\psi \end{aligned}$$

(iv) the right-invariant 1-forms f^a, given by the decomposition

$$(dA)A^{-1} = (dA)A^{+} \overset{!}{=} f^a E_a \equiv -\frac{i}{2}\begin{pmatrix} f^3 & f^1 - if^2 \\ f^1 + if^2 & -f^3 \end{pmatrix}$$

and the dual right-invariant vector fields f_a turn out to be

$$\begin{aligned} f^1 &= -\sin\varphi\, d\vartheta + \sin\vartheta\cos\varphi\, d\psi & f_1 &= -\sin\varphi\,\partial_\vartheta - \cot\vartheta\cos\varphi\,\partial_\varphi + \frac{\cos\varphi}{\sin\vartheta}\partial_\psi \\ f^2 &= \cos\varphi\, d\vartheta + \sin\vartheta\sin\varphi\, d\psi & f_2 &= \cos\varphi\,\partial_\vartheta - \cot\vartheta\sin\varphi\,\partial_\varphi + \frac{\sin\varphi}{\sin\vartheta}\partial_\psi \\ f^3 &= d\varphi + \cos\vartheta\, d\psi & f_3 &= \partial_\varphi \end{aligned}$$

Hint: see (11.7.12) and (10.1.12). □

Summary of Chapter 11

Differential geometry turns out to provide an effective tool for studying such sophisticated objects as Lie groups represent. Their rich geometry stems from the compatibility of the two structures involved. A much simpler object, the Lie *algebra* (it is a finite-dimensional vector space), may be associated canonically with each Lie group with the help of the *left-invariant* vector fields. In spite of its simplicity the Lie algebra of a group encodes a great deal (the essential part) of information concerning the group itself. The exponential map from the Lie algebra to the group is introduced.

$L_g h := gh,\ R_g h := hg$	Left translation, right translation	(11.1.1)
$L_g^* T = T$	T is left-invariant tensor field on G	(11.1.4)
$e_a(g) = L_{g*}E_a,\ E_a = e_a(e)$	Left-invariant frame field generated by E_a	(11.1.6)
$(x^{-1})^i_k\, dx^k_j \equiv (x^{-1}\, dx)^i_j$	Left-invariant 1-forms on $GL(n,\mathbb{R})$	(11.1.9)
$x^i_k \partial^k_j \equiv (x\partial)^i_j$	Left-invariant vector fields on $GL(n,\mathbb{R})$	(11.1.10)
$[E_a, E_b] = c^c_{ab} E_c$	Structure constants with respect to E_a	(11.2.2)
$de^a + \frac{1}{2}c^a_{bc} e^b \wedge e^c = 0$	Maurer–Cartan formula in terms of e^a	(11.2.3)
$\langle\theta, L_X\rangle := X,\ \theta = e^a E_a$	Canonical (Maurer–Cartan) 1-form θ on G	(11.2.6)
$d\theta + \frac{1}{2}[\theta\wedge\theta] = 0$	Maurer–Cartan formula in terms of θ	(11.2.6)
$\gamma(t+s) = \gamma(t)\gamma(s),\ \gamma(0) = e$	One-parameter subgroup on G	Sec. 11.3
$\gamma^X(t) = e^{tX}$	One-parameter subgroup in terms of exp	(11.4.1)
$f(e^X) = e^{f'(X)}$	Derived homomorphism f'	(11.5.3)
$x^{-1}\, dx$	Canonical 1-form on $GL(n,\mathbb{R})$	(11.7.19)
$j^*(x^{-1}\, dx) = x^{-1}(z)\, dx(z)$	Canonical 1-form on matrix groups	(11.7.21)

12

Representations of Lie groups and Lie algebras

• As we already mentioned at the beginning of Section 10.1, groups always occur as groups of transformations of something, through their *action* on a set (usually endowed with some additional structure). Thus, there exists a rule which assigns to each element g of a group a transformation L_g of some set M. A study of group theory thus naturally incorporates[181] besides knowledge of the groups themselves also the question of where and how a given group may act.

In this chapter we will systematically treat a particular, but very important, class of actions, which are called *representations*. From the perspective of general actions, to be discussed in more detail in Chapter 13, they are singled out by operating in *linear spaces* and, moreover, *linearly*. Such a distinguished position of just this class of actions within the scope of all actions is simply the reflection of the distinguished position of linear spaces within the scope of various mathematical structures. Representations may be found wherever symmetries and linearity meet in one place.

12.1 Basic concepts

• If a symmetry group is to act in a linear space V, it is natural to ask for the compatibility of the symmetry operations with the linear structure. This means that to each group element g we should assign a *linear* operator $\rho(g)$, i.e. $\rho(g) \in \text{End}\, V$. Moreover, these maps should also "reproduce" the behavior of the abstract group G itself, i.e. to be *homomorphisms* from G to $\text{End}\, V$. Recall, however, that *all* the linear maps (i.e. $\text{End}\, V$) *do not constitute* a group so that the concept of a homomorphism from G to $\text{End}\, V$ formally (as a group homomorphism) makes no sense.

12.1.1 Be sure to understand that

(i) $\text{Aut}\, V \subset \text{End} V \equiv \text{Hom}\,(V, V)$

[181] Especially in those despicable cases when we put our mind to this with, from the very beginning, the view of *using* our knowledge somewhere.

(ii) if ρ is to be a homomorphism, its image $\text{Im}\,\rho \equiv \rho(G)$ has to lie in the part $\text{Aut}\,V$, i.e. each operator $\rho(g)$ is to be *invertible*; in general, we may write

$$\rho(G) \subset \text{Aut}\,V \subset \text{End}\,V$$

Hint: (ii) to each g there is a g^{-1}. □

• Symmetries in V are thus described by homomorphisms

$$\rho : G \to \text{Aut}\,V \equiv GL(V) \qquad \rho(g\tilde{g}) = \rho(g)\rho(\tilde{g})$$

Such a homomorphism[182] is called a *representation of a group G* in a vector space V. The dimension of the space V is called the *dimension of the representation* ρ. The representation is thus a *left linear* action.[183]

There is an alternative "matrix" point of view on representations: to each group element g a representation ρ assigns a linear (invertible) operator $\rho(g)$ and thus, after introducing a basis E_a in V, also the *matrix* of the operator

$$\rho(g)E_a =: (\rho(g))^b_a E_b \qquad g \mapsto (\rho(g))^b_a$$

Then we obtain an assignment "group element $\mapsto$ matrix," in which a product of elements is mapped to the product of the corresponding *matrices*.

Consider, for example, the group consisting of just three elements $G = \{e, a, b\}$ from problem (11.1.3). Then one easily checks that the assignment

$$e \mapsto \begin{pmatrix} 1 & 0 & 0 \\ 0 & 1 & 0 \\ 0 & 0 & 1 \end{pmatrix} \quad a \mapsto \begin{pmatrix} 0 & 1 & 0 \\ 0 & 0 & 1 \\ 1 & 0 & 0 \end{pmatrix} \quad b \mapsto \begin{pmatrix} 0 & 0 & 1 \\ 1 & 0 & 0 \\ 0 & 1 & 0 \end{pmatrix}$$

is a representation of the group in $\mathbb{R}^3$. Let us look at two further simple (and important) examples.

12.1.2 Given $G \subset GL(n, \mathbb{R})$, a *matrix* group, check that the prescription

$$A \mapsto \rho(A) \equiv A$$

is a representation of the group in $\mathbb{R}^n$ (the matrix A, regarded as an element of the group G, is represented by the *matrix A itself*, interpreted, however, as a linear operator in $\mathbb{R}^n$). We will denote this representation as $\rho = \text{id}$ ("identity"; sometimes it is also called "tautological"). □

12.1.3 Given a non-singular square matrix A (i.e. $A \in GL(n, \mathbb{R})$) check that

(i) the prescription $A \mapsto \det A$ may be regarded as a (one-dimensional) representation of the group $GL(n, \mathbb{R})$

[182] When we speak about *Lie* groups, it is natural to add the requirement of a *smooth* dependence on g, i.e. of the smoothness of the map $G \times V \to V$, $(g, v) \mapsto \rho(g)v$.

[183] The (more general) concept of *left action* is introduced in Section 13.1.

(ii) if H is a subgroup of G then the representation of G also automatically gives the representation of H (the *restriction on a subgroup*); in particular, $A \mapsto \det A$ is the representation of an arbitrary *sub*group of $GL(n, \mathbb{R})$ (= matrix group)
(iii) the generalization to $A \mapsto (\det A)^\lambda$, $\lambda \in \mathbb{R}$ is also a representation. □

• Recall that the idea of a representation of a group stems from the simple observation that *invertible* linear operators are naturally endowed with the structure of a *group*. More careful observations, however, result in a more general statement; namely, that *particular subsets* of linear operators are naturally endowed with various useful algebraic structures. Representations are then homomorphisms (in order to preserve the operations) of the structure under consideration to the corresponding class of operators. Two (or three) important cases are of particular interest for us: the *invertible* operators, i.e. $\operatorname{Aut} V \equiv GL(V)$, constitute a *group* (the case we started with) whereas *all* linear operators, i.e. $\operatorname{End} V$, constitute an *associative algebra* and consequently also a *Lie* algebra (the commutator being realized as $[A, B] := AB - BA$ – we learned this in (11.7.1); this will be used (now) to represent Lie algebras and (later) Clifford algebras). Namely, the *representation of a Lie algebra* $\mathcal{G}$ in a vector space V is a Lie algebra homomorphism

$$f : \mathcal{G} \to \operatorname{End} V$$

i.e. we assign to an arbitrary element X of the Lie algebra $\mathcal{G}$ a linear operator $f(X)$ so that

$$f(X + \lambda Y) = f(X) + \lambda f(Y) \qquad \text{depends linearly on } X$$

$$f([X, Y]) = [f(X), f(Y)] \equiv f(X)f(Y) - f(Y)f(X) \qquad \text{preserves the commutator}$$

Note that (as many as) *two*[184] linearities occur here: $f(X)$ is a *linear* operator in V, depending moreover *linearly* on X.

12.1.4 Let $f : \mathcal{G} \to \operatorname{End} V$ be a representation of a Lie algebra $\mathcal{G}$ in V, E_j a basis in $\mathcal{G}$ and let $\mathcal{E}_j := f(E_j)$ be the *generators of the representation*, i.e. the linear operators in V which represent the *basis* elements of the Lie algebra. Check that the operators $\mathcal{E}_j$ satisfy "the same commutation relations" as the basis elements E_j do for the Lie algebra itself, i.e.

$$[E_i, E_j] = c^k_{ij} E_k \quad \Rightarrow \quad [\mathcal{E}_i, \mathcal{E}_j] = c^k_{ij} \mathcal{E}_k$$

To find a representation of a Lie algebra thus (also) means finding a set of operators $\mathcal{E}_j$, whose commutation relations coincide with the commutation relations of the basis E_j.
Hint: f is a homomorphism. □

12.1.5 Let $\mathcal{G}, \tilde{\mathcal{G}}$ be two Lie algebras of the same dimension. Check that

(i) they are isomorphic if and only if their structure constants happen to coincide in appropriately chosen bases
(ii) the Lie algebras $so(3)$ and $su(2)$ are isomorphic, moreover they are isomorphic to the Lie algebra $(E^3, \times)$ (common three-dimensional vectors, commutator := the vector product)

[184] In a more general case of *action* on a manifold (see Section 13.4) the first linearity dies away, but the second one still survives.

(iii) the correspondence is

$$\mathbf{a}\cdot\mathbf{l} \leftrightarrow -\frac{i}{2}\mathbf{a}\cdot\boldsymbol{\sigma} \leftrightarrow \mathbf{a}$$

(iv) the commutators of the general elements turn out to be

$$[\mathbf{a}\cdot\mathbf{l},\mathbf{b}\cdot\mathbf{l}] = (\mathbf{a}\times\mathbf{b})\cdot\mathbf{l} \qquad \left[-\frac{i}{2}\mathbf{a}\cdot\boldsymbol{\sigma}, -\frac{i}{2}\mathbf{b}\cdot\boldsymbol{\sigma}\right] = -\frac{i}{2}(\mathbf{a}\times\mathbf{b})\cdot\boldsymbol{\sigma} \qquad [\mathbf{a},\mathbf{b}] = \mathbf{a}\times\mathbf{b}$$

(v) these isomorphisms may also be regarded as either a two-dimensional complex ($\equiv$ in the space $\mathbb{C}^2$) representation of the algebra $so(3)$ or a three-dimensional real ($\equiv$ in the space $\mathbb{R}^3$) representation of the algebra $su(2)$

$$\mathbf{a}\cdot\mathbf{l} \mapsto \rho_1(\mathbf{a}\cdot\mathbf{l}) := -\frac{i}{2}\mathbf{a}\cdot\boldsymbol{\sigma} \qquad -\frac{i}{2}\mathbf{a}\cdot\boldsymbol{\sigma} \mapsto \rho_2\left(-\frac{i}{2}\mathbf{a}\cdot\boldsymbol{\sigma}\right) := \mathbf{a}\cdot\mathbf{l}$$

or on the bases

$$l_j \mapsto \rho_1(l_j) := -\frac{i}{2}\sigma_j \qquad -\frac{i}{2}\sigma_j \mapsto \rho_2\left(-\frac{i}{2}\sigma_j\right) := l_j$$

(vi) similarly, any homomorphism of two matrix Lie algebras may be regarded as a representation.

Hint: (i) $E_j \leftrightarrow \tilde{E}_j \equiv f(E_j)$; (iii) see (11.7.12) and (11.7.13). □

• The relation between Lie groups and Lie algebras is reflected in a corresponding relation between their representations. The unique and trouble-free route $G \mapsto \mathcal{G}$ (a Lie group G induces *its* Lie algebra $\mathcal{G}$) corresponds to the unique and trouble-free route $\rho \mapsto \rho'$ at the level of their representations: a representation ρ of a Lie group G induces a unique representation ρ' of the Lie algebra $\mathcal{G}$; it is called the *derived* representation (12.1.6). However, just as the route $\mathcal{G} \mapsto G$ need not be so simple (two isomorphic algebras may have "above them" non-isomorphic groups, e.g. $su(2) = so(3)$, but $SU(2) \neq SO(3)$), in general also the route $\rho' \mapsto \rho$ may be more involved.

The explicit form of the derived representation of $\mathcal{G}$ is but a special case of the derived homomorphism from Section 11.5.

12.1.6 Given ρ a representation of a Lie group G in V, show that

(i) the Lie algebra of the group $H = \text{Aut}\,V \equiv GL(V)$ is $\mathcal{H} = \text{End}\,V$ (in the matrix version $H = GL(n,\mathbb{R})$ leads to $\mathcal{H} = gl(n,\mathbb{R}) =$ *all* real $n\times n$ matrices)

(ii) the derived homomorphism ρ', given by the commutative diagram

$$\begin{array}{ccc} G & \xrightarrow{\ \rho\ } & \text{Aut}\,V \\ \uparrow{\scriptstyle\exp} & & \uparrow{\scriptstyle\exp} \\ \mathcal{G} & \xrightarrow[\ \rho'\ ]{} & \text{End}\,V \end{array} \qquad \text{i.e.}\quad \rho(e^X) = e^{\rho'(X)}$$

and by the explicit formula

$$\rho'(X) = \left.\frac{d}{dt}\right|_0 \rho(e^{tX})$$

is indeed a representation of $\mathcal{G}$ in V; it is called the *derived representation*

(iii) the formula given above for the computation of ρ' may be rewritten in the convenient form

$$\rho(1+\epsilon X) = 1 + \epsilon\rho'(X) \qquad \text{within order } \epsilon$$

(iv) if a basis E_a is fixed in V then the derived representation ρ' of the Lie algebra is encoded in the matrix elements ρ^b_{ai} of the generators $\mathcal{E}_i \equiv \rho'(E_i)$ of the representation, given by

$$\rho'(E_i)E_a =: \rho^b_{ai}E_b$$

Hint: (i) the operator $1+\epsilon X$ has the inverse $1-\epsilon X$ (i.e. it has an inverse operator for *any* linear map X); (ii) see (11.5.3). □

12.1.7 Let $A \mapsto \det A$, or more generally $A \mapsto (\det A)^\lambda$, be representations of the group $GL(n,\mathbb{R})$. Show that

(i) for their derived representations we get

$$\begin{aligned} \rho : A \mapsto \det A \quad &\Rightarrow \quad \rho' : C \mapsto \operatorname{Tr} C \\ \rho : A \mapsto (\det A)^\lambda \quad &\Rightarrow \quad \rho' : C \mapsto \lambda \operatorname{Tr} C \end{aligned}$$

(ii) the maps ρ' are indeed linear and (trivially) preserve the commutator.

Hint: (i) see (5.6.5), (12.1.3) and (12.1.6). □

12.1.8 Let $\rho : G \to \operatorname{Aut} V$ be a representation of G. Show that

(i) the prescription

$$\langle \check\rho(g)\alpha, v\rangle := \langle \alpha, \rho(g^{-1})v\rangle \qquad \alpha \in V^*, v \in V$$

defines the representation

$$\check\rho : G \to \operatorname{Aut} V^*$$

of the group G in the *dual* space V^*, called the *contragredient representation* (or sometimes the *dual representation*)

(ii) if we adopt the dual bases in V and V^* we get

$$\rho(g)E_a = A^b_a E_b \quad \Rightarrow \quad \check\rho(g)E^a = (A^{-1})^a_b E^b$$

so that at the level of matrices the transition from a given representation to the contragredient is

$$\rho : g \leftrightarrow A \quad \check\rho : g \leftrightarrow (A^{-1})^{\mathrm{T}}$$

(iii) the canonical pairing is invariant[185] with respect to the *simultaneous* action of ρ in V and $\check\rho$ in V^* (that is the point about $\check\rho$)

$$\langle \check\rho(g)\alpha, \rho(g)v\rangle = \langle \alpha, v\rangle$$

or at the level of components

$$v^a \mapsto v'^a \equiv A^a_b v^b \qquad \alpha_a \mapsto \alpha'_a \equiv (A^{-1})^b_a \alpha_b \quad \Rightarrow \quad \alpha_a v^a \mapsto \alpha'_a v'^a = \alpha_a v^a$$

Hint: a straightforward check. □

[185] This invariance should not be confused with the concept of invariant scalar product in V, to be discussed in (12.1.10) and beyond.

12.1.9 Let $\rho : G \to \text{Aut}\, V$ be a representation of G, ρ' its derived representation of $\mathcal{G}$ and let ρ^b_{ai} be the corresponding matrix elements. Show that

(i) the matrix elements of the (derived) contragredient representation (with respect to the dual basis) are $-\rho^b_{ai}$

$$\rho'(E_i)E_a = \rho^b_{ai}E_b \quad \Rightarrow \quad \check{\rho}'(E_i)E^b = -\rho^b_{ai}E^a$$

so that at the level of the matrices the transition from a given representation to the contragredient one (for the derived representation ρ') consists in $C \mapsto -C^{\text{T}}$

(ii) arbitrary commutation relations of matrices are invariant with respect to this substitution, so that

$$[\mathcal{E}_i, \mathcal{E}_j] = c^k_{ij}\mathcal{E}_k \quad \Rightarrow \quad [(-\mathcal{E}_i)^{\text{T}}, (-\mathcal{E}_j)^{\text{T}}] = c^k_{ij}(-\mathcal{E}_k)^{\text{T}}$$

Hint: (12.1.6), (12.1.8). □

• A representation ρ of a group G is called a *faithful representation* if the map $\rho : G \to \text{Aut}\, V$ is injective (and the same holds for representations of Lie algebras). The (abstract) group itself is then isomorphic to the subgroup $\text{Im}\, \rho \subset \text{Aut}\, V$ of operators (matrices), the isomorphism being $g \leftrightarrow \rho(g)$. For example, the embedding of the affine group $GA(n, \mathbb{R})$ into $GL(n+1, \mathbb{R})$ from (10.1.15) may also be regarded as a faithful representation of $GA(n, \mathbb{R})$ in $\mathbb{R}^{n+1}$. (We will encounter faithful representations of Clifford algebras in Chapter 22.)

Consider a representation of a group G in a linear space V, i.e. a pair (V, ρ). The same linear space may also be endowed with a scalar product h; thus we also have (V, h). If both the structures happen to occur *simultaneously* in some situation, as a rule they are *compatible* – we have a compatible triple (V, h, ρ). This means that the scalar product of any two vectors does not change if an arbitrary group element g is applied (by means of the representation ρ) simultaneously on both of the vectors. If this is the case we speak about a *ρ-invariant scalar product.*

12.1.10 Let h be a ρ-invariant scalar product of the type (r, s) in V. So we have[186]

$$h(\rho(g)v, \rho(g)w) = h(v, w) \qquad v, w \in V; g \in G$$

Check that

(i) the operators $\rho(g)$ are *(pseudo-)orthogonal*, i.e.

$$\rho : G \to O(r, s) \equiv \text{Aut}\,(V, h) \subset \text{Aut}\, V \qquad r + s = \dim V$$

(ii) the operators of the derived representation satisfy

$$h(\rho'(X)v, w) = -h(v, \rho'(X)w)$$

so that they are (pseudo-)antisymmetric;[187] $\rho' : \mathcal{G} \to o(r, s)$

[186] If the representation is fixed and it cannot be confused with some other possible representation, a more compact (and lucid) way of writing is preferred, namely $\rho(g)v \equiv gv$ (this is also the way it is often written in the general case of a left action, see Section 13.1). Then the invariance of the scalar product is simply $h(gv, gw) = h(v, w)$.

[187] A linear operator A is said to be (pseudo-)antisymmetric (with respect to h) if the bilinear form $B(v, w) := h(v, Aw)$ turns out to be antisymmetric; in components $B_{ab} := h_{ac}A^c_b$ and $B_{ab} = -B_{ba}$.

(iii) the matrix elements ρ^b_{ai} of the derived representation of the Lie algebra $\mathcal{G}$ (defined by $\rho'(E_i)E_a = \rho^b_{ai}E_b$) satisfy

$$h_{bc}\rho^c_{ai} + h_{ac}\rho^c_{bi} = 0 \quad \text{or} \quad \rho_{bai} = -\rho_{abi} \quad \text{for} \quad \rho_{abi} := h_{ac}\rho^c_{bi}$$

Hint: (i) see (10.1.4) and (10.1.5); (ii) set $g = e^{tX}$ and differentiate in $t = 0$; (iii) see (12.1.6). □

• In this situation we speak about an *orthogonal* representation of a group. More frequently, however, a complex version of this concept occurs, the *unitary representation* of a group (in a complex space endowed with the scalar product of the type (10.1.12), which happens to be ρ-invariant) and (derived) *antihermitean* representation of its Lie algebra.

Sometimes a situation offers invariant scalar products naturally (as in the case of the Killing metric in $\mathcal{G}$, see (12.3.8)). Another time it might be fairly useful to know when and how one can construct such a product if it is not available and we need it. Now we describe a simple construction of a ρ-invariant scalar product, which is applicable for *finite* groups and then we will contemplate what is to be improved in the method in order for it to be viable for *Lie* groups as well and what problems possibly may occur.

12.1.11 Let $G \equiv \{e \equiv g_1, \dots, g_n\}$ be a *finite group*, ρ its representation in V and h_0 *any positive-definite* scalar product in V. Show that

(i) the new scalar product

$$h(v, w) := \frac{1}{n}\sum_{j=1}^{n} h_0(g_j v, g_j w)$$

already turns out to be ρ-invariant

(ii) if h_0 *is* already invariant, then $h_0 \mapsto h = h_0$, so that a "bad" scalar product gets repaired and a "good" one remains unchanged (this is *the only* role of the factor $1/n$ in front; h also is clearly invariant without it).

Hint: $nh(gv, gw) := \sum_{j=1}^{n} h_0(g_j g v, g_j g w) \equiv n \sum_{j=1}^{n} h_0((R_g g_j)v, (R_g g_j)w)$; since R_g is a *bijection*, we perform the summation effectively again over the whole group so that we only get $nh(v, w)$. □

• We see that the invariance of the product is accomplished by *averaging* over the group (one performs the arithmetic mean of the scalar products[188] with the arguments being transformed step by step by all the elements of the group). It is clear that for a continuous (Lie) group one should replace the sum by the *integral*. For the integral over the group we need a *volume form* but there are an infinite number of volume forms on the group (multiples of a single one by an arbitrary non-vanishing function). Which volume form should we choose? An approved mentor, the computation, will be consulted again.

12.1.12 Let ω be a (not yet specified) volume form on a *compact* Lie group G, so that the volume of the group vol $G := \int_G \omega$ turns out to be finite. Let further h_0 be an arbitrary

[188] The procedure of performing the linear combination of the bilinear forms (which is done in the averaging) may spoil the non-degeneracy; this does not happen, however, in the positive-definite case. In general, it can happen that the averaging of h_0 gives rise to degenerate (although invariant) h, so that it cannot then be used as a scalar product.

(positive definite) scalar product in V, where the representation ρ of the group G acts. For fixed vectors $v, w \in V$ we define the function on the group

$$f_{v,w}(g) := h_0(gv, gw)$$

(the scalar product of g-transformed vectors) and introduce in V a *new* scalar product by the prescription (stolen from finite groups)

$$h(v, w) := \frac{1}{\text{vol } G} \int_G f_{v,w}\omega \equiv \langle f_{v,w} \rangle_G$$

The operation $\langle(\,\cdot\,)\rangle_G$ is again (see the end of Section 7.7) the *average over the group*, so that $h(v, w)$ is actually the mean value of the scalar product of g-transformed vectors. Check that

(i) the function $f_{v,w}$ on G satisfies

$$f_{gv,gw} = f_{v,w} \circ R_g \equiv R_g^* f_{v,w}$$

(ii) if we take ω to be the *right-invariant volume form* ($R_g^*\omega = \omega$), then h turns out to be ρ-invariant

(iii) if h_0 already *is* invariant, then $h_0 \mapsto h = h_0$, so that a "bad" scalar product gets repaired and a "good" one remains unchanged (this is *the only* role of the factor $1/\text{vol } G$ in front; h is also clearly invariant without it).

Hint: (i) ρ is a homomorphism, from which $f_{gv,gw}(k) = f_{v,w}(kg)$; (ii)

$$h(gv, gw) = \frac{1}{\text{vol } G} \int_G (R_g^* f_{v,w})\omega = \frac{1}{\text{vol } G} \int_G R_g^*(f_{v,w}\omega) = \frac{1}{\text{vol } G} \int_{R_g(G) \equiv G} f_{v,w}\omega$$
$$\equiv h(v, w)$$

(iii) $f_{v,w}$ is then a constant function and it may be put in front of the integral. □

12.1.13 Consider the standard representation of $SO(2)$ in $\mathbb{R}^2$, i.e. ordinary rotations of the Euclidean plane $v \mapsto Av$, $A \in SO(2)$. Check that

(i) after averaging over $SO(2)$ the "bad" (= rotation non-invariant) scalar product h_0 changes to the "good" one, which looks as anticipated:

$$h_0(v, w) \equiv (v^1, v^2)\begin{pmatrix} 1 & 0 \\ 0 & 2 \end{pmatrix}\begin{pmatrix} w^1 \\ w^2 \end{pmatrix} \quad \mapsto \quad h\,(v, w) := (v^1, v^2)\begin{pmatrix} k & 0 \\ 0 & k \end{pmatrix}\begin{pmatrix} w^1 \\ w^2 \end{pmatrix}$$

(ii) after the averaging over $SO(2)$ of the (also rotation non-invariant) scalar product $\hat{h}_0$ of type $(1, -1)$, the computation gives

$$\hat{h}_0(v, w) \equiv (v^1, v^2)\begin{pmatrix} 1 & 0 \\ 0 & -1 \end{pmatrix}\begin{pmatrix} w^1 \\ w^2 \end{pmatrix} \quad \mapsto \quad \hat{h}_0(v, w) := (v^1, v^2)\begin{pmatrix} 0 & 0 \\ 0 & 0 \end{pmatrix}\begin{pmatrix} w^1 \\ w^2 \end{pmatrix}$$

so that the averaging spoils non-degeneracy (even in an extreme way, since the resulting invariant "scalar product" is useless: it simply vanishes).

Hint: see (12.1.12), the right- (as well as left-) invariant volume form on $SO(2)$ is $d\varphi$ (11.1.12). □

• We see that for *compact Lie groups* a more or less straightforward modification of the method used for finite groups may be applied with the same positive result – an invariant scalar product *always exists* (and we even know the algorithm of its construction). For non-compact groups the situation differs significantly since the crucial integral $\int_G f\omega$ *need not exist.* An invariant scalar product then cannot be constructed in this way and its existence is not guaranteed in general. This is one of the reasons why the theory of representations of non-compact groups gets much more complicated in comparison with compact (or finite) groups.

12.2 Irreducible and equivalent representations, Schur's lemma

• Sometimes in a given representation (V, ρ) a *smaller* representation (of the same group G) is hidden. This happens when there exists an *invariant subspace* $W \subset V$ in V, i.e. such that the vectors which belong there also remain in it after a transformation by an *arbitrary* operator of the representation (10.1.13); formally

$$\rho(G)W \subset W \qquad \text{i.e.} \quad w \in W \;\Rightarrow\; \rho(g)w \in W \quad g \in G$$

We emphasize that we are speaking about a *simultaneous* invariant subspace of *all* operators $\rho(g)$.

Two such subspaces are always available: $W =$ the whole V and $W = \{0\}$. They are said to be *trivial.* If a representation ρ has *no other* invariant subspaces, it is called an *irreducible representation* (so that it cannot be reduced to a smaller representation (a *subrepresentation*), which we would get by *restriction* of the initial one on the subspace W, i.e. by confining a domain of $\rho(g)$ to W alone). A *reducible representation,* in contrast, has at least one non-trivial invariant subspace.

12.2.1 Consider the representation of $SO(2)$ in $\mathbb{R}^2$ from (12.1.13), i.e. the ordinary rotations of the Euclidean plane $v \mapsto Av$, $A \in SO(2)$. Show that this representation is irreducible.

Hint: a non-trivial subspace would have to be one-dimensional, i.e. a line passing through the origin; no such line is, however, invariant with respect to rotations. □

• If we find in V a non-trivial subspace W, we feel fairly encouraged and start to seek its *invariant complement* $\hat{W}$, i.e. a subspace which is (also) invariant and which together with W constitutes the whole space V, so that $V = W \oplus \hat{W}$. If such a complement exists, the representation is said to be *completely reducible*; it essentially decomposes into two smaller representations: the first in W and the second in $\hat{W}$ (compare the concept of a sum of representations in problem (12.4.10)). It turns out, however, that the invariant complement need not always exist, i.e. a reducible[189] representation need not always also be completely reducible.

[189] For example the two-dimensional representation $(x, y) \mapsto (x + ay, y)$ of the group $G = (\mathbb{R}, +)$ has *only a single* one-dimensional invariant subspace, given by the vectors of the form $(x, 0)$. It is reducible, but not decomposable.

12.2.2 Let ρ be a reducible or completely reducible representation of a Lie group G. Show that

(i) the reducibility as well as the complete reducibility automatically carries over to the *derived* representation ρ' of the Lie algebra $\mathcal{G}$ (i.e. to operators $\rho'(X)$ for all $X \in \mathcal{G}$)
(ii) in an appropriately chosen basis in V the matrices of all the operators $\rho(g)$ and $\rho'(X)$ have the following block structure:

$$\begin{pmatrix} A & C \\ 0 & B \end{pmatrix} \quad \text{or} \quad \begin{pmatrix} A & 0 \\ 0 & B \end{pmatrix}$$

(iii) if there is an *invariant* scalar product h in V, then each reducible representation is already necessarily also completely reducible
(iv) for *compact* (as well as finite) groups each reducible representation is already necessarily also completely reducible.

Hint: (i) the definition of the derived representation; (ii) adapted bases (10.1.13) and (10.1.14); for example, for a reducible one

$$\text{if } v = x^i e_i + y^\alpha e_\alpha \leftrightarrow \begin{pmatrix} x \\ y \end{pmatrix} \qquad \text{then} \quad \rho(g) : \begin{pmatrix} x \\ y \end{pmatrix} \mapsto \begin{pmatrix} A & C \\ 0 & B \end{pmatrix} \begin{pmatrix} x \\ y \end{pmatrix}$$

(iii) the *orthogonal complement* $W^\perp$ is invariant; (iv) there exists an invariant scalar product (12.1.12). □

• It is convenient to introduce an equivalence relation among representations, since the difference is irrelevant between some of them. Given two representations (ρ_1, V_1) and (ρ_2, V_2) of a group G, a linear map A is said to be *equivariant* if it "commutes" with the corresponding operators of the representations, i.e. if the following diagram commutes:

$$\begin{array}{ccc} V_1 & \xrightarrow{\rho_1(g)} & V_1 \\ {\scriptstyle A}\downarrow & & \downarrow{\scriptstyle A} \\ V_2 & \xrightarrow[\rho_2(g)]{} & V_2 \end{array} \qquad \text{i.e.} \quad \rho_2(g)A = A\rho_1(g) \quad \forall g \in G$$

Such a map A is alternatively called an *intertwining operator* for these two representations.[190] Note that the same operator then also "intertwines" the *derived* representations of $\mathcal{G}$

$$\begin{array}{ccc} V_1 & \xrightarrow{\rho_1'(X)} & V_1 \\ {\scriptstyle A}\downarrow & & \downarrow{\scriptstyle A} \\ V_2 & \xrightarrow[\rho_2'(X)]{} & V_2 \end{array} \qquad \text{i.e.} \quad \rho_2'(X)A = A\rho_1'(X) \quad \forall X \in \mathcal{G}$$

[190] They are also studied in Section 12.5. The concept of equivariance itself turns out to be useful also in the broader context of *actions* (linearity drops out), see (13.1.13).

One can easily check that the intertwining operators for the fixed two representations are naturally endowed with the structure of a linear space and in the particular case of *equal* representations they even constitute an associative (unital) *algebra.*

12.2.3 Representations (ρ_1, V_1) and (ρ_2, V_2) of a group G are called *equivalent* if there exists an *equivariant isomorphism* between V_1 and V_2; the intertwining operator is thus an *isomorphism* and

$$\rho_2(g) = A\rho_1(g)A^{-1}$$

Show that

(i) this indeed introduces an equivalence on the set of representations of G
(ii) in appropriately chosen bases equivalent representations have equal matrices of the corresponding operators (i.e. $\rho_1(g)$ and $\rho_2(g)$ for the same g)
(iii) for derived representations then $\forall X \in \mathcal{G}$

$$\rho_2'(X) = A\rho_1'(X)A^{-1}$$

(iv) if there is a ρ-invariant scalar product h in (V, ρ), the contragredient representation $\check{\rho}$ is equivalent to the representation ρ itself.

Hint: (ii) if $E_a \in V_1$, take $A(E_a) \equiv \tilde{E}_a \in V_2$; (iii) set $g = e^{tX}$ and differentiate with respect to t in $t = 0$; (iv) $v \mapsto h(v, \ \cdot\ ) \equiv \flat_h v$ (lowering of the index by h) is an intertwining isomorphism. □

12.2.4 The prescription $l_j \mapsto -\frac{1}{2}i\sigma_j \equiv S_j$ defines a (complex) representation of $so(3)$ in $\mathbb{C}^2$ (12.1.5). Show that

(i) the complex conjugate matrices S_i^* also provide a representation
(ii) S_i and S_i^* are (in this particular case) equivalent.

Hint: (ii) see (12.2.3) with $A = \sigma_2$ (which *is* a linear isomorphism $\mathbb{C}^2 \to \mathbb{C}^2$, since $\det\sigma_2 \neq 0$); $\sigma_j^* \equiv \sigma_j^{\mathrm{T}} = -\sigma_2\sigma_j\sigma_2$ may come in handy. □

12.2.5 Check that

(i)

$$S_j \equiv -\frac{i}{2}\sigma_j \quad \text{and} \quad N_j \equiv \frac{1}{2}\sigma_j$$

is a representation of $\mathcal{G} = so(1,3)$ (= the Lie algebra of the *Lorentz group*)
(ii) the complex conjugate matrices S_j^*, N_j^* also provide a representation of $so(1,3)$
(iii) the two representations are now (contrary to (12.2.4)) *inequivalent.*[191]

Hint: according to (12.2.3) one is to find a non-singular complex 2×2 matrix A such that

$$\left(-\frac{i}{2}\sigma_j\right)^* = A\left(-\frac{i}{2}\sigma_j\right)A^{-1} \qquad \text{and at the same time} \qquad \left(\frac{1}{2}\sigma_j\right)^* = A\left(\frac{1}{2}\sigma_j\right)A^{-1}$$

See that this is a bit too much for a single matrix A. □

[191] This turns out to be important for the theory of "*dotted* and *undotted* spinors" (see the text after (13.3.15)).

• *Schur's lemma* is a standard formal tool used in connection with irreducible representations. We mention *two* Schur lemmas here; the first one deals with the intertwining operators between *two* irreducible representations and the second one is then a simple consequence for a special case of two *equal* representations, i.e. it is a statement about the intertwining maps for *a single irreducible* representation. When we read a reference to "Schur's lemma" in physics texts, as a rule the *second* one is meant.

12.2.6 Let (ρ_1, V_1) and (ρ_2, V_2) be two irreducible representations of a group G. Consider all intertwining operators A between them, i.e. all linear maps satisfying

$$A : V_1 \to V_2 \qquad A\rho_1(g) = \rho_2(g)A$$

Prove the ("first") *Schur lemma*: then,

(i) there are (only) two possibilities: either ρ_1, ρ_2 are *inequivalent* and then the only A is $A = 0$, or ρ_1, ρ_2 are *equivalent* and then A is an isomorphism (put together, an intertwining operator A between two *irreducible* representations may be either zero or an isomorphism)

(ii) if ρ_1, ρ_2 are equivalent *complex representations*, then the isomorphism A is *unique* up to the freedom (clear in advance) $A \mapsto \lambda A \quad (\lambda \in \mathbb{C})$.

Hint: (i) Ker A and Im A are invariant subspaces (check) $\Rightarrow$ they are necessarily trivial. Analyze the four possibilities available (three of them give $A = 0$, the fourth, Ker $A = 0$, Im $A = V_2$, says that A is an isomorphism so that they are equivalent); (ii) let B be another candidate, then (recall the linear structure) also $C \equiv A - \lambda B$ for any $\lambda \in \mathbb{C}$ is good $\Rightarrow$ also for the root of the equation $\det(A - \lambda B) = 0$ (a *complex* root exists for sure[192] and it is *acceptable here* (check; see (12.2.8)). In this case, however, C fails to be invertible $\Rightarrow$ it is not an isomorphism $\Rightarrow$ by item (i) it is necessarily zero, $C = 0$. □

12.2.7 Let (ρ, V) be an *irreducible complex* representation of a group G and let A be its intertwining operator, or equivalently (!) a linear operator in V, which *commutes with all* the operators of the representation: $\forall g \in G$ we have

$$\rho(g)A = A\rho(g) \qquad \text{i.e.} \quad [A, \rho(g)] = 0 \quad \forall g \in G$$

Prove the ("second") *Schur lemma*: then

$$A = \lambda \mathbf{1} \qquad \lambda \in \mathbb{C}$$

i.e. A is necessarily only a *multiple of the identity operator* (of course, any multiple of the identity operator commutes with all $\rho(g)$; we assert that *no other* operator does).

Hint: for $\rho_1 = \rho_2$ in (12.2.6) item (ii) is true; since $A = \mathbf{1}$ clearly *is* acceptable, the most general A is $A = \lambda\mathbf{1}$. □

12.2.8 Convince yourself by means of the following counterexample that the assumption of the *complexity* of the representation in Schur's lemma is essential (and not a specific

[192] $\mathbb{C}$ is an *algebraically closed field*, i.e. such that all the solutions of the equations "polynomial with coefficients from $\mathbb{C} = 0$" belong to the field as well; we are not forced to extend the field because of them (contrary to $\mathbb{R}$, where the equation $x^2 + 1 = 0$ forces us to sit down and invent the imaginary number $i \notin \mathbb{R}$).

feature of the proof, which may be bypassed by some other proof): consider again the "identical" representation of $SO(2)$ (12.1.13), (12.2.1) $A \mapsto \rho(A) = A$. We already know that (when regarded as a *real* representation, i.e. in $\mathbb{R}^2$) it is irreducible. Check that

(i) in spite of being irreducible, all of its operators $\rho(A)$ commute (besides the multiples of the identity matrix) with an independent non-singular matrix

$$B = \begin{pmatrix} 0 & 1 \\ -1 & 0 \end{pmatrix}$$

(but already with no other independent matrices)

(ii) geometrically this stems from the fact that in addition to the standard metric tensor there is also a rotationally invariant *volume form* in E^2 (in *two-dimensional* space its components form a 2×2 matrix)

(iii) the proof from (12.2.6) fails in that the equation $\det(A - \lambda B) \equiv \det(\mathbf{1} - \lambda B) = 0$ has no *real* solution.

Hint: (iii) $\det(\mathbf{1} - \lambda B) = 1 + \lambda^2 = 0$ gives only $\lambda = \pm i$; this is not acceptable for a real representation, since $\mathbf{1} \pm iB$ is not a linear map $\mathbb{R}^2 \to \mathbb{R}^2$. □

12.2.9 Let G be a *commutative group*. Show that all its complex irreducible representations are necessarily one-dimensional.

Hint: for fixed g the operator $\rho(g)$ commutes with all $\rho(g')$, $g' \in G$; therefore $\rho(g) = \lambda(g)\mathbf{1}$; if the space (V, ρ) is more than one-dimensional, it contradicts irreducibility (*each* subspace is invariant for $\lambda(g)\mathbf{1}$). □

12.2.10 Give a complete classification of all irreducible complex representations of the Lie group $U(1)$. What do the corresponding derived representations of the algebra $u(1)$ look like?

Solution: from (12.2.9) they are one-dimensional, i.e. we have $z \mapsto w(z)$, $||z|| = 1$, $w \in \mathbb{C}$; the requirement of *homomorphism* gives (within first-order accuracy in ϵ)

$$\begin{aligned} \rho(z(1 + i\epsilon)) &\stackrel{1}{=} \rho(z)\rho[1 + \epsilon i] = \rho(z)[1 + \epsilon\rho'(i)] && \text{definition of } \rho' \\ &\stackrel{2}{=} w(z + \epsilon i z) = w(z) + \epsilon i z \frac{dw(z)}{dz} && \text{Taylor expansion of } w(z) \end{aligned}$$

so that $izw'(z) = \lambda w(z)$, $\lambda \equiv \rho'(i) \in \mathbb{C}$; *continuity* restricts λ to $\lambda = in$, $n \in \mathbb{N}$; so $z \mapsto \rho_n(z) = z^n$, where $n \in \mathbb{N}$ and the derived representation is then $\rho'_n(i) = in$; equivalently

$$e^{i\alpha} \mapsto \rho_n(e^{i\alpha}) := e^{in\alpha} \qquad \rho'_n(i) = in \quad n \in \mathbb{N}$$

The integer n, which completely characterizes a given irreducible representation, is in applications in physics connected with the *charge* of the object transforming according to this representation. What kind of charge it is (electrical, etc.) depends on the particular context; in general, it is some $U(1)$-*charge*. □

12.2.11 If $\rho(g)$ in (12.2.8) is regarded as an operator in $\mathbb{C}^2$ (instead of $\mathbb{R}^2$; one speaks about the *complexification* of the representation), the representation becomes reducible (it is two-dimensional, see (12.2.9)).

(i) Explain why it is even *completely* reducible, so that there exist in $\mathbb{C}^2$ two invariant (complex) one-dimensional subspaces.
(ii) Find the corresponding invariant subspaces and thus obtain the form of the matrices from problem (12.2.2).
(iii) Find out which irreducible representations (in the sense of (12.2.10)) act in the one-dimensional invariant subspaces (i.e. they appear as the diagonal elements of the matrices in the form (12.2.2)).

Solution: (i) the group $SO(2)$ is compact (12.1.12), (12.2.2); (ii), (iii) first we find the intertwining operator $A : \mathbb{C}^1 \to \mathbb{C}^2$ between the nth irreducible representation ($e^{i\alpha} \mapsto e^{in\alpha}$) and the two-dimensional (complex) representation under consideration from (12.2.8). Because of the freedom $A \mapsto \lambda A$ we may seek it in the form $1 \mapsto \binom{1}{w}$, where $w \in \mathbb{C}$ is to be determined. For the corresponding *derived* representations (which is a bit simpler) we get the equivariance condition (the text before (12.2.3))

$$\begin{pmatrix} 1 \\ w \end{pmatrix} in = \begin{pmatrix} 0 & 1 \\ -1 & 0 \end{pmatrix} \begin{pmatrix} 1 \\ w \end{pmatrix} \qquad \text{from where} \qquad n^2 = 1, w = in$$

So the non-trivial (non-zero) intertwining operators exist only for the representations labelled by $n = \pm 1$ and their matrices explicitly read

$$A = \begin{pmatrix} 1 \\ i \end{pmatrix} \quad \text{for} \quad n = 1 \qquad A = \begin{pmatrix} 1 \\ -i \end{pmatrix} \quad \text{for} \quad n = -1$$

The images of these operators constitute in $\mathbb{C}^2$ two one-dimensional (searched) subspaces, in which the representations ρ_n with $n = 1$ and $n = -1$ operate. Thus they are spanned by the vectors $\begin{pmatrix} 1 \\ i \end{pmatrix}$ and $\begin{pmatrix} 1 \\ -i \end{pmatrix}$. A change of basis in $\mathbb{C}^2$ resulting in just these two vectors is given by the matrix $\hat{A} = \begin{pmatrix} 1 & 1 \\ i & -i \end{pmatrix}$; one checks that it indeed satisfies

$$\begin{pmatrix} 1 & 1 \\ i & -i \end{pmatrix} \begin{pmatrix} e^{i\alpha} & 0 \\ 0 & e^{-i\alpha} \end{pmatrix} \begin{pmatrix} 1 & 1 \\ i & -i \end{pmatrix}^{-1} = \begin{pmatrix} \cos\alpha & \sin\alpha \\ -\sin\alpha & \cos\alpha \end{pmatrix}$$

so that $\hat{A}$ is an intertwining *isomorphism* between $\rho_1 \oplus \rho_{-1}$ and the given representation from problem (12.2.8). □

• In (12.2.11) we found the invariant subspaces by comparison (through the construction of the intertwining operator) with *known* irreducible representations. There are also other methods available. We may use, for example, the result of the problem (12.2.8), where we learned that there are (in this particular case) only two independent operators which happen to commute with all the operators of the representation, namely $\mathbf{1}$ and B. As we have already mentioned before, the operators commuting with all the operators $\rho(g)$ of the representation (i.e. the intertwining operators for this representation) are endowed with the structure of an algebra and it turns out that this algebra encodes the invariant subspaces of the representation.

12.2.12* Let ρ be a representation of G in V. Consider all linear isomorphisms $A : V \to V$ which commute with all operators of the representation. Check that

(i) these operators are closed with respect to linear combinations as well as *product* (composition), so that they constitute a (unital) associative *algebra* (see Appendix A.2)
(ii) for the representation of $SO(2)$ from (12.2.8) this algebra is *two-dimensional* and it is spanned by the basis $\mathbf{1}, B$ ($B^2 = -\mathbf{1}$), so that a general element is $a1 + bB$ and the product reads

$$(a\mathbf{1} + bB)(a'\mathbf{1} + b'B) = (aa' - bb')\mathbf{1} + (ab' + a'b)B$$

(iii) if we consider this algebra *over* $\mathbb{C}$ (i.e. the coefficients may be from $\mathbb{C}$), one can form the linear combinations $P_\pm$, which behave like *projectors* and they realize the *decomposition of unity*

$$P_+^2 = P_+ \qquad P_-^2 = P_- \qquad P_+P_- = 0 = P_-P_+ \qquad P_+ + P_- = \mathbf{1}$$

(iv) the space of the representation $V \equiv \mathbb{C}^2$ may be then decomposed into the sum of *invariant* (one-dimensional) subspaces $P_\pm V$

$$V = (P_+ + P_-)V \equiv P_+V \oplus P_-V \equiv V_+ \oplus V_-$$

in which the restrictions of the operators $\rho(g)$ act
(v) in this way we get exactly the decomposition from (12.2.11)
(vi) if a representation is irreducible, the algebra is only one-dimensional and we can project out no non-trivial invariant subspace.

Hint: (iii) write down the requirements with general coefficients and solve. ($P_\pm = \frac{1}{2}(\mathbf{1} \pm iB)$; $\rho(g) = \rho(g)\mathbf{1} = \rho(g)P_+ + \rho(g)P_- \equiv \rho_\pm(g)$, where $\rho_\pm(g)$ effectively acts (only) in $V_\pm$.) □

• Let us look more closely at the structure of the intertwining operators between two representations whose spaces may be decomposed into the sum of invariant subspaces.

12.2.13* Let $(\rho_1, V \equiv \oplus_i V_i)$ and $(\rho_2, W \equiv \oplus_\alpha W_\alpha)$ be two complex representations of G, the subspaces V_i and W_α being invariant and not containing already smaller (non-trivial) invariant subspaces (so that the restrictions of the representations to these subspaces are irreducible). Let P_i and P_α be the projectors onto these subspaces. Check that

(i) if $A : V \to W$ is an intertwining operator between ρ_1 and ρ_2, then the operators

$$A_{\alpha i} : V_i \to W_\alpha \qquad A_{\alpha i} := P_\alpha A P_i, \; A = \sum_{i,\alpha} A_{\alpha i}$$

are the intertwining operators for the irreducible subrepresentations which arise by the restriction to V_i and W_α
(ii) each operator $A_{\alpha i}$ is either zero or an isomorphism given uniquely up to multiplication by a constant

(iii) all the intertwining operators between ρ_1 and ρ_2 constitute a linear space (it is denoted by $\mathcal{C}(\rho_1, \rho_2)$ or $\mathrm{Hom}_G(V_1, V_2)$)
(iv) the *dimension* of the space $\mathcal{C}(\rho_1, \rho_2)$ expresses the total number of *pairs* of mutually equivalent irreducible subrepresentations (one member of the pair being in ρ_1 and the other one in ρ_2)
(v) if some particular irreducible representation $\hat{\rho}$ occurs k times in ρ_1 and l times in ρ_2, it contributes to the dimension of the space $\mathcal{C}(\rho_1, \rho_2)$ by a term kl.

Hint: (ii) use Schur's lemma. □

• Let us remark at the end of this section that in the proof of Schur's lemma we actually never used the fact that the operators $\rho(g)$ (or $\rho'(X)$) *represent a group* (or a Lie algebra). As a matter of fact, the lemma tells something about *any* family of operators which act in a common linear space: what are the properties of an operator commuting with the whole family, what happens when the family acts irreducibly (i.e. there is no simultaneous non-trivial invariant subspace), etc.? A practical consequence of this observation is that the lemma may be used (and in fact *is* used) in connection with representations of other algebraic structures, like *associative* algebras and, in particular, *Clifford* algebras.

12.3 Adjoint representation, Killing–Cartan metric

Both the adjoint representation Ad of a group G as well as its derived representation ad of the Lie algebra $\mathcal{G}$ are frequently encountered in various applications. The group does not worry too much about finding a vector space V to carry the representation. It simply uses its own Lie algebra to do this. So (V, ρ, ρ') becomes in this particular case $(\mathcal{G}, \mathrm{Ad}, \mathrm{ad})$.

Although this may be regarded as an admirably economical behavior of G (instead of two structures $\mathcal{G}, V$ to be paid from taxpayers' money a single one makes do), it might at the same time make it a bit harder to grasp these ideas quickly (one should always be careful to understand clearly whether a given $X \in \mathcal{G}$ stands in the role of the Lie algebra element to be represented or in the role of an element of the carrier space $V \equiv \mathcal{G}$).

12.3.1 Consider the *conjugation* I_g on a Lie group G, i.e. (see also (13.1.3)) the map

$$I_g : G \to G \qquad k \mapsto gkg^{-1}$$

Check that

(i) I_g is (for fixed $g \in G$) a bijective homomorphism, i.e. an *automorphism of the group* G; it is also called an *inner automorphism*
(ii) if the derived homomorphism to I_g is denoted by $\mathrm{Ad}_g := I'_g \equiv I_{g*}$, the following commutative diagram holds:

$$\begin{array}{ccc} G & \xrightarrow{I_g} & G \\ \uparrow{\scriptstyle \exp} & & \uparrow{\scriptstyle \exp} \\ \mathcal{G} & \xrightarrow[\mathrm{Ad}_g \equiv I'_g]{} & \mathcal{G} \end{array} \qquad \text{i.e.} \quad ge^X g^{-1} = e^{\mathrm{Ad}_g X}$$

(this diagram (or formula) actually may be regarded as *defining* the map Ad_g)

(iii) for *matrix* groups the map Ad_A ($g \equiv A$ being a matrix from G) may be explicitly written as follows:[193]

$$\text{Ad}_A X = AXA^{-1}$$

Hint: (ii) see (11.5.3); (iii) $Be^X B^{-1} = B(\mathbf{1} + X + \frac{1}{2}XX + \cdots)B^{-1} = \mathbf{1} + BXB^{-1} + \cdots = e^{BXB^{-1}}$. □

• *For each* $g \in G$ we thus have the automorphism I_g of the group G as well as the automorphism Ad_g of the Lie algebra $\mathcal{G}$ (the bijectivity of the latter will be checked in a while). Let us concentrate now on their behavior with respect to its "parameter" g.

12.3.2 Check that

(i) both the maps I_g and Ad_g behave with respect to the "parameter" g as *left actions*, i.e.

$$I_{gh} = I_g \circ I_h \qquad \text{Ad}_{gh} = \text{Ad}_g \circ \text{Ad}_h$$

(ii) for each g the map Ad_g is a bijection so that, when put together, it is an *automorphism of the Lie algebra* $\mathcal{G}$

(iii) the map

$$\text{Ad} : G \to \text{Aut}\,\mathcal{G} \qquad g \mapsto \rho(g) \equiv \text{Ad}_g$$

is a *representation*[194] of G in $\mathcal{G}$; it is called the *adjoint representation* of the group G.

Hint: (i) I_{gh} trivial, $\text{Ad}_{gh} = I'_{gh} = (I_g \circ I_h)' = \cdots$ (11.5.3); (ii) the preimage for X is $\text{Ad}_{g^{-1}} X$; if $\text{Ad}_g X = \text{Ad}_g Y$, then $X = Y$ (apply $\text{Ad}_{g^{-1}}$); (iii) the definition and the preceding results here. □

• We now mention two geometrical situations in which this representation occurs. The first is the transformation of the *left*-invariant objects with respect to the *right* translations on a group, the second one concerns the relation between the right- and left-invariant fields.

12.3.3* Let L_X be the left-invariant vector field on G which is given by its value X at the point $e \in G$, i.e. $L_X(e) = X$. Also let E_i be a basis in $\mathcal{G} \equiv T_e G$, E^i the dual basis in $\mathcal{G}^* \equiv T_e^* G$ and e_i, e^i the corresponding left-invariant fields on G and finally let L_g and R_g be the left and right translation on G. Check that

(i) the left-invariant fields behave with respect to L_g and R_g as follows:[195]

$$L_g^* L_X = L_X \qquad R_g^* L_X = L_{\text{Ad}_g X}$$

[193] We emphasize that this (simple) expression is meaningful *only* for matrix groups, since in general the product of a group element with a Lie algebra element is *not defined* at all. In the case of the matrix groups, however, both objects are matrices and they *may* be multiplied (as matrices). Note that in the general formula $ge^X g^{-1} = e^{\text{Ad}_g X}$ the objects get multiplied correctly also from an "abstract" point of view.

[194] It is actually even *more than* a representation, since the operators from $\text{Aut}\,\mathcal{G}$ are not only linear, but they also happen to preserve free of charge (beyond their legal duties resulting from being a representation of a group) the *commutator* in $\mathcal{G}$ (recall that Ad_g is defined as a derived *homomorphism* so that $\text{Ad}_g[X, Y] = [\text{Ad}_g X, \text{Ad}_g Y]$); automorphism of a Lie algebra $\mathcal{G}$ means more than an automorphism of a vector space $\mathcal{G}$, being a *part* of the structure of the Lie algebra $\mathcal{G}$.

[195] Recall that both L_g and R_g are diffeomorphisms of G on itself so that it makes sense (3.1.6) to apply both maps L_g^* and L_{g*} to arbitrary tensor fields. In particular, one can perform the pull-back of vector fields and $L_g^* = L_{g^{-1}*}$.

(ii) in particular, for the frame and coframe fields e_i, e^i

$$L_g^* e_i = e_i \qquad R_g^* e_i = (\mathrm{Ad}_g)_i^j e_j$$
$$L_g^* e^i = e^i \qquad R_g^* e^i = (\mathrm{Ad}_{g^{-1}})_j^i e^j$$

We see that *left*-invariant frame fields e_i transform under *right* translations on G into one another by the Ad_g-representation and they remain unchanged under the left translation, since they remember well from schooldays what the word "invariant" means in plain English

(iii) the *values* themselves of the right- and left-invariant fields L_X and R_X corresponding to the same X are related in the point g by

$$L_X(g) = R_{\mathrm{Ad}_g X}(g) \qquad R_X(g) = L_{\mathrm{Ad}_{g^{-1}} X}(g)$$

or for the bases

$$e_i(g) = (\mathrm{Ad}_g)_i^j f_j(g) \qquad f_i(g) = (\mathrm{Ad}_{g^{-1}})_i^j e_j(g)$$
$$e^i(g) = (\mathrm{Ad}_{g^{-1}})_j^i f^j(g) \qquad f^i(g) = (\mathrm{Ad}_g)_j^i e^j(g)$$

Hint: (i) according to (11.1.1), (11.1.4), (12.3.1) we have

$$R_{g*} L_X(h) = R_{g*} L_{h*} X = L_{h*} R_{g*} X = L_{h*} L_{g*} L_{g^{-1}*} R_{g*} X = L_{h*} L_{g*} I_{g^{-1}*} X$$
$$= L_{h*} L_{g*} \mathrm{Ad}_{g^{-1}} X = L_{hg*} \mathrm{Ad}_{g^{-1}} X \equiv L_{\mathrm{Ad}_{g^{-1}} X}(hg)$$

(ii) it should be $\langle e^i, e_j \rangle = \langle R_g^* e^i, R_g^* e_j \rangle$; (iii) $L_X(g) = L_{g*} X = R_{g*} R_{g^{-1}*} L_{g*} X = R_{g*} I_{g*} X \equiv R_{g*} \mathrm{Ad}_g X = R_{\mathrm{Ad}_g X}(g)$. □

12.3.4 Let θ be the *canonical* 1-*form on* G. Check that it behaves under the left and right translation on G as follows:

$$L_g^* \theta = \theta \qquad \text{i.e.} \qquad (L_g^* \theta^i) E_i = \theta^i E_i$$
$$R_g^* \theta = \mathrm{Ad}_{g^{-1}} \theta \qquad \text{i.e.} \qquad (R_g^* \theta^i) E_i = \big((\mathrm{Ad}_{g^{-1}})_j^i \theta^j \big) E_i$$

The form θ is thus left-invariant and with respect to right translations it is (in the terminology of Section 13.5) "of type Ad."

Hint: left-invariance is already known from (11.2.6), with respect to R_g: the component forms of θ happen to coincide (11.2.6) with e^i, (12.3.3). □

• Let us look more closely at how the adjoint representation appears in these formulas. If we apply, for example, R_g^* on e_i, the result is to be some constant linear combination of the fields e_j (since $R_g^* e_i$ constitute a left-invariant frame field, see (11.1.1iv) and (11.1.6v)). One checks that the matrices by which e_i are combined realize a representation of G, which may be regarded as acting in the Lie algebra $\mathcal{G}$. *Only at this stage* does the true question arise: *which* representation of G in $\mathcal{G}$ comes out there? There are just two natural possibilities: the adjoint and the trivial one. The computation showed us that it is the adjoint representation (the trivial one is realized on *right*-invariant fields f_i).

Now we look at the derived representation of the adjoint one.

12.3.5 Let $\rho = \mathrm{Ad}$ be the adjoint representation of G in $\mathcal{G}$. Its derived representation will be denoted by $\rho' \equiv \mathrm{Ad}' =: \mathrm{ad}$. Check that

(i) the commutative diagram holds

$$\begin{array}{ccc} G & \xrightarrow{\mathrm{Ad}} & \mathrm{Aut}\,\mathcal{G} \\ {\scriptstyle \exp}\uparrow & & \uparrow{\scriptstyle \exp} \\ \mathcal{G} & \xrightarrow[\mathrm{ad}\equiv\mathrm{Ad}']{} & \mathrm{End}\,\mathcal{G} \end{array} \qquad \text{i.e.} \quad \mathrm{Ad}_{\exp X} = e^{\mathrm{ad}_X}$$

(ii) the explicit formula for ad is[196]

$$\mathrm{ad}_X Y = [X, Y]$$

(iii) this formula indeed defines a representation of $\mathcal{G}$ in $\mathcal{G}$, i.e.

$$\mathrm{ad}_{X+\lambda Y} = \mathrm{ad}_X + \lambda \mathrm{ad}_Y$$
$$\mathrm{ad}_{[X,Y]} = \mathrm{ad}_X \mathrm{ad}_Y - \mathrm{ad}_Y \mathrm{ad}_X \equiv [\mathrm{ad}_X, \mathrm{ad}_Y]$$

(iv) if c_{ij}^k are the structure constants with respect to a basis $E_i \in \mathcal{G}$, then the matrix elements of the operator ad_{E_i} with respect to the basis E_j (a special case of ρ_{ai}^b; the indices of the type $a, b, \dots$ and $i, j, \dots$ coincide here) are

$$(\mathrm{ad}_{E_i})_j^k = c_{ij}^k \qquad \text{i.e.} \quad \mathrm{ad}_{E_i} E_j = (\mathrm{ad}_{E_i})_j^k E_k = c_{ij}^k E_k$$

This is usually referred to as that in the adjoint representation the "generators are given directly in terms of the structure constants"

(v) check directly that these matrices satisfy appropriate commutation relations.

Hint: (i) (12.1.6); (ii) in the *matrix* case it is easily computed from (12.3.1); in general, consider on G the flow $\Phi_t^{L_X} := R_{\exp tX}$, generated by the field L_X (11.4.6); the result (12.3.3) says that $R^*_{\exp tX} L_Y = L_{(\mathrm{Ad}_{\exp tX} Y)}$ and its differentiation with respect to t at $t = 0$ gives $\mathcal{L}_{L_X} L_Y = L_{(\mathrm{ad}_X Y)}$ ($\mathcal{L}_{L_X}$ is the Lie derivative along the field L_X) and since $\mathcal{L}_{L_X} L_Y = [L_X, L_Y] = L_{[X,Y]}$ and $X \leftrightarrow L_X$ is a bijection, we get $\mathrm{ad}_X Y = [X, Y]$; (v) the Jacobi identity $[E_i, [E_j, E_k]] + \cdots = 0$. □

12.3.6 Write down explicitly the matrices ad_{E_i} with respect to the basis E_i for the Lie algebras $so(3)$, $su(2)$ and $ga(1, \mathbb{R})$.

Hint: (11.7.12), (11.7.13) and (11.7.10) (for both $su(2)$ and $so(3)$ it turns out that $\mathrm{ad}_{E_j} = l_j$; thus, in particular, $\mathrm{ad} = \mathrm{id}$ for $so(3)$). □

• In a Lie algebra $\mathcal{G}$ there is often a distinguished scalar product, given by the *Killing–Cartan* metric tensor. Its importance lies in the fact that it is invariant with respect to the adjoint representation, i.e. it is Ad-*invariant*. When we spread this metric tensor by left

[196] Notice that we write $\mathrm{ad}(X)$ as ad_X, just like $\mathrm{Ad}(g)$ is written as Ad_g (so that the *argument* of the maps is written as if it were an *index*). It is more convenient, since in the usual form the left-hand side of the formula to be proved, as an example, would be $\mathrm{ad}\,(X)(Y)$, which is rather obscure.

translations over the whole group, we gain a metric tensor on G (already as a field on a manifold G), which is (trivially) left-invariant, but (surprisingly = non-trivially) it is also right-invariant. In a series of problems we now learn more details about this important idea.

12.3.7 Let ρ be a representation of a Lie group G in V and let ρ' be the derived representation of $\mathcal{G}$. Check that

(i) the operators $\rho(g)$ and $\rho'(X)$ (where $g \in G$, $X \in \mathcal{G}$) satisfy

$$\rho(g)\rho'(X)\rho(g^{-1}) = \rho'(\mathrm{Ad}_g X) \qquad \text{and on the basis} \qquad \rho(g)\mathcal{E}_i\rho(g^{-1}) = (\mathrm{Ad}_g)_i^j \mathcal{E}_j$$

(ii) in the Lie algebra $\mathcal{G}$ we may define a symmetric bilinear form

$$B(X, Y) := \mathrm{Tr}\,(\rho'(X)\rho'(Y)) \equiv \langle E^a, \rho'(X)\rho'(Y)E_a\rangle$$

which is Ad-invariant

(iii) its matrix of components with respect to the basis $E_i \in \mathcal{G}$ is

$$B_{ij} \equiv B(E_i, E_j) = \rho^a_{bi}\rho^b_{aj}$$

and it does not depend on the choice of a basis E_a in V (even though the parts from which it is pieced together, the matrix elements ρ^a_{bi}, do depend on the choice).

Hint: (i) apply ρ to $ge^{\epsilon X}g^{-1} = e^{\epsilon \mathrm{Ad}_g X}$, (12.1.6); (ii) symmetry: under the trace symbol the operators do commute; invariance:

$$B(\mathrm{Ad}_g X, \mathrm{Ad}_g Y) = \mathrm{Tr}\,(\rho(g)\rho'(X)\rho'(Y)\rho(g^{-1})) = B(X, Y)$$

due to the cyclic invariance of a trace; (iii) ρ^a_{bi} behave as the components of a *tensor* of the type $\binom{1}{1}$ in V in the pair of indices $(\,)^a_b$. □

12.3.8 Check that

(i) in the Lie algebra $\mathcal{G}$ we may define a symmetric bilinear form

$$K(X, Y) := \mathrm{Tr}\,(\mathrm{ad}_X \mathrm{ad}_Y) \equiv \langle E^i, \mathrm{ad}_X \mathrm{ad}_Y E_i\rangle$$

which is Ad-invariant; it is called the *Killing–Cartan form*

(ii) its matrix of components with respect to the basis $E_i \in \mathcal{G}$ is

$$k_{ij} \equiv K(E_i, E_j) = c^k_{il}c^l_{jk} \qquad \text{and it satisfies} \qquad k_{ij}(\mathrm{Ad}_g)^i_r(\mathrm{Ad}_g)^j_s = k_{rs}$$

(iii) the form[197] $K(X, Y)$ is invariant with respect to *all* the automorphisms of $\mathcal{G}$ (not only with respect to inner automorphisms, i.e. Ad_g).

Hint: (i), (ii) see (12.3.7) for $\rho = \mathrm{Ad}$; (iii) rewrite the definition of the automorphism $A([X, Y]) = [A(X), A(Y)]$ as $\mathrm{ad}_{A(X)} = A \circ \mathrm{ad}_X \circ A^{-1}$. □

[197] The word “form” does not mean a form in the sense of Chapter 5 (an “exterior” form) here, since K is *symmetric* (bilinear form) and the exterior forms are, in contrast, *antisymmetric*.

12.3.9 Let K be the Killing–Cartan form in a Lie algebra $\mathcal{G}$. Check that

(i) the infinitesimal version of its Ad-invariance is

$$K(\mathrm{ad}_Z X, Y) + K(X, \mathrm{ad}_Z Y) = 0$$
$$\text{i.e.} \quad K([Z, X], Y) + K(X, [Z, Y]) = 0$$

(ii) on a basis it gives

$$c_{ijk} + c_{jik} = 0 \qquad c_{ijk} := k_{il} c^l_{jk}$$

(iii) this condition gives that c_{ijk} are the components of some (truly "exterior") 3-*form* in the Lie algebra (see also (12.6.5)); its explicit expression reads

$$c(X, Y, Z) := K(X, [Y, Z]) \equiv c_{ijk} X^i Y^j Z^k$$

(iv) the 3-form $c(X, Y, Z)$ is Ad-invariant

$$c(\mathrm{Ad}_g X, \mathrm{Ad}_g Y, \mathrm{Ad}_g Z) = c(X, Y, Z).$$

Hint: (i), (ii): (12.1.10) for $\rho = \mathrm{Ad}$; (iii) antisymmetry of c_{ijk} in the first and the last pair results automatically in the antisymmetry of the pair (ik), so that c_{ijk} is a *completely* antisymmetric tensor of the type $\binom{0}{3}$; (iv)

$$c(\mathrm{Ad}_g X, \mathrm{Ad}_g Y, \mathrm{Ad}_g Z) = K(\mathrm{Ad}_g X, [\mathrm{Ad}_g Y, \mathrm{Ad}_g Z]) = \cdots \qquad \square$$

12.3.10 Write down explicitly the matrices of the components of the Killing–Cartan forms k_{ij} for the Lie algebras $so(3)$, $su(2)$ and $ga(1, \mathbb{R})$. Check that for $su(2) = so(3)$ the bilinear form $-\frac{1}{2}K(X, Y)$ is positive definite and Ad-invariant (so that it gives an invariant scalar product in $\mathcal{G}$)

Solution: according to (12.3.8), (11.7.12), (11.7.13) and (11.7.10) we get for $su(2)$ and $so(3)$ the result $k_{ij} = -2\delta_{ij}$; for $ga(1, \mathbb{R})$ we get $k_{ij} = \begin{pmatrix} 1 & 0 \\ 0 & 0 \end{pmatrix}$. $\square$

12.3.11 Consider the matrix Lie algebras $gl(n, \mathbb{R})$ and $sl(n, \mathbb{R})$. Show

(i) that the Killing–Cartan form for $\mathcal{G} = gl(n, \mathbb{R})$ turns out to be

$$K(X, Y) = 2n\mathrm{Tr}\,(XY) - 2(\mathrm{Tr}\,X)(\mathrm{Tr}\,Y)$$

(ii) explicitly its Ad-invariance
(iii) that it is (on $gl(n, \mathbb{R})$) degenerate
(iv) that its restriction to $sl(n, \mathbb{R})$ gives

$$K(X, Y) = 2n\mathrm{Tr}\,(XY)$$

Hint: (i) the trace in the definition of K may be computed with the help of the Weyl's basis $(E^i_j)^k_l = \delta^i_l \delta^k_j$; check that if $\hat{A} E^i_j = A^{ik}_{jl} E^l_k$, then $\mathrm{Tr}\hat{A} = A^{ij}_{ji}$; (ii) (12.3.1) and the cyclic invariance of a trace; (iii) the elements of the form $\lambda\mathbf{1}$ are orthogonal to the whole of

$gl(n, \mathbb{R})$ (this is visible right from the form of K, but also from the fact that these elements commute with all elements so that $\text{ad}_X = 0$ for them); (iv) $\text{Tr}X = 0$. □

• We can learn from these results that the Killing–Cartan form may not always be used as an invariant scalar product in $\mathcal{G}$ – sometimes it is degenerate. In the case of $so(3) = su(2)$ it turns out to be negative definite, so that if we help it a bit by multiplying it by a negative number, we get a beautiful positive definite invariant scalar product. For the algebras $ga(1, \mathbb{R})$ and $gl(n, \mathbb{R})$ the form K turns out, however, to be *degenerate* and there is no chance in this case to make a scalar product (even indefinite) from it. It turns out that the form K is non-degenerate just for a particular class of Lie algebras, which are called the *semi-simple Lie algebras*. By definition they are the algebras which do not contain non-zero commutative ideals[198] (see Appendix A.3). For our particular algebras under consideration we easily verify that this statement indeed holds.

12.3.12 Check that

(i) the algebras $ga(1, \mathbb{R})$ and $gl(n, \mathbb{R})$ indeed *do have* a non-zero commutative ideal
(ii) the algebra $so(3) = su(2)$ indeed *does not have* a non-zero commutative ideal.

Hint: (i) in $ga(1, \mathbb{R})$ the needed ideal is spanned by E_2 (in the notation of (11.7.10)), in $gl(n, \mathbb{R})$ by the identity matrix $\mathbb{I}$; (ii) the only non-zero ideal is given here by the *whole* algebra $\mathcal{G}$ and this, in turn, is non-commutative (if $\mathcal{I}$ is an ideal, $[\mathcal{G}, \mathcal{I}] \subset \mathcal{I}$ should hold; this means that $\mathcal{I}$ is an invariant subspace for the ad-representation of $\mathcal{G}$. In the language of (12.1.5) the subspace is constituted by the vectors $\mathbf{b}$ such that they remain there when multiplied "vectorially" by *any* vector $\mathbf{a}$). □

• The operators in the representation space, which are formed from the generators of the representation and which in turn happen to commute with all the generators, play an important role in the representation theory of Lie algebras. They are called the *Casimir operators*. By Schur's lemma (12.2.7) they are necessarily just scalar multiples of the identity operator (for an irreducible representation). The value of the multiple (the number λ in $\lambda\mathbf{1}$) depends on the representation, so it may in turn be used to characterize it. There exists a theory[199] which says how many independent Casimir operators a given Lie algebra has (and so how many numbers are to be specified to fix an irreducible representation of this algebra) and how they are constructed from the generators (they form "polynomials" in the generators). It turns out that this issue is closely related with the structure of the Ad-invariant tensors in Lie algebra. As an illustration (which is the case of greatest interest in physics) we mention the construction of the *quadratic* Casimir operator. For the Lie algebra $so(3)$, this used to be the starting point of the construction of all irreducible representations, which is worked out in every textbook on quantum mechanics in chapters devoted to the *theory of angular momentum*.

[198] Any such algebra is, in turn, the direct sum (see (12.4.2)) of *simple* algebras (i.e. non-commutative algebras which have only trivial ideals – zero and the whole algebra) and the simple algebras are classified (i.e. all of them are known), so that in this sense the structure of all the semi-simple Lie algebras is known. We shall not be concerned with the structure and the classification of the Lie algebras in this book; this may be found in almost every book on Lie algebras.

[199] We shall not be concerned with this part of representation theory in this book.

12.3.13 Let $\mathcal{G}$ be a Lie algebra such that an Ad-invariant (non-degenerate) metric tensor k exists in it. Let ρ' be a representation of $\mathcal{G}$ with generators $\mathcal{E}_i \equiv \rho'(E_i)$, k_{ij} the matrix of k and k^{ij} the inverse matrix to k_{ij}. Check that

(i) the *quadratic Casimir operator*

$$\hat{C} \equiv \hat{C}_2 := k^{ij}\mathcal{E}_i\mathcal{E}_j$$

commutes with all generators of the representation

$$[\hat{C}, \mathcal{E}_j] = 0 \qquad j = 1, \ldots, \dim \mathcal{G}$$

(ii) for any *irreducible* representation

$$\hat{C} = \lambda \mathbf{1}$$

(iii) for $so(3) = su(2)$ it is (when multiplied by an appropriate constant) the standard quantum-mechanical *operator of the square of the angular momentum*

$$-\hbar^2\hat{C} = \mathbf{J}^2 \equiv \mathbf{J}\cdot\mathbf{J} \equiv J_1J_1 + J_2J_2 + J_3J_3 \qquad J_j \equiv -i\hbar\mathcal{E}_j \equiv -i\hbar\rho'(E_j)$$

Hint: $[\hat{C}, \mathcal{E}_i] = k^{jk}[\mathcal{E}_j\mathcal{E}_k, \mathcal{E}_i] = \cdots = c^{jk}{}_i(\mathcal{E}_j\mathcal{E}_k + \mathcal{E}_k\mathcal{E}_j)$, where $c^{jk}{}_i := k^{kr}c^j_{ri}$; this vanishes if $c^{jk}{}_i = -c^{kj}{}_i$, which is, however, a consequence of the Ad-invariance of k (c_{ijk} is completely antisymmetric by (12.3.9)); (ii) Schur's lemma; (iii) (12.3.10), take $-\frac{1}{2}K$. □

• Now we would like to carry these objects from the Lie algebra $\mathcal{G}$ onto the corresponding Lie group G. We will assume that the form K is non-degenerate. By the standard technique from (11.1.4) and (11.1.8) we may then assign to the tensor K in the Lie algebra $\mathcal{G} \equiv T_eG$ the left-invariant tensor *field* $\mathcal{K}$ of type $\binom{0}{2}$ on the group G. Since it is non-degenerate in $\mathcal{G}$ and the left translation is bijective, it will also be non-degenerate at each point $g \in G$, so that we obtain the *left-invariant metric tensor* on the group. This field would be, however, left-invariant (by the construction) for *any* choice of the metric tensor in $\mathcal{G}$. A natural question then arises, what then is the reward for a particular choice of K?

12.3.14 Let K be the Killing–Cartan metric tensor in $\mathcal{G}$, and let $\mathcal{K}$ be the corresponding left-invariant metric tensor on G. Check that

(i) it may be written in terms of the left-invariant or right-invariant coframe fields $e^i \leftrightarrow E^i \leftrightarrow f^i$ as follows:

$$\mathcal{K} = k_{ij}e^i \otimes e^j = k_{ij}f^i \otimes f^j \qquad k_{ij} \equiv K(E_i, E_j) = \text{constant}$$

(ii) the metric tensor $\mathcal{K}$ is actually even *bi-invariant*, i.e. invariant with respect to both *left as well as right* translations on G

$$L_g^*\mathcal{K} = \mathcal{K} \qquad R_g^*\mathcal{K} = \mathcal{K}$$

Hint: (i) (11.1.8); according to (12.3.3) and (12.3.8) we have

$$\begin{aligned}\mathcal{K}(g) &= k_{ij}e^i(g) \otimes e^j(g) = k_{ij}(\mathrm{Ad}_{g^{-1}})^i_r(\mathrm{Ad}_{g^{-1}})^j_s f^r(g) \otimes f^s(g)\\ &= k_{rs}f^r(g) \otimes f^s(g)\end{aligned}$$

(ii) left-invariance by means of the presentation in terms of e^i ($L_g^* e^i = e^i \Rightarrow L_g^*(k_{ij} e^i \otimes e^j) = \cdots = k_{ij} e^i \otimes e^j$), right-invariance in full analogy by means of the presentation in terms of f^i. □

12.3.15 Express the (improved = positive definite) Killing–Cartan metric tensor $-\frac{1}{2}\mathcal{K}$ on the group $SU(2)$ in coordinates $(\varphi, \vartheta, \psi)$ (the Euler angles). Convince yourself that the expression containing the left-invariant 1-forms e^i indeed gives the same result as with the right-invariant 1-forms f^i (i.e. that indeed $k_{ij} e^i \otimes e^j = k_{ij} f^i \otimes f^j$).

Hint: according to (12.3.14) and the explicit expressions in (11.7.23)

$$-\frac{1}{2}\mathcal{K} = e^1 \otimes e^1 + e^2 \otimes e^2 + e^3 \otimes e^3$$

$$= (d\vartheta \quad d\varphi \quad d\psi) \otimes \begin{pmatrix} 1 & 0 & \cos\vartheta \\ 0 & 1 & 0 \\ \cos\vartheta & 0 & 1 \end{pmatrix} \begin{pmatrix} d\vartheta \\ d\varphi \\ d\psi \end{pmatrix}$$ □

• Let $\omega_L \equiv e^1 \wedge \cdots \wedge e^n$ and $\omega_R \equiv f^1 \wedge \cdots \wedge f^n$ be the left- and right-invariant volume forms on a Lie group G. As we already mentioned in Section 11.6, in general they differ. Let us see more closely by "how much" these forms differ.

12.3.16 Check that

(i) in general, the forms ω_R and ω_L are related as[200]

$$\omega_R(g) = (\det \mathrm{Ad}_g)\, \omega_L(g)$$

(ii) for the group $GA(1, \mathbb{R})$ it turns out that

$$\mathrm{Ad}_{(x,y)} = \begin{pmatrix} 1 & 0 \\ -y & x \end{pmatrix} \qquad \omega_R = x\, \omega_L$$

which perfectly matches the general result from item (i)

(iii) for the group $SO(3)$ we have

$$\mathrm{Ad}_A = \text{the rotation given by the matrix } A \qquad \omega_R = \omega_L$$

once more perfectly matching the general result from item (i); this means that on $SO(3)$ the right-invariant volume form happens to be at the same time also left-invariant.

Hint: (i) according to (12.3.3) we have $\omega_R(g) \equiv f^1(g) \wedge \cdots \wedge f^n(g) = \cdots = (\det \mathrm{Ad}_g)\, e^1(g) \wedge \cdots \wedge e^n(g) \equiv (\det \mathrm{Ad}_g)\, \omega_L(g)$; (ii) (11.1.15) and

$$\mathrm{Ad}_{(x,y)}(aE_1 + bE_2) = \begin{pmatrix} x & y \\ 0 & 1 \end{pmatrix} \begin{pmatrix} a & b \\ 0 & 0 \end{pmatrix} \begin{pmatrix} x^{-1} & -yx^{-1} \\ 0 & 1 \end{pmatrix} = \begin{pmatrix} a & xb - ya \\ 0 & 0 \end{pmatrix}$$

$$\equiv aE_1 + (xb - ya)E_2$$

[200] We mean the relation between the forms, which arose on G by the right and left translation of *the same* form $E^1 \wedge \cdots \wedge E^n$ in the tangent space of the unit element (i.e. the Lie algebra).

(iii) by (12.3.6) we have (for $SO(3)$!) $\mathrm{Ad}_{\exp X} = e^{(\mathrm{ad}X)} = e^{(\mathrm{id}X)} = e^X$, i.e. $\mathrm{Ad}_A = A \Rightarrow \det \mathrm{Ad}_A = \det A = 1$; by (11.7.23) $e^1 \wedge e^2 \wedge e^3 = f^1 \wedge f^2 \wedge f^3 = \sin\vartheta \, d\vartheta \wedge d\varphi \wedge d\psi$. □

• We obtained a fairly remarkable result on the group $SO(3)$, namely that the left- and the right-invariant volume forms coincide; in other words, that we have the *bi-invariant* volume form $\omega_L = \omega_R$. In the following problem we convince ourselves that it turns out likewise on an arbitrary *compact* Lie group (and trivially also on a commutative one, where $e^a = f^a$; on a non-commutative one non-trivially, since there $e^a \neq f^a$ in general and only their *product* happens to coincide).

12.3.17 Prove that on each *compact* Lie group there exists a *bi-invariant volume form*, namely that each right-invariant volume form turns out to be at the same time also left-invariant (and vice versa).

Hint: since R_g is a diffeomorphism of G onto itself, $R_g^*\omega_L$ is also a volume form; this form is left-invariant ($L_h^*(R_g^*\omega_L) = R_g^* L_h^* \omega_L = R_g^*\omega_L$), so that it may be at most a constant multiple of ω_L, the "constant" being possibly dependent on g, $R_g^*\omega_L = \Phi(g)\omega_L$. Then

$$\mathrm{vol}\, G = \int_G \omega_L = \int_{R_g(G)} \omega_L = \int_G R_g^*\omega_L = \int_G \Phi(g)\omega_L = \Phi(g)\int_G \omega_L = \Phi(g)\, \mathrm{vol}\, G$$

so that $\Phi(g) = 1$ and $R_g^*\omega_L = \omega_L$; the left-invariant volume form ω_L is at the same time also right-invariant. □

12.3.18 Prove that there exists a *positive-definite* Ad-invariant metric tensor in the Lie algebra of each *compact* Lie group[201]

$$B(\mathrm{Ad}_g X, \mathrm{Ad}_g Y) = B(X, Y)$$

so that there also holds

$$B(\mathrm{ad}_Z X, Y) + B(X, \mathrm{ad}_Z Y) \equiv B([Z, X], Y) + B(X, [Z, Y]) = 0$$

Any Lie algebra in which there exists a positive-definite scalar product satisfying $B([Z, X], Y) + B(X, [Z, Y]) = 0$ is called a *compact Lie algebra*. We see that compact Lie groups have compact Lie algebras.

Hint: average over the group (12.1.12) the *arbitrary positive-definite* initial scalar product B_0 (δ_{ij} in some fixed basis in $\mathcal{G}$); the relation for ad is the infinitesimal version of the relation for Ad. □

• This result is useful, for example, for the structure of the action integrals in the theory of gauge fields (21.5.6).

To close the section let us have a look at an important representation of G which takes place in the *dual* space $\mathcal{G}^*$ to the Lie algebra $\mathcal{G}$. (The reason why this representation is fairly interesting will be clarified in Section 14.6).

[201] For example, in $su(2) = so(3)$ it is minus the *Killing–Cartan* metric tensor.

12.3.19 The *coadjoint representation* Ad_g^* of a Lie group G is the contragredient (12.1.8) representation to the adjoint one Ad_g. It thus acts in the linear space $\mathcal{G}^*$ by the prescription

$$\langle \mathrm{Ad}_g^* X^*, Y\rangle := \langle X^*, \mathrm{Ad}_{g^{-1}} Y\rangle \qquad X^* \in \mathcal{G}^*, Y \in \mathcal{G}$$

Check that

(i) if there is an Ad-invariant (non-degenerate) scalar product in $\mathcal{G}$, then the coadjoint representation Ad_g^* is equivalent to the adjoint one Ad_g
(ii) it acts on the dual basis via inverse and transpose matrices of the adjoint representation

$$\mathrm{Ad}_g^* E^i = E^j (\mathrm{Ad}_{g^{-1}})^i_j$$

(iii) in the derived representation ad_X^* the "generators are given (just like for ad_X) directly in terms of the structure constants"

$$\langle \mathrm{ad}_X^* Z^*, Y\rangle = -\langle Z^*, [X, Y]\rangle \qquad \mathrm{ad}_{E_i}^* E^j \equiv (\mathrm{ad}_{E_i}^*)^j_k E^k = -c^j_{ik} E^k$$

Hint: (i) see (12.2.3); (ii) see (12.1.8); (iii) see (12.3.5). □

Note: sometimes the coadjoint representation is introduced without "re-doing the right action to the left one" (the trick $g \mapsto g^{-1}$), i.e. by the relation $\langle \mathrm{Ad}_g^* X^*, Y\rangle := \langle X^*, \mathrm{Ad}_g Y\rangle$. This clearly results in the *right* linear action ("*anti*representation").

12.4 Basic constructions with groups, Lie algebras and their representations

• In this section we will speak about the simplest (and at the same time the most frequent and important) constructions which are routinely performed with groups, Lie algebras and their representations. Let us start with the direct and semidirect product of groups and the corresponding direct and semidirect sum of Lie algebras.

12.4.1 Given two groups G and H, consider the Cartesian product of the sets $G \times H$ (the elements being ordered pairs (g, h)) and introduce the multiplication "by components" into it

$$(g_1, h_1) \circ (g_2, h_2) := (g_1 g_2, h_1 h_2)$$

Check that

(i) this multiplication indeed satisfies the axioms of a group; thus we get a new group, the *direct product of the groups* G and H; it is denoted by $G \times H$
(ii) if *matrix* groups are concerned, a simple realization of their direct product reads

$$(g, h) \leftrightarrow \begin{pmatrix} g & 0 \\ 0 & h \end{pmatrix}$$

that is to say, the group of matrices of the block-diagonal form displayed above (endowed with the operation of a product realized by the common matrix multiplication) *is isomorphic* to the "official" group $G \times H$

(iii) within the group $G \times H$ there are hidden the subgroups $\tilde{G}$ and $\tilde{H}$, which are isomorphic to the original building blocks G and H; the elements of $\tilde{G}$ and $\tilde{H}$ commute with each other and moreover each element from $G \times H$ may be uniquely expressed in the form of a product $\tilde{g}\tilde{h}$, where $\tilde{g} \in \tilde{G}, \tilde{h} \in \tilde{H}$ (this will be written as $\tilde{G} \cdot \tilde{H}$)

(iv) the opposite statement also holds: if a group K has two mutually commuting subgroups G and H such that $K = G \cdot H$ (i.e. if each element form K may be uniquely expressed in the form of a product $k = gh$, where $g \in G, h \in H$), then K is isomorphic to the direct product $G \times H$ in the sense of the definition from item (i).

Hint: (iii) $\tilde{G} \leftrightarrow (g, e_H)$, $\tilde{H} \leftrightarrow (e_G, h)$, $e_{\dots}$ being the identity elements in the corresponding groups; (iv) $k \equiv gh \leftrightarrow (g, h)$. □

12.4.2 Given two Lie algebras $\mathcal{G}$ and $\mathcal{H}$, consider the Cartesian product of the sets $\mathcal{G} \times \mathcal{H}$ (the elements being ordered pairs (X, Y)) and introduce the linear structure and commutator "by components" into it

$$(X_1, Y_1) + \lambda(X_2, Y_2) := (X_1 + \lambda X_2, Y_1 + \lambda Y_2)$$
$$[(X_1, Y_1), (X_2, Y_2)] := ([X_1, X_2], [Y_1, Y_2])$$

Check that

(i) the axioms of a Lie algebra are indeed satisfied; thus we get a new Lie algebra, the *direct sum of the Lie algebras* $\mathcal{G}$ and $\mathcal{H}$; it is denoted by $\mathcal{G} + \mathcal{H}$

(ii) if *matrix* Lie algebras are concerned, a simple realization of their direct sum reads

$$(X, Y) \leftrightarrow \begin{pmatrix} X & 0 \\ 0 & Y \end{pmatrix}$$

that is to say, the Lie algebra of matrices of the block-diagonal form displayed above (endowed with the operation of the commutator realized by the common matrix commutator) *is isomorphic* to the Lie algebra $\mathcal{G} + \mathcal{H}$

(iii) within the Lie algebra $\mathcal{G} + \mathcal{H}$ there are hidden the subalgebras $\tilde{\mathcal{G}}$ and $\tilde{\mathcal{H}}$, which are isomorphic to the original building blocks $\mathcal{G}$ and $\mathcal{H}$; the elements of $\tilde{\mathcal{G}}$ and $\tilde{\mathcal{H}}$ commute with each other and moreover each element from $\mathcal{G} + \mathcal{H}$ may be uniquely expressed in the form of a sum $\tilde{X} + \tilde{Y}$, where $\tilde{X} \in \tilde{\mathcal{G}}, \tilde{Y} \in \tilde{\mathcal{H}}$ (so that the Lie algebra *as a linear space* is the direct sum of these commuting subalgebras)

(iv) the opposite statement also holds: if an algebra $\mathcal{K}$ has two mutually commuting subalgebras $\mathcal{G}$ and $\mathcal{H}$ such that it is as a linear space a direct sum of these subalgebras (each element from $\mathcal{K}$ may be uniquely expressed in the form of a sum $Z = X + Y$, where $X \in \mathcal{G}, Y \in \mathcal{H}$), then $\mathcal{K}$ is isomorphic to the direct sum $\mathcal{G} + \mathcal{H}$ (in the sense of the definition from item (i)).

Hint: (iii) $\tilde{\mathcal{G}} \leftrightarrow (X, 0)$, $\tilde{\mathcal{H}} \leftrightarrow (0, Y)$; (iv) $Z \equiv X + Y \leftrightarrow (X, Y)$. □

12.4.3 Let $K = G \times H$ be a direct product of Lie groups G and H and let $\mathcal{K}, \mathcal{G}$ and $\mathcal{H}$ be the corresponding Lie algebras. Show that $\mathcal{K}$ is isomorphic to the direct sum of $\mathcal{G}$ and

$\mathcal{H}$, so that "under" a direct product of groups there is the direct sum of their Lie algebras[202]

$$K = G \times H \quad \Rightarrow \quad \mathcal{K} = \mathcal{G} + \mathcal{H}$$

Hint: in matrix language it is elementary: close to the unit element of $G \times H$ is

$$\begin{pmatrix} g(\epsilon) & 0 \\ 0 & h(\epsilon) \end{pmatrix} = \begin{pmatrix} 1+\epsilon X & 0 \\ 0 & 1+\epsilon Y \end{pmatrix} \equiv \begin{pmatrix} 1 & 0 \\ 0 & 1 \end{pmatrix} + \epsilon \begin{pmatrix} X & 0 \\ 0 & Y \end{pmatrix}$$

so that $\mathcal{K}$ is given by the matrices $\begin{pmatrix} X & 0 \\ 0 & Y \end{pmatrix}$, being just $\mathcal{G} + \mathcal{H}$. "Scientifically": $\mathcal{K}$ is the tangent space in $(e_G, e_H) \in G \times H$; the latter is decomposed by (2.2.14) into $T_{e_G} G \oplus T_{e_H} H \equiv \mathcal{G} \oplus \mathcal{H}$, so that $Z = X + Y$. Left-invariant fields arise standardly by the left translation (11.1.4) from the unit element, $L_Z(g, h) = L_{(g,h)*} Z$. Since L_X and L_Y generate, according to (11.4.6), the flows $(g, h) \mapsto (ge^{tX}, h)$ and $(g, h) \mapsto (g, he^{tY})$ respectively and the *flows commute*, also their generators, namely L_X and L_Y, commute and consequently also X and Y commute. Within their own "halves" (two Xs or two Ys) life goes on as before, which gives altogether the rule from problem (12.4.2). □

• The fact that a group (Lie algebra) happens to be the direct product (sum) of its subgroups (subalgebras) need not be evident at first sight. This is illustrated by the following simple examples.

12.4.4 Consider the Lie group $GL_+(n, \mathbb{R})$ and its Lie algebra $gl(n, \mathbb{R})$. Check that

(i) the group $GL_+(n, \mathbb{R})$ is a direct product of the groups $GL_+(1, \mathbb{R})$ and $SL(n, \mathbb{R})$

$$GL_+(n, \mathbb{R}) = GL_+(1, \mathbb{R}) \times SL(n, \mathbb{R})$$

(ii) the Lie algebra $gl(n, \mathbb{R})$ is a direct sum of the algebras $gl(1, \mathbb{R}) \sim \mathbb{R}$ and $sl(n, \mathbb{R})$

$$gl(n, \mathbb{R}) = \mathbb{R} + sl(n, \mathbb{R})$$

Hint: (i) if $A \in GL_+(n, \mathbb{R})$, then $A = (\det A)^{1/n} B \equiv \lambda B \leftrightarrow (\lambda, B)$; (ii) if $X \in gl(n, \mathbb{R})$, then $X = (1/n)(\text{Tr}\, X)\mathbb{I} + Y \leftrightarrow ((1/n)(\text{Tr}\, X), Y)$. □

12.4.5 Show that the Lie algebra $u(n)$ is a direct sum of its subalgebras (isomorphic to) $su(n)$ and $u(1)$, but the corresponding "capitals version" of the statement *does not* hold, i.e. the group $U(n)$ *is not* a direct product of $SU(n)$ and $U(1)$,

$$u(n) = su(n) + u(1) \qquad U(n) \neq SU(n) \times U(1)$$

Hint: once more one should extract[203] a trace: $X = (1/n)(\text{Tr}\, X)\mathbb{I} + Y \leftrightarrow (Y, (1/n)(\text{Tr}\, X))$; the decomposition $A = \lambda B$ with $\lambda \in U(1)$ and $B \in SU(n)$ is now *ambiguous* (there are n

[202] The opposite statement *does not* hold in general: "over" a direct sum $\mathcal{G} + \mathcal{H}$ need not be *only* $G \times H$ (the latter *is* clearly there), but the issue gets complicated topologically. In general, it turns out that *several* groups related through a *covering* (being a discrete version of a *bundle*, a concept to be found in Chapter 17 and beyond) may have isomorphic Lie algebras or conversely through a *factorization* (projection in the language of bundles); see an example in (12.4.5).

[203] The "scalar" matrices $\lambda\mathbb{I}$ commute with all other matrices and they constitute a *one-dimensional* subalgebra; such an algebra is (up to an isomorphism) *unique*; here we denote it by $u(1)$, and in the preceding problem it was written as $gl(1, \mathbb{R})$ or $\mathbb{R}$.

possibilities, $A = \lambda_1 B_1 = \lambda_2 B_2 = \cdots = \lambda_n B_n$, where λ_k is the kth copy of the nth root (we have n pieces of nth roots) of $\det A \equiv e^{i\alpha}$; see also (13.2.13), where it is shown that the product turns out to be isomorphic to the group $U(n)$ *after being factorized* by the subgroup $\mathbb{Z}_n$). □

12.4.6 Check that the Lie algebra $so(4)$ is a direct sum of two copies of $so(3)$

$$so(4) = so(3) + so(3)$$

Hint: first introduce the antisymmetrized Weyl basis $\tilde{E}_{ab} := e_{ab} - e_{ba}$ $((e_{ab})_{cd} = \delta_{ac}\delta_{bd})$, $a, \ldots = 1, \ldots, 4$; then set $E_i := \frac{1}{2}\epsilon_{ijk}\tilde{E}_{jk}$, $F_i := \tilde{E}_{i4}$ $(i, \ldots = 1, 2, 3)$ and finally $e_i = E_i + F_i$, $f_i = E_i - F_i$; convince yourself that the commutation relations of the basis (e_i, f_i) correspond to a direct sum of the type mentioned above. □

• Now we will introduce a slightly more involved product for the case of groups and sum for Lie algebras, corresponding to a *semidirect* product and sum. In order to do this, however, it is not enough to have two groups G, H (or the algebras $\mathcal{G}, \mathcal{H}$). Instead we need some more structure, namely an action of G by automorphisms of H (and of $\mathcal{G}$ by derivations of $\mathcal{H}$). Well-known examples of this construction are provided by the semidirect product in the affine group and its subgroups (the Poincaré group and the Euclidean group).

12.4.7* Let G and H be two groups and let G act from the left by automorphisms on H, i.e. there exists a map

$$\phi : G \to \text{Aut}\, H \qquad \text{such that} \qquad \phi_{g\tilde{g}} = \phi_g \circ \phi_{\tilde{g}}$$
$$g \mapsto \phi_g \qquad\qquad \phi_g(h\tilde{h}) = \phi_g(h)\phi_g(\tilde{h})$$

On the Cartesian product of the sets $G \times H$ we introduce a multiplication by the prescription

$$(g_1, h_1) \circ (g_2, h_2) := (g_1 g_2, h_1 \phi_{g_1}(h_2))$$

(in the right slot there is the product (in the group H) of the element h_1 and the ϕ_{g_1}-image of the element h_2). Check that

(i) this multiplication satisfies the axioms of a group; thus we get a new group, the *semidirect product of the groups* G and H; it will be denoted by $G \ltimes H$ or $G \times_\phi H$
(ii) for the *trivial* automorphism $\phi_g = \text{id}$ (i.e. $\phi_g(h) = h$) the semidirect product reduces to the direct one
(iii) the elements of the form (e_G, h) constitute a normal (= invariant) subgroup (the concept to be defined in (13.2.10)) isomorphic to H. □

12.4.8* Let G be a group and (V, ρ) a representation. Check that

(i) any vector space V may be regarded as a (commutative = Abelian) Lie group
(ii) the representation ρ in V provides a left action of G by the automorphisms of (a group) V, so that the assumptions for introducing a semidirect product from (12.4.4) are satisfied

(iii) in this particular case the product explicitly reads

$$(g, v) \circ (\tilde{g}, \tilde{v}) := (g\tilde{g}, v + \rho(g)\tilde{v})$$

(iv) the product in the affine, Poincaré or Euclidean group (see (4.6.10) and (10.1.15))

$$(A, a) \circ (B, b) := (AB, a + Ab)$$

(where a, b are the columns from $\mathbb{R}^n$, A, B belong to $GL(n, \mathbb{R})$ for the *affine group*, to $SO(1, 3)$ for the *Poincaré group* and to $SO(n)$ for the *Euclidean group*) is of just that type, so that the affine (etc.) group is a semidirect product of its (linear and translational) subgroups.

Hint: (i) multiplication is $v + w$; (ii) $\rho(g)$ are *linear*; (iv) $\rho = \text{id}$, so that $\rho(A)b = Ab$. □

12.4.9* Let $\mathcal{G}$ and $\mathcal{H}$ be two Lie algebras and let $\mathcal{G}$ be represented by derivations of $\mathcal{H}$, i.e. there exists a map

$$\begin{array}{lll} D : \mathcal{G} \to \text{Der}\, \mathcal{H} & D_{X+\lambda\tilde{X}} = D_X + \lambda D_{\tilde{X}} & D_X(Y + \lambda\tilde{Y}) = D_X Y + \lambda D_X \tilde{Y} \\ \quad X \mapsto D_X & D_{[X,\tilde{X}]} = D_X D_{\tilde{X}} - D_{\tilde{X}} D_X & D_X[Y, \tilde{Y}] = [D_X Y, \tilde{Y}] + [Y, D_X \tilde{Y}] \end{array}$$

On the Cartesian product of the sets $\mathcal{G} \times \mathcal{H}$ we introduce a structure of a Lie algebra by the prescriptions

$$(X_1, Y_1) + \lambda(X_2, Y_2) := (X_1 + \lambda X_2, Y_1 + \lambda Y_2)$$
$$[(X_1, Y_1), (X_2, Y_2)] := ([X_1, X_2], [Y_1, Y_2] + D_{X_1} Y_2 - D_{X_2} Y_1)$$

Check that

(i) the axioms of a Lie algebra are satisfied; thus we get a new Lie algebra, the *semidirect sum of the Lie algebras* $\mathcal{G}$ and $\mathcal{H}$; it will be denoted by $\mathcal{G} \ltimes \mathcal{H}$ (i.e. in the same way as the semidirect *product* of groups; it should be clear what is actually meant from what *objects* are around the symbol) or $\mathcal{G} +_D \mathcal{H}$
(ii) the alternative way to define this commutator is the following: inside the original Lie algebras life goes on as before and the "mixed" commutator is defined as

$$[X, Y] := D_X(Y)$$

(iii) for the trivial (= zero) derivation the semidirect sum reduces to the direct one (the mixed commutator is now defined *to vanish*)
(iv) the elements of the form $(0, Y)$ constitute an *ideal* of this algebra
(v) the derivation D_X corresponding to the construction from (12.4.8) is $D_X x = \rho'(X)x$ (ρ' being the derived representation of the representation ρ, x is an element of the Abelian (commutative) Lie algebra of the group V), so that the commutator reads

$$[(X, x), (Y, y)] = ([X, Y], \rho'(X)y - \rho'(Y)x)$$

and in the particular case of the affine (Poincaré, Euclidean) Lie algebra we get

$$[(X, x), (Y, y)] = ([X, Y], Xy - Yx)$$

which matches the result obtained from the embedding of $ga(n, \mathbb{R})$ into $gl(n+1, \mathbb{R})$, resulting from the embedding of the corresponding groups (10.1.15)

$$(A, a) \mapsto \begin{pmatrix} A & a \\ 0 & 1 \end{pmatrix} \quad \Rightarrow \quad (X, x) \mapsto \begin{pmatrix} X & x \\ 0 & 0 \end{pmatrix}$$

Hint: (ii) write down the particular cases $[(X_1, 0), (X_2, 0)]$, $[(0, Y_1), (0, Y_2)]$ and $[(X, 0), (0, Y)]$. □

• Let us turn our attention to the constructions leading to the direct sum and direct product of *representations*. Assume that we are given two representations *of the same* group G, namely (ρ_1, V_1) and (ρ_2, V_2). Then, as is described in detail in Appendix A.1, two "new" vector spaces are automatically available, the direct sum $V_1 \oplus V_2$ and the direct product $V_1 \otimes V_2$. If we are also given some linear operators in the initial spaces (A in V_1 and B in V_2), we may consider new operators $A \oplus B$ in $V_1 \oplus V_2$ and $A \otimes B$ in $V_1 \otimes V_2$. Simple algebraic properties of these particular operators guarantee that if the initial operators corresponded to the *representations* of a group, the resulting operators define a representation as well. These representations are called the *direct sum* $\rho_1 \oplus \rho_2$ and the *direct product* $\rho_1 \otimes \rho_2$ of the initial representations.

12.4.10 Let (ρ_1, V_1) and (ρ_2, V_2) be two representations of the same group G. Check that

(i) by the prescription

$$(\rho_1 \oplus \rho_2)(g) := \rho_1(g) \oplus \rho_2(g)$$

(the right-hand side, i.e. the operator of the structure $A \oplus B$, being in the sense of Appendix A.1) one indeed defines a representation of the group G

(ii) its dimension is $n_1 + n_2$ (if $n_1 \equiv \dim V_1$ and $n_2 \equiv \dim V_2$)

(iii) in matrix language this reads

$$(\rho_1 \oplus \rho_2)(g) \quad \leftrightarrow \quad \begin{pmatrix} \rho_1(g) & 0 \\ 0 & \rho_2(g) \end{pmatrix}$$

(iv) the derived representation of the Lie algebra is the direct sum of the initial derived representations

$$(\rho_1 \oplus \rho_2)'(X) = \rho_1'(X) \oplus \rho_2'(X)$$

Hint: (i) according to Appendix A.1 the following identity holds: $(A \oplus B)(C \oplus D) = AC \oplus BD$; (iv)

$$\begin{aligned}
(\rho_1 \oplus \rho_2)(e^{\epsilon X})(v_1, v_2) &= \rho_1(e^{\epsilon X}) \oplus \rho_2(e^{\epsilon X})(v_1, v_2) = (\rho_1(e^{\epsilon X})v_1, \rho_2(e^{\epsilon X})v_2) \\
&= (v_1 + \epsilon\rho_1'(X)v_1 + \cdots, v_2 + \epsilon\rho_2'(X)v_2 + \cdots) \\
&\equiv (v_1, v_2) + \epsilon(\rho_1'(X) \oplus \rho_2'(X))(v_1, v_2) + \cdots
\end{aligned}$$

□

12.4.11 Let (ρ_1, V_1) and (ρ_2, V_2) be two representations of the same group G. Check that

(i) by the prescription[204]

$$(\rho_1 \otimes \rho_2)(g) := \rho_1(g) \otimes \rho_2(g)$$

(the right-hand side, i.e. the operator of the structure $A \otimes B$, being in the sense of Appendix A.1) one indeed defines a representation of the group G

(ii) its dimension is $n_1 \cdot n_2$ (if $n_1 \equiv \dim V_1, n_2 \equiv \dim V_2$)

(iii) the expression of the derived representation of the Lie algebra in terms of the initial derived representations turns out to be

$$(\rho_1 \otimes \rho_2)'(X) = \rho_1'(X) \otimes \hat{1} + \hat{1} \otimes \rho_2'(X)$$

Hint: (i) according to Appendix A.1 the identity $(A \otimes B)(C \otimes D) = AC \otimes BD$ holds; (iii) first check the bilinearity of the direct product of operators, i.e. $(A + \lambda C) \otimes B = A \otimes B + \lambda C \otimes B$ and similarly on the right; then $(\rho_1 \otimes \rho_2)(e^{\epsilon X}) = \rho_1(e^{\epsilon X}) \otimes \rho_2(e^{\epsilon X}) = (\hat{1} + \epsilon\rho_1'(X) + \cdots) \otimes (\hat{1} + \epsilon\rho_2'(X) + \cdots) = \hat{1} \otimes \hat{1} + \epsilon(\hat{1} \otimes \rho_1'(X) + \rho_2'(X) \otimes \hat{1}) = \cdots$. □

• We often encounter representations which are sums or products of simpler ones in the index language. Recall how these look.

12.4.12 According to Appendix A.1 a general vector $u \in V_1 \oplus V_2$ may be written as $u = u^i E_i + u^\alpha E_\alpha$. Be sure to understand that

(i) (the components of) vectors from $u \in V_1 \oplus V_2$ carry only *a single* index, taking values "of two types" (i or α)

(ii) the indices of type i transform under the action of $\rho_1 \oplus \rho_2$ by the matrix $(\rho_1(g))^i_{\ j}$, the indices of type α transform by the matrix $(\rho_2(g))^\alpha_\beta$, i.e.

$$u^i \mapsto (\rho_1(g))^i_{\ j} u^j \qquad u^\alpha \mapsto (\rho_2(g))^\alpha_\beta u^\beta$$

(iii) for the derived representation $(\rho_1 \oplus \rho_2)'$ in full analogy

$$u^i \mapsto (\rho_1'(X))^i_{\ j} u^j \qquad u^\alpha \mapsto (\rho_2'(X))^\alpha_\beta u^\beta$$

Hint: according to Appendix A.1 the operators $A \oplus B$ act on a component column (u^i, u^α) as follows:

$$\begin{pmatrix} u^i \\ u^\alpha \end{pmatrix} \mapsto \begin{pmatrix} A^i_{\ j} & 0 \\ 0 & B^\alpha_\beta \end{pmatrix} \begin{pmatrix} u^j \\ u^\beta \end{pmatrix}$$

□

12.4.13 According to Appendix A.1 a general vector $u \in V_1 \otimes V_2$ may be written as $u = u^{i\alpha} E_i \otimes E_\alpha$. Be sure to understand that

[204] Note that the construction of the direct sum of representations $\rho_1 \oplus \rho_2$ may be obtained amazingly simply (and surprisingly no literature mentions this fact!) from the direct product $\rho_1 \otimes \rho_2$ by the well-known operation "turning a bulb by an angle $\pm\pi/4$" (its iteration is usually applied when the bulb $\otimes$ is blown and we replace it by a new one).

(i) (the components of) vectors from $u \in V_1 \otimes V_2$ carry *two* indices, one "of the type i" and the other "of the type α"

(ii) the index of type i is transformed under the action of $(\rho_1 \otimes \rho_2)(g)$ by the matrix $(\rho_1(g))^i_{\ j}$ and *at the same time* the index of the type α is transformed by the matrix $(\rho_2(g))^\alpha_\beta$, i.e.

$$u^{i\alpha} \mapsto (\rho_1(g))^i_{\ j}(\rho_2(g))^\alpha_\beta u^{j\beta}$$

A general rule (valid for the direct product of an *arbitrary number* of representations) thus reads: each index is transformed by "its" matrix, this occurring *all at once*: under $\rho_1 \otimes \rho_2 \otimes \cdots \otimes \rho_n$ the components transform according to the rule

$$u^{i\alpha\ldots a} \mapsto (\rho_1(g))^i_{\ j}(\rho_2(g))^\alpha_\beta \ldots (\rho_n(g))^a_b u^{j\beta\ldots b}$$

(iii) for the derived representation $(\rho_1 \otimes \rho_2)'$ we get

$$u^{i\alpha} \mapsto (\rho_1'(X))^i_{\ j} u^{j\alpha} + (\rho_2'(X))^\alpha_\beta u^{i\beta}$$

(and in full analogy for the derived representation of the product of *more than two* representations we get a *sum of several* such terms).

Hint: (ii) according to Appendix A.1 the operators $A \otimes B$ act on the components $u^{i\alpha}$ as follows: $u^{i\alpha} \mapsto A^i_{\ j} B^\alpha_\beta u^{j\beta}$; (iii) see (12.4.11). □

12.4.14 Consider tensors of the type $\binom{p}{q}$ in a linear space L. Check that

(i) in the formula for the transformation of components $t \equiv t^{a\ldots b}_{c\ldots d}$ under the change of a basis $e_a \mapsto A^b_{\ a} e_b$ in L a representation ρ of the group $GL(n, \mathbb{R})$ occurs[205]

$$t(eA) = \rho(A^{-1})t(e)$$

($t(e)$ being the *components* of the tensor with respect to the basis e) which is a tensor product of several representations of the type $\rho^1_0 \equiv \mathrm{id}$ ($A \mapsto A$) and several representations of the type ρ^0_1, contragredient to the former

$$\rho \equiv \rho^p_q = \underbrace{\rho^1_0 \otimes \cdots \otimes \rho^1_0}_{p} \otimes \underbrace{\rho^0_1 \otimes \cdots \otimes \rho^0_1}_{q}$$

i.e.

$$(\rho(A)t)^{a\ldots b}_{c\ldots d} = A^a_{\ i} \ldots A^b_{\ j}(A^{-1})^k_{\ c} \ldots (A^{-1})^l_{\ d} t^{i\ldots j}_{k\ldots l}$$

(so that there is one ρ^1_0 for each upper index and one ρ^0_1 for each lower index)

(ii) its derived representation turns out to be

$$(\rho'(C)t)^{a\ldots b}_{c\ldots d} = C^a_{\ i} t^{i\ldots b}_{c\ldots d} + \cdots + C^b_{\ i} t^{a\ldots i}_{c\ldots d} - C^i_{\ c} t^{a\ldots b}_{i\ldots d} - \cdots - C^i_{\ d} t^{a\ldots b}_{c\ldots i}$$

(iii) as an example, for a *metric* tensor we get

$$(\rho(A)g)_{ab} = (A^{-1})^k_{\ a}(A^{-1})^l_{\ b} g_{kl} \qquad (\rho'(C)g)_{ab} = -C^i_{\ a} g_{ib} - C^i_{\ b} g_{ai}$$

Hint: see (2.4.6). □

[205] A *right* action is present in this formula due to the argument A^{-1} in ρ (13.1.1); this is all *right*, since also $e \mapsto eA$ is a right action.

12.4.15 We denote by (V^p_q, ρ^p_q) the representation of the group $GL(n, \mathbb{R})$ in the space of components of tensors of type $\binom{p}{q}$ in the linear space L in the sense of (12.4.14); thus it acts by the formula

$$\left(\rho^p_q(A)t\right)^{a\dots b}_{c\dots d} = A^a_i \dots A^b_j (A^{-1})^k_c \dots (A^{-1})^l_d t^{i\dots j}_{k\dots l}$$

(there are p matrices A and q matrices A^{-1}). Check that an arbitrary *contraction* C is an *intertwining operator* between the representations (V^p_q, ρ^p_q) and $\left(V^{p-1}_{q-1}, \rho^{p-1}_{q-1}\right)$, i.e. that

$$C : \left(V^p_q, \rho^p_q\right) \to \left(V^{p-1}_{q-1}, \rho^{p-1}_{q-1}\right) \qquad C \circ \rho^p_q(A) = \rho^{p-1}_{q-1}(A) \circ C$$

Hint: see (2.4.8). □

• In connection with a gauge fields theory another simple construction of a representation might come in handy, namely the representation of a direct product of groups $G_1 \times G_2$ in the tensor product $V_1 \otimes V_2$ from two given representations ρ_1 of the group G_1 in V_1 and ρ_2 of G_2 in V_2, as well as a little thing concerning an Ad-invariant metric tensor in the Lie algebra of a product of groups.

12.4.16* Let ρ_1 be a representation of a group G_1 in V_1 and ρ_2 a representation of G_2 in V_2. Denote by E_a a basis in V_1 and E_A a basis in V_2, so that a basis in the tensor product $V_1 \otimes V_2$ is $E_a \otimes E_A$. Recall that the Lie algebra of the product $G_1 \times G_2$ is the direct sum $\mathcal{G}_1 \oplus \mathcal{G}_2$, so that its general element may be written as $X = X_1 + X_2$. Check that

(i) the prescription

$$\rho(g_1, g_2) := \rho_1(g_1) \otimes \rho_2(g_2)$$

provides a representation of $G_1 \times G_2$ in $V_1 \otimes V_2$ (compare with the tensor product of representations in (12.4.11); note that the latter may be obtained from the representation considered here if we represent the "diagonal" in $G \times G$, i.e. the elements of the form (g, g))

(ii) its derived representation turns out to be

$$\rho'(X_1 + X_2) = \rho'_1(X_1) \otimes \hat{1} + \hat{1} \otimes \rho'_2(X_2)$$

(iii) if E_i is a basis of the Lie algebra $\mathcal{G}_1$ and E_I is a basis of the Lie algebra $\mathcal{G}_2$, then for the matrix elements ρ^{bB}_{aAi} and ρ^{bB}_{aAI} of the derived representation we get

$$\rho^{bB}_{aAi} = \rho^b_{ai}\delta^B_A \qquad \rho^{bB}_{aAI} = \delta^b_a \rho^B_{AI}$$

where ρ^b_{ai} and ρ^B_{AI} are the matrix elements of the initial derived representations (12.1.6)

(iv) if h_1 is a ρ_1-invariant scalar product in V_1 and h_2 is a ρ_2-invariant scalar product in V_2, then

$$h := h_1 \otimes h_2 \qquad (h_1 \otimes h_2)(E_a \otimes E_\alpha, E_b \otimes E_\beta) := h_1(E_a, E_b) h_2(E_\alpha, E_\beta)$$

turns out to be ρ-invariant in $V_1 \otimes V_2$.

Hint: (ii) $\rho_1(e^{\epsilon X})E_a \otimes \rho_2(e^{\epsilon Y})E_A = \cdots$. □

12.4.17* We now look at a simple fact concerning Ad-invariant scalar products in Lie algebras for the case of a direct product of two groups

$$G = G_1 \times G_2 \qquad \mathcal{G} = \mathcal{G}_1 + \mathcal{G}_2$$

Namely, verify that if h_1 is an Ad_{G_1}-invariant scalar product in $\mathcal{G}_1$ and h_2 is an Ad_{G_2}-invariant scalar product in $\mathcal{G}_2$, then

$$h \equiv \lambda_1 h_1 \oplus \lambda_2 h_2$$

is G-invariant *for arbitrary* λ_1, λ_2

Hint: $h(X_1 + X_2, \tilde{X}_1 + \tilde{X}_2) := \lambda_1 h_1(X_1, \tilde{X}_1) + \lambda_2 h_2(X_2, \tilde{X}_2)$, $\mathrm{Ad}_{(g_1,g_2)}(X_1 + X_2) = \mathrm{Ad}_{g_1} X_1 + \mathrm{Ad}_{g_2} X_2$. □

• If, for example, the scalar products h_1 and h_2 were given uniquely up to a constant factor, we see that the freedom in $\mathcal{G}_1 + \mathcal{G}_2$ is already *bigger*, namely it is as large as a *two-parameter* class of invariant scalar products, now (the freedom up to a factor would be only a one-parameter class). This simple fact is reflected in the structure of the action integrals in the gauge theories: if a gauge group happens to be a product, the corresponding action contains *more free constants* (resulting in the lower "predictive power" of the theory).

To close the section let us say a few words on irreducibility. If the initial representations (V_1, ρ_1) and (V_2, ρ_2) happen to be irreducible, then their direct sum clearly fails to be an irreducible representation (both V_1 and V_2 being invariant subspaces). In general, neither is the direct *product* irreducible. For certain classes of groups the product may be decomposed into a direct *sum* of irreducible representations; this is called the *Clebsch–Gordan series*. The decomposition procedure may be rephrased as finding another basis in $V_1 \otimes V_2$ (instead of $E_i \otimes E_\alpha$, which is adapted to the *initial* representations ρ_1 and ρ_2), namely a basis which is adapted to the structure of invariant subspaces with respect to the action of the *resulting* representation $\rho_1 \otimes \rho_2$. This (important and extensive) subject will not be treated in this book. Elements of the technique for the representations of $SU(2)$ may be found in textbooks on quantum mechanics ("addition of angular momenta").

12.5 Invariant tensors and intertwining operators

• Each *equivariant* map $A : (V_1, \rho_1) \to (V_2, \rho_2)$ provides us with a wand which enables us to reach the wishful thinking of whole generations of alchemists, namely a "transmutation of a quantity of type ρ_1 into a quantity of type ρ_2." It is enough to pretend deep concentration for a while, to mutter mysteriously abracadabra and (not forget) at the same time to assign to a vector $v_1 \in V_1$ in an unobtrusive way[206] the vector $Av_1 =: v_2 \in V_2$, since the action of g then gives

$$v_1 \mapsto \rho_1(g)v_1 \quad \Rightarrow \quad v_2 \equiv Av_1 \mapsto A(\rho_1(g)v_1) = \rho_2(g)(Av_1) \equiv \rho_2(g)v_2$$

[206] The words *in an unobtrusive way* should be emphasized. Sometimes small children in the audience succeed in seeing through the trick and then they shout "ha ha, he applied the equivariant map $A : (\rho_1, V_1) \to (\rho_2, V_2)$!"

so that the effect of the equivariant map A consists in a loss of cultural heritage of V_1 and the complete assimilation to the novel milieu of V_2. Now we are going to learn that the same thing may also be achieved with the help of an *invariant* element in the space $V_1^* \otimes V_2$.

12.5.1 Check that there exists a close relation between intertwining operators (= *equivariant* maps $V \to W$) and *invariant* elements in $V^* \otimes W$. In more detail, let A be an intertwining operator between the representations (ρ_1, V_1) and (ρ_2, V_2) of a group G, E_i a basis in V_1 and E_a a basis in V_2

$$A : V_1 \to V_2 \qquad \rho_2(g)A = A\rho_1(g) \qquad AE_i =: A_i^a E_a$$

Check that then

(i) the element $\hat{A}$ of the space $V_1^* \otimes V_2$

$$\hat{A} := A_i^a E^i \otimes E_a \qquad \text{i.e.} \quad \hat{A}(v_1, v_2^*) := \langle v_2^*, Av_1 \rangle$$

is *invariant* with respect to the representation $\check{\rho}_1 \otimes \rho_2$

$$(\check{\rho}_1 \otimes \rho_2)(g)\hat{A} = \hat{A}$$

(ii) in the opposite direction, an element $\hat{B} \in V_1 \otimes V_2$ which is invariant with respect to the representation $\rho_1 \otimes \rho_2$ induces the intertwining operator B between $(V_1^*, \check{\rho}_1)$ and (V_2, ρ_2)

$$(\rho_1 \otimes \rho_2)(g)\hat{B} = \hat{B} \quad \Rightarrow \quad B : V_1^* \to V_2 \quad \hat{B}(v_1^*, v_2^*) =: \langle v_2^*, Bv_1^* \rangle$$
$$\rho_2(g)B = B\check{\rho}_1(g)$$

Hint: (i) $A \leftrightarrow \hat{A}$ for $\hat{A}(v_1, v_2^*) := \langle v_2^*, Av_1 \rangle$; then

$$\begin{aligned} ((\check{\rho}_1 \otimes \rho_2)(g)\hat{A})(v_1, v_2^*) &= \langle \check{\rho}_2(g)v_2^*, A\rho_1(g)v_1 \rangle = \langle \check{\rho}_2(g)v_2^*, \rho_2(g)Av_1 \rangle \\ &= \langle v_2^*, \rho_2(g^{-1})\rho_2(g)Av_1 \rangle = \hat{A}(v_1, v_2^*) \end{aligned}$$

or $(\check{\rho}_1 \otimes \rho_2)(g)\hat{A} = A_i^a \check{\rho}_1(g)E^i \otimes \rho_2(g)E_a = \cdots$. □

• Such invariant elements are often called *invariant tensors*; in general they are elements $\hat{A}$ of a multiple tensor product $V_1 \otimes \cdots \otimes V_n$, which happen to be invariant with respect to the action of the natural representation in this space, namely the tensor product of the individual representations ρ_i

$$(\rho_1 \otimes \cdots \otimes \rho_n)(g)\hat{A} = \hat{A} \quad \hat{A} = A^{i \dots a} E_i \otimes \cdots \otimes E_a \in V_1 \otimes \cdots \otimes V_n$$

Note that a *single* invariant tensor induces[207] *several* intertwining operators in general; for example, an invariant tensor $\hat{A}$ for the product of two representations $V_1 \otimes V_2$

$$\hat{A} = A^{ia} E_i \otimes E_a$$

[207] Just as one may interpret in several ways a tensor in L; see, for example, the three ways in which one may regard a tensor of type $(1, 1)$ in (2.4.5).

defines as many as four intertwining operators:

$$\begin{aligned} A_{(1)} : V_1^* \to V_2 \qquad & E^i \mapsto A^{ia} E_a \\ A_{(2)} : V_2^* \to V_1 \qquad & E^a \mapsto A^{ia} E_i \\ A_{(3)} : V_1^* \otimes V_2^* \to \mathbb{R} \qquad & E^a \otimes E^i \mapsto A^{ia} \\ A_{(4)} : \mathbb{R} \to V_1 \otimes V_2 \qquad & 1 \mapsto A^{ia} E_a \otimes E_i \end{aligned}$$

($\mathbb{R}$ being the *trivial* one-dimensional representation of a group G). A single invariant tensor may thus serve for several "transmutations" of the quantities of different types.

Let us look at some concrete examples of important invariant tensors.

12.5.2 Consider the "constants" ρ^a_{bi}, which emerge as the matrix elements of the represented basis of a Lie algebra $\mathcal{G}$, i.e. in the formula $\rho'(E_i)E_a =: \rho^b_{ai} E_b$ (12.1.6). Check that

(i) in these constants, there is information about an *invariant tensor* $\hat{A}$ with respect to the representation $\text{Ad}^* \otimes \check{\rho} \otimes \rho$ in the space $\mathcal{G}^* \otimes V^* \otimes V$, defined by

$$\hat{A} := \rho^a_{bi} E^i \otimes E^b \otimes E_a \qquad \hat{A}(X, v_1, v_2^*) := \langle v_2^*, \rho'(X) v_1 \rangle$$

(ii) this tensor induces (for example) the intertwining operators[208]

$$\begin{aligned} A_{(1)} : \mathcal{G} \to V^* \otimes V \qquad & E_i \mapsto \rho^a_{bi} E^b \otimes E_a \qquad \text{i.e.} \quad X^i \mapsto \rho^a_{bi} X^i =: X^a_b \\ A_{(2)} : V^* \otimes V \to \mathcal{G}^* \qquad & E^a \otimes E_b \mapsto \rho^a_{bi} E^i \qquad \text{i.e.} \quad v^b_a \mapsto \rho^a_{bi} v^b_a =: v_i \end{aligned}$$

The objects "of type Ad" are thus converted to the objects "of type $\check{\rho} \otimes \rho$" and vice versa

(iii) the structure constants of a Lie algebra $\mathcal{G}$ provide an invariant tensor $\hat{A}$ with respect to the representation $\text{Ad}^* \otimes \text{Ad}^* \otimes \text{Ad}$ in the space $\mathcal{G}^* \otimes \mathcal{G}^* \otimes \mathcal{G}$, defined by

$$\hat{A} := c^i_{jk} E^j \otimes E^k \otimes E_i \ (E_i \in \mathcal{G}) \qquad \hat{A}(X, Y, Z^*) := \langle Z^*, [X, Y] \rangle$$

Hint: (i) according to (12.1.8) and (12.3.7)

$$\begin{aligned} ((\text{Ad}^* \otimes \check{\rho} \otimes \rho)(g)\hat{A})(X, v_1, v_2^*) &= \hat{A}(\text{Ad}_g X, \rho(g) v_1, \check{\rho}(g) v_2^*) \\ &= \langle \check{\rho}(g) v_2^*, \rho'(\text{Ad}_g X) \rho(g) v_1 \rangle \\ &= \langle v_2^*, \rho(g^{-1}) \rho'(\text{Ad}_g X) \rho(g) v_1 \rangle = \langle v_2^*, \rho'(X) v_1 \rangle \\ &\equiv \hat{A}(X, v_1, v_2^*) \end{aligned}$$

(ii) the text before (12.5.2); (iii) the special case for $\rho = \text{Ad}$. □

12.5.3 In quantum mechanics one works with a two-component complex wave function $\psi(\mathbf{r})$ in order to describe a particle with spin $\frac{1}{2}$; this function transforms under rotations (by α around $\mathbf{n}$) according to the rule[209]

$$\psi \mapsto A\psi \equiv e^{-\frac{1}{2} i\alpha \mathbf{n} \cdot \boldsymbol{\sigma}} \psi \qquad A \in SU(2)$$

[208] The first of them will be used, for example, to introduce the "represented connection form" (and the gauge potential) $\omega^a_b := \rho^a_{bi} \omega^i$, $A^a_b := \rho^a_{bi} A^i$ in (20.4.6) and (21.2.4), the second one in a while (12.5.3) for creating a vector from two spinors in quantum mechanics as well as in (21.5.4), where the current $\mathcal{J}^i = k^{ij} \rho_{abj} \phi^a (\mathcal{D}\phi)^b$ is introduced.

[209] It is a representation treated in general in (13.4.11). In addition to the transformation of components the argument is altered, too, $\mathbf{r} \mapsto R(-\alpha, \mathbf{n})\mathbf{r}$; however, this is not relevant now.

Often one makes use of the statement that it is possible to form a *vector* from two such wave functions ψ, χ

$$\mathbf{w} = \psi^+\boldsymbol{\sigma}\chi \qquad \text{or, in particular, from } \psi \text{ alone} \quad \psi^+\boldsymbol{\sigma}\psi$$

Check that

(i) this statement stems from the fact that the matrix elements of *Pauli matrices* provide an invariant $SU(2)$-tensor[210]

$$(\sigma_j)^a{}_b$$

(the indices of type $(\ )^a$ correspond to the representation $\rho = \text{id}$ in $\mathbb{C}^2$, the indices $(\ )^i$ correspond to $\rho(A) = R =$ a common rotation in the sense of covering from (13.3.3))

(ii) also, the two-dimensional Levi-Civita symbols ϵ_{ab} and ϵ^{ab} are invariant $SU(2)$-tensors – this is the reason one can canonically raise and lower indices of the type $a, b, \dots$ by them.

Hint: (i) like in (12.3.16) we check that for $SU(2)$ the Ad-representation reduces to ordinary *rotation* (Ad_A performs a rotation in $su(2) = E^3$ by the matrix $R(A)$, covered by A in the sense of (13.3.3)); the lower index on σ_i then indicates Ad^* (in the dual space $su(2)^*$), i.e. the transformation by the inverse matrix R^{-1}. The question thus is whether

$$(R^{-1})^i_j A^a_b (A^{-1})^c_d (\sigma_i)^b{}_c \stackrel{?}{=} (\sigma_j)^a{}_d \qquad \text{i.e.} \quad A\sigma_i A^+ \stackrel{?}{=} R^j_i \sigma_j$$

The positive answer results from (13.3.3); alternatively the same: the question is whether $\psi \mapsto A\psi$, $\chi \mapsto A\chi$ ($\Rightarrow \chi^+ \mapsto \chi^+ A^+ \equiv \chi^+ A^{-1}$, so that χ^+ has *lower* index) and $\mathbf{a} \mapsto R\mathbf{a}$ lead to

$$\chi^+(\mathbf{a}\cdot\boldsymbol{\sigma})\psi \mapsto \chi^+ A^+ (R\mathbf{a}\cdot\boldsymbol{\sigma}) A\psi \equiv R\mathbf{a}\cdot \chi^+(A^+\boldsymbol{\sigma}A)\psi \stackrel{?}{=} \chi^+(\mathbf{a}\cdot\boldsymbol{\sigma})\psi$$

The positive answer once again comes from (13.3.3); (ii) $A^a_b A^c_d \epsilon_{ac} = \det A \epsilon_{bd} = \epsilon_{bd}$ (5.6.5), similarly for ϵ^{ab} use A^{-1}; thus the raising and lowering of an index $(\ )_a \mapsto \epsilon^{ab}(\ )_b$ and $(\ )^a \mapsto \epsilon_{ab}(\ )^b$ are indeed *equivariant* maps, which "transmute the quantities of type id" to the "quantities of type $\check{\text{id}}$" and vice versa. □

12.5.4 Check that

(i) for *any* representation (V, ρ) of a group G the *identity* tensor is an invariant tensor with respect to $\check{\rho}\otimes\rho$

$$\hat{1} := \delta^a_b E^b \otimes E_a \equiv E^a \otimes E_a \qquad \text{i.e.} \quad \hat{1}(v, w^*) := \langle w^*, v\rangle \quad (\check{\rho}\otimes\rho)(g)\hat{1} - \hat{1}$$

(ii) for the (pseudo-)orthogonal group $G = O(r, s)$ the tensors

$$\eta_{ab} \qquad \text{as well as} \qquad \eta^{ab}$$

are invariant and for $SO(r, s)$ we can *furthermore* add[211]

$$\epsilon_{a\dots b} \qquad \text{as well as} \qquad \epsilon^{a\dots b}$$

[210] In a similar way *Dirac* matrices provide (22.5.9) Lorentzian tensors $\bar{\psi}\gamma^a \dots \gamma^b\psi$ composed of Dirac spinors.

[211] That is to say, if we assigned to the Levi-Civita symbol *upper* indices (corresponding to the representation $\rho = \text{id}$ of the group $SO(r, s)$), we did it just as it should be. In the same way, if we assigned to *the same* Levi-Civita symbol *lower* indices (corresponding to the representation $\rho = \check{\text{id}}$ of the group $SO(r, s)$), we did it just as it should be again. One checks easily, however, that *only these two* possibilities actually work – the indices should be *either* all upper *or* all lower.

(iii) for the symplectic groups $G = \mathrm{Sp}(n, \mathbb{R})$ the tensors

$$\omega_{ab} \qquad \text{as well as} \qquad \omega^{ab}$$

are invariant, $\omega \equiv \epsilon \otimes \mathbb{I}$ being the canonical form of a non-degenerate antisymmetric matrix (10.1.6).

Hint: (i) the definition of $\check{\rho}$ (12.1.8); (ii) and (iii) the *definition* of the groups; for example, the (pseudo-)orthogonal matrices satisfy $A^{\mathrm{T}}\eta A = \eta$, being just the invariance of η_{ab} (with respect to $\check{\rho} \otimes \check{\rho}$), the same relation for $A \mapsto A^{-1}$ says that also η^{ab} is an invariant tensor (with respect to $\rho \otimes \rho$); the restriction "determinant $= 1$" gives $A^a_b \dots A^c_d \epsilon_{a\dots c} = (\det A)\ \epsilon_{b\dots d} = \epsilon_{b\dots d}$, so that the indices of $\epsilon_{a\dots b}$ are justified; similarly (use $\det A^{\mathrm{T}} = \det A$) $A^a_b \dots A^c_d \epsilon^{b\dots d} = (\det A)\ \epsilon^{a\dots c} = \epsilon^{a\dots c}$, so that the indices $\epsilon^{a\dots b}$ are justified, too. □

12.5.5 Check that

(i) the invariant tensors are naturally endowed with the structure of an *algebra* (with respect to the *tensor* product)
(ii) a contraction (if it is meaningful) does not spoil the invariance of a tensor
(iii) therefore, as an example

$$\rho^a_{bi}\rho^c_{dj} \qquad \rho^a_{bk}c^k_{ij} \qquad \delta^a_b\delta^c_d$$

are also invariant tensors (and enable one to "transmute the type of objects").

Hint: (i)

$$\{\rho_1(g)\hat{A} = \hat{A},\ \rho_2(g)\hat{B} = \hat{B}\} \ \Rightarrow\ (\rho_1 \otimes \rho_2)(g)(\hat{A} \otimes \hat{B}) = \rho_1(g)\hat{A} \otimes \rho_2(g)\hat{B} = \hat{A} \otimes \hat{B}$$

□

12.6* Lie algebra cohomologies

• The concept of cohomology, which we introduced in Chapter 9, has numerous applications in modern mathematics and mathematical physics. Here we will speak about a complex canonically induced by a triple $(\mathcal{G}, \rho, V)$, i.e. by a Lie algebra $\mathcal{G}$ and its representation ρ in the vector space V. This complex appears naturally in its various forms (with various $\mathcal{G}$, ρ and V) in a number of unrelated contexts. For example, we will encounter it in Chapter 14, when treating the Poisson actions of groups.

Let us have a look at the *Cartan formulas* from problem (6.2.13) from a different point of view. It turns out that these formulas may be naturally reinterpreted in the language of Lie algebras and their representations.

12.6.1 Let $\mathcal{G} := \mathfrak{X}(M)$ be a (∞-dimensional) Lie algebra of the vector fields on a manifold M, $X \in \mathfrak{X}(M)$ and let $V := \mathcal{F}(M)$ be the (∞-dimensional) algebra (thus in particular a vector space) of (smooth) functions on M. Show that

(i) the prescription

$$\rho(X)f := \mathcal{L}_X f \equiv Xf$$

defines a representation of $\mathcal{G}$ in V

(ii) Cartan formulas may be rewritten in the form

$$d\alpha(X_1, \dots, X_{p+1}) = \sum_{j=1}^{p+1}(-1)^{j+1}\rho(X_j)\alpha(X_1, \dots, \hat{X}_j, \dots, X_{p+1}) + \sum_{i<j}(-1)^{i+j}\alpha([X_i, X_j], \dots, \hat{X}_i, \dots, \hat{X}_j, \dots, X_{p+1})$$

Hint: see (6.2.13). □

• This form suggests strongly that the operation d might also be introduced in a more general context than on "ordinary" differential forms on manifolds. In the new context, nevertheless, all the properties will be preserved which are essential for introducing a complex and its cohomologies (i.e. a new d will operate as a differential in an appropriate complex, which will substitute for the deRham complex of differential forms).

Consider (V, ρ)-valued p-forms in $\mathcal{G}$, ρ being a representation of $\mathcal{G}$ in V. Thus what we have in mind are polylinear completely antisymmetric maps

$$\alpha : \underbrace{\mathcal{G} \times \dots \times \mathcal{G}}_{p} \to (V, \rho) \qquad \underbrace{(v, \dots, w)}_{p} \mapsto \alpha(v, \dots, w) \in (V, \rho)$$

Denote the space of such forms by $\Lambda^p(\mathcal{G}^*, V)$ (see Section 6.4). Now regard this space as a subspace of degree p of the $\mathbb{Z}$-graded space $\Lambda(\mathcal{G}^*, V) := \oplus_{p=0}^{n} \Lambda^p(\mathcal{G}^*, V)$. The last stage of the construction of a complex we look for will consist in the definition of an appropriate differential $\hat{d}$.

12.6.2 Define a linear operator

$$\hat{d} : \Lambda^p(\mathcal{G}^*, V) \to \Lambda^{p+1}(\mathcal{G}^*, V) \qquad \alpha \mapsto \hat{d}\alpha$$

by the prescription[212] inspired by (12.6.1)

$$\hat{d}\alpha(X_1, \dots, X_{p+1}) = \sum_{j=1}^{p+1}(-1)^{j+1}\rho(X_j)\alpha(X_1, \dots, \hat{X}_j, \dots, X_{p+1}) + \sum_{i<j}(-1)^{i+j}\alpha([X_i, X_j], \dots, \hat{X}_i, \dots, \hat{X}_j, \dots, X_{p+1}) \quad X_j \in \mathcal{G}$$

Show that

[212] A view based on components might also be instructive: we have $\alpha^A_{i\dots j}$ and we want to get $(\hat{d}\alpha)^A_{i\dots jk}$, so that we need somehow to add a single lower index of type "i." Now the structure constants c^i_{jk} as well as the matrices of the generators of the representation ρ^A_{Bi} (given by $\rho(e_i)e_A =: \rho^B_{Ai}e_B$) are available. If we set

$$(\hat{d}\alpha)^A_{ijk\dots l} := \lambda(p)c^r_{[ij}\alpha^A_{k\dots l]r} + \mu(p)\rho^A_{B[i}\alpha^B_{jk\dots l]}$$

then for appropriate $\lambda(p)$, $\mu(p)$ we may eventually obtain (making use of Jacobi identity as well as of the fact that ρ is a representation) $\hat{d}\hat{d} = 0$.

(i) in the first three degrees the formulas read

$$\begin{aligned}\hat{d}v(X) &= \rho(X)v \\ \hat{d}\alpha(X,Y) &= \rho(X)(\alpha(Y)) - \rho(Y)(\alpha(X)) - \alpha([X,Y]) \\ \hat{d}\beta(X,Y,Z) &= \rho(X)(\beta(Y,Z)) - \rho(Y)(\beta(X,Z)) + \rho(Z)(\beta(X,Y)) \\ &\quad - \beta([X,Y],Z) + \beta([X,Z],Y) - \beta([Y,Z],X)\end{aligned}$$

(ii) the operator $\hat{d}$ is *nilpotent*, i.e. $\hat{d}\hat{d} = 0$

(iii) the diagram

$$\Lambda^0(\mathcal{G}^*, V) \xrightarrow{\hat{d}} \Lambda^1(\mathcal{G}^*, V) \xrightarrow{\hat{d}} \Lambda^2(\mathcal{G}^*, V) \xrightarrow{\hat{d}} \cdots \xrightarrow{\hat{d}} \Lambda^n(\mathcal{G}^*, V)$$

represents a *complex*; its cohomologies

$$H^p(\mathcal{G}; V) := Z^p(\mathcal{G}; V)/B^p(\mathcal{G}; V) \equiv \text{Ker}\, \hat{d}_p / \text{Im}\, \hat{d}_{p-1}$$

are called *Lie algebra cohomologies* (or, in more detail, the cohomologies of $\mathcal{G}$ with respect to the representation (ρ, V))

(iv) the particular formulas from item (i) are indeed consistent with $\hat{d}\hat{d} = 0$, i.e. that

$$\hat{d}\hat{d}v(X,Y) = 0 \qquad \hat{d}\hat{d}\alpha(X,Y,Z) = 0$$

Hint: (iv) $\hat{d}\hat{d}v(X,Y) = \rho(X)(\hat{d}v)(Y) - \rho(Y)(\hat{d}v)(X) - (\hat{d}v)([X,Y]) = \cdots$. □

• For the construction of the complex we need, as we can see, not only the Lie algebra $\mathcal{G}$ itself, but also a representation ρ in V. The corresponding cohomologies depend in general on both constituents involved. Let us have a look in more detail at two particular representations, which are always available (free of charge), namely the *trivial representation* ($\rho(X) = 0$) in $\mathbb{R}$ and the *adjoint representation* ($\rho(X) = \text{ad}_X$) in the Lie algebra $\mathcal{G}$ itself.

For the trivial representation we actually work with *ordinary* ($\mathbb{R}$-valued) forms in $\mathcal{G}$, i.e. $\Lambda^p(\mathcal{G}^*, V) \equiv \Lambda^p\mathcal{G}^*$.

12.6.3 Let (V, ρ) be the trivial representation in $\mathbb{R}$. Check that

(i) the operator $\hat{d}$ then simplifies to[213]

$$\hat{d}\alpha(X_1, \dots, X_{p+1}) = \sum_{i<j} (-1)^{i+j} \alpha([X_i, X_j], \dots, \hat{X}_i, \dots, \hat{X}_j, \dots, X_{p+1}) \qquad X_j \in \mathcal{G}$$

or in the first three degrees explicitly

$$\begin{aligned}\hat{d}\lambda(X) &= 0 \qquad \lambda \in \mathbb{R} \\ \hat{d}\alpha(X,Y) &= -\alpha([X,Y]) \\ \hat{d}\beta(X,Y,Z) &= -\beta([X,Y],Z) + \beta([X,Z],Y) - \beta([Y,Z],X) \\ &\equiv \beta(X,[Y,Z]) + \text{ cycl.}\end{aligned}$$

(ii) the corresponding complex reads

$$\mathbb{R} \xrightarrow{0} \mathcal{G}^* \xrightarrow{\hat{d}} \Lambda^2\mathcal{G}^* \xrightarrow{\hat{d}} \cdots \xrightarrow{\hat{d}} \Lambda^n\mathcal{G}^*$$ □

[213] In components to $(\hat{d}\alpha)_{ijk\dots l} = \lambda(p) c^r_{[ij}\alpha_{k\dots l]r}$.

• For the Poisson actions of groups mentioned above (to be introduced after (14.5.6)) the essential information is encoded in the *second* cohomology class of this very complex, $H^2(\mathcal{G}^*, \mathbb{R})$. The crucial question to be answered is thus whether there exist *2-cocycles* which are not coboundaries. It turns out (we do not give a proof here) that, in general, such 2-cocycles *do* exist; for *semi-simple* Lie algebras, however, "Whitehead's lemmas" state that this is *not* possible

$$H^1(\mathcal{G}^*, \mathbb{R}) = 0 \qquad H^2(\mathcal{G}^*, \mathbb{R}) = 0$$

12.6.4 "Smell" the cohomologies of this type for the Lie algebra $\mathcal{G} = so(3)$, i.e. compute explicitly $H(so(3)^*, \mathbb{R})$. In particular, check that

(i) the general forms of the relevant degrees in $so(3)$ look like ($\lambda, \mu \in \mathbb{R}$)

$$\lambda \quad \alpha(X) = \mathbf{a}\cdot\mathbf{x} \qquad \beta(X, Y) = \mathbf{b}\cdot(\mathbf{x}\times\mathbf{y}) \qquad \sigma(X, Y, Z) = \mu\mathbf{x}\cdot(\mathbf{y}\times\mathbf{z})$$

(ii) the operator $\hat{d}$ acts as follows:

$$\lambda \mapsto 0 \qquad \mathbf{a}\mapsto\mathbf{a} \qquad \mathbf{b}\mapsto 0 \qquad \mu\mapsto 0$$

(μ is trivial due to the dimension), i.e. in detail

$$\hat{d}\lambda(X) = 0 \qquad \hat{d}\alpha(X, Y) = \mathbf{a}\cdot(\mathbf{x}\times\mathbf{y}) \qquad \hat{d}\beta(X, Y, Z) = 0$$

(iii) this means that

$$\begin{array}{llll} \Lambda^0 \sim \mathbb{R} & \Lambda^1 \sim \mathbb{R}^3 & \Lambda^2 \sim \mathbb{R}^3 & \Lambda^3 \sim \mathbb{R} \\ Z^0 \sim \mathbb{R} & Z^1 \sim 0 & Z^2 \sim \mathbb{R}^3 & Z^3 \sim \mathbb{R} \\ B^0 \sim 0 & B^1 \sim 0 & B^2 \sim \mathbb{R}^3 & B^3 \sim 0 \\ H^0 \sim \mathbb{R} & H^1 \sim 0 & H^2 \sim 0 & H^3 \sim \mathbb{R} \end{array}$$

so that

$$H(so(3)^*, \mathbb{R}) \equiv H^0 \oplus H^1 \oplus H^2 \oplus H^3 \sim \mathbb{R}\oplus 0\oplus 0\oplus\mathbb{R}$$

(iv) these results are consistent with Whitehead's lemmas.

Hint: (i) see (12.1.5); (ii) see (12.6.3), $c^i_{jk} = \epsilon_{ijk}$, $(\mathbf{a}\times\mathbf{b})_i = \epsilon_{ijk}a_jb_k$. □

12.6.5 Let $c \in \Lambda^3\mathcal{G}^*$ be the 3-form which arises by the lowering of the index on the tensor of structure constants by means of the Killing–Cartan form, i.e.

$$c = \frac{1}{3!}c_{ijk}e^i \wedge e^j \wedge e^k \qquad c_{ijk} := k_{ir}c^r_{jk}$$

or (12.3.9)

$$c(X, Y, Z) := k(X, [Y, Z])$$

Show that

(i) this form is always *closed* (= a cocycle, $\hat{d}c = 0$)

(ii) for $\mathcal{G} = so(3)$ *it is not* exact ($\neq$ a coboundary), so that $H^3(so(3)^*, \mathbb{R}) \neq 0$.

Hint: (i) Jacobi identity; (ii) $\varepsilon_{ijk} \overset{?}{=} \varepsilon_{r[ij}\beta_{k]r} \Rightarrow$ (multiply both sides by ε_{ijk}) $6 \overset{?}{=} 2\beta_{jj} \equiv 0$, answer truly *no!* □

- Let us look at the second *gratis* representation, now, $(V, \rho) = (\mathcal{G}, \mathrm{ad})$.

12.6.6 Let $(V, \rho) = (\mathcal{G}, \mathrm{ad})$. Show that

(i) the corresponding complex reads

$$\mathcal{G} \xrightarrow{\hat{d}} \mathcal{L}(\mathcal{G}, \mathcal{G}) \xrightarrow{\hat{d}} \Lambda^2(\mathcal{G}^*, \mathcal{G}) \xrightarrow{\hat{d}} \cdots \xrightarrow{\hat{d}} \Lambda^n(\mathcal{G}^*, \mathcal{G})$$

(ii) the operator $\hat{d}$ acts as[214]

$$\hat{d}\alpha(X_1, \dots, X_{p+1}) = \sum_{j=1}^{p+1}(-1)^{j+1}[X_j, \alpha(X_1, \dots, \hat{X}_j, \dots, X_{p+1})] + \sum_{i<j}(-1)^{i+j}\alpha([X_i, X_j], \dots, \hat{X}_i, \dots, \hat{X}_j, \dots, X_{p+1}) \quad X_j \in \mathcal{G}$$

or in the first three degrees explicitly

$$\begin{aligned}
\hat{d}X(Y) &= [X, Y] \\
\hat{d}\alpha(X, Y) &= [X, \alpha(Y)] - [Y, \alpha(X)] - \alpha([X, Y]) \\
\hat{d}\beta(X, Y, Z) &= [X, \beta(Y, Z)] - [Y, \beta(X, Z)] + [Z, \beta(X, Y)] \\
&\quad - \beta([X, Y], Z) + \beta([X, Z], Y) - \beta([Y, Z], X) \\
&\equiv [X, \beta(Y, Z)] + \beta(X, [Y, Z]) + \text{cycl.}
\end{aligned}$$

(iii) closed 1-forms coincide with *derivations of the Lie algebra*, the exact 1-forms being the *inner* derivations (so that vanishing of $H^1(\mathcal{G}^*, \mathcal{G})$ means that in the Lie algebra under consideration there are no other derivations than the inner ones)

(iv) the identity tensor $\hat{1}$ in $\mathcal{G}$ may be regarded as $\hat{1} \in \Lambda^1(\mathcal{G}^*, \mathcal{G})$, the tensor of structure constants $\hat{c}$ (with components c^i_{jk}) as $\hat{c} \in \Lambda^2(\mathcal{G}^*, \mathcal{G})$, the latter being the differential of the former

$$\hat{d}\hat{1} = \hat{c} \qquad (\Rightarrow \hat{d}\hat{c} = 0)$$

Hint: (iii) $\hat{d}\alpha = 0 \Leftrightarrow \alpha([X, Y]) = [\alpha(X), Y] + [X, \alpha(Y)]$; (iv) according to (i) for $\alpha = \hat{1}$ we have $\hat{d}\hat{1}(X, Y) = [X, Y] \equiv \hat{c}(X, Y)$. □

12.6.7 Show directly that for the Lie algebra $so(3)$ any derivation is necessarily inner and deduce from this that $H^1(so(3)^*, so(3)) = 0$.

Hint: look for linear maps $D : so(3) \to so(3)$ such that $D([x, y]) = [D(x), y] + [x, D(y)]$; if $D(e_i) = D^j_i e_j$, then check (on a basis, for $c^i_{jk} = \epsilon_{ijk}$) that $D^i_j = \epsilon_{ijk}z^k$, so that always $D = \mathrm{ad}_z$ for some $z \in so(3)$. □

[214] In components $\rho^A_{Bi} \mapsto c^r_{ji}$.

• Let us mention, to close the section, that the general differential $\hat{d}$ from the problem (12.6.2) may also be expressed in an alternative way, with no use of "catalytic" arguments. Instead the operators i_v and j_v known from (5.8.6) are used.

12.6.8* Show that the differential $\hat{d}$, being the key element of the section, may also be written in the form

$$\hat{d} = -\frac{1}{2}c^k_{mn} j^m j^n i_k \otimes \hat{1} + j^k \otimes \rho(E_k)$$

where E_k is a basis in the Lie algebra, $j^k(\alpha^A E_A) = (E^k \wedge \alpha^A)E_A$, $i_k(\alpha^A E_A) = (i_{E_k}\alpha^A)E_A$ and $(C \otimes D)(\alpha^A E_A) = (C\alpha^A)(DE_A)$. Check its nilpotence explicitly. □

Summary of Chapter 12

A Lie group often shows its presence via its *representations*, i.e. there exists a homomorphism of the former to the group of invertible linear operators in a vector space and in a given situation we encounter only the image of the group with respect to this homomorphism. Each representation of a Lie group induces automatically a representation of its Lie *algebra* (called the derived representation), the latter meaning in general a homomorphism of a Lie algebra into the Lie algebra of (all) linear operators (in a fixed vector space). If a representation admits a non-trivial invariant subspace it is called *reducible*, since it may be reduced (by restriction) to a (smaller) representation in this subspace. *Irreducible* representations cannot be reduced in this way. Schur's lemma provides a useful criterion of irreducibility. If the invariant subspace admits an invariant *complement* as well, the representation is equivalent to a direct sum of two simpler ones. Such a complement sometimes happens to be *orthogonal* with respect to an *invariant* scalar product (if it does exist; on compact groups its existence is guaranteed and the procedure for its construction is given here). One can perform some standard constructions with representations, such as the dual (contragredient) one and the direct sum and the direct product; combining these two with the restriction to invariant subspaces in the resulting spaces, a lot of new representations may be obtained from a small number of them at the beginning (like all "tensor" representations ρ^p_q from a single "vector" one ρ^1_0; in some cases even all irreducible representations from just a single one, see Section 13.3). Invariant tensors and related intertwining operators enable one to "transmute the type" of quantities, i.e. associate with vectors acted by a representation ρ_1 vectors acted by a representation ρ_2 (of the same group). A representation of a Lie algebra induces a complex; we study its cohomologies for a while.

$\rho(1+\epsilon X) = 1 + \epsilon\rho'(X)$	Computation of the derived representation ρ'	(12.1.6)
$\rho'(E_i)E_a =: \rho^b_{ai} E_b$	Matrix elements of generators	(12.1.6)
$\langle\check{\rho}(g)\alpha, v\rangle := \langle\alpha, \rho(g^{-1})v\rangle$	Contragredient (dual) representation $\check{\rho}$	(12.1.8)
$h(\rho(g)v, \rho(g)w) = h(v, w)$	Scalar product h is ρ-invariant	(12.1.10)
$h_{bc}\rho^c_{ai} + h_{ac}\rho^c_{bi} = 0$	Component expression of the same fact	(12.1.10)

$\rho_2(g)A = A\rho_1(g)$	A is intertwining operator for ρ_1 and ρ_2	Sec. 12.2
$ge^X g^{-1} = e^{\mathrm{Ad}_g X}$	Adjoint representation Ad of G	(12.3.1, 2)
$\mathrm{Ad}_A X = AXA^{-1}$	Explicit expression of Ad for matrix groups	(12.3.1)
$\mathrm{ad}_X Y = [X, Y]$ $(\mathrm{ad} \equiv \mathrm{Ad}')$	Adjoint representation ad of $\mathcal{G}$	(12.3.5)
$\mathrm{ad}_{E_i} E_j = c^k_{ij} E_k$	Component expression of ad	(12.3.5)
$K(X, Y) := \mathrm{Tr}\,(\mathrm{ad}_X \mathrm{ad}_Y)$	Killing–Cartan form on $\mathcal{G}$	(12.3.8)
$\hat{C}_2 := k^{ij}\rho'(E_i)\rho'(E_j)$	Quadratic Casimir operator	(12.3.13)
$(g_1, h_1) \circ (g_2, h_2) := (g_1 g_2, h_1 h_2)$	Direct product of groups	(12.4.7)
$(\rho_1 \otimes \rho_2)(g) := \rho_1(g) \otimes \rho_2(g)$	Direct product of representations of G	(12.4.11)
$(\rho_1 \otimes \rho_2)' = \rho_1' \otimes \hat{1} + \hat{1} \otimes \rho_2'$	Derived representation for $\rho_1 \otimes \rho_2$	(12.4.11)

13

Actions of Lie groups and Lie algebras on manifolds

• In the last chapter we discussed the particular cases of actions, the representations (the actions in linear spaces). Here we will treat more general actions of Lie groups and algebras, namely the actions on *manifolds*.[215] It might be useful to mention, however, that the validity of some concepts and results discussed in this chapter is actually even broader. We will use "narrower" language speaking about the actions of Lie groups on manifolds, but we recommend always to think about the particular situation whether the group which acts indeed needs to be a *Lie* group and whether the set where the group acts indeed needs to be a *manifold*.

13.1 Action of a group, orbit and stabilizer

• Given a Lie group G and a smooth manifold M we say that G *acts from the left* (or *acts from the right*)[216] on M if for each group element $g \in G$ a diffeomorphism L_g (or R_g) is given, which satisfies ($g, h \in G$)

$L_g : M \to M$	$L_{gh} = L_g \circ L_h$	$L_e = \mathrm{id}_M$	G acts from the left on M
$R_g : M \to M$	$R_{gh} = R_h \circ R_g$	$R_e = \mathrm{id}_M$	G acts from the right on M

For a given action (left or right) the shorthand notation is often used ($x \in M$)

$$x \mapsto L_g x =: gx \quad \text{or} \quad x \mapsto R_g x =: xg$$

so that the property of being a left or right action looks like

$$(gh)x = g(hx) \quad \text{or} \quad x(gh) = (xg)h$$

We emphasize that gx does not denote a product,[217] it is nothing but a shorthand notation for $L_g x$.

A manifold M on which a left action of a group G is available is called *left G-space* (and *right G-space* for a right action).

[215] Of course, linear spaces are also manifolds, so that now we begin to study the *more general* situation, relaxing the assumption of linearity; sometimes one finds the terminology *non-linear realizations* in a physics literature, which we will, however, not use. Our actions will be simply "general," they *need not* (but may) be linear.

[216] Alternatively, G has a *left action* (or *right action*) on M.

[217] Although in particular cases it may be realized in terms of appropriate products; see, for example, (13.1.3).

13.1.1 Check that

(i) $(L_g)^{-1} = L_{g^{-1}}$, $(R_g)^{-1} = R_{g^{-1}}$ holds
(ii) if L_g is a *left* action, then the prescription $\hat{R}_g := L_{g^{-1}}$ provides a *right* action
(iii) if R_g is a *right* action, then the prescription $\hat{L}_g := R_{g^{-1}}$ provides a *left* action. □

13.1.2 Often the actions mentioned above are introduced in terms of the maps

$$L : G \times M \to M \qquad L(g, x) := L_g x \equiv gx$$
$$R : M \times G \to M \qquad R(x, g) := R_g x \equiv xg$$

the *smooth action* being defined in terms of smoothness of these maps. Write down the conditions which L and R are to satisfy in order to indeed correspond to actions. ($L(gh, x) = L(g, L(h, x))$, $R(x, gh) = R(R(x, g), h)$).) □

• On the same manifold sometimes several actions of the same group may be defined; we then regard them as *different* G-spaces. A simple albeit very important example provides the group G itself, regarded as (its own) G-space (so that we speak about the case in which $M = G$).

13.1.3 Check that a group G may be regarded in as many as three ways as a left G-space and moreover in (as many as) three ways as a right G-space:

left actions	$h \mapsto gh$	$h \mapsto hg^{-1}$	$h \mapsto I_g h := ghg^{-1}$
right actions	$h \mapsto hg$	$h \mapsto g^{-1}h$	$h \mapsto I_{g^{-1}} h := g^{-1}hg$

In the first column we recognize the good old left and right *translations* (11.1.1), the second column is obtained from the first one by the trick from (13.1.1) and the third column (being actually a combination of the two) is the (good old as well, see (12.3.1)) *conjugation* I_g (by the elements g and g^{-1}). □

13.1.4 Check that

(i) $\mathbb{R}^n$ may be regarded in two ways as a left $GL(n, \mathbb{R})$-space and also in two ways as a right $GL(n, \mathbb{R})$-space ($A \in GL(n, \mathbb{R})$, $x \in \mathbb{R}^n$):

left actions	$x \mapsto Ax$	$x \mapsto (A^{\mathrm{T}})^{-1}x$
right actions	$x \mapsto A^{\mathrm{T}}x$	$x \mapsto A^{-1}x$

The prescription $x \mapsto A^{\mathrm{T}}x$ may also be equivalently written as $x^{\mathrm{T}} \mapsto x^{\mathrm{T}}A$ (both of them actually mean $x_i \mapsto A_{ji}x_j$, whereas $x \mapsto Ax$ corresponds to $x_i \mapsto A_{ij}x_j$)
(ii) the same is true for the complex analog, the space $\mathbb{C}^n$ and the group $GL(n, \mathbb{C})$. □

13.1.5 Consider the structure (X, s) (in the sense of Section 10.1) to be realized as a group $G \equiv X$ together with a right action $R_g k := kg$ (the right translation). Check that the group of automorphisms of (X, s) is isomorphic to the group G itself.

Solution: let $f : G \to G$ be from the group to be determined. In order to preserve the structure "s," f has to commute with R_g for each $g \in G$. Then it turns out, however, that f

is necessarily a *left translation* $f = L_h$ for appropriate $h \in G$. Indeed, $R_g f(k) \equiv f(k)g \overset{!}{=} f(R_g k) \equiv f(kg)$. For $k = e$ this gives $f(g) = f(e)g \equiv L_{f(e)}g$, so that $f = L_{f(e)}$. □

• If a group G acts on M, both players in the relation become marked for the rest of their lives; on both of them, willy-nilly, an additional structure arises. Namely, on M a system of subsets occurs, called the *orbits*, whereas on G subgroups arise, called *stabilizers*.

13.1.6 The *orbit* $\mathcal{O}_x$ of a point $x \in M$ is the set of points in M which may be reached from x by an action of an appropriate element of the group, so that

$$M \supset \mathcal{O}_x := \{y \in M \mid \exists g \in G \text{ such that } y = gx\}$$

Be sure to understand that

(i) an orbit is fully given by any of its points

$$y \in \mathcal{O}_x \quad \Leftrightarrow \quad \mathcal{O}_x = \mathcal{O}_y$$

(ii) "to share an orbit" is an equivalence $\sim$ on M.

A G-space thus, in general, falls into several orbits, which actually do not communicate (by the action of the group) with each other. □

• It may happen sometimes that a G-space consists of *a single orbit*; this means that any two points may be connected by the action of the group. We then speak about a *transitive action* and the corresponding G-space is called a *homogeneous space*. The opposite extreme is provided by the case when an orbit $\mathcal{O}_x$ consists of *a single point* x; we then speak about a *fixed point* of the action.

13.1.7 Find the orbits of the action of $GL(n, \mathbb{R})$ in $\mathbb{R}^n$ (as well as $GL(n, \mathbb{C})$ in $\mathbb{C}^n$), treated in problem (13.1.4). Decide whether $\mathbb{R}^n$ (and $\mathbb{C}^n$) is a homogeneous space.

Hint: the point $0 \in \mathbb{R}^n$ represents an orbit in its own right (it is a fixed point); the rest of $\mathbb{R}^n$ is an orbit, too, which may be generated from (say) the point $x_0 := (0, \dots, 1)$ (so that $\mathbb{R}^n$ fails to be a homogeneous space). Indeed, the equation $x = Ax_0$ for $x \neq 0$ fixes the last column of the matrix A: $A_{in} = x_i$; complete arbitrarily to a basis in $\mathbb{R}^n$ and locate the basis vectors as columns of the matrix A. The matrix A belongs to $GL(n, \mathbb{R})$ and it sends x_0 to x. The complex case follows in full analogy. □

13.1.8 Consider the *restriction of the action* $x \mapsto Ax$ of the group $GL(n, \mathbb{R})$ in $\mathbb{R}^n$ to the subgroup $SO(n)$. Show that

(i) the orbits of the new action coincide with the *spheres* S_r^{n-1} of all possible radii r and centered at the origin

$$S_r^{n-1} := \{x \in \mathbb{R}^n \mid \eta(x, x) \equiv (x^1)^2 + \cdots + (x^n)^2 = r^2\}$$

(ii) the new action is indeed transitive on each sphere; an arbitrary point on the sphere of radius r may be generated from (say) the point $x_0 := (0, \dots, r)$.

Hint: $\eta(Ax, Ax) = \eta(x, x)$ according to (10.1.5); the equation $x = Ax_0$ (for x from the sphere of radius r) fixes the last column[218] of the matrix A: $A_{in} = r^{-1}x_i$ (it is normalized to unity); complete arbitrarily to a right-handed orthonormal basis in E^n and locate the basis vectors as columns of the matrix A. The matrix A belongs to $SO(n)$ and it sends x_0 to x. □

13.1.9 Let V be a *complete* vector field on M and consider its flow Φ_t. Check that

(i) the map $x \mapsto \Phi_t x$ may be regarded as an action of the Lie group $(\mathbb{R}, +)$ on M
(ii) the orbits of the action are just the *integral curves* (or rather "integral *paths*," since the parametrization is irrelevant) of the field V.

Hint: $R_t x := \Phi_t(x)$, $R_{t+s}x = \Phi_{t+s}(x) = (\Phi_t \circ \Phi_s)(x) = (R_t \circ R_s)x$. □

13.1.10 Let a group G act (say, from the left) on M and let $x \in M$. The *stabilizer*[219] of the point x is the subgroup $G_x \subset G$ which contains only those elements g of the group G, which leave the point x fixed

$$G_x := \{g \in G \mid gx = x\}$$

Check that

(i) it is indeed a sub*group* (not only a sub*set*)
(ii) the stabilizers of all points from the same orbit happen to be isomorphic; in more detail, if $y = \hat{g}x$, then G_y may be obtained from G_x by *conjugation* by the element $\hat{g}$

$$y = \hat{g}x \ (\Rightarrow y \in \mathcal{O}_x) \quad \Rightarrow \quad G_y = I_{\hat{g}}(G_x) \equiv \hat{g}G_x\hat{g}^{-1} \equiv \{\tilde{g} \in G \mid \tilde{g} = I_{\hat{g}}g \text{ for some } g \in G_x\}$$

(iii) in general, if H is a subgroup of a group G, then for any $g \in G$ also $I_g(H) \equiv gHg^{-1}$ turns out to be a subgroup of the group G, which is moreover isomorphic to the group H (it is called a *conjugate subgroup*); the conjugation is an equivalence relation on the set of subgroups of the group G, so that all the subgroups fall into the classes of subgroups which are conjugate to each other
(iv) each orbit fixes such a class.

Hint: $gy = y$ gives $(g\hat{g})x = \hat{g}x$, so that $I_{\hat{g}^{-1}}g \in G_x$. □

13.1.11 Show that the stabilizer of any point on the sphere S_r^{n-1}, regarded as an orbit of the action (13.1.8) of the group $SO(n)$ in E^n, is a subgroup isomorphic to $SO(n-1)$.

Hint: according to (13.1.10) the subgroup does not depend (up to an isomorphism given by the conjugation) on the choice of a point on the sphere; choose the north pole. For this particular point we have

$$Ax_0 \equiv \begin{pmatrix} A^1_1 & \dots & A^1_n \\ \vdots & \ddots & \vdots \\ A^n_1 & \dots & A^n_n \end{pmatrix} \begin{pmatrix} 0 \\ \vdots \\ r \end{pmatrix} = \begin{pmatrix} rA^1_n \\ \vdots \\ rA^n_n \end{pmatrix} \overset{!}{=} \begin{pmatrix} 0 \\ \vdots \\ r \end{pmatrix} \Rightarrow \begin{pmatrix} A^1_n \\ \vdots \\ A^n_n \end{pmatrix} = \begin{pmatrix} 0 \\ \vdots \\ 1 \end{pmatrix}$$

[218] A watchful reader of hints might feel this reasoning to be familiar.

[219] The other frequently used terms for the stabilizer of the point x are the *stationary subgroup* or the *little group* of x.

The orthogonality conditions $A^T A = AA^T = \mathbb{I}$ fulfilled by the matrix A (saying that both the columns as well as the rows of the matrix constitute an orthonormal system) result in that its bottom row is necessarily $(0, \dots, 1)$ and that the remaining $(n-1) \times (n-1)$ block B has to be an orthogonal matrix. The condition $\det A = 1$ gives $\det B = 1$, so that $B \in SO(n-1)$. Together

$$A = \begin{pmatrix} B & 0 \\ 0 & 1 \end{pmatrix}, \qquad B \in SO(n-1) \qquad \text{so that} \quad B \leftrightarrow \begin{pmatrix} B & 0 \\ 0 & 1 \end{pmatrix}$$

is an isomorphism of the group $SO(n-1)$ and $G_{x_0} \equiv$ the stabilizer of the north pole. □

13.1.12 Given two (left) G-spaces (M, L_g) and $(N, \hat{L}_g)$, check that the prescription

$$(x, y) \mapsto (gx, gy) \equiv (L_g x, \hat{L}_g y) =: \tilde{L}_g(x, y)$$

defines a new (left) G-space (their product $(M \times N, \tilde{L}_g)$). In particular, also for some of the actions being trivial (the identity transformation for each g). □

13.1.13 Given two G-spaces (M, R_g) and $(N, \hat{R}_g)$, check that

(i) the prescription

$$\mathcal{L}_g : \mathcal{F}(M, N) \to \mathcal{F}(M, N) \qquad \mathcal{L}_g f := \hat{R}_{g^{-1}} \circ f \circ R_g$$

defines a left action of G on $\mathcal{F}(M, N)$, the *set of all maps* from M to N; it is a simultaneous *right* action on arguments (the domain) of f and a *left* action (by the same group element) on values (the image set) of the map f

(ii) the stable points of the action are given just by the *equivariant maps* $M \to N$, being the maps which "commute with the action" in the sense of

$$\begin{array}{ccc} M & \xrightarrow{f} & N \\ {\scriptstyle R_g}\downarrow & & \downarrow{\scriptstyle \hat{R}_g} \\ M & \xrightarrow[f]{} & N \end{array} \qquad \text{i.e.} \quad f \circ R_g = \hat{R}_g \circ f$$

Such maps may also be regarded as the maps *compatible* with the (right) actions on M and N and they constitute morphisms for the category of (here right) G-spaces (see Appendix A.6).

Hint: $\mathcal{L}_{gh} f = \hat{R}_{(gh)^{-1}} \circ f \circ R_{gh} = \hat{R}_{g^{-1}} \circ (\hat{R}_{h^{-1}} \circ f \circ R_h) \circ R_g \equiv (\mathcal{L}_g \circ \mathcal{L}_h) f.$ □

13.1.14 Consider a subgroup H of $SO(3)$ which rotates vectors around a fixed unit vector $\mathbf{n}$ (it is a one-parameter subgroup containing the matrices $R(\alpha, \mathbf{n})$ mentioned in (11.7.13)). Prove that

(i) the elements of $SO(3)$, which are conjugate to $R(\alpha, \mathbf{n})$, are just $R(\alpha, \mathbf{m})$, i.e. they rotate vectors by *the same angle* around an arbitrary unit vector $\mathbf{m}$

(ii) any subgroup conjugate to H consists of the rotations around arbitrary fixed axes (so the subgroups are labeled by unit vectors $\mathbf{m}$ or, alternatively, by the points of the unit sphere).

Hint: (i) let $R \equiv R(\alpha, \mathbf{m})$ and let $Re_a = R^b_a e_b$ in some basis; then for any $A \in SO(3)$ we have $(ARA^{-1})(Ae_a) = R^b_a(Ae_b)$, so if we denote $E_a := Ae_a$, we have $(I_A R)E_a = R^b_a E_b$; thus the (conjugate) matrix $I_A R$ acts on the basis E_a in the same way as R acts on the basis e_a; in particular, if $e_3 = \mathbf{n}$, then $A\mathbf{n} =: \mathbf{m}$ is the new axis of rotation ($(I_A R)\mathbf{m} = \mathbf{m}$) and the angle of rotation is the same ($Re_1 = e_1 \cos\alpha + e_2 \sin\alpha \Rightarrow (I_A R)E_1 = E_1 \cos\alpha + E_2 \sin\alpha$, similarly for e_2). □

13.2 The structure of homogeneous spaces, G/H

• It turns out that any homogeneous space M of a group G may be replaced by its isomorphic copy, which is constructed directly in terms of the group. First, we will look at these particular homogeneous spaces and then we learn how exactly such a space may be assigned to a given homogeneous space.

13.2.1 Let G be a group, $H \subset G$ a subgroup, $g \in G$. The *left coset* $gH \subset G$ is the set of all elements of the form gh for h from the subgroup H, i.e.

$$gH = \{k \mid k = gh, h \in H\}$$

Show that

(i) the cosets realize a *decomposition* of G, i.e. each element $k \in G$ belongs in just one coset
(ii) all cosets have the same cardinality ("number of elements"), i.e. there exists a bijection $g_1 H \leftrightarrow g_2 H$ for any two cosets
(iii) the relation on G, defined as

$$g_1 \sim g_2 \Leftrightarrow g_1 = g_2 h \qquad \text{for some } h \in H$$

is an equivalence and the equivalence classes just coincide with the left cosets.

Hint: (ii) $g_1 h \leftrightarrow g_2 h$ (with the same $h \in H$). □

13.2.2 Let $H \subset G$ be a subgroup of G. Define (yet nothing more than) a set G/H as a factor-set of G with respect to the equivalence from (13.2.1), i.e. a *point* $[g] \in G/H$ is the equivalence class of the point g in $(G, \sim)$. Be sure to understand that

(i) the *canonical projection*

$$\pi : G \to G/H \qquad g \mapsto [g]$$

is a surjective map
(ii) the preimage of a point $[g] \in G/H$ is just the coset $gH \subset G$. □

13.2.3 Let $G \equiv (\mathbb{Z}, +)$ be the group of integers[220] with respect to addition and consider the subgroup $H \equiv n\mathbb{Z}$ containing all the (integer) multiples of a fixed number $n \in \mathbb{Z}$.

[220] So that it is not a *Lie* group.

(i) Check that $n\mathbb{Z}$ is indeed a subgroup of $\mathbb{Z}$
(ii) how many elements does the set $G/H \equiv \mathbb{Z}/n\mathbb{Z}$ contain?
(iii) what is $\pi(7)$ for $n = 4$?
(iv) what is $\pi^{-1}([0])$ and $\pi^{-1}([1])$ for $n = 2$?
((ii) n; (iii) [3]; (iv) even and odd numbers.) □

13.2.4 Consider a *finite* group G and prove the validity of

(i) the *Lagrange theorem*: for any subgroup $H \subset G$ we have

$$o(G) = k \cdot o(H) \qquad k \in \mathbb{N}$$

($o(G)$ is the *order* of the group, i.e. the number of elements). So the theorem simply says that the order of a group is always divisible by the order of *any* of its subgroups
(ii) a corollary: if $o(G) =$ a *prime number*, then G has *only trivial* subgroups ($H = G$ or $H = \{e\}$)
(iii) another corollary: given any element g in a finite group, there holds

$$g^{o(G)} \equiv \underbrace{g \cdot g \cdots g}_{o(G) \text{ entries}} = e$$

Hint: (i) see (13.2.1), $k =$ the number of cosets, i.e. points of G/H; (iii) the order of G is to be divisible (also) by the order of the *cyclic* subgroup $G_g \subset G$ generated by the element g (all powers of g), $o(G) = k \cdot o(G_g)$ so that $g^{o(G)} = g^{k \cdot o(G_g)} = (g^{o(G_g)})^k = e^k = e$, since $g^{o(G_g)} = e$ (due to the definition of the order of a cyclic group). □

• Now we learn that there is a natural left transitive action of the group G on G/H so that G/H is naturally endowed with the structure of a *homogeneous space*.

13.2.5 Check that

(i) by the prescription

$$L_{\hat{g}}[g] \equiv \hat{g}[g] := [\hat{g}g]$$

a left action of the group G on G/H is well defined (it is a prescription of the type "by means of representatives," check that it does not depend on their choice)
(ii) the action is *transitive*, so that G/H is a natural *homogeneous space* of the group G
(iii) if we denote by $\mathcal{L}_g$ the left translation on G, then for the relation between L_g and $\mathcal{L}_g$ the following commutative diagram holds:

$$\begin{array}{ccc} G & \xrightarrow{\mathcal{L}_g} & G \\ \pi \downarrow & & \downarrow \pi \\ G/H & \xrightarrow[L_g]{} & G/H \end{array} \qquad \text{i.e.} \quad \pi \circ \mathcal{L}_g = L_g \circ \pi$$

so that the action L_g on G/H may actually be regarded as the projection of the action $\mathcal{L}_g$ on the group G itself (and the projection π is then an equivariant map)
(iv) in problem (13.2.3) we have

$$L_n[m] = [n + m]$$

(v) for the two extreme cases, $H = G$ and $H = \{e\}$, we get $L_g = \text{id}$ ($G/H \equiv G/G$ then reduces to a single point) and $L_g = \mathcal{L}_g$ respectively.

Hint: (ii) the left translation *is* transitive on G. □

• Now we have at our disposal a tool for a simple construction of the class of homogeneous spaces of a group G – it is enough to find its subgroups H. Our pleasure, being already far from negligible, grows to a true rapture when we learn that (up to isomorphism) there are actually no other homogeneous spaces at all except this class so that this construction exhausts as a matter of fact *all* homogeneous spaces.[221]

13.2.6 Two (left) homogeneous spaces (M, L_g) and $(N, \tilde{L}_g)$ of the same group G are called *isomorphic* (equivalent) if there exists an *equivariant* bijective map between them.[222] Check that this indeed introduces an equivalence among homogeneous spaces. □

13.2.7 Prove that

(i) if H, H' are conjugate subgroups of the group G, then the canonical homogeneous spaces G/H and G/H' are isomorphic
(ii) *each* homogeneous space of the group G is isomorphic to an appropriate *canonical* homogeneous space G/H (i.e. there are, in fact, no other homogeneous spaces than G/H)
(iii) the biggest homogeneous space of the group G is the *principal homogeneous space*, the space isomorphic to the group G itself (regarded as a homogeneous space with an action given by the left or right translation)
(iv) the action of $GL(n, \mathbb{R})$ on the bases $e \mapsto eA \equiv R_A$ from problem (5.7.2) makes from the space of bases $E(L)$ just a *principal* homogeneous space; note that $E(L)$ *is not a group* (two bases cannot be multiplied, there is no unit basis), but still it is diffeomorphic to the group $GL(n, \mathbb{R})$ and from the algebraic point of view it is "only" a (principal) homogeneous space of the group.

Hint: (i) if π, π' are the canonical projections of G onto G/H and G/H', then $f : G/H \to G/H'$ is given by $\pi(g) \mapsto \pi'(g)$ (i.e. $f \circ \pi = \pi'$); (ii) let $(M, \hat{L}_g)$ be a given homogeneous space, $x \in M$ and $G_x \subset G$ the stabilizer of the point x; then the map f from the commutative diagram

$$\begin{array}{ccc} M & \xrightarrow{f} & G/G_x \\ \hat{L}_g \downarrow & & \downarrow L_g \\ M & \xrightarrow[f]{} & G/G_x \end{array} \qquad f : y \equiv gx \mapsto [g]$$

is an isomorphism $M \to G/G_x$ (check), so that the given space *is* indeed isomorphic to the space G/H, namely for the subgroup $H \equiv G_x$; another choice of the point x only leads (according to item (i) and (13.1.10)) to an isomorphic space; (iii) the biggest homogeneous space results from the smallest subgroup. □

[221] Including the homogeneous spaces, which any missions from other planets, "solar systems" or even other galaxies will carry sometime in the future (with a view to investigating it in laboratories under the microscope). Sometimes the strength of our slender earthly mathematics indeed takes the breath away.
[222] The bijectivity gives the "isomorphism of sets," the equivariance adds the "isomorphism of the actions."

13.2.8 Consider the standard spheres in Cartesian spaces (radius 1, centered at the origin). Check that there are the following isomorphisms of homogeneous spaces (denoted by "="):

(i)
$$S^n = SO(n+1)/SO(n) = O(n+1)/O(n)$$

(ii)
$$S^{2n-1} = SU(n)/SU(n-1) = U(n)/U(n-1)$$

(iii)
$$S^1 = (\mathbb{R}, +)/\mathbb{Z}$$

(iv)
$$T^n = (\mathbb{R}^n, +)/\mathbb{Z}^n$$

Hint: (i) (13.1.11) and its modification for $O(n)$; (ii) a similar action of $U(n)$ and $SU(n)$ in $\mathbb{C}^n \sim \mathbb{R}^{2n}$, $|z_1|^2 + \cdots + |z_n|^2 = 1$ is a sphere S^{2n-1}; (iii) the action of $\mathbb{R}$ on S^1 of the form $z \mapsto L_x z := e^{i2\pi x} z$ $(|z| = 1)$; (iv) the torus is a product of circles, a similar action to the preceding item. □

• The general method common to all of these examples (as well as for numerous further situations) might be summed up in terms of the following steps:

1. find an action L_g of the group G on a set X
2. find the orbit $\mathcal{O}_x =: M$ of a point $x \in X$; this is already a homogeneous space
3. find the stabilizer $H \equiv G_x$ of the point x
4. write down the result in the ("deep science") form $M = G/H$, i.e. $\mathcal{O}_x = G/G_x$.

For matrix groups it is useful to keep in mind that they naturally act in the space of columns of appropriate dimension (equal to the dimension of the matrices), realizing thus item 1; the remaining items should now be just straightforwardly computed. One more possibility for 2×2 matrices adds the following observation.

13.2.9 Consider the (extended) complex plane and an element $A \in GL(2, \mathbb{C})$. Show that

(i) the *Möbius transformation* of the plane (also known as the *linear-fractional transformation*) given by the formula

$$z \mapsto \frac{az+b}{cz+d} \equiv L_A z \qquad A \equiv \begin{pmatrix} a & b \\ c & d \end{pmatrix} \in GL(2, \mathbb{C})$$

defines a left action of $GL(2, \mathbb{C})$ (as well as any of its subgroups) in the (extended) complex plane

(ii) for $0 \neq \lambda \in \mathbb{C}$ we have

$$L_{\lambda A} = L_A$$

so that from the class of matrices λA we can always choose an unimodular representative, $A \in SL(2, \mathbb{C})$. □

• So far we have learned that the set G/H is *always* endowed with the structure of a (left) *homogeneous space* (and that there are no other homogeneous spaces). It turns out that this object is *sometimes* also endowed with another structure, namely with the structure of a *group*. If this is the case, it is called the *factor* group. Let us investigate when exactly this is the case. The idea is fairly standard (it has proved to be useful many times, the last time when we defined the action of G in G/H), namely we try to define a multiplication in G/H by means of the representatives (i.e. by a projection of the product in the group).

13.2.10 Define a multiplication in G/H by the prescription

$$[g][\tilde{g}] := [g\tilde{g}] \qquad \text{i.e.} \quad \pi(g)\pi(\tilde{g}) := \pi(g\tilde{g})$$

Check that

(i) the multiplication, in general, *depends* on the choice of the representatives (and is thus useless)
(ii) if one requires the multiplication to be independent of the representative, this imposes a condition on the subgroup H, namely

$$I_g H \equiv gHg^{-1} = H \qquad \text{for all} \quad g \in G$$

i.e. the subgroup is to be *invariant* with respect to conjugation by an arbitrary element of G. Such a subgroup is usually called a *normal subgroup*, or sometimes also an *invariant subgroup*
(iii) the prescription

$$H \mapsto I_g H \equiv gHg^{-1}$$

defines a *left action* of G on the set of its subgroups; the orbit $\mathcal{O}_H$ is given by the class of the subgroups conjugate to the subgroup H and the fixed points of the action are represented just by the normal subgroups.

Hint: (ii) if $g\tilde{g} = \hat{g}$ (so that $[g][\tilde{g}] = [\hat{g}]$), then if we require the product to be independent of the representative, it needs $(gh)(\tilde{g}\tilde{h}) = g\tilde{g}h'$ for some $h' \in H$. This may be written as $gHg^{-1} = H$ (i.e. $ghg^{-1} =$ some h'' again $\in H$). □

13.2.11 Check that

(i) in a *commutative* group *each* subgroup is normal
(ii) the subgroup $n\mathbb{Z}$ is a normal subgroup in $\mathbb{Z}$
(iii) the product in $\mathbb{Z}_n := \mathbb{Z}/n\mathbb{Z}$ reduces to the good old addition of numbers "modulo n."

Hint: see (13.2.3); the multiplication (addition here) is $[k] + [l] := [k + l]$. □

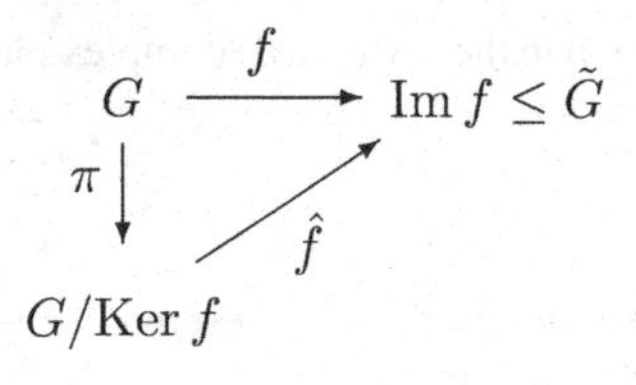

• Now we introduce a theorem which generates numerous interesting isomorphisms of groups. The input datum consists of an arbitrary *homo*morphism of groups, the output is then a certain *iso*morphism of (modified) groups.

13.2.12 Let $f : G \to \tilde{G}$ be a homomorphism of groups. Check that

(i) the kernel of the homomorphism $\text{Ker} f$ (i.e. the elements which are mapped to the unit element of $\tilde{G}$) is a *normal* subgroup of the group G

(ii) the image of the homomorphism $\operatorname{Im} f$ (those elements in $\tilde{G}$ which have a preimage in G) is a subgroup in $\tilde{G}$

(iii) the canonical projection $\pi : G \to G/\operatorname{Ker} f$ is a surjective homomorphism

(iv) the following *homomorphism theorem* holds: the map $\hat{f} : G/\operatorname{Ker} f \to \operatorname{Im} f$, $\hat{f}(\pi(g)) := f(g)$ is an isomorphism of groups; thus

$$G/\operatorname{Ker} f = \operatorname{Im} f \qquad (= \text{ is an isomorphism of groups})$$

Hint: (iii) preservation of the operation: $\pi(g) = [g]$, (13.2.10), surjection by means of (13.2.2); (iv) a straightforward (easy) check. □

• The patient f may complain at most of being non-injective and non-surjective (if compared with an isomorphism). The non-injectivity is healed by factorization and the non-surjectivity by ignoring that part of $\tilde{G}$ which is outside the image of f. The patient (becoming $\hat{f}$ after the medical help) may then be discharged from the hospital and (bubbling over with sound health) left to plunge into the whirl of life again.

13.2.13 Check that the homomorphisms on the left give the isomorphisms on the right:[223]

$$\begin{array}{rccr}
GL(n,\mathbb{R}) \to GL(1,\mathbb{R}) & A \mapsto \det A & \Rightarrow & GL(n,\mathbb{R})/SL(n,\mathbb{R}) = GL(1,\mathbb{R}) \\
GL(1,\mathbb{C}) \to GL(1,\mathbb{R}) & z \mapsto |z| & \Rightarrow & GL(1,\mathbb{C})/U(1) = GL_+(1,\mathbb{R}) \\
GL(1,\mathbb{C}) \to GL(1,\mathbb{C}) & z \mapsto z^2 & \Rightarrow & GL(1,\mathbb{C})/\mathbb{Z}_2 = GL(1,\mathbb{C}) \\
SU(n) \times U(1) \to U(n) & (A, e^{i\alpha}) \mapsto e^{i\alpha} A & \Rightarrow & SU(n) \times U(1)/\mathbb{Z}_n = U(n) \\
GL_+(1,\mathbb{R}) \to (\mathbb{R},+) & x \mapsto \ln x & \Rightarrow & GL_+(1,\mathbb{R}) = (\mathbb{R},+)
\end{array}$$

Hint: $A \mapsto \det A$: an equivalence class (left coset) in $GL(n,\mathbb{R})/SL(n,\mathbb{R})$ is specified by the value of the determinant of matrices. Since the classes are multiplied by means of representatives and since the determinant of the product of matrices is the product of their determinants, the multiplication of classes exactly matches the multiplication of ordinary real numbers, i.e. the multiplication in $GL(1,\mathbb{R})$, giving rise to the isomorphism $GL(n,\mathbb{R})/SL(n,\mathbb{R}) = GL(1,\mathbb{R})$. $z \mapsto |z|$: an equivalence class (left coset) in $GL(1,\mathbb{C})/U(1)$ is specified by the modulus of complex numbers (the radius of the corresponding circle). Since the modulus of the product of complex numbers is the product of their moduli, the multiplication of classes exactly matches the multiplication of ordinary (positive) real numbers, i.e. the multiplication in $GL_+(1,\mathbb{R})$, giving rise to the isomorphism $GL(1,\mathbb{C})/U(1) = GL_+(1,\mathbb{R})$. The remaining cases can be treated along similar lines. □

13.3 Covering homomorphism, coverings $SU(2) \to SO(3)$ and $SL(2,\mathbb{C}) \to L_+^\uparrow$

• An important particular case of a homomorphism of Lie groups $f : G \to \tilde{G}$ (13.2.12) is provided by a situation where the image $\operatorname{Im} f$ turns out to be the *whole* group $\tilde{G}$ and the kernel $\operatorname{Ker} f$ is a *discrete* (normal, possibly infinite) subgroup. If we denote the elements of

[223] The last isomorphism (given by the logarithm) is the heart of how the *slide rule* functions, some years ago an essential piece of equipment for any true engineer. It converts a product into a sum, the latter then being realized mechanically. Rulers based on the remaining isomorphisms still await a producer.

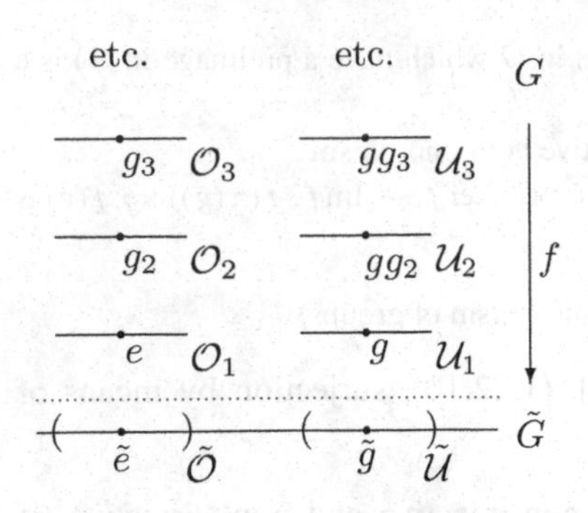

the subgroup by $(g_1 \equiv e, g_2, \ldots, g_k, \ldots)$, then (by definition) they constitute the preimage of the unit element $\tilde{e} \in \tilde{G}$. It results from the continuity of f that the pre-image of a (sufficiently small) *neighborhood* $\tilde{\mathcal{O}}$ of the unit element $\tilde{e} \in \tilde{G}$ is given by some neighborhoods $\mathcal{O}_k$ of the elements g_k, each of them being mapped bijectively onto $\tilde{\mathcal{O}}$.[224] This does not occur, however, only "over" the unit element $\tilde{e}$, but rather this picture may be found over *each* element $\tilde{g} = f(g)$ in $\tilde{G}$ (the preimages of $f(g)$ being $(gg_1 \equiv g, gg_2, \ldots, gg_k, \ldots)$). In general, a map of manifolds which has these properties is called a *covering* or a *local homeomorphism*.[225] If the number of preimages is n, we speak about an *n-sheeted covering*. If the covering of Lie groups $f : G \to \tilde{G}$ moreover happens to be a homomorphism (as is the case in the situation we started with at the beginning of the section), it is called a *covering homomorphism*. And finally if G is connected as well as *simply connected* (i.e. one can shrink to a point each loop (lasso) in this space), G is called the *universal covering group* of the group $\tilde{G}$ (one can show that for a given connected Lie group $\tilde{G}$ there is up to isomorphism a unique such group). Let us have a look at how some objects are related which are connected with the group $\tilde{G}$ and the covering group G, namely their Lie algebras and representations.

13.3.1 Let G and $\tilde{G}$ be Lie groups and let $f : G \to \tilde{G}$ be a covering homomorphism. Check that

$$\begin{array}{ccc} G & \xrightarrow{\ \rho\ } & \text{Aut}\, V \\ {\scriptstyle f}\downarrow & \nearrow{\scriptstyle \tilde{\rho}} & \\ \tilde{G} & & \end{array}$$

(i) their Lie algebras are *isomorphic* (even if the groups themselves are only homomorphic)

$$f : G \to \tilde{G} \ \text{ is a covering homomorphism} \quad \Rightarrow \quad \mathcal{G} = \tilde{\mathcal{G}}$$

(ii) each representation of the group $\tilde{G}$ "is" at the same time also a representation (i.e. induces a representation) of the covering group G (this may not be true in the opposite direction, however, so that the covering group may have "more" representations in general – see problem (13.3.8)).

Hint: (i) the Lie algebras are hidden in the *infinitesimal neighborhoods* of the unit elements of the groups; *these neighborhoods* are mapped *iso*morphically one onto another; (ii) if $\tilde{\rho} : \tilde{G} \to \text{Aut}\, V$ is a representation of $\tilde{G}$, we may "lift" it to the representation $\rho \equiv \tilde{\rho} \circ f : G \to \text{Aut}\, V$; what obstructs the opposite assignment $\rho \mapsto \tilde{\rho}$ is the fact that f may not be invertible (since it is non-injective, nor in general does a canonical choice of one of the possible preimages exist.)[226] □

[224] So there are *several* preimages of $\tilde{\mathcal{O}}$ with respect to π (their number coinciding with the number of elements of the subgroup $\text{Ker} f$), each of them being homeomorphic to $\tilde{\mathcal{O}}$ itself.

[225] The restriction of f to *a single* homeomorphic copy mentioned above is already a "true" homeomorphism.

[226] A covering is a special case of the concept of a *bundle* (see Section 17.2), when the fiber happens to be discrete; $f : G \to G/\text{Ker} f$ is a *principal bundle* (see Section 20.1). The canonical choice of one of the preimages would provide a *global section* of the bundle, resulting (20.1.3) in its triviality, $G = (G/\text{Ker} f) \times \text{Ker} f$.

13.3.2 Check that the map (13.2.8)

$$f : (\mathbb{R}, +) \to U(1) \qquad x \mapsto e^{2\pi i x}$$

(visually a winding of the real axis onto the unit circle in the complex plane) is a covering homomorphism and that moreover it is the *universal* covering. Verify the isomorphism of the Lie algebras and check that the "lift" of the representation $\tilde{\rho}_n : e^{i\alpha} \mapsto e^{in\alpha}$ (12.2.10) of the group $U(1)$ to a representation ρ_n of the covering group reads

$$\rho_n : x \mapsto e^{2\pi i n x} \qquad (\Rightarrow \rho_n(x+y) = \rho_n(x)\rho_n(y))$$

Hint: the real axis is connected as well as simply connected. □

• At first sight the figure at the beginning of the section may give the false impression that it is possible somehow to link together the individual "pancakes" in the upper part and end with several *copies* of the group $\tilde{G}$ (the number of copies coinciding with the number of leaves; that, for example, $\mathcal{O}_1$ and $\mathcal{U}_1$ are parts of a first copy, etc.). The elementary problem (13.3.2) shows, however, that the matter *is not* so simple (and that the concept of the covering might be interesting and non-trivial): there is a single circle $S^1 = U(1)$ in the lower part, but in the upper part we do not have infinitely many circles but rather *only a single* long (infinite) line $\mathbb{R}$.

(Imagine we covered the circle in the lower part by several "pancakes" (neighborhoods) of the type $\tilde{\mathcal{U}}$. Choose over one of them, say $\tilde{\mathcal{U}}_1$, a fixed preimage $\mathcal{U}_1$. Now glue a neighboring "pancake" $\tilde{\mathcal{U}}_2$ to $\tilde{\mathcal{U}}_1$ (such that they have a non-empty overlap) and choose that particular preimage $\mathcal{U}_2$ over $\tilde{\mathcal{U}}_2$, which has non-empty overlap with $\mathcal{U}_1$ (for sufficiently small domains it is unique). What we have obtained are two "bigger pancakes," one of them in G and the other one (its projection) in $\tilde{G}$ (in the example (13.3.2) the pancakes have the form of short lines and these lines are glued into longer lines). If we continue with this procedure, in the lower part, on $U(1)$, we will eventually glue together the whole circle (a *closed* line), whereas in the upper part an *open* line still continues to carry on which may be visually regarded also as part of an infinite *helix* produced over the circle.)

A nice, non-trivial and important illustration of a covering homomorphism is the *two-sheeted covering* of the rotational group in three dimensions $SO(3)$ by the group $SU(2)$. (This is closely related to *spinors* and we will return to the subject at a more sophisticated level in Chapter 22.) Since $SU(2)$ is a sphere S^3 (so that it is connected as well as simply connected), this is an example of a *universal* covering.[227] Let us have a look at how it happens.

13.3.3 Denote by $H_0(2,\mathbb{C})$ the 2×2 Hermitian traceless matrices. Check that

(i) it is a three-dimensional (real) vector space with a basis σ_a and that the prescription

$$\psi : \mathbb{R}^3 \to H_0(2,\mathbb{C}) \qquad x \mapsto \psi(x) \equiv \tilde{x} := x_a \sigma_a \equiv \begin{pmatrix} x_3 & x_1 - i x_2 \\ x_1 + i x_2 & -x_3 \end{pmatrix}$$

is a linear isomorphism

[227] And at the same time a reaffirmation of the warning mentioned in the preceding paragraph: $SU(2)$ regarded as a manifold *differs* from two copies of the manifold $SO(3)$ (the latter being non-connected).

(ii) the standard Euclidean structure is in $H_0(2, \mathbb{C})$ incorporated by

$$\det \tilde{x} = -x_a x_a \equiv -\mathbf{x} \cdot \mathbf{x} \equiv -||\mathbf{x}||^2 \qquad \tilde{x}\tilde{x} = ||\mathbf{x}||^2 \mathbb{I}_2$$

(iii) the prescription

$$\tilde{x} \mapsto L_A \tilde{x} := A\tilde{x}A^+$$

gives a left linear action (i.e. a representation) of $SU(2)$ in $H_0(2, \mathbb{C})$, which moreover preserves the Euclidean norm $||\mathbf{x}||^2$; thus it is a representation by means of *orthogonal* operators and we get the map (homomorphism) $f : SU(2) \to O(3)$ being sought (so far we (only) see that it is to $O(3)$; as a matter of fact it will be clear in a moment that its image is in $SO(3)$)

(iv) on the basis σ_a this gives the matrix elements of the orthogonal operator

$$\sigma_a \mapsto A\sigma_a A^+ =: R_{ba}\sigma_b \qquad R_{ab} = \frac{1}{2}\mathrm{Tr}\,(\sigma_a A \sigma_b A^+)$$

so that effectively in coordinates

$$x_a \mapsto x'_a \equiv R_{ab} x_b$$

Hint: (iii) $\det(L_A\tilde{x}) = \det \tilde{x}$, $\mathrm{Tr}\,(L_A\tilde{x}) = 0$; (iv) the formula for R_{ab} with the help of (13.3.4). □

13.3.4 Prove the validity of the following useful identities for the traces of Pauli matrices:

$$\mathrm{Tr}\,\sigma_a = 0 \qquad \mathrm{Tr}\,(\sigma_a\sigma_b\sigma_c) = 2i\epsilon_{abc}$$
$$\mathrm{Tr}\,(\sigma_a\sigma_b) = 2\delta_{ab} \qquad \mathrm{Tr}\,(\sigma_a\sigma_b\sigma_c\sigma_d) = 2(\delta_{ab}\delta_{cd} - \delta_{ac}\delta_{bd} + \delta_{ad}\delta_{cb})$$

Hint: $\sigma_a\sigma_b = \delta_{ab}\mathbb{I}_2 + i\epsilon_{abc}\sigma_c$. □

13.3.5 Show that if we parametrize the matrices $A \in SU(2)$ according to (11.7.12)

$$A = e^{-\frac{1}{2}i\alpha\mathbf{n}\cdot\boldsymbol{\sigma}} \equiv \mathbb{I}_2 \cos\frac{\alpha}{2} - i(\mathbf{n}\cdot\boldsymbol{\sigma})\sin\frac{\alpha}{2}$$

then the matrix R_{ab} from (13.3.3) turns out to be

$$R_{ab} = \delta_{ab}\cos\alpha + (1-\cos\alpha)n_a n_b - \epsilon_{abc}n_c \sin\alpha \equiv (e^{\alpha\mathbf{n}\cdot\mathbf{l}})_{ab}$$

which is by (11.7.13) *the most general* matrix from $SO(3)$. Thus, the image of $SU(2)$ is not given by the whole orthogonal group $O(3)$, but rather[228] by the (whole) subgroup $SO(3)$. □

• Thus for each rotation (a matrix $R_{ab} \in SO(3)$) we are in a position to find a matrix $A \in SU(2)$ which is mapped to the rotation. It even turns out that the matrix A is not unique, but there are actually two of them. This fact results from the examination of the *kernel* of the homomorphism $f : SU(2) \to SO(3)$.

[228] Since the map $SU(2) \to O(3)$ is continuous and $SU(2)$ is connected, the image is necessarily a connected group, i.e. (the unit element goes to the unit element) the image necessarily lies in $SO(3)$. It is not clear from this argument, however, that it is the *whole* group $SO(3)$.

13.3.6 Prove that the map under consideration

$$f : SU(2) \to SO(3)$$

given explicitly by the prescription[229]

$$(f(A)x)\tilde{} := A\tilde{x}A^+$$

$$\text{i.e.} \quad f(A) := \psi^{-1} \circ L_A \circ \psi \qquad \text{or also} \qquad e^{-\frac{1}{2}i\alpha\mathbf{n}\cdot\boldsymbol{\sigma}} \mapsto e^{\alpha\mathbf{n}\cdot\mathbf{l}}$$

is a surjective homomorphism with the kernel

$$\operatorname{Ker} f = (\mathbb{I}_2, -\mathbb{I}_2) \equiv \mathbb{Z}_2 \subset SU(2)$$

so that each rotational matrix $R \equiv f(A)$ has *just two* preimages, A and $-A$.

Hint: the homomorphism is clear, surjectivity is from (13.3.5). The matrices from the kernel are to satisfy $A\sigma_a A^+ = \sigma_a$, i.e. $[A, \sigma_a] = 0$, from where $A = \lambda\mathbb{I}_2$ (a direct computation, or certain multiples of σ_a realize a complex irreducible representation of $su(2)$, Schur's lemma). This belongs to $SU(2)$ only for $\lambda = \pm 1$. □

13.3.7 Be sure to understand that the map f from (13.3.6) is a *universal two-sheeted* covering of the group $SO(3)$ by the group $SU(2)$ and confirm for this case the result concerning the relation between the Lie algebras of the original and the covering groups from (13.3.1)

$$SO(3) = SU(2)/\mathbb{Z}_2 \qquad \text{but} \quad so(3) = su(2)$$

Hint: (13.2.12) for f from (13.3.6); $SU(2) \sim S^3$ is connected as well as simply connected; Lie algebras:

$$f : e^{-\frac{1}{2}i\alpha\mathbf{n}\cdot\boldsymbol{\sigma}} \mapsto e^{\alpha\mathbf{n}\cdot\mathbf{l}} \qquad \Rightarrow \qquad f' : -\frac{i}{2}\alpha\mathbf{n}\cdot\boldsymbol{\sigma} \mapsto \alpha\mathbf{n}\cdot\mathbf{l}$$

(12.1.5). □

13.3.8 Also confirm for this case the second result (concerning the relation between representations) from (13.3.1). Namely prove that the simplest representation of the covering group $SU(2)$, the representation $\rho = \mathrm{id}$ $(A \mapsto \rho(A) = A)$, *is not a lift* of any representation $\tilde{\rho}$ of the group $SO(3)$.

Hint: if it were a lift, there would hold $\mathrm{id} \equiv \rho = \tilde{\rho} \circ f$, i.e. $A \equiv \rho(A) = \tilde{\rho}(f(A)) \equiv \tilde{\rho}(R)$; for the matrix of the rotation by α around the z-axis this means

$$\tilde{\rho} : R_z(\alpha) \mapsto \operatorname{diag}\left(e^{-i\alpha/2}, e^{i\alpha/2}\right)$$

This gives, however, $R_z(2\pi) \mapsto \operatorname{diag}(-1,-1) \equiv -\mathbb{I}$, which is bad, since the element with $\alpha = 2\pi$ in $SO(3)$ is the *identity element* of the group and it is necessarily mapped into the (plus) identity operator (*in each* representation; $\Rightarrow \tilde{\rho}$ does not exist). □

[229] So that $A\tilde{x}A^+ = x_b A\sigma_b A^+ = (R_{ab}x_b)\sigma_a = (f(A)x)_a\sigma_a$.

• This shows that there are *more* representations of the group $SU(2)$ than of $SO(3)$. The analysis of the situation[230] leads to the result that the finite-dimensional complex irreducible representations of $SU(2)$ are labelled by a single non-negative integer or half-integer $j = 0, \frac{1}{2}, 1, \frac{3}{2}, \dots$ (in quantum mechanics it is related to the angular momentum of the object, in particular to its spin). It turns out that the representations labelled by integers j are lifted from $SO(3)$, whereas the half-integer representations are specific for $SU(2)$, i.e. one cannot obtain them by means of the lift from $SO(3)$. For example, the representation $\rho = \text{id}$ has spin $j = \frac{1}{2}$, so that it is not a lift (in accordance with the result of problem (13.3.8)). It is called the *spinor representation* of the group $SU(2)$ and the objects (the two-component complex columns), transforming according to this representation are known as ("non-relativistic") *spinors*. Spinors in a general context are treated in more detail in Chapter 22 (in particular, see (22.3.7)).

It is of practical importance to know that the covering $f : SU(2) \to SO(3)$ is a part of the covering of larger groups, namely the (two-sheeted) covering $f : SL(2, \mathbb{C}) \to L_+^\uparrow$, where $L_+^\uparrow$, the *proper orthochronous Lorentz* group, is a subgroup of $SO(1, 3)$.

13.3.9 Show that

(i) the *Lorentz group* $L \equiv O(1, 3)$ has (at least) four connected components

$$L = L_+^\uparrow \cup L_-^\uparrow \cup L_+^\downarrow \cup L_-^\downarrow \qquad L^\uparrow \cap L^\downarrow = \varnothing = L_+ \cap L_-$$

(ii) the multiplication by the matrices

$$\begin{aligned} T &:= \text{diag}\,(-1, \mathbb{I}_3) \in L_-^\downarrow \\ P &:= \text{diag}\,(1, -\mathbb{I}_3) \in L_-^\uparrow \\ PT &:= \text{diag}\,(-1, -\mathbb{I}_3) \in L_+^\downarrow \end{aligned}$$

realizes diffeomorphisms between them; the four connected components thus happen to be "topologically equal." One can prove that the connected component of the *identity element*, i.e. the subgroup $L_+^\uparrow$, is already connected,[231] so that $O(1, 3)$ has *just four* connected components.

Hint: $\pm$ because of $\det \Lambda = \pm 1$, $\uparrow\downarrow$ because of $\Lambda^0_0 \geq 1$ or ≤ -1 (resulting from the $(\ )_{00}$ component of the defining equations of $O(1, 3)$: according to (10.1.5) we have $(\Lambda^T \eta \Lambda)_{00} \equiv (\Lambda^0_0)^2 - \Lambda^j_0 \Lambda^j_0 \stackrel{!}{=} \eta_{00} \Rightarrow (\Lambda^0_0)^2 \geq 1$). □

• The condition $\Lambda^0_0 \geq 1$ says (through $x'^0 = \Lambda^0_0 x^0 + \cdots$) that the transformation does not invert the direction of time (the primed time thus flows "correctly," hence the word "orthochronous") and together with the condition on the determinant being $+1$ we get that Λ does not contain the space reflections; we thus study the Lorentz transformations which can be realized by a transition to the frame of reference which is (only) rotated and it moves

[230] This will not be performed in this book (the representations of the Lie *algebra* $su(2)$ used to be studied in textbooks on quantum mechanics, in the chapters devoted to rotations and angular momentum).

[231] Namely in problem (20.1.6) we learn that $L_+^\uparrow \sim \mathbb{R}^3 \times SO(3)$ (being at the same time $\mathbb{R}^3 \times \mathbb{R}P^3$). But $\mathbb{R}^3$ as well as $SO(3)$ *are* connected ($SO(3)$ is exponential, see the text after (11.7.18)).

uniformly along a straight line (with respect to a fiducial one). It is clear that such "proper" Lorentz transformations form a subgroup.

13.3.10 Denote by $H(2,\mathbb{C})$ the 2×2 Hermitian matrices (we do not require tracelessness in comparison with (13.3.3)). Show that

(i) it is a four-dimensional (real) vector space with a basis $\sigma_\mu \equiv (\mathbb{I}_2, \boldsymbol{\sigma})$ or alternatively $\bar{\sigma}_\mu \equiv (\mathbb{I}_2, -\boldsymbol{\sigma})$ ($\mu = 0, 1, 2, 3$; in both cases one defines the zeroth matrix to be the *identity* 2×2 matrix) and that the prescriptions

$$\psi : \mathbb{R}^4 \to H(2,\mathbb{C}) \qquad x \mapsto \psi(x) \equiv \underset{\sim}{x} = x^\mu \sigma_\mu$$
$$\chi : \mathbb{R}^4 \to H(2,\mathbb{C}) \qquad x \mapsto \chi(x) \equiv \tilde{x} = x^\mu \bar{\sigma}_\mu$$

are linear isomorphisms

(ii) the standard pseudo-Euclidean structure is in $H(2,\mathbb{C})$ incorporated by

$$\det \tilde{x} = \det \underset{\sim}{x} = (x, x) \equiv \eta_{\mu\nu} x^\mu x^\nu \qquad \underset{\sim}{x}\tilde{x} = \tilde{x}\underset{\sim}{x} = (x, x)\,\mathbb{I}_2$$

and that this results in the following identities:

$$\sigma_\mu \bar{\sigma}_\nu + \sigma_\nu \bar{\sigma}_\mu = 2\eta_{\mu\nu}\,\mathbb{I}_2$$
$$\bar{\sigma}_\mu \sigma_\nu + \bar{\sigma}_\nu \sigma_\mu = 2\eta_{\mu\nu}\,\mathbb{I}_2$$
$$\mathrm{Tr}\,(\bar{\sigma}_\mu \sigma_\nu) = \mathrm{Tr}\,(\sigma_\mu \bar{\sigma}_\nu) = 2\eta_{\mu\nu}$$

(iii) the prescription

$$\underset{\sim}{x} \mapsto L_A \underset{\sim}{x} := A \underset{\sim}{x} A^+$$

is a left linear action (i.e. a representation) of $SL(2,\mathbb{C})$ in $H(2,\mathbb{C})$, which moreover preserves the pseudo-Euclidean norm $||x||^2$; thus it is a representation by means of *pseudo-orthogonal* operators and we get the map (homomorphism) $f : SL(2,\mathbb{C}) \to O(1,3)$ being sought (so far we (only) see that it is to $O(1,3)$; as a matter of fact it will be clear in a moment that its image is in $L_+^\uparrow$)

(iv) on the basis σ_ν this gives the matrix elements of the pseudo-orthogonal operator

$$\sigma_\nu \mapsto A\sigma_\nu A^+ =: \Lambda^\mu_\nu \sigma_\mu \qquad \Lambda^\mu_\nu = \eta^{\mu\alpha} \frac{1}{2} \mathrm{Tr}\,(\bar{\sigma}_\alpha A \sigma_\nu A^+)$$

so that effectively in coordinates

$$x^\mu \mapsto \Lambda^\mu_\nu x^\nu$$

Hint: (iii) $\det(L_A \underset{\sim}{x}) = \det \underset{\sim}{x} = (x, x)$. □

• Now we would like to see again which Lorentz transformation is the f-image of a given matrix A (an analog of (13.3.5)). In order to do this we need to find a suitable parametrization of the matrices from $SL(2,\mathbb{C})$ and this is, in turn, conveniently done making use of the *polar decomposition*.

13.3.11 Show that any matrix $A \in SL(2,\mathbb{C})$ may be written in the form of the product of a special unitary matrix and a Hermitian unimodular matrix

$$A = UH \qquad U \in SU(2) \quad H^+ = H \quad \det H = 1$$

Hint: A^+A is a positive definite Hermitian matrix ($\langle v, A^+Av\rangle = \langle Av, Av\rangle \geq 0$), so that it has a (positive definite Hermitian) square root $H := \sqrt{A^+A}$; one checks that $\det H = 1$; then we get for U the result $U = AH^{-1}$ – check that it is unitary and realize that $\det U = 1$. □

• Since f is a homomorphism, we have $f(A) = f(UH) = f(U)f(H)$ and so the image of an arbitrary matrix A is a *composition of two* Lorentz transformations, the first being the image of U and the second being the image of H. Thus, it is enough to find out how these special types of matrices are mapped.

13.3.12 Check that for these special types of matrices U and H we get:

(i) they may be parametrized in the form

$$U = e^{-\frac{1}{2}i\alpha\mathbf{n}\cdot\boldsymbol{\sigma}} \qquad H = \pm\cosh\frac{\alpha}{2}\mathbb{I}_2 + (\mathbf{n}\cdot\boldsymbol{\sigma})\sinh\frac{\alpha}{2} = \pm e^{\pm\frac{1}{2}\alpha\mathbf{n}\cdot\boldsymbol{\sigma}} \quad \alpha \in \mathbb{R} \; ||\mathbf{n}|| = 1$$

(ii) for U parametrized in this way we get from (13.3.10) as $\Lambda \equiv f(U)$ explicitly

$$\Lambda^0_0 = 1 \qquad \Lambda^0_i = 0 = \Lambda^i_0 \qquad \Lambda^i_j = \cos\alpha\,\delta_{ij} + (1-\cos\alpha)n_in_j - \sin\alpha\,\varepsilon_{ijk}n_k$$

and so $x^\mu \mapsto x'^\mu \equiv \Lambda^\mu_\nu x^\nu$ reads

$$\begin{aligned} x'^0 &= x^0 \\ \mathbf{r}' &= \mathbf{r}\cos\alpha + (\mathbf{n}\cdot\mathbf{r})\mathbf{n}(1-\cos\alpha) + (\mathbf{n}\times\mathbf{r})\sin\alpha \end{aligned}$$

which is by (11.7.13) *the most general rotation* (by α around $\mathbf{n}$)

(iii) for H parametrized in this way we get from (13.3.10) as $\Lambda \equiv f(H)$ explicitly

$$\Lambda^0_0 = \cosh\alpha \qquad \Lambda^i_0 = \Lambda^0_i = \pm n_i\sinh\alpha \qquad \Lambda^i_j = \delta^i_j + (\cosh\alpha - 1)n_in_j$$

and so $x^\mu \mapsto x'^\mu \equiv \Lambda^\mu_\nu x^\nu$ reads

$$\begin{aligned} x'^0 &= x^0\cosh\alpha \pm (\mathbf{n}\cdot\mathbf{r})\sinh\alpha \\ x'^i &= \pm n_i\sinh\alpha\, x^0 + x^i + (\cosh\alpha - 1)(\mathbf{r}\cdot\mathbf{n})n_i \end{aligned}$$

If $\mathbf{r}$ is decomposed into its longitudinal and transversal parts (with respect to $\mathbf{n}$)

$$\mathbf{r} = \mathbf{r}_{\parallel} + \mathbf{r}_{\perp} \qquad \begin{aligned} \mathbf{r}_{\parallel} &= (\mathbf{n}\cdot\mathbf{r})\mathbf{n} \equiv r_{\parallel}\mathbf{n} \\ \mathbf{r}_{\perp} &= \mathbf{r} - (\mathbf{n}\cdot\mathbf{r})\mathbf{n} \end{aligned}$$

then

$$\begin{aligned} x'^0 &= x^0\cosh\alpha \pm r_{\parallel}\sinh\alpha \\ r'_{\parallel} &= \pm x^0\sinh\alpha + r_{\parallel}\cosh\alpha \\ \mathbf{r}'_{\perp} &= \mathbf{r}_{\perp} \end{aligned}$$

which is just a *general boost* (in the direction of **n**). The image of the whole group $SL(2,\mathbb{C})$ is thus the composition of the most general rotation and the most general boost, i.e. it is the most general element of the group $L_+^\uparrow$. □

13.3.13 Prove that the map under consideration

$$f : SL(2,\mathbb{C}) \to L_+^\uparrow$$

given explicitly by the prescription[232]

$$(f(A)x)_{\sim} := A\,\underset{\sim}{x}\,A^+ \qquad \text{i.e.} \quad f(A) = \psi^{-1} \circ L_A \circ \psi$$

is a surjective homomorphism with the kernel

$$\text{Ker}\, f = (\mathbb{I}_2, -\mathbb{I}_2) \equiv \mathbb{Z}_2 \subset SL(2,\mathbb{C})$$

so that each (proper and orthochronous) Lorentz matrix $\Lambda \equiv f(A)$ has *just two* preimages, A and $-A$.

Hint: surjectivity from (13.3.12), the kernel as in (13.3.6). □

13.3.14 Check that the map f from (13.3.13) is a *universal two-sheeted* covering of the group $L_+^\uparrow$ by the group $SL(2,\mathbb{C})$, a part being the covering of $SO(3)$ by the group $SU(2)$ from (13.3.7). Confirm also here the result concerning the relation between the Lie algebras of an initial and a covering group from (13.3.1)

$$L_+^\uparrow = SL(2,\mathbb{C})/\mathbb{Z}_2 \qquad \text{but} \qquad so(1,3) = sl(2,\mathbb{C})$$

Hint: (13.2.12) for f from (13.3.13); S^3 (and consequently also $SL(2,\mathbb{C}) \sim \mathbb{R}^3 \times S^3$, see (20.1.5)) is connected as well as simply connected; restriction of L_A from (13.3.10) to the subgroup $SU(2)$ preserves the subspace $H_0(2,\mathbb{C}) \subset H(2,\mathbb{C})$ and coincides on it with L_A from (13.3.3), so that the restriction of f to the subgroup $SU(2)$ coincides with f from (13.3.6); the isomorphism of Lie algebras explicitly reads as

$$\begin{pmatrix} 0 & 0 \\ 0 & l_j \end{pmatrix} \leftrightarrow -\frac{i}{2}\sigma_j \qquad \begin{pmatrix} 0 & n_j \\ n_j & 0_3 \end{pmatrix} \leftrightarrow \frac{1}{2}\sigma_j$$

$$(l_j)_{km} \equiv -\epsilon_{jkm} \qquad\qquad (n_j)_m \equiv \delta_{jm}$$

□

13.3.15 Confirm also for this case the second result (concerning the relation between representations) from (13.3.1). Namely, prove that here the simplest representation of the covering group $SL(2,\mathbb{C})$, the representation $\rho = \text{id}$ ($A \mapsto \rho(A) = A$), *is not a lift* of any representation $\tilde{\rho}$ of the group $L_+^\uparrow$.

Hint: if it were a lift, it would be so for its restriction to the subgroup $SU(2)$ as well; we already learned in (13.3.8) that this is not the case. □

[232] So that $A\,\underset{\sim}{x}\,A^+ = x_A^\nu \sigma_\nu A^+ = (\Lambda_\nu^\mu x^\mu)\sigma_\nu \equiv (f(A)x)^\nu \sigma_\nu$.

• This shows that there are *more* representations of the group $SL(2, \mathbb{C})$ than those of $L^{\uparrow}_{+}$. An analysis of the situation[233] leads to the result that the finite-dimensional complex irreducible representations of $SL(2, \mathbb{C})$ are labeled by *a pair* of non-negative integers or half-integers $(j, j') = 0, \frac{1}{2}, 1, \frac{3}{2}, \dots$. It turns out that the representations with integer *sum* $j + j'$ are lifted from those of $L^{\uparrow}_{+}$, whereas the half-integer sum $j + j'$ corresponds to the representations which are specific for $SL(2, \mathbb{C})$, i.e. one cannot obtain them by means of the lift from $L^{\uparrow}_{+}$.

For example, the representation $\rho = \text{id}$ has $(j, j') = (\frac{1}{2}, 0)$, so that it is not a lift (as we already learned in (13.3.15)). It is one of its two *spinor representations*, the second one being $(j, j') = (0, \frac{1}{2})$ and it turns out to be complex conjugate to $(\frac{1}{2}, 0)$ so that it is also two-dimensional. One can show that these two representations are inequivalent (12.2.5) and that by tensor products and restrictions to invariant subspaces *all* the finite-dimensional complex irreducible representations of $SL(2, \mathbb{C})$ may be obtained; this is the reason why they are called the *fundamental representations*.

Since the two fundamental representations are inequivalent, in the component approach we need to use *different* types of indices in order to distinguish them. A standard convention is to use *dotted* indices for $(j, j') = (0, \frac{1}{2})$ and *undotted* indices for $(j, j') = (\frac{1}{2}, 0)$ (so that, for example, $u^a \mapsto A^a_b u^b$ and $w^{\dot{a}} \mapsto (A^*)^{\dot{a}}_{\dot{b}} w^{\dot{b}}$). The translation of the properties mentioned above to the language of indices is that a general representation occurs on *tensors* carrying several dotted and several undotted indices (*on such* tensors the tensor products of the fundamental representations act) with a particular type of symmetry (in *this way* the invariant subspaces are singled out). It turns out in more detail that the representation of the type (j, j') takes place on tensors with $2j$ undotted and $2j'$ dotted indices, which are *completely symmetric* with respect to both types of indices separately (for example, on $t^{(ab)(\dot{c}\dot{d}\dot{e})}$ the representation $(j, j') = (1, \frac{3}{2})$ is realized).

For $SU(2)$ there is only *a single* fundamental representation, namely $j = \frac{1}{2}$ (or $\rho = \text{id}$; according to (12.2.4), the complex conjugate representation is equivalent to ρ). A general representation of type j is realized on completely symmetric tensors with $2j$ indices and its dimension is $2j + 1$.

We know from problem (13.3.1) that any representation of the "covered" group "is" automatically also a representation of the covering group, but the opposite is not true in general. In order to soften the mental trauma which this unpleasant result causes to the covered[234] groups $\tilde{G}$, modern mental hygiene introduced the concept of *multi-valued representations*. How does it work?

As already mentioned in the hint to (13.3.1), the reason why the converse assignment $\rho \mapsto \tilde{\rho}$ is problematic is that the map f cannot be inverted, since it is not injective. Nor does the canonical choice of one of the preimages, in general, exist. If there is no distinguished choice of a preimage, the most fair decision is to take *all* the preimages. (The other equally

[233] This analysis will not be performed in this book (the representations of the Lie *algebra* $so(1, 3) = sl(2, \mathbb{C})$ used to be studied in textbooks of relativistic quantum mechanics (or the quantum field theory)).

[234] According to (13.3.1) the groups $\tilde{G}$ readily and voluntarily lend *all* of their representations to their covering groups G. Fairly often, however, there are serious problems with the reciprocity.

fair decision is to accept *no* preimage, i.e. to take a conservative stand that there is ("we are sorry") no representation, there's an end to it.) This means that we first assign to the group element $\tilde{g} \in \tilde{G}$ all the preimages $(g_1, g_2, \ldots)$, then we represent them (by means of ρ) and finally we declare *all* the resulting operators $(\rho(g_1), \rho(g_2), \ldots)$ (their order not being important) to constitute the image of the element $\tilde{g}$ with respect to a multi-valued representation

$$\tilde{g} \mapsto (\rho(g_1), \rho(g_2), \ldots)$$

As an example, for the two-sheeted covering $f : SU(2) \to SO(3)$, each rotational matrix $R \in SO(3)$ has just *two* preimages $(A, -A)$, so that we get a *two-valued representation* of the group $SO(3)$ in this way

$$R \mapsto (\rho(A), \rho(-A))$$

Such unordered n-tuples (for the n-sheeted covering) may be multiplied correctly (each operator from the first n-tuple is multiplied by each operator from the second one and then we throw away those results which are already there) and this multiplication[235] "copies" the multiplication in the group $\tilde{G}$, i.e. these n-tuples of operators in this sense indeed represent[236] the group $\tilde{G}$.

13.3.16 Consider the two-sheeted covering $f : SU(2) \to SO(3)$ and the representation $A \to \rho(A) = A$ of the group $SU(2)$ (according to the result (13.3.8) this is one of the representations not shared by the covering group $SU(2)$ with the group $SO(3)$). Contemplate how the two-valued representation works in this particular case (i.e. how the group $SO(3)$ can use it as *at least a two-valued* representation).

Hint: if $f(A) = f(-A) = R \in SO(3)$, we have the assignment $R \mapsto (A, -A)$ in the two-valued assignment, the order being irrelevant. The multiplication of pairs reads

$$(A_1, -A_1) \circ (A_2, -A_2) = (A_1A_2, -A_1A_2)$$

We get as many as *four* operators as the result of the multiplication

$$A_1A_2, \; A_1(-A_2), \; (-A_1)A_2, \; (-A_1)(-A_2)$$

but then we throw away those which are already there once and we end with a *pair*

$$A_1A_2, \; -A_1A_2$$

just copying the multiplication of the orthogonal matrices,

$$R_1R_2 \leftrightarrow (A_1, -A_1) \circ (A_2, -A_2) \qquad \square$$

[235] What is left behind is actually the multiplication of the *cosets* $(g_1H)(g_2H) = (g_1g_2)H$, which *is* well defined, since H is a (discrete) *normal* subgroup (13.2.10).

[236] Note, however, that the concept of multi-valued representation is not a special case of a representation, but rather an *extension* of the concept of a representation. The multi-valued representation is thus *not a representation* as understood so far.

Although there is no canonical choice of one of the preimages of $\tilde{g}$ in general, there *is* such a choice in a (sufficiently small) neighborhood of the *identity element* $\tilde{e} \in \tilde{G}$. Namely, we choose the preimage which is also from the neighborhood of the *identity* element $e \in G$ (in particular, we take e to be *the* preimage of $\tilde{e}$). This means that each multi-valued representation of $\tilde{G}$ induces an "ordinary" representation of the "local group" (a small neighborhood of the group in which the multiplication is defined (only) for those pairs of elements for which the result once again belongs to the neighborhood). This shows that the *derived* representation of the multi-valued representation of the group is well defined, too, being an "ordinary" representation of the Lie algebra $\tilde{\mathcal{G}}$. Put another way, there is no asymmetry at the level of the *derived* representations (recall that the Lie algebras themselves are isomorphic!), *all* representations ρ' of Lie algebra $\mathcal{G}$ of the covering group G may be used as ("ordinary") representations $\tilde{\rho}'$ of the Lie algebra $\tilde{\mathcal{G}}$ of the "covered" group $\tilde{G}$. So, for example, the Lie algebras $su(2) = so(3)$ share the same set of representations in spite of the fact that this is more complicated at the level of the representations of the groups $SU(2)$ and $SO(3)$.

13.4 Representations of G and $\mathcal{G}$ in the space of functions on a G-space, fundamental fields

• Consider a manifold M which is a right G-space of a Lie group G. Thus by definition there exists the right *action* R_g of the group G on M. Since M need not be a linear space, the action need not be in general a representation. A closer inspection reveals, however, that the input data which are available nevertheless enable one to easily construct a *linear space* in which G acts *linearly*, i.e. to obtain a *representation* of the group G. The linear space is $\mathcal{F}(M)$, the space of (smooth) *functions*[237] on M and the representation is provided by the action *on arguments* of the functions.

13.4.1 Let R_g be the right action of G on a manifold M. Check that

(i) the prescription

$$\psi \mapsto \rho(g)\psi := \psi \circ R_g \qquad \text{i.e.} \quad (\rho(g)\psi)(m) := \psi(R_g m) \equiv \psi(mg)$$

defines a (∞-dimensional) *representation* of G in $\mathcal{F}(M)$

(ii) the right action on arguments (i.e. on M) results in the left action on functions and vice versa (this is the reason why the representation needs the *right* action on arguments)

(iii) this representation may also be expressed in terms of the *pull-back* of the action R_g as

$$\psi \mapsto \rho(g)\psi := R_g^*\psi \qquad \text{i.e} \quad \rho(g) = R_g^*$$

(iv) this $\rho(g)$ is actually *more* than a representation, since it also respects the structure of an *algebra* on $\mathcal{F}(M)$: in an effort to win customers it offers a "freebie" property, it preserves the *product*

$$\rho(g)(\psi\phi) = (\rho(g)\psi)(\rho(g)\phi)$$

[237] This is the simplest version of the construction. We will also study its (still simple) far-reaching generalizations.

This may be expressed as the fact that it is a representation of G by *automorphisms* of the algebra $\mathcal{F}(M)$ (compare with the situation in (12.3.2)).

Hint: (i) a straightforward check that $\rho(g)$ is a left linear action; (iv) similarly, or recall the general properties of the pull-back f^*. □

13.4.2 If we take in particular in (13.4.1) the group G itself as M and the right *translation* on G as R_g, we get the *right regular representation* of the group G. Thus it is a representation on functions on the group itself, which is realized by the right[238] translation of the argument:

$$(\text{Reg}\,(g)\psi)(k) := \psi(kg)$$

Check that each finite-dimensional irreducible representation $\rho : G \to \text{Aut}\, V$ is equivalent to some subrepresentation of the right regular representation. In other words, for an arbitrary given irreducible representation ρ in V there exists in $\mathcal{F}(G)$ (the representation space of the regular representation) an invariant subspace such that the restriction of the regular representation to the subspace is equivalent to the given representation.

Hint: on the basis $E_b \in V$,

$$\rho(g)E_b = r^c_b(g)E_c$$

For the functions on the group

$$\psi_b : G \to \mathbb{R} \qquad \psi_b(g) := r^a_b(g) \qquad a \text{ fixed } (a\text{th row of the matrix } r^a_b)$$

we have

$$(\text{Reg}\,(g)\psi_b)(k) = \psi_b(kg) = r^a_b(kg) = r^a_c(k)r^c_b(g) = r^c_b(g)\psi_c(k)$$

$\Rightarrow$ in the subspace spanned[239] by $(\psi_1, \ldots, \psi_{\dim V})$ the right regular representation gives the subrepresentation which is equivalent to the representation ρ in V (the bases $E_a \leftrightarrow \psi_a$ are transformed by the same matrices $r^a_b(g)$, see (12.2.3)).

Note: *Each* row of the matrix r^a_b thus yields an invariant subspace with respect to the right regular representation $\Rightarrow$ (each irreducible) representation ρ is "hidden" in the representation Reg exactly $n = \dim V$ times (we have n rows). An analogous result is easily verified for the columns for the left regular representation. □

• One of the lessons from this problem is that the construction of representations by means of (13.4.1) might also be useful if we are interested in the *finite-dimensional* representations: although the whole representation in the space $\mathcal{F}(M)$ of functions on M is *infinite*-dimensional, it is in general *reducible*, containing finite-dimensional invariant subspaces (spanned by several functions on M, which transform only within themselves under the action on the arguments) and *on these functions* we obtain finite-dimensional irreducible

[238] In full analogy the *other right* action $k \mapsto g^{-1}k$ yields the *left regular representation*.

[239] If the representation in V *is not* irreducible, the correspondence $E_a \leftrightarrow \psi_a$ would not be an isomorphism, since for an appropriate choice of the basis E_a (if it is adapted to the invariant subspace) some functions ψ_a would *vanish*.

representations of G. Since the whole representation ρ is given by a simple and universal formula $(\rho(g)\psi)(m) := \psi(mg)$, all we need for particular situations is to be able to find the invariant subspaces mentioned above. If we know the explicit expressions for the generators $\mathcal{E}_i := \rho'(E_i)$ of the *derived* representation, we might use the *Casimir operators* mentioned before problem (12.3.13).

Let us return to the general case of ρ from (13.4.1) and focus our attention to its derived representation. The representation of G under consideration is a homomorphism of G to the group Aut $\mathcal{F}(M)$, the group of automorphisms of the associative algebra $\mathcal{F}(M)$. Its Lie algebra (see Appendix A.2) is the (Lie) algebra Der $\mathcal{F}(M)$ of *derivations* of the (associative) algebra $\mathcal{F}(M)$; according to (2.2.8), this is the Lie algebra of vector fields $\mathfrak{X}(M)$ with the standard operation of the commutator. Then the derived representation ρ' is now to be a homomorphism

$$\mathcal{G} \to \mathfrak{X}(M) \qquad X \mapsto \xi_X = \text{ a vector field on } M$$

Let us have a look at the construction and the general properties of such vector fields.

If we fix a point m on M and apply the one-parameter subgroup $g(t) = \exp tX$ to this point (by means of the action R_g), the *curve* $\gamma(t) := R_{\exp tX}m \equiv me^{tX}$ arises on M which starts at $\gamma(0) = m$. Since the one-parameter group is completely given by the element $X \in \mathcal{G}$, we may denote the tangent vector at zero to this curve by $\xi_X(m)$. Repeating the same procedure at all points on M we get a vector *field* on M, called the *fundamental field of the action* R_g. It is clear from the construction of the fundamental field that it provides an aid for visualizing the action itself: it displays the directions of the possible motions at a given point by means of the action of the group (this may be used, for example, to determine the form of the orbit in the vicinity of m).

13.4.3 Define the *fundamental field* of the action R_g on M (known also as the *generator of the action*) by the formula

$$\xi_X(m) := \left.\frac{d}{dt}\right|_0 R_{\exp tX}m \qquad \text{i.e.} \qquad \xi_X(m) = \dot{\gamma}(0) \quad \text{for} \quad \gamma(t) := me^{tX}$$

and, in particular, for the basis elements $E_i \in \mathcal{G}$ we will sometimes use the condensed notation $\xi_i \equiv \xi_{E_i}$ (note that this is the *field itself*, not a component). Show that

(i) ξ_X behaves with respect to the action R_g as follows:

$$R_g^*\xi_X = \xi_{\mathrm{Ad}_g X} \qquad \text{and, in particular, on the basis} \qquad R_g^*\xi_i = (\mathrm{Ad}_g)_i^j\, \xi_j$$

so that the fields ξ_i are linearly combined with one another by the matrices of the *adjoint* representation

(ii) the infinitesimal version of this transformation law reads

$$[\xi_X, \xi_Y] = \xi_{[X,Y]} \qquad \text{and, in particular, on the basis} \qquad [\xi_i, \xi_j] = c_{ij}^k \xi_k$$

(iii) the prescription

$$X \mapsto \rho'(X) \equiv \xi_X$$

is a representation of the Lie algebra $\mathcal{G}$ in $\mathcal{F}(M)$, being just the derived representation ρ' of the representation of the group considered in (13.4.1), so that

$$\xi_{X+\lambda Y} = \xi_X + \lambda \xi_Y$$
$$\xi_{[X,Y]} = [\xi_X, \xi_Y]$$

Hint: (i) making use of (3.1.2) and (12.3.1) we get

$$R_{g*}\xi_X(m) := R_{g*} \left.\frac{d}{dt}\right|_0 R_{\exp tX} m = \left.\frac{d}{dt}\right|_0 R_g R_{\exp tX} m$$
$$= \left.\frac{d}{dt}\right|_0 mg(g^{-1}e^{tX}g) = \left.\frac{d}{dt}\right|_0 (mg)e^{t\mathrm{Ad}_{g^{-1}}X} \equiv \xi_{\mathrm{Ad}_{g^{-1}}X}(mg)$$

(ii) set $g = e^{tY}$ in (i) and differentiate in $t = 0$; realize that $R_{\exp tY}$ is a *flow* on M, generated by the field ξ_Y, that $\mathcal{L}_{\xi_Y}\xi_X = [\xi_Y, \xi_X]$ and that $\mathrm{Ad}' = \mathrm{ad}$; (iii) compare the definition of ξ_X with the formula for $\rho'(X)$ in (12.1.6). □

• Before we proceed further let us learn which modifications result from considering the *left* actions on M instead of right actions.

13.4.4 Let L_g be a *left* action on M. Its *generator* is defined by (compare with (13.4.3))

$$\hat{\xi}_X(m) := \left.\frac{d}{dt}\right|_0 L_{\exp tX} m \qquad \text{i.e.} \qquad \hat{\xi}_X(m) = \dot{\gamma}(0) \quad \text{for} \quad \gamma(t) := e^{tX} m$$

and, in particular, for the basis elements $E_i \in \mathcal{G}$ we use the condensed notation $\hat{\xi}_i \equiv \hat{\xi}_{E_i}$. Check that

(i) $\hat{\xi}_X$ behaves with respect to the action L_g as follows:

$$L_g^* \hat{\xi}_X = \hat{\xi}_{\mathrm{Ad}_{g^{-1}}X} \qquad \text{and, in particular, on the basis} \qquad L_g^* \hat{\xi}_i = (\mathrm{Ad}_{g^{-1}})_i^j \hat{\xi}_j$$

(ii) the infinitesimal version of this transformation law reads

$$[\hat{\xi}_X, \hat{\xi}_Y] = -\hat{\xi}_{[X,Y]} \qquad \text{and, in particular, on the basis} \qquad [\hat{\xi}_i, \hat{\xi}_j] = -c_{ij}^k \hat{\xi}_k$$

(note the change of sign; *this* is the main difference compared to the generators of the right action)

(iii) the prescription

$$X \mapsto \hat{\xi}_X$$

is an *anti*representation of the Lie algebra $\mathcal{G}$ in $\mathcal{F}(M)$ (i.e. the commutator has the *opposite sign* compared to a representation).

Hint: just like in (13.4.3): $L_{g*}\hat{\xi}_X(m) := L_{g*} \left.\frac{d}{dt}\right|_0 L_{\exp tX} m = \cdots$. □

• Let us now compute some concrete simple fundamental fields which will come in handy later, being interesting, however, also in their own right. In the first problem we realize that we have already encountered particular cases of the fundamental fields before.

13.4.5 Let R_g be the *right translation* on a group. Since it is a right action (13.1.3), there should be some generators. Check that

(i) there holds

$$\xi_X = L_X$$

so that the generators (fundamental fields) of the right translations on G coincide with our good old left-invariant fields on G

(ii) the vector fields

$$V_C = x^i_k C^k_j \partial^j_i \equiv \mathrm{Tr}\,(xC\partial)$$

from (11.1.10) are just the fundamental fields of the right translation on $GL(n, \mathbb{R})$

(iii) the generators of the *left translation* L_g on the group are the *right-invariant* fields

$$\hat{\xi}_X = R_X$$

Hint: see (11.4.6) and (11.1.10). □

13.4.6 Consider the standard (right) action $\mathbf{r} \mapsto A^{-1}\mathbf{r} \equiv R_A\mathbf{r}$ of the group $SO(3)$ in $\mathbb{R}^3$. Show that

(i) the fundamental fields ξ_{l_j} corresponding to the basis $l_j \in so(3)$ explicitly read

$$\xi_{l_j} = -\epsilon_{jkm} x_k \partial_m \equiv (-\mathbf{r} \times \boldsymbol{\nabla})_j$$

and thus confirm the folklore knowledge that the operators of the *orbital angular momentum* in quantum mechanics are given (up to a constant multiple) by the generators of the rotations in $\mathbb{R}^3$

$$\hat{L}_j = i\hbar\xi_{l_j}$$

(ii) if expressed in the *spherical* polar coordinates in (part of) $\mathbb{R}^3$, they read

$$\begin{aligned} \xi_{l_1} &= -\sin\varphi\partial_\vartheta - \cot\vartheta\cos\varphi\partial_\varphi \\ \xi_{l_2} &= \cos\varphi\partial_\vartheta - \cot\vartheta\sin\varphi\partial_\varphi \\ \xi_{l_3} &= \partial_\varphi \end{aligned}$$

explain their independence of the "radial" coordinate r

(iii) this action may be restricted to the unit *sphere* centered at the origin of $\mathbb{R}^3$, $\mathbf{n} \mapsto A^{-1}\mathbf{n} \equiv R_A\mathbf{n}$ and that the generators of this action look equally as they come out in item (ii)

(iv) the fields ξ_{l_j} are particular cases (part of a basis) of the *Killing vectors* of the (pseudo-)Euclidean space $M_{ij} \equiv -M_{ji} \equiv x_i\partial_j - x_j\partial_i$ from (4.6.10) and their restrictions to the sphere coincide with our good old Killing vectors on the sphere from (4.6.11).

Hint: (i) for $A(\epsilon) \equiv e^{\epsilon l_i}$ we have $A^{-1}(\epsilon) \doteq 1 - \epsilon l_i \Rightarrow$

$$x_j \mapsto (R_A x)_j \doteq x_j - \epsilon(l_i x)_j = x_j + \epsilon\epsilon_{ijm} x_m \equiv x_j(\epsilon).$$

Then,

$$\xi_{l_j} = \dot{x}_m(0)\partial_m = -\epsilon_{jkm} x_k \partial_m \equiv (-\mathbf{r} \times \boldsymbol{\nabla})_j \equiv \left(-\frac{i}{\hbar}\vec{r} \times \vec{p}\right)_j \equiv -\frac{i}{\hbar}\hat{L}_j$$

(ii) the orbits are spheres (therefore ∂_r is absent) and the action rotates all the points on a "ray" emanating from the origin by the same angle (therefore r in components is absent); (iv) $\xi_j = \frac{1}{2}\,\epsilon_{jkl} M_{kl}$; the action respects the ordinary metric tensor in $\mathbb{R}^3$ and consequently also its restriction to the *orbit* (i.e. the sphere). □

13.4.7 Find the fundamental field ξ_X for the natural (right) action in $\mathbb{R}^n$ for

(i) the group $GL(n, \mathbb{R})$ (linear transformations)

$$\mathbb{R}^n \ni x \mapsto A^{-1}x \equiv R_A x \qquad A \in GL(n, \mathbb{R})$$

(ii) the group $GA(n, \mathbb{R})$ (affine transformations)

$$\mathbb{R}^n \ni x \mapsto (A, a)^{-1}x \equiv R_{(A,a)}x \qquad (A, a)x := Ax + a$$

(iii) obtain from this as a special case again the solution of (13.4.6)
(iv) obtain from this as a special case again the generators of the *Euclidean transformations* in E^n (translations and rotations).

Hint: (i) for $A(t) = e^{tC}$ we have $x \mapsto A(-t)x = x - tCx + \cdots \equiv x(t)$; then,

$$\xi_C \equiv \dot{x}^j(0)\partial_j = -x^k C^j_k \partial_j$$

One verifies easily that $C \mapsto \xi_C$ is indeed a representation: for the "linear" vector fields of the form $V_A := x_j A_{ji}\partial_i$ there holds $[V_A, V_B] = V_{[A,B]}$; here this then reduces to checking whether $C \mapsto -C^{\mathrm{T}}$ is a representation, see (12.1.9); (ii) a convenient form of the (so far left) action reads

$$\begin{pmatrix} x \\ 1 \end{pmatrix} \mapsto \begin{pmatrix} A & a \\ 0 & 1 \end{pmatrix}\begin{pmatrix} x \\ 1 \end{pmatrix} \quad \Rightarrow \quad \text{for} \quad \begin{pmatrix} A & a \\ 0 & 1 \end{pmatrix}^{-1} = \begin{pmatrix} 1 & 0 \\ 0 & 1 \end{pmatrix} - \epsilon\begin{pmatrix} C & c \\ 0 & 0 \end{pmatrix}$$

we get $x(\epsilon) = x - \epsilon(Cx + c)$ so that $\xi_{(C,c)} \equiv \dot{x}^i(0)\partial_i = -x^j C^i_j \partial_i - c^i \partial_i$; (iii) plug the matrix $l_j \in so(3)$ for C; (iv) restrict C to the antisymmetric matrices (since A are orthogonal). □

13.4.8 Let ρ be a representation of G in V, $\check{\rho}$ the contragredient representation in V^*. Then $R_g = \rho(g^{-1})$ and $\check{R}_g = \check{\rho}(g^{-1})$ are right actions on the manifolds V and V^*. Introduce canonical coordinates x^a and y_a on these manifolds by $v = x^a E_a \in V$, $\alpha = y_a E^a \in V^*$. Check that the fundamental fields of the right actions read

$$\xi_i = -x^b \rho^a_{bi}\partial_a \quad \text{and} \quad \check{\xi}_i = y_b \rho^b_{ai}\partial^a \qquad \partial_a \equiv \frac{\partial}{\partial x^a} \quad \partial^a \equiv \frac{\partial}{\partial y_a}$$

Hint: if $v = x^a E_a \in V$ then,

$$R_{\exp \epsilon E_i} v = x^a \rho(\exp(-\epsilon E_i))E_a = x^a E_a - \epsilon x^a \rho^b_{ai} E_b \equiv x^a(\epsilon)E_a \quad \Rightarrow$$
$$\xi_i \equiv \dot{x}^a(0)\partial_a = -x^b \rho^a_{bi}\partial_a$$

and similarly $\check{R}_{\exp \epsilon E_i}\alpha = \cdots$ (or use the relation between $\check{\rho}^a_{bi}$ and ρ^a_{bi} from (12.1.9)); note that we get the "linear vector fields" mentioned in the particular case of (13.4.7). □

13.4.9 Find the fundamental fields for the adjoint as well as coadjoint actions (both regarded as right actions in the sense of (13.4.8)).

Hint: let $x = x^i E_i \in \mathcal{G}$, $y = y_i E^i \in \mathcal{G}^*$; (12.3.19) and (13.4.8) yield: for Ad we have $\rho^j_{ki} = -c^j_{ki}$ and for Ad* $\check{\rho}^j_{ki} = +c^j_{ki}$ so that

$$\xi_i = x^k c^j_{ki}\partial_j \qquad \text{and} \qquad \check{\xi}_i = -y_k c^k_{ji}\partial^j \quad \partial_k \equiv \frac{\partial}{\partial x^k}, \; \partial^k \equiv \frac{\partial}{\partial y_k}$$

□

13.4.10 Find the fundamental fields of the adjoint and the coadjoint actions for the group $GA(1,\mathbb{R})$ and use them to determine the orbits of the actions. Confirm that all orbits of the coadjoint action are even-dimensional (as the general theory of such orbits asserts, see (14.6.4)), whereas for the adjoint action here we also have odd-dimensional orbits (which shows clearly at the same time that the representations Ad and Ad* are *inequivalent* for this group, which is consistent with the degeneracy of the Killing–Cartan metric, (12.2.3) and (12.3.10)).

Hint: from (13.4.9) and (11.2.4) one should obtain

$$\begin{aligned} \text{Ad} \;:\; & \xi_1 = -x^2\partial_2 & \qquad \text{Ad}^* \;:\; & \xi_1 = y_2\partial^2 \\ & \xi_2 = x^1\partial_2 & & \xi_2 = -y_2\partial^1 \end{aligned}$$

Visualize these vectors in all the possible points of the (x^1, x^2) and (y_1, y_2) planes and infer that the orbits in the (x^1, x^2) plane (i.e. those of the Ad-action) are: (i) the point $(0, 0)$, (ii) the remaining parts of the x^2 axis and (iii) all lines parallel to the x^2 axis (so there are zero- and one-dimensional orbits), whereas in the (y_1, y_2) plane we have: (i) each point on the y^1 axis and (ii) the upper as well as the lower half-planes (so there are zero- and two-dimensional orbits). □

• Let us now have a look at how we can significantly (albeit surprisingly simply) generalize our construction of the representations on $\mathcal{F}(M)$. The first generalization stems from the observation that the "ordinary" $\mathbb{R}$-valued functions may be replaced by the functions on M with values in an arbitrary *linear space* $(V, \hat{\rho})$, which carries some representation $\hat{\rho}$ of the group G. The second direction of the generalization consists in replacing *functions* on M by arbitrary *tensor fields* (in particular, differential forms) on M. Combining these two ideas we eventually come to a fairly general and useful concept of the representation of G on the $(V, \hat{\rho})$-valued tensor fields on M.

13.4.11 Let a manifold M be a right G-space and let $\hat{\rho}$ be a representation of G in V. Denote by $\mathcal{F}(M, V)$ the space of V-valued functions on M.[240] Check that

[240] The elements of $\mathcal{F}(M, V)$ are nothing but the *several-component fields* frequently used in physics. Recall (see Section 6.4, the functions being particular forms) that introducing a basis E_a into V enables one to write uniquely each $\psi \in \mathcal{F}(M, V)$ in the form $\psi = \psi^a E_a$, where ψ^a are the *component functions* with respect to the basis E_a and that they may, in turn, be written as an n-component column, $n \equiv \dim V$.

(i) it is a linear space

(ii) the prescription $\psi \equiv \psi^a E_a \mapsto \rho(g)\psi$, where

$$\rho(g)\psi := \hat{\rho}(g) \circ \psi \circ R_g \equiv (\hat{\rho}(g) \circ R_g^*)\psi \equiv (R_g^* \psi^a)(\hat{\rho}(g)E_a)$$
$$(\rho(g)\psi)(m) := \hat{\rho}(g)\psi(mg) \equiv \psi^a(mg)(\hat{\rho}(g)E_a)$$

defines a *representation* of G in $\mathcal{F}(M, V)$

(iii) it is a particular case of (13.1.13), where the action on the maps between two G-spaces is treated

(iv) the representation from (13.4.1) is a special case.

Hint: (ii) $\hat{\rho}(gh) \circ \psi \circ R_{gh} = \hat{\rho}(g) \circ (\hat{\rho}(h) \circ \psi \circ R_h) \circ R_g = \rho(g)(\hat{\rho}(h) \circ \psi \circ R_h) = \rho(g)\rho(h)\psi$; (iii) for $N = V$, $\hat{R}_g = \hat{\rho}(g^{-1})$; (iv) for $V = \mathbb{R}$, $\hat{\rho} =$ the trivial representation. □

13.4.12 Let ρ be the representation of G in $\mathcal{F}(M, V)$ introduced in (13.4.11), E_a be a basis in V and $\psi \equiv \psi^a E_a \in \mathcal{F}(M, V)$. Check that

(i) the derived representation ρ' reads

$$\rho'(X) = \xi_X + \hat{\rho}'(X) \qquad \text{i.e.} \quad \rho'(X)(\psi^a E_a) = (\xi_X \psi^a)E_a + \psi^a(\hat{\rho}'(X)E_a)$$

(ii) for the generators $\mathcal{E}_i \equiv \rho'(E_i)$ we get

$$(\rho'(E_i)\psi)^a = \xi_i \psi^a + \rho^a_{bi}\psi^b$$

so that the generator is a sum of two parts: the first one *only differentiates* the component functions ψ^a whereas the second one, in contrast, *only "scrambles"* the components (but does not differentiate them; at the level of the component functions, thus the whole operator is a sum of a differential operator and a matrix).

Hint: (i) for $g(\epsilon) = e^{\epsilon X}$ we have

$$\rho(g(\epsilon))\psi \overset{1}{=} \hat{\rho}(g(\epsilon))R^*_{g(\epsilon)}\psi = (1 + \epsilon\hat{\rho}'(X))(1 + \epsilon\mathcal{L}_{\xi_X})\psi = \cdots$$
$$\overset{2}{=} (1 + \epsilon\rho'(X))\psi$$

□

13.4.13 Consider a wave function of a single particle with spin s in quantum mechanics. In this theory it is explained that a $\mathbb{C}^n$-valued ("wave") function on $\mathbb{R}^3$ (being the configuration space of a classical particle) corresponds to such a particle, where $n = 2s + 1$, i.e. an element ψ from $\mathcal{F}(\mathbb{R}^3, \mathbb{C}^n)$. The group of rotations $SO(3)$ is represented on these functions in a way described in (13.4.11)

$$\psi(\mathbf{r}) \mapsto \psi_A(\mathbf{r}) \equiv (\rho(A)\psi)(\mathbf{r}) = \hat{\rho}(A)\psi(A^{-1}\mathbf{r})$$

i.e. there is a simultaneous rotation of the argument $\mathbf{r}$ and "mixing" of the components by a $2s + 1 \equiv n$-dimensional complex representation $\hat{\rho}$ of the group $SO(3)$. In physical terms this corresponds to the transition to the *rotated state* $\psi \mapsto \rho(A)\psi \equiv \psi_A$ (an "active" interpretation of the rotation).[241] Show that if we identify (up to a factor of $i\hbar$) the operators

[241] A "passive" rotation corresponds to the transition to the rotated *observer* (reference frame), giving the same formulas up to the replacement $A \mapsto A^{-1}$ (the observer rotated by A sees the world as if being rotated by A^{-1}), thus resulting in the *right* linear action of the group and consequently in a *change of the sign* of the generator (13.4.4).

of the *total angular momentum* J_i with the generators of the rotations, we get the good old expression

$$\mathbf{J} = \mathbf{L} + \mathbf{s} \qquad \text{i.e.} \quad J_j = L_j + s_j$$

where L_j are the (genuinely differential) operators of the orbital (part of the) angular momentum and s_j, known as the *operators of spin*, are, in contrast, genuinely matrix operators (they only "mix" the components).

Hint: see (13.4.12). □

• Let us mention now two particular types of action of groups, namely free and effective actions, as well as the reflection of these properties of the action in the generators (fundamental fields).

13.4.14 An action of a group on M is called a *free action* if the stabilizer of each point happens to be trivial, $G_x = e$ for all $x \in M$, and an *effective action* if the stabilizer is trivial at least somewhere. The visual meaning of the freedom is that *each* point $x \in M$ "indeed moves away" from its original place after the action of any (non-identity) element of the group (so that all the non-identity group elements really work everywhere) and the effectiveness means that *at least a single* point $x \in M$ "indeed moves away," so that there are no elements in the group (except the identity) which "do nothing anywhere" (misusing the social system in this way). Be sure that you understand that

(i) a free action is necessarily effective
(ii) the action of $GL(n, \mathbb{R})$ on the bases in L (see problems (5.7.2) and (13.2.7)) is free; in general, G acts freely on the *principal* homogeneous space
(iii) the fundamental fields ξ_X of a free action are everywhere non-zero and for an effective action they are non-zero at least somewhere (for each $0 \neq X \in \mathcal{G}$). □

• To close the section let us make clear the relation between the generators of the left translation on a group G and the generators of the "projected" left action of G on the homogeneous space G/H.

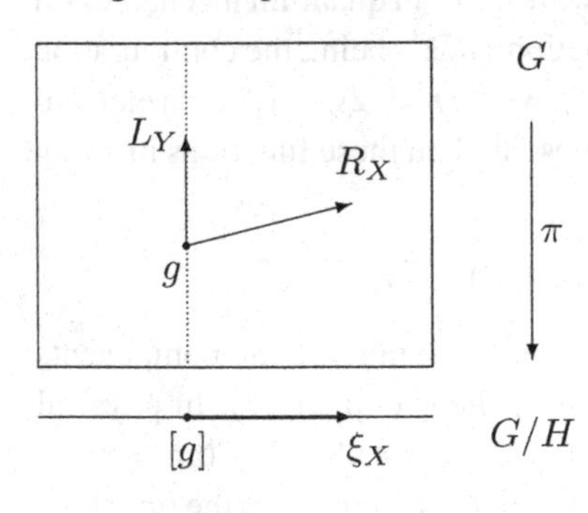

13.4.15* Consider a Lie group G and the homogeneous space G/H. On G we consider[242] the right action $\mathcal{R}_h$ of the subgroup H along the coset gH (the right translation $g \mapsto gh$) as well as the left translation $\mathcal{L}_g$, $g \in G$. We know that the projection of the left translation $\mathcal{L}_g$ on the group G yields the canonical left action L_g of the group G on the homogeneous space G/H (13.2.5), i.e. there holds

$$\pi \circ \mathcal{L}_g = L_g \circ \pi \qquad \pi \circ \mathcal{R}_h = \pi$$

[242] A useful view of the same situation is (in the language of "fiber bundles") presented in (20.1.2).

Show that

(i) the generators ξ_X of the action L_g happen to be the π_*-images of the generators of the left translation $\mathcal{L}_g$, i.e. the π_*-images of the right-invariant fields R_X on the group G

$$\xi_X = \pi_* R_X$$

(ii) they obey the commutation relations

$$[\xi_X, \xi_{X'}] = -\xi_{[X,X']}$$

(iii) the generators of the right action $\mathcal{R}_h$ of the subgroup H along the coset gH are the left-invariant fields L_Y, $Y \in \mathcal{H}$.

Hint: (i) in $\pi \circ \mathcal{L}_g = L_g \circ \pi$ set g equal to the one-parameter subgroup $g(t) = \exp(tX)$; the flow $L_{g(t)}$ is generated by the field ξ_X on G/H we are looking for; recall that the flow $\mathcal{L}_{g(t)}$ is generated by the right-invariant field R_X (13.4.5); then differentiation in $t = 0$ gives immediately $\pi_* R_X = \xi_X$; (ii) see (13.4.4); (iii) the action is $g \mapsto ge^{tY} \in gH$, (11.4.6). □

13.5 Representations of G and $\mathcal{G}$ in the space of tensor fields of type $\hat{\rho}$

• We mentioned in the text before problem (13.4.11) that the second direction of the generalization of the construction of representations on functions on G-space consists in replacing *functions* on M by arbitrary *tensor fields* (in particular, differential forms) on M. We investigate this possibility more closely in this section.

According to the results of problem (13.4.1) we know that the representation studied there may be succinctly written as

$$g \mapsto R_g^* : \mathcal{F}(M) \to \mathcal{F}(M)$$

However, the pull-back map R_g^* is defined not only on functions but also on arbitrary tensor fields (recall that R_g is a diffeomorphism). One may check easily that by this simple rule we indeed obtain a representation of G in the space $\mathcal{T}_s^r(M)$ of tensor fields of *arbitrary rank* on a manifold (G-space) M.

13.5.1 Let $A \in \mathcal{T}_s^r(M)$ be a tensor field on a right G-space M. Check that the prescriptions

$$A \mapsto \rho(g)A := R_g^* A \qquad A \mapsto \rho'(X)A := \mathcal{L}_{\xi_X} A$$

(where ξ_X is the fundamental field of the action and $\mathcal{L}_{\xi_X}$ is the Lie derivative along this field) define a representation of the group G in $\mathcal{T}_s^r(M)$ and the corresponding derived representation of its Lie algebra $\mathcal{G}$ (the situation treated in (13.4.1) is just a particular case).

Hint: $R_{gh}^* = (R_h \circ R_g)^* = R_g^* R_h^*$; the flow $\Phi_t := R_{\exp tX}$ is (by definition, see (13.4.3)) generated by the fundamental field ξ_X and the required "time" derivative in zero just gives (by definition, see Section 4.2) the Lie derivative. □

• In the text after problem (13.4.2) we already mentioned that although we obtain an ∞-dimensional representation in the space $\mathcal{F}(M)$, it is highly reducible and by restriction to appropriate finite-dimensional invariant subspaces we get finite-dimensional subrepresentations. All this is still true. Imagine there is an N-dimensional G-invariant subspace in $\mathcal{T}^r_s(M)$, i.e. we have a set of *tensor fields* A^a, $a = 1, \ldots, N$ (each A^a being a *whole* tensor *field, not* a component of the latter), such that they are "scrambled" only within the set under the action R^*_g of the group. Then for each g we have the matrix $r^a_b(g)$ given by $R^*_g A^a = r^a_b(g) A^b$.

13.5.2 Check that

(i) the prescription

$$g \mapsto r^a_b(g)$$

is an *anti*representation (*right* linear action) of the group G

(ii) the following statement holds:

$$\text{if} \quad \hat{\rho}^a_b(g) := r^a_b(g^{-1}) \quad \text{i.e.} \quad R^*_g A^a = \hat{\rho}^a_b(g^{-1}) A^b$$

$$\text{then} \quad g \mapsto \hat{\rho}^a_b(g) \quad \text{is a representation}$$

(iii) if we regard the matrices $\hat{\rho}^a_b(g)$ as corresponding to an abstract (finite-dimensional) representation $\hat{\rho}$ in a space V with respect to a basis E_a, i.e. as given by the standard relations

$$\hat{\rho}(g) E_a = \hat{\rho}^b_a(g) E_b$$

and if we regard the tensor fields A^a as the *component fields* of a V*-valued* tensor field (in the sense[243] of Section 6.4), then the component expression from item (ii) corresponds to the formula

$$R^*_g A = \hat{\rho}(g^{-1}) A$$

$$\text{i.e.} \quad (R^*_g A^a) E_a = A^a (\hat{\rho}(g^{-1}) E_a) \quad \leftrightarrow \quad R^*_g A^a = \hat{\rho}^a_b(g^{-1}) A^b$$

Hint: (i) hint to (13.5.1); (ii) see (13.1.1). □

• The finite-dimensional subrepresentations hidden in the representation $g \mapsto R^*_g$ on tensor fields bring us naturally to particular kinds of tensor fields, namely to those $(V, \hat{\rho})$-valued tensor fields which respond to the action of the group according to the rule

$$R^*_g A = \hat{\rho}(g^{-1}) A$$

We will call them *tensor fields of type* $\hat{\rho}$ and, in particular, *differential forms of type* $\hat{\rho}$. Such fields are fairly frequent in geometry; they provide a useful tool in all those situations when tensor fields (in particular, forms) are treated on manifolds on which group G acts.

(Up to now, we have been considering tensor fields of type $\binom{r}{s}$ as those belonging to $\mathcal{T}^r_s(M)$; no role of a group G acting on M was mentioned. The "type" introduced here is clearly something completely different; it says that some *set* of tensors (all of them being of "type $\binom{r}{s}$" as understood up to now) transforms only within the set under the *action of the group G on M*, namely that it is "mixed" by the representation $\hat{\rho}$. A potential confusion

[243] Although we treated only V-valued *forms* explicitly there, the reader is undoubtedly able to extend the idea to arbitrary V-valued *tensors* as well.

from these two concepts is fortunately excluded in the *most important* case of interest, for the *differential forms*: if we say "a 2-form of type Ad," it is clear that the "standard" type is $\binom{0}{2}$ now (plus the antisymmetry), so that the "type Ad" necessarily means the "new" one – we speak about a Lie algebra valued 2-form (since *this* is the space where the representation Ad operates) satisfying moreover $R_g^*\alpha = \text{Ad}_{g^{-1}}\alpha$).

13.5.3 Consider a generalization of the representation of G on $(V, \hat{\rho})$-valued functions on M (13.4.11), which consists in replacing functions by general tensor fields. This means that the representation space is $\mathcal{T}_s^r(M, V)$, the space of V-valued tensor fields on M, and we define a representation ρ on it by the prescription

$$\rho(g) := \hat{\rho}(g) \circ R_g^* \qquad \text{i.e.} \quad \rho(g)(A^a E_a) := (R_g^* A^a)(\hat{\rho}(g) E_a)$$

Check that

(i) it is a representation of G, which is moreover equivalent to the *tensor product* $R_g^* \otimes \hat{\rho}(g)$
(ii) (13.4.11) is its special case
(iii) the tensor fields A, which happen to be *invariant* with respect to *this* representation, are nothing but the tensor fields of *type* $\hat{\rho}$ introduced above

$$\rho(g)A = A \quad \Leftrightarrow \quad R_g^* A = \hat{\rho}(g^{-1})A \quad \Leftrightarrow \quad \text{the fields of type } \hat{\rho}$$

Hint: (i) if $A \equiv A^a E_a \leftrightarrow A^a \otimes E_a$, then the fact that $\rho(g)(A^a E_a) := (R_g^* A^a)(\hat{\rho}(g) E_a)$ corresponds just to the tensor product (12.4.11):

$$(R_g^* A^a) \otimes (\hat{\rho}(g) E_a) =: (R_g^* \otimes \hat{\rho}(g))(A^a \otimes E_a)$$

(ii) $r = s = 0$. □

13.5.4 Consider the representation $\rho(g)$ on $\mathcal{T}_s^r(M, V)$ from problem (13.5.3). Check that

(i) the derived representation ρ' reads

$$\rho'(X) = \mathcal{L}_{\xi_X} + \hat{\rho}'(X) \qquad \text{i.e.} \quad \rho'(X)(A^a E_a) = (\mathcal{L}_{\xi_X} A^a) E_a + A^a (\hat{\rho}'(X) E_a)$$

(ii) we get for the generators $\mathcal{E}_i \equiv \rho'(E_i)$

$$(\rho'(E_i)A)^a = \mathcal{L}_{\xi_i} A^a + \rho^a_{bi} A^b$$

so that the generator is a sum of two parts, the first of them containing *only* the (Lie) *differentiation* (but not a "scrambling") of the component tensor fields A^a, whereas the second of them (in contrast) *only "scrambles"* the components (but it does not differentiate them; at the level of the component tensor fields the whole is the sum of a "differential" operator (the *Lie* derivative operator) and a matrix)
(iii) the condition expressing the fact that $A \equiv A^a E_a$ is of type $\hat{\rho}$ may be rewritten in the *infinitesimal* version as

$$\mathcal{L}_{\xi_X} A = -\hat{\rho}_X A \qquad \text{or in components} \qquad \mathcal{L}_{\xi_i} A^a = -\rho^a_{bi} A^b$$

Hint: see (13.4.12), (13.5.1) and (13.5.3). □

• This construction combines the two directions of generalization of the representation on $\mathcal{F}(M)$ (mentioned before (13.4.11)). We came to $\mathcal{T}^r_s(M, V)$, so that we generalized functions to arbitrary tensor fields and at the same time the $\mathbb{R}$-valued objects to the V-valued ones.

Let us mention two simple examples illustrating the objects introduced above. We will see from them that actually the tensor fields of type $\hat{\rho}$ are by no means rare and dangerous beasts living in virgin forests far from here, but rather they are fairly frequent, useful and good-natured pets living in our immediate environment.

13.5.5 Consider M to be the circle S^1 together with a standard action (by rotations) of the group $G = U(1) = SO(2)$

$$e^{i\alpha} : \varphi \mapsto \varphi + \alpha$$

(φ being the common polar angle). Check that

(i) the fundamental field $\xi \equiv \xi_{E_1}$ of the action, corresponding to the basis element $X \equiv E_1 = i \in u(1)$, is

$$\xi = \partial_\varphi$$

(ii) if we consider $\hat{\rho}$ to be the irreducible representation $\hat{\rho}_n$ from problem (12.2.10), then the (infinitesimal) condition which singles out a *function of type* $\hat{\rho}_n$ reads

$$\partial_\varphi \Phi_n(\varphi) = -in\Phi_n(\varphi) \quad \Rightarrow \quad \Phi_n(\varphi) = \Phi_n(0)e^{-in\varphi}$$

(iii) the *Fourier expansion* of a function on the circle (the expansion to the series with respect to the basis functions $\sim e^{-in\varphi}$) may be regarded as an expansion with respect to the *functions of type* $\hat{\rho}$ for *all the possible irreducible* representations of the group $U(1)$

(iv) the differentials $d\Phi_n$ of the functions Φ_n are 1-forms of type $\hat{\rho}_n$ on the circle:

$$\mathcal{L}_\xi(d\Phi_n) = -in(d\Phi_n)$$

Hint: (iv) (6.2.10). □

13.5.6 Consider M to be the sphere S^2 together with a standard (right) action of the group $G = SO(3)$

$$\mathbf{n} \mapsto A^{-1}\mathbf{n} \equiv R_A\mathbf{n}$$

(rotations from (13.4.6)). From a course on quantum mechanics we know[244] the *spherical harmonics* $Y^l_m(\mathbf{n})$. They may be regarded as (complex-valued) functions on the sphere, $Y^l_m : S^2 \to \mathbb{C}$. In connection with the theory of angular momentum in quantum mechanics their important property – the behavior under the rotation of the argument – is mentioned

$$Y^l_m(A^{-1}\mathbf{n}) = \sum_{m'=-l}^{l} D^l_{m'm}(A)\, Y^l_{m'}(\mathbf{n})$$

[244] Everybody who has seen the standard solution of the Schrödinger equation for the hydrogen atom has to be familiar with them. These functions are also widely used for more advanced computations in various branches of classical physics, such as in electromagnetism, etc.

The functions $D^l_{m'm}(A)$ on the group $SO(3)$ are called the *rotational matrices* or alternatively the *Wigner rotational functions* (l fixed, the matrix indices being m, m'). Check that

(i) the rotational matrices satisfy the condition

$$D^l_{m'm}(AB) = \sum_{m''=-l}^{l} D^l_{m'm''}(A) D^l_{m''m}(B) \qquad \text{i.e.} \quad D^l(AB) = D^l(A) D^l(B)$$

so that they provide a *representation*[245] of the group $SO(3)$

(ii) also the prescription

$$A \mapsto \hat{\rho}^l(A) := (D^l)^{\mathrm{T}}(A^{-1})$$

provides a representation

(iii) the behavior of the functions Y^l_m with respect to the rotation of the argument may also be understood so that they constitute the component functions of *a single* $\mathbb{C}^{2l+1}$-valued function, which is in turn *of type* $\hat{\rho}$: if a natural basis E_m is introduced in $\mathbb{C}^{2l+1}$ (the columns with all entries vanishing except for the mth place, being 1) then the function $Y^l := Y^l_m E_m$ (to be summed over m) obeys

$$R^*_A Y^l = \hat{\rho}^l(A^{-1}) Y^l \qquad Y^l : S^2 \to \mathbb{C}^{2l+1}$$

(iv) the expansion of a function on the sphere into the series

$$f(\mathbf{n}) = \sum_{l=0}^{\infty} \sum_{m=-l}^{l} c^m_l Y^l_m(\mathbf{n})$$

may be thus again interpreted as the expansion with respect to the *functions of type* $\hat{\rho}$ for *all the possible irreducible* representations of the group $SO(3)$.

Hint: (i) $Y^l_m((AB)^{-1}\mathbf{n}) = Y^l_m(B^{-1}(A^{-1}\mathbf{n}))$. □

13.5.7 Consider again M to be the sphere (S^2, g) with the natural action of $G = SO(3)$ and, moreover, endowed with the standard "round" metric tensor g. As we have already learned in (13.5.6), the *functions* Y^l_m may be regarded as the component functions of a single function $Y^l : S^2 \to \mathbb{C}^{2l+1}$ of type $\hat{\rho}^l$. From this function, however, we can easily obtain (making use of the standard operations on forms) two 1-*forms* of type $\hat{\rho}^l$ as well as two *vector fields* of type $\hat{\rho}^l$ (and in principle, combining them appropriately, also "higher" tensor fields, if needed). Namely, let ψ^l_m represent any of the following fields on the sphere:

functions on S^2:	Y^l_m	
one-forms on S^2:	dY^l_m and	$* dY^l_m$
vector fields on S^2:	$\sharp dY^l_m$ and	$\sharp * dY^l_m$

[245] The representation is $(2l+1)$-dimensional, complex and (as one can also show) irreducible. These representations are studied in detail in textbooks of quantum mechanics. In the sense of the remark in the text after (13.3.8), they correspond to *integer* $j = l$. A closer analysis shows that these representations (for all $l = 0, 1, 2, \ldots$) actually exhaust *all* irreducible (unitary, finite-dimensional) representations of the group $SO(3)$.

and introduce (for each particular object) the corresponding $\mathbb{C}^n$-valued quantity

$$\psi^l := \sum_{m=-l}^{l} \psi^l_m E_m$$

Then check that

(i) each ψ^l is indeed of type $\rho = \hat{\rho}^l$

$$R_A^* \psi^l_m = \sum_{m'=-l}^{l} D^l_{m'm}(A)\, \psi^l_{m'} \qquad \text{i.e.} \quad R_A^* \psi^l = \hat{\rho}^l(A^{-1})\psi^l$$

(ii) each ψ^l_m happens to represent the simultaneous eigenstate of the operators $\mathbf{J}^2$ and J_3, the operator of the *square of the angular momentum*[246] (12.3.13) and the third component of the angular momentum

$$\mathbf{J}^2 \psi^l_m = \hbar^2 l(l+1)\psi^l_m \qquad J_3\psi^l_m = m\psi^l_m$$

where[247]

$$\mathbf{J}^2 \equiv \mathbf{J}\cdot\mathbf{J} \equiv J_1J_1 + J_2J_2 + J_3J_3 \qquad J_j \equiv -i\hbar\mathcal{E}_j \equiv -i\hbar\rho'(E_j) \equiv -i\hbar\mathcal{L}_{\xi_{E_j}}$$

Hint: for the *function* Y^l_m (where $\mathcal{L}_{\xi_{E_j}} Y^l_m$ simplifies to $\xi_{E_j} Y^l_m$) both statements are assumed to be known (from, say, a course on quantum mechanics). For higher rank fields recall that R_A is an isometry, and consequently all generators ξ_{E_j} are Killing fields. Then it is enough to realize that R_A^* commutes with $d, *_g$ and $\sharp_g$ (see (6.2.11) and (8.3.8)), and the same is then also true for the Lie derivative along the Killing fields ξ_{E_j} (8.3.14). □

• Note that we actually also met a form of type $\hat{\rho} = \mathrm{Ad}$ in (12.3.4) (the canonical 1-form on a Lie group) and that such forms will play a prominent role in connection theory (in Chapter 19 and beyond).

To close the section let us have a look at two simple properties of forms of type $\hat{\rho}$.

13.5.8* Let (V_1, ρ_1) and (V_2, ρ_2) be two representations of the group G, E_i be a basis in V_1, E_a a basis in V_2, $\alpha \equiv \alpha^i E_i$ be a form of type ρ_1, $\beta \equiv \beta^a E_a$ a form of type ρ_2 and let A be an intertwining operator between (V_1, ρ_1) and (V_2, ρ_2)

$$A : V_1 \to V_2 \qquad \rho_2(g) \circ A = A \circ \rho_1(g)$$

Show that then

(i) the form $A \circ \alpha$

$$A \circ \alpha := \alpha^i A(E_i) \equiv \left(A^a_i \alpha^i\right) E_a$$

is already a form of type ρ_2

$$R_g^*\alpha = \rho_1(g^{-1})\alpha \quad \Rightarrow \quad R_g^*(A\circ\alpha) = \rho_2(g^{-1})(A\circ\alpha)$$

[246] It turns out that this operator coincides with (a multiple of) the Laplace–deRham operator on the sphere, $\mathbf{J}^2 = -\hbar^2\Delta_g$.

[247] Such states (normalized to unity) are denoted by $|l, m\rangle$ in quantum-mechanical treatments of rotations and the angular momentum.

so that if α^i deserves the index $(\ )^i$, then $A^a_i\alpha^i$ *indeed* deserves the index $(\ )^a$, as indicated by the summation over i; the matrix A^a_i of an *intertwining operator* is thus able to "transmute" an index of type "i" into an index of type "a" (see also Section 12.5)

(ii) the form $\alpha \overset{\otimes}{\wedge} \beta$

$$\alpha \overset{\otimes}{\wedge} \beta := (\alpha^i \wedge \beta^a) E_i \otimes E_a$$

is already a form of type $\rho_1 \otimes \rho_2$

$$\begin{aligned} R^*_g\alpha &= \rho_1(g^{-1})\alpha \\ R^*_g\beta &= \rho_2(g^{-1})\beta \quad \Rightarrow \quad R^*_g(\alpha \overset{\otimes}{\wedge} \beta) = (\rho_1 \otimes \rho_2)(g^{-1})\alpha \overset{\otimes}{\wedge} \beta \end{aligned}$$

Solution: (i)

$$\begin{aligned} R^*_g(A \circ \alpha) &= R^*_g(A^a_i\alpha^i)E_a = A^a_i(R^*_g\alpha^i)E_a = A^a_i(\rho_1(g^{-1}))^i_j\alpha^j E_a \\ &= (A\rho_1(g^{-1}))^a_j\alpha^j E_a = (\rho_2(g^{-1})A)^a_j\alpha^j E_a = (\rho_2(g^{-1}))^a_b(A \circ \alpha)^b E_a \\ &\equiv \rho_2(g^{-1})(A \circ \alpha) \end{aligned}$$

(ii) $R^*_g(\alpha \overset{\otimes}{\wedge} \beta) = R^*_g(\alpha^i \wedge \beta^a)E_i \otimes E_a = (R^*_g\alpha^i \wedge R^*_g\beta^a)E_i \otimes E_a = \cdots$ □

Summary of Chapter 13

From the point of view of differential geometry, the most interesting actions of groups are their smooth actions on manifolds. As a rule, there is some additional structure on the manifold and the action preserves this structure (e.g. actions via isometries on Riemannian manifolds or symplectic actions on symplectic manifolds). An action of a Lie group induces at the infinitesimal level an action of its Lie algebra, generated by the *fundamental* (vector) *fields* (the generators). An action on points of a manifold results (using the tools of Section 3.1) in an action on functions on the manifold (and more generally on tensor fields on the latter). By this simple method we obtain a construction of (∞-dimensional) representations of groups and their algebras (tensor fields, in particular functions, are naturally endowed with a linear space structure). Upon restriction to invariant subspaces finite-dimensional representations are also obtained by this method. Restriction to a G-invariant subspace of functions (tensor fields) is a standard useful way to solve complicated differential equations (an ansatz with some symmetry properties).

$L_{gh} = L_g \circ L_h,\ R_{gh} = R_h \circ R_g$	Left action, right action of G on M	Sec. 13.1
$L_{\hat g}[g] := [\hat g g]$	Left action of G on the homogeneous space G/H	(13.2.5)
$[g][\tilde g] := [g\tilde g]$	Multiplication in the factor group G/H	(13.2.10)
$gHg^{-1} = H$	H is a normal (invariant) subgroup of G	(13.2.10)
$G/\mathrm{Ker} f = \mathrm{Im} f$	Homomorphism theorem	(13.2.12)
$e^{-\frac{1}{2}i\alpha\mathbf{n}\cdot\boldsymbol{\sigma}} \mapsto e^{\alpha\mathbf{n}\cdot\mathbf{l}}$	Universal 2-sheet covering of $SO(3)$ by $SU(2)$	(13.3.6)
$\rho(g)\psi := \psi \circ R_g \equiv R^*_g\psi$	Representation of G in $\mathcal{F}(M)$	(13.4.1)
$\xi_X(m) := (d/dt)\vert_0\, R_{\exp tX} m$	Fundamental field (generator) of the action R_g	(13.4.3)

$\rho'(X) = \xi_X$	Derived representation in $\mathcal{F}(M)$	(13.4.3)
$\xi_j = -\epsilon_{jkm} x_k \partial_m \equiv (-\mathbf{r} \times \nabla)_j$	Generators of the rotations in $\mathbb{R}^3$	(13.4.6)
$\rho(g)\psi := \hat{\rho}(g) \circ \psi \circ R_g$	Representation of G in $\mathcal{F}(M, V)$	(13.4.11)
$\rho'(X) = \xi_X + \hat{\rho}'(X)$	Derived representation in $\mathcal{F}(M, V)$	(13.4.12)
$R_g^* A = \hat{\rho}(g^{-1}) A$	A is a tensor field of type $\hat{\rho}$	(13.5.2)
$\rho(g) := \hat{\rho}(g) \circ R_g^*$	Representation of G in $\mathcal{T}_s^r(M, V)$	(13.5.3)

14
Hamiltonian mechanics and symplectic manifolds

• Analysis of the dynamics governed by the Hamiltonian equations reveals that classical Hamiltonian mechanics has an elegant geometric interpretation in terms of the *symplectic structure* on a manifold. This structure allows one to gain a deeper insight into the essential properties of the solutions of the equations. In the course of the whole chapter we will assume that the Hamiltonian does not depend explicitly on time (so that the system of equations is autonomous); otherwise the formalism requires a modification, which will be mentioned in Section 18.5.

14.1 Poisson and symplectic structure on a manifold

• As early as in problem (2.3.7), being true beginners in the field, we learned that the Hamilton equations

$$\dot{q}^a = \frac{\partial H}{\partial p_a} \qquad \dot{p}_a = -\frac{\partial H}{\partial q^a} \quad a = 1, \ldots, n$$

may be regarded as the equations for the integral curves of a certain vector field on $\mathbb{R}^{2n}[q^a, p_a]$, namely the field

$$\zeta_H = \frac{\partial H}{\partial p_a}\frac{\partial}{\partial q^a} - \frac{\partial H}{\partial q^a}\frac{\partial}{\partial p_a}$$

(The indices on the coordinates p_a indeed are to be written as *subscripts*. There is a natural motivation for this, a somewhat unusual convention in the treatment of Hamiltonian mechanics on the *cotangent bundle* T^*M (to be discussed in Chapter 17 and beyond; although these words might raise fears, actually this is the *most common* situation where everybody uses the Hamilton equations, namely the phase space of a configuration space M). One should realize that this interchange of subscripts and superscripts on (some) coordinates results in the corresponding interchange of the position of indices on the (corresponding) components of tensors (for example, the corresponding components of vectors carry *lower* indices, etc.). The purpose of this chapter, however, is to introduce a *coordinate-free* formalism, after all, thus escaping from "problems" of this sort completely.)

Now, after learning in numerous examples that progress often starts when we are able to write down equations in a coordinate-free way, we try to see some structure behind the Hamilton equations (or the vector field ζ_H). We will come quickly to the conclusion that an antisymmetric tensor field of type $\binom{2}{0}$ waits for us restlessly there.

14.1.1 In the $2n$-dimensional phase space $\mathbb{R}^{2n}[q, p]$ rename the coordinates as follows:

$$\begin{aligned} z^i \equiv (z^1, \dots, z^n, z^{n+1}, \dots, z^{2n}) &:= (q^1, \dots, q^n, p_1, \dots, p_n) \qquad i = 1, \dots, n, \dots, 2n \\ &\equiv (q^a, p_a) \qquad\qquad\qquad\qquad\quad a = 1, \dots, n \end{aligned}$$

Check that

(i) the Hamilton equations may be written in these coordinates in the form

$$\dot{z}^i = \zeta_H^i(z) =: (\partial_j H)\mathcal{P}^{ji} \equiv (dH)_j \mathcal{P}^{ji}$$

i.e. when retold in the coordinate-free language, as the equations for integral curves $\gamma(t)$ of the vector field ζ_H, given in terms of a *bivector field* $\mathcal{P}$ (i.e. an antisymmetric tensor field of type $\binom{2}{0}$) and the gradient of a function H

$$\dot{\gamma} = \zeta_H \qquad \zeta_H = \mathcal{P}(dH, \,\cdot\,)$$

In the coordinates z we have explicitly

$$\mathcal{P}^{ij}(z) = \begin{pmatrix} \mathbb{O}_n & -\mathbb{I}_n \\ \mathbb{I}_n & \mathbb{O}_n \end{pmatrix} = -\mathcal{P}^{ji}(z)$$

$$\mathcal{P} = \mathcal{P}^{ij}(z)\partial_i \otimes \partial_j = \frac{1}{2}\mathcal{P}^{ij}(z)\partial_i \wedge \partial_j = \frac{\partial}{\partial p_a} \wedge \frac{\partial}{\partial q^a}$$

(ii) the matrix $\mathcal{P}^{ij}$ is *non-singular* (namely $\det \mathcal{P}^{ij} = 1$)

(iii) the standard *Poisson bracket* of two functions $f(z)$, $g(z)$ may be succinctly written in terms of the field $\mathcal{P}$ as

$$\{f, g\} \equiv \frac{\partial f}{\partial p_a}\frac{\partial g}{\partial q^a} - \frac{\partial f}{\partial q^a}\frac{\partial g}{\partial p_a} = \mathcal{P}^{ij}(df)_i(dg)_j \equiv \mathcal{P}(df, dg)$$

□

• The bivector field $\mathcal{P}$, which emerged from the Hamilton equations, satisfies a differential identity, which guarantees the validity of the *Jacobi identity* for the Poisson bracket. We postpone its derivation until later (see (14.1.8)), since the direct computation here would be unnecessarily tedious and it will also be more instructive later.

We obtained a new field on a particular manifold $\mathbb{R}^{2n}$ and in particular coordinates on it. It turns out to be convenient to introduce this field in a general setting as a new useful geometrical structure. Namely, an arbitrary bivector field $\mathcal{P}$ on a manifold M, which happens to satisfy the identity mentioned above, is called a *Poisson tensor* and a manifold $(M, \mathcal{P})$ endowed with such a tensor is a *Poisson manifold*. The vector field of the form $\zeta_f := \mathcal{P}(df, \,\cdot\,)$ is known as the *Hamiltonian field* generated by the function f.

Note that we did not include the requirement of *non-degeneracy* (i.e. vanishing of the kernel of the map $\alpha \mapsto \mathcal{P}(\alpha, \,\cdot\,)$ or non-singularity of the matrix $\mathcal{P}^{ij}$) into the general definition of $\mathcal{P}$ (although "our" tensor $\mathcal{P}$ *was* non-degenerate). The reason is that there are interesting and non-trivial applications of Poisson manifolds with a *singular* matrix $\mathcal{P}^{ij}$. In what follows we restrict, however, to the most important special case[248] where $\det \mathcal{P}^{ij} \neq 0$. Then the situation may be studied by means of its "mirror image," the whole story may be retold in the language of *differential forms*, which greatly simplifies the matter.

From now on we will thus *automatically* understand the *non-degenerate* case when speaking about the Poisson tensor.

Our coordinate-free way of expressing the Hamilton equations in the sense of problem (14.1.1) was based on the particular coordinate presentation of the Poisson tensor in the form $\mathcal{P} = \frac{\partial}{\partial p_a} \wedge \frac{\partial}{\partial q^a}$. In Section 14.2 we will learn, however, that the defining properties of the tensor (non-degenerate bivector field, which satisfies the differential identity from (14.1.8)) already guarantee the possibility of just this local presentation (it is a "canonical form" of a non-degenerate Poisson tensor). This means that the study of the dynamics given locally by the system of the Hamilton equations

$$\dot{q}^a = \frac{\partial H}{\partial p_a} \qquad \dot{p}_a = -\frac{\partial H}{\partial q^a} \qquad a = 1, \ldots, n$$

is the same thing as the study of the dynamics given in a global and coordinate-free form by

$$\dot{\gamma} = \zeta_H \equiv \mathcal{P}(dH, \cdot)$$

where $\mathcal{P}$ is a non-degenerate Poisson tensor on a manifold M and H is a distinguished function on this manifold, the *Hamiltonian*.

Now let us have a look at how we can pass to the "mirror image" of the (non-degenerate) Poisson dynamics, which is the *symplectic* dynamics.

14.1.2 Define on the phase space $\mathbb{R}^{2n}[q^a, p_a] \equiv \mathbb{R}^{2n}[z^i]$ the tensor field ω of type $\binom{0}{2}$, whose matrix is (up to the sign) inverse to the matrix $\mathcal{P}^{ij}$

$$\mathcal{P} \circ \omega = -\hat{1} \qquad \text{i.e.} \qquad \mathcal{P}^{ik}\omega_{kj} := -\delta^i_j$$

$$\Rightarrow \qquad \omega_{ij}(z) = \begin{pmatrix} \mathbb{O}_n & -\mathbb{I}_n \\ \mathbb{I}_n & \mathbb{O}_n \end{pmatrix} = -\omega_{ji}(z)$$

Check that

(i) this matrix happens to be antisymmetric, $\omega_{ij} = -\omega_{ji}$, so that actually ω is a *2-form*
(ii) in the coordinates z^i and (q^a, p_a) it reads

$$\omega = \frac{1}{2}\omega_{ij}\, dz^i \wedge dz^j = dp_a \wedge dq^a$$

(iii) the form ω is *closed* ($d\omega = 0$), even *exact* ($\omega = d\theta$).

Hint: (iii) $\omega \equiv dp_a \wedge dq^a = d(p_a dq^a)$. □

[248] The general case contains in a sense a family of these special situations. We will see an illustration of a degenerate $\mathcal{P}$ in problem (14.6.8).

[14.1.3] Define (see also (5.8.15)) the concepts of the *rank of a 2-form* $\alpha \in \Lambda^2 L^*$ and the *rank of a bivector* $b \in \Lambda^2 L$ to be the ranks of the linear maps

$$\begin{array}{llll} \hat{\alpha} : L \to L^* & v \mapsto \alpha(v, \cdot\,) \equiv i_v \alpha & \text{i.e.} & v^i \mapsto v^j \alpha_{ji} \\ \hat{b} : L^* \to L & \beta \mapsto b(\beta, \cdot\,) & \text{i.e.} & \beta_i \mapsto \beta_j b^{ji} \end{array}$$

Check that the bivector $\mathcal{P}$ from (14.1.1) as well as the 2-form ω from (14.1.2) happen to be *non-degenerate*, i.e. they have (at each point) *maximum* rank ($= 2n$).

Hint: the matrices of their components are (at each point) non-singular; see (2.4.17) and (5.8.14). □

• An arbitrary closed and non-degenerate 2-form ω on a manifold M is called a *symplectic form* and the pair (M, ω) is a *symplectic manifold*. If ω is exact, it is called an *exact symplectic form*. In this book we will study in more detail two inexhaustible reservoirs of symplectic manifolds. The first of them is provided by *cotangent bundles* T^*M of *arbitrary* manifolds M (see (17.6.7); we mentioned this already at the beginning of this section). The second class is given by *coadjoint orbits* of an arbitrary Lie group G, see (14.6.3).

In many respects the symplectic form ω resembles a metric tensor g on M. Both of them are non-degenerate tensors of type $\binom{0}{2}$. They differ, however, in that ω satisfies the differential identity $d\omega = 0$ as well as in the symmetry properties: the metric tensor is symmetric ($g_{ij} = g_{ji}$), whereas the symplectic form is *anti*symmetric ($\omega_{ij} = -\omega_{ji}$). These differences result in a considerable dissimilarity of some properties of these structures. As a trivial example let us mention the following simple observation.

[14.1.4] A symplectic form may live only on an even-dimensional manifold.

Hint: the fact from linear algebra: non-singular antisymmetric matrices exist only in even-dimensional space (for $n \times n$ matrices $A^{\mathrm{T}} = -A \Rightarrow \det A^{\mathrm{T}} \equiv \det A = (-1)^n \det A$). □

• On the other hand, there are also properties of the structures under consideration where the symmetry/antisymmetry of the corresponding tensors does not matter. For example, the non-degeneracy alone (shared by both ω and $\mathcal{P}$) is enough (see the text before (2.4.15)) for introducing the operations of lowering and raising of indices $\flat_\omega$ and $\sharp_{\mathcal{P}}$ just as was done with the help of the metric tensor.

[14.1.5] Let (M, ω) be a symplectic manifold and $\mathcal{P}$ the corresponding (non-degenerate) Poisson tensor (so that $\mathcal{P} \circ \omega = -\hat{1}$, i.e. $\mathcal{P}^{ik}\omega_{kj} = -\delta^i_j$). Define the maps (the lowering and raising of indices)

$$\begin{array}{llll} \flat_\omega : \mathfrak{X}(M) \to \Omega^1(M) & V \mapsto \omega(\,\cdot\,, V) & \text{i.e.} & V^i \mapsto \omega_{ij} V^j \\ \sharp_{\mathcal{P}} : \Omega^1(M) \to \mathfrak{X}(M) & \alpha \mapsto \mathcal{P}(\alpha, \cdot\,) & \text{i.e.} & \alpha_i \mapsto \alpha_j \mathcal{P}^{ji} \end{array}$$

Check that

(i) they are canonical isomorphisms, inverse to each other (they enable the canonical identification of vector and covector fields)

(ii) the lowering of an index by ω may also be expressed in terms of the *interior product* as

$$\flat_\omega V = -i_V \omega$$ □

• An especially important position in the geometry on symplectic manifolds is occupied by the vector fields, which we obtain by raising the index[249] on the *gradient* df of a function f. It turns out that by this construction we get just the *Hamiltonian fields* ζ_f generated by the function f which we already defined in terms of the Poisson tensor (recall that the dynamics behind the Hamilton equations is generated by the flow of the Hamiltonian field ζ_H); the collection of all Hamiltonian fields on M will be denoted by Ham (M). Let us study in more detail some of the most important general properties shared by all of these remarkable vector fields.

14.1.6 Let $\zeta_f \in \text{Ham}\,(M)$, so that it is a Hamiltonian field generated by the function f. Check that

(i) the following definitions turn out to be equivalent:

$$i_{\zeta_f}\omega = -df \quad \Leftrightarrow \quad \zeta_f = \mathcal{P}(df,\ \cdot\ ) \equiv \sharp_{\mathcal{P}}\, df$$

(ii) there holds

$$\zeta_{(f+\text{const})} = \zeta_f$$

(iii) Hamiltonian fields may also be regarded as analogues of the Killing vectors from Riemannian geometry, since they preserve ω in just the same way as Killing vectors preserve g

$$\mathcal{L}_{\zeta_f}\omega = 0$$

(iv) the collection of all Hamiltonian fields is closed with respect to linear combinations (over $\mathbb{R}$) as well as the commutator; namely,

$$\zeta_f + \lambda\zeta_g = \zeta_{f+\lambda g} \qquad [\zeta_f, \zeta_g] = \zeta_{\{f,g\}}$$

so that they constitute an (∞-dimensional) *Lie algebra* Ham $(M) \subset \mathfrak{X}(M)$.

Hint: (iii) (6.2.8); (iv) making use of (6.2.9) and the preceding items here we find

$$i_{[\zeta_f,\zeta_g]}\omega = \mathcal{L}_{\zeta_f} i_{\zeta_g}\omega - i_{\zeta_g}\mathcal{L}_{\zeta_f}\omega = d(i_{\zeta_f} i_{\zeta_g}\omega) = \dots (14.1.8) \dots = -d\{f, g\}$$ □

• We learned that Hamiltonian fields preserve the symplectic structure in the sense of $\mathcal{L}_{\zeta_f}\omega = 0$. The vector fields with this property will be called *symplectic fields* and denoted by Symp (M)

$$W \in \text{Symp}\,(M) \quad \Leftrightarrow \quad \mathcal{L}_W\omega = 0$$

[249] It may thus be regarded as a "symplectic" counterpart of the standard (see Section 2.6) "metric" object, the vector field $\text{grad} f \equiv \nabla f := \sharp_g\, df$.

14.1.7 Check that

(i) Symp (M) is a Lie algebra and

$$\text{Ham}(M) \subset \text{Symp}(M) \subset \mathfrak{X}(M)$$

where $\subset$ means to be a Lie subalgebra, here

(ii) locally Ham $(\mathcal{U})$ = Symp $(\mathcal{U})$, so that the difference (if any) may occur only at the global level (symplectic = at least locally Hamiltonian)

(iii) there holds

$$[\text{Symp}(M), \text{Symp}(M)] \subset \text{Ham}(M)$$

i.e. the commutator of two (possibly only) symplectic fields is already a Hamiltonian field ($\Rightarrow$ Ham (M) is an *ideal* in Symp (M)).

Hint: (i) (4.3.8); (ii) $\mathcal{L}_W\omega = d(i_W\omega)$; (iii) let $V, W \in \text{Symp}(M)$, then $i_{[V,W]}\omega = \dots$ (6.2.9), (6.2.8) $\dots = -d(\omega(V, W))$ so that $[V, W] = \zeta_{\omega(V,W)} \in \text{Ham}(M)$. □

• Now we turn our attention to the Poisson bracket and its relation to the Poisson tensor $\mathcal{P}$ and the symplectic form ω.

14.1.8 Check that

(i) all the following coordinate-free definitions of the Poisson bracket of the functions f, g are equivalent to each other

$$\{f, g\} \overset{1}{=} \omega(\zeta_f, \zeta_g) \overset{2}{=} \mathcal{P}(df, dg) \overset{3}{=} \zeta_f g = -\zeta_g f$$

$$\omega^n\{f, g\} \overset{4}{=} n\omega^{(n-1)} \wedge df \wedge dg$$

(the kth power of a form is its k-fold *exterior* product, see (5.6.9))

(ii) when expressed in the language of the symplectic form ω, the Jacobi identity turns out to be equivalent to its *closedness*

$$\{f, \{g, h\}\} + \{h, \{f, g\}\} + \{g, \{h, f\}\} = 0 \quad \Leftrightarrow \quad d\omega = 0$$

(iii) when expressed in the language of the Poisson tensor $\mathcal{P}$, the Jacobi identity turns out to be equivalent to its *invariance* with respect to an *arbitrary Hamiltonian* field[250]

$$\{f, \{g, h\}\} + \{h, \{f, g\}\} + \{g, \{h, f\}\} = 0 \quad \Leftrightarrow \quad \mathcal{L}_{\zeta_\psi}\mathcal{P} = 0, \ \psi \in \mathcal{F}(M)$$

and in components this gives the differential identity

$$\mathcal{L}_{\zeta_f}\mathcal{P} = 0 \ \text{ for all } f \in \mathcal{F}(M) \quad \Leftrightarrow \quad \mathcal{P}^{r[i}\mathcal{P}^{jk]}{}_{,r} = 0$$

So this is the identity (mentioned above) which each Poisson tensor is to satisfy.

Hint: (i) $\{f, g\}\omega^n = (i_{\zeta_g} i_{\zeta_f}\omega) \wedge \omega^n = \dots$, (5.4.2); (ii) write down explicitly the expression $d\omega(\zeta_f, \zeta_g, \zeta_h)$ by means of the Cartan formula (6.2.13) and show that

$$d\omega(\zeta_f, \zeta_g, \zeta_h) = 2(\{f, \{g, h\}\} + \{h, \{f, g\}\} + \{g, \{h, f\}\})$$

[250] See also (14.1.10).

Explain then that vanishing of the form on arbitrary Hamiltonian fields already implies its vanishing as such (as a form); (iii) $0 = \mathcal{L}_{\zeta_\psi}(\mathcal{P} \circ \omega) = (\mathcal{L}_{\zeta_\psi}\mathcal{P}) \circ \omega + \mathcal{P} \circ (\mathcal{L}_{\zeta_\psi}\omega) = (\mathcal{L}_{\zeta_\psi}\mathcal{P}) \circ \omega + \mathcal{P} \circ (i_{\zeta_\psi} d\omega + d i_{\zeta_\psi}\omega) = (\mathcal{L}_{\zeta_\psi}\mathcal{P}) \circ \omega + \mathcal{P} \circ (i_{\zeta_\psi} d\omega)$ so that $(\mathcal{L}_{\zeta_\psi}\mathcal{P}) \circ \omega = -\mathcal{P} \circ (i_{\zeta_\psi} d\omega)$; because of the non-degeneracy of $\mathcal{P}$ and ω we then get

$$\mathcal{L}_{\zeta_\psi}\mathcal{P} = 0 \quad \Leftrightarrow \quad i_{\zeta_\psi} d\omega = 0 \quad \Leftrightarrow \quad d\omega = 0$$ □

• The Poisson bracket adds an extra structure to the (*associative*) algebra of functions $\mathcal{F}(M)$ on a symplectic manifold, namely the structure of a *Lie* algebra. We get a combined *algebra of observables* of the classical mechanics $\mathcal{A}(M)$ (the manifold M serves as a *phase space* of the classical mechanics). Its elements, the *observables*, are the functions $f \in \mathcal{F}(M)$ on the phase space; one can form their linear combinations as well as pointwise products (⇒ so far it is an associative algebra), but also their Poisson brackets (⇒ a Lie algebra).

The name "observable" for a function on a phase space corresponds to the interpretation that these functions represent (in classical mechanics) the objects of the theory, which correspond to measurable quantities and enable a comparison with the results of real measurements. The measurements are performed on the "states," which are in turn represented by the *points* of the phase space M in the theory. The prediction of the theory is that if an observable f is measured in the state $p \in M$, the result of the measurement will be the *number* $f(p)$, the *value* of f at the point p.

As an example, consider as a physical system a single point mass moving in the space $\mathbb{R}^3$. Then we associate with it the phase space $M = \mathbb{R}^6[\mathbf{r}, \mathbf{p}]$. A possible state is (say) the point $(\mathbf{r}, \mathbf{p}) = (\mathbf{0}, \mathbf{0})$ (the point stands still at the origin). If we intend to measure (say) the z-component of the angular momentum in this state, we should first introduce an observable $f = L_z = xp_y - yp_x$ (being a function on the phase space) for this quantity and then evaluate f at the point $(\mathbf{r}, \mathbf{p}) = (\mathbf{0}, \mathbf{0})$. We obtain $L_z(\mathbf{0}, \mathbf{0}) = 0$. According to the theory we will measure the number 0.

Let us remark that the points of the phase space actually correspond to so-called *pure states*. In (classical) *statistical* mechanics it is convenient to consider more general states, known as *mixed states*. These states are represented by the "probabilistic measures" on M (we know the positions and momenta only with some probabilities) and the prediction of the theory is that the result of the measurement of the quantity represented by f will be the *integral* of f in the sense of the measure which corresponds to the state. The pure states are the special cases for which the measure is "concentrated at the point p" – then the integral reduces just to the evaluation of f in p.

14.1.9 Consider the algebra of observables of the classical mechanics $\mathcal{A}(M)$. Check that

(i) the two "products" $\mathcal{A}(M) \times \mathcal{A}(M) \to \mathcal{A}(M)$ involved are interconnected by the identity

$$\{f, gh\} = \{f, g\}h + g\{f, h\}$$

(ii) the prescription

$$\zeta : \mathcal{A}(M) \to \text{Ham}\,(M) \quad f \mapsto \zeta_f$$

is a homomorphism of Lie algebras, its kernel being constituted by the constant functions on M.

Hint: (i) $\{f, \cdot\} = \zeta_f(\cdot) \Rightarrow$ it is a vector field, i.e. a *derivation* of the (associative) algebra $\mathcal{F}(M)$; (ii) see (14.1.6). □

14.1.10 Let $\mathcal{A}(M)$ be the algebra of observables of classical mechanics. Since its elements (observables) are the *functions* on the phase space M, there is a natural action of the group of diffeomorphisms on the algebra ($f \mapsto \Phi^* f$). Check that

(i) the structure of the algebra is preserved just by the *symplectomorphisms* of (M, ω), i.e. by such diffeomorphisms of M to itself, which preserve the symplectic form ω (or equivalently the Poisson tensor $\mathcal{P}$)

$$\Phi^* \omega = \omega$$

(ii) the *flows* of such transformations are generated by the symplectic (in particular, by the Hamiltonian) fields

(iii) the action of the flow of a Hamiltonian field on the algebra $\mathcal{A}(M)$

$$U_t^f : \mathcal{A}(M) \to \mathcal{A}(M) \qquad U_t^f := \left(\Phi_t^f\right)^* \qquad \Phi_t^f \leftrightarrow \zeta_f, \ f \in \mathcal{A}(M)$$

can also be expressed in the form of the series

$$U_t^f g = g + t\{f, g\} + \frac{t^2}{2!}\{f, \{f, g\}\} + \frac{t^3}{3!}\{f, \{f, \{f, g\}\}\} + \cdots$$

(iv) the *Jacobi identity* for the Poisson bracket is just the infinitesimal version of the condition that the Poisson bracket (of two arbitrary functions) is *preserved* by the flow of an arbitrary Hamiltonian field, i.e. of the condition

$$U_t^f \{g, h\} = \left\{U_t^f g, U_t^f h\right\} \qquad f, g, h \in \mathcal{A}(M)$$

(v) the map

$$U_t^f : \mathcal{A}(M) \to \mathcal{A}(M)$$

is for each t an *automorphism* of the algebra of observables $\mathcal{A}(M)$ (it preserves its linear structure as well as *both products*) and the prescription $t \mapsto U_t^f$ is the one-parameter group of such automorphisms.

Hint: (i) $\Phi^*\{f, g\} \equiv \Phi^*(\mathcal{P}(df, dg)) = (\Phi^*\mathcal{P})(d\Phi^* f, d\Phi^* g) \overset{!}{=} \mathcal{P}(d\Phi^* f, d\Phi^* g)$ so that $\Phi^*\mathcal{P} \overset{!}{=} \mathcal{P}$ and consequently $\Phi^*\omega \overset{!}{=} \omega$; (ii) in the standard way $\Phi_t^*\omega \overset{!}{=} \omega \ \Rightarrow \mathcal{L}_W \omega = 0$; (iii) by definition $\frac{d}{dt}\big|_0 U_t^f = \mathcal{L}_{\zeta_f}$, (4.4.2) and $\mathcal{L}_{\zeta_f} g \equiv \zeta_f g = \{f, g\}$; (iv) the differentiation of $U_t^f\{g, h\} = \{U_t^f g, U_t^f h\}$ with respect to t in $t = 0$ gives $\mathcal{L}_{\zeta_f}\{g, h\} = \{\mathcal{L}_{\zeta_f} g, h\} + \{g, \mathcal{L}_{\zeta_f} h\}$; (v) preserving the pointwise product and linear combinations is trivial (this holds for each Φ^*), preserving of the Poisson bracket solves item (iv). □

• From this new point of view (and using a new terminology) the situation connected with the dynamics may be summarized as follows. Behind the Hamilton equations there is a vector field of a very special structure on a very special manifold, namely the Hamiltonian field ζ_H generated by a Hamiltonian (function) H on a symplectic manifold (M, ω). The flow Φ_t^H corresponding to the field (called the *phase flow* or also the *Hamiltonian flow*) moves the points of M, which is interpreted as the *time development of the states* of the classical mechanics, $p \mapsto \Phi_t^H(p)$. This flow also induces the one-parameter group of maps $U_t^H \equiv (\Phi_t^H)^*$, which acts on the observables and may be interpreted as the *time development of the observables*, $f \mapsto U_t^H f$.

Now, there are two main approaches to the issue of time development. Within the most common one, which corresponds to the *Schrödinger picture* in quantum mechanics, the states undergo a time development whereas the observables do not, $(p, f) \mapsto (\Phi_t^H(p), f)$. There is also a "dual" point of view (which, in turn, corresponds in quantum mechanics to the *Heisenberg picture*), in which states remain still whereas the observables undergo a time development, $(p, f) \mapsto (p, U_t^H f)$. From the definition of the pull-back of functions it is clear that both approaches yield equal predictions concerning the *results of measurements*, since $f(\Phi_t^H(p)) = (U_t^H f)(p)$.

A more general concept than the symplectomorphism is the *symplectic map* between two symplectic manifolds. It is a map f, which is compatible with the symplectic structures involved, i.e. such that

$$f : (M, \omega_M) \to (N, \omega_N) \qquad f^*\omega_N = \omega_M$$

(As a special case we have transformations of a single manifold

$$f : (M, \omega) \to (M, \omega) \qquad f^*\omega = \omega$$

and thus we come back to a *symplectomorphism*.) They are clearly the analogs of the Riemannian concept of isometry, where in general we have $f : (M, g) \to (N, h)$ such that $f^*h = g$ and, in particular, for a single manifold this reduces to $f : (M, g) \to (M, g)$ such that $f^*g = g$. As we already mentioned in (14.1.6), the analogs of Killing vectors from Riemannian geometry (the generators of isometries) are the Hamiltonian fields (the generators of symplectomorphisms). Both types of fields (Killing as well as Hamiltonian) constitute Lie algebras. In Section 4.6 we mentioned that the Lie algebra of Killing fields is always finite-dimensional. In the symplectic case it is just the opposite – the Lie algebra of Hamiltonian fields is always ∞-dimensional.

14.1.11 Check that the Hamiltonian fields indeed constitute an ∞-dimensional Lie algebra.

Hint: according to (14.1.9) they are given as the homomorphic image of the (∞-dimensional) Lie algebra $(\mathcal{F}(M), \{\,\cdot\,,\,\cdot\,\})$, the kernel being given (only) by constant functions. Check that

$$c_1\zeta_f + c_2\zeta_g = 0 \;\Rightarrow\; c_1 f + c_2 g = \text{ constant } \;\Rightarrow\; f, g, 1 \text{ are linearly dependent}$$

so that we can construct an arbitrary number of linearly independent Hamiltonian fields (or $\flat_\omega$ gives a bijection $\zeta_f \leftrightarrow df$, so that there are "just so many" independent Hamiltonian fields as we have independent differentials df). □

• We close the section by showing a useful way in which the Poisson bracket of arbitrary functions may be written in terms of the Poisson bracket of coordinates on M.

14.1.12 Let $\mathcal{P}$ be a Poisson tensor and $\{\cdot, \cdot\}$ the corresponding Poisson bracket. Check that

(i) in arbitrary coordinates we have

$$\mathcal{P}^{ij} = \{x^i, x^j\} \qquad \{f, g\} = (\partial_i f)\{x^i, x^j\}(\partial_j g)$$

(ii) in particular, in canonical coordinates it gives the well-known expressions

$$\{q^a, q^b\} = 0 = \{p_a, p_b\} \qquad \{p_a, q^b\} = \delta^b_a \qquad \{f, g\} = \frac{\partial f}{\partial p_a}\frac{\partial g}{\partial q^a} - \frac{\partial f}{\partial q^a}\frac{\partial g}{\partial p_a}$$

Hint: $\{f, g\} = \mathcal{P}(df, dg)$. □

14.2 Darboux theorem, canonical transformations and symplectomorphisms

• Often it is convenient to use the coordinates tailored to the structure under consideration, since some facts, which are obscure in general coordinates, may become evident in these *adapted coordinates*.[251] It is fairly instructive to compare from this point of view the possibilities of the optimal choice of coordinates for a metric tensor and a symplectic form. By an appropriate choice of the *frame field* (possibly not coordinate) one can transform an arbitrary metric tensor in some domain (say, in a finite neighborhood of each point) to the canonical form

$$g(e_a, e_b) = \eta_{ab} \equiv \text{diag}\,(1, \dots, 1, -1, \dots, -1)$$

(the non-trivial part of the statement being a well-known result from linear *algebra*). One can see easily, however, that it is not possible to reach this form in a *coordinate* frame field in general: there are (only) n levers (the choices of new coordinates as functions of the old ones) to satisfy (as many as) $n(n+1)/2$ conditions (the values of independent components $g_{ij}(x) \equiv g(\partial_i, \partial_j)(x) = \pm 1$ or 0).

However, the situation with the symplectic form is different. It is closed, so that it is locally exact and this is the reason it is actually enough to optimize an object with fewer degrees of freedom – a potential θ (locally $\omega = d\theta$), being *only a* 1*-form*, which has (only) n components, i.e. the same number as the number of levers mentioned above. It should not be surprising, then, that the component matrix of the symplectic form can *always* be brought *in coordinates* (in a finite domain – a finite neighborhood of an arbitrary point) to the canonical form, which is (as we again know from linear algebra, see also (5.6.8)) the form from problem (14.1.2).

All these are of course only estimations obtained by a "rule of thumb," but they are actually confirmed by the following important theorem.

[251] For example, in the *Cartesian* coordinates in E^n the translational invariance of the metric tensor is evident.

14.2.1 The canonical form of a closed 2-form is described by the *Darboux theorem.*[252] It claims that if a *closed 2-form* β on an n-dimensional manifold M has constant rank in a neighborhood of a point P, then there are local coordinates $x^1, \dots, x^n$ such that β looks like

$$\beta = dx^1 \wedge dx^2 + \dots + dx^{2k-1} \wedge dx^{2k} \equiv d(x^1 dx^2 + \dots + x^{2k-1} dx^{2k}) \quad 2k \le n$$

Check that

(i) the rank of this form is $2k$ (thus always *even*)
(ii) if $2k < n$, β is degenerate.

Hint: (i) find the images of the linear map $v \mapsto i_v \beta$ for $v = \partial_i$, $i = 1, \dots, n$; (ii) the vectors ∂_i for $i > 2k$ map to zero $\Rightarrow$ the last $n - 2k$ rows of the matrix β_{ij} vanish. □

- For us the most important consequence of the Darboux theorem is the canonical coordinate presentation of the *symplectic* form.

14.2.2 Let (M, ω) be an $n = 2m$-dimensional symplectic manifold. Check that

(i) in appropriate local coordinates x^i the symplectic form ω may be written as

$$\omega \equiv \frac{1}{2} \omega_{ij}\, dx^i \wedge dx^j = dx^1 \wedge dx^2 + \dots + dx^{2m-1} \wedge dx^{2m}$$
$$\equiv d(x^1\, dx^2 + \dots + x^{2m-1}\, dx^{2m})$$

(ii) the matrix ω_{ij} then looks like

$$\omega_{ij} = \text{diag}\,(\varepsilon, \dots, \varepsilon) \qquad \varepsilon = \begin{pmatrix} 0 & 1 \\ -1 & 0 \end{pmatrix}$$

(so that it is a block-diagonal matrix with 2×2 blocks ε)

(iii) in *other* appropriate local coordinates (p_a, q^a), $a = 1, \dots, m$, ω may be written as in (14.1.2), i.e.

$$\omega = dp_a \wedge dq^a$$

The coordinates (q^a, p_a), in which the symplectic form ω looks like this and whose existence is guaranteed by the Darboux theorem, are called the *canonical coordinates*[253]

(iv) in canonical coordinates the corresponding Poisson tensor is

$$\mathcal{P} \equiv -\omega^{-1} = \frac{\partial}{\partial p_a} \wedge \frac{\partial}{\partial q^a} := \frac{\partial}{\partial p_a} \otimes \frac{\partial}{\partial q^a} - \frac{\partial}{\partial q^a} \otimes \frac{\partial}{\partial p_a}$$

(v) in canonical coordinates the Hamiltonian field is

$$\zeta_f = (\partial_j f)\mathcal{P}^{ji} \partial_i \equiv \frac{\partial f}{\partial p_a} \frac{\partial}{\partial q^a} - \frac{\partial f}{\partial q^a} \frac{\partial}{\partial p_a}$$

[252] Its proof may be based, for example, on Frobenius' theorem concerning integrable distributions, which we encounter in the treatment of connection theory (19.3.6).

[253] Our route to the revelation of the symplectic structure led just through the canonical coordinates (14.1.2), which were, however, *global* in the space under consideration, thus making the situation rather specific. Here it is asserted that *locally* such coordinates may be used even in the general case.

(vi) in canonical coordinates the equations for the integral curve of a Hamiltonian field ζ_f are

$$\dot{\gamma} = \zeta_f \quad \Leftrightarrow \quad \dot{q}^a = \frac{\partial f}{\partial p_a} \qquad \dot{p}_a = -\frac{\partial f}{\partial q^a}$$

(vii) Hamilton's equations may be presented geometrically as

$$\dot{\gamma} = \zeta_H \qquad i_{\zeta_H}\omega = -dH$$

Hint: (i) see (14.2.1), ω is non-degenerate; (iii) just rename the coordinates $x^{2a-1} = p_a, x^{2a} = q^a, a = 1, \dots, m$; (vii) see (14.1.1) and (14.1.6). □

14.2.3 The sphere (S^2, ω) along with a common (metric) volume form (6.3.9) *is a symplectic* manifold (check). Find out

(i) whether it is also exact symplectic and whether the coordinates ϑ, φ are canonical
(ii) what the Hamilton equations for a general Hamiltonian $H(\vartheta, \varphi)$ in the variables ϑ, φ look like
(iii) what the Hamiltonian $H(\vartheta, \varphi)$ looks like, which generates as the time development the uniform circulation along the parallel line directed eastwards

$$\varphi(t) = \varphi_0 + t \qquad \vartheta(t) = \vartheta_0$$

Hint: (i) see (9.1.1); you might be confused by the coordinate result (6.3.10) $\omega = \sin\vartheta\, d\vartheta \wedge d\varphi = d(-\cos\vartheta\, d\varphi)$; remember, however, that this does not work *globally* (!); is $\omega = \pm d\vartheta \wedge d\varphi$?; check that $p = -\cos\vartheta, q = \varphi$ *are* canonical; (ii) either solve in canonical coordinates and then rewrite the result in the coordinates demanded here, or (better): for a general $H(\vartheta, \varphi)$ find the field i_{ζ_H} from the equation $i_{\zeta_H}\omega = -dH$ and write down the equations for its integral curves, i.e. express in coordinates ϑ, φ the equations from the last item in (14.2.2); (iii) we need $\zeta_H = \partial_\varphi$; plug this into $i_{\zeta_H}\omega = -dH$ and find H. ((ii) $\dot{\vartheta}\sin\vartheta = -\partial_\varphi H,\ \dot{\varphi}\sin\vartheta = \partial_\vartheta H$; (iii) $H(\vartheta, \varphi) = -\cos\vartheta$.) □

- We see that in order to formulate the Hamiltonian dynamics we need just two things:

1. a *phase space*, which is a symplectic manifold M
2. an exact 1-form dH on M.

A triple (M, ω, dH) is therefore called a *Hamiltonian system* and it enables one to introduce the standard time development $x \mapsto x(t) := \Phi_t^H(x)$, where Φ_t^H is the flow generated by the field ζ_H. It turns out that the time development given just by the *Hamiltonian* field has some remarkable special properties (if compared with a general field), which hold universally, being independent of the particular choice of Hamiltonian.

Recall that the Hamilton equations have their well-known ("canonical") form

$$\dot{q}^a = \frac{\partial H}{\partial p_a} \qquad \dot{p}_a = -\frac{\partial H}{\partial q^a} \quad a = 1, \dots, n$$

only in canonical coordinates; in general coordinates they look like

$$\dot{x}^i = (\partial_j H(x))\mathcal{P}^{ji}(x)$$

where the matrix $\mathcal{P}^{ij}(x)$ does not have the simple structure from (14.1.2) (depending non-trivially on x in general). On the other hand, the canonical coordinates are far from being unique and in the textbooks of classical mechanics coordinate transformations are studied in which we pass from one system of canonical coordinates to another such system – the *canonical transformations*. Such transformations may thus be characterized either as the coordinate transformations

$$f : (q^a, p_a) \mapsto (Q^a(q, p), P_a(q, p))$$

preserving the canonical form of the Hamilton equations,

$$\dot{q}^a = \frac{\partial H}{\partial p_a} \quad \dot{p}_a = -\frac{\partial H}{\partial q^a} \quad \Rightarrow \quad \dot{Q}^a = \frac{\partial \mathcal{H}}{\partial P_a} \quad \dot{P}_a = -\frac{\partial \mathcal{H}}{\partial Q^a}$$

where $H(q, p) = \mathcal{H}(Q(q, p), P(q, p))$ (i.e. $H = f^*\mathcal{H}$), or equivalently (and more simply) as the coordinate transformations preserving the canonical form of the *symplectic form*:

$$f : (q^a, p_a) \mapsto (Q^a(q, p), P_a(q, p)) \qquad \omega = dp_a \wedge dq^a = f^*(dP_a \wedge dQ^a)$$

Usually the pull-back f^* is omitted so that the condition looks simply like

$$dp_a \wedge dq^a \stackrel{!}{=} dP_a \wedge dQ^a$$

There are two standard techniques of finding and describing the canonical transformations. The first one operates with the concept of a *generating function* of the transformation, the second one is based on the concept of the *generator* of the transformation.

14.2.4 Let (q^a, p_a) as well as (Q^a, P_a) be canonical coordinates in a domain $\mathcal{U}$. From the requirement that the transformation of coordinates $(q^a, p_a) \mapsto (Q^a(q, p), P_a(q, p))$ is canonical

$$dp_a \wedge dq^a \stackrel{!}{=} dP_a \wedge dQ^a \equiv dP_a(q, p) \wedge dQ^a(q, p)$$

there follows

$$d(p_a dq^a - P_a dQ^a) \equiv d\sigma = 0 \quad \Rightarrow \quad \sigma = d\Phi \quad \Phi \in \mathcal{F}(\mathcal{U})$$

The function Φ on $\mathcal{U}$ (which is given up to an additive constant) fully characterizes the canonical transformation and it is called its *generating function*. Check that

(i) if it is possible to choose as independent coordinates on $\mathcal{U}$ the set (q^a, Q^a),[254] then the formulas describing the transformation are (implicitly) given by

$$p_a = \frac{\partial \Phi}{\partial q^a} \qquad P_a = -\frac{\partial \Phi}{\partial Q^a}$$

[254] Which set of $2n$ functions is independent (so that we may choose them as coordinates) depends on the transformation itself. For example, if we take the *identity* transformation, the set mentioned above is evidently *not suitable*.

(ii) if it is possible to choose as independent coordinates on $\mathcal{U}$ the set (q^a, P_a), the corresponding relations are

$$p_a = \frac{\partial \hat{\Phi}}{\partial q^a} \qquad Q^a = \frac{\partial \hat{\Phi}}{\partial P_a} \qquad \hat{\Phi} := \Phi + P_a Q^a$$

(iii) the change of coordinates

$$(q^a, p_a) \mapsto (Q^a(q, p), P_a(q, p)) = (p_a, -q^a)$$

is canonical; find its generating function

(iv) the identity transformation

$$(q^a, p_a) \mapsto (Q^a(q, p), P_a(q, p)) = (q^a, p_a)$$

is canonical; find its generating function.

Hint: (i) write down $d\Phi$ in these coordinates; (ii) $p_a\, dq^a - P_a\, dQ^a = p_a\, dq^a - d(P_a Q^a) + Q^a\, dP_a$; (iii) $\Phi(q^a, Q^a) = q^a Q^a$; (iv) $\hat{\Phi}(q^a, P_a) = q^a P_a$. □

• The second approach to canonical transformations may be regarded as a "symplectic version" of the result (4.6.26) from Riemannian geometry. Recall that if there is a flow Φ_t of *isometries* of (M, g), we can introduce new coordinates $x_t^i := \Phi_t^* x^i$ in which the functional expression of the components of the metric tensor will be the same as it was in the initial coordinates. If, in particular, the initial coordinates were in some sense *canonical* (tailored to the metric, like the Cartesian coordinates are for the Euclidean metric), the new coordinates will be canonical as well.

The same technique may also be repeated here. If we have a flow Φ_t which preserves the symplectic form, the transformation to the coordinates $x_t^i := \Phi_t^* x^i$ preserves the functional form of its components and, in particular, also its canonical form. This means that $x^i \mapsto x_t^i := \Phi_t^* x^i$ is a canonical transformation in the sense understood here. Now realize that it is extremely simple to find such a flow Φ_t, for we learned in (14.1.6) that the symplectic form is preserved by the flow generated by an *arbitrary Hamiltonian* field ζ_f. Put another way, *each function* on the phase space induces a one-parameter class of canonical transformations.

14.2.5 Let ζ_f be a Hamiltonian field, $\Phi_t \leftrightarrow \zeta_f$ its flow and let (q^a, p_a) be canonical coordinates. Check that if the coordinate presentation of the flow is

$$(q^a, p_a) \mapsto (q^a(t), p_a(t))$$

i.e. if $(q^a(t), p_a(t))$ are the solutions of the Hamilton equations with the initial conditions $(q^a(0), p_a(0)) = (q^a, p_a)$ and the Hamiltonian f, then the change of coordinates

$$(q^a, p_a) \mapsto (Q^a(q, p; t) := q^a(t), P_a(q, p; t) := p_a(t))$$

(t being fixed, but arbitrary) is a canonical transformation, so that we get a *one-parameter class* of canonical transformations, starting with the identity (for $t = 0$) and little by little (as t grows) differing increasingly from it.

Hint: we have $(q^a(t), p_a(t)) = (q^a \circ \Phi_t, p_a \circ \Phi_t) \equiv (\Phi_t^* q^a, \Phi_t^* p_a)$, so that

$$dp_a \wedge dq^a = \Phi_t^*(dp_a \wedge dq^a) = d(\Phi_t^* p_a) \wedge d(\Phi_t^* q^a) = dp_a(t) \wedge dq^a(t) \equiv dP_a \wedge dQ^a$$

$\Rightarrow Q^a, P_a$ *indeed are* the canonical coordinates. □

14.2.6 Consider the (simplest possible) phase space $\mathbb{R}^2[q, p]$ and choose the function $f(q, p)$ as $f(q, p) = qp$.

(i) Find the Hamiltonian field ζ_f and its flow Φ_t
(ii) check explicitly that $(q, p) \equiv (q(0), p(0)) \mapsto (Q, P) \equiv (q(t), p(t))$ (for any fixed t) is indeed a canonical transformation
(iii) find the generating function of this canonical transformation.

Hint: $(Q(q, p; t), P(q, p; t)) = \cdots = (e^t q, e^{-t} p) \Rightarrow dP \wedge dQ = (e^{-t} dp) \wedge (e^t dq) = (e^{-t} e^t) dp \wedge dq = dp \wedge dq$; (iii) $\hat{\Phi}(q, P) = e^t q P$. □

• So we see that any function f on a phase space (M, ω) induces a one-parameter class of canonical transformations. It is given by the flow Φ_t of the Hamiltonian field ζ_f corresponding to the function f.

14.2.7 Check that the time dependence of the coordinates resulting from the solution of the Hamilton equations may be regarded as a sequence (one-parameter family, the parameter being the time) of canonical transformations.

Hint: see (14.1.12) and (14.2.5), $f = H$. □

14.3 Poincaré–Cartan integral invariants and Liouville's theorem

• The geometrical formulation (14.2.2) of the Hamilton equations enables one to handle in an elegant and simple way some issues which turn out to be much more complicated without geometry. The Poincaré–Cartan integral invariants may serve as a nice example. As the nomenclature indicates, they describe some integrals which are preserved under some transformations. Since both the concepts (the integral as well as the invariance) are standardly and successfully treated by geometry, we should not be surprised to learn that this stuff is indeed mastered effectively and easily by geometrical methods.

Let us first introduce some important concepts. Recall (see Section 4.2) that a tensor field T is said to be invariant (or Lie constant) with respect to a vector field W (and its flow $\Phi_t \leftrightarrow W$) if

$$\mathcal{L}_W T = 0 \quad \leftrightarrow \quad \Phi_t^* T = T$$

This is clearly true also in the particular case of forms: an *invariant form* (with respect to W or Φ_t) is a form for which

$$\mathcal{L}_W \alpha = 0 \quad \leftrightarrow \quad \Phi_t^* \alpha = \alpha$$

In the case of forms we automatically get, however, some further consequences.

14.3.1 Let the forms α, β be invariant with respect to the field W (i.e. $\mathcal{L}_W\alpha = 0 = \mathcal{L}_W\beta$). Check that then the forms

$$d\alpha \qquad \alpha \wedge \beta \qquad i_W\alpha$$

are also invariant. This means that the space of invariant forms (with respect to W) is closed with respect to some important operations on forms, namely d, $\wedge$ and i_W.

Hint: $i_W\alpha$ from (6.2.9) for $W = V$. □

• A slightly weaker concept is the *relative invariance* of forms: a form τ is said to be relative invariant (with respect to W) if $d\tau$ (rather then τ itself) is invariant ($\mathcal{L}_W(d\tau) \equiv d(\mathcal{L}_W\tau) = 0$).

14.3.2 Check that there holds

$$\{\text{invariant forms}\} \subset \{\text{relative invariant forms}\}$$

Hint: see (14.3.1). □

14.3.3 Check that

(i) the symplectic form ω is invariant with respect to an arbitrary Hamiltonian field ζ_f
(ii) any "exterior power" (i.e. a $2k$-form $\omega^k := \underbrace{\omega \wedge \cdots \wedge \omega}_{k \text{ entries}}$) of the symplectic form ω is invariant with respect to an arbitrary Hamiltonian field ζ_f
(iii) if the symplectic form ω happens to be exact and $\omega = d\theta$, then θ is relative invariant with respect to an arbitrary Hamiltonian field ζ_f.

Hint: see (14.1.10) and (14.3.1). □

• Since forms may be regarded as expressions under the integral sign (see Section 7.1), these properties together with the behavior of integrals with respect to maps of manifolds (see Section 7.8) immediately lead to certain "integral" consequences – integral invariants arise.

14.3.4 Let $\mathcal{D} \subset M$ be a $2k$-dimensional domain on (M, ω), ζ_f a Hamiltonian field, Φ_t its flow and $\Phi_t(\mathcal{D})$ the image of $\mathcal{D}$ with respect to the flow. Define the expressions (the *Poincaré–Cartan integral invariants*)

$$I^k \equiv I^k[\mathcal{D}] := \int_{\mathcal{D}} \underbrace{\omega \wedge \cdots \wedge \omega}_{k \text{ entries}}$$

Prove that they indeed deserve their name; in particular[255] prove that

$$I^k[\Phi_t(\mathcal{D})] = I^k[\mathcal{D}]$$

[255] For the proof of the "Poincaré–Cartan" part use an appropriate source on the history of mathematics; the term "integral" is clear; here we only concentrate on the word "invariant," namely invariant *with respect to the flow* Φ_t.

Hint: according to (7.8.1) and (14.3.3)

$$\int_{\Phi_t(\mathcal{D})} \omega \wedge \cdots \wedge \omega = \int_{\mathcal{D}} \Phi_t^*(\omega \wedge \cdots \wedge \omega) = \int_{\mathcal{D}} \Phi_t^*\omega \wedge \cdots \wedge \Phi_t^*\omega = \int_{\mathcal{D}} \omega \wedge \cdots \wedge \omega.$$

□

14.3.5 Show that the statement about the invariance of the expressions I^k is true, in particular, for the *time development* in the Hamiltonian system (M, ω, H), i.e. that these integrals do not depend on time t.

Hint: according to (14.2.2) for $f = H$ the map Φ_t is just the time development of the phase points (pure states of the system); then $\Phi_t(\mathcal{D})$ is the domain $\mathcal{D}$ time developed by t and the result (14.3.4) in this particular case says that the integral of the form ω^k over a (arbitrary) $2k$-dimensional domain $\mathcal{D}$ gives the same result, as given by the integral over the domain which results from the time development of all the points from $\mathcal{D}$ by t (according to the Hamilton equations). □

- The best known as well as the most important is the last of the invariants, i.e. the case where $k = n$; the corresponding statement is known as Liouville's theorem.

14.3.6 Let (M, ω) be a symplectic manifold, $\dim M = 2n$. Check that

(i) the n-fold product of the form ω (as well as its arbitrary non-zero multiple) defines on M the *volume form*

(ii) in general coordinates $z^1, \ldots, z^{2n}$ there holds

$$\tilde{\Omega}_\omega \equiv \tilde{\Omega} := \underbrace{\omega \wedge \cdots \wedge \omega}_{n \text{ entries}} = n!\,\mathrm{Pf}(\omega_{ij})\, dz^1 \wedge \cdots \wedge dz^{2n}$$
$$= \pm n! \sqrt{|\det \omega_{ij}|}\, dz^1 \wedge \cdots \wedge dz^{2n}$$

(iii) in particular, in canonical coordinates (q^a, p_a) it gives

$$\omega_{ij} \leftrightarrow \begin{pmatrix} 0 & -\mathbb{I} \\ \mathbb{I} & 0 \end{pmatrix} \qquad \mathrm{Pf}(\omega_{ij}) = (-1)^{n(n+1)/2}$$

(the blocks being $n \times n$) so that the form $\tilde{\Omega}$ reads

$$\tilde{\Omega} = (-1)^{n(n+1)/2} n!\, dq^1 \wedge \cdots \wedge dq^n \wedge dp_1 \wedge \cdots \wedge dp_n \equiv (-1)^{n(n+1)/2} n!\, dq\, dp$$

Therefore one usually adopts its appropriate constant multiple, the *Liouville form*

$$\Omega_\omega \equiv \Omega := (-1)^{n(n+1)/2} \frac{1}{n!} \tilde{\Omega} = dq\, dp$$

as *the volume form* on a symplectic manifold and as the *phase volume* of the domain $\mathcal{D} \subset M$ we mean the expression $\int_{\mathcal{D}} \Omega$ (i.e. the volume of the domain $\mathcal{D}$ in the sense of the Liouville volume form).

(iv) *Liouville's theorem* holds: the phase volume of an arbitrary ($2n$-dimensional) domain $\mathcal{D}$ is preserved under the time development Φ_t of the phase points (more generally under the flow of an arbitrary Hamiltonian field ζ_f)[256]

$$\Phi_t \leftrightarrow \zeta_H \quad \Rightarrow \quad \int_{\Phi_t(\mathcal{D})} \Omega = \int_{\mathcal{D}} \Omega$$

(v) any symplectic manifold is orientable.

Hint: (i) the fact that $\omega \wedge \cdots \wedge \omega$ is everywhere non-zero is clear from its coordinate presentation; (ii) see (5.6.8); (iv) see (14.3.4); (v) see (6.3.5). □

• Liouville's theorem about preserving the phase volume may be contemplated also in the broader setting of preserving volumes in general, where we naturally encounter the concept of Ω-divergence.

14.3.7 Consider a manifold M endowed with a distinguished volume form Ω and define the Ω-*divergence* of a vector field V by

$$\mathcal{L}_V \Omega = (\text{div}_\Omega V)\Omega$$

Check that

(i) Ω is invariant with respect to the flow $\Phi_t \leftrightarrow V$ if and only if the Ω-divergence of the field V *vanishes*, i.e.

$$\Phi_t^* \Omega = \Omega \quad \Leftrightarrow \quad \text{div}_\Omega V = 0$$

(ii) for the flows generated by vector fields with *vanishing Ω-divergence* "Liouville's theorem" holds (what is the difference between the figure presented here and that in problem (8.2.2)?)

$$\text{div}_\Omega V = 0 \quad \Rightarrow \quad \text{the volume of } D = \text{the volume of } \Phi_t D \qquad \text{vol } D := \int_D \Omega$$

(iii) if Ω locally looks like $\Omega = f\, dx^1 \wedge \cdots \wedge dx^n$, then

$$\text{div}_\Omega V = \frac{1}{f}(fV^i)_{,i}$$

(iv) an arbitrary *Hamiltonian* field *has* vanishing Ω-divergence for the Liouville volume form $\Omega =$ constant ω^n (and "that's why" Liouville's theorem holds)

(v) for the *metric* volume form $\Omega \equiv \omega_g$ the ω_g-divergence coincides with the "ordinary" divergence (i.e. with $*_g^{-1} d *_g \flat_g V$) as well as with the "covariant" divergence (see (8.2.1) and (15.6.18))

(vi) if σ is a (non-zero) function, then

$$\text{div}_{\sigma\Omega} V = \frac{1}{\sigma}\text{div}_\Omega(\sigma V)$$

(vii) let M be a symplectic manifold, (q^a, p_a) the canonical local coordinates,

$$V = A^a(q, p)\partial/\partial q^a + B_a(q, p)\partial/\partial p_a$$

[256] If we regard the flow Φ_t as the flow of a fluid, then the result says that the fluid is *incompressible*.

Write down the component expression for $\operatorname{div}_\Omega V$ (Ω as in item (iv)) and convince yourself once again that

$$V \text{ is a } \textit{Hamiltonian} \text{ field} \quad \Rightarrow \quad \operatorname{div}_\Omega V = 0$$

and locally also the converse is true.

Hint: (i) see (4.4.2); (ii) see (8.2.3); (iii)

$$\mathcal{L}_V(f\,dx^1 \wedge \cdots \wedge dx^n) = (Vf)\,dx^1 \wedge \cdots \wedge dx^n + f d(Vx^1) \wedge \cdots \wedge dx^n + \cdots$$
$$+ f dx^1 \wedge \cdots \wedge d(Vx^n) = \cdots;$$

(iv) see (14.3.3); (v) see (8.2.1); (vi) see (6.2.14). □

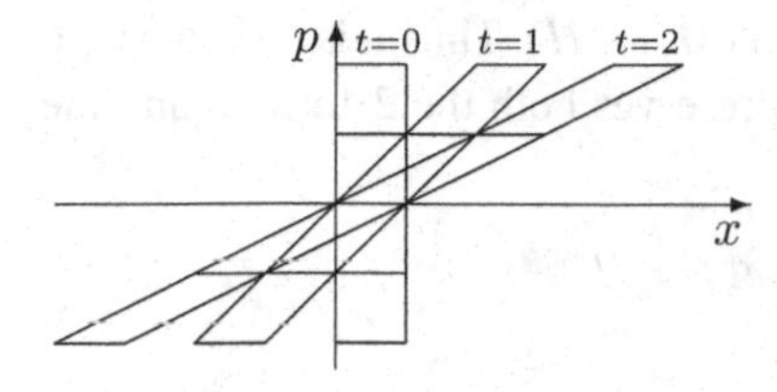

14.3.8 The simplest manifestation of Liouville's theorem is provided by the dynamics in *two-dimensional* phase spaces. Consider three simple systems with a single degree of freedom x (consequently with two-dimensional phase spaces xp), namely the free particle, the linear harmonic oscillator and a particle on which a constant force is applied. Check that

(i) the Liouville volume form Ω_ω reduces here to the expression $dx \wedge dp$ (which coincides up to a sign with the symplectic form ω itself), so that the phase volume of a domain is simply its ordinary area in the xp-plane
(ii) the phase flows Φ_t given by these dynamics explicitly read

free motion	$H(x,p) = p^2/2$	$(x,p) \mapsto (x+pt, p)$
harmonic oscillator	$H(x,p) = (p^2+x^2)/2$	$(x,p) \mapsto (x\cos t + p\sin t, -x\sin t + p\cos t)$
constant force	$H(x,p) = p^2/2 - x$	$(x,p) \mapsto (x+pt+t^2/2, p+t)$

(iii) for the case of free motion the system of unit squares in the phase space maps after 1 and 2 seconds as displayed on the figure (what do the corresponding figures look like for the remaining two cases as well as for the Hamiltonian $H(x,p) = xp$ from problem (14.2.6)?)
(iv) it is clear from the figure that the flow under consideration preserves the areas, i.e. that Liouville's theorem holds here (and the same is true for the remaining figures worked out a moment ago).

Hint: (iii) the formulas for the flows determine the motion of *all* points, in particular (for example) also the vertices of the squares. □

• If we indeed draw the pictures for at least the four two-dimensional dynamics considered above, we see clearly that to preserve *only the areas* (phase volumes) is a considerably weaker requirement than to preserve both the shapes and dimensions; the given figures may be deformed in various ways (the "distances" of individual points may be changed), but always only to the extent that their area after the deformation is the same as it was before it.

14.4 Symmetries and conservation laws

• The useful physics folklore also contains a clever exploitation of the close relation between the symmetries and the conservation laws. In this section we learn how this works in the context of the Hamiltonian system (M, ω, H). (An illuminating point of view to see the same ideas is also provided by Noether's theorem, see (21.6.7) and (21.6.8).)

The Hamiltonian system (M, ω, H) consists, as we see, of three parts and its automorphisms (i.e. symmetries) should preserve according to standard rules all three parts. Preservation of M gives a diffeomorphism (10.1.2) and preservation of (ω, H) means the restriction to such diffeomorphisms whose pull-back does not alter (ω, H).[257] In quite the same manner as we descended to the infinitesimal level when studying isometries and we came to the notion of Killing vectors as generators of *one-parameter groups* of isometries, also here we introduce the vector fields which generate the one-parameter groups of automorphisms of the Hamiltonian system. Such a field[258] is called a *Cartan symmetry* (hereafter in this section CS) of the Hamiltonian system (M, ω, H). Thus it is a vector field V on M such that the corresponding flow $\Phi_t \leftrightarrow V$ preserves both the 2-form ω and the 0-form H

$$\Phi_t^* \omega = \omega \qquad \Phi_t^* H \equiv H \circ \Phi_t = H$$

or infinitesimally

$$\mathcal{L}_V \omega = 0 \qquad \mathcal{L}_V H \equiv VH = 0$$

14.4.1 Check that

(i) for $V \in$ CS the 1-form $i_V \omega$ happens to be closed
(ii) if V, W are CS then their linear combinations (over $\mathbb{R}$) as well as the commutator are CS, too, i.e. Cartan symmetries constitute a *Lie algebra* (subalgebra of the algebra of all vector fields on M)
(iii) the corresponding maps $M \to M$ constitute a group (a subgroup of the group of all diffeomorphisms of M).

Hint: (i) see (6.2.8); (ii) see (4.3.8). □

• The conservation laws are directly related to a particular subclass of Cartan symmetries, called the *exact Cartan symmetries* (hereinafter in this section ECS). They are characterized by the condition that the 1-form $i_V \omega$ is not only closed, but is *exact*.[259]

14.4.2 Consider *exact* Cartan symmetries (ECS). Check that

(i) they are just those *Hamiltonian* fields, with respect to which the Hamiltonian is moreover invariant
(ii) they constitute a Lie algebra which is a subalgebra of the algebra of all Cartan symmetries (ECS $\subset$ CS), moreover it is an *ideal* in this algebra (so that [ECS,CS] $\subset$ ECS).

[257] Compare with the isometries (i.e. automorphisms of (M, g)) – see the text before (4.6.2).
[258] Maybe it would be more natural to call the map Φ_t itself a "symmetry" (rather than its generator V).
[259] As we learned in Chapter 9, *locally* these concepts coincide.

Hint: (i) if $i_V\omega = d(-F_V)$ (the function $-F_V$ being a potential), then according to the definition in (14.1.6) $V = \zeta_{F_V}$; (ii) see (6.2.9). □

• Now we explain what it means when one speaks about a conserved quantity of the Hamiltonian system (M, ω, H). Let $\gamma(t)$ be an arbitrary integral curve of the generator of the time development, the Hamiltonian field ζ_H (a motion along $\gamma(t)$ thus corresponds to the time development of the phase point $\gamma(0) \in M$, or in coordinates a solution of the Hamilton equations). A function F on M is said to be a *conserved quantity*, if it is constant on $\gamma(t)$ (the function $F(\gamma(t))$ does not depend on t, or equivalently $F \circ \Phi_t \equiv \Phi_t^* F = F$).[260]

14.4.3 Check that

(i) for the conserved quantity F there holds

$$\zeta_H F \equiv \{H, F\} = 0$$

(so that F is *in involution*[261] with the Hamiltonian H)

(ii) conserved quantities are closed with respect to the operations in the algebra of the observables $\mathcal{A}(M)$, thus constituting a *subalgebra* of this (combined, see (14.1.7)) algebra

(iii) if F is a conserved quantity, then so is $F +$ constant.

Hint: (i) (14.1.6); (ii) Jacobi identity. □

• We learned that each exact Cartan symmetry V is necessarily a Hamiltonian field, being thus related to a *function* F_V (its generator), which is given up to an additive constant. It turns out that it is just this function, which is the conserved quantity corresponding to the symmetry V. Conversely, if F is a conserved quantity, then the resulting Hamiltonian field ζ_F happens to be an exact Cartan symmetry.

14.4.4 Show that there is a one-to-one correspondence between the *exact Cartan symmetries* and the *conserved quantities*. In particular, that

(i) if V is ECS, then F_V (defined by $i_V\omega = -dF_V$) is a conserved quantity

(ii) if F is a conserved quantity, then ζ_F is ECS.

Hint: (i) $\dot{F}_V \equiv \zeta_H F_V = \langle dF_V, \zeta_H\rangle = -\omega(V, \zeta_H) = -VH = 0$; (ii) $i_{\zeta_F}\omega$ is exact due to ζ_F being Hamiltonian, $\zeta_F H = -\{H, F\} = -\zeta_H F = 0$ due to the conservation of F. □

14.4.5 Check that

(i) the generator of the time development ζ_H is ECS (so that the "shift in time" is a symmetry of *each* Hamiltonian system), the corresponding conserved quantity being the Hamiltonian H

(ii) the Hamiltonian field ζ_f is ECS if and only if $\{f, H\} = 0$ (when f is in involution with the Hamiltonian H). □

[260] On a pragmatic coordinate level it is such a combination of the variables q^a, p_a that *does not* depend on t in spite of the fact that the "components" from which it is "put together" *do*: $F(t) := F(q^a(t), p_a(t)) = $ constant.

[261] We say that two functions f, g on (M, ω) are in involution if they happen to "commute" in the sense of Poisson brackets, i.e. if $\{f, g\} = 0$.

• Symmetries may be used for a construction of the new solutions of the equations of motion from given ("old") solutions. If, for example, a physical situation happens to be translationally invariant, our intuition suggests that if we are given a solution of the equations of motion (i.e. a possible motion), we can obtain a new solution (another possible motion) by applying a translation to the known solution. In general, this construction looks for the Hamiltonian systems as follows.

14.4.6 Let V be a Cartan symmetry of a Hamiltonian system (M, ω, H). Check that

(i) the field V commutes with the field ζ_H (generating the time development)

$$[V, \zeta_H] = 0$$

(ii) the flow Φ_t^V commutes with the time development (i.e. with the flow $\Phi_t \leftrightarrow \zeta_H$)

$$\Phi_s^V \circ \Phi_t = \Phi_t \circ \Phi_s^V$$

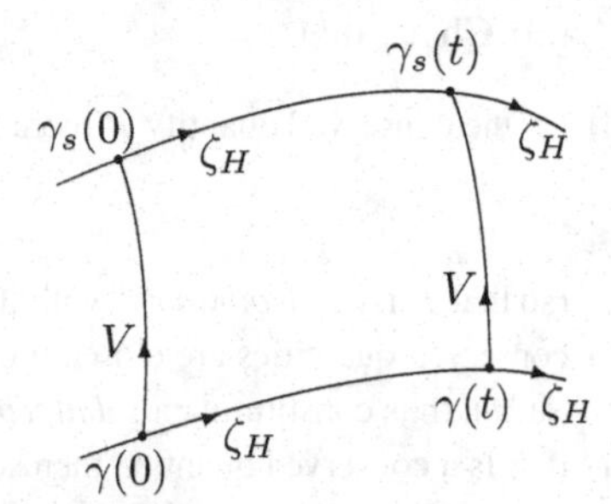

(iii) if $\gamma(t)$ is a possible motion of the system (i.e. the coordinate presentation of $\gamma(t)$ is a solution of the Hamilton equations of motion) and if we denote by $\gamma_s(t) := \Phi_s^V(\gamma(t))$ the image of the solution with respect to the flow of the symmetry Φ_s^V, then the curve $\gamma_s(t)$ *also* happens to be a possible motion of the system (for each s), so that we are able to generate from a single solution $\gamma(t)$ by means of the symmetry V the whole one-parameter class of new solutions $\gamma_s(t)$

(iv) if $V = \zeta_F$ is an *exact* Cartan symmetry corresponding to the conserved quantity F, then the value of F for the initial motion coincides with the value of F on the new (class of) solutions[262]

$$F(\gamma_s(t)) = F(\gamma(t)) = \text{constant}$$

$F(\gamma_s(t))$ depends neither on t (since F is conserved on all γ_s) *nor on* s (since the value of the conserved quantity happens to be the same on all γ_s); F is thus constant on the two-dimensional sheet $\gamma_s(t) \subset M$).

Hint: (i) by definition $i_{\zeta_H}\omega = -dH$; if we apply $\mathcal{L}_V$ on both sides of the equation (and use (6.2.9) and $VH = 0$), we get $i_{[V,\zeta_H]}\omega = 0$; non-degeneracy of ω then immediately yields the result; (ii) (4.5.8); (iii) if $x := \gamma(0)$ is the initial value of the solution $\gamma(t)$, then $\gamma(t) = \Phi_t(x)$; but $\gamma_s(t) := \Phi_s^V(\gamma(t)) = \Phi_s^V \circ \Phi_t(x) = \Phi_t(\Phi_s^V(x)) \equiv \Phi_t(y)$, so this is the solution with the initial value $y := \Phi_s^V(\gamma(0))$; (iv) $\zeta_F F \equiv \{F, F\} = 0$ $(= \zeta_H F \equiv \{H, F\}$ by (14.4.3)). □

14.4.7 Check that for $V = \zeta_H$ we get as the new solutions $\gamma_s(t) = \gamma(t + s)$ (the same motion which takes place *sooner* by s; this class of the "new solutions" has the same *energy*, since $F = H$ now).

Hint: (14.4.5) and (14.4.6). □

[262] This means, for example (as will be clear after reading (18.4.3)), that if the new solution is obtained by a translation, it has the same *momentum* as before and the new solution obtained by a rotation has the same *angular* momentum.

• We will elaborate on various more involved aspects of this in the remaining three sections of this chapter; we then return to this important topic in Section 18.4. There we will make more explicit the results obtained here for the case where the manifold carrying the Hamiltonian system is realized as the *(co)tangent bundle* of a *configuration* space (being the standard *phase* space of analytic mechanics).

14.5* Moment map

• The moment map plays an important role in various contexts in modern mathematical physics. It appears wherever a Lie group acts on a symplectic manifold so that it preserves the symplectic structure. It thus puts in an appearance, for example, in the context of the symmetries of Hamiltonian systems treated in the last section (where, however, the preservation of H was demanded in addition). Since its occurrence is not restricted, however, to Hamiltonian systems alone, its field of applications being broader, it might be useful to devote slightly more time to grasping this topic than its applications to the conservation laws in Hamiltonian systems alone would need.

Let (M, ω, R_g) be a symplectic manifold along with a right action R_g of a Lie group G and suppose that the action is compatible with the symplectic structure; this means that each map R_g is a symplectomorphism

$$R_g : M \to M \qquad R_g^* \omega = \omega$$

It turns out that each such *symplectic action* automatically gives rise to a closed 1-form.

14.5.1 Check that

(i) the generators (fundamental fields) ξ_X of the action R_g are *symplectic* fields

$$\mathcal{L}_{\xi_X} \omega = 0 \qquad (\Rightarrow \quad \xi_X \in \text{Symp}\,(M))$$

(ii) the 1-forms α_X, which are obtained from the generators ξ_X by the lowering of the index with the help of the form ω, are *closed*

$$\alpha_X := \flat_\omega \xi_X \quad \text{i.e.} \quad i_{\xi_X} \omega =: -\alpha_X \quad \Rightarrow \quad d\alpha_X = 0$$

(iii) the 1-form α_X depends on $X \in \mathcal{G}$ linearly

$$\alpha_{X+\lambda Y} = \alpha_X + \lambda \alpha_Y \qquad \lambda \in \mathbb{R}$$

so that if E_i is a basis in $\mathcal{G}$ and $X = X^i E_i$, then

$$\alpha_X = X^i \alpha_i \qquad \alpha_i := \alpha_{E_i}$$

(iv) under the action R_g the form α_X behaves as follows:

$$R_g^* \alpha_X = \alpha_{\text{Ad}_g X} \qquad \text{i.e.} \quad R_g^* \alpha_i = (\text{Ad}_g)_i^j \alpha_j$$

so that the 1-forms α_i, $i = 1, \ldots, \dim \mathcal{G}$ are "scrambled" by the Ad-representation of the group G

(v) the infinitesimal version of the transformation of α_X with respect to R_g looks like

$$\mathcal{L}_{\xi_X}\alpha_Y = \alpha_{[X,Y]} \qquad \text{or also} \qquad \mathcal{L}_{\xi_i}\alpha_j = c_{ij}^k\alpha_k \qquad \xi_i := \xi_{E_i}$$

Hint: (i) see Section 4.2; (ii) (6.2.8); (iii) (13.4.3); (iv) (13.4.3); (v) set $g = e^{tX}$ and differentiate in $t = 0$. □

• So the forms $\alpha_X \equiv \flat_\omega \xi_X$ inherited some user-friendly properties from the generators ξ_X, like the linearity with respect to X and "scrambling" by the Ad-representation (these properties are preserved by the isomorphism $\flat_\omega$). They also have, however, a new specific property (not being directly inherited from ξ_X), namely the *closedness* (which may be traced back to being partly due to the closedness of the form ω). The most interesting case occurs when the form α_X actually happens to be *exact* (rather than only closed), so when there exists a global potential, a function $P_X \in \mathcal{F}(M)$ such that $\alpha_X = dP_X$.[263] Then we speak about a *globally Hamiltonian action*, since the field ξ_X is then a Hamiltonian field generated by the function P_X

$$i_{\xi_X}\omega = -dP_X \qquad \text{so that} \qquad \xi_X = \zeta_{P_X} \qquad \text{or also} \qquad \xi_X f = \{P_X, f\}$$

Assume that the potential exists. A natural question then arises as to whether the properties mentioned above are passed on from the 1-form α_X also to the potential (function) P_X. It turns out that the general answer is "not straightforwardly." Namely, the linearity may always be achieved by a simple trick, but the Ad-behavior with respect to the action of G may sometimes turn out to be beyond one's reach.

Let us take the linearity first. As always happens with potentials, the function P_X is not fixed uniquely by the defining equation $\alpha_X = dP_X$, but rather only up to an *additive constant* (function). One has to realize, however, that we are actually speaking about an *infinite number* of 1-forms α_X and of potentials P_X (one for each $X \in \mathcal{G}$) and one thus has to make a choice of a potential P_X *for each* $X \in \mathcal{G}$. Then it is clear that if the choice is made "at random," the linearity in X will hardly be valid.[264] Linearity may be easily achieved, however, by a coordinated choice of the additive constants.

14.5.2 Let E_i be a basis of the Lie algebra $\mathcal{G}$. Fix the functions P_X for the basis elements[265] E_i and then define P_X for an arbitrary element $X \equiv X^i E_i$ by

$$P_X := X^i P_i \qquad P_i \equiv P_{E_i}$$

Check that

(i) the functions defined in this way indeed depend on X linearly

$$P_{X+\lambda Y} = P_X + \lambda P_Y$$

[263] This is locally always true, as we know. Globally it occurs, for example, if the first cohomology group of the manifold M is trivial (i.e. $H^1(M,\mathbb{R}) = 0$), but this is not necessary, however, since in spite of $H^1(M,\mathbb{R}) \neq 0$, the particular 1-form α_X may represent the trivial class (it depends on the action) even if there are also non-trivial classes on M.

[264] If the potentials P_X, P_Y and P_{X+Y} are fixed *randomly*, there will hardly hold $P_X + P_Y = P_{X+Y}$, to say nothing of this *for each* X, Y.

[265] Since the potentials are functions (0-forms), the freedom is fixed by assigning values of the functions at an arbitrary point, i.e. in the same way as standardly done with the scalar potential in electrostatics, the only difference being in that we have to fix $n \equiv \dim \mathcal{G}$ such potentials P_i here.

This may also be regarded as the fact that the map

$$\tilde{P} : \mathcal{G} \to \mathcal{A}(M) \qquad X \mapsto P_X$$

is linear. The preservation of the commutator (both spaces actually being Lie algebras) turns out, however, to be a much more delicate issue; see (14.5.4))

(ii) the freedom in the fixation on the basis E_i may also be rephrased as the freedom

$$P_i \mapsto \hat{P}_i := P_i + p_i \quad p_i \in \mathbb{R} \qquad \Leftrightarrow \qquad P_X \mapsto \hat{P}_X := P_X + \langle p, X \rangle \quad p \in \mathcal{G}^*$$

where $\mathcal{G}^*$ denotes the linear space which is dual to $\mathcal{G}$. □

• The linearity of the function P_X with respect to $X \in \mathcal{G}$ enables one to encode the information carried by P_X in an equivalent object, namely a map $P : M \to \mathcal{G}^*$.

14.5.3 Define the map P by the formula

$$P : M \to \mathcal{G}^* \qquad \langle P(x), X \rangle := P_X(x), \ x \in M$$

It may also be regarded as a $\mathcal{G}^*$-valued 0-form on M, i.e. as $P \in \Omega^0(M, \mathcal{G}^*)$. Check that

(i) the freedom in P_X from problem (14.5.2) is reflected on P as the freedom

$$P \mapsto \hat{P} := P + p \qquad p \in \mathcal{G}^*$$

so that $\hat{P}$ differs from P by a *constant* shift by p in the target space $\mathcal{G}^*$ (this enables one to prescribe to the point x an arbitrary value in $\mathcal{G}^*$)

(ii) if E^i is the basis in $\mathcal{G}^*$ (which is dual to E_i in $\mathcal{G}$), then

$$P = P_i E^i$$

so that the functions P_i from problem (14.5.2) are just the *component forms* (actually functions, here) of the form P.

Hint: (i) $\hat{P}_X(x) = P_X(x) + \langle p, X \rangle = \langle P(x), X \rangle + \langle p, X \rangle \equiv \langle P(x) + p, X \rangle$. □

• The map P interconnects two manifolds, on which the right action of the group G is defined – on M it is (by assumption) R_g, whereas in $\mathcal{G}^*$ it is the *coadjoint action* Ad_g^* (as we learned in (12.3.18)). Therefore it is natural to address the question of whether P is *equivariant* with respect to these actions. It turns out that in general it is not. But the issue is more subtle (and interesting) than this plain answer might indicate. Namely, the analysis shows that sometimes the equivariance can be achieved (making use of the freedom available), but there are situations when this is simply not possible. We will see that the answer may be most conveniently expressed in *cohomological* terms, meaning here, however, the Lie algebra cohomologies introduced in Section 12.8 rather than deRham cohomologies of differential forms. Moreover we will see that the reformulation of the question in the language of the function P_X turns out to be just the issue of preserving the commutator by the map $X \mapsto P_X$, which we mentioned in problem (14.5.2).

14.5.4 Check that

(i) the simple behavior of α_X under the action of the group G from (14.5.1) gets more involved at the level of the potential P_X, namely

$$R_g^* P_X = P_{\mathrm{Ad}_g X} + k(g, X) \qquad k(g, X) \in \mathbb{R}$$

(ii) the infinitesimal version of the same behavior reads

$$\xi_X P_Y = P_{[X,Y]} + \beta(X, Y) \quad \text{or equivalently} \quad \{P_X, P_Y\} = P_{[X,Y]} + \beta(X, Y)$$

The term $\beta(X, Y)$ thus measures the *deviation* from a homomorphism for the map $X \mapsto P_X$ from (14.5.2) (actually of the commutator alone; the linearity is all right).

Hint: (i) if $\alpha_X = dP_X$, then $R_g^* \alpha_X = \alpha_{\mathrm{Ad}_g X}$ from (14.5.1) gives $d(R_g^* P_X - P_{\mathrm{Ad}_g X}) = 0$; (ii) $\mathcal{L}_{\xi_X} \alpha_Y = \alpha_{[X,Y]}$ similarly leads to $d(\xi_X P_Y - P_{[X,Y]}) = 0$. □

14.5.5 Check that

(i) the map

$$\beta : \mathcal{G} \times \mathcal{G} \to \mathbb{R}$$

defined by the relation (see (14.5.4))

$$\beta(X, Y) := \{P_X, P_Y\} - P_{[X,Y]}$$

is bilinear and antisymmetric, so that it is a *2-form* on $\mathcal{G}$[266]

(ii) the 2-form $\beta \in \Lambda^2 \mathcal{G}^*$ happens to be *closed* (a *2-cocycle*)

$$\hat{d}\beta = 0 \qquad \text{i.e.} \quad \beta(X, [Y, Z]) + \text{ cycl.} = 0$$

(iii) the change of P_X (caused by the freedom of choice of P_X on a basis) results in the following change of the 2-form β:

$$P_X \mapsto \hat{P}_X := P_X + \langle p, X\rangle \quad \Rightarrow \quad \beta \mapsto \hat{\beta} := \beta + \hat{d}p \qquad p \in \mathcal{G}^* \equiv \Lambda^1 \mathcal{G}^*$$

(iv)

$$[\beta] = 0 \quad \Leftrightarrow \quad \text{one can obtain} \quad \{\hat{P}_X, \hat{P}_Y\} = \hat{P}_{[X,Y]}$$

so that the (undesirable) additive constant $\beta(X, Y)$ in (14.5.4) may be eliminated, making use of the freedom if and only if β represents the *trivial* cohomological class, i.e. if it is a *coboundary*. This is clearly guaranteed when there are no other classes except the trivial one, i.e. when $H^2(\mathcal{G}^*, \mathbb{R}) = 0$, which is fulfilled, as we mentioned in (11.8), by the important class of *semi-simple* Lie algebras

(v)

$$\beta(X, Y) = 0 \quad \Leftrightarrow \quad k(g, X) = 0$$

[266] Actually $\beta(X, Y)$ is the constant *function* on M, so, strictly speaking, $\beta \in \Omega^0(M, \Lambda^2 \mathcal{G}^*)$; since it is, however, a *constant*, there is the *same* element from $\Lambda^2 \mathcal{G}^*$ at each point, which may be regarded simply as an element $\beta \in \Lambda^2 \mathcal{G}^*$.

so that the elimination of the first unpleasant term also results in the elimination of the second one.

Hint: (ii) (12.6.3), the Jacobi identity for $\{\cdot\,,\cdot\}$ and $[\cdot\,,\cdot]$; (iv) $\beta \equiv \hat{d}\tau \mapsto \hat{d}\tau + \hat{d}p \Rightarrow$ one should choose a new P_X so that $p = -\tau$, i.e. $\hat{P} = P - \tau$; (v) if $g(t) = e^{tX}$ is a one-parameter subgroup, then

$$\begin{aligned} \xi_X P_Y = P_{[X,Y]} \;&\Rightarrow\; \mathcal{L}_{\xi_X} P_Y = P_{\mathrm{ad}_X Y} \\ &\Rightarrow\; e^{t\mathcal{L}_{\xi_X}} P_Y = P_{\exp(t\,\mathrm{ad}_X)Y} \equiv P_{\mathrm{Ad}_{g(t)}Y} \\ &\Rightarrow\; R^*_{g(t)} P_Y = P_{\mathrm{Ad}_{g(t)}Y} \end{aligned}$$

so that the equality holds for arbitrary elements of the form $g = e^X$. □

• A reformulation of these properties of the function P_X in the language of the map $P : M \to \mathcal{G}^*$ answers the question of under what conditions does P happen to be equivariant.

14.5.6 Check that the equivariance of P, i.e. the validity of the commutative diagram

$$\begin{array}{ccc} M & \xrightarrow{\;P\;} & \mathcal{G}^* \\ {\scriptstyle R_g}\downarrow & & \downarrow{\scriptstyle \mathrm{Ad}^*_g} \\ P & \xrightarrow[\;P\;]{} & \mathcal{G}^* \end{array}$$

is equivalent to

(i) the condition

$$R^*_g P_X = P_{\mathrm{Ad}_g X} \qquad \text{(or alternatively } k(g, X) = 0\text{)}$$

so that there holds

$$P \text{ can be improved to become equivariant} \quad \Leftrightarrow \quad [\beta] = 0$$

(ii) the fact that the (already improved) map $\tilde{P} : \mathcal{G} \to \mathcal{A}(M)$, $X \mapsto P_X$ from (14.5.2) is a homomorphism of Lie algebras. We may then summarize it as follows:

$$P \text{ “is” equivariant} \quad \Leftrightarrow \quad [\beta] = 0 \quad \Leftrightarrow \quad \tilde{P} \;\text{ is a homomorphism of Lie algebras}$$

(“is” = “can be improved to become”; by $\tilde{P}$ an already improved P is meant).

Hint:

$$\langle (P \circ R_g)(x), X\rangle = P_X(R_g x) = P_{\mathrm{Ad}_g X}(x) = \langle P(x), \mathrm{Ad}_g X\rangle = \langle (\mathrm{Ad}^*_g \circ P)(x), X\rangle$$

□

• The map $P : M \to \mathcal{G}^*$, which happens to be equivariant (so that it has $\beta = 0$), is called the *moment map* corresponding to the action of G on (M, ω) and such a special Hamiltonian action R_g is then in turn called a *Poisson action*.[267] In terms of the function P_X then the map $\tilde{P}: \mathcal{G} \to \mathcal{A}(M)$, $X \mapsto P_X$, is a Lie algebra homomorphism and it is called the *comoment map*. Recall that both maps carry exactly the same information, being related by

$$\langle P(x), X \rangle = \tilde{P}(X)(x) \equiv P_X(x)$$

14.5.7 Consider now the situation which is most interesting for us, namely when the globally Hamiltonian action of the Lie group G also preserves the *Hamiltonian* H (so that we have a symmetry of the whole Hamiltonian system (M, ω, H)). Check that

(i) then the function P_X is for each $X \in \mathcal{G}$ a conserved quantity of the Hamiltonian system
(ii) if E_i is a basis of $\mathcal{G}$, then all the functions $P_i(x) = P_{E_i}(x)$ are conserved, so that the number of conserved quantities resulting from the symmetry group G coincides with the dimension of the group G (and consequently of the Lie algebra $\mathcal{G}$)
(iii) an equivalent way of expressing this result is to say that "the moment map is conserved," meaning $\dot{P} \equiv \dot{P}_i E^i = 0$, where P is the moment map.

Hint: the fundamental fields $\xi_X \equiv \zeta_{P_X}$ are then exact Cartan symmetries, see (14.4.4). □

14.6* Orbits of the coadjoint action

• The orbits of the coadjoint action Ad* play an important role in various questions related to the actions of Lie groups and symplectic geometry. It turns out that they carry a canonical symplectic structure.

Recall from (12.3.18) that the *coadjoint action* Ad^*_g of a Lie group G takes place on the linear space $\mathcal{G}^*$ dual to the Lie algebra $\mathcal{G}$ of the group G and it is defined[268] by the prescription

$$\langle \mathrm{Ad}^*_g X^*, Y \rangle := \langle X^*, \mathrm{Ad}_g Y \rangle \qquad X^* \in \mathcal{G}^*, \quad Y \in \mathcal{G}$$

So it is the contragredient (anti)representation[269] to the adjoint representation Ad.

Denote by ξ_X its fundamental fields; for an arbitrary function Φ on $\mathcal{G}^*$ we have by definition

$$\xi_X(Y^*)\Phi = \left.\frac{d}{dt}\right|_0 \Phi(\mathrm{Ad}^*_{\exp tX} Y^*)$$

14.6.1 On a manifold L, which is at the same time a linear space, we may define *linear* functions, i.e. the functions obeying $\Phi(v + \lambda w) = \Phi(v) + \lambda\Phi(w)$. Be sure to understand that

[267] It arises, for example, when an arbitrary action is *lifted* from the base to the total space of the (co)tangent bundle (see (18.4.1)) or in the case of the *coadjoint* action Ad^*_g on $\mathcal{G}^*$ (see (12.3.19) and (14.6.5)).

[268] We will denote by $X, Y, \ldots$ the elements of $\mathcal{G}$ and by $X^*, Y^*, \ldots$ the elements of $\mathcal{G}^*$. Thus the star on X^* *does not denote* the result of some operation performed on the element X.

[269] This is a *right* action. The representation is given by $\mathrm{Ad}^*_{g^{-1}}$.

(i) these functions are nothing but the covectors on L, i.e. the elements of the dual space L^*

$$\Phi_{w^*}(v) = \langle w^*, v\rangle$$

(ii) if we know how a vector field W acts on *these* particular functions, we already know *everything* about W.

Hint: (ii) if e_i is a basis in L and e^i the dual basis in L^*, then $\Phi_{w^*}(x) = w_i x^i$ and, in particular, $\Phi_{e^i}(x) = x^i$; then for $W = W^i\partial_i$ we get $(W\Phi_{e^i})(x) = W^i(x)$, so that from the action on linear functions we get the components of the field W. □

14.6.2 The linear functions on the manifold $\mathcal{G}^*$ are thus uniquely parametrized by the elements of the Lie algebra $\mathcal{G}$

$$\Phi_X(Y^*) = \langle Y^*, X\rangle$$

Check that

(i) the action of the group G on these functions (by pull-back) gives

$$R_g^*\Phi_X = \Phi_{\mathrm{Ad}_g X} \qquad R_g \equiv \mathrm{Ad}_g^*$$

(ii) the infinitesimal version of this action reads

$$\xi_X\Phi_Y = \Phi_{[X,Y]}$$

Hint: (i) $(R_g^*\Phi_X)(Y^*) = \Phi_X(\mathrm{Ad}_g^* Y^*) = \langle \mathrm{Ad}_g^* Y^*, X\rangle = \langle Y^*, \mathrm{Ad}_g X\rangle = \Phi_{\mathrm{Ad}_g X}(Y^*)$; (ii) set $g = \exp tZ$ and differentiate in $t = 0$; you get $\xi_Z\Phi_X = \Phi_{\mathrm{ad}_Z X} = \Phi_{[Z,X]}$. □

14.6.3 Consider now the *orbit* $\mathcal{O}_{Z^*}$ generated by a point $Z^* \in \mathcal{G}^*$. The tangent space *to the orbit* in the point Z^* is spanned (as is always the case for orbits) by the values of the fundamental fields at this point. Therefore, if we want to define a differential form *on the orbit*, it is enough to specify its values on the fundamental fields. Define in the tangent space $T_{Z^*}\mathcal{O}_{Z^*}$ *to the orbit* in the point Z^* the 2-form ω_{Z^*} by the formula

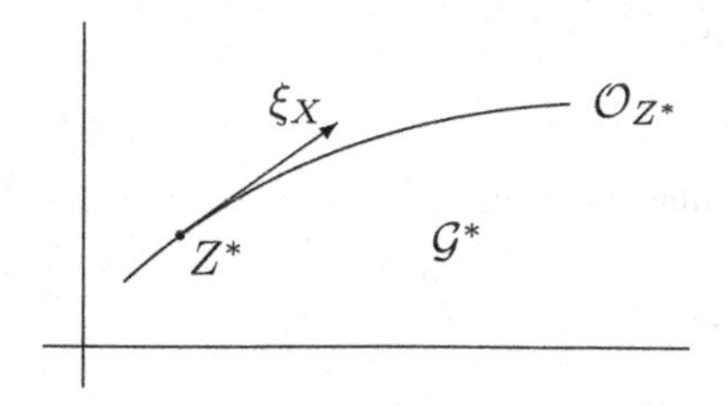

$$\omega_{Z^*}(\xi_X, \xi_Y) := \langle Z^*, [X, Y]\rangle \equiv \Phi_{[X,Y]}(Z^*)$$

In this way (by the pointwise construction) we get a smooth differential 2-form ω on the manifold (the orbit) $\mathcal{O}_{Z^*}$. Check that ω happens to be a *symplectic form*.

Solution: the closedness: in an arbitrary point on the orbit the Cartan formula (6.2.13) yields

$$\begin{aligned}
d\omega(\xi_X, \xi_Y, \xi_Z) &= (\xi_X\omega(\xi_Y, \xi_Z) - \omega([\xi_X, \xi_Y], \xi_Z)) + \text{ cycl.} && \text{Cartan formula}\\
&= (\xi_X\Phi_{[Y,Z]} - \Phi_{[[X,Y],Z]}) + \text{ cycl.} && \text{the definition } \omega\\
&= 2\Phi_{[[X,Y],Z]+\text{ cycl.}} && (14.6.2)\\
&= 2\Phi_0 = 0 && \text{Jacobi identity}
\end{aligned}$$

the non-degeneracy: we have $\omega_{Z^*}(\xi_X, \xi_Y) := \Phi_{[X,Y]}(Z^*) = \xi_X(Z^*)\Phi_Y$; so if $\omega_{Z^*}(\xi_X, \xi_Y) = 0$ for all ξ_Y, the vector $\xi_X(Z^*)$ yields zero by the action on *all linear* functions, but then also (14.6.2) necessarily on *all* functions, so that the vector ξ_X *itself* vanishes at the point Z^*. □

14.6.4 Show that each orbit of the coadjoint action is an *even-dimensional* manifold.

Hint: this belongs to the life-style of *symplectic* manifolds, see (14.1.4). □

• So we learned that a rich source of symplectic manifolds is available, namely *all* the orbits[270] of the coadjoint action. The input data consist of (only) a Lie group G and a point in the dual space $\mathcal{G}^*$ to the Lie algebra $\mathcal{G}$.

We already met a particular example of such orbits in problem (13.4.10), where we found all orbits of the actions Ad and Ad* for the group $GA(1, \mathbb{R})$. This case is fairly instructive, since it shows that the orbit structure may be substantially different for these two actions (recall that in this particular case Ad* has, as is proper, only zero- and two-dimensional orbits, whereas Ad has also one-dimensional orbits), so that the star on Ad* or $\mathcal{G}^*$ is in general[271] to be taken seriously.

14.6.5 The orbits of the action Ad* are canonical symplectic manifolds. At the same time there is, however, a transitive action of a Lie group G on them. This means that they actually represent a fairly interesting "combined" geometrical object, the *homogeneous symplectic spaces* of the group G. The natural question arises as to whether these two structures are compatible. Considering the naturalness of both the structures involved one should expect that the action Ad* respects the symplectic structure ω. Verify that this is indeed the case. Namely check that

(i) the fundamental fields ξ_X of the action Ad* happen to be *Hamiltonian* fields, their generators ("Hamiltonians") being just the *linear* functions on $\mathcal{G}^*$ (or more precisely their *restrictions* to the orbit)

$$i_{\xi_X}\omega = -d\Phi_X$$

(ii) the action Ad* is *globally Hamiltonian*, where in particular $P_X = \Phi_X$

(iii) the action Ad* even happens to be a *Poisson action*

(iv) the moment and comoment maps for Ad* explicitly read

$$P : \mathcal{O} \to \mathcal{G}^* \qquad X^* \mapsto X^*$$
$$\tilde{P} : \mathcal{G} \to \mathcal{A}(\mathcal{O}) \qquad X \mapsto \Phi_X$$

so that the moment map is, in fact, the *identity* map (more precisely it is the canonical embedding of the orbit $\mathcal{O}$ into $\mathcal{G}^*$) and the comoment map assigns to the element X *its linear* function

(v) the action Ad* preserves the symplectic structure ω.

[270] Except for zero-dimensional ones, the isolated points. All orbits which are more than zero-dimensional are already interesting (being *two-*, *four-*, ... dimensional).

[271] Sometimes, however, it is *not* relevant; for example, on semi-simple Lie algebras the Killing form provides an equivariant isomorphism between $\mathcal{G}$ and $\mathcal{G}^*$ and the orbits then look equal.

Hint: (i) $\langle i_{\xi_X}\omega, \xi_Y\rangle \equiv \omega(\xi_X, \xi_Y) = \Phi_{[X,Y]} = -\Phi_{[Y,X]} = -\xi_Y\Phi_X = \langle -d\Phi_X, \xi_Y\rangle$; (ii) the result of item (i) plus the definition in Section 14.5; (iii) $\{P_X, P_Y\} \equiv \{\Phi_X, \Phi_Y\} = \omega(\xi_X, \xi_Y) = \Phi_{[X,Y]} \equiv P_{[X,Y]}$ so that $\beta(X, Y) = 0$; (iv) $P_X(Z^*) = \Phi_X(Z^*) = \langle Z^*, X\rangle \overset{!}{=} \langle P(Z^*), X\rangle$; (v) the flows of *Hamiltonian* fields *do* preserve ω. □

14.6.6 Find an explicit expression of the form ω for the case $G = GA(1,\mathbb{R})$ and check its Ad*-invariance.

Hint: (13.4.10); if a general element from $ga(1,\mathbb{R})^*$ is $Z^* = xE^1 + yE^2$, then $\xi_{E_1} = y\partial_y$, $\xi_{E_2} = -y\partial_x$. The 2-form which is to be found is $a(x, y)\, dx \wedge dy$, the condition $\omega(\xi_{E_1}, \xi_{E_2}) \equiv y^2 a(x, y) \overset{!}{=} \langle xE^1 + yE^2, [E_1, E_2]\rangle \equiv y$ gives $a(x, y) = 1/y$. The coadjoint action reads $\mathrm{Ad}^*_{(a,b)}(x, y) = (x - by, ay)$ and $(1/ay)\; d(x - by) \wedge d(ay) = (1/y)\, dx \wedge dy$. □

14.6.7 Check that the Ad*-action may be identified for $SU(2)$ as well as for $SO(3)$ with the ordinary rotations in E^3, so that the orbits reduce to ordinary spheres centered at the origin and the form ω coincides (up to a constant multiple) with the ordinary "round" (rotationally invariant) volume form on the sphere.

Hint: for $A = \mathbb{I} + \epsilon n_j l_j$ the Ad action gives $\mathrm{Ad}_A(x_i l_i) = A(x_i l_i)A^{-1} = \cdots = (x_i + \epsilon\epsilon_{ijk} n_j x_k) l_i$, i.e. $\mathbf{r} \mapsto \mathbf{r} + \epsilon\mathbf{n} \times \mathbf{r}$, so that the Ad action indeed reduces to rotations; since the Killing form is $\sim \delta_{ij}$, the same is true also for Ad*; the (constant multiple of the) ordinary "round" volume form (6.3.10) is *the only* volume form which is rotationally invariant. □

• On the *orbits* of the coadjoint action in $\mathcal{G}^*$ we found the canonical *symplectic* structure. This means that the orbits may also be regarded as *non-degenerate Poisson* manifolds (see Section 14.1). It turns out, however, that the Poisson tensor actually exists canonically *on the whole* manifold $\mathcal{G}^*$. Why then have we bothered about some orbits instead of taking simply $\mathcal{G}^*$ itself? The reason is that the Poisson tensor happens to be *degenerate*, when regarded on the whole $\mathcal{G}^*$ and we are forced to *restrict it to the orbits* in order to obtain a non-degenerate tensor (and to "invert" it then to obtain a symplectic form).

Consider the manifold $\mathcal{G}^*$ along with the Cartesian coordinates x_i given by the decomposition[272] of X^* with respect to an arbitrary basis $X^* = x_i E^i$. Recall that the *linear functions* on $\mathcal{G}^*$ have the form $\Phi_Y(X^*) = \langle X^*, Y\rangle \equiv y^i x_i$; in particular, the functions $\Phi_i \equiv \Phi_{E_i} = x_i$ turn out to be directly the coordinates x_i. This means (see (14.1.12)) that it is enough to specify the Poisson bracket on *linear* functions.

14.6.8 Define the Poisson bracket of two linear functions on a manifold $\mathcal{G}^*$ by the formula

$$\{\Phi_Y, \Phi_Z\} := \Phi_{[Y,Z]} \equiv \xi_Y \Phi_Z$$

i.e.

$$\{\Phi_Y, \Phi_Z\}(X^*) := \Phi_{[Y,Z]}(X^*) \equiv \langle X^*, [Y, Z]\rangle \equiv (\mathcal{P}(d\Phi_Y, d\Phi_Z))(X^*)$$

[272] There are natural *lower* indices on the coordinates, so that the whole index machinery gets "inverted": for example, the Poisson tensor also has *lower* indices $\mathcal{P}_{ij}(x)$, etc.

Check that

(i) it satisfies all the requirements imposed in general on a Poisson bracket
(ii) for the coordinates x_i themselves we get, in particular,

$$\{x_i, x_j\} = c_{ij}^k x_k$$

(c_{ij}^k being the structure constants of the Lie algebra $\mathcal{G}$ with respect to the basis E_i) so that the formula for arbitrary functions f, g reads

$$\{f, g\} = (\partial^i f) x_k c_{ij}^k (\partial^j g)$$

(iii) the Poisson bracket is related to the symplectic form on the orbit $\mathcal{O}_{X^*}$ by

$$\{\Phi_Y, \Phi_Z\} = \omega(\xi_Y, \xi_Z)$$

(iv) for an arbitrary function f on $\mathcal{G}^*$ there holds

$$\{\Phi_Y, f\} = \xi_Y f$$

so that this bracket (i.e. the corresponding Poisson tensor) is *degenerate*.

Hint: (i) the Jacobi identity for $\{\cdot\,,\cdot\}$ is a direct consequence of the Jacobi identity for $[\cdot\,,\cdot]$ in $\mathcal{G}$, since $\{\{\Phi_Y, \Phi_Z\}, \Phi_W\} = \Phi_{[[Y,Z],W]}$; (iii) both expressions in the point X^* yield $\langle X^*, [Y, Z]\rangle$; (iv) it vanishes for all functions f which are constant on orbits. □

- Let us compute this bracket explicitly on $\mathcal{G}^* = (so(3))^*$ and $ga(1, \mathbb{R})$.

14.6.9 Check that

(i) if we identify $(so(3))^*$ with an ordinary three-dimensional Euclidean space and adopt the usual vector notation, then

$$\{f, g\} = \mathbf{r} \cdot (\nabla f \times \nabla g) \equiv ((\mathbf{r} \times \nabla f) \cdot \nabla) g \equiv \zeta_f g$$

(ii) the Poisson bracket vanishes for the functions which depend on r alone (i.e. constant on the spheres centered at the origin, i.e. on the orbits of the action)
(iii) on $\mathcal{G}^* = (ga(1, \mathbb{R}))^*$ we get in the coordinates (x, y) from (14.6.6)

$$\{x, y\} = y \qquad \text{i.e.} \quad \{f, g\} = y(\partial_x f \partial_y g - \partial_y f \partial_x g)$$

Hint: (i) (14.6.8) for $c_{ij}^k = \varepsilon_{kij}$; (ii) $\nabla f(r) \sim \mathbf{r}$, (14.6.7). □

14.6.10 The Lie algebra $so(3)$ along with the Killing metric may be regarded as an ordinary Euclidean space E^3 and consequently the same is true for its dual space $(so(3))^*$. We will denote as x_i the coordinates with respect to some orthonormal basis in this $E^3 \equiv (so(3))^*$ (as is possible and common in E^3, we need not distinguish the upper/lower indices; we will write *all* of them as *lower*). The orbits of the coadjoint action are the spheres of all possible radii centered at the origin. On this dual space we have according to (14.6.9) now also the

Poisson tensor $\mathcal{P}$, hidden in the Poisson bracket

$$\{x_i, x_j\} \equiv \mathcal{P}_{ij} = \epsilon_{ijk}x_k \qquad \text{i.e.} \qquad \{f, g\} = \mathbf{r} \cdot (\nabla f \times \nabla g) = \zeta_f g$$
$$\text{where} \qquad \zeta_f \equiv (\mathbf{r} \times \nabla f) \cdot \nabla$$

Check that

(i) the restriction of $\mathcal{P}$ to the unit sphere $x_i x_i \equiv |\mathbf{r}|^2 = 1$ (one of the coadjoint orbits) is already non-degenerate and the corresponding symplectic form is just the ordinary ("round" = rotationally invariant) volume form on the sphere,[273] mentioned in (14.2.3) and (14.6.7)

(ii) the Poisson dynamics in E^3 generated by a general Hamiltonian field ζ_H reads

$$\dot{\mathbf{r}} = \boldsymbol{\Omega} \times \mathbf{r} \qquad \boldsymbol{\Omega}(\mathbf{r}) := -\nabla H(\mathbf{r})$$

i.e. the point $\mathbf{r}$ rotates with angular velocity $\boldsymbol{\Omega}$, which in turn *depends on* $\mathbf{r}$, so that (in general) the whole space does not rotate as a rigid body (and neither does the fixed orbit)

(iii) for a Hamiltonian which is *linear in* $\mathbf{r}$, the angular velocity $\boldsymbol{\Omega}$ happens to be constant, so that the time development reduces to a uniform rotation of the points in E^3 around $\boldsymbol{\Omega}$ with angular velocity $|\boldsymbol{\Omega}|$,

$$H(\mathbf{r}) = -\boldsymbol{\Omega} \cdot \mathbf{r} \qquad \Rightarrow \qquad \dot{\mathbf{r}} = \boldsymbol{\Omega} \times \mathbf{r} \quad \boldsymbol{\Omega} = \text{ constant}$$

Hint: (i) if $\mathbf{r} \equiv \mathbf{n}$ is a point on the unit sphere, then the value of the symplectic form ω in the point $\mathbf{n}$ on a pair of vectors $\mathbf{a}, \mathbf{b}$ is $\omega_{\mathbf{n}}(\mathbf{a}, \mathbf{b}) = \omega_{ij}(\mathbf{n})a_i b_j = \epsilon_{ijk}n_k a_i b_j = \mathbf{n} \cdot (\mathbf{a} \times \mathbf{b}) \Rightarrow$ on an arbitrary right-handed orthonormal basis in $\mathbf{n} \in S^2$ there holds $\omega_{\mathbf{n}}(\mathbf{e}_1, \mathbf{e}_2) = \mathbf{n} \cdot (\mathbf{e}_1 \times \mathbf{e}_2) = 1$, so that ω is just the metric volume form on the sphere; (ii) $\dot{x}_i = \{H, x_i\} = \epsilon_{jkl}x_l(\partial_j H)(\partial_k x_i)$. □

14.6.11 Recall that the state of a quantum-mechanical system is in general given by the *density operator* $\hat{\rho}$, which is a positive Hermitian operator with unit trace. For a *two-level system* (like the spin $\frac{1}{2}$) it is an operator in the Hilbert space $\mathbb{C}^2$; to summarize,

$$\hat{\rho}^+ = \hat{\rho} \qquad \operatorname{Tr} \hat{\rho} = 1 \qquad \xi^+ \hat{\rho} \xi \geq 0, \quad \xi \in \mathbb{C}^2$$

Its dynamics is governed by the *Liouville equation*

$$i\hbar\dot{\hat{\rho}} = [\hat{H}, \hat{\rho}]$$

where $\hat{H}$ is the (quantum-mechanical) *Hamiltonian*, being in general a Hermitian operator (in $\mathbb{C}^2$, here). Check that

(i) the most general density operator and the Hamiltonian may be parametrized as

$$\hat{\rho} = \frac{1}{2}(\mathbb{I} + \mathbf{P} \cdot \boldsymbol{\sigma})$$
$$\hat{H} = \frac{\hbar}{2}\mathbf{B} \cdot \boldsymbol{\sigma} \qquad \mathbf{B}, \mathbf{P} = \text{constant}, \ |\mathbf{P}| \leq 1$$

[273] Also the restriction of $\mathcal{P}$ to the *other* orbits (the spheres with different radii) gives the symplectic forms proportional to the round volume forms on the orbits.

The vector $\mathbf{P}$, which (as we see) fully characterizes the state of the (two-level) system, is called the *polarization vector*. If $|\mathbf{P}| = 1$, we actually have a *pure state* (the vector $\mathbf{P} \equiv \mathbf{n}$ is then called the *vector of spin*); for $|\mathbf{P}| < 1$ we speak about the *mixed state*

(ii) when expressed in the language of the vectors $\mathbf{P}$ and $\mathbf{B}$, the dynamics governed by the Liouville equation manifests itself through the equation of motion

$$\dot{\mathbf{P}} = \mathbf{B} \times \mathbf{P}$$

(iii) this equation corresponds to a ("classical"!) *Poisson* dynamics with the *linear* Hamiltonian

$$\mathcal{P} = \frac{1}{2} P_k \epsilon_{kij} \frac{\partial}{\partial P_i} \wedge \frac{\partial}{\partial P_j} \qquad H(\mathbf{P}) = -\mathbf{B} \cdot \mathbf{P}$$

and thus the solution is a uniform rotation of the vector $\mathbf{P}$ around the vector $\mathbf{B}$

(iv) in particular, the dynamics of the *pure* states is *Hamiltonian* dynamics on the unit sphere, the symplectic form being the ordinary metric (round) volume form and the Hamiltonian being linear in $\mathbf{n}$

$$\dot{\mathbf{n}} = \mathbf{B} \times \mathbf{n} \qquad \mathbf{n}(\vartheta, \varphi) \equiv (\sin\vartheta \cos\varphi, \sin\vartheta \sin\varphi, \cos\vartheta)$$

$$\omega = \omega_g = \sin\vartheta \, d\vartheta \wedge d\varphi \qquad H(\vartheta, \varphi) = -\mathbf{B} \cdot \mathbf{n}(\vartheta, \varphi)$$

(v) this result is consistent with (14.2.3).

Hint: (i) the hermiticity gives $\hat{\rho} = a\mathbb{I} + \mathbf{b} \cdot \boldsymbol{\sigma}$ ($a, \mathbf{b}$ real), the unit trace $\hat{\rho} = \frac{1}{2}\mathbb{I} + \mathbf{b} \cdot \boldsymbol{\sigma}$; we have $(\lambda\xi)^+ \hat{\rho}(\lambda\xi) = |\lambda|^2 \xi^+ \hat{\rho}\xi$, so that the positivity suffices for normalized ξ; then $\xi^+ \hat{\rho}\xi = \frac{1}{2}(1 + 2\mathbf{b} \cdot \mathbf{c})$, where $\mathbf{c} := \xi^+ \boldsymbol{\sigma}\xi$; since $|\mathbf{c}| = 1$ for normalized ξ, the condition $2\mathbf{b} \cdot \mathbf{c} \geq -1$ needs $2|\mathbf{b}| \leq 1$; the term of the form $a\mathbb{I}$ is irrelevant in the Hamiltonian (it commutes with all operators, thus having no influence on the time development of the density operator); (ii) $i\hbar\dot{\hat{\rho}} = i\hbar/2(\dot{\mathbf{P}} \cdot \boldsymbol{\sigma})$, $[\hat{H}, \hat{\rho}] = \hbar/4[\mathbf{B} \cdot \boldsymbol{\sigma}, \mathbf{P} \cdot \boldsymbol{\sigma}] = i\hbar/2(\mathbf{B} \times \mathbf{P}) \cdot \boldsymbol{\sigma}$; (iii) then the Hamiltonian field is $\zeta_H = \mathcal{P}(dH, \cdot) = (\mathbf{B} \times \mathbf{P})_i \partial/\partial P_i$ as needed; (v) $\mathbf{B} = (0, 0, 1)$. □

14.7* Symplectic reduction

• One of the valuable lessons from elementary mechanics is that the analysis of numerous situations is greatly simplified by passing to the center of mass system, since this procedure eliminates the *irrelevant* degrees of freedom connected with the motion of the center of mass and the remaining contemplation concerns then only the *relevant* degrees of freedom; namely, those of the relative motion *with respect to* the center of mass.

To illustrate, we may take the two-body problem. At the beginning we have the variables describing the initial point masses, $(m_1, \mathbf{r}_1)$ and $(m_2, \mathbf{r}_2)$. At the first step we pass to the variables describing *two other* (fictitious) point masses, $(M, \mathbf{R})$ and $(\mu, \mathbf{r})$ (M being situated at the center of mass, μ is connected with the relative vector $\mathbf{r}$). At the second step we notice that the dynamics of the part $(M, \mathbf{R})$ is separated out (and moreover it turns out to be trivial). Therefore it is ignored at the third step and we focus on the problem concerning the part $(\mu, \mathbf{r})$ alone, the latter then being finalized with some effort. It is important for us to notice that we started with a 12-dimensional phase space $\mathbb{R}^{12}[\mathbf{r}_1, \mathbf{r}_2, \mathbf{p}_1, \mathbf{p}_2]$ and a Hamiltonian system in this large space, but after the third step we ultimately work only in a "reduced" six-dimensional phase space $\mathbb{R}^6[\mathbf{r}, \mathbf{p}]$ and in this *reduced phase space* we analyze a new (less dimensional, but still) *Hamiltonian* dynamics. It is well known that the grey eminence behind the possibility of the elimination of the (irrelevant) motion of the center of mass is the translational *invariance* of the initial problem.[274]

The procedure of reduction may be described in a fairly general setting, when a Hamiltonian system (M, ω, H) has a symmetry G. It turns out that the most interesting part of the dynamics corresponds in this situation effectively to a *smaller Hamiltonian* system $(\hat{M}, \hat{\omega}, \hat{H})$. In this section we first show how all this happens for the *symplectic* part of the problem alone (thus ignoring temporarily the Hamiltonian), i.e. we will be concerned with the construction of a smaller (reduced) symplectic manifold $(\hat{M}, \hat{\omega})$ from the initial bigger one (M, ω), making use of the symmetry available. (This procedure turns out to be interesting enough in its own right in recent mathematical physics.) Then we focus on the dynamics and show how the new Hamiltonian may be obtained ($H \mapsto \hat{H}$) on the new phase space, thus completing the reduction of the whole Hamiltonian system.

Consider a $2n$-dimensional symplectic manifold (M, ω) along with a *free* and *Poisson* action R_g of a Lie group G. Then there is the corresponding equivariant moment map $P : M \to \mathcal{G}^*$ (14.5.6). Recall that if G also happens to be the symmetry of the Hamiltonian (i.e. it is a symmetry of the whole Hamiltonian system (M, ω, H)), then the components P_i of the moment map P are conserved (14.5.7), so that the whole trajectory necessarily lies in a *subset* of the phase space M, namely in the part which P maps into a single point $p \in \mathcal{G}^*$. We therefore fix a point $p \in \mathcal{G}^*$ and denote by M_p the preimage of p with respect to P

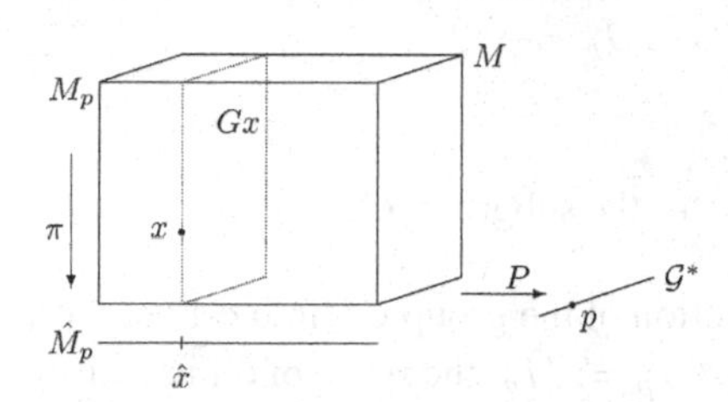

$$M_p := \{x \in M \mid P(x) = p\}$$

(in the figure it is represented by the front face of the cuboid M). It is a submanifold of dimension $2n - \dim G$, which is defined implicitly by the equations

$$P_i(x) = p_i \equiv \text{ constant}, \quad i = 1, \ldots, \dim G$$

14.7.1 Let x be an arbitrary point in M_p and let Gx be its orbit with respect to the action of G (the dotted plane on the figure). In the tangent space at the point x two natural subspaces then arise: the subspace tangent to the orbit (spanned by the generators of the action ξ_X) and that tangent to the submanifold M_p. Check that the two subspaces happen to be *symplectic orthogonal complements* to each other, meaning that if some vector has vanishing "scalar product" *in the sense of* ω with all vectors from one of them, then it is from the complement.

Hint: first check that (for a non-degenerate bilinear form) the complement of the complement turns out to be the initial subspace once again, so that it is enough to prove the statement

[274] The *homogeneity* of the empty space. Note that the decisive factor in the finalization of the remaining problem for $(\mu, \mathbf{r})$ is again a symmetry, this time the rotational symmetry (the *isotropy* of the space). The general technique which is described in this section is illustrated on the two-body problem at the end of Section 18.4.

in one direction (preferably the easier one, then). The generators of the action R_g are at the same time also Hamiltonian fields generated by the functions P_X (i.e. $\xi_X = \zeta_{P_X}$, see the text after (14.5.1)), so that

$$\omega(u, \xi_X) = (-i_{\xi_X}\omega)(u) = \langle dP_X, u\rangle = uP_X$$

Thus vanishing of the "scalar product" $\omega(u, \xi_X)$ for all ξ_X means that the vector u annihilates all the functions P_i and, consequently, that it is tangent to M_p. We see that the subspace tangent to M_p is just the complement of the subspace tangent to the orbit. □

14.7.2 Check that

(i) the *sets* M_p are transformed ("permuted") by the action R_g according to the formula

$$R_g M_p = M_{\mathrm{Ad}^*_g p}$$

(ii) if $G_p \subset G$ is the stabilizer of the point p (with respect to the coadjoint action Ad* on $\mathcal{G}^*$), then the restriction of the action R_g to G_p already leaves M_p fixed (as a set; the points inside the set may move)

$$R_g M_p = M_p \qquad g \in G_p$$

(iii) the group G_p acts on M_p *freely*.

Hint: (iii) R_g is free for the whole G and, a fortiori, for the subgroup G_p. □

14.7.3 The manifold M_p decomposes under the action of the group G_p into orbits[275] $\mathcal{O}_x$ and a natural projection onto the factor manifold $M_p/G_p =: \hat{M}_p$ (the space of orbits) arises, $\pi : M_p \to \hat{M}_p$ $(x \mapsto \hat{x} \equiv [x])$. Check that

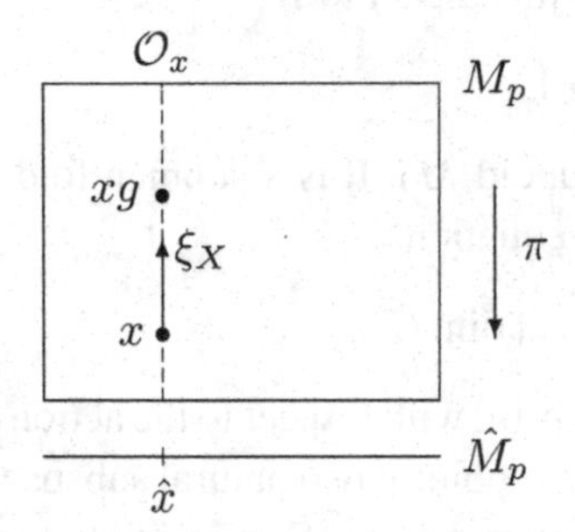

(i) the action of the group G_p on M_p is *vertical*, meaning that

$$\pi \circ R_g = \pi$$

so that it moves the points in the figure only in the vertical direction

(ii) the generators (fundamental fields) ξ_X of the action are *vertical*

$$\pi_* \xi_X = 0$$

so that one should draw ξ_X on the figure as a vertical arrow (tangent to the orbit $\mathcal{O}_x$ of the group G_p)

(iii) *each* vertical vector may be uniquely written as ξ_X for an appropriate element X from the Lie algebra of the group G_p

$$\pi_* w = 0 \quad \Rightarrow \quad w = \xi_X$$

[275] The orbit of the *whole* group G is denoted by Gx in the figure before problem (14.7.1), the orbit of the subgroup G_p by $\mathcal{O}_x$ here. There holds $\mathcal{O}_x = Gx \cap M_p$.

Hint: (i) we factorize just by means of this action; (iii) $\pi_* w = 0$ means that it is tangent to the orbit; *by definition* the whole orbit is, however, generated by the action of the group; the uniqueness stems from the *freedom* of the action. □

• Take a step forward, now, and pay attention to the symplectic form ω. This form lives on the initial manifold M. We may consider its *restriction* $\omega|_{M_p}$ to the submanifold M_p; it will be denoted by $\tilde{\omega}$. This 2-form on M_p has some remarkable properties, which enable one to use it for the construction of a *symplectic* form on $\hat{M}_p$.

14.7.4 Let $\tilde{\omega} \equiv \omega|_{M_p}$ be the restriction of the symplectic form ω to the submanifold M_p. Check that

(i) the form $\tilde{\omega}$ is G_p-invariant, i.e.

$$R_g^* \tilde{\omega} = \tilde{\omega} \qquad g \in G_p$$

(ii) the form $\tilde{\omega}$ is *horizontal*, i.e. it is annihilated by (even a single) *vertical* argument

$$\tilde{\omega}(w, \cdot\ ) = 0 \qquad \text{for any vertical vector } w$$

Hint: (i) if $j : M_p \to M$ is the canonical embedding, then $\tilde{\omega} = j^* \omega$, so that $R_g^* \tilde{\omega} = j^* R_g^* \omega = \tilde{\omega}$ due to the invariance of ω; (ii) for $w = \xi_X \equiv \zeta_{P_X}$ we get $\omega|_{M_p}(w, \cdot\ ) = (i_{\xi_X} \omega)|_{M_p} = -\ (dP_X)|_{M_p} = 0$, since P_X is constant on M_p. □

• It turns out, however, that whenever a form on M_p happens to be horizontal and at the same time G_p-invariant, it corresponds to a unique form on $\hat{M}_p$. This enables us to assign a unique form $\hat{\omega}$ on $\hat{M}_P$ to $\tilde{\omega}$ and a simple check then reveals that actually a *symplectic* form is obtained by this construction.

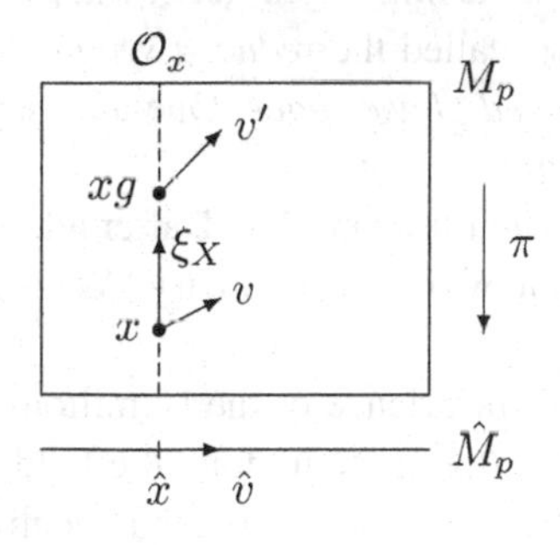

14.7.5 Check that forms on $\hat{M}_p$ are in one-to-one correspondence with G_p-invariant horizontal forms on M_p; namely, that each G_p-invariant horizontal form α on M_p may be regarded as the *pull-back* of a unique form $\hat{\alpha}$ on $\hat{M}_p$

$$R_g^* \alpha = \alpha \qquad i_{\xi_X} \alpha = 0 \qquad \Rightarrow \qquad \alpha = \pi^* \hat{\alpha}$$

Hint: for a given $\hat{\alpha}$ on $\hat{M}_p$ the form $\pi^* \hat{\alpha}$ *is* horizontal as well as G_p-invariant; now let α be a horizontal and G_p-invariant form on M_p; we assign to it the form $\hat{\alpha}$, which is defined in $\hat{x}$ by the prescription

$$\hat{\alpha}(\pi_* v, \dots) := \alpha(v, \dots) \qquad (\Rightarrow \alpha = \pi^* \hat{\alpha})$$

All the vectors in $\hat{x}$ are of the form $\pi_* v$, where v is a vector at a point over $\hat{x}$, for example in x or xg (the vectors v and v' project to the same image). Since R_{g*} is a linear isomorphism and $\pi \circ R_g = \pi$, we may write $v' = R_{g*} v + w$ for some *vertical* vector w in the point xg.

Then due to G_p-invariance of α we get

$$\begin{aligned}\alpha_{xg}(v', \dots) &= \alpha_{xg}(R_{g*}v + w, \dots) = \alpha_{xg}(R_{g*}v, \dots) \\ &= (R_g^*\alpha)_x(v, \dots) = \alpha_x(v, \dots)\end{aligned}$$

so that $\hat{\alpha}$ is well defined (it does not depend on arbitrary choices). □

• Now we apply this result to the form $\tilde{\omega}$. This form, being invariant as well as horizontal, *can be* uniquely "projected" on $\hat{M}_p$. Moreover it is also *closed* and we will see that this is transferred to the form $\hat{\omega}$ on $\hat{M}_p$, too. Finally, one checks that $\hat{\omega}$ is also non-degenerate, so that a *symplectic* form on $\hat{M}_p$ is actually obtained by this construction.

14.7.6 Let $\tilde{\omega} \equiv \omega|_{M_p}$ be the restriction of the symplectic form ω to the submanifold M_p. Since it is G_p-invariant and horizontal, a unique form $\hat{\omega}$ may be assigned to it on $\hat{M}_p$. Check that the 2-form $\hat{\omega}$ is a *symplectic form*, i.e. verify that it is closed and non-degenerate.

Hint: closedness: if $j : M_p \to M$ is the canonical embedding, then

$$\tilde{\omega} = j^*\omega = \pi^*\hat{\omega} \qquad \text{so that} \qquad 0 = j^*d\omega = \pi^*(d\hat{\omega}) \Rightarrow d\hat{\omega} = 0$$

(π^* is injective on forms); non-degeneracy: let $\hat{\omega}(\pi_*u, \pi_*v) = 0$ for arbitrary π_*v, i.e. $\tilde{\omega}(u, v) = 0$ for arbitrary v, which is tangent to M_p. The vector u is thus "symplectic orthogonal" to an arbitrary vector, which is tangent to M_p; then, due to (14.7.1), it is tangent to the orbit Gx and since it moreover lives on M_p, it is *vertical*, i.e. of the form $u = \xi_X$ for some X from the Lie algebra of the group G_p. Such u projects, however, to zero, $\pi_*u = 0$. □

• So far we treated the "symplectic part" of the Hamiltonian system alone and we came to the end of this story – after a series of steps we obtained from the initial symplectic manifold (M, ω) a new (smaller) symplectic manifold $(\hat{M}_p, \hat{\omega})$. It is called the *reduced symplectic manifold* and within the mechanical context also the *reduced phase space*. One also says that $(\hat{M}_p, \hat{\omega})$ results from (M, ω) by *reduction by the group G*.

And what about the rest of the Hamiltonian system, the Hamiltonian H and, after all, the Hamiltonian (dynamical) field ζ_H? Is it also reduced somehow by the symmetry? We may check easily that this indeed *is the case*.

The G-invariance of ω and H immediately yields the G-invariance of the Hamiltonian field ζ_H on M. This field is tangent to the submanifold M_p (the motion along the field = the time development and the latter preserves, according to (14.5.7), all P_is) so that we may consider its restriction to this submanifold as well as to the subgroup which acts on M_p; we thus get on M_p a G_p-invariant vector field. Such fields, however, "project" uniquely onto vector fields on $\hat{M}_p$; we declare *our* vector field obtained in this way to be *the* dynamical field on $\hat{M}_p$ (the time development := motion along its integral curves), call it the *reduced field* and denote it by $\hat{\zeta}_H$.

14.7.7 Check that the dynamical field $\hat{\zeta}_H$ is actually a *Hamiltonian* field, its Hamiltonian $\hat{H}$ being the "restriction" of the original Hamiltonian H to $\hat{M}_p$.

Hint: the question is whether $\hat{\omega}(\hat{\zeta}_H, \hat{W}) \overset{?}{=} \langle -d\hat{H}, \hat{W}\rangle \equiv -\hat{W}\hat{H}$ for each vector field $\hat{W}$ on $\hat{M}_p$. All vector fields on $\hat{M}_p$ may be regarded as projections of some G_p-invariant fields on M_p; let $\hat{W}$ stem from W (i.e. $\pi_* W = \hat{W}$). Then the question is $\tilde{\omega}(\zeta_H, W) \overset{?}{=} -WH$, which *is true*, since on the *whole* manifold M there holds $\omega(\zeta_H, W) = -WH$ for each W. □

• So we finally verified that the procedure of the reduction by the group G results in a new (and smaller) *reduced Hamiltonian system* $(\hat{M}_p, \hat{\omega}, \hat{H})$ starting from the original system (M, ω, H). This procedure is illustrated in detail on concrete examples in problems (18.4.11)–(18.4.17) (for the particular symmetries "lifted from the configuration space").

As we learned in Section 14.1, any symplectic form corresponds to a unique non-degenerate Poisson tensor and eventually to the Poisson bracket. This means that the symplectic reduction treated above should have an equivalent counterpart in terms of the Poisson bracket

$$(M, \omega) \mapsto (\hat{M}_p, \hat{\omega}) \quad \Rightarrow \quad (M, \mathcal{P}) \mapsto (\hat{M}_p, \hat{\mathcal{P}})$$

(where $\hat{\mathcal{P}}$ is "inverse" to $\hat{\omega}$) or, alternatively, in terms of the algebra of observables

$$(M, \omega) \mapsto (\hat{M}_p, \hat{\omega}) \quad \Rightarrow \quad \mathcal{A} \equiv (\mathcal{F}(M), \{\cdot\,,\cdot\}) \mapsto (\mathcal{F}(\hat{M}_p), \{\cdot\,,\cdot\}^\wedge) \equiv \hat{\mathcal{A}}_p$$

The reduced algebra of observables (i.e. effectively the whole procedure of reduction) may be equivalently described also in terms of the initial algebra of observables. We concentrate first on the construction of an algebra which is isomorphic to the algebra of *functions* $\mathcal{F}(\hat{M}_p)$ (ignoring thus temporarily the new Poisson bracket $\{\cdot\,,\cdot\}^\wedge$ and focusing attention on an algebraic description of the transition to the resulting *manifold* $M \mapsto M_p$ alone). For the sake of simplicity, we restrict in what follows to a *one-dimensional* symmetry group (so that the reduced phase space has *two* dimensions fewer than the initial space).

The manifold $\hat{M}_p$ arises in two steps from M, $M \mapsto M_p \mapsto \hat{M}_p$. The first step, $M \mapsto M_p$, consists in restriction to a *subset* and the second step, $M_p \mapsto \hat{M}_p$, is based on a *factorization*. It turns out that both of these procedures have a parallel at the level of the algebra of functions, being *reversed*, however: the algebra of functions on a submanifold may be obtained naturally by a *factorization* of the original algebra of functions on the whole manifold, whereas the algebra of functions on a factorized manifold is realized as a *subset* of the original algebra.

14.7.8 Let N be a submanifold of a manifold M, $\mathcal{F}(M)$ and $\mathcal{F}(N)$ denote their algebras of functions and let $\mathcal{I}_N \subset \mathcal{F}(M)$ be the set of functions on M *vanishing on* N

$$\mathcal{I}_N := \{f \in \mathcal{F}(M) \mid f(x) = 0,\ x \in N \subset M\}$$

Check that

(i) the set $\mathcal{I}_N$ is an *ideal* in the algebra $\mathcal{F}(M)$ (see Appendix A.2)

(ii) the factorization of the algebra $\mathcal{F}(M)$ with respect to $\mathcal{I}_N$ results in an algebra which is isomorphic to the algebra $\mathcal{F}(N)$

$$\mathcal{F}(M)/\mathcal{I}_N \approx \mathcal{F}(N)$$

(iii) if the submanifold N is singled out by the equations $\Phi_\alpha(x) = 0$ (the differentials $d\Phi_\alpha$ being independent throughout the set $\Phi_\alpha(x) = 0$) then for the *differentials* of the elements of the ideal $\mathcal{I}_N$ there holds

$$di(x) = u^\alpha(x)\, d\Phi_\alpha(x)$$

Hint: (ii) introduce in $\mathcal{F}(M)$ the equivalence $f \sim g \Leftrightarrow f$ and g coincide on N (thus f, g are equivalent if they may be regarded as smooth extensions of a *single* function from N to the whole of M). It is clear that there is a bijection between equivalence classes and functions on N. The equivalence classes may be regarded at the same time as the elements of the factor-algebra mentioned above: note that $f - g$ vanishes on N, thus being from the ideal $\mathcal{I}_N$, or $f = g + i, i \in \mathcal{I}_N$; (iii) in the tangent space $T_x M$ there is a subspace $T_x N \subset T_x M$ of vectors which are tangent to N; the subspace induces a subspace $W \subset T_x^* M$ in the cotangent space (the *annihilator* of $T_x N$; it consists of the covectors which annihilate the vectors from $T_x N$, see (2.4.19)), spanned by a basis $d\Phi_\alpha$ (Φ_α vanish (i.e. are constant) on N); since the function $i(x)$ is constant (vanishes) on N, its differential *is* from W and therefore it *may be* decomposed with respect to the basis $d\Phi_\alpha$. □

• Let us apply all this to the case of our interest, where the manifold M_p is singled out by the equation

$$P - p = 0 \qquad p \in \mathcal{G}^* \cong \mathbb{R}$$

This means that the algebra of functions on M_p is isomorphic to the factor-algebra of the algebra of functions on M with respect to the ideal I of functions $\psi(x)$ vanishing in the points x such that $P(x) = p$; for the differentials of these functions there holds

$$d\psi(x) = u(x)\, dP(x)$$

and so *at the level of differentials* we have an equivalence[276]

$$df \sim df + u(x)\, dP(x)$$

Let us turn our attention now to the case of the factorization of manifolds.

14.7.9 Let $\pi : M \to N$ be a projection of M onto the factor manifold $N \equiv M/\sim$ (where $\sim$ denotes equivalence, with respect to which the manifold is factorized; if it arises, for example, by the action of a group on M, then N is the manifold of orbits). Denote by $\mathcal{F}_\pi(M) \subset \mathcal{F}(M)$ the set of functions on M which are *constant on the preimages* of the map π, i.e.

$$\mathcal{F}_\pi(M) := \{f \in \mathcal{F}(M) \mid \pi(x) = \pi(x') \Rightarrow f(x) = f(x')\}$$

[276] *This* particular equivalence will come in handy in a moment in (14.7.11) where the Poisson bracket will be examined.

Check that $\mathcal{F}_\pi(M)$ is an algebra (a subalgebra of $\mathcal{F}(M)$), which is isomorphic to the algebra $\mathcal{F}(N)$

$$\mathcal{F}_\pi(M) \approx \mathcal{F}(N)$$

Hint: $\hat{f} \leftrightarrow f$ for $\hat{f}(\pi(x)) = f(x)$. □

• In the case of our interest this yields the requirement of constancy *on the orbits* $\mathcal{O}_x$. This condition may also be written in terms of Poisson brackets.

14.7.10 Check that the functions which are constant on the orbits coincide with the functions which are in involution with the moment map[277] P

$$f \text{ is constant on orbits} \quad \Leftrightarrow \quad \{f, P\} = 0$$

Hint: we consider the Poisson action (its generator is the Hamiltonian field generated by $P(x)$). □

14.7.11 Let $\mathcal{A}$ be the algebra of observables on the original symplectic manifold (M, ω). Denote by $\hat{\mathcal{A}}$ the algebra which results from $\mathcal{F}(M)$ by the combination of both procedures mentioned above, i.e.

1. we consider only those functions which satisfy

$$\{f, P\} = 0$$

2. we introduce the equivalence among them

$$f(x) \sim f(x) + \psi(x) \qquad \psi(x) = 0, \quad x \in M_p$$

The algebra $\hat{\mathcal{A}}$ is isomorphic (regarded as an associative algebra) to the algebra $\mathcal{F}(\hat{M}_p)$ of functions on the manifold $\hat{M}_p$. Check that by the prescription

$$\{[f], [g]\} := [\{f, g\}]$$

(i.e. by means of representatives) *also the Poisson bracket* is correctly inherited from $\mathcal{A}$ onto $\hat{\mathcal{A}}$, so that $\hat{\mathcal{A}}$ is isomorphic to the (already combined, both associative and Lie) *algebra of observables* corresponding to the reduced phase space $(\hat{M}_p, \hat{\omega})$.

Hint: for the representatives $f + \psi$ and $g + \chi$ (where ψ and χ vanish on M_p, thus being from the ideal) we get (making use of $\{f, P\} = 0 = \{g, P\}$)

$$\begin{aligned}
\{f + \psi, g + \chi\} &= \mathcal{P}(df + d\psi, dg + d\chi) = \mathcal{P}(df + u\,dP, dg + v\,dP) \\
&= \{f, g\} + v\mathcal{P}(df, dP) + u\mathcal{P}(dP, dg) + uv\mathcal{P}(dP, dP) \\
&= \{f, g\} + v\{f, P\} + u\{P, g\} + uv\{P, P\} \\
&= \{f, g\}
\end{aligned}$$

□

[277] The moment map, being in general $\mathcal{G}^*$-valued, reduces now to an "ordinary function," since we consider a *one-dimensional* Lie algebra $\mathcal{G}$, which may be identified with $\mathbb{R}$.

Summary of Chapter 14

An appropriate relabeling of coordinates reveals that there is an elegant geometrical structure hidden behind the Hamilton canonical equations. Its essential part is a closed non-degenerate 2-form ω on the phase space of the system, the *symplectic* form. It enables us to raise and lower the indices in a similar manner as we did before with the metric tensor. The vector field which is the counterpart of the gradient field in the Riemannian case (i.e. which is obtained by raising an index on the gradient of a function f understood as a *co*vector field) is called the *Hamiltonian* field generated by the function f. The Hamilton equations turn out to be simply the equations for the integral curves of the Hamiltonian field generated by a distinguished function H, the *Hamiltonian* of the system. Thus we come to the notion of the Hamiltonian *system* (M, ω, H). The vector fields which generate automorphisms of a Hamiltonian system (they preserve the symplectic form as well as the Hamiltonian) are called Cartan symmetries and, those obeying a specific additional property, *exact* Cartan symmetries. There is a one-to-one correspondence between the exact Cartan symmetries and the conserved quantities of the system. More details can be found in sections devoted to the moment map and symplectic reduction. The orbits of the coadjoint action (which is an action of the group G on the dual space $\mathcal{G}^*$ of its own Lie algebra $\mathcal{G}$) provide a rich source of interesting (G-invariant) symplectic manifolds (there is a canonical symplectic structure on them).

$\zeta_f = \mathcal{P}(df, \cdot\,)$	Hamiltonian field in terms of $\mathcal{P}$	(14.1.1)
$\{f, g\} = \mathcal{P}(df, dg)$	Poisson bracket in terms of the Poisson tensor $\mathcal{P}$	(14.1.1)
$\dot{\gamma} = \zeta_H$	Hamilton equations – coordinate-free version	(14.1.1)
$i_{\zeta_f}\omega = -df$	Hamiltonian field in terms of symplectic form ω	(14.1.6)
$\omega = dp_a \wedge dq^a$	Symplectic form in canonical (Darboux) coordinates	(14.2.2)
$\Omega_\omega := \text{constant}\ \omega \wedge \cdots \wedge \omega$	Liouville volume form on (M, ω)	(14.3.6)
$\int_{\Phi_t(\mathcal{D})} \Omega_\omega = \int_{\mathcal{D}} \Omega_\omega$	Liouville's theorem	(14.3.6)
$i_V\omega = -dF, \ \ VH = 0$	V is exact Cartan symmetry of (M, ω, H)	(14.4.2)
$\gamma_s(t) := \Phi_s^V(\gamma(t))$	A new solution generated by a symmetry flow	(14.4.6)
$\langle P(x), X\rangle := P_X(x)$	Moment map corresponding to the Poisson action	(14.5.3)
$\omega_{Z^*}(\xi_X, \xi_Y) := \langle Z^*, [X, Y]\rangle$	Canonical symplectic form on coadjoint orbits	(14.6.3)

15
Parallel transport and linear connection on M

• In Chapter 4 we already encountered the possibility of transporting vectors, namely Lie transport and the related Lie derivative. Here we introduce another type of transport, the so-called *parallel* transport. The corresponding derivative is called the *covariant derivative*. Although the two transports share some common features, in many respects their geometrical meaning differs and one should understand which one is appropriate for use in a concrete application.

15.1 Acceleration and parallel transport

• Recall (see Section 2.2) that it makes no direct sense to perform a linear combination of a vector at point x with a vector at point y since the tangent spaces at different points $x, y \in M$ are vector spaces which are not at all related (except for the dimension).

15.1.1 Let $B : W \to V$ be an isomorphism of linear spaces V and W. Check that the rule

$$v + \lambda w := v + \lambda B(w) \qquad v \in V, \quad w \in W$$

gives a sense to the linear combination of two vectors from *different* spaces.

Hint: B enables one to identify V with W. □

• So if a *distinguished* (canonical, independent of arbitrary choices) isomorphism $B : T_yM \to T_xM$ existed, we might define the combination by the trick $v + \lambda w := v + \lambda B(w)$, $v \in T_xM$, $w \in T_yM$. However, already in Section 2.2 we warned that although on a "bare" manifold the spaces T_xM and T_yM are isomorphic, the isomorphism is *not* canonical.

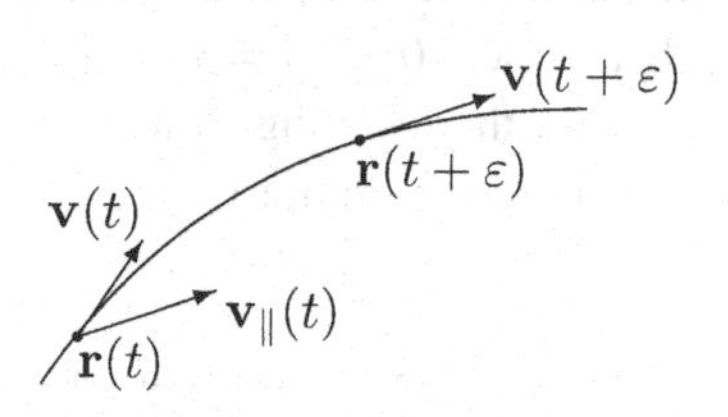

After these general considerations let us have a look at how it is related to the definition of the concept of *acceleration* of a point mass in elementary mechanics. Given $\mathbf{r}(t)$, the trajectory of the point, its (instantaneous) velocity is $\mathbf{v}(t) := \dot{\mathbf{r}}(t)$ and the (instantaneous) acceleration is $\mathbf{a}(t) := \dot{\mathbf{v}}(t)$. The conception of the velocity of the point mass helped us to introduce the key concept of a vector on a manifold as early as in Chapter 2. The acceleration turns out to be equally inspiring (although it had to wait patiently for its chance until Chapter 15) – it brings us to the concept of the parallel transport of the vector as well as the covariant derivative.

In order to compute $\mathbf{a}(t)$ at the point $\mathbf{r}(t)$ one has by definition to perform the *difference* (i.e. a linear combination) of the vectors $\mathbf{v}(t+\varepsilon)$ and $\mathbf{v}(t)$, i.e. just the procedure which makes no sense officially. It is clear, however, what is meant by this (and often even explicitly stated) in mechanics: the "obvious" fact is used that in E^3 the vectors may be shifted (not altering either their length or direction) around and then used at the point where we need them to sit.[278] In the case of acceleration one is namely to shift the vector $\mathbf{v}(t+\varepsilon)$ from the point $\mathbf{r}(t+\varepsilon)$ back to the point $\mathbf{r}(t)$ (thus obtaining $\mathbf{v}_{\parallel}(t)$) and only this vector may be compared with the vector $\mathbf{v}(t)$. So it is nothing but the trick from (15.1.1), the role of the isomorphism B being played by an appropriate shift. Everything is so clear *here* that one might even be abashed at why an issue should be made of all this.[279]

Nevertheless, there is a tiny cloudlet in the blue sky, namely a slightly conspicuous significance of the Cartesian coordinates in the technical realization of the shifts (on a general manifold all the local coordinates should be equivalent; the fact that this is *not the case* here indicates that E^3 is exceptional from this point of view).

15.1.2 Verify that the operation of the shift of a vector in the Euclidean space E^3 (or, even simpler, in E^2) happens to be technically trivial in Cartesian coordinates *alone* (if we accept as Cartesian coordinates also those with a shifted origin and a modified direction of axes, so that they are related through an *affine* transformation to some "true" Cartesian coordinates).

Hint: in Cartesian coordinates the components of the shifted vector remain *the same*; transform this explicitly to polar, spherical polar, cylindrical, etc. coordinates and check that it becomes complicated. □

- The particular trajectories with *vanishing* acceleration are closely related to the concept of acceleration. They correspond to *uniform straight-line* motion of the point mass. By definition, vanishing of the acceleration means that on this trajectory the velocity $\mathbf{v}(t+\varepsilon)$ is equal to the velocity $\mathbf{v}(t)$. Or, more precisely (since they sit at different points), the velocity vector $\mathbf{v}(t+\varepsilon)$ arises by a *shift* alone (not changing either its length or direction) of the vector $\mathbf{v}(t)$ from the point $\mathbf{r}(t)$ to the point $\mathbf{r}(t+\varepsilon)$ (so that we set $\mathbf{v}(t+\epsilon)=\mathbf{v}_{\parallel}(t+\epsilon)$).

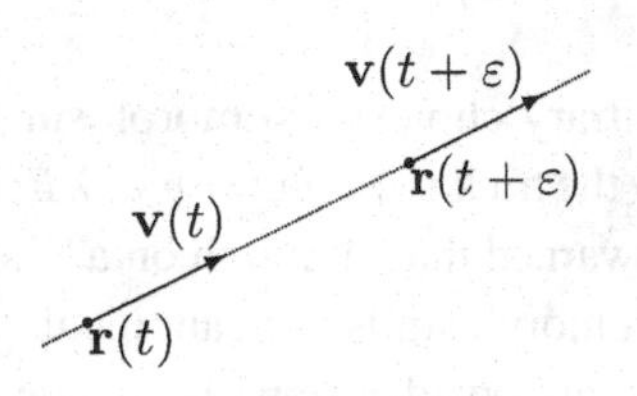

By an iteration of such infinitesimal shifts of the velocity vector the resulting straight line arises, being the trajectory of the point mass (the uniform straight-line motion).

278 One speaks about *free* and *bounded* vectors there.

279 Mathematical physics is sometimes blamed for "making an issue" of quite "simple" things. There is a perfect consensus in that this blame is indeed legitimate in p percent of concrete cases, a bit less concord takes place in the numerical value of the number p. Extensive research (based on elaborate questionnaires) revealed that the distribution of p over the world population is actually uniform, bounded by the values $p=0$ and $p=100$.

Now, let us try to repeat the same procedure on a different manifold, for example on the sphere S^2. Imagine that an ambitious technocratic ideal was accomplished at last – throughout the Earth, first all the irregularities were straightened out by bulldozers (they were, one should admit, both impractical and unaesthetic) and then the whole surface of the Earth was nicely covered by a neat asphalt. If we now roll a ball along such a smooth surface,[280] it has to roll, according to the laws of mechanics, uniformly along a straight line, since the only force available is the gravitational force, directed everywhere downwards. This force constrains the ball to remain on the two-dimensional *surface* of the Earth (it keeps the ball from flying away along a "truly" straight-line trajectory and escaping into space); the ball gets accustomed to this status quo and it does not regard it as a restriction.[281] It considers pragmatically the sphere S^2 to be its living space and it does not care whether the sphere actually is or is not a subset of any larger ambient space. Since the projection of the gravitational force on to the plane which is tangent to the sphere always vanishes, the ball feels[282] *no force* acting on it and it thus has no reason to change its velocity (neither length nor direction); it therefore moves with vanishing acceleration along a straight line. Note, however, that from the point of view of the ambient space E^3 this is by no means an ordinary straight line, but rather it is a circle (with maximum possible radius), which encircles the whole Earth. The uniform motion along this circle which arises by the iteration of the (infinitesimal) shifts of the velocity vector is the straightest possible motion on the sphere. The shift of the velocity vector keeping its length as well as direction unaltered *in the sense of the sphere* is, as we see from the resulting trajectory, something considerably different from the same procedure performed *in the sense of* E^3 – from the point of view of E^3, in the course of the shifts the vector also continually *rotates* a bit in order to remain tangent to the sphere.

The lesson from this as well as numerous similar particular cases resulted in the following picture: the definition of the concept of *acceleration* as well as *uniform straight-line motion* (i.e. motion with zero acceleration) which is based on it requires the ability to *transport* the velocity *vector* (at least by infinitesimal distances) along a given trajectory. In the space E^3 there is a natural rule of transport and this rule is indeed used in elementary mechanics in E^3. However, in general the matter may not be so simple. It turns out that the most fruitful point of view is to regard the rule of transporting vectors on a manifold as an *independent structure*, which is *a priori not available* on a general manifold, although in particular cases (like in E^3) there may exist most natural realizations.

If such a rule (satisfying some requirements) is introduced on a manifold, we say that a *parallel transport* (or an associated concept – *linear connection*) is defined on M, denoted by (M, ∇). For example, the natural parallel transport of vectors in E^3 is realized as an ordinary shift, but if we introduced another connection into E^3 (which *is* perfectly possible), the parallel transport would be performed in a different way. The straight lines which result

[280] We also ensure zero air resistance and a couple of similar technical details.

[281] This is confidential information from one such ball; for reasons of protection of privacy it has no wish to make either its center or radius public.

[282] See the previous footnote.

from the iteration of the infinitesimal parallel transport of the velocity vector (the trajectories with zero acceleration) are called the (affinely parametrized) *geodesics* on (M, ∇).

The concept of a linear connection is very important in physics, although its presence is fairly obscure in many applications (like in acceleration in elementary mechanics).

15.1.3 Estimate (or evaluate exactly) the fraction

$$f = lc/a$$

where a denotes the number of people on Earth who understand what the acceleration is (including the formula which enables one to compute it) and lc denotes the number of people on Earth who are aware that the linear connection is used in this formula.

Hint: ask all of them and then divide the two numbers; (1.1.1)–(22.5.12). □

• However, there are also disciplines like the general theory of relativity, in which the linear connection lies at the very heart of the mathematical formulation, being explicitly present in the fundamental equations of the theory.

15.1.4 Estimate (or evaluate exactly) the fraction

$$f = lc/gr$$

where gr denotes the number of people on Earth who understand elements of general relativity (including the basic formulas) and lc denotes the number of people on Earth who are aware that the linear connection is used in these formulas.

Hint: see the hint in (15.1.3). □

• The far-reaching generalization of the linear connection, to be explained in more detail in Chapter 20 and beyond, is the basis of the formalism of modern gauge field theories.

15.2 Parallel transport and covariant derivative

• We convinced ourselves that the introduction of the concept of acceleration requires the ability to transport velocity vectors along curves (the trajectories of a point mass). A similar requirement also occurs in numerous other contexts. We say that a *rule of parallel transport* is given on a manifold M, if, for an arbitrary curve γ on M and two points x, y on the curve, there is a prescription which assigns uniquely to vectors in x vectors in y, i.e. a map

$$\tau^{\gamma}_{y,x} : T_x M \to T_y M \qquad v \mapsto \tau^{\gamma}_{y,x} v$$

Clearly, one can think out lots of such rules, but if they are to be useful in the contexts from which the motivation for their introduction came, they should satisfy some restrictive conditions. For the moment we mention the two most important of them.

First, it is natural to ask that the transport of a sum of vectors or a multiple of a vector by a constant should yield the sum of the results of the transport of the individual vectors or the multiple of the transported vector, i.e. to ask for the *linearity* of the map $\tau^{\gamma}_{y,x}$. Secondly, if there are *three* points x, y, z on a curve γ, we expect the parallel transport from x to y followed by the transport (of the vector just brought to y) from y to z to yield the same result as the direct transport (without a moment's rest in y) from x to z would yield. This may be written as the composition property of the maps $\tau^{\gamma}_{y,x}$

$$\tau^{\gamma}_{z,y} \circ \tau^{\gamma}_{y,x} = \tau^{\gamma}_{z,x} \qquad x, y, z \text{ on the curve } \gamma \text{ (otherwise arbitrary)}$$

and, in particular,

$$\tau^{\gamma}_{x,x} = \text{identity} \qquad \left(\tau^{\gamma}_{y,x}\right)^{-1} = \tau^{\gamma}_{x,y}$$

Note that the rule of parallel transport needs as an input not only the edge points x, y, but also a *path* connecting them.[283] So if we are given at the point x a vector v and a path from x to y, we are able to transport v uniquely to the point y; given another path, the transport is unique as well, but the resulting transported vector may be different in general. We will see that the path dependence of the parallel transport is an important and typical phenomenon in the situations where the connection is applied and it enables one to speak about the *curvature* of the manifold (M, ∇).[284]

Suppose we have some particular fixed rule of a parallel transport of vectors. This rule then enables one to introduce a derivative, which is based on it. Namely, let $\gamma(t)$ be a curve and let $V(t) \equiv V_{\gamma(t)}$ be a vector field defined on the curve.[285] If we intend to differentiate the vector field V along the curve γ, in order to find out whether (and how much) it varies in this direction), we are to compare the vectors $V(t+\varepsilon)$ and $V(t)$. However, these two vectors sit at different points and it means that their difference has no *direct* meaning. Still, the difference of the vectors may be legalized by making use of the rule of parallel transport. Namely, we first *transport* the vector $V(t+\varepsilon)$ along the curve γ from the point $\gamma(t+\varepsilon)$ backwards to the point $\gamma(t)$ and then we compare (subtract) the vector transported back with the initial vector $V(t)$.

Denote the vector transported backwards by $V^{\parallel}_{\varepsilon}(t)$. Then the corresponding derivative, which is called the *absolute derivative* of the vector field V along the curve γ, is defined as

$$\frac{DV(t)}{Dt} := \lim_{\varepsilon \to 0} \frac{V^{\parallel}_{\varepsilon}(t) - V(t)}{\varepsilon}$$

Let us contemplate some immediate consequences of the definition. First, it is clear that the derivative depends on the particular rule of parallel transport.

Next, note that it uses only the behavior of the objects on the curve γ – the field V may (but need not) be defined also outside the curve, but nothing from outside the curve has any

[283] We mentioned a *curve* a minute ago, here we speak about (only) a *path*, i.e. a non-parametrized curve. The transport to be studied here actually depends only on the path alone (see (15.2.6) and (15.2.12)).

[284] This does not mean that the parallel transport always *indeed* depends on a path, but rather that *in general it may* depend on it. For example, the ordinary shifts of vectors in E^3 are evidently path-independent, whereas the transport of the vectors on the sphere, which we discussed in Section 15.1 really depends on a path (15.3.9).

[285] The vector $V(t)$ is an element of the tangent space $T_{\gamma(t)}M$ and it may not be directed along the curve; i.e. we contemplate n-dimensional vectors which need not exist at each point of an n-dimensional domain, as is the case when we treat vector fields on a domain, but they are instead defined only on a one-dimensional domain, at the points of the curve γ.

influence on the value of the derivative along the curve. This differs essentially from the *Lie* derivative. If the curve γ were the integral curve of a vector field W and if both the fields, W and V, were defined in some neighborhood of the curve γ, then the Lie derivative of the field V along W (which corresponds to the derivative of V along the curve γ) would also depend on the values of the field W *outside* the curve γ (15.2.4), so that the transport of the field V along the curve γ actually depends on the structure of *additional curves* aside from the curve γ (the neighboring integral curves of the field W; note that they are indeed necessary since the Lie derivative may be performed only on the fields defined (at least) in a domain).

Realize finally that the *vanishing* of the absolute derivative on some (part of a) curve means that the field $V(t)$ may then be regarded as that its values everywhere on γ arose by (only) a parallel transport of its value at a single fixed point into all the points of the (above mentioned part of the) curve. Such a field on a curve (one might say that it is *constant* on the curve) is called an *autoparallel field.* Thus the absolute derivative informs us exactly about the *deviation from being autoparallel.*

The relation between the absolute derivative of a vector field and the rule of the parallel transport may be used to reverse the roles of what is a "primary" concept and what is a "secondary" one: if we were technically able to perform the derivative, it would allow us in turn to reconstruct the rule of parallel transport. Namely, the rule says: do the transport so *as to make* the derivative *vanish.* This is exactly the way one usually introduces the concept of the linear connection on a manifold. Instead of specifying in detail the requirements which the parallel transport should satisfy, one postulates, on the contrary, the properties of the derivative and the parallel transport is then in turn defined by the simple equation "the derivative should vanish." The corresponding defining properties of the derivative are to be chosen so as to be clear and brief and so as not to contradict any particular case of the transport, which served as the source of inspiration for the general theory (like E^3, the sphere, etc.; i.e. so as to guarantee that all the useful known cases might be regarded as "particular cases of a general theory").

Before we write down the resulting requirements regarding the derivative, we realize that we also have to contemplate the issue of the parallel transport (as well as the derivative) of general *tensors* (just like we did for the Lie derivative).[286] For the Lie stuff, where the "primary" concept was the (Lie) transport (being realized technically as the pull-back Φ_t^* of the flow $\Phi_t \leftrightarrow W$), this issue was simply computed and it *turned out* that the transport preserves the degree and commutes with the tensor product and the contraction, so that the (Lie) derivative *turned out* to be the derivation of the tensor algebra, which preserves the degree and commutes with contraction.

Here it is necessary to *postulate* the properties either at the level of the (parallel) transport, or at the level of the (covariant) derivative. The standard definition says that in this respect we simply copy the properties in the Lie case: one postulates that the (parallel) transport

[286] Note that there are the same problems with the linear combinations of tensors of type $\binom{p}{q}$ at different points x, y, as with vectors (being the special case $p = 1, q = 0$), *except for the case of scalars* ($p = q = 0$): the numbers in x and in y are "canonically" combined without any problems.

preserves the degree and commutes with the tensor product and contraction.[287] The (covariant) derivative, which corresponds to the (parallel) transport, is then necessarily a derivation of the tensor algebra, which preserves the degree and commutes with contractions.

So at last we now state the official definition of the concept of *linear connection* on a manifold M. It says that with each vector field W on M one may associate an operator ∇_W, the *covariant derivative* along the field W, enjoying the following properties:

1. it is a linear operator on the tensor algebra, which preserves the degree
$$\nabla_W : \mathcal{T}^p_q(M) \to \mathcal{T}^p_q(M)$$
$$\nabla_W(A + \lambda B) = \nabla_W A + \lambda \nabla_W B \qquad A, B \in \mathcal{T}^p_q(M), \quad \lambda \in \mathbb{R}$$
2. on a tensor product it behaves according to the Leibniz rule
$$\nabla_W(A \otimes B) = (\nabla_W A) \otimes B + A \otimes (\nabla_W B) \qquad A \in \mathcal{T}^p_q(M), B \in \mathcal{T}^{p'}_{q'}(M)$$
3. in degree $\binom{0}{0}$ (i.e. the functions) it gives
$$\nabla_W \psi = W\psi \equiv \mathcal{L}_W \psi \qquad \psi \in \mathcal{F}(M) \equiv \mathcal{T}^0_0(M)$$
4. it commutes with contractions
$$\nabla_W \circ C = C \circ \nabla_W \qquad C = \text{ (any) contraction}$$
5. it is $\mathcal{F}$*-linear* with respect to W, i.e.[288]
$$\nabla_{V+fW} = \nabla_V + f\nabla_W \qquad V, W \in \mathfrak{X}(M), \quad f \in \mathcal{F}(M)$$

Now, let us have a look at how such a connection may be technically determined.

15.2.1 Show that the covariant derivative is uniquely specified by the *coefficients of linear connection* $\Gamma^a_{bc}(x)$ with respect to a frame field e_a, which are the functions defined by
$$\nabla_a e_b =: \Gamma^c_{ba} e_c \qquad \nabla_a := \nabla_{e_a}$$

Solution:[289] we have (the numbers indicate the property used)
$$\nabla_W \big(A^{a\ldots b}_{c\ldots d} e^c \otimes \cdots \otimes e_b\big) = \text{making use of } 1, 2, 3$$
$$= \big(W A^{a\ldots b}_{c\ldots d}\big) e^c \otimes \cdots \otimes e_b + A^{a\ldots b}_{c\ldots d} (\nabla_W e^c) \otimes \cdots \otimes e_b + \cdots + A^{a\ldots b}_{c\ldots d} e^c \otimes \cdots \otimes (\nabla_W e_b)$$

Thus, one needs to be able to compute $\nabla_W e_a$ and $\nabla_W e^b$. If $W = W^b e_b$ then
$$\nabla_W e_a \equiv \nabla_{(W^b e_b)} e_a \overset{5}{=} W^b \nabla_b e_a \equiv \big(\Gamma^c_{ab} W^b\big) e_c$$

[287] Preserving of a degree is clear, commuting with the contraction in plain English says that the transported tensor yields the same number on the transported arguments as the original tensor did on the original arguments.

[288] This is the only property in which the operator of the covariant derivative ∇_W *differs* from the operator of the Lie derivative $\mathcal{L}_W$ ($\mathcal{L}_W$ happens to be only $\mathbb{R}$-linear with respect to W); it turns out that it reflects the requirement mentioned above, so as the parallel transport does not depend (in contrast with the Lie transport) on objects outside the curve.

[289] It may be briefly summarized as follows: ∇_W is a derivation of the tensor algebra $\Rightarrow$ one comes to $\binom{0}{0}$ and the bases of $\binom{1}{0}$ and $\binom{0}{1}$. The case of $\binom{0}{0}$ is handled by property 3, the commuting with contractions enables one to reduce $\binom{0}{1}$ to $\binom{1}{0}$.

and since

$$\begin{aligned}
0 &\overset{3}{=} \nabla_W \delta^a_b = \nabla_W \langle e^a, e_b \rangle = \nabla_W (C(e^a \otimes e_b)) \\
&\overset{4,2}{=} C((\nabla_W e^a) \otimes e_b + e^a \otimes (\nabla_W e_b)) = \langle \nabla_W e^a, e_b \rangle + \langle e^a, \nabla_W e_b \rangle \\
&= \langle \nabla_W e^a, e_b \rangle + \langle e^a, \Gamma^d_{bc} W^c e_d \rangle = (\nabla_W e^a)_b + \Gamma^a_{bc} W^c
\end{aligned}$$

we obtain

$$\nabla_W e^a = -(\Gamma^a_{bc} W^c) e^b \qquad \text{and in particular} \quad \nabla_b e^a = -\Gamma^a_{cb} e^c$$

The knowledge of the coefficients of linear connection $\Gamma^c_{ab}(x)$ with respect to a frame field $e_a(x)$ thus indeed enables one to compute $\nabla_W A$ for arbitrary W and A, i.e. there is complete information about the connection in them. □

• The coefficients of the linear connection have one upper index and two lower indices. One might anticipate from this that they form the components of a tensor field of type $\binom{1}{2}$. A computation yields something different, however.

15.2.2 Let $e_a \mapsto e'_a = A^b_a(x) e_b$ be a change of a frame field. Check that the primed coefficients of linear connection (given by the prescription $\nabla_{e'_a} e'_b =: \Gamma'^c_{ba} e'_c$) are related to the unprimed coefficients by

$$\Gamma'^c_{ab} = \Gamma^d_{ef} (A^{-1})^c_d A^e_a A^f_b + (A^{-1})^c_d A^f_b (e_f A^d_a)$$

so that in addition to the first term, corresponding (if it were alone) to a tensor of type $\binom{1}{2}$, there is also the "non-tensorial" second term (which does not contain the unprimed coefficients Γ at all; one speaks about an *inhomogeneous* transformation rule).

Hint: $\nabla_{e'_a} e'_b = A^c_a \nabla_{e_c} (A^d_b e_d) = \dots$; use the properties of the covariant derivative and the definition of the initial coefficients themselves. □

15.2.3 The coefficients of linear connection $\Gamma^k_{ij}(x)$ with respect to the *coordinate* frame field $e_i = \partial_i$ are called the *Christoffel symbols* of the *second kind*. Thus, they are defined by

$$\nabla_j \partial_i =: \Gamma^k_{ij} \partial_k \qquad \nabla_i := \nabla_{\partial_i}$$

Check that under the change of coordinates $x^i \mapsto x'^i(x)$ the following transformation rule holds:

$$\Gamma'^i_{jk} = \frac{\partial x'^i}{\partial x^r} \frac{\partial x^s}{\partial x'^j} \frac{\partial x^m}{\partial x'^k} \Gamma^r_{sm} + \frac{\partial x'^i}{\partial x^r} \frac{\partial^2 x^r}{\partial x'^j \partial x'^k}$$

Hint: (15.2.2), $A^i_j = (J^{-1})^i_j = \partial x^i / \partial x'^j$, $A^f_b (e_f A^d_a) = (e'_b A^d_a)$. □

• Since the connection is a global structure on a manifold, this fairly complicated transformation rule for the Christoffel symbols should necessarily have the correct composition properties on a triple overlap of charts (see Section 2.5). This may be verified "by brute force" here, but one can do this more easily after learning how to encode the coefficients of linear connection into so-called connection 1-forms (15.6.2).

Making use of the covariant derivative we are now in a position to realize the program outlined at the beginning of the section: to express the absolute derivative and then to describe the parallel transport.

15.2.4 Let (M, ∇) be a manifold endowed with a connection, ∇_W the corresponding covariant derivative operator, $\gamma(t)$ a curve and V a vector field. The *absolute derivative* of the field V along γ is defined as[290]

$$\frac{DV(t)}{Dt} := \nabla_{\dot{\gamma}} V$$

Assume that the curve γ happens to be the integral curve of the field W (i.e. $\dot{\gamma} = W$ on γ) and that both fields V, W as well as the curve γ are given in a coordinate patch. Check that

(i) in local coordinates we get on the curve γ

$$\nabla_W V = \left(\dot{V}^i + \Gamma^i_{jk}\dot{x}^k V^j\right)\partial_i$$
$$\mathcal{L}_W V = \left(\dot{V}^i - V^j W^i_{\ ,j}\right)\partial_i$$

where $V^i(t) := V^i(\gamma(t))$ are the components of the field V, regarded as the functions on the curve alone

(ii) from the expression of the covariant derivative we can infer that no knowledge of the field W outside the curve is necessary for its computation and thus the formula $\nabla_{\dot{\gamma}} := \nabla_W$ is indeed correct (recall that the connection officially defines only the notion of the covariant derivative along the vector *field* W) and, consequently, also the definition of the absolute derivative is all right; the expression of the Lie derivative, on the contrary, shows that the behavior of W in a neighborhood of the curve has an influence on this object.

Hint: (i) $\nabla_W V = (\nabla_W V^i)\partial_i + V^i W^j \nabla_j \partial_i = (\dot{\gamma} V^i)\partial_i + \Gamma^k_{ij}\dot{x}^j V^i \partial_k$; here $\Gamma^k_{ij}\dot{x}^j \equiv \Gamma^k_{ij}(\gamma(t))\dot{x}^j(t)$ is a known function of t on the curve; (ii) $\mathcal{L}_W V = (WV^i - VW^i)\partial_i \equiv (\dot{V}^i - V^j W^i_{\ ,j})\partial_i$; to compute $VW^i \equiv V^j W^i_{\ ,j}$ we also need to know W in a neighborhood of γ. □

15.2.5 A vector field V on γ will be called *autoparallel* if its absolute derivative along γ vanishes, i.e. if

$$\frac{DV(t)}{Dt} \equiv \nabla_{\dot{\gamma}} V = 0$$

Check that

(i) the components $V^i(t) := V^i(\gamma(t))$ then satisfy the equations

$$\dot{V}^i + \Gamma^i_{jk}\dot{x}^k V^j = 0$$

[290] The absolute derivative was defined before in terms of the parallel transport (assumed to be known), here it is defined from the opposite point of view, namely in terms of the (known) covariant derivative.

(ii) these equations may be written in the form

$$\dot{V}^i(t) = S^i_j(t)V^j(t) \qquad S^i_j(t) \text{ known functions of } t$$

so that they form an autonomous system of $n = \dim M$ ordinary first-order linear differential equations with *non*-constant coefficients.

Hint: (i) (15.2.4); (ii) $S^i_j(t) = -\Gamma^i_{jk}(\gamma(t))\dot{x}^k(t)$, which is a known function, provided that the connection (represented by $\Gamma^i_{jk}(x)$) and the curve (in the form of $x^i(t)$) are given. □

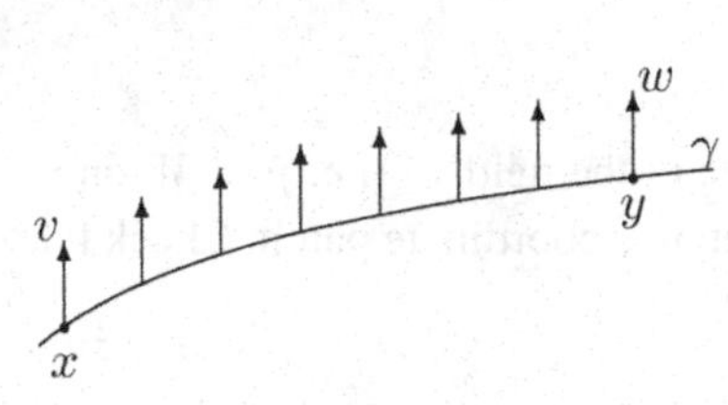

15.2.6 In terms of the covariant derivative we may now introduce the operation of the *parallel transport* of a vector along a curve as follows: if there is a vector v at a point x on a curve γ and we want to transport the vector to a point y, then we first construct the autoparallel field generated by the vector v and then take its value at the point y; this value w will be regarded as the result of the parallel transport of the vector, $w = \tau^\gamma_{y,x} v$. Show that

(i) if v has the components (with respect to a coordinate basis in x) $v^i \in \mathbb{R}$, then the components $w^i \in \mathbb{R}$ of the transported vector w (with respect to the coordinate basis in y) are obtained by solving the *equations of parallel transport*

$$\dot{V}^i + \Gamma^i_{jk}\dot{x}^k V^j = 0 \qquad \text{i.e.} \quad \dot{V}^i(t) = S^i_j(t)V^j(t)$$

(for $S^i_j(t) = -\Gamma^i_{jk}(\gamma(t))\dot{x}^k(t)$) with the initial condition $V^i(t_1) = v^i$ (if $x = \gamma(t_1)$); then w^i are obtained as the value of the solution for $t = t_2$ (if $y = \gamma(t_2)$); so in brief

$$\dot{V}^i(t) = S^i_j(t)V^j(t) \qquad V^i(t_1) = v^i \quad w^i := V^i(t_2)$$

(ii) if the assignment $v \mapsto w$ is interpreted as a map $\tau^\gamma_{y,x} : T_x M \to T_y M$, then the map (the *operator of the parallel transport*) is *linear* and it satisfies the requirement

$$\tau^\gamma_{z,y} \circ \tau^\gamma_{y,x} = \tau^\gamma_{z,x} \qquad x, y, z \text{ on the curve } \gamma \text{ (otherwise arbitrary)}$$

(iii) the parallel transport does not feel the parametrization of the curve, i.e. it depends on the *path* rather than on the *curve*.

Hint: (i) according to (15.2.5) the equations $\dot{V}^i + \Gamma^i_{jk}\dot{x}^j V^k = 0$ along with the initial condition $V^i(t_1) = v^i$ yield the unique autoparallel field $V(t)$ generated by the vector v; (ii) the composition property of the operator $\tau^\gamma_{y,x}$ results immediately from the fact that the solution of the equations is *unique* – its linearity stems from the linearity of the equations (a solution linearly depends on the initial conditions); (iii) from the form of the equations of parallel transport

$$\frac{dV^i}{dt} + \Gamma^i_{jk}\frac{dx^k}{dt}V^j = 0$$

we see the reparametrization invariance of the latter (there is the relation $\delta V^i = -\Gamma^i_{jk} V^k \delta x^j$ between the infinitesimal increments irrespective of a parametrization of $x^i(t)$, or alternatively if $V^i(t)$ is a solution for $x^i(t)$, then the solution for $x^i(\sigma(t))$ reads $V^i(\sigma(t))$).

□

• Before we embark on developing a technique of transporting general tensor fields, we need to derive the coordinate component formulas for the computation of the covariant derivatives of an arbitrary tensor field, i.e. to finalize the computation from (15.2.1).

15.2.7 Check that

(i) for the covariant derivative of the coordinate frame and coframe fields there holds

$$\nabla_j \partial_i = +\Gamma^k_{ij}\partial_k \qquad \nabla_W \partial_i = +\Gamma^k_{ij} W^j \partial_k$$
$$\nabla_j dx^i = -\Gamma^i_{kj} dx^k \qquad \nabla_W dx^i = -\Gamma^i_{kj} W^j dx^k$$

(ii) the component formula for the covariant derivative of a general tensor field reads

$$(\nabla_W A)^{i\dots j}_{k\dots l} = W^m A^{i\dots j}_{k\dots l,m} - \Gamma^n_{km} W^m A^{i\dots j}_{n\dots l} - \dots + \Gamma^j_{nm} W^m A^{i\dots n}_{k\dots l}$$

(iii) this result may be concisely summarized in the form of a table – a recipe for cooking the house speciality $(\nabla_W A)^{i\dots j}_{k\dots l}$ (the recipe for the Lie derivative is also repeated for the convenience of gourmets)

	for preparation of $\nabla_W A$	for $\mathcal{L}_W A$
put on the bottom of a pan	$WA^{\dots}_{\dots} \equiv W^m A^{\dots}_{\dots,m}$	$WA^{\dots}_{\dots} \equiv W^m A^{\dots}_{\dots,m}$
plus for each $A^{\dots i \dots}$ add	$+W^m \Gamma^i_{nm} A^{\dots n \dots}$	$-W^i_{\ ,m} A^{\dots m \dots}$
plus for each $A_{\dots i \dots}$ add	$-W^m \Gamma^n_{im} A_{\dots n \dots}$	$+W^m_{\ ,i} A_{\dots m \dots}$

Hint: (15.2.1) and (15.2.3); compare with (4.3.4). □

15.2.8 Compute the components of the tensor $\nabla_W g$, the covariant derivative of the *metric* tensor along the vector field W.

Hint: the table yields (there is the lump part plus two terms for two lower indices) $(\nabla_W g)_{ij} = W^m g_{ij,m} - W^m \Gamma^n_{im} g_{nj} - W^m \Gamma^n_{jm} g_{in} \equiv W^m (g_{ij,m} - \Gamma^n_{im} g_{nj} - \Gamma^n_{jm} g_{in})$. □

15.2.9 The $\mathcal{F}$-linearity of the operator ∇_W with respect to W enables one to introduce the operation of the *covariant gradient* by

$$\nabla : \mathcal{T}^p_q(M) \to \mathcal{T}^p_{q+1}(M) \qquad (\nabla A)(V, \dots, W; \alpha, \dots) := (\nabla_W A)(V, \dots; \alpha \dots)$$

Check that

(i) ∇A is indeed a tensor field of the type stated above, so that ∇ is a *tensor* operation
(ii) in components (with respect to the coordinate basis) it gives

$$(\nabla A)^{i\dots j}_{k\dots lm} = (\nabla_m A)^{i\dots j}_{k\dots l} =: A^{i\dots j}_{k\dots l;m}$$

where

$$A^{i\dots j}_{k\dots l;m} := A^{i\dots j}_{k\dots l,m} - \Gamma^n_{km}A^{i\dots j}_{n\dots l} - \dots + \Gamma^j_{nm}A^{i\dots n}_{k\dots l}$$

(iii) the computation of $A^{i\dots j}_{k\dots l;m}$ is performed according to the recipe

	for preparation of $A^{\dots}_{\dots;m}$
first put on the bottom of a pan	$\partial_m A^{\dots}_{\dots} \equiv A^{\dots}_{\dots,m}$
plus for each $A^{\dots i \dots}$ add	$+\Gamma^i_{nm}A^{\dots n \dots}$
plus for each $A_{\dots i \dots}$ add	$-\Gamma^n_{im}A_{\dots n \dots}$

(iv) for a general (possibly non-coordinate) frame field we have

$$(\nabla A)^{a\dots b}_{c\dots df} \equiv (\nabla_f A)^{a\dots b}_{c\dots d} =: A^{a\dots b}_{c\dots d;f}$$

where

$$A^{a\dots b}_{c\dots d;f} := e_f A^{a\dots b}_{c\dots d} - \Gamma^n_{cf}A^{a\dots b}_{n\dots d} - \dots + \Gamma^b_{nf}A^{a\dots n}_{c\dots d}$$

(v) for $p = q = 0$ (on functions) the covariant gradient coincides with the "ordinary" gradient (regarded as a *co*vector field)

$$\nabla f = df \qquad f \in \mathcal{F}(M)$$

(vi) the covariant derivative along W may be written in terms of the covariant gradient as

$$(\nabla_W A)^{i\dots j}_{k\dots l} = W^m A^{i\dots j}_{k\dots l;m}$$

Hint: (15.2.7). □

• So it holds that "semicolon" = comma *plus* a term containing the Christoffel symbols added for each index; the expression $A^{i\dots j}_{k\dots l;m}$ is usually called in short the "*covariant derivative* of $A^{i\dots j}_{k\dots l}$ *by m*" and it consists of the partial derivative by m plus the terms with Christoffel symbols.

The covariant gradient may be regarded as a "derivative in an unspecified direction"; if then one intends to compute the (covariant) derivative along a particular vector W, a "scalar product" of the vector is to be performed with the "semi-finished product" ∇A (this immediately results from the $\mathcal{F}$-linearity: $\nabla_W A = W^m \nabla_m A$). If the covariant gradient of a tensor field happens to vanish in some domain, the field is said to be *covariantly constant.* Then the covariant derivative of the field along *any direction* vanishes and so the field may be regarded as being transported into all the points within the domain from its value at a single point (just like the value of a constant *function* in an arbitrary point is known as long as its value at a single point is known).

15.2.10 Evaluate the components of the tensor ∇g, the covariant gradient of the *metric* tensor.

Hint: the table yields (there is the lump part plus two terms for two lower indices) $(\nabla g)_{ijk} \equiv g_{ij;k} = g_{ij,k} - \Gamma^l_{ik}g_{lj} - \Gamma^l_{jk}g_{il}$. □

• Now we may return to the parallel transport of tensors. The concepts of the absolute derivative, the autoparallel field and parallel transport may be extended in a straightforward way from vector fields to arbitrary tensor fields.

15.2.11 The *absolute derivative* of a tensor field A along γ is defined as

$$\frac{DA(t)}{Dt} := \nabla_{\dot{\gamma}} A$$

and the field A on γ is called *autoparallel* if its absolute derivative along γ vanishes. Check that

(i) the concept of the absolute derivative is well defined (if on γ there holds $\dot{\gamma} = W$, then the derivative does not depend on W outside γ)

(ii) in local coordinates the condition for $A^{i\ldots j}_{k\ldots l}(t) := A^{i\ldots j}_{k\ldots l}(\gamma(t))$ being autoparallel reads

$$\dot{A}^{i\ldots j}_{k\ldots l} + \dot{x}^m \left(\Gamma^i_{nm} A^{n\ldots j}_{k\ldots l} + \cdots - \Gamma^n_{lm} A^{i\ldots j}_{k\ldots n}\right) = 0$$

(iii) this equation may also be written as

$$\dot{A}^{i\ldots j}_{k\ldots l}(t) = S^{i\ldots jc\ldots d}_{k\ldots la\ldots b}(t) A^{a\ldots b}_{c\ldots d}(t) \qquad S^{i\ldots jc\ldots d}_{k\ldots la\ldots b}(t) \text{ known functions of } t$$

so that (if $n = \dim M$) they form an autonomous system of n^{p+q} ordinary first-order linear differential equations with *non*-constant coefficients.

Hint: (15.2.4), (15.2.5) and (15.2.7). □

15.2.12 The operation of *parallel transport* of a tensor field A along γ is introduced as follows: if there is a tensor $\hat{a}$ at a point x on a curve γ and we want to transport the tensor to a point y, then we first construct the autoparallel field generated by the tensor $\hat{a}$ and then take its value at the point y; this value $\hat{b}$ will be regarded as the result of the parallel transport of the tensor, $\hat{b} = \tau^{\gamma}_{y,x}\hat{a}$. Show that

(i) the components $\hat{b}^{i\ldots j}_{k\ldots l} \in \mathbb{R}$ of the transported tensor $\hat{b}$ (with respect to the coordinate basis in y) are obtained by solving the *equations of parallel transport*

$$\dot{A}^{i\ldots j}_{k\ldots l} + \dot{x}^m \left(\Gamma^i_{nm} A^{n\ldots j}_{k\ldots l} + \cdots - \Gamma^n_{lm} A^{i\ldots j}_{k\ldots n}\right) = 0$$

with the initial condition $A^{i\ldots j}_{k\ldots l}(t_1) = \hat{a}^{i\ldots j}_{k\ldots l}$ and $\hat{b}^{i\ldots j}_{k\ldots l}$ are obtained as the value of the solution for $t = t_2$; so in brief

$$\dot{A}^{i\ldots j}_{k\ldots l}(t) = S^{i\ldots jc\ldots d}_{k\ldots la\ldots b}(t) A^{a\ldots b}_{c\ldots d}(t) \qquad A^{i\ldots j}_{k\ldots l}(t_1) = \hat{a}^{i\ldots j}_{k\ldots l} \qquad \hat{b}^{i\ldots j}_{k\ldots l} := A^{i\ldots j}_{k\ldots l}(t_2)$$

(ii) if the assignment $\hat{a} \mapsto \hat{b}$ is interpreted as a map $\tau^{\gamma}_{y,x} : T^p_{qx} M \to T^p_{qy} M$, then the map (the *operator of parallel transport*) is *linear* and it satisfies the requirement

$$\tau^{\gamma}_{z,y} \circ \tau^{\gamma}_{y,x} = \tau^{\gamma}_{z,x} \qquad x, y, z \text{ on the curve } \gamma \text{ (otherwise arbitrary)}$$

(iii) the parallel transport depends on the *path* rather than on the *curve*.

Hint: just like in (15.2.6). □

• In Section 4.4 we learned that the operator of Lie transport Φ_t^* may be expressed in the form of the exponent of the Lie derivative, $\Phi_t^* = e^{t\mathcal{L}_W}$. There is a similar possibility also for the parallel transport and the covariant derivative, since the formula stems from the composition property of the transport, being valid in *both* cases under consideration.

15.2.13 Let $\gamma(t)$ be the integral curve of a field W. Denote by τ_t^γ the operator of parallel transport *backwards* along γ by the parametric distance t, i.e.

$$\tau_t^\gamma := \tau_{\gamma(s),\gamma(s+t)}^\gamma$$

for any s. Show that

(i) τ_t^γ has the composition property

$$\tau_{t+s}^\gamma = \tau_t^\gamma \circ \tau_s^\gamma$$

(ii) the covariant derivative may be expressed as

$$\nabla_W A = \left.\frac{d}{ds}\right|_0 \tau_s^\gamma A$$

(iii) for the derivative of τ_t^γ with respect to t there holds

$$\frac{d}{dt}\tau_t^\gamma = \tau_t^\gamma \circ \nabla_W$$

(iv) for C^ω tensor fields we may write

$$\tau_t^\gamma = e^{t\nabla_W} \equiv 1 + t\nabla_W + \frac{t^2}{2!}\nabla_W\nabla_W + \cdots$$

(v) the ordinary Taylor expansion of a function

$$\psi(x+t) = \psi(x) + t\psi'(x) + \frac{t^2}{2!}\psi''(x) + \cdots$$

may be regarded as a special case for $(M, \nabla) = (\mathbb{R}[x]$, *arbitrary* connection on $\mathbb{R})$, $W = \partial_x$.

Hint: (i) (15.2.6) and (15.2.12); (iii) $\frac{d}{dt}\tau_t^\gamma = \frac{d}{ds}\big|_{s=0}\tau_{t+s}^\gamma$; (iv) $(\frac{d}{dt})^n\tau_t^\gamma = \cdots = \tau_t^\gamma(\nabla_W)^n$, (4.4.2); (v) (4.4.1). □

• This expression enables one, just like in the case of the Lie derivative, to perform a systematic expansion of the operator of *infinitesimal* parallel transport τ_ε^γ in terms of powers of ε; for example, to within second-order accuracy in ε we have $\tau_\varepsilon^\gamma = e^{\varepsilon\nabla_W} \equiv 1 + \varepsilon\nabla_W + \frac{\varepsilon^2}{2!}\nabla_W\nabla_W$. This will be used for the study of the relation between the *curvature* and the dependence of the parallel transport on a path in Section 15.5.

15.3 Compatibility with metric, RLC connection

• All the particular examples of parallel transport which we mentioned in Section 15.1, namely in E^2 and E^3 as well as on the sphere S^2, shared a common property: the vectors preserve the *length* under the transport. This means, however, that we actually treat the manifolds (M, g, ∇) endowed with a *pair* of structures, the metric tensor g, which enables us to measure the lengths of the vectors and the linear connection ∇, which enables us to transport the vectors along paths. The invariance of the length of vectors under parallel transport means that the connection is *compatible* with the metric, or in short that we treat the *metric connection*. In the component language this may also be stated as that for some particular Christoffel symbols $\Gamma^i_{jk}(x)$, being dependent on given $g_{ij}(x)$, a computation of the change of the length of an arbitrary vector under parallel transport yields zero. Let us focus our attention on this fact in more detail, now.

15.3.1 Consider a vector v at a point x on a curve γ. Starting from v, generate an autoparallel field V ($\nabla_{\dot{\gamma}} V = 0$). Check that

(i) the requirement of preservation of the length of v by parallel transport may be stated as

$$\nabla_{\dot{\gamma}}(g(V, V)) = 0 \qquad \text{if} \quad \nabla_{\dot{\gamma}} V = 0$$

(ii) if this is to be true for an *arbitrary* curve γ and an *arbitrary* initial vector v, then for any *two* vector fields W, V one should demand

$$\nabla_W(g(V, V)) = 0 \qquad \text{if} \quad \nabla_W V = 0$$

(iii) if this is to be true for any two *equal* arguments V, V, it should also be true for *any* two (possibly different) arguments,[291] i.e. for any *three* vector fields W, V, U the covariant derivative should obey

$$\nabla_W(g(V, U)) = 0 \qquad \text{if} \quad \nabla_W V = 0 = \nabla_W U$$

(iv) this condition is equivalent to the requirement

$$\nabla g = 0 \qquad \text{or in local coordinates} \qquad g_{ij;k} = 0$$

A connection ∇ which satisfies this equation is called the *metric connection*.

Hint: (i) the expression $f(t) := g(V(t), V(t))$ is a function on the curve and $\dot{f} = \dot{\gamma} f = \nabla_{\dot{\gamma}} f$; (iii) $g(U + V, U + V) = g(U, U) + g(V, V) + 2g(U, V)$; (iv) $(\nabla g)(V, U, W) = (\nabla_W g)(V, U)$ and $g(V, U) = CC(g \otimes V \otimes U) \Rightarrow \nabla_W(g(V, U)) = (\nabla_W g)(V, U) + g(\nabla_W V, U) + g(V, \nabla_W U)$. □

15.3.2 Check that the requirement

$$g_{ij;k} = 0$$

represents $n^2(n + 1)/2$ constraints imposed on n^3 functions (the Christoffel symbols $\Gamma^i_{jk}(x)$), so that it is very promising; it even seems that one could satisfy an *additional* $n^2(n - 1)/2$ constraints.

Hint: $g_{ij} = g_{ji}$. □

[291] Preserving of all lengths under the parallel transport thus also automatically leads to the preserving of all angles between the vectors.

• It turns out that this naive counting of the "degrees of freedom" indeed leads to a true conclusion. A metric indeed induces a lot of compatible linear connections. Moreover, if one *adds one more* condition, the vanishing *torsion* of the connection, the solution is even unique. So let us first explain what exactly the torsion is and then prove the main result indicated above.

15.3.3 Let (M, ∇) be a manifold with a linear connection. Check that

(i) the map

$$T : \mathfrak{X}(M) \times \mathfrak{X}(M) \to \mathfrak{X}(M) \qquad T(U, V) := \nabla_U V - \nabla_V U - [U, V]$$

is $\mathcal{F}(M)$-linear in both arguments, so that actually a tensor field of type $\binom{1}{2}$, which is associated with the connection ∇, is defined by this rule; it is called the *torsion tensor*[292] (or briefly the *torsion*) connection ∇

(ii) the tensor is antisymmetric (in the lower indices)

$$T(U, V) = -T(V, U) \qquad \text{i.e.} \quad T^i_{jk} = -T^i_{kj}$$

and so it has $n^2(n-1)/2$ independent components

(iii) for its (coordinate) components one obtains the expression

$$\langle dx^i, T(\partial_j, \partial_k)\rangle \equiv T^i_{jk} = \Gamma^i_{kj} - \Gamma^i_{jk} \equiv -2\Gamma^i_{[jk]} \qquad \text{i.e.} \quad \Gamma^i_{jk} = \Gamma^i_{(jk)} - \frac{1}{2}T^i_{jk}$$

(iv) if the torsion of the connection *vanishes*, i.e. if

$$\nabla_U V - \nabla_V U = [U, V]$$

then the Christoffel symbols are *symmetric* in the lower indices

$$\Gamma^i_{jk} = \Gamma^i_{kj}$$

this is the motivation to call it the *symmetric connection*

(v) the coefficients of a symmetric connection Γ^a_{bc} with respect to a *non-holonomic* basis e_a *are not* symmetric in the lower indices.[293]

Hint: (v) (non-vanishing) coefficients of *anholonomy* (see (9.2.10)) enter the formula. □

• Each linear connection is thus characterized (also) by its torsion and, in particular, the torsion of the symmetric connection (by definition) vanishes (the connection is then said to be *torsion-free*). If the connection is required to be at the same time metric and symmetric, it imposes altogether n^3 constraints on n^3 functions Γ^i_{jk}. This "rule of thumb" calculation indicates that the connection with this property might be unique.

15.3.4 Show that there is a unique connection which is simultaneously metric and symmetric. In order to do this check step by step that

[292] Geometrical meaning of the torsion is studied in (15.8.1).

[293] Since the coefficients of a connection *do not* constitute the components of a tensor, it *may* happen that they are symmetric with respect to one basis, but they are not symmetric in another one. It may even happen, just as is true for the anholonomy coefficients, that they *vanish* in one basis, but they do not vanish in another one; see, for example, (15.3.5).

(i) the Christoffel symbols of the *first kind*[294]

$$\Gamma_{ijk} := g_{il}\Gamma^l_{jk}$$

of the connection which is metric and symmetric satisfy

$$\begin{aligned} \Gamma_{ijk} + \Gamma_{jik} &= g_{ij,k} && \text{since it is metric} \\ \Gamma_{ijk} - \Gamma_{ikj} &= 0 && \text{since it is symmetric} \end{aligned}$$

(ii) the two relations result in

$$g_{ij,k} + g_{ik,j} - g_{jk,i} = 2\Gamma_{ijk}$$

and eventually

$$\Gamma^i_{jk} = \frac{1}{2}g^{il}(g_{lj,k} + g_{lk,j} - g_{jk,l})$$

so that the requirement of being metric and symmetric indeed leads to the unique result for the Christoffel symbols of the connection.[295] This distinguished linear connection on a Riemannian manifold is usually called the *Riemann connection* or the *Levi-Civita connection*; we will therefore use the abbreviation *RLC connection*[296]

(iii) the non-coordinate definition of the RLC connection reads

$$\begin{aligned} g(\nabla_U V, W) := \frac{1}{2}\{&Ug(V, W) + Vg(U, W) - Wg(U, V) \\ &+ g([U, V], W) - g([U, W], V) - g(U, [V, W])\} \end{aligned}$$

which is to be regarded as a definition of the expression $\nabla_U V$ in terms of the right-hand side, where no covariant derivative occurs.

Hint: (i) (15.3.1), (15.2.10) and (15.3.3); (iii) check the $\mathcal{F}$-linearity of the right-hand side and set the coordinate basis as U, V, W. □

• Since we already created a stockpile of the manifolds endowed with the metric, the formulas obtained in this problem enable us to examine everything concerning connections on real examples.

15.3.5 Compute the Christoffel symbols of the RLC connection in E^n directly from the formula in (15.3.4) and check that

(i) for arbitrary n we obtain in Cartesian coordinates $\Gamma^i_{jk} = 0$
(ii) for $n = 2$ in polar coordinates the only non-vanishing gammas read[297]

$$\Gamma^r_{\varphi\varphi} = -r \qquad \Gamma^\varphi_{r\varphi} = 1/r$$

(iii) the same result for polar coordinates may be also obtained by transforming the Cartesian Christoffel symbols (which are zero according to item (i)) to the polar coordinates by means of (15.2.3)

[294] Since $\Gamma^i_{jk}(x)$ *do not* constitute the components of a tensor, this *is not* the operation of the lowering of the index; it is indeed a *definition*.
[295] One may check that the transformational properties of $g_{ij}(x)$ under the change of coordinates yield the proper (15.2.3) transformational properties of $\Gamma^i_{jk}(x)$.
[296] Its role in the analysis of RLC circuits in electronics still remains obscure.
[297] Due to the symmetry we do not list Γ^i_{kj} explicitly, if Γ^i_{jk} is already there.

(iv) for $n = 3$ in spherical polar coordinates the only non-vanishing gammas are

$$\Gamma^r_{\vartheta\vartheta} = -r \qquad \Gamma^\vartheta_{r\vartheta} = 1/r \qquad \Gamma^\varphi_{r\varphi} = 1/r$$
$$\Gamma^r_{\varphi\varphi} = -r\sin^2\vartheta \qquad \Gamma^\vartheta_{\varphi\varphi} = -\sin\vartheta\cos\vartheta \qquad \Gamma^\varphi_{\vartheta\varphi} = \cot\vartheta$$

and in cylindrical coordinates one gets (just like in polar coordinates in the plane)

$$\Gamma^r_{\varphi\varphi} = -r \qquad \Gamma^\varphi_{r\varphi} = 1/r.$$

□

15.3.6 Check that these expressions for Christoffel symbols in E^n result in a common concept of the parallel transport rule of vectors in the Euclidean space E^n (the vectors are just shifted with no change either of length or direction).

Hint: according to (15.2.6) and (15.3.5) the equations of parallel transport along *any* curve read in *Cartesian* coordinates $\dot{V}^i = 0$, whence $V^i(t) = \text{constant}$. □

15.3.7 Use the formula obtained in (15.3.4) to compute the Christoffel symbols of the RLC connection on the sphere S^2 with the standard metric; check that in coordinates ϑ, φ the only non-vanishing symbols are

$$\Gamma^\vartheta_{\varphi\varphi} = -\sin\vartheta\cos\vartheta \qquad \Gamma^\varphi_{\vartheta\varphi} = \frac{\cos\vartheta}{\sin\vartheta}$$

Hint: see (3.2.4). □

15.3.8 Check that on the sphere S^2 with the standard metric the equations of parallel transport of a vector V read as follows:

(i) along a general curve $\vartheta(t), \varphi(t)$

$$\dot{V}^\vartheta - \dot{\varphi}\sin\vartheta\cos\vartheta\, V^\varphi = 0 \qquad \dot{V}^\varphi + \frac{\cos\vartheta}{\sin\vartheta}(\dot{\vartheta}V^\varphi + \dot{\varphi}V^\vartheta) = 0$$

(ii) along a parallel (of latitude) parametrized as $\vartheta(t) = \vartheta_0, \varphi(t) = t$

$$\dot{V}^\vartheta - \sin\vartheta_0\cos\vartheta_0\, V^\varphi = 0 \qquad \dot{V}^\varphi + \frac{\cos\vartheta_0}{\sin\vartheta_0}V^\vartheta = 0$$

and, in particular, along the equator

$$\dot{V}^\vartheta = 0 \qquad \dot{V}^\varphi = 0$$

(iii) along a meridian parametrized as $\vartheta(t) = t, \varphi(t) = \varphi_0$

$$\dot{V}^\vartheta = 0 \qquad \dot{V}^\varphi + \frac{\cos t}{\sin t}V^\varphi = 0$$

Hint: (i) see (15.2.6). □

15.3.9 Let us test the equations from (15.3.8) and check the result, which is well known from the pictures in popular books trying to illustrate the subtleties of parallel transport in "curved spaces." Namely, the parallel transport of a vector around a *right spherical* triangle.

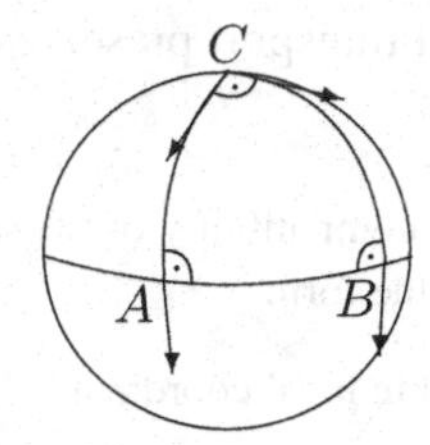

So let ABC be a right triangle on the sphere S^2 with the vertices being the north pole ($= C$) and two points at the equator, the point B lying a quarter of the equator's perimeter eastwards from A. Check that if we transport (in the sense of the RLC connection on the sphere) along the route $C \to A \to B \to C$ a vector which is directed to the point A at the beginning, the transported vector is rotated by $\pi/2$ counterclockwise with respect to the initial one (so that it has the same length and is directed to the point B) and if we traversed the same route in the opposite direction (along $C \to B \to A \to C$), then it is rotated by the same angle clockwise.

Hint: the result itself is clear immediately *without any computation* from the *metric compatibility* of the RLC connection: the transported vector must not change its length as well as the angle it makes with the line along which it is transported. Concerning the equations, the "singularity" in a neighborhood of the point C (the coordinates happen to be defective there) may be "healed" by a substitution of the edge of the triangle by an infinitesimal quarter-circle $\vartheta = \epsilon$. The three "long" parts are computed trivially, on this short one the equations linearize (due to $\epsilon \ll 1$) to $\dot{V}^\vartheta = -\epsilon V^\varphi$, $\dot{V}^\varphi = (1/\epsilon)V^\vartheta$ and they are easily solved; altogether one obtains that $k\partial_\vartheta \mapsto \frac{k}{\epsilon}\partial_\varphi$, which is just what is needed. □

15.3.10 Check that if a vector is transported around the parallel line $\vartheta(t) = \vartheta_0$ on the sphere[298] S^2, it ends up rotated by the angle $\beta_\parallel = 2\pi \cos\vartheta_0$ with respect to its initial direction. Around this parallel of the *fictitious non-rotating* globe an arbitrary object passes in just one day, which is at rest on the *real rotating* globe. In particular, this also holds for a *Foucault pendulum* that is observed somewhere on the Earth. Check that the angle β_{Fouc} of the rotation of the plane in which it swings *coincides* with $\beta_\parallel$

$$\beta_\parallel = \beta_{\text{Fouc}} = 2\pi \cos\vartheta_0$$

(and also in detail the angle of the rotation of the Foucault pendulum due to the shift along the parallel line by a small angle $\delta\varphi$ is $\delta\varphi \cos\vartheta_0$, coinciding with the angle of the rotation of a vector due to the parallel transport along the same trajectory). What does this result say about the Foucault pendulum?

Hint: see (15.3.8); (a vector in the direction of the swinging undergoes parallel[299] transport). □

15.3.11 Let (M, ω, ∇) be an n-dimensional manifold endowed with a volume form and a linear connection. Given n vectors $v, \dots, w$ at a point, the volume of the parallelepiped spanned by them is $\omega(v, \dots, w)$. A parallel transport of the vectors results in n new vectors $\hat{v}, \dots, \hat{w}$ (at a different point) and the corresponding new volume $\omega(\hat{v}, \dots, \hat{w})$. The two

[298] The angle is measured *from the z-axis*, the standard *latitude* α is measured *from the equator*; therefore $\sin\alpha = \cos\vartheta_0$.
[299] "The vector of swinging" tries to remain parallel in the "ambient" space E^3, but the situation continually forces it to "project" into the tangent plane to the sphere ≡ the Earth; one can prove that this is exactly the way in which the RLC connection works with respect to the *induced* metric.

structures are said to be compatible if an arbitrary parallel transport preserves the volume of each such parallelepiped. Show that

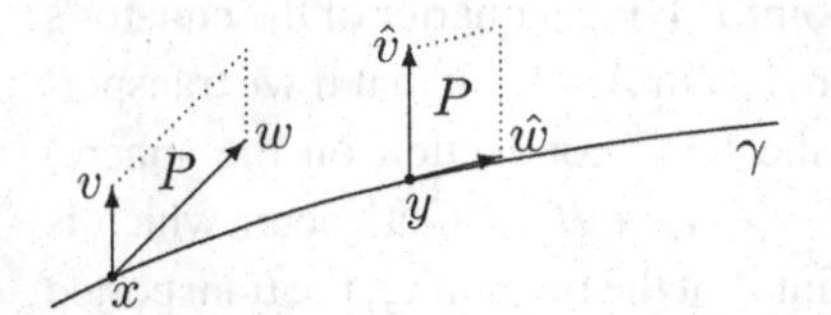

(i) the condition of compatibility of the structures may be expressed in the form

$$\nabla\omega = 0 \qquad \text{and in local coordinates} \qquad \omega_{i\dots j;k} = 0$$

(ii) if in local coordinates $\omega = f dx^1 \wedge \dots \wedge dx^n$ (i.e. $\omega_{i\dots j} = f\epsilon_{i\dots j}$), then the condition from item (i) relates f and Γ^i_{jk} by

$$f_{,k} = f\Gamma^i_{ik} \qquad \text{i.e.} \quad (\ln f)_{,k} = \Gamma^i_{ik}$$

(iii) the RLC connection *is* compatible with the metric volume form ω_g, i.e. the RLC Christoffel symbols obey[300]

$$(\ln\sqrt{|g|})_{,k} = \Gamma^i_{ik}$$

Hint: (i) like in (15.3.1); (ii) write down explicitly $\omega_{i\dots j;k} = 0$ and use (5.6.4); (iii) according to (5.6.7) we have $\partial g/\partial g_{ij} = gg^{ji}$ ($g \equiv \det g$); it is convenient to use the machinery of connection forms (see Section 15.6) and to write in an orthonormal basis $\nabla_V\omega_g = \nabla_V(e^1 \wedge \dots \wedge e^n) = (\nabla_V e^1) \wedge \dots \wedge e^n + \dots + e^1 \wedge \dots \wedge (\nabla_V e^n) = \dots = -\omega^a_a(V)(e^1 \wedge \dots \wedge e^n) = 0$ due to $\omega_{ab} = -\omega_{ba}$ (15.6.6). □

• Let us have a look at some practical manipulations with the coordinate expressions containing covariant derivatives.[301]

15.3.12 Check that

(i) the "semicolon" operation (just like the "colon" operation, the ordinary partial derivative) is linear and on a product it behaves according to the Leibniz rule; so, for example,

$$\left(A^i_{jk} + \lambda B^i_{jk}\right)_{;l} = A^i_{jk;l} + \lambda B^i_{jk;l} \qquad \left(A^i_j B^k_{lm}\right)_{;n} = A^i_{j;n} B^k_{lm} + A^i_j B^k_{lm;n}$$

(ii) also the inverse metric tensor g^{ij} is covariantly constant with respect to the metric connection, i.e.

$$g_{ij;k} = 0 \quad \Rightarrow \quad g^{ij}{}_{;k} = 0$$

(iii) the semicolon operation in the sense of the *metric* connection (in particular, RLC) commutes with the raising and lowering of indices; e.g.

$$(g^{ij}A_{jl})_{;k} = g^{ij}A_{jl;k}$$

Hint: (i) this is the behavior of ∇_i, see (15.2.9); (ii) $\nabla\hat{1} = 0$ (commuting with contractions), i.e. $\delta^i_{j;k} = 0$; (iii) both $\sharp_g$ and $\flat_g$ are combinations of the tensor product with the (covariantly

[300] It is clear intuitively that if parallel transport preserves the scalar products (consequently, also the unit cube) then it also preserves the volume, since the volume form ω_g is just "tuned" to the unit cube (see Section 5.7). The compatibility of the pairs metric ↔ volume form and metric ↔ connection thus results automatically in the compatibility of the pair volume form ↔ connection.

[301] This may be regarded as a continuation of the exercises of the *index gymnastics* from (2.4.14) and (5.2.6) (see footnote 50), which is made possible by the addition of a further popular gymnastic apparatus, the semicolon.

constant) tensors g or g^{-1} and contractions; e.g. $\nabla_W(\flat_g V) = \nabla_W(C(g \otimes V)) = C((\nabla_W g) \otimes V) + C(g \otimes \nabla_W V) = \flat_g(\nabla_W V)$. □

• For the computations of the covariant derivatives of *forms* it is sometimes fairly useful to realize how the operator ∇_V behaves with respect to the Hodge star $*_g$. For the *metric* connection the simplest possible behavior takes place.

15.3.13 Check that the operator of the covariant derivative ∇_V with respect to the metric connection ∇ *commutes* with the operator of dualization $*_g$

$$[\nabla_V, *_g] = 0$$

Hint: realize that $*_g$ is composed from the operations of the raising of indices, contractions and the tensor product with the (covariantly constant, $\nabla_V \omega_g = 0$) volume form: $\nabla_V *_g \alpha \sim \nabla_V\{C \dots C((\sharp_g \dots \sharp_g \alpha) \otimes \omega_g)\} = C \dots C((\sharp_g \dots \sharp_g \nabla_V \alpha) \otimes \omega_g) \sim *_g \nabla_V \alpha$. □

15.3.14* Consider a connection which is metric, yet not necessarily symmetric. Generalize the results of problem (15.3.4) for this case. In particular, check that

(i) the Christoffel symbols of the first kind of the connection with a given torsion satisfy

$$\Gamma_{ijk} + \Gamma_{jik} = g_{ij,k} \qquad \text{since it is metric}$$
$$\Gamma_{ijk} - \Gamma_{ikj} = -T_{ijk} \qquad \text{from the definition of the torsion}$$

(ii) the two relations yield

$$g_{ij,k} + g_{ik,j} - g_{jk,i} = 2\Gamma_{i(jk)} + (T_{jki} + T_{kji})$$

from where

$$\Gamma_{i(jk)} = \frac{1}{2}(g_{ij,k} + g_{ik,j} - g_{jk,i}) - \frac{1}{2}(T_{jki} + T_{kji})$$

and eventually

$$\Gamma^i_{jk} = \Gamma^i_{(jk)} - \frac{1}{2}T^i{}_{jk} = \frac{1}{2}g^{il}(g_{lj,k} + g_{lk,j} - g_{jk,l}) - \frac{1}{2}\left(T_{jk}{}^i + T_{kj}{}^i + T^i{}_{jk}\right)$$

The metricity plus the prescribed torsion thus result in a unique expression for the Christoffel symbols of the sought connection.[302] (The torsion being *zero*, we return to the RLC connection.)

Hint: (i) (15.3.1), (15.2.10) and (15.3.3); (iii) set the coordinate basis for U, V, W. □

15.4 Geodesics

• Now, having been equipped with the machinery of the linear connection, we may return to the concept which opened the chapter, the concept of acceleration. If we realize what is actually performed with the velocity field defined on a curve in order to compute the acceleration, we can immediately conclude that the acceleration at a given point on the

[302] Note that the *symmetric* part of the Christoffel symbols is not given by the expression for the RLC connection *alone*, but rather it contains in addition a part composed of (the tensor) $-T_{(jk)}^{\ \ i}$.

curve is nothing but the *absolute* derivative of the velocity field $\dot{\gamma}$ along the curve, or in terms of (15.2.4) the *covariant* derivative of the velocity along the velocity itself[303]

$$a = \nabla_{\dot{\gamma}} \dot{\gamma} = \nabla_v v \qquad v := \dot{\gamma} = \text{the velocity vector}$$

A case of particular interest arises when the acceleration *vanishes*. This is good old *uniform straight-line* motion. The corresponding curve on (M, ∇) is thus characterized by the equation $\nabla_{\dot{\gamma}} \dot{\gamma} = 0$; it is called an (affinely parametrized) *geodesic* and it represents the most reasonable realization of the concept of a "straight line" (together with a particular "speed" of the motion along the line) on a general manifold with a linear connection.

15.4.1 Let γ be an affinely parametrized geodesic on (M, ∇). Show that

(i) in local coordinates we have

$$\nabla_{\dot{\gamma}} \dot{\gamma} = 0 \quad \leftrightarrow \quad \ddot{x}^i + \Gamma^i_{jk} \dot{x}^j \dot{x}^k = 0 \quad \text{the } \textit{geodesic equation}$$

so that we get a system of n ordinary quasi-linear second-order differential equations for the unknown functions $x^i(t)$

(ii) the geodesic feels only the symmetric part of the Christoffel symbols[304]

(iii) the geodesics in E^n, when expressed in Cartesian coordinates, are just the curves

$$x^i(t) = x^i_0 + v^i t \qquad v^i \equiv \dot{x}^i(0), \, x^i_0 \equiv x^i(0)$$

(iv) in the general case the first two terms of the expansion in t of the coordinate presentation of a geodesic read

$$x^i(t) = x^i_0 + v^i t - \frac{1}{2} \Gamma^i_{jk} v^j v^k t^2 + \cdots$$
$$v^i \equiv \dot{x}^i(0), \quad x^i_0 \equiv x^i(0), \quad \Gamma^i_{jk} \equiv \Gamma^i_{jk}(x^i_0)$$

Hint: (i) see (15.2.5) for $V = \dot{\gamma}$; (iii) see (15.3.5); (iv) $x^i(t) = x^i(0) + \dot{x}^i(0)t + \frac{1}{2}\ddot{x}^i(0)t^2 + \cdots$. □

15.4.2 Let $(M, \nabla) = (S^2, \nabla_{\text{RLC}})$. Check that

(i) the acceleration corresponding to the uniform motion along a meridian is zero

(ii) the acceleration corresponding to the uniform motion along a parallel is *not* zero (even $a \not\parallel v$), except for the longest parallel = *the equator* (and trivially also for the opposite extreme "parallel," staying still at any pole)

(iii) all the meridians are geodesics, the only parallel which happens to be a geodesic is the equator; in general the only geodesics on the sphere are the *great circles* (the circles with the maximum possible radius; trivially also the curve which represents standing still at any point).

Hint: according to the results from (15.3.7) we get: (i) $a \sim \nabla_\vartheta \partial_\vartheta = 0$; (ii) $a \sim \nabla_\varphi \partial_\varphi = -\sin\vartheta \cos\vartheta \, \partial_\vartheta$. □

[303] So *the dot* in the expression $\mathbf{a} = \dot{\mathbf{v}}$, corresponding to the rate of change of the vector $\mathbf{v}$, implicitly contains its parallel transport; thus it is actually the *covariant* derivative.

[304] If the torsion of the connection does not vanish, it contributes to the symmetric part (15.3.14). Then if there are two connections, both of them being metric (with respect to the same g), differing, however, in the torsion, they will generate in general *different* families of geodesics – see an example in (15.8.3).

• Now let us concentrate on the issue of the parametrization. One may also traverse the path which corresponds to the uniform straight-line motion non-uniformly. Although the acceleration does not vanish in this case, it is rather specific, being at each point tangent to the curve, i.e. proportional to the velocity $\nabla_{\dot\gamma}\dot\gamma \sim \dot\gamma$. From the opposite point of view, a curve which satisfies the equation $\nabla_{\dot\gamma}\dot\gamma = f(t)\dot\gamma$ is also straight if regarded as a path, only the motion along this path fails to be uniform. Therefore such curves are called geodesics as well, being, however, not affinely parametrized.

15.4.3 Let γ be an affinely parametrized geodesic and let $\hat\gamma := \gamma \circ \sigma$ be a reparametrized curve, $\hat\gamma(t) := \gamma(\sigma(t))$, $\sigma'(t) > 0$. Check that

(i)

$$\nabla_{\dot{\hat\gamma}}\dot{\hat\gamma} = \sigma''\dot{\hat\gamma}$$

(ii) the affine reparametrization $\sigma(t) = at + b$ (and no other one) does not spoil the affine parametrization of a geodesic

(iii) by means of a unique affine reparametrization one can "tune" a given geodesic to two given points P, Q on it (not too far from each other) so as to satisfy $P = \gamma(0)$ and $Q = \gamma(1)$; now the points P, Q are at a *parametric* distance $= 1$ from each other and the point P is its "origin"

(iv) if a geodesic $\gamma(t)$ turns out to be "badly" (non-affinely) parametrized, one can always make a reparametrization such that the new geodesic $\bar\gamma$ is already affinely parametrized.

Hint: (i) (2.3.5) $\Rightarrow \nabla_{\dot{\hat\gamma}}\dot{\hat\gamma} = \sigma'(\sigma''\dot\gamma + \sigma'\nabla_{\dot\gamma}\dot\gamma)$; (ii) we need $\sigma'' = 0$; (iii) we need to map $(t_1, t_2) \mapsto (0, 1)$ by means of $t \mapsto at + b$; (iv) if $\nabla_{\dot\gamma}\dot\gamma = f(t)\dot\gamma$, then for $\gamma := \bar\gamma \circ s$ ($s = s(t) \Rightarrow \dot\gamma = s'\dot{\bar\gamma}$) we get $\nabla_{\dot\gamma}\dot\gamma = s''\dot{\bar\gamma} + (s')^2\nabla_{\dot{\bar\gamma}}\dot{\bar\gamma} = fs'\dot{\bar\gamma}$ so that to reach $\nabla_{\dot{\bar\gamma}}\dot{\bar\gamma} = 0$ it is enough to solve $s'' = s'f$ (to find $s(t)$ for given $f(t)$), which is easy. □

• So we see that the parametrization which is optimal from the point of view of the simplicity of the equations (the affine one) can always be achieved. Therefore we will automatically understand an *affinely parametrized* geodesic when speaking about a geodesic from now on and we will specially point out if this will not be the case.

The procedure of finding geodesics of the *RLC connection* as we learned up to now is fairly lengthy and laborious. Fortunately, there is a convenient alternative way, which is based on the Lagrange equations from analytical mechanics. The steps performed in both approaches may be summarized as follows:

$$g \mapsto \Gamma \mapsto \ddot x + \Gamma\dot x\dot x = 0 \mapsto x(t) \qquad \text{straightforward approach}$$
$$g \mapsto L \mapsto \mathcal{E} \mapsto x(t) \qquad \text{Lagrangian approach}$$

(see the problem). For a given g, the Lagrangian approach actually turns out to be the easiest and quickest way for

1. finding the explicit form of the equations for geodesics
2. finding the solutions of the equations
3. finding the explicit expressions for the Christoffel symbols themselves

(the main source of the power of the new approach concerning item 2 lies in making use of well-known tricks from the Lagrangian machinery – a relation between the cyclic coordinates and the conservation laws, see (18.4.2)).

15.4.4 Let (M, g) be a Riemannian manifold, γ a curve on M and $\nabla = \nabla_{\text{RLC}}$ the RLC connection corresponding to g. Then the functional

$$S[\gamma] := \frac{1}{2}\int g(\dot{\gamma}, \dot{\gamma})\,dt \equiv \int L\,dt \qquad L(x, \dot{x}) = \frac{1}{2}g_{ij}(x)\dot{x}^i\dot{x}^j$$

may be regarded as an *action integral* for the *free* motion $\gamma(t)$, since the Lagrangian L contains the kinetic energy alone ($L = T$). Check that

(i) the *Euler–Lagrange expression* $\mathcal{E}_i$ corresponding to this Lagrangian is[305]

$$\mathcal{E}_i(x, \dot{x}) \equiv \frac{\partial L}{\partial x^i} - \frac{d}{dt}\frac{\partial L}{\partial \dot{x}^i} = -g_{ij}\left(\ddot{x}^j + \Gamma^j_{kl}\dot{x}^k\dot{x}^l\right) \equiv -g_{ij}(\nabla_{\dot{\gamma}}\dot{\gamma})^j$$

(ii) the Lagrange equations are equivalent to the geodesic equations

$$\mathcal{E}_i(x, \dot{x}) = 0 \quad \Leftrightarrow \quad \nabla_{\dot{\gamma}}\dot{\gamma} = 0$$

(iii) the explicit form of the Lagrange equations for this Lagrangian enables one to immediately read off the Christoffel symbols Γ^i_{jk}.

Hint: (i) see (15.3.4); (ii) g_{ij} is non-singular; (iii) the antisymmetric part of Γ^i_{jk} vanishes for the RLC connection and the symmetric part may be read off from $\Gamma^j_{kl}\dot{x}^k\dot{x}^l$. □

15.4.5 Let (T^2, g) be the torus in E^3 with the induced metric. Check by plugging into the equations that the following curves happen to be geodesics and draw the corresponding pictures:

(i) $\psi(t) = kt$, $\varphi(t) = \varphi_0$
(ii) $\psi(t) = \psi_0$, $\varphi(t) = kt$ for particular values of ψ_0 (which ones?).

Hint: see (3.2.2), $L = \frac{1}{2}[(a + b\sin\psi)^2\dot{\varphi}^2 + b^2\dot{\psi}^2]$. □

• Consider now a more general Lagrangian, also containing the potential energy, $L = T - U$. The motion deviates from a "straight line" due to the force corresponding to U.

15.4.6 Let the action integral be

$$S[\gamma] := \int L\,dt \qquad L(x, \dot{x}) = \frac{1}{2}g_{ij}(x)\dot{x}^i\dot{x}^j - U(x) \equiv T(x, \dot{x}) - U(x)$$

Check that

(i) the Euler–Lagrange expression, corresponding to this Lagrangian, comes out as

$$\mathcal{E}_i(x, \dot{x}) \equiv \frac{\partial L}{\partial x^i} - \frac{d}{dt}\frac{\partial L}{\partial \dot{x}^i} = -g_{ij}(\nabla_{\dot{\gamma}}\dot{\gamma})^j - U_{,i}$$

[305] For a coordinate-free derivation see (15.4.16).

(ii) the Lagrange equations are equivalent to

$$\mathcal{E}_i(x,\dot{x}) = 0 \quad \Leftrightarrow \quad \nabla_{\dot{\gamma}}\dot{\gamma} = -\sharp_g dU$$

so that the motion is no longer along a geodesic in general, but rather it has an acceleration of $-\text{grad}\, U$.

Hint: see (15.4.4). □

15.4.7 We know from analytical mechanics that the *Lagrange equations* (of the second kind) in general (even if there does not exist any potential energy) read

$$\frac{d}{dt}\frac{\partial T}{\partial \dot{x}^i} - \frac{\partial T}{\partial x^i} = Q_i \qquad T = \frac{1}{2}g_{ij}(x)\dot{x}^i\dot{x}^j$$

where Q_i is the ith *generalized force* and T is the kinetic energy of a system.

(i) Check that their coordinate-free version is

$$a \equiv \nabla_{\dot{\gamma}}\dot{\gamma} = Q$$

with Q being a force (vector) field $Q = Q^i\partial_i$, $Q^i = g^{ij}Q_j$, so that actually they represent the "Newton" equation[306]

"acceleration = force"

on a configuration manifold $(M, g, \nabla_{\text{RLC}})$

(ii) according to textbooks of analytical mechanics the generalized force is computed by the formula

$$Q_i := \sum_{k=1}^{N} \mathbf{F}_k \cdot \frac{\partial \mathbf{r}_k}{\partial x^i}$$

where $\mathbf{r}_k(x^1,\dots x^n)$ represent the parametrization of the positions of individual point masses in terms of the generalized coordinates x^i and $\mathbf{F}_k$ is the force acting on the kth point mass. Check that if this parametrization is regarded as a map $f : M \to \mathbb{R}^{3N}$, then the expression for Q_i is nothing but a component expression of the *pull-back* of the force (as a covector field) from $\mathbb{R}^{3N}$ to the configuration space M (see also (3.2.9)). □

• Our lifelong experience results in a clear feeling that the *shortest* path connecting two points is the *straight* path. This experience stems from the particular spaces E^3 or E^2. Now we are in a position, however, to investigate the relation between the straight and the shortest lines on an arbitrary Riemannian manifold. Since Section 2.6 we can compute the lengths of curves and now we have learned that the straight lines are the geodesics. So the question is whether geodesics (regarded as straight lines) happen to also be at the same time the shortest paths.

Right at the beginning we should realize that the length of a curve does not depend on the parametrization (2.6.9), so that the shortest path is indeed only a path (without

[306] With unit mass; recall that if a system of particles with various masses is under consideration, it is formally described as the motion of *a single* particle in a many-dimensional Riemannian (configuration) space, the masses being hidden in the *metric tensor* g; see (2.6.7) and (3.2.9).

parametrization). This means that even if we find that the shortest path turns out to be a geodesic, the result certainly may not come out as the *affinely parametrized* geodesic.

15.4.8 The *functional of the length* of a curve (see (2.6.9) and (4.6.1)) may be regarded as an action integral with the Lagrangian[307]

$$L(x,\dot{x}) := \sqrt{g_{ij}(x)\,\dot{x}^i\dot{x}^j}$$

Check that

(i) the Lagrange equations for this Lagrangian are

$$\ddot{x}^i + \Gamma^i_{jk}\dot{x}^j\dot{x}^k = \frac{\dot{L}}{L}\dot{x}^i \qquad \text{i.e. in a coordinate-free form} \qquad \nabla_{\dot\gamma}\dot\gamma = \chi\dot\gamma$$

where $\dot{L}$ is the "total" time derivative[308] of the Lagrangian L, Γ^i_{jk} correspond to the RLC connection and $\chi \equiv \dot{L}/L$

(ii) the shortest path is a geodesic

(iii) the affine parametrization of this geodesic is achieved in the *natural parameter* s, being a parameter such that its increment coincides with the increment of the actual length of the curve (i.e. the length of the path between the points $\gamma(s_1)$ and $\gamma(s_2)$ is simply $s_2 - s_1$)

(iv) if ψ is a non-zero function, then

$$\nabla_{\dot\gamma}\dot\gamma = \frac{\dot\psi}{\psi}\dot\gamma \quad\Leftrightarrow\quad \nabla_{\dot\gamma}\left(\frac{\dot\gamma}{\psi}\right) = 0$$

so that the equation for the shortest path may also be written in the form

$$\nabla_{\dot\gamma}\left(\frac{\dot\gamma}{L}\right) \equiv \nabla_{\dot\gamma}\left(\frac{\dot\gamma}{\sqrt{g_{ij}(x)\,\dot{x}^i\dot{x}^j}}\right) \equiv \nabla_{\dot\gamma}\left(\frac{\dot\gamma}{||\dot\gamma||}\right) = 0$$

(in the natural parameter we have $L \equiv \sqrt{g_{ij}(x)\,\dot{x}^i\dot{x}^j} \equiv ||\dot\gamma|| = 1$, so that $\nabla_{\dot\gamma}\dot\gamma = 0$ then).

Hint: (i) see (15.3.4); (ii) see (15.4.3); (iii) if $\nabla_{\dot\gamma}\dot\gamma = (\dot{L}/L)\dot\gamma$, then the improving procedure from (15.4.3) yields the equation $s'' = s'L'/L$ with a solution $L = s'$, or $ds = L\,dt$; hence $s_2 - s_1 = \int_{t_1}^{t_2} L\,dt =$ *the length* of the curve. □

• By variation of the functional of the length of the curve we investigate in principle only its *stationary points* (*local* extrema or saddle points). In numerous simple particular cases it is intuitively clear what the situation looks like "globally." For example, consider two points A, B on a sphere, which do not happen to be opposite one another. If we join them by the shorter part of a great circle, we get the path with *minimal* length, whereas the complementary longer part of the great circle turns out to be only a saddle point of the functional of length, since any warp evidently results in its prolongation and its "rotation" along the sphere (with a view to deforming it step by step to the shorter part) makes it shorter

[307] Note that this (reparametrization invariant) Lagrangian is (up to a factor of 2) a *square root* of the Lagrangian from problem (15.4.4), which *was not* reparametrization invariant and therefore it *could* yield as extremals the curves with a particular parametrization; see also (15.4.16).

[308] The derivative of the function $L(x(t), \dot{x}(t))$ with respect to time, which takes into account the fact that the time enters through both $x(t)$ and $\dot{x}(t)$; it may be written in detail as $\dot{L} = \frac{\partial L}{\partial x^i}\dot{x}^i + \frac{\partial L}{\partial \dot{x}^i}\ddot{x}^i$, but here it is more convenient to leave it as it is.

(as our intuition signals and an elementary computation confirms). In general, however, the issue of the global properties of the critical points of the length functional may be fairly complicated and they are beyond the scope of this book.

Additional complication refers to certain particular pairs of points (they are called *conjugate* points). If we were to choose, for example, the north and south poles as A, B on the sphere, there would be an infinite number of (globally) shortest paths (each meridian). This occurs only for certain exceptional pairs of points, being always "very far" to each other. It turns out that for each point $x \in M$ there is a neighborhood $\mathcal{O}$ (called the *geodesic neighborhood*), in which there is just one[309] shortest path leading to each $y \in \mathcal{O}$ from x (it is clearly a geodesic).

In this neighborhood one can introduce extremely useful coordinates, tailored to the linear connection; they are known as the *normal* coordinates. For their construction recall that a geodesic may be uniquely fixed either by a point where it is in some "time" and a tangent vector at that point (the position and the velocity at a *single time*) or by the positions at *two* instants of time.[310]

15.4.9 Denote by $\gamma_v(t)$ the geodesic starting at time zero from the point P with velocity v,

$$\gamma_v(0) = P \in M \qquad \dot{\gamma}_v(0) = v \in T_P M$$

Check that a simple relation holds

$$\gamma_v(bt) = \gamma_{bv}(t) \qquad b \in \mathbb{R}$$

or, put another way, a b-fold increase of the *initial* velocity results in a b-fold increase of the velocity along the *whole* trajectory (the motion takes place along the same path).

Hint: the curve $\hat{\gamma}(t) := \gamma_v(bt)$ is a geodesic ($\dot{\hat{\gamma}} = b\dot{\gamma}_v \Rightarrow \nabla_{\dot{\hat{\gamma}}}\dot{\hat{\gamma}} = \cdots = 0$) which satisfies $\hat{\gamma}(0) = P$, $\dot{\hat{\gamma}}(0) = bv \Rightarrow$ due to the uniqueness it necessarily coincides with $\gamma_{bv}(t)$; or alternatively: the operator of the parallel transport (of the velocity vector along the geodesic) is linear. □

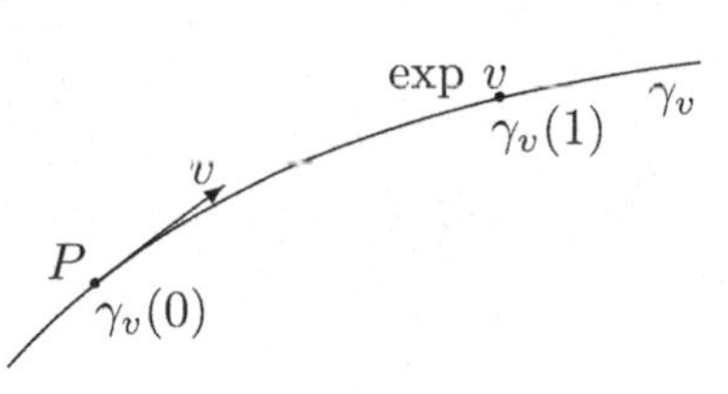

15.4.10 On a manifold with connection (M, ∇) define the *exponential map* (centered at $P \in M$)

$$\exp : T_P M \to M \qquad v \mapsto \exp v := \gamma_v(1)$$

where $\gamma_v(t)$ is the geodesic from problem (15.4.9). So one assigns to a vector v the point from M to which we arrive at $t = 1$, if at time $t = 0$ we start from the point P with initial velocity v and all the while the motion is uniform and straight-line (i.e. along a geodesic). Check that

[309] This may be obtained as a result of an analysis of differential equations governing a geodesic. It is a second-order equation and a contemplation of additional conditions leading to a unique solution yields the conclusion mentioned above.

[310] Each of these input data evidently fixes the uniform straight-line motion; from a formal point of view a second-order system $\ddot{x} + \Gamma\dot{x}\dot{x} = 0$ needs either $x(t_0)$ and $\dot{x}(t_0)$, or $x(t_0)$ and $x(t_1)$, the second possibility being trouble-free only in the geodesic neighborhood of the point $\gamma(t_0)$.

(i)

$$\exp(v = 0) = P$$

(ii) the coordinate presentation of the exponential map reads

$$\exp : v^i \mapsto x^i(v^1, \dots, v^n) \equiv x^i(P) + v^i - \frac{1}{2}\Gamma^i_{jk}(P)v^j v^k + \cdots$$

(iii) $\exp_{*0}$ is a *non-degenerate* (i.e. its kernel vanishes) linear map, so that exp maps bijectively (*diffeomorphically*) some neighborhood of zero in $T_P M$ to some neighborhood of the point P

(iv) the uniform straight-line motion in the *tangent* space is mapped to the uniform straight-line motion *on a manifold* (i.e. along a geodesic)

$$\exp(vt) = \gamma_v(t)$$

Hint: (ii) $t = 1$ in the expression of a geodesic (15.4.1); (iii) the Jacobian matrix at zero is $J^i_j(0) = \delta^i_j$ (for small v^i it reduces to a *translation* $v^i \mapsto x^i_0 + v^i$); (iv) see (15.4.9). □

• The fact that a neighborhood of a point P may be diffeomorphically mapped on a neighborhood of the zero in a *linear* space $T_P M$ means in practice that we obtain *local coordinates* in the neighborhood of the point P. The most important property of the coordinates constructed in this particular way is the vanishing of all Christoffel symbols *in the point* P. This fact greatly simplifies numerous computations and proofs (actually all that is needed in doing so is only to be aware of their existence, there is no need to construct them explicitly).[311]

15.4.11 Let exp be the exponential map centered at $P \in M$. If in $T_P M$ an (arbitrary) basis e_i is fixed, we may introduce in a neighborhood of the point P *Riemann normal coordinates* by the prescription

$$x^i \leftrightarrow Q \quad \Leftrightarrow \quad Q = \exp(v) \equiv \exp(x^i e_i)$$

So a geodesic is constructed starting ($t = 0$) in P and passing at $t = 1$ through *the* point Q, which is to be assigned coordinates. The geodesic has the unique initial velocity v and this velocity in turn has components with respect to e_i; *these components* are declared (by definition) as the coordinates x^i. Check that in these coordinates

(i) the geodesic $\gamma_v(t)$ reads

$$x^i(t) = v^i t \qquad \text{if} \quad v = v^i e_i$$

(ii) for any *symmetric* connection (in particular, also RLC)

$$\Gamma^i_{jk}(P) = 0$$

(iii) in these coordinates there holds

$$g_{ij,k}(P) = 0 \qquad \text{so that in a neighborhood of } P \qquad g_{ij}(x) \doteq g_{ij}(P) + \frac{1}{2}g_{ij,kl}(P)x^k x^l$$

[311] Let us mention also that in general relativity these coordinates have a direct physical meaning as the coordinates with respect to a frame of reference which freely falls in a gravitational field (locally inertial frame), so that the action of the force due to the gravitational field (locally) vanishes.

(i.e. a linear term is missing in the expansion) and, in particular, for an *orthonormal* basis e_i the coordinate expression of the metric tensor in a neighborhood of the point P is

$$g_{ij}(x) = \eta_{ij} + K_{ijkl}x^k x^l + \cdots \qquad K_{ijkl} = K_{(ij)(kl)} = \text{constant}$$

Hint: (i) according to (15.4.10) we have $\exp((v^i t)e_i) = \gamma_v(t)$; (ii) the general equation for a geodesic $\ddot{x}^i + \Gamma^i_{jk}\dot{x}^j\dot{x}^k = 0$ and the fact that here $x^i(t) = v^i t$ yields $\Gamma^i_{jk}(x(t))v^j v^k = 0$ on the whole geodesic $x(t)$; for $t = 0$ this gives $\Gamma^i_{jk}(P)v^j v^k = 0$ for *all* $v^i \Rightarrow \Gamma^i_{(jk)}(P) = 0$ (for $t \neq 0$ $\Gamma^i_{jk}(x(t))$ depends on v^i via $x(t)$ so that on different v^i actually different quadratic forms vanish, which *does not* allow one to deduce the vanishing of the forms); (iii) for an RLC connection it yields item (ii) and (15.3.4):

$$\begin{aligned}
\Gamma_{ijk} + \Gamma_{jik} &= g_{ij,k} && \text{metric connection (holds in arbitrary coordinates)}\\
\Gamma_{ijk} - \Gamma_{ikj} &= 0 && \text{symmetric connection (holds in arbitrary coordinates)}\\
\Gamma_{ijk}(P) + \Gamma_{ikj}(P) &= 0 && \text{holds in normal coordinates centered at } P
\end{aligned}$$

so that all $\Gamma_{jik}(P) = 0$ and then also $g_{ij,k}(P) = 0$. □

• As a simple illustration (see also (15.5.8)) of the use of these coordinates let us mention the following useful technicality, which holds for the coordinate computation of Lie and exterior derivatives.

15.4.12 Let α be a 1-form and V a vector field. Then in (arbitrary) coordinates we have

$$(\mathcal{L}_V\alpha)_i = \alpha_{i,j}V^j + V^j_{\ ,i}\alpha_j \qquad (d\alpha)_{ij} = -2\alpha_{[i,j]}$$

Check by a direct computation that

(i) if we substitute in these expressions *each* comma by a semicolon (the partial derivative by the covariant) in the sense of an arbitrary *symmetric* connection (in particular, also RLC), the expression actually does not change (the new terms pairwise cancel)
(ii) the same rule holds in general when the *Lie* derivative of an *arbitrary* tensor field as well as the *exterior* derivative of an *arbitrary* form are computed.

Hint: see (5.2.6), (6.2.5), (4.3.4), (15.2.9) and the symmetry $\Gamma^i_{jk} = \Gamma^i_{kj}$ (15.3.3). □

15.4.13 Check the validity of the general statement from (15.4.12) making use of the normal coordinates.

Hint: both expressions to be compared (with commas versus semicolons) are (a priori different) *tensor fields*,[312] the expression with semicolons containing additional terms with Christoffel symbols; in normal coordinates centered at $P \in M$ they coincide at the point P for any *symmetric* connection (15.4.11), so that at *this point* (being arbitrary) and in *these particular* coordinates the two tensors are indeed equal; the equality of two tensors does not, however, depend on the choice of the coordinates. □

[312] Consider, for example, $\alpha_{i,j}V^j + V^j_{\ ,i}\alpha_j$. This *is* a tensor field, since it is $(\mathcal{L}_V\alpha)_i$ (although neither of the two terms by itself is a tensor). After replacing commas by semicolons both terms become tensor fields even by themselves ($\alpha_{i;j}V^j = (\nabla_V\alpha)_i$ and $V^j_{\ ;i}\alpha_j = (\nabla V(\alpha))_i$) so that also their sum is all right.

15.4.14 Show that the *Killing equations* may also be written in terms of covariant derivatives in the sense of RLC connection and then take a form

$$\xi_{i;j} + \xi_{j;i} = 0 \qquad \xi_i := g_{ij}\xi^j$$

Hint: see (4.6.6), (15.3.1) and (15.4.13). □

• Recall that we have already encountered the exponential map when speaking about Lie groups; namely in Section 11.4 we studied the map

$$\exp : T_e G \equiv \mathcal{G} \to G \qquad X \mapsto \exp X := \gamma^X(1)$$

We see that this definition coincides with the definition introduced here (15.4.10), provided that $M = G$, $P = e$ (so that it is centered at the unit element of the group), $v = X$ and *if* the one-parameter subgroup $\gamma^X(t)$ *were a geodesic* on the group G in the sense of some linear connection on G. It turns out that such a connection may indeed be easily constructed so that the "group" exponential map actually reduces to be a particular case of the "geodesic" one.

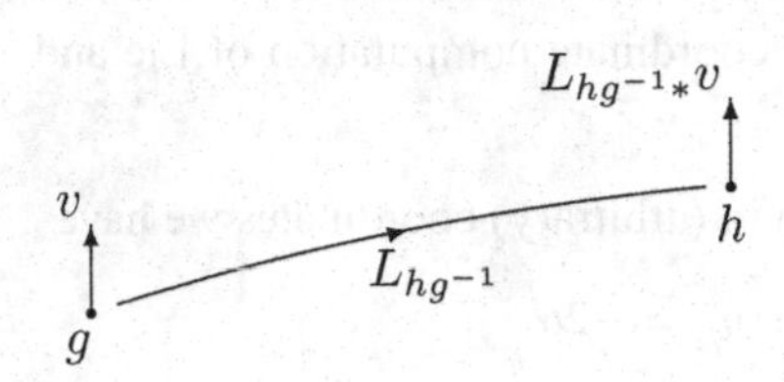

15.4.15* Define on a Lie group the parallel transport of vectors by means of the left translation, i.e. declare the operator

$$\tau_{h,g} := L_{hg^{-1}*} : T_g G \to T_h G \qquad g, h \in G$$

to be the operator of parallel transport (it does not depend on the path between the points). Check that

(i) it is linear and satisfies

$$\tau_{h,k} \circ \tau_{k,g} = \tau_{h,g} \qquad h, k, g \in G$$

(it may indeed serve as a parallel transport operator)

(ii) the covariant derivative corresponding to this parallel transport is defined by

$$\nabla_V W = 0$$

for any *left-invariant* field W (and arbitrary V)

(iii) the coefficients of the connection Γ^a_{bc} with respect to the *left-invariant* frame field e_a vanish

(iv) the tensor of the torsion reads (in the left-invariant basis)

$$T(e_a, e_b) = -[e_a, e_b] \equiv -c^c_{ab} e_c \qquad \text{i.e.} \quad T^a_{bc} = -c^a_{bc}$$

so that this (simple) connection has (for non-Abelian groups) non-vanishing torsion

(v) the geodesics emanating from the unit element of the group happen to coincide with one-parameter subgroups.

Hint: (ii) the covariant derivative measures a *deviation* from the parallel transport and the field W, being left-invariant, satisfies $\tau_{h,g} W(g) = W(h)$, so that it is invariant with respect to the parallel transport along any curve; (iv) $\nabla_a e_b = 0$; (v) see (11.3.3). □

• Now, let us have a look at how the equation of the *geodesics* of the RLC connection may be derived from the functional (15.4.4) in a coordinate-free way.

15.4.16* Let (M, g) be a Riemannian manifold, γ a curve on M and ∇ the RLC connection corresponding to g. Perform an infinitesimal variation of the curve γ by means of the flow of a "deforming" vector field W, i.e. pass to the curve $\gamma_\epsilon(t) \equiv \Phi_\epsilon \circ \gamma(t)$, where Φ_s is the flow generated by the field W (the field W should vanish at the points $\gamma(t_1)$ and $\gamma(t_2)$ since the endpoints of the curve are to be kept fixed in the course of the variation). Check that

(i) the functional $S[\gamma]$ from (15.4.4) responds to the change of argument as follows:

$$S[\gamma] := \frac{1}{2}\int g(\dot\gamma, \dot\gamma)\, dt \mapsto S[\gamma_\epsilon] = S[\gamma] + \epsilon \int \langle \mathcal{E}, W\rangle\, dt + \cdots$$

where the *Euler–Lagrange 1-form* $\mathcal{E}$ reads

$$\mathcal{E} := -g(\cdot\, , \nabla_{\dot\gamma}\dot\gamma) \equiv -\flat_g \nabla_{\dot\gamma}\dot\gamma$$

so that the critical points of the functional $S[\gamma]$ coincide with the (affinely parametrized) *geodesics* $(\nabla_{\dot\gamma}\dot\gamma = 0)$

(ii) if a potential energy is added to the action (15.4.6), i.e. we add the term $-\int U(\gamma(t))\, dt$, the Euler–Lagrange 1-form undergoes a change to

$$\mathcal{E} = -\flat_g \nabla_{\dot\gamma}\dot\gamma - dU$$

so that the critical points of the functional $S[\gamma]$ turn out to be the solutions of the (actually "Newton") equation (see (15.4.7))

$$\nabla_{\dot\gamma}\dot\gamma = -\sharp_g dU \equiv -\operatorname{grad} U$$

$\Rightarrow$ we no longer move along the geodesics, but there is the non-vanishing acceleration $-\operatorname{grad} U$

(iii) for the "square root" action $\hat{S}[\gamma] := \int \sqrt{g(\dot\gamma, \dot\gamma)}\, dt$ (the reparametrization invariant *functional of the length*) we similarly get[313]

$$\hat{\mathcal{E}} := -\flat_g \nabla_{\dot\gamma}\left(\frac{\dot\gamma}{\sqrt{g(\dot\gamma, \dot\gamma)}}\right) \equiv -\flat_g \nabla_{\dot\gamma}\left(\frac{\dot\gamma}{\|\dot\gamma\|}\right)$$

so that the critical points of the functional $\hat{S}[\gamma]$ again turn out to be the *geodesics* (this time parametrized arbitrarily, $\nabla_{\dot\gamma}(\dot\gamma/\|\dot\gamma\|) = 0$).

Hint: (i) $\gamma \mapsto \Phi_\epsilon \circ \gamma \;\Rightarrow\; \dot\gamma \mapsto \Phi_{\epsilon *}\dot\gamma$. Then,

$$\begin{aligned} S[\Phi_\epsilon \circ \gamma] &= \frac{1}{2}\int g(\Phi_{\epsilon *}\dot\gamma, \Phi_{\epsilon *}\dot\gamma)\, dt = \frac{1}{2}\int (\Phi_\epsilon^* g)(\dot\gamma, \dot\gamma)\, dt \\ &= S[\gamma] + \epsilon \frac{1}{2}\int (\mathcal{L}_W g)(\dot\gamma, \dot\gamma)\, dt + \cdots \end{aligned}$$

Disentangling $\mathcal{L}_W\{g(U, V)\} = \nabla_W\{g(U, V)\}$ for the RLC connection gives

$$(\mathcal{L}_W g)(U, V) = g(\nabla_V W, U) + g(\nabla_U W, V)$$

[313] This reduces to $\mathcal{E}$ after the choice of the "natural parameter," in which $g(\dot\gamma, \dot\gamma) = 1$.

(in coordinates it is the identity $(\mathcal{L}_W g)_{ij} = W_{i;j} + W_{j;i}$ from (15.4.14)), from where

$$(\mathcal{L}_W g)(\dot\gamma, \dot\gamma) = 2g(\nabla_{\dot\gamma} W, \dot\gamma) = 2\dot\gamma g(W, \dot\gamma) - 2g(W, \nabla_{\dot\gamma}\dot\gamma)$$

Since $\dot\gamma$ in the first term is actually the derivative with respect to t of the function standing on the right, under the integral sign it may be omitted (we get $g(W, \dot\gamma)$ evaluated at the boundary of the interval $\langle t_1, t_2\rangle$, which is zero since there $W = 0$). One is left with

$$S[\Phi_\epsilon \circ \gamma] = S[\gamma] - \epsilon \int g(W, \nabla_{\dot\gamma}\dot\gamma)\, dt + \cdots \equiv S[\gamma] + \epsilon \int \langle \mathcal{E}, W\rangle\, dt + \cdots$$

from where we immediately get (W is arbitrary, g is non-degenerate) the equation $\nabla_{\dot\gamma}\dot\gamma = 0$. (ii) $\int (U \circ \gamma)\, dt \mapsto \int (U \circ \Phi_\epsilon \circ \gamma)\, dt = \int (\Phi_\epsilon^* U \circ \gamma)\, dt = \int (U \circ \gamma)\, dt + \epsilon \int (WU \circ \gamma)\, dt + \cdots = \int (U \circ \gamma)\, dt + \epsilon \int (\langle dU, W\rangle \circ \gamma)\, dt + \cdots$; (iii) $\delta\sqrt{2u} = (2u)^{-1/2}\delta u$ and (15.4.8). □

15.4.17 Consider two highways, both of them starting from Bratislava (or any other town of your choice) going in a westward direction. The first highway proceeds all the time straight forward (i.e. it is a geodesic) and the second one is always directed westward (i.e. it keeps track along a parallel line). Find out the (approximate) distance between the two highways after 1, 10 and 100 km from the starting point.

Hint: for the geodesic highway there holds $a = \nabla_v v = 0$ (where $v \equiv \dot\gamma$ is the velocity of the motion and a is its acceleration); for the "parallel line" highway choose $v = -e_\varphi$ (e_ϑ, e_φ being the standard orthonormal frame field on the sphere, i.e. the Earth), so that we run like mad at constant unit speed $||v|| = 1$ westwards; the acceleration is

$$a = \nabla_v v = \nabla_{(-e_\varphi)}(-e_\varphi) = \cdots = -\frac{1}{R}\frac{\cos\vartheta}{\sin\vartheta} e_\vartheta \equiv -||a|| e_\vartheta$$

The motion with unit speed along the parallel line highway thus has an acceleration directed to the north (perpendicular to the driving direction) with magnitude $||a||$. The same acceleration corresponds to the motion along a bend with radius $r = 1/||a||$ (recall that for motion on a circle $||a|| = ||v||^2/r$ holds, here we have $||v|| = 1$). The geodesic highway is thus regarded by the driver as being straight and the one running along parallel in turn as a right-hand *bend of radius* $r \equiv R\tan\vartheta$.[314] One can easily check that if we perform a motion by ϵ along the *tangent* to a circle of radius r, we move off the circle by $\Delta l \equiv \epsilon^2/2r$. In our case the distance between the highways is thus $\Delta l \equiv \epsilon^2/2R\tan\vartheta$. Since in Bratislava $\vartheta \sim 42°$ (and the radius of the Earth $R \sim 6378$ km), we get approximately $\Delta l = 10^{-4}\,\text{km}^{-1}\epsilon^2$, so that for $\epsilon = 1$, 10 and 100 km we get the distances around 10 cm, 10 m and 1 km respectively. □

[314] This fact (and thus the result of the whole problem as well) may be also seen by an elementary consideration: $\hat r \equiv R\sin\vartheta$ is the radius of the parallel (its center lying on the Earth's axis); the motion along this circle has an acceleration $v^2/\hat r$. From the acceleration, however, the driver feels as the acceleration "due to the bend" only its projection onto the plane of the road (the remaining part raises the car up), producing a factor of $\cos\vartheta$ and this may be reformulated as an effective radius $r = \hat r/\cos\vartheta \equiv R\tan\vartheta$.

15.5 The curvature tensor

• The parallel transport of a vector (as well as an arbitrary tensor) in general depends on the (oriented) path along which it is performed. An alternative formulation of the same fact is that if the tensor is transported along a *closed* path (a loop), the resulting tensor may *differ* from the initial one. We already convinced ourselves that this phenomenon is indeed real in the case of the transport of a vector around a particular spherical triangle (15.3.9), where the change consisted in a *rotation* by $\pi/2$. The fact that the resulting vector had the same length as the initial one (so that the net change consisted *only* in the rotation) is a particular feature of the RLC connection (actually its metricity). In general, only the *linearity* of the operator of the parallel transport along the closed path is guaranteed.

It turns out that an immensely important piece of information about the local dependence of the parallel transport on the path is stored in the further tensor field characterizing the connection, the curvature tensor. In order to motivate its formal definition, let us first compute what the operator of the parallel transport along a particular infinitesimal loop looks like, namely the loop we already encountered in Chapter 4, when studying the geometrical meaning of the commutator of vector fields.

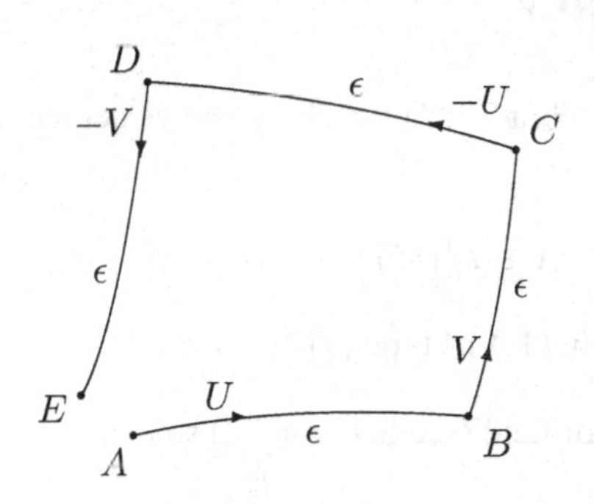

15.5.1 Consider two vector fields U, V. We saw in problem (4.5.3) how an infinitesimal loop

$$A \overset{\Phi^U_\varepsilon}{\to} B \overset{\Phi^V_\varepsilon}{\to} C \overset{\Phi^U_{-\varepsilon}}{\to} D \overset{\Phi^V_{-\varepsilon}}{\to} E \overset{\Phi^{[U,V]}_{-\varepsilon^2}}{\to} A$$

is generated by the fields composed of four pieces of parametric length ε and the fifth "closing" piece of parametric length ε^2 ($E \mapsto A$). Check that the operator $\tau_{A,A} \equiv \tau_{A\mapsto B\mapsto C\mapsto D\mapsto E\mapsto A}$ of the parallel transport of an arbitrary tensor along this loop may be expressed within second-order accuracy in ϵ as

$$\tau_{A,A} = \hat{1} - \varepsilon^2 R(U, V) + \cdots$$

where the *curvature operator* $R(U, V)$ is the expression

$$R(U, V) := \nabla_U \nabla_V - \nabla_V \nabla_U - \nabla_{[U,V]} \equiv [\nabla_U, \nabla_V] - \nabla_{[U,V]}$$

Hint: according to (15.2.13) the transport from A to B is performed by the operator $\tau_{B,A} = e^{-\varepsilon\nabla_U} = \hat{1} - \varepsilon\nabla_U + \frac{1}{2}\varepsilon^2\nabla_U^2 + \cdots$; so one should multiply out the product

$$\tau_{A,A} = e^{\varepsilon^2\nabla_{[U,V]}} e^{\varepsilon\nabla_V} e^{\varepsilon\nabla_U} e^{-\varepsilon\nabla_V} e^{-\varepsilon\nabla_U} = \cdots$$

up to order ϵ^2 □

• The curvature operator $R(U, V)$, which we obtained in this way, has some fairly remarkable properties.

15.5.2 Show that for the curvature operator $R(U,V)$ there holds

(i) it is a *derivation* of the tensor algebra $\mathcal{T}(M)$, which commutes with contractions
(ii) it *vanishes* on degree $\binom{0}{0}$, i.e. on $\mathcal{F}(M)$

$$R(U,V)f = 0 \qquad f \in \mathcal{F}(M)$$

(iii) it depends $\mathcal{F}(M)$-linearly on both U and V.

Hint: (i) it has the structure $[D_1, D_2] + D_3$, where D_1, D_2, D_3 *are* such according to the definition of ∇_W and (4.3.7); or alternatively: (each) operator of parallel transport should have certain properties and the operator treated here has the form $\tau_{A,A} = \hat{1} - \varepsilon^2 R(U,V)$, resulting in some properties of $R(U,V)$; (iii) the properties of the covariant derivative. □

• The derivations of the tensor algebra which commute with contractions and vanish on $\mathcal{F}(M)$ turn out to have a fairly simple structure – they are completely given by a certain *tensor* (field) of type $\binom{1}{1}$. Let us investigate this useful fact from a slightly more general perspective.

15.5.3 Show that each derivation D of the tensor algebra $\mathcal{T}(M)$ which preserves degree and commutes with contractions has the form

$$D = \mathcal{L}_V + A \qquad V \in \mathfrak{X}(M), \quad A \in \mathcal{T}^1_1(M)$$

i.e. it is parametrized by a vector field V and a tensor field A of type $\binom{1}{1}$.

Solution: on $\mathcal{F}(M)$, each derivation is given by a vector field (see Section 2.2) so that

$$Df = Vf \equiv \mathcal{L}_V f \qquad \text{for some } V \in \mathfrak{X}(M)$$

Set $\hat{D} := D - \mathcal{L}_V$. According to (4.3.7) it is a derivation of $\mathcal{T}(M)$ commuting with contractions, moreover it is by construction zero on $\mathcal{F}(M)$. Then it is enough (see (4.3.1) and the text after the problem) to specify it on vector fields. There we have $W \mapsto \hat{D}(W) =$ a vector field again (due to preserving of degree) and

$$\hat{D}(fW) \equiv \hat{D}(f \otimes W) = (\hat{D}f)W + f\hat{D}(W) = f\hat{D}(W) \qquad f \in \mathcal{F}(M)$$

so that $W \mapsto \hat{D}(W)$ is $\mathcal{F}(M)$-linear $\Rightarrow$ it is a tensor field $\hat{D} = A$ of type $\binom{1}{1}$, namely $A(W,\alpha) := \langle \alpha, \hat{D}(W) \rangle$. □

15.5.4 Let us see in more detail how the operator A (regarded as a part of a general derivation D from (15.5.3) corresponding to a tensor of type $\binom{1}{1}$) acts on tensors. Check that

(i) the action of the operator A on a type-$\binom{1}{1}$ tensor B, on a volume form ω and on a general[315] tensor T of type $\binom{p}{q}$ looks in components as follows:

$$B^a_b \mapsto B^c_b A^a_c - B^a_c A^c_b \equiv [A, B]^a_b$$
$$\omega \mapsto (-\text{Tr } A)\, \omega$$
$$T^{a\ldots b}_{c\ldots d} \mapsto T^{e\ldots b}_{c\ldots d} A^a_e + \cdots - T^{a\ldots b}_{c\ldots e} A^e_d$$

(ii) in particular, for the *covariant* derivative $D = \nabla_V$ we have $A = \nabla V$ so that

$$\nabla_V = \mathcal{L}_V + (\nabla V)$$

Hint: (i) if $Ae_a = A^b_a e_b$, then $Ae^a = -A^a_b e^b$ (since $0 = A\langle e^a, e_b\rangle = A(C(e^a \otimes e_b)) = \cdots$); then

$$A\big(B^b_a e^a \otimes e_b\big) = \big(AB^b_a\big)e^a \otimes e_b + B^b_a(Ae^a) \otimes e_b + B^b_a e^a \otimes (Ae_b) = \cdots$$
$$\begin{aligned} A(fe^1 \wedge \cdots \wedge e^n) &= (Af)e^1 \wedge \cdots \wedge e^n + f(Ae^1) \wedge \cdots \wedge e^n + \cdots + fe^1 \wedge \cdots \wedge (Ae^n) \\ &= f\big(-A^1_1 e^1\big) \wedge \cdots \wedge e^n + \cdots + fe^1 \wedge \cdots \wedge \big(-A^n_n e^n\big) \\ &= \big(-A^a_a\big) fe^1 \wedge \cdots \wedge e^n \equiv (-\text{Tr } A)\, \omega \end{aligned}$$

□

• The operator $R(U, V)$ thus has the form (15.5.3) with the *missing* part $\mathcal{L}_W$, so that all the information about the operator is stored in its action (as a tensor field of type $\binom{1}{1}$) on *vector* fields, i.e. in the expression $R(U, V)W$ (a vector field) or, alternatively, in the expression $\langle\alpha, R(U, V)W\rangle$ (a function).

15.5.5 Check that

(i) the expression

$$R(W, U, V; \alpha) := \langle\alpha, R(U, V)W\rangle \equiv \langle\alpha, ([\nabla_U, \nabla_V] - \nabla_{[U,V]})W\rangle$$

is $\mathcal{F}(M)$-linear in all four arguments $U, V, W \in \mathfrak{X}(M)$, $\alpha \in \Omega^1(M)$ so that a tensor field of type $\binom{1}{3}$ is defined by this formula; this important tensor field is called the *curvature tensor*, or also the *Riemann tensor*; in components

$$R^a{}_{bcd} = \langle e^a, R(e_c, e_d)e_b\rangle = \langle e^a, (\nabla_c\nabla_d - \nabla_d\nabla_c - \nabla_{[e_c, e_d]})e_b\rangle$$

(ii) from the definition we have

$$R(U, V)W = \big(R^a{}_{bcd} U^c V^d W^b\big)\, e_a$$

so that the value of the expression at the point $P \in M$ depends *only* on the values of the quantities *at the point* P (in spite of the fact that there are *derivatives* in the detailed expression of $R(U, V)W$ and therefore one could expect that the values of the objects in some infinitesimal *neighborhood* of the point might be necessary)

[315] The first two objects are clearly particular cases of the general one, but on B one sees most easily how it works and ω illustrates specificity of forms (it comes in handy in (15.6.18)).

(iii) the curvature tensor is antisymmetric in the last pair of indices

$$R^a{}_{bcd} = -R^a{}_{bdc}$$

(iv) in the coordinate basis it may be expressed in terms of the Christoffel symbols by the formula

$$R^i{}_{jkl} = \Gamma^i_{jl,k} - \Gamma^i_{jk,l} + \Gamma^m_{jl}\Gamma^i_{mk} - \Gamma^m_{jk}\Gamma^i_{ml}$$

Hint: (i) see (15.5.2); (iv) $R^i{}_{jkl} = \langle e^i, R(e_k, e_l)e_j\rangle = \langle dx^i, (\nabla_k\nabla_l - \nabla_l\nabla_k)\partial_j\rangle = \cdots$. □

- If we take the *coordinate* basis vectors ∂_i, ∂_j as the fields U, V in (15.5.1), the loop contains only four steps of parametric length ε along the coordinate curves (the commutator term is not needed for its closure) and it lies completely on the ijth coordinate two-dimensional surface (the remaining coordinates being constant there).

15.5.6 Check that

(i) for $U = \partial_i$, $V = \partial_j$ the curvature operator reduces to the *commutator of the covariant derivatives* in the ith and jth coordinate directions

$$R(\partial_i, \partial_j) = \nabla_i\nabla_j - \nabla_j\nabla_i \equiv [\nabla_i, \nabla_j]$$

so that the components of the Riemann tensor enter the result of the computation of the commutator of the "coordinate" covariant derivatives on the coordinate basis as follows:

$$[\nabla_i, \nabla_j]\partial_k = R^l{}_{kij}\partial_l \qquad [\nabla_i, \nabla_j]\,dx^k = -R^k{}_{lij}\,dx^l$$

(ii) for an arbitrary tensor field there holds

$$A^{k\ldots l}_{r\ldots s;i;j} - A^{k\ldots l}_{r\ldots s;j;i} = A^{m\ldots l}_{r\ldots s}R^k_{mji} + \cdots + A^{k\ldots m}_{r\ldots s}R^l_{mji} - A^{k\ldots l}_{m\ldots s}R^m_{rji} - \cdots - A^{k\ldots l}_{r\ldots m}R^m_{sji}$$

Hint: (i) see (15.5.5) and $0 = R(U, V)\langle\alpha, W\rangle = \langle R(U, V)\alpha, W\rangle + \langle\alpha, R(U, V)W\rangle$; (ii)

$$A^{k\ldots l}_{r\ldots s;i;j} - A^{k\ldots l}_{r\ldots s;j;i} = \{[\nabla_j, \nabla_i]A\}^{k\ldots l}_{r\ldots s} = \{R(\partial_j, \partial_i)A\}^{k\ldots l}_{r\ldots s}$$
$$R(\partial_j, \partial_i)A = R(\partial_j, \partial_i)\{A^{k\ldots l}_{r\ldots s}dx^r \otimes \cdots \otimes \partial_l\} = \cdots$$

□

- The curvature tensor admits (as a $\binom{1}{3}$-type tensor) three contractions

$$R^c{}_{cab} \qquad R^c{}_{acb} \qquad R^c{}_{abc}$$

all the resulting tensors being of type $\binom{0}{2}$. It follows from the antisymmetry in the last pair of indices that the second contraction differs from the third one only in a sign and it turns out that the first one vanishes *for the RLC connection*, so that it is usually ignored. In the case of a *Riemannian* manifold a further contraction is possible (a tool for raising the index on the tensor of type $\binom{0}{2}$ is available) and one can define a further tensor field (being already of type $\binom{0}{0}$, a function). The definitions read

$$R_{ab} := R^c{}_{acb} \qquad \text{Ricci tensor}$$
$$R := R^a{}_a \equiv g^{ab}R_{ba} \equiv R^{ab}{}_{ab} \qquad \text{scalar curvature}$$

Clearly these tensors carry less information in general then the whole Riemann tensor; this may not be true, however, for manifolds with very low dimensions (for example, we will see in (15.6.11) and (15.6.12) that on two-dimensional manifolds the whole Riemann tensor of the RLC connection may be reconstructed from the scalar curvature). A highly effective way of computing the curvature tensor consists in using the machinery of differential forms to be discussed in the next section. We will also mention some of its further properties there.

15.5.7 Consider two Riemannian manifolds. Show that if the scalar curvature of the first manifold vanishes and this is not the case for the second one then the two manifolds cannot be isometric to each other. Infer from this that the sphere S^2 is not (locally) isometric to the Euclidean plane (we have already proved the same result before, referring to different Killing algebras, see (4.6.13)).

Hint: let $f : (M, g) \to (N, h)$ be an isometry; if y^a are the coordinates on N and $h = h_{ab}\, dy^a \otimes dy^b$, then in coordinates $x^a := f^* y^a$ on M we get $g \equiv f^* h = h_{ab}(x)\, dx^a \otimes dx^b$ (with *the same* functions h_{ab}); then according to (15.3.4) also $\Gamma^a_{bc}, \ldots, R$, will be the same, i.e. the scalar curvature on M arises by the substitution $y \mapsto x \equiv f^* y$ in the expression of the scalar curvature on N (being a pull-back, $R_M = f^* R_N$). □

15.5.8 In a neighborhood of a point P consider Riemann normal coordinates centered at P, corresponding to an orthonormal basis e_i in P (15.4.11). Check that

(i) the components of the Riemann tensor of the RLC connection *in the point* P in these coordinates read

$$R^{ns}_{ijkl} = -(g_{i[k,l]j} - g_{j[k,l]i}) \qquad \text{at the point } P$$

(ii) the tensor has the symmetries (being already valid *in arbitrary* coordinates)

$$R_{ijkl} = -R_{jikl} = -R_{ijlk}$$

Hint: (i) according to (15.5.5) and (15.4.11) we have $R^i{}_{jkl}(P) = \Gamma^i_{jl,k}(P) - \Gamma^i_{jk,l}(P)$; using (15.3.4) $2\Gamma^i_{jl,k}(P) = \cdots = \eta^{ir}(g_{rj,lk} + g_{rl,jk} - g_{jl,rk})(P)$, so that

$$R^i{}_{jkl}(P) = \Gamma^i_{jl,k}(P) - \Gamma^i_{jk,l}(P) \equiv 2\Gamma^i_{j[l,k]}(P) = -\eta^{ir}(g_{r[k,l]j} - g_{j[k,l]r})(P)$$

(ii) they are explicit in these particular coordinates; the symmetries, however, do not depend on the choice of coordinates. □

• We close the section with a few words about an important concept, which is based on the path dependence of the parallel transport.

On a manifold (M, ∇) with a connection consider a point x and a *loop* c, which starts and ends at the point x. If we take a vector in x and perform the parallel transport along the loop, in general we arrive (according to the result of (15.5.1)) at a *different* vector. However, since the operator of parallel transport is always a *linear isomorphism*, the transported vector may be obtained from the initial one by the action of a certain *linear invertible* operator $T_x M \to T_x M$, i.e. an element of the group $G \equiv GL(T_x M) \cong GL(n, \mathbb{R})$.

Contemplate now all the possible loops emanating from the same point x. To each of them a group element may be assigned, so eventually we get a map "loop $\mapsto$ element of the group." This group element is said to be the *holonomy* (corresponding to the pair (x, c)) and the group itself in which the group elements lie the *holonomy group*. For a linear connection this is not necessarily the *whole* group $GL(n, \mathbb{R})$, but rather it may be only a *subgroup*. This happens when the parallel transport "preserves something"; the preserving of something is thus reflected in a restriction of the resulting group (the automorphism group of a stronger structure is smaller, see the concrete results in Section 10.1). For example, the connection, which is *metric*, assigns to each loop some *rotation*,[316] so that here the holonomy group is at best the rotation group; in reality it may be even smaller sometimes, as the example of the "ordinary" connection in E^3 shows, where the parallel transport does not depend on the path and *each* loop gives the *identity* map (the holonomy group being trivial, containing a single element).[317]

15.6 Connection forms and Cartan structure equations

• The formalism of differential forms is also very efficient in the theory of linear connections. Let us have a look, first, at how information about the connection may be encoded into appropriate 1-forms.

15.6.1 Let e_a be a frame field on $\mathcal{O} \subset (M, \nabla)$. Check that

(i) the relations

$$\nabla_V e_a = \omega^b_a(V) e_b \qquad \omega^b_a \in \Omega^1(\mathcal{O})$$

define a set of 1-forms ω^b_a on $\mathcal{O}$; they are known as *connection forms* with respect to e_a

(ii) these forms are related to the coefficients of the connection via

$$\omega^a_b = \Gamma^a_{bc} e^c$$

and, in particular, for the *coordinate* frame $e_i = \partial_i$ they can be written in terms of Christoffel symbols of the second kind as

$$\omega^i_j = \Gamma^i_{jk} \, dx^k$$

(iii) a different choice of frame field $e_a \mapsto A^b_a(x) e_b$, where $A^b_a(x) \in GL(n, \mathbb{R})$, results in a transformation of connection forms according to the rule

$$\omega'^a_b = (A^{-1})^a_c \omega^c_d A^d_b + (A^{-1})^a_c \, dA^c_b$$

(iv) this general rule contains (as a special case) the correct transformation law for Christoffel symbols

(v) for a *coframe* field one has

$$\nabla_V e^a = -\omega^a_b(V) e^b$$

[316] We learned in problems (15.3.9) and (15.3.10) that to the loop = the spherical right triangle, the group element = the rotation by $\pi/2$ is assigned and similarly to the loop = the parallel line, the group element = the rotation by the *Foucault angle* β_{Fouc} is assigned.

[317] The structure behind this is *complete parallelism* (see Section 15.8).

Hint: (i) according to the axioms of the covariant derivative, $\nabla_V e_a$ is a vector (field) which depends in an $\mathcal{F}(M)$-linear way on V; (ii) see (15.2.1); (iv) see (15.2.3); (v) $0 = \nabla_V \langle e^a, e_b \rangle = \cdots$. □

15.6.2 It is convenient to interpret the 1-forms ω^a_b as well as the functions (0-forms) A^a_b as component forms of *matrix algebra-valued* forms ω and A (in the sense of Section 6.4); if E^b_a is the usual Weyl basis in the matrix algebra $M_n(\mathbb{R})$, then

$$\omega = \omega^a_b E^b_a \in \Omega^1(\mathcal{O}, M_n(\mathbb{R})) \qquad A = A^a_b E^b_a \in \Omega^0(\mathcal{O}, M_n(\mathbb{R}))$$

Show that

(i) the results of (15.6.1) may be written as

$$e' = eA \quad \Rightarrow \quad \omega' = A^{-1}\omega A + A^{-1}\, dA$$

where the operations of multiplication and exterior derivative of forms are to be understood *in the sense of* (6.4.2) and (6.4.4)

(ii) this result is consistent on a *triple* overlap,

$$e \mapsto e' = eA \mapsto e'' = e'B \\ \equiv e(AB)$$

$$\Rightarrow$$

$$\omega \mapsto \omega' = A^{-1}\omega A + A^{-1}\, dA \mapsto \omega'' = B^{-1}\omega' B + B^{-1}\, dB \\ = (AB)^{-1}\omega(AB) + (AB)^{-1} d(AB)$$

which says (cf. the end of Section 2.5) that a *global* structure on a manifold (linear connection ∇) is actually defined by means of *local* quantities (the forms ω on domains $\mathcal{O}$, where the frame fields e_a dwell). □

• If one also encodes tensors related to the connection, namely the curvature and torsion tensor, into suitable forms, their definitions result, after translation into the language of forms, in Cartan structure equations.

15.6.3 On a domain $\mathcal{O}$ with a frame field e_a, let us define *torsion forms* T^a and *curvature forms* Ω^a_b (both of them being 2-forms) with respect to this frame by

$$T^a(U, V) := \langle e^a, T(U, V)\rangle \equiv \langle e^a, \nabla_U V - \nabla_V U - [U, V]\rangle \qquad \text{or} \quad T^a = \frac{1}{2} T^a_{bc} e^b \wedge e^c$$

$$\Omega^a_b(U, V) := \langle e^a, R(U, V) e_b\rangle \equiv \langle e^a, ([\nabla_U, \nabla_V] - \nabla_{[U,V]}) e_b\rangle \qquad \text{or} \quad \Omega^a_b = \frac{1}{2} R^a_{bcd} e^c \wedge e^d$$

where T^a_{bc} and R^a_{bcd} are the components of the *tensors* of torsion and curvature with respect to e_a. Check that

(i) they are indeed 2-forms

(ii) under the transformation $e_a \mapsto e'_a = A^b_a(x) e_b$ of the frame field the forms transform as follows:

$$\Omega^a_b \mapsto \Omega'^a_b = (A^{-1})^a_c \Omega^c_d A^d_b \qquad T^a \mapsto T'^a = (A^{-1})^a_b T^b$$

(iii) the encoding described above is bijective.

Hint: (ii) $e_a \mapsto A^b_a e_b$, $e^a \mapsto (A^{-1})^a_b e^b$; $R(U,V)$ acts as a derivation which vanishes on functions so that we have $R(U,V)(A^c_b e_c) = A^c_b R(U,V) e_c$. □

15.6.4 One may also interpret the 2-forms Ω^a_b as component forms of *a single* $M_n(\mathbb{R})$-valued 2-form Ω and similarly T^a as component forms of *a single* $\mathbb{R}^n$-valued 2-form T.[318] Check that the transformation rules for Ω and T under the change $e \mapsto eA(x)$ of a frame field then read

$$\Omega \mapsto \Omega' = A^{-1}\Omega A \qquad T \mapsto T' = A^{-1}T$$

Hint: see (15.6.3). □

• Let us now have a look at the geometrical meaning of the connection and curvature forms introduced above. It should not be too surprising to hear that ω^a_b carries information about parallel transport of a frame field e_a in an arbitrary direction and Ω^a_b informs us about the result of such transport along an infinitesimal loop. A more detailed discussion of these topics is in order, however, since it may help the reader to develop some intuition for the work with both forms and, moreover, it paves the way for the theory of general connections and gauge fields, to be developed in Chapter 21.

15.6.5 Let ω^a_b be connection 1-forms on $(\mathcal{O}, e_a)$ and Ω^a_b the corresponding curvature 2-forms. Verify that a parallel transport of the frame e_a by ϵ along V and around the pentagon-shaped ϵ-loop spanned by the vectors V, W (see (15.5.1)), respectively, results in

translation along V	$e_a \mapsto e_a - \epsilon\omega^b_a(V)e_b$
translation around a loop based on V, W	$e_a \mapsto e_a - \epsilon^2\Omega^b_a(V,W)e_b$

or in an index-free version

$$e \mapsto (\hat{1} - \epsilon\omega(V))e \qquad e \mapsto (\hat{1} - \epsilon^2\Omega(V,W))e$$

Hint: according to the definition of the covariant derivative

$$\nabla_V e_a = \frac{e^{\parallel}_a - e_a}{\epsilon} \qquad \text{so that} \qquad e^{\parallel}_a = e_a + \epsilon\nabla_V e_a \equiv \left(\delta^b_a + \epsilon\omega^b_a(V)\right)e_b$$

where $e^{\parallel}_a$ is the vector e_a parallel transported *against* the direction of V by a parameter ϵ; then the translation *along* V (against the vector $-V$) gives an additional minus sign. Curvature: the role of the curvature *operator* $R(U,V)$ within the context of translation along a loop (15.5.1). □

• The situation can be described as follows: if on a domain $\mathcal{O}$ one has a frame field e, one has also, in particular, the frames $e(x)$ and $e(x+\epsilon V)$ at the points x and "$x+\epsilon V$"[319]

[318] $\mathbb{R}^n$ serves as an $M_n(\mathbb{R})$-module here: columns = elements of $\mathbb{R}^n$ can be multiplied by matrices = elements of $M_n(\mathbb{R})$ (the result being again a column, cf. Appendix A.4). According to Section 6.4 we can then introduce the exterior product of two $M_n(\mathbb{R})$-valued forms as well as the product of an $M_n(\mathbb{R})$-valued form with an $\mathbb{R}^n$-valued one.

[319] "$x+\epsilon V$" denotes in a compact way the point at which we arrive when moving by a parameter ϵ from the point x along *any* curve of the equivalence class defining V.

(the values of the field e at the two points). The result of a parallel transport of a frame $e(x)$ by ϵ along the vector V is some particular frame at the point $x + \epsilon V$; each frame there can be, however, obtained by an appropriate "mixing" of the elements of the frame $e(x + \epsilon V)$. We see from the result of (15.6.5) that, as a matter of fact, mixing by the matrix $\hat{1} - \epsilon\omega(V)$ takes place. This matrix is non-singular and infinitesimally close to the identity matrix. Thus, we see that it is natural to treat the matrix $X := \omega(V)$ as an element of the *Lie algebra* $gl(n, \mathbb{R})$ and the frame is then mixed by the element of the *Lie group* $GL(n, \mathbb{R})$ of the form $\hat{1} + \epsilon X \equiv e^{\epsilon X}$. This point of view reveals with no computation at all, as an example, that for the *metric* connection the matrix of the connection 1-forms with respect to an *orthonormal* frame field has to be (pseudo-)antisymmetric. In order for the parallel transport not to spoil orthonormality of a frame, the group element has to belong to a (pseudo-)orthogonal subgroup and, consequently, the element of the Lie algebra $X \equiv \omega(V)$ has to belong to a (pseudo-)orthogonal subalgebra which is, according to (11.7.6), just the algebra of all (pseudo-)antisymmetric matrices. This should be valid for all V, hence the matrix ω itself must be (pseudo-)antisymmetric. The same reasoning is valid for the curvature forms as well; since the matrix $Y \equiv \Omega(V, W)$ has to be (pseudo-)antisymmetric for all V, W, the matrix 2-form Ω must be (pseudo-)antisymmetric, too.

15.6.6 Verify these statements by formal computation; namely check that

(i) for the matrices of connection 1-forms and curvature 2-forms of a *metric* connection with respect to a *general* (that means, including coordinate) frame field there holds

$$dg_{ij} = \omega_{ij} + \omega_{ji} \qquad \omega_{ij} := g_{ik}\omega^k_j$$
$$0 = \Omega_{ij} + \Omega_{ji} \qquad \Omega_{ij} := g_{ik}\Omega^k_j$$

(ii) this results in (anti)symmetry of the Riemann tensor

$$R_{ijkl} = -R_{jikl} \qquad R_{ijkl} := g_{im}R^m{}_{jkl}$$

(iii) for an *orthonormal* frame field the matrices of these forms are (pseudo-)antisymmetric:

$$\omega_{ab} + \omega_{ba} = 0 \qquad \omega_{ab} := \eta_{ac}\omega^c_b$$
$$\Omega_{ab} + \Omega_{ba} = 0 \qquad \Omega_{ab} := \eta_{ac}\Omega^c_b$$

In matrix notation thus $(\eta\omega)^{\mathrm{T}} = -\eta\omega$, which means, according to (11.7.6), that $\omega \in so(r, s)$.

Hint: (i) for a *general* frame field and arbitrary connection there holds

$$Vg_{ij} = \nabla_V(g(e_i, e_j)) = (\nabla_V g)(e_i, e_j) + g(\nabla_V e_i, e_j) + g(e_i, \nabla_V e_j)$$

so that for the *metric* connection $Vg_{ij} = g(\nabla_V e_i, e_j) + g(e_i, \nabla_V e_j)$, which gives just $dg_{ij} = \omega_{ij} + \omega_{ji}$; for the curvature forms, replace $\nabla_V \mapsto R(V, W)$ and use the fact that for the metric connection $R(V, W)g = 0$; (ii) use the relation between Ω^i_j and $R^i{}_{jkl}$ (15.6.3). □

• The forms ω, Ω and T are not independent. Full information about a connection for a given (co)frame field e is encoded in ω. Since the connection determines the torsion and curvature tensors, it determines the forms Ω and T as well. Consequently, there should exist equations relating these forms. This is the way in which the Cartan structure equations are obtained.

15.6.7 Let ω^a_b be connection 1-forms on $(\mathcal{O}, e_a)$, Ω^a_b and T^a the corresponding curvature and torsion 2-forms. Check the validity of the *Cartan structure equations*

$$de^a + \omega^a_b \wedge e^b = T^a$$
$$d\omega^a_b + \omega^a_c \wedge \omega^c_b = \Omega^a_b$$

or in index-free version (i.e. if one regards e, ω, Ω, T as the forms with values in the algebra $M_n(\mathbb{R})$ or the $M_n(\mathbb{R})$-module $\mathbb{R}^n$ respectively)

$$de + \omega \wedge e = T$$
$$d\omega + \omega \wedge \omega = \Omega$$

Hint: a straightforward computation using general arguments, definitions of objects and Cartan formulas (6.2.13) for the exterior derivative, e.g.

$$\begin{aligned} de^a(U,V) &= U\langle e^a, V\rangle - V\langle e^a, U\rangle - \langle e^a, [U,V]\rangle \\ &= \langle \nabla_U e^a, V\rangle + \langle e^a, \nabla_U V\rangle - \langle \nabla_V e^a, U\rangle - \langle e^a, \nabla_V U\rangle - \langle e^a, [U,V]\rangle \\ &= \dots \ (15.6.1) \\ &= \langle e^a, T(U,V)\rangle - \big(\omega^a_b \wedge e^b\big)(U,V) \\ &= \big(T^a - \omega^a_b \wedge e^b\big)(U,V) \end{aligned}$$

The second relation in full analogy: start with $d\omega^a_b(U,V) = \cdots$. □

• As we know, an important role among linear connections is played by the RLC connection (see Section 15.3). Let us have a look at the modifications of the Cartan structure equations in this particular case.

15.6.8 Let ω^a_b be RLC connection 1-forms with respect to an *orthonormal* frame field e_a, Ω^a_b and T^a the corresponding curvature and torsion 2-forms. Check that the structure equations in this case read

$$\omega_{ab} + \omega_{ba} = 0 \qquad \omega_{ab} := \eta_{ac}\omega^c_b$$
$$de^a + \omega^a_b \wedge e^b = 0$$
$$d\omega^a_b + \omega^a_c \wedge \omega^c_b = \Omega^a_b$$

or in index-free notation

$$(\eta\omega)^{\mathrm{T}} + \eta\omega = 0$$
$$de + \omega \wedge e = 0$$
$$d\omega + \omega \wedge \omega = \Omega$$

Hint: see (15.6.7). Also consider vanishing of the torsion and ω being a (pseudo-) antisymmetric matrix due to the metricity of the connection (15.6.6). □

• For a given metric tensor g, the application of these equations consists of the following three-step procedure:

1. one finds an orthonormal coframe field e^a (so that $g = \eta_{ab} e^a \otimes e^b$)
2. the *first two* equations from (15.6.8) are written down and solved, i.e. one looks for a set of 1-forms ω^a_b such that they satisfy the second equation and at the same time the matrix $\omega_{ab} \equiv \eta_{ac}\omega^c_b$ is *antisymmetric* (due to this condition there exist only $n(n-1)/2$ unknown 1-forms instead of n^2)
3. if we already do have connection 1-forms ω^a_b, we plug them into the third equation and find curvature 2-forms Ω^a_b and maybe, depending on what we actually need, also the components of the curvature tensor, Ricci tensor and scalar curvature from the relations

$$\Omega^a_b = \frac{1}{2} R^a{}_{bcd} e^c \wedge e^d \qquad R_{ab} = R^c{}_{acb} \qquad R = R^a_a \equiv \eta^{ab} R_{ab}$$

Recall (15.3.4) that the computation of Christoffel symbols (and components of all remaining objects then) for the RLC connection is not a creative procedure, one has simply to plug g_{ij} into the corresponding formulas (and to make no mistake in the course of the computation of all the necessary partial derivatives, the number of which increases rapidly with the dimension of this manifold). That is why the solution of Cartan structure equations should not be a creative procedure either. Step 1 amounts to the diagonalization of the matrix of the metric tensor; in real life situations, however, one often obtains the required frame field without the formal procedure of diagonalization. For step 2, there should exist some formulas expressing ω^a_b in terms of the (already known) 1-forms e^a.

15.6.9 Let e_a be an *orthonormal* frame field, c^a_{bc} its coefficients of anholonomy and let ω^a_b be the RLC connection 1-forms. Check that

(i) the following relations are valid (it is useful to compare them with their coordinate counterparts displayed in (15.3.4)):

$$\omega_{ab}(e_c) + \omega_{ba}(e_c) \equiv \Gamma_{abc} + \Gamma_{bac} = 0 \qquad \text{metric connection}$$

$$\omega_{ab}(e_c) - \omega_{ac}(e_b) \equiv \Gamma_{abc} - \Gamma_{acb} = -c_{abc} \qquad \text{symmetric connection}$$

(ii) the RLC connection 1-forms may be expressed explicitly as

$$\omega^a_b = \eta^{ac}\omega_{cb} = \eta^{ac}\Gamma_{cbd} e^d \equiv \frac{1}{2}\eta^{ac}(c_{dcb} + c_{bcd} - c_{cbd})e^d$$

Hint: (i) see (9.2.10), $0 = T(e_a, e_b) = \nabla_a e_b - \nabla_b e_a - [e_a, e_b]$. □

• One often obtains, however, the solution by plugging an appropriate *ansatz* into the structure equations and solving the rest by "trial and error." This is illustrated most easily on *two-dimensional* manifolds.

15.6.10 From (15.6.6) it follows that if we are given an *orthonormal* frame field on a *two-dimensional* manifold (a surface) (M, g), there is only a *single independent* connection 1-form; we will denote it by $\alpha \equiv \omega_{12} = -\omega_{21}$. By the same reasoning, there is only a single independent curvature form; let us denote it by $\beta \equiv \Omega_{12} = -\Omega_{21}$. This form (as is the case for *any* 2-form) is necessarily a scalar multiple of the metric volume form $e^1 \wedge e^2$; then we can write

$$\omega_{ab} = \epsilon_{ab}\alpha \qquad \Omega_{ab} = \epsilon_{ab}\beta \equiv \epsilon_{ab} K e^1 \wedge e^2$$

The function $K(x)$ is called the *Gaussian curvature* of the surface. Show that

(i) the symmetries of the Riemann tensor lead to the conclusion that the complete Riemann (curvature) tensor may be reconstructed from the *scalar* curvature R alone (and the same then clearly holds for the Ricci tensor), the latter being simply twice the *Gaussian* curvature; in the case of signature $(+,+)$ (the other ones need minor modifications here and there) we may namely write

$$R_{abcd} = K(x)\epsilon_{ab}\epsilon_{cd} \qquad R_{ab} = K(x)\delta_{ab} \qquad K = R/2$$

(ii) the structure equations for the RLC connection (15.6.8) reduce to the simple system

$$de^1 + \alpha \wedge e^2 = 0$$
$$de^2 - \alpha \wedge e^1 = 0$$
$$d\alpha = \beta \equiv K e^1 \wedge e^2$$

The computation of all relevant quantities thus consists in the solution of the *first two* (very simple) equations for the unknown 1-form α. Differentiation of α then results in β, the latter being necessarily of the form $Ke^1 \wedge e^2$; eventually K doubled gives R.

Hint: (i) $R_{abcd} = -R_{bacd} = -R_{abdc}$; (ii) $\omega^1_a \wedge \omega^a_2 = 0 \wedge 0 + \alpha \wedge (-\alpha) = 0$. □

15.6.11 Solve the system (15.6.10) for $(S^2_\rho, g) =$ the two-dimensional sphere of radius ρ endowed with the standard "round" metric (3.2.4). Compute the Gaussian curvature (show that it is *constant* $\cdot K(x) = 1/\rho^2$), Ricci tensor and the scalar curvature and check that

$$R_{abcd} = \frac{1}{\rho^2}\epsilon_{ab}\epsilon_{cd} \qquad R_{ab} = \frac{1}{\rho^2}\delta_{ab} \qquad R(x) = \frac{2}{\rho^2} \equiv 2K(x)$$

so that the scalar curvature is constant, inversely proportional to the square of the radius of the sphere (which matches the intuitive notion of the curvature of the sphere: it is everywhere the same and the bigger the sphere the less it is "curved").

Solution: for $e^1 = \rho\, d\vartheta$, $e^2 = \rho \sin\vartheta\, d\varphi$ we have the equations $\alpha \wedge e^2 = 0$, $\rho \cos\vartheta\, d\vartheta \wedge d\varphi - \alpha \wedge e^1 = 0$ from which one easily gets $\alpha = -\cos\vartheta\, d\varphi$, so that $\beta = \sin\vartheta d\vartheta \wedge d\varphi \equiv \rho^{-2} e^1 \wedge e^2$. □

15.6.12 Solve the system (15.6.10) for $(T^2, g) =$ the two-dimensional torus with the metric induced from the embedding into E^3 (3.2.2). Compute its Gaussian curvature, Ricci tensor and the scalar curvature and check that

$$R_{abcd} = \frac{\sin\psi}{b(a + b\sin\psi)}\epsilon_{ab}\epsilon_{cd} \qquad R_{ab} = \frac{\sin\psi}{b(a + b\sin\psi)}\delta_{ab}$$

$$R(x) = \frac{2\sin\psi}{b(a + b\sin\psi)} \equiv 2K(x)$$

so that the scalar curvature is no longer constant, rather it depends on the coordinate ψ (in particular, it *vanishes* on the two circles, where the torus touches the slices of bread when eaten for lunch, it is positive on the part seen by the consumer from outside and negative on the part which is not visible).

Hint: following the lines of (15.6.11) for $e^1 = (a + b \sin \psi) \, d\varphi$, $e^2 = b \, d\psi$ one quickly gets $\alpha = \cos \psi \, d\varphi$ and $\beta = - \sin \psi \, d\psi \wedge d\varphi \equiv (R/2) e^1 \wedge e^2$. □

15.6.13 We mention without proof a quick method of obtaining the scalar curvature of two-dimensional surfaces (like a sphere or a torus). At any given point, imagine two mutually perpendicular circles of appropriate radii such that they match optimally the surface in the neighborhood of the point. Let their radii be r_1, r_2. Then let the Gaussian curvature be the product of the inverse values of the radii, $K = (r_1 r_2)^{-1}$. If the circles lie on the opposite sides of the tangent plane at the given point (so that there is a "saddle" in the neighborhood of this point), the curvature is negative. Verify that this algorithm is consistent with the results we obtained for the sphere and the torus.

Hint: the sphere: the radii of the circles coincide, being $r_1 = r_2 = \rho$; the torus: on the outer perimeter (say) there holds $r_1 = a + b$, $r_2 = b$, on the inner perimeter $r_1 = a - b$, $r_2 = b$ (and with a saddle point there) and on the upper as well as the bottom circle one has $r_1 = \infty$, $r_2 = b$. □

• Important examples[320] of working with the Cartan structure equations on higher than two-dimensional manifolds are provided by ordinary three-dimensional Euclidean space E^3 and four-dimensional Minkowski space $E^{1,3}$, when non-Cartesian frame fields are used; in particular, for orthonormal frame fields generated by the cylindrical and spherical polar coordinates.

15.6.14 Consider the three-dimensional Euclidean space E^3 endowed with the cylindrical and spherical polar orthonormal (co)frame fields (2.6.4),

$$\begin{array}{llll} e^1 = dr & e^2 = r \, d\varphi & e^3 = dz & \text{cylindrical} \\ e^1 = dr & e^2 = r \, d\vartheta & e^3 = r \sin \vartheta \, d\varphi & \text{spherical polar} \end{array}$$

Check that the Cartan structure equations for the RLC connection lead in these two cases

(i) to the following connection forms:[321]

$$\begin{array}{llll} r\omega_{12} = -e^2 & \omega_{13} = 0 & \omega_{23} = 0 & \text{cylindrical} \\ r\omega_{12} = -e^2 & r\omega_{13} = -e^3 & r\omega_{23} = -(\cot \vartheta) \, e^3 & \text{spherical polar} \end{array}$$

and that these forms are consistent with the Christoffel symbols obtained in (15.3.5)

(ii) to vanishing curvature forms (as should be the case in *Euclidean* space).

[320] Moreover, their results will turn out to be useful later.

[321] The forms, obtained trivially from the antisymmetry $\omega_{ab} = -\omega_{ba}$ are omitted. Recall that $g_{ab} = \delta_{ab}$ so that $\omega_{ab} = \omega^a_b$.

Hint: (i) for example, for the spherical polar case one obtains in notation $\sigma^a_b \equiv r\omega^a_b$ the equations

$$\sigma^1_2 \wedge e^2 + \sigma^1_3 \wedge e^3 = 0$$
$$\sigma^1_2 \wedge e^1 - \sigma^2_3 \wedge e^3 = e^1 \wedge e^2$$
$$\sigma^1_3 \wedge e^1 + \sigma^2_3 \wedge e^2 = e^1 \wedge e^3 + (\cot \vartheta)\, e^2 \wedge e^3$$

Since we know that the solution is unique, we try to find the *simplest possible* forms satisfying all the equations: e.g. the second equation suggests that (maybe) $\sigma^2_3 \sim e^3$ (e^3 is missing on the right-hand side) and $\sigma^1_2 = -e^2$ (there might be a term $\sim e^1$ there, but we try the simplest ansatz first[322]); the result $\omega^1_2 = -d\vartheta$ gives $\nabla_{e_2} e_2 = \omega^1_2(e_2)e_1 = -(1/r)\partial_r$, at the same time it should be $(1/r^2)\Gamma^i_{\vartheta\vartheta}\partial_i$, from where we get $\Gamma^r_{\vartheta\vartheta} = -r$ and (the remaining) $\Gamma^i_{\vartheta\vartheta} = 0$, which is in agreement with (15.3.5). □

15.6.15 Consider the four-dimensional Minkowski space $E^{1,3}$ endowed with the orthonormal (co)frame fields of the form $e^a \equiv (e^0, e^j)$, where $e^0 = dt$ and e^j, $j = 1, 2, 3$, are the cylindrical and spherical polar frames treated in (15.6.14). Check that the Cartan structure equations for the RLC connection lead in these two cases to

(i) the common result

$$\omega^0_j = 0 \qquad \omega^i_j = \text{ as in } E^3$$

i.e.

$$\omega_{0j} = 0 \qquad \omega_{ij} = -\text{ as in } E^3$$

(ii) vanishing curvature forms (as should be the case in *Minkowski* space).

Hint: (i) in detail the equations take the form

$$de^0 + \omega^0_j \wedge e^j = 0$$
$$de^j + \omega^j_0 \wedge e^0 + \omega^j_k \wedge e^k = 0$$

with the evident solution (recall that it is unique)

$$\omega^0_j = 0 \qquad \omega^j_k = \text{the solutions of the system } \; de^j + \omega^j_k \wedge e^k = 0$$

This system coincides, however, with the system met in the case of E^3. Since now $g_{ab} = \eta_{ab}$, we have $\omega_{0j} = \eta_{00}\omega^0_j = \omega^0_j$ and (no summation) $\omega_{ij} = \eta_{ii}\omega^i_j = -\omega^i_j$. □

• Let us now return to general manifolds with the connection (M, ∇). The curvature and torsion tensors enter important identities (Bianchi and Ricci), which are most easily derived, and even formulated, in the language of forms.

[322] Recall the "Ockham's razor" (law of parsimony) principle, which advises us: "*Pluralitas non est ponenda sine necessitate,*" i.e. plurality should not be posited without necessity. Fortunately, there is no "necessity" for "positing plurality," here.

15.6.16 Let ω^a_b be connection 1-forms on $(\mathcal{O}, e_a)$, Ω^a_b and T^a the corresponding curvature and torsion 2-forms. Check that

(i) the following identities hold[323]

$$d\Omega + \omega \wedge \Omega - \Omega \wedge \omega = 0 \qquad \text{Bianchi identity}$$
$$dT + \omega \wedge T = \Omega \wedge e \qquad \text{Ricci identity}$$

(ii) they are equivalent to

$$\{(\nabla_U R)(V, W) - R(U, T(V, W))\} + \text{cycl.} = 0 \qquad \text{Bianchi}$$
$$(\nabla_U T)(V, W) + T(T(U, V), W) + \text{cycl.} = R(U, V)W + \text{cycl.} \qquad \text{Ricci}$$

(iii) and in components also to

$$R^i{}_{j[kl;m]} + R^i{}_{jr[m}T^r_{kl]} = 0 \qquad \text{Bianchi}$$
$$T^i_{[jk;l]} + T^i_{m[j}T^m_{kl]} = R^i{}_{[jkl]} \qquad \text{Ricci}$$

(iv) in particular, for the RLC connection the identities simplify (in the three different versions mentioned above) to

$$d\Omega + \omega \wedge \Omega - \Omega \wedge \omega = 0 \quad (\nabla_U R)(V, W) + \text{cycl.} = 0 \quad R^i{}_{j[kl;m]} = 0 \qquad \text{Bianchi identity}$$
$$\Omega \wedge e = 0 \quad R(U, V)W + \text{cycl.} = 0 \quad R^i{}_{[jkl]} = 0 \qquad \text{Ricci identity}$$

Hint: (i) apply d on the Cartan structure equations (15.6.7); (ii) insert arguments U, V, W and use Cartan formulas (6.2.13) (for $p = 2$ in the form with "+ cycl."); (iii) replace the (general) fields U, V, W by the coordinate basis fields. □

• Let us have a look, next, at how the basic differential operators on forms, the exterior derivative ("differential") d and the codifferential δ, may be expressed in terms of the covariant derivatives.

15.6.17* Let i_a, $j^a \equiv g^{ab} j_b$ be the operators on forms introduced in (5.8.6), (5.8.10) (the fields e_a, e^a are supposed to be dual to each other, but they need not be orthonormal) and T^a the torsion forms. Check that

(i) the exterior derivative of forms may be expressed in terms of covariant derivatives $\nabla_a \equiv \nabla_{e_a}$ as

$$d = j^a \nabla_a + T^a i_a \qquad \text{i.e.} \quad d\alpha = e^a \wedge \nabla_a \alpha + T^a \wedge i_a \alpha$$

(ii) in particular, for the RLC connection this simplifies to

$$d = j^a \nabla_a \qquad \text{i.e.} \quad d\alpha = e^a \wedge \nabla_a \alpha$$

and for the components of the exterior derivative of p-forms in a coordinate basis one obtains

$$(d\alpha)_{i \dots jk} = (-1)^p (p+1) \alpha_{[i \dots j;k]}$$

(iii) relate the result of (ii) to (6.2.5)

[323] We will also encounter these identities later in a more general context of connections on principal bundles, see (20.4.4)–(20.4.8). With the help of the exterior *covariant* derivative $\mathcal{D}$ which will be introduced there, they even simplify to $\mathcal{D}\Omega \equiv \mathcal{D}\mathcal{D}\omega = 0$ and $\mathcal{D}T \equiv \mathcal{D}\mathcal{D}e = \Omega \wedge e$ (the Cartan structure equations themselves read $\mathcal{D}e = T$ and $\mathcal{D}\omega = \Omega$, see (21.7.4)).

(iv) for the RLC connection, the *codifferential* may be expressed in terms of covariant derivatives as

$$\delta_g \alpha = -i^a \nabla_a \alpha$$

(v) coordinate expression of the codifferential then reads (compare with (8.3.5))

$$(\delta\alpha)_{i\dots j} = -\alpha_{ki\dots j;}{}^{k} \qquad \text{i.e.} \quad (\delta\alpha)^{i\dots j} = -\alpha^{ki\dots j}{}_{;k}$$

(vi) and, in particular, the *Laplace–Beltrami operator* reads

$$\Delta f \equiv -\delta d f = f_{;k}{}^{;k}$$

Hint: (i) according to (6.1.7) and (6.1.8), the right-hand side is a degree $+1$ derivation of $\Omega(M)$, so that it is enough to check this on $\Omega^0(M)$ (trivial) and on $e^a \in \Omega^1(M)$ (the first Cartan equation); (ii) $(d\alpha)_{i\dots jk} = (dx^r \wedge \nabla_r \alpha)_{i\dots jk} = \cdots$ (iii) see (15.4.12); (iv) using (5.8.9) and (15.3.13) we get $*^{-1} d * \hat{\eta} = *^{-1} j^a \nabla_a * \hat{\eta} = *^{-1} j^a * \hat{\eta} \nabla_a = - *^{-1} * i^a \hat{\eta}\hat{\eta} \nabla_a \equiv -i^a \nabla_a$; (v) $(\delta_g \alpha)_{i\dots j} = -(i^k \nabla_k \alpha)_{i\dots j} = -g^{kr} (\nabla_k \alpha)_{ri\dots j} = -g^{kr} \alpha_{ri\dots j;k} = -\alpha_{ki\dots j;}{}^{k}$. □

15.6.18* Check that for the RLC connection the two apparently different definitions of the divergence (the first one, based on the metric volume form (8.2.1), and the *covariant divergence*, which is defined to be the trace of the covariant gradient of the field V)

$$\mathcal{L}_V \omega_g = (\mathrm{div}_{(1)} V) \omega_g \qquad \mathrm{div}_{(2)} V = \mathrm{Tr}\,(\nabla V) \equiv V^i{}_{;\,i}$$

actually lead to the same result.

Hint: according to (15.5.4) there holds

$$\mathcal{L}_V \omega_g = \nabla_V \omega_g - (\nabla V) \cdot \omega_g$$

where the action of the tensor $\nabla V \equiv A$ of type $\binom{1}{1}$ on the volume form ω_g is

$$(\nabla V) \cdot \omega_g \equiv A \cdot \omega_g = (-\mathrm{Tr}\, A)\, \omega_g \equiv (-\mathrm{div}_{(2)} V) \omega_g$$

Since according to (15.3.11) for the RLC connection $\nabla_V \omega_g = 0$, we can finally conclude that $\mathrm{div}_{(1)} V = \mathrm{div}_{(2)} V$. □

- In the following exercises we will be concerned with several simple facts one usually encounters when studying *spinor fields* on Riemannian manifolds (M, g) (in "curved spaces"; the spinor fields are treated in more detail in Chapter 22).

15.6.19* Let $\omega^a_b \equiv \omega^a_{b\mu}\, dx^\mu$ be connection 1-forms with respect to a (co)frame field $e^a \equiv e^a_\mu\, dx^\mu$ and let $\Gamma^\mu_{\nu\rho}$ be the Christoffel symbols *of the same* (!) connection ∇ with respect to local coordinates x^μ. Verify that

(i) $\omega^a_{b\mu}$ and $\Gamma^\mu_{\nu\rho}$ are related as follows:

$$\partial_\mu e^a_\nu - \Gamma^\rho_{\mu\nu} e^a_\rho + \omega^a_{b\mu} e^b_\nu = 0$$

(ii) if this is regarded as the prescription for finding $\omega^a_{b\mu}$ in terms of given $\Gamma^\rho_{\mu\nu}$ and e^a_μ, we may rewrite it as

$$\omega^a_{b\mu} = e^a_\rho e^\nu_b \Gamma^\rho_{\mu\nu} + e^a_\rho \partial_\mu e^\rho_b$$

If the (co)frame field happens to be *orthonormal* and the connection ∇ is *metric* (possibly not symmetric, however), then the fields[324] $\omega^a_{b\mu}(x)$ are known as the *spin connection*. (They enter the formulas for the covariant derivative of spinor fields, cf. (22.4.8) and (22.5.1), as well as the explicit expression of the Dirac operator (22.5.4). The formula obtained above may sometimes be found in the literature on spinors in general relativity under the noble name of the *tetrad postulate*. The fields e^a_μ and e^μ_a are usually called *vielbein fields* (in the four-dimensional case *tetrad fields*, see Section 4.5).)

Hint: the particular case of (15.6.1) for the change of a frame field $e_a \mapsto \partial_\mu \equiv e^a_\mu e_a$ (so that $A \leftrightarrow e^a_\mu$, $A^{-1} \leftrightarrow e^\mu_a$, see (4.5.3)). □

15.6.20* Let e_a be an *orthonormal* frame field. In the general theory of relativity the following objects are often introduced when working within the *tetrad formalism*:

$$\gamma_{abc} := e^\mu_a e_{b\mu;\nu} e^\nu_c \qquad \text{Ricci coefficients of rotation}$$

(in particular, for computations with spinors, cf. (22.5.4)). Verify that

(i) they may be expressed as follows:

$$\begin{aligned} \gamma_{abc} &= (\nabla_c g)_{ab} + g(e_a, \nabla_c e_b) \\ &\equiv g_{ab;c} + \Gamma_{abc} \qquad\qquad \Gamma_{abc} := \eta_{ad}\Gamma^d_{bc} \end{aligned}$$

(where Γ^a_{bc} are the coefficients of the connection (15.6.1) with respect to e_a), so that for the *metric* connection (the case notably interesting for the general theory of relativity and spinors) we get that the Ricci coefficients of rotation simply coincide with the coefficients of connection (with respect to an orthonormal frame) with a "lowered index"

$$\gamma_{abc} = \Gamma_{abc} = \langle \omega_{ab}, e_c \rangle \qquad \text{i.e.} \quad \omega_{ab} = \Gamma_{abc} e^c = \gamma_{abc} e^c$$

(ii) they are antisymmetric with respect to the first pair of indices

$$\gamma_{abc} = -\gamma_{bac}$$

(iii) one can also express these coefficients in terms of *coefficients of anholonomy* (9.2.10) of the frame field e_a and this gives (for the metric connection)

$$\gamma_{abc} = \frac{1}{2}(c_{cab} + c_{bac} - c_{abc})$$

Hint: (i)

$$\begin{aligned} e^\mu_a e_{b\mu;\nu} e^\nu_c &= e^\mu_a e^\nu_c \left(g_{\mu\sigma} e^\sigma_b\right)_{;\nu} = e^\mu_a e^\nu_c \left(g_{\mu\sigma;\nu} e^\sigma_b + g_{\mu\sigma} e^\sigma_{b;\nu}\right) \\ &= e^\mu_a e^\nu_c \left\{(\nabla_\nu g)_{\mu\sigma} e^\sigma_b + g_{\mu\sigma}(\nabla_\nu e_b)^\sigma\right\} = (\nabla_c g)_{ab} + g(e_a, \nabla_c e_b) \end{aligned}$$

(ii) $\omega_{ab} = -\omega_{ba}$ because of metricity; (iii) see (15.6.9). □

• Let us have a look at how one can write down the parallel transport equations in terms of connection forms.

[324] That is, the *coordinate* components of the *metric* connection 1-forms with respect to an *orthonormal* frame field.

15.6.21 Let ω^a_b be the connection forms with respect to a frame field e_a, $V = V^a(t)e_a$ a vector field defined on a curve $\gamma(t)$ and $A = A^{a\dots b}_{c\dots d}(t)e^c \otimes \dots \otimes e_b$ a tensor field of type (r, s) at the same curve. Check that

(i) the parallel transport equations of the vector V and the tensor A take the form

$$\dot{V}^a = S^a_b(t)V^b \qquad\qquad S^a_b(t) := -\omega^a_b(\dot{\gamma}(t))$$
$$\dot{A}^{a\dots b}_{c\dots d} = S^a_f(t)A^{f\dots b}_{c\dots d} + \dots - \dots - S^f_d(t)A^{a\dots b}_{c\dots f}$$

(ii) the equations from (15.2.6) and (15.2.12) are the special cases for the *coordinate* frame field

(iii) for an *orthonormal* frame field the matrix S^a_b is (pseudo-)antisymmetric

(iv) if S^a_b does not depend on time, the explicit solution (for the vector) may be written in the form

$$V^a(t) = (e^{tS})^a_b V^b(0) \equiv V^a(0) + tS^a_b V^b(0) + \frac{t^2}{2} S^a_c S^c_b V^b(0) + \cdots$$

and the matrix $(e^{tS})^a_b$ is (pseudo-)orthogonal.

Hint: (i) $0 = \nabla_{\dot{\gamma}}(V^a e_a) = \cdots$; (ii) $\omega^i_j = \Gamma^i_{jk}\,dx^k$ (15.6.1). □

15.6.22 Check that for the case of the two-dimensional sphere from problem (15.6.11) the result of (15.6.21) gives

$$S^a_b(t) = \epsilon_{ab}\dot{\varphi}\cos\vartheta$$

and, in particular, for the motion along a parallel $\vartheta = \vartheta_0$, $\varphi = t$ we get

$$S^a_b(t) = S^a_b = \epsilon_{ab}\cos\vartheta_0 \qquad \text{i.e.} \quad e^{tS} \equiv e^{\varphi S} = \begin{pmatrix} \cos(\varphi\cos\vartheta_0) & \sin(\varphi\cos\vartheta_0) \\ -\sin(\varphi\cos\vartheta_0) & \cos(\varphi\cos\vartheta_0) \end{pmatrix}$$

This means that the motion along the parallel with $\vartheta = \vartheta_0$ results in a (clockwise) *uniform rotation* of the vector which is parallel transported. The net effect of the transport by the angle φ (directed toward the east) consists in the rotation of the vector by the angle $\varphi\cos\vartheta_0$; in particular the transport around the entire parallel gives just the *Foucault angle* $2\pi\cos\vartheta_0 \equiv 2\pi\sin\alpha$ from (15.3.10).

Hint: $\omega^a_b = \epsilon_{ab}\alpha = -\epsilon_{ab}\cos\vartheta\,d\varphi$. □

15.7 Geodesic deviation equation (Jacobi's equation)

• Imagine two boats sailing across a lake, their motion being uniform and along a straight line. We may write down their trajectories as

$$\mathbf{r}_1(t) = \mathbf{r}_1(0) + \mathbf{v}_1 t$$
$$\mathbf{r}_2(t) = \mathbf{r}_2(0) + \mathbf{v}_2 t$$

and they represent (affinely parametrized) *geodesics* in E^2. For their *relative* position vector, relative velocity and relative acceleration we get

$$\mathbf{r}(t) \equiv \mathbf{r}_2(t) - \mathbf{r}_1(t) = (\mathbf{r}_2(0) - \mathbf{r}_1(0)) + (\mathbf{v}_2 - \mathbf{v}_1)t \equiv \mathbf{r}(0) + \mathbf{v}t$$
$$\dot{\mathbf{r}}(t) = \mathbf{v}$$
$$\ddot{\mathbf{r}}(t) = \mathbf{0}$$

These equations say that also from the point of view of a man sitting in the first boat the motion of the second boat is *uniform* and *along a straight line*. This fact is so evident (we knew it in advance and no computation was needed for it) that the reader might be astonished as to why this trivial stuff should enter Section 15.7 of the chapter devoted to linear connection.

Let's try to have a look at what happens when our freshwater beginners are substituted by fearless sea wolves, moving at the scale of the whole globe. Imagine they start their sails simultaneously, being (say) 100 m from one another (the second one eastwards from the first one). Both of them move *uniformly along a straight line* again (with the same speed) to the south, each one along their meridian. Their trajectories thus also represent (affinely parametrized) *geodesics*, but this time with geodesics *on the sphere* S^2. The behavior of a "relative vector," however, essentially differs in this case: the trajectories of the boats first diverge from one another and then (after passing the equator) they start to converge! From this "oscillation" it is clear that their "relative motion" is no longer "uniform," even though the motion of either of the boats *is* uniform and along a straight line.

Now we will try to discuss all of this in a more general setting, on a *manifold with a connection* (M,∇). It turns out that the phenomenon already occurs at the *local* level and it is a manifestation of the behavior of *nearby geodesics*.

Contemplate a geodesic $\gamma(t)$. We may construct the whole *one-parameter class* of geodesics from the *single* geodesic $\gamma(t)$ as follows: at the point $P = \gamma(0)$ we consider a vector ξ (which is not directed along $\dot{\gamma}$) and we construct an arbitrary curve $\sigma(s)$ such that it is tangent to ξ at the point P, so that

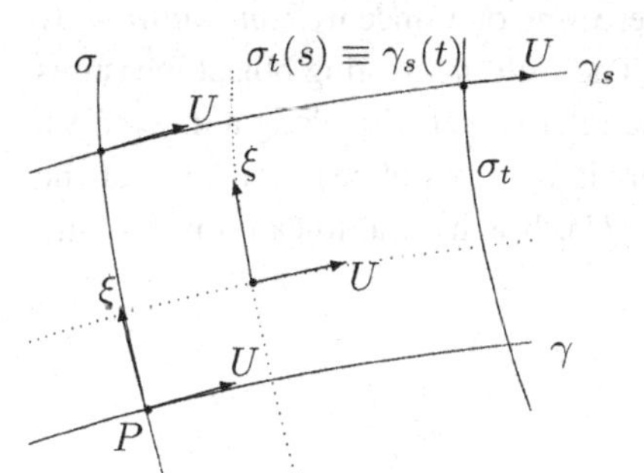

$$P = \sigma(0) = \gamma(0) \qquad \dot{\sigma}(0) = \xi$$

Now consider any vector field $U(s)$ on (a small piece of) the curve $\sigma(s)$ such that it is smooth and that for $s = 0$, i.e. at P, it coincides with the velocity vector $\dot{\gamma}(0)$ of the initial geodesic. The points of the curve $\sigma(s)$ plus the vectors $U(s)$ at these points induce unique[325] geodesics (see the text before (15.4.9)) $\gamma_s(t)$: there holds

$$\gamma_s(t=0) = \sigma(s) \qquad \dot{\gamma}_s(t=0) = U(s)$$

[325] What this construction (for $s = \epsilon \ll 1$) actually does is a small *variation of the initial conditions* of the original geodesic: we perform the "variation" of the initial point $P \equiv \sigma(0)$ to $\sigma(\epsilon)$ and the variation of the initial velocity $\dot{\gamma}(P) \equiv U(0)$ to $U(\epsilon)$ at the point $\sigma(\epsilon)$. The aim is then to learn what effect the small variation of the initial conditions has on the future course of the geodesic. Put another way, what is the variation of the *rest* of the geodesic for a given variation of its *initial conditions?*

The curve $\sigma(s)$ by construction binds the *initial points* (the points $\gamma_s(0)$) of the *one-parameter class* of the geodesics γ_s. Now define a similar curve $\sigma_t(s)$ *for each* t, i.e. so that the curve $\sigma_t(s) \equiv \sigma(t,s)$ with fixed t binds the points on the geodesics γ_s which share the same value of the parameter t. Thus, there holds

$$\sigma(t,0) = \gamma(t) \equiv \gamma_{s=0}(t) \qquad \sigma(0,s) = \sigma(s) \equiv \sigma_{t=0}(s) \qquad \sigma(t,s) = \gamma_s(t) = \sigma_t(s)$$

The parameter s thus "labels" the geodesics whereas the parameter t "runs in" them. It is intuitively clear that this one-parameter class of geodesics forms a *two-dimensional surface* $\mathcal{S}$ (which is parametrized by $\sigma(t,s)$). There are two natural *vector fields* on the surface (it is enough to consider the fields in an infinitesimal neighborhood of the initial geodesic): the velocity field U of the motion along the individual geodesics (it may be regarded as an extension to the surface $\mathcal{S}$ of its values $U(s)$ on the curve $\sigma(s)$) and the field ξ which "links the individual neighboring geodesics" or, more exactly, whose integral curves are (by definition) the curves which bind the points with the same value of "the time" t, i.e. the curves $\sigma_t(s) \equiv \sigma(t,s)$ with fixed t (for its flow Φ_s^ξ we may write $\Phi_s^\xi \gamma(t) = \gamma_s(t)$).

15.7.1 Check that the vector fields U and ξ on $\mathcal{S}$ *commute*

$$[U,\xi] = 0$$

Hint: they generate the flows $(t,s) \mapsto (t+\lambda, s)$ and $(t,s) \mapsto (t, s+\lambda)$, see (4.5.8). □

• Let us now have a look at how objects in this construction correspond to objects in the situation with the boats. The relative velocity of the boats $\mathbf{v} \equiv \mathbf{v}_2 - \mathbf{v}_1$ is the difference of two vectors sitting at two distinct points. In order to make the comparison we need first to (parallel) transport one of the vectors along the line joining the two points – the *relative velocity* of nearby geodesics is thus the covariant derivative of the velocity vector along the joining line, $\nabla_\xi U$. This object is still not the most interesting one since *we can control it* by means of the choice of its value at the time zero.[326] The truly interesting object measures the *rate* of the relative velocity (i.e. the change of the relative velocity along a trajectory), i.e. we are to study the (covariant, *parallel* transport is again implicit) derivative of the relative velocity along the (ordinary) velocity, $\nabla_U(\nabla_\xi U)$. It is natural to call this quantity the *relative acceleration* of "neighboring" geodesics

relative velocity	$\leftrightarrow$	$\nabla_\xi U$
relative acceleration	$\leftrightarrow$	$\nabla_U(\nabla_\xi U)$

This object, the relative acceleration, is already out of our control (by means of any choices in time $t=0$) and as Jacobi's equation (to be introduced in a while) shows, the relative acceleration feels the *curvature* of a manifold (M,∇) where the geodesics are studied.

[326] When we had chosen arbitrarily $U(s)$ on $\sigma(s)$ and also we have already fixed implicitly $\nabla_{\dot\sigma} U \equiv \nabla_\xi U$. In particular, if we had chosen as $U(s)$ on $\sigma(s)$ the *autoparallel* field given by the vector $\dot\gamma(0)$, it should mean that we perform the variation of the *position alone* leaving the initial value of the velocity "the same" or, put another way, that we study a free motion of two objects which start from neighboring points with the same speed, their initial velocities being directed parallel to each other.

15.7.2 Let U and ξ be the fields introduced above and let the connection ∇ be *torsion-free*. Show that

(i) the relative acceleration may also be expressed as $\nabla_U^2 \xi$, since

$$\nabla_U \xi = \nabla_\xi U$$

(ii) there holds the identity

$$\nabla_U^2 \xi = R(U, \xi)U$$

(iii) for the field ξ on the initial geodesic γ this results in *Jacobi's equation* for *geodesic deviation*

$$\frac{D^2\xi}{Dt^2} \equiv \nabla_{\dot\gamma}^2 \xi = R(\dot\gamma, \xi)\dot\gamma \qquad \text{or briefly} \quad \ddot\xi = R(\dot\gamma, \xi)\dot\gamma$$

Notice that in Jacobi's equation the only quantities that occur are (i.e. it is *enough* if they are) defined on the geodesic γ alone (for $\nabla_\xi U$ we need the *field* U also in a neighborhood of γ, whereas for $\nabla_U \xi = \nabla_{\dot\gamma} \xi$ we make do with ξ on the curve γ itself).

Hint: (i) in general, $\nabla_U \xi - \nabla_\xi U - [U, \xi] = T(U, \xi)$; (ii) since U is a "geodesic field," we have $\nabla_U U = 0$; then $\nabla_U \nabla_U \xi = \nabla_U \nabla_\xi U = \nabla_U \nabla_\xi U - \nabla_\xi \nabla_U U - \nabla_{[U,\xi]} U = R(U, \xi)U$. □

15.7.3 Be sure to understand that on the right-hand side of Jacobi's equation there is a *linear operator* (depending quadratically on $\dot\gamma$) applied on the vector ξ

$$\xi \mapsto B(\xi) \equiv R(\dot\gamma, \xi)\dot\gamma \qquad \xi^i \mapsto B^i_l \xi^l \quad B^i_l := R^i{}_{jkl}\dot x^j \dot x^k$$

so that in components the equation reads

$$\nabla_{\dot\gamma} \nabla_{\dot\gamma} \xi = \left(R^i{}_{jkl}\dot x^j \dot x^k \xi^l\right)\partial_i \qquad \text{i.e.} \quad (\nabla_{\dot\gamma} \nabla_{\dot\gamma} \xi)^i = R^i{}_{jkl}\dot x^j \dot x^k \xi^l$$

Hint: $R(U, V)W$ is $\mathcal{F}$-linear in *all* arguments (15.5.5). □

15.7.4 Let us examine how this equation works on the ordinary sphere S^2. Here we may consider *meridians* as a one-parameter class of geodesics (15.4.2). Then we take as the fields U and ξ simply the coordinate basis fields ∂_ϑ and ∂_φ respectively. Check that

(i) a direct computation of the left-hand side of Jacobi's equation gives

$$\nabla_{\partial_\vartheta}^2 \partial_\varphi = \cdots = -\partial_\varphi$$

(ii) by comparison with what the right-hand side of the equation should give we obtain

$$-\partial_\varphi \overset{!}{=} R(\partial_\vartheta, \partial_\varphi)\partial_\vartheta \equiv R^\vartheta{}_{\vartheta\vartheta\varphi}\partial_\vartheta + R^\varphi{}_{\vartheta\vartheta\varphi}\partial_\varphi$$

(iii) from there we can read the values of the components of the Riemann curvature tensor

$$R^\vartheta{}_{\vartheta\vartheta\varphi} = 0 \qquad R^\varphi{}_{\vartheta\vartheta\varphi} = -1$$

(iv) these results coincide with those obtained by a direct computation from the formula for components of the curvature tensor in terms of Christoffel symbols (15.5.5) or by expressing the result from (15.6.11) ("orthonormal" components obtained from the Cartan structure equations) in coordinate components.

Hint: see (15.3.7). □

15.8* Torsion, complete parallelism and flat connection

• We encountered the concept of (the tensor of) torsion in the section devoted to the RLC connection (15.3.3), where we learned that the requirement of *vanishing* torsion leads in combination with metricity to a unique (i.e. RLC) connection. So in this particular connection the torsion is *by definition* completely "disabled." On the other hand, exactly this particular connection is by far the most frequent linear connection met by most people (say, in general relativity). This results in the torsion mostly remaining hidden in the shadow of its much more popular sibling, the curvature.[327]

The torsion must appreciate then (even be touched to the heart) knowing that we did not forget about it. In this section we will learn in which geometrical situation the (non-vanishing) torsion manifests its presence. Namely it turns out that it causes "disclosure of a geodesic parallelogram."

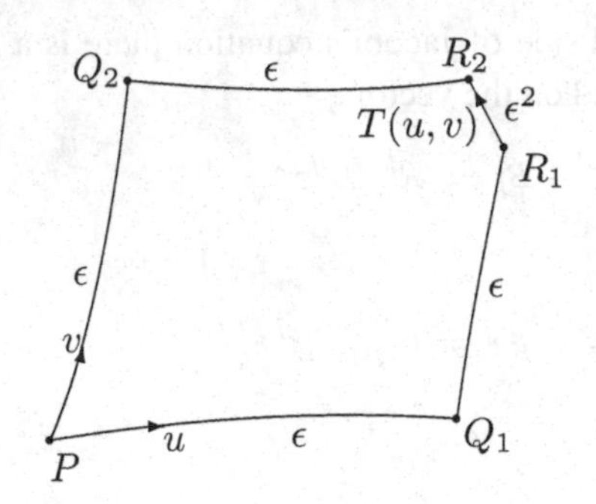

15.8.1 At a point P consider two vectors u, v. The point P and the vector u define a unique (affinely parametrized) geodesic $\gamma_u(t)$. We parallel transport v to the point $Q_1 \equiv \gamma_u(\epsilon)$ along the geodesic; this results in $v_\parallel$ in Q_1. The point Q_1 and the vector $v_\parallel$ define in turn a geodesic $\gamma_{v_\parallel}(t)$. The point $\gamma_{v_\parallel}(\epsilon)$ will be labeled R_1. Now we perform the same steps with the vectors $u \leftrightarrow v$ being interchanged. In this way we obtain the points Q_2 and R_2. It is clear that in the ordinary plane we should draw a parallelogram with vertices P, Q_1, $R_1 \equiv R_2$, Q_2. It turns out, however, that on a general manifold with connection (M, ∇) there holds $R_1 \neq R_2$, and the step by which we get from R_1 to R_2 up to second-order accuracy in ϵ is realized by the vector $T(u, v)$, where T is the *torsion* of the connection ∇. One may then say that the vector $T(u, v)$ encloses within order ϵ^2 the infinitesimal geodesic parallelogram given by the vectors u and v. Check this statement by a computation.

Hint: for example, in coordinates: according to (15.4.1) the point Q_1 has the coordinates $x^i(Q_1) = x^i(P) + u^i\epsilon - \frac{1}{2}\Gamma^i_{jk}u^j u^k \epsilon^2 + \cdots$. Components of the transported vector v are

[327] As scientists recently discovered (under microscopes, I expect) this spectacular astronomical phenomenon was already pretty well known to Mayan civilization. Mayan astronomers compiled precise tables of positions for the Moon, Venus, Curvature and Torsion and were able to predict with astonishing accuracy torsion eclipses (caused by the curvature; their prediction namely stated that it always happens).

by (15.2.6) $v^i_{\parallel} = v^i - \epsilon\Gamma^i_{jk}u^k v^j + \cdots$ (since $\dot{x}^k = u^k$). Within order ϵ^2 then the coordinates of the point R_1 are

$$\begin{aligned} x^i(R_1) &= x^i(Q_1) + v^i_{\parallel}\epsilon - \frac{1}{2}\Gamma^i_{jk}v^j_{\parallel}v^k_{\parallel}\epsilon^2 \\ &\equiv x^i(P) + \epsilon(u^i + v^i) + \frac{1}{2}\epsilon^2\Gamma^i_{jk}(-u^j u^k - v^j v^k - 2v^j u^k) \end{aligned}$$

The corresponding result for $x^i(R_2)$ is obtained by $u \leftrightarrow v$ so that

$$\begin{aligned} x^i(R_2) - x^i(R_1) &= \frac{1}{2}\epsilon^2\Gamma^i_{jk}2(v^j u^k - v^k u^j) = \epsilon^2\big(-2\Gamma^i_{[jk]}\big)u^j v^k \equiv \epsilon^2 T^i_{jk}u^j v^k \\ &\equiv (\epsilon^2 T(u, v))^i \end{aligned}$$

□

• These results probably reminded the reader of a similar computation in Chapter 4 where the geometrical meaning of the *commutator* $[U, V]$ of two vector fields U and V was discussed (4.5.3). In what way do these constructions actually differ?

In Chapter 4 we managed without any connection, here we definitely need it. In particular, there we moved along *integral curves* of the vector fields involved, whereas here we move along *geodesics*. There we needed the fields U, V *also in a neighborhood* of the point P, whereas here we make do with the vectors u, v *at the point P alone.*

In both cases unclosed parallelograms arose; then due to non-vanishing $[U, V]$, now due to non-vanishing $T(U, V)$. As an enclosing piece (up to order ϵ^2) one had to add then $-\epsilon^2[U, V]$, now it is $+\epsilon^2 T(U, V)$.

There is also an equivalent way of expressing the effect of torsion. Contemplate vectors u, v at the point P. Extend them to vector *fields* U, V in a small *neighborhood* of the point P as follows: if Q is a point in the neighborhood, we construct a geodesic from P to Q and parallel transport the vectors u, v to Q along the geodesic (recall that a parametrization of the geodesic does not matter). All the transported vectors then constitute the vector fields U, V. By construction their covariant derivative in any direction vanishes *at the point P* so that we get for the tensor of torsion at that point $T_P(U, V) = (\nabla_U V)_P - (\nabla_V U)_P - [U, V]_P = -[U, V]_P$. The effect of torsion thus happens to coincide with the effect of (minus) the commutator *of these* vector fields. The latter manifests itself when traveling along their *integral curves*, coinciding here in turn just with geodesics (a geodesic given by a vector v arises by the parallel transport of v "along itself"; that is, however, exactly the way in which the values of the field V arise), so that the above-mentioned "geodesic" construction actually matches the construction in terms of the integral curves used here.

Recall also that Section 15.5, describing curvature, starred yet another important "non-closure phenomenon" which is related to connections (this time, however, not at the level of the points along which we travel, but rather at the level of tensors being transported; a tensor suffers a change due to the parallel transport along a loop). Let us illustrate non-vanishing torsion with the example of a simple connection where the effect of the torsion may be easily grasped visually.

Consider as a manifold the two-dimensional sphere S^2 with both the north and south poles removed, endowed with the common "round" metric tensor. If it is as big as the surface of the Earth, it may easily happen we actually do not recognize it is a sphere (it took some time for mankind, too) and we believe we walk on a Euclidean plane. Then it is natural to perform the parallel transport of vectors as follows.

First, we measure the length of the vector to be transported and arrange the *length* to be *the same* after the transport. Then the only issue which remains is its direction. In order to fix the direction we use a compass and measure the azimuth of the initial vector (i.e. the angle clockwise from due north; this does not work at the poles, but recall they were removed from the manifold at the very beginning with wondrous foresight). We then prescribe the *same azimuth* to the transported vector. If we believe we walk on a Euclidean plane (endowed with a distinguished "north" direction) we have a clear conscience that we did our best to realize parallel transport in the *most common* intuitive sense.[328]

15.8.2 Check that the connection on the sphere with removed poles which was introduced above is metric, it has vanishing curvature and *non-vanishing torsion.*

Hint: by construction it is evident that the scalar product of vectors is preserved (so that it is metric) and that parallel transport does not depend on the path ($\Rightarrow$ vanishing curvature). We also see that the standard orthonormal frame field (e_ϑ, e_φ) on the sphere *is parallel*, i.e. that for *any* V there holds $\nabla_V e_\vartheta = 0 = \nabla_V e_\varphi$ (it is enough to realize how parallel transport of *these* vectors to another point turns out). Then (on the sphere with unit radius),

$$T(e_\vartheta, e_\varphi) \equiv \nabla_{e_\vartheta} e_\varphi - \nabla_{e_\varphi} e_\vartheta - [e_\vartheta, e_\varphi] = -[e_\vartheta, e_\varphi] = -\left[\partial_\vartheta, \frac{1}{\sin\vartheta}\partial_\varphi\right]$$
$$= \frac{\cos\vartheta}{\sin^2\vartheta}\partial_\varphi \equiv \frac{\cos\vartheta}{\sin\vartheta} e_\varphi \neq 0$$

□

15.8.3 Check that all meridians *as well* (in contrast to RLC) *as parallel lines* (and even in general all *loxodromes*) turn out to be geodesics of this connection.

Hint: $\nabla_U e_\vartheta = 0 = \nabla_U e_\varphi$ (for arbitrary U) results in $\nabla_V V = 0$ for $V = k_1 e_\vartheta + k_2 e_\varphi$; in particular, $\nabla_{e_\varphi} e_\varphi = 0$ says that integral curves of the field e_φ (parallel lines) are geodesics; for general k_1, k_2 the integral curves happen to coincide with loxodromes (3.2.8). □

• The fact that in this particular case there holds $T(e_\vartheta, e_\varphi) = -[e_\vartheta, e_\varphi] \neq 0$ means that the vector which encloses a *geodesic* parallelogram coincides with the vector enclosing the parallelogram made from *integral curves*. This is not an accident. Both procedures of construction of the parallelograms eventually lead to the same result: if we take e_ϑ, e_φ as U, V, a motion along integral curves is the same as the motion along geodesics (the first halves of the construction thus coincide), the parallel transport of the second vector along

[328] This technique can be safely used at the scale of a town, say; as a preparation the reader is invited to use it at a copy-book scale.

the geodesic given by the first one results in the value of the second field at a new place so that also the second halves of the construction give the same result. The effect of torsion thus coincides here with the effect of non-commutativity of the fields e_ϑ and e_φ. However, here the latter is *very clear* visually.

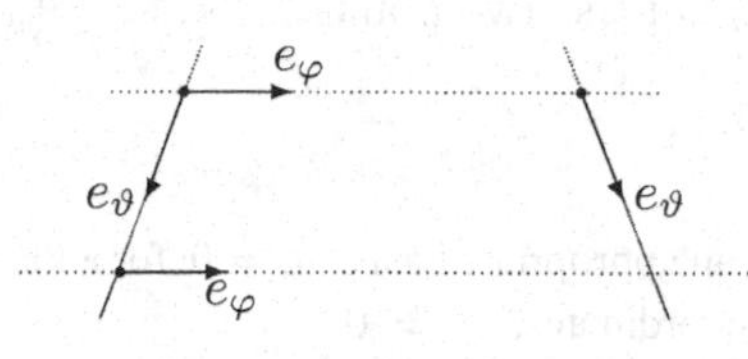

15.8.4 Be sure you understand that the effect of non-commutativity of the fields e_φ and e_ϑ (and consequently also of the non-vanishing torsion of the connection under consideration) consists of the elementary fact that if we move a small distance eastwards and then the same distance southwards, we do not reach (exactly) the same point as if we did the same steps in the opposite order. Try to obtain (by an elementary computation) the difference and check that you get the same result as you get by "scientific" consideration, i.e. by the computation of the term $\epsilon^2 T(e_\varphi, e_\vartheta) \equiv -\epsilon^2 [e_\varphi, e_\vartheta]$.

Hint: the distance between meridians gets shorter when we start to move in a direction toward the poles. □

• In the example discussed above an important class of connections has been illustrated, called a *complete parallelism*. This comes into being when in a domain on a manifold there is a *covariantly constant frame field* e_a (alternatively it is known as a *parallel* frame field), i.e. a frame field for which the covariant derivative in an arbitrary direction vanishes, $\nabla_V e_a = 0$ for each V, so that also

$$\nabla e_a = 0 \quad \text{or} \quad e_{a;\mu} = 0$$

Such a field may in general be non-holonomic. If it happens to be holonomic (i.e. coordinate), we speak about a *flat connection*.

Yet another name used for a complete parallelism is *teleparallelism*, i.e. parallelism "at a distance." The origin of this terminology will be clear from the next problem.

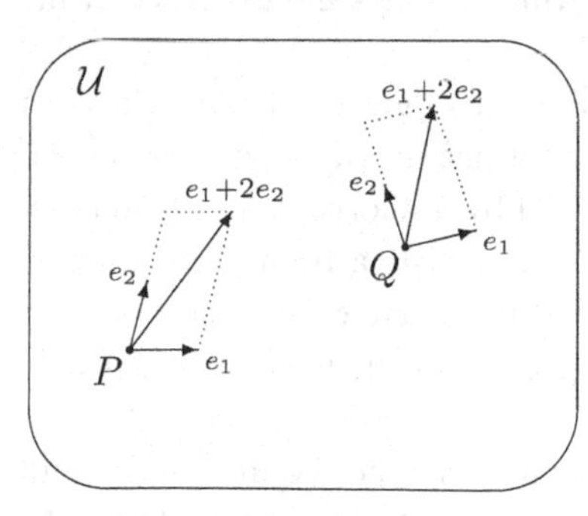

15.8.5 Let e_a be a covariantly constant frame field in a domain $\mathcal{U}$. Be sure to understand that parallel transport in this domain *does not depend on the path*, so that there exists a natural identification of any two tangent spaces in $\mathcal{U}$. Consequently a comparison of vectors sitting in different (possibly fairly remote) points in $\mathcal{U}$ now *makes* "absolute" *sense* (see the motivation of the concept of a connection at the very beginning of the chapter).

Hint: covariant constancy of the frame field gives the following equation of the parallel transport of (say) a vector field: $0 = \nabla_{\dot\gamma}(A^a e_a) = \dot A^a e_a$, i.e. $A^a(t) = k^a \equiv$ constant, *regardless of the path* along which the vector is transported $\Rightarrow$

the transport (within the domain where the frame field e_a operates) from a point P to any point Q looks like $k^a e_a(P) \mapsto k^a e_a(Q)$ (the transport thus consists in decomposing the vector at the point P with respect to the basis e_a and thereafter in composing it back *with the same coefficients* at the final point Q; so it works as if we made an "immediate leap" to Q (i.e. transport "at a distance")). Two vectors at different points are regarded as being "equal" if they have equal coefficients with respect to e_a at these two points. □

15.8.6 Check that

(i) for a complete parallelism the connection forms vanish at an appropriate basis, $\omega^a_b = 0$, for a flat connection the Christoffel symbols vanish at appropriate coordinates, $\Gamma^i_{jk} = 0$

(ii) for a complete parallelism the curvature vanishes and for a flat connection both the curvature and torsion vanish

$$\begin{aligned} \nabla_{e_a} e_b = 0 \ \text{(complete parallelism)} \quad &\Rightarrow \quad R^a{}_{bcd} = 0 \\ \nabla_{\partial_a} \partial_b = 0 \ \text{(flat connection)} \quad &\Rightarrow \quad R^a{}_{bcd} = 0 = T^a_{bc} \end{aligned}$$

Hint: (i) for a covariantly constant frame we have $\nabla_V e_a = \omega^b_a(V) e_b = 0$ or $\nabla_{\partial_a} \partial_b = \Gamma^c_{ba} \partial_c = 0$; (ii) from the Cartan structure equations (15.6.7); or since $\nabla_{e_a} e_b = 0$, we have $R(e_a, e_b) e_c \equiv (\nabla_{e_a} \nabla_{e_b} - \nabla_{e_b} \nabla_{e_a} - \nabla_{[e_a, e_b]}) e_c = 0$; $T(e_a, e_b) \equiv \nabla_{e_a} e_b - \nabla_{e_b} e_a - [e_a, e_b] = -[e_a, e_b]$, which vanishes exactly for $e_a = \partial_a$. □

• Note that both statements in problem (15.8.6) had the form of a one-way implication. The opposite implication

$$\begin{aligned} R^a{}_{bcd} = 0 \quad &\overset{?}{\Rightarrow} \quad \exists e_a : \nabla_{e_a} e_b = 0 \ \text{(complete parallelism)} \\ R^a{}_{bcd} = 0 = T^a_{bc} \quad &\overset{?}{\Rightarrow} \quad \exists x^a : \nabla_{\partial_a} \partial_b = 0 \ \text{(flat connection)} \end{aligned}$$

is a priori not clear and the issue needs a special analysis. One line of thought might be based on the way in which the curvature tensor occurred: its vanishing guarantees the triviality of parallel transport around particular infinitesimal loops. This can also be extended to bigger (finite) loops and one may infer from that the possibility of transport of a frame from the point P to its neighborhood *independent* of path, which implies the existence of the covariantly constant frame field being sought.

A different line of thought goes as follows: vanishing curvature means that *for each* frame field we have $0 = \Omega \equiv d\omega + \omega \wedge \omega$. This does not necessarily also mean $\omega = 0$ (covariantly constant e_a). A general change of frame field by a matrix A results in $\omega \mapsto \hat{\omega} = A^{-1}(\omega A + dA)$, so that the question of whether there exists a frame field $\hat{e}_a$ such that $\hat{\omega} = 0$ leads to the formulation of the problem of whether there exists a non-singular matrix field $A(x)$ which obeys a system of (partial differential) equations $\omega A + dA = 0$. The problem may be solved and the answer is *yes*.

The third possibility (which will be adopted here) is to postpone the discussion until Chapter 20, see (20.4.11), to the general context of connection theory. In this theory the notion of a covariantly constant frame field takes an interesting geometrical interpretation in terms of "integrable distributions"; by referring to the "Frobenius integrability condition" we learn that vanishing curvature is indeed also a *sufficient* condition for the existence of a covariantly constant frame field[329] (so the first implication *holds*), from which then already the *validity of the second* implication follows immediately.[330] Vanishing of both the curvature and torsion tensors thus means that the connection is flat.

15.8.7 Check that

(i) the ordinary RLC connection in (pseudo-)Euclidean space $E^{r,s}$ is flat
(ii) the connection on a Lie group G which we mentioned in problem (15.4.15) (parallel transport being given by left translation) is a complete parallelism, but in general it is not flat.

Hint: (i) $e_a = \partial_a =$ *Cartesian* frame field; (ii) $e_a =$ *left-invariant* basis; the latter fails to be holonomic for non-Abelian groups. □

15.8.8 * Define a connection on a Lie group G by the formula

$$\nabla_{L_X} L_Y := \lambda [L_X, L_Y] \equiv \lambda L_{[X,Y]} \qquad L_X, L_Y \text{ left-invariant fields}$$

(λ being a real parameter to be specified later). Check that the explicit expressions for the curvature and torsion of this connection read

$$R(L_X, L_Y)L_Z = \lambda(\lambda - 1)L_{[[X,Y],Z]} \qquad \text{i.e.} \quad R^a{}_{bcd} = \lambda(\lambda - 1)c^a_{fb}c^f_{cd}$$
$$T(L_X, L_Y) = (2\lambda - 1)L_{[X,Y]} \qquad \text{i.e.} \quad T^a_{bc} = (2\lambda - 1)c^a_{bc}$$

(e_a is a left-invariant frame field) and we may identify the following special cases:

$\lambda = \frac{1}{2}$	$R^a{}_{bcd} \neq 0$	$T^a_{bc} = 0$	RLC connection for Killing metric $\mathcal{K}$ on G
$\lambda = 0$	$R^a{}_{bcd} = 0$	$T^a_{bc} \neq 0$	parallel transport is *left* translation (15.4.15)
$\lambda = 1$	$R^a{}_{bcd} = 0$	$T^a_{bc} \neq 0$	parallel transport is *right* translation

This connection is not flat, but for $\lambda = 0$ as well as $\lambda = 1$ we have complete parallelism.

Hint: computation of R and T right from the definitions; for any λ the connection turns out to be metric with respect to the Killing metric:

$$\begin{aligned}\nabla_{L_Z}\{\mathcal{K}(L_X, L_Y)\} &\overset{1}{=} L_Z K(X, Y) = 0 \\ &\overset{2}{=} (\nabla_{L_Z}\mathcal{K})(L_X, L_Y) + \mathcal{K}(\nabla_{L_Z}L_X, L_Y) + \mathcal{K}(L_X, \nabla_{L_Z}L_Y) \\ &= (\nabla_{L_Z}\mathcal{K})(L_X, L_Y) + \lambda\{K([Z, X], Y) + K(X, [Z, Y])\} \\ &= (\nabla_{L_Z}\mathcal{K})(L_X, L_Y) \quad \text{due to (12.3.9)}\end{aligned}$$

Then (15.8.7) and (15.4.15). □

[329] The notion of curvature itself leans heavily on the integrability condition mentioned above. Namely the curvature is introduced *so that* integrability would (by definition) mean *vanishing* curvature. It turns out that a covariantly constant frame field corresponds to a "horizontal section" and that the latter exists if and only if a horizontal distribution happens to be integrable.

[330] If $R^a{}_{bcd} = 0$ gives $\nabla_{e_a} e_b = 0$, then *with respect to this* frame field we have $0 = T^a_{bc} = \cdots = -\langle e^a, [e_b, e_c]\rangle \Rightarrow [e_b, e_c] = 0$, i.e. $e_a = \partial_a$.

Summary of Chapter 15

In many applications (e.g. in the computation of acceleration of a point mass in elementary mechanics) one performs linear combinations (in particular, the difference in the case of acceleration) of vectors (or more generally tensors) sitting at different points of a manifold. This is not possible on a "bare" manifold. The structure which makes it legal is a (linear) *connection* ∇ on M. The connection enables one to transport vectors along a given path (the transport being path-dependent in general) and consequently to perform the above-mentioned comparison (vector in x is compared with the one being *transported to* x from y). This transport is *by definition* called *parallel* (in the sense of the connection ∇). A connection is frequently defined by postulating the properties of a derived object, the *covariant derivative*. One can introduce the concept of a straight line (geodesic) on (M, ∇). Two tensor fields are associated with a linear connection, the curvature and torsion tensors. It is shown that the requirements of compatibility of a connection with the metric (conservation of any scalar product upon any parallel transport) together with vanishing of its torsion result in a unique connection, the Riemannian or Levi-Civita (RLC) connection. The curvature tensor encodes the local information of "how much" (if ever) the parallel transport (along infinitesimal paths) is path-dependent; it also displays itself in the behavior of nearby geodesics, causing their deviation (Jacobi's equation). A non-zero torsion implies non-closure of a geodesic parallelogram. An efficient tool for working with a connection is provided by the machinery of differential forms. Basic objects are encoded into forms and relations between them are given by the Cartan structure equations. A connection is called a complete parallelism if there exists a covariantly constant frame field. Then the curvature tensor turns out to vanish and moreover a comparison of vectors (as well as tensors) in different (possibly remote) points makes sense. A connection is said to be flat if the covariantly constant frame field happens to be holonomic (coordinate). Then both the curvature and torsion tensors turn out to vanish.

$\nabla_a e_b =: \Gamma^c_{ba} e_c$	Coefficients of connection with respect to e_a	(15.2.1)
$\nabla_j \partial_i =: \Gamma^k_{ij} \partial_k$	Christoffel symbols of the second kind	(15.2.3)
$\dot V^i + \Gamma^i_{jk} \dot x^k V^j = 0$	Equations of parallel transport of vector	(15.2.6)
$\nabla g = 0 \quad (g_{ij;k} = 0)$	Connection ∇ is metric	(15.3.1)
$T(U, V) := \nabla_U V - \nabla_V U - [U, V]$	Torsion tensor induced by ∇	(15.3.3)
$\Gamma^i_{jk} = \frac{1}{2} g^{il}(g_{lj,k} + g_{lk,j} - g_{jk,l})$	Riemann/Levi-Civita connection (RLC)	(15.3.4)
$\nabla_{\dot\gamma} \dot\gamma = 0 \quad \left(\ddot x^i + \Gamma^i_{jk} \dot x^j \dot x^k = 0\right)$	Geodesic equation	(15.4.1)
$\exp v := \gamma_v(1), \quad \dot\gamma_v(0) = v \in T_P M$	Exponential map centered at $P \in M$	(15.4.10)
$\langle \alpha, ([\nabla_U, \nabla_V] - \nabla_{[U,V]}) W \rangle$	Riemann curvature tensor	(15.5.5)
$R_{ab} := R^c{}_{acb}, \quad R := R^a{}_a \equiv R^{ab}{}_{ab}$	Ricci tensor and scalar curvature	Sec. 15.5
$\nabla_V e_a = \omega^b_a(V) e_b \quad \left(\omega^a_b = \Gamma^a_{bc} e^c\right)$	Connection forms ω^b_a with respect to e_a	(15.6.1)
$\omega' = A^{-1} \omega A + A^{-1} dA$	Transformation law for ω under $e' = eA$	(15.6.2)
$de + \omega \wedge e = T, \quad d\omega + \omega \wedge \omega = \Omega$	Cartan structure equations	(15.6.7)
$d\Omega + \omega \wedge \Omega - \Omega \wedge \omega = 0, \quad \Omega \wedge e = 0$	Bianchi and Ricci identities (for RLC)	(15.6.16)
$\nabla^2_{\dot\gamma} \xi = R(\dot\gamma, \xi) \dot\gamma$	Jacobi's equation for geodesic deviation	(15.7.2)
$R^a{}_{bcd} = 0 = T^a{}_{bc}$	Flat connection	(15.8.6)

16

Field theory and the language of forms

• A standard machinery developed in courses on the special theory of relativity consists of a mathematics of 4-vectors and 4-tensors in Minkowski space. In this language one achieves explicit Lorentz covariance of all equations. It turns out, for example, that from a four-dimensional perspective the electric and magnetic fields turn out to be parts of a single object, the tensor of the electromagnetic field with components $F_{\mu\nu}$. This tensor is, however, not "general," but rather, *antisymmetric*

$$F_{\mu\nu} = -F_{\nu\mu}$$

This fact is a clear signal for us "to switch over to another channel" in contemplation on these ideas, the new "channel" being the language of *differential forms*. And if we call to mind that in the 4-tensor formalism Maxwell's equations look like

$$F^{\mu\nu}{}_{,\nu} = j^{\mu} \qquad F_{[\mu\nu,\rho]} = 0$$

then our experienced eye[331] readily reports that it noticed the component expression of *codifferential and differential of a 2-form F* on the left-hand sides of the equations. Thus, the fundamental equations of the electromagnetic theory may be written in terms of fundamental objects and operations of the theory of differential forms.

In this chapter we derive the four-dimensional expressions of field theory equations in the language of differential forms and we outline several advantages of this approach. In the first section we investigate thoroughly a structure of differential forms in Minkowski space along with their expression in a ("(1+3)-decomposed") form which facilitates immediate contact with the three-dimensional expressions typically encountered in the usual vector analysis approach. We also compute the results of all standard operations on such forms. By doing this we reveal that some of them contain just those combinations of terms we need for Maxwell's equations.[332] This enables us eventually to write down the equations in terms of forms on a four-dimensional space-time and to admire their amazingly simple and lucid structure.

[331] The left eye for right-handers and the right eye for left-handers (recall the well-known crossing of neural pathways).

[332] The reader is expected to be familiar with the standard three-dimensional vector analysis form of Maxwell's equations ($\operatorname{div} \mathbf{E} = \rho$, etc.).

16.1 Differential forms in the Minkowski space $E^{1,3}$

• *Minkowski space* (2.6.6) results from a "symbiosis" of the time axis $E^1[t]$ with ordinary three-dimensional space $E^3[\mathbf{r}]$ into a single entity, a four-dimensional *space-time* of the special theory of relativity. So first there is a Cartesian product of manifolds $\mathbb{R}^1 \times \mathbb{R}^3 \sim \mathbb{R}^4$, in the second step it is endowed with a metric tensor[333] η with signature $(1, 3)$. Recall that splitting of space-time into a "time" and a "space" does not have an absolute character in the theory of relativity, but rather it is *observer-dependent*: another observer may regard another direction[334] (one-dimensional subspace) in the ("absolute") tangent space at a given point as a "time" direction and consequently also another three-dimensional subspace interpreted as a "genuine space" (since the latter should be perpendicular to the "time" dimension).

So in Minkowski space there is an important inherent structure of $(1 + 3)$-*decomposition* in each tangent space which is usually realized by a choice of particular Cartesian coordinates $(t, \mathbf{r})$. If the choice is made, each vector may be unambiguously written as a sum of its "time" and "space" parts. Since vectors serve as arguments for forms, it should not surprise us too much that the decomposition of vectors has a direct influence on *forms* on the manifold $E^{1,3}$. We will see that forms on $E^{1,3}$ may also be uniquely written in a decomposed way. The decomposition incorporates *two* forms, which are *as if* they live only in E^3; on such forms one can apply the standard operations of *three-dimensional vector analysis* which we discussed in detail in Section 8.5. The possibility of decomposition of vectors thus has an important implication for the formalism of forms in Minkowski space: the same statement may be made either "objectively," in a language of genuinely four-dimensional forms, or "subjectively," in a language of three-dimensional observer-dependent forms. Both languages are very important: the four-dimensional language of a "noble science" often reveals an extremely simple formal expression of a physical law (however, at the expense of passing to higher-dimensional space, which is not "directly visible" by us, mere mortals), the three-dimensional "language of common people" is a language of "directly visible" quantities (vector of the electric field, etc.). Therefore it is worth putting in the time and effort to develop such a version of the machinery of forms in Minkowski space which offers a simple transition between the two languages. Note that in *both* languages the calculus of *differential forms* plays a central role, thus providing an excellent example of its remarkable power combined with amazing simplicity.

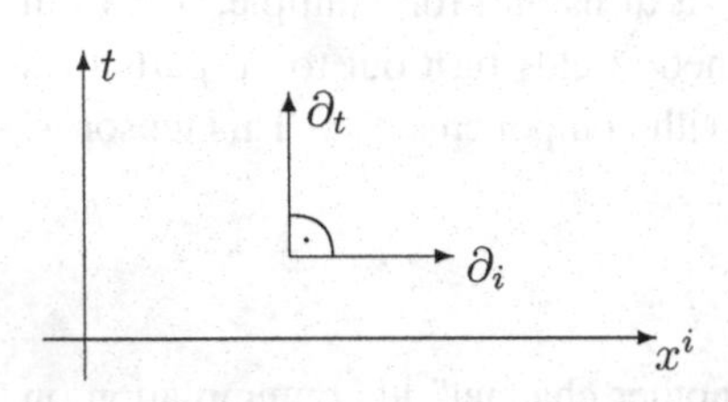

[333] This text by no means has the ambition to replace textbooks on relativity, where one finds a *physical motivation* for the choice of just this metric tensor on the resulting manifold $\mathbb{R}^4$, i.e. why we do not combine E^1 and E^3 to, say, E^4, but instead $E^{1,3}$; a word of caution is in order: roughly equally frequently the *opposite convention* is adopted, in which η has the signature $(3, 1)$.

[334] A time direction corresponds to the choice of the vector e_0, which obeys $\eta(e_0, e_0) = +1$. This vector is not unique, any other possibility looks like $e'_0 = \Lambda^a_0 e_a \equiv \Lambda^0_0 e_0 + \Lambda^i_0 e_i$ with a Lorentzian matrix Λ. This produces a "mixture" of the previous time and space.

16.1.1 Consider Minkowski space $E^{1,3}$ with global Cartesian coordinates $x^\mu \equiv (x^0, x^i) \equiv (t, x^i)$. By definition in these coordinates the metric tensor takes the form (2.6.6)

$$\eta = dt \otimes dt - h \equiv dt \otimes dt - (dx \otimes dx + dy \otimes dy + dz \otimes dz) \equiv g^{(1)} \oplus (-g^{(3)})$$

and an orientation is introduced (in a standard way, see the end of Section 5.5) so that the coordinate frame field $\partial_\mu \equiv (\partial_0, \partial_1, \partial_2, \partial_3) \equiv (\partial_t, \partial_x, \partial_y, \partial_z)$ is declared to be right-handed. Check that

(i) each differential form on $E^{1,3}$ may be uniquely written in the form

$$\alpha = dt \wedge \hat{s} + \hat{r}$$

where the forms $\hat{s}, \hat{r}$ already *do not contain* the differential dt; they decompose with respect to a "spatial" part dx^i alone of the coordinate coframe field $dx^\mu \equiv (dt, dx^i) \equiv (dt, dx, dy, dz)$; we will call such forms *spatial forms*

(ii) if α is a p-form, then $\hat{s}$ is a $(p-1)$-form and $\hat{r}$ is a p-form

(iii) $\alpha = 0$ is equivalent to $\hat{s} = 0, \hat{r} = 0$

(iv) a form α is spatial if and only if $i_{\partial_t}\alpha = 0$

(v) explicit expressions of the forms $\hat{s}, \hat{r}$ read

$$\alpha = \frac{1}{p!}\alpha_{\mu\dots\nu}\, dx^\mu \wedge \dots \wedge dx^\nu \quad \Rightarrow \quad \hat{s} = \frac{1}{(p-1)!}\alpha_{0i\dots j}\, dx^i \wedge \dots \wedge dx^j$$

$$\hat{r} = \frac{1}{p!}\alpha_{i\dots j}\, dx^i \wedge \dots \wedge dx^j$$

so that

$$\hat{s}_{i\dots j}(t, \mathbf{r}) = \alpha_{0i\dots j}(t, \mathbf{r}) \qquad \hat{r}_{i\dots j}(t, \mathbf{r}) = \alpha_{i\dots j}(t, \mathbf{r})$$

Hint: (i) trivial: according to the rules for manipulating forms, in each term the differential dt may occur *at most* once, i.e. once or not at all; $\hat{s}$ is obtained by collecting all terms in which dt occurs *once*, $\hat{r}$ collects all terms in which dt is *absent*; (iii) the two terms are linearly independent; (iv) $i_{\partial_t}\alpha = \hat{s}$; (v) a sum over μ = a sum over 0 (one term) + a sum over i (three terms); for example, for a 2-form

$$\begin{aligned}
\alpha &= \frac{1}{2}\alpha_{\mu\nu}\, dx^\mu \wedge dx^\nu \\
&= \frac{1}{2}\alpha_{00}\, dx^0 \wedge dx^0 + \frac{1}{2}\alpha_{0i}\, dx^0 \wedge dx^i + \frac{1}{2}\alpha_{i0}\, dx^i \wedge dx^0 + \frac{1}{2}\alpha_{ij}\, dx^i \wedge dx^j \\
&= dt \wedge (\alpha_{0i}\, dx^i) + \frac{1}{2}\alpha_{ij}\, dx^i \wedge dx^j \\
&= dt \wedge (\hat{s}_i\, dx^i) + \frac{1}{2}\hat{r}_{ij}\, dx^i \wedge dx^j \equiv dt \wedge \hat{s} + \hat{r}
\end{aligned}$$

□

16.1.2 Check that a general decomposition of differential forms in Minkowski space $\alpha = dt \wedge \hat{s} + \hat{r}$ gives for the concrete relevant degrees the following explicit "parametrizations" in the language of three-dimensional vector analysis:

$$\begin{array}{llll}
\text{0-forms:} & \alpha = f & \text{i.e.} & (\hat{s},\hat{r}) = (0, f) \\
\text{1-forms:} & \alpha = f\,dt + \mathbf{a}\cdot d\mathbf{r} & \text{i.e.} & (\hat{s},\hat{r}) = (f, \mathbf{a}\cdot d\mathbf{r}) \\
\text{2-forms:} & \alpha = dt\wedge \mathbf{a}\cdot d\mathbf{r} + \mathbf{b}\cdot d\mathbf{S} & \text{i.e.} & (\hat{s},\hat{r}) = (\mathbf{a}\cdot d\mathbf{r}, \mathbf{b}\cdot d\mathbf{S}) \\
\text{3-forms:} & \alpha = dt\wedge \mathbf{a}\cdot d\mathbf{S} + f\,dV & \text{i.e.} & (\hat{s},\hat{r}) = (\mathbf{a}\cdot d\mathbf{S}, f\,dV) \\
\text{4-forms:} & \alpha = dt\wedge f\,dV & \text{i.e.} & (\hat{s},\hat{r}) = (f\,dV, 0)
\end{array}$$

where the basis spatial forms $d\mathbf{r}$, $d\mathbf{S}$, dV are given by the expressions introduced in (8.5.2).

Hint: see (16.1.1) and (8.5.2). □

• Although the spatial forms f, $\mathbf{a}\cdot d\mathbf{r}$, $\mathbf{a}\cdot d\mathbf{S}$ or $f\,dV$ are at first sight indistinguishable from the true forms on E^3, a closer look reveals an important difference, namely the time dependence of their *components*.[335] This results in an additional term in the expression of their exterior derivative.

16.1.3 Check that

(i) for the exterior derivative of *spatial* forms we have the following general expression:

$$d\hat{r} = dt\wedge \partial_t\hat{r} + \hat{d}\hat{r} \qquad \partial_t\hat{r} \equiv \mathcal{L}_{\partial_t}\hat{r}$$

i.e. we may write an operator identity

$$d = dt\wedge\mathcal{L}_{\partial_t} + \hat{d}$$

where $\hat{d}$ stands for a *spatial exterior derivative*, which is performed as if the spatial form under consideration *indeed* lived only in E^3 (its components did not depend on t), i.e. according to the results discussed in the section on vector analysis (8.5.4)

(ii) this explicitly gives for particular degrees

$$\begin{array}{lrl}
\text{0-forms:} & df &= (\partial_t f)dt + \text{grad} f\cdot d\mathbf{r} \\
\text{1-forms:} & d(\mathbf{a}\cdot d\mathbf{r}) &= dt\wedge(\partial_t\mathbf{a})\cdot d\mathbf{r} + (\text{curl}\,\mathbf{a})\cdot d\mathbf{S} \\
\text{2-forms:} & d(\mathbf{a}\cdot d\mathbf{S}) &= dt\wedge(\partial_t\mathbf{a})\cdot d\mathbf{S} + (\text{div}\,\mathbf{a})dV \\
\text{3-forms:} & d(f dV) &= dt\wedge(\partial_t f)dV
\end{array}$$

Hint: (i) for a 1-form, say, we have

$$\begin{aligned}
d(\hat{r}_i(t,\mathbf{r})\,dx^i) &= d\hat{r}_i\wedge dx^i = ((\partial_t\hat{r}_i)\,dt + \hat{r}_{i,j}\,dx^j)\wedge dx^i \\
&= dt\wedge\mathcal{L}_{\partial_t}(\hat{r}_i dx^i) + \frac{1}{2}(-2\hat{r}_{[i,j]})dx^i\wedge dx^j \equiv dt\wedge\partial_t\hat{r} + \hat{d}\hat{r}
\end{aligned}$$

(ii) see (8.5.4); for $\hat{d}$ *here* the formulas for d *there* are valid. □

• Now it is already a simple matter to write down the action of an exterior derivative on arbitrary (not only spatial) forms.

[335] One should keep in mind that these forms live in $E^{1,3}$ rather than in E^3, so that we may write in detail, for example, $\hat{r} = \hat{r}_i(t,\mathbf{r})\,dx^i$.

16.1.4 Check that

(i) for the exterior derivative of a general form in $E^{1,3}$ we have the following formula:

$$d\alpha \equiv d(dt \wedge \hat{s} + \hat{r}) = dt \wedge (\partial_t \hat{r} - \hat{d}\hat{s}) + \hat{d}\hat{r}$$

(ii) this gives explicitly for particular degrees

$$\begin{aligned}
&\text{0-forms:} & df &= (\partial_t f)dt + \text{grad} f \cdot d\mathbf{r} \\
&\text{1-forms:} & d(f dt + \mathbf{a} \cdot d\mathbf{r}) &= dt \wedge (\partial_t \mathbf{a} - \text{grad} f) \cdot d\mathbf{r} + (\text{curl}\,\mathbf{a}) \cdot d\mathbf{S} \\
&\text{2-forms:} & d(dt \wedge \mathbf{a} \cdot d\mathbf{r} + \mathbf{b} \cdot d\mathbf{S}) &= dt \wedge (\partial_t \mathbf{b} - \text{curl}\,\mathbf{a}) \cdot d\mathbf{S} + (\text{div}\,\mathbf{b}) dV \\
&\text{3-forms:} & d(dt \wedge \mathbf{a} \cdot d\mathbf{S} + f dV) &= dt \wedge (\partial_t f - \text{div}\,\mathbf{a}) dV
\end{aligned}$$

Hint: see (16.1.3). □

• The second important operator on forms to be tackled is the Hodge star operator $*$. Just as the concept of the spatial exterior derivative $\hat{d}$ (which acts only on spatial forms) appears naturally in the course of the computation of d, it is convenient to introduce the *spatial Hodge operator* $\hat{*}$, which also acts only on spatial forms. This operator is defined in a standard way with respect to the (positive definite!) metric tensor $g^{(3)}$ and standard orientation $\hat{o}$ in E^3 (the basis $(\partial_x, \partial_y, \partial_z)$ is declared to be right-handed), i.e. using the notation from (5.8.1)

$$\hat{*} := *_{g^{(3)},\hat{o}} \qquad * := *_{\eta,o}$$

(i.e. $\hat{*}$ is an ordinary star operator for the subspace corresponding to E^3, endowed with its "original" positive definite[336] metric $g^{(3)}$). It turns out that if we apply the ("total") operator $*$ on a decomposed form, the result may be written (again in a decomposed form) in terms of the spatial operator $\hat{*}$.

16.1.5 Check that

(i) the Hodge operator of dualization gives on a general form in $E^{1,3}$

$$*\alpha \equiv *(dt \wedge \hat{s} + \hat{r}) = dt \wedge (\hat{*}\hat{r}) + \hat{*}\hat{\eta}\hat{s}$$

(ii) for particular degrees this explicitly reads

$$\begin{aligned}
&\text{0-forms:} & *f &= dt \wedge f\, dV \\
&\text{1-forms:} & *(f\, dt + \mathbf{a} \cdot d\mathbf{r}) &= dt \wedge \mathbf{a} \cdot d\mathbf{S} + f\, dV \\
&\text{2-forms:} & *(dt \wedge \mathbf{a} \cdot d\mathbf{r} + \mathbf{b} \cdot d\mathbf{S}) &= dt \wedge \mathbf{b} \cdot d\mathbf{r} - \mathbf{a} \cdot d\mathbf{S} \\
&\text{3-forms:} & *(dt \wedge \mathbf{a} \cdot d\mathbf{S} + f\, dV) &= f\, dt + \mathbf{a} \cdot d\mathbf{r} \\
&\text{4-forms:} & *(dt \wedge f\, dV) &= -f
\end{aligned}$$

(iii) these formulas are consistent with the result $** = -\hat{\eta}$, which we obtain for the case $E^{1,3}$ from (5.8.2)

(iv) the operator $\hat{\eta}$ (the main automorphism) acts as follows:

$$\hat{\eta}(dt \wedge \hat{s} + \hat{r}) = dt \wedge (-\hat{\eta}\hat{s}) + \hat{\eta}\hat{r}$$

[336] *This* choice of a sign then enables one to express hatted quantities in the language of vector analysis, i.e. the hatted quantities then behave exactly as the objects treated in Section 8.5.

Hint: (i) with the help of (5.8.1) straightforwardly check that on a "mixed" and a genuine spatial basis p-form there holds ($e^0 \equiv dt, e^i \equiv dx^i$)

$$*(e^0 \wedge e^i \wedge \cdots \wedge e^j) = \hat{*}\hat{\eta}(e^i \wedge \cdots \wedge e^j)$$
$$*(e^k \wedge \cdots \wedge e^l) = e^0 \wedge \hat{*}(e^k \wedge \cdots \wedge e^l)$$

(ii) see (8.5.3); (iii) see (5.3.3). □

• Finally, let us have a look at the last two operators needed, the codifferential δ and the Laplace–deRham operator Δ. Since they are only composed from operators which we have already mastered, the only thing we are to do is to bring the parts together. As is commonly done, we will denote the Laplace–deRham operator for the case $E^{1,3}$ by $\Box$ rather than Δ (one should keep in mind, however, that this operator in general acts on forms, not only on functions; in particular, on functions it is also called the *d'Alembert operator* or the *wave operator*).[337]

16.1.6 Check that

(i) the codifferential $\delta \equiv *^{-1}d * \hat{\eta}$ (on $E^{1,3}$ this comes out as $*d*$) and the Laplace–deRham operator $\Box = -(\delta d + d\delta)$ act on a general form on $E^{1,3}$ as follows:

$$\delta\alpha \equiv \delta(dt \wedge \hat{s} + \hat{r}) = dt \wedge (\hat{\delta}\hat{s}) + (-\partial_t \hat{s} - \hat{\delta}\hat{r})$$
$$\Box\alpha \equiv \Box(dt \wedge \hat{s} + \hat{r}) = dt \wedge \left[\left(\partial_t^2 - \hat{\Delta}\right)\hat{s}\right] + \left[\left(\partial_t^2 - \hat{\Delta}\right)\hat{r}\right]$$

where we introduced the *spatial codifferential*

$$\hat{\delta} := \hat{*}^{-1}\hat{d}\hat{*}\hat{\eta} \qquad \text{(on } E^3 \text{ this is } \hat{*}\hat{d}\hat{*}\hat{\eta}\text{)}$$

and *spatial Laplace–deRham operator*

$$\hat{\Delta} = -(\hat{\delta}\hat{d} + \hat{d}\hat{\delta})$$

(ii) for particular degrees this explicitly yields

$$\begin{aligned}
\text{0-forms:} \quad & \delta f = 0 \\
& \Box f = \left(\partial_t^2 - \hat{\Delta}\right) f \\
\text{1-forms:} \quad & \delta(f\, dt + \mathbf{a}\cdot d\mathbf{r}) = -\partial_t f + \operatorname{div}\mathbf{a} \\
& \Box(f\, dt + \mathbf{a}\cdot d\mathbf{r}) = \left(\partial_t^2 - \hat{\Delta}\right) f\, dt + \left(\partial_t^2 - \hat{\Delta}\right)\mathbf{a}\cdot d\mathbf{r} \\
\text{2-forms:} \quad & \delta(dt \wedge \mathbf{a}\cdot d\mathbf{r} + \mathbf{b}\cdot d\mathbf{S}) = (-\operatorname{div}\mathbf{a})dt + (-\partial_t\mathbf{a} - \operatorname{curl}\mathbf{b})\cdot d\mathbf{r} \\
& \Box(dt \wedge \mathbf{a}\cdot d\mathbf{r} + \mathbf{b}\cdot d\mathbf{S}) = dt \wedge \left(\partial_t^2 - \hat{\Delta}\right)\mathbf{a}\cdot d\mathbf{r} + \left(\partial_t^2 - \hat{\Delta}\right)\mathbf{b}\cdot d\mathbf{S} \\
\text{3-forms:} \quad & \delta(dt \wedge \mathbf{a}\cdot d\mathbf{S} + f\, dV) = dt \wedge (\operatorname{curl}\mathbf{a}\cdot d\mathbf{r}) + (-\partial_t\mathbf{a} + \operatorname{grad} f)\cdot d\mathbf{S} \\
& \Box(dt \wedge \mathbf{a}\cdot d\mathbf{S} + f\, dV) = dt \wedge \left(\partial_t^2 - \hat{\Delta}\right)\mathbf{a}\cdot d\mathbf{S} + \left(\partial_t^2 - \hat{\Delta}\right) f\, dV \\
\text{4-forms:} \quad & \delta(dt \wedge f\, dV) = dt \wedge (-\operatorname{grad} f\cdot d\mathbf{S}) + (-\partial_t f)dV \\
& \Box(dt \wedge f\, dV) = dt \wedge \left(\partial_t^2 - \hat{\Delta}\right) f\, dV
\end{aligned}$$

Hint: this is a bit easier with the help of (16.1.7). □

[337] Exceptional foresight was definitely not the strong point of the people who introduced this notation long ago; they seem to have missed the elementary fact that, in this book, the same symbol denotes the end of problems.

• One can manipulate such combinations of operators more simply using the following matrix formalism: assign a column with components $\hat{s}$ and $\hat{r}$ to a form $\alpha = dt \wedge \hat{s} + \hat{r}$. According to (16.1.4) then the action of the operator d (to give an example) gives

$$\begin{pmatrix} \hat{s} \\ \hat{r} \end{pmatrix} \stackrel{d}{\mapsto} \begin{pmatrix} \partial_t \hat{r} - \hat{d}\hat{s} \\ \hat{d}\hat{r} \end{pmatrix} \equiv \begin{pmatrix} -\hat{d} & \partial_t \\ 0 & \hat{d} \end{pmatrix} \begin{pmatrix} \hat{s} \\ \hat{r} \end{pmatrix} \quad \Rightarrow \quad d \leftrightarrow \begin{pmatrix} -\hat{d} & \partial_t \\ 0 & \hat{d} \end{pmatrix}$$

All the other operators which we have encountered may also be written in this way, reducing the calculation of their combinations to a mechanical matrix multiplication.

16.1.7 Check that the following matrix operators are to be assigned to the operators discussed so far:

$$d \leftrightarrow \begin{pmatrix} -\hat{d} & \partial_t \\ 0 & \hat{d} \end{pmatrix} \qquad * \leftrightarrow \begin{pmatrix} 0 & \hat{*} \\ \hat{*}\hat{\eta} & 0 \end{pmatrix} \qquad *^{-1} \leftrightarrow \begin{pmatrix} 0 & -\hat{*}\hat{\eta} \\ \hat{*} & 0 \end{pmatrix}$$

$$\hat{\eta} \leftrightarrow \begin{pmatrix} -\hat{\eta} & 0 \\ 0 & \hat{\eta} \end{pmatrix} \qquad \delta \leftrightarrow \begin{pmatrix} \hat{\delta} & 0 \\ -\partial_t & -\hat{\delta} \end{pmatrix} \qquad \Box \leftrightarrow \begin{pmatrix} \partial_t^2 - \hat{\Delta} & 0 \\ 0 & \partial_t^2 - \hat{\Delta} \end{pmatrix}$$

Hint: see (16.1.4), (16.1.5), (5.8.2) and (16.1.6). □

• Also Stokes' theorem takes a specific form if it concerns strictly spatial domains. By a *spatial domain* we mean a domain $\hat{\mathcal{D}}$ in Minkowski space such that if γ is an arbitrary curve lying *in* $\hat{\mathcal{D}}$ then the tangent vector $\dot{\gamma}$ is strictly spatial (i.e. perpendicular to the vector ∂_t at this point). Put another way, its parametrization has a form

$$(u, \dots, v) \mapsto (t, x^i) \equiv (t_0 = \text{constant}, x^i(u, \dots, v))$$

(for example, a ball in E^3 at time $t = t_0$ is a three-dimensional spatial domain; such domains exist with one, two and three dimensions).

16.1.8 Let $\hat{\mathcal{D}}$ be a *spatial* domain of dimension $p = 1, 2$ or 3 in Minkowski space $E^{1,3}$ and let $\hat{r}$ be any *spatial* $(p-1)$-form. Show that the *spatial Stokes' theorem* holds:

$$\int_{\hat{\mathcal{D}}} \hat{d}\hat{r} = \int_{\partial\hat{\mathcal{D}}} \hat{r}$$

i.e. we may substitute d in this case by $\hat{d}$ if we wish (reducing then to grad, curl or div; thus, for both spatial forms and domains, Stokes' theorem presented here is at first sight indistinguishable from one of three integral theorems discussed in vector analysis (8.5.6)).

Hint: according to (16.1.3) we have $d\hat{r} = dt \wedge (\cdots) + \hat{d}\hat{r}$; the part which contains dt drops out upon restriction on the spatial domain. □

• Let us sum up for later convenience all the general relations we have encountered in this section. We learned that any differential form on Minkowski space $E^{1,3}$ may be written as

$$\alpha = dt \wedge \hat{s} + \hat{r}$$

and that the basic operations performed on forms which are decomposed in this way are given by the formulas

$$\begin{aligned}
d\alpha &\equiv d(dt\wedge\hat{s}+\hat{r}) &&= dt\wedge(\partial_t\hat{r}-\hat{d}\hat{s})+\hat{d}\hat{r}\\
*\alpha &\equiv *(dt\wedge\hat{s}+\hat{r}) &&= dt\wedge(\hat{*}\hat{r})+\hat{*}\hat{\eta}\hat{s}\\
\hat{\eta}\alpha &\equiv \hat{\eta}(dt\wedge\hat{s}+\hat{r}) &&= dt\wedge(-\hat{\eta}\hat{s})+\hat{\eta}\hat{r}\\
\delta\alpha &\equiv \delta(dt\wedge\hat{s}+\hat{r}) &&= dt\wedge(\hat{\delta}\hat{s})+(-\partial_t\hat{s}-\hat{\delta}\hat{r})\\
\Box\alpha &\equiv \Box(dt\wedge\hat{s}+\hat{r}) &&= dt\wedge\left(\partial_t^2-\hat{\Delta}\right)\hat{s}+\left(\partial_t^2-\hat{\Delta}\right)\hat{r}
\end{aligned}$$

One should think of typical expressions from vector analysis f, $\mathbf{a}\cdot d\mathbf{r}$, $\mathbf{b}\cdot d\mathbf{S}$, $h\,dV$ (depending on degree) representing the spatial forms and standard operators grad, curl and div (depending again on degree) representing the spatial differential and codifferential $\hat{d}$, $\hat{\delta}$.

A spatial Stokes' theorem holds in which formally only the spatial exterior derivative takes part.

16.2 Maxwell's equations in terms of differential forms

- Write down the result of an exterior derivative of a general 2-form in Minkowski space

$$d(dt\wedge\mathbf{a}\cdot d\mathbf{r}+\mathbf{b}\cdot d\mathbf{S}) = dt\wedge(\partial_t\mathbf{b}-\operatorname{curl}\mathbf{a})\cdot d\mathbf{S}+(\operatorname{div}\mathbf{b})dV$$

and compare it with a standard three-dimensional version of the second series of Maxwell's equations in vacuum (i.e. the homogeneous part of the equations)

$$\operatorname{curl}\mathbf{E}+\partial_t\mathbf{B}=\mathbf{0}\qquad \operatorname{div}\mathbf{B}=0$$

An overwhelming majority of readers would probably agree that there are some common features detectable between the expression above and the last two equations.

16.2.1 Check that

(i) the second series (homogeneous half) of Maxwell's equations may be written in the form

$$dF=0\qquad\text{where}\quad F:=dt\wedge\mathbf{E}\cdot d\mathbf{r}-\mathbf{B}\cdot d\mathbf{S}$$

is a 2-*form of the electromagnetic field*

(ii) an explicit expression of its (Cartesian) components in terms of (Cartesian) components of vectors of electric and magnetic field reads

$$F_{0i}=E_i$$

$$F_{ij}=-\epsilon_{ijk}B_k\qquad\text{i.e.}\quad F_{\mu\nu}=\begin{pmatrix}0 & E_x & E_y & E_z\\ -E_x & 0 & -B_z & B_y\\ -E_y & B_z & 0 & -B_x\\ -E_z & -B_y & B_x & 0\end{pmatrix}$$

(iii) a transition to the dual form may be expressed in terms of the fields $\mathbf{E}$, $\mathbf{B}$ as

$$F\mapsto *F\quad\Leftrightarrow\quad(\mathbf{E},\mathbf{B})\mapsto(-\mathbf{B},\mathbf{E})$$

Hint: see (16.1.4). □

• So we see that a remarkably convenient way of encoding complete information about electric and magnetic fields consists in building them into a particular 2-form F in Minkowski space. Then the homogeneous half of Maxwell's equations reduces to an amazingly simple statement saying that *the form is closed.*

Now, let's talk about the first series (inhomogeneous half) of Maxwell's equations. They are first-order partial differential equations for the fields **E**, **B** as well, so we expect some first-order differential operator on forms in Minkowski space to act on F. Since d was already used, the only remaining (known) candidate is the codifferential δ. Now when we write down the result of the action of δ or $d*$ on the form F (using the general formulas derived in the preceding section)

$$\delta(dt \wedge \mathbf{E} \cdot d\mathbf{r} - \mathbf{B} \cdot d\mathbf{S}) = (-\operatorname{div} \mathbf{E})dt + (-\partial_t \mathbf{E} + \operatorname{curl} \mathbf{B}) \cdot d\mathbf{r}$$
$$d * (dt \wedge \mathbf{E} \cdot d\mathbf{r} - \mathbf{B} \cdot d\mathbf{S}) = dt \wedge (-\partial_t \mathbf{E} + \operatorname{curl} \mathbf{B}) \cdot d\mathbf{S} - (\operatorname{div} \mathbf{E}) dV$$

and confront it with the first series[338]

$$\operatorname{div} \mathbf{E} = \rho \qquad \operatorname{curl} \mathbf{B} - \partial_t \mathbf{E} = \mathbf{j}$$

we see that we were lucky again – just the right combinations occur.

16.2.2 Check that

the first series (inhomogeneous half) of Maxwell's equations may be written in the form

$$\delta F = -j \quad \text{or equivalently} \quad d * F = -J \equiv - * j$$

where the three-dimensional quantities ρ (electric *charge density*) and **j** (electric *current density*) are built into a single object living in Minkowski space, the 1-*form of current* or alternatively its dual 3-*form of current*

$$j = \rho\, dt - \mathbf{j} \cdot d\mathbf{r} \equiv j_\mu\, dx^\mu$$
$$J = dt \wedge (-\mathbf{j} \cdot d\mathbf{S}) + \rho\, dV \equiv j^\mu\, d\Sigma_\mu \equiv *j$$

Hint: see (16.1.5) and (16.1.6). □

16.2.3 Check that the total electric charge in a spatial domain $\hat{\mathcal{D}}_3$ is given by the integral

$$Q = \int_{\hat{\mathcal{D}}_3} J \equiv \int_{\hat{\mathcal{D}}_3} *j$$

Hint: according to (16.2.2) we have $J = dt \wedge (-\mathbf{j} \cdot d\mathbf{S}) + \rho\, dV$ and the first term does not contribute to the integral over $\hat{\mathcal{D}}_3$ due to the factor of dt. □

[338] This holds in *Heaviside's* system of units and setting $c = 1$; otherwise various constants may enter the equations which are, however, irrelevant for us.

16.2.4 Check that

(i) the system of *Maxwell's equations*

$$\delta F = -j$$
$$dF = 0$$

is consistent only for forms of the current j that obey the *continuity equation*, i.e. any of the following (equivalent) conditions:

$$\delta j = 0 \qquad dJ = 0 \qquad j^{\mu}{}_{;\mu} = 0 \qquad \partial_t \rho + \operatorname{div} \mathbf{j} = 0$$

(ii) this condition is a local formulation of *electric charge conservation*

(iii) the conservation of electric charge, in contrast, *necessarily needs* the existence of a 2-form F which obeys the inhomogeneous part of Maxwell's equations, i.e. the equation

$$\delta F = -j$$

Put another way, the conserved charge turns out to be the *source of a field of type* F (a 2-form field or, in $1+3$ language, of the fields $\mathbf{E}, \mathbf{B}$); the *homogeneous* half of Maxwell's equations is not fixed, however, by the conserved charge.

Solution: (i) according to (8.3.12) $\delta\delta = 0$; by (16.1.6) and (16.2.2) $\delta j = -(\partial_t \rho + \operatorname{div} \mathbf{j})$; (ii) if $\mathcal{D}_4$ is a four-dimensional domain, then

$$0 = \int_{\mathcal{D}_4} dJ = \int_{\partial\mathcal{D}_4} J = \int_{\partial\mathcal{D}_4} \rho\, dV - \int_{\partial\mathcal{D}_4} dt \wedge \mathbf{j} \cdot d\mathbf{S} \qquad \text{i.e.} \qquad \int_{\partial\mathcal{D}_4} \rho\, dV = \int_{\partial\mathcal{D}_4} dt \wedge \mathbf{j} \cdot d\mathbf{S}$$

For $\mathcal{D}_4$ of the form of an infinitesimal cylinder over a spatial domain $\hat{\mathcal{D}}_3$, i.e. $\mathcal{D}_4 = I \times \hat{\mathcal{D}}_3$, $I = \langle t_0, t_0 + \epsilon \rangle$ we have $\partial\mathcal{D}_4 = \partial I \times \hat{\mathcal{D}}_3 - I \times \partial\hat{\mathcal{D}}_3 = \{t_0 + \epsilon\} \times \hat{\mathcal{D}}_3 - \{t_0\} \times \hat{\mathcal{D}}_3 - I \times \partial\hat{\mathcal{D}}_3$ so that we get

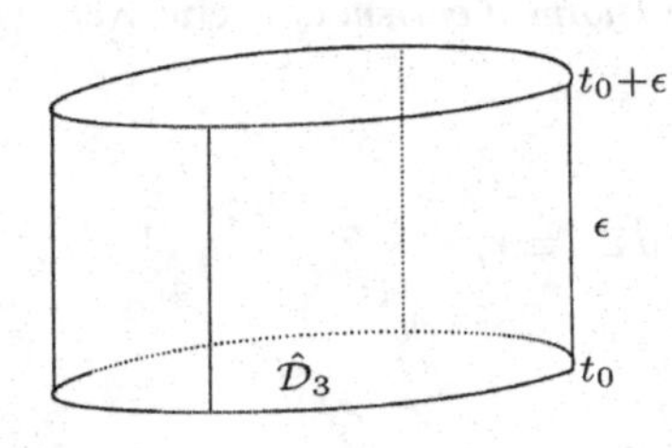

$$\int_{\hat{\mathcal{D}}_3} \rho(t_0+\epsilon, \mathbf{r})\, dV - \int_{\hat{\mathcal{D}}_3} \rho(t_0, \mathbf{r})\, dV = - \int_{I \times \partial\hat{\mathcal{D}}_3} dt \wedge \mathbf{j} \cdot d\mathbf{S}$$

$$= - \int_{t_0}^{t_0+\epsilon} dt \int_{\partial\hat{\mathcal{D}}_3} \mathbf{j}(t, \mathbf{r}) \cdot d\mathbf{S} = -\epsilon \oint_{\partial\hat{\mathcal{D}}_3} j(t_0, \mathbf{r}) \cdot d\mathbf{S}$$

leading to

$$\frac{d}{dt} Q(t) = - \oint_{\partial\hat{\mathcal{D}}_3} \mathbf{j}(t, \mathbf{r}) \cdot d\mathbf{S} \qquad Q(t) := \int_{\hat{\mathcal{D}}_3} \rho(t, \mathbf{r})\, dV$$

so that the increase of the total charge $Q(t)$ in a spatial domain $\hat{\mathcal{D}}_3$ is (only) given by the negative flux of the current density for the boundary of the domain; this is, however, exactly a conservation law (charge is neither produced nor destroyed, it is only transferred from place to place); (iii) co-version of the Poincaré lemma (9.2.12): $\delta j = 0 \Rightarrow$ locally $j = \delta(-F)$. □

• Maxwell's equations represent a *local* (differential) formulation of the laws of electromagnetism. By means of an appropriate integration one obtains their integral versions, which express the laws in terms of directly measurable quantities. For example, although an object which enters the local Maxwell's equations is the charge *density* ρ, it is only the *integral* $Q = \int_{\hat{\mathcal{D}}_3} \rho\, dV$ of the density ρ over a spatial domain $\hat{\mathcal{D}}_3$ rather than the density itself, which is a measurable quantity. Since we learned in Chapter 7 that the quantities under the integral sign are differential forms, it is clear that from the point of view of the relation between differential and integral formulations of a physical law the most convenient differential formulation is that directly in terms of differential forms.

16.2.5 Check that the following integral laws are represented by the local equation $dF = 0$.

$$\frac{d}{dt}\int_{\hat{\mathcal{D}}_2} \mathbf{B}\cdot d\mathbf{S} = -\oint_{\partial\hat{\mathcal{D}}_2} \mathbf{E}\cdot d\mathbf{r} \qquad \text{Faraday's law of induction}$$

$$\oint_{\partial\hat{\mathcal{D}}_3} \mathbf{B}\cdot d\mathbf{S} = 0 \qquad \text{there are no isolated magnetic poles (monopoles)}$$

Hint: for an *arbitrary* three-dimensional domain there holds

$$0 = \int_{\mathcal{D}_3} dF = \int_{\partial\mathcal{D}_3} F = \int_{\partial\mathcal{D}_3} (dt \wedge \mathbf{E}\cdot d\mathbf{r} - \mathbf{B}\cdot d\mathbf{S})$$

$$\text{i.e.} \quad \int_{\partial\mathcal{D}_3} dt \wedge \mathbf{E}\cdot d\mathbf{r} = \int_{\partial\mathcal{D}_3} \mathbf{B}\cdot d\mathbf{S}$$

For a *spatial* domain $\mathcal{D}_3 = \hat{\mathcal{D}}_3$ the integral containing dt vanishes and we get the second statement; for a domain of the form of an infinitesimal cylinder over a two-dimensional spatial domain, i.e. for $\mathcal{D}_3 = I \times \hat{\mathcal{D}}_2$, $I = \langle t_0, t_0 + \epsilon\rangle$ we proceed in analogy with (16.2.4) and get the first statement. □

16.2.6 Check that the following integral laws are represented by the local equation $d * F = -J$:

$$\oint_{\partial\hat{\mathcal{D}}_3} \mathbf{E}\cdot d\mathbf{S} = Q \equiv \int_{\hat{\mathcal{D}}_3} \rho\, dV \qquad \text{Gauss' law}$$

$$\oint_{\partial\hat{\mathcal{D}}_2} \mathbf{B}\cdot d\mathbf{r} = \int_{\hat{\mathcal{D}}_2} (\mathbf{j} + \partial_t \mathbf{E})\cdot d\mathbf{S} \qquad \text{Ampère's law plus Maxwell's displacement current}$$

Hint: for an *arbitrary* three-dimensional domain there holds

$$\int_{\mathcal{D}_3} d * F = \int_{\partial\mathcal{D}_3} *F = -\int_{\mathcal{D}_3} J$$

$$\text{i.e.} \quad \int_{\partial\mathcal{D}_3} (dt \wedge +\mathbf{B}\cdot d\mathbf{r} + \mathbf{E}\cdot d\mathbf{S}) = \int_{\mathcal{D}_3} (\rho\, dV - dt \wedge \mathbf{j}\cdot d\mathbf{S})$$

For a *spatial* domain $\mathcal{D}_3 = \hat{\mathcal{D}}_3$ we get the first statement; for $\mathcal{D}_3 = I \times \hat{\mathcal{D}}_2, I = \langle t_0, t_0 + \epsilon \rangle$ we proceed in analogy with (16.2.4), (16.2.5) and get

$$\oint_{\partial \hat{\mathcal{D}}_2} \mathbf{B} \cdot d\mathbf{r} = \int_{\hat{\mathcal{D}}_2} \mathbf{j} \cdot d\mathbf{S} + \frac{d}{dt} \int_{\hat{\mathcal{D}}_2} \mathbf{E}(t, \mathbf{r}) \cdot d\mathbf{S} \equiv \int_{\hat{\mathcal{D}}_2} (\mathbf{j} + \partial_t \mathbf{E}) \cdot d\mathbf{S}$$

□

• The theory of the electromagnetic field historically arose from the uniting of two "statics," electro*statics* and magneto*statics*. The two parts are still inherent in the theory as particular cases in which ∂_t (of anything) $= 0$ (i.e. for static sources and fields). Then Maxwell's equations decouple into two subsystems:

electrostatics	*magnetostatics*
$\operatorname{div} \mathbf{E} = \rho$	$\operatorname{curl} \mathbf{B} = \mathbf{j}$
$\operatorname{curl} \mathbf{E} = \mathbf{0}$	$\operatorname{div} \mathbf{B} = 0$

Each statics is thus governed by a pair of equations. One of them is inhomogeneous, where the sources of the fields stand on the right-hand side, and the other is homogeneous, where no sources occur. Yet the two equations themselves differ a bit at first sight. It turns out, however, that they become as similar as two peas in a pod,[339] when written in terms of differential forms.

16.2.7 Let us assemble the vector fields **E**, **B** and **j** into appropriate spatial forms

$$\hat{E} = \mathbf{E} \cdot d\mathbf{r} \qquad \hat{B} = \mathbf{B} \cdot d\mathbf{S} \qquad \hat{j} = \mathbf{j} \cdot d\mathbf{r}$$

so that

$$F = dt \wedge \hat{E} - \hat{B} \qquad j = \rho\, dt - \hat{j}$$

Check that

(i) in terms of these quantities Maxwell's equations are as follows

inhomogeneous half	homogeneous half
$\hat{\delta}\hat{E} = -\rho$	$\partial_t \hat{B} + \hat{d}\hat{E} = 0$
$-\partial_t \hat{E} + \hat{\delta}\hat{B} = \hat{j}$	$\hat{d}\hat{B} = 0$

(ii) in the static case the system of equations decouples into independent electrostatics and magnetostatics halves; their equations read

electrostatics	magnetostatics	electromagnetism
$\hat{\delta}\hat{E} = -\rho$	$\hat{\delta}\hat{B} = \hat{j}$	$\delta F = -j$
$\hat{d}\hat{E} = 0$	$\hat{d}\hat{B} = 0$	$dF = 0$

(for convenience, Maxwell's electrodynamic equations are also displayed).

Hint: (ii) $\partial_t(\cdots) \mapsto 0$. □

[339] Recently scientists found that actually the equations are as similar as one oak leaf to another.

• We see that the structure of equations of both the subsystems (electrostatics and magnetostatics) is exactly the same and it even coincides with that of the complete electrodynamics. One equation relates a source to the codifferential of a form being sought and the other one expresses its closedness (vanishing of its differential). Put another way, if we look for fields generated by given sources, we are to find in all three cases a closed form with a prescribed value of the codifferential.

Of course, there are also differences: in electrostatics a 1-form is sought whereas in magnetostatics as well as electrodynamics it is a 2-form. Notice also that $\hat{E}$ and $\hat{B}$ may be regarded as forms in (only) three-dimensional Euclidean space E^3 (they are spatial and neither of their components depend on time), whereas F is necessarily a form in four-dimensional Minkowski space $E^{1,3}$.

16.3 Gauge transformations, action integral

• Maxwell's equation $dF = 0$ says that the 2-form F is closed. Due to the Poincaré lemma it is then (at least locally) also exact.

16.3.1 Check that

(i) the structure of Maxwell's equations immediately leads (at least locally) to the *existence of a potential*[340] A of the form F

$$F = dA \qquad \text{i.e.} \quad F_{\mu\nu} = \partial_\mu A_\nu - \partial_\nu A_\mu$$

(ii) in three-dimensional language it reads

$$A = \phi\, dt - \mathbf{a} \cdot d\mathbf{r}$$
$$\mathbf{E} = -\operatorname{grad}\phi - \partial_t \mathbf{A}$$
$$\mathbf{B} = \operatorname{curl}\mathbf{A}$$

where ϕ, $\mathbf{A}$ are three-dimensional quantities known as the *scalar potential* and *vector potential* respectively

(iii) the potential A is not uniquely fixed by F, but rather there is a freedom given by *gauge transformations*

$$A \mapsto A' = A + d\chi \qquad \text{or in three-dimensional language} \qquad \phi \mapsto \phi' = \phi + \partial_t \chi$$
$$A_\mu \mapsto A'_\mu = A_\mu + \partial_\mu \chi \qquad \qquad \mathbf{A} \mapsto \mathbf{A}' = \mathbf{A} - \operatorname{grad}\chi$$

Hint: (i) see (9.2.4); (ii) we should have $dt \wedge \mathbf{E} \cdot d\mathbf{r} - \mathbf{B} \cdot d\mathbf{S} = d(\phi\, dt - \mathbf{a} \cdot d\mathbf{r})$, (16.1.4); (iii) $F' \equiv dA' = dA + dd\chi = F$, so that also $\mathbf{E}$ and $\mathbf{B}$ remain unchanged. □

• These transformations represent the simplest special case of a far-reaching generalization, which is realized in *gauge theories*; we will discuss this stuff in more detail in Chapter 21.

It is a folklore piece of knowledge in theoretical physics that various technical advantages result from the possibility of deriving a differential equation under consideration from a

[340] In physics, the components A_μ of the 1-form A are called the 4-*potential* (since it is a (co)vector field in *four*-dimensional space), in order to distinguish it from the (vector) potential A_i in *three*-dimensional space.

variational principle (see, for example, (15.4.4) or Sections 16.4 and 21.6). Therefore we turn our attention now to the issue of finding an appropriate action integral for Maxwell's equations (which will be further generalized to an action for more involved gauge theories in Section 21.5). We will see that the machinery of differential forms offers both a simple expression for the action itself and an equally simple way of performing the variational procedure needed for the derivation of the equations of motion (Maxwell's equations in this particular case).

It turns out that the action, being a (4-)volume integral, is to be regarded as a functional of the 1-form of the *potential* A (rather than the 2-form of the electromagnetic *field* F, which is unknown in Maxwell's equations). Put another way, the potential A serves as an independent variable of the action and a variation is to be performed with respect to this very quantity in order to derive Maxwell's equations.

16.3.2 In Minkowski space $E^{1,3}$ consider a four-dimensional domain $\mathcal{D}$, in which A is a potential for the field F. By a *variation* of the potential we mean a replacement

$$A \mapsto A + \epsilon\alpha$$

where α is an arbitrary 1-form which *vanishes* on the boundary $\partial\mathcal{D}$ of $\mathcal{D}$; the potential is thus altered arbitrarily (though infinitesimally) inside the domain $\mathcal{D}$, being, however, kept fixed on the boundary. Check that

(i) the condition for the functional

$$S_0[A] := -\frac{1}{2}\langle dA, dA\rangle \equiv -\frac{1}{2}\langle F, F\rangle \equiv -\frac{1}{2}\int_{\mathcal{D}} F \wedge *F$$

to be stationary is the equation

$$\delta dA \equiv \delta F = 0$$

i.e. the source-free Maxwell's equation

(ii) if j is a fixed 1-form, then the condition for the functional

$$S[A] := -\frac{1}{2}\langle dA, dA\rangle - \langle A, j\rangle \equiv -\int_{\mathcal{D}} \left(\frac{1}{2} dA \wedge *dA + A \wedge *j\right)$$

to be stationary is already the fully fledged Maxwell's equation

$$\delta F = -j$$

The first term (containing derivatives of A) is called the *kinetic term* and the second one is the *interaction term* of the field with a given source j

(iii) the action integral may also be written in terms of a *Lagrangian density* $L(A, dA)$,

$$S[A] = \int_{\mathcal{D}} L\,\omega_\eta \qquad L(A, dA) \equiv -\frac{1}{2}(F, F) - (A, j) \equiv -\frac{1}{4}F_{\mu\nu}F^{\mu\nu} - A_\mu j^\mu$$

with $\omega_\eta \equiv dt \wedge dV$ being the metric volume form in $E^{1,3}$.

Hint: (i) if $A \mapsto A + \epsilon\alpha$, then $F \mapsto F + \epsilon\, d\alpha$ and so $\langle F, F\rangle \mapsto \langle F, F\rangle + 2\epsilon\langle F, d\alpha\rangle$ so that an extremum requires $\langle F, d\alpha\rangle = 0$ for arbitrary α. According to (8.3.2) this is equivalent to $\langle \delta F, \alpha \rangle = 0$ and since the bilinear form (8.3.1) is non-degenerate (and α is arbitrary), we get $\delta F = 0$; (ii) $S[A + \epsilon\alpha] = S[A] - \epsilon\langle \delta F + j, \alpha \rangle$; (iii) see (8.3.1). □

• In this context it might be useful to mention a concept which is often referred to, the *variational derivative* of an action functional with respect to its argument (if there are more arguments the derivative exists for each of them). The meaning of this concept for functionals is the same as that of an ordinary derivative for functions (or the *partial* derivative of a function, if there are more arguments), i.e. it informs us about the sensitivity of a functional with respect to small changes of an argument.[341] When the variation of an action functional is performed, its linear increment (the first variation) has the structure of a volume integral, in which the expression under the integral sign depends linearly on the varied argument; the factor standing by this variation is (by definition) just the variational derivative. For example, the result[342]

$$S[A + \epsilon\alpha] = S[A] - \langle \delta F + j, \epsilon\alpha \rangle + \cdots \equiv S[A] - \epsilon \int_{\mathcal{D}} \alpha^\mu (\delta F + j)_\mu \, d^4x$$

is rewritten as

$$\frac{\delta S[A]}{\delta A^\mu(x)} = -(\delta F + j)_\mu(x)$$

16.3.3 It is already an open secret that the salary which letters receive for their performance in mathematics and physics are scandalously poor indeed. We should not be surprised then to hear that many of them try to earn a little extra so that they put a signature to a contract for more than a single role (⇒ more salaries). Neither are they discouraged enough in awkward situations when they have to perform *two* roles in a *single* equation! Find out where in this equation δ performs the role of a *variation* and where it denotes[343] the *codifferential*. □

• From the point of view of deriving Maxwell's equations from a variational principle we see that one equation ($dF = 0$) is a trivial consequence of the choice of an independent variable in the action functional (the potential A is independent, the field F is *defined* by the equation $F := dA$ and so $dF = 0$ holds trivially), the other equation ($\delta F = -j$) is a truly variational (Euler–Lagrange) equation, expressing the condition for extremizing the particular functional $S[A]$.

16.3.4 Check that the action functional $S[A]$ (with a conserved current j) is *gauge invariant*. Be sure to understand that one can deduce from this that also the equations derived from the action are necessarily gauge invariant. Also verify this result straightforwardly.

Hint: F is gauge invariant so that $\langle F, F \rangle$ is all right; $\langle A, j \rangle \mapsto \langle A, j \rangle + \langle d\chi, j \rangle = \langle A, j \rangle + \langle \chi, \delta j \rangle$; the equations contain (invariant) F alone. □

[341] Here we have already entered a realm where another large mathematical discipline reigns with an iron hand, *functional analysis*. Our approach here is from this point of view only "formal," making use of the geometric machinery we learned. (This is, however, fairly standard in ordinary theoretical physics.)

[342] In books on physics the quantity $\epsilon\alpha$ is often written as δA (increment = variation of A), so that $S[A + \delta A] = S[A] - \int_{\mathcal{D}} \delta A^\mu (\delta F + j)_\mu \, d^4x$.

[343] Note that it is even a *juvenile* δ. Evidently the problem has already developed insomuch that although the adult Δ by no means idles and denotes everything possible, it is still not enough to reasonably support a family.

16.3.5 Check that the three-dimensional version of the action $S[A] \equiv S[\phi, \mathbf{A}]$ reads as

$$S[\phi, \mathbf{A}] = \int_{\mathcal{D}} L(\phi, \mathbf{A})\, dt \wedge dV \qquad L(\phi, \mathbf{A}) = \frac{1}{2}(\mathbf{E}^2 - \mathbf{B}^2) - (\rho\phi - \mathbf{j}\cdot\mathbf{A})$$

where $\mathbf{E}(\phi, \mathbf{A})$ and $\mathbf{B}(\phi, \mathbf{A})$ are given by their expressions from (16.3.1). □

16.3.6 Let $\alpha \equiv dt \wedge \hat{s} + \hat{r}$ and $\beta \equiv dt \wedge \hat{S} + \hat{R}$ be two p-forms in Minkowski space. Check that

(i) their scalar product $\langle\alpha, \beta\rangle$ may be written in terms of spatial forms as

$$\langle\alpha, \beta\rangle \equiv \int_{\mathcal{D}} \alpha \wedge *\beta \equiv \int_{\mathcal{D}} (\alpha, \beta)\, dt \wedge dV = \int_{\mathcal{D}} [(\hat{\eta}\hat{s}, \hat{S}) + (\hat{\eta}\hat{r}, \hat{R})]\, dt \wedge dV$$

i.e. that there holds

$$(\alpha, \beta)_{E^{1,3}} = (\hat{\eta}\hat{s}, \hat{S})_{E^3} + (\hat{\eta}\hat{r}, \hat{R})_{E^3}$$

where the indices on the parentheses mean "in the sense of $E^{1,3}$" or "in the sense of E^3"

(ii) in particular, for $\alpha = \beta$ we have a *difference* of "squares" (positive definite expressions) in the square bracket

(iii) for concrete parametrizations we get explicitly for various degrees

0-forms:	$\alpha = f$	$(\alpha, \alpha)_{E^{1,3}} = f^2$
1-forms:	$\alpha = f\, dt + \mathbf{a}\cdot d\mathbf{r}$	$(\alpha, \alpha)_{E^{1,3}} = f^2 - \mathbf{a}^2$
2-forms:	$\alpha = dt \wedge \mathbf{a}\cdot d\mathbf{r} + \mathbf{b}\cdot d\mathbf{S}$	$(\alpha, \alpha)_{E^{1,3}} = \mathbf{b}^2 - \mathbf{a}^2$
3-forms:	$\alpha = dt \wedge \mathbf{a}\cdot d\mathbf{S} + f\, dV$	$(\alpha, \alpha)_{E^{1,3}} = \mathbf{a}^2 - f^2$
4-forms:	$\alpha = dt \wedge f\, dV$	$(\alpha, \alpha)_{E^{1,3}} = -f^2$

(iv) (16.3.5) is a special case and the results mentioned there and here are consistent.

Hint: (i) using formulas from Section 16.1 and (8.3.1) the relevant product gives

$$\begin{aligned}(dt \wedge \hat{s} + \hat{r}) \wedge *(dt \wedge \hat{S} + \hat{R}) = \cdots &= dt \wedge [(\hat{\eta}\hat{s}) \wedge \hat{*}\hat{S} + (\hat{\eta}\hat{r}) \wedge \hat{*}\hat{R}] \\ &\equiv dt \wedge [(\hat{\eta}\hat{s}, \hat{S})_{E^3} + (\hat{\eta}\hat{r}, \hat{R})_{E^3}]dV\end{aligned}$$

(ii) the forms $\hat{s}$ and $\hat{r}$ have neighboring degrees and expressions of the type $(\hat{r}, \hat{r})_{E^3}$ are positive definite; (iii) see (8.5.8) and (16.1.2). □

• It turns out that the electromagnetic field shares a general structure $\int \alpha \wedge *\beta$ of the action with some other physical fields (one may even say that this structure is *typical*). Here we will discuss, for example, the scalar and (co)vector fields and we will also encounter such actions in Chapter 21, when speaking about gauge fields.

A *scalar field* in Minkowski space is simply a (real-valued) *function* ϕ in $E^{1,3}$ (*complex-valued* scalar fields also occur frequently in physics).

16.3.7 The action of a (free) scalar field ϕ is defined by the expression

$$S[\phi] := \frac{1}{2}\langle d\phi, d\phi\rangle - \frac{m^2}{2}\langle \phi, \phi\rangle$$
$$\equiv \int_{\mathcal{D}} \frac{1}{2} d\phi \wedge *d\phi - \frac{m^2}{2}\int_{\mathcal{D}} \phi \wedge *\phi \equiv \int_{\mathcal{D}} L(\phi, d\phi)\, dt \wedge dV$$

i.e. the Lagrangian density reads

$$L(\phi, d\phi) = \frac{1}{2}(d\phi, d\phi) - \frac{m^2}{2}(\phi, \phi) \equiv \frac{1}{2}(\partial_\mu\phi)(\partial^\mu\phi) - \frac{m^2}{2}\phi^2$$

The first term (containing derivatives) is called the *kinetic term*, the second term (with no derivatives, containing the mass m) is the *mass term*. Check that

(i) the equation of motion (condition for an extremum of the action functional) is the *Klein–Gordon equation*

$$(-\delta d + m^2)\phi \equiv (\Box + m^2)\phi = 0$$

(ii) in Cartesian coordinates in $E^{1,3}$ this equation reads

$$(\partial_\mu \partial^\mu + m^2)\phi = 0$$

(iii) in $1+3$ perspective the action and the equation are

$$S[\phi] = \frac{1}{2}\int_{\mathcal{D}} [(\partial_t\phi)^2 - (\operatorname{grad}\phi)^2 - m^2\phi^2]\, dt \wedge dV \qquad (\partial_t^2 - \hat{\Delta} + m^2)\,\phi = 0$$

Hint: (i) $\phi \mapsto \phi + \epsilon\chi \Rightarrow \cdots$ like in (16.3.2); (ii) $\Box f \equiv -\delta df = \partial_\mu\partial^\mu f$ according to (8.3.5) and (16.1.6). □

16.3.8 The action of a (free co)*vector field* W is defined by the expression

$$S[W] := -\frac{1}{2}\langle dW, dW\rangle + \frac{m^2}{2}\langle W, W\rangle$$
$$\equiv -\frac{1}{2}\langle F, F\rangle + \frac{m^2}{2}\langle W, W\rangle \equiv -\frac{1}{2}\int_{\mathcal{D}} (F \wedge *F - m^2 W \wedge *W)$$

(where $F := dW$, $F_{\mu\nu} = \partial_\mu W_\nu - \partial_\nu W_\mu$), i.e. the Lagrangian density is

$$L(W, dW) = -\frac{1}{2}(F, F) + \frac{m^2}{2}(W, W) \equiv -\frac{1}{4}F_{\mu\nu}F^{\mu\nu} + \frac{m^2}{2}W_\mu W^\mu$$

The first term (containing derivatives) is called the *kinetic term* and the second[344] term (with no derivatives, containing the mass m) is the *mass term*. Check that

(i) the equation of motion (condition of an extremum of the action functional) is the *Proca equation*

$$(-\delta d + m^2)W \equiv -\delta F + m^2 W = 0$$

[344] We see that the action for the case $m = 0$ reduces to that for the electromagnetic field.

(ii) the *Lorentz gauge condition*

$$\delta W = 0$$

turns out to be a direct *consequence*[345] of the equation (for $m \neq 0$) so that the equations of motion may, in turn, also be written in the form

$$(\Box + m^2)W = 0 \qquad \delta W = 0 \qquad \Box \equiv -(\delta d + d\delta)$$

(iii) in Cartesian coordinates in $E^{1,3}$ this leads to the system

$$(\partial_\mu \partial^\mu + m^2)W_\nu = 0 \qquad \partial^\mu W_\mu = 0$$

(iv) the action (for $m \neq 0$) *lacks* the usual electromagnetic gauge symmetry (freedom) $W \mapsto W + d\chi$.

Hint: (i) $W \mapsto W + \epsilon w \Rightarrow \cdots$ like in (16.3.2); (ii) $\delta\delta = 0$; (iii) see (8.3.5); (iv) the symmetry is brought just by the *mass* term (for $m = 0$ we return back to a free electromagnetic field and the gauge freedom is once more recovered). □

• Now let us turn our attention to the action and equations of motion of a point charge (e, m), which moves in a *given external* electromagnetic field with a potential A.

Consider a Lorentzian manifold (space-time) (M, g) and time-like curves (*world lines* of the charge) $\gamma(\tau)$, so that $g(\dot\gamma, \dot\gamma) > 0$. Recall (15.4.8) that the variation of the action (a functional of the length) $\int_{\tau_1}^{\tau_2} d\tau \sqrt{g(\dot\gamma, \dot\gamma)}$ leads to the equation of (an arbitrarily parametrized) geodesic and also that an affine parametrization is achieved by the choice of a *natural parameter*, where the increment of the parameter equals the corresponding length moved. In the context of world lines such a parameter is called the *proper time* and it will be denoted by s (it is the time displayed by a clock of the observer for which γ is the world line); there holds $g(\dot\gamma, \dot\gamma) = 1$ (if the dot corresponds to the derivative with respect to s) and thus $\sqrt{g(\dot\gamma, \dot\gamma)}\, d\tau = ds$.

16.3.9 An interaction with the electromagnetic field given by a potential A is arranged by addition of an *interaction term* $S_{\text{int}}[\gamma; A]$ to the action, being proportional to the integral of A along γ; namely the resulting complete action[346] takes the form

$$\begin{aligned} S[\gamma; A] := -m \int_{s_1}^{s_2} ds - e \int_\gamma A &\equiv -m \int_{\tau_1}^{\tau_2} \sqrt{g(\dot\gamma, \dot\gamma)}\, d\tau + S_{\text{int}}[\gamma; A] \\ &\equiv -m \int_{\tau_1}^{\tau_2} \sqrt{g(\dot\gamma, \dot\gamma)}\, d\tau - e \int_{\tau_1}^{\tau_2} \langle A, \dot\gamma \rangle d\tau \end{aligned}$$

Check that the variation of this (reparametrization invariant) action reads ($\gamma_\epsilon \equiv \Phi_\epsilon(\gamma)$ and $\Phi_\epsilon \leftrightarrow W$)

$$S[\gamma; A] \mapsto S[\gamma_\epsilon; A] = S[\gamma] + \epsilon \int \langle \mathcal{E}, W \rangle\, d\tau + \cdots$$

[345] For the massless case ($m = 0$, e.g. in electromagnetism) the Lorentz gauge condition is *no longer* a consequence of the equations of motion. Still it is often used as a convenient condition reducing the gauge freedom of the theory.

[346] This action is regarded as a functional of the curve γ (alone), keeping the field A fixed (it is an "external," i.e. prescribed field; it would acquire its own dynamics if we also added *its* kinetic term $\sim \langle dA, dA \rangle$ (16.3.2) and we performed a variation with respect to A as well). Note that the interaction term has just the structure $-\langle A, j \rangle$ of the interaction term from (16.3.2). One should realize, however, that now j does not vanish only along the world line and is proportional there to $e\dot\gamma$ (so that j is not a smooth 1-form on M, it may be (as a distribution) explicitly expressed in terms of the "Dirac δ-function").

where the *Euler–Lagrange 1-form* $\mathcal{E}$ turns out to be

$$\mathcal{E} := m\flat_g \nabla_{\dot\gamma}\left(\frac{\dot\gamma}{||\dot\gamma||}\right) + e i_{\dot\gamma} F \qquad ||\dot\gamma|| \equiv \sqrt{g(\dot\gamma,\dot\gamma)}$$

so that the equation of motion resulting from the variation of this action is (in arbitrary parametrization of $\gamma(\tau)$)

$$m\nabla_{\dot\gamma}\left(\frac{\dot\gamma}{||\dot\gamma||}\right) = -e\sharp_g i_{\dot\gamma} F$$

or in the proper time parametrization (where $\tau = s$)

$$ma = f \qquad a := \nabla_{\dot\gamma}\dot\gamma = \text{(4-)acceleration}$$

where the *Lorentz (4-)force* by which the electromagnetic field acts on the point charge is

$$f = \sharp_g \tilde f \qquad \tilde f := -e i_{\dot\gamma} F \equiv -eF(\dot\gamma,\cdot) \equiv +eF(\cdot,\dot\gamma)$$
$$\tilde f_\mu = eF_{\mu\nu}\dot x^\nu$$

Hint: the variation of the first term was computed in (15.4.16); the second term gives

$$\begin{aligned} S_{\text{int}}[\gamma;A] \mapsto S_{\text{int}}[\Phi_{\epsilon *}\gamma;A] &= S_{\text{int}}[\gamma;\Phi^*_\epsilon A] = S_{\text{int}}[\gamma;A+\epsilon\mathcal{L}_W A] \\ &= S_{\text{int}}[\gamma;A+\epsilon(i_W d + d i_W)A] \\ &= S_{\text{int}}[\gamma;A] - \epsilon e\int F(W,\dot\gamma)\,d\tau - \epsilon e\int_\gamma d\langle A,W\rangle \\ &= S_{\text{int}}[\gamma;A] + \epsilon\int \langle e i_{\dot\gamma}F, W\rangle\,d\tau - \epsilon e[\langle A,W\rangle]^{\tau_2}_{\tau_1} \end{aligned}$$

The last term vanishes since we do not vary the curve at the ends ($W(\tau_1) = W(\tau_2) = 0$). □

16.3.10 Check that if a world line $\gamma(\tau)$ in Minkowski space is parametrized by an "ordinary" coordinate time $\tau = t = x^0$, then the (1+3)-version of the Lorentz 4-force reads

$$\tilde f \equiv -e i_{\dot\gamma} F = e(\mathbf{E}\cdot\mathbf{v})\,dt - e(\mathbf{E}+\mathbf{v}\times\mathbf{B})\cdot d\mathbf{r}$$

where[347] the "ordinary" *Lorentz force* $\mathbf{f}_{\text{Lor}} = e(\mathbf{E}+\mathbf{v}\times\mathbf{B})$ may be recognized.

Hint: in the parametrization $\tau = t$ we have $\gamma(t) \leftrightarrow x^\mu(t) = (t, x^i(t))$, so that $\dot\gamma = \partial_t + v^i\partial_i$, $v^i = \dot x^i \equiv dx^i/dt$; then (16.2.1) and (8.5.8) yield

$$i_{\dot\gamma}F = i_{(\partial_t + v^i\partial_i)}(dt\wedge\mathbf{E}\cdot d\mathbf{r} - \mathbf{B}\cdot d\mathbf{S}) = \cdots = (\mathbf{E}+\mathbf{v}\times\mathbf{B})\cdot d\mathbf{r} - (\mathbf{E}\cdot\mathbf{v})dt$$

□

- Maxwell's equations may also be derived from the following, less standard, action.

[347] The term $e\mathbf{E}\cdot\mathbf{v}$ corresponds to the *power of the electric field* (the work produced by Lorentz force in unit time $(\mathbf{f}_{\text{Lor}}\cdot d\mathbf{r})/dt$; the magnetic field does not perform work, so its power vanishes).

16.3.11* Consider the action $S[A, F]$, where the fields A, F are regarded as being *independent* and which has the form

$$S[A, F] := \frac{1}{2}\langle F, F\rangle - \langle F, dA\rangle - \langle A, j\rangle$$

Check that

(i) conditions for the functional to be stationary with respect to F and A are

$$F = dA \qquad \delta F = -j$$

i.e. first the "correct" relation $F(A) = dA$ between F and A (with the consequence $dF = 0$, the homogeneous part of Maxwell's equations) and then the inhomogeneous part of Maxwell's equations $\delta F = -j$

(ii) if we plug the *extremal* $F(A) = dA$ as F into the action, we return to the standard action from (16.3.2)

$$S[A, F(A)] = -\frac{1}{2}\langle F(A), F(A)\rangle - \langle A, j\rangle = -\frac{1}{2}\langle dA, dA\rangle - \langle A, j\rangle \equiv S[A] \text{ from (16.3.2)}$$

We thus see that a variation of an appropriate action within a *wider class* of fields (F, A independent) results eventually *in the same* system of equations

$$dF = 0 \quad \delta F = -j \quad F = dA$$

as a variation (of a different action) within a narrower class of fields (F is no longer independent of A, but rather it is expressed in terms of *a single independent* field A). □

16.4 Energy–momentum tensor, space-time symmetries and the conservation laws due to them

• The energy–momentum tensor is a concept of the highest importance in physics. It encodes, as follows from the nomenclature, the energy and momentum of a physical object under consideration, but there is also information about "transfers" of these quantities in space. Here we will first define the tensor in terms of the variation of an action with respect to the metric tensor (this is not the only possibility for defining it) and we show how its key property, vanishing of the "covariant divergence," follows from the definition. This property then enables us to construct conserved quantities, which correspond to Killing vectors (i.e. to space-time *symmetries*; see also Section 21.6 about Noether's theorem). We will also learn how to compute it in practice for a certain (important) class of action integrals and, in particular, we will compute the tensor explicitly for electromagnetic, scalar and vector fields.

16.4.1 Consider a field theory on (M, g) with an action which is *natural with respect to diffeomorphisms* in the following sense (see also (8.3.6), (8.3.7), etc.): it is the (4-)volume integral of a 4-form Ω, which depends on some tensor fields ψ as well as the metric tensor g

$$S[\psi, g] := \int_{\mathcal{D}} \Omega[\psi, g] \equiv \int_{\mathcal{D}} L(\psi, g)\omega_g$$

(ω_g being the metric volume form) and for any diffeomorphism $f : M \to M$ it satisfies

$$f^*(\Omega[\psi, g]) = \Omega[f^*\psi, f^*g]$$

Now in the domain $\mathcal{D}$ consider an arbitrary vector field V such that it vanishes on the boundary $\partial\mathcal{D}$; consequently its flow Φ_t is "arbitrary" inside $\mathcal{D}$, but it does not move the points on the boundary. Check that then the *energy–momentum tensor* of the system, defined as[348] (minus) variational derivative of action $S[\psi, g]$ with respect to the metric tensor g

$$T^{\mu\nu}(x) := -\frac{\delta S[\psi, g]}{\delta g_{\mu\nu}(x)} \qquad \text{i.e.} \quad S[\psi, g + \epsilon h] =: S[\psi, g] - \epsilon \int_{\mathcal{D}} \frac{1}{2} h_{\mu\nu} T^{\mu\nu} \omega_g$$

has vanishing *covariant divergence*

$$T^{\mu\nu}{}_{;\nu} = 0$$

when computed for the fields ψ which *obey the equations of motion.*

Solution: since V vanishes on $\partial\mathcal{D}$, we have $\Phi_\epsilon(\mathcal{D}) = \mathcal{D}$; then,

$$\begin{aligned} S[\psi, g] &= \int_{\Phi_\epsilon(\mathcal{D})} \Omega[\psi, g] = \int_{\mathcal{D}} \Phi_\epsilon^* \Omega[\psi, g] = \int_{\mathcal{D}} \Omega[\Phi_\epsilon^* \psi, \Phi_\epsilon^* g] \\ &= \int_{\mathcal{D}} \Omega[\psi + \epsilon \mathcal{L}_V \psi, g + \epsilon \mathcal{L}_V g] = \int_{\mathcal{D}} \Omega[\psi, g + \epsilon \mathcal{L}_V g] \end{aligned}$$

where for the last equality we used the fact that ψ extremizes S (obeys the equations of motion). Then, by the definition of $T^{\mu\nu}$ we have

$$S[\psi, g] = S[\psi, g] - \epsilon \int_{\mathcal{D}} \frac{1}{2} (\mathcal{L}_V g)_{\mu\nu} T^{\mu\nu} \omega_g \quad \Rightarrow \quad \frac{1}{2} \int_{\mathcal{D}} (\mathcal{L}_V g)_{\mu\nu} T^{\mu\nu} \omega_g = 0$$

for arbitrary V (which vanishes on the boundary). If we use the expression $(\mathcal{L}_V g)_{\mu\nu} = V_{\mu;\nu} + V_{\nu;\mu}$ from (15.4.14), we get

$$\begin{aligned} \frac{1}{2} (\mathcal{L}_V g)_{\mu\nu} T^{\mu\nu} &= V_{\mu;\nu} T^{\mu\nu} = (V_\mu T^{\mu\nu})_{;\nu} - V_\mu \left(T^{\mu\nu}{}_{;\nu}\right) \\ &= W^\nu{}_{;\nu} - V_\mu \left(T^{\mu\nu}{}_{;\nu}\right) \qquad W^\nu := V_\mu T^{\mu\nu} \end{aligned}$$

The first term (divergence) drops out via Gauss' theorem (8.2.8), vanishing of the second term for V which is arbitrary in $\mathcal{D}$ yields the result being sought. □

[348] Actually this definition fixes its *symmetric part* alone; according to a standard choice the tensor *is symmetric*, $T^{\mu\nu} = T^{\nu\mu}$.

• Now we will show that the energy–momentum tensor may be used for obtaining conserved quantities corresponding to space-time symmetries. By these symmetries we mean symmetries of the manifold (M, g) regarded as a (pseudo-)Riemannian manifold, i.e. its *isometries*, or in some specific cases (as it turns out) also *conformal* transformations.

16.4.2 Let ξ be a *Killing vector* of the manifold (M, g) and let $T_{\mu\nu}$ be the energy–momentum tensor introduced in (16.4.1). Denote by $\tilde{\mathcal{J}}$ the 1-form

$$\tilde{\mathcal{J}} := T(\xi, \cdot) \qquad \text{i.e.} \quad \tilde{\mathcal{J}}_\mu := \xi^\nu T_{\nu\mu}$$

and by $\mathcal{J}$ the corresponding vector field (obtained by raising of indices by means of g). Check that

(i) the vector field $\mathcal{J}$ has vanishing (covariant) divergence or, equivalently, that the form $\tilde{\mathcal{J}}$ is coclosed

$$\mathcal{J}^\mu{}_{;\mu} \equiv (\xi_\nu T^{\nu\mu})_{;\mu} = 0 \qquad \text{i.e.} \quad \delta\tilde{\mathcal{J}} = 0$$

(ii) this may be reformulated as the *closedness* of the 3-*form* $*\tilde{\mathcal{J}}$

$$*\tilde{\mathcal{J}} \equiv *(\xi^\nu T_{\nu\mu}\, dx^\mu) \equiv \xi_\nu T^{\nu\mu}\, d\Sigma_\mu \equiv \mathcal{J}^\mu\, d\Sigma_\mu \quad \Rightarrow \quad d * \tilde{\mathcal{J}} = 0$$

(iii) for Minkowski space the integral of the 3-form $*\tilde{\mathcal{J}}$ over the whole three-dimensional space V does not depend on time; so it represents a *conserved quantity*, which may be assigned to the Killing vector ξ,

$$\begin{aligned} Q(t) &:= \int_V *\tilde{\mathcal{J}} \\ &\equiv \int_V \mathcal{J}^\mu\, d\Sigma_\mu \equiv \int_V \xi_\nu T^{\nu 0}\, d\Sigma_0 \equiv \int_V \xi_\nu T^{\nu 0}\, dV \quad \Rightarrow \quad \frac{d}{dt} Q(t) = 0 \end{aligned}$$

(the integral is over the whole 3-space V in time t).

Hint: (i) $(\xi_\nu T^{\nu\mu})_{;\mu} = \xi_{\nu;\mu} T^{\nu\mu} + \xi_\nu T^{\nu\mu}{}_{;\mu} = \frac{1}{2}(\mathcal{L}_\xi g)_{\mu\nu} T^{\nu\mu} + \xi_\nu T^{\nu\mu}{}_{;\mu} = 0$; (ii) according to (8.2.1) and (14.3.7) we have $(\mathcal{J}^\mu{}_{;\mu})\omega_g = (\text{div}\, \mathcal{J})\omega_g = d(i_{\mathcal{J}}\omega_g) = d(*\tilde{\mathcal{J}})$; (iii) $0 = \int_{\mathcal{D}_4} d * \tilde{\mathcal{J}} = \int_{\partial\mathcal{D}_4} *\tilde{\mathcal{J}}$; if we take as $\mathcal{D}_4$ a four-dimensional domain $\langle t_1, t_2\rangle \times V$ and we use the fact that the fields (and consequently also $T^{\mu\nu}$) vanish in (spatial) infinity ∂V, we get $Q(t_2) - Q(t_1) = 0$. □

• A conservation law itself is thus hidden in the existence of a *closed* 3-*form* and exactly such a closed 3-form[349] may be obtained from the energy–momentum tensor $T_{\mu\nu}$ and a Killing vector ξ. (An instructive point of view in terms of symmetries of the action and Noether's theorem is presented in problem (21.6.6).)

16.4.3 Take, in particular, Minkowski space $E^{1,3}$. In (4.6.10) we already found all independent Killing vectors for this space and we learned that a general Killing vector may be written in the form (in Cartesian coordinates)

$$\xi_{(A,a)} = \xi^\mu_{(A,a)} \partial_\mu \qquad \xi^\mu_{(A,a)} = (A\eta)^{\mu\nu} x_\nu + a^\mu \qquad (A\eta) + (A\eta)^{\text{T}} = 0$$

[349] In $(1 + d)$-dimensional space it is a closed d-*form* $*\mathcal{J} \equiv \mathcal{J}^\mu\, d\Sigma_\mu$.

It is thus parametrized in terms of a (pseudo-)antisymmetric matrix A and a column vector a. Check that

(i) then the conserved quantity Q is also parametrized in terms of a pair (A, a) and it has the form

$$Q_{(A,a)} = (\eta A)_{\mu\nu} M^{\mu\nu} + a_\mu P^\mu$$

where

$$P^\mu = \int_V T^{\mu 0}\, dV \qquad M^{\mu\nu} = \int_V x^{[\mu} T^{\nu]0}\, dV$$

(ii) in particular, due to the invariance with respect to time and space translations as well as rotations in the (spatial) plane (ij) the conserved quantities turn out to be the total *energy of the field*, total *momentum of the field* and the total *angular momentum of the field* (all of them computed for the whole 3-volume V)

$$E = \int_V T^{00}\, dV \equiv P^0 \qquad P^i = \int_V T^{i0}\, dV \qquad M^{ij} = \int_V x^{[i} T^{j]0}\, dV$$

Hint: (i) according to (16.4.3) we have $Q = \int_V \xi_\nu T^{\nu 0}\, dV$. □

• Let us look more closely at the *balance* of the energy and momentum in a *finite* 3-*volume* $\hat{\mathcal{D}}_3$. Conservation laws for the quantities in question then admit a visual interpretation of the type "the amount a quantity increased, since it came across the boundary (rather than being created inside the volume) or decreased, since it escaped across the boundary (rather than being destroyed inside the volume)." This enables us at the same time to clarify in more detail the meaning of particular components of the energy–momentum tensor.

16.4.4 Consider Minkowski space $E^{1,3}$ and the Killing vector ξ, which corresponds to space-time *translations*, i.e. $\xi = \partial_0$ (translation in time) or $\xi = \partial_j$ (translation along the jth spatial axis). Check that

(i) the closed 3-form $*\mathcal{J} \equiv \xi_\mu T^{\mu\nu}\, d\Sigma_\nu$ then looks like

$$*\mathcal{J} = T^{00}\, dV - dt \wedge T^{0i}\, dS_i \qquad \text{for} \quad \xi = \partial_0$$
$$*\mathcal{J} = T^{j0}\, dV - dt \wedge T^{ji}\, dS_i \qquad \text{for} \quad \xi = \partial_j$$

(ii) if we integrate the (vanishing) 4-form $d * \mathcal{J}$ over a four-dimensional domain $\mathcal{D}_4$ given as an infinitesimal cylinder over a spatial domain $\hat{\mathcal{D}}_3$, i.e. over $\mathcal{D}_4 = I \times \hat{\mathcal{D}}_3$, $I = \langle t_0, t_0 + \epsilon \rangle$, we get for the two cases the equations

$$\dot{E} = -\oint_{\partial\hat{\mathcal{D}}_3} T^{0i}\, dS_i \qquad \dot{P}^j = -\oint_{\partial\hat{\mathcal{D}}_3} T^{ji}\, dS_i$$

where

$$E(t) := \int_{\hat{\mathcal{D}}_3} T^{00}\, dV \qquad P^i(t) := \int_{\hat{\mathcal{D}}_3} T^{i0}\, dV$$

represent the total energy and the ith component of the momentum contained *in the domain* $\hat{\mathcal{D}}_3$

(iii) we can deduce from this that there is the following information in the particular components of the energy–momentum tensor:

$$T^{00}\,dV = \text{energy contained in volume element } dV$$
$$T^{i0}\,dV = i\text{th component of momentum contained in volume } dV$$
$$T^{0i}\,dS_i = \text{energy which flows out per unit time across the surface element } d\mathbf{S}$$
$$T^{ij}\,dS_j = i\text{th component of momentum which flows out per unit time across } d\mathbf{S}$$

(iv) so that the components $T^{\mu\nu}$ themselves we identify as

$$T^{00} = \text{energy density}$$
$$T^{i0} = \text{density of the } i\text{th component of momentum}$$
$$T^{0i} = i\text{th component of the density of the current of energy}$$
$$T^{ij} = j\text{th component of the density of the current of the } i\text{th component of momentum.}$$

□

• How can one compute the energy–momentum tensor effectively for an important class of action integrals? Consider, as an example, the expression under the integral sign in the action functional for the electromagnetic field

$$-\frac{1}{2}F\wedge *F - A\wedge *j$$

We are to compute the variational derivative of (an integral of) this expression with respect to the metric tensor g. So we embark on seeking where g actually is. What we find out is that the only place where g occurs is in the Hodge operator $* = *_g$. So we are to determine how exactly the expression of the structure

$$\alpha\wedge *_g\beta$$

responds to a small change of metric tensor $g\mapsto g+\epsilon h$, where h is an arbitrary symmetric tensor of type $\binom{0}{2}$.

16.4.5 Consider the form

$$\Omega(\alpha,\beta,g) := \alpha\wedge *_g\beta \equiv (\alpha,\beta)_g\omega_g$$

Compute a *variation of* Ω *with respect to* g; namely check that

$$\Omega(\alpha,\beta,g+\epsilon h) = \Omega(\alpha,\beta,g) - \epsilon\frac{1}{2}h^{ab}t_{ab}\omega_g$$

where

$$t_{ab}\omega_g := 2\Omega(i_a\alpha, i_b\beta, g) - g_{ab}\Omega(\alpha,\beta,g) \equiv \{2(i_a\alpha, i_b\beta)_g - g_{ab}(\alpha,\beta)_g\}\omega_g$$

i.e.

$$t_{ab} = 2(i_a\alpha, i_b\beta)_g - g_{ab}(\alpha,\beta)_g$$

Solution: (i) if e_a is an arbitrary basis and $h^{ab} := g^{ac}g^{bd}h_{cd}$, then

$$g_{ab} \mapsto \hat{g}_{ab} \equiv g_{ab} + \epsilon h_{ab} \quad \Rightarrow \quad g^{ab} \mapsto \hat{g}^{ab} \equiv g^{ab} - \epsilon h^{ab}$$

Then if $i_a \equiv i_{e_a}$, we get

$$\begin{aligned}
(\alpha, \beta)_{g+\epsilon h} &= \frac{1}{p!}\alpha_{a\ldots b}(g^{ai} - \epsilon h^{ai})\ldots(g^{bj} - \epsilon h^{bj})\beta_{i\ldots j} \\
&= (\alpha, \beta)_g - \epsilon \frac{1}{p!}\underbrace{\alpha_{a\ldots b}\left(h^a_i \beta^{i\ldots b} + \cdots + h^b_j \beta^{a\ldots j}\right)}_{p\alpha_{a\ldots b}h^a_i\beta^{i\ldots b}} \\
&= (\alpha, \beta)_g - \epsilon \underbrace{\frac{1}{(p-1)!}\alpha_{ar\ldots b}h^{ac}\beta_c{}^{r\ldots b}}_{h^{ac}(i_a\alpha, i_c\beta)} \\
&= (\alpha, \beta)_g - \epsilon h^{ab}(i_a\alpha, i_b\beta)
\end{aligned}$$

According to (5.7.3) in an arbitrary *right-handed* basis $\omega_g = \sqrt{|\det g_{ab}|}\, e^1 \wedge \cdots \wedge e^n \equiv \sqrt{|g|}\, e^1 \wedge \cdots \wedge e^n$; then,

$$|g + \epsilon h| = |g|\underbrace{|1 + \epsilon g^{-1}h|}_{1+\epsilon \operatorname{Tr} g^{-1}h} = |g|(1 + \epsilon g^{ab}h_{ba}) = |g|(1 + \epsilon h^a_a)$$

$$\Rightarrow \quad \omega_{g+\epsilon h} = \omega_g + \frac{1}{2}\epsilon h^{ab} g_{ab}\omega_g$$

and finally

$$\begin{aligned}
\alpha \wedge *_{g+\epsilon h}\beta = (\alpha, \beta)_{g+\epsilon h}\omega_{g+\epsilon h} &= \{(\alpha, \beta)_g - \epsilon h^{ab}(i_a\alpha, i_b\beta)_g\}\left\{\omega_g + \frac{1}{2}\epsilon h^{ab} g_{ab}\omega_g\right\} \\
&= \alpha \wedge *_g\beta - \frac{1}{2}\epsilon h^{ab}\underbrace{(2(i_a\alpha, i_b\beta)\omega_g - g_{ab}(\alpha, \beta)_g\omega_g)}_{t_{ab}\omega_g}
\end{aligned}$$

Note: since $h^{ab} = h^{ba}$, we see that t_{ab} *is not* determined *uniquely* – its *symmetric part* alone is fixed (cf. also (16.4.1))

$$t_{(ab)} = \{(i_a\alpha, i_b\beta) + (i_b\alpha, i_a\beta)\} - g_{ab}(\alpha, \beta)_g$$

□

• This result shows that the computation of the energy–momentum tensor for action integrals which are sums of terms of the structure $\langle\alpha, \alpha\rangle \equiv \int(\alpha, \alpha)\omega_g$ turns out to be very simple.

16.4.6 Check that if an action consists of terms of the form

$$S = \langle\alpha, \alpha\rangle \equiv \int(\alpha, \alpha)\omega_g$$

then the energy–momentum tensor is, in turn, the sum of terms

$$T_{ab} = 2(i_a\alpha, i_b\alpha) - g_{ab}(\alpha, \alpha)$$

(for each term in the action one gets a corresponding term T_{ab}).

Hint: (16.4.1) and (16.4.5). □

16.4.7 Compute explicitly the energy–momentum tensors for the electromagnetic, (massive) vector as well as scalar fields. In particular, check that

(i) from the action integrals (16.3.2), (16.3.7) and (16.3.8) we get (in Cartesian coordinates) the expressions

$$T_{\mu\nu} = \frac{1}{4}\eta_{\mu\nu}F_{\rho\sigma}F^{\rho\sigma} - F_{\mu\rho}F_{\nu}{}^{\rho} \qquad \text{electromagnetic field}$$

$$T_{\mu\nu} = \frac{1}{4}\eta_{\mu\nu}F_{\rho\sigma}F^{\rho\sigma} - F_{\mu\rho}F_{\nu}{}^{\rho} + m^2\left(W_\mu W_\nu - \frac{1}{2}\eta_{\mu\nu}W^\rho W_\rho\right) \qquad \text{massive vector field}$$

$$T_{\mu\nu} = (\partial_\mu\phi)(\partial_\nu\phi) - \frac{1}{2}\eta_{\mu\nu}(\partial_\rho\phi\partial^\rho\phi) + \frac{m^2}{2}\eta_{\mu\nu}\phi^2 \qquad \text{scalar field}$$

(ii) the energy and momentum of an electromagnetic field turn out to be

$$E = \frac{1}{2}\int_V (\mathbf{E}^2 + \mathbf{B}^2)\,dV \qquad \mathbf{P} = \int_V (\mathbf{E}\times\mathbf{B})\,dV$$

(iii) the energy and momentum of a massive vector field are

$$E = \frac{1}{2}\int_V \left[\mathbf{E}^2 + \mathbf{B}^2 + m^2\left(W_0^2 + \mathbf{W}^2\right)\right]dV \qquad \mathbf{P} = \int_V (\mathbf{E}\times\mathbf{B} + m^2 W^0\mathbf{W})\,dV$$

(iv) the energy and momentum of a scalar field are

$$E = \int_V [(\partial_t\phi)^2 + (\nabla\phi)^2 + m^2\phi^2]\,dV \qquad \mathbf{P} = -\int_V (\partial_t\phi)(\nabla\phi)\,dV$$

Hint: (16.4.3), (16.4.5) and (16.4.6). □

• As we have already mentioned before problem (16.4.2), sometimes a *conformal* Killing vector is enough for construction of a conserved quantity. This happens for actions leading to *traceless* energy–momentum tensors, $T^\mu_\mu \equiv g^{\mu\nu}T_{\mu\nu} = 0$. It turns out that the vanishing trace in turn occurs for actions which happen to be invariant with respect to *conformal rescaling of the metric* (5.8.3); an important example is provided by the electromagnetic field.

16.4.8 Let ξ be a *conformal Killing vector* of a manifold (M, g) and let $T_{\mu\nu}$ be the *traceless* energy–momentum tensor. Check that the quantity Q which is constructed in the same way as we did for a Killing vector in (16.4.2) is still conserved.

Hint: like in (16.4.2), but now $(\mathcal{L}_\xi g)_{\mu\nu} = f g_{\mu\nu}$, so that $(\mathcal{L}_\xi g)_{\mu\nu}T^{\nu\mu} = f T^\mu_\mu$, which *still* vanishes, so that everything proceeds as before. □

16.4.9 Consider an infinitesimal conformal rescaling of the metric

$$g_{\mu\nu}(x) \mapsto g_{\mu\nu}(x) + \epsilon\sigma(x)g_{\mu\nu}(x)$$

Check that the invariance of action with respect to arbitrary transformations of this type (i.e. with an arbitrary function $\sigma(x)$) is equivalent to tracelessness of the energy–momentum tensor.

Hint: from the definition in (16.4.1) we have $S[\psi, g + \epsilon\sigma g] =: S[\psi, g] - \epsilon \int_{\mathcal{D}} \frac{1}{2}\sigma(x) T^{\mu}_{\mu}\omega_g$. □

16.4.10 Let α be a p-form on a manifold (M, g) of dimension n and consider a term in the action

$$S = \langle \alpha, \alpha \rangle \equiv \int (\alpha, \alpha)\omega_g$$

Check that

(i) the trace of the energy–momentum tensor corresponding to this particular term in the action is

$$T^a_a = (2p - n)(\alpha, \alpha)$$

so whenever $n = 2p$, the trace *vanishes*

(ii) this occurs exactly for a pure electromagnetic field (in four-dimensional space-time)

(iii) we get the vanishing trace in electromagnetism also directly from (16.4.7)

(iv) also, the tensor which corresponds to the kinetic term of a *scalar* field in *two-dimensional* space-time[350] is traceless.

Hint: (i) according to (5.8.6) and (5.8.11) we have

$$\begin{aligned} T^a_a &= 2(i^a\alpha, i_a\alpha) - g^a_a(\alpha, \alpha) = 2(\alpha, j^a i_a\alpha) - g^a_a(\alpha, \alpha) = 2(\alpha, p\alpha) - g^a_a(\alpha, \alpha) \\ &= (2p - n)(\alpha, \alpha) \end{aligned}$$

(ii) $2p - n = 2.2 - 4$; (iv) $2p - n = 2.1 - 2$. □

• A technical issue to be answered in the early history of general relativity was to settle on a rule by which physical laws should be generalized so that they were also valid in a "curved space-time" (M, g).

(The new theory of gravitation has replaced Minkowski *flat* space-time $E^{1,3}$ by a more general space-time, a *Lorentzian manifold* (M, g), i.e. a four-dimensional manifold M endowed with a metric tensor g with signature $(+---)$; the effects of the gravitational field are explained in the theory in terms of curving of the space-time (the reader is recommended to study the specialized literature in order to learn more).)

It turns out that the standard rule for the generalization reads: *replace all commas by semicolons*, i.e. if there is a *partial* derivative in a physical equation (written in Cartesian

[350] As well as the mass term of a vector field in two-dimensional space-time; notice, however, that the tracelessness is spoiled by the kinetic term in this case.

coordinates in Minkowski space), it is to be replaced by the *covariant* derivative *in the sense of the RLC connection*.

(This is closely related to the *equivalence principle*, which demands that physics in a "falling lift," i.e. in a reference frame freely falling in a gravitational field, should be the same as it is with *no* gravitational field (the gravitational field may be locally transformed away by the free fall); the details should again be found in the specialized literature.)

For Maxwell's equations there is, however, another natural possibility for how to generalize them to a "curved" space-time. Namely, if we write them *in terms of forms* in Minkowski space(-time), the only place from which we know we are in Minkowski space is the metric tensor η *on* the *codifferential* (or, equivalently, on the Hodge star operator). The most natural generalization to a Lorentzian manifold is then clearly a formal replacement $\eta \mapsto g$ on the codifferential:[351]

$$\delta_\eta F = -j \quad dF = 0 \quad \mapsto \quad \delta_g F = -j \quad dF = 0$$

So we have as many as two natural ways for how to pass to (M, g). Which is then to be preferred? Fortunately, it turns out that it is actually not a hard choice since both ways eventually yield the same result.

16.4.11 Let (M, g) be a Lorentzian manifold. Check that the two prescriptions indeed yield the same result, i.e. that

$$\delta_g F = -j \quad dF = 0 \quad \Leftrightarrow \quad F^{\mu\nu}{}_{;\nu} = j^\mu \quad F_{[\mu\nu;\rho]} = 0$$

Hint: see (8.3.5), (15.2.9), (15.4.12) and (15.6.17). □

• The machinery of forms enables us to express particularly clearly transformations of fields under symmetries. From a purely technical point of view this is due to the naturalness of the operators d and δ.

16.4.12 Let $f : M \to M$ be an *isometry* of a Lorentzian manifold (M, g), i.e. $f^* g = g$. Check that if a pair $(F \equiv dA,\ j)$ satisfies Maxwell's equations (i.e. the current j generates the field $F \equiv dA$), then they are also satisfied by the pair $(f^* F \equiv df^* A,\ f^* j)$ (so that the current $f^* j$ then generates the field $f^* F$, given by[352] the potential $f^* A$):

$$\begin{aligned} &f \text{ is an isometry of } (M, g) \\ \Rightarrow \quad &\{(j \text{ generates } F \leftrightarrow A) \Rightarrow (f^* j \text{ generates } f^* F \leftrightarrow f^* A)\} \end{aligned}$$

Hint: apply f^* to the equations $\delta_g F = -j, dF = 0$ and use the naturalness of the differential and codifferential ($f^* d = df^*$, $f^* \delta_g = \delta_{f^* g} f^*$, (6.2.11) and (8.3.8)).

16.4.13 Consider Minkowski space $E^{1,3}$ and an isometry given by an element of the Poincaré group

$$f : x^\mu \mapsto \Lambda^\mu_\nu x^\nu + a^\mu$$

[351] Note also that there is *nothing at all* to be generalized in the homogeneous part of Maxwell's equations. The homogeneous equations do not depend on any particular characteristic of a manifold, so that everywhere it is the same (this is sometimes expressed as it being "purely geometrical").

[352] Of course, the potential $f^* A$ still has its usual gauge freedom, so that $f^* A$ itself represents only a possible choice.

Be sure to understand that $F \mapsto f^*F$, $A \mapsto f^*A$ and $j \mapsto f^*j$ yields here (the well-known "4-tensor") component relations

$$\begin{aligned} F_{\mu\nu}(x) &\mapsto \Lambda^\rho_\mu \Lambda^\sigma_\nu F_{\rho\sigma}(\Lambda x + a) \\ A_\mu(x) &\mapsto \Lambda^\rho_\mu A_\rho(\Lambda x + a) \\ j_\mu(x) &\mapsto \Lambda^\rho_\mu j_\rho(\Lambda x + a) \end{aligned}$$

Hint: a standard form computation; e.g. $f^*(j_\rho(x)\,dx^\rho) = j_\rho(\Lambda x + a)\,d(\Lambda^\rho_\mu x^\mu + a^\mu) = \cdots$. □

• This enables one, as we know from a course on electrodynamics, to obtain (say) the field of a charge which moves uniformly along a straight line from a (simpler) field of a charge being *at rest* (by mere transformation, without repeated *solution* of differential equations with a new source), since passing to a reference frame which moves uniformly along a line *is an isometry* of Minkowski space.

Also time and space *inversions* happen to be isometries of Minkowski space.[353] Both of them *reverse the orientation* of Minkowski space, which manifests itself in a subtlety in using the rule $(F, j) \mapsto (f^*F, f^*j)$ for computation of new sources and fields. Consider the result of the procedure $j \mapsto f^*j$ for time inversion $f_T : (t, \mathbf{r}) \mapsto (-t, \mathbf{r})$. We get

$$j \equiv \rho(t, \mathbf{r})\,dt - \mathbf{j}(t, \mathbf{r}) \cdot d\mathbf{r} \mapsto f_T^* j \equiv -\rho(-t, \mathbf{r})\,dt - \mathbf{j}(-t, \mathbf{r}) \cdot d\mathbf{r}$$

so that

$$(\rho, \mathbf{j})(t, \mathbf{r}) \mapsto (-\rho, \mathbf{j})(-t, \mathbf{r})$$

However, this transformation corresponds physically not only to "T-inversion," but rather to a combined "CT-inversion" (also a change of sign of the *charge* occurs).[354] Then if we indeed want to realize only T-inversion, the transformation $j \mapsto j_T \equiv -f_T^* j$ of the source is needed (or, as one checks easily, $J \mapsto J_T \equiv f_T^* J$, where $J \equiv *j$ is the 3-*form* of a current).[355] Then according to the result of problem (16.4.12) (which still holds, since δ_g *does not feel* orientation) the corresponding field will be $F_T \equiv -f_T^* F$. The cases of P and C transformations are similar.

16.4.14 Summarize the behavior of the quantities occurring in the theory of the electromagnetic field, (F, A, j) and $(\mathbf{E}, \mathbf{B}, \phi, \mathbf{A}, \rho, \mathbf{j})$, under the time, space and charge inversions. In particular, check that if $f_T : (t, \mathbf{r}) \mapsto (-t, \mathbf{r})$ and $f_P : (t, \mathbf{r}) \mapsto (t, -\mathbf{r})$, then

$$\begin{aligned} (F, A, j)_T &= (-f_T^* F, -f_T^* A, -f_T^* j) \\ (F, A, j)_P &= (f_P^* F, f_P^* A, f_P^* j) \\ (F, A, j)_C &= (-F, -A, -j) \end{aligned}$$

[353] The result of time inversion may be seen when a film is played backwards (mistakenly or just for fun; in a popular sequence slivers of glass lying on the floor suddenly start to recombine and in a moment they form a jar standing on a table). The result of space inversion may be seen by an appropriate combination of a rotation and viewing the object in a mirror, e.g. by first turning it by 180° and then viewing it in a mirror whose plane is perpendicular to the axis of rotation.

[354] Standard notation for the three inversions in physics is C $(Q \mapsto -Q)$, P $(\mathbf{r} \mapsto -\mathbf{r})$ and T $(t \mapsto -t)$.

[355] If an isometry reverses orientation, it also *reverses the sign* of the Hodge operator: $f^* *_{g,o} = *_{g,-o} f^* = - *_{g,o} f^*$.

or in three-dimensional notation

$$(\mathbf{E}, \mathbf{B}, \phi, \mathbf{A}, \rho, \mathbf{j})_T(t, \mathbf{r}) = (\mathbf{E}, -\mathbf{B}, \phi, -\mathbf{A}, \rho, -\mathbf{j})(-t, \mathbf{r})$$
$$(\mathbf{E}, \mathbf{B}, \phi, \mathbf{A}, \rho, \mathbf{j})_P(t, \mathbf{r}) = (-\mathbf{E}, \mathbf{B}, \phi, -\mathbf{A}, \rho, -\mathbf{j})(t, -\mathbf{r})$$
$$(\mathbf{E}, \mathbf{B}, \phi, \mathbf{A}, \rho, \mathbf{j})_C(t, \mathbf{r}) = (-\mathbf{E}, -\mathbf{B}, -\phi, -\mathbf{A}, -\rho, -\mathbf{j})(t, \mathbf{r})$$

Hint: see (16.4.12). □

• The differential d commutes *with all* diffeomorphisms. What then constrains the possibilities of transformations of fields is the *codifferential* in Maxwell's equations. This operator *depends on* g and restricts diffeomorphisms to those which behave nicely with respect to g. The nicest possible behavior, isometry, was already discussed. The second nicest behavior is *conformal transformation* and it turns out that also here some result may be obtained.

16.4.15* Let $f : M \to M$ be a *conformal transformation* of a *four-dimensional* Lorentzian manifold (M, g) and let $f^*g = \sigma^2(x)g, \sigma^2(x) > 0$. Check that if a pair $(F = dA, j)$ satisfies Maxwell's equations (i.e. the current j generates the field $F = dA$), then they are also satisfied by the pair $(f^*F, f^*A, \sigma^2 f^*j)$ (so that the current $\sigma^2 f^*j$ then generates the field f^*F, given by the potential f^*A):

$$f \text{ is a conformal transformation of } (M, g)$$
$$\Rightarrow \quad \{(j \text{ generates } F \leftrightarrow A) \Rightarrow (\sigma^2 f^* j \text{ generates } f^*F \leftrightarrow f^*A)\}$$

(In particular, the *sourceless* theory ($j = 0$) is *conformally invariant*: if F is a solution, then also f^*F is a solution.)

Hint: like in (16.4.12): $f^*(d *_g F) = f^*(- *_g j)$, but now (5.8.3) $f^* *_g = *_{\sigma^2 g} f^* = \sigma^{4-2p} *_g f^*$. □

16.5* Einstein gravitational field equations, Hilbert and Cartan action

• As we learned in Section 15.5, given a curvature tensor $R^a{}_{bcd}$ we can obtain by contractions two simpler objects, the Ricci tensor R_{ab} and (if a metric tensor g_{ab} is also available) the scalar curvature R. Combining these tensors one gets another important tensor which plays a prominent role in the general theory of relativity entering the equations of the gravitational field[356]

$$G_{ab} \equiv R_{ab} - \frac{1}{2}Rg_{ab} \qquad \text{Einstein tensor}$$
$$G_{ab} = 8\pi T_{ab} \qquad \text{Einstein equations}$$

(nothing less than *space-time* (M, g) itself is "computed from them"; T_{ab} denotes the *energy–momentum tensor* (16.4.1) of the matter which generates the gravitational field). The reason why the field equations look exactly like this and, as a matter of fact, how the gravitational

[356] This holds in a system of units where $c = 1 = G$; otherwise there is a constant $8\pi G/c^4$.

field itself is actually encoded into the metric tensor g_{ab}, is discussed at length in textbooks on the general theory of relativity.[357]

Here we only bring to the reader's notice some technical facts which may be seen very quickly and which might turn out to be fairly useful in the course of study of the theory (like what exactly is interesting in just this combination of R_{ab}, g_{ab} and R and so on).

16.5.1 Let R_{ab} be the Ricci tensor. Check that

(i) the tensor may also be encoded into *Ricci forms* R_a and that these forms are simply related to the curvature forms Ω^a_b and the scalar curvature R

$$R_a := R_{ab}e^b \qquad R_a = i_b \Omega^b_a \qquad i^a R_a \equiv i^a i_b \Omega^b_a = R$$

(ii) for the RLC connection the Ricci (and then also Einstein) tensor is symmetric

$$\text{RLC connection} \quad \Rightarrow \quad R_{ab} = R_{ba} \quad G_{ab} = G_{ba}$$

Hint: (i) $i_b\Omega^b_a = \frac{1}{2}R^b{}_{acd}i_b(e^c \wedge e^d) = R^b{}_{abd}e^d = R_{ad}e^d = R_a$; (ii) for the RLC connection the Ricci identity (15.6.16) gives $\Omega \wedge e = 0$; then $0 = i_b(\Omega^b_a \wedge e^a) = (i_b\Omega^b_a) \wedge e^a + \Omega^b_a \wedge (i_b e^a) = R_{ac}e^c \wedge e^a + \Omega^a_a$; in an orthonormal basis $\Omega^a_a = 0$, so that R_{ac} is symmetric; this fact, however, does not depend on a choice of basis. □

16.5.2 Consider the RLC connection on (M, g) and let $G^a_b \equiv R^a_b - \frac{1}{2}R\delta^a_b$ be the Einstein tensor. Check that this tensor

(i) can be obtained from the curvature tensor by contraction with 3-delta

$$G^a_b = -\frac{3}{2}\delta^{ars}_{bcd}R^{cd}{}_{rs} \qquad \delta^{ars}_{bcd} \equiv \delta^a_{[b}\delta^r{}_c\delta^s{}_{d]}$$

(ii) has vanishing *covariant divergence*

$$G^{ab}{}_{;b} = 0$$

Hint: (i) (5.6.2), straightforward check; (ii) generalized Kronecker symbols are covariantly constant (being combined from "ordinary" Kronecker symbols *which are* constant due to commuting of ∇_V with contractions), so that

$$G^a_{b;a} = -\frac{3}{2}\delta^{ars}_{bcd;a}R^{cd}{}_{rs} - \frac{3}{2}\delta^{ars}_{bcd}R^{cd}{}_{rs;a} = -\frac{3}{2}\delta^{ars}_{bcd}R^{cd}{}_{rs;a} = -\frac{3}{2}R^{cd}{}_{[cd;b]} = 0$$

The last equality is due to the Bianchi identity (15.6.16). □

• From problem (16.4.1) we know that the energy–momentum tensor also has vanishing covariant divergence, so that vanishing of the divergence of the Einstein tensor turns out to be its key property which guarantees consistency[358] of the field equations $G_{ab} \sim T_{ab}$.

In the same problem (16.4.1) we also learned that the energy–momentum tensor may be computed by means of variation of the action integral with respect to the metric tensor g.

[357] And we will not walk in their shoes.

[358] This consistency is similar to vanishing of the codifferential of both sides of Maxwell's equations $\delta F = -j$ (16.2.4).

Here we will see that also the Einstein tensor may be obtained by means of variation of an appropriate action with respect to the metric tensor. The action turns out to be the volume integral of the *scalar curvature*.

16.5.3 The *Hilbert action* of a gravitational field is (up to a conventional factor $(-1/16\pi)$) the volume integral of the scalar curvature, regarded as a functional of the metric tensor

$$S_H \equiv S_H[g] := -\frac{1}{16\pi} \int R_g \omega_g$$

where $R_g \equiv R$ is the scalar curvature of the RLC connection on (M, g) and ω_g is the metric volume form. Check that

(i) it may be rewritten in the form

$$S_H = -\frac{1}{16\pi} \int \Omega_{ab} \wedge *_g(e^a \wedge e^b)$$

with e^a being an arbitrary (co)frame field; Ω_{ab} then denotes curvature forms with respect to e^a

(ii) in *two-dimensional* space-time the form under the integral sign turns out to be *exact* (so that the action may then be rewritten (due to Stokes' theorem) as an integral *over the boundary* of the space-time volume under consideration and it is completely insensitive to variations of g *inside* the volume $\Rightarrow$ it gives no equations of motion).

Hint: (i) according to (16.5.1), (5.8.4), (5.8.9) and (5.8.10) we have

$$\begin{aligned} R_g \omega_g &= (i^a i_b \Omega^b_{\ a})\omega_g = (i^a i_b \Omega^b_{\ a}, 1)_g \omega_g = (\Omega_{ba}, j^b j^a 1)_g \omega_g \\ &= (\Omega_{ba}, e^b \wedge e^a)_g \omega_g = \Omega_{ba} \wedge *_g(e^b \wedge e^a) \end{aligned}$$

(ii) according to (15.6.10) we then have with respect to an *orthonormal* frame field $\Omega_{ab} = \epsilon_{ab}\, d\alpha$, $*(e^a \wedge e^b) = \pm\epsilon^{ab}$ so that $\Omega_{ab} \wedge *_g(e^a \wedge e^b) = \pm 2 d\alpha$. □

• It turns out that the variational derivative of this functional is just the Einstein tensor G_{ab}, so that this very part of the action gives as "equations of motion" just $G_{ab} = 0$ ("vacuum" or "sourceless" Einstein equations). In the following problem we confirm this statement by a computation (postponing part of the proof until later, when we learn more appropriate machinery (see Section 21.7)).

16.5.4 The metric tensor g_{ab}, with respect to which we would like to perform variation of the functional $S_H[g]$ occurs at *three* places in the form under the integral sign:

$$S_H[g] = -\frac{1}{16\pi} \int g_{ac} \Omega^c_{\ b}(g) \wedge *_g(e^a \wedge e^b)$$

A contribution to the form under the integral sign from a variation $g \mapsto g + \epsilon h$ in the curvature forms $\Omega^c_{\ b}(g)$ happens to be an *exact* form, so that this term is of no interest for the computation of the variational derivative.[359] Compute the change of the form under the

[359] In (21.7.8) we will see that to within first-order accuracy in ϵ there holds $g_{ac}\Omega^c_{\ b}(g + \epsilon h) \wedge *_g(e^a \wedge e^b) = g_{ac}\Omega^c_{\ b}(g) \wedge *_g(e^a \wedge e^b) + \epsilon d(\text{something})$, so that upon integration over a domain $\mathcal{U}$ Stokes' theorem gives an integral over the *boundary* $\partial\mathcal{U}$ which vanishes, since $h = 0$ there.

integral sign due to a change $g \mapsto g + \epsilon h$ in the *remaining two* places; convince yourself that it is (within order ϵ)

$$(g_{ac} + \epsilon h_{ac})\Omega^c_b(g) \wedge *_{g+\epsilon h}(e^a \wedge e^b) = g_{ac}\Omega^c_b(g) \wedge *_g(e^a \wedge e^b) - \epsilon h^{ab} G_{ab}\omega_g$$

so that in combination with the comment about the (sad) fate of the variation $\Omega^c_b(g + \epsilon h)$ we indeed obtain as a total result

$$S_H[g + \epsilon h] = S_H[g] + \epsilon \int h^{ab} \left(\frac{1}{16\pi} G_{ab}\right) \omega_g + \cdots \qquad h^{ab} := g^{ac} g^{bd} h_{cd}$$

Solution: in order to get an increment due to a change of g_{ab} in the first place one has, just as in problem (16.5.3), to compute that

$$\begin{aligned} h_{ac}\Omega^c_b \wedge *_g(e^a \wedge e^b) &= h_{ac}\big(\Omega^c_b, e^a \wedge e^b\big)\omega_g = h_{ac}\big(\Omega^c_b, j^a j^b 1\big)\omega_g = h_{ac}\big(i^b i^a \Omega^c_b\big)\omega_g \\ &= h^{ac}(i_a R_c)\omega_g = h^{ab} R_{ab}\omega_g \end{aligned}$$

The change of the star operator due to a variation of g_{ab} is discussed in problem (16.4.5); here it gives

$$\begin{aligned} & g_{ac}\Omega^c_b \wedge *_{g+\epsilon h}(e^a \wedge e^b) \\ &= \Omega_{ab} \wedge *_g(e^a \wedge e^b) + \epsilon h^{rs} \left\{ \frac{1}{2} g_{rs}(\Omega_{ab}, e^a \wedge e^b) - (i_r\Omega_{ab}, i_s(e^a \wedge e^b)) \right\} \omega_g \\ &= \Omega_{ab} \wedge *_g(e^a \wedge e^b) + \cdots \\ &= \Omega_{ab} \wedge *_g(e^a \wedge e^b) + \epsilon h^{ab} \left\{ \frac{1}{2} g_{ab} R - 2R_{ab} \right\} \omega_g \end{aligned}$$

These two terms thus indeed add up to

$$\begin{aligned} (g_{ac} + \epsilon h_{ac})\Omega^c_b \wedge *_{g+\epsilon h}(e^a \wedge e^b) &= \Omega_{ab} \wedge *_g(e^a \wedge e^b) + \epsilon h^{ab} \left\{ R_{ab} + \frac{1}{2} g_{ab} R - 2R_{ab} \right\} \omega_g \\ &= \Omega_{ab} \wedge *_g(e^a \wedge e^b) + \epsilon h^{ab}(-G_{ab})\omega_g \end{aligned}$$

□

• Now everything is already prepared for the derivation of Einstein's field equations from an action principle. We are virtually finished with the equations for the gravitational field "in vacuum," i.e. equations which are valid in regions of space-time where there is nothing else except for a gravitational field. Then one takes the action to be $S_H[g]$ *alone.*

16.5.5 Check that

(i) from the Hilbert action

$$S_H \equiv S_H[g] := -\frac{1}{16\pi} \int R_g \omega_g$$

the following equations of motion result:

$$G_{ab} = 0 \qquad \text{Einstein's vacuum equations}$$

(ii) these equations are equivalent to the condition for a space-time to be *Ricci-flat*, i.e. the condition of vanishing *Ricci* tensor itself

$$G_{ab} = 0 \quad \Leftrightarrow \quad R_{ab} = 0$$

Hint: (i) (16.5.4); (ii) contraction of the equation $R_{ab} - \frac{1}{2}Rg_{ab} = 0$ by the tensor g^{ab} gives in n-dimensional space-time $(2-n)R = 0$, whence[360] $R = 0$ and consequently $0 = G_{ab} = R_{ab}$. □

• Now, if we found a term in the action which would yield as a variational derivative with respect to g_{ab} just the energy–momentum tensor of the "matter" generating the gravitational field, we would be able to derive from the variational principle the *complete* Einstein equations (including the sources = their right-hand side).

This is, however, very easy since the energy–momentum tensor was defined in (16.4.1) just in the way we need: for a field ψ with action $S[\psi, g]$ it was introduced by the relation

$$S[\psi, g + \epsilon h] =: S[\psi, g] - \epsilon \int_{\mathcal{D}} \frac{1}{2} h^{ab} T_{ab} \omega_g$$

i.e. exactly as the variational derivative of the action of the field ψ with respect to the metric tensor g. So if we represent the "matter" by such a field and the action integral of the field is denoted by S_{matter}, then the sum of the Hilbert action and the action of the "matter" will necessarily lead to the complete Einstein equations.

16.5.6 Consider an action which is a combination of the Hilbert action for the gravitational field and an action of (non-gravitational) "matter" represented by a field ψ

$$S[g, \psi] := S_H[g] + S_{\text{matter}}[\psi, g] \equiv -\frac{1}{16\pi} \int_{\mathcal{D}} R_g \omega_g + \int_{\mathcal{D}} L(\psi, g) \omega_g$$

Check that its variation with respect to g gives (within order ϵ)

$$S[g + \epsilon h, \psi] = S[g, \psi] + \epsilon \int_{\mathcal{D}} \frac{1}{2} h^{ab} \left(\frac{1}{8\pi} G_{ab} - T_{ab} \right) \omega_g$$

so that the equations of motion resulting from this action coincide with Einstein's equations

$$G_{ab} = 8\pi T_{ab}$$

Hint: see (16.4.1) and (16.5.4). □

• One can also arrive at Einstein's equations from another action, in a way proposed by Cartan. This modification of the original Einstein theory of gravitation is usually called *Einstein–Cartan theory*. In general it is indeed a *modification*, i.e. the theory is in general inequivalent to the theory based on the Hilbert action, which we discussed until now. Namely, sometimes the *torsion* of the connection occurs here (it is generated by spinor fields), which

[360] For $n \neq 2$; recall that a "two-dimensional gravitation" needs a completely *different action* since then the *Hilbert* action "does not work," see (16.5.3).

was clearly impossible in the approach based *a priori on the RLC connection*, induced[361] by the metric tensor g. Again, we will mention only some simple ideas of Cartan's approach, leaving the details to more specialized texts.

In a domain $\mathcal{O}$ one considers as *basic independent variables*

1. a coframe field e^a (or *vielbein fields* e^a_μ given by a decomposition $e^a = e^a_\mu \, dx^\mu$)
2. 1-forms $\omega_{ab} = -\omega_{ba}$

   (Recall that the vielbein field is usually called the *tetrad field* in four dimensions.) These two basic objects then *secondarily* induce in domain $\mathcal{O}$ further (already derived) important quantities:
3. metric tensor

$$g := \eta_{ab} e^a \otimes e^b \equiv g(e)$$

   i.e. $g \equiv g(e)$ is (uniquely) defined *so as* the field e^a were *orthonormal*
4. linear connection ∇; it is defined as follows:

$$\nabla_V e^a := -\omega^a_b(V) e^b \qquad \omega^a_b := \eta^{ac} \omega_{cb}$$

   i.e. so that the forms ω_{ab} were (after due modification of the position of the index) connection forms of *the* connection[362] with respect to *the* (already orthonormal) basis e^a.
5. volume form and orientation

$$\omega_{abcd} := \epsilon_{abcd} \quad \text{or equivalently} \quad \omega := e^0 \wedge e^1 \wedge e^2 \wedge e^3$$

   i.e. the orientation is (uniquely) defined so that the frame field e^a is *right-handed* and the volume form is then a standard form compatible with the metric and orientation $\omega_{g(e),o(e)}$.

16.5.7 Be sure to understand that

(i) if a particular (co)frame field is declared to be orthonormal, the metric tensor is indeed uniquely defined by this
(ii) a linear connection is indeed defined by the prescription given above; the connection is moreover *metric with respect to* $g(e)$ from item 3 (its torsion being not restricted by the prescription, however; it may turn out to be non-vanishing)
(iii) the *curvature forms* Ω_{ab} depend on the forms ω_{ab} *alone*, $\Omega(\omega)$.

Solution: (ii)

$$\begin{aligned} 0 &= V\eta_{ab} = \nabla_V\{g(e_a, e_b)\} = (\nabla_V g)(e_a, e_b) + g(\nabla_V e_a, e_b) + g(e_a, \nabla_V e_b) \\ &= (\nabla_V g)(e_a, e_b) + (\omega_{ab} + \omega_{ba})(V) = (\nabla_V g)(e_a, e_b) \\ &\Rightarrow \nabla g = 0 \end{aligned}$$

(iii) standard relations $\Omega_{ab} = d\omega_{ab} + \omega_{ac} \wedge \omega^c_b$. □

[361] In physical parlance the fact of vanishing torsion was put in "by hand."

[362] Notice that the "objective" connection ∇ depends on the choice of *both* primary objects, e^a as well as ω^b_a (a connection is given by forms ω with respect to the (co)frame field e), whereas the metric tensor depends on the choice of e^a alone.

• Let us now have a look at what the action looks like. At first glance the new (Cartan) action for the gravitational field coincides with the old (Hilbert) one – it is a $(-1/16\pi)$-multiple of an integral of the form

$$\tau = \Omega_{ab} \wedge *(e^a \wedge e^b)$$

However, the action is to be regarded differently (for the purpose of performing the variation), since the basic independent variables have changed.[363] The Hodge star operator is implicitly understood to be defined with respect to $g \equiv g(e)$ and $o(e)$, so that (e^a being *by definition* right-handed and orthonormal) it actually gives

$$*(e^a \wedge e^b) = \frac{1}{2}\omega^{ab}{}_{cd} e^c \wedge e^d = \frac{1}{2}\eta^{ar}\eta^{bs}\epsilon_{rscd} e^c \wedge e^d$$

($\omega^{ab}{}_{cd} := \eta^{ar}\eta^{bs}\omega_{rscd}$), so the explicit dependence of $*$ on g drops out and one obtains

$$\frac{1}{2}\Omega_{ab} \wedge \eta^{ar}\eta^{bs}\epsilon_{rscd} e^c \wedge e^d \equiv \frac{1}{2}\Omega^{ab}(\omega) \wedge \epsilon_{abcd} e^c \wedge e^d$$

Then Ω_{ab} depends on ω_{ab} (alone), the rest depends on the frame field e^a (alone) and one is to perform variations of the 4-form under the integral sign *with respect to these two independent* objects

$$\tau[e, \omega] = \frac{1}{2}\epsilon_{abcd}\Omega^{ab}(\omega) \wedge e^c \wedge e^d$$

16.5.8 The *Cartan action* of the gravitational field is the integral

$$S_C \equiv S_C[e, \omega] := -\frac{1}{16\pi}\int_{\mathcal{D}} \tau[e, \omega] \equiv -\frac{1}{32\pi}\int_{\mathcal{D}} \epsilon_{abcd}\Omega^{ab}(\omega) \wedge e^c \wedge e^d$$

regarded as a functional of the frame field e and 1-forms ω. Check that an infinitesimal change of the frame field $e^a \mapsto e^a + \epsilon f^a$ (f^a being arbitrary 1-forms) results in a variation of the functional

$$S_C[e + \epsilon f, \omega] = S_C[e, \omega] + \epsilon \int_{\mathcal{D}} f^a \wedge \left(\frac{1}{8\pi} * G_a\right)$$

where we introduced

$$G_a := G_{ba} e^b \qquad \text{Einstein 1-forms}$$

Solution:

$$\begin{aligned} \tau[e + \epsilon f, \omega] &= \frac{1}{2}\epsilon_{abcd}\Omega^{ab}(\omega) \wedge (e^c + \epsilon f^c) \wedge (e^d + \epsilon f^d) \\ &= \tau[e, \omega] + \epsilon f^c \wedge \{\epsilon_{abcd}\Omega^{ab}(\omega) \wedge e^d\} \\ &=: \tau[e, \omega] + \epsilon f^c \wedge \sigma_c \end{aligned}$$

[363] It resembles a situation in mechanics where we consider the action $S[q] = \int L(q, \dot{q})\, dt$ as well as $S[q, p] = \int (p\dot{q} - H(q, p))\, dt$ (see (18.5.5) and (18.5.6)): since H and L are related by $H = p\dot{q} - L$, at first sight "the same" object is under the integral sign in both cases. The actions nevertheless differ, since the variations are performed with respect to curves in different spaces (the first one in configuration space, whereas the second one in a broader phase space).

But

$$\sigma_c \equiv \epsilon_{abcd}\Omega^{ab} \wedge e^d = \frac{1}{2}\epsilon_{abcd}R^{ab}{}_{\alpha\beta}e^\alpha \wedge e^\beta \wedge e^d = \frac{1}{2}\epsilon_{abcd}R^{ab}{}_{\alpha\beta}\omega^{\alpha\beta d\rho}\eta_{\rho\kappa} * e^\kappa$$
$$= \frac{1}{2}\epsilon_{dabc}\epsilon^{d\alpha\beta\rho}R^{ab}{}_{\alpha\beta}\eta_{\rho\kappa} * e^\kappa = 3\delta^{\alpha\beta\rho}_{abc}R^{ab}{}_{\alpha\beta}\eta_{\rho\kappa} * e^\kappa = -2G^\rho_c\eta_{\rho\kappa} * e^\kappa$$
$$= -2G_{bc} * e^b = -2 * G_{bc}e^b \equiv -2 * G_c$$

(since $e^a \wedge e^b \wedge e^c = \omega^{abc}{}_d * e^d$ and $\omega^{abcd} = (\det \eta)\epsilon_{abcd} = -\epsilon_{abcd} \equiv -\epsilon^{abcd}$). □

• Let us now have a look at the variation with respect to the forms ω_{ab}. We need to know how the form $\tau[e, \omega]$ under the integral sign responds to an infinitesimal change of connection forms

$$\omega_{ab} \mapsto \omega_{ab} + \epsilon\sigma_{ab} \qquad \sigma_{ab} = -\sigma_{ba}$$

Also here we prefer to postpone the computation itself to Section 19.6 (as we already did in (16.5.4)), where we will learn how to use the exterior *covariant* derivative $\mathcal{D}$. The result of the computation reads

$$\tau[e, \omega + \epsilon\sigma] = \tau[e, \omega] + d\{\cdots\} - \epsilon\sigma^{ab} \wedge \epsilon_{abcd}T^c \wedge e^d$$

16.5.9 Be sure to understand that our knowledge concerning the variation of the Cartan action may be summarized as follows:

$$S_C[e + \epsilon_1 f, \omega + \epsilon_2\sigma] = S_C[e, \omega] + \epsilon_1 \int_{\mathcal{D}} f^a \wedge \left(\frac{1}{8\pi} * G_a\right)$$
$$+ \epsilon_2 \int_{\mathcal{D}} \sigma^{ab} \wedge \left(\frac{1}{16\pi}\epsilon_{abcd}T^c \wedge e^d\right)$$

Conditions for the functional to be stationary with respect to e and ω thus read

$$G_a = 0 \qquad \epsilon_{abcd}T^c \wedge e^d = 0$$

Check that they happen to be equivalent to

$G_{ab} = 0$ Einstein vacuum (sourceless) equations
$T^a = 0$ connection is symmetric

so that we get just the same as the Einstein–Hilbert approach gives: we have vanishing torsion[364] and the Einstein vacuum equations hold.

Solution: variations according to (16.5.8), (21.7.8) and (21.7.9); vanishing torsion:

$$\epsilon_{abcd}T^c \wedge e^d = 0 \Rightarrow \epsilon_{abcd}T^b \wedge e^c \wedge e^d = 0 \Rightarrow T_b \wedge *(e^a \wedge e^b) = 0 \Rightarrow (T_b, j^a j^b 1) = 0$$
$$\Rightarrow (i^b i^a T_b, 1) = 0 \Rightarrow i^a(i^b T_b) \equiv i^a\hat{T} = 0 \Rightarrow \hat{T} \equiv i^b T_b = 0$$

[364] Also, from the very beginning, we have metricity imposed "by hand" ⇒ RLC connection altogether.

and then

$$\begin{aligned}&\epsilon_{abcd}T^c\wedge e^d=0\quad\Rightarrow T^b\wedge e^c=T^c\wedge e^b\quad\Rightarrow\quad j^aT^b=j^bT^a\Rightarrow i_aj^aT^b=i_aj^bT^a\\ \Rightarrow\quad&(\delta^b_a-j^bi_a)T^a=2T^b\Rightarrow\quad 2T^b=T^b-j^bi_aT^a\Rightarrow T^b=-j^b\hat{T}=j^b0\quad\Rightarrow\quad T^b=0\end{aligned}$$

□

• In order to derive the *complete* Einstein equations (including sources = their right-hand side), we again need to add to the action a term S_{matter} representing "non-gravitational matter."

Inspection of the concrete actions for fields which we have encountered (electromagnetic, scalar, vector) shows that they have the structure $S_{\text{matter}}[\psi, g]$. Thus in addition to the fields themselves they depend on the metric tensor g, its only place of occurrence being the Hodge operator $*_g$ (16.4.5). Now we are to express the action integrals of non-gravitational matter in terms of new variables, namely $S_{\text{matter}}[\psi, e, \omega]$. We see that for the fields under consideration it is *enough to express g in terms of e* and set

$$S_{\text{matter}}[\psi, e, \omega] := S_{\text{matter}}[\psi, g(e)]$$

This has two consequences. First, variation with respect to e effectively reduces to variation with respect to g, leading again to the energy–momentum tensor (and thus to the complete Einstein equations).

Secondly, *these* actions *do not depend* at all on ω, so that variation of the *complete* action (sum of the terms corresponding to the gravitational field and to non-gravitational matter) will eventually be the same as the variation of the gravitational field part alone. This means that we again obtain the condition of *vanishing torsion*. Put another way, we end up with exactly *the same results* as we obtained before from the original (Hilbert) action in terms of g rather than e, ω. Let us have a look at these two points in more detail.

16.5.10 Parametrize the variation of the coframe field by a symmetric matrix $S_{ab}(x)$ and an antisymmetric matrix $A_{ab}(x)$

$$\begin{aligned}e^a\mapsto e^a+\epsilon f^a&=e^a+\epsilon B^a_{\ b}(x)e^b=e^a+\epsilon\eta^{ab}\big(B_{(bc)}+B_{[bc]}\big)e^c\\&\equiv e^a+\epsilon\eta^{ab}(S_{bc}(x)+A_{bc}(x))e^c\end{aligned}$$

Check that

(i) the resulting change of the metric tensor $g(e)$ reads

$$g\mapsto g+\epsilon h\qquad h_{ab}=2S_{ab}$$

so that the antisymmetric part $A_{ab}(x)$ of the general matrix $B(x)$ only generates transitions to other *orthonormal* coframe fields[365] (so that $g = g(e)$ does not change; these transformations are called *local Lorentz transformations*) and the symmetric part $S_{ab}(x)$ generates the variations which *indeed alter* g

[365] They are orthonormal in the sense of the *initial* metric $g \equiv g(e)$. Recall that *any* new coframe field $e + \epsilon f$ is *by definition* regarded as orthonormal in the sense of the *new* metric. Any antisymmetric matrix A_{ab} generates fields $e^a + \epsilon f^a$, which remain orthonormal with *no change* of g.

(ii) for actions of the type discussed above there holds

$$S_{\text{matter}}[\psi, e + \epsilon f, \omega] = S_{\text{matter}}[\psi, e, \omega] + \epsilon \int_{\mathcal{D}} f^a \wedge (- * \tau_a) \qquad \tau_a := T_{ba} e^b$$

where T_{ab} is the energy–momentum tensor

(iii) for the total action

$$S[e, \omega, \psi] = S_C[e, \omega] + S_{\text{matter}}[\psi, e, \omega]$$

we then have

$$S[e + \epsilon f, \omega, \psi] = S[e, \omega, \psi] + \epsilon \int_{\mathcal{D}} f^a \wedge * \left(\frac{1}{8\pi} G_a - \tau_a \right)$$

so that the extremum yields the equation of motion

$$G_a = 8\pi \tau_a \qquad \text{i.e.} \quad G_{ab} = 8\pi T_{ab}$$

Hint: (i) $g \mapsto \eta_{ab}(e^a + \epsilon f^a) \otimes (e^b + \epsilon f^b) = \cdots = g + \epsilon (B_{ab} + B_{ba}) e^a \otimes e^b \overset{!}{=} g + \epsilon h$ (this is clear also without any computation since we know that the matrices $(1 + \epsilon B)$ for anti-symmetric (ηB) turn out to be (pseudo-)orthogonal, so that only the *symmetric* part of (ηB) is able to alter g);

(ii)

$$\begin{aligned} S_{\text{matter}}[\psi, e + \epsilon f, \omega] &= S_{\text{matter}}[\psi, g(e + \epsilon f)] = S_{\text{matter}}[\psi, g + \epsilon 2S] \\ &= S_{\text{matter}}[\psi, g] - \epsilon \int_{\mathcal{D}} \frac{1}{2} 2S^{ab} T_{ab} \omega_{g(e)} \end{aligned}$$

At the same time $f^a \wedge * \tau_a = B^a{}_b e^b \wedge T_{ca} * e^c = B^a{}_b T_{ca} e^b \wedge * e^c = B^a{}_b T_{ca} \eta^{bc} \omega_g = B^{ab} T_{ba} \omega_g = S^{ab} T_{ab} \omega_g$. □

There are also fields, however, for which this comes out *differently*. Namely, for *spinor fields* the action contains explicitly[366] the connection forms ω, thus *contributing* non-trivially to the variation with respect to ω; this results in *non-vanishing torsion* $T^a \neq 0$ (Einstein–Cartan theory then indeed differs from the Einstein–Hilbert theory). Things become more complicated: Einstein as well as energy–momentum tensors turn out to be neither symmetric nor "conserved" (they have non-vanishing covariant divergence), etc.

16.6* Non-linear sigma models and harmonic maps

• In mathematical physics a *non-linear sigma model* denotes a field theory in which it is a "*non-linear field*" that is the subject of a dynamics. This is simply a *map*

$$f : M \to N$$

between two manifolds, where the "target manifold" N *fails to be* a linear space. Such maps, unlike the majority of common fields in physics (most often tensor fields, possibly with values in linear spaces), *do not* constitute a linear space.

[366] Its kinetic term turns out to be proportional to $(\bar{\psi} \gamma^a i_{e_a} \mathcal{D}\psi) \omega_{g(e)} \equiv (\bar{\psi} \gamma_a \mathcal{D}\psi) \wedge *_{g(e)} e^a$ and in $\mathcal{D}\psi \equiv d\psi + \rho'(\omega)\psi = \cdots$ is hidden ω. An explicit expression for $\mathcal{D}\psi$ will be derived in (22.5.1).

The dynamics of a field f is introduced by postulating an appropriate action integral. A valuable hint for the possible structure of the action may be provided by convenient rewriting of two particular actions we have already encountered previously. The first one is the kinetic term of the action of an ordinary (linear) *scalar field* (16.3.7) and the second source of inspiration is the action for a *free* motion[367] ($U = 0$) in Lagrangian *mechanics* with a finite number of degrees of freedom (3.2.9).

16.6.1 Consider the kinetic term of the action of a (real) scalar field (16.3.7) on a Riemannian manifold (M, g),

$$S[\phi; g] := \frac{1}{2} \int_{\mathcal{U}} (\partial_\mu \phi)(\partial^\mu \phi)\omega_g \equiv \int_{\mathcal{U}} g^{\mu\nu}(\partial_\mu \phi)(\partial_\nu \phi)\omega_g$$

Check that

(i) if ϕ is regarded as a (global Cartesian) *coordinate* (function) on a "target" space $\mathbb{R}[\phi]$ and the scalar field as a *map* $f : M[x^\mu] \to \mathbb{R}[\phi]$, then $\phi(x)$ is the coordinate presentation of the map (and at the same time the pull-back $f^*\phi$ of the coordinate function ϕ on M)

(ii) the expression

$$\partial_\mu \phi \equiv J^1_\mu$$

is actually the *Jacobian matrix* of the map f, so that the action may also be written as

$$S[f; g] := \frac{1}{2} \int_{\mathcal{U}} g^{\mu\nu} J^1_\mu J^1_\nu \omega_g$$

(iii) if we introduce the ordinary Euclidean metric on the target $\mathbb{R}[\phi]$

$$h := d\phi \otimes d\phi \qquad \text{i.e.} \quad h_{11}(\phi) := 1$$

the action takes the form

$$S[f; g, h] := \frac{1}{2} \int_{\mathcal{U}} g^{\mu\nu}(f^*h)_{\mu\nu}\omega_g$$

Hint: (iii) $g^{\mu\nu} J^1_\mu J^1_\nu = g^{\mu\nu} J^1_\mu J^1_\nu h_{11} = g^{\mu\nu}(f^*h)_{\mu\nu}$. □

• Under the integral sign the expression of the structure $A^{\mu\nu} B_{\mu\nu}$ occurred which reminds us of the scalar product of *forms* $\alpha \wedge *_g \beta \sim \alpha^{\mu\ldots\nu}\beta_{\mu\ldots\nu}\omega_g$ introduced in (5.8.4) and (8.3.1). The (only) difference consists in the fact that our tensors happen to be *symmetric*, so that the expression under the integral sign containing their scalar product

$$A \cdot B \equiv (A, B)_g := \frac{1}{2} A^{\mu\nu} B_{\mu\nu} \equiv \frac{1}{2} g^{\mu\alpha} g^{\nu\beta} A_{\alpha\beta} B_{\mu\nu}$$

cannot be written in terms of the exterior product and the star operator, but only in terms of components of the tensors.

16.6.2 Check that the kinetic term in the action of a real scalar field

$$f : (M, g) \to (\mathbb{R}^1, h)$$

[367] This functional corresponds at the same time (as we know from (15.4.4)) to geodesics of the RLC connection.

may be written in terms of the scalar product of tensors in a simple and transparent form

$$S[f; g, h] := \int_{\mathcal{U}} (g, f^*h)_g \omega_g$$

□

• We see that the action simply represents a volume integral (over a domain on a manifold M) of the scalar product of an initial metric tensor g on M with the pull-back f^*h of the metric tensor h from $\mathbb{R}$ to M. Now, let us have a look at the second particular action integral mentioned above.

16.6.3 Consider the standard kinetic term of the action integral of a mechanical system with a finite number of degrees of freedom (3.2.9)

$$S[\gamma; h] := \frac{1}{2}\int_{t_1}^{t_2} h(\dot{\gamma}, \dot{\gamma})\, dt \equiv \frac{1}{2}\int_{t_1}^{t_2} h_{ab}\dot{q}^a\dot{q}^b dt$$

Check that

(i) if the curve γ is regarded (in accordance with an official definition) as a map $\gamma : \mathbb{R}[t] \to M[q^a]$, then $\dot{q}^a \equiv J_1^a$ is actually the Jacobian matrix of the map γ, so that the action may also be written as

$$S[\gamma; h] := \frac{1}{2}\int_{t_1}^{t_2} J_1^a J_1^b h_{ab}\, dt$$

(ii) if we introduce the ordinary Euclidean metric on the "parameter" real axis $\mathbb{R}[t]$

$$g := dt \otimes dt \qquad \text{i.e.} \quad g_{11}(t) := 1$$

then the action takes the form

$$S[\gamma; g, h] := \frac{1}{2}\int_{\mathcal{U}} g^{\mu\nu}(f^*h)_{\mu\nu}\omega_g \equiv \int_{\mathcal{U}} (g, f^*h)_g \omega_g \qquad \mathcal{U} = \langle t_1, t_2\rangle$$

so that the action considered here and that of the scalar field from (16.6.2) happen to be precisely alike, not merely having a close resemblance as would two peas in a pod. □

• Compare the results we obtained so far:

scalar field: $f : (M, g) \to (\mathbb{R}^1, h)$ $\quad S[f; g, h] = \int_{\mathcal{U}} (g, f^*h)_g \omega_g \quad h \equiv d\phi \otimes d\phi$

mechanics: $\gamma : (\mathbb{R}^1, g) \to (N, h)$ $\quad S[\gamma; g, h] = \int_{\mathcal{U}} (g, f^*h)_g \omega_g \quad g \equiv dt \otimes dt$

The form of the action, shared by the two systems under consideration, strongly suggests a way that we might easily generalize the concept to maps between *arbitrary Riemannian* manifolds

$$f : (M, g) \to (N, h)$$

It suffices to replace the particular manifold $\mathbb{R}$ by a general Riemannian manifold *and leave* the form of the action derived in (16.6.2) and (16.6.3) *unchanged.*

16.6.4 Consider a map f between two Riemannian manifolds

$$f : (M, g) \to (N, h)$$

and introduce as an action integral the expression[368]

$$S[f; g, h] := \int_{\mathcal{U}} (g, f^* h)_g \omega_g$$

Check that

(i) if x^μ, y^a are coordinates on M and N respectively, then the coordinate presentation of the action reads

$$S[f; g, h] \leftrightarrow S[y^a(x), y^a_{,\mu}(x), x] = \frac{1}{2}\int_{\mathcal{U}} g^{\mu\nu} \frac{\partial y^a}{\partial x^\mu}\frac{\partial y^b}{\partial x^\nu} h_{ab}\sqrt{|g|}\, d^m x \equiv \frac{1}{2}\int_{\mathcal{U}} g^{\mu\nu} y^a_{,\mu} y^b_{,\nu} h_{ab}\omega_g$$

(ii) this action may also be rewritten in terms of differential forms and the standard scalar product of forms

$$S[f; g, h] \leftrightarrow S[y^a, dy^a, x] = \frac{1}{2}\int_{\mathcal{U}} dy^a \wedge *_g h_{ab}\, dy^b \equiv \frac{1}{2}\langle dy^a, h_{ab}\, dy^b\rangle$$

Note that in this expression only the coordinates y^a on N take part, whereas it is "coordinate-free" with respect to M[369]

(iii) the condition for the extremum of this action with respect to $f \leftrightarrow y^a(x)$ gives the equations of motion

$$\Delta_g y^a + (dy^b, dy^c)_g \Gamma^a_{bc} = 0$$

where Γ^a_{bc} are the Christoffel symbols of the RLC connection corresponding to h

(iv) so that if the RLC connection for the manifold (N, h) happens to be *flat*, the fields $y^a(x)$ (for a suitable choice[370] of y^a on N) turn out to be *harmonic functions* on M

(v) in two particular cases which served as sources of inspiration concerning the form of the action this general equation of motion indeed reduces to the equations to be expected, i.e. the massless Klein–Gordon equation (16.3.7) and the equation of a geodesic (15.4.1)

$$\Box_g \phi = 0 \quad \text{and} \quad \ddot{y}^a + \Gamma^a_{bc}\dot{y}^a\dot{y}^b = 0$$

Hint: (iii) if $y^a(x) \mapsto y^a(x) + \epsilon z^a(x)$ (z^a being arbitrary, vanishing on the boundary ∂U), then

$$S[y] \mapsto S[y + \epsilon z] \equiv \frac{1}{2}\langle d(y^a + \epsilon z^a), h_{ab}(y + \epsilon z) d(y^b + \epsilon z^a)\rangle = S[y] + \epsilon\alpha$$

[368] It is interpreted as a functional of f; the metric tensors g, h are given, so that they may be regarded as "parameters."

[369] Coordinates y^a may be regarded in this context as a "multiplet" of *scalar fields* on M. Actually they are, as we have already mentioned, *pull-backs* $y^a(x) \equiv f^* y^a$ of the coordinate functions on M, i.e. *functions on* M. In the same way (as pull-back) one should also understand h_{ab} – it depends on x through y, $h_{ab}(y(x))$.

[370] Namely for ("Cartesian") coordinates *adapted* to the flat connection, i.e. such that $\Gamma^a_{bc} = 0$.

where

$$\alpha = \langle dz^a, h_{ab}\, dy^b\rangle + \frac{1}{2}\langle dy^a, h_{ab,c} z^c\, dy^b\rangle$$
$$= \langle z^a, \delta_g(h_{ab}\, dy^b)\rangle + \frac{1}{2}\langle z^a, (dy^b, dy^c)_g h_{bc,a}\rangle \equiv \langle z^a, \delta_g(h_{ab}\, dy^b) + \frac{1}{2}(dy^b, dy^c)_g h_{bc,a}\rangle$$

Vanishing of the first variation (that means $\alpha \stackrel{!}{=} 0$ for arbitrary z^a) thus amounts to the condition $\delta_g(h_{ab}\, dy^b) + \frac{1}{2}(dy^b, dy^c)_g h_{bc,a} = 0$. Since the codifferential leads to

$$\delta_g(h_{ab}\, dy^b) \equiv *^{-1}\, d * \hat{\eta}(h_{ab}\, dy^b) = - *^{-1}\, d(h_{ab} * dy^b) = \cdots$$
$$= -h_{ab,c}(dy^b, dy^c) - h_{ab}\Delta y^b$$

we may eventually write

$$0 = (dy^b, dy^c)\left(\frac{1}{2}h_{bc,a} - h_{a(b,c)}\right) - h_{ab}\Delta y^b \equiv -h_{ab}\left(\Delta y^b + \Gamma^b_{cd}(dy^c, dy^d)\right)$$

□

• Maps of Riemannian manifolds $f : (M, g) \to (N, h)$ which extremize the action presented above (i.e. which satisfy local differential equations $\Delta y^a + \Gamma^a_{bc}(dy^a, dy^b) = 0$) are called in the mathematically oriented literature *harmonic maps* and they have numerous applications. We saw, for example, that a (real massless) scalar field (obeying equations of motion) as well as a curve describing the uniform motion along a straight line on (N, h) (a geodesic) realize particular harmonic maps (some further examples will be mentioned later on).

Now contemplate symmetries of the action. Since two *Riemannian* manifolds are involved, we may expect that groups of automorphisms of the Riemannian structures might be relevant, i.e. groups of *isometries* of the manifolds. Then we should not be overly surprised to hear that the action actually has the symmetry $G_1 \times G_2$ (with G_1 and G_2 being the isometry groups of (M, g) and (N, h) respectively).

16.6.5 Let $f : M \to N$ be a non-linear field and let $\psi : M \to M$ and $\chi : N \to N$ be diffeomorphisms. By composition we get a new non-linear field (map $M \to N$)

$$M \xrightarrow{\psi} M \xrightarrow{f} N \xrightarrow{\chi} N$$

In particular, by composition with *isometries* of the corresponding Riemannian manifolds we get a right action of the direct product $G_1 \times G_2$ of the groups of isometries on the space of fields

$$f \mapsto R_{(k_1,k_2)} f \equiv R_{k_2} \circ f \circ R^{-1}_{k_1} \qquad (k_1, k_2) \in G_1 \times G_2$$

Check that

(i) the following identity holds:

$$\psi^*((A, B)_g \omega_g) = (\psi^* A, \psi^* B)_{\psi^* g}\, \omega_{\psi^* g}$$

(ii) for the action integral under consideration

$$S[f] \equiv S[f; g, h, \mathcal{U}] := \int_{\mathcal{U}} (g, f^*h)_g \omega_g$$

we may write

$$S[\chi \circ f \circ \psi; g, h, \mathcal{U}] = S[f; (\psi^{-1})^* g, \chi^* h, \psi(\mathcal{U})]$$

(iii) in particular, the action integral remains *invariant* with respect to the action of the group $G_1 \times G_2$

$$S[R_{(k_1,k_2)} f; g, h, M] = S[f; g, h, M)]$$

Hint: (i) $(A, B)_g = CC\sharp_g \sharp_g A \otimes B, \omega_g = *_g 1$, see (3.1.7) and (8.3.8). □

• Up to now, we have been considering both metric tensors in the action as given (i.e. as "parameters"). Now we look at a variation procedure with respect to the metric tensor g on M (so that we regard g as an additional "variable"). In doing so we generalize the action a bit (the reason becomes clear at the end of the computations) by adding a "cosmological term" $\sim$ constant $\int \omega_g$: so let us consider from now on the action integral

$$S[f; g, h, \lambda] := \int_{\mathcal{U}} \{(g, f^*h)_g + \lambda/2\}\omega_g \equiv \frac{1}{2}\int_{\mathcal{U}} \{g^{\mu\nu}(f^*h)_{\mu\nu} + \lambda\}\omega_g \equiv \int_{\mathcal{U}} L(g, \hat{G}, \lambda)\omega_g$$

where the notation $\hat{G} \equiv f^*h$ was introduced and $\lambda \in \mathbb{R}$ is so far arbitrary.

16.6.6 Consider a variation $g_{\mu\nu} \mapsto g_{\mu\nu} + \epsilon\sigma_{\mu\nu}, \sigma_{\mu\nu} = \sigma_{\nu\mu}$ in the action integral

$$S[g, \hat{G}, \lambda] := \int_{\mathcal{U}} L(g, \hat{G}, \lambda)\omega_g \qquad L(g, \hat{G}, \lambda) \equiv \frac{1}{2}\{g^{\mu\nu}\hat{G}_{\mu\nu} + \lambda\} \qquad \hat{G} \equiv f^*h$$

Check that

(i) under the variation of g the variation of the form under the integral sign reads

$$L(g + \epsilon\sigma, \hat{G}, \lambda)\omega_{g+\epsilon\sigma} = L(g, \hat{G}, \lambda)\omega_g + 2\epsilon(\sigma, Lg - \hat{G})_g \omega_g$$

so that the critical $g = \hat{g}$ is given by the condition

$$\hat{L}\hat{g} = \hat{G} \quad \text{i.e.} \quad \hat{L}\hat{g}_{\mu\nu} = \hat{G}_{\mu\nu} \qquad \hat{L} \equiv L(\hat{g}, \hat{G}, \lambda)$$

Note that this is an *algebraic* equation (a "constraint" rather than a differential equation) for $\hat{g}$

(ii) a necessary condition for obeying this relation is non-degeneracy of $\hat{G}$ and consequently also $\dim M \leq \dim N$

(iii) the equation $\hat{L}\hat{g} = \hat{G}$ leads to the condition

$$(2 - m)\hat{L} = \lambda \qquad m \equiv \dim M$$

Thus the requirement of existence of critical $g = \hat{g}$ selects for $m = 2$ the action with $\lambda = 0$ and for $m \neq 2$ in turn $\lambda \neq 0$

(iv) for $m = 2$ the solution of the equation $\hat{L}\hat{g} = \hat{G}$ consists of exactly the *whole conformal class* given by the representative $\hat{G}$, i.e. it contains all metric tensors of the form $\hat{g} = \chi\hat{G}$, where χ is an *arbitrary* nowhere vanishing function; for $m \neq 2$ we have the solution

$$\hat{g} = \hat{L}^{-1}\hat{G} \equiv \frac{\lambda}{2-m}\hat{G}$$

and, in particular, for the choice $\lambda = 2 - m$ we have the simplest solution[371] $\hat{g} = \hat{G}$.

Hint: (i) $g^{\mu\nu} \mapsto g^{\mu\nu} - \epsilon\sigma^{\mu\nu}$ (where $\sigma^{\mu\nu} = g^{\mu\alpha}g^{\nu\beta}\sigma_{\alpha\beta}$), $\omega_g \mapsto (1 + \epsilon\frac{1}{2}\sigma^{\mu}_{\mu})\omega_g$; (ii) $\hat{g}$ is a metric tensor, so that *it is* non-degenerate; (iii) multiply by $g^{\mu\nu}$ the equation $\hat{L}\hat{g}_{\mu\nu} = \hat{G}_{\mu\nu}$. □

• Now let us investigate what we obtain if the *critical* $\hat{g}$ is *inserted back* into the action from which it was obtained (the action then becomes a functional of a "single variable," the map f, as it was before). We already set the parameter λ to be $\lambda = 2 - m$ (since this turned out to be the most convenient choice).

16.6.7 Let $\hat{G}$ be non-degenerate (i.e. it is also a *metric tensor* on M). Consider the action from (16.6.6) for $\lambda = 2 - m$, i.e.

$$S[g, \hat{G}] \equiv S[g, \hat{G}, 2-m] \equiv \int_{\mathcal{U}} L(g, \hat{G}, 2-m)\omega_g \equiv \frac{1}{2}\int_{\mathcal{U}} \{g^{\mu\nu}\hat{G}_{\mu\nu} + (2-m)\}\omega_g$$

where $m \equiv \dim M = 1, 2, \dots$. Check that if we replace g with the critical[372] value $g = \hat{g}$, we get the action[373]

$$S[f; h] \equiv S[\hat{g}, \hat{G}] \equiv \int_{\mathcal{U}} \hat{L}\omega_{\hat{g}} = \int_{\mathcal{U}} \omega_{\hat{G}}$$

i.e.

$$S[f; h] = \int_{\mathcal{U}} \omega_{f^*h}$$

Hint: according to (16.6.6) for $m \neq 2$ we have $\hat{L} = 1$ and $\hat{g} = \hat{G}$; for $m = 2$ with the help of (5.8.3) we have

$$\hat{L}\omega_{\hat{g}} = (\chi\hat{G}, \hat{G})_{\chi\hat{G}}\omega_{\chi\hat{G}} = \frac{1}{2}\chi^{-1}\hat{G}^{\mu\nu}\hat{G}_{\mu\nu}\chi\omega_{\hat{G}} = \frac{1}{2}\chi^{-1}\delta^{\mu}_{\mu}\chi\omega_{\hat{G}} = \omega_{\hat{G}}$$

so in both cases there holds $\hat{L}\omega_{\hat{g}} = \omega_{\hat{G}}$. □

• However, the functional given by the last integral already has a clear geometrical meaning: it represents the *volume of the domain* $\mathcal{U}$ in the sense of the metric tensor f^*h *induced* on M from (N, h) or, alternatively[374] the volume of the *image* $f(\mathcal{U})$ in the manifold N with respect to the metric tensor h (restricted to $f(\mathcal{U})$). So we end up with the good old

[371] Note that $\hat{g} = \hat{G}$ is also a solution for the choice $\lambda = 2 - m$ *in the case* $m = 2$.
[372] For $m = 2$ we are free to choose *any representative* of the conformal class given by the tensor $\hat{G}$.
[373] This action is to be regarded as a functional of a *map* f alone, keeping the metric tensor h on the target manifold N fixed.
[374] For injective f.

functionals of length, area, etc., expressed technically by "square root" Lagrangians (7.7.5),

$$\text{length of } \gamma = \int \sqrt{h_{ab}\dot{y}^a \dot{y}^b}\, dt \equiv \int \sqrt{h(\dot{\gamma},\dot{\gamma})}\, dt$$

$$\text{area of } S = \int \sqrt{\det\left(h_{ab} y^a{}_{,\mu} y^b{}_{,\nu}\right)}\, dx^1 \wedge dx^2 \qquad y^a{}_{,\mu} \equiv \frac{\partial y^a}{\partial x^\mu}$$

etc.

which are used for finding the shortest paths, minimal surfaces, etc. What is then the point of the result of problem (16.6.7)? What did we actually learn from it? The problem showed that the functionals for finding the shortest paths, minimal surfaces, etc. may also be written in a different way than we knew up to now. Namely, instead of $S[f;h]$ one takes the functional $S[f;g,h]$, which corresponds to a non-linear *sigma model* and which contains an "auxiliary" metric tensor g on M. If we now regard $S[f;g,h]$ as a functional of "*two* variables" f, g and if we, in addition to f (as we intended to do for $S[f;h]$), vary it *also with respect to* g, we achieve by means of *both variations* the same effect as we would achieve by means of a variation of the initial action $S[f;h]$ with respect to f *alone*.

Well, and what might then be the use of such an "extension" of the problem (which looks more like a complication than a simplification)? The reason is that this is a cheap way (with a minimum grant-in-aid) of how to *get rid of the square root* under the integral sign in the action.[375] Let us have a look explicitly at the particular cases $\dim M = 1$ (minimal curves) and $\dim M = 2$ (minimal surfaces).

16.6.8 Be sure to understand that the particular case $\dim M = 1$ of the scheme mentioned above enables one to get rid of the square root in the action (16.3.9) corresponding to a charged particle[376] (m, e) which moves in a space-time (N, h) in an external electromagnetic field given by a potential A; in particular, check that

(i) if we consider the (quadratic) action

$$S[\gamma,\beta;A] := -\frac{m}{2}\int_{\tau_1}^{\tau_2}\left(\frac{h(\dot{\gamma},\dot{\gamma})}{\beta} + \beta\right)d\tau + S_{\text{int}}[\gamma;A]$$
$$\equiv -\frac{m}{2}\int_{\tau_1}^{\tau_2}\left(\frac{h_{ab}\dot{y}^a\dot{y}^b}{\beta} + \beta\right)d\tau - e\int_{\tau_1}^{\tau_2}\langle A,\dot{\gamma}\rangle\, d\tau$$

then its variation with respect to an "auxiliary function" $\beta(\tau)$ plus inserting the critical value $\hat{\beta}(\tau)$ back leads to the original (square root) action (16.3.9)

$$S[\gamma;A] := -m\int_{\tau_1}^{\tau_2}\sqrt{h(\dot{\gamma},\dot{\gamma})}\, d\tau + S_{\text{int}}[\gamma;A] \equiv -m\int_{\tau_1}^{\tau_2}\sqrt{h_{ab}\dot{y}^a\dot{y}^b}\, d\tau - e\int_{\tau_1}^{\tau_2}\langle A,\dot{\gamma}\rangle\, d\tau$$

[375] This turns out to be very desirable for a *quantization* of the theory (one obtains a nice *quadratic* action by this trick); we will not pursue this topic at all in this book.

[376] Here m *does not* denote the dimension of the manifold M, but rather the *mass* of the charged particle.

(ii) the new action is also *reparametrization invariant*: namely β *transforms* under reparametrization according to the prescription

$$\tau \mapsto \tau'(\tau) \quad \Rightarrow \quad \beta \mapsto \beta' = \frac{d\tau}{d\tau'}\beta$$

Hint: (i) if $g = \beta(\tau)\,d\tau \otimes \beta(\tau)\,d\tau \equiv e^1 \otimes e^1$, then $\omega_g = e^1$ and $((g, f^*h)_g + \frac{1}{2})\omega_g$ reduces to $\frac{1}{2}(h_{ab}\dot{y}^a\dot{y}^b/\beta + \beta)\,d\tau$; (ii) the "function" $\beta(\tau)$ is actually a *component of a 1-form*[377] e^1 with respect to the coordinate basis $d\tau$; then the transformation rule under a change of the coordinate follows immediately ($\beta\,d\tau \stackrel{!}{=} \beta'\,d\tau'$ should hold). □

- The case of a *two*-dimensional manifold M (two-dimensional minimal surfaces in (N, h)) turns out to be of particular interest for two groups of the civilian population.

The first group is represented by small children, who are fascinated by *soap bubbles* while playing in a bathtub. The surface tension forces the bubble to take a form with minimum area under the given additional conditions; these conditions may be realized, say, by a wire rim, to which the boundary of the bubble should be attached or by the pressure of the air enclosed by the bubble (if it is attached to nothing and hovers in the form of S^2 featherlight in the air). Some children continue with this fascination until adulthood, they write complicated papers and (complicated) monographs, in which they do not hesitate to attack the (complicated) problems of the theory of soap bubbles using "heavy artillery" of differential geometry and algebraic topology.

16.6.9 Consider a soap bubble whose shape may be expressed in terms of a function $u(x, y)$ in a domain $\mathcal{U}$ in the xy-plane. Check that the requirement of minimality of area of the bubble leads to a (non-linear) second-order partial differential equation for u

$$(1 + u_{,y}^2)u_{,xx} + (1 + u_{,x}^2)u_{,yy} - 2u_{,x}u_{,y}u_{,xy} = 0 \qquad (x, y) \in \mathcal{D}$$
$$u|_{\partial\mathcal{D}} = \Phi$$

The function Φ, which is defined on the boundary $\partial\mathcal{D}$ (and serves as a boundary condition for the differential equation), carries information about the form of the wire rim on which the bubble was formed.

Hint: if we induce a metric tensor on $\mathcal{D}$ from $E^3 \equiv (\mathbb{R}^3, h)$ by means of the map

$$f : \mathcal{D} \to E^3 \qquad (x, y) \mapsto (x, y, z = u(x, y))$$

we get

$$f^*h = (1 + u_{,x}^2)\,dx \otimes dx + \left(1 + u_{,y}^2\right) dy \otimes dy + u_{,x}u_{,y}(dx \otimes dy + dy \otimes dx)$$
$$\omega_{f^*h} = \sqrt{1 + u_{,x}^2 + u_{,y}^2}\,dx \wedge dy$$

[377] According to the general definition $e^a = e^a_\mu\,dx^\mu$ the function β is often called the *vielbein field* ($e^1 = e^1_\tau(\tau)\,d\tau \equiv \beta\,d\tau$), although a defence of the word *viel* (many) in this (*one*-dimensional) case might be hard work even for an experienced lawyer.

so that we are to write down the Euler–Lagrange equation

$$\partial_x \frac{\partial L}{\partial u_{,x}} + \partial_y \frac{\partial L}{\partial u_{,y}} = \frac{\partial L}{\partial u} \qquad L = \sqrt{1 + u_{,x}^2 + u_{,y}^2}$$

An alternative way is to use the "quadratic" technique instead of the "square root" formalism, i.e. to use the equation $\Delta_g y^a + \Gamma^a_{bc}(dy^a, dy^b)_g = 0$ from (16.6.4): we set $\Gamma^a_{bc} = 0, g = \hat{g} \equiv f^*h$ and write down the Laplace equation $\Delta_{f^*h} u(x, y) = 0$ (we are to compute $d * du = 0$; it is a bit laborious, since although we are in the ordinary xy-plane, we have there a *non-standard* metric tensor f^*h); we check (inserting $u \mapsto x,\ u \mapsto y$) that the equations $\Delta_{f^*h} x = 0$ and $\Delta_{f^*h} y = 0$ *are* satisfied as well. □

- Other children continuously diffuse in full age from being fascinated by soap bubbles to being (even more) fascinated by *string theory*; this theory tries to reach the ambitious goal of explaining all the physics in the universe from a minimal number of first principles. Instead of considering a *world-line* $y^a(\tau)$ of a point particle it introduces a *world-sheet* $y^a(\tau, \sigma)$ of a (one-dimensional) string. The action of the string[378] is introduced by a natural modification of the action of a point particle. Instead of the extremal *length* of a world-line (in the sense of the space-time (N, h)) *Nambu and Goto* proposed a principle of least action for the string, which requires extremal *area* of the world-sheet. The square root present in the action causes, as we have already mentioned before, technical problems; making use of the trick (16.6.7) based on an auxiliary metric g on M it was then rewritten in a "quadratic" form by *Polyakov*; the two actions are then proportional to the integrals

$$\int_{\mathcal{U}} \sqrt{\det\left(h_{ab} y^a_{,\mu} y^b_{,\nu}\right)}\, d\tau \wedge d\sigma \qquad \text{Nambu, Goto}$$

$$\frac{1}{2} \int_{\mathcal{U}} g^{\mu\nu} y^a_{,\mu} y^b_{,\nu} h_{ab} \sqrt{|g|}\, d\tau \wedge d\sigma \qquad \text{Polyakov}$$

($(x^1, x^2) \equiv (\tau, \sigma)$). The reader anxious to learn more about strings is recommended to read, just to start somewhere (best this very day!), several thousands of papers, waiting patiently in the electronic preprint library at the site http://arxiv.org/.

Summary of Chapter 16

The (4-)tensor version of Maxwell's equations in Minkowski space(-time) reveals that the tensors involved are rather special – they may actually be regarded as *differential forms*. That is why the most natural way of formulating four-dimensional electrodynamics is provided by the language of differential forms. Forms in Minkowski space exhibit additional particular structure (as a consequence of the splitting of the space-time into "time" and "space"): one can express any form (in an observer-dependent way) in terms of a pair of *spatial* forms. Such an expression of forms (as well as of operations on them) offers a convenient

[378] Actually the actions discussed here correspond (only) to the "bosonic" string; for a "superstring" one considers on the world-sheet additional odd ("anticommuting") variables (in the sense of $\mathbb{Z}_2$-grading of supermathematics, see the end of Section 5.3) and the action has a "local supersymmetry," i.e. a local symmetry mixing even and odd variables.

bridge between a four-dimensional and the (original) three-dimensional formulation of electrodynamics. Forms are not only useful in electrodynamics, but rather in field theory in general. The *action integrals* are simply expressed (since the objects under the integral sign are always forms) and their extrema, providing the equations of motion, are simply computed, too (the codifferential appears naturally). There is a deep link between the space-time symmetries and the *energy–momentum tensor* of the field, which may be defined via variation of the action functional with respect to the metric tensor. The energy–momentum tensor of matter occurs (as a source) in the Einstein equations of the gravitational field, too. Both the Hilbert and Cartan approaches to the derivation of the Einstein equations from a variational principle are discussed. In the former approach, the metric tensor is the key independent field variable (with respect to which small variations are to be performed); the latter approach makes use of (co)frame (tetrad) fields and connection forms. In non-linear sigma models mappings of two Riemannian manifolds are regarded as field variables. There is a natural action integral for such mappings. Harmonic maps are extremals of this action. They correspond to "minimal surfaces," representing, for example, soap bubbles, but also the world-sheets in string theory. There is a technical trick enabling one to get rid of a "square root" action by means of a variation of the "quadratic" one with respect to one of two metric tensors (then called "auxiliary").

$\alpha = dt \wedge \hat{s} + \hat{r}$	Decomposition of forms in Minkowski space	(16.1.1)
$d\alpha = dt \wedge (\partial_t \hat{r} - \hat{d}\hat{s}) + \hat{d}\hat{r}$	Action of d on a decomposed form	(16.1.4)
$*\alpha = dt \wedge (\hat{*}\hat{r}) + \hat{*}\hat{\eta}\hat{s}$	Action of the Hodge star $*$ on a decomposed form	(16.1.5)
$\delta\alpha = dt \wedge (\hat{\delta}\hat{s}) + (-\partial_t \hat{s} - \hat{\delta}\hat{r})$	Action of the codifferential δ on a decomposed form	(16.1.6)
$F := dt \wedge \mathbf{E} \cdot d\mathbf{r} - \mathbf{B} \cdot d\mathbf{S}$	2-form of the electromagnetic field	(16.2.1)
$j = \rho dt - \mathbf{j} \cdot d\mathbf{r} \equiv j_\mu dx^\mu$	1-form of current	(16.2.2)
$\delta F = -j, \ \ dF = 0$	Maxwell's equations	(16.2.1, 2)
$F = dA$	A is a potential for F	(16.3.1)
$-\frac{1}{2}\langle dA, dA\rangle - \langle A, j\rangle$	Action integral $S[A]$ for an electromagnetic field	(16.3.2)
$\frac{1}{2}\langle d\phi, d\phi\rangle - (m^2/2)\langle \phi, \phi\rangle$	Action integral $S[\phi]$ for a free scalar field	(16.3.7)
$T^{\mu\nu}{}_{;\nu} = 0$	Energy–momentum tensor is divergence-free	(16.4.1)
$R_{ab} - \frac{1}{2}Rg_{ab} = 8\pi T_{ab}$	Einstein equations	Sec. 16.5

17

Differential geometry on TM and T^*M

• In this chapter we begin the part of the book in which the concept of the *fiber bundle* enters the story. Bundles play a significant role in modern geometry and their language as well as techniques are widely used in modern theoretical and mathematical physics. That is why the strategy of ignoring them, although in principle possible, would be fairly short-sighted. In the forthcoming two chapters we will look in some detail at two particular bundles closely associated with Lagrangian and Hamiltonian mechanics, the tangent and cotangent bundle. These (as well as numerous further) bundles may be *canonically* constructed for an arbitrary manifold M and one can find a fairly rich geometry on them, resulting "free of charge" directly from the way they are defined. Moreover, this additional geometrical structure turns out to be just what is needed for the formulation of the two versions (Lagrangian and Hamiltonian) of classical mechanics. Later on (in Chapter 19) another bundle will be introduced, which may be canonically assigned to an arbitrary manifold M, the *frame bundle*. It provides a novel view of a linear connection on M. Generalizing the three bundles we will then introduce the concepts of the *principal G-bundle* and the associated bundle, which turn out to be the essential ingredients needed for the development of the theory of connections and gauge fields.

From a didactic point of view it is convenient to begin to study some particular bundle and to notice its relevant features, which then enter the official abstract definition of the concept of a bundle. The two particular bundles which suit our purpose well are the tangent and cotangent bundles. Their points have a simple visual meaning and they are closely related to analytical mechanics.

17.1 Tangent bundle TM and cotangent bundle T^*M

• Let M be a smooth manifold and let T_xM be the tangent space at a point $x \in M$. Define (for the moment only) the set TM as the collection (union) of all tangent spaces at all points of M

$$TM := \bigcup_{x \in M} T_xM$$

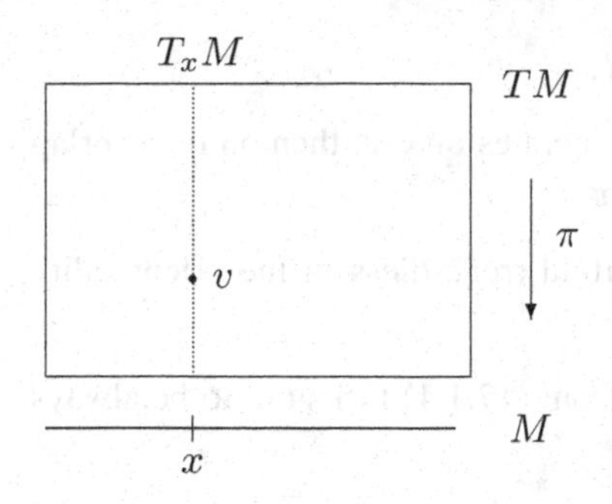

i.e. all vectors at all points $x \in M$ are regarded as points of a new set TM. If we assign to a vector $v \in T_xM$ its point of tangency x, we get a surjective map called the *canonical projection*

$$\pi : TM \to M \qquad T_xM \ni v \mapsto x$$

It turns out that the set TM has the natural structure of a smooth manifold. A convenient atlas on TM may be introduced making use of the atlas on the manifold M. Let x^a be local coordinates in a neighborhood $\mathcal{O}$ of a point x, i.e. let

$$\psi : \mathcal{O} \to \mathbb{R}^n[x^1, \dots, x^n]$$

be a chart. Consider the domain

$$\hat{\mathcal{O}} := \pi^{-1}(\mathcal{O}) \subset TM$$

Then one can introduce on $\hat{\mathcal{O}}$ *canonical coordinates* as follows: if $v \in \hat{\mathcal{O}} \Rightarrow v \in T_xM$ for some $x \in \mathcal{O}$ then

$$v = v^a \left.\frac{\partial}{\partial x^a}\right|_x \qquad (v^1, \dots, v^n) \in \mathbb{R}^n$$

Then it is clear that the $2n$-tuple of numbers $(x^1, \dots, x^n, v^1, \dots, v^n)$ uniquely corresponds to a point $v \in \hat{\mathcal{O}}$ ($(x^1, \dots, x^n)$ shows where the vector v resides and $(v^1, \dots, v^n)$ provides its decomposition with respect to a coordinate basis in T_xM). Put another way,

$$\hat{\psi} : \hat{\mathcal{O}} \to \mathbb{R}^{2n}[x^1, \dots, x^n, v^1, \dots, v^n]$$

is a chart on $\hat{\mathcal{O}} \subset TM$, which is induced by the chart ψ on $\mathcal{O} \subset TM$. If $\{\mathcal{O}_\alpha, \psi_\alpha\}$ is an atlas on M, $\{\hat{\mathcal{O}}_\alpha, \hat{\psi}_\alpha\}$ is an atlas on TM.

17.1.1 Check that a change of coordinates on M

$$x^a \mapsto x'^a(x)$$

induces the change of (canonical) coordinates

$$(x^a, v^a) \mapsto (x'^a(x), J^a_b(x)v^b) \qquad J^a_b \equiv \frac{\partial x'^a}{\partial x^b}$$

on TM, i.e.

$$x'^a(x, v) = x'^a(x)$$
$$v'^a(x, v) = J^a_b(x)v^b$$

Hint: $v = v^a \partial_a = v'^a \partial'_a$. □

17.1.2 Check that $\{\hat{\mathcal{O}}_\alpha, \hat{\psi}_\alpha\}$ is a smooth atlas on TM.

Hint: if (x^a, v^a) operates on $\hat{\mathcal{O}}_\alpha$ and (x'^a, v'^a) in turn operates on $\hat{\mathcal{O}}_\beta$, then on the overlap (17.1.1) holds. Justify the smoothness of these relations. □

17.1.3 Prove that TM is always an orientable manifold (regardless of the orientability of M).

Hint: show that the Jacobian of a coordinate change from (17.1.1) turns out to be always positive

$$\hat{J} = \frac{\partial(x', v')}{\partial(x, v)} = \cdots = J^2 > 0$$

so that $(\hat{\mathcal{O}}_\alpha, \hat{\psi}_\alpha)$ is an oriented atlas regardless of whether $(\mathcal{O}_\alpha, \psi_\alpha)$ was oriented. □

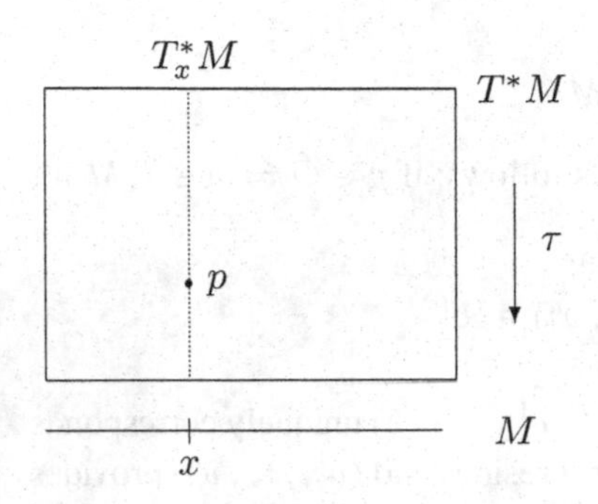

• In a very similar way another important manifold, denoted by T^*M, is introduced. As a set it is the collection (union) of all *cotangent* spaces at all points of M

$$T^*M := \bigcup_{x \in M} T_x^*M$$

i.e. all *covectors* at all points $x \in M$ are regarded as points of the set T^*M. In this case we will denote the corresponding canonical projection by the letter τ

$$\tau : T^*M \to M \qquad T_x^*M \ni p \mapsto x$$

An atlas on M naturally induces an atlas on T^*M. If $p \in T_x^*M$ and if its decomposition with respect to the coordinate basis reads

$$p = p_a \, dx^a|_x \qquad (p_1, \ldots, p_n) \in \mathbb{R}^n$$

then the $2n$-tuple $(x^1, \ldots, x^n, p_1, \ldots, p_n)$ uniquely corresponds to a point $p \in \hat{\mathcal{O}} \equiv \tau^{-1}(\mathcal{O})$, so that

$$\hat{\psi} : \hat{\mathcal{O}} \to \mathbb{R}^{2n}[x^1, \ldots, x^n, p_1, \ldots, p_n]$$

is a chart on $\hat{\mathcal{O}} \subset T^*M$ induced by the chart ψ on $\mathcal{O} \subset M$ (*canonical coordinates* on T^*M).

17.1.4 Check that a change of coordinates on M

$$x^a \mapsto x'^a(x)$$

induces the change of (canonical) coordinates on T^*M,

$$(x^a, p_a) \mapsto \left(x'^a(x), (J^{-1})^b_a(x) p_b\right) \qquad \text{i.e.} \qquad \begin{aligned} x'^a(x, p) &= x'^a(x) \\ p'_a(x, p) &= (J^{-1})^b_a(x) p_b \end{aligned}$$

Hint: $p = p_a \, dx^a = p'_a \, dx'^a$. □

17.1.5 Check that $\{\hat{\mathcal{O}}_\alpha, \hat{\psi}_\alpha\}$ represents a smooth atlas on T^*M.

Hint: see (17.1.2) and (17.1.4). □

17.1.6 Prove that T^*M is always an orientable manifold (regardless of the orientability of M).

Hint: see (17.1.3). □

17.1.7 Check that for the projections π and τ the following formulas hold:

(i)

$$\pi : (x^a, v^a) \mapsto x^a \qquad \tau : (x^a, p_a) \mapsto x^a$$

(ii)

$$\pi_*(\partial/\partial x^a) = \partial/\partial x^a \qquad \pi_*(\partial/\partial v^a) = 0$$

(iii)

$$\tau_*(\partial/\partial x^a) = \partial/\partial x^a \qquad \tau_*(\partial/\partial p_a) = 0$$

Note: realize that actually there are *three sorts* of vectors $\partial/\partial x^a$ in these formulas: those on TM, T^*M and M. □

17.1.8 Show that the map χ defined below is a diffeomorphism

$$\chi : \pi^{-1}(\mathcal{O}) \to \mathcal{O} \times \mathbb{R}^n[v^1, \dots, v^n] \qquad v \mapsto (\pi(v), (v^1, \dots, v^n))$$

and that it moreover satisfies

$$\pi_1 \circ \chi = \pi$$

□

17.1.9 Let M be a part of the plane $\mathbb{R}^2$, in which both Cartesian coordinates (x, y) and polar coordinates (r, φ) operate. Then on TM two sets of canonical coordinates emerge, (x, y, p_x, p_y) and $(r, \varphi, p_r, p_\varphi)$. Check that they are then related by[379]

$$x = r\cos\varphi \quad p_x = p_r \cos\varphi - p_\varphi \frac{\sin\varphi}{r} \qquad r = \sqrt{x^2 + y^2} \quad r p_r = x p_x + y p_y$$

$$y = r\sin\varphi \quad p_y = p_r \sin\varphi + p_\varphi \frac{\cos\varphi}{r} \qquad \varphi = \arctan \tfrac{y}{x} \quad p_\varphi = x p_y - y p_x$$

Hint: according to (17.1.4) $p_x\, dx + p_y\, dy = p_r\, dr + p_\varphi\, d\varphi$. □

• The manifolds TM and T^*M which we introduced in this section represent part (total spaces) of structures called the *tangent bundle* and the *cotangent bundle*. A motivation for this terminology should be clear in the following section (see (17.2.5)), where we explain what is a bundle in general.

[379] From the future development of these ideas we will see that these formulas may be regarded as transformational relations between *canonical momenta* in *phase* space corresponding to the transformations of coordinates in *configuration* space (they are called *point transformations* in textbooks on analytical mechanics).

17.2 Concept of a fiber bundle

• The manifolds TM and T^*M provide paradigmatic examples of a useful object in modern differential geometry, a so-called fiber bundle. Note that in both cases (TM as well as T^*M) we have at each point x of the manifold M quasi-hidden another manifold, namely the vector space T_xM or T_x^*M, so that both "hidden" manifolds are diffeomorphic to $\mathbb{R}^n$. A natural generalization then would be to study a situation in which we "paste" at each point $x \in M$ a manifold F_x, all of the manifolds being diffeomorphic to a common manifold F (i.e. if $x, x' \in M$, then $F_x \sim F_{x'} \sim F$). The manifold F is called a *typical fiber*, F_x is the *fiber over a point* x, M is the *base* and

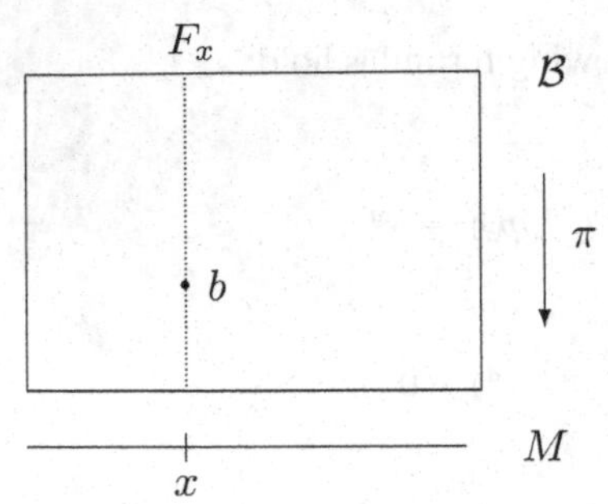

$$\mathcal{B} := \bigcup_{x \in M} F_x$$

is the *total space*. All of these elements, when taken together, constitute a structure called a *fiber bundle* (also a *fibered manifold* or a *bundle*). The concept of a fiber bundle actually is comprised of *two* manifolds $\mathcal{B}$ and M and *a surjective map*

$$\pi : \mathcal{B} \to M$$

(called the *canonical projection*). All the preimages $F_x \equiv \pi^{-1}(x)$ are required to be diffeomorphic to a common manifold F and in addition each F_x is to be a submanifold in $\mathcal{B}$ (so it is to be "nicely placed" in $\mathcal{B}$). The last item of the definition is the requirement of *local product structure*: there exists a covering $\mathcal{O}_\alpha$ of the base M and a system of diffeomorphisms

$$\psi_\alpha : \pi^{-1}(\mathcal{O}_\alpha) \to \mathcal{O}_\alpha \times F$$

(the map ψ_α is called a *local trivialization*) such that

$$\pi_1 \circ \psi_\alpha = \pi$$

(see problem (17.1.8)). An alternative notation for a fiber bundle $\pi : \mathcal{B} \to M$, fairly often used in the literature, is $(\mathcal{B}, M, \pi, F)$.

The simplest fiber bundle is a *product bundle*, where the total space is simply the Cartesian product of the base and the fiber and the projection is realized as the projection onto the first factor of the product.

17.2.1 Check that

$$\pi_1 : M \times F \to M$$

is indeed a fiber bundle. □

• Fiber bundles are often mapped to one another. Clearly those maps of total spaces which preserve the structure of fibers (if b_1, b_2 are in the same fiber, the same is true for their images) are distinguished. They are formalized in the concept of a *bundle map* (or a *fibered map*): let

$$\pi : \mathcal{B} \to M \qquad \pi' : \mathcal{B}' \to M'$$

be two bundles. Then a bundle map is actually a *pair* of maps $f, \hat{f}$, for which the following diagram commutes:

$$\begin{array}{ccc} \mathcal{B} & \xrightarrow{f} & \mathcal{B}' \\ \pi \downarrow & & \downarrow \pi' \\ M & \xrightarrow[\hat{f}]{} & M' \end{array} \qquad \text{i.e. for which} \quad \pi' \circ f = \hat{f} \circ \pi$$

For *equivalent bundles* f should be a diffeomorphism and $\hat{f}$ the identity map. Finally, a *trivial bundle* is equivalent to a product bundle: a diffeomorphism

$$f : \mathcal{B} \to M \times F$$

exists which obeys

$$\pi_1 \circ f = \pi$$

Such a map f is called a *global trivialization.*

In terms of these concepts one can say something about the structure of a general bundle. We see from the definition that on sufficiently small pieces (on $\mathcal{O}_\alpha$) any bundle is *trivial*: the restriction to

$$\pi : \pi^{-1}(\mathcal{O}_\alpha) \to \mathcal{O}_\alpha$$

is (for each α) trivial (ψ_α is a "global trivialization" of this piece), but the pieces are glued together in such a way that the bundle itself

$$\pi : \mathcal{B} \to M$$

need not be trivial. One usually says that fiber bundles are in general only *locally trivial.* This means that the knowledge of a base M and a typical fiber F is not enough to reproduce the global structure of a bundle; in general, there may exist several essentially (i.e. topologically) different ways of gluing the pieces $\mathcal{O}_\alpha \times F$ into a single whole $\mathcal{B}$.

17.2.2 The surface of a cylinder may be regarded as $\mathcal{B} \equiv S^1 \times I,\ I = \langle a, b \rangle$, i.e. as the total space of the product bundle $\pi : S^1 \times I \to S^1$. Check that even though the Möbius band M^2 (6.3.2) shares the base (a circle) and the typical fiber (a line) with the surface of the cylinder, the whole $\mathcal{B}' \equiv M^2$ is glued differently and a global trivialization evidently cannot exist. □

• The concept of a fiber bundle thus may be regarded as a generalization of the concept of a Cartesian product. It turns out that the generalization is highly non-trivial, with numerous important applications in various branches of mathematics and mathematical physics. In this book, however, we will restrict ourselves to *differential geometry* on (some particular) fibered manifolds and we will not discuss *topological* aspects of these manifolds.

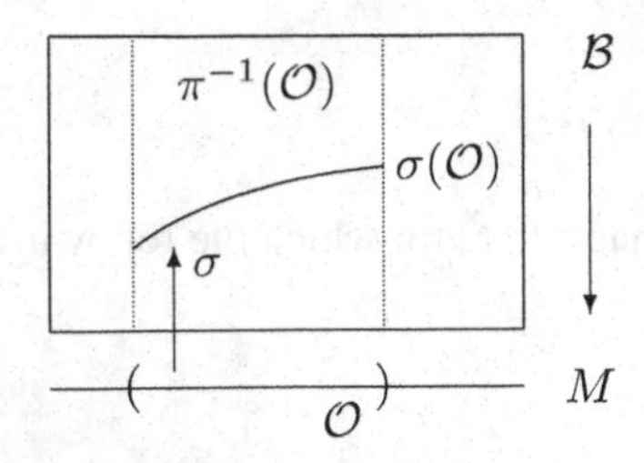

The next important concept is that of a *local section* of a bundle $\pi : \mathcal{B} \to M$. It is a smooth map

$$\sigma : \mathcal{O} \to \mathcal{B} \qquad \mathcal{O} \subset M$$

such that

$$\pi \circ \sigma = id_{\mathcal{O}}$$

If $\mathcal{O}$ = the whole M, the section is said to be global.

17.2.3 Check that a section maps a point from M to "its own" fiber (so that x is mapped to the fiber $\pi^{-1}(x)$). □

17.2.4 Be sure to understand that sections of a product bundle $\pi : M \times F \to M$ are in one-to-one correspondence with maps from M to F.

Hint: see (1.4.12). □

• What is the point in distinguishing between local and global sections (in making life difficult for considering local sections)? Well, it turns out that whereas local sections always exist (for sufficiently small domains on the base), a global section may sometimes not exist at all (for example, in the case of *principal* bundles, playing a key role in connection theory, the existence of a global section turns out to be equivalent to triviality of the bundle, see (20.1.3)).

Bundles we encounter in real-life situations are mostly endowed with an additional structure in fibers, like that of a linear or homogeneous space. A *vector bundle* is a fibered manifold which has the structure of a linear space in each fiber, i.e. for each $x \in M$ a linear combination

$$b_1(x) + \lambda b_2(x) \qquad b_1(x), b_2 \in \pi^{-1}(x), \quad \lambda \in \mathbb{R}$$

is defined.

17.2.5 Check that both of the objects we introduced in Section 17.1,

$$\pi : TM \to M \qquad \text{tangent bundle}$$
$$\tau : T^*M \to M \qquad \text{cotangent bundle}$$

are vector bundles with base M and standard fiber $F = \mathbb{R}^n$.

Hint: see (1.4.12). □

17.2.6 Be sure to understand that local sections

$$\sigma : \mathcal{O} \to TM \qquad \text{and} \qquad \tilde{\sigma} : \mathcal{O} \to T^*M$$

are in one-to-one correspondence with vector and covector *fields on* $\mathcal{O} \subset M$ respectively. □

17.2.7 Given local coordinates x^i on $\mathcal{O} \subset M$, consider a vector field V and a covector field α on $\mathcal{O}$

$$V = V^i(x)\partial_i \qquad \text{and} \qquad \alpha = \alpha_i(x)\,dx^i$$

Check that the coordinate presentation of the corresponding sections (in the sense of (17.2.6)) reads

$$\sigma : x^i \mapsto (x^i(x), v^i(x)) = (x^i, V^i(x)) \qquad \tilde{\sigma} : x^i \mapsto (x^i(x), p_i(x)) = (x^i, \alpha_i(x))$$

□

17.2.8 Consider a vector bundle $\pi : \mathcal{B} \to M$. Check that local sections over a domain $\mathcal{U}$ constitute an $\mathcal{F}(\mathcal{U})$-module, i.e. one can perform linear combinations with coefficients being smooth *functions* on $\mathcal{U}$ (if σ_1, σ_2 are sections and f is a smooth function, then $\sigma_1 + f\sigma_2$ is a section as well). Which fact (known since long ago) gives this for TM and T^*M?

Hint: see (2.2.12). □

17.3 The maps Tf and T^*f

- Let

$$f : M \to N$$

be a (smooth) map of manifolds and let

$$\pi_M : TM \to M \qquad \pi_N : TN \to N$$

be the corresponding tangent bundles of M and N respectively. We learned in Section 3.1 that the map f induces a map of vectors

$$f_* : T_x M \to T_{f(x)} N$$

Vectors on M may be, however, regarded as points on TM. Consequently a further map[380]

$$Tf : TM \to TN$$

is induced, given by the prescription

$$(Tf)(v) := f_* v$$

Here, v on the left is to be interpreted as a *point* on TM whereas on the right as the corresponding *vector* in the point $x \in M$. The map Tf may therefore be regarded as the collection of *all* f_* maps (for all $x \in M$).

[380] It is a kind of *lift* of the map f from the base M to the total space TM in the sense of Section 17.5.

17.3.1 Check that Tf "closes" the commutative diagram

$$\begin{array}{ccc} TM & \xrightarrow{Tf} & TN \\ \pi_M \downarrow & & \downarrow \pi_N \\ M & \xrightarrow[f]{} & N \end{array} \qquad \text{i.e. that} \quad f \circ \pi_M = \pi_N \circ Tf$$

□

17.3.2 Check that for the composition of maps one has the simple formula

$$T(f \circ g) = Tf \circ Tg$$

Hint: see (3.1.2). □

17.3.3 Check that the map Tf is a *morphism of vector bundles*, i.e. it preserves fibers and is linear on them. □

• These properties of Tf say that T itself may be regarded as a *covariant functor* (see Appendix A.6) from the category of smooth manifolds (the objects being smooth manifolds, morphisms smooth maps between them) to the category of vector bundles (where objects are vector bundles and morphisms are those bundle maps which are linear on fibers). The *tangent functor* T thus assigns to a manifold M the vector bundle TM ("its" tangent bundle) and to a map $f : M \to N$ the map $Tf : TM \to TN$ (the lift of f to the tangent bundle).

In full analogy a similar map is also defined for T^*M: if

$$\tau_M \; : T^*M \to M \qquad \tau_N \; : T^*N \to N$$

are cotangent bundles and $f : M \to N$ is an *injective* map, we define a map

$$T^*f \; : T^*N \to T^*M$$

by

$$(T^*f)(\alpha) = f^*\alpha$$

f^* being the pull-back[381] of the covector α.

17.3.4 Check the validity of the commutative diagram

$$\begin{array}{ccc} T^*M & \xleftarrow{T^*f} & T^*N \\ \tau_M \downarrow & & \downarrow \tau_N \\ M & \xrightarrow[f]{} & N \end{array} \qquad \text{i.e.} \quad f \circ \tau_M \circ T^*f = \tau_N$$

□

[381] On the left α is a point from T^*N, on the right it is the corresponding covector in $f(x) \in N$. If f fails to be surjective, T^*f is only a map from $T^*(f(M)) \subset T^*N$; if f failed to be injective, T^*f would not be a map at all (a single point would be "mapped" to several images).

17.3.5 Check that for a composition of maps one has a simple formula

$$T^*(f \circ g) = T^* g \circ T^* f$$

Hint: see (3.1.4). □

17.3.6 Check that $T^* f$ is a morphism of vector bundles.

Hint: see (17.3.3). □

17.3.7 Let (x^i, v^i) and (y^a, w^a) be canonical coordinates on TM and TN, (x^i, p_i) and (y^a, q_a) canonical coordinates on T^*M and T^*N and consider a map f with a coordinate presentation

$$f : M \to N \qquad x^i \mapsto y^a(x^1, \dots, x^m) \qquad i = 1, \dots, m \quad a = 1, \dots, n$$

Check that then the coordinate presentations of Tf and $T^* f$ respectively read

$$Tf : TM \to TN \qquad (x^i, v^i) \mapsto (y^a(x, v), w^a(x, v)) = \big(y^a(x), J^a_i(x) v^i\big)$$
$$T^* f : T^*N \to T^*M \qquad (y^a(x), q_a) \mapsto (x^i(y, q), p_i(y, q)) = \big(x^i, J^a_i(x) q_a\big)$$

and if f happens to be a diffeomorphism ($\Rightarrow$ there exists f^{-1} and then also $T^* f^{-1}$)

$$T^* f^{-1} : T^*M \to T^*N \qquad (x^i, p_i) \mapsto (y^a(x, p), q_a(x, p)) = \big(y^a(x), (J^{-1})^i_a p_i\big)$$

Note: we actually used these formulas in the particular case of $M = N = \mathcal{O}_\alpha \cap \mathcal{O}_\beta$ and f = a change of coordinates. □

17.4 Vertical subspace, vertical vectors

• Let $\pi : \mathcal{B} \to M$ be a fiber bundle. The very existence of the projection π singles out in the tangent space of any point $b \in \mathcal{B}$ a *vertical subspace*

$$\text{Ver}_b \mathcal{B} \leq T_b \mathcal{B} \qquad \qquad \text{Ver}_b := \text{Ker}\, \pi_{*b}$$

A vector $w \in T_b\mathcal{B}$ is thus said to be *vertical* if it projects (via π_*) to zero. Visually this means that w is tangent to a fiber, i.e. an arbitrary curve γ which represents the vector w ($[\gamma] \equiv W$) goes from b along the fiber in which b resides.

Since the canonical coordinates (x^i, v^i) and (x^i, p_i) on TM and T^*M respectively are adapted to the structure of the fibration, one should not be too surprised that vertical vectors may be easily identified in these coordinates.

17.4.1 Check that the most general vertical vector fields on TM and T^*M respectively read

$$V = V^i(x, v) \frac{\partial}{\partial v^i} \qquad \text{and} \qquad W = W_i(x, p) \frac{\partial}{\partial p_i}$$

Hint: see (17.1.7). □

• Problems (17.1.7) and (17.4.1) show that vertical subspaces on TM and T^*M respectively span the vectors $(\frac{\partial}{\partial v^1}, \dots, \frac{\partial}{\partial v^n})$ or $(\frac{\partial}{\partial p_1}, \dots, \frac{\partial}{\partial p_n})$. It might seem from these computations that in a similar way we can define a "horizontal" subspace to be the span of $(\frac{\partial}{\partial x^1}, \dots, \frac{\partial}{\partial x^n})$. A computation shows, however, that this is only an illusion.

17.4.2 Check that the "horizontal" subspace discussed above is (contrary to the vertical one) non-canonical, i.e. it depends on the choice of coordinates x^i.

Hint: Check that if $x^i \mapsto x'^i(x)$, then the vectors $\frac{\partial}{\partial x'^i}$ also contain terms $\frac{\partial}{\partial v^i}$, whereas $\frac{\partial}{\partial v'^i}$ do not contain $\frac{\partial}{\partial x^i}$. □

• This makes the (canonical) decomposition of a vector $w \in T_v TM$ into vertical and horizontal parts impossible (i.e. one cannot even separate in a coordinate-independent way the *vertical part* of the vector; only the statement that the *complete vector* is vertical makes sense).[382]

17.4.3 Imitating $TM \equiv T_0^1 M$ and $T^*M \equiv T_1^0 M$ related to vectors and covectors on M, describe the bundle $T_1^1 M$ which is related to tensors of type $\binom{1}{1}$ on M. Construct a smooth atlas (induced by an atlas on M), find a typical fiber F as well as the dimension of the total space, write down explicitly (in canonical coordinates) the projection and a general *vertical* vector field. □

17.5 Lifts on TM and T^*M

• Within the context of fibre bundles a *lift* is in general a procedure which assigns to a geometrical object on the base M a geometrical object (possibly of another type) on the total space $\mathcal{B}$ of the bundle $\pi : \mathcal{B} \to M$. In this section we will introduce some lifts from M to TM as well as to T^*M which will prove to be useful later.

We begin with a lift of curves from M to TM. Let

$$\gamma : \mathbb{R} \to M \qquad t \mapsto \gamma(t)$$

be a curve on M. Then the curve

$$\hat{\gamma} : \mathbb{R} \to TM \qquad t \mapsto \dot{\gamma}(t)$$

is called the *natural lift of the curve* γ from M to TM.

17.5.1 Check that

(i) the lifted curve $\hat{\gamma}$ is always exactly "over" the curve γ

$$\pi \circ \hat{\gamma} = \gamma$$

[382] The above-mentioned decomposition is possible provided a *linear connection* is given on M (the ambitious reader is invited to find a way in which this may then be achieved); here we study TM for a "bare" manifold M.

(ii) if $x^a(t)$ is the coordinate presentation of the curve γ, then the coordinate presentation of the lifted curve is $(x^a(t), \dot{x}^a(t))$

$$\gamma(t) \leftrightarrow x^a(t)$$
$$\hat{\gamma}(t) \leftrightarrow (x^a(t), v^a(t)) = (x^a(t), \dot{x}^a(t))$$

• The Lagrangian function in analytical mechanics is introduced as a function $L(q^i(t), \dot{q}^i(t))$, i.e. a function of generalized coordinates q^i and generalized velocities $\dot{q}^i(t)$. From the perspective of TM, the most natural geometrical interpretation is as follows: there is a Lagrangian L as a *function on* TM

$$L : TM \to \mathbb{R}$$

and this function is, in turn, evaluated on the natural lift $\hat{\gamma}$ of the trajectory γ in the configuration space M:

$$L(t) := L(\hat{\gamma}(t))$$

Now consider a vector u on M, $u \in T_xM \equiv \pi^{-1}(x)$. We may associate a curve in the fiber $\pi^{-1}(x)$ over x with this vector

$$\sigma(t) := v + tu \qquad v \in \pi^{-1}(x)$$

The tangent vector at zero of the curve is a vector at the point $v \in TM$

$$u^{\uparrow} := \dot{\sigma}(0) \equiv \left.\frac{d}{dt}\right|_0 (v + tu)$$

This vector is called a *vertical lift* of the vector u to the point $v \in TM$.

17.5.2 Check that

(i) the coordinate presentation of the curve $\sigma(t)$ reads

$$x^a(t) = x^a \qquad v^a(t) = v^a + tu^a$$

(ii) the resulting vector $u^{\uparrow}$ is *vertical*

$$u^{\uparrow} \equiv \dot{\sigma}(0) \in \text{Ver}_v TM \leq T_v TM$$

(iii) the coordinate expression of the resulting vector is

$$u = u^a \frac{\partial}{\partial x^a}, \qquad v = v^a \frac{\partial}{\partial x^a} \qquad \Rightarrow \qquad u^{\uparrow} = u^a \frac{\partial}{\partial v^a}$$

Hint: (iii) $(u^{\uparrow}\phi)(x^a, v^a) := \left.\frac{d}{dt}\right|_0 \phi(x^a(t), v^a(t)) = \cdots$. □

• Note that a single vector u may be lifted in this way to *each* point in the fiber π^{-1} over x, giving rise to a (vertical) vector *field* defined on the fiber. If there is a vector *field* $u = u^a(x)\partial_a$ available on M, the vertical lift (to each point of $\pi^{-1}(x)$, for all $x \in M$) generates a vector *field* on TM, which is called the *vertical lift of the field* u.

17.5.3 Check that

(i) in coordinates the operation of the vertical lift consists in

$$u = u^a(x)\frac{\partial}{\partial x^a} \quad \Rightarrow \quad u^\uparrow = u^a(x)\frac{\partial}{\partial v^a}$$

(ii) the operation of the vertical lift of fields

$$()^\uparrow : \mathfrak{X}(M) \to \mathfrak{X}(TM)$$

is an $\mathcal{F}(M)$-linear map $\Rightarrow$ it is enough to know it on basis vector fields

(iii) on a coordinate basis field the map reads

$$\frac{\partial}{\partial x^a} \mapsto \left(\frac{\partial}{\partial x^a}\right)^\uparrow = \frac{\partial}{\partial v^a}$$

(iv) a general vector field on TM

$$W \equiv A^a(x,v)\frac{\partial}{\partial x^a} + B^a(x,v)\frac{\partial}{\partial v^a}$$

may be regarded as the result of a vertical lift $W = V^\uparrow$ if and only if

$$A^a(x,v) = 0 \qquad B^a(x,v) = B^a(x)$$

• Consider next a $\binom{1}{p}$-type tensor A at $x \in M$. We may assign to A a *vertical lift* $A^\uparrow$, which is a $\binom{1}{p}$-type tensor in $v \in T_xM$:

$$A^\uparrow(U,\dots,V) := (A(\pi_* U,\dots,\pi_* V))^\uparrow \qquad U,\dots,V \in T_vTM$$

($A^\uparrow$ is indeed a tensor since it is in each argument the composition of three linear mappings, $\pi_*, A, (\ )^\uparrow$.) If there is a $\binom{1}{p}$-type tensor *field* available on M, the vertical lift (to each point of $\pi^{-1}(x)$, for all $x \in M$) generates a $\binom{1}{p}$-type tensor *field* $A^\uparrow$ on TM, giving rise to a map[383]

$$(\ )^\uparrow \ : \ T^1_p(M) \to T^1_p(TM)$$

17.5.4 Check that

(i) this map is $\mathcal{F}(M)$-linear in the sense of $(A + fB)^\uparrow = A^\uparrow + (\pi^* f)B^\uparrow$ ($\Rightarrow$ it is enough to know it on basis tensor fields)

(ii) in coordinates the map reads

$$A \equiv A^a_{b\dots c}(x)dx^b \otimes \cdots \otimes dx^c \otimes \frac{\partial}{\partial x^a} \quad \mapsto \quad A^\uparrow = A^a_{b\dots c}(x)\, dx^b \otimes \cdots \otimes dx^c \otimes \frac{\partial}{\partial v^a}$$

(iii) the tensor $A^\uparrow$ is *horizontal*, i.e. it is annihilated by (even a single) *vertical* argument.

Hint: (i) evaluate both sides on (the same) arguments. □

[383] We will actually use this lift only for the case where $p = 1$ (see (17.6.4)).

• The definitions of the lifts of *fields* treated up to now were *pointwise*, i.e. we first defined a lift of an individual object sitting at the point and then "adding" of these lifts produced the lift of the corresponding field. The following (important) example shows that it need not always be so.

Consider a vector field V on M. This field induces an (at least local) flow

$$\Phi_t : M \to M$$

It turns out that the flow ($\Rightarrow$ also its generator V) may be lifted to a flow on TM, making use of the *tangent functor* T (see the text after (17.3.3)):

$$\begin{array}{ccc} TM & \xrightarrow{T\Phi_t} & TM \\ {\scriptstyle \pi_M}\downarrow & & \downarrow{\scriptstyle \pi_M} \\ M & \xrightarrow[\Phi_t]{} & M \end{array}$$

17.5.5 Check that $T\Phi_t$ is indeed an (at least local) flow on TM and that it is "exactly over" the flow Φ_t, i.e. that it obeys

$$\pi \circ T\Phi_t = \Phi_t \circ \pi$$

Hint: apply T to $\Phi_{t+s} = \Phi_t \circ \Phi_s$, (17.3.2). □

• Since $T\Phi_t$ is a flow, it is necessarily generated by some vector field $\tilde{V}$ on TM; the field $\tilde{V}$ is called a *complete lift* of the field V.

17.5.6 Consider a flow on M

$$\Phi_t : M \to M$$

which is generated by a vector field $V \equiv V^a(x)\partial_a$. Check that the coordinate presentation of the infinitesimal flows Φ_ϵ and $T\Phi_\epsilon$ reads

$$\begin{array}{ll} \Phi_\epsilon : & x^a \mapsto x^a(\epsilon) = x^a + \epsilon V^a(x) \\ T\Phi_\epsilon : & (x^a, v^a) \mapsto (x^a(\epsilon), v^a(\epsilon)) = (x^a + \epsilon V^a(x), v^a + \epsilon V^a{}_{,b}(x) v^b) \end{array}$$

Hint: see (17.3.7). □

17.5.7 Check that the coordinate expression of the lifted field $\tilde{V}$ is

$$V = V^a(x)\frac{\partial}{\partial x^a} \quad \Rightarrow \quad \tilde{V} = V^a(x)\frac{\partial}{\partial x^a} + V^a{}_{,b}(x) v^b \frac{\partial}{\partial v^a}$$

Hint: $(\tilde{V}\Phi)(x, v) = \frac{d}{d\epsilon}\big|_0 \Phi(x(\epsilon), v(\epsilon)) = \cdots$; (17.5.6). □

17.5.8 Check that the complete lift, when regarded as a map

$$\tilde{(\ )} : \mathfrak{X}(M) \to \mathfrak{X}(TM) \qquad V \mapsto \tilde{V}$$

is $\mathbb{R}$-linear, but *fails* to be $\mathcal{F}(M)$-linear.[384] □

[384] This is a consequence of the fact that the construction is not pointwise (see the discussion in Section 2.5) – there is no complete lift of an individual vector.

• On T^*M we will discuss two lifts, the vertical lift of a covector and the complete lift of a vector field.

Consider a covector α residing in a point x on M. We may assign to this covector a curve in the fiber $\tau^{-1}(x)$ over x

$$\sigma(t) := p + t\alpha \qquad p \in \tau^{-1}(x)$$

The tangent vector at zero of the curve is a vector in the point $p \in T^*M$

$$\alpha^{\uparrow} := \dot{\sigma}(0) \equiv \left.\frac{d}{dt}\right|_0 (p + t\alpha) \in T_pT^*M$$

This vector is called the *vertical lift of the covector* of the covector α to the point $p \in T^*M$ (the lift of a *covector* is thus a *vector*).

17.5.9 Check that

(i) the coordinate presentation of the curve $\sigma(t)$ is

$$x^a(t) = x^a$$
$$p_a(t) = p_a + t\alpha_a$$

(ii) the resulting vector $\alpha^{\uparrow}$ is *vertical*

$$\alpha^{\uparrow} = \dot{\sigma}(0) \in \mathrm{Ver}_p T^*M \leq T_pT^*M$$

(iii) in coordinates the lift reads

$$\alpha = \alpha_a\, dx^a \quad \Rightarrow \quad \alpha^{\uparrow} = \alpha_a \frac{\partial}{\partial p_a}$$

Hint: (iii) $(\alpha^{\uparrow}\Phi)(x, p) := \frac{d}{dt}\big|_0 \Phi(x^a(t), p_a(t)) = \cdots$. □

• If there is a covector *field* $\alpha = \alpha_a(x)\, dx^a$ available on M, the vertical lift (to each point of $\tau^{-1}(x)$, for all $x \in M$) generates a vector *field* on T^*M, which is called the *vertical lift of the field* α.

17.5.10 Check that

(i) the coordinate formula for lifting a field α is

$$\alpha = \alpha_a(x)\, dx^a \quad \Rightarrow \quad \alpha^{\uparrow} = \alpha_a(x) \frac{\partial}{\partial p_a}$$

(ii) the operation of vertical lift on fields

$$(\ )^{\uparrow} : \mathcal{T}^0_1(M) \to \mathcal{T}^1_0(T^*M) \qquad \alpha \mapsto \alpha^{\uparrow}$$

is an $\mathcal{F}(M)$-linear map $\Rightarrow$ it is enough to know it on basis fields

(iii) on a coordinate basis we have

$$dx^a \mapsto (dx^a)^{\uparrow} = \frac{\partial}{\partial p_a}$$

(iv) a general vector field W on T^*M

$$W \equiv A^a(x,p)\frac{\partial}{\partial x^a} + B_a(x,p)\frac{\partial}{\partial p_a}$$

may be regarded as the result of the vertical lift $W = \alpha^\uparrow$ if and only if

$$A^a(x,p) = 0 \qquad B_a(x,p) = B_a(x)$$

• The procedure of complete lift of a vector field on M to TM may also be repeated with minor modification on T^*M. The difference stems from the fact that T^* is a *contravariant* functor (it reverses arrows). Consequently, we have to use the *inverse* map when lifting the flow: the lift of Φ_t on M is actually $T^*(\Phi_t^{-1}) \equiv T^*\Phi_{-t}$, i.e. the following commutative diagram is used:

$$\begin{array}{ccc} T^*M & \xrightarrow{T^*\Phi_{-t}} & T^*M \\ \tau\downarrow & & \downarrow\tau \\ M & \xrightarrow[\Phi_t]{} & M \end{array}$$

17.5.11 Check that $T^*\Phi_{-t}$ is indeed an (at least local) flow on T^*M and that it is "exactly over" the flow Φ_t (i.e. it obeys $\tau \circ T^*\Phi_{-t} = \Phi_t \circ \tau$).

Hint: apply T^* to $\Phi_{-(t+s)} = \Phi_{-t} \circ \Phi_{-s}$, (17.3.5) and (17.5.5). □

• Since $T^*\Phi_{-t}$ is a flow, it is necessarily generated by some vector field $\widetilde{V}$ on T^*M; the field $\widetilde{V}$ is called the *complete lift* of the field V.

17.5.12 Consider a flow on M

$$\Phi_t : M \to M$$

which is generated by a vector field $V \equiv V^a(x)\partial_a$. Check that the coordinate presentation of the infinitesimal flow $T^*\Phi_{-\epsilon}$ reads

$$T^*\Phi_{-\epsilon} : \ (x^a, p_a) \mapsto (x^a(\epsilon), p_a(\epsilon)) = (x^a + \epsilon V^a(x), p_a - \epsilon V^b{}_{,a}(x)p_b)$$

Hint: see (17.3.7) and (17.5.6). □

17.5.13 Check that the coordinate expression of the lifted field $\widetilde{V}$ is

$$V = V^a(x)\frac{\partial}{\partial x^a} \qquad \Rightarrow \qquad \widetilde{V} = V^a(x)\frac{\partial}{\partial x^a} - V^b{}_{,a}(x)p_b\frac{\partial}{\partial p_a}$$

Hint: see (17.5.7) and (17.5.12). □

17.5.14 Check that the complete lift to T^*M, when regarded as a map

$$\widetilde{(\ )} : \mathfrak{X}(M) \to \mathfrak{X}(T^*M) \qquad V \mapsto \widetilde{V}$$

is $\mathbb{R}$-linear, but *fails* to be $\mathcal{F}(M)$-linear.

Hint: see (17.5.8). □

17.6 Canonical tensor fields on TM and T^*M

• Specific features of total spaces of TM and T^*M (fibration, linearity in fibers and a close relation of the points of the total spaces with the geometry of the base) result in the existence of important *canonical* tensor fields on these manifolds, i.e. fields which may be introduced "objectively" ≡ with no arbitrary choices. In particular, on TM we will encounter a vector field Δ (the *Liouville field* or *dilation field*) and a $\binom{1}{1}$-type tensor field S (vertical endomorphism) and on T^*M it will be (once again) the Liouville field Δ, the canonical 1-form θ and the canonical exact symplectic form $\omega = d\theta$.

17.6.1 Check that

(i) by means of the formulas

$$\Phi_t : \; v \mapsto e^t v \qquad \text{and} \qquad p \mapsto e^t p$$

(canonical) flows on TM and T^*M respectively are introduced

(ii) their coordinate presentations read

$$(x^a(t), v^a(t)) = (x^a, e^t v^a) \qquad \text{and} \qquad (x^a(t), p_a(t)) = (x^a, e^t p_a)$$

(iii) their generators $\Delta \in \mathfrak{X}(TM)$ and $\Delta \in \mathfrak{X}(T^*M)$ are given by coordinate formulas

$$\Delta = v^a \frac{\partial}{\partial v^a} \qquad \text{and} \qquad \Delta = p_a \frac{\partial}{\partial p_a}$$

(iv) the fields Δ are (in both cases) vertical

(v) check the fact that Δ is a canonical field in (canonical) coordinates: if $x^a \mapsto x'^a (\Rightarrow v^a \mapsto v'^a = \cdots)$, then

$$v^a \frac{\partial}{\partial v^a} = v'^a \frac{\partial}{\partial v'^a} \qquad p'_a \frac{\partial}{\partial p'_a} = p_a \frac{\partial}{\partial p_a}$$

□

17.6.2 Consider a tensor field A on TM (or on T^*M) which is *homogeneous of degree k* in fiber coordinates. This means that if we express the field in terms of canonical coordinates and then substitute $(x^a, v^a) \mapsto (x^a, \lambda v^a)$ or $(x^a, p_a) \mapsto (x^a, \lambda p_a)$, we get

$$A \mapsto \lambda^k A$$

As an example, the fields

$$\widetilde{V} = V^a(x) \frac{\partial}{\partial x^a} + V^a{}_{,b}(x) v^b \frac{\partial}{\partial v^a} \qquad \alpha^\uparrow = \alpha_a(x) \frac{\partial}{\partial p_a} \qquad \hat{g} = g_{ab}(x) v^a v^b$$

are homogeneous of degree 0, -1 and 2 respectively. Check that A then satisfies[385]

$$\mathcal{L}_\Delta A = kA$$

[385] This means that the operators $\mathcal{L}_\Delta$ for $\Delta = v^a \partial_{v^a}$ or $\Delta = p_a \partial_{p_a}$ may serve as *measuring instruments* of degree of homogeneity in fiber coordinates.

In particular, for $A = V =$ a vector field and $A = f =$ a function we get

$$[\Delta, V] = kV \qquad \text{and} \qquad \Delta f = kf$$

Hint: first note that $\psi^* A = \lambda^k A$ for $\psi : (x^a, v^a) \mapsto (x^a, \lambda v^a)$ or $(x^a, p_a) \mapsto (x^a, \lambda p_a)$; then set $\lambda = e^t$ and differentiate the equation

$$\Phi_t^* A = e^{kt} A \qquad \Phi_t \leftrightarrow \Delta$$

with respect to t in $t = 0$. □

17.6.3 Let $\tilde{V}$ be the complete lift of a field V from M to TM or T^*M respectively, $V^\uparrow$ its vertical lift to TM and $\alpha^\uparrow$ denote the vertical lift of a 1-form α to T^*M. Check the relations

$$[\Delta, \tilde{V}] = 0 \qquad [\Delta, V^\uparrow] = -\,V^\uparrow \qquad [\Delta, \alpha^\uparrow] = -\alpha^\uparrow$$

Hint: see (17.5.3), (17.5.7), (17.5.10), (17.5.13) and (17.6.2). □

- The next canonical field on TM which we will introduce is a $\binom{1}{1}$-type tensor field which is called a[386] *vertical endomorphism* $S \in \mathcal{T}_1^1(TM)$. It is defined as

$$S := 1^\uparrow$$

where 1 is the identity tensor (of type $\binom{1}{1}$) on M and $(\ )^\uparrow$ denotes the operation of vertical lift from (17.5.4).

17.6.4 Check that

(i) in canonical coordinates the tensor field S reads

$$S = dx^a \otimes \frac{\partial}{\partial v^a}$$

(ii) if S is regarded as a prescription vector $\mapsto$ vector or covector $\mapsto$ covector, then

$$S\left(\frac{\partial}{\partial x^a}\right) = \frac{\partial}{\partial v^a} \qquad S(dx^a) = 0$$

$$S\left(\frac{\partial}{\partial v^a}\right) = 0 \qquad S(dv^a) = dx^a$$

(iii) both the kernel and the image of S_v coincide with the vertical subspace

$$\operatorname{Ker} S_v = \operatorname{Im} S_v = \operatorname{Ver}_v TM$$

(iv) this results in nilpotence of S_v

$$S_v \circ S_v = 0$$

(v) if $\tilde{V}$ and $V^\uparrow$ are the complete and vertical lift of a field $V \in \mathfrak{X}(M)$ respectively, then

$$S(\tilde{V}) = V^\uparrow$$

386 "Endomorphism" expresses the fact that S_v is a linear map $S_v : T_v TM \to T_v TM$, "vertical" in turn means that its image is (only) the vertical subspace (see item (iii) in problem (17.6.4)).

(vi)

$$\mathcal{L}_\triangle S = -S$$

Hint: (v) canonical coordinates, (vi) (17.6.2). □

• The most important canonical object on T^*M is the *canonical 1-form* θ (and its exterior derivative, the *canonical (exact) symplectic form* $\omega = d\theta$). Its pointwise definition reads: let $p \in T^*M,\ W \in T_pT^*M$. Then,

$$\langle \theta, W\rangle := \langle p, \tau_* W\rangle$$

We thus first project the vector W to $x \equiv \tau(p) \in M$ and then insert it into the 1-form $p \in T_x^*M \equiv \tau^{-1}(x)$, which corresponds to the point $p \in T^*M$.

17.6.5 Check that

(i) θ_p is indeed a 1-form in the point $p \in T^*M$, i.e. a linear map $T_pT^*M \to \mathbb{R}$
(ii) in canonical coordinates on T^*M it is

$$\theta = p_a\, dx^a$$

□

Hint: (ii) let $W = A^a\partial/\partial x^a + B^a\partial/\partial p_a$ and $\theta = C_a\, dx^a + D^a dp_a$. Then,

$$\tau_* W = A^a \frac{\partial}{\partial x^a} \Rightarrow \langle p, \tau_* W\rangle = \left\langle p_a\, dx^a, A^b \frac{\partial}{\partial x^b}\right\rangle = \cdots \overset{!}{=} \langle \theta, W\rangle$$

17.6.6 It turns out that the canonical 1-form θ on T^*M may be regarded as the "Platonic eternal Idea" of a differential form on M in the following sense: let α be a 1-form on $\mathcal{O} \subset M$ and let $\sigma : \mathcal{O} \to T^*M$ be the corresponding section of the cotangent bundle $\tau : T^*M \to M$ (17.2.6). Check that

$$\sigma^*\theta = \alpha$$

so that *any* differential form on M may be viewed as the result of an appropriate pull-back of "the 1-form θ" on T^*M. The 1-form θ, living in the "real world of eternal Ideas" T^*M, is then "the Platonic Idea of a differential form" whereas α, living in the "apparent world of material objects" M is its "immersion in the material world." □

17.6.7 Consider the 2-form $\omega = d\theta$. Check that

(i) in canonical coordinates (x^a, p_a) it has automatically the canonical (Darboux) form[387]

$$\omega = dp_a \wedge dx^a$$

(ii) ω turns out to be an *exact symplectic* form on T^*M (so that T^*M is always a symplectic manifold).

Hint: (i) (17.6.5); (ii) non-degeneracy (14.1.2) and (14.3.6). □

[387] Canonical coordinates (x^a, p_a) were tailored to the structure of T^*M itself (not knowing anything about ω then). It turns out that they are free of charge also tailored to the structure of the form ω, i.e. that they happen to be "canonical" in the sense of Darboux's theorem (14.2.1).

17.6.8 Let $f : M \to M$ be a diffeomorphism of M; then $T^* f$ is a diffeomorphism of T^*M. Check that the canonical 1-form θ (hence also the symplectic form ω) is *invariant* with respect to *each* $T^* f$ (for *arbitrary* f)

$$(T^* f)^* \theta = \theta \qquad (\Rightarrow \text{ also } (T^* f)^* \omega = \omega)$$

Solution:

$$\begin{aligned} \langle ((T^* f)^* \theta)_p, w \rangle &= \langle \theta_{f^* p}, (T^* f)_* w \rangle \\ &= \langle f^* p, \tau_* (T^* f)_* w \rangle \\ &= \langle p, f_* \tau_* (T^* f)_* w \rangle \\ &= \ldots (17.3.4) \ldots \\ &= \langle p, \tau_* w \rangle \\ &\equiv \langle \theta_p, w \rangle \end{aligned}$$

□

17.7 Identities between the tensor fields introduced here

• The objects lifted (in various ways) from M to TM or T^*M as well as the canonical tensor fields on TM (T^*M) are related by numerous useful identities. Let us mention explicitly at least the following ones:

17.7.1 Prove the following identities on TM and T^*M:

$$\begin{array}{cccc} TM & T^*M & & \\ [V^\uparrow, W^\uparrow] = 0 & [\alpha^\uparrow, \beta^\uparrow] = 0 & & \\ [\widetilde{V}, \widetilde{W}] = \widetilde{[V, W]} & [\widetilde{V}, \widetilde{W}] = \widetilde{[V, W]} & & \\ [\widetilde{V}, W^\uparrow] = [V, W]^\uparrow & [\widetilde{V}, \alpha^\uparrow] = (\mathcal{L}_V \alpha)^\uparrow & & \\ & & & \\ [\Delta, V^\uparrow] = -V^\uparrow & [\Delta, \alpha^\uparrow] = -\alpha^\uparrow & & \\ [\Delta, \widetilde{V}] = 0 & [\Delta, \widetilde{V}] = 0 & & \\ & & & \\ \mathcal{L}_{W^\uparrow} S = 0 & \mathcal{L}_{\alpha^\uparrow} \theta = \tau^* \alpha & \Rightarrow & \mathcal{L}_{\alpha^\uparrow} \omega = \tau^* d\alpha \\ \mathcal{L}_{\widetilde{W}} S = 0 & \mathcal{L}_{\widetilde{W}} \theta = 0 & \Rightarrow & \mathcal{L}_{\widetilde{W}} \omega = 0 \\ \mathcal{L}_\Delta S = -S & \mathcal{L}_\Delta \theta = \theta & \Rightarrow & \mathcal{L}_\Delta \omega = \omega \end{array}$$

Hint: (for example) canonical coordinates; we have already mentioned some of them (those in which Δ appears) in (17.6.3) and (17.6.4).

Summary of Chapter 17

In this chapter the concept of a fiber bundle is introduced. Rather than develop a general theory at the very beginning, instead we begin with a fairly detailed treatment of two paradigmatic examples of fiber bundles, in order to motivate the definition. Namely, we show that with each manifold M two other manifolds of double the dimension, TM and T^*M, may be canonically associated. Both of them are endowed with a remarkable geometrical structure even if M happens to be just a "bare" smooth manifold. For example, they turn out to represent the total spaces of vector bundles over M and carry various canonical tensor fields (in particular, T^*M always carries a symplectic structure), several objects may be lifted from M to the total spaces, etc. In analytical mechanics they serve as the playing fields for Lagrangian and Hamiltonian formulation of the dynamics respectively; this will be discussed in more detail in the following chapter, this one provides the necessary preliminaries.

$\pi : (x^a, v^a) \mapsto x^a, \quad \tau : (x^a, p_a) \mapsto x^a$	Canonical projections on TM and T^*M	(17.1.7)
$T(f \circ g) = Tf \circ Tg$	A property of the tangent map Tf	(17.3.2)
$\gamma(t) \mapsto \dot{\gamma}(t)$	Natural lift of a curve from M to TM	(17.5.1)
$\Phi_t \mapsto T\Phi_t$	Lift of a flow from M to TM	(17.5.5)
$\Delta = v^a \partial/\partial v^a \;\; (\Delta = p_a \partial/\partial p_a)$	Liouville dilation field on TM (T^*M)	(17.6.1)
$S := 1^\uparrow = dx^a \otimes \partial/\partial v^a$	Vertical endomorphism on TM	(17.6.4)
$\langle \theta_p, W \rangle := \langle p, \tau_* W \rangle$	Canonical 1-form $\theta = p_a\, dx^a$ on T^*M	(17.6.5)
$\omega = d\theta = dp_a \wedge dx^a$	Canonical symplectic form on T^*M	(17.6.7)

18

Hamiltonian and Lagrangian equations

• Now that we already know the ropes concerning TM and T^*M and we are also aware of the essential canonical geometrical objects living on these manifolds, we embark on an examination of how they are related to analytical mechanics. We will learn that *Lagrangian* mechanics may be naturally formulated on TM whereas the *Hamiltonian* formulation turns out to be natural on T^*M. In the case of a regular Lagrangian (or Hamiltonian) a standard relation between the two formulations will be discussed. This is usually based on the Legendre transformation in analytical mechanics, here it will be presented as the Legendre *map* between the two manifolds under consideration.

18.1 Second-order differential equation fields

• Consider a system of ordinary second-order quasi-linear ($\equiv$ linear in the *highest* derivative) autonomous differential equations, which is already solved with respect to terms containing second derivatives, i.e. a system

$$\ddot{x}^a = \Gamma^a(x, \dot{x}) \qquad a = 1, \dots, n$$

By introducing a new variable

$$v^a(t) := \dot{x}^a(t)$$

this may be standardly rewritten as a system of $2n$ first-order equations

$$\dot{x}^a = v^a \qquad \dot{v}^a = \Gamma^a(x, v)$$

We may regard them as the equations for integral curves of the vector field

$$\Gamma = v^a \frac{\partial}{\partial x^a} + \Gamma^a(x, v) \frac{\partial}{\partial v^a}$$

Note that if x^a are treated as (local) coordinates on a manifold M, then it is natural to treat the vector field Γ as living on TM. We begin to understand that also second-order equations are closely related to vector fields, albeit the fields do not live on the same manifold M where the equations do, but rather on the tangent bundle TM of the manifold M. Not all fields on TM are, however, relevant in this context.

[18.1.1] Check that

(i) if

$$W \equiv A^a(x,v)\frac{\partial}{\partial x^a} + B^a(x,v)\frac{\partial}{\partial v^a}$$

is a general vector field on TM, then the constraints on its components imposed by demanding it to be a field of type Γ read

$$A^a(x,v) = v^a \qquad B^a(x,v) = \Gamma^a(x,v)$$

i.e. it restricts the form of the components $A^a(x,v)$ (alone)

(ii) these restrictions may be succinctly characterized by the (coordinate-free) equation

$$S(\Gamma) = \Delta$$

($S =$ the vertical endomorphism, $\Delta =$ the Liouville field).

Hint: see (17.6.1) and (17.6.4). □

[18.1.2] Consider a vector field Γ on TM which satisfies the equation $S(\Gamma) = \Delta$. Check that

(i) each integral curve of the field Γ is the natural lift $\hat{\gamma}$ of a curve γ on M

(ii) the coordinate presentation $x^a(t)$ of the curve γ on M satisfies the system of second-order equations

$$\ddot{x}^a = \Gamma^a(x,\dot{x})$$ □

• Second-order differential equations on M are in one-to-one correspondence with a class of vector fields on TM, namely with fields which satisfy the equation

$$S(\Gamma) = \Delta$$

Such fields are therefore called *second-order differential equation fields.*

[18.1.3] Write down Γ explicitly for the equation of motion of the linear harmonic oscillator and sketch the corresponding integral curves. Be sure to recognize that your drawing basically coincides with a *phase portrait* of the oscillator. □

18.2 Euler–Lagrange field

• A particular class of second-order equations under consideration is given by Lagrange's equations ("of the second kind")

$$\frac{d}{dt}\frac{\partial L}{\partial \dot{x}^a} - \frac{\partial L}{\partial x^a} = 0 \qquad \text{Lagrange's equations}$$

It is known that *not all* second-order equations may be written in this form; necessary and sufficient conditions for the existence of a Lagrangian were already found by (Hermann Ludwig Ferdinand von) Helmholtz.

18.2.1 Find out when (Hermann Ludwig Ferdinand von) Helmholtz lived and estimate then how long the Helmholtz criterion[388] of the existence of a Lagrangian for a given second-order ordinary differential equations has been known.

Hint: see Appendix B or Google. □

18.2.2 Check that if $L(x, \dot{x})$ is a Lagrangian and if

$$A_{ab}(x,\dot{x}) := \frac{\partial^2 L}{\partial \dot{x}^a \partial \dot{x}^b} \qquad B_{ab}(x,\dot{x}) := \frac{\partial^2 L}{\partial \dot{x}^a \partial x^b} \qquad C_a(x,\dot{x}) := \frac{\partial L}{\partial x^a}$$

then Lagrange's equations read

$$\ddot{x}^a = \Gamma^a(x,\dot{x}) \qquad \Gamma^a = -(A^{-1})^{ab} B_{bc}\dot{x}^c + (A^{-1})^{ab} C_b$$

(provided that A^{-1} exists; if it does not, what do they look like?) □

• Now we show that if on TM an appropriate function L is available, one can introduce a symplectic structure there and consequently also Hamiltonian fields and the corresponding dynamics. For an appropriate Hamiltonian the dynamics turns out to coincide with that given by Lagrange's equations on M.

18.2.3 Consider a function L on TM and define *Cartan forms* by

$$\theta_L := S(dL) \qquad \text{Cartan 1-form}$$
$$\omega_L := d\theta_L \qquad \text{Cartan 2-form}$$

where S denotes the *vertical endomorphism* from (17.6.4), regarded as a linear map, which sends 1-forms to 1-forms, $\alpha \mapsto S(\,\cdot\,, \alpha)$. Check that

(i) in canonical coordinates we get

$$\theta_L = \frac{\partial L}{\partial v^a} dx^a \qquad \omega_L = -\frac{\partial^2 L}{\partial v^a \partial v^b} dx^a \wedge dv^b + \frac{\partial^2 L}{\partial x^a \partial v^b} dx^a \wedge dx^b$$

(ii) the form ω_L is closed

$$d\omega_L = 0$$

(iii) we may characterize Lagrangians leading to non-degenerate ω_L by

$$\omega_L \text{ is non-degenerate} \quad \Leftrightarrow \quad \det\left(\frac{\partial^2 L}{\partial v^a \partial v^b}\right) \neq 0$$

(iv) the condition on the Lagrangian from (iii) does not depend on the choice of coordinates x^a on M (it is an intrinsic property of the function L).

[388] For the sake of incompleteness we do not mention an explicit form of the criterion here.

Hint: (iii) recall that non-degeneracy of ω_L is equivalent to $\overbrace{\omega_L \wedge \cdots \wedge \omega_L}^{n \text{ items}} \neq 0$ (5.6.8). Show that

$$\underbrace{\omega_L \wedge \cdots \wedge \omega_L}_{n \text{ items}} \propto \det\left(\frac{\partial^2 L}{\partial v^a \partial v^b}\right) \underbrace{dx^1 \wedge \cdots \wedge dx^n \wedge dv^1 \wedge \cdots \wedge dv^n}_{\neq 0}$$

□

• A Lagrangian which satisfies the condition from item (iii) is called *non-singular* ($\equiv$ *non-degenerate* $\equiv$ *regular*).

18.2.4 Check that

$$L \text{ is non-singular} \quad \Leftrightarrow \quad \omega_L \text{ is a symplectic form}$$

Hint: see (18.2.3). □

• Thus any *regular* Lagrangian L makes a *symplectic manifold* (TM, ω_L) from TM. Contrary to T^*M, which is a symplectic manifold *by itself* (needing no structure to be added), TM becomes a symplectic manifold only in combination with the appropriate Lagrangian L.

18.2.5 Check that a standard Lagrangian encountered in analytical mechanics

$$L = T - \mathcal{U} \equiv \frac{1}{2} g_{ab}(x) v^a v^b - \mathcal{U}(x, v) \qquad \mathcal{U}(x, v) = \phi(x) + A_a(x) v^a$$

i.e. the difference of the kinetic and (possibly *generalized*) potential energy, is regular (see also (18.4.9)).

Hint: here the relevant matrix from (18.2.3) turns out to be g_{ab}; this in turn defines a *metric tensor* on M, since the kinetic energy is *positive* for any true motion (i.e. g_{ab} is positive definite). □

• Now, when TM became a symplectic manifold (TM, ω_L), we may already proceed to perform standard steps: introduce a *Hamiltonian field* ζ_f corresponding to an arbitrary generator $f \in \mathcal{F}(TM)$ by

$$i_{\zeta_f} \omega_L = -df$$

and consequently a *Hamiltonian system* as a triple $(TM, \omega_L, \mathcal{H})$ by the choice of a distinguished function $f \equiv \mathcal{H}$ (Hamiltonian). The motion along integral curves $\gamma(t)$ of the field $\zeta_{\mathcal{H}}$ is then regarded as the dynamics (= *time development*) of the system:

$$\dot{\gamma} = \zeta_{\mathcal{H}} \quad \text{i.e.} \quad x \equiv \gamma(0) \mapsto \Phi_t(x) \equiv \gamma(t) \qquad \Phi_t \leftrightarrow \zeta_{\mathcal{H}}$$

It turns out, finally, that for the appropriate choice of the Hamiltonian $\mathcal{H}$, namely for

$$\mathcal{H} = E_L := \Delta L - L$$

(E_L being the *energy* corresponding to the Lagrangian L), the dynamics of the Hamiltonian system (TM, ω_L, E_L) just coincides with a standard dynamics generated by Lagrange's equations (of the second kind). Put another way, for a non-degenerate Lagrangian the content of the system

$$\dot{\gamma} = \zeta_{E_L} \qquad i_{\zeta_{E_L}} \omega_L = -dE_L$$

is the same as that of the standard Lagrange equations.

18.2.6 Check this statement by expressing explicitly the defining equation for the *Euler–Lagrange field* Γ

$$i_\Gamma \omega_L = -dE_L \qquad \text{i.e.} \quad \Gamma \equiv \zeta_{E_L}$$

in canonical coordinates.

Hint: sufficiently large piece of paper, (18.2.2) and (18.2.3). □

• For an alternative coordinate-free expression of Lagrange's equations, we first verify in (18.2.7)–(18.2.9) that Γ is indeed a second-order differential equation field (fortunately, it turns out well) and then in (18.2.11) we will see at last that the equations derived there indeed happen to coincide with Lagrange's equations.

18.2.7* For an arbitrary $\binom{1}{1}$-type tensor field A on $\mathcal{M}$ define a $\binom{1}{2}$-type tensor field N_A on $\mathcal{M}$ by the prescription

$$N_A(V, W) := A^2([V, W]) + [A(V), A(W)] - A([A(V), W]) - A([V, A(W)])$$

The field N_A is called the *Nijenhuis tensor* associated with A. Check that

(i) N_A is indeed a $\binom{1}{2}$-type tensor field (i.e. $\mathcal{F}(M)$-linearity)
(ii) in particular, for $A = S$ on $\mathcal{M} = TM$ (S being the vertical endomorphism) the corresponding Nijenhuis tensor *vanishes*, $N_S = 0$. □

18.2.8* Prove the identity

$$\omega_L(S(V), W) = -\,\omega_L(V, S(W)) \qquad V, W \text{ arbitrary}$$

(i.e. the operator S is "antisymmetric with respect to ω_L").

Hint: $\omega_L(S(V), W) - d\theta_L(S(V), W) = \cdots$; then use the Cartan formulas (6.2.13) for computation of d, definitions of the quantities involved and the result of (18.2.7). □

18.2.9* Show that

$$i_\Gamma \omega_L = -dE_L \quad \Rightarrow \quad S(\Gamma) = \Delta$$

i.e. that the Euler–Lagrange field is a second-order differential equation field.

Hint: set $\Gamma = V$ in (18.2.8) and climb step by step to the equation

$$\omega_L(S(\Gamma), W) = \cdots = \omega_L(\Delta, W) \qquad \text{for arbitrary } W$$

juggling with definitions and relevant identities from Section 17.7. Furthermore, consider non-degeneracy of ω_L. □

18.2.10 * Define a map

$$\mathcal{E}^L : (V, \Gamma) \mapsto \Gamma(V^\uparrow L) - \widetilde{V} L$$

where $V^\uparrow$ and $\widetilde{V}$ denote the vertical and complete lift of a field V on M respectively, L is a Lagrangian and Γ is a second-order differential equation field. Check that

(i) the map is $\mathcal{F}(M)$-linear with respect to V, so that

$$\mathcal{E}^L(V, \Gamma) = V^a(x)\mathcal{E}^L_a(x, v) \qquad \mathcal{E}^L_a \equiv \mathcal{E}^L(\partial_a, \Gamma)$$

(ii) for *Euler–Lagrange expression* $\mathcal{E}^L_a$ we get in canonical coordinates

$$\mathcal{E}^L_a(x, v) = \Gamma \frac{\partial L}{\partial v^a} - \frac{\partial L}{\partial x^a}$$

(iii) the function $\mathcal{E}^L_a$, when evaluated on an integral curve of the field Γ (= on the natural lift $\hat{\gamma}$ of a curve γ on M), gives

$$\mathcal{E}^L_a(\hat{\gamma}) = \frac{d}{dt} \frac{\partial L}{\partial \dot{x}^a} - \frac{\partial L}{\partial x^a}$$

Hint: a direct computation in canonical coordinates. Just like in (18.2.9) use $v^a = \dot{x}^a$. □

18.2.11 * Show that the map $\mathcal{E}^L$ from (18.2.10) *vanishes* just for a particular second-order differential equation field, namely for the *Euler–Lagrange* field, i.e. then

$$\Gamma(V^\uparrow L) - \widetilde{V} L = 0 \qquad \text{for arbitrary } V$$

This may then be regarded as a coordinate-free version of Lagrange's equations. Put another way

$$i_\Gamma \omega_L = -dE_L \quad \Rightarrow \quad \mathcal{E}^L_a(\hat{\gamma}) = 0$$

so that a curve γ on M, whose natural lift $\hat{\gamma}$ to TM happens to be an integral curve of the dynamical Hamiltonian field Γ, satisfies the standard Lagrange equations.

Hint: evaluate (18.2.6) on the complete lift $\widetilde{V}$ of an arbitrary field V on M; i.e.

$$\begin{aligned} 0 &= \langle i_\Gamma \omega_L + dE_L, \widetilde{V} \rangle = \omega_L(\Gamma, \widetilde{V}) + \widetilde{V} E_L = (d\theta_L)(\Gamma, \widetilde{V}) + \widetilde{V} \Delta L - \widetilde{V} L = \cdots \\ &= \Gamma(V^\uparrow L) - \widetilde{V} L \end{aligned}$$

The process denoted by $= \cdots =$ uses definitions, Cartan formulas (6.2.13), identities from (17.7.1) and the property (18.2.9) of the field Γ. □

18.3 Connection between Lagrangian and Hamiltonian mechanics, Legendre map

• Up to now, we have learned that if a regular Lagrangian $L : TM \to \mathbb{R}$ is available on TM, we get naturally a Hamiltonian dynamics there. Namely, we first construct a symplectic form ω_L from L and then a dynamical vector field $\Gamma_L \in \mathfrak{X}(TM)$ defined by

$$i_{\Gamma_L}\omega_L = -dE_L$$

i.e. as the Hamiltonian field generated by the function E_L. Eventually the time development is identified with a motion along Γ_L.

All this is even simpler on T^*M. There is a *canonical* symplectic form ω there and if we fix a Hamiltonian $H : T^*M \to \mathbb{R}$, we immediately get a dynamical vector field $\Gamma \in \mathfrak{X}(T^*M)$ defined by the similar relation

$$i_{\Gamma_H}\omega = -dH$$

i.e. as the Hamiltonian field generated by the function H.

This means that we now have *two* Hamiltonian systems, (TM, ω_L, E_L) and (T^*M, ω, H). We will see in what follows that provided some conditions are satisfied (a proper relation between L and H), there exists a diffeomorphism $TM \leftrightarrow T^*M$, which realizes an equivalence of the two Hamiltonian systems under consideration. The corresponding diffeomorphism represents a global version of the Legendre transformation, well known from ordinary analytical mechanics. Let us start by a description of the diffeomorphism in the direction $TM \to T^*M$.

Let

$$L : TM \to \mathbb{R}$$

be a Lagrangian on TM. Define the *Legendre map*

$$\hat{L} : TM \to T^*M$$

by the relations

$$\tau \circ \hat{L} = \pi \qquad \langle \hat{L}(v), w \rangle := w_v^\uparrow L \equiv \left.\frac{d}{dt}\right|_0 L(v + tw)$$

where $v, w \in \pi^{-1}(x)$ and $w_v^\uparrow$ is the vertical lift to v.

18.3.1 Check that

(i) the first condition simply says that a vector sitting at $x \in M$ is mapped to a covector sitting *at the same point* $x \in M$

(ii) the second condition indeed defines a covector at x (i.e. the linearity with respect to w)

(iii) in canonical coordinates on TM and T^*M the map reads

$$\hat{L} : (x^a, v^a) \mapsto (x^a(x, v), p_a(x, v)) \equiv \left(x^a, \frac{\partial L}{\partial v^a}\right)$$

i.e. it coincides with standard formulas for the Legendre transformation

(iv) $\hat{L}$ is a (local) diffeomorphism $\Leftrightarrow$ L is regular.

Hint: (iii) $\langle \hat{L}(v), w\rangle = \hat{L}(v)_a w^a \overset{!}{=} w^{\uparrow}L$, (17.5.2); (iv) the relevant Jacobian is

$$\frac{\partial(x,p)}{\partial(x,v)} = \det\left(\frac{\partial p_a}{\partial v^b}\right) = \det\left(\frac{\partial^2 L}{\partial v^a \partial v^b}\right)$$

see the definition of regularity in (18.2.3). □

• Lessons:

(i) regularity of L is not only important for non-degeneracy of ω_L, but also for local invertibility of the map $\hat{L}$

(ii) what in analytical mechanics is regarded as a change of coordinates is interpreted here as a coordinate presentation of a map between *two different* manifolds.

Now we will investigate how various important tensor fields are mapped under $\hat{L}$ and eventually how the complete dynamics is related.

18.3.2 Check that

$$\theta_L \equiv S(dL) = \hat{L}^*\theta$$

Hint: $\langle (\hat{L}^*\theta)_v, w\rangle = \langle \theta_{\hat{L}(v)}, \hat{L}_* w\rangle = \cdots$ definition $\cdots = \langle \theta_L, w\rangle$. □

18.3.3 Check that

$$\omega_L = \hat{L}^*\omega$$

Hint: see (18.3.2). □

• This means that the symplectic structure on TM may also be regarded as the $\hat{L}^*$-image of the canonical symplectic structure on T^*M (recall that ω_L was introduced in (18.2.3) *independently* of T^*M).

Now if we compare the defining equations for dynamical fields Γ_L on TM and Γ_H on T^*M and take into account (18.3.3), we get

$$i_{\Gamma_L}(\hat{L}^*\omega) = -dE_L \qquad i_{\Gamma_H}\omega = -dH$$

Since the interior product i_V is natural with respect to diffeomorphisms f^* (8.3.6) one can see that a proper correspondence between the Lagrangian L and the Hamiltonian H results in equivalence of the two dynamics involved.

18.3.4 Check that if L and H are related by

$$\hat{L}^*H = E_L \qquad \text{i.e.} \quad H = (\hat{L}^{-1})^*E_L \equiv E_L \circ \hat{L}^{-1}$$

then

(i) also the dynamical fields are related

$$\Gamma_H = \hat{L}_*\Gamma_L$$

(ii) if

$$\Phi_t^L \leftrightarrow \Gamma_L \quad \text{and} \quad \Phi_t^H \leftrightarrow \Gamma_H$$

denote the flows corresponding to the time development in (TM, ω_L, E_L) and (T^*M, ω, H), then the flows are related by

$$\hat{L} \circ \Phi_t^L = \Phi_t^H$$

Hint: (i) application of $\hat{L}^*$ to $i_{\Gamma_H}\omega = -dH$ gives (6.2.11)

$$i_{\hat{L}^*\Gamma_H}\omega_L = -d(\hat{L}^* H)$$

$\Rightarrow$ for $\hat{L}^* H = E_L$ we necessarily have $\hat{L}^*\Gamma_H = \Gamma_L$ (due to non-degeneracy of ω_L). □

18.3.5 Check that the relation $H = (\hat{L}^{-1})^* E_L$ we just obtained is nothing but the good old formula which we know from analytical mechanics, where it is usually written in the form

$$H(x, p) = \dot{x}^a \frac{\partial L}{\partial \dot{x}^a} - L \qquad \text{for} \quad p_a(x, \dot{x}) = \frac{\partial L(x, \dot{x})}{\partial \dot{x}^a}$$

Hint: $H = E_L \circ \hat{L}^{-1} \Rightarrow$ it is actually the function $E_L = v^a \frac{\partial L}{\partial v^a} - L$, expressed in terms of the variables (x, p) according to (18.3.1). □

• The Legendre map may be also defined in the opposite direction and expressed in terms of the Hamiltonian. Let

$$\hat{H} : T^*M \to TM$$

be defined by the relations

$$\pi \circ \hat{H} = \tau \qquad \langle \alpha, \hat{H}(p) \rangle := \alpha_p^{\uparrow} H \equiv \frac{d}{dt}\bigg|_0 H(p + t\alpha)$$

where $\alpha, p \in \tau^{-1}(x)$ and $\alpha^{\uparrow}$ is the vertical lift to p.

18.3.6 Check that

(i) in canonical coordinates

$$\hat{H} : (x^a, p_a) \to (x^a(x, p), v^a(x, p)) \equiv \left(x^a, \frac{\partial H(x, p)}{\partial p_a} \right)$$

(ii) $\hat{H}$ is a (local) diffeomorphism $\leftrightarrow$ H is regular (i.e. $\det \left(\frac{\partial^2 H(x,p)}{\partial p_a \partial p_b} \right) \neq 0$)
(iii) if L and H match in the sense of (18.3.4), then

$$\hat{L} \circ \hat{H} = id_{T^*M} \quad \hat{H} \circ \hat{L} = id_{TM} \qquad \text{i.e.} \quad \hat{H} = \hat{L}^{-1}$$

□

18.4 Symmetries lifted from the base manifold (configuration space)

In Section 14.4 we learned that there is a one-to-one correspondence between conserved quantities of a Hamiltonian system $(\mathcal{M}, \omega, H)$ and *exact Cartan symmetries*, i.e. vector fields V on $\mathcal{M}$ which satisfy $i_V\omega = d(-F_V)$ for some function F_V (so that they are actually *Hamiltonian fields* $V = \zeta_{F_V}$) and $VH = 0$.

Here we will look more closely at an important particular case, where

1. we take as symplectic manifolds the total spaces of the (co)tangent bundle, $(TM, d\theta_L)$ or $(T^*M, d\theta)$
2. Cartan symmetries happen to be *complete lifts* of vector fields from the base M.

In physical terms this corresponds to a situation in which there is a symmetry of a mechanical system already present in the *configuration space* (= the base M; this symmetry may thus be "directly seen") and it is only lifted to the *phase space* T^*M (or TM). Both the motion *and the symmetry* thus "actually" occur on M, transition to the space of double dimension, however, brings technical advantages – it opens the possibility of using the full strength of symplectic machinery.

Let us find out, first, what the *moment map* from Section 14.5 looks like in this specific case.

18.4.1 Let R_g be a right action of a Lie group G on M and denote by ξ_X fundamental fields of the action. Check that

(i) then also the maps TR_g and $T^*R_{g^{-1}}$ provide actions of G;

$$R_g \text{ is an action of } G \text{ on } M \quad \Rightarrow \quad \begin{array}{l} TR_g \text{ is an action of } G \text{ on } TM \\ T^*R_{g^{-1}} \text{ is an action of } G \text{ on } T^*M \end{array}$$

They are called *lifts of the action of the group* G from M to TM and T^*M respectively

(ii) the generators of the lifted actions turn out to be just the *complete lifts* of the generators of the initial action

$$\xi_X \text{ generates } R_g \text{ on } M \quad \Rightarrow \quad \begin{array}{l} \tilde{\xi}_X \text{ generates } TR_g \text{ on } TM \\ \tilde{\xi}_X \text{ generates } T^*R_{g^{-1}} \text{ on } T^*M \end{array}$$

(iii) the lifted actions are *globally Hamiltonian* (see Section 14.5), their "Hamiltonians" P_X being given by simple formulas[389]

$$P_X = \langle \theta_L, \tilde{\xi}_X \rangle \text{ on } TM \qquad \text{and} \qquad P_X = \langle \theta, \tilde{\xi}_X \rangle \text{ on } T^*M$$

as well as *Poisson*, so that

$$\{P_X, P_Y\} = P_{[X,Y]}$$

(on the right a potentially possible non-trivial 2-cocycle $\beta(X, Y)$ from (14.5.4) *does not* appear)

(iv) if $\xi_X = X^i \xi_i^a(x)\partial_a$, the function P_X explicitly reads as

$$P_X \equiv X^i P_i = X^i \frac{\partial L}{\partial v^a} \xi_i^a(x) \qquad \text{on } TM$$
$$\text{and} \quad P_X \equiv X^i P_i = X^i p_a \xi_i^a(x) \qquad \text{on } T^*M$$

389 We assume that a *G-invariant Lagrangian* is given on TM, i.e. such that $L \circ TR_g = L$, or consequently $\tilde{\xi}_X L = 0$.

(Recall (14.5.7) that if G is a symmetry of the complete Hamiltonian system, the functions P_i are conserved.)

Hint: (i) (17.3.2), (17.3.5); (ii) (17.5.6), (17.5.12); (iii) for example, on T^*M (on TM in full analogy)

$$i_{\tilde{\xi}_X} d\theta = \mathcal{L}_{\tilde{\xi}_X}\theta - d i_{\tilde{\xi}_X}\theta = -d\langle\theta, \tilde{\xi}_X\rangle \equiv -dP_X$$
$$\{P_X, P_Y\} = \tilde{\xi}_X P_Y = \mathcal{L}_{\tilde{\xi}_X}\langle\theta, \tilde{\xi}_Y\rangle = \langle\mathcal{L}_{\tilde{\xi}_X}\theta, \tilde{\xi}_Y\rangle + \langle\theta, \mathcal{L}_{\tilde{\xi}_X}\tilde{\xi}_Y\rangle = \langle\theta, [\tilde{\xi}_X, \tilde{\xi}_Y]\rangle = \langle\theta, \tilde{\xi}_{[X,Y]}\rangle$$
$$\equiv P_{[X,Y]}$$

where we used $\mathcal{L}_{\tilde{V}}\theta = 0$ from (17.7.1); (iv) (17.5.7), (17.5.13), (17.6.5) and (18.2.3). □

18.4.2 Consider a Lagrangian $L(x, v)$. In analytical mechanics a particular coordinate x^a is called a *cyclic coordinate* if it does not actually enter L (where it might be in principle) and the same nomenclature will be used for a Hamiltonian $H(x, p)$. Show how standard (and useful) results about cyclic coordinates follow[390] from the result of the preceding problem:

Lagrangian mechanics: x^a is cyclic $\Rightarrow$ conserved quantity is $p_a(x, v) := \dfrac{\partial L(x, v)}{\partial v^a}$

Hamiltonian mechanics: x^a is cyclic $\Rightarrow$ conserved quantity is p_a

The function $p_a(x, v)$ on TM is called the ath *canonical momentum* corresponding to the ath coordinate x^a. (The reader can easily check that for regular Lagrangian L the $2n$ coordinates $(x^a, p_a(x, v))$ are indeed canonical in the sense of the Darboux theorem (14.2.2). Note also that the ath canonical momentum may alternatively be regarded as the pull-back of the canonical coordinate p_a on T^*M to TM with respect to the Legendre map (18.3.1).) In the Hamiltonian case the conserved quantity is just p_a = the ath "fiber" canonical coordinate on T^*M.

Hint: if x^a is cyclic, the dynamics has a *symmetry lifted from* M, which is generated on M by the field $\partial_a \Rightarrow$ according to (iv) in (18.4.1) we get $P_X = \partial L/\partial v^a$ in the Lagrangian formalism and p_a in the Hamiltonian one. From (18.2.3) we know that $\theta_L = p_a(x, v)\, dx^a$ so that $\omega_L = dp_a(x, v) \wedge dx^a$. □

18.4.3 Consider the standard (left) action of the *Euclidean group* $E(3)$ in the configuration space $M = E^3$ of a single point mass

$$\mathbf{r} \mapsto L_{(A,\mathbf{a})}\mathbf{r} \equiv A\mathbf{r} + \mathbf{a} \qquad (A, \mathbf{a}) \in E(3)$$

Check that

(i) if E_j and l_j is the standard basis of the Lie algebra $e(3)$ corresponding to translations and rotations respectively, then the fundamental fields ξ_X (of the right action $L_{(A,\mathbf{a})^{-1}}$) on M are

$$\xi_{E_j} = -\partial_j \qquad \xi_{l_j} = -\epsilon_{jik} x_i \partial_k$$

[390] Both conservation laws are easily seen directly from Lagrange's and Hamilton's equations. This way of revealing the conserved quantities is, however, strongly dependent on a "lucky" choice of coordinates in configuration space (leading to an extremely simple form of $\xi_i^a(x)$ in P_X), whereas (18.4.1) works for any coordinates.

(ii) their (complete) lifts to the *phase space* $T^*M[\mathbf{r},\mathbf{p}]$ read

$$\tilde{\xi}_{E_j} = -\partial_j \qquad \tilde{\xi}_{l_j} = -\epsilon_{jik}\left(x_i \frac{\partial}{\partial x_k} + p_i \frac{\partial}{\partial p_k}\right)$$

(iii) the functions $P_X = \langle \theta, \tilde{\xi}_X \rangle$ for $X = E_j$ and l_j come out as

$$P_{E_j}(\mathbf{r},\mathbf{p}) = -p_j \qquad P_{l_j}(\mathbf{r},\mathbf{p}) = -\epsilon_{jkl} x_k p_l \equiv -(\mathbf{r}\times\mathbf{p})_j \equiv -L_j$$

(iv) the property $\{P_X, P_Y\} = P_{[X,Y]}$ is realized here as the validity of the well-known Poisson brackets between the observables $\mathbf{p}$, $\mathbf{L}$

$$\{p_i, p_j\} = 0 \qquad \{L_i, L_j\} = -\epsilon_{ijk} L_k \qquad \{L_i, p_j\} = -\epsilon_{ijk} p_k$$

Hint: (13.4.6), (13.4.7), $\theta = p_i\, dx^i \equiv \mathbf{p}\cdot d\mathbf{r}$. □

• Now we restrict to the most important class of Lagrangians (and corresponding Hamiltonians), containing kinetic and potential energy. We are speaking about ordinary Lagrangians (and Hamiltonians) encountered in analytical mechanics, with the structure $L = T - U$ and $H = T + U$. First we develop a convenient formalism for treating such objects.

18.4.4 Let B be a $\binom{0}{k}$-type ("strictly covariant") *tensor field* on M. We may associate a *function* $\overset{\circ}{B}$ on TM with B as follows:

$$\overset{\circ}{B}(v) := B_x(v,\dots,v)$$

Here, v on the left denotes a *point* on TM whereas v on the right is the (corresponding) *vector* at the point $x \equiv \pi(v)$. In particular, for a function ($k = 0$) this is to be understood as $\overset{\circ}{B}(v) := B(\pi(v)) \equiv (\pi^* B)(v)$, i.e. $\overset{\circ}{B} := \pi^* B$. Check that

(i) in canonical coordinates this gives

$$B \equiv B_{a\dots b}(x)\, dx^a \otimes \cdots \otimes dx^b \quad\Rightarrow\quad \overset{\circ}{B}(x,v) = B_{a\dots b}(x) v^a \dots v^b$$

(ii) in full analogy we may associate a function $\overset{\circ}{B}$ on T^*M with a $\binom{k}{0}$-type ("strictly contravariant") tensor field B on M (the same notation is used for both cases under consideration) by

$$\overset{\circ}{B}(p) := B_x(p,\dots,p)$$

where p on the left denotes a *point* on T^*M whereas p on the right is the (corresponding) *covector* at the point $x \equiv \tau(p)$ (again for a function we set $\overset{\circ}{B} := \pi^* B$) and in canonical coordinates we get

$$B \equiv B^{a\dots b}(x)\partial_a \otimes \cdots \otimes \partial_b \quad\Rightarrow\quad \overset{\circ}{B}(x,p) = B^{a\dots b}(x) p_a \dots p_b$$

(iii) the maps

$$f : T^0_k(M) \to \mathcal{F}(TM) \qquad B \mapsto \overset{\circ}{B}$$
$$f : T^k_0(M) \to \mathcal{F}(T^*M) \qquad B \mapsto \overset{\circ}{B}$$

are injective (no information is lost) when restricted to *fully symmetric* tensor fields on M

(iv) in both cases (TM as well as T^*M) the following statements hold:

$$\mathcal{L}_{\tilde{V}} \overset{\circ}{B} \equiv \tilde{V} \overset{\circ}{B} = \overset{\circ}{(\mathcal{L}_V B)} \qquad \text{i.e.} \quad \mathcal{L}_{\tilde{V}} \circ f = f \circ \mathcal{L}_V$$
$$\triangle \overset{\circ}{B} = k \overset{\circ}{B} \qquad \text{i.e.} \quad \overset{\circ}{B} \text{ is homogeneous of degree } k \text{ in fiber coordinates}$$

where $\tilde{V}$ is the *complete lift* of a field V and $\triangle$ is the *Liouville field* from (17.6.1).

Hint: (iv) for example, on TM we get for $\Phi_t \leftrightarrow V$

$$(\tilde{V} \overset{\circ}{B})(v) = \frac{d}{dt}\bigg|_0 \overset{\circ}{B}(T\Phi_t(v)) = \frac{d}{dt}\bigg|_0 B(\Phi_{t*}v, \dots) = \frac{d}{dt}\bigg|_0 (\Phi_t^* B)(v, \dots) = (\mathcal{L}_V B)(v, \dots)$$
$$= \overset{\circ}{(\mathcal{L}_V B)}(v)$$

□

18.4.5 Test the computation of the Lie derivative by this unusual method on the example of a vector field and a metric tensor.

Solution: if $W = W^a \partial_a$ and $g = g_{ab}\, dx^a \otimes dx^b$, then $\overset{\circ}{W} = W^a p_a$ and $\overset{\circ}{g} = g_{ab} v^a v^b$. According to (17.5.7) and (17.5.13) we then get for $V = V^a \partial_a$

$$\tilde{V} \overset{\circ}{W} = \left(V^a(x) \frac{\partial}{\partial x^a} - V^b{}_{,a}(x) p_b \frac{\partial}{\partial p_a} \right) (W^c p_c) = \cdots = (V W^a - W V^a) p_a$$
$$\tilde{V} \overset{\circ}{g} = \left(V^a(x) \frac{\partial}{\partial x^a} + V^a{}_{,b}(x) v^b \frac{\partial}{\partial v^a} \right) (g_{cd} v^c v^d) = \cdots = (V g_{ab} + V^c{}_{,a} g_{cb} + V^c{}_{,b} g_{ac}) v^a v^b$$

The terms gathered in the last brackets indeed coincide with the component expressions of the Lie derivatives $\mathcal{L}_V W$ and $\mathcal{L}_V g$ (4.3.4). □

• Now recall that the standard Lagrangians we usually encounter in analytical mechanics contain two parts, the *kinetic energy* T, which is a quadratic function of (generalized) velocities and the *potential energy* U, which depends on coordinates alone:

$$L(q, \dot{q}) = T(q, \dot{q}) - U(q) \equiv \frac{1}{2} T_{ab}(q) \dot{q}^a \dot{q}^b - U(q)$$

Notice that both of them are just expressions of the type under consideration. A function L of this structure is called a *natural Lagrangian*.

18.4.6 Check that

(i) a natural Lagrangian may be written in a simple form

$$L = \frac{1}{2} \overset{\circ}{g} - \overset{\circ}{\phi}$$

where g represents a *metric tensor* on M and ϕ is a *function* on M

(ii) the Legendre map for such L explicitly reads

$$\hat{L} : (x^a, v^a) \mapsto (x^a(x, v), p_a(x, v)) = (x^a, g_{ab} v^b)$$

(iii) a Hamiltonian which corresponds to this Lagrangian via the Legendre map is

$$H = \frac{1}{2}\overset{\circ}{g}{}^{-1} + \overset{\circ}{\phi} \qquad \text{or in detail} \qquad H(x, p) = \frac{1}{2}g^{ab}(x)p_a p_b + \phi(x)$$

Hint: (i) see (2.6.7) and (3.2.9); (ii) $E_L = \Delta L - L = \Delta(\frac{1}{2}\overset{\circ}{g} - \overset{\circ}{\phi}) - (\frac{1}{2}\overset{\circ}{g} - \overset{\circ}{\phi}) = \frac{1}{2}\overset{\circ}{g} + \overset{\circ}{\phi}$ (due to 18.4.4)); see (18.3.4). □

• In what follows we are going to characterize *all* exact Cartan symmetries which are *lifted* from the base for this type of Lagrangian (there still may be additional "hidden" symmetries involved, which are not lifted from the base).

18.4.7 Consider a natural Lagrangian $L = \frac{1}{2}\overset{\circ}{g} - \overset{\circ}{\phi}$. Check that

(i) for exact Cartan symmetries of the form of the complete lift there holds

$$\widetilde{V} = \text{exact Cartan symmetry} \qquad \Leftrightarrow \qquad \begin{aligned} \mathcal{L}_V g &= 0 \\ V\phi \equiv \mathcal{L}_V \phi &= 0 \end{aligned}$$

i.e. the complete lift $\widetilde{V}$ of a field V is the exact Cartan symmetry if and only if the field V itself happens to be the *Killing vector* of the metric tensor (associated with the kinetic energy) and moreover if the potential energy is invariant with respect to V

(ii) the corresponding conserved quantity turns out to be

$$F_{\widetilde{V}} \equiv F = g(v, V) \equiv v \cdot V \equiv g_{ab}(x)v^a V^b(x) = p_a V^a$$

The quantity $g(\dot{\gamma}, V) \equiv \dot{\gamma} \cdot V$, the scalar product of the instantaneous velocity and the Killing vector V, thus remains unchanged in the course of the motion

(iii) if a coordinate x^a does not enter any component of the metric tensor and also does not enter the function ϕ, we may take ∂_a to be the field V and F then becomes just p_a.

Hint: our Hamiltonian system is $(TM, d\theta_L, E_L \equiv \frac{1}{2}\overset{\circ}{g} + \overset{\circ}{\phi})$; from (18.4.1) we know that $\widetilde{V}$ is a Hamiltonian field on $(TM, d\theta_L)$, its Hamiltonian being $F_{\widetilde{V}} = \langle \theta_L, \widetilde{V} \rangle = \cdots = g(v, V)$. Moreover, $\widetilde{V}E_L = \frac{1}{2}\widetilde{V}\overset{\circ}{g} + \widetilde{V}\overset{\circ}{\phi} = \frac{1}{2}(\overset{\circ}{\mathcal{L}_V g}) + (\overset{\circ}{V\phi})$. □

18.4.8 Let V be a Killing vector on (M, g). We already know from problem (15.4.4) that if we consider the Lagrangian

$$L = \frac{1}{2}\overset{\circ}{g} \equiv \frac{1}{2}g_{ab}(x)v^a v^b$$

(containing the *kinetic* energy alone), then the solutions of Lagrange's equations coincide with *geodesics* of RLC connection on (M, g). Be sure to understand that

(i) in the course of motion along a geodesic γ the quantity $F = \dot{\gamma} \cdot V \equiv g(\dot{\gamma}, V)$ remains constant[391] (if we choose an affine parameter so that the velocity is normed to unity, this becomes just a *projection* of the Killing vector onto the direction of motion); also prove this fact with no use of our symplectic tools at all (use covariant derivatives instead)

[391] This is useful in general relativity, where free test particles move along geodesics of the space-time.

(ii) the result from (18.4.7) generalizes the fact of the conservation of this quantity also to motion with non-vanishing acceleration; then it is, however, restricted to the case where the force field obeys $V\phi = 0$; put another way, the vector field V is required to be a symmetry of the (potential) force field as well.

Hint: (i) $\dot{F} = \dot{\gamma} F = \nabla_{\dot{\gamma}} g(\dot{\gamma}, V) = (\nabla_{\dot{\gamma}} g)(\dot{\gamma}, V) + g(\nabla_{\dot{\gamma}} \dot{\gamma}, V) + g(\dot{\gamma}, \nabla_{\dot{\gamma}} V) = g(\dot{\gamma}, \nabla_{\dot{\gamma}} V) = V_{i;j}\dot{x}^i \dot{x}^j = 0$, since due to (15.4.14) $V_{(i;j)} = 0$. □

• Recall that in analytical mechanics one also introduces the concept of *generalized potential energy*, which depends (linearly) on velocities as well. In order to incorporate this case in our formalism, we add still another term to L. The class of Lagrangians under consideration is namely extended to

$$L = \frac{1}{2} \overset{\circ}{g} - (\overset{\circ}{\phi} + \overset{\circ}{A})$$

where A is a *covector* field on M. A complete Lagrangian is thus given (parametrized) by three independent objects living on M, namely g, A, ϕ. Let us study in more detail how each of the three terms in L manifests itself on equations of motion.

18.4.9 Consider a Lagrangian of the form

$$L = \frac{1}{2} \overset{\circ}{g} - (\overset{\circ}{\phi} + \overset{\circ}{A}) \equiv L_g + L_\phi + L_A$$

Since the relevant elements of the equation $i_\Gamma \omega_L = -dE_L$ depend on L additively, each of them may be written as a sum of three terms, which correspond to the individual terms of L. Check that

(i) explicitly it looks as follows:

$$\begin{aligned} \theta_L &= \theta_{L_g} + \theta_{L_\phi} + \theta_{L_A} = \theta_{L_g} - \pi^* A \\ \omega_L &= \omega_{L_g} + \omega_{L_\phi} + \omega_{L_A} = \omega_{L_g} - \pi^* dA \\ E_L &= E_{L_g} + E_{L_\phi} + E_{L_A} = E_{L_g} + \pi^* \phi \end{aligned}$$

(ii) the complete equation for the dynamical field $\Gamma \equiv \zeta_{E_L}$ has the form

$$i_\Gamma(\omega_{L_g} - \pi^* dA) = -d(E_{L_g} + \pi^* \phi)$$

(iii) this may be interpreted as the fact that the dynamics with the Lagrangian containing all three terms differs from the free motion (with the Lagrangian given by the kinetic energy alone) as follows:

the term $\overset{\circ}{A}$ modifies the symplectic form	$\omega_{L_g} \mapsto \omega_{L_g} - \pi^* dA$
the term $\overset{\circ}{\phi}$ modifies the energy	$E_{L_g} \mapsto E_{L_g} + \pi^* \phi$

(iv) the equation of motion (and consequently, also the motion itself) feels the presence of A and ϕ through their (exterior) *derivatives* dA and $d\phi$ *alone*; this means that the motion remains *the same*, if the substitution[392]

$$A \mapsto A + d\chi \qquad \phi \mapsto \phi + \text{constant}$$

is performed in the original Lagrangian L.

Hint: see (18.2.3) and (18.2.6). □

• Let us examine next what is the characterization of all exact Cartan symmetries lifted from the base for the Lagrangian $L = L_g + L_\phi + L_A$ (i.e. how the result of (18.4.7) is to be modified, if we add the term $\overset{\circ}{A}$).

18.4.10 Consider the Lagrangian of the form

$$L = \frac{1}{2}\overset{\circ}{g} - (\overset{\circ}{\phi} + \overset{\circ}{A})$$

where A is a *covector* field on M. Check that

(i) for exact Cartan symmetries of the form of the complete lift there holds

$$\tilde{V} = \text{exact Cartan symmetry} \quad \Leftrightarrow \quad \begin{aligned} \mathcal{L}_V g &= 0 \\ V\phi \equiv \mathcal{L}_V \phi &= 0 \\ \mathcal{L}_V A &= d\chi \end{aligned}$$

i.e. except for the requirements on V that it should be a *Killing vector* of the metric tensor associated with the kinetic energy and that the function ϕ should be invariant with respect to this field, we are to add a new requirement: the Lie derivative of a "vector potential" A should be *exact*, $\mathcal{L}_V A = d\chi$ for some function χ

(ii) the corresponding conserved quantity is

$$\begin{aligned} F_{\tilde{V}} = \langle g(v, \cdot) - A, V\rangle + \chi &\equiv V^a(x)(g_{ab}(x)v^b - A_a(x)) + \chi(x) \\ &\equiv V^a(x)p_a(x, v) + \chi(x) \end{aligned}$$

where $p_a(x, v)$ is the *canonical momentum* introduced in (18.4.2)

(iii) this expression *is* "gauge invariant" (with respect to the freedom $A \mapsto A + df$ from (18.4.9)).

Hint: the energy function turns out to be the same as in (18.4.7), $E_L \equiv \frac{1}{2}\overset{\circ}{g} + \overset{\circ}{\phi}$, so that the condition $\tilde{V}E_L \overset{!}{=} 0$ again yields $\mathcal{L}_V g = 0$ and $V\phi = 0$. A computation then gives $i_{\tilde{V}} d\theta_L = -d(g(v, V)) - \pi^*(i_V dA) \overset{!}{=} -d(\text{conserved quantity})$; if $A \mapsto A + df$, then $\chi \mapsto \chi + Vf$ and the combination $\langle A, V\rangle - \chi$ remains unchanged. □

• In Section 14.7 we studied a *symplectic reduction* of a Hamiltonian system by a symmetry G, i.e. a separation of the less interesting part of the dynamics, which was made possible by the symplectic action of a symmetry group. As a motivating example we mentioned there a standard separation of a (trivial) motion of the center of mass in the two-body problem. This separation is due to translational symmetry of the *space*, so that the corresponding

[392] In electromagnetic interpretation this is a gauge transformation of (static) potentials; dA corresponds to the magnetic field and $d\phi$ to the electric field.

symmetry in *phase space* is just a lift of the symmetry from the base, which we study in this section. Now we will show explicitly what the individual steps of the reduction procedure look like in this simple case.

18.4.11 Consider a motion of two point masses in E^3. Their dynamics is described by the Hamiltonian

$$H(\mathbf{r}_1, \mathbf{r}_2, \mathbf{p}_1, \mathbf{p}_2) = \frac{\mathbf{p}_1^2}{2m_1} + \frac{\mathbf{p}_2^2}{2m_2} + U(|\mathbf{r}_1 - \mathbf{r}_2|)$$

which we regard as a function in the *phase* space $\mathbb{R}^{12}[\mathbf{r}_1, \mathbf{r}_2, \mathbf{p}_1, \mathbf{p}_2] \equiv T^*\mathbb{R}^6[\mathbf{r}_1, \mathbf{r}_2]$, i.e. on the total space of the cotangent bundle of the configuration space $\mathbb{R}^6[\mathbf{r}_1, \mathbf{r}_2]$. In the *configuration* space there is a standard action of the (three-dimensional) *translational* group $\mathbb{R}^3[\mathbf{a}]$ given by the prescription

$$\hat{R}_{\mathbf{a}} : (\mathbf{r}_1, \mathbf{r}_2) \mapsto (\mathbf{r}_1 + \mathbf{a}, \mathbf{r}_2 + \mathbf{a})$$

Check that

(i) the lift $R_{\mathbf{a}} := T^*\hat{R}_{-\mathbf{a}}$ of this action to $T^*\mathbb{R}^6[\mathbf{r}_1, \mathbf{r}_2]$ reads

$$R_{\mathbf{a}} : (\mathbf{r}_1, \mathbf{r}_2, \mathbf{p}_1, \mathbf{p}_2) \mapsto (\mathbf{r}_1 + \mathbf{a}, \mathbf{r}_2 + \mathbf{a}, \mathbf{p}_1, \mathbf{p}_2)$$

(ii) if we pass in the configuration space to the coordinates of the center of mass $\mathbf{R}$ and the relative vector $\mathbf{r}$ instead of $(\mathbf{r}_1, \mathbf{r}_2)$, the complete (induced) change of coordinates in the phase space is

$$(\mathbf{r}_1, \mathbf{r}_2, \mathbf{p}_1, \mathbf{p}_2) \mapsto (\mathbf{R}, \mathbf{r}, \mathbf{P}, \mathbf{p})$$

where

$$\mathbf{R} = \frac{m_1\mathbf{r}_1 + m_2\mathbf{r}_2}{m_1 + m_2} \qquad \mathbf{P} = \mathbf{p}_1 + \mathbf{p}_2$$

$$\mathbf{r} = \mathbf{r}_1 - \mathbf{r}_2 \qquad \mathbf{p} = \frac{m_2}{m_1 + m_2}\mathbf{p}_1 - \frac{m_1}{m_1 + m_2}\mathbf{p}_2$$

and moreover the change turns out to be *canonical*[393]

$$\omega = d\mathbf{p}_1 . \wedge d\mathbf{r}_1 + d\mathbf{p}_2 . \wedge d\mathbf{r}_2 = d\mathbf{P} . \wedge d\mathbf{R} + d\mathbf{p} . \wedge d\mathbf{r}$$

(iii) the action $R_{\mathbf{a}}$ in the phase space looks even simpler when expressed in these coordinates, since now the action concerns the center of mass coordinates alone; it is

$$R_{\mathbf{a}} : (\mathbf{R}, \mathbf{r}, \mathbf{P}, \mathbf{p}) \mapsto (\mathbf{R} + \mathbf{a}, \mathbf{r}, \mathbf{P}, \mathbf{p})$$

it is *free* and the symplectic form ω is invariant with respect to it

(iv) in new coordinates the Hamiltonian reads

$$H(\mathbf{R}, \mathbf{r}, \mathbf{P}, \mathbf{p}) = \frac{\mathbf{P}^2}{2M} + \frac{\mathbf{p}^2}{2\mu} + U(|\mathbf{r}|) \qquad M \equiv m_1 + m_2 \quad \mu \equiv \frac{m_1 m_2}{m_1 + m_2}$$

and it is *invariant* with respect to the action of $\mathbb{R}^3$ under consideration.

[393] $d\mathbf{p} . \wedge d\mathbf{r}$ is a shortcut notation for $dp_x \wedge dx + dp_y \wedge dy + dp_z \wedge dz$.

Hint: (ii) (17.1.4); a change of coordinates induced in this way is *always* canonical (ω on T^*M is invariant with respect to T^*f (17.6.8) for an arbitrary diffeomorphism $f : M \to M$; or in coordinates $dp'_a(x, p) \wedge dx'^a(x, p) = d\{(J^{-1})^b_a(x)p_b\} \wedge dx'^a(x) = \cdots = dp_a \wedge dx^a$; (iii) the argument (17.6.8) again, or explicitly $d(\mathbf{P} + \mathbf{a}). \wedge d\mathbf{R} + d\mathbf{p}. \wedge d\mathbf{r} = d\mathbf{P}. \wedge d\mathbf{R} + d\mathbf{p}. \wedge d\mathbf{r}$. □

18.4.12 Check that the moment map for the action under consideration of the translational group may be written as

$$P : (\mathbf{R}, \mathbf{r}, \mathbf{P}, \mathbf{p}) \mapsto \mathbf{P}$$

so that the manifold M_p (the level surface of the moment for the value p) from the general construction (14.7.1) corresponds here to a part of the phase space with a *fixed* value of the *total momentum* $\mathbf{P}$; we may thus identify it with $\mathbb{R}^9[\mathbf{R}, \mathbf{r}, \mathbf{p}]$.

Hint: see (18.4.1), the canonical 1-form is $\theta = \mathbf{P} \cdot d\mathbf{R} + \mathbf{p} \cdot d\mathbf{r}$, generators of the (already lifted) action of the translational group are due to (18.4.11) $\partial_{\mathbf{R}}$. □

18.4.13 Check that the group G_p from the general formalism turns out to be *the whole* translational group $\mathbb{R}^3[\mathbf{a}]$, so that the resulting manifold $\hat{M}_p$ may be identified here with $\mathbb{R}^6[\mathbf{r}, \mathbf{p}]$.

Hint: the Ad-representation is trivial for any *commutative* group (and the same consequently holds for the Ad*-action) $\Rightarrow$ the condition $R_g M_p = M_p$ (14.7.2) then gives $G_p = G$; the factorization is to be performed in the sense of the equivalence $(\mathbf{R}, \mathbf{r}, \mathbf{p}) \sim (\mathbf{R} + \mathbf{a}, \mathbf{r}, \mathbf{p})$. □

18.4.14 Check that

(i) restriction of the symplectic form ω to the submanifold M_p (the form $\tilde{\omega}$ from (14.7.4)) reads $\tilde{\omega} = d\mathbf{p}. \wedge d\mathbf{r}$ (it lives on $\mathbb{R}^9[\mathbf{R}, \mathbf{r}, \mathbf{p}]$)
(ii) this 2-form is indeed horizontal and $\mathbb{R}^3[\mathbf{a}]$-invariant
(iii) the "reduced" symplectic form $\hat{\omega}$ is the form $\hat{\omega} = d\mathbf{p}. \wedge d\mathbf{r}$ (it looks exactly like $\tilde{\omega}$, but it already lives only on $\mathbb{R}^6[\mathbf{r}, \mathbf{p}]$).

Hint: (ii) vertical fields span the generators of translations, i.e. $\partial_{\mathbf{R}}$; (iii) the projection $\pi : M_p \to \hat{M}_p$ is $(\mathbf{R}, \mathbf{r}, \mathbf{p}) \mapsto (\mathbf{r}, \mathbf{p})$. □

• The *translational* symmetry of the original *two-body* problem thus brought us (by means of symplectic reduction) to a simpler problem concerning a *single* body. The reduced phase space is already only $\mathbb{R}^6[\mathbf{r}, \mathbf{p}]$ (the phase space of a fictitious point with a "position vector" $\mathbf{r}$) endowed with a symplectic form $\hat{\omega} = d\mathbf{p}. \wedge d\mathbf{r}$. In this phase space the dynamics is generated by the Hamiltonian

$$\hat{H} = \hat{H}(\mathbf{r}, \mathbf{p}) = \frac{\mathbf{p}^2}{2\mu} + U(r)$$

In order to obtain $\hat{H}$ from the complete $H(\mathbf{R}, \mathbf{r}, \mathbf{P}, \mathbf{p})$, one sets $\mathbf{P} =$ constant (due to the restriction to M_p; an irrelevant additive constant in H may be ignored); it does not actually depend on $\mathbf{R}$ (due to the invariance) and so passing to $\hat{M}_p$ (ignoring of $\mathbf{R}$) formally manifests itself in no way.

The ("single-body") problem which remains may be, however, reduced further, since it still has *rotational* symmetry.

18.4.15 On the configuration space $\mathbb{R}^3[\mathbf{r}]$ there is a natural action of the rotational group. Check that *the lift* of the action to the phase space $\mathbb{R}^6[\mathbf{r}, \mathbf{p}]$ is a symmetry of the Hamiltonian system under consideration.

Hint: it is a symmetry of (M, ω) due to a mere fact of the lift (17.6.8); since the action is $(\mathbf{r}, \mathbf{p}) \mapsto (A^{-1}\mathbf{r}, A^{-1}\mathbf{p})$, the symmetry of $H(\mathbf{r}, \mathbf{p}) = \mathbf{p}^2/2\mu + U(r)$ is evident. □

18.4.16 Check that the symplectic reduction by the rotation group leads for the problem of a motion in the *central field* $U(r)$ to the dynamics of the "radial" degree of freedom r, i.e. the dynamics in the phase space with coordinates (r, p_r), the symplectic form $\hat{\omega} = dp_r \wedge dr$ and the Hamiltonian

$$\hat{H}(r, p_r) = \frac{p_r^2}{2\mu} + U_{\text{eff}}(r) \qquad U_{\text{eff}}(r) := U(r) + \frac{L^2}{2\mu r^2} \equiv \text{effective potential energy}$$

Hint: according to (18.4.3), the moment map reads $(\mathbf{r}, \mathbf{p}) \mapsto -\mathbf{L}(\mathbf{r}, \mathbf{p}) \equiv -\mathbf{r} \times \mathbf{p}$, so that the level surface M_p of the moment is the submanifold $M_{\mathbf{L}} \subset \mathbb{R}^6[\mathbf{r}, \mathbf{p}]$ of the points with a fixed value of the angular momentum vector $\mathbf{L} \equiv \mathbf{r} \times \mathbf{p} =$ constant. Scalar multiplication by the vectors $\mathbf{r}$ and $\mathbf{p}$ gives $\mathbf{L} \cdot \mathbf{r} = 0 = \mathbf{L} \cdot \mathbf{p}$, so if we choose the z-axis directed along $\mathbf{L}$, both the vectors $\mathbf{r}$ and $\mathbf{p}$ will be situated in the "xy-plane." Using *polar* coordinates in these planes we thus have so far the coordinates $(r, \varphi, p_r, p_\varphi)$ (being coordinates on the total space of $T^*\mathbb{R}^2[r, \varphi]$). A condition fixing the *length* of the vector $\mathbf{L}$ gives in addition (due to (17.1.9)) $L_z = xp_y - yp_x = p_\varphi = L =$ constant, so that in coordinates what remains for $M_{\mathbf{L}}$ is only $\mathbb{R}^3(r, \varphi, p_r)$. The role of G_p is played by rotations, which preserve $\mathbf{L}$, i.e. rotations about the z-axis alone ($SO(2) \subset SO(3)$). The equivalence on orbits of the action is $(r, \varphi, p_r) \sim (r, \varphi + a, p_r)$ so that the factorization gives $\hat{M}_p \equiv \hat{M}_{\mathbf{L}} \cong \mathbb{R}^2(r, p_r)$. The symplectic form in the initial $\mathbb{R}^6$ was $\omega = dp_r \wedge dr + dp_\vartheta \wedge d\vartheta + dp_\varphi \wedge d\varphi$; its restriction to $M_{\mathbf{L}}$ is then $\tilde{\omega} = dp_r \wedge dr$, being invariant as well as horizontal with respect to the generator ∂_φ of the group $SO(2)$; on $\hat{M}_{\mathbf{L}}$ we have at last $\hat{\omega} = dp_r \wedge dr$. Finally, the restriction of the Hamiltonian is $(p_\vartheta \mapsto 0,\ p_\varphi \mapsto L \equiv \text{constant})$

$$H(\mathbf{r}, \mathbf{p}) = \frac{\mathbf{p}^2}{2\mu} + U(r) \equiv \frac{p_r^2}{2\mu} + \frac{p_\vartheta^2}{2\mu r^2 \sin^2\vartheta} + \frac{p_\varphi^2}{2\mu r^2} + U(r) \mapsto \frac{p_r^2}{2\mu} + \frac{L^2}{2\mu r^2} + U(r)$$

□

18.4.17 Consider a Hamiltonian system given by the phase space $\mathbb{R}^6[\mathbf{r}, \mathbf{p}]$ (a single point with a "position vector" $\mathbf{r}$) with the symplectic form $\omega = d\mathbf{p}. \wedge d\mathbf{r}$ and the Hamiltonian

$H(\mathbf{r}, \mathbf{p}) = \mathbf{p}^2/2m + mgz$ (so we are speaking about a motion of a point mass in a homogeneous gravitational field). Check that

(i) the translations in horizontal directions turn out to be symmetries of the system
(ii) the reduced Hamiltonian system (with respect to this symmetry) is $\mathbb{R}^2[z, p]$ with the symplectic form $\hat{\omega} = dp \wedge dz$ and the Hamiltonian $H(z, p) = p^2/2m + mgz$; so this is a motion of a point mass in the homogeneous gravitational field, where the (irrelevant = horizontal) *degrees of freedom* (x, y) are *ignored* and only the relevant (= vertical) part of the problem survives.

Hint: the lifted action and the moment map read:

$$\text{lifted action} \qquad (x, y, z, p_x, p_y, p_z) \mapsto (x + a_x, y + a_y, z, p_x, p_y, p_z)$$
$$\text{moment map} \qquad (x, y, z, p_x, p_y, p_z) \mapsto (p_x, p_y) \in \mathbb{R}^2 \sim \mathcal{G}^*$$

so that $M_p \sim \mathbb{R}^4[x, y, z, p_z]$. The group is commutative, consequently G_p is *the whole* group $G = \mathbb{R}^2[\mathbf{a}]$ and the factorization is performed according to the equivalence given by $(x, y, z, p_z) \sim (x + a_x, y + a_y, z, p_z)$; what remains is $\hat{M}_p \sim \mathbb{R}^2[z, p_z]$. Restriction of ω to $M_p \sim \mathbb{R}^4[x, y, z, p_z]$ is $\tilde{\omega} = dp_z \wedge dz$ and it gives $\hat{\omega} = dp_z \wedge dz$. Finally, the restriction of H leads to $\hat{H}(z, p_z) = p_z^2/2m + mgz$. □

18.5 Time-dependent Hamiltonian, action integral

• The whole time we treated Hamiltonian dynamics (in Chapters 14, 17 and 18) we tacitly assumed that the Hamiltonian (the generator of a dynamical field) *does not depend on time*. It was namely a function on a *symplectic* manifold, or in canonical coordinates a function of the variables (q, p) and not of t. From a course on analytical mechanics we know, however, that in Hamilton's (as well as Lagrange's) equations one *does not* assume a priori that a Hamiltonian (Lagrangian) does not depend on time; in general Hamilton's equations read

$$\dot{q}^a = \frac{\partial H(q, p, t)}{\partial p_a} \qquad \dot{p}_a = -\frac{\partial H(q, p, t)}{\partial q^a} \qquad a = 1, \dots n$$

and similarly Lagrange's equations

$$\frac{d}{dt}\frac{\partial L(q, \dot{q}, t)}{\partial \dot{q}^a} - \frac{\partial L(q, \dot{q}, t)}{\partial q^a} = 0 \quad a = 1, \dots n$$

are not attempting at all to hide a possible dependence of $L(q, \dot{q}, t)$ on time. Then how should this possibility be incorporated into our geometrical contemplations?

First of all, we need to say on *which manifold* the Hamiltonian actually is defined, if its coordinate presentation is $H(q, p, t)$. A minimal modification of the approach adopted up to now consists in an assumption[394] that a mere extension of the original symplectic manifold by the time axis takes place, i.e. that we consider as an *extended phase space* the Cartesian *product* of manifolds $T^*M \times \mathbb{R}[t]$; the needed set (q^a, p_a, t) then clearly may be used as local coordinates on such a direct product.

[394] There is also an alternative approach based on so-called *jet manifolds*.

18.5.1 According to (2.2.14), each vector field on $T^*M \times \mathbb{R}$ may be regarded as a sum of two particular vector fields, a field "along T^*M" and a field "along $\mathbb{R}$"; since our local co-ordinates (q^a, p_a, t) happen to be adapted to the product structure of the resulting manifold, the decomposition in coordinates reads

$$W = A^a(q, p, t)\frac{\partial}{\partial x^a} + B_a(q, p, t)\frac{\partial}{\partial p_a} + C(q, p, t)\frac{\partial}{\partial t} \equiv W_0 + C(q, p, t)\partial_t$$

Be sure to understand that if we are now interested in a dynamical vector field Γ whose integral curves are of the form $t \mapsto (q^a(t), p_a(t), t)$, where in addition

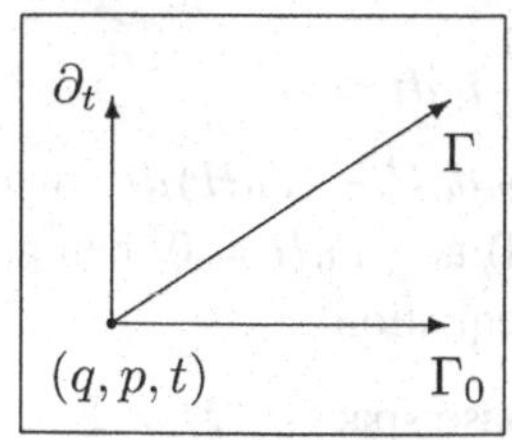

$$\dot{q}^a = \frac{\partial H(q, p, t)}{\partial p_a} \qquad \dot{p}_a = -\frac{\partial H(q, p, t)}{\partial q^a}$$

holds, i.e. they are curves which are parametrized by the *time* t (being one of the coordinates) and $q^a(t)$, $p_a(t)$ satisfy Hamilton's equations, then the field Γ is

$$\Gamma = \frac{\partial H(q, p, t)}{\partial p_a}\frac{\partial}{\partial x^a} - \frac{\partial H(q, p, t)}{\partial q^a}\frac{\partial}{\partial p_a} + \partial_t \equiv \Gamma_0 + \partial_t$$

so that the first part, denoted by Γ_0, looks like a Hamiltonian field generated by the function H (its components depending, however, on time) plus there is a simple term ∂_t added.

Hint: in general, the parameter of an integral curve of W is some s and we have (among others) the equation $\dot{t} = C$ (where the dot denotes a derivative with respect to the parameter, i.e. s); since we now have $s = t$, the equation is $\dot{t} = dt/dt = 1$ and $C = 1$ (note that this may also be written as $i_\Gamma\, dt = 0$). □

18.5.2 Denote by $\pi_1 : T^*M \times \mathbb{R} \to T^*M$ projection to the first factor of the product and by $\hat{\omega} \equiv d\hat{\theta} := \pi_1^*\omega \equiv d\pi_1^*\theta$ the pull-back of the canonical symplectic form ω to the extended phase space $T^*M \times \mathbb{R}$. Check that

(i) the form $\hat{\omega}$ has a coordinate expression

$$\hat{\omega} = dp_a \wedge dq^a \qquad \text{i.e. formally the same as } \omega \text{ has on } T^*M$$

(ii) the projection Γ_0 of the vector field Γ satisfies the equation for the "Hamiltonian" field generated by the function H

$$i_{\Gamma_0}\hat{\omega} = -d_0 H$$

where d_0 is a part of the exterior derivative d which "does not see the coordinate t" (as if d were performed on the factor T^*M alone), so that the full dH is given as $dH \equiv d_0 H + (\partial_t H)\, dt$; d_0 is a counterpart of the *spatial* exterior derivative $\hat{d}$ in Minkowski space (16.1.3).

Hint: coordinates. □

• So we know already more or less what a geometrical expression of Hamilton's equations might look like; all that is needed is just to rewrite it in terms of "complete" objects Γ, d instead of Γ_0, d_0.

18.5.3 Check that integral curves of the dynamical field $\Gamma \equiv \Gamma_0 + \partial_t$ correspond to solutions of Hamilton's equations if and only if Γ satisfies the equation

$$i_\Gamma(\hat{\omega} - dH \wedge dt) \equiv i_\Gamma d(\hat{\theta} - H\,dt) = 0$$

so that we may finally write the geometrical version of *Hamilton's equations* (for the Hamiltonian possibly depending on time) in the form

$$\dot{\gamma} = \Gamma \qquad i_\Gamma(\hat{\omega} - dH \wedge dt) \equiv i_\Gamma d(\hat{\theta} - H\,dt) = 0 \quad i_\Gamma dt = 1$$

Hint: $0 = i_\Gamma(\hat{\omega} - dH \wedge dt) = i_{\Gamma_0}\hat{\omega} - (\Gamma H)dt + dH = i_{\Gamma_0}\hat{\omega} + d_0 H - (\Gamma_0 H)dt$; *two* equations result from this: $i_{\Gamma_0}\hat{\omega} = -d_0 H$ (exactly what we need) and $\Gamma_0 H = 0$, being, however, just a *consequence* of the first one (apply i_{Γ_0} on the first equation). □

• We see that from behind Hamilton's equations on extended phase space $T^*M \times \mathbb{R}$ all of a sudden the (apparently important) 1-form

$$\hat{\theta} - H\,dt \equiv p_a\,dq^a - H(q,p,t)\,dt \equiv \text{“}p\,dq - H\,dt\text{”}$$

unexpectedly arose.

18.5.4 Check that the exterior derivative of the 1-form $\hat{\theta} - H\,dt$ has maximum rank (namely $2n$).

Hint: for example, write it down in components and examine how the coordinate basis is mapped. □

• A 1-form in odd-dimensional space (dim $= 2n + 1$), whose exterior derivative has maximum rank ($2n$) is called a *contact form* and it defines a *contact structure* on a manifold. This turns out to be an important structure with numerous applications. We will, however, no longer encounter it in this book and it is mentioned just for completeness.[395]

There is yet another reason for being interested in the 1-form $p\,dq - H\,dt$. Imagine we would like to derive Hamilton's (or Lagrange's) equations from a variational principle. An action integral should be a *line* integral (an extremal *curve* is to be found), consequently some (yet unknown) distinguished *1-form,* closely related to the resulting equations, has to be under the integral sign. The form (18.5.3) of the equations offers, however, *just a single* 1-form, namely $p\,dq - H\,dt$. So the idea that it is just $p\,dq - H\,dt$ which is to be integrated along γ in order to obtain an action integral for the Hamilton equations looks fairly promising. Come on, let's have a look.

18.5.5 Define a functional on curves on the extended phase space

$$S[\gamma] = \int_\gamma (\hat{\theta} - H\,dt) \equiv \int_\gamma (p\,dq - H\,dt) \equiv \int_{t_1}^{t_2} \{p_a(t)\dot{q}^a(t) - H(q(t), p(t), t)\}\,dt$$

[395] Arnold's book is highly recommended for learning more.

Check that appropriate extremals of this functional indeed coincide with solutions of Hamilton's equations and find out what exactly should be kept fixed in order make the statement true.

Solution: a standard procedure in analytical mechanics looks as follows:

$$\delta \int_{t_1}^{t_2} (p_a \dot{q}^a - H)\,dt = \int_{t_1}^{t_2} (\delta p_a \dot{q}^a + p_a \delta \dot{q}^a - \delta H)\,dt$$

$$= \int_{t_1}^{t_2} \left\{ \delta p_a \left(\dot{q}^a - \frac{\partial H}{\partial p_a} \right) + \delta q^a \left(-\dot{p}_a - \frac{\partial H}{\partial q^a} \right) \right\} dt + [p_a \delta q^a]_{t_1}^{t_2}$$

so if q^a in times t_1 and t_2 are fixed in the course of variation (note that nothing special is required though from p_a), the condition of extremum results in the validity of Hamilton's equations; geometrically a procedure of variation might look (for injective γ) as follows:[396] a new curve is produced by an infinitesimal shift realized by means of a *flow* of a *vector field* ξ: $\gamma \mapsto \gamma_\epsilon \equiv \Phi_\epsilon \circ \gamma$. The field ξ should be everywhere tangent to T^*M (it is to produce changes of q, p at a given time instant, so that $\xi = \xi^a(q,p)\partial_a + \xi_a(q,p)\partial^a + 0\partial_t$). Then,

$$S[\gamma_\epsilon] = \int_{\Phi_\epsilon \circ \gamma} (\hat{\theta} - H\,dt) = \int_\gamma \Phi_\epsilon^*(\hat{\theta} - H\,dt)$$

$$= S[\gamma] + \epsilon \int_\gamma \mathcal{L}_\xi(\hat{\theta} - H\,dt) = S[\gamma] + \epsilon \int_\gamma (i_\xi d + d i_\xi)(\hat{\theta} - H\,dt)$$

$$= S[\gamma] + \epsilon \int_\gamma i_\xi(\hat{\omega} - dH \wedge dt) + \int_{\partial\gamma} i_\xi \hat{\theta}$$

$$= S[\gamma] + \epsilon \int_\gamma i_\xi(\hat{\omega} - dH \wedge dt) + \epsilon [p_a \xi^a(q,p)]_{t_1}^{t_2}$$

Now if we ensure $\xi^a(q,p)|_{t_1} = \xi^a(q,p)|_{t_2} = 0$ (so that q^a are not varied at the ends) and we require vanishing of the first-order contribution, we get $i_{\dot{\gamma}} i_\xi(\hat{\omega} - dH \wedge dt) = 0$ for each $\xi \Rightarrow i_{\dot{\gamma}}(\hat{\omega} - dH \wedge dt) = 0$, which are already Hamilton equations due to (18.5.3). □

• Now let us have a look at how the corresponding action might work in the Lagrangian case.

18.5.6 Consider a curve $\gamma(t)$ in *configuration* space M. Let $\hat{\gamma}(t)$ be its *natural lift* to TM (17.5.1) and denote by the same letter also its lift to the *extended* "phase" space $TM \times \mathbb{R}$, i.e. $t \mapsto (\hat{\gamma}(t), t) \equiv \hat{\gamma}(t)$. Denote further by hats pull-backs of θ_L and E_L with respect to the natural projection $\pi_1 : TM \times \mathbb{R} \to TM$ and finally define a functional on curves γ in the *configuration* space as

$$S[\gamma] := \int_{\hat{\gamma}} (\hat{\theta}_L - \hat{E}_L\,dt)$$

[396] Notice that no particular technical simplification occurs in the geometrical version here. This method of reasoning is, however, often useful for variations of *more complicated* functionals.

Put another way, we use the well-established model for the "Hamiltonian" action (18.5.4), but we consider this action only for those curves *which are lifts*[397] from M. Check that if we work out the action in more detail, we get the standard expression for the action in Lagrangian mechanics

$$S[\gamma] \equiv \int_{\hat{\gamma}} (\hat{\theta}_L - \hat{E}_L\, dt) = \int_{t_1}^{t_2} L(q(t), \dot{q}(t), t)\, dt$$

Solution: the hatted objects have the same coordinate expressions as they had on TM; then

$$S[\gamma] = \int_{\hat{\gamma}} \left\{ \frac{\partial L}{\partial v^a} dx^a - \left(v^a \frac{\partial L}{\partial v^a} - L \right) dt \right\} = \int_{\hat{\gamma}} \frac{\partial L}{\partial v^a}(dx^a - v^a dt) + \int_{\hat{\gamma}} L dt$$
$$= \int_{\hat{\gamma}} L\, dt \equiv \int_{t_1}^{t_2} L(q(t), \dot{q}(t), t)\, dt$$

since the 1-form $(dx^a - v^a dt)$ vanishes *on any natural lift.* □

Summary of Chapter 18

It is shown how classical mechanics may be formulated on TM and T^*M. In the non-degenerate case, both dynamics turn out to be completely equivalent geometrically: they realize a standard "symplectic" dynamics we studied in Chapter 14, i.e. a motion along integral curves of the Hamiltonian field. On T^*M the canonical symplectic structure is available from the outset so that the choice of a function H is the only step to be made. On TM the situation is a bit more complicated; rather than a symplectic form there is a canonical $\binom{1}{1}$-type tensor field available and the required symplectic structure is given only after the latter is combined with the (non-degenerate) Lagrangian, regarded as a function on TM. The standard Lagrange equations result by the projection of the symplectic dynamics onto the base M. The projection adds one order, so that the final equations are second-order differential equations on M. Hamilton's equations live directly on the total space T^*M and that is why they are (as is always the case for equations for integral curves) only first-order differential equations. Making use of the Lagrangian L one may construct the Legendre map $TM \to T^*M$, which serves as a bridge between the two dynamics. If the Hamiltonian (or Lagrangian) depends explicitly on time, a modification of the formalism is needed since the carrier manifold is now odd-dimensional. The distinguished 1-form $p\,dq - H\,dt$ enters the equations and it turns out that this form also plays a decisive role in a construction of the action functional.

$\theta_L := S(dL),\ \ \omega_L := d\theta_L$	Cartan 1-form, Cartan 2-form	(18.2.3)
$E_L := \Delta L - L$	Energy corresponding to the Lagrangian L	Sec. 18.2
$\dot{\gamma} = \zeta_{E_L},\ \ i_{\zeta_{E_L}}\omega_L = -dE_L$	Lagrange's equations (on TM yet)	(18.2.6)

[397] This was *not possible* in T^*M since there is *no natural way* to lift curves from M to T^*M.

$\langle \hat{L}(v), w \rangle := (d/dt)\vert_0 \, L(v + tw)$	Legendre map $\hat{L} : TM \to T^*M$	(18.3.1)
$\hat{L} \circ \Phi_t^L = \Phi_t^H$	Lagrangian and Hamiltonian flows related	(18.3.4)
$TR_g, \;\; T^*R_{g^{-1}}$	Lifts of action R_g on M to TM and T^*M	(18.4.1)
$\tilde{\xi}_X$	Generators of the lifted actions	(18.4.1)
$P_X = \langle \theta_L, \tilde{\xi}_X \rangle, \;\; P_X = \langle \theta, \tilde{\xi}_X \rangle$	"Hamiltonians" of the lifted actions	(18.4.1)
$L = \frac{1}{2} \overset{\circ}{g} - \overset{\circ}{\phi}$	Natural Lagrangian on TM	(18.4.6)
$\int_\gamma (\hat{\theta} - H\,dt) \equiv \int_\gamma (p\,dq - H\,dt)$	Action integral for the Hamiltonian dynamics	(18.5.6)
$\int_{\hat{\gamma}} (\hat{\theta}_L - \hat{E}_L dt) \equiv \int_{t_1}^{t_2} L(\hat{\gamma}(t))\,dt$	Action integral for the Lagrangian dynamics	(18.5.6)

19

Linear connection and the frame bundle

• In this chapter we begin with a systematic study of connections in principal bundles, which have numerous important applications in modern theoretical physics (*gauge field* theory representing the most prominent application). This theory deals with concepts which are fairly simple, yet not sufficiently motivated for a beginner in the field. Therefore in our presentation a whole chapter is devoted first to one particular case – the *linear* connection. Although this topic is already well known from Chapter 15, here we adopt a brand new approach. It turns out that the novel point of view on the good old linear connection clearly indicates a direction towards a straightforward, albeit far-reaching generalization. This results in an elegant and powerful conceptual framework unifying such seemingly different structures as those represented by linear connection and gauge fields.

19.1 Frame bundle $\pi : LM \to M$

• A novel point of view on the good old linear connection on (M, ∇) consists in expressing the fundamental concepts of the theory in terms of a larger manifold, which is denoted by LM and which may be canonically associated with any manifold M. This manifold is automatically endowed with some structure (due to the way it is constructed; recall a similar situation for TM and T^*M). If there is a connection on M, however, the structure becomes even richer and it affords the opportunity of the complete reformulation of the concept of connection on M in terms of the new structure on LM. (This reformulation turns out to be particularly convenient from the perspective of a generalization.) We will see that contrary to the description of a connection on M, where we are forced to describe it locally (on appropriate pieces and set the rules for how to glue the pieces on overlaps), on LM connections may be described explicitly globally, in terms of a (single global) connection form.

So the first step to be done is to introduce the concept of a frame bundle $\pi : LM \to M$ and to learn the structures which are available there already *prior to* adding a connection on M.

19.1.1 Consider (so far only) the set LM of all frames $e(x)$ at all points x of a manifold M

$$LM := \bigcup_{x \in M, e} e(x) \quad e(x) \leftrightarrow e_a(x) - \text{a frame in } T_x M$$

Show that this set may be naturally endowed with the structure of a smooth manifold of dimension $n + n^2$ (provided that M has dimension n). This manifold will be called the *manifold of frames* and denoted by LM.

Hint: let x^i be local coordinates on $\mathcal{O} \subset M$ and $e(x)$ a frame field defined on the same $\mathcal{O}$. Then for an arbitrary frame E in x we may write

$$E = e(x)y \qquad \text{i.e.} \quad E_a = e_b(x)y_a^b$$

for a unique non-singular matrix $y \in GL(n, \mathbb{R}) \Rightarrow (x^i, y_b^a)$ may serve as coordinates in the domain $\hat{\mathcal{O}} := \bigcup_{x \in \mathcal{O}, e} e(x) \equiv L\mathcal{O}$. If in $\mathcal{O}' \subset M$ we have local coordinates x'^i and a frame field $e'(x)$, such that on the overlap $\mathcal{O} \cap \mathcal{O}'$ there holds $x' = x'(x)$ and $e' = eA$ (i.e. $e'_a(x) = A_a^b(x)e_b(x)$), then

$$E = ey = e'y' \quad \Rightarrow \quad (x, y) \mapsto (x'(x), y'(x, y) \equiv A^{-1}(x)y)$$

$\Rightarrow$ we have smooth relations. Check the non-vanishing Jacobian $\partial(x', y')/\partial(x, y)$; it is useful to compare all of this with problems (17.1.2) and (17.1.5), where TM and T^*M were constructed. □

• We thus see that in our list of larger manifolds which may be canonically assigned to a manifold M, we may add to TM and T^*M from Chapter 17 our largest trophy up to now, the $(n + n^2)$-dimensional manifold LM. And in like manner as we revealed additional structure on TM and T^*M, we will find it here, too.

19.1.2 Define a map

$$\pi : LM \to M \qquad e(x) \mapsto x$$

i.e. we assign to a frame $e(x)$ in (the tangent space of) a point x just the point x itself. Check that

(i) it is a smooth map with a coordinate presentation[398]

$$\pi : \left(x^i, y_b^a\right) \mapsto x^i$$

(ii) for arbitrary x the preimage $\pi^{-1}(x)$ is diffeomorphic to $GL(n, \mathbb{R})$ (so that for any two points $x, x' \in M$, $\pi^{-1}(x)$ and $\pi^{-1}(x')$ are diffeomorphic to each other).

[398] Recall that a single symbol x^i actually denotes as many as *two* distinct objects, local coordinates on M as well as (a part of) the coordinates on LM; thus the functions x^i on the right differ from those on the left in this expression.

We see that we have obtained a *fiber bundle* (see Section 17.2) with total space LM, base M, canonical projection π and typical fiber $GL(n,\mathbb{R})$. The bundle $\pi : LM \to M$ is called the *frame bundle*.

Hint: (ii) a diffeomorphism is $(x^i, y^a_b) \leftrightarrow y^a_b$ (with fixed x^i). □

• A part of the structure on LM thus simply consists in the bundle structure. There is, however, more structure than this. Namely there is a natural action of the group $GL(n,\mathbb{R})$.

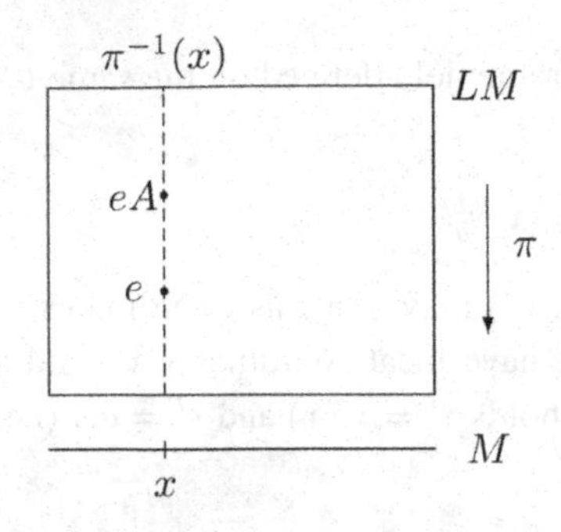

19.1.3 Let $\pi : LM \to M$ be the frame bundle. Show that on the manifold LM there is a natural structure of a right $GL(n,\mathbb{R})$-space which is moreover compatible with the fibration, i.e. in more detail that

(i) if $A \in GL(n,\mathbb{R})$, then the map

$$R_A : LM \to LM \qquad e \mapsto R_A e = eA$$

is a right action of $GL(n,\mathbb{R})$ on LM

(ii) in coordinates (x^i, y^a_b) it reads

$$R_A : (x^i, y^a_b) \mapsto \left(x'^i, y'^a_b\right) \equiv (x^i, y^a_c A^c_b)$$
$$\text{i.e.} \quad (x', y') = (x, yA)$$

(iii) the action is *free* ((13.4.14), all stationary subgroups happen to be trivial) and *transitive in a fiber*
(iv) the action is *vertical*,

$$\pi \circ R_A = \pi$$

i.e. it always transforms the points of LM within their own fiber (in the picture this is the vertical dashed line). □

• A fiber bundle whose total space is a right G-space, fibers are diffeomorphic to a group G, the action is vertical, free and in a fiber transitive (so that each fiber represents a *principal homogeneous space* of the group G, see (13.2.7)), is called a *principal G-bundle* or a *principal fibered manifold with a group G*. The frame bundle is a paradigmatic example of a *principal $GL(n,\mathbb{R})$-bundle*.

In the total space of an arbitrary fiber bundle the natural concept of *vertical* vectors and a vertical subspace in each tangent space arises. Recall (see Section 17.4) that a vector is said to be vertical if it is projected to the zero vector (the vertical subspace thus coincides with the kernel of the linear map π_*). This concept has some specific features on LM (as well as on an arbitrary principal G-bundle, see (20.2.2)) resulting from a relation to the action of the group.

19.1.4 Denote by ξ_C the fundamental field of the action R_A, which corresponds to an element $C \in gl(n,\mathbb{R})$. Check that

(i) in coordinates (x^i, y^a_b) it explicitly reads

$$\xi_C = (yC)^a_b \partial^b_a \equiv C^a_b \xi_{E^b_a} \qquad \xi_{E^a_b} = y^c_b \partial^a_c \qquad \partial^a_c \equiv \frac{\partial}{\partial y^c_a}$$

and that the map $C \mapsto \xi_C$ is indeed a representation of the Lie algebra $gl(n, \mathbb{R})$, i.e. that

$$\xi_{C+\lambda D} = \xi_C + \lambda \xi_D \qquad \xi_{[C,D]} = [\xi_C, \xi_D]$$

(ii) there holds

$$\pi_* \partial_i = \partial_i \qquad \pi_* \partial^a_b = 0$$

(in the first equation ∂_i on the left are vectors on LM and on the right on M)

(iii) at each point $e \in LM$ the fundamental fields $\xi_{E^a_b}$ constitute a basis of the *vertical subspace*

$$\text{Ver}_e LM := \text{Ker}\, \pi_*$$

(iv) an arbitrary vertical vector in $e \in LM$ may be written in the form ξ_C for an appropriate *unique* $C \in gl(n, \mathbb{R})$.

Hint: (iii) we have $\partial^a_b = (y^{-1})^c_b \xi_{E^a_c}$ and ∂^a_b *is* (according to (ii)) a basis ($\Rightarrow$ the number of vectors $\xi_{E^a_c}$ is sufficient); (iv) a consequence of items (i) and (iii). □

19.2 Connection form on LM

• Recall deep into the past how we described a linear connection on M. If a frame field $e(x)$ was available in a domain $\mathcal{O} \subset M$, we characterized the connection in this domain by a set of connection 1-forms[399] $\hat{\omega}^a_b$ with respect to the field $e_a(x)$ (15.6.1). In order to describe a global connection on the manifold M we thus need a covering of the manifold by open domains $\mathcal{O}_\alpha$ along with locally defined connection 1-forms; on each overlap $\mathcal{O} \cap \mathcal{O}'$ the compatibility of the system is ensured by the relations (15.6.2)

$$e' = eA \quad \Rightarrow \quad \hat{\omega}' = A^{-1} \hat{\omega} A + A^{-1}\, dA$$

So the geometrical object under consideration, a linear connection ∇ on M, is global, but its description turns out to be local (on patches and gluing rules). Here we will learn how by passing to a larger space, to a manifold LM, we can describe a connection in terms of *a single global* 1-form.

19.2.1 Let $\hat{\omega}$ be connection forms of a connection ∇ with respect to a frame field $e(x)$ living in a domain $\mathcal{O}$ and let (x^i, y^a_b) be local coordinates in $\hat{\mathcal{O}} \equiv \pi^{-1}(\mathcal{O})$ which are tailored to the frame field $e(x)$ (19.1.1). In $\hat{\mathcal{O}}$ define a square matrix with 1-form entries as follows:

$$\omega_{\mathcal{O}} \equiv \omega := y^{-1}(\pi^* \hat{\omega}) y + y^{-1}\, dy$$

$$\text{or in detail} \quad \omega^a_b := (y^{-1})^a_c (\pi^* \hat{\omega}^c_d) y^d_b + (y^{-1})^a_c\, dy^c_b$$

[399] Connection forms ω^a_b (on M) introduced in Chapter 15 will be denoted by $\hat{\omega}^a_b$ here, whereas the hats will be omitted on corresponding forms on LM (to be introduced in a while, see (19.2.1)).

Check that

(i) if the same is repeated in a domain $\mathcal{O}'$ with a frame field e' and coordinates (x'^i, y'^a_b) tailored to the primed frame field, then on the overlap $\hat{\mathcal{O}} \cap \hat{\mathcal{O}}'$ we have (assuming that $e' = eA(x)$ holds there)

$$\omega_{\mathcal{O}} = \omega_{\mathcal{O}'}$$

(ii) this result actually says that there is a *global* matrix[400] 1-form on LM

$$\omega \equiv \omega^a_b E^b_a \in \Omega^1(LM, gl(n, \mathbb{R}))$$

We will call this global $gl(n, \mathbb{R})$-valued 1-form ω on LM the *connection form.*

Hint: (i) making use of (19.1.3) and (15.6.2) we get

$$\begin{aligned}\omega_{\mathcal{O}'} &:= y'^{-1}(\pi^*\hat{\omega}')y' + y'^{-1}dy' \\ &= (A^{-1}y)^{-1}(\pi^*(A^{-1}\hat{\omega}A + A^{-1}dA))A^{-1}y + (A^{-1}y)^{-1}d(A^{-1}y) = \cdots \\ &= \omega_{\mathcal{O}}\end{aligned}$$

(notice that the coordinate expressions of $\pi^*\hat{\omega}$ and π^*A coincide with $\hat{\omega}$ and A respectively); (ii) in each particular domain $\hat{\mathcal{O}}$ we have $\omega := \omega_{\mathcal{O}}$; a computation confirms that the definition does not depend on the choice of coordinates and a frame field on $\mathcal{O}$. □

• Making use of (in general several) local 1-forms $\hat{\omega}_{\mathcal{O}}$ on (parts of) M we succeeded in constructing a single global 1-form ω on (the whole) LM. It is moreover always possible to pass (in both directions) between the two descriptions of connection in case of need. We will see now how the forms $\hat{\omega}_{\mathcal{O}}$ may be reobtained, if ω is available. The concept of a *local section* introduced in Section 17.2 proves to be useful for this purpose.

19.2.2 Check that

(i) local sections

$$\sigma : \mathcal{O} \to LM$$

of the frame bundle $\pi : LM \to M$ are in one-to-one correspondence with frame fields on $\mathcal{O} \subset M$

(ii) if $\sigma \leftrightarrow e(x) \leftrightarrow (x^i, y^a_b)$, then the coordinate presentation of the section itself is

$$x^i \mapsto (x^i, y^a_b = \delta^a_b) \qquad \text{i.e.} \quad x^i(x) = x^i \qquad y^a_b(x) = \delta^a_b$$

Hint: (i) $e(x) = \sigma(x)$; see (17.3.8); (ii) the definition of coordinates (x^i, y^a_b) and the concept of a section. □

19.2.3 Let $\sigma : \mathcal{O} \to LM$ be a local section, $e(x)$ the corresponding frame field on $\mathcal{O}$, $\hat{\omega}$ a connection form (on $\mathcal{O} \subset M$) with respect to $e(x)$ and $\omega \in \Omega^1(LM, gl(n, \mathbb{R}))$ the

[400] We see from the notation $\Omega^1(LM, gl(n, \mathbb{R}))$ that the space of $n \times n$ matrices, in which the form has values (and in which also the form $\hat{\omega}$ on $\mathcal{O}$ has its values), is regarded as the *Lie algebra* of such matrices. Although nothing forces us to this interpretation so far, it will prove to be convenient later on.

connection form on LM. Check that

$$\hat{\omega} = \sigma^* \omega$$

Hint: in $\pi^{-1}(\mathcal{O})$ we have $\omega = y^{-1}(\pi^*\hat{\omega})y + y^{-1}\,dy$ and the section reads in coordinates (19.2.2) $\sigma : x^i \mapsto (x^i, y^a_b = \delta^a_b)$; hence $\sigma^*\omega = \sigma^*\pi^*\hat{\omega} = (\pi \circ \sigma)^*\hat{\omega} \equiv \hat{\omega}$. □

• So we see that the information content in a single (matrix-valued) form ω on LM is the same as it is in the totality of all local (matrix-valued) forms $\hat{\omega}$ on M. The "translation" in the "upward" direction (from M to LM) may be performed according to (19.2.1), the "downward" direction (from LM to M) is performed via pull-back with respect to a local section according to (19.2.3).

Now let us have a look at two important properties of a connection form ω.

19.2.4 Check that a connection form ω

(i) is of type Ad, i.e. it behaves with respect to the action R_A of the group $GL(n, \mathbb{R})$ on LM as follows:

$$R_A^*\omega = \mathrm{Ad}_{A^{-1}}\omega \equiv A^{-1}\omega A$$

or in more detail

$$\left(R_A^*\omega^a_b\right) E^b_a = \omega^a_b \left(A^{-1} E^b_a A\right) \qquad \text{so that} \quad R_A^*\omega^a_b = (A^{-1})^a_c \omega^c_d A^d_b$$

(ii) satisfies the identity

$$\langle \omega, \xi_C \rangle = C$$

where ξ_C is the fundamental field of the action R_A corresponding to the element $C \in gl(n, \mathbb{R})$

(iii) also satisfies the derived identities

$$\mathcal{L}_{\xi_C}\omega = -\mathrm{ad}_C\omega \equiv -[C, \omega] \qquad i_{\xi_C}\omega = C$$
$$i_{\xi_C} d\omega = -[C, \omega]$$

Hint: (i), (ii) a direct computation in coordinates, (19.1.3), (19.1.4) and (19.2.1); (iii) the first is an infinitesimal version of item (i) (set $A(t) = \exp(tC)$ and differentiate at zero) and the third is a combination of the first two (6.2.8). □

• So far we have learned that a connection on a manifold M induces on LM a global $gl(n, \mathbb{R})$-valued 1-form ω. As we will see shortly this form may be interpreted visually in terms of a "horizontal distribution" on LM. The concept of a distribution itself (along with its integrability) may be ranked among the most important geometrical notions, being an extremely useful instrument in numerous contexts (curvature of a connection, integrability conditions for systems of differential equations, various kinds of adapted coordinates, thermodynamics, etc.). The next section may be regarded as a short digression worked into the main narrative, in which basic notions of the theory of smooth distributions[401] on manifolds will be presented.

[401] Caution: the distributions treated here are something *completely different* from the distributions also known as *generalized functions* (such as the Dirac δ-function).

19.3 k-dimensional distribution $\mathcal{D}$ on a manifold $\mathcal{M}$

• Consider an n-dimensional manifold $\mathcal{M}$. At each point there is an n-dimensional linear space $T_x M$, the tangent space at x. Imagine we fix a k-dimensional subspace in each of these spaces (k being the same for all the points of the manifold)

$$\mathcal{D}_x \subset T_x M \qquad \dim \mathcal{D}_x = k$$

If these subspaces depend smoothly on x (we will see in a moment what this means precisely), we say that a *k-dimensional smooth distribution* was defined on M.

For example, this occurs when the manifold is "stratified" into k-dimensional submanifolds: we may then take as $\mathcal{D}_x$ the space of those vectors in x which are *tangent* to the submanifold (they are represented by curves along the submanifold). If, say, the air in a room is stratified into two-dimensional surfaces of constant temperature, at each point x in the room we get a two-dimensional subspace of vectors which are tangent to the surface passing through x; we thus obtain a two-dimensional distribution in a three-dimensional domain. For applications (in particular, in the theory of connections), however, the opposite situation turns out to be much more interesting: a k-dimensional distribution on $\mathcal{M}$ is given and we look for such a stratification of $\mathcal{M}$ to k-dimensional submanifolds (exactly one submanifold $\mathcal{S}_x$ is to pass through each point $x \in \mathcal{M}$), which "interlock" (integrate) neighboring subspaces into a single whole (called an integral submanifold). If the distribution happens to be smooth (the choice of a subspace smoothly depends on the point), one might expect that integral submanifolds should necessarily exist. It turns out, however, that life is not so easy (and boring). There are two classes of distributions, integrable (when the integral submanifolds do exist) and non-integrable (when they do not exist) and a simple (Frobenius) criterion exists, which enables us to test the integrability of each particular distribution. In order to formulate the criterion we have first to learn two basic approaches for the description of distributions themselves.

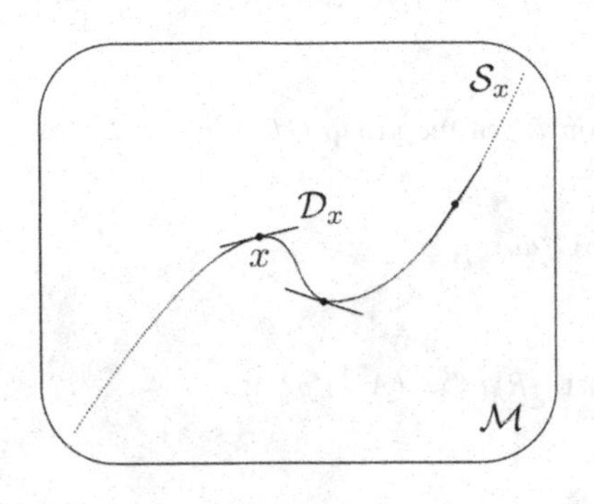

We begin with a discussion of subspaces in a given n-dimensional linear space L. Imagine we want to characterize technically a k-dimensional subspace $W \subset L$. The first idea might be to fix a basis in the subspace, i.e. to choose k linearly independent *vectors* in W. There is, however, also another possibility, in which (linearly independent) *covectors* are used. A subspace W may be namely thought of as consisting of those vectors in L which *annihilate* all the covectors under consideration, i.e. which *vanish* when inserted into the covectors.

19.3.1 Let L be an n-dimensional linear space and $\alpha \in L^*$ be a covector (1-form). Check that

(i) the set $\tilde{W}$ of vectors which *annihilate* the covector α, i.e.

$$\tilde{W} := \{w \in L \mid \langle \alpha, w \rangle = 0\}$$

is actually an $(n-1)$-dimensional *subspace* of the space L; the covector α is then called a *constraint 1-form* for $\tilde{W}$

(ii) if β is another covector in L which is linearly independent of α, then the set $\hat{W}$ of vectors which annihilate *both* the covectors is actually an $(n-2)$-dimensional subspace of L and moreover

$$\hat{W} \subset \tilde{W} \subset L$$

(iii) if there are q linearly independent covectors in L, then their *annihilator* (the set of vectors which annihilate all of these covectors) turns out to be an $(n-q)$-dimensional subspace in L.

Hint: (i) set $\alpha = e^1$ and complete *arbitrarily* to a basis in L^*; if e_a is the dual basis in L, then $\tilde{W}$ consists of exactly those vectors $w = w^a e_a$ for which $w^1 = 0$; (ii) set $\alpha = e^1$, $\beta = e^2$, etc. and complete arbitrarily; vectors from $\hat{W}$ then have $w^1 = w^2 = 0$; (iii) we call them $e^1, \dots, e^q$, etc. □

19.3.2 Consider an n-dimensional linear space L with a k-dimensional subspace $W \subset L$. Let $e_a = (e_\alpha, e_i)$ be a basis in L which is *adapted to the subspace* W, i.e. $e_\alpha \in W$, $\alpha = 1, \dots, k$. Check that

(i) if (e^α, e^i) is the dual basis, then the covectors e^i $(i = k+1, \dots, n)$ define (as constraint 1-forms) *the same* subspace W

(ii) the subspace W is not changed if we scramble the vectors e_α by an arbitrary non-singular $k \times k$-matrix A^α_β or the covectors e^i by an arbitrary non-singular $(n-k) \times (n-k)$-matrix B^i_j.

Hint: (i) from the definition of the dual basis $\langle e^i, e_\alpha \rangle = 0$; (ii) the subspace spanned by $e'_\alpha \equiv A^\beta_\alpha e_\beta$ coincides with the original subspace W, the covectors $e'^i \equiv B^i_j e^j$ annihilate just linear combinations of the vectors e_α. □

• We see that there is a considerable freedom (given by a matrix $A \in GL(k, \mathbb{R})$ or $B \in GL(n-k, \mathbb{R})$) in fixing a subspace using either the system of k vectors or $(n-k)$ covectors. The freedom may be greatly reduced if a *single* special (decomposable) $(n-k)$-form (related to the exterior product of the initial 1-forms) is used instead of $(n-k)$ 1-forms. We will not, however, pursue this possibility in more detail, since we will not need it in the future work with distributions.

Now it is already clear how a k-dimensional distribution $\mathcal{D}$ on a domain $\mathcal{O}$ *on a manifold* $\mathcal{M}$ might be described. In the first approach one should fix k linearly independent vectors at each point of the domain, i.e. specify k vector *fields* e_α, $\alpha = 1, \dots, k$ in $\mathcal{O}$, such that their values are linearly independent at each point. Provided that the fields moreover happen to be *smooth*, we speak about a *smooth distribution in a domain* $\mathcal{O}$. The other possibility is to specify $(n-k)$ constraint 1-forms θ^i, $i = k+1, \dots, n$; again, they should be smooth and at each point linearly independent. We say that a vector field V *belongs to the distribution* $\mathcal{D}$ (and write $V \in \mathcal{D}$), if at each point of the domain $\mathcal{O}$ the value V_x of the field belongs to the subspace $\mathcal{D}_x$, given by the distribution.

19.3.3 Let a k-dimensional distribution $\mathcal{D}$ in a domain $\mathcal{O}$ be given by vector fields e_α or, alternatively, by constraint 1-forms θ^i. Be sure to understand that then $\langle\theta^i, e_\alpha\rangle = 0$ and

$$V \in \mathcal{D} \quad \Leftrightarrow \quad V = V^\alpha(x)e_\alpha \quad \Leftrightarrow \quad \langle\theta^i, V\rangle = 0 \qquad \square$$

19.3.4 Consider a two-dimensional smooth distribution $\mathcal{D}$ in $\mathbb{R}^3$ given by the constraint 1-form

$$\theta^3 \equiv \theta := dz + x\,dy - y\,dx \equiv dz + r^2\,d\varphi$$

(the expression in cylindrical coordinates (r, φ, z) holds, of course, only at their points of applicability). Check that

(i) if $a, b, c \in \mathcal{F}(\mathbb{R}^3)$ and $V \equiv a\partial_x + b\partial_y + c\partial_z$, then

$$V \in \mathcal{D} \quad \Leftrightarrow \quad c = ya - xb \quad \Leftrightarrow \quad V = a(\partial_x + y\partial_z) + b(\partial_y - x\partial_z)$$

(ii) we may choose as e_α the fields

$$e_1 = \partial_x + y\partial_z \qquad e_2 = \partial_y - x\partial_z$$

Hint: (i) solve $\langle dz + x\,dy - y\,dx, a\partial_x + b\partial_y + c\partial_z\rangle = 0$. $\square$

• It turns out that this distribution is actually more interesting than we might expect at first sight; it is namely *non-integrable*. In the following problem we will first check this fact in a direct way; after introducing a formal Frobenius criterion we will confirm this result once again.

19.3.5 Consider the two-dimensional distribution $\mathcal{D}$ from problem (19.3.4). Verify that this distribution is not integrable. In order to do this check that

(i) the subspaces $\mathcal{D}_x$ are nowhere "vertical," so that any potential integral submanifold $\mathcal{S}$ would be necessarily a two-dimensional surface which may be expressed in the form $z = f(x, y)$
(ii) if we contemplate a curve $\gamma(t)$, which lies in such a surface and whose *projection* to the (x, y)-plane happens to be a *loop* (closed curve) $\hat{\gamma}(t)$, then we can deduce that also the *curve* $\gamma(t)$ *itself* is a *loop*
(iii) if $\pi : (x, y, z) \mapsto (x, y)$ is the projection of $\mathbb{R}^3$ to the (x, y)-plane, then the curve γ may be reconstructed from the projection $\hat{\gamma}$ by the requirements

$$\pi \circ \gamma = \hat{\gamma} \qquad \langle\theta, \dot{\gamma}\rangle = 0 \qquad z(t_0) = z_0$$

(iv) if we reconstruct in this way the curve $\gamma(t)$, whose projection is the *circle* (i.e. a loop)

$$x(t) = R\cos t \qquad y(t) = R\sin t$$

we get the *helix* ($\neq$ a loop)

$$x(t) = R\cos t \qquad y(t) = R\sin t \qquad z(t) = z(0) - R^2 t$$

(v) this shows (by contradiction) that there is actually *no integral submanifold* $\mathcal{S}$ of $\mathcal{D} \Rightarrow \mathcal{D}$ is *non-integrable*.

Hint: (i) $\partial_z \notin \mathcal{D}$; (ii) trivial; (iii) the definition of an integral submanifold and constraint forms. □

• Now we will learn (omitting the proof), what *Frobenius' theorem* (criterion) says about the integrability of a smooth distribution. The theorem exists in two equivalent versions, reflecting the two ways of specifying distributions. In terms of vector fields it says that a distribution $\mathcal{D}$ is integrable if and only if the commutator of arbitrary vector fields from $\mathcal{D}$ also belongs to $\mathcal{D}$:

$$\mathcal{D} \text{ is integrable} \quad \Leftrightarrow \quad \{U, V \in \mathcal{D} \Rightarrow [U, V] \in \mathcal{D}\}$$

19.3.6 Check that if $\mathcal{D}$ is given by vector fields e_α, $\alpha = 1, \ldots, k$, then we also have

$$\mathcal{D} \text{ is integrable} \quad \Leftrightarrow \quad [e_\alpha, e_\beta] = c^\rho_{\alpha\beta}(x) e_\rho \qquad \alpha, \beta, \rho = 1, \ldots, k$$

i.e. if and only if each commutator of the fields e_α may also be expressed in terms of the fields themselves.

Hint: each commutator $[e_\alpha, e_\beta]$ should belong to $\mathcal{D}$; see (19.3.3). □

19.3.7 Check that (also) according to Frobenius' integrability theorem the distribution $\mathcal{D}$ from (19.3.4) and (19.3.5) turns out to be non-integrable.

Hint: $[e_1, e_2] = -2\partial_z \neq a e_1 + b e_2$. □

• In terms of constraint 1-forms θ^i Frobenius' theorem says that a distribution $\mathcal{D}$ is integrable if and only if for arbitrary vectors $U, V \in \mathcal{D}$ there holds $d\theta^i(U, V) = 0$, i.e. if the *restriction* of all the 2-forms $d\theta^i$ to the distribution $\mathcal{D}$ vanish:

$$\mathcal{D} \text{ is integrable} \quad \Leftrightarrow \quad \{\theta^i|_{\mathcal{D}} = 0 \Rightarrow d\theta^i|_{\mathcal{D}} = 0\} \quad \text{i.e.} \quad \{U, V \in \mathcal{D} \Rightarrow d\theta^i(U, V) = 0\}$$

19.3.8 Check that this formulation of Frobenius' theorem is equivalent to the formulation in terms of vector fields given above.

Hint: according to Cartan's formulas (6.2.13) we have

$$d\theta^i(U, V) = U\langle\theta^i, V\rangle - V\langle\theta^i, U\rangle - \langle\theta^i, [U, V]\rangle$$

so, in particular, for $U, V \in \mathcal{D}$ this yields $d\theta^i(U, V) = -\langle\theta^i, [U, V]\rangle$. □

19.3.9 Check that (also) according to this version of Frobenius' integrability theorem the distribution $\mathcal{D}$ from (19.3.4) is non-integrable.

Hint: $d\theta(e_1, e_2) = 2\ (\neq 0)$. □

• We also mention that the form version of Frobenius' theorem may also be found as an (equivalent) statement that $\mathcal{D}$ is integrable if and only if there are $(n-k)^2$ 1-forms σ^i_j such that $d\theta^i = \sigma^i_j \wedge \theta^j$

$$\mathcal{D} \text{ is integrable} \quad \Leftrightarrow \quad \left\{\exists \sigma^i_j : d\theta^i = \sigma^i_j \wedge \theta^j\right\}$$

19.3.10 Check that (also) according to this version of Frobenius' integrability theorem the distribution $\mathcal{D}$ from (19.3.4) is non-integrable.

Hint: $d\theta = 2dx \wedge dy \neq \sigma \wedge \theta$ for any 1-form σ (try $\sigma = a\,dx + b\,dy + c\,dz$ and get a contradiction). □

• We may illustrate Frobenius' theorem by applying it to a derivation of the integrability conditions of a system of first-order partial differential equations of the form

$$\partial_\alpha y^i = f^i_\alpha(x, y) \qquad i = 1, \dots, m; \quad \alpha = 1, \dots, n$$

for m unknown functions $y^i \equiv y^1, \dots, y^m$, depending on n coordinates $x^\alpha \equiv x^1, \dots, x^n$; the functions f^i_α on the right-hand side are regarded as being given and they may depend on all x^α as well as y^i. Since we have altogether mn equations for m unknown functions, the system is overdetermined (for $n \geq 2$, i.e. if we indeed contemplate *partial* differential equations) and it may possess no solution at all (the equations may turn out to be contradictory). The exact form of consistency conditions for such a system may be obtained by first expressing it in terms of distributions (we reformulate the problem as an integrability problem for an appropriate distribution) and then solve it with the help of Frobenius' theorem.

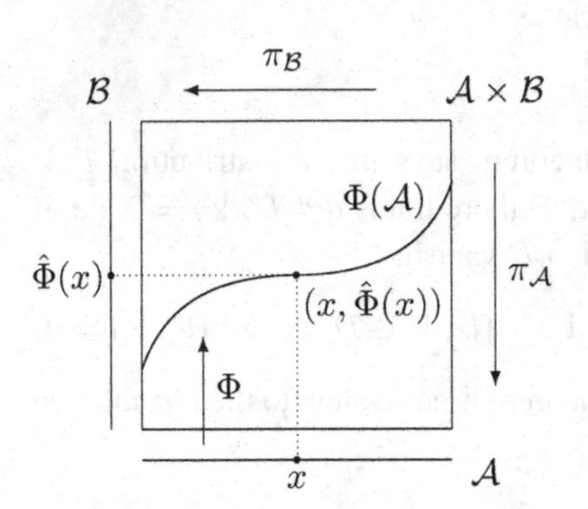

19.3.11* Consider maps $\hat{\Phi} : \mathcal{A} \to \mathcal{B}$, where $\mathcal{A}$ and $\mathcal{B}$ are open domains in $\mathbb{R}^n[x^\alpha]$ and $\mathbb{R}^m[y^i]$ respectively. In coordinates they are given by functions $y^i(x^\alpha)$, which are just unknowns in the equations under consideration. Next, define the map

$$\Phi : \mathcal{A} \to \mathcal{A} \times \mathcal{B} \qquad x \mapsto (x, \hat{\Phi}(x))$$
$$\text{i.e.} \qquad x^\alpha \mapsto (x^\alpha, y^i(x^\alpha))$$

The *graph* of the map $\hat{\Phi}$, i.e. the image $\Phi(\mathcal{A})$, is according to (1.4.10) a submanifold in $\mathcal{A} \times \mathcal{B}$. Notice also that actually we are dealing with a *product bundle* $\pi : \mathcal{A} \times \mathcal{B} \to \mathcal{A}$ (17.2.1), the map Φ being its *section* (17.2.4) and the submanifold $\Phi(\mathcal{A})$ being the image (with respect to the section) of the base in the total space of the bundle. Finally define the key element of the construction, a distribution $\mathcal{D}$ on $\mathcal{A} \times \mathcal{B}$, given by constraint 1-forms

$$\theta^i := dy^i - f^i_\alpha dx^\alpha \qquad f^i_\alpha(x, y) \text{ being the right-hand sides of the system of equations}$$

Check that

(i) the tangent space in an arbitrary point of the submanifold $\Phi(\mathcal{A})$ is spanned by a basis which is constituted by the vectors

$$\Phi_* \partial_\alpha = \partial_\alpha + y^i_{\ ,\alpha} \partial_i$$

(ii) the tangent space coincides with a subspace given by the distribution $\mathcal{D}$ if and only if the functions $y^i(x^\alpha)$ happen to be *solutions* of the system of differential equations under consideration, $y^i{}_{,\alpha} = f^i_\alpha$

$$\Phi_* T_x \mathcal{A} = \mathcal{D}_{\Phi(x)} \quad \Leftrightarrow \quad y^i(x^\alpha) \text{ are solutions of } \quad y^i{}_{,\alpha} = f^i_\alpha$$

(iii) to find a solution of the system is thus the same thing as to construct an *integral submanifold* $\mathcal{S}$ of the distribution $\mathcal{D}$ (then $\Phi(\mathcal{A}) = \mathcal{S}$)

(iv) the system of equations has a solution if and only if the distribution $\mathcal{D}$ is integrable.

Hint: (i) the definition of Φ_* and $\Phi(\mathcal{A})$; (ii) check that $\langle \theta^i, \Phi_* \partial_\alpha \rangle = y^i{}_{,\alpha} - f^i_\alpha(x, y)$. □

• The question of whether there is a solution of the system of equations $y^i{}_{,\alpha} = f^i_\alpha$ is thus replaced by an equivalent question concerning the integrability of the distribution $\mathcal{D}$. The latter question may be, however, answered with the help of Frobenius' theorem.

19.3.12* Check that

(i) the distribution $\mathcal{D}$ from (19.3.11) is integrable (so that the system of differential equations $y^i{}_{,\alpha} = f^i_\alpha$ has a unique solution) if and only if the functions $f^i_\alpha(x, y)$ satisfy the *integrability condition*

$$f^i_{[\alpha,\beta]} = f^j_{[\alpha} f^i_{\beta],j}$$

(ii) for $n = 1$ (for *ordinary* differential equations) the condition is automatically satisfied (so that individual equations of the system of ordinary differential equations cannot be contradictory to each other)

(iii) the system of two equations

$$\frac{\partial f}{\partial x} = f \sin y \quad \frac{\partial f}{\partial y} = \lambda f x \cos y \qquad \lambda \in \mathbb{R}$$

for the unknown function $f(x, y)$ has a solution only for $\lambda = 1$.

Hint: (i) a straightforward computation gives

$$d\theta^i = \left(f^i_{\alpha,\beta} + f^j_\beta f^i_{\alpha,j}\right) dx^\alpha \wedge dx^\beta + f^i_{\alpha,j}\, dx^\alpha \wedge \theta^j$$

which should be equal, according to Frobenius' theorem, to $\sigma^i_j \wedge \theta^j$; since the forms dx^α and θ^i constitute a basis, the decomposition of σ^i_j with respect to this basis and a comparison gives the needed result (plus an explicit form $\sigma^i_j = f^i_{\alpha,j}\, dx^\alpha$); (ii) indices of the type α, β take only a *single* value (so that the antisymmetrization *trivially* gives zero on both sides); (iii) use the criterion from item (i).[402] □

• In a particular case of a two-dimensional distribution in ordinary three-dimensional Euclidean space E^3 the integrability conditions may be expressed in the notation of vector analysis.

19.3.13 A constraint 1-form θ of a two-dimensional distribution in ordinary three-dimensional Euclidean space E^3 has the form (as is the case for each 1-form in E^3)

$$\theta = \mathbf{A} \cdot d\mathbf{r},$$

[402] An elementary method: the system may be written as $F_{,x} = \sin y$, $F_{,y} = \lambda x \cos y$ for $F = \ln f$; clearly $F_{,xy} = F_{,yx}$ holds.

so that it may be parametrized by a *vector field* $\mathbf{A}$. Check that

(i) the distribution $\mathcal{D}$ given by this 1-form consists in each point of vectors which are *perpendicular to* $\mathbf{A}$
(ii) Frobenius' integrability condition for $\mathcal{D}$ may be written, using the standard vector analysis notations, in the form

$$\mathbf{A} \cdot \text{curl}\, \mathbf{A} = 0$$

Hint: (i) according to (8.5.8) there holds $i_{\mathbf{B}}(\mathbf{A} \cdot d\mathbf{r}) = \mathbf{B} \cdot \mathbf{A}$; (ii) according to the text after (19.3.9) integrability is equivalent to the existence of a 1-form σ such that $d\theta = \sigma \wedge \theta$; if $\sigma = \mathbf{C} \cdot d\mathbf{r}$, then due to (8.5.4) and (8.5.8) we obtain

$$\text{curl}\, \mathbf{A} \cdot d\mathbf{S} = (\mathbf{C} \times \mathbf{A}) \cdot d\mathbf{S} \qquad \text{whence} \quad \text{curl}\, \mathbf{A} = \mathbf{C} \times \mathbf{A}$$

Justify that $\mathbf{A} \cdot \mathbf{G} = 0$ is equivalent to $\mathbf{G} = \mathbf{C} \times \mathbf{A}$ for some (not unique) $\mathbf{C}$. □

• We can see from the examples (19.3.4) and (19.3.13) that on three-dimensional manifolds a two-dimensional distribution may not be integrable. One easily finds that the lowest dimension of a manifold for which there is a chance for non-integrability is three and that moreover at least the two-dimensional distribution is needed for that.

19.3.14 Be sure to understand that

(i) any one-dimensional distribution is necessarily integrable
(ii) on a two-dimensional manifold any distribution is necessarily integrable.

Hint: (i) restriction of $d\theta^i$ to $\mathcal{D}$ vanishes by virtue of the *dimension* (restriction of a 2-form to the one-dimensional space); (ii) potentially interesting are only one-dimensional distributions. □

• An important application of integrable distributions is also encountered in *thermodynamics*. Recall that the state of a thermodynamic system is characterized by several "generalized coordinates" $(x^1, \dots, x^n)$ and the temperature T (also serving as a coordinate x^{n+1}). Various processes may be regarded as *curves* in the space with coordinates $(x^1, \dots, x^n, T)$. In the processes the heat Q may be absorbed and work A may be done. A computation of these two quantities consists in performing *line integrals* of appropriate 1-forms along the curves, which correspond to the processes, namely a *heat 1-form* $đQ$ and a *work 1-form* $đA$,

$$Q = \int_\gamma đQ \quad A = \int_\gamma đA \qquad đA = X_i\, dx^i$$

(X_i are called "generalized forces" corresponding to generalized coordinates x^i). For example, for an ideal gas we have $n = 1$ (such systems are called *simple*), the coordinate is the volume V and the force is the pressure p, so that the work is given by the line integral

$$A \equiv \int_\gamma đA = \int_\gamma p\, dV$$

A small bar on $đ$ on 1-forms of heat and work indicates the important fact that the forms are *not exact* – there are no *state quantities* (i.e. functions on a manifold with the above-mentioned coordinates), which correspond to heat or work – both the heat and work correspond to *processes* rather than states.

(For historical reasons, 1-forms are often called *Pfaffian forms* (or linear differential forms) in thermodynamics and, in particular, exact 1-forms df are called "exact differentials" (f being then a state quantity); the symbol $đ$ present in $đQ$ and $đA$ is related to the infinitesimal nature of corresponding quantities (just as was the case with surface elements dS_i in integration theory) rather than with the action of the exterior derivative (gradient) d (by means of *integration* of these 1-forms along infinitesimal curves we get infinitesimal amounts of heat and work).)

According to the *first law of thermodynamics*, the 1-forms under consideration are related by

$$đQ = dE + đA$$

(where E is a state quantity called the "internal energy"), i.e. upon integrating along a curve γ which corresponds to the process, we get a condition expressing the balance of heat, work and energy in the process

$$Q = E(\gamma(1)) - E(\gamma(0)) + A$$

(the heat absorbed in the process is used to change the internal energy and for performing work).

A process in which no heat is absorbed is called an *adiabatic process*. Curves which correspond to such processes (i.e. an *adiabatic curve*) thus have the property

$$i_{\dot\gamma} đQ = 0$$

i.e. tangent vectors to the curves *annihilate* the heat 1-form. The subspaces of such vectors constitute an n-dimensional distribution $\mathcal{D}$ on an $(n+1)$-dimensional manifold, where the life of the thermodynamic system under consideration is enacted. Integrability of the distribution turns out to be an important issue. Namely, one of the formulations of the *second law of thermodynamics* (originated by Carathéodory) says that this distribution *is integrable*. Put another way, an arbitrarily small neighborhood of a point (i.e. state) contains points which cannot be reached by an adiabatic process – the process namely "runs" (by definition) within a single integral submanifold of the distribution, so that the (nearby) points which *do not reside* in the same integral submanifold *cannot* be reached in this way. In order to reach points on different integral submanifolds a process is needed in which a non-vanishing amount of heat is absorbed (or released). One can show[403] that from integrability of the "adiabatic" distribution, i.e. from validity of Frobenius' condition

$$d(đQ) = \sigma \wedge đQ$$

[403] The statement is that if a 1-form α satisfies the condition $d\alpha = \sigma \wedge \alpha$ (i.e. the distribution defined by α is integrable), then (locally) functions f, g exist such that $\alpha = f\,dg$ (in particular, for $f =$ constant the form α is exact). Integral submanifolds then evidently coincide with the surfaces $g =$ constant.

the possibility of introducing a function (state quantity) S follows (called *entropy*), such that the heat 1-form $đQ$ may be written as

$$đQ = T\, dS$$

so that integral submanifolds are given by the "level surfaces" of the function S – they are just submanifolds of *constant entropy*

$$S(x, T) = \text{constant}$$

Two nearby points cannot be reached by an adiabatic (i.e. isoentropic) process, if the value of the entropy at the two points turns out to be different. In this approach the existence of the entropy is a mathematical consequence of a physical postulate – (the Carathéodory formulation of) the second law of thermodynamics.

19.3.15 Be sure to understand that for *simple* systems this postulate is not needed for the existence of S.

Hint: the thermodynamics takes place on a *two-dimensional* manifold, so that the "adiabatic" distribution is *one-dimensional*; see (19.3.14). □

19.4 Geometrical interpretation of a connection form: horizontal distribution on LM

• Once we become world-acclaimed authorities in the theory and applications of distributions and their integrability, we immediately embark on searching for them on LM (from where we just made a short digression). What we are looking for is a distinguished subspace $\mathcal{D}_e \subset T_e LM$ in each tangent space of the manifold. We will actually find *two* interesting subspaces in $T_e LM$: the first (vertical) subspace is always available (it follows from the mere existence of fibers, thus being available already for the "bare" manifold M), the second one (horizontal) enters into play only for manifolds endowed *with connection* (M, ∇).

Let us start with the ubiquitous vertical distribution. We already encountered the concept of a vertical subspace in Section 17.4 in the context of the tangent and cotangent bundles TM and T^*M and on LM it was mentioned at the end of Section 19.1.

19.4.1 A *vertical distribution* $\mathcal{D}^v$ may be defined on LM so that the *vertical subspace* $\text{Ver}_e LM$ is declared to be the subspace which the distribution singles out in each tangent space

$$\mathcal{D}^v_e := \text{Ver}_e LM \equiv \text{Ker}\, \pi_* \subset T_e LM \qquad \text{so that} \quad W \in \mathcal{D}^v_e \quad \Leftrightarrow \quad \pi_* W = 0$$

Check that the distribution

(i) is at each point spanned by a basis of the *fundamental* fields $\xi_{E^a_b}$ of the action R_A, so that a general vertical vector field has the form

$$W = W^b_a(x, y)\xi_{E^a_b}$$

(ii) has dimension n^2
(iii) is *integrable* (its integral submanifold through the point e over x being just the fiber $\pi^{-1}(x)$ over x)
(iv) is also integrable according to the formal Frobenius criterion.

Hint: (i) see (19.1.4); (iii) the reason is mentioned in the bracket; (iv) the property $[\xi_X, \xi_Y] = \xi_{[X,Y]}$ (13.4.3) of fundamental fields ξ_X and the Frobenius' criterion (the version from (19.3.6)). □

• If there is an action of a group G on a manifold $\mathcal{M}$, an action on *distributions* on the manifold is naturally induced: if R_g shifts points, then R_{g*} shifts vectors and, consequently, also subspaces

$$\mathcal{D}_x \mapsto R_{g*}\mathcal{D}_x =: (R_g\mathcal{D})_{xg}$$

It may happen, in particular, that a distribution is *G-invariant* ($R_g\mathcal{D} = \mathcal{D}$), i.e. the shifted subspace always happens to coincide with the subspace residing originally at the shifted point. If the distribution is moreover integrable, it results in G-invariance of integral submanifolds. All of this is true for the vertical distribution on LM.

19.4.2 Check that the vertical distribution $\mathcal{D}^v$ on LM is $GL(n,\mathbb{R})$-invariant, i.e. that for each element $A \in GL(n,\mathbb{R})$ there holds

$$R_A\mathcal{D}^v = \mathcal{D}^v \quad \text{or in more detail} \quad R_{A*}\left(\mathcal{D}^v_e\right) = \mathcal{D}^v_{eA}$$

Hint: verticality (19.1.3) of the action yields $\pi_* R_{A*} = \pi_*$, whence we get

$$w \in \text{Ver}_e LM \quad \Rightarrow \quad R_{A*}w \in \text{Ver}_{eA} LM$$

Alternatively making use of the property

$$R_g^*\xi_X = \xi_{\text{Ad}_g X}$$

(13.4.3) of the fundamental fields ξ_X. □

• We see that the construction of the vertical distribution indeed needs no connection; actually all that is needed is the fiber structure of the total space.

Just the opposite is true, however, for the horizontal distribution (to be defined presently). Its construction is inconceivable without the connection; actually it may even be regarded as a convenient (alternative) way of expressing what exactly the connection itself *is*.

A connection form ω on LM is a 1-form with values in an n^2-dimensional linear space $gl(n,\mathbb{R})$. Then it may be regarded as a collection of n^2 "ordinary" (component) 1-forms ω^a_b (see Section 6.4). If these 1-forms were linearly independent at each point, they would define (according to (19.3.1) and (19.3.3)) a smooth n-dimensional distribution on the manifold LM. We will check that this is indeed the case.

19.4.3 Let $\omega \equiv \omega^a_b E^b_a$ be a connection form from problem (19.2.1). Check that

(i) its component 1-forms ω^a_b are linearly independent at each point

(ii) by the condition

$$V \in \mathcal{D}^h \quad \Leftrightarrow \quad \langle \omega, V \rangle = 0$$

a smooth n-dimensional distribution $\mathcal{D}^h$ on LM is defined; it is called the *horizontal distribution*, the subspace singled out (at each point $e \in LM$) by the distribution is called the *horizontal subspace*

$$\text{Hor}_e LM \equiv \mathcal{D}^h_e \subset T_e LM$$

and vectors which belong to the subspace ($V \in \mathcal{D}^h$) are *horizontal vectors.*

Hint: (i) let $k^a_b \omega^b_a = 0$; then, according to (19.2.4), for arbitrary $C \in gl(n, \mathbb{R})$ we have

$$0 = \langle k^a_b \omega^b_a, \xi_C \rangle = k^a_b C^b_a$$

so that $k^a_b = 0$ □

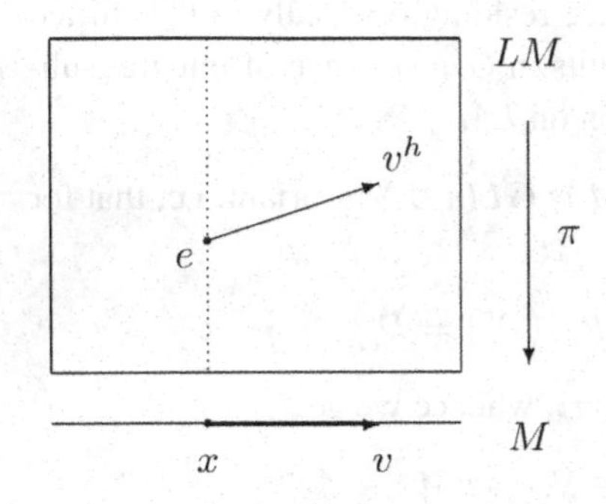

19.4.4 Let $\mathcal{D}^h$ be the horizontal distribution on LM, given by a connection form ω. Check that

(i) if $v \in T_x M$ is an arbitrary vector at the point $x \in M$, then at each point $e \in \pi^{-1}(x)$ there exists its *horizontal lift*, i.e. a unique vector $v^h \in T_e LM$ such that

$$\pi_* v^h = v \qquad \text{(it projects to } v\text{)}$$

$$v^h \in \text{Hor}_e LM \qquad \text{(it is horizontal)}$$

(ii) the distribution $\mathcal{D}^h$ is spanned by (a basis consisting of) vector fields

$$H_i \equiv \partial^h_i := \partial_i - \langle \omega^a_b, \partial_i \rangle y^b_c \partial^c_a \equiv \partial_i - \langle \omega^a_b, \partial_i \rangle \xi_{E^b_a}$$

where $\xi_{E^b_a}$ are the fundamental fields of the action R_A on LM; moreover the fields $H_i \equiv \partial^h_i$ are *horizontal lifts of the coordinate*[404] *basis* ∂_i on M

(iii) a general horizontal vector field V on LM may be written in the form

$$V \in \mathcal{D}^h \quad \Leftrightarrow \quad V = V^i(x, y) H_i \equiv V^i(x, y) \partial^h_i$$

(iv) the operation of the horizontal lift $v \mapsto v^h$ is a *linear isomorphism* of the (whole) tangent space in x and the *horizontal* subspace in e.

Hint: (i), (iii): ansatz $v^h = v^i \partial_i + v^a_b \partial^b_a$, the explicit expression of ω^a_b from (19.2.1) and the horizontality condition $\langle \omega^a_b, v^h \rangle = 0$; (ii) see (19.3.3); (iv) $v^i \partial_i \leftrightarrow v^i H_i$. □

• The nomenclature "*vertical* distribution" is fairly clear from the conventions concerning drawing of pictures of the type (19.4.4): fibers used to be drawn vertically. Now why is the distribution given by the connection called horizontal? If we took an opinion poll about what

[404] Recall that ∂_i actually denotes *two* fields, namely on M as well as on LM; it should always be clear from a context which one is relevant.

precise meaning people actually associate with the word horizontal, we would probably learn sort of "flat," "level" or maybe "contrary (perhaps perpendicular?) to vertical." In more official sources we read[405] *horizontal* = "flat or level; parallel to the ground or to the bottom or top edge of something" or[406] we first learn that *horizon* is the "line at which the earth or sea and sky seem to meet" and then *horizontal* = "parallel to the horizon; flat or level" and finally[407] "parallel to, in the plane of, or operating in a plane parallel to the horizon or to a base line." And, by the way, *horizon* = "the apparent junction of earth and sky," *level* = "having no part higher than another: conforming to the curvature of the liquid parts of the earth's surface."

What part of this piece of wisdom concerns *our* notion of horizontality? If we look at the figure in problem (19.4.4), we can see that the horizontal vector v^h is (on purpose) not displayed as being horizontal in the sense of "liquid parts of the earth's surface," since those "parts" (say, the surface of a lake) used to be *perpendicular* to the truly "vertical" direction (given by, say, a plumbline at rest). If we, however, adopt the definition which refers to a "line at which the earth ... and sky seem to meet" and by "earth" we understand the beautiful scenery of a national park with a marvellous chain of mountains afar, then the horizon need not be necessarily "flat" or "level" and the vector v^h actually may be tangent to the horizon. So in this broader sense "horizontality" need not necessarily mean orthogonality with respect to the vertical direction (not all of us happen to be mariners), but rather *complementarity* to the latter. By this we mean that the vertical *plus* horizontal is already enough to produce *any* direction whatsoever. (Actually *any* direction, which is *not vertical*, may be declared to be horizontal – it suffices to find a place on the mountains afar with a slope just steep enough).

Returning to connections, we will check that the concept of horizontality is indeed based on complementarity rather than on orthogonality (actually no metric tensor on LM was even mentioned to express orthogonality). The fact that the horizontal subspace is complementary to the vertical one is vital for the possibility of a unique decomposition of vectors into horizontal and vertical parts with all the consequences of this decomposition (see (19.4.5), Section 20.2, etc.).

19.4.5 At a point $e \in LM$, consider the vertical and horizontal subspaces. Check that

(i) if a vector turns out to be at the same time horizontal and vertical, it is necessarily zero; we may write

$$\text{Hor}_e LM \cap \text{Ver}_e LM = 0$$

(ii) the (direct) sum of the two subspaces already gives the whole tangent space

$$T_e LM = \text{Ver}_e LM \oplus \text{Hor}_e LM$$

so that the horizontal subspace is indeed complementary with respect to the vertical one

[405] *Cambridge Advanced Learner's Dictionary*, Cambridge University Press, 2003.
[406] A. S. Hornby, *Oxford Advanced Learner's Dictionary of Current English*, Oxford University Press, 1974.
[407] *Merriam-Webster's Collegiate Dictionary*.

(iii) a decomposition of a general vector into its vertical and horizontal parts may be written in the form

$$v = \text{ver } v + \text{hor } v \equiv \xi_C + v^i H_i$$

i.e. the vertical part may be parametrized in terms of a matrix $C \in gl(n, \mathbb{R})$ and the horizontal part by an n-tuple of numbers v^i (the coefficients with respect to the horizontal lift of the coordinate basis, i.e. with respect to $H_i \equiv \partial_i^h$).

Hint: (i) according to (19.1.4) the vertical part is of the form ξ_C, (19.2.4) adds $\langle \omega, \xi_C \rangle = C$, and by (19.4.3) this should be 0; (ii) the dimensions of the subspaces are n^2 and n respectively and the dimension of the whole space $T_e LM$ is just $n^2 + n$ (19.1.1). □

• Behavior of the horizontal distribution $\mathcal{D}^h$ with respect to the action of the group $GL(n, \mathbb{R})$ turns out to be, like it was in the case of the vertical one $\mathcal{D}^v$, the simplest possible: it is invariant with respect to the action.

19.4.6 Check that the horizontal distribution $\mathcal{D}^h$ on LM is $GL(n, \mathbb{R})$-invariant, i.e. that for each element $A \in GL(n, \mathbb{R})$ there holds

$$R_A \mathcal{D}^h = \mathcal{D}^h \quad \text{or in more detail} \quad R_{A*}(\mathcal{D}^h_e) = \mathcal{D}^h_{eA}$$

Hint: the property $R_A^* \omega = \text{Ad}_{A^{-1}} \omega \equiv A^{-1} \omega A$ (19.2.4) of connection form ω is behind: if $v \in \mathcal{D}^h_e \equiv \text{Hor}_e LM$, i.e. $\langle \omega, v \rangle = 0$, then $\langle \omega, R_{A*} v \rangle = \langle R_A^* \omega, v \rangle = A^{-1} \langle \omega, v \rangle A = 0$, so that $R_{A*} v \in \mathcal{D}^h_{eA} \equiv \text{Hor}_{eA} LM$ (it is also horizontal). □

• Let's summarize. There are two relevant distributions on LM: vertical, which is always available and horizontal, for which a connection is needed on M. Both of them are invariant with respect to the action of the group $GL(n, \mathbb{R})$ and they are complementary to each other (a direct sum of the vertical and the horizontal subspaces at each point equals the whole tangent space). The vertical distribution is integrable. In the next section we will show that the horizontal distribution provides a brand new view on the procedure of parallel transport. The fact that in general the horizontal distribution *fails* to be integrable turns out to be closely related to path dependence of parallel transport; this stuff will only be discussed, however, in the context of general connections in Chapter 20 (in Section 20.4).

19.4.7 Consider once more the situation from (19.4.4). Check that

(i) if a vector v is horizontally lifted to all points in the fiber over x, we get a vector field v^h in the fiber, which is (right, horizontal, but also) $GL(n, \mathbb{R})$-*invariant*: $R_{A*} v^h = v^h$
(ii) if a vector *field* V on M is lifted in this pointwise way, we get on LM a vector field V^h, which is (sure, horizontal, but also) $GL(n, \mathbb{R})$-invariant:[408]

$$R_{A*} V^h = V^h \qquad \langle \omega, V^h \rangle = 0$$

(iii) conversely, *each* $GL(n, \mathbb{R})$-invariant and horizontal vector field W on LM may be regarded as the horizontal lift $W = V^h$ of a field V on M.

[408] The fields $H_i \equiv \partial_i^h$ from (19.4.4) represent particular cases.

Hint: (i) $R_{A*}v^h_e$ satisfies both requirements (19.4.4) on v^h_{eA} (because of $\pi \circ R_A = \pi$ and $GL(n, \mathbb{R})$-invariance of the horizontal distribution (19.4.6)), consequently this is nothing but v^h_{eA}. □

We end this section with a remarkable basis of horizontal vector fields on LM. Contrary to $H_i \equiv \partial^h_i$ from (19.4.4), which exist only locally (in $\pi^{-1}(\mathcal{O})$ over a coordinate patch $\mathcal{O}$), depend on the choice of coordinates x^i in $\mathcal{O}$ and are invariant with respect to R_A, the fields to be discussed here are global, canonical and they transform non-trivially under the action R_A.

19.4.8* Consider vector fields $\hat{\mathcal{E}}_a$, $a = 1, \dots, n \equiv \dim M$ on LM, defined as follows: a point $e \in LM$ corresponds to the frame e_a in $x \equiv \pi(e)$; each vector of the frame may be horizontally lifted to e; by definition, the lifts provide the values of the fields $\hat{\mathcal{E}}_a$ at the point e

$$\hat{\mathcal{E}}_a(e) := e^h_a(e) \qquad \text{(the lift is to the point } e)$$

Check that the fields $\hat{\mathcal{E}}_a$ (known as the *standard horizontal fields*)

(i) constitute at each point a basis of the horizontal subspace
(ii) are canonically associated with each manifold with linear connection (M, ∇)
(iii) are transformed under the action R_A of the group $GL(n, \mathbb{R})$ on LM[409] according to the rule

$$R^*_A \hat{\mathcal{E}}_a = A^b_a \hat{\mathcal{E}}_b$$

(so they are scrambled by the matrix A).

Hint: (i) according to (19.4.4) the lift is an isomorphism; (ii) the lift depends on the connection; (iii) $\hat{\mathcal{E}}_a(eA) = (eA)^h_a(eA) = A^b_a e^h_b(eA) = A^b_a R_{A*} e^h_b(e)$ (the last equality because of the $GL(n, \mathbb{R})$-invariance of e^h_a), so that $R^*_A \hat{\mathcal{E}}_a(eA) = A^b_a e^h_b(e)$, i.e. $R^*_A \hat{\mathcal{E}}_a = A^b_a \hat{\mathcal{E}}_b$. □

19.4.9* Be sure to understand that the standard horizontal vector fields $\hat{\mathcal{E}}_a$ along with the fundamental fields $\xi_{E^a_b}$ of the action R_A are global vector fields on LM, whose values at each point constitute a basis of the tangent space; it then follows that LM is always *parallelizable*, and consequently also an *orientable* manifold (even if M itself were not).

Hint: see (19.4.1) and (19.4.8); parallelizable by definition of the concept, orientable: the exterior product of the dual basis provides a global volume form (see also (21.7.3)). □

19.5 Horizontal distribution on LM and parallel transport on M

• Imagine we have a curve $\gamma(t)$ on M and a field of frames $e(t)$ on the curve (that is to say a frame field $e_a(\gamma(t))$ defined at the points of the *curve alone*). Such a field of frames then induces naturally a curve $\hat{\gamma}$ on LM: one assigns a frame $e_a(\gamma(t))$ to the parameter t, interpreted now, however, as a *point on LM*. This curve clearly projects onto the original curve γ via π, i.e. $\pi \circ \hat{\gamma} = \gamma$.

409 Contrary to ∂^h_i, which are *invariant* with respect to R_A (as is each horizontal lift, (19.4.7)).

Now fix a frame E at a point of the curve γ. Making use of the connection, we may generate an *autoparallel* frame field $e^{\|}(t)$ on γ (each vector E_a generates the autoparallel field $e_a^{\|}$ on γ; since the parallel transport operator is a linear isomorphism, the linear independence of the vectors $e_a^{\|}$ is guaranteed at each point of γ). This particular frame field also induces on LM the corresponding curve $\hat{\gamma}$. However, since we have associated it with a very special frame field, we expect it should exhibit some very special features (in addition to the fact that it projects onto γ, which always holds). A computation reveals that the special property of this particular curve is *horizontality*, i.e. the fact that its *tangent vector* is horizontal at each point.

19.5.1 Verify that an *autoparallel* frame field $e^{\|}$ on a curve $\gamma(t)$ on M induces the *horizontal curve* $\hat{\gamma}(t)$ on LM

$$\langle \omega, \dot{\hat{\gamma}} \rangle = 0$$

Hint: let $\hat{\gamma}(t)$ be represented by $(x^i(t), y_b^a(t))$, with (x^i, y_b^a) being coordinates with respect to an arbitrary frame field $e(x)$ in a domain $\mathcal{O}$ (19.1.1); then $e_a^{\|} = y_a^b e_b$. From the condition that $e^{\|}$ is autoparallel we get

$$0 = \nabla_{\dot{\gamma}} e_a^{\|} = \nabla_{\dot{\gamma}} (y_a^b e_b) = \dot{y}_a^b e_b + y_a^b \nabla_{\dot{\gamma}} e_b = \dot{y}_a^b e_b + y_a^b \langle \hat{\omega}_b^c, \dot{\gamma} \rangle e_c = (\dot{y}_a^b + y_a^c \langle \hat{\omega}_c^b, \dot{\gamma} \rangle) e_b$$

so that (for $\dot{\gamma} = \dot{x}^i \partial_i$)

$$\dot{\hat{\gamma}} \equiv \dot{x}^i \partial_i + \dot{y}_a^b \partial_b^a = \dot{x}^i (\partial_i - \langle \hat{\omega}_c^b, \partial_i \rangle y_a^c \partial_b^a) = \dot{x}^i H_i \equiv \dot{x}^i (\partial_i)^h = (\dot{\gamma})^h$$

Thus the tangent vector $\dot{\hat{\gamma}}$ to the curve $\hat{\gamma}$ is the *horizontal lift* of the tangent vector $\dot{\gamma}$ to the curve γ. □

19.5.2 Let γ be a curve on M and e an arbitrary point in the fiber over $\gamma(0)$. Check that

(i) there is a unique curve γ^h on LM, specified by the conditions

$$\pi \circ \gamma^h = \gamma \qquad \gamma^h(0) = e \qquad \langle \omega, (\dot{\gamma^h}) \rangle = 0$$

i.e. γ^h is everywhere "exactly over" γ (it projects onto γ), it passes through a prescribed point e and its tangent vector is *horizontal* at each point. The curve γ^h is called the *horizontal lift of the curve* γ

(ii) tangent vectors to the curve γ^h are *horizontal lifts* of tangent vectors to the original curve γ

$$(\dot{\gamma^h}) = (\dot{\gamma})^h$$

(iii) the curve γ^h coincides with the curve $\hat{\gamma}$ discussed in (19.5.1).

Hint: (i) if $\gamma \leftrightarrow x^i(t)$, the first condition gives $\gamma^h \leftrightarrow (x^i(t), y_b^a(t))$ (with $x^i(t)$ being the same functions as $x^i(t) \leftrightarrow \gamma$ and $y_b^a(t)$ are so far arbitrary); the horizontality condition leads (in analogy with (19.5.1)) to a first-order differential equation for $y_b^a(t) \Rightarrow$ together with the initial condition for $y_b^a(0)$ (from $\gamma^h(0) = e$) we have a unique solution $y_b^a(t)$; (ii) see (19.4.4). □

• The lesson from exercises (19.5.1) and (19.5.2) is that the horizontal lift γ^h of a curve γ carries exactly the same information as the autoparallel frame field on γ, so that if we are able to construct the horizontal lift of a curve, we are then also able to perform the parallel transport of a frame and vice versa. This simple observation enables us to express the operation of parallel transport on M (so far only of frames[410] alone) entirely in terms of an appropriate construction on LM, based on the concept of the horizontal distribution $\mathcal{D}^h$ on this manifold. A complementary point of view is, however, of vital importance for the prospect of development of a general concept of a connection: the distribution $\mathcal{D}^h$ carries the *full information* needed for the procedure of parallel transport on M.

19.6 Tensors on M in the language of LM and their parallel transport

• As we have already noticed in the previous section, the formalism using the manifold LM is tailored to treat frames but the description of more elementary objects (vectors, say) remains obscure. Let us first have a look at this problem within the simpler context of linear algebra.

Suppose we have a (finite-dimensional) linear space L and let $E(L)$ denote the set of all frames in L. If $e \equiv e_a$ is a fixed frame, then an arbitrary vector $v \in L$ may be written as $v = \hat{v}^a e_a$, and an ordered *pair* $(e, \hat{v}) \in E(L) \times \mathbb{R}^n$ may be assigned to this decomposition $(\hat{v} \equiv \hat{v}^a)$. For our purposes the following point of view will be most useful: for each vector $v \in L$ consider *a map* $\Phi^v : E(L) \to \mathbb{R}^n$, which assigns to a frame e the components of the vector v with respect to this particular frame, $e \mapsto \Phi^v(e) \equiv \hat{v}$. Then there holds $v \leftrightarrow (e, \hat{v}) = (e, \Phi^v(e))$. A choice of a pair "frame + components" is, however, far from unique: infinitely many pairs $(\tilde{e}, \tilde{\hat{v}}) \in E(L) \times \mathbb{R}^n$ may be assigned to the same vector, corresponding to decompositions of the vector with respect to all possible frames in L. This results in a severe restriction on the map Φ^v.

19.6.1 Consider a map $\Phi^v : E(L) \to \mathbb{R}^n$, which assigns to a frame e the components of a (fixed) vector v with respect to this frame

$$\Phi^v : E(L) \to \mathbb{R}^n \qquad e \mapsto \Phi^v(e) \equiv \hat{v}$$

Check that

(i) the transition to another frame $e \mapsto eA \equiv R_1(A)e$ is a right action of $GL(n, \mathbb{R})$ on $E(L)$
(ii) the transition to corresponding new components of the vector $\hat{v} \mapsto R_2(A)\hat{v} \equiv A^{-1}\hat{v}$ (with respect to a frame eA instead of e) is, in turn, a right action of $GL(n, \mathbb{R})$ on $\mathbb{R}^n$
(iii) Φ^v is an *equivariant map* between two (right) $GL(n, \mathbb{R})$-spaces, i.e.

$$\Phi^v : E(L) \to \mathbb{R}^n \qquad \Phi^v(eA) = A^{-1}\Phi^v(e)$$

[410] Frames on M are described naturally in the language of the manifold LM (it was after all invented for this very reason), but we realized just now that we do not see any possibility to incorporate *vectors themselves* into this scheme (let alone tensors). *Points* on LM are already *n-tuples* of vectors and the idea to identify *one* vector with some *part of a point* (the nth part) does not raise too much hope. The problem has an amazingly simple and elegant solution, which is described in Section 19.6. The solution even survives the generalization of the whole context and we will return to this issue in Section 20.3.

and it is completely given by its value at a single (arbitrary) point $e \in E(L)$ (it is then extended to all other points by the equivariance property)

(iv) the linear space of all such equivariant maps is (canonically) isomorphic to the space L itself (and can thus fully replace L as an equivalent alternative); there holds

$$\Phi^{v+\lambda w} = \Phi^v + \lambda \Phi^w$$

Hint: (i) see (5.7.2); (ii) see (13.1.4); (iii) $v \leftrightarrow (e, \hat{v}) \leftrightarrow (eA, A^{-1}\hat{v})$; since R_1 is transitive and free, it is enough to specify the value of Φ^v at a single point; (iv) $v \leftrightarrow \Phi^v$. □

• In complete analogy the *dual* space L^* may also be replaced by an appropriate map space. Namely, one should decompose a covector α with respect to the dual frame as $\alpha = \hat{\alpha}_a e^a$ and then associate an ordered *pair* $(e, \hat{\alpha}) \in E(L) \times \mathbb{R}^n$ ($\hat{\alpha} \equiv \hat{\alpha}^a$) as well as a *map*

$$\Phi^\alpha : E(L) \to \mathbb{R}^n \qquad e \mapsto \Phi^\alpha(e) \equiv \hat{\alpha}$$

with this decomposition. Note that the resulting space of pairs actually *coincides as a set* with that constructed for vectors (since both vector and covector components are elements of $\mathbb{R}^n$); there is a difference, however, in the action of the group $GL(n, \mathbb{R})$ in the $\mathbb{R}^n$-part of it.

19.6.2 Consider a map $\Phi^\alpha : E(L) \to \mathbb{R}^n$ which assigns to a frame e the components of a (fixed) covector α with respect to this particular frame

$$\Phi^\alpha : E(L) \to \mathbb{R}^n \qquad e \mapsto \Phi^\alpha(e) \equiv \hat{\alpha}$$

Check that

(i) the transition to new components of the covector (with respect to the frame eA instead of e), namely $\hat{\alpha} \mapsto R_2(A)\hat{\alpha} \equiv A^{\mathrm{T}}\hat{\alpha}$, is a right action of $GL(n, \mathbb{R})$ on $\mathbb{R}^n$

(ii) Φ^α is an *equivariant map* between two (right) $GL(n, \mathbb{R})$-spaces, i.e.

$$\Phi^\alpha : E(L) \to \mathbb{R}^n \qquad \Phi^\alpha(eA) = A^{\mathrm{T}}\Phi^\alpha(e)$$

(iii) the linear space of all such equivariant maps is (canonically) isomorphic to the space L^*; there holds

$$\Phi^{\alpha+\lambda\beta} = \Phi^\alpha + \lambda\Phi^\beta \quad \langle \Phi_\alpha, \Phi^v \rangle := \langle \Phi_\alpha(e), \Phi^v(e) \rangle \equiv \hat{\alpha}_a \hat{v}^a \qquad e \in E(L)$$

Hint: see (19.6.1); the pairing does not depend on the choice of e. □

• Note that in both cases (for vectors as well as covectors) the following general scheme was applied:

(i) one creates a space of *pairs* (frame, components)

(ii) there is an *action of the group* $GL(n, \mathbb{R})$ both on the first and the second part of the pair

(iii) the action on the first element is always the same; on the second element $GL(n, \mathbb{R})$ acts by means of *different representations*

(iv) the type of the resulting object is given by the action (representation, here) on the *second* part of a pair

(v) elements of the spaces L and L^* may be identified with *equivariant maps* from the set of frames to the set of components.

It is easily seen that this scheme can also be extended in a straightforward way from vectors and covectors to *arbitrary* tensors on L. The only thing to do is to substitute the "one-index" real numbers (the space $\mathbb{R}^n$) by the multi-index ones $\hat{B} \equiv \hat{B}^{a...b}_{c...d}$, and the representations $A \mapsto \rho^1_0(A) = A$ (for vectors) and $A \mapsto \rho^0_1(A) = (A^{-1})^{\mathrm{T}}$ (for covectors)[411] by the corresponding tensor representations $A \mapsto \rho^r_s(A)$, i.e. by tensor products of r vector and s covector representations, in order to obtain tensors of type $\binom{r}{s}$; the resulting representation space (module) will be denoted by (V, ρ^r_s)). The corresponding equivariant map, which encodes a tensor B of type $\binom{r}{s}$ in L, will be

$$\Phi^B : E(L) \to \left(V, \rho^r_s\right) \qquad \Phi^B(eA) = \rho^r_s(A^{-1})\Phi^B(e)$$

As an example, let us see how this works for tensors of type $\binom{0}{2}$.

19.6.3 Consider a representation $\rho^0_2 = \rho^0_1 \otimes \rho^0_1$, where ρ^0_1 is the "covector" representation $A \mapsto \rho^0_1(A) = (A^{-1})^{\mathrm{T}}$. It acts on $V \equiv \mathbb{R}^n \otimes \mathbb{R}^n$ as

$$\hat{B} \mapsto \rho^0_2(A)\hat{B} \qquad \hat{B}_{ab} \mapsto \left(\left(\rho^0_1 \otimes \rho^0_1\right)(A)\hat{B}\right)_{ab} := (A^{-1})^c_a (A^{-1})^d_b \hat{B}_{cd}$$

Check that the scheme explained above results here in equivariant maps which may be identified with *tensors of type* $\binom{0}{2}$ on L, i.e.

$$T^0_2(L) \approx (E(L) \times V)/GL(n, \mathbb{R}) \qquad V = \left(\mathbb{R}^n \otimes \mathbb{R}^n, \rho^0_1 \otimes \rho^0_1\right)$$

Hint: $B \equiv \hat{B}_{ab} e^a \otimes e^b \leftrightarrow \Phi^B, \; e \mapsto \Phi^B(e) = \hat{B}; \; (\Phi^B(eA))_{ab} = A^c_a A^d_b (\Phi^B(e))_{cd}$. □

• Note that within this approach we succeeded in describing the *tensors* on L (and, in particular, also the space L itself) in terms of the set of *frames* in the space L and "component" spaces (V, ρ); the tensor is identified with an equivariant map Φ from the space of frames to the space of components (it is reconstructed from a frame and its image with respect to Φ; equivariance of Φ guarantees that proper pairs are combined). The type of tensor, $\binom{r}{s}$, is completely given by the choice of a representation ρ^r_s in the space of components V. In this language it is quite natural to introduce a general notion of a *quantity of type* ρ (in a linear space L) as an equivariant map

$$\Phi : E(L) \to (V, \rho) \qquad \Phi \circ R_A = \rho(A^{-1}) \circ \Phi$$

where R_A is the standard right action ($e \mapsto eA$) of $GL(n, \mathbb{R})$ in the space of frames $E(L)$ from (5.7.2) and (19.6.1). All *tensors* in L provide basic examples of such quantities (for $\rho = \rho^r_s$), but there are actually more of them, e.g. *tensor densities of weight* λ for the representation $\rho(A) = (\det A)^\lambda \rho^r_s(A)$, in particular $\rho(A) = \det A$ for the ordinary *scalar density* (of weight 1; see also (21.7.10)). The notion of a quantity of type ρ can be straightforwardly

[411] One should realize that the rules $\hat{v} \mapsto A^{-1}\hat{v}$ and $\hat{\alpha} \mapsto A^{\mathrm{T}}\hat{\alpha}$ in the examples mentioned above are *right* actions, so if we need to express them in terms of *representations* (which are *left* actions), the trick $A \mapsto A^{-1}$ from (13.1.1) is to be used.

generalized from the group $GL(n, \mathbb{R})$ to a general Lie group G (we will do this later, in Section 20.3).

Now the situation is clear at the level of linear algebra and we can repeat the idea on a manifold. The role of the space $E(L)$ will be played by a fiber $\pi^{-1}(x)$ over x on a manifold LM, a set of all frames at the point $x \in M$. A vector v in x may be identified with the map $\Phi^v_x : \pi^{-1}(x) \to \mathbb{R}^n$ satisfying the equivariance condition $\Phi^v_x \circ R_A = \rho^1_0(A^{-1}) \circ \Phi^v_x$, a tensor of type ρ^r_s at the point x is obtained by a substitution of the representation ρ^1_0 by ρ^r_s, a *tensor field* of type $\binom{r}{s}$ on M may be identified with an equivariant map

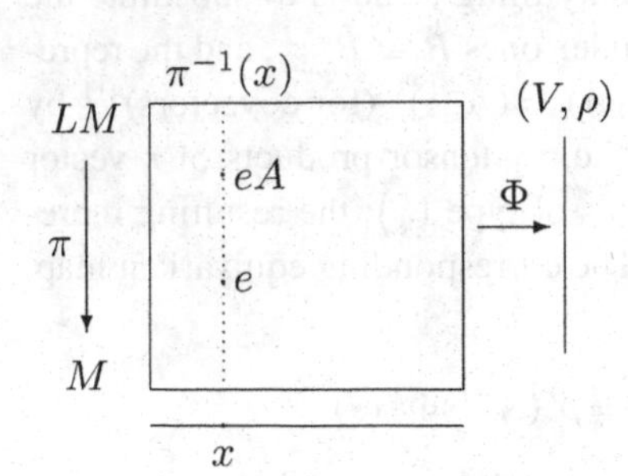

$$\Phi : LM \to \left(V, \rho^r_s\right) \qquad \Phi \circ R_A = \rho^r_s(A^{-1}) \circ \Phi$$

and finally a general *field of type* ρ on M may be identified with an equivariant map

$$\Phi : LM \to (V, \rho) \qquad \Phi \circ R_A = \rho(A^{-1}) \circ \Phi$$

where R_A is the canonical action of $GL(n, \mathbb{R})$ on LM (19.1.3) and (V, ρ) is an arbitrary representation module of the group $GL(n, \mathbb{R})$.

We see that the frame bundle LM enables us to introduce a novel global, *unified* and surprisingly *simple* formal treatment of a wide class of geometrical objects on a base manifold M: each object is identified with an equivariant V-valued *function*[412] on LM, various types of objects *only* differing by the choice of a particular module (V, ρ) (e.g. the choice $(\mathbb{R}^n, \rho^1_0)$ corresponds to *vector* fields).

Let us now discuss the procedure of *parallel transport* of quantities of type ρ. First, observe an elementary fact concerning the parallel transport of tensors in the good old language of Chapter 15.

19.6.4 Let $B = \hat{B}^{a\dots b}_{c\dots d} e^c \otimes \dots \otimes e_b$ be the expression of a tensor field B on (M, ∇) with respect to a frame field e, and suppose that e is *autoparallel* on a curve γ. Check that

(i) the covariant derivative $\nabla_{\dot\gamma} B$ along $\gamma(t)$ actually reduces to the *ordinary* derivative of the *components* of the tensor field

(ii) B is autoparallel on γ if and only if it has *constant* components[413] (with respect to e) on the curve γ

Solution: since (by assumption) $\nabla_{\dot\gamma} e_a = 0 = \nabla_{\dot\gamma} e^a$, we have

$$\nabla_{\dot\gamma} B \equiv \nabla_{\dot\gamma} \left(\hat{B}^{a\dots b}_{c\dots d} e^c \otimes \dots \otimes e_b\right) = \dots = \dot{\hat{B}}^{a\dots b}_{c\dots d} e^c \otimes \dots \otimes e_b$$

so that $\nabla_{\dot\gamma} B = 0 \Leftrightarrow \dot{\hat{B}}^{a\dots b}_{c\dots d} = 0$. □

[412] (V, ρ) is a representation module of $GL(n, \mathbb{R})$.

[413] Recall that a similar fact holds on Lie groups: the components of left-invariant fields with respect to the left-invariant frame field are constant (11.1.6).

• As a result, any autoparallel tensor field has constant components with respect to an autoparallel field of frames. This fact might serve as a motivation for the definition of basic concepts related to the parallel transport in the language of the manifold LM.

In this formalism a tensor at a point x is given by a *pair* which consists of a frame in x plus components (an element of a module (V, ρ)). If we want to move along a curve γ and parallel transport the tensor along it, at each point of the curve such a pair is needed. We also know that the pairs are not unique (there is a freedom given by the action of $GL(n, \mathbb{R})$), and actually any particular representative is sufficient. There is a preferred frame field on the curve, given by the parallel transport of the frame at the starting point; this frame field corresponds (as we learned in (19.5.2)) to the *horizontal lift* γ^h of the curve γ on M. Now, what components should be combined with *these particular* frames $\gamma^h(t)$ in order to obtain the *autoparallel* tensor field? The result of exercise (19.6.4) offers a clear hint: take the *constant* element of (V, ρ). Thus if we want to construct the autoparallel tensor field $B(t)$ on a curve $\gamma(t)$, we should combine the pairs $(\gamma^h(t), \hat{B} \equiv \text{constant})$. In the language of an equivariant function Φ this means that all the points of $\gamma^h(t)$ (lying in the fibers over $\gamma(t)$) should be mapped into *a single* element $\hat{B} \in V$ (independent of t). (This in turn fixes the function Φ completely at all points of all fibers which $\gamma^h(t)$ intersects, see (19.6.1).)

19.6.5 Consider an equivariant function Φ which is defined on the fibers over the curve $\gamma(t)$ and which moreover corresponds to the *autoparallel* field of quantities of type ρ (e.g. an autoparallel tensor field B) on γ. Be sure to understand that the function is constant on a horizontal lift, i.e. that

$$\Phi(\gamma^h(t)) = \text{constant} \qquad \text{i.e.} \quad (\dot{\gamma})^h \Phi = 0$$

Hint: according to (19.5.1) $\dot{\gamma^h} = (\dot{\gamma})^h$; here we have $\frac{d}{dt}\Phi(\gamma^h(t)) \equiv \dot{\gamma^h}\Phi = 0$ and therefore also $(\dot{\gamma})^h \Phi = 0$. □

• If the derivative of the function Φ along the horizontal lift of a curve $\gamma(t)$ *does not vanish*, it means that the components $\hat{B}$ with respect to the autoparallel frame field *are not* constant and consequently this field *is not* autoparallel on $\gamma(t)$. What piece of information is then carried by the derivative? Clearly, it informs us about the *covariant* derivative of the corresponding tensor field B along the curve γ.

19.6.6 Consider an equivariant function Φ which is defined on the fibers over the curve $\gamma(t)$ and which corresponds to a field of quantities of type ρ (e.g. a tensor field B) on γ. Check that its derivative along the horizontal lift γ^h of the curve γ corresponds (just in the sense that Φ corresponds to B) to the *covariant* derivative $\nabla_{\dot{\gamma}} B$ of the field B

$$\Phi \leftrightarrow B \quad \Rightarrow \quad (\dot{\gamma})^h \Phi \leftrightarrow \nabla_{\dot{\gamma}} B$$

Hint: see (19.6.4). □

• We can see that the extension of a manifold M to LM leads to a notable *simplification* of the description of various objects. Frames on M become *points* of LM, tensor fields on M

become *functions* on LM and covariant derivatives of tensor fields on M become *ordinary* directional derivatives of these functions.

Summary of Chapter 19

In order to pave the way for a possible generalization of the theory of linear connection well known from Chapter 15 (to be done in the next chapter) we reformulate it in a new language. The new description takes place on a new playing field, a manifold LM which may be canonically assigned to any manifold M. The points of LM are all frames at all points of M. There is a fairly rich structure on LM even prior to introducing the connection on M: the manifold LM namely turns out to be a total space of a principal $GL(n, \mathbb{R})$-bundle over M. A connection on M adds more structure on LM, a $GL(n, \mathbb{R})$-invariant horizontal distribution. We may reformulate the procedure of parallel transport of a frame along a curve γ on M in terms of the horizontal lift γ^h of the curve γ. There is also an interesting possibility of treating a wide class of geometrical objects on M (in particular tensor fields and more generally fields of type ρ) in terms of equivariant functions Φ on LM. Their parallel transport is discussed and it is shown that an appropriate directional derivative of Φ corresponds to the covariant derivative on M of the geometrical object described by Φ.

$\omega \equiv \omega^a_b E^b_a$	Connection form on the frame bundle LM	(19.2.1)
$R^*_A\omega = A^{-1}\omega A,\ \langle \omega, \xi_C \rangle = C$	Crucial properties of the connection form	(19.2.4)
$U, V \in \mathcal{D} \Rightarrow [U, V] \in \mathcal{D}$	$\mathcal{D}$ is integrable (Frobenius' criterion)	Sec. 19.3
$\theta^i\vert_{\mathcal{D}} = 0 \ \Rightarrow\ d\theta^i\vert_{\mathcal{D}} = 0$	Alternative formulation of the criterion	Sec. 19.3
$V \in \mathcal{D}^h \ \Leftrightarrow\ \langle \omega, V \rangle = 0$	Horizontal distribution on LM	(19.4.3)
$T_e LM = \text{Ver}_e LM \oplus \text{Hor}_e LM$	Decomposition induced by a connection	(19.4.5)
$\langle \omega, \dot{\hat{\gamma}} \rangle = 0$	$\hat{\gamma}$ corresponds to autoparallel frame field	(19.5.1)
$\Phi \circ R_A = \rho(A^{-1}) \circ \Phi$	Φ is a quantity of type ρ	Sec. 19.6
$\Phi(\gamma^h(t)) =$ constant	Autoparallel field of quantities of type ρ	(19.6.5)

20

Connection on a principal G-bundle

• In the previous chapter we learned that the concept of a linear connection on M may be encoded into a horizontal distribution $\mathcal{D}^h$ on a manifold LM. Here we accomplish a simple, albeit far-reaching generalization. The novelty may be briefly described as an acceptance of a more general stage where, however, an old dramatic piece is performed. Clearly, there are some restrictions on possible new stages (in order that the old piece *can be* performed on it in principle). From a variety of coulisses and properties we only insist on retaining truly unthinkable elements; all others, in truth not essential for the drama, may be altered by a director and replaced by anything else, more congenial to his taste and artistic intent.

20.1 Principal G-bundles

• In Section 19.1 we learned that the stage used for the linear connection, the manifold LM, is the total space of a *principal* $GL(n, \mathbb{R})$*-bundle*. A natural step therefore consists in considering a general *principal G-bundle* over M instead of just a particular case $\pi : LM \to M$. The change thus consists in a replacement of the group $GL(n, \mathbb{R})$ by a general Lie group G. This is, however, not all.

The bundle $\pi : LM \to M$ actually turns out to be a *very specific* principal $GL(n, \mathbb{R})$-bundle, which may be canonically associated with an arbitrary manifold M. Points of the total space (of the manifold LM) are namely closely related[414] to certain objects (i.e. frames) on the base manifold M (this is analogous to TM and T^*M, where points of the total spaces are vectors and covectors on the base M). For a general principal G-bundle

$$\pi : P \to M$$

however, *no* such relation of points of the total space and whatever objects on the base M is required. The points of the manifold P may have nothing to do with the manifold M.

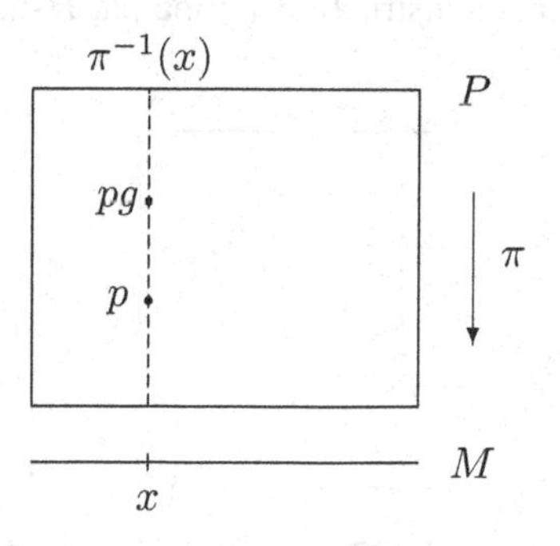

[414] The relation is known as *soldering*. The fact that the total space is "soldered" to the base results in the existence of various canonical geometrical objects on LM, TM and T^*M, see Sections 17.6 and 21.7.

Recall then what *is*, in contrast, required from a principal G-bundle (see Section 17.2 and the text after (19.1.3)). First, it is to be a *bundle*; so two manifolds P and M are to be given along with a smooth surjective map $\pi : P \to M$ (projection), all preimages being submanifolds of P diffeomorphic to each other. Moreover, since we speak of a *principal* bundle, a right vertical action of a Lie group G in the total space P is to be added

$$R_g : P \to P \qquad R_{gh} = R_h \circ R_g \qquad \pi \circ R_g = \pi$$

The action is *free* (i.e. all stabilizers trivial) and *transitive* in fibers (any two points in a single fiber can be joined by the action; the fiber thus becomes a *principal homogeneous space* of the group G). The local product structure also takes a specific character; namely, local trivializations are to be maps

$$\psi_\alpha : \pi^{-1}(\mathcal{O}_\alpha) \to \mathcal{O}_\alpha \times G$$

which in addition to the general requirement

$$\pi_1 \circ \psi_\alpha = \pi$$

should also satisfy

$$\psi_\alpha : p \mapsto (m, h) \;\Rightarrow\; pg \mapsto (m, hg)$$

($h \mapsto hg \equiv \mathcal{R}_g h$ is the right *translation* on G); the trivialization thus transforms (on sufficiently small pieces) a principal bundle to a *product principal bundle*

$$\pi : M \times G \to M \qquad \pi : (m, g) \mapsto m \qquad R_{\tilde{g}} : (m, g) \mapsto (m, g\tilde{g})$$

20.1.1 Check that for any two points p, p' residing in a common fiber ($\pi(p) = \pi(p')$) there is a *unique* group element $g \in G$, which links the points in the sense that $p' = pg$.

Hint: in general they are linked by the set HgH', where H, H' are stabilizers of the points p and p' respectively; both of them are, however, trivial (the group acts freely). □

• In Section 13.2 we examined the object G/H as a homogeneous space, or perhaps as a group (provided that H happens to be a *normal* subgroup). Now we realize that by the same construction a principal H-bundle is also obtained.

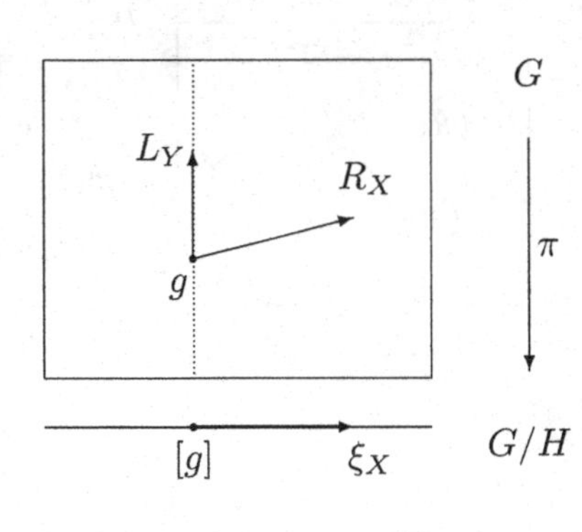

20.1.2 Let G be a Lie group, H a (closed) Lie subgroup. It then turns out that each homogeneous space $M \sim G/H$ is the base of a principal H-bundle $\pi : G \to M \sim G/H$. A specific feature of exactly *this* principal bundle is that the total space is a *group* and the base is a *homogeneous space* of the group. Let us see how this works in more detail. Check that

(i)

$$\pi : G \to G/H \qquad g \mapsto [g]$$

is indeed a principal H-bundle

(ii) apart from "compulsory figures," there is an important *additional* structure on H-bundles of this type, namely a fiber-preserving action $\mathcal{L}_g$ of a larger group G on the total space, as well as its projection L_g, being a left action of G on the base M (and satisfying $\pi \circ \mathcal{L}_g = L_g \circ \pi$)
(iii) the generators of the action of H along fibers are the left-invariant fields L_Y on G, corresponding to elements Y which belong to the *subalgebra* $\mathcal{H} \subset \mathcal{G}$
(iv) the generators of the action $\mathcal{L}_g$ in the total space G are (all) right-invariant fields R_X on G, $X \in \mathcal{G}$.

Hint: (i) the fiber over $[g]$ is gH and the action is $R_h g := gh$; check its verticality, freedom and transitivity in fibers; (ii) it preserves fibers, since it commutes with the right action of H (left translations commute with right ones), see (13.2.5); (iii) and (iv) see (13.4.15). □

• In Section 17.2 we introduced the concepts of *equivalent bundles* and a *trivial bundle* (as one which is equivalent to the product bundle). For *principal* bundles we have to add certain requirements of equivariance (they relate to the action of the group): we say that two principal G-bundles over the same base, $\pi : P \to M$ and $\tilde{\pi} : \tilde{P} \to M$, are equivalent, if there exists an *equivariant* diffeomorphism of total spaces which preserves fibers, i.e.

$$\psi : P \to \tilde{P} \qquad \tilde{R}_g \circ \psi = \psi \circ R_g \qquad \tilde{\pi} \circ \psi = \pi$$

A principal bundle is called *trivial* if it is equivalent (already in the new sense) to a product one; then the map $\psi : P \to M \times G$ is called a *global trivialization.*

20.1.3 There is a simple and useful criterion for triviality of a *principal* bundle. Namely, check that

$$\pi : P \to M \text{ is trivial} \quad \Leftrightarrow \quad \text{there exists its } \textit{global} \text{ section}$$

(Put another way, a global *trivialization* exists if and only if a global *section* does.)

Hint: $\rightarrow$: a global trivialization $\psi : P \to M \times G$ exists; a section is $\sigma : m \mapsto \psi^{-1}(m, e)$; $\leftarrow$: if a section σ exists, in each fiber we get a distinguished point[415] ($\sigma(m)$ over m). All the points in the fiber may be now related to $\sigma(m)$: they are associated with a unique group element g such that $p = R_g\sigma(m) \equiv \sigma(m)g$; a global trivialization is $p \mapsto (m \equiv \pi(p), g)$. □

20.1.4 Check that a *vector bundle always* has a (global) section, so that the criterion for triviality from (20.1.3) may not hold for bundles, which are not principal (a vector bundle is not principal).

Hint: for example, the *zero section* $m \mapsto 0 \in \pi^{-1}(m)$; zero is the only distinguished point in a linear space; notice that the (infinite) Möbius band is the total space of a *non-trivial* vector bundle $\pi : M^2 \to S^1$, since $M^2 \neq S^1 \times \mathbb{R}$. □

• Let us look at a concrete example of a principal H-bundle, the triviality of which is not immediately evident, but may be revealed by an explicit construction of a global section.

[415] We already feel that we have won: Archimedes (would) have moved the Earth, as soon he (would) have been given a fixed point, we (indeed) move forward with the proof, since we do (indeed) have a distinguished point.

20.1.5 The group $G \equiv SL(2, \mathbb{C})$ acts naturally from the left in $\mathbb{C}^2$. Let H be the stabilizer of the point $\binom{1}{0} \in \mathbb{C}^2$. Check that

(i) $H \sim (\mathbb{C}, +)$
(ii) the orbit of the point $\binom{1}{0}$ is a homogeneous space $M \equiv \mathbb{C}^2 \setminus \binom{0}{0} \sim G/H$
(iii) the principal H-bundle $\pi : G \to G/H \sim M$ turns out to be *trivial* in this case
(iv) the Lie group $SL(2, \mathbb{C})$ is diffeomorphic as a manifold to $\mathbb{R}^3 \times S^3$.

Hint: (i) if $g = \begin{pmatrix} a & b \\ c & d \end{pmatrix} \in SL(2, \mathbb{C})$, then $H \ni h = \begin{pmatrix} 1 & b \\ 0 & 1 \end{pmatrix}$; (ii)

$$\begin{pmatrix} a & b \\ c & d \end{pmatrix}\begin{pmatrix} 1 \\ 0 \end{pmatrix} = \begin{pmatrix} a \\ c \end{pmatrix} \neq \begin{pmatrix} 0 \\ 0 \end{pmatrix}$$

(otherwise det $= 0$); accessibility of all non-vanishing points is clear from the form of a section (see below); $M \sim G/H$ from general considerations in Section 13.2; (iii) the bundle has a global section

$$\begin{pmatrix} a \\ c \end{pmatrix} \mapsto \begin{pmatrix} a & -\bar{c}/\kappa \\ c & \bar{a}/\kappa \end{pmatrix} \qquad \kappa \equiv |a|^2 + |c|^2 \neq 0$$

(iv) $M \equiv \mathbb{C}^2 \setminus \binom{0}{0} \sim \mathbb{R}^4 \setminus \{0\} \sim \mathbb{R} \times S^3$ ($\mathbf{r} \equiv r\mathbf{n} \leftrightarrow (r, \mathbf{n}), \mathbf{n}^2 = 1$); the triviality of the bundle gives

$$P \equiv SL(2, \mathbb{C}) \sim M \times H = (\mathbb{R} \times S^3) \times (\mathbb{C} \sim \mathbb{R}^2) \sim \mathbb{R}^3 \times S^3$$

□

• A further useful criterion for triviality of any (not only principal) bundle reads (we mention it without proof)

$$\text{contractible base} \quad \Rightarrow \quad \text{trivial bundle}$$

Notice that, contrary to (20.1.3), the criterion does not provide us with a constructive recipe saying how a global trivialization is to be found; it just asserts that the latter certainly exists.

Any *non*-trivial bundle thus *needs* to have a *non*-contractible base. On the other hand, the non-contractibility is not enough for non-triviality, as an example of a *product* bundle with a non-contractible base shows. Let us see how this works for a more interesting example.

20.1.6 The *proper orthochronous Lorentz group* $G \equiv L_+^\uparrow$ (all $\Lambda \in SO(1,3)$ obeying $\Lambda^0_0 \geq 1$, so that they do not reverse the direction of time) naturally acts on the left on columns $x \equiv (x^0, x^1, x^2, x^3) \equiv (x^0, \mathbf{x})$ from $E^{1,3}$ (points of Minkowski space)

$$x \mapsto \Lambda x$$

Let H be the stabilizer of the point $\hat{x} \equiv (1, 0, 0, 0)$. Check that

(i) the orbit of the point $\hat{x}$ is (a homogeneous space)

$$M = \text{the upper hyperboloid} = \{x \in \mathbb{E}^{1,3} \mid \eta(x, x) = 1,\ x^0 > 0\}$$

(ii) as a manifold, $M \sim \mathbb{R}^3$
(iii) the group H is isomorphic to $SO(3)$
(iv)

$$\pi : L_+^\uparrow \to M$$

is a *trivial* principal $SO(3)$-bundle, so that (as a manifold)

$$L_+^\uparrow \sim \mathbb{R}^3 \times SO(3)$$

Hint: (i) see (11.1.5) and (11.1.8); (ii) $(x^0, \mathbf{x}) \mapsto \mathbf{x}$; (iii) see (13.1.11); (iv) $M \sim \mathbb{R}^3$ is contractible. □

• Yet another interesting principal bundle is the *Hopf bundle* $\pi : S^3 \to S^2$. It appears in various realizations, which seem fairly different at first sight. For example, a description presented in (20.1.8) is related to a quantum-mechanical equation $(\mathbf{n} \cdot \boldsymbol{\sigma})\zeta = \zeta$ for a "spin $\frac{1}{2}$ along $\mathbf{n}$," another one (20.1.9) places emphasis on the fact that it is a bundle of the type $\pi : G \to G/H$ (20.1.2) for $G = SU(2)$ and $H = U(1)$ and yet another one (20.1.11) uses the idea of a projective space.

20.1.7* Consider the space $\mathbb{C}^2$ with elements χ (two-component complex unnormalized columns) and the space $\mathbb{R}^3$ with elements $\mathbf{r} \leftrightarrow x_a$ (three-component real unnormalized columns). There is a natural action of the group $SU(2)$ on both of these manifolds: on $\mathbb{C}^2$ it acts directly ($\chi \mapsto A\chi \equiv L_A\chi$), on $\mathbb{R}^3$ through the (two-sheeted covering) homomorphism $f : SU(2) \to SO(3)$ introduced in problem (13.3.6) ($\mathbf{r} \mapsto R\mathbf{r} \equiv f(A)\mathbf{r} \equiv \hat{L}_{f(A)}\mathbf{r}$). Columns normalized to unity, which we will denote by ζ and $\mathbf{n} \leftrightarrow n_a$ (there holds $\zeta^+\zeta = 1 = \mathbf{n}^2$) may be regarded as points on the unit spheres $S^3 \subset \mathbb{C}^2$ and $S^2 \subset \mathbb{R}^3$ respectively (they are orbits of the actions mentioned above). Next, we define a (non-linear) map

$$\pi : \mathbb{C}^2 \to \mathbb{R}^3 \qquad \chi \mapsto \mathbf{r} \qquad \mathbf{r} := \chi^+\boldsymbol{\sigma}\chi$$

Check that

(i) for an arbitrary $\chi \in \mathbb{C}^2$ the 2×2 (Hermitian) matrix $\chi\chi^+$ may be parametrized in the form

$$\chi\chi^+ = \frac{1}{2}(r\mathbb{I}_2 + \mathbf{r} \cdot \boldsymbol{\sigma}) \qquad \text{where} \quad r := \chi^+\chi$$

(ii) restriction of π to the 3-sphere of radius $\sqrt{r}$ in $\mathbb{C}^2$ (i.e. to columns χ which satisfy $\chi^+\chi = r$) has as the image the 2-sphere of the radius r in $\mathbb{R}^3$ (so that r and $\mathbf{r}$ in the parametrization mentioned above are related by $r = |\mathbf{r}|$); for $r = 1$ we thus also have a map $\pi : S^3 \to S^2$, $\zeta \mapsto \mathbf{n}$ (we will denote it by the same letter as the original map π) and the parametrization

$$\zeta\zeta^+ = \frac{1}{2}(\mathbb{I}_2 + \mathbf{n} \cdot \boldsymbol{\sigma}) \qquad \text{where} \quad \mathbf{n} := \zeta^+\boldsymbol{\sigma}\zeta, \ |\mathbf{n}| = 1$$

(iii) the map π is $SU(2)$-*equivariant* in the sense that

$$\pi \circ L_A = \hat{L}_{f(A)} \circ \pi \qquad \text{i.e.} \quad \{\zeta \mapsto A\zeta\} \Rightarrow \{\mathbf{n} \mapsto R\mathbf{n}\}$$

(iv) the map π is surjective (each vector $\mathbf{n}$ has a preimage ζ)
(v) if ζ happens to be a preimage of $\mathbf{n}$, then $e^{i\alpha}\zeta$ is also a preimage for all $e^{i\alpha} \in U(1)$

$$\zeta \mapsto \mathbf{n} \quad \Rightarrow \quad e^{i\alpha}\zeta \mapsto \mathbf{n}$$

(vi) if the map π is expressed in coordinates $(\vartheta, \varphi, \psi)$ from problem (11.7.23) on $SU(2) = S^3$ (Euler angles) and the standard "spherical" angles (ϑ, φ) on S^2 (spherical coordinates in $\mathbb{R}^3$ for $r = 1$), it reads

$$\pi : (\vartheta, \varphi, \psi) \mapsto (\vartheta, \varphi)$$

Hint: (i) Pauli matrices plus the identity matrix constitute a basis of 2×2 Hermitian matrices (13.3.10), coefficients with respect to them using (13.3.4); (ii) if $\chi^+\chi = r$ (the sphere of radius $\sqrt{r}$), then the parametrization mentioned above (as well as the identity $\sigma_a\sigma_b = \delta_{ab}\mathbb{I} + i\epsilon_{abc}\sigma_c$) gives

$$|\mathbf{r}|^2 \equiv x_a x_a = (\chi^+\sigma_a)(\chi\chi^+)(\sigma_a\chi) = \frac{1}{2}(\chi^+\sigma_a)(r\mathbb{I}_2 + \mathbf{r}\cdot\boldsymbol{\sigma})(\sigma_a\chi) = \cdots$$
$$= \frac{3}{2}r^2 - \frac{1}{2}|\mathbf{r}|^2 \quad \Rightarrow \quad x_a x_a = r^2$$

($\Rightarrow$ the sphere of radius r); (iii) if $\chi \mapsto A\chi$, then according to (13.3.3)

$$x_a \mapsto (A\chi)^+\sigma_a(A\chi) = \chi^+(A^{-1}\sigma_a(A^{-1})^+)\chi = \chi^+(R^{-1}_{ba}\sigma_b)\chi = \chi^+(R_{ab}\sigma_b)\chi = R_{ab}x_b$$

and after the restriction $n_a \mapsto R_{ab}n_b$; (iv) a consequence of equivariance of π and transitivity of the action $SU(2)$ on S^3 (π connects two *orbits*); (v) and (vi) a direct calculation. □

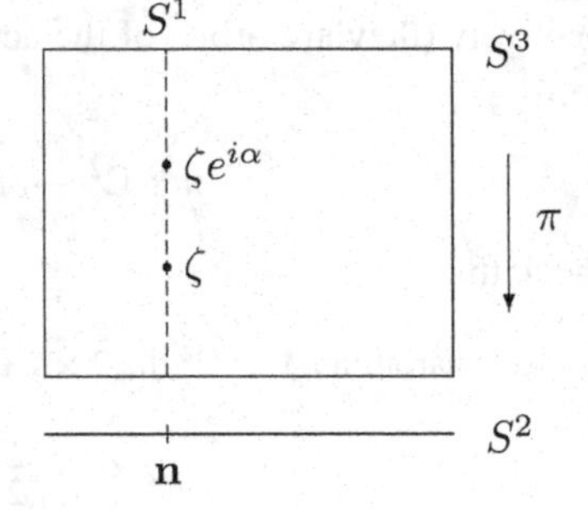

20.1.8 * Consider the situation from problem (20.1.7). Let π map a normalized column ζ to $\mathbf{n}$, i.e. $\zeta^+\boldsymbol{\sigma}\zeta = \mathbf{n}$ (since according to (20.1.7) π is surjective, so, certainly, some ζ for a given $\mathbf{n}$ exists). Check that

(i) the column ζ is a (normalized) solution of the equation

$$(\mathbf{n}\cdot\boldsymbol{\sigma})\zeta = \zeta$$

(ii) *all* normalized solutions $\hat{\zeta}$ of the equation just coincide with all preimages of $\mathbf{n}$ with respect to π and at the same time with the columns $\hat{\zeta} = e^{i\alpha}\zeta$ for all possible $e^{i\alpha} \in U(1)$

$$\{(\mathbf{n}\cdot\boldsymbol{\sigma})\hat{\zeta} = \hat{\zeta} \quad \text{and} \quad \hat{\zeta}^+\hat{\zeta} = 1\} \Leftrightarrow \pi(\hat{\zeta}) = \mathbf{n} \Leftrightarrow \hat{\zeta} = e^{i\alpha}\zeta$$

(iii) a preimage with respect to π for a given point $\mathbf{n} \in S^2$ is a *circle* $S^1 \subset S^3 \subset \mathbb{C}^2$; the three-dimensional sphere may thus be regarded as a union of one-dimensional spheres (i.e. circles)[416]

[416] The ambitious reader may check that all the circles have the *same length* (which is a bit different from the *two-dimensional* sphere being regarded as a union of circles (say, parallel lines).

(iv) the map π is the projection of a non-trivial principal $U(1)$-bundle

$$\pi : S^3 \to S^2 \qquad \text{Hopf bundle}$$

The corresponding action of $U(1)$ on S^3 reads

$$\zeta \mapsto e^{i\alpha}\zeta$$

Hint: (i) according to (20.1.7) we have $(\mathbf{n}\cdot\boldsymbol{\sigma})\zeta = (2\zeta\zeta^+ - \mathbb{I}_2)\zeta = 2\zeta(\zeta^+\zeta) - \zeta = \zeta$; (ii) let ζ be mapped to $\mathbf{n}$, we look for *all* preimages of $\mathbf{n}$; since $SU(2)$ transitively acts on normalized columns, we may parametrize them in the form $\hat{\zeta} = B\zeta$ for $B \in SU(2)$. But $B\zeta$ maps to $\hat{R}\mathbf{n}$ (where B covers $\hat{R} \equiv f(B)$), so that $\hat{R}$ may be at most a rotation (by an arbitrary angle) *about* $\mathbf{n}$, corresponding (13.3.6) to the one-parameter subgroup $B(t) = \exp(-\frac{i}{2}t(\mathbf{n}\cdot\boldsymbol{\sigma})) \in SU(2)$, which is isomorphic to $U(1)$; the whole preimage of $\mathbf{n}$ is then its orbit $B(t)\zeta \sim S^1$; from the expression for $\zeta\zeta^+$ we obtain that the action of $U(1)$ on the fiber over $\mathbf{n}$ is

$$\begin{aligned}\zeta \mapsto B(t)\zeta &= \exp\left(-\frac{i}{2}t(\mathbf{n}\cdot\boldsymbol{\sigma})\right)\zeta = \left(\cos\frac{t}{2}\mathbb{I}_2 - i\sin\frac{t}{2}(\mathbf{n}\cdot\boldsymbol{\sigma})\right)\zeta \\ &= \left(\cos\frac{t}{2} - i\sin\frac{t}{2}\right)\zeta = e^{-\frac{i}{2}t}\zeta \equiv e^{i\alpha}\zeta\end{aligned}$$

So all preimages of $\mathbf{n}$ are given just by the columns $e^{i\alpha}\zeta$ for all possible α; (iv) a trivial bundle should have the total space diffeomorphic to $S^2 \times S^1$, which is not simply connected (a loop "around S^1" is not contractible), whereas S^3 is simply connected. □

• In this approach, a fiber of the Hopf bundle emerges as a set of all normalized solutions of the equation $(\mathbf{n}\cdot\boldsymbol{\sigma})\zeta = \zeta$. In textbooks on quantum mechanics we may read that if $\mathbf{n}$ is an arbitrary unit vector, then on solving this equation we get a (normalized) spinor ζ (an element of $\mathbb{C}^2$), which corresponds to a particle with "spin $\frac{1}{2}$ along $\mathbf{n}$." All "directions" (unit vectors $\mathbf{n}$) form the sphere S^2 (base of the bundle), normalized "spinors" ζ in turn constitute the sphere S^3 (the total space of the bundle) and the fiber over $\mathbf{n}$ is the "sphere" S^1, since we learned that the *only* freedom in the solutions of the equation mentioned above (for a given $\mathbf{n}$) is $\zeta \mapsto e^{i\alpha}\zeta$.

Let us look at the realization of the bundle as a bundle of type $G \to G/H$.

20.1.9* Recall that the group $SU(2)$, when regarded as a manifold, is the three-dimensional sphere S^3. Its embedding into $\mathbb{C}^2$ is given by

$$A(\zeta) \equiv \begin{pmatrix} z & -\bar{w} \\ w & \bar{z}\end{pmatrix} \leftrightarrow \zeta \equiv \begin{pmatrix} z \\ w\end{pmatrix} \in \mathbb{C}^2 \qquad \zeta^+\zeta \equiv |z|^2 + |w|^2 = 1 \quad (z, w \in \mathbb{C})$$

Check that

(i) the right translation of the subgroup $U(1) \subset SU(2)$, generated by the basis element $E_3 \equiv -\frac{i}{2}\sigma_3 \in su(2)$ (i.e. of the subgroup of matrices $e^{\alpha E_3} = \operatorname{diag}(e^{-\frac{i}{2}\alpha}, e^{\frac{i}{2}\alpha})$), on $SU(2)$ explicitly reads

$$A \mapsto Ae^{\alpha E_3} \qquad \Leftrightarrow \qquad \zeta \mapsto e^{-\frac{i}{2}\alpha}\zeta \qquad \Leftrightarrow \qquad (\vartheta, \varphi, \psi) \mapsto (\vartheta, \varphi, \psi + \alpha)$$

(ii) a standard construction of principal H-bundle $\pi : G \to G/H$ from (20.1.2) gives for this particular case ($G = SU(2)$, $e^{\alpha E_3} \in H = U(1)$) just the Hopf bundle from problem (20.1.8).

Hint: (i) see (11.7.12) and the parametrization from (11.7.23); we have

$$A(\zeta) \equiv \begin{pmatrix} z & -\bar{w} \\ w & \bar{z} \end{pmatrix} \mapsto \begin{pmatrix} z & -\bar{w} \\ w & \bar{z} \end{pmatrix} \begin{pmatrix} e^{-\frac{i}{2}\alpha} & 0 \\ 0 & e^{\frac{i}{2}\alpha} \end{pmatrix} = \cdots = A\left(e^{-\frac{i}{2}\alpha}\zeta\right)$$

so that the right coset gH for $g = \zeta$ contains just elements $e^{-\frac{i}{2}\alpha}\zeta$; (ii) according to (20.1.8), exactly the whole coset is mapped to $\mathbf{n} \in S^2$, so that we have a bijection $gH \leftrightarrow [g] \leftrightarrow \mathbf{n}$; this means that an abstract scheme $\pi : SU(2) \to SU(2)/U(1)$ may be here equivalently replaced by (identified with) the concrete bundle $\pi : S^3 \to S^2$ discussed in (20.1.8). □

20.1.10* In problem (13.4.15) we studied generators of the left action of G on a homogeneous space G/H (we learned there that they may be obtained by the projection of the right-invariant fields R_X on the group). Here, we will treat the special case where $G = SU(2)$, $H = U(1)$ generated by the element $E_3 = -\frac{i}{2}\sigma_3 \in su(2)$ and we get as G/H (20.1.9) ordinary sphere S^2. Check that

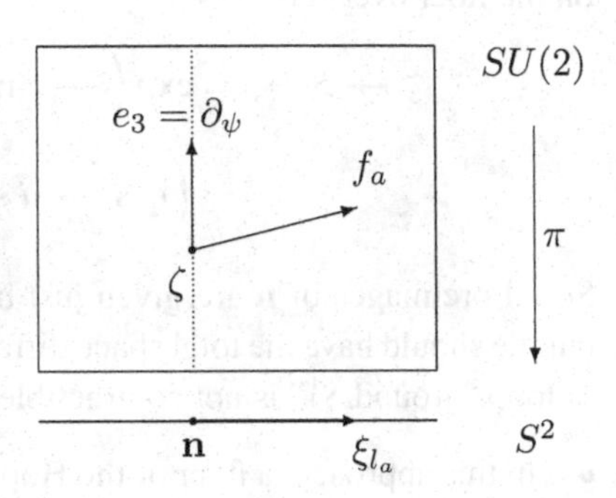

(i) the canonical projection π in G/H in coordinates reads

$$\pi : (\vartheta, \varphi, \psi) \mapsto (\vartheta, \varphi)$$

(ii) generators of rotations ξ_{l_a} on the sphere S^2 from (13.4.6) may be obtained in this way by the projection of the right-invariant basis f_a on $SU(2)$ from (11.7.23)

$$\xi_{l_a} = \pi_* f_a \qquad a = 1, 2, 3$$

(iii) because of that they satisfy "equal commutation relations"

$$[f_a, f_b] = -\epsilon_{abc} f_c \qquad [\xi_{l_a}, \xi_{l_b}] = -\epsilon_{abc}\xi_{l_c} \qquad a = 1, 2, 3$$

(iv) the generator of the right action $\mathcal{R}_h$ of the subgroup $U(1)$ along cosets is the left-invariant field $e_3 = \partial_\psi$.

Hint: (i) see (20.1.7) and the identification of projections π from the hint to (20.1.9); (ii) use $\pi \circ L_A = \hat{L}_{f(A)} \circ \pi$ (20.1.7) in a similar way as $\pi \circ \mathcal{L}_g = L_g \circ \pi$ was used in (13.4.15); the difference lies in that here we have an *additional* homomorphism (covering) f, which results in a general relation $\xi_{f'(X)} = \pi_* R_X$, $X \in su(2)$ (instead of $\xi_X = \pi_* R_X$ in (13.4.15)); for the basis $E_a = -\frac{i}{2}\sigma_a$ we have $f'(E_a) = l_a \in so(3)$ (13.3.7) and (12.1.5), so that, in particular, $\xi_{f'(E_a)} = \xi_{l_a} = \pi_* R_{E_a} = \pi_* f_a$; in coordinates $\pi_*(\partial_\vartheta, \partial_\varphi, \partial_\psi) \mapsto (\partial_\vartheta, \partial_\varphi, 0)$

and so

$$\pi_* \left(-\sin\varphi \partial_\vartheta - \cot\vartheta \cos\varphi \partial_\varphi + \frac{\cos\varphi}{\sin\vartheta} \partial_\psi \right) = -\sin\varphi \partial_\vartheta - \cot\vartheta \cos\varphi \partial_\varphi$$

$$\pi_* \left(\cos\varphi \partial_\vartheta - \cot\vartheta \sin\varphi \partial_\varphi + \frac{\sin\varphi}{\sin\vartheta} \partial_\psi \right) = \cos\varphi \partial_\vartheta - \cot\vartheta \sin\varphi \partial_\varphi$$

$$\pi_* \partial_\varphi = \partial_\varphi$$

(iii) and (iv) see (13.4.4) and (13.4.15). □

• Yet another way of looking at the Hopf bundle is provided by the language of projective spaces.

20.1.11* Consider the complex space $\mathbb{C}^{n+1}$ with points χ and the corresponding projective space $\mathbb{C}P^n$ (1.3.2) with points $[\chi]$. On each ray in $\mathbb{C}^{n+1}$ the group $GL(1, \mathbb{C})$ acts naturally ($\chi \mapsto \lambda\chi$; the points of $\mathbb{C}P^n$ may be identified just with *orbits* of the action). Check that

(i) a ray may also be characterized by a *normalized* representative ζ (such that $\zeta^+\zeta = 1$ holds)
(ii) the normalized representative is still not unique, the freedom being $\zeta \mapsto e^{i\alpha}\zeta$, so that on normalized vectors (i.e. on the sphere S^{2n+1}) the group $U(1)$ acts
(iii) in this way a principal $U(1)$-bundle arises

$$\pi : S^{2n+1} \to \mathbb{C}P^n$$

(iv) the case $n = 1$ just corresponds to the *Hopf bundle* discussed in (20.1.8) and (20.1.9)
(v) at the same time this corresponds, for $n = 1$, to the map described in (1.4.6).

Hint: (i) the norm of a vector changes as $\chi^+\chi \mapsto |\lambda|^2\chi^+\chi$; (ii) $|\lambda|^2 = 1$; (iii) the orbit space is (by definition) just $\mathbb{C}P^n$; (iv) $\mathbb{C}P^1 = S^2$; (v) see (1.4.6). □

20.2 Connection form $\omega \in \Omega^1(P, \mathrm{Ad})$

• In Chapter 19 we learned that the linear connection on M may be encoded into a horizontal $GL(n, \mathbb{R})$-invariant distribution $\mathcal{D}^h$ on LM. Here, we will promote this result to a starting point for our discussion on connections on arbitrary principal bundles, i.e. we *define* a connection in terms of the distribution itself:

Definition A *connection on a principal G-bundle* $\pi : P \to M$ is a(n arbitrary) horizontal G-invariant distribution $\mathcal{D}^h$ on the total space P (or anything which is equivalent to this object).[417]

This means that in each tangent space T_pP of the manifold P we have, in addition to the canonical *vertical subspace* (whose existence reflects the fiber structure on the manifold P), also a distinguished complementary subspace, which is called a *horizontal subspace*; each

[417] The bracket is a loophole for the future, where we will see that a connection in this sense may also be encoded differently, say, into the connection *form*, and it would be inconvenient to prevent such alternative methods of description from having a chance to serve as a *definition* (we already encountered a similar situation with the definition of the concept of a vector on a manifold in Section 2.2).

vector v on the manifold P may thus be *uniquely* decomposed into the sum of its vertical and horizontal parts

$$T_pP = \text{Ver}\,_pP \oplus \text{Hor}\,_pP \qquad v = \text{ver}\,v + \text{hor}\,v$$
$$\text{ver}\,v \in \text{Ver}\,_pP, \ \text{hor}\,v \in \text{Hor}\,_pP$$

20.2.1 Be sure to understand that

(i) all horizontal subspaces within a single fiber may be linked to each other by the action of the group G and the same is also true for vertical subspaces

$$R_{g*}\text{Hor}\,_pP = \text{Hor}\,_{pg}P \qquad R_{g*}\text{Ver}\,_pP = \text{Ver}\,_{pg}P$$

(ii) the decomposition of a vector into its horizontal and vertical parts commutes with the action of the group

$$\text{hor}\,R_{g*}v = R_{g*}\text{hor}\,v \qquad \text{ver}\,R_{g*}v = R_{g*}\text{ver}\,v$$

(iii) in order to specify a horizontal distribution it is enough to (smoothly) fix the horizontal subspace at a single (arbitrary) point in each fiber.

Hint: (i) horizontal: the definition of the concept G-invariance of $\mathcal{D}^h$, i.e. $R_g^*\mathcal{D}^h = \mathcal{D}^h$; vertical: verticality of the action; (ii) and (iii) a direct consequence of (i). □

• We saw in (19.2.1) that a horizontal distribution on LM (i.e. for a linear connection) may be concisely specified in terms of a connection form, which is a matrix-valued 1-form on LM; its component forms served as constraint forms of the distribution. Here we will learn that also in the general case of a connection in $\pi : P \to M$, the horizontal distribution may be conveniently expressed in terms of an appropriate 1-form. It turns out that it is natural to regard the values of the form to lie in the Lie algebra $\mathcal{G}$ of the group G. For the construction of the form it is useful first to formalize a well-known property of fundamental fields by introducing a map Ψ_p.

20.2.2 Given $X \in \mathcal{G}$, consider the fundamental field ξ_X of the action R_g on P. Define a map

$$\Psi_p : \mathcal{G} \to \text{Ver}\,_pP \qquad X \mapsto \xi_X(p)$$

Check that

(i) Ψ_p is a linear isomorphism
(ii) an arbitrary *vertical* vector at the point $p \in P$ may be *uniquely* written as a certain fundamental field in the point, i.e. in the form of $v \equiv \text{ver}\,v = \xi_X(p)$ for a unique element X.

Hint: (i) $X \mapsto \xi_X$ is a representation of the Lie algebra $\mathcal{G}$ (13.4.3) $\Rightarrow$ it is linear. The bijectivity is still needed. Injectivity is due to the freedom of the action (if $\xi_X(p) = \xi_Y(p)$, then $\xi_{X-Y}(p) = 0$, so that $X - Y$ is (4.1.6) from the Lie subalgebra corresponding to the stabilizer; the latter is, however, trivial), surjectivity is due to the transitivity in a fiber (there is no direction which cannot be produced by the action); (ii) $X = \Psi_p^{-1}(v)$. □

[20.2.3] Check the following behavior of the map Ψ_p from (20.2.2) with respect to the action of the group

$$R_{g*} \circ \Psi_p = \Psi_{pg} \circ \mathrm{Ad}_{g^{-1}} \qquad \left(\Rightarrow \text{ also } \Psi_{pg}^{-1} \circ R_{g*} = \mathrm{Ad}_{g^{-1}} \circ \Psi_p^{-1}\right)$$

Hint: see (13.4.3); the second one by applying $(\ )^{-1}$ to the first one plus the replacements $p \mapsto pg,\ g \mapsto g^{-1}$. □

- Now we are prepared to introduce a connection form and to learn its essential properties.

[20.2.4] Define at the point $p \in P$ a 1-form ω_p with values in the Lie algebra $\mathcal{G}$ by the prescription: if v_p is an arbitrary vector in the point p, then

$$\langle \omega_p, v_p \rangle := \Psi_p^{-1}(\text{ver } v_p) \qquad \text{i.e.} \quad \omega_p = \Psi_p^{-1} \circ \text{ver} : T_p P \to \mathcal{G}$$

So, we first project out the *vertical* part[418] of the vector v_p and express it, in the sense of (20.2.2), as a fundamental field ξ_X at the point p. The fundamental field is parametrized by a *unique* element X from the Lie algebra $\mathcal{G}$. *This* X is declared to be the image of the vector v_p with respect to the map ω_p. Check that

(i) it is indeed a ($\mathcal{G}$-valued) 1-form
(ii) *horizontal vectors* are just those vectors which are annihilated by the form ω

$$v_p \in \mathrm{Hor}_p P \Leftrightarrow \langle \omega_p, v_p \rangle = 0 \qquad \text{i.e.} \quad \mathrm{Hor}_p P = \mathrm{Ker}\, \omega_p$$

(iii) if E_i is an arbitrary basis of the Lie algebra $\mathcal{G}$, then

$$\omega_p = \omega_p^i E_i$$

and (ordinary, i.e. $\mathbb{R}$-valued) component 1-forms ω_p^i may serve as constraint 1-forms of the horizontal subspace $\mathrm{Hor}_p P \subset T_p P$.

Hint: (i) it is a composition of two *linear* maps; (ii) the vertical part of horizontal vectors vanishes; (iii) see (6.4.1) and (19.3.1). □

- If we define a form ω_p for *each* $p \in P$ (so that it varies smoothly with p), we get a *connection 1-form* $\omega \in \Omega^1(P, \mathcal{G})$, i.e. already a 1-form on a manifold P with values in the Lie algebra $\mathcal{G}$. In this simple object, a (global) Lie algebra valued 1-form, there is (by construction) encoded full information concerning the horizontal distribution $\mathcal{D}^h$ on P, i.e. concerning connection in the principal G-bundle $\pi : P \to M$.

In particular, the fact that the distribution $\mathcal{D}^h$ is G-invariant (in the sense that $R_{g*}\mathcal{D}^h = \mathcal{D}^h$) should be somehow reflected in properties of ω. In order to see this in detail, recall that the concept of differential forms of *type* ρ was introduced in Section 13.5. In our situation

[418] Note that the *vertical* part of a vector depends on how the *horizontal* (complementary) subspace is defined, i.e. *on a connection*; consider, for example, a basis e_1, e_2 in a two-dimensional space and let $v = ae_1 + be_2$. The projection onto the subspace spanned by e_1 (if the complementary subspace is spanned on e_2) is ae_1. If, however, we chose the complementary subspace to be spanned on $\tilde{e}_2 \equiv e_1 + e_2$, then we would write $v = \tilde{a}e_1 + \tilde{b}\tilde{e}_2 \equiv (a-b)e_1 + b\tilde{e}_2$ and the projection on e_1 would be $\tilde{a}e_1 = (a-b)e_1 \neq ae_1$. A change of e_2 thus results in a change of the projection onto e_1.

(when we have an action R_g on a manifold P) a p-form α on P with values in (V, ρ) is called a *form of type* ρ (we write $\alpha \in \Omega^p(P, \rho)$), if it satisfies

$$R_g^*\alpha = \rho(g^{-1})\alpha$$

Alternatively, introducing component forms by

$$\alpha = \alpha^a E_a \qquad E_a \text{ – a basis in } V$$

there holds

$$(R_g^*\alpha^a)E_a = \alpha^a \rho(g^{-1})E_a \quad \text{or} \quad R_g^*\alpha^a = (\rho(g^{-1}))^a_b \alpha^b$$

20.2.5 Check that a connection form ω has the following two important properties:

$$R_g^*\omega = \text{Ad}_{g^{-1}}\omega \qquad \text{i.e.} \quad \omega \in \Omega^1(P, \text{Ad}) \qquad (\text{it is } \textit{of type } \text{Ad})$$
$$\langle \omega, \xi_X \rangle = X$$

Solution: using (20.2.1) and (20.2.3) we get

$$\begin{aligned}\langle R_g^*\omega_{pg}, v_p \rangle &= \langle \omega_{pg}, R_{g*}v_p \rangle = \Psi_{pg}^{-1}(\text{ver } R_{g*}v_p) = \Psi_{pg}^{-1} \circ R_{g*} \circ \text{ver } v_p \\ &= \text{Ad}_{g^{-1}} \circ \Psi_p^{-1} \circ \text{ver } v_p = \text{Ad}_{g^{-1}} \langle \omega_p, v_p \rangle \\ \langle \omega_p, \xi_X \rangle &= \Psi_p^{-1} \text{ver } \xi_X(p) = \Psi_p^{-1} \xi_X(p) = X\end{aligned}$$

□

20.2.6 Check that if, on the contrary, there is a $\mathcal{G}$-valued 1-form ω on P, which has the two properties from problem (20.2.5), then the prescription

$$\text{Hor}_p P := \text{Ker } \omega_p$$

defines on P a G-invariant distribution $\mathcal{D}^h$, which is complementary to the vertical distribution $\mathcal{D}^v$, i.e. a *connection* in the G-bundle $\pi : P \to M$. (This gives the possibility of an alternative (equivalent) *definition* of connection simply as such a 1-form.)

Hint: $\langle \omega_{pg}, R_{g*}v_p \rangle = \langle R_g^*\omega_{pg}, v_p \rangle = \dots (20.2.5) = \dots \text{Ad}_{g^{-1}} \langle \omega_p, v_p \rangle$, so that $R_{g*}\text{Hor}_p = \text{Hor}_{pg}$ (the horizontal distribution is G-invariant). The second equation says that the fundamental fields *are not* horizontal and guarantees that component 1-forms ω^i are at each point linearly independent (if $c_i\omega^i_p = 0$, then evaluation of both sides on the vectors ξ_{E_j} yields $c_j = 0$), so that the dimension of the horizontal distribution is just complementary to the dimension of the vertical one (the dimension of the manifold P is a sum of the dimensions of a base and a group, and the vertical distribution has the dimension of the group, so that the horizontal one has the dimension of the base). □

20.2.7 Check that by passing to an infinitesimal version of (20.2.5) we get the identities

$$\begin{aligned}\mathcal{L}_{\xi_X}\omega &= -\text{ad}_X\omega \equiv -[X, \omega] \\ i_{\xi_X}\omega &= X\end{aligned}$$

and, combining them, also

$$i_{\xi_X} d\omega = -[X, \omega]$$

Hint: see (19.4.2); $di_{\xi_X}\omega = dX = 0$, since X is a constant (function) on P. □

20.3 Parallel transport and the exterior covariant derivative D

• In Section 19.5 we learned that the procedure of parallel transport of a frame along a curve γ on the base M may be equivalently described as a construction of the curve γ^h in the total space LM, the horizontal lift of the original curve γ. The lift is specified by two conditions:

1. it projects to the original curve γ on M
2. it is horizontal at each point.

If, in the original approach discussed in Chapter 15, we performed the transport of a frame $e(x)$ from a point x to a point y (along the curve γ), in the novel approach the lift begins at the point $e(x)$ in the fiber over x and we then declare the point on γ^h in the fiber over y to be the parallel transported frame.

In Section 19.6 we then learned that, surprisingly, in the formalism based on the manifold of frames LM also tensor fields on M may be treated remarkably simply, namely as equivariant functions on LM. We also learned how parallel transport of tensor fields may be performed in this approach as well as a computation of their covariant derivatives.

Here we will show how all of these concepts and procedures may be easily extended to the case of a *general* principal bundle.

Recall that points p of the total space of a general principal bundle $\pi : P \to M$ have no interpretation in terms of the base (there is no counterpart of the fact that a point of LM corresponds to a frame on M). So if we intend somehow to define their parallel transport, we cannot base the method (contrary to Chapter 19) upon our knowledge of how it is done on the base manifold M. A natural possibility suggests itself, however, to repeat almost verbatim the idea used in the special case of a principal bundle, the frame bundle $\pi : LM \to M$. A point p of the total space P of a general principal bundle may be regarded in this context as a generalization of a frame. We learned that the parallel transport of a frame could be interpreted *there* as a construction of horizontal lift and that (only) the horizontal distribution is needed for that. However, just this structure is available on a general principal bundle, too (*this is* namely a connection), so that the horizontal lift of a curve can also be constructed in the general setting.

20.3.1 Let $\mathcal{D}^h$ be a horizontal distribution on P, given by a connection form ω. Check that

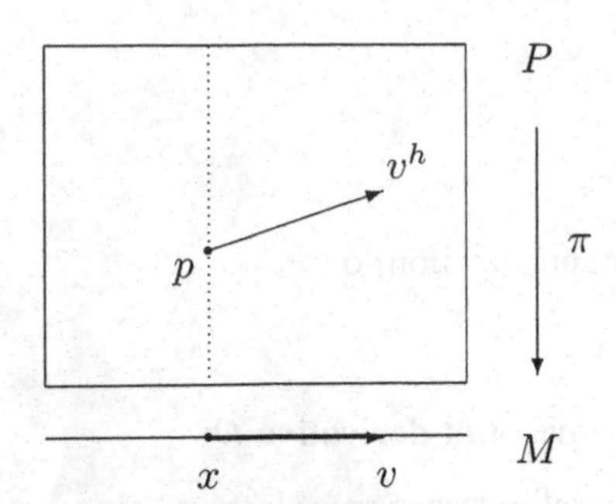

(i) if $v \in T_x M$ is an arbitrary vector at a point $x \in M$, then at each point $p \in \pi^{-1}(x)$ in the fiber over x there exists the unique *horizontal lift*, i.e. a vector $v^h \in T_p P$ such that

$$\pi_* v^h = v \qquad \text{(it projects onto } v\text{)}$$
$$v^h \in \text{Hor}_{\,p} P \qquad \text{(it is horizontal)}$$

(ii) the operation of the horizontal lift $v \mapsto v^h$ is a *linear isomorphism* of the (whole) tangent space in x and the *horizontal* subspace in p

(iii) if we lift v in this way to all points of the fiber over x, we get a vector field v^h in the fiber which is *G-invariant*: $R_{g*} v^h = v^h$

(iv) if we lift in this pointwise way a vector *field* V from M, we get on P a vector field V^h, which is *G-invariant* and *horizontal*:

$$R_{g*} V^h = V^h \qquad \langle \omega, V^h \rangle = 0$$

(v) on the other hand, *each* G-invariant and horizontal vector field W on P may be regarded as a horizontal lift V^h of an appropriate field V on M.

Hint: the projection induces a surjective linear map $\pi_* : T_p P \to T_x M$; the distribution $\mathcal{D}^h$ gives *by definition* the decomposition $T_p P = \text{Ver}_{\,p} P \oplus \text{Hor}_{\,p} P$, where $\text{Ver}_{\,p} P := \text{Ker}\, \pi_*$ has the dimension of the group G and $\text{Hor}_{\,p} P$ is complementary, so it is *canonically isomorphic* to the target space $T_x M$; the isomorphism is in one direction (the restriction of) π_*, in the opposite one the lift (being moreover just the *inverse map* to the restriction of π_*); (iii) $R_{g*} v_p^h$ satisfies both the properties of v_{pg}^h (due to $\pi \circ R_g = \pi$ and the G-invariance of the horizontal distribution (20.2.1)), so that *it is* v_{pg}^h. □

20.3.2 Let γ be a curve on the base M of a principal bundle $\pi : P \to M$ with connection and p an arbitrary point from the fiber over $\gamma(0)$. Check that

(i) there exists a unique curve γ^h on P given by the conditions

$$\pi \circ \gamma^h = \gamma \qquad \text{(it projects to } \gamma\text{)}$$
$$\gamma^h(0) = p \qquad \text{(it starts in the point } p\text{)}$$
$$\langle \omega, (\dot{\gamma^h}) \rangle = 0 \qquad \text{(it is horizontal)}$$

i.e. γ^h is everywhere "exactly above" γ, it passes through a given point p and the tangent vector to this curve is at each point *horizontal*. The curve γ^h is called the *horizontal lift of the curve* γ

(ii) tangent vectors to the curve γ^h are *horizontal lifts* of tangent vectors to the initial curve γ

$$(\dot{\gamma^h}) = (\dot{\gamma})^h$$

(iii) the horizontal lift of a reparametrized curve is the (equally) reparametrized horizontal lift of the initial curve; this means that also the horizontal lift of a (non-parametrized) *path* (corresponding to a curve γ) is well defined,

$$(\gamma \circ \sigma)^h = \gamma^h \circ \sigma \qquad \sigma : t \mapsto \sigma(t) \in \mathbb{R}$$

Hint: (i) we start in $\gamma^h(0) = p$, make a step by ϵ along $(\dot{\gamma})^h$ (being just over $\gamma(t)$ at this small piece of curve and moving horizontally), etc.; (ii) from the construction; (iii) the tangent vector of a reparametrized curve turns out (2.3.5) to be just a multiple of the initial one, the procedure of the lift of a vector is, however, linear so that the new lifted tangent vector is the same multiple of the old lifted one. □

• We will interpret points of the horizontal lift $p(t) \equiv \gamma^h(t)$ *by definition* (motivated by the special case (19.5.1) and (19.5.2) for LM) as a *parallel transported* "generalized frame."

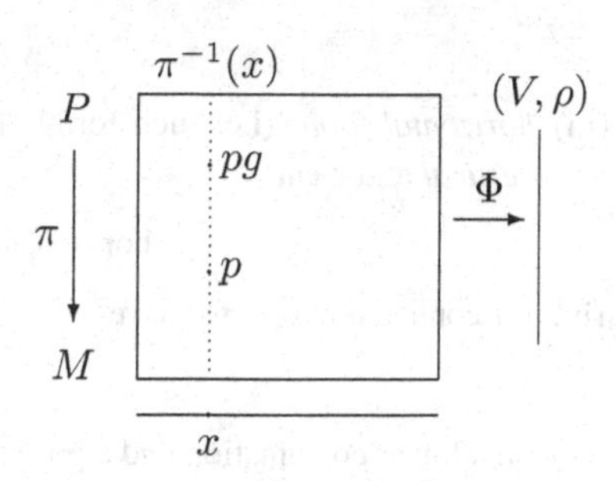

If we also wish to introduce parallel transport of objects other than generalized frames, we need first to define the objects themselves. In the case of LM we first studied tensor fields, but it then turned out that their expression in terms of LM also offers free of charge even more general objects (19.6.5), quantities of *type* ρ. The good news is that these objects may also be naturally introduced on a general principal G-bundle $\pi : P \to M$. Again, they are *equivariant functions* on P, i.e. maps

$$\Phi : P \to (V, \rho) \qquad \Phi \circ R_g = \rho(g^{-1}) \circ \Phi$$
$$\text{i.e.} \quad \Phi(pg) = \rho(g^{-1})\Phi(p)$$

where (V, ρ) is a linear space (its elements play the role of *components* of the quantity Φ with respect to the "frame" p), in which a representation ρ of the group G is available. A *quantity of type* ρ at a point $x \in M$ is introduced as an equivariant function Φ from the fiber over x to V; its value $\Phi(p)$ is regarded as "components" of the quantity with respect to the "basis" p. If we change the "basis" according to $p \mapsto pg$, the equivariance of Φ results in a transformation of "components" $\Phi(p) \mapsto \Phi(pg) \equiv \rho(g^{-1})\Phi(p)$, i.e. "components" are mixed up according to the (anti)representation $\rho(g^{-1})$.

Now, the parallel transport of such an object may be defined in the same way, as we did for the particular case of LM (see (19.6.4) and the text further on). Namely, if we *parallel* transport a "frame" along a curve γ (i.e. we move along the *horizontal lift* γ^h), the corresponding "components" are to be kept (by definition) *constant*

$$\Phi \text{ is parallel transported along } \gamma \quad \Leftrightarrow \quad \Phi(\gamma^h(t)) = \text{constant}$$

20.3.3 Check that this condition may also be written in the form

$$\langle d\Phi, (\dot{\gamma})^h \rangle = 0$$

Hint: item (ii) in (20.3.2). □

• Yet another way of expressing the fact that Φ corresponds to an *autoparallel* quantity of type ρ on a curve γ may be written in terms of special and important operations on forms, namely projection on the horizontal part and a combination of such a projection with the exterior derivative.

20.3.4 For an arbitrary p-form α on P, define a new p-form $\text{hor}\,\alpha$ (the *horizontal part* of α) by the prescription

$$(\text{hor}\,\alpha)(U,\ldots,V) := \alpha(\text{hor}\,U,\ldots,\text{hor}\,V)$$

Check that

(i) it is well defined (the result is indeed a p-form)
(ii) the map $\text{hor}: \Omega^p(P) \to \Omega^p(P)$ is a projection, i.e.

$$\text{hor} \circ \text{hor} = \text{hor}$$

(iii) *horizontal forms* (i.e. such forms α for which $\text{hor}\,\alpha = \alpha$) are annihilated by (even a single) *vertical* argument

$$\text{hor}\,\alpha = \alpha \quad \Leftrightarrow \quad i_W\alpha = 0 \text{ for vertical } W$$

(iv) for connection form we have

$$\text{hor}\,\omega = 0$$

(v) on a linear combination and a product there holds

$$\text{hor}\,(\alpha + \lambda\beta) = \text{hor}\,\alpha + \lambda\,\text{hor}\,\beta$$
$$\text{hor}\,(\alpha \wedge \beta) = (\text{hor}\,\alpha) \wedge (\text{hor}\,\beta)$$

so that the operator hor is an (endo)*morphism* of the Cartan algebra $\Omega(P)$ of differential forms on P; the image of the morphism is a subalgebra of the Cartan algebra

$$\bar{\Omega}(P) := \text{Im hor} \subset \Omega(P) \qquad \textit{the algebra of horizontal forms on } P$$

(vi) if the operator hor is applied in the standard way on forms of type ρ (i.e. with values in (V,ρ), $\text{hor}\,(\alpha^A E_A) := (\text{hor}\,\alpha^A)E_A$, (6.4.4)), then hor preserves the type ρ of the forms, on which it acts

$$\alpha \text{ is of type } \rho \quad \Rightarrow \quad \text{hor}\,\alpha \text{ is of type } \rho$$

Hint: (i) $V \mapsto \text{hor}\,V$ is $\mathcal{F}$-linear operation; (ii) evaluate both sides on general arguments; (vi) $R_g^* \circ \text{hor} = \text{hor} \circ R_g^*$ according to (20.2.1). □

20.3.5 Consider differential forms on P and define their *exterior covariant derivative* by the prescription

$$D\alpha := \text{hor}\,d\alpha$$

i.e. as the horizontal part of their (ordinary) exterior derivative. Check that

(i) it is a map

$$D: \Omega^p(P) \to \bar{\Omega}^{p+1}(P)$$

(ii) on a linear combination and a product it behaves analogously to the "ordinary" exterior derivative

$$D(\alpha + \lambda\beta) = D\alpha + \lambda D\beta$$
$$D(\alpha \wedge \beta) = (D\alpha) \wedge \text{hor}\,\beta + (\hat{\eta}\,\text{hor}\,\alpha) \wedge D\beta$$

so that the operator D behaves on the Cartan subalgebra $\tilde{\Omega}(P)$ of *horizontal* forms on P as a *derivation of degree* $+1$

(iii) if we apply the operator D in the standard way on forms of type ρ (i.e. on forms with values in (V, ρ), the prescription being $D(\alpha^A E_A) := (D\alpha^A)E_A$, (6.4.4)), then D preserves the type ρ of forms on which it acts

$$\alpha \text{ is of type } \rho \quad \Rightarrow \quad D\alpha \text{ is of type } \rho$$

Hint: (ii) standard properties of d and hor; (iii) $R_g^* \circ D = D \circ R_g^*$ according to (20.3.4). □

20.3.6 Consider a quantity Φ of type ρ (an equivariant function $\Phi : P \to (V, \rho)$) which satisfies the condition

$$D\Phi = 0$$

over some domain $\mathcal{U}$ on M (i.e. in $\pi^{-1}(\mathcal{U}) \subset P$). Check that if γ is any curve passing through $\mathcal{U}$, then the restriction of Φ to fibers over γ corresponds to an *autoparallel* quantity of type ρ defined on the curve γ.

Hint: see (20.3.3) and $\langle d\Phi, (\dot{\gamma})^h \rangle = \langle d\Phi, \mathrm{hor}\,(\dot{\gamma})^h \rangle = \langle \mathrm{hor}\, d\Phi, (\dot{\gamma})^h \rangle = \langle D\Phi, (\dot{\gamma})^h \rangle$. □

• The mere fact of writing the property of a quantity Φ as being autoparallel in terms of D actually represents no technical progress, since the computation right from the definition is too complicated (one has to perform explicitly horizontal projections of arguments). The good news is that for truly important objects, horizontal forms of type ρ (in particular, functions of type ρ) and a connection form ω, the result may be written in an amazingly simple explicit form (see (20.4.3) and (20.4.6)).

20.3.7 Be sure to understand that

(i) parallel transport *does not depend on the parametrization* of a curve γ; it is completely given by an oriented *path* (non-parametrized "curve"), along which an object is transported

(ii) the net effect of a transport from x to y and then returning back from y to x along *the same* path is zero, i.e. the complete transport is equivalent to doing nothing (the actions of the transports there and back along the same path "cancel" one another).

Hint: (i) we can see from the result of (20.3.2) that the final point on the lifted path does not depend on a parametrization of the path itself; (ii) we return back along the *same* horizontal lift. □

20.4 Curvature form $\Omega \in \Omega^2(P, \mathrm{Ad})$ and explicit expressions of D

• As we learned in Sections 19.5 and 20.3, parallel transport involves the construction of the horizontal lift of a path from the base to the total space of a bundle. The lift, however, may *depend on the path*. Namely, consider two different paths connecting the points x, y on the base. Then, the lifts of the paths (starting from the same point p over x) do not necessarily end at the same *point*; actually what is only guaranteed is that the lifts end in the same *fiber*

$\pi^{-1}(y)$ over the common endpoint y of the paths under consideration. This fact may be alternatively reformulated as that parallel transport around a *loop* may be non-trivial, i.e. the values of the geometrical quantity before and after the transport around the loop may differ. Whether or not the transport *indeed* depends on the path in any particular case depends (locally) eventually on *integrability* of the horizontal distribution. Indeed, if the horizontal distribution happens to be integrable, then the lift of a small enough loop lies entirely within the integral submanifold passing through the point x (being the start as well as the endpoint of the loop). Then, however, also the start and endpoint *of the lift* necessarily coincide, since there is *exactly one* integral submanifold of the horizontal distribution passing through any point of the fiber over x.

On the other hand, if the distribution is non-integrable over some neighborhood of x, we can construct a small loop based at x such that the lift of the loop already fails to be a loop (the start point differs from the endpoint; we encountered a similar situation in problem (19.3.5)), i.e. the parallel transport turns out to be non-trivial. This shows that it is just the examination of integrability of the horizontal distribution which reveals whether or not a parallel transport is (locally) path-dependent.[419]

In this way we naturally come to the task of finding an object which reflects a measure of non-integrability of the horizontal distribution (if the latter happens to be integrable the object should vanish, if it is not integrable it should be non-zero and "the more" non-integrable the distribution is, "the larger value" the object should have). It turns out that this piece of information may be conveniently carried by an appropriate 2-form on P.

Recall that one of the formulations of Frobenius' integrability condition of a distribution (19.3.8) states that the distribution is integrable if and only if the restriction of the exterior derivative of all constraint 1-forms to $\mathcal{D}$ vanishes ($d\theta^i = 0$ on $\mathcal{D}$). In our particular case this amounts to saying that

$$\mathcal{D}^h \text{is integrable} \quad \Leftrightarrow \quad \{U, V \in \mathcal{D}^h \Rightarrow d\omega(U, V) = 0\}$$

i.e. the restriction of the 2-form $d\omega$ to the horizontal subspace results in the zero 2-form at each point. This criterion is not very convenient so far since, in addition to a (simple) computation of the exterior derivative of the connection form ω, it requires the (more complicated) restriction of the result to the horizontal subspace. Before we progress to a truly convenient formulation, we will express it in terms of operations which we introduced in (20.3.5).

20.4.1 Check that the horizontal distribution $\mathcal{D}^h$ is integrable if and only if the 2-form $\Omega \equiv \text{hor}\, d\omega$ vanishes

$$\mathcal{D}^h \text{ is integrable} \quad \Leftrightarrow \quad \Omega := D\omega = 0$$

This (immensely important) 2-form Ω (with values in the Lie algebra $\mathcal{G}$) is called the *curvature form*.

[419] This issue gets more involved for "large" loops (or for "remote" points x and y); we will return to this after problem (20.4.8).

Hint: this is nothing but a rewriting of the criterion mentioned before (20.4.1) in terms of the operation hor; for general vector fields U, V we have $\Omega(U, V) = d\omega(\text{hor } U, \text{hor } V)$. □

Although this criterion of integrability seems to be simple and elegant, it still lacks perfection since so far a *simple* algorithm for the computation of the curvature form is not available (a simple realization of the "hor" procedure). This last defect is eventually removed by a surprisingly simple formula presented in problem (20.4.4).

However, in order to understand the formula it is necessary first to introduce a specific exterior product of Lie algebra valued forms, which will be denoted by $[\alpha \wedge \beta]$ (we mentioned it briefly in Section 6.4 in the context of forms with values in a general *algebra*).

20.4.2 Let α and β be differential forms with values in a Lie algebra $\mathcal{G}$. Check that

(i) if $\alpha = \alpha^i E_i$, $\beta = \beta^i E_i$, then the prescription

$$[\alpha \wedge \beta] := \alpha^i \wedge \beta^j [E_i, E_j] \equiv \left(c^k_{ij} \alpha^i \wedge \beta^j\right) E_k$$

provides a well-defined product (being the Lie algebra valued $(p+q)$-form, if α and β are p-form and q-form respectively)

(ii) the product behaves under the exchange of the factors as follows:

$$[\alpha \wedge \beta] = -(-1)^{pq} [\beta \wedge \alpha]$$

so that the rule contains an *extra minus sign* compared to ordinary forms (5.2.4)

(iii) for two 1-*forms* this may also be written as[420]

$$[\alpha \wedge \beta](U, V) := [\alpha(U), \beta(V)] + [\beta(U), \alpha(V)]$$

and, in particular, for two *equal* 1-forms we have

$$[\alpha \wedge \alpha](U, V) := 2[\alpha(U), \alpha(V)]$$

(iv) the exterior derivative of such a product (in the sense of Section 6.4) turns out to be in accordance with expectations

$$d[\alpha \wedge \beta] = [d\alpha \wedge \beta] + [\hat{\eta}\alpha \wedge d\beta]$$

(v) also the action of the operator hor does not bring any annoying surprises

$$\text{hor } [\alpha \wedge \beta] = [\text{hor } \alpha \wedge \text{hor } \beta]$$

Hint: (i) check that it does not depend on the choice of basis in $\mathcal{G}$; (ii) the extra minus sign is due to the commutator in the Lie algebra; (iii)

$$\begin{aligned}
[\alpha \wedge \beta](U, V) &= (\alpha^i \wedge \beta^j)(U, V)[E_i, E_j] \\
&= (\alpha^i(U)\beta^j(V) - \beta^j(U)\alpha^i(V))[E_i, E_j] \\
&= [\alpha(U), \beta(V)] + [\beta(U), \alpha(V)]
\end{aligned}$$

(iv) see (20.3.4). □

420 The product may be written in the same spirit also for higher degree forms, but we will not need it.

• Now we are already familiar with all the operations needed for a simple and easily applicable expression of the curvature form Ω in terms of a given connection form ω.

20.4.3 Check that the curvature form $\Omega \equiv D\omega$ may be expressed in the form of the

Cartan structure equation $$\Omega = d\omega + \frac{1}{2}[\omega \wedge \omega] \qquad \Omega := D\omega \equiv \text{hor}\, d\omega$$

so that for the component forms Ω^i we have[421]

$$\Omega^i = d\omega^i + \frac{1}{2}c^i_{jk}\omega^j \wedge \omega^k \qquad \Omega = \Omega^i E_i, \quad \omega = \omega^i E_i$$

Solution: according to (20.4.2) and (20.3.4) we are to prove that for arbitrary vector fields U, V there holds

$$d\omega(\text{hor}\, U, \text{hor}\, V) = d\omega(U, V) + [\omega(U), \omega(V)]$$

This relation is $\mathcal{F}(P)$-linear (equality of two tensors), so that it is enough to prove it on an appropriate "basis." According to (20.2.2), at an arbitrary point p a general vector U (as well as V) may be written as $U_p = \hat{U}_p + \xi_X(p)$, where $\hat{U}_p$ is already horizontal and $\xi_X(p)$ is the fundamental field (with appropriate X). This means that it is enough to consider three particular cases, namely $U, V = \hat{U}, \hat{V}$ or $\hat{U}, \xi_X$ or ξ_X, ξ_Y. The first one is trivial, the second and the third one are

$$d\omega(\xi_X, \hat{U}) \equiv i_{\hat{U}} i_{\xi_X} d\omega = 0$$
$$\text{or} \quad d\omega(\xi_X, \xi_Y) \equiv i_{\xi_Y} i_{\xi_X} d\omega = -[X, Y]$$

Both equalities are evident from the result $i_{\xi_X} d\omega = -\text{ad}_X \omega \equiv -[X, \omega] \equiv -c^k_{ij} X^i \omega^j E_k$, proved in (20.2.7). □

• This expression of the curvature form Ω in terms of a given connection form ω is indeed "user friendly": the operation hor is realized as a simple addition of the term $\frac{1}{2}[\omega \wedge \omega]$, so that we do not need to bother at all with any horizontal directions, instead we make do perfectly with an elementary manipulation with the form ω itself. Thus, let us look at the basic properties of the curvature form Ω, which may be easily found either right from the definition or from the derived explicit expression.

20.4.4 Check that the curvature form Ω always enjoys the following properties:

(i) it is a horizontal 2-form of type Ad, i.e.

$$\text{hor}\, \Omega = \Omega \qquad R^*_g \Omega = \text{Ad}_{g^{-1}} \Omega$$

(ii) it satisfies the *Bianchi identity* (see also (20.4.7))

$$D\Omega \equiv DD\omega = 0$$

[421] The relation of the Cartan structure equation to the equations of the same name introduced in (15.6.7) is discussed in (21.2.7) and (21.7.4).

Hint: (i) each form which results from applying the hor operator to another form is horizontal; Ad-behavior due to the commutation of R_g^* with D (20.3.5) and Ad-behavior of ω itself; (ii)

$$DD\omega = \text{hor}\, d\left(d\omega + \frac{1}{2}[\omega \wedge \omega]\right) = \text{hor}\,[d\omega \wedge \omega] = [D\omega \wedge \text{hor}\,\omega] = [\Omega \wedge 0] = 0$$

where the results of (20.3.4) and (20.4.2) were used. □

• As we have already mentioned, the operation D may be simply expressed in two other relevant cases, namely for functions of type ρ and for horizontal forms of type ρ. In order to understand the formulas we need to manage yet another type of product of forms.

20.4.5 On the total space P, consider a differential p-form α with values in the Lie algebra $\mathcal{G}$ and a q-form β with values in a *representation* space (W, ρ') of the Lie algebra $\mathcal{G}$. Check that

(i) if $\alpha = \alpha^i E_i$, $\beta = \beta^a E_a$ (E_a being a basis in W), then by the prescription

$$\rho'(\alpha) \dot{\wedge} \beta := \alpha^i \wedge \beta^a \rho'(E_i) E_a \equiv \left(\rho^a_{bi}\alpha^i \wedge \beta^b\right) E_a \equiv \left(\alpha^a_b \wedge \beta^b\right) E_a$$

a well-defined (exterior) product is defined, resulting in a $(p+q)$-form with values in W; in particular, if β is a *function* ($\beta \equiv \Phi$, $q = 0$), the (trivial) symbol of the exterior product may be *omitted* and we simply write

$$\rho'(\alpha) \dot{\wedge} \Phi = \rho'(\alpha)\Phi = \alpha^i \Phi^a \rho'(E_i) E_a = \left(\rho^a_{bi}\alpha^i \Phi^b\right) E_a \equiv \left(\alpha^a_b \Phi^b\right) E_a$$

(ii) the product of two $\mathcal{G}$-valued forms from (20.4.2) is just a special case for $\rho' = \mathrm{ad}$

$$\rho'(\alpha) \dot{\wedge} \beta \mapsto \mathrm{ad}\,(\alpha) \dot{\wedge} \beta = [\alpha \wedge \beta]$$

(iii) the exterior derivative of this product (in the sense of Section 6.4) gives the standard result

$$d\{\rho'(\alpha) \dot{\wedge} \beta\} = \rho'(d\alpha) \dot{\wedge} \beta + \rho'(\hat{\eta}\alpha) \dot{\wedge} d\beta$$

(iv) the operator hor also acts according to our expectations

$$\text{hor}\,\{\rho'(\alpha) \dot{\wedge} \beta\} = \rho'(\text{hor}\,\alpha)\ \dot{\wedge}\ \text{hor}\,\beta$$

(v) if α is of type Ad and β is of type ρ, then $\rho'(\alpha) \dot{\wedge} \beta$ is of type ρ.

Hint: (i) check that it is independent of the choice of bases in $\mathcal{G}$ and W; (iv) see (20.3.4); (v)

$$\begin{aligned} R_g^*\{\rho'(\alpha) \dot{\wedge} \beta\} &= \rho'(R_g^*\alpha) \dot{\wedge} R_g^*\beta = \rho'(\mathrm{Ad}_{g^{-1}}\alpha) \dot{\wedge} \rho(g^{-1})\beta \\ &= \text{according to (12.3.7)} \\ &= \rho(g^{-1})\rho'(\alpha)\rho(g) \dot{\wedge} \rho(g^{-1})\beta = \rho(g^{-1})\{\rho'(\alpha) \dot{\wedge} \beta\} \end{aligned}$$

□

20.4.6 Let $\pi : P \to M$ be a principal G-bundle, $\omega \equiv \omega^i E_i$ a connection form, $\alpha \equiv \alpha^a E_a$ a horizontal p-form of type ρ, $\Phi \equiv \Phi^a E_a$ a function of type ρ and denote $\omega^a_b \equiv \rho^a_{bi}\omega^i$.

Check that for exterior covariant derivatives $D\alpha$ and $D\Phi$ the following explicit expressions hold:

$$D\alpha = d\alpha + \rho'(\omega) \dot{\wedge} \alpha \qquad \text{i.e.} \qquad D\alpha^a = d\alpha^a + \omega^a_b \wedge \alpha^b$$
$$D\Phi = d\Phi + \rho'(\omega)\Phi \qquad \text{i.e.} \qquad D\Phi^a = d\Phi^a + \omega^a_b \Phi^b$$

(recall that according to Section 6.4 we have $D\alpha^a \equiv (D\alpha)^a$ and $D\Phi^a \equiv (D\Phi)^a$).

Hint: proceed similarly as in (20.4.3): we are to check

$$d\alpha(\text{hor } U, \text{hor } V, \dots) = d\alpha(U, V, \dots) + \rho'(\omega) \dot{\wedge} \alpha(U, V, \dots)$$

A decomposition $U_p = \hat{U}_p + \xi_X(p)$, $V_p = \hat{V}_p + \xi_Y(p), \dots$ reduces the proof to particular cases, namely to $U, V, \dots = \hat{U}, \hat{V}, \dots,\ \xi_X, \hat{U}, \dots,\ \xi_X, \xi_Y, \dots$. For all arguments horizontal it is trivial, for more than one fundamental field we obtain the identity $0 = 0$ (the term $d\alpha(\xi_X, \xi_Y, \dots)$ due to Cartan formulas (6.2.13) and horizontality of α, the term $\rho'(\omega) \dot{\wedge} \alpha(\xi_X, \xi_Y, \dots)$ due to horizontality of α, since at least one fundamental field necessarily ends in α) and for a combination $\xi_X, \hat{U}, \dots$ we are to check the validity of

$$d\alpha(\xi_X, \hat{U}, \dots) + \rho'(\omega) \dot{\wedge} \alpha(\xi_X, \hat{U}, \dots) \equiv (i_{\xi_X} d\alpha)(\hat{U}, \dots) + \rho'(X)\alpha(\hat{U}, \dots) = 0$$

This may be seen from the result $i_{\xi_X} d\alpha = -\rho'(X)\alpha$ (being an infinitesimal version of $R^*_g \alpha = \rho(g^{-1})\alpha$). For the function Φ we are to verify

$$d\Phi(\text{hor } U) = d\Phi(U) + \rho'(\omega(U))\Phi)$$

For $U = \hat{U}$ it is trivial, for $U = \xi_X$ the question is whether $\xi_X \Phi = -\rho'(X)\Phi$, which once more results from the fact that Φ is of type ρ. □

• Having gone through this computation let us reward ourselves by deriving further useful results. First, we will write down a more detailed version of the Bianchi identity and then we will investigate whether the operator D is also nilpotent (as was d), or whether this property is lost.

20.4.7 Check that the *Bianchi identity* may also be written in the form

$$d\Omega + [\omega \wedge \Omega] = 0$$
$$\text{or equivalently} \qquad d\Omega^i + c^i_{jk} \omega^j \wedge \Omega^k = 0$$

Hint: since according to (20.4.4) Ω is a horizontal form type Ad, (20.4.6) gives

$$D\Omega = d\Omega + [\omega \wedge \Omega]$$

□

20.4.8 Check that for the *square* of the exterior covariant derivative we get on a horizontal form α of type ρ and on a function Φ of type ρ the results (also known as the *Ricci identity*)

$$DD\alpha = \rho'(\Omega) \dot{\wedge} \alpha \qquad \text{i.e.} \qquad DD\alpha^a = \Omega^a_b \wedge \alpha^b$$
$$DD\Phi = \rho'(\Omega)\Phi \qquad \text{i.e.} \qquad DD\Phi^a = \Omega^a_b \Phi^b$$

We see that the square DD in general *does not* vanish, but rather it is proportional to the *curvature* form. Due to the derived formula, the square nevertheless *does* vanish in the case of zero curvature as well as for $\rho' = 0$ (in the case of the trivial representation).

Hint: according to (20.4.6)

$$DD\alpha = \text{hor}\, d\{d\alpha + \rho'(\omega) \stackrel{\wedge}{} \alpha\} = \rho'(\text{hor}\, d\omega) \stackrel{\wedge}{} \alpha - \rho'(\text{hor}\, \omega) \stackrel{\wedge}{} \text{hor}\, d\alpha = \rho'(\Omega) \stackrel{\wedge}{} \alpha$$

□

• Now, let us look in some detail at the path (in)dependence of parallel transport for the case where the *curvature vanishes* (so that the horizontal distribution is *integrable*), but the beginning and the endpoint of a path happen to be "remote" from one another (or, equivalently, the loop is "large"). The argument about the path independence (mentioned at the beginning of Section 20.4) is based on the fact that the *lifts* of both paths lie in *a single common* integral submanifold (then the endpoints of both lifts necessarily coincide) and this is only guaranteed locally by the integrability of the distribution for *some neighborhood* of a given point. The question is then: consider two paths c_0 and c_1 such that they both begin in x and terminate in y ($c_0(0) = c_1(0) = x$, $c_0(1) = c_1(1) = y$); do the lifts lie in a single common integral submanifold even if the points x and y fail to be close to each other (so that *a priori* it is not clear whether the integral submanifold passing through x coincides with that passing through y)? We can see intuitively that *if* c_0 and c_1 happen to be *homotopic*,[422] then their lifts actually *do lie* in a common integral submanifold. Namely, divide the two-dimensional surface given by the homotopy (we may visualize it as a *soap bubble*, stretched at a contour given by the paths c_0, c_1) into *small enough* "rectangles" (as is shown schematically in the figure), with small enough meaning that they already lie entirely in *a single* integral submanifold (given, say, by any vertex of the rectangle). Then we can see easily that the *lift* of this "spider web" turns out to be again a spider web which necessarily lies entirely in *a single common large* integral submanifold S, so that the lifts of the *endpoints* of the paths c_0 and c_1 necessarily *coincide*.

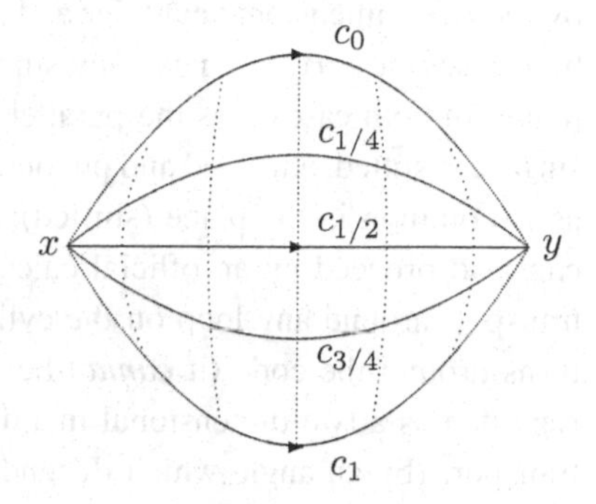

20.4.9 Contemplate this argument until it is clear, since it is a fairly useful result.

Hint: let an integral submanifold be given by the left bottom corner of the rectangle; then also the right bottom corner lies in the same submanifold (since the rectangle is small), the former coinciding, however, with the *left* bottom corner of the *neighboring* rectangle (at the right),

[422] That is, such that one can be continuously deformed to the other. Thus for each τ from 0 to 1 there is a path c_τ (a one-parameter class of paths altogether; it is called a *homotopy*) such that for $\tau = 0$ it is just (our original) c_0 and for $\tau = 1$ we get (our original) c_1. More formally, let T_1, T_2 be topological spaces. Two continuous maps $f, g : T_1 \to T_2$ are called homotopic, if there exists a continuous map $F : [0, 1] \times T_1 \to T_2$ (homotopy) such that $F(0, x) = f(x)$ and $F(1, x) = g(x)$ for all $x \in T_1$.

so that it also defines an integral submanifold, in which the entire neighboring rectangle lies, but, taking into account *uniqueness* of an integral submanifold passing through a given point, this necessarily means that *both* rectangles actually lie in a *single common* larger integral submanifold and eventually (by repeating this procedure) also that *all* the rectangles lie in a *single common* large integral submanifold. □

• If the paths c_0 and c_1 *fail to be* homotopic to each other (or, speaking in terms of loops, if the "large" loop fails to be *homotopic to zero*, i.e. it cannot be *continuously shrunk* into its beginning = endpoint), the argument mentioned above cannot be used and it turns out that in this case parallel transport *may* (but does not need to) depend on the path (it may be non-trivial around a non-contractible loop). Simple and instructive examples are provided by the two-dimensional cylinder and cone endowed with the standard RLC connection. In both cases the *curvature vanishes*, since they happen to be *locally isometric* to the Euclidean plane. In both cases thus the parallel transport of vectors may be realized as follows: the surface is slitted, unfolded and put onto the plane and a vector is then translated "ordinarily" as is common in the plane (shifted); eventually the surface is glued back (of course, we can also proceed by an official calculation). In this way one can see that while parallel transport around any loop on the cylinder is trivial, the transport around the loop, which turns around the cone (it *cannot* be shrunk, since the cusp *does not* belong to the cone regarded as a two-dimensional manifold), is *non-trivial* – the vector is *turned* due to the transport (by an angle which depends on the "steepness" of the cone or, put differently, on its "angular deficit" (the angle to be cut off from the plane in order to produce the cone).

20.4.10 If we "repaired" the cone so that we smoothed its cusp replacing it by a nice two-dimensional small cap, we already *would* be able to shrink the loop under consideration (through the cap). In doing so, the region outside the cap did not change at all. Consequently, the parallel transport around the loop encircling the cap should remain *non-trivial*. Is it still possible to use the argument with the lift of the "spider net" (and we have a contradiction), or it is not (and the matter is all right)?

Hint: the cap necessarily has non-zero curvature (it should be rounded off in order to close the cone) and so at least a single rectangle lies in a region with *non-integrable* distribution (where the gluing of the rectangles into a larger whole collapses). □

• Toward the end of Section 15.8, in the context of linear connections, we discussed (and did not settle) the question of whether vanishing curvature also represents a sufficient condition of complete parallelism. Now the (positive) answer may be easily seen. Recall that by *complete parallelism* we mean a linear connection for which there exists a covariantly constant frame field e_a, i.e. a field such that $\nabla_V e_a = 0$ for each V. In terms of the connection form this gives $\omega^a_b = 0$ and, consequently, also $\Omega^a_b \equiv d\omega^a_b + \omega^a_c \wedge \omega^c_d = 0$. The opposite implication is, however, still unclear: does $\Omega^a_b = 0$ result in existence of a frame field e_a, in which $\omega^a_b = 0$? Well, we will be wiser after the next problem.

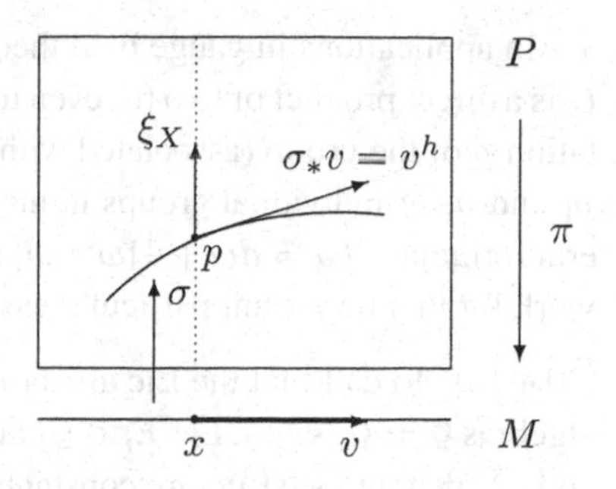

20.4.11 Check that if the curvature form of a linear connection vanishes, there exists a covariantly constant frame field, i.e. we have the complete parallelism

$$\Omega^a_b = 0 \quad \Rightarrow \quad \exists e_a \; : \; \hat{\omega}^a_b = 0$$

For a *general* connection this may be rephrased so that the vanishing curvature form guarantees the existence of a (local) section such that the pull-back of the connection form by means of the section vanishes

$$\Omega = 0 \quad \Rightarrow \quad \exists \sigma \; : \; \sigma^* \omega = 0$$

Hint: $\Omega = 0$ says that the horizontal distribution is integrable; consider an integral submanifold $\mathcal{S}$, which passes through a point $p \in P$ and introduce a section σ such that $\mathcal{S}$ is just the image of a domain on M with respect to the section. In this way we obtained a *horizontal section*, i.e. a section such that all vectors tangent to its image turn out to be horizontal. Now, if we pull back the connection form ω to the base manifold by *this* section, we get $\hat{\omega} \equiv \sigma^* \omega$, which *vanishes* ($\langle \sigma^* \omega, v \rangle = \langle \omega, \sigma_* v \rangle = 0$, since $\sigma_* v$ is horizontal). □

• Now, let us look at a very simple connection which is available free of charge in each *product* principal bundle; it is called *flat*, since its curvature vanishes.

20.4.12 Consider a *product* principal G-bundle $\pi : M \times G \to M$

$$\pi : M \times G \to M \qquad \pi : (m, g) \mapsto m \qquad R_{\tilde{g}} : (m, g) \mapsto (m, g\tilde{g})$$

and let $\pi_G : M \times G \to G$, $(m, g) \mapsto g$, be a natural projection onto the factor G. Check that

(i) if we declare vectors on $P \equiv M \times G$ tangent to M (tangent to curves $t \mapsto (m(t), g)$) to be horizontal, we get a connection (vertical vectors are tangent to G so they correspond to curves $t \mapsto (m, g(t))$)
(ii) even without doing any calculations we can see that the connection has *vanishing curvature*, $\Omega = 0$; it is called the *canonical flat connection*
(iii) its connection form is $\omega = \pi^*_G \theta$, where θ denotes the canonical 1-form on the group G; this form singles out the correct horizontal subspace and it also fulfills all the formal requirements which any connection form is to satisfy
(iv) also a direct computation of Ω from this connection form gives $\Omega \equiv D\omega = 0$.

Hint: (i) G-invariance evident from the action $(m(t), g) \mapsto (m(t), g\tilde{g})$; (ii) the horizontal distribution is integrable (the integral submanifold passing through (m, g) is (M, g)); (iii) the form $\pi^*_G \theta \equiv \theta \circ \pi_{G*}$ annihilates part of a vector, which is directed along M; one should check (20.2.5); if R_g is the action on $M \times G$ and $\hat{R}_g$ is the right translation on G, then $\pi_G \circ R_g = \hat{R}_g \circ \pi_G$; if ξ_X is a generator of R_g on P, then $\pi_{G*} \xi_X = L_X$; see (11.2.6), (12.3.4) and (20.2.4); (iv) apply π^* to the Maurer–Cartan relations (11.2.6). □

• In applications in gauge field theories one often encounters a situation when the group G is a direct product of two (or even more) smaller groups, $G = G_1 \times G_2$ and the representation ρ of the group (associated with a function of type ρ) is composed of representations ρ_1 and ρ_2 of individual groups in the way described in (12.4.16). Let us see how the general formulas $D\omega = d\omega + \frac{1}{2}[\omega \wedge \omega]$ from (20.4.3) and $D\Phi = d\Phi + \rho'(\omega)\Phi$ from (20.4.6) work for this important particular case.

20.4.13* Recall that the Lie algebra of a direct product $G_1 \times G_2$ is the direct sum of Lie algebras $\mathcal{G} = \mathcal{G}_1 + \mathcal{G}_2$. Let $E_i \in \mathcal{G}_1$ and $E_I \in \mathcal{G}_2$ denote bases in the initial algebras and c^k_{ij} and c^K_{IJ} denote the structure constants with respect to these bases. Check that

(i) the connection form has a unique decomposition $\omega = \omega_1 + \omega_2$ to a $\mathcal{G}_1$-valued part and a $\mathcal{G}_2$-valued part respectively

(ii) the curvature form has a unique decomposition $\Omega = \Omega_1 + \Omega_2$ to a $\mathcal{G}_1$-valued part and a $\mathcal{G}_2$-valued part, where

$$\Omega_1 = d\omega_1 + \frac{1}{2}[\omega_1 \wedge \omega_1] \quad \text{i.e.} \quad \Omega^i_1 = d\omega^i_1 + \frac{1}{2}c^i_{jk}\omega^j_1 \wedge \omega^k_1$$

$$\Omega_2 = d\omega_2 + \frac{1}{2}[\omega_2 \wedge \omega_2] \quad \text{i.e.} \quad \Omega^I_2 = d\omega^I_2 + \frac{1}{2}c^I_{JK}\omega^J_2 \wedge \omega^K_2$$

so that the first (second) part of the curvature form is to be computed by a standard formula applied to the first (second) part of the connection form *alone*, as if the second (first) part did not exist at all.

Hint: (i) $\omega = \omega^i E_i + \omega^I E_I \equiv \omega_1 + \omega_2$ and the decomposition does not depend on the choice of bases in $\mathcal{G}_1$ and $\mathcal{G}_2$; (ii) $\Omega = d(\omega_1 + \omega_2) + \frac{1}{2}[(\omega_1 + \omega_2) \wedge (\omega_1 + \omega_2)] = \cdots$, $[\mathcal{G}_1, \mathcal{G}_2] = 0$. □

20.4.14* Consider still $G = G_1 \times G_2$ and let Φ be a function of type ρ for the representation $\rho(g_1, g_2) = \rho(g_1) \otimes \rho(g_2)$ in $V = V_1 \otimes V_2$ from (12.4.16). Denote bases in the initial spaces by $E_a \in V_1$ and $E_\alpha \in V_2$ respectively so that $E_a \otimes E_\alpha$ is a basis in $V_1 \otimes V_2$ and $\Phi = \Phi^{a\alpha} E_a \otimes E_\alpha$. Check that the formula for computing the exterior covariant derivative of Φ may be written in the form of

$$D\Phi^{a\alpha} = d\Phi^{a\alpha} + (\omega_1)^a_b \Phi^{b\alpha} + (\omega_2)^\alpha_\beta \Phi^{a\beta} \qquad (\omega_1)^a_b = \rho^a_{bi}\omega^i_1$$

$$(\omega_2)^\alpha_\beta = \rho^\alpha_{\beta I}\omega^I_2$$

so that both types of indices are separately "processed" by "their own" connection form.

Hint: according to (12.4.16) we have $\rho'(X_1 + X_2) = \rho'_1(X_1) \otimes \hat{1} + \hat{1} \otimes \rho'_2(X_2)$ so that $\rho'(\omega_1 + \omega_2)\Phi = \cdots$. □

20.5* Restriction of the structure group and connection

• Consider a principal G-bundle $\pi : P \to M$ and a function $\Phi : P \to V$ of type ρ. We will show that, provided that some requirements are fulfilled, we automatically get a submanifold $Q \subset P$ "inside" P, which moreover turns out to be the total space of a "smaller" principal H-bundle $\hat{\pi} : Q \to M$ (where H is a subgroup of G) and if there was a connection in P, it is also available in Q.

[20.5.1] Let $\pi : P \to M$ be a principal G-bundle and consider a function $\Phi : P \to V$ of type ρ. Be sure to understand that the whole fiber $\pi^{-1}(x)$ is mapped by Φ into a *single* orbit of the action (representation) ρ in V.

Hint: Φ is equivariant, $\Phi(pg) = \rho(g^{-1})\Phi(p)$, so that it preserves orbits (any fiber is just a whole orbit). □

• In general, different fibers may be mapped into different orbits (i.e. orbits $\mathcal{O}_{\Phi(p)}$ may depend on $x \in M$). In what follows we shall restrict ourselves to the case where the Φ-image of the whole P lies *in a single orbit* W of the representation ρ, so that actually $\Phi : P \to W \subset V$. Let us see how this works for some simple examples (what the requirement may mean in concrete situations).

[20.5.2] Consider the frame bundle $\pi : LM \to M$ and take as Φ a function of type ρ_0^1 with values in $\mathbb{R}^n$, corresponding to a *vector field* on M (see the text before (19.6.4)). Be sure you understand that if Φ is to map the whole LM into a single orbit of the representation ρ_0^1, then the corresponding vector field on M either vanishes everywhere or it is, on the contrary, everywhere non-zero.

Hint: the representation ρ_0^1 of the group $GL(n, \mathbb{R})$ in R^n ($x \mapsto Ax$) possesses just two orbits: 0 and everything else. □

[20.5.3] Consider again the frame bundle $\pi : LM \to M$ and now take Φ to be a function of type ρ_2^0 with values in $\mathbb{R}^n \otimes \mathbb{R}^n$, which corresponds to a field of *bilinear forms* on M (19.6.3). Check that if Φ is to map the whole LM into the (single) orbit of the representation ρ_2^0 generated by the *identity* matrix, then the corresponding object is a *symmetric positive definite* bilinear form, i.e. the field of a ("true") *metric tensor* on M. A field of the more general metric tensor with signature (r, s) on M arises in like manner from the orbit of the matrix η with signature (r, s).

Hint: the representation ρ_2^0 of the group $GL(n, \mathbb{R})$ is given by $\rho(A)g := (A^{\mathrm{T}})^{-1} g A^{-1}$; then $g_{ab}(eA) = (\rho(A^{-1})g)_{ab}(e)$ gives $g_{ab}(eA) = A^c_a g_{cd}(e) A^d_b$. The identity matrix happens to be the canonical form of any symmetric and positive definite bilinear form.[423] □

• Now, choose a point w_0 on the orbit W and define a submanifold $Q \subset P$ as the preimage of w_0 with respect to Φ (i.e. $Q := \Phi^{-1}(w_0)$) and the subgroup $H \subset G$ as the stabilizer of the point w_0 (i.e. $Hw_0 = w_0$):

$$Q := \{p \in P \mid \Phi(p) = w_0\}$$
$$H := \{h \in G \mid \rho(h)w_0 = w_0\}$$

[423] The procedure $\hat{B} \mapsto (A^{\mathrm{T}})^{-1}\hat{B}A^{-1} \equiv \rho(A)\hat{B}$ may be regarded as the change of the matrix of a bilinear form due to a change of basis. Here it is, however, regarded as a transition to another matrix on the *same orbit* and, consequently, orbits themselves may be labeled by *canonical forms* of symmetric matrices and the orbit under consideration which is given by the canonical form $= \delta_{ab}$ contains just all *positive definite* symmetric matrices.

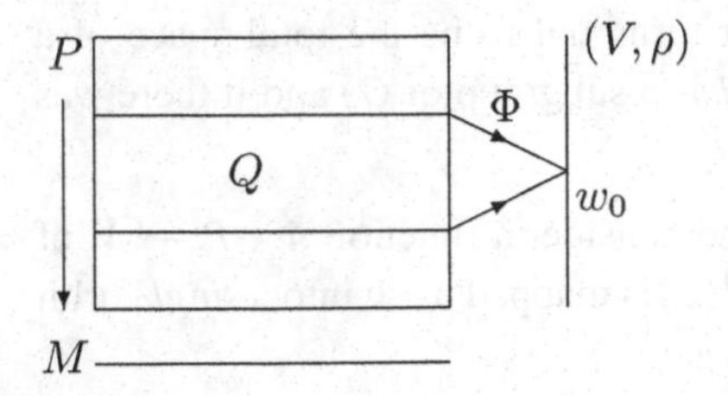

It turns out that if we restrict the projection from P to the submanifold Q and the action from G to the subgroup H, we get a new principal bundle (a *subbundle* of the bundle $\pi : P \to M$), namely the principal H-bundle $\pi : Q \to M$.

20.5.4 Check that $\pi : Q \to M$ is indeed a principal H-bundle over M; it is called the *restriction of the bundle* $\pi : P \to M$. This procedure is also known as a *restriction of the structure group* (from G to a subgroup H).

Hint: in each fiber of the manifold P there exists at least one point from Q: if some $p' \mapsto w \equiv \Phi(p')$, then $w = \rho(g)w_0$ for some $g \in G$ and consequently $\Phi(p'g) = \rho(g^{-1})\Phi(p') = \rho(g^{-1})\rho(g)w_0 = w_0$, so that $p \equiv p'g \mapsto w_0$. In addition to this particular point there are also exactly all points ph for $h \in H$ – these points constitute the fiber $\pi^{-1}(x)$ for $\pi : Q \to M$. The action of H on Q is just the restriction of the action of G on P, so that it satisfies all the needed properties. □

20.5.5 Consider the frame bundle $\pi : LM \to M$ and the function Φ discussed in problem (20.5.3). Take as the point w_0 the matrix η with signature (r, s). Check that if we apply the procedure described above, we get as $\pi : Q \to M$ the *bundle of orthonormal frames* $\pi : OM \to M$ (points of the total space are all *orthonormal* frames with respect to a metric tensor g on M), which is a principal $O(r, s)$-bundle, canonically associated to any *Riemannian* manifold (M, g).

Hint: $\Phi(e)$ is regarded as the matrix of the metric tensor g with respect to the basis e, i.e. $\Phi_{ab}(e) := g(e_a, e_b) \equiv g_{ab}$; then the condition $\Phi_{ab}(e) \overset{!}{=} \eta_{ab}$ picks up just orthonormal frames. This works in both directions: if we have (M, g), we define a function Φ of type ρ_2^0 on LM by the formula $\Phi_{ab}(e) := g(e_a, e_b) \equiv g_{ab}$; if, on the other hand, we have such a function Φ on LM, we can define *orthonormal* frames by $\Phi_{ab}(e) \overset{!}{=} \eta_{ab}$; this is equivalent to introducing g on M (if we know that a basis e_a is orthonormal, we put $g = \eta_{ab}e^a \otimes e^b$). □

• Although the H-subbundle $\pi : Q \to M$, strictly speaking, depends on the choice of a point w_0 on the W, this dependence is (as all of us feel in consideration of the equivalence of points on orbits) irrelevant: another choice just leads to an equivalent principal H-bundle.

20.5.6 Let w_1 be another point on the orbit W. Check that the principal H-bundle which arises from this point is isomorphic to the initial H-bundle, which is associated with w_0.

Hint: if $w_1 = \rho(\hat{g})w_0$, then we get the stabilizer H_1 of the point w_1 by conjugation of H_0 by the element $\hat{g}$ (being an isomorphism, see (13.1.10) and (13.2.7)) and the submanifold Q_1 is evidently the image of the submanifold Q_0 with respect to the right action by $\hat{g}^{-1}$ $(Q_1 = R_{\hat{g}^{-1}}Q_0,\ H_1 = \hat{g}H_0\hat{g}^{-1})$. □

• We introduced the restriction of a principal bundle in terms of an equivariant function Φ on P. The example of the bundle of orthonormal frames, however, showed that sometimes it is also possible to describe directly, with no reference to any Φ (to say, for example, which points of LM constitute OM). It turns out, however, that even if this is the case, the function Φ still may be always introduced (in a sense specified in more detail after the next problem), i.e. there actually exists a bijection between restrictions of a principal bundle and equivariant functions on the total space such that they map the whole P into a single orbit. How may such a function Φ be constructed for a given restriction?

We may take inspiration from the fact discussed in problem (13.2.7), that the orbit W is isomorphic (as a homogeneous space) to the canonical homogeneous space G/H and, in particular, that $w_0 \leftrightarrow [e] \leftrightarrow eH$.

20.5.7 Check that the field Φ is given by the prescription

$$\Phi(q) := [e] \leftrightarrow eH \equiv H \qquad q \in Q \subset P$$

Hint: if, by definition, the whole Q maps into the point $[e] \leftrightarrow H$, its value at other points of P is already given by equivariance, $\Phi(p) = \Phi(qg) = \rho(g^{-1})\Phi(q) = \rho(g^{-1})[e] = [g^{-1}] \in G/H$. □

• Note that the map Φ is not defined here as a field with values in a vector space V (as a usual *linear field*) with a given *representation* ρ, but rather as a map with values in some manifold (a *non-linear field*, to throw the physical jargon around a bit, see Section 16.6), namely in the homogeneous space G/H endowed with the canonical *action* $\rho(g)[g'] = [gg']$. Put another way, here the orbit is not realized as a subset of a vector space (as was the case in the approach discussed at the beginning of the section), but rather it is given as an abstract manifold "in its own right," not being embedded into anything.

Now we turn our attention to connections. If there was any on P, we also would like to inherit it on Q. How to ensure this? First we should realize what kind of problems may actually occur within a "probate process."

A connection on P may be realized in terms of connection 1-form ω (with values in $\mathcal{G}$). It seems then that it should be sufficient to restrict it to the submanifold Q and the problem is solved. However, if the restriction is to serve as a connection form on Q, where already only the *subgroup* $H \subset G$ acts, its values must only lie (by definition of a connection form) in the Lie *subalgebra* $\mathcal{H} \subset \mathcal{G}$. In addition to this there is a natural requirement with respect to a parallel transport: if we "parallel transport" a point q (i.e. we perform the horizontal lift of a curve which starts from the point $q \in Q$), we must also end in Q (for a general connection on P the endpoint of the lift may lie outside Q in spite of the fact that the beginning of the lift is in Q). We will see in what follows that these two requirements are actually only *a single* requirement with respect to the original connection in $\pi : P \to M$ (the requirement of compatibility of the original connection with the procedure of the restriction of the principal bundle).

20.5.8 Consider a principal G-bundle $\pi : P \to M$ with a connection form ω and let $\pi : Q \to M$ be a restricted principal H-bundle, which arose with the help of a function Φ of type ρ. Check that if the connection is to be inherited by Q (in the sense that ω regarded as a form living only on Q is a connection form in $\pi : Q \to M$), then

(i) the exterior covariant derivative of the field Φ should vanish on (the whole) P

$$D\Phi \equiv d\Phi + \rho'(\omega)\Phi = 0$$

(ii) equivalently, *upon restriction to the manifold* Q, the connection form ω should have values only[424] in the Lie *subalgebra* $\mathcal{H} \subset \mathcal{G}$.

Hint: (i) *each lifted* curve should remain in Q (provided that it begins in Q); therefore $\Phi(\gamma^h(t)) = w_0$, so that $0 = \dot{\gamma}^h\Phi = \langle d\Phi, \dot{\gamma}^h\rangle = \langle \text{hor}\, d\Phi, \dot{\gamma}^h\rangle \equiv \langle D\Phi, \dot{\gamma}^h\rangle \;\Rightarrow\; D\Phi = 0$ ($\dot{\gamma}^h$ is actually an *arbitrary horizontal* vector, any vertical one annihilates $D\Phi$, so that $D\Phi$ vanishes on *arbitrary* vectors); this holds at all points from Q, but due to the equation $R_g^* D\Phi = \rho(g^{-1})D\Phi$ it even holds on the whole of P; (ii) on Q we have $\Phi(Q) = w_0$, so $d\Phi(Q) = 0$; then from $D\Phi(Q) = 0$ one can deduce $\rho'(\omega(Q))\Phi(Q) = 0 = \rho'(\omega(Q))w_0$, and consequently $\omega(Q)$ has values only in $\mathcal{H}$. □

20.5.9 Let us look at a statement concerning the relation between the connection on $\pi : P \to M$ and $\pi : Q \to M$ in a more general setting. Assume that our Lie algebra $\mathcal{G}$ admits a decomposition $\mathcal{G} = \mathcal{H} \dot{+} \mathcal{L}$ (by $\dot{+}$ we denote the direct sum of vector spaces, but not necessarily of Lie algebras), where $\mathcal{L}$ is so far an arbitrary complementary subspace to $\mathcal{H}$. Then also the connection form has the unique decomposition into two parts, the first one being $\mathcal{H}$-valued and the second one being $\mathcal{L}$-valued. If we choose in $\mathcal{G}$ a basis which is adapted to the two subspaces, $E_\alpha \in \mathcal{H}$, $E_i \in \mathcal{L}$, then

$$\begin{aligned} \omega &= \omega_{\mathcal{H}} + \omega_{\mathcal{L}} \\ &\equiv \omega^\alpha E_\alpha + \omega^i E_i \end{aligned}$$

Check that

(i) if (also) $\mathcal{L}$ happens to be Ad_H-invariant,[425] i.e. if

$$\text{Ad}_H \mathcal{L} \subset \mathcal{L}$$

then the first part $\omega_{\mathcal{H}}$ itself of the form ω represents a connection form on Q (i.e. we can then "project out" a connection on Q from a connection on P)

(ii) in the case discussed in (20.5.8) we have

$$\omega_{\mathcal{L}} = 0 \quad \text{on } Q$$

[424] Recall that the restriction to the (sub)manifold $Q \subset P$ concerns points in which we consider the form ω, but implicitly also vectors, which are from now allowed to be inserted into ω (only those vectors which are tangent to Q, see (7.6.8)). For example, if we insert to ω (after restriction to points of Q) the vector of the fundamental field generated by an element $Y \notin \mathcal{H}$ (it is *forbidden* now, since, moving in this direction, we leave Q), we get, according to the basic property (20.2.5) of the connection form, just Y; this, however, contradicts the statement that ω has values only in $\mathcal{H}$.

[425] Clearly $\text{Ad}_H \mathcal{H} \subset \mathcal{H}$, since $\mathcal{H}$ is a sub*algebra*. Thus *both* subspaces are supposed to be Ad_H-invariant.

i.e. if the restriction of a bundle with connection $\pi : P \to M$ to a subbundle $\pi : Q \to M$ is realized by means of a function Φ *such that* $D\Phi = 0$, nothing (no $\omega_{\mathcal{L}}$) is already to be thrown away on Q in the course of the projection (ω on Q is directly a connection form, nothing is to be projected out).[426]

Hint: (i) for a vector $v \in T_q Q$ and a group element $h \in H$ we may write

$$\langle \omega, R_{h*}v \rangle = \langle \omega_{\mathcal{H}}, R_{h*}v \rangle + \langle \omega_{\mathcal{L}}, R_{h*}v \rangle$$
$$\text{Ad}_{h^{-1}} \langle \omega, v \rangle = \text{Ad}_{h^{-1}} \langle \omega_{\mathcal{H}}, v \rangle + \text{Ad}_{h^{-1}} \langle \omega_{\mathcal{L}}, v \rangle$$

The left-hand sides are equal (due to $R_g^* \omega = \text{Ad}_{g^{-1}} \omega$) and both right-hand sides are of the form $X + Y, X \in \mathcal{H}, Y \in \mathcal{L}$ (in the term of type Y in the second equation the assumption of Ad_H-invariance of $\mathcal{L}$ is crucial). Equating the terms of type X gives the required property of a connection form $R_h^* \omega_{\mathcal{H}} = \text{Ad}_{h^{-1}} \omega_{\mathcal{H}}$ (20.2.5), the second needed property $\langle \omega_{\mathcal{H}}, \xi_X \rangle = X$ (for arbitrary $X \in \mathcal{H}$) is evident from $\omega_{\mathcal{H}} = \omega^\alpha E_\alpha$ and its validity for ω as a whole. □

20.5.10 Consider once again the function Φ of type ρ_2^0 on LM, which we contemplated in problems (20.5.3) and (20.5.5). Check that

(i) the condition $D\Phi = 0$ on $P \equiv LM$ gives here explicitly

$$D\Phi_{ab} = 0 \quad \Leftrightarrow \quad d\Phi_{ab} = \omega_a^c \Phi_{cb} + \omega_a^c \Phi_{cb}$$

and upon restriction to $Q \equiv OM$ it reduces to

$$\omega_{ab} + \omega_{ba} = 0 \qquad \omega_{ab} := \Phi_{ac} \omega_b^c$$

(ii) this says that the connection is *metric*
(iii) here, ω restricted to Q indeed has its values only in $\mathcal{H}$
(iv) here, the needed complement $\mathcal{L}$ *does exist*, so if only a "general" (non-metric) linear connection is available, we can project out its "metric part."

Hint: (i) according to (19.6.3), $(\rho(A)\hat{B})_{ab} = (A^{-1})_a^c (A^{-1})_b^d \hat{B}_{cd}$, from where $(\rho'(C)\hat{B})_{ab} = -C_a^c \hat{B}_{cb} - C_b^c \hat{B}_{ac}$; (ii) see (15.6.6); (iii) on OM we have $\omega_{ab} + \omega_{ba} = 0$, i.e. ω has values only in $\mathcal{H} \equiv so(r, s) \subset gl(r + s, \mathbb{R}) \equiv \mathcal{G}$; (iv) an appropriate complement to $o(n) \subset gl(n, \mathbb{R})$ (antisymmetric matrices) is the space of *symmetric* matrices; one easily checks its invariance (if $Y = Y^{\text{T}}$ is a symmetric matrix and $A \in O(n)$, then $(AYA^{-1})^{\text{T}} = \cdots = AYA^{-1}$); so in order to project out a metric connection one should write down ω_{ab} with respect to an orthonormal frame field and then perform $\omega_{ab} \mapsto \omega_{[ab]}$. □

• Let us look at the restriction of a principal bundle from still another point of view, namely that of *morphisms of principal bundles*. We have already introduced the concept of *bundle maps* in Section 17.2, playing the role of morphisms in the category of fibre bundles. Now we need to relate two *principal* bundles, so we have to add some requirement concerning the actions of the groups. We will discuss only the case when both the bundles share a *common base* $M_1 = M_2 \equiv M$.

[426] So if $D\Phi \neq 0$, there is still a chance (if an appropriate $\mathcal{L}$ exists) to produce a connection on Q from that on P, but we need to throw away a part of ω on Q.

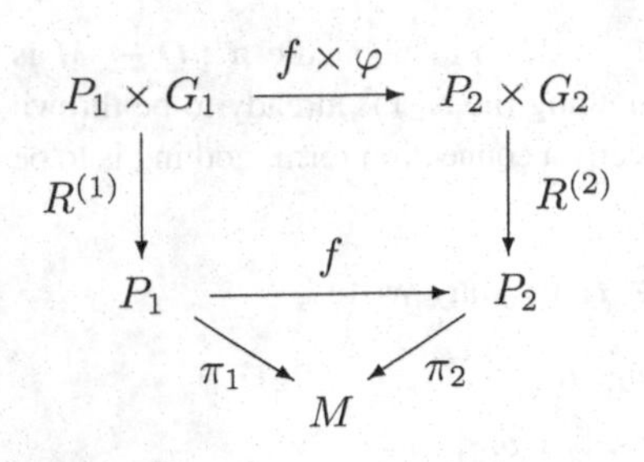

20.5.11 Consider then two *principal* bundles over the same base, $\pi_1 : P_1 \to M$ with group G_1 and $\pi_2 : P_2 \to M$ with group G_2. A morphism of the bundles will be the map f, discussed[427] after (17.2.1), but, except for preserving fibers, the requirement of *preserving actions* in the total spaces is added: a homomorphism of groups $\varphi : G_1 \to G_2$ has to exist such that the actions $R^{(1)}$ and $R^{(2)}$ on P_1 and P_2 "commute in the sense of φ," i.e. it does not matter whether we first move (on P_1) from $p_1 \in P_1$ by $g_1 \in G_1$ and then we map the result by means of f, or we first map p_1 by means of f and then we move (on P_2) from the resulting point by $\varphi(g_1) \in G_2$:

$$f \circ R^{(1)}_{g_1} = R^{(2)}_{\varphi(g_1)} \circ f$$

Check that all of these requirements turn out to be equivalent to the validity of the commutative diagram.

Hint: the action $R_g : P \to P$ may also be written as $R : P \times G \to P,\ (p, g) \mapsto R_g p$ (13.1.2); the commutativity of the rectangle in the diagram says that $(p_1, g_1) \mapsto R^{(1)}_{g_1} p_1 \mapsto f(R^{(1)}_{g_1} p_1)$ and at the same time $(p_1, g_1) \mapsto (f(p_1), \varphi(g_1)) \mapsto R^{(2)}_{\varphi(g_1)} f(p_1)$. □

• For the most interesting morphisms, the maps f and φ are injective[428] (then $\pi_1 : P_1 \to M$ is called the *restriction of the bundle* $\pi_2 : P_2 \to M$ and $\pi_2 : P_2 \to M$ is in turn an *extension of the bundle* $\pi_1 : P_1 \to M$, or surjective (then $\pi_1 : P_1 \to M$ is a *prolongation of the bundle* $\pi_2 : P_2 \to M$,[429] and $\pi_2 : P_2 \to M$ is in turn a *reduction of the bundle* $\pi_1 : P_1 \to M$).

20.5.12 Be sure you understand that the restriction of a bundle which we discussed in (20.5.4) is also a restriction in the sense we introduced here; in particular, that the bundle of orthonormal frames $\pi : OM \to M$ is a restriction of the frame bundle $\pi : LM \to M$.

Hint: $P_1 = Q$, $P_2 = P$, f is the embedding of Q into P, the action $R^{(1)}$ of the group H is just a restriction of the action $R^{(2)}$ of the larger group G; in particular for OM: the action of $O(p, q)$ is just the restriction of the action of $GL(n, \mathbb{R})$ to a subgroup, forced by the need to connect only orthonormal frames; set in (20.5.11) $P_1 = OM$, $P_2 = LM$, $f =$ the canonical embedding of OM into LM (we recall that an *orthonormal* frame is at the same time also a *frame*), $\varphi =$ the canonical embedding of $O(p, q)$ into $GL(n \equiv p + q, \mathbb{R})$ (we recall that an *orthogonal* transformation is at the same time *linear*). □

• Clearly, as an example of extension we can take LM with respect to OM; a more interesting one (we will, however, not examine it in any detail, here) is the bundle of

[427] The map $\hat{f}$ mentioned there is now the identity map, since there is only a single (common) base.

[428] If f is injective, also its restriction to the fiber over $p_1 \in P_1$ is injective. The fiber is mapped into the fiber over $f(p_1) \in P_2$; since both fibers are diffeomorphic to groups, also the homomorphism $\varphi : G_1 \to G_2$ is necessarily injective. The situation is similar with surjectivity.

[429] This terminology is not unique. For example, the concepts of reduction and restriction are sometimes interchanged in the literature with respect to the definitions mentioned here.

affine frames $\pi : AM \to M$, regarded as an extension of $\pi : LM \to M$; one considers all "affine frames" in each $x \in M$ being somewhat richer objects than good old ordinary "linear" frames, which we frequently encountered here.[430]

As an interesting example of a surjective morphism f, i.e. of the relation prolongation/reduction, we mention the *spin bundle* $\pi : SM \to M$ with respect to the bundle of orthonormal frames $OM \to M$; this important point will be discussed in Section 22.4.

Let us return once more to the relation between various structures on M and the procedure of restriction of a principal bundle with the base M. If we fix on M a metric tensor g, we single out by this a class of *distinguished frames* – namely the orthonormal frames with respect to this g – and then also a particular subset (submanifold) $OM \subset LM$. A different g would single out a different subset $OM \subset LM$. On both of them, however, the orthogonal group would naturally act and we would get two different (but isomorphic) realizations of an abstract principal $O(p, q)$-bundle $\pi : OM \to M$. Introduction of g on M thus leads to a "concrete version" of an $O(p, q)$-bundle $\pi : OM \to M$, which is a restriction of the $GL(n, \mathbb{R})$-bundle of (all) frames $\pi : LM \to M$. If we were to, the other way round, single out a submanifold $P \subset LM$, such that it served as the total space of a principal $O(p, q)$-bundle $\pi : P \to M$ (so in each fiber of LM we chose a submanifold where the group $O(p, q)$ acted freely and transitively, etc. – we would call its points distinguished frames), we could introduce, *with the help of* this submanifold, on M a metric tensor g: we would simply declare the points of $P \subset LM$ as *orthonormal* frames (the defining properties of a bundle then guarantee the consistency of the definition). We thus see that the introduction of a metric tensor g on M with signature (p, q) and the restriction of the frame bundle $\pi : LM \to M$ to an $O(p, q)$-subbundle $\pi : OM \to M$ "are actually the same thing" (this fact was already mentioned in the language of a function Φ at the end of the hint to (20.5.5)).

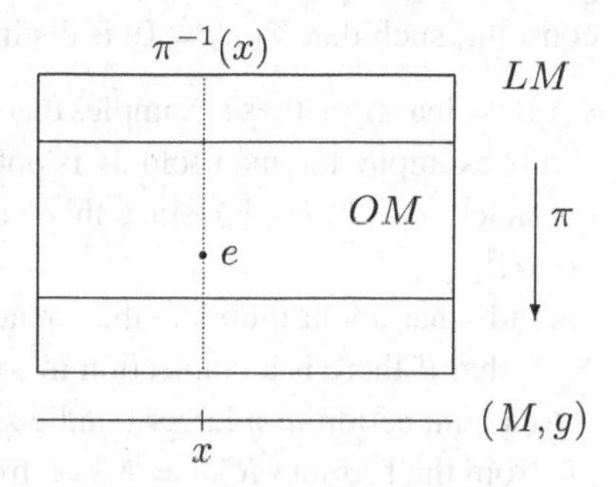

One can proceed in this way when also introducing other geometrical structures – a structure may be realized in terms of an appropriate *restriction* of an initial principal bundle to a subbundle (for example, making use of maps Φ of type ρ),[431] in which already only a *subgroup* $H \subset G$ of the initial group G acts.

20.5.13 Introduction of a metric tensor with signature (p, q) may be realized, as we learned, by a restriction of the structure group in the bundle $\pi : LM \to M$ from $GL(n, \mathbb{R})$ to a subgroup $O(p, q)$. Contemplate which restrictions of $GL(n, \mathbb{R})$, as well as distinguished

[430] A connection in the principal bundle $\pi : AM \to M$ is called a "generalized" *affine connection* and it turns out that any such connection may be parametrized in terms of a linear connection (in $LM \to M$) plus a $(1, 1)$-type tensor field on M. If the tensor field happens to be the *identity* tensor, the attribute "generalized" is omitted. We see from this that there is a one-to-one relation between affine and linear connections and they are often interchanged because of this fact.

[431] This is consistent with a description of various geometrical quantities (like tensor fields) as functions of type ρ, see Sections 19.6 and 20.3.

frames, correspond to introduction of the following structures:

(i) an orientation on M
(ii) a volume form on M
(iii) the complete parallelism on M.

Hint: (i) $H = GL_+(n, \mathbb{R})$ (10.1.9), right-handed frames are distinguished; (ii) $H = SL(n, \mathbb{R})$ (10.1.7), *unimodular frames* are distinguished, i.e. such which span a parallelepiped with volume 1 (then also $\omega = e^1 \wedge \cdots \wedge e^n$); (iii) $H = \{e\}$ (the trivial group = a group containing just a single element e), "parallel" frame field (= covariantly constant; such that $\nabla_V e_a = 0$) is distinguished,[432] (15.8.4). □

• It is clear from these examples that in general the restriction to a subgroup may not exist. If, for example, the manifold M is not orientable, the structure group $GL(n, \mathbb{R})$ cannot be restricted to $GL_+(n, \mathbb{R})$, since this would just be equivalent to introducing an orientation on M.[433]

And what about inducing the connection in the opposite direction (to a larger bundle)? Note that if there is a connection in a smaller bundle $\pi : Q \to M$, we can trivially extend it to a connection in a larger bundle $\pi : P \to M$; in terms of the connection form we get ω_p from the formula $R_g^* \omega = \mathrm{Ad}_g \omega$, from where $\omega_{pg} = \mathrm{Ad}_g R_{g*} \omega_p$, so the connection form may be uniquely extended from points $q \in Q$ also outside Q as $\omega_{qg} = \mathrm{Ad}_g R_{g*} \omega_q, qg \in P$. (Even before this, however, the form is to be "extended" at the point q itself so as to also allow the insertion of vectors which are not tangent to Q; this is unique since these new directions in the tangent space $T_q P$ are just the directions of the fundamental field $\xi_Y, Y \notin \mathcal{H}$ and there is a strictly specified way of how an arbitrary connection form should respond to such vectors according to the constitution (namely, it should give Y); the extension in q thus takes place so that the form responds to "old" vectors "as before" and to "new" vectors (also) by course of the constitution.)

This connection does not, however, represent the most general connection in the larger bundle: namely on Q it has values only in a subalgebra $\mathcal{H}$, and there is no reason for that for a general connection. The specific feature of this particular connection is that it is compatible with just the structure which leads to the restriction to the subbundle (horizontal lifts which start in Q already lie completely in Q, meaning that parallel transport preserves distinguished "frames").

If there is, for example, an *arbitrary* connection form in the bundle $\pi : OM \to M$, we also automatically get a unique connection in the bundle $\pi : LM \to M$. (If we know the results of $\nabla_U e_a = \ldots$ for an *orthonormal* frame field, the general properties of the covariant derivative already enable us to compute this expression also for a "general" frame field.)

[432] If e_a is a distinguished frame field (a section σ with the image in $P \subset LM$), then also $\hat{e}_a := A_a^b(x) e_b$ should be distinguished for an arbitrary matrix $A_a^b(x)$, which *for each x belongs to H*. A new frame field should also be parallel, which gives $\nabla_V \hat{e}_a = (V A_a^b) e_b \overset{!}{=} 0$ for each $V \Rightarrow A_a^b$ is a constant matrix $\Rightarrow A_a^b = \delta_a^b$ (the only subgroup of $GL(n, \mathbb{R})$ containing a single element is $H = \{\mathbb{I}\}$).

[433] Whether structures may or may not exist is often determined by appropriate topological objects, like some so-called characteristic classes; non-triviality of a concrete characteristic class represents an *obstacle* to introducing a concrete structure, i.e. to a concrete restriction of the structure group.

The latter is, however, not "general," but rather it carries an imprint of its history (extension from the subbundle): it is *metric*. For if we parallel transport *orthonormal* frames along curves on M, we construct horizontal lifts of the curves from M. Since the connection was originally defined as a connection in the subbundle $\pi : OM \to M$, the lifts which start in OM necessarily lie *completely* in OM. Extension of the connection outside OM clearly cannot have any impact on these lifts, so that also after the extension, when the connection becomes a connection on the whole LM, the lifts remain only in OM if they start in OM. This just means, however, that the connection is metric. In this particular case we thus get the following result: if a subbundle results from introducing a metric tensor g on M, then a connection extended from the subbundle is special in that it is compatible with g (it is metric). This also holds in general (in particular for an orientation, a volume form, etc. from problem (20.5.13)).

If a structure leads to a restriction of a bundle to a subbundle, then a connection in the bundle which arises by extension from the subbundle to the whole bundle preserves the structure.

Summary of Chapter 20

The translation of the concepts related to the *linear* connection into the language of the frame bundle, which was performed in Chapter 19, clearly indicates a possible generalization. One should simply replace $\pi : LM \to M$ by $\pi : P \to M$, a general principal G-bundle. A connection in this bundle is then (by definition) any G-invariant horizontal distribution on the total space P. The distribution may be conveniently encoded in a connection form ω, a $\mathcal{G}$-valued 1-form on P ($\mathcal{G}$ being the Lie algebra of G). The points of P are now the natural counterparts of frames, and their parallel transport along a curve on M is defined as the horizontal lift of the curve to P. The (local) dependence of parallel transport on a path may be rephrased in terms of integrability of the horizontal distribution and the $\mathcal{G}$-valued *curvature* 2-form Ω enters the scene (via the Frobenius integrability condition) just as a measure of this non-integrability. As a convenient formal tool for the explicit computation of the curvature form one introduces the exterior *covariant* derivative D; we then prove the formula $\Omega = D\omega = d\omega + \frac{1}{2}[\omega \wedge \omega]$. Similarly we compute the action of D on another important class of objects, horizontal forms of type ρ, where we get $D\alpha = d\alpha + \rho'(\omega) \dot{\wedge} \alpha$. Applying D twice, Bianchi and Ricci identities arise. The last, starred, section is devoted to an interesting relation between subbundles, structures on M and connections compatible with the structures. It explains, for example, how special connections in $\pi : P \to M$ (say, metric connection in $\pi : LM \to M$) may be regarded as being extended from appropriate subbundles $\pi : Q \to M, Q \subset P$ (from $\pi : OM \to M$ in the case of the metric connection).

$R_{g*}\text{Hor}_{\,p} P = \text{Hor}_{\,pg} P$	Horizontal distribution is G-invariant	(20.2.1)
$\omega_p := \Psi_p^{-1} \circ \text{ver} : \ T_p P \to \mathcal{G}$	Connection 1-form in $p \in P$	(20.2.4)
$R_g^* \omega = \text{Ad}_{g^{-1}} \omega, \ \langle \omega, \xi_X \rangle = X$	Crucial properties of the connection form	(20.2.5)

$\pi \circ \gamma^h = \gamma, \gamma^h(0) = p, \langle \omega, \dot{(\gamma^h)} \rangle = 0$	Horizontal lift of γ starting in $p \in P$	(20.3.2)
$(\text{hor}\,\alpha)(U, \dots) := \alpha(\text{hor}\,U, \dots)$	Horizontal part of a form	(20.3.4)
$D\alpha := \text{hor}\,d\alpha$	Exterior covariant derivative of a form	(20.3.5)
$\Omega := D\omega = \Omega^i E_i$	Curvature 2-form on P	(20.4.1, 3)
$\Omega = d\omega + \frac{1}{2}[\omega \wedge \omega]$	Cartan structure equation	(20.4.3)
$D\alpha = d\alpha + \rho'(\omega) \dot{\wedge} \alpha$	Action of D on horizontal forms of type ρ	(20.4.6)
$DD\omega \equiv D\Omega = d\Omega + [\omega \wedge \Omega] = 0$	Bianchi identity	(20.4.4, 7)
$DD\alpha = \rho'(\Omega) \dot{\wedge} \alpha$	Ricci identity	(20.4.8)
$\Omega = 0 \;\Rightarrow\; \exists \sigma \;:\; \sigma^* \omega = 0$	Zero curvature $\Rightarrow$ complete parallelism	(20.4.11)

21
Gauge theories and connections

• We saw in the preceding chapter how the concept of linear connection from Chapter 15 may be generalized to the connection in an arbitrary principal G-bundle. So far it is, however, not clear at all what is the use of all these new ideas. Here we will show that the key concepts from this exactly match up[434] to corresponding key concepts of the formal scheme, which is known in the physics literature as a *gauge field theory*. There the group G is called the *gauge group*.

21.1 Local gauge invariance: "conventional" approach

• The best known gauge theory is "ordinary" electromagnetism. Long ago people realized that the introduction of potentials is a useful technical trick (see the beginning of Chapter 9), which greatly simplifies some issues. Instead of *measurable* objects, vectors of electric and magnetic *fields* **E**, **B**, one introduces auxiliary[435] quantities Φ, **A**, the electromagnetic *potentials*:

$$\operatorname{div} \mathbf{B} = 0 \quad \Rightarrow \quad \exists \mathbf{A} \text{ such that } \mathbf{B} = \operatorname{curl} \mathbf{A}$$
$$\operatorname{curl}(\mathbf{E} + \partial_t \mathbf{A}) = \mathbf{0} \quad \Rightarrow \quad \exists \Phi \text{ such that } \mathbf{E} = -\nabla \Phi - \partial_t \mathbf{A}$$

We can see right from the definition that potentials are not given uniquely; instead of $(\Phi, \mathbf{A})$ we can use another pair $(\Phi', \mathbf{A}')$ such that

$$\Phi' = \Phi + \partial_t \chi \qquad \mathbf{A}' = \mathbf{A} - \operatorname{grad} \chi \quad \text{for a function } \chi$$

The most convenient way of speaking about this freedom (as we learned in Chapter 16) provides the language of differential forms in space-time: there $(\mathbf{E}, \mathbf{B}) \leftrightarrow F$, $(\Phi, \mathbf{A}) \leftrightarrow A$ and the argumentation reads

$$dF = 0 \Rightarrow \exists A \text{ such that } F = dA, \text{ and moreover } A \sim A' := A + d\chi$$

The transition from initial potentials to new (equivalent) ones, $A \mapsto A' \equiv A + d\chi$, is known as a *gauge transformation* of the potentials.

[434] There are also some other applications in physics, but this one is certainly the best known as well as the most important.
[435] In *quantum* theory potentials turn out to play a much more important role and the use of the word "auxiliary" might actually be inadequate.

There is, however, also another, at first sight completely unrelated situation in physics, where one naturally encounters a (co)vector field A with a transformation law $A \mapsto A' \equiv A + d\chi$. It namely occurs when the action integral of a field theory which is *globally* gauge invariant is required to become *locally* gauge invariant. Let us see how this works for the simple case of the theory of a single complex scalar field ϕ.

21.1.1 Consider a complex scalar field ϕ on (M, g) with the standard action integral (16.3.6)

$$S[\phi] = \int \mathcal{L}(\phi, d\phi)\omega_g \qquad \mathcal{L}(\phi, d\phi) = (d\phi^*, d\phi) - m^2(\phi^*, \phi)$$
$$\equiv \partial_\mu \phi^* \partial^\mu \phi - m^2 \phi^* \phi$$

Check that

(i) the Lagrangian (density) $\mathcal{L}$ (as well as the action S) is invariant with respect to a "global" change of "phase" of the field ϕ

$$\{\phi(x) \mapsto e^{-i\alpha}\phi(x),\ \phi^*(x) \mapsto e^{i\alpha}\phi^*(x)\} \quad \Rightarrow \quad \mathcal{L} \mapsto \mathcal{L} \quad (\alpha = \text{constant})$$

(ii) if the function $\alpha(x)$ *depends* on x, the action *fails* to be invariant; a "local" change of phase thus *spoils* the invariance

(iii) the invariance of the action with respect to a *local* change of phase of the field ϕ may be restored[436] by the following simple recipe: the original action is *modified* by means of the substitution $d \mapsto \mathcal{D}$, defined as

$$\mathcal{D}\phi := d\phi + iA\phi \qquad \mathcal{D}\phi^* := d\phi^* - iA\phi^*$$

where A is a new field, a 1-form subject to the transformation law $A \mapsto A + d\alpha$. So the statement is that the (new, modified) action

$$S[\phi, A] = \int \mathcal{L}(\phi, d\phi, A)\omega_g \qquad \mathcal{L}(\phi, d\phi, A) = (\mathcal{D}\phi^*, \mathcal{D}\phi) - m^2(\phi^*, \phi)$$
$$\equiv (\partial_\mu - iA_\mu)\phi^*(\partial^\mu + iA^\mu)\phi - m^2\phi^*\phi$$

happens to be invariant with respect to *simultaneous* change of the fields

$$\phi \mapsto e^{-i\alpha(x)}\phi \qquad \phi^* \mapsto e^{i\alpha(x)}\phi^* \qquad A \mapsto A + d\alpha(x)$$

The type as well as the transformation law of the field A is thus chosen *so as* to *compensate* undesirable *changes* in the original action.

Hint: (ii) $d\phi \mapsto d(e^{-i\alpha(x)}\phi) = e^{-i\alpha(x)}(d\phi - i\phi d\alpha) \neq e^{-i\alpha(x)}d\phi$; (iii) if a transformation law $A \mapsto A'$ of the field A existed such that $\mathcal{D}'\phi' \equiv (d + iA')\phi' \overset{!}{=} e^{-i\alpha(x)}\mathcal{D}\phi$ were true, we could check the invariance of the new action in exactly the same way,[437] as was done for the original one for *constant* α; now

$$\mathcal{D}'\phi' \equiv (d + iA')\phi' = e^{-i\alpha(x)}(d\phi - i\phi d\alpha + iA'\phi) \overset{!}{=} e^{-i\alpha(x)}(d\phi + iA\phi)$$

[436] Note, however, that actually we speak about a *different* (modified) action.

[437] For $\alpha =$ constant the operator d is able to "jump over" $e^{-i\alpha}$ and this factor then cancels with $e^{+i\alpha}$, coming from the left; for $\alpha \neq$ constant $\mathcal{D}$ has the same ability with respect to the dangerous factor $e^{-i\alpha(x)}$ and the phase factors again cancel.

from where $A' = A + d\alpha$. So the needed transformation law indeed exists (and it even turns out to be remarkably simple). □

• So far the "dynamics" of the field A itself has been absent, since there are no derivatives of A in the action. This shortcoming may, however, be eliminated (not spoiling, at the same time, the local gauge invariance).

21.1.2 Check that

(i) any expression which only depends on A through $F := dA$ is invariant with respect to a transformation $A \mapsto A + d\alpha$, so that a possible dynamical term in the action reads

$$-\frac{1}{2}\langle F, F\rangle = -\frac{1}{2}\int F \wedge *F \equiv -\frac{1}{2}\int (F, F)\omega_g \equiv -\frac{1}{4}\int F_{\mu\nu}F^{\mu\nu}\omega_g$$

(ii) the complete action

$$S[\phi, A] = \int \mathcal{L}(\phi, d\phi, A, dA)\omega_g$$

$$\mathcal{L}(\phi, d\phi, A, dA) = (\mathcal{D}\phi^*, \mathcal{D}\phi) - m^2(\phi^*, \phi) - \frac{1}{2}(F, F)$$

$$\equiv (\partial_\mu - iA_\mu)\phi^*(\partial^\mu + iA^\mu)\phi - m^2\phi^*\phi - \partial_{[\mu}A_{\nu]}\partial^{[\mu}A^{\nu]}$$

is invariant with respect to $U(1)$-*local gauge transformations*, i.e. a simultaneous change of fields

$$\phi \mapsto e^{-i\alpha(x)}\phi \qquad \phi^* \mapsto e^{i\alpha(x)}\phi^* \qquad A \mapsto A + d\alpha(x)$$

(iii) the phase factor $e^{i\alpha(x)}$ may be regarded as a position-dependent element of the group $U(1)$ and the transformations of the fields may also be written in the form

$$\phi \mapsto B^{-1}\phi \qquad \phi^* \mapsto B\phi^* \qquad iA \mapsto iA + B^{-1}dB \qquad B(x) \equiv e^{i\alpha(x)} \in U(1)$$

which explains the occurrence of "$U(1)$" in the nomenclature used for these transformations. □

• What is the lesson to take away from this problem? We learned that if we start from the action of an "*isolated*" complex scalar field and we try to modify the theory *so as* to make it $U(1)$-*locally* invariant, we "unwittingly" come to the action of the *scalar electrodynamics*, i.e. the theory of *two interacting* fields, the scalar field itself and a new field A, possessing all the characteristics of the *electromagnetic* field.

The resulting action may also be rewritten in an instructive (real) matrix form, in which the group $U(1)$ formally arises as the *isomorphic* group $SO(2)$.

21.1.3 Let us look at the results of the preceding problem in the real language. Let $\phi =: \frac{1}{\sqrt{2}}(\phi^1 + i\phi^2)$ and introduce the following notation:[438]

$$\phi := \begin{pmatrix}\phi^1\\ \phi^2\end{pmatrix} \qquad \mathcal{A} := \begin{pmatrix}0 & -A\\ A & 0\end{pmatrix}$$

$$\mathcal{F} := \begin{pmatrix}0 & -F\\ F & 0\end{pmatrix} \qquad B(x) := \begin{pmatrix}\cos\alpha(x) & -\sin\alpha(x)\\ \sin\alpha(x) & \cos\alpha(x)\end{pmatrix}$$

[438] Caution: the same notation is used for the new real column $\phi = (\phi_1, \phi_2)$ as for the initial complex field; there is, however, an *additional factor* of $\sqrt{2}$ involved (see the relevant formula).

Check that the formulas from the preceding problem are modified as follows:

(i) the original action is

$$S[\phi] \leftrightarrow S[\phi^1, \phi^2] = \int \mathcal{L}(\phi^a, d\phi^a)\omega_g \qquad a = 1, 2$$

$$\mathcal{L}(\phi, d\phi) \leftrightarrow \mathcal{L}(\phi^a, d\phi^a) = \frac{1}{2}h_{ab}(d\phi^a, d\phi^b) - \frac{1}{2}m^2 h_{ab}(\phi^a, \phi^b) \qquad h_{ab} = \delta_{ab}$$

$$\equiv \frac{1}{2}h_{ab}\partial_\mu\phi^a\partial^\mu\phi^b - \frac{1}{2}m^2 h_{ab}\phi^a\phi^b$$

(ii) the action is invariant with respect to the transformations $\phi \mapsto B^{-1}\phi$ with a *constant* matrix B (i.e. α = constant)

(iii) for position-dependent matrices $B(x)$ invariance is obtained by the replacement $d\phi \mapsto \mathcal{D}\phi \equiv d\phi + \mathcal{A}\phi$, i.e.

$$\begin{pmatrix} d\phi^1 \\ d\phi^2 \end{pmatrix} \mapsto \begin{pmatrix} \mathcal{D}\phi^1 \\ \mathcal{D}\phi^2 \end{pmatrix} \equiv \begin{pmatrix} d\phi^1 \\ d\phi^2 \end{pmatrix} + \begin{pmatrix} 0 & -A \\ A & 0 \end{pmatrix}\begin{pmatrix} \phi^1 \\ \phi^2 \end{pmatrix}$$

(iv) the complete action now reads

$$S[\phi, \mathcal{A}] = \int \mathcal{L}(\phi^a, d\phi^a, A, dA)\omega_g$$

$$\mathcal{L}(\phi^a, d\phi^a, A, dA) = \frac{1}{2}h_{ab}(\mathcal{D}\phi^a, \mathcal{D}\phi^b) - \frac{1}{2}m^2 h_{ab}(\phi^a, \phi^b) - \frac{1}{2}(F, F)$$

$$\equiv \frac{1}{2}h_{ab}(\partial_\mu - iA_\mu)\phi^a(\partial^\mu + iA^\mu)\phi^b - \frac{1}{2}m^2 h_{ab}\phi^a\phi^b - \partial_{[\mu}A_{\nu]}\partial^{[\mu}A^{\nu]}$$

and it is invariant with respect to $SO(2)$-*local gauge transformations*

$$\begin{pmatrix} \phi^1 \\ \phi^2 \end{pmatrix} \mapsto \begin{pmatrix} \cos\alpha(x) & -\sin\alpha(x) \\ \sin\alpha(x) & \cos\alpha(x) \end{pmatrix}\begin{pmatrix} \phi^1 \\ \phi^2 \end{pmatrix}$$

$$\begin{pmatrix} 0 & -A \\ A & 0 \end{pmatrix} \mapsto \begin{pmatrix} 0 & -A \\ A & 0 \end{pmatrix} + \begin{pmatrix} \cos\alpha(x) & -\sin\alpha(x) \\ \sin\alpha(x) & \cos\alpha(x) \end{pmatrix}^{-1} d\begin{pmatrix} \cos\alpha(x) & -\sin\alpha(x) \\ \sin\alpha(x) & \cos\alpha(x) \end{pmatrix}$$

$$\equiv \begin{pmatrix} 0 & -A \\ A & 0 \end{pmatrix} + \begin{pmatrix} 0 & -d\alpha \\ d\alpha & 0 \end{pmatrix}$$

i.e.

$$\phi \mapsto B^{-1}\phi \qquad \mathcal{A} \mapsto \mathcal{A} + B^{-1}dB \qquad B(x) \in SO(2)$$

□

• Now, a way of generalizing the scheme from $SO(2) \approx U(1)$ to an arbitrary matrix group G should be clear.

21.1.4 Consider a column $\phi(x)$ with components $\phi^a(x)$ and let a group G act on such columns via an appropriate matrix representation, so that $\phi \mapsto B^{-1}\phi$. Let h be an invariant scalar product with respect to the representation, so that $B^{\mathrm{T}}hB = h$. Check that

(i) the action

$$S[\phi^1, \dots, \phi^n] = \int \mathcal{L}(\phi^a, d\phi^a)\omega_g$$

$$\mathcal{L}(\phi^a, d\phi^a) = \frac{1}{2}h_{ab}(d\phi^a, d\phi^b) - \frac{1}{2}m^2 h_{ab}(\phi^a, \phi^b) \qquad B^c_a h_{cd} B^d_b = h_{ab}$$

$$\equiv \frac{1}{2}h_{ab}\partial_\mu\phi^a\partial^\mu\phi^b - \frac{1}{2}m^2 h_{ab}\phi^a\phi^b$$

is invariant with respect to *G-global gauge transformations*, i.e. with respect to $\phi \mapsto B^{-1}\phi$ with *constant* matrices from the corresponding representation of the group G

(ii) the action is *not* invariant with respect to *G-local gauge transformations*, i.e. with respect to $\phi \mapsto B^{-1}(x)\phi$ with *position-dependent* matrices from the corresponding representation of the group G

(iii) if the *gauge potential*, i.e. a covector field $\mathcal{A}$ with values in the (represented) Lie algebra $\mathcal{G}$, transforms under a local gauge transformation with a matrix $B(x)$ according to the rule

$$\mathcal{A} \mapsto B^{-1}\mathcal{A}B + B^{-1}dB$$

then the *covariant derivative*[439] of the field ϕ

$$\mathcal{D}\phi := d\phi + \mathcal{A}\phi$$

transforms (contrary to the ordinary derivative $d\phi$) in the same way as ϕ itself does, i.e. there holds

$$\phi \mapsto B^{-1}(x)\phi \qquad \Rightarrow \qquad \mathcal{D}\phi \mapsto B^{-1}(x)\mathcal{D}\phi$$

(iv) as a consequence of this fact the replacement $d \mapsto \mathcal{D}$ in the original action extends its invariance to *local* gauge transformations: the action

$$S\left[\phi^a, A^a_{b\mu}\right] = \int \mathcal{L}\left(\phi^a, d\phi^a, A^a_{b\mu}\right)\omega_g$$

$$\mathcal{L}\left(\phi^a, d\phi^a, A^a_{b\mu}\right) = \frac{1}{2}h_{ab}(\mathcal{D}\phi^a, \mathcal{D}\phi^b) - \frac{1}{2}m^2 h_{ab}(\phi^a, \phi^b) \qquad B^c_a h_{cd} B^d_b = h_{ab}$$

$$\equiv \frac{1}{2}h_{ab}\left(\partial_\mu\phi^a + A^a_{c\mu}\phi^c\right)\left(\partial^\mu\phi^b + A^{b\mu}_c\phi^c\right) - \frac{1}{2}m^2 h_{ab}\phi^a\phi^b$$

is invariant with respect to *local gauge transformations*, i.e. to a simultaneous replacement

$$\phi \mapsto B^{-1}(x)\phi \qquad \mathcal{A} \mapsto B^{-1}\mathcal{A}B + B^{-1}dB$$

Hint: (iii)

$$\mathcal{D}'\phi' \equiv (d + \mathcal{A}')(B^{-1}\phi) = (dB^{-1})\phi + B^{-1}d\phi + \mathcal{A}'B^{-1}\phi \overset{!}{=} B^{-1}\mathcal{D}\phi \equiv B^{-1}d\phi + B^{-1}\mathcal{A}\phi$$

which gives $\mathcal{A}' = B^{-1}\mathcal{A}B + B^{-1}dB$. □

• We see that the requirement of local gauge invariance of the action of the field ϕ dragged into play a new field $\mathcal{A}$, the gauge potential. We were able to find a transformation law for

[439] For the moment one should ignore the coincidence of this nomenclature (or perhaps even the letter denoting the concept) with any covariant derivative encountered so far. The justification of the coincidence will be clarified later (in Section 21.5).

$\mathcal{A}$ such that it compensates all the undesirable changes produced by the initial fields, which otherwise would spoil the invariance of the action. Because of that, it is also known as a *compensating field.*

The field $\mathcal{A}$ still lacks its own dynamics, since the derivatives of its components are still missing in the action. (If there are no derivatives of $\mathcal{A}$, the variation with respect to $\mathcal{A}$ does not result in a *differential* equation for $\mathcal{A}$ as the "equation of motion," but rather it produces only an algebraic one, a "constraint.") However, we can once more add an appropriate (although a bit more involved) term, which contains derivatives and, at the same time, is locally gauge invariant.

21.1.5 Consider the gauge potential $\mathcal{A}$ from problem (21.1.4). Then the prescription

$$\mathcal{F} := d\mathcal{A} + \mathcal{A} \wedge \mathcal{A}$$

defines a new field, the *gauge field strength.* It is a 2-form with values in the represented Lie algebra $\mathcal{G}$, i.e. an $n \times n$ matrix with 2-form entries; $\mathcal{A} \wedge \mathcal{A}$ denotes the matrix product in which the matrix elements (1-forms) are multiplied in the sense of the *exterior* product, so that $(\mathcal{A} \wedge \mathcal{A})^a_b = \mathcal{A}^a_c \wedge \mathcal{A}^c_b$. Check that

(i) the transformation law for the field $\mathcal{F}$ is (contrary to the first term alone, the ordinary exterior derivative $d\mathcal{A}$) very simple (even simpler than that of the field $\mathcal{A}$ itself), namely

$$\mathcal{F} \mapsto B^{-1}\mathcal{F}B$$

(ii) this is also the way in which $*\mathcal{F}$ transforms

$$*\mathcal{F} \mapsto B^{-1}(*\mathcal{F})B$$

(iii) and also how $\mathcal{F} \wedge *\mathcal{F}$ transforms (this is a 4-form with values in the represented Lie algebra $\mathcal{G}$, i.e. an $n \times n$ matrix with 4-form entries)

$$\mathcal{F} \wedge *\mathcal{F} \mapsto B^{-1}(\mathcal{F} \wedge *\mathcal{F})B$$

(iv) the *trace* of this expression (an *ordinary* 4-form) is already gauge *invariant*

$$\mathrm{Tr}\,\{\mathcal{F} \wedge *\mathcal{F}\} \mapsto \mathrm{Tr}\,\{B^{-1}(\mathcal{F} \wedge *\mathcal{F})B\} = \mathrm{Tr}\,\{\mathcal{F} \wedge *\mathcal{F}\}$$

(v) the integral of the trace may be used as a (locally gauge invariant) term into the action

$$S[\mathcal{A}] \sim \int \mathrm{Tr}\,\{\mathcal{F} \wedge *\mathcal{F}\} \qquad S[\mathcal{A}] = S[\mathcal{A}']$$

Hint: (i) a direct computation $\mathcal{F}' = d\mathcal{A}' + \mathcal{A}' \wedge \mathcal{A}' = \cdots = B^{-1}\mathcal{F}B$; (ii) $*$ is only related to the "form" part of $\mathcal{F}$ and B^{-1} and B are 0-forms, so that $*(B^{-1}\mathcal{F}B) = B^{-1}(*\mathcal{F})B$; (iii) $\mathcal{F}' \wedge *\mathcal{F}' = B^{-1}\mathcal{F}B \wedge B^{-1} * \mathcal{F}B = B^{-1}(\mathcal{F} \wedge *\mathcal{F})B$. □

• The standard nomenclature in physics is as follows: the field ϕ is called the *matter field*, $\mathcal{A}$ is the *gauge potential* and $\mathcal{F}$ is the *gauge field strength.* The basic idea is that the matter fields (their quanta) describe various kinds of "charged" particles and they in turn interact by means of the quanta of corresponding gauge fields (potentials). For example, in scalar electrodynamics the charged pions[440] may be associated with the field ϕ and the photons with the field $\mathcal{A}$. If the interaction term which couples the field ϕ with $\mathcal{A}$ (i.e. the term containing the *product* of the fields) arises in the Lagrangian as a consequence of the replacement $d \mapsto \mathcal{D} \equiv d + \mathcal{A}$, we speak about *minimal coupling*. Notice that only the fields ϕ which transform according to *non-trivial* representation ρ interact with the field $\mathcal{A}$ (otherwise $\rho' = 0$, $\mathcal{D} = d$ and the relevant term which contains the product is actually absent); just those fields (and corresponding particles) are *charged.*

Now let us see what is in fact the reason why this section introducing the formalism of the gauge fields is in chapters devoted to connections. This is a bit strange at first sight since the starting points of the two theories seem to be completely different. However, some formulas derived in this section look oddly familiar... Let us concentrate our attention on the similarity by explicitly comparing the relevant expressions from gauge field theory with the corresponding ones we encountered in the theory of linear connections.

21.1.6 Let $\hat{\omega}$ be the 1-form of the (*linear*) connection, which we introduced in Section 15.6 and let $\hat{\Omega}$ denote the corresponding curvature 2-form. Compare

(i) the transformation law of $\hat{\omega}$ (under the change of a frame field $e_a \mapsto B^b_a e_b$) with the transformation law of the field $\mathcal{A}$ discussed here

$$\hat{\omega}' = B^{-1}\hat{\omega}B + B^{-1}dB$$
$$\mathcal{A}' = B^{-1}\mathcal{A}B + B^{-1}dB$$

and express (by raised eyebrows) your sincere amazement at the unexpected similarity

(ii) the explicit formulas for the computation of $\mathcal{F}$ from $\mathcal{A}$ and $\hat{\Omega}$ from $\hat{\omega}$

$$\hat{\Omega} = d\hat{\omega} + \hat{\omega} \wedge \hat{\omega}$$
$$\mathcal{F} = d\mathcal{A} + \mathcal{A} \wedge \mathcal{A}$$

and do not forget, please, to look stunned and raise your eyebrows even higher; actually you should be struck by the similarity and we warmly recommend letting your "eyebrows rise so high that they are in danger of disappearing into your hair" – the situation indeed deserves strong emotions

(iii) the transformation laws of $\hat{\Omega}$ and $\mathcal{F}$

$$\hat{\Omega}' = B^{-1}\hat{\Omega}B$$
$$\mathcal{F}' = B^{-1}\mathcal{F}B$$

Hint: see (15.6.2) and (21.1.5). □

• We learned in this section that the basic relations of the formalism of the *gauge fields* and the basic relations of the theory of (so far only) the *linear connection* become (when appropriately written) as similar as two peas in a pod. In the next section we will see that the similarity is actually not limited to the linear connection, but rather it also concerns

[440] As well as the electrons in the *spinor electrodynamics*; although the structure of the action is *different* for electrons (ϕ is no longer a *scalar* field there), the way in which the potential $\mathcal{A}$ arises is basically the same.

the *general* connection introduced in Chapter 20. Moreover, we will learn that all the basic objects and relations encountered in the theory of gauge fields[441] admit a simple and natural geometrical interpretation in terms of the theory of connections.

21.2 Change of section and a gauge transformation

• At the end of the last section we realized the striking similarity between the scheme of the gauge fields and the theory of the *linear* connection. Recall that the corresponding objects are

$$\mathcal{A} \leftrightarrow \hat{\omega} \qquad \mathcal{F} \leftrightarrow \hat{\Omega}$$

and the similarity consists in the relations

$$\begin{array}{lll} \hat{\omega}' = B^{-1}\hat{\omega}B + B^{-1}dB & \hat{\Omega} = d\hat{\omega} + \hat{\omega} \wedge \hat{\omega} & \hat{\Omega}' = B^{-1}\hat{\Omega}B \\ \mathcal{A}' = B^{-1}\mathcal{A}B + B^{-1}dB & \mathcal{F} = d\mathcal{A} + \mathcal{A} \wedge \mathcal{A} & \mathcal{F}' = B^{-1}\mathcal{F}B \end{array}$$

In this section we first convince ourselves that the similarity naturally extends from the *linear* connection to the *general* connection as well and then we start to contemplate what all this resemblance means. The contemplation will result in the joyful conclusion[442] that the formalism of the gauge fields and the theory of connections actually speak "about the same thing in different words."

Let us start with the question of how to pass from the linear connection to the general one. Recall that the transformation law $\hat{\omega} \mapsto \hat{\omega}' = B^{-1}\hat{\omega}B + B^{-1}dB$ speaks about the change of the forms of the (linear) connection under the *change of the frame field* $e_a \mapsto B^b_a e_b$. On the other hand, the choice of a frame field, from the point of view of the formulation on LM (see (19.2.2)), may be regarded as the choice of a *local section* $\sigma : \mathcal{U} \to LM$. If we, moreover, recall (19.2.3) that $\hat{\omega} = \sigma^*\omega$, where ω represents the connection form on LM (which is already independent of any sections), we see that the change $\hat{\omega} \mapsto \hat{\omega}'$ may also be interpreted as passing to the form pulled back by *another local section*:

$$\sigma^*\omega \mapsto \hat{\sigma}^*\omega$$

This formulation already turns out to be convenient for treating the situation in the general case. We will compute how the pull-back $\sigma^*\omega$ of the connection form on a *general* principal bundle transforms under the change of the section $\sigma \mapsto \hat{\sigma}$. We will also compute the transformation of the pull-back of an arbitrary horizontal form of type ρ. This then gives the transformation of the pull-back $\sigma^*\Omega$ of the curvature form but it is also closely related to the transformation of the matter field ϕ from the theory of gauge fields.

So consider an arbitrary principal G-bundle $\pi : P \to M$. Let $\omega \in \Omega^1(P, \text{Ad})$ be a connection form, $\Omega \in \bar{\Omega}^2(P, \text{Ad})$ the curvature form, $\Lambda \in \bar{\Omega}^p(P, \rho)$ a horizontal p-form of

[441] We mean explicitly the *classical* (as opposed to the quantum) theory of gauge fields. On the other hand, the quantum theory stems from the classical one and this explains why the connection theory is of importance to (some parts of) the quantum field theory as well.

[442] A lot of our brain capacity is saved for other interesting facts if the same thing does not need to be stored twice.

type ρ and $\Phi \in \Omega^0(P, \rho)$ a function of type ρ. If σ and $\hat{\sigma}$ represent two local sections, we may pull back the forms ω, Ω, Λ and Φ by means of both of these sections and get

$$A := \sigma^*\omega \qquad F := \sigma^*\Omega \qquad \lambda := \sigma^*\Lambda \qquad \phi := \sigma^*\Phi$$
$$\hat{A} := \hat{\sigma}^*\omega \qquad \hat{F} := \hat{\sigma}^*\Omega \qquad \hat{\lambda} := \hat{\sigma}^*\Lambda \qquad \hat{\phi} := \hat{\sigma}^*\Phi$$

Passing from the unhatted quantity to the hatted one is called the *gauge transformation* of the corresponding quantity. In this section, our goal is to obtain the explicit formulas for these transformations. As the first step we need to find a convenient way of describing quantitatively the relation between two sections themselves.

21.2.1 Consider two sections which share the same domain $\mathcal{U}$

$$\sigma : \mathcal{U} \to P \qquad \hat{\sigma} : \mathcal{U} \to P$$

(if the domains were different, everything should be referred to their *intersection*). Check that

(i) there is a *unique* map

$$S : \mathcal{U} \to G \qquad \text{given by} \quad \hat{\sigma}(x) = \sigma(x)S(x) \equiv R_{S(x)}\sigma(x)$$

(ii) if, the other way round, there is a section σ and a map $S : \mathcal{U} \to G$, a unique second section $\hat{\sigma}$ is then defined by the same formula; this means that there is a one-to-one correspondence between such maps and transformations from one section to another and the maps under consideration may thus be used as a quantitative measure of the relation of two sections

(iii) the maps $S : \mathcal{U} \to G$ are naturally endowed with the structure of a group, if we define the product as

$$(S \circ \tilde{S})(x) := S(x)\tilde{S}(x)$$

(this (infinite-dimensional) group is called the *group* of *local gauge transformations* and we will denote[443] it by $G^{\mathcal{U}}$)

(iv) *constant* elements of the group $G^{\mathcal{U}}$ (i.e. the maps which assign to each point from $\mathcal{U}$ *the same* group element) constitute a subgroup $\hat{G} \subset G^{\mathcal{U}}$, which is isomorphic to the group G itself (so it is finite-dimensional); this subgroup is called (in the context of gauge theories)[444] the group of *global gauge transformations.*

Hint: (i) the action R_g is free and transitive in each fiber; (iv) if $S(x) = g =$ constant, then the isomorphism is $S \mapsto g$. □

• Our goal is to compare the pull-backs $\sigma^*\omega$ and $\hat{\sigma}^*\omega$ (and the same for the form Λ and the function Φ). Inserting arguments this means to compare

$$\langle \sigma^*\omega, v\rangle \equiv \langle \omega, \sigma_* v\rangle \qquad \text{and} \qquad \langle \hat{\sigma}^*\omega, v\rangle \equiv \langle \omega, \hat{\sigma}_* v\rangle$$

[443] In general, A^B denotes all maps from B to A (since the maps behave in a sense like the ordinary power of numbers a^b).
[444] We see that the meaning of the words *global* and *local* differs here from its usual meaning in differential geometry: namely "global" means constancy within the whole domain $\mathcal{U}$, here (being still locality from the point of view of differential geometry), local in turn means non-constancy in $\mathcal{U}$.

From these expressions we see that we are to clarify the issue of how two "lifts" of the same vector v in terms of two different sections are related, i.e. of the relation of the vectors $\sigma_* v$ and $\hat{\sigma}_* v$.

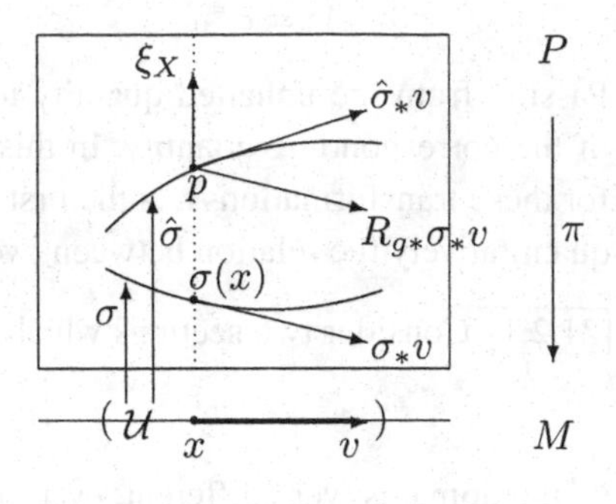

21.2.2 Let v be a vector at the point $x \in \mathcal{U} \subset M$ and let σ and $\hat{\sigma}$ be two local sections $\mathcal{U} \to P$. The sections are related by the function $S : \mathcal{U} \to G$ in the sense that $\hat{\sigma}(x) = \sigma(x)S(x)$. Check that the images of the vector v with respect to these two sections are then related by

$$\hat{\sigma}_* v = R_{g*}(\sigma_* v) + \xi_X(p)$$

$$\text{for} \quad g = S(x), \quad X = \langle S^*\theta, v\rangle, \quad p = \sigma(x)S(x)$$

where $\xi_X(p)$ is the value of the fundamental field (generator) of the action R_g in the point $p \in P$ and θ is the canonical 1-form on the group G (11.2.6).

Solution: let $v = \frac{d}{dt}\big|_0 \gamma(t)$; then

$$\begin{aligned}
\hat{\sigma}_* v = \hat{\sigma}_* \frac{d}{dt}\bigg|_0 \gamma(t) &= \frac{d}{dt}\bigg|_0 \hat{\sigma}(\gamma(t)) = \frac{d}{dt}\bigg|_0 R_{S(\gamma(t))}\sigma(\gamma(t)) \\
&= \frac{d}{dt}\bigg|_0 R_{S(\gamma(0))}\sigma(\gamma(t)) + \frac{d}{dt}\bigg|_0 R_{S(\gamma(t))}\sigma(\gamma(0)) \\
&= R_{S(x)*}\frac{d}{dt}\bigg|_0 \sigma(\gamma(t)) + \frac{d}{dt}\bigg|_0 R_{S(\gamma(t))}\sigma(\gamma(0)) \\
&= R_{g*}(\sigma_* v) + \frac{d}{dt}\bigg|_0 R_{S(\gamma(t))}\sigma(x)
\end{aligned}$$

So we are to show that the second term gives just $\xi_X(p)$ for $X = \langle S^*\theta, v\rangle$. In order to do that, we proceed as follows: the curve (in the group) $S(\gamma(t))$ begins in $S(x)$; for infinitesimal $t \equiv \epsilon$ it may be written as the right translation $S(\gamma(\epsilon)) = S(x)e^{\epsilon X}$ for a unique element $X \in \mathcal{G}$. So we need the element X; the infinitesimal right translation by $\exp \epsilon X$ under consideration is generated by the *left*-invariant field $L_X(g)$ (11.4.6), so that

$$\frac{d}{dt}\bigg|_0 S(\gamma(t)) \equiv S_* v = L_X(S(x)) \equiv L_X(g)$$

and consequently

$$X = L_X(e) = L_{g^{-1}*}L_X(g) = L_{g^{-1}*}S_* v \equiv \langle\theta, S_* v\rangle = \langle S^*\theta, v\rangle$$

Then

$$R_{S(\gamma(\epsilon))} = R_{S(x)\exp\epsilon X} = R_{\exp\epsilon X}R_{S(x)}$$

and

$$\frac{d}{d\epsilon}\bigg|_0 R_{S(\gamma(\epsilon))}\sigma(x) = \frac{d}{d\epsilon}\bigg|_0 R_{\exp \epsilon X}(\sigma(x)S(x)) \equiv \frac{d}{d\epsilon}\bigg|_0 R_{\exp \epsilon X}(p) \equiv \xi_X(p)$$

□

• So we are already prepared to answer the motivating question of how $\sigma^*\omega$ and $\hat{\sigma}^*\omega$ (as well as $\sigma^*\Lambda$ and $\hat{\sigma}^*\Lambda$) are related.

21.2.3 Let ω be a connection form, Ω the (corresponding) curvature form, Λ a horizontal p-form of type ρ, Φ a function of type ρ, σ and $\hat{\sigma}$ two local sections $\mathcal{U} \to P$, which are related by the function $S : \mathcal{U} \to G$, so that $\hat{\sigma}(x) = \sigma(x)S(x)$. Denote further

$$A := \sigma^*\omega \qquad F := \sigma^*\Omega \qquad \lambda := \sigma^*\Lambda \qquad \phi := \sigma^*\Phi$$
$$\hat{A} := \hat{\sigma}^*\omega \qquad \hat{F} := \hat{\sigma}^*\Omega \qquad \hat{\lambda} := \hat{\sigma}^*\Lambda \qquad \hat{\phi} := \hat{\sigma}^*\Phi$$

We may regard all of this as that the quantities $\hat{A}, \hat{F}, \hat{\lambda}, \hat{\phi}$ result from the *gauge transformation* of the "original" quantities A, F, λ, ϕ, the transformation being quantitatively given by the function $S(x)$. Check[445] that the following formulas hold:

$$\begin{aligned}
\hat{A} &= \mathrm{Ad}_{S^{-1}} A + S^*\theta \qquad \text{i.e.} \quad & \hat{A}(x) &= \mathrm{Ad}_{(S(x))^{-1}} A(x) + S^*\theta(x) \\
\hat{F} &= \mathrm{Ad}_{S^{-1}} F & \hat{F}(x) &= \mathrm{Ad}_{(S(x))^{-1}} F(x) \\
\hat{\lambda} &= \rho(S^{-1})\lambda & \hat{\lambda}(x) &= \rho((S(x))^{-1})\lambda(x) \\
\hat{\phi} &= \rho(S^{-1})\phi & \hat{\phi}(x) &= \rho((S(x))^{-1})\phi(x)
\end{aligned}$$

Solution: let $g = S(x)$, $X = \langle S^*\theta, v\rangle$, $p = \sigma(x)$; then,

$$\begin{aligned}
(\hat{\sigma}^*\omega)(v) &= \langle \omega, \hat{\sigma}_* v\rangle \\
&= \langle \omega, R_{g*}(\sigma_* v)\rangle + \langle \omega, \xi_X(pg)\rangle && (21.2.2) \\
&= \langle R_g^*\omega, \sigma_* v\rangle + X && (20.2.5) \\
&= \langle \mathrm{Ad}_{g^{-1}}\omega, \sigma_* v\rangle + X && (20.2.5) \\
&= \langle \mathrm{Ad}_{g^{-1}} A, v\rangle + X \equiv \langle \mathrm{Ad}_{g^{-1}} A + S^*\theta, v\rangle
\end{aligned}$$

Similarly,

$$\begin{aligned}
(\hat{\sigma}^*\Lambda)(v, \dots) &= \Lambda(\hat{\sigma}_* v, \dots) \\
&= \Lambda(R_{g*}(\sigma_* v, \dots) + \Lambda(\xi_X(pg), \dots) && (21.2.2) \\
&= R_g^*\Lambda(\sigma_* v, \dots) + 0 && \Lambda \text{ is horizontal} \\
&= (\rho(g^{-1})\Lambda)(\sigma_* v, \dots) && \Lambda \text{ is of type } \rho \\
&= (\rho(g^{-1})\lambda)(v, \dots)
\end{aligned}$$

and for the 0-form $\Lambda \equiv \Phi$ it is even simpler

$$\begin{aligned}
\hat{\phi}(x) &\equiv (\hat{\sigma}^*\Phi)(x) = \Phi(\hat{\sigma}(x)) = \Phi(R_{S(x)}\sigma(x)) = \rho((S(x))^{-1})\Phi(\sigma(x)) \\
&= \rho((S(x))^{-1})\Phi(\sigma(x)) \equiv \rho((S(x))^{-1})\phi(x)
\end{aligned}$$

The curvature form is a horizontal 2-form of type Ad (i.e. the special case of Λ). □

[445] The *function* Φ of type ρ is clearly the special case of p-forms of type ρ (for $p = 0$), but since the horizontality makes no sense in this particular case, the slight difference in the derivation of the corresponding formula (the result turns out to be *the same*) should be at least realized.

• For the ultimate confrontation of the two formalisms (connections versus gauge fields) we still need to compute how the expressions obtained above behave after a *representation* and also to introduce the exterior covariant derivative $\mathcal{D}$ *on the base* M.

21.2.4 Let ρ be a representation of G in W (i.e. the space where Λ, the horizontal p-form of type ρ, and Φ, the function of type ρ, have their values). Let E_a and E_i be bases in W and $\mathcal{G}$ respectively and let ρ' be the derived representation of $\mathcal{G}$, so that

$$\omega = \omega^i E_i \qquad A \equiv \sigma^*\omega = A^i E_i \qquad \lambda \equiv \sigma^*\Lambda = \lambda^a E_a$$
$$\Omega = \Omega^i E_i \qquad F \equiv \sigma^*\Omega = F^i E_i \qquad \phi \equiv \sigma^*\Phi = \phi^a E_a$$

Denote

$$\mathcal{A} := \rho'(A) \equiv A^i \rho'(E_i) \qquad \mathcal{A}E_b = A^i \rho^c_{bi} E_c =: \mathcal{A}^c_b E_c$$
$$\mathcal{F} := \rho'(F) \equiv F^i \rho'(E_i) \qquad \mathcal{F}E_b = F^i \rho^c_{bi} E_c =: \mathcal{F}^c_b E_c$$

and introduce $\mathcal{D}$, the "exterior covariant derivative on the base" M, by the relation

$$\mathcal{D} \circ \sigma^* := \sigma^* \circ D$$

Check that then

(i) the *Cartan structure equations* on the base for the represented objects read

$$\mathcal{F} = \mathcal{D}\mathcal{A} = d\mathcal{A} + \mathcal{A} \wedge \mathcal{A} \qquad \text{i.e.} \quad \mathcal{F}^a_b = \mathcal{D}\mathcal{A}^a_b = d\mathcal{A}^a_b + \mathcal{A}^a_c \wedge \mathcal{A}^c_b$$

(ii) the formula for the exterior covariant derivative of a horizontal p-form of type ρ and a function of type ρ is

$$\mathcal{D}\lambda = d\lambda + \mathcal{A} \dot{\wedge} \lambda \qquad \text{i.e.} \quad \mathcal{D}\lambda^b = d\lambda^b + \mathcal{A}^b_c \wedge \lambda^c$$
$$\mathcal{D}\phi = d\phi + \mathcal{A}\phi \qquad \text{i.e.} \quad \mathcal{D}\phi^b = d\phi^b + \mathcal{A}^b_c \phi^c$$

(iii) the *Bianchi identity* now reads

$$\mathcal{D}\mathcal{F} \equiv d\mathcal{F} + \mathcal{A} \wedge \mathcal{F} - \mathcal{F} \wedge \mathcal{A} = 0$$
$$\text{i.e.} \quad \mathcal{D}\mathcal{F}^a_b \equiv d\mathcal{F}^a_b + \mathcal{A}^a_c \wedge \mathcal{F}^c_b - \mathcal{F}^a_c \wedge \mathcal{A}^c_b = 0$$

(iv) and the *Ricci identity* acquires the form

$$\mathcal{D}\mathcal{D}\lambda = \mathcal{F} \dot{\wedge} \lambda \qquad \text{i.e.} \quad \mathcal{D}\mathcal{D}\lambda^a = \mathcal{F}^a_b \wedge \lambda^b$$
$$\mathcal{D}\mathcal{D}\phi = \mathcal{F}\phi \qquad \text{i.e.} \quad \mathcal{D}\mathcal{D}\phi^a = \mathcal{F}^a_b \phi^b$$

Hint: a typical computation: (i)

$$\begin{aligned}\rho'(\sigma^*[\omega \wedge \omega]) &= \rho'([A \wedge A]) = A^i \wedge A^j \rho'[E_i, E_j] \\ &= A^i \wedge A^j (\rho'(E_i)\rho'(E_j) - \rho'(E_j)\rho'(E_i)) \\ &= 2A^i \wedge A^j \rho'(E_i)\rho'(E_j) \equiv 2\mathcal{A} \wedge \mathcal{A}\end{aligned}$$

so that in the representation the commutator becomes *twice the product*; then the application of $\rho'\sigma^*$ to the Cartan structure equation $\Omega = D\omega = d\omega + \frac{1}{2}[\omega \wedge \omega]$ yields $\mathcal{F} = \mathcal{D}\mathcal{A} = d\mathcal{A} + \mathcal{A} \wedge \mathcal{A}$, and the remaining expressions similarly. □

• And what do the gauge transformations of these represented objects eventually look like?

21.2.5 Denote by $B(x)$ the *matrix representation* of the function $S(x)$, which encodes (by means of the relation $\hat{\sigma}(x) = \sigma(x)S(x)$) the gauge transformation

$$B := \rho \circ S : \mathcal{U} \to GL(N, \mathbb{R}) \qquad \text{i.e.} \quad B(x) := \rho(S(x))$$

($N \equiv \dim\ (W, \rho)$). So $B(x)$ is an $N \times N$ matrix valued function defined in the domain $\mathcal{U}$. Likewise, also $\mathcal{A}$ and $\mathcal{F}$ are $N \times N$ matrices (with 1-form and 2-form entries respectively) and λ is an N-component column vector. Check that the gauge transformations now read as follows:

$$\hat{\mathcal{A}} = B^{-1}\mathcal{A}B + B^{-1}dB \qquad \hat{\lambda} = B^{-1}\lambda$$
$$\hat{\mathcal{F}} = B^{-1}\mathcal{F}B \qquad \hat{\phi} = B^{-1}\phi$$

Hint: apply ρ' to the results $\hat{A} = \dots$, $\hat{F} = \dots$ from (21.2.3), use the formula $\rho'(\mathrm{Ad}_g X) = \rho(g)\rho'(X)(\rho(g))^{-1}$ from (12.3.7) and, in the case of $\mathcal{A}$, also the fact that on $GL(n, \mathbb{R})$ we have $\theta = x^{-1}dx$, see (11.7.19). □

• What exactly did all of these computations show? We learned that if the basic objects of the theory of connections, ω, Ω and Φ, are pulled back to the base by means of a section, they behave *exactly the same* as the basic quantities $\mathcal{A}$, $\mathcal{F}$ and ϕ of the physical formalism of gauge fields. The simplest and most natural conclusion one can make from this indisputable technical fact is to *identify* the two groups of objects. Namely, we assume that the quantities encountered in a gauge theory with the group G on a manifold M (usually the space-time) *are* the results of the pull-back of global invariant objects living in the total space of a principal G-bundle $\pi : P \to M$ and gauge transformations of these quantities just reflect the relations between the results of pull-back by different local sections. A useful vocabulary to be used in order to relate the connection theory with the gauge theory then reads

connection form ω	$\leftrightarrow$	gauge potential $\mathcal{A}$
curvature form Ω	$\leftrightarrow$	gauge field strength $\mathcal{F}$
function Φ of type ρ	$\leftrightarrow$	matter field ϕ of type ρ
choice of a section σ	$\leftrightarrow$	gauge fixing

(In the spirit of problem (17.6.6), the objects on the left live in the "real world of eternal Ideas" P, whereas the objects on the right live in the "apparent world of material objects" M. Each object on the right represents a particular "immersion in the material world" of the corresponding object on the left.)

21.2.6* Sometimes we encounter gauge transformations in an *infinitesimal* version. If $B(x)$ encodes the gauge transformation for represented quantities in the sense of (21.2.5), then the infinitesimal gauge transformation is given by the matrix

$$B(x) = \mathbb{I} + \epsilon b(x)$$

Indeed, if we write $S(x) = \exp\{\epsilon s(x)\}$, where $s : \mathcal{U} \to \mathcal{G}$, then we get within first-order accuracy in ϵ

$$B(x) \equiv \rho(S(x)) = \exp\{\epsilon\rho'(s(x))\} \equiv \mathbb{I} + \epsilon b(x)$$

for $b(x) := \rho'(s(x))$. Check that

(i) the gauge transformations may then be written in the form

$$\hat{\mathcal{A}} = \mathcal{A} + \epsilon\{[\mathcal{A}, b] + db\} \qquad \hat{\lambda} = \lambda - \epsilon b\lambda$$
$$\hat{\mathcal{F}} = \mathcal{F} + \epsilon\{[\mathcal{F}, b]\} \qquad \hat{\phi} = \phi - \epsilon b\phi$$

(ii) in particular, for the gauge group $SU(2)$ the infinitesimal gauge transformation of the potential $\mathcal{A}$ and the field strength $\mathcal{F}$ is

$$\hat{\mathbf{A}} = \mathbf{A} + \epsilon\{\mathbf{A} \times \mathbf{b} + d\mathbf{b}\} \qquad \mathcal{A} =: \mathbf{A} \cdot \left(-\frac{i}{2}\sigma\right) \equiv A^j_\mu \, dx^\mu \left(-\frac{i}{2}\sigma_j\right)$$

$$\hat{\mathbf{F}} = \mathbf{F} + \epsilon\{\mathbf{F} \times \mathbf{b}\} \qquad \mathcal{F} =: \mathbf{F} \cdot \left(-\frac{i}{2}\sigma\right) \equiv \frac{1}{2} F^j_{\mu\nu} \, dx^\mu \wedge dx^\nu \left(-\frac{i}{2}\sigma_j\right)$$

$$b =: \mathbf{b} \cdot \left(-\frac{i}{2}\sigma\right) \equiv b^j \left(-\frac{i}{2}\sigma_j\right)$$

Hint: (ii) see (12.1.5). □

21.2.7 Consider the *frame* bundle $\pi : LM \to M$. Here, the group G becomes $GL(n, \mathbb{R})$ and we may choose the Weyl basis E^a_b in the Lie algebra $gl(n, \mathbb{R})$. Recall also that a section of this bundle corresponds to a frame field on (part of) M (19.2.2). Be sure to understand that in this particular case

(i) $A \mapsto \hat{\omega} = \hat{\omega}^a_b E^b_a$, i.e. the gauge potential with respect to σ amounts to good old connection forms from (15.6.1)
(ii) $F \mapsto \hat{\Omega} = \hat{\Omega}^a_b E^b_a$, i.e. the field strength with respect to σ amounts to the curvature forms from (15.6.3)
(iii) the general transformation rules for A and F match the corresponding formulas from Chapter 15
(iv) the pull-back σ^* of the *general* Cartan structure equation from (20.4.3) yields just the good old *second* Cartan structure equation from (15.6.7) (if you are concerned about the *first* structure equation, consult Section 21.7, in particular problem (21.7.4)).

Hint: (i) see (21.2.3) and (19.2.3). □

21.3 Parallel transport equations for an object of type ρ in a gauge σ

• In Section 20.3 we introduced the concept of the parallel transport of the quantity of type ρ along a curve γ (or, actually, along an oriented path c, since according to (20.3.7) it does not depend on the parametrization). Recall what is going on. A quantity of type ρ on a curve γ is given by the equivariant map $\Phi : \pi^{-1}(\gamma) \to (V, \rho)$; this provides the pairs $(p, \Phi(p)) \in P \times V$, which may be regarded as the pairs ("a frame," "components with respect to the frame"). Due to the equivariance property it is enough to define Φ at a single point in each fiber over γ and we get the quantity of type ρ on γ. In order to make this quantity *autoparallel*, we have (by definition) to proceed as follows:

1. first we horizontally lift the curve γ on the base M to the total space P (the lift being γ^h); this procedure results in the "autoparallel frame field" on γ
2. the parallel transported quantity has the pairs $(\gamma^h(t), \Phi(\gamma^h(t)) = \text{constant})$.

So one combines *constant* "components" with the *horizontal* lift. All the non-trivial work is performed on P and it consists in the computation of the horizontal lift γ^h.

However, *the same* object may also be described by *other pairs*, provided that they happen to be equivalent (in the sense of $(p, v) \sim (pg, \rho(g^{-1})v)$) with those we mentioned above. What other pair might then represent a more convenient choice than $(\gamma^h(t), v \equiv \text{ constant})$? Recall that when we studied the parallel transport of vectors in Chapter 15, we never wrote the (parallel transport) equations for components with respect to an autoparallel (i.e. highly specific) field on the curve (this was first done as late as in problem (19.6.4)), but rather with respect to a *general* frame field on the curve (at least a general *coordinate* frame field at the beginning). From the point of view of the present section this means that we should not lift the curve γ to P horizontally, but rather by making use of an *arbitrary section*, $\gamma(t) \mapsto \sigma(\gamma(t))$. So we are to solve the following technical problem: what representatives $v \in V$ are to be combined with the points of the curve lifted with the help of an arbitrary *section*, so that the resulting pairs were equivalent to the pairs $(\gamma^h(t), v \equiv \text{constant})$, for which the curve is lifted horizontally? The representative $v(t)$ already necessarily depends on t, since constant v only combines with the horizontal lift. Our goal is to derive a *differential equation* for $v(t)$.

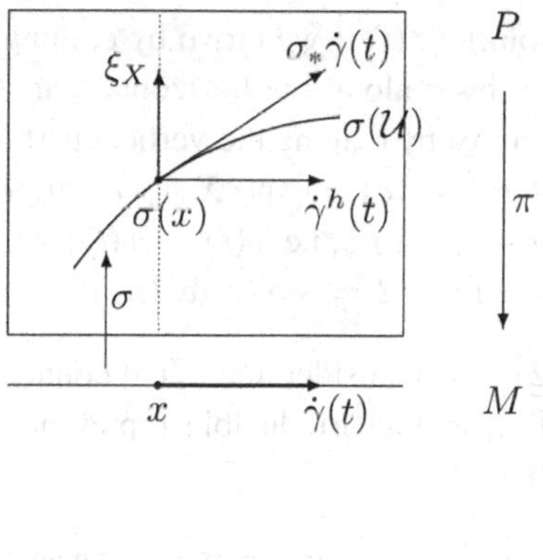

21.3.1 Let $\gamma(t)$ be a curve on the base, $\dot\gamma(t)$ the tangent vector at the point $\gamma(t)$, $\dot\gamma^h(t)$ the horizontal lift and $\sigma_*\dot\gamma(t)$ the lift with the help of the section $\sigma : \mathcal{U} \to P$. Check that

(i) the *horizontal* part of the vector $\sigma_*\dot\gamma(t)$ coincides with the horizontal lift $\dot\gamma^h(t)$

(ii) the *vertical* part of the vector $\sigma_*\dot\gamma(t)$ has the form ξ_X for $X = \langle A, \dot\gamma\rangle$, where $A \equiv \sigma^*\omega$ is the *gauge potential* in the gauge σ; putting it all together the decomposition is

$$\begin{aligned}\sigma_*\dot\gamma(t) &= \text{hor } \sigma_*\dot\gamma(t) + \text{ver } \sigma_*\dot\gamma(t)\\ &= \dot\gamma^h(t) + \xi_X(p) \qquad\qquad p = \sigma(\gamma(t))\\ &\qquad\qquad\qquad\qquad\quad X = \langle A, \dot\gamma\rangle\end{aligned}$$

Hint: (i) both vectors are horizontal and both project to $\dot\gamma(t)$; (ii) according to (20.2.4) we may write $\text{ver}_p\sigma_*\dot\gamma(t) = (\Psi_p \circ \omega_p)\sigma_*\dot\gamma(t) = \Psi_p\langle\sigma^*\omega, \dot\gamma\rangle = \Psi_p\langle A, \dot\gamma\rangle = \Psi_p X = \xi_X(p)$, where $p \equiv \sigma(\gamma(t))$. □

• By this, as we will see in a moment, a substantial part of the derivation of the parallel transport equations is in fact done.

21.3.2 Let $\Phi : \pi^{-1}(\gamma) \to (V, \rho)$ be a function of type ρ, which is defined over a curve γ, σ a local section (a choice of the gauge), $\phi = \sigma^*\Phi \equiv \Phi \circ \sigma$ the pull-back of Φ to the base, $A = \sigma^*\omega$ the gauge potential in the gauge σ, $\mathcal{A} = \rho'(A)$ the represented gauge potential, $v(t) = \phi(\gamma(t))$ the "components" (in the gauge σ) of the quantity of type ρ, defined (by the function Φ) on the curve with respect to the "frame field" $\sigma(\gamma(t))$ on the curve. Check that

(i) the quantity of type ρ is *autoparallel* on the curve if and only if the vector $v(t)$ satisfies the differential *equation of parallel transport*

$$\dot{v} + \langle \mathcal{A}, \dot{\gamma} \rangle v = 0$$

(ii) if $v(t) = v^a(t) E_a$ and $\mathcal{A}^a_b \equiv A^i \rho^a_{bi}$ are the components of the represented gauge potential (see (21.2.4)), the parallel transport equation may also be written in components as the system of ordinary linear differential equations with non-constant coefficients

$$\dot{v}^a(t) = S^a_b(t) v^b(t) \qquad \text{where} \quad S^a_b(t) := -\left\langle \mathcal{A}^a_b, \dot{\gamma} \right\rangle = -\dot{x}^\mu(t) \mathcal{A}^a_{b\mu}(x(t))$$

(iii) the quantities v and S, which arise in the parallel transport equation $\dot{v} = Sv$, change under a gauge transformation as follows:

$$v \mapsto \hat{v} = B^{-1} v \qquad S \mapsto \hat{S} = B^{-1} S B + B^{-1} \dot{B}$$

and the equation *is invariant* with respect to the changes (as it behoves, since the equation is expected to describe *objective* parallel transport).

Solution: (i) if we move by ϵ along the curve $\sigma(\gamma(t))$, we proceed along the vector $\sigma_* \dot{\gamma}$, i.e. by ϵ along the horizontal part (here the "components" do not change) and then (or before) by ϵ along the vertical part ξ_X; this is, however, the right action on P by the group element $g(\epsilon) = \exp(\epsilon X)$ $(p \mapsto pg(\epsilon))$, which causes the compensation $v \mapsto \rho(g(\epsilon)^{-1}) v = (1 - \epsilon \rho'(X)) v$, i.e. $v(t) \mapsto v(t + \epsilon) = v(t) - \epsilon \rho'(X) v(t)$. The differential equation for $v(t)$ thus reads $\dot{v} = -\rho'(X) v$; (ii) $\dot{v}^a + \langle \mathcal{A}^a_b, \dot{\gamma} \rangle v^b = 0$. □

21.3.3 Consider the $U(1)$-connection in $\pi : P \to M$ and choose as (V, ρ) the one-dimensional irreducible representation $(\mathbb{C}, \rho_n)$, where $\rho_n : e^{i\alpha} \mapsto e^{in\alpha}$ (12.2.10). Check that

(i) the parallel transport equation now reads

$$\dot{z} = Sz \qquad S(\lambda) \equiv -in \langle A, \dot{\gamma} \rangle$$

$$\text{i.e. in more detail} \quad \dot{z}(\lambda) = -in \dot{x}^\mu(\lambda) A_\mu(x(\lambda)) z(\lambda)$$

(ii) the solution may be written in the form of the integral

$$z(\lambda) = z(0) e^{-in \int_0^\lambda A_\mu(x(s)) \dot{x}^\mu(s)\, ds} \equiv z(0) e^{-in \int_\gamma A_\mu\, dx^\mu} \equiv z(0) e^{-in \int_\gamma A}$$

(iii) if A is regarded as the *electromagnetic* 4-potential, $\gamma(s)$ as the world-line of a charged particle and n as its electric charge, then the parallel transport from $\gamma(0)$ to $\gamma(1)$ may also be written as

$$z(1) = z(0)e^{-iS_{\text{int}}} \qquad S_{\text{int}}[\gamma; A] := n \int_\gamma A \equiv n \int_\gamma A_\mu \, dx^\mu$$

According to (16.3.9) $S_{\text{int}}[\gamma; A]$ represents the interaction part of the action of a charged particle moving along the world-line $\gamma(s)$ in the electromagnetic field given by the potential A.

Hint: (i) if a basis in $u(1)$ is $E_1 = i$, then $\omega^j E_j \mapsto i\omega$, $A^j E_j \mapsto iA$ (A being already an ordinary 1-form on M), $\mathcal{A} \equiv \rho'_n(A) = inA$, so that $\langle \mathcal{A}, \dot\gamma \rangle = in\langle A, \dot\gamma \rangle = in\dot x^\mu A_\mu$; (ii) $\dot z = Sz \Rightarrow z^{-1}dz \equiv d\ln z = S(\lambda)\, d\lambda \Rightarrow \ln z(\lambda) - \ln z(0) = \int_0^\lambda S(s)\, ds$. □

21.3.4 Consider the "ordinary" *linear* connection (i.e. the connection in $\pi : LM \to M$) and the tensor representation of type $\binom{r}{s}$ of the group $GL(n, \mathbb{R})$. Check that the concise equation from (21.3.1)

$$\dot v + \langle \mathcal{A}, \dot\gamma \rangle v = 0 \quad \mathcal{A} \equiv \rho'(A)$$

may then be "unzipped" to the well-known form (15.2.12) from Chapter 15

$$\dot v^{i\ldots j}_{k\ldots l} + \dot x^m \left(\Gamma^i_{nm} v^{n\ldots j}_{k\ldots l} + \cdots - \Gamma^n_{lm} v^{i\ldots j}_{k\ldots n} \right) = 0$$

Hint: first realize (15.6.1) that $\dot x^m \Gamma^i_{nm} = \langle \hat\omega^i_n, \dot\gamma \rangle \equiv \langle A^i_n, \dot\gamma \rangle$. For the representation under consideration we have $(\rho(B)v)^{i\ldots j}_{k\ldots l} := B^i_m \ldots (B^{-1})^n_l v^{m\ldots}_{\ldots n}$, so that for the derived one we get $(\rho'(C)v)^{i\ldots j}_{k\ldots l} = C^i_m v^{m\ldots j}_{k\ldots l} + \cdots - C^m_l v^{i\ldots j}_{k\ldots m}$ (12.4.14). Then

$$\begin{aligned}
0 &= \dot v^{i\ldots j}_{k\ldots l} + \dot x^m \left(\Gamma^i_{nm} v^{n\ldots j}_{k\ldots l} + \cdots - \Gamma^n_{lm} v^{i\ldots j}_{k\ldots n} \right) \\
&= \dot v^{i\ldots j}_{k\ldots l} + \langle A^i_n, \dot\gamma \rangle v^{n\ldots j}_{k\ldots l} + \cdots - \langle A^n_l, \dot\gamma \rangle v^{i\ldots j}_{k\ldots n} \\
&= \dot v^{i\ldots j}_{k\ldots l} + (\langle \mathcal{A}, \dot\gamma \rangle v)^{i\ldots j}_{k\ldots l} \\
&\equiv (\dot v + \langle \mathcal{A}, \dot\gamma \rangle v)^{i\ldots j}_{k\ldots l}
\end{aligned}$$

□

• As we learned in Section 15.2, the *deviation* of a given field defined on a curve from being autoparallel is measured by the *absolute* derivative along the curve (i.e. the covariant derivative along the tangent vector to the curve, see (15.2.4) and (15.2.11)). If $v(t)$ is a field of type ρ in a gauge σ, defined on a curve γ on M (in the sense of problem (21.3.2)), then the absolute derivative of the field at the point $\gamma(t)$ measures the difference between the result of the backward parallel transport from the point $\gamma(t + \epsilon)$ to $\gamma(t)$ and the actual value in $\gamma(t)$

$$\frac{Dv(t)}{Dt} := \lim_{\epsilon \to 0} \frac{v^\parallel_\epsilon(t) - v(t)}{\epsilon}$$

21.3.5 Check that the explicit formula for the absolute derivative is

$$\frac{Dv(t)}{Dt} = \frac{dv(t)}{dt} + \langle \mathcal{A}, \dot\gamma \rangle v \equiv \dot v + \langle \mathcal{A}, \dot\gamma \rangle v$$

so that the parallel transport equation just expresses the condition for *vanishing* of the absolute derivative

$$\frac{Dv^{\|}(t)}{Dt} = 0$$

(as it should – it says that the values of the autoparallel field $v^{\|}(t)$ on the curve γ may be regarded as (only) *parallel transported* from some (arbitrary) point; the difference between the "transported" and the "original" value thus always vanishes for such a field).

Hint: if the value of v at the point $\gamma(t+\epsilon)$ is $v(t+\epsilon)$, then, according to (21.3.2), the backward parallel transport to $\gamma(t)$ gives (within first-order accuracy in ϵ) $v^{\|}(t) \equiv v^{\|}((t+\epsilon)-\epsilon) = v(t+\epsilon) - \epsilon\dot{v}^{\|}(t+\epsilon) = v(t+\epsilon) + \epsilon\langle\mathcal{A}, \dot{\gamma}\rangle v = v(t) + \epsilon(\dot{v} + \langle\mathcal{A}, \dot{\gamma}\rangle v)$ so that

$$\lim_{\epsilon\to 0} \frac{v^{\|}_{\epsilon}(t) - v(t)}{\epsilon} = \dot{v} + \langle\mathcal{A}, \dot{\gamma}\rangle v$$

□

21.3.6 Consider a quantity of type ρ which is also defined *in a neighborhood* of the curve γ, in some (open) *domain* $\mathcal{O}$ (rather than only on the curve itself). In a gauge σ we may then write $v^a(x)E_a \equiv v(x) \equiv \phi(x) := \Phi(\sigma(x)) \equiv (\sigma^*\Phi)(x)$ and we may also introduce its *exterior covariant* derivative *on the base* (21.2.4)

$$\mathcal{D}v = dv + \mathcal{A}v \quad \text{i.e.} \quad \mathcal{D}v^a = dv^a + \mathcal{A}^a_b v^b \qquad \mathcal{D}\circ\sigma^* := \sigma^*\circ D$$

(it is a 1-form on $\mathcal{O}\subset M$ with values in (V,ρ), resulting from pull-back of the horizontal 1-form $D\Phi$ from P to M by the section σ). Check that the absolute derivative along the curve may then be written in the following useful form:

$$\frac{Dv(t)}{Dt} \equiv \dot{v} + \langle\mathcal{A}, \dot{\gamma}\rangle v = i_{\dot{\gamma}}\mathcal{D}v$$

This form naturally leads to the definition of the *covariant* derivative of the field $v(x)$ along a *vector field* W on $\mathcal{O}\subset M$

$$\nabla_W v := i_W\mathcal{D}v \equiv Wv + \langle\mathcal{A}, W\rangle v$$

which satisfies (it might be useful to compare it with (15.2.11))

$$\frac{Dv(t)}{Dt} = \nabla_{\dot{\gamma}}v(t)$$

Hint: $\dot{v} + \langle\mathcal{A}, \dot{\gamma}\rangle v = i_{\dot{\gamma}}(dv + \mathcal{A}v) = i_{\dot{\gamma}}(d\sigma^*\Phi + \rho'(\sigma^*\omega)\sigma^*\Phi) = i_{\dot{\gamma}}\sigma^*(d\Phi + \rho'(\omega)\Phi) = i_{\dot{\gamma}}\sigma^*D\Phi = i_{\dot{\gamma}}\mathcal{D}\sigma^*\Phi \equiv i_{\dot{\gamma}}\mathcal{D}v.$ □

21.3.7 Introduce the following standard coordinate notation related to the covariant derivatives: the 1-form $\mathcal{D}v^a$ may be expressed in terms of (say) the coordinate frame field dx^μ

on $\mathcal{O} \subset M$; we will denote the corresponding components as[446]

$$\mathcal{D}v^a =: (\mathcal{D}_\mu v^a)\, dx^\mu \equiv v^a_{\ ;\mu}\, dx^\mu$$

Check that then

(i) a more detailed expression of $\mathcal{D}_\mu v^a$ is

$$\mathcal{D}_\mu v^a = \partial_\mu v^a + \mathcal{A}^a_{b\mu} v^b \quad \text{i.e.} \quad v^a_{\ ;\mu} = v^a_{\ ,\mu} + \mathcal{A}^a_{b\mu} v^b \qquad \mathcal{A}^a_b =: \mathcal{A}^a_{b\mu} dx^\mu$$

(ii) the parallel transport equation may also be written as

$$\dot{x}^\mu \mathcal{D}_\mu v^a \equiv \dot{x}^\mu v^a_{\ ;\mu} = 0$$

(iii) the covariant derivative may also be expressed as

$$(\nabla_W v)^a = W^\mu \mathcal{D}_\mu v^a \equiv W^\mu v^a_{\ ;\mu}$$

Hint: see (21.3.6). □

• Finally, let us look at the generalization of the parallel transport equation in a gauge σ for the case when the representation $(V, \rho(g))$ is replaced by a *general action* on a manifold (N, r_g) (so we do not require the *linearity* of the action). A *quantity of type* ρ is still regarded as an equivariant map

$$\Phi : P \to N \qquad \Phi \circ R_g = r_g \circ \Phi$$

The quantity is said to be *parallel transported* if the parallel transported "frame field," i.e. the horizontal lift $\gamma^h(t)$, is (again) combined with *constant* "components" (the image $\Phi(\gamma^h(t))$; recall that the "components" *are not* even the elements of a linear space, now, but rather only the points of a general manifold endowed with an action of a group).

21.3.8* Let $\pi : P \to M$ be a principal bundle with connection, $\Phi : \pi^{-1}(\gamma) \to (N, r(g))$ a quantity of type $r(g)$ defined on a curve γ, i.e. the equivariant map

$$\Phi : \pi^{-1}(\gamma) \to N \qquad \Phi \circ R_g = r_g \circ \Phi$$

σ a local section (a choice of the gauge), $A = \sigma^*\omega$ the gauge potential in the gauge σ, $n(t) = (\Phi \circ \sigma)(\gamma(t))$ the "components" (in the gauge σ) of the quantity of type $r(g)$ which is defined (in terms of the function Φ) on the curve, with respect to the "frame field" $\sigma(\gamma(t))$ on the curve (compare with problem (21.3.2)). Check that

(i) the quantity of type $r(g)$ is *autoparallel* on the curve if and only if the curve $n(t)$ on the manifold N satisfies the differential *equation of parallel transport*

$$\dot{n}(t) = \hat{\xi}_{X(t)} \qquad X(t) \equiv \langle A, \dot{\gamma} \rangle \in \mathcal{G}$$

i.e. if it is an "integral curve" of the *fundamental field* $\hat{\xi}_X$ of the action $r(g)$ on N (the meaning of the quotation marks is that the field actually *varies* with "time" t, since X depends on t; so we are to find the integral curve for a vector field which *depends on time*)

[446] In the same way a general frame field e_a may be used. If a basis in V is denoted by E_α (the "frame" indices a should not be confused with the indices α in V), we define $\mathcal{D}v^\alpha =: (\mathcal{D}_a v^\alpha)e^a = v^\alpha_{\ ,a} e^a$ and it then comes out as $\mathcal{D}_a v^\alpha = e_a v^\alpha + \mathcal{A}^\alpha_{\beta a} v^\beta$.

(ii) if the action eventually grows wise and again becomes a representation (i.e. becomes linear), the more general equation derived here reduces (as is proper) to the linear equation from (21.3.2).

Hint: (i) we can proceed in full analogy with (21.3.2), but the required compensation is realized by means of the action rather than a representation, i.e. we move by ϵ *along the generator*, being just $\hat{\xi}_X$; (ii) if $N = V$, then $n(t) = v(t) = v^a(t)E_a$. According to (13.4.8) the fundamental field reads $\hat{\xi}_X = -X^i v^b \rho^a_{bi} \partial_a$, so that the equation $\dot{v} = \hat{\xi}_X$ becomes

$$\dot{v}^a = (\hat{\xi}_X)^a = -X^i \rho^a_{bi} v^b = -\langle A^i \rho^a_{bi}, \dot{\gamma} \rangle v^b = -\langle \mathcal{A}^a_b, \dot{\gamma} \rangle v^b$$

in agreement with (21.3.2). □

21.4 Bundle $P \times_\rho V$ associated to a principal bundle $\pi : P \to M$

In Section 20.3 we introduced the quantities of type ρ as equivariant functions $\Phi : P \to (V, \rho)$. For example, an ordinary vector field on M (see Section 19.6) in this formalism "becomes" an equivariant function on LM with $(V, \rho) = (\mathbb{R}^n, \text{id})$ and a covector field has $(V, \rho) = (\mathbb{R}^n, \check{\text{id}})$ (where id denotes the "identity" representation $A \mapsto A$ and $\check{\text{id}}$ is the contragredient one). Recall, however, that we learned in (17.2.6) that vector and covector fields on M may also be described differently, as *sections* of appropriate vector bundles over M (a section of TM encodes a vector field and a section of T^*M encodes a covector field). This is not an accident. It turns out that one may associate a vector bundle[447] over the base M with a quantity of type ρ such that *sections* of the bundle may, in turn, be canonically identified with the equivariant functions, i.e. with the quantities of type ρ as treated up to now. The bundle is called an *associated bundle* and it is denoted as $P \times_\rho V$. In this section the total space of the bundle will be denoted by E and the projection by $\hat{\pi}$.

21.4.1 Let $\pi : P \to M$ be a principal G-bundle and (V, ρ) a representation of G. The points of the total space E of the associated bundle $\hat{\pi} : E \to M$ are the *orbits* of the action of the group G on the Cartesian product $P \times V$, i.e. the equivalence classes $[(p, v)] \sim [(pg, g^{-1}v)]$. Check that the bundle is indeed well defined, i.e. be sure to understand that

(i) $P \times V$ is a right G-space with respect to the action $(p, v) \mapsto (pg, g^{-1}v)$
(ii) there is a natural projection

$$\hat{\pi} : E \to M \qquad [(p, v)] \mapsto \pi(p) \equiv x$$

(iii) there is a natural linear structure in each fiber $\hat{\pi}^{-1}(x)$ given by

$$[(p, v)] + \lambda[(p, w)] := [(p, v + \lambda w)]$$

(p is to be the *same* (although arbitrary); by appropriate $p \mapsto pg$ the representatives with the same p may be always chosen and *then* the linear combination of *the second* components of the pairs should be performed)

[447] We will not use it *explicitly* in what follows, but it is worth being familiar with the concept (even though only a little bit) since it is fairly frequent in the geometrical literature.

(iv) each fiber is a linear space isomorphic to V; so the procedure of the (Cartesian) multiplication by V and subsequent factorization over the group action effectively *replaces* the fiber of the principal bundle P by a copy of the vector space V, giving rise to a *vector bundle* with fibers isomorphic to V.

Hint: (iii) check the independence on the choice of p; (iv) a (non-canonical) isomorphism is $v \leftrightarrow [(p, v)]$. □

• So far the quantities of type ρ were described by equivariant functions $\Phi : P \to V$. It turns out that such *functions* may be canonically identified with *sections* of the vector bundle introduced above.

21.4.2 Check that there is a one-to-one correspondence between equivariant functions $\Phi : P \to V$ and sections of the associated bundle $\hat{\pi} : E \to M$.

Hint: for $\Phi : P \to V$ the section $\sigma : M \to E$ is given by $\sigma(x) := [(p, \Phi(p))]$ (it does not depend on the choice of p over x); if a section σ is available and $\sigma(x) = [(p, v)]$, we set $\Phi(p) = v$. □

• At first sight the construction of the associated bundle looks a bit too abstract. Then we might be glad to realize that actually we are already familiar with some particular examples of such objects (albeit their method of introduction was different): the tangent and cotangent bundles turn out to be associated with the frame bundle.

21.4.3 Be sure to understand that TM and T^*M are indeed associated bundles with the principal bundle LM; more generally the *tensor bundle* $T^r_s M$ of type (r, s) (the fiber over x consists of all tensors of type (r, s) in the point $x \in M$; in particular, $T^1_0 M = TM$, $T^0_1 = T^*M$) is associated with LM.

Hint: the description of vectors and tensors in Section 19.6 (before (19.6.4)). □

• A short comment might be in order. If contemplated over sufficiently small patches $\mathcal{O}$ on M, both the principal and the associated bundles are trivial ($\mathcal{O} \times G$ and $\mathcal{O} \times V$ respectively). So the essential information consists in specifying the way in which the trivial pieces are *glued together* to form the whole P and E. Now it turns out that the principal bundle and *all* the bundles which are associated with it are *glued equally* (in a sense we will not describe here). This means that if we know the global topology of (the total space of) a principal bundle, the global topology of (the total spaces of) all the associated bundles is already fixed (in particular, if the principal bundle happens to be trivial, all the associated bundles are necessarily trivial as well).

21.5 Gauge invariant action and the equations of motion

In Section 21.1 we already encountered the typical actions of the theory of gauge fields in the traditional "physical" language. Here we will discuss the general structure of the actions, making use of the machinery of the connection theory. This also greatly simplifies the derivation of the equations of motion.

Recall that a natural scalar product in the space of differential forms on a Riemannian manifold was introduced in problem (8.3.1) by the formula

$$\langle \alpha, \beta \rangle := \int_{\mathcal{U}} \alpha \wedge *_g \beta$$

Later, in (16.3.2) and (16.3.6), we were able to write down the action functionals for the electromagnetic and the scalar field in terms of the scalar product in the form of

$$S[A] = -\frac{1}{2}\langle dA, dA \rangle - \langle A, j \rangle \qquad S[\phi] = \frac{1}{2}\langle d\phi, d\phi \rangle - \frac{m^2}{2}\langle \phi, \phi \rangle$$

This particular form proved to be remarkably convenient for the manipulations that are necessary in order to derive the corresponding equations of motion

$$\delta F = -j \qquad (-\delta d + m^2)\phi \equiv (\Box + m^2)\phi = 0$$

and it stems from the simple fact that the operator of the exterior derivative d converts to the *codifferential* (8.3.2) when "reshuffled" to the other side of the scalar product, $d^+ \equiv \delta$; this is how the codifferential δ arises naturally in the equations of motion.

In the theory of gauge fields we work with forms with values in various *linear spaces*. We already mentioned in (8.3.1) that the generalization of the scalar product to vector-valued forms is straightforward and it reads

$$\langle \alpha, \beta \rangle := \int_{\mathcal{U}} h_{ab} \alpha^a \wedge *\beta^b$$

In the action integrals to be studied we will encounter the *covariant* exterior derivative $\mathcal{D}$ in the scalar product and the issue of the "reshuffling" of such $\mathcal{D}$ to the other side of the scalar product arises naturally; put another way, we are interested in the explicit form of the conjugated operator $\mathcal{D}^+$. If we succeeded in solving this problem, the equations of motion might be derived from the action integral as easily as we did it in the case of "ordinary" ($\mathbb{R}$-valued) forms.

So let us concentrate for a while on the structure of (locally) gauge invariant action integrals. The actions are usually expressed as integrals of forms defined *on the base M* (space-time), i.e. the relevant *fields* depend on the choice of the gauge; i.e. on the section σ. The *action* itself should be locally *gauge invariant*, however, i.e. it should *not depend* on the choice of the section σ. Two technical tools are to be used in order to reach this goal: the exterior *covariant* derivative $\mathcal{D}$ (rather than d) and an *invariant* scalar product in the vector space, where the forms take their values.

21.5.1 Consider $\alpha, \beta \in \tilde{\Omega}^p(P, \rho)$, i.e. horizontal p-forms of type ρ. Let e_a be a basis in (V, ρ) (so that $\alpha = \alpha^a e_a$, $\beta = \beta^a e_a$), $h \leftrightarrow h_{ab}$ the ρ-invariant scalar product in (V, ρ) and $\sigma : \mathcal{U} \to P$ a local section (a choice of the gauge). Denote by $\hat{\alpha}$, $\hat{\beta}$ the pull-backs of the forms α, β to $\mathcal{U}$ with respect to the section σ and by $\hat{\alpha}', \hat{\beta}'$ the pull-back of the forms α, β to $\mathcal{U}$ with respect to another section σ', where $\sigma'(x) = \sigma(x)S(x)$. Check that the expression

$$\langle \hat{\alpha}, \hat{\beta} \rangle_h := \int_{\mathcal{U}} h_{ab}\hat{\alpha}^a \wedge *\hat{\beta}^b \equiv h_{ab}\langle \hat{\alpha}^a, \hat{\beta}^b \rangle$$

is *locally gauge invariant*, i.e. it does not depend on the section σ

$$\langle \hat{\alpha}, \hat{\beta} \rangle_h = \langle \hat{\alpha}', \hat{\beta}' \rangle_h \equiv \langle \alpha, \beta \rangle_h$$

(consequently it may indeed be written without the hats, as is done on the right).

Hint: let $\hat{\alpha}' \equiv \sigma'^*\alpha$, $\hat{\beta}' \equiv \sigma'^*\beta$; then according to (21.2.3) $\hat{\alpha}'^a = \rho(S^{-1}(x))^a_c\hat{\alpha}^c$, $\hat{\beta}'^b = \rho(S^{-1}(x))^b_d\hat{\beta}^d$, so that

$$h_{ab}\hat{\alpha}'^a \wedge *\hat{\beta}'^b = \underbrace{h_{ab}\rho(S^{-1}(x))^a_c\rho(S^{-1}(x))^b_d}_{h_{cd}}\hat{\alpha}^c \wedge *\hat{\beta}^d = h_{ab}\hat{\alpha}^a \wedge *\hat{\beta}^b$$

□

It is worth pointing out the essential elements of the construction. They are

1. the simple transformation law

$$\hat{\alpha}'^a = \rho(S^{-1}(x))^a_c\hat{\alpha}^c \qquad \hat{\beta}'^b = \rho(S^{-1}(x))^b_d\hat{\beta}^d$$

this is due to the *horizontality* (and type ρ) of the forms α, β on P

2. the ρ-*invariance* of the scalar product h in V; exactly this property ensures the "disarmament" of the operators $\rho(S(x))$ (being potentially dangerous due to their x dependence).

The first of the two facts provides the key for understanding of how *derivatives* of fields should enter any locally gauge invariant action: it is not enough to use the *exterior* derivative alone, but one should also take care of the *horizontality* of the result of applying d, i.e. one should make[448] the correction $d \mapsto \text{hor}\, d \equiv D$. Actually exactly this procedure, the replacing of the exterior derivative d by the exterior *covariant* derivative $\mathcal{D}$

$$d \mapsto \mathcal{D} \equiv \text{“}d + \mathcal{A}\text{”} = \textit{minimal coupling} \text{ prescription}$$

(or introducing the *minimal interaction*) provides the local gauge invariance of the action integral. Notice that this rule produces the terms containing *concrete products* of the field $\mathcal{A}$ with the field being differentiated in the action, thus specifying the *interactions*. This is the typical feature of gauge theories: the requirement of local gauge invariance severely constrains possible individual terms in the action.

Now that we have constructed the locally gauge invariant scalar product let us keep in mind our goal of constructing locally gauge invariant actions. Concentrate first on the structure of the action of the gauge field $\mathcal{A}$ itself. If the field is to possess its own dynamics, derivatives are needed. As we have just learned, not $d\mathcal{A}$ but rather $\mathcal{D}\mathcal{A} \equiv \mathcal{F}$; we recognize

[448] The correction is already to be done on forms "upstairs" *on* P, prior to the pull-back to the base. So one should pull back forms containing $D\alpha$ rather than $d\alpha$, resulting in $\mathcal{D}\hat{\alpha}$ rather than $d\hat{\alpha}$ "downstairs" on M.

the good old *gauge field strength* (the pull-back of the curvature form to the base). This 2-form has values in the Lie algebra $\mathcal{G}$, so we need an invariant scalar product in $\mathcal{G}$. Recall we already encountered such a product in (12.3.8), namely the Killing–Cartan metric k_{ij} in $\mathcal{G}$. So a good candidate for the action might be[449] an appropriate multiple of the integral

$$\langle \mathcal{F}, \mathcal{F} \rangle_k \equiv \langle \mathcal{DA}, \mathcal{DA} \rangle_k \equiv k_{ij} \langle \mathcal{F}^i, \mathcal{F}^j \rangle \equiv \int_{\mathcal{U}} k_{ij} \mathcal{F}^i \wedge *_g \mathcal{F}^j$$

We take the action of the gauge field itself in the (standard) form

$$S[\mathcal{A}] = -\frac{1}{2} \langle \mathcal{F}, \mathcal{F} \rangle_k$$

In order to derive the corresponding equations of motion we can proceed in full analogy with (16.3.2): consider the increment of $S[\mathcal{A}]$ resulting from the change $\mathcal{A} \mapsto \mathcal{A} + \epsilon a$, the quantity $\mathcal{A} + \epsilon a$ being of the "same type" as $\mathcal{A}$, so that a itself should be a 1-form with values in $\mathcal{G}$. We see that the linear part of the increment turns out to be $2\epsilon \langle \mathcal{D}a, \mathcal{DA} \rangle_k$ and if we want to use the fact that a is arbitrary, we need to get rid of $\mathcal{D}$ on the left by its "reshuffling" to the right, where it converts, however, to the *adjoint* operator $\mathcal{D}^+$. It should be very useful to understand the technical side of this "reshuffling" in a general setting. Let us therefore look at how the operator $\mathcal{D}^+$ with respect to the scalar product $\langle \alpha, \beta \rangle_h$ of forms with values in (V, ρ, h) may be written explicitly.

21.5.2 Consider $\alpha \in \bar{\Omega}^p(P, \rho)$, $\beta \in \bar{\Omega}^{p+1}(P, \rho)$ and h a ρ-invariant scalar product in (V, ρ) and denote

$$\rho_{abi} := h_{ac} \rho^c_{bi} \qquad \mathcal{A}_{ab} := \rho_{abi} \mathcal{A}^i \equiv h_{ac} \mathcal{A}^c_b$$

Check that

(i) the matrix $\mathcal{A}_{ab}$ is antisymmetric

$$\mathcal{A}_{ab} = -\mathcal{A}_{ba}$$

(ii) the following identity holds:

$$h_{ab} (\mathcal{D}\hat{\alpha}^a) \wedge *\hat{\beta}^b = h_{ab} \hat{\alpha}^a \wedge *(*^{-1} \mathcal{D} * \hat{\eta} \hat{\beta}^b) + d(h_{ab} \hat{\alpha}^a \wedge *\hat{\beta}^b)$$

(iii) for the scalar product of forms we then get

$$\langle \mathcal{D}\hat{\alpha}, \hat{\beta} \rangle_h = \langle \hat{\alpha}, \mathcal{D}^+ \hat{\beta} \rangle_h + \int_{\partial \mathcal{U}} h_{ab} \hat{\alpha}^a \wedge *\hat{\beta}^b \qquad \mathcal{D}^+ := *^{-1} \mathcal{D} * \hat{\eta}$$

(it is instructive to compare it with (8.3.2)). The operator $\mathcal{D}^+$ is called the *covariant codifferential*

[449] The scalar product k is often realized in terms of the *trace* of matrices. If we work with the *represented* potential $\mathcal{A} = \rho'(A) \equiv A^i \rho'(E_i)$ (entering the covariant derivative of the matter fields $\mathcal{D} = d\phi + \mathcal{A}\phi$), the scalar product is of type (12.3.7), if we speak directly about $A = A^i E_i$, the scalar product also usually reduces to the trace since the Killing metric of *matrix* groups is realized by the trace (12.3.11). This sheds light upon the structure of (21.1.5).

(iv) if, for any reason whatsoever,[450] the integral over the boundary $\partial\mathcal{U}$ of the domain $\mathcal{U}$ happens to vanish, then

$$\langle \mathcal{D}\hat{\alpha}, \hat{\beta}\rangle_h = \langle \hat{\alpha}, \mathcal{D}^+\hat{\beta}\rangle_h$$

so that $\mathcal{D}^+$ becomes the *adjoint of* $\mathcal{D}$ (in the sense of $\langle\cdot,\cdot\rangle_h$).

Hint: (i) see (12.1.10); (ii) the straightforward calculation goes as follows: $\mathcal{D}\hat{\alpha}^a = d\hat{\alpha}^a + \mathcal{A}^a_b \wedge \hat{\alpha}^b \Rightarrow$

$$\begin{aligned}
h_{ab}(\mathcal{D}\hat{\alpha}^a)\wedge *\hat{\beta}^b &= h_{ab}d\hat{\alpha}^a \wedge *\hat{\beta}^b + \mathcal{A}_{bc}\wedge\hat{\alpha}^c\wedge *\hat{\beta}^b \\
&= d(h_{ab}\hat{\alpha}^a\wedge *\hat{\beta}^b) - h_{ab}(\hat{\eta}\hat{\alpha}^a)\wedge d*\hat{\beta}^b - (\hat{\eta}\hat{\alpha}^c)\wedge\mathcal{A}_{cb}\wedge *\hat{\beta}^b \\
&= h_{ab}\hat{\alpha}^a\wedge d*(\hat{\eta}\hat{\beta}^b) + \hat{\alpha}^c\wedge\mathcal{A}_{cb}\wedge *(\hat{\eta}\hat{\beta}^b) + d(h_{ab}\hat{\alpha}^a\wedge *\hat{\beta}^b) \\
&= h_{ab}\hat{\alpha}^a\wedge *(*^{-1}d*\hat{\eta}\hat{\beta}^b) + h_{ab}\hat{\alpha}^a\wedge\mathcal{A}^b_c\wedge *\hat{\eta}\hat{\beta}^c + d(h_{ab}\hat{\alpha}^a\wedge *\hat{\beta}^b) \\
&= h_{ab}\hat{\alpha}^a\wedge *(*^{-1}\mathcal{D}*\hat{\eta}\hat{\beta}^b) + d(h_{ab}\hat{\alpha}^a\wedge *\hat{\beta}^b)
\end{aligned}$$

Another useful (simpler as well as more instructive) way: the expression $h_{ab}\hat{\alpha}^a\wedge *\hat{\beta}^b$ is a locally gauge invariant form and so we may put $\mathcal{D} = d$ on this form ($\rho' = 0$ for invariant quantities); it has the structure $f^a \wedge g_a$, where f, g themselves are forms of type ρ and $\check{\rho}$ (the latter being contragredient to ρ). Then $d(f^a\wedge g_a) = \mathcal{D}(f^a\wedge g_a) = (\mathcal{D}f^a)\wedge g_a + (\hat{\eta}f^a)\wedge(\mathcal{D}g_a)$; this (plus the "trick" $\hat{1} = *^{-1}*$) already yields directly the required identity; thus it is enough to realize that *also capital* $\mathcal{D}$ behaves as a derivation (20.3.5). □

• Since we already know how to "reshuffle" $\mathcal{D}$ to the other side of the scalar product, the path towards a simple derivation of the corresponding equations of motion is already free.

21.5.3 Consider the gauge field $\mathcal{A}$ alone. Its (locally gauge invariant) action has the form

$$S[\mathcal{A}] = -\frac{1}{2}\langle\mathcal{F},\mathcal{F}\rangle_k \qquad \mathcal{F}\equiv\mathcal{D}\mathcal{A}$$

Check that

(i) the equations of motion which correspond to (the extremum of) this action are

$$\mathcal{D}^+\mathcal{F} = 0 \qquad \text{i.e.} \quad \mathcal{D}*\mathcal{F} = 0$$

(ii) if contemplated along with the Bianchi identity (note, however, that the latter *does not* result from the extremizing of the action), the complete system of the equations of motion of the gauge field itself reads

$$\mathcal{D}*\mathcal{F} = 0 \qquad \mathcal{D}\mathcal{F} = 0$$

Solution: (i) imagine we perform the variation "upstairs" (prior to the pull-back), i.e. we vary $\omega\mapsto\omega+\epsilon\alpha$; then α should be the *horizontal* 1-form of *type* Ad (then the new whole $\omega+\epsilon\alpha$ has the compulsory properties of a connection form, see (20.2.5) and (20.2.6)), so that $D\alpha = d\alpha + [\omega\wedge\alpha]$; then $\Omega\equiv d\omega+\frac{1}{2}[\omega\wedge\omega]\mapsto\Omega+\epsilon(d\alpha+$

[450] The reasons may be the same as those mentioned in the text following (8.3.2). Now the reason is "the calculus of variations."

$[\omega \wedge \alpha]) \equiv \Omega + \epsilon D\alpha$; after the pull-back ("downstairs") this means that if we perform the variation $\mathcal{A} \mapsto \mathcal{A} + \epsilon a$, then $\mathcal{F} \equiv \mathcal{D}\mathcal{A} \mapsto \mathcal{F} + \epsilon \mathcal{D}a$; with the help of (21.5.2) we then get $\langle \mathcal{F}, \mathcal{F} \rangle_k \mapsto \langle \mathcal{F}, \mathcal{F} \rangle_k + 2\epsilon \langle \mathcal{D}a, \mathcal{F} \rangle_k = \langle \mathcal{F}, \mathcal{F} \rangle_k + 2\epsilon \langle a, \mathcal{D}^+ \mathcal{F} \rangle_k$, so that $S[\mathcal{A} + \epsilon a] = S[\mathcal{A}] - \epsilon \langle a, \mathcal{D}^+ \mathcal{F} \rangle_k \Rightarrow \mathcal{D}^+ \mathcal{F} = 0$. □

• There is a remarkable point concerning the dynamics of the fields under consideration. Namely, the structure of the action reveals that for the non-commutative (i.e. *non-Abelian*) gauge group G the dynamics exhibits the *self-interaction* of the field $\mathcal{A}$ (already making the theory of the "free" gauge field fairly non-trivial). If we write down the action more explicitly, we get

$$S[\mathcal{A}] = -\frac{1}{2} \langle \mathcal{F}, \mathcal{F} \rangle_k = -\frac{1}{2} \langle d\mathcal{A} + \mathcal{A} \wedge \mathcal{A}, d\mathcal{A} + \mathcal{A} \wedge \mathcal{A} \rangle_k$$

from which we see that it contains the products of as many as four $\mathcal{A}$s, which indicates the self-interaction[451] (for an Abelian group, like $U(1)$, the field $\mathcal{F}$ reduces to just $d\mathcal{A}$ and there is no self-interaction in the theory).

We now pass to a slightly more complicated action by adding a *matter field* ϕ of type ρ (the pull-back of an equivariant[452] function $\Phi : P \to (V, \rho)$). For the sake of simplicity, let us begin with the kinetic term alone, yet already modified by the "correction" $d \mapsto \mathcal{D}$.

21.5.4 Consider the action of the coupled system containing the matter field ϕ of type ρ and the gauge field $\mathcal{A}$

$$S[\phi, \mathcal{A}] = -\frac{1}{2} \langle \mathcal{D}\mathcal{A}, \mathcal{D}\mathcal{A} \rangle_k + \frac{1}{2} \langle \mathcal{D}\phi, \mathcal{D}\phi \rangle_h$$

Check that

(i) the response of the *new* term to the variation $\mathcal{A} \mapsto \mathcal{A} + \epsilon a$ may be written in terms of the "current" $\mathcal{J} \equiv \mathcal{J}(\phi, \mathcal{A})$ as

$$\mathcal{A} \mapsto \mathcal{A} + \epsilon a \Rightarrow \frac{1}{2} \langle \mathcal{D}\phi, \mathcal{D}\phi \rangle_h \mapsto \frac{1}{2} \langle \mathcal{D}\phi, \mathcal{D}\phi \rangle_h - \epsilon \langle a, \mathcal{J} \rangle_k \qquad \mathcal{J}^i = -k^{ij} \rho_{abj} \phi^a (\mathcal{D}\phi)^b$$

where the 1-form of the "current" $\mathcal{J} = \mathcal{J}^i(\phi, \mathcal{A}) e_i$ (with values in the Lie algebra[453] $\mathcal{G}$) is of type Ad

(ii) the variation of the complete action with respect to both fields $\mathcal{A}$ and ϕ turns out to be

$$S[\phi + \epsilon_1 \psi, \mathcal{A} + \epsilon_2 a] = S[\phi, \mathcal{A}] + \epsilon_1 \langle \psi, \mathcal{D}^+ \mathcal{D}\phi \rangle - \epsilon_2 \langle a, \mathcal{D}^+ \mathcal{F} + \mathcal{J} \rangle$$

(iii) the equations of motion corresponding to the action are

$$\mathcal{D}^+ \mathcal{F} = -\mathcal{J} \qquad \mathcal{D}^+ \mathcal{D}\phi = 0$$

[451] In quantum field theory the term containing four $\mathcal{A}$s corresponds to a "vertex" with four lines of type $\mathcal{A}$, i.e. there exists an elementary interaction, in which four particles described by the field $\mathcal{A}$ (like gluons in chromodynamics) are involved.

[452] In physical parlance this is the "multiplet of scalar fields, which transform according to the representation ρ of the gauge group G."

[453] Note that the *change* of the value of the scalar product $\langle \cdot, \cdot \rangle_h$ may be written in terms of a *different* scalar product $\langle \cdot, \cdot \rangle_k$.

Hint: (i) $\mathcal{A} \mapsto \mathcal{A} + \epsilon a \Rightarrow \mathcal{D}\phi \mapsto \mathcal{D}\phi + \epsilon a\phi \Rightarrow \langle \mathcal{D}\phi, \mathcal{D}\phi \rangle_h \mapsto \langle \mathcal{D}\phi, \mathcal{D}\phi \rangle_h + 2\epsilon \langle a\phi, \mathcal{D}\phi \rangle_h$; now $a\phi \equiv (a\phi)^a e_a = a^i \phi^b \rho'(e_i) e_b \equiv (\rho^a_{bi} a^i \phi^b) e_a$ so that

$$\langle a\phi, \mathcal{D}\phi \rangle_h = \int h_{ab} (a\phi)^a \wedge *(\mathcal{D}\phi)^b = \int h_{ab} \rho^a_{ci} a^i \phi^c \wedge *(\mathcal{D}\phi)^b$$

$$= \int a^i \wedge *\rho_{bci} \phi^c (\mathcal{D}\phi)^b \equiv -\int a^i \wedge *k_{ij} \mathcal{J}^j \equiv -\langle a, \mathcal{J} \rangle_k$$

the type Ad: either use the fact that the expression $\langle a, \mathcal{J} \rangle_k$ should be invariant and $a \equiv a^i e_i$ is of type Ad (since $\mathcal{A} \mapsto \mathcal{A} + \epsilon a$ should preserve the type Ad), or directly: if $\phi \mapsto \rho(S^{-1}(x))\phi \equiv B^{-1}(x)\phi$, then $\mathcal{D}\phi \mapsto B^{-1}(x)\mathcal{D}\phi$; consequently

$$\mathcal{J}^i \mapsto -k^{ij} \rho_{abj} (B^{-1})^a_c (B^{-1})^b_d \phi^a (\mathcal{D}\phi)^b = (\mathrm{Ad}_S)^i_j \mathcal{J}^j$$

according to (12.5.2); (ii) $\phi \mapsto \phi + \epsilon\psi \Rightarrow \mathcal{D}\phi \mapsto \mathcal{D}\phi + \epsilon\mathcal{D}\psi \Rightarrow \langle \mathcal{D}\phi, \mathcal{D}\phi \rangle \mapsto \cdots$. □

- The last term to be added to the action is the "potential energy" $U(|\phi|)$ of the matter field ϕ (in particular, the mass term).

21.5.5 Consider the matter field ϕ of type ρ. Check that

(i) the expression

$$|\phi|^2 \equiv h(\phi, \phi) \equiv h_{ab} \phi^a \phi^b$$

is a locally gauge invariant function

(ii) for any function $U(|\phi|)$ we may use the invariant term $\int U(|\phi|)\omega_g$ in the action; the response of this term to the variation of the field ϕ reads

$$\phi \mapsto \phi + \epsilon\psi \quad \Rightarrow \quad U(|\phi|)\omega_g \mapsto U(|\phi|)\omega_g + \epsilon \left\langle \psi, U'(|\phi|) \frac{\phi}{|\phi|} \right\rangle_h$$

(iii) the choice $U(a) = -m^2 a^2/2$ gives rise to the mass terms of the fields ϕ^a, the masses of all the component fields ϕ^a being *equal* $(= m)$.

Hint: (ii)

$$\begin{aligned} \phi \mapsto \phi + \epsilon\psi \quad &\Rightarrow \quad |\phi|^2 \equiv h(\phi, \phi) \mapsto |\phi|^2 (1 + 2\epsilon h(\phi, \psi)/h(\phi, \phi)) \\ &\Rightarrow \quad |\phi| \mapsto |\phi|(1 + \epsilon h(\phi, \psi)/h(\phi, \phi)) \equiv |\phi| + \epsilon h(\phi/|\phi|, \psi) \\ &\Rightarrow \quad U(|\phi|) \mapsto U(|\phi|) + \epsilon h(U'(|\phi|)\phi/|\phi|, \psi) \end{aligned}$$

(iii) if e_a represents an *orthonormal* basis in (V, ρ) $(h_{ab} = \delta_{ab})$, we get standard kinetic terms of the fields ϕ^a and $U(|\phi|) = -(m^2/2)((\phi^1)^2 + \cdots + (\phi^n)^2)$. □

21.5.6 Consider the action of a coupled system consisting of a matter field ϕ of type ρ and a gauge field $\mathcal{A}$

$$S[\phi, \mathcal{A}] = -\frac{1}{2} \langle \mathcal{D}\mathcal{A}, \mathcal{D}\mathcal{A} \rangle_k + \frac{1}{2} \langle \mathcal{D}\phi, \mathcal{D}\phi \rangle_h + \int_{\mathcal{U}} U(|\phi|)\omega$$

(In order to obtain correct signs in (both) *kinetic* terms, both scalar products should be positive definite. This is guaranteed for *compact* groups (12.3.18) and (trivially) also for *commutative* groups G.) Check that

(i) it is locally gauge invariant
(ii) the variation with respect to the fields $\mathcal{A}$ and ϕ gives

$$S[\phi + \epsilon_1 \psi, \mathcal{A} + \epsilon_2 a] = S[\phi, \mathcal{A}] + \epsilon_1 \left\langle \psi, \mathcal{D}^+ \mathcal{D}\phi + U'(|\phi|)\frac{\phi}{|\phi|} \right\rangle - \epsilon_2 \langle a, \mathcal{D}^+ \mathcal{F} + \mathcal{J} \rangle$$

(iii) the corresponding equations of motion (including the "non-variational" Bianchi identity) read[454]

$$\mathcal{D}^+ \mathcal{F} = -\mathcal{J}(\phi, \mathcal{A}) \qquad \mathcal{D}\mathcal{F} = 0 \qquad \left(\mathcal{D}^+ \mathcal{D} + U'(|\phi|)\frac{1}{|\phi|}\right)\phi = 0$$

(iv) in particular, for $U(a) = -m^2 a^2/2$ the action is

$$S[\phi, \mathcal{A}] = -\frac{1}{2}\langle \mathcal{D}\mathcal{A}, \mathcal{D}\mathcal{A}\rangle_k + \frac{1}{2}\langle \mathcal{D}\phi, \mathcal{D}\phi\rangle_h - \frac{m^2}{2}\langle \phi, \phi\rangle_h$$

and the corresponding equations of motion are

$$\mathcal{D}^+ \mathcal{F} = -\mathcal{J}(\phi, \mathcal{A}) \qquad \mathcal{D}\mathcal{F} = 0 \qquad (\mathcal{D}^+ \mathcal{D} - m^2)\phi = 0$$

(v) the equations of motion imply the equation

$$d * j \equiv d * (\mathcal{J} + *^{-1}[\mathcal{A} \wedge *\mathcal{F}]) = 0 \qquad \text{(i.e. } \delta j = 0)$$

where j is the "conserved" ("Noether," see (21.6.4)) current; making use of the current we get the conservation of the *charges*

$$Q^i := \int_{\mathcal{U}_3} *j^i \qquad j^i := \mathcal{J}^i + c^i_{jk} \mathcal{A}^j \wedge *\mathcal{F}^k$$

which may also be regarded as the conservation of *a single* element $Q \equiv Q^i e_i$ from the *Lie algebra* $\mathcal{G}$.

Hint: see (21.5.4) and (21.5.5); (v) the equation of motion $\mathcal{D}^+\mathcal{F} = -\mathcal{J}$ gives $\mathcal{D} * \mathcal{F} = d * \mathcal{F} + [\mathcal{A} \wedge *\mathcal{F}] = -*\mathcal{J}$, i.e. $d * \mathcal{F} = - * \mathcal{J} - [\mathcal{A} \wedge *\mathcal{F}]$; see (16.2.4). □

• The equations of motion are derived from a locally gauge invariant action and, consequently, the equations themselves are necessarily locally gauge invariant. It is easy to verify this directly.

21.5.7 Check *directly* that the equations of motion obtained above are indeed locally gauge invariant.

Hint: $\mathcal{D}\mathcal{F} = 0$ is an equality of two 3-forms of type Ad; $*$ does not change type, so that $\mathcal{D}^+\mathcal{F} = \mathcal{J}$ is an equality of two 1-forms of type Ad ($*^{-1}\mathcal{D} * \eta\mathcal{F}$ transforms just as $\mathcal{F}$); by the same argument $(-\mathcal{D}^+\mathcal{D} + m^2)\phi = 0$ is an equality of two 0-forms of type ρ. □

[454] If we compare the first equation $\mathcal{D}^+\mathcal{F} = -\mathcal{J}(\phi, \mathcal{A})$ with the "ordinary" Maxwell equation $\delta F = -j$ (16.2.2), we can see that the "current" $\mathcal{J}$, which formally "generates" the field $\mathcal{A}$ (hidden in $\mathcal{F}$) depends itself on $\mathcal{A}$.

• As we have already mentioned, in real physical theories the group G is often a direct product of two (or more) subgroups, $G = G_1 \times G_2$ and the representation ρ of the group (according to which the matter field transforms) is composed from the representations ρ_1 and ρ_2 of the individual groups in a way described in (12.4.16).[455] According to the results of problems (20.4.13) and (20.4.14) the connection form then naturally decomposes into two parts, $\omega = \omega_1 + \omega_2$, and each of them "takes care of its own index" in the computation of the exterior covariant derivative of the function $\Phi = \Phi^{a\alpha} E_a \times E_\alpha$. It may be instructive to look at the action in this case in more detail.

21.5.8* Check that the part of the action which corresponds to a pure gauge field takes in the case $G = G_1 \times G_2$ the form

$$S[\mathcal{A}] = -\frac{1}{2}\langle \mathcal{F}, \mathcal{F}\rangle_k \qquad \mathcal{F}_1 =: \rho'(\sigma^*\Omega_1)$$
$$= -\frac{1}{2}\lambda_1\langle \mathcal{F}_1, \mathcal{F}_1\rangle_{k_1} - \frac{1}{2}\lambda_2\langle \mathcal{F}_2, \mathcal{F}_2\rangle_{k_2} \qquad \mathcal{F}_2 =: \rho'(\sigma^*\Omega_2)$$

where k_1 and k_2 are Ad-invariant scalar products in the Lie algebras $\mathcal{G}_1$ and $\mathcal{G}_2$ and

$$k = \lambda_1 k_1 \oplus \lambda_2 k_2 \qquad \lambda_1, \lambda_2 \in \mathbb{R}$$

is the Ad-invariant scalar product on the whole Lie algebra $\mathcal{G} = \mathcal{G}_1 \oplus \mathcal{G}_2$.

Hint: see (20.4.13), (21.5.3) and (12.4.17). □

21.5.9* Still consider $G = G_1 \times G_2$ and let $\phi = \sigma^*\Phi$ be a matter field which results from the pull-back to the base of a function of type ρ for the representation $\rho(g_1, g_2) = \rho(g_1) \otimes \rho(g_2)$ in the space $V = V_1 \otimes V_2$ discussed in (12.4.16). Check that

(i) the formula for the exterior covariant derivative of a matter field $\mathcal{D}\phi$ may now be written in the form

$$\mathcal{D}\phi = d\phi + \mathcal{A}_1\phi + \mathcal{A}_2\phi \qquad \mathcal{A}_1 =: \rho'(\sigma^*\omega_1)$$
$$\mathcal{A}_2 =: \rho'(\sigma^*\omega_2)$$

or, using indices (21.3.7)

$$\phi^{a\alpha}{}_{;\mu} \equiv \mathcal{D}_\mu \phi^{a\alpha} \qquad (\mathcal{A}_1)^a_b =: (\mathcal{A}_1)^a_{b\mu}\, dx^\mu$$
$$= \partial_\mu \phi^{a\alpha} + (\mathcal{A}_1)^a_{b\mu}\phi^{b\alpha} + (\mathcal{A}_2)^\alpha_{\beta\mu}\phi^{a\beta} \qquad (\mathcal{A}_2)^\alpha_\beta =: (\mathcal{A}_2)^\alpha_{\beta\mu}\, dx^\mu$$

so that each index of the field ϕ is "managed" (additively) by "its own gauge potential"

(ii) for the corresponding part of the action we get in more detail

$$S[\phi, \mathcal{A}] = \frac{1}{2}\langle \mathcal{D}\phi, \mathcal{D}\phi\rangle_h \equiv \frac{1}{2}(h_1)_{ab}(h_2)_{\alpha\beta}\langle \mathcal{D}\phi^{a\alpha}, \mathcal{D}\phi^{b\beta}\rangle$$

where h_1 is a ρ_1-invariant scalar product in V_1 (and similarly h_2), and $h = h_1 \otimes h_2$ is ρ-invariant in $V_1 \otimes V_2$.

[455] For example, in the standard model of electroweak interactions we have $G = SU(2) \times U(1)$ and, if the chromodynamics is incorporated, it is even $G = SU(3) \times SU(2) \times U(1)$; the matter field $\Phi^{a\alpha n}$ then has as many as *three* indices, each of them acted on by another group.

Hint: (i) see (21.2.4); the application of σ^* to (20.4.14) gives $\mathcal{D}\phi^{a\alpha} = d\phi^{a\alpha} + (\mathcal{A}_1)^a_b\phi^{b\alpha} + (\mathcal{A}_2)^\alpha_\beta\phi^{a\beta}$; (ii) see (21.5.4) and (12.4.16). □

21.5.10* Consider the gauge group $G = SU(2) \times U(1)$ and the matter field ϕ of type $\rho_1(g_1) \otimes \rho_2(g_2)$, where $\rho_1 = \text{id}$ (the "fundamental" representation of $SU(2)$) and ρ_2 is the nth irreducible representation of $U(1)$. Check that

(i) the gauge potential may be written as

$$A = -\frac{i}{2}\begin{pmatrix} A_3 & A_1 - iA_2 \\ A_1 + iA_2 & -A_3 \end{pmatrix} + iA_4$$

(ii) the exterior covariant derivative of the field ϕ is

$$\mathcal{D}\begin{pmatrix} \phi^1 \\ \phi^2 \end{pmatrix} \equiv \begin{pmatrix} \mathcal{D}\phi^1 \\ \mathcal{D}\phi^2 \end{pmatrix} = \begin{pmatrix} d\phi^1 \\ d\phi^2 \end{pmatrix} - \frac{i}{2}\begin{pmatrix} A_3 & A_1 - iA_2 \\ A_1 + iA_2 & -A_3 \end{pmatrix}\begin{pmatrix} \phi^1 \\ \phi^2 \end{pmatrix} + \text{in}\, A_4 \begin{pmatrix} \phi^1 \\ \phi^2 \end{pmatrix}$$

Hint: (i) a basis in $su\,(2) \oplus u(1)$ is given by $(-i/2\,\sigma_j, i)$, so that $A = -i/2\; A_j\sigma_j + iA_4$; (ii) $V_1 \otimes V_2 = \mathbb{C}^2 \otimes \mathbb{C} \sim \mathbb{C}^2$, see (21.2.6); according to (21.5.9) and (12.4.16)

$$\mathcal{D}\phi = d\phi + \rho'(A)\phi = d\phi - \frac{i}{2}A_j\sigma_j\phi + in\; A_4\phi$$

□

• Let us conclude with a short comment. If we compare the action integral from problem (21.5.4)

$$S[\phi, \mathcal{A}] = -\frac{1}{2}\langle \mathcal{D}\mathcal{A}, \mathcal{D}\mathcal{A}\rangle_k + \frac{1}{2}\langle \mathcal{D}\phi, \mathcal{D}\phi\rangle_h$$

with a corresponding expression found in the majority of the physical literature, we will find that a *coupling constant* g is used in physics. What is going on? As we have already mentioned, there is a freedom in the scalar product k in the action: if some particular k is acceptable (it is Ad-invariant and positive definite), then so is an arbitrary (positive) *multiple* of k. So if the scalar product is fixed by (say) the relation $\langle\cdot\rangle_k := \text{Tr}\,(\cdot)$, the freedom may be written in terms of a free parameter g as

$$-\frac{1}{2g^2}\langle \mathcal{D}\mathcal{A}, \mathcal{D}\mathcal{A}\rangle_k + \frac{1}{2}\langle \mathcal{D}\phi, \mathcal{D}\phi\rangle_h \qquad \mathcal{A} := \rho'(\sigma^*\omega)$$

Now, this g may be *reshuffled to a different place* from the first term of the action by introducing a new gauge potential $\mathcal{A}^{\text{new}}$, which is defined (with a view to pulling $1/g^2$ into the scalar product) as an appropriate *multiple* of the original represented form $\sigma^*\omega$:

$$\mathcal{A}^{\text{new}} := \frac{1}{g}\rho'(\sigma^*\omega) \equiv \frac{1}{g}\mathcal{A}$$

Then in terms of *the new* potential $\mathcal{A}^{\text{new}}$ we evidently obtain

$$S[\phi, \mathcal{A}^{\text{new}}] = -\frac{1}{2}\langle \mathcal{D}\mathcal{A}^{\text{new}}, \mathcal{D}\mathcal{A}^{\text{new}}\rangle_k + \frac{1}{2}\langle \mathcal{D}\phi, \mathcal{D}\phi\rangle_h$$

The coupling constant does not disappear completely, it just arises in *another place.*

[21.5.11]* Check that in terms of the new potential the following formulas are modified (the original formulas correspond to $g = 1$):

(i) the exterior covariant derivative of $\mathcal{A}^{\text{new}}$ and ϕ read

$$\mathcal{F}^{\text{new}} \equiv \mathcal{D}\mathcal{A}^{\text{new}} = d\mathcal{A}^{\text{new}} + g\mathcal{A}^{\text{new}} \wedge \mathcal{A}^{\text{new}}$$
$$\mathcal{D}\phi = d\phi + g\mathcal{A}^{\text{new}}\phi$$

(ii) the gauge transformation of the potential is given as

$$\mathcal{A}^{\text{new}} \mapsto B^{-1}(x)\mathcal{A}^{\text{new}}B(x) + \frac{1}{g}B^{-1}(x)\,dB(x)$$

Hint: simply insert $\mathcal{A} = g\mathcal{A}^{\text{new}}$ into the corresponding good old formulas from (21.2.4) and (21.2.5). □

[21.5.12]* Check that in the particular case of $G = G_1 \times G_2$ we get in this way as many as *two* coupling constants g_1 and g_2 and the corresponding formulas read (for example)

$$\mathcal{F}_1^{\text{new}} = d\mathcal{A}_1^{\text{new}} + g_1\mathcal{A}_1^{\text{new}} \wedge \mathcal{A}_1^{\text{new}}$$
$$\mathcal{F}_2^{\text{new}} = d\mathcal{A}_2^{\text{new}} + g_2\mathcal{A}_2^{\text{new}} \wedge \mathcal{A}_2^{\text{new}}$$
$$\mathcal{D}\phi = d\phi + g_1\mathcal{A}_1^{\text{new}}\phi + g_2\mathcal{A}_2^{\text{new}}\phi$$

Hint: according to (21.5.8) in terms of the original fields the action is

$$S[\phi, \mathcal{A}_1, \mathcal{A}_2] = -\frac{1}{2g_1^2}\langle \mathcal{F}_1, \mathcal{F}_1\rangle_{k_1} - \frac{1}{2g_2^2}\langle \mathcal{F}_2, \mathcal{F}_2\rangle_{k_2} + \frac{1}{2}\langle \mathcal{D}\phi, \mathcal{D}\phi\rangle_h$$

Again we should pull g_1, g_2 into the new fields $\mathcal{A}_1/g_1 \equiv \mathcal{A}_1^{\text{new}}$ and $\mathcal{A}_2/g_2 \equiv \mathcal{A}_2^{\text{new}}$ and express the covariant derivative $\mathcal{D}\phi$ according to (21.5.9) in terms of the new fields. □

(In the standard model of elementary particles with the gauge group $SU(3) \times SU(2) \times U(1)$ we thus obtain as many as *three* coupling constants; since the constants are free parameters in the theory, increasing the number of coupling constants results in decreasing of the "predictive power" of the theory. The idea behind "grand unification theories" consists in regarding the model as being embedded in a *bigger* model, a gauge theory based on a bigger *simple* Lie group G, into which $SU(3) \times SU(2) \times U(1)$ may be embedded as a subgroup (in particular, $SU(5)$ or $SO(10)$). Since now G is simple (rather than a product), only *a single* coupling constant g enters the action. The three coupling constants of the theory based on $SU(3) \times SU(2) \times U(1)$ are no longer free parameters since they may be expressed in terms of g.)

21.6 Noether currents and Noether's theorem

As we already mentioned in (21.5.6), the "conserved" current j is also called the "Noether" current. It turns out that such a Noether current (a 1-form j with vanishing codifferential, $\delta j = 0$) and the corresponding conserved charge Q may be associated with *each global* (continuous) symmetry in field theory.[456] Let us look at the statement, which is definitely one of the most profound observations in theoretical physics. The main general statement is in problem (21.6.1). In (21.6.4) we check that the above-mentioned current from (21.5.6) is indeed of this type. Then we find out that the conserved quantities for fields, which we constructed in Section 16.4 with the help of the energy-momentum tensor, may also be regarded as particular cases of the formalism treated here. Finally, we discuss from this useful point of view the conservation laws in Hamiltonian mechanics which we treated in Chapter 14.

Consider an action integral $S[\psi]$ for a field ψ in the domain $\mathcal{U}$ of the space-time M (there are possibly several fields, ψ denotes all of them). Assume there is a "global" action $\psi(x) \mapsto \rho(g)\psi(x)$ of the group G on the fields ψ (so that g does not depend on x; at each point x the same g acts), such that the action integral is invariant with respect to the "global" action of the group

$$S[\rho(g)\psi] = S[\psi]$$

In particular, for an infinitesimal global transformation we have

$$S[\rho(e^{\epsilon X})\psi] = S[\psi] \qquad X \in \mathcal{G}$$

Consider now an infinitesimal "*local*" transformation

$$\psi(x) \mapsto \rho(e^{\epsilon s(x)})\psi(x)$$

given in terms of the function $s : \mathcal{U} \to \mathcal{G}, x \mapsto s(x) \equiv s^i(x)E_i \in \mathcal{G}$. The action integral is no longer invariant in general.

[21.6.1] Since for a *constant* function the action integral is supposed to remain unchanged, the variation is now expected to be proportional to the 1-form $\epsilon\, ds = \epsilon\, ds^i\, E_i$ (it is a 1-form on $\mathcal{U}$ with values in the Lie algebra $\mathcal{G}$)

$$S[\rho(e^{\epsilon s(x)})\psi] = S[\psi] + \epsilon \int_{\mathcal{U}} ds^i \wedge J_i(\psi) + o(\epsilon)$$

where $(n-1)$-forms $J_i(\psi)$, constructed from the fields ψ, depend on the detailed structure of the action (n is the dimension of the space-time M). Check that *Noether's theorem* holds, i.e. that

(i) the forms $J_i(\psi)$, when evaluated *on the solutions of the equations of motion*, are *closed*

$$\psi \text{ is a solution of the equations of motion} \quad \Rightarrow \quad dJ_i(\psi) = 0$$

[456] It is named after the distinguished German mathematician ("whose innovations in higher algebra," according to Encyclopaedia Britannica, "gained her recognition as the most creative abstract algebraist of modern times") Emmy Noether; she published the celebrated result in 1918.

(ii) this results in conservation of the *spatial* integrals of the forms J_i, known as the *Noether charges*

$$Q_i := \int_{\text{the whole space}} J_i \quad \Rightarrow \quad \dot{Q}_i = 0 \qquad i = 1, \dots, \dim G$$

this may also be regarded as conservation of a *single* element $Q \equiv Q_i E^i$ of the *dual of the Lie algebra* $\mathcal{G}^*$

(iii) if we parametrize the forms J_i in terms of *Noether currents*, the dual 1-forms $j^i(\psi)$

$$J_j(\psi) = k_{ji} * j^i(\psi) \qquad \text{i.e. if} \quad \int_{\mathcal{U}} ds^i \wedge J_i(\psi) = k_{ji} \int_{\mathcal{U}} ds^j \wedge * j^i(\psi) \equiv \langle ds, j \rangle_k$$

(k is the scalar product in the Lie algebra $\mathcal{G}$), then the currents (1-forms) are coclosed

$$\psi \text{ is a solution of the equations of motion} \quad \Rightarrow \quad \delta j^i(\psi) = 0$$

(and consequently the charges $\int * j^i \equiv Q^i = k^{ij} Q_j$ are conserved).

Hint: (i) if ψ is a solution of the equations of motion, then it extremizes the action, so that the action $S[\psi]$ does not change within first-order accuracy with respect to an *arbitrary* small change of the field ψ, which vanishes on $\partial\mathcal{U}$; in particular, for functions $s^i(x)$ vanishing on $\partial\mathcal{U}$ (and arbitrary inside $\mathcal{U}$) we get $S[\rho(e^{\epsilon s(x)})\psi] = S[\psi]$, i.e.

$$\psi \text{ is a solution of the equations of motion} \Rightarrow \int_{\mathcal{U}} ds^i \wedge J_i(\psi) = 0$$

Stokes' theorem and the vanishing of s^i on $\partial\mathcal{U}$ then gives

$$\begin{aligned} 0 &= \int_{\mathcal{U}} ds^i \wedge J_i(\psi) = \int_{\mathcal{U}} d(s^i J_i(\psi)) - \int_{\mathcal{U}} s^i \, dJ_i(\psi) \\ &= \int_{\partial\mathcal{U}} s^i J_i(\psi)) - \int_{\mathcal{U}} s^i \, dJ_i(\psi) = \int_{\mathcal{U}} s^i dJ_i(\psi) \end{aligned}$$

and since $s^i(x)$ are arbitrary inside $\mathcal{U}$, this already leads to $dJ_i(\psi) = 0$; (ii) see (16.2.4). □

- We can see immediately from the derivation that the currents are not given uniquely.

21.6.2 Check that neither the $(n-1)$-forms J_i nor Noether currents j^i are given uniquely, but instead only up to the freedom

$$J_i \mapsto J_i + d\,(\text{something})_i \qquad j^i \mapsto j^i + \delta\,(\text{something'})^i$$

i.e. only up to additive *exact* or *coexact* forms respectively; the conserved quantities (charges) Q are, however, already unique.

Hint: the relevant expression $\int_{\mathcal{U}} ds^i \wedge J_i(\psi)$ does not change under the replacement $J_i \mapsto J_i + d\sigma_i$, since $\int_{\mathcal{U}} ds^i \wedge d\sigma_i(\psi) = \int_{\mathcal{U}} d(s^i \, d\sigma_i(\psi)) = \int_{\partial\mathcal{U}} s^i \, d\sigma_i(\psi) = 0$; the conservation law is obtained from $0 = \int dJ_i = \cdots$ and $J_i \mapsto J_i + d\,(\text{something})_i$ clearly has no effect on this expression. □

• In order to illustrate the method, let us work out the Noether current explicitly for the "global" $U(1)$-symmetry of the action of the complex scalar field.

21.6.3 Consider the complex scalar field with the standard action integral

$$S[\phi] = \langle d\phi^*, d\phi\rangle - m^2\langle\phi^*, \phi\rangle$$

and the "global" action of the group $U(1)$ on the fields according to the nth irreducible representation, i.e.

$$\phi \mapsto \rho_n(e^{-i\alpha})\phi = e^{-in\alpha}\phi \qquad \phi^* \mapsto e^{in\alpha}\phi^*$$

Check that

(i) the corresponding Noether current j and the charge Q read

$$j(\phi, \phi^*) = in(\phi^* d\phi - \phi d\phi^*) \equiv in\phi^* \overleftrightarrow{d}\, \phi \qquad Q = \int j_0\, dV = in \int \phi^* \overleftrightarrow{\partial_0}\, \phi$$

(ii) the equations of motion indeed guarantee the coexactness of j

(iii) if we also add the electromagnetic field, i.e. if we extend the action to

$$S[\phi, A] = \langle \mathcal{D}\phi^*, \mathcal{D}\phi\rangle - m^2\langle\phi^*, \phi\rangle - \tfrac{1}{2}\langle F, F\rangle \qquad \begin{aligned}\mathcal{D}\phi &= d\phi + inA\phi \\ \mathcal{D}\phi^* &= d\phi^* - inA\phi^*\end{aligned}$$

then the *global* $U(1)$ symmetry is

$$\phi \mapsto e^{-in\alpha}\phi \qquad \phi^* \mapsto e^{in\alpha}\phi^* \qquad A \mapsto A$$

(it arises from the local gauge symmetry discussed in (21.1.2) for α = constant) and the corresponding Noether current and charge then read

$$j(\phi, \phi^*, A) = in(\phi^*\mathcal{D}\phi - \phi\mathcal{D}\phi^*) \equiv in\phi^* \overleftrightarrow{\mathcal{D}}\, \phi \qquad Q = \int j_0\, dV = in \int \phi^* \overleftrightarrow{\mathcal{D}_0}\, \phi$$

(iv) also here the equations of motion result in the coexactness of j (and consequently the conservation of the charge Q).

Hint: (i) the infinitesimal local transformation is $\phi \mapsto \phi - i\epsilon n\alpha(x)\phi$, $\phi^* \mapsto \phi^* + i\epsilon n\alpha(x)\phi^*$; the action $S[\phi] = \langle d\phi^*, d\phi\rangle - m^2\langle\phi^*, \phi\rangle$ changes within first-order accuracy by the term $\epsilon\langle d\alpha, j\rangle$, where $j(\phi, \phi^*) = in(\phi^*\, d\phi - \phi\, d\phi^*)$; (ii) the equation of motion is $\delta d\phi = m^2\phi$, i.e. $d * d\phi = -m^2 * \phi$ (plus the complex conjugated one), so that $d * (\phi^* d\phi - \phi d\phi^*) = \cdots = 0$; (iii) only the term $\langle \mathcal{D}\phi^*, \mathcal{D}\phi\rangle$ changes; we have $\mathcal{D}\phi \mapsto \mathcal{D}\phi - i\epsilon n(d\alpha(x))\phi$ plus an expression independent of $d\alpha$, similarly $\mathcal{D}\phi^*$. □

21.6.4* We now come back to the action integral

$$S[\phi, \mathcal{A}] = \frac{1}{2}\langle \mathcal{D}\phi, \mathcal{D}\phi\rangle_h - \frac{m^2}{2}\langle\phi, \phi\rangle - \frac{1}{2}\langle \mathcal{D}\mathcal{A}, \mathcal{D}\mathcal{A}\rangle_k$$

from (21.5.6). The action is locally gauge invariant with respect to the transformations of the fields

$$\phi \mapsto \hat{\phi} \equiv B^{-1}(x)\phi \qquad \mathcal{A} \mapsto \hat{\mathcal{A}} \equiv B^{-1}(x)\mathcal{A}B(x) + B^{-1}(x)dB(x)$$

and thus, in particular, also with respect to the same change with a *constant* matrix B; then,

$$\phi \mapsto B^{-1}\phi \qquad \mathcal{A} \mapsto B^{-1}\mathcal{A}B \qquad B = \text{constant}$$

This, in turn, may be regarded as a global symmetry with the corresponding Noether current. Check that the current is given by the expression

$$j = \mathcal{J}(\phi, \mathcal{A}) + *^{-1}[\mathcal{A} \wedge *\mathcal{F}]$$

we encountered in (21.5.6).

Hint: if we take B depending on x in the formulas for the *global* transformation, we get $\phi \mapsto \hat{\phi}$, $\mathcal{A} \mapsto \hat{\mathcal{A}} - B^{-1}(x)\,dB(x)$; the action is invariant with respect to the local transformation $S[\hat{\phi}, \hat{\mathcal{A}}] = S[\phi, \mathcal{A}]$, we need to evaluate it for the values $S[\hat{\phi}, \hat{\mathcal{A}} - B^{-1}(x)\,dB(x)]$, but in fact only for $B(x) = \mathbb{I} + \epsilon b(x)$; this gives $B^{-1}(x)\,dB(x) = \epsilon\, db(x)$; we thus need $S[\hat{\phi}, \hat{\mathcal{A}} - \epsilon db(x)]$; from the derivation of the equations of motion we know (21.5.6) that the variation of the action with respect to $\mathcal{A}$ gives the expression $\mathcal{J} + \mathcal{D}\mathcal{F}$; although it is true for hatted fields, but in the view of the order of ϵ before db a correction to unhatted is not needed, so that we have within first-order accuracy in ϵ

$$\begin{aligned} S[\hat{\phi}, \hat{\mathcal{A}} - \epsilon\, db(x)] &= S[\hat{\phi}, \hat{\mathcal{A}}] + \epsilon \langle db(x), \mathcal{J} + \mathcal{D}^{+}\mathcal{F}\rangle \\ &= S[\phi, \mathcal{A}] + \epsilon \langle db(x), \mathcal{J} + \mathcal{D}^{+}\mathcal{F}\rangle \\ &\equiv S[\phi, \mathcal{A}] + \epsilon \langle db(x), j\rangle \end{aligned}$$

Then the Noether current turns out to be

$$j(\phi, \mathcal{A}) = \mathcal{J} + \mathcal{D}^{+}\mathcal{F} \equiv \mathcal{J} + \delta\mathcal{F} + *^{-1}[\mathcal{A} \wedge *\mathcal{F}]$$

which differs only by the (completely harmless = coexact) term $\delta\mathcal{F}$ from[457] the required one. □

• In Section 16.4 we saw how the conserved quantities (energy and momentum; also angular momentum may be computed in this way) for various fields may be obtained from isometries (sometimes also conformal transformations) of the space-time. The explicit expressions of the conserved quantities were obtained by combining the generators of isometries (Killing vectors) with the *energy–momentum tensor* – the conserved current turns out to be (for a somewhat miraculous reason) given by $T(\xi, \cdot)$ (16.4.2). Let us see what this looks like from the point of view of Noether's theorem.

Contemplate a simple mechanical system and study its motion in the time interval from $t = 0$ to $t = 3$. For definiteness, think of a pebble thrown upwards and let this interval represent the time the pebble just moves upwards and downwards. Common sense then says that if we threw the pebble (in the same way) 5 seconds later, it would move in exactly the same way. In scientific parlance we appeal to the homogeneity of time, i.e. to

[457] Note that this j is *vanishing* on *solutions* of the equations of motion (the equations read $\mathcal{D}^{+}\mathcal{F} = -\mathcal{J}$), whereas for the equivalent $j = \mathcal{J} + *^{-1}[\mathcal{A} \wedge *\mathcal{F}]$ only *the codifferential* vanishes.

the invariance of mechanics with respect to shifts in time. When expressed in terms of the action integral this might mean the following:

If γ denotes the original trajectory, the new one is γ_5, where $\gamma_5(t) := \gamma(t-5)$. Introduce the map $\Phi_5 : t \mapsto t+5$ (the shift of time by 5 seconds). If $I \equiv \langle 0, 3 \rangle$ is the original interval, the new one is $\Phi_5(I)$ and the new trajectory is $\gamma_5 \equiv \gamma \circ \Phi_{-5}$. Thus, the expected invariance of the action integral (and consequently the dynamics) may be written in the noble form

$$S[\gamma \circ \Phi_{-5}; \Phi_5(I)] = S[\gamma; I]$$

The action is to be invariant with respect to the simultaneous shift of the integration domain by Φ_5 and the solution by Φ_{-5}.

Now we pass to the field theory. Contemplate a field ψ living in the space-time (M, g) and let the action integral be of the form $S[\psi, g; \mathcal{U}]$. Assume a symmetry group G acts on M, the generators (fundamental fields) being ξ_X. Move the domain $\mathcal{U}$ by means of the flow $\Phi_t : M \to M$ of the field ξ_X (this mimics the time shift of I from the mechanical example) and at the same time pull back the field ψ with respect to the flow, $\psi \mapsto \Phi^*_{-t}\psi \equiv \psi \circ \Phi_{-t}$ (this mimics the shift of the argument of the trajectory in the mechanical example). We expect that the action integral is the same for the two configurations:

$$S[\Phi^*_{-t}\psi, g; \Phi_t(\mathcal{U})] = S[\psi, g; \mathcal{U})]$$

In what follows we restrict the action integrals to be discussed to the physically most important class of actions which are *natural with respect to diffeomorphisms*. Recall (16.4.1) that the action is given as the integral of a form Ω, which in turn depends on some fields ψ as well as on the metric tensor g

$$S[\psi, g; \mathcal{U}] := \int_{\mathcal{U}} \Omega[\psi, g] \equiv \int_{\mathcal{U}} L(\psi, g)\omega_g$$

(ω_g being the metric volume form) and for an arbitrary diffeomorphism $f : M \to M$ it satisfies the condition

$$f^*(\Omega[\psi, g]) = \Omega[f^*\psi, f^*g]$$

21.6.5 Consider an action integral which is natural with respect to diffeomorphisms. Check that the condition under consideration

$$S[\Phi^*_{-t}\psi, g; \Phi_t(\mathcal{U})] = S[\psi, g; \mathcal{U})]$$

restricts the transformations Φ_t to those which satisfy

$$\Omega[\psi, \Phi^*_t g] = \Omega[\psi, g]$$

i.e. virtually to *isometries*, or in particular cases[458] also *conformal transformations*.

[458] For example, for the action integral of the pure electromagnetic field $S[A, g; \mathcal{U})] = \int_{\mathcal{U}} \Omega[A, g] = \int_{\mathcal{U}} dA \wedge *_g dA$.

Solution:

$$S[\Phi^*_{-t}\psi, g; \Phi_t(\mathcal{U})] = \int_{\Phi_t(\mathcal{U})} \Omega[\Phi^*_{-t}\psi, g] = \int_{\mathcal{U}} \Phi^*_t\{\Omega[\Phi^*_{-t}\psi, g]\}$$
$$= \int_{\mathcal{U}} \Omega[\Phi^*_t\Phi^*_{-t}\psi, \Phi^*_t g] = S[\psi, \Phi^*_t g; \mathcal{U}]$$

□

• Let us now understand the simple meaning behind this result. The field ψ lives and evolves (according to the equations of motion) on a certain playing field, the (pseudo-) *Riemannian* manifold (M, g) (space-time). If we require a "shift" of the configuration (the field ψ as well as the domain $\mathcal{U}$) by a transformation f to be possible – so that the new field in the new domain behaves "equally" as the old field in the old domain, we are to create in the new domain "equal conditions" to those the old field had in the old domain. The "conditions" are, however, encoded in the action integral and the latter in turn contains g; thus we are to create equal "*metric*" conditions. Actually this is to be interpreted just from the opposite side (since g is given): we have only the moral right to require "equal behavior" from the field for such diffeomorphisms f which preserve the metric conditions. These are as a rule isometries, but for some actions it may be even extended to conformal transformations.

So consider a Killing vector ξ_X, where $X \in \mathcal{G}$ belongs to the Lie algebra of the group of isometries. The corresponding flow is Φ_t. The action is invariant with respect to the simultaneous transformations of the field ψ and the domain $\mathcal{U}$

$$\psi \mapsto \Phi^*_{-t}\psi \qquad \mathcal{U} \mapsto \Phi_t(\mathcal{U})$$

Now this may be regarded as a *global symmetry* of the action integral (the field ψ is transformed at each point x *by the same* group element $\exp tX \in G$). This means that Noether's theorem may be used and we can compute the corresponding Noether current j and the (conserved) charge Q. Doing all the necessary steps we find that the current turns out to be just $j = T(\xi, \cdot)$.

[21.6.6] Check that this is indeed the case.

Hint: according to (21.6.5) the global symmetry of the action integral with respect to the transformation $\psi \mapsto \Phi^*_{-t}\psi$ of the field ψ may be replaced by the (global) symmetry with respect to the transformation $g \mapsto \Phi^*_t g$ of the metric tensor g. For the computation of the Noether current j we first need to write down the infinitesimal *global* symmetry, which is $S[\psi, \Phi^*_\epsilon g; \mathcal{U}] = S[\psi, g; \mathcal{U}]$ and then to evaluate the change under the *local* transformation, i.e. the expression $S[\psi, g + \epsilon\mathcal{L}_{\xi_{X(x)}} g; \mathcal{U}]$ for $X(x)$ dependent on the point x (the field $\xi_{X(x)}$ is *no longer* a Killing vector!); denote the change[459] of g by $g + \epsilon h := g + \epsilon\mathcal{L}_{\xi_{X(x)}} g$ and realize that the very existence of h is due to the *non-constancy* of $X(x) = X^i(x)E_i$ (the

[459] We should realize at the latest just now that the energy–momentum tensor necessarily enters the result, since this is (by definition) exactly the quantity which arises when the infinitesimal change $g \mapsto g + \epsilon h$ is performed in the action (16.4.1).

terms containing X^i itself (rather than the derivatives of X^i) *vanish*, since ξ_{E_i} are Killing vectors); we then have

$$h_{\mu\nu} = (\mathcal{L}_{\xi_{X(x)}} g)_{\mu\nu} = (X^i(x)\xi_{E_i})_{\mu;\nu} + (X^i(x)\xi_{E_i})_{\nu;\mu} = X^i{}_{,\nu}(\xi_{E_i})_\mu + (\mu \leftrightarrow \nu)$$

and consequently

$$\begin{aligned} S[\psi, g + \epsilon \mathcal{L}_{\xi_{X(x)}} g; \mathcal{U}] &= S[\psi, g; \mathcal{U}] - \epsilon \int_{\mathcal{U}} \frac{1}{2} h_{\mu\nu} T^{\mu\nu} \omega_g \\ &= S[\psi, g; \mathcal{U}] - \epsilon \int_{\mathcal{U}} X^i{}_{,\nu}(\xi_{E_i})_\mu T^{\mu\nu} \omega_g \\ &= S[\psi, g; \mathcal{U}] - \epsilon \int_{\mathcal{U}} (dX^i)_\nu T(\xi_{E_i}, \cdot)^\nu \omega_g \\ &= S[\psi, g; \mathcal{U}] - \epsilon \int_{\mathcal{U}} dX^i \wedge *_g T(\xi_{E_i}, \cdot) \end{aligned}$$

This shows that for each basis Killing vector ξ_{E_i} there is a Noether current

$$j_i = T(\xi_{E_i}, \cdot)$$

in full agreement with the result (16.4.1). □

• And if our admiration for the wide range of Noether's theorem were still not fervent enough, we might also add the conservation laws in Hamiltonian mechanics. Recall that they are closely related to exact Cartan symmetries (see Section 14.4) or, alternatively, to the moment map (see Section 14.5). Here we will look at them from the perspective of the symmetries of the *action integral*.

A trajectory $\gamma(t)$ in mechanics may be regarded as a (non-linear) *field* (it was already mentioned in Section 16.6 concerning non-linear sigma models). The "space-time" of this "field" is just the time axis[460] and it has values in the phase space, i.e. in a symplectic manifold (M, ω). We know from Section 18.5 that if we want to write down an action integral, we need to add still another dimension (time), so that the curves actually have their values in the extended phase space $M \times \mathbb{R}[t]$ (they read $t \mapsto (\gamma(t), t)$; we will also denote the "extended" curves by γ).

The action $S[\gamma]$ may be expressed as the integral of an appropriate 1-form along the curve

$$S[\gamma] = \int_\gamma (\theta - H\,dt) \equiv \int_\gamma (\text{“}p\,dq - H\,dt\text{”}) \equiv \int_\gamma \sigma$$

Now consider a Lie group G acting on M and let Φ_s be the flow generated by the fundamental field ξ_X of the action.[461] The flow also transforms curves (by the prescription $\gamma \mapsto \Phi_s \circ \gamma \equiv \gamma_s$) and we get *global transformations* of the "field" γ (each point $\gamma(t)$ is transformed by

[460] Field theorists like to speak about the mechanics as a "zero-dimensional field theory," meaning that the fields live in a space-time with a single time dimension and zero spatial dimensions.
[461] More precisely, the flow (as well as the generators) is to be (trivially) extended from M to $M \times \mathbb{R}[t]$.

the same group element $\exp sX \in G$). The action integral of the infinitesimally transformed curve is

$$S[\Phi_\epsilon \circ \gamma] = \int_{\Phi_\epsilon \circ \gamma} \sigma = \int_\gamma \Phi_\epsilon^* \sigma = S[\gamma] + \epsilon \int_\gamma \mathcal{L}_{\xi_X} \sigma + o(\epsilon)$$
$$= S[\gamma] + \epsilon \int_\gamma (i_{\xi_X} d\sigma + d i_{\xi_X} \sigma) + o(\epsilon)$$

Since the action of the group is expected to be a symmetry of the dynamics under consideration, the increment of order ϵ should actually vanish. Let us see what restrictions this imposes on the action of the group (on the generators ξ_X).

[21.6.7] Check that the requirement of the invariance of the action integral imposes the following restrictions on the generators ξ_X of the action of G:

$$i_{\xi_X} \omega = -d\langle \theta, \xi_X \rangle \qquad \xi_X H = 0$$

or, in plain English, ξ_X is to be an *exact Cartan symmetry* (the Hamiltonian field which moreover preserves the Hamiltonian, (14.4.2)).

Hint: vanishing of the increment of order ϵ (i.e. of the integral $\int (i_{\xi_X} d\sigma + d i_{\xi_X} \sigma)$) for each curve needs to fulfill $i_{\xi_X} d\sigma + d i_{\xi_X} \sigma = 0$); since ξ_X does not contain ∂_t, we get

$$0 = i_{\xi_X} d\sigma + d i_{\xi_X} \sigma = i_{\xi_X} (\omega - dH \wedge dt) + d i_{\xi_X} (\theta - H\, dt)$$
$$= i_{\xi_X} \omega - (\xi_X H)\, dt + d\langle \theta, \xi_X \rangle$$

The second term alone contains the basis form dt, so it should vanish in its own right. □

- The exact Cartan symmetries thus act as generators of *global symmetries* of a "field" theory. So they necessarily amount to corresponding Noether currents and charges. What do they look like?

[21.6.8] Consider the symmetry given by the field ξ_X. For constant X then the variation of order ϵ vanishes. Check that

(i) the variation of order ϵ for $X(t)$ which depends on t (for the "local" transformation) reads

$$\epsilon \int_\gamma dX^i \langle \theta, \xi_{E_i} \rangle$$

(ii) so that the closed $(n-1)$-forms J_i from the general formalism are realized as *constant functions*

$$J_i = \langle \theta, \xi_{E_i} \rangle$$

(iii) the same functions serve at the same time as the conserved charges

$$Q_i = \langle \theta, \xi_{E_i} \rangle$$

which agrees with the result from Section 14.4 obtained for exact Cartan symmetries.

Hint: (i) the term proportional to dX^i arises from $\int_\gamma di_{\xi_X}\sigma = \int_\gamma d(X^j i_{\xi_{E_j}}(\theta - H dt)) = \cdots$; (ii) $n = 1$; (iii) in general the conservation laws are obtained by integration of the condition $dJ_i = 0$, i.e. here from

$$0 = \int_{t_1}^{t_2} d\langle\theta, \xi_{E_i}\rangle = \langle\theta, \xi_{E_i}\rangle(t_2) - \langle\theta, \xi_{E_i}\rangle(t_1) \equiv Q_i(t_2) - Q_i(t_1)$$

thus the "spatial integral" in a fixed moment of time is here represented simply by the evaluation of the function at this time (recall how the integral of 0-forms over a zero-dimensional chain is computed); according to (14.4.4) the "Hamiltonian" F_V of the exact Cartan symmetry V is conserved; for ξ_{E_i} this is (according to (21.6.7)) just $\langle\theta, \xi_{E_i}\rangle$. □

21.7* Once more (for a while) on LM

We began a whole group of chapters treating connections by Chapter 19, introducing LM and describing the linear connection in terms of this manifold. Then we extended the concept of the connection to an arbitrary principal bundle $\pi : P \to M$ and treated various facts already in the general setting. Clearly everything which is said about the general case is also true for the particular case $\pi : LM \to M$. Nevertheless, because of the exceptional importance as well as some specific features of the good old frame bundle it is worth returning there for a while.[462] For example, a careful reader probably did not miss the fact that while the concept of curvature naturally generalizes to P, the *torsion* did not appear at all on P. How, as a matter of fact, does the torsion arise from the point of view of principal bundles? It turns out[463] that the torsion is indeed a specific feature of the bundle $\pi : LM \to M$ and it is closely related to the existence of another *canonical object* living on LM, the canonical 1-form θ with values in $\mathbb{R}^n$.

21.7.1 After some time, welcome again to LM! Introduce at the point $e \in LM$ the 1-forms θ^a_e, $a = 1, \ldots, n \equiv \dim M$ by the prescription

$$\langle\theta^a_e, w\rangle := \langle e^a, \pi_* w\rangle \equiv (\pi_* w)^a \qquad e \equiv e_a \leftrightarrow e^a \text{ (the dual basis)}$$

The vector w is first projected from e to $x = \pi(e)$ and then decomposed with respect to the basis e_a corresponding to the point e; the components are then regarded as the θ^a-images of w. Put differently, if the projection of the vector is $w_e \mapsto \pi_* w \equiv (\pi_* w)^a e_a$, then θ^a_e performs the map $w_e \mapsto (\pi_* w)^a$. Check that

(i) if σ is the section which corresponds to a frame field e_a in a domain $\mathcal{U} \subset M$, then

$$\sigma^*\theta^a = e^a$$

[462] We know from detective stories that culprits like to return to the scene of the crime; perhaps we have already committed enough on LM in Chapter 19 so as to return there for a while.

[463] Another natural route leading to the torsion is based on the concept of *affine connection*. This is a (particular) connection in another principal bundle which may be canonically associated with a manifold M, the *bundle of affine frames* (we will not discuss it in this book). The *affine* group $GA(n, \mathbb{R})$ (10.1.15) acts naturally in the fibers of the bundle. Since the group happens to be the semidirect product of $GL(n, \mathbb{R})$ and $\mathbb{R}^n$ (12.4.8), the *curvature* form of the *affine* connection decomposes to the part with values in $gl(n, \mathbb{R})$ (this part is closely related to the "ordinary" *curvature* form of the *linear* connection) and the part with values in $\mathbb{R}^n$ – and this very part corresponds to the *torsion* form of the *linear* connection.

(so that in the spirit of (17.6.6) the canonical 1-form θ on LM may be regarded as "Platonic eternal Idea" of a coframe field on M)

(ii) if θ^a are regarded as the component forms of *a single* 1-form $\theta = \theta^a E_a$ with values in $\mathbb{R}^n$, then θ is a horizontal 1-form of type id, i.e.

$$\text{hor}\,\theta = \theta \qquad R_A^*\theta = A^{-1}\theta \equiv \rho(A^{-1})\theta \qquad \rho(A) = A \equiv \text{id}_A$$

It is called the *canonical 1-form on LM* (with values in $\mathbb{R}^n$)

(iii) the forms θ^a are pointwise linearly independent, so that they may be used as a (global) basis for *horizontal* forms on LM

(iv) in particular, the decomposition of the *curvature 2-forms* Ω^a_b on LM (recall that they *are* horizontal) gives

$$\Omega^a_b = \frac{1}{2} R^a{}_{bcd}\theta^c \wedge \theta^d$$

The components $R^a{}_{bcd}(e)$ constitute a (global) *function* of type $\text{id} \otimes \check{\text{id}} \otimes \check{\text{id}} \otimes \check{\text{id}}$ on LM and by the pull-back to the base M (by means of the section from item (i)) we get "ordinary" components $\hat{R}^a{}_{bcd}(x)$ of the curvature *tensor* with respect to the frame field e_a (introduced in Chapter 15; they are only defined locally)

$$\hat{R}^a{}_{bcd} := \sigma^* R^a{}_{bcd}$$

Hint: (i) for $\sigma(x) \equiv e \equiv e_a$ we have $\langle \sigma^*\theta^a_e, v_x\rangle = \langle \theta^a_{\sigma(x)}, \sigma_* v_x\rangle = \langle e^a, \pi_*\sigma_* v_x\rangle = \langle e^a, v_x\rangle$; (ii) horizontality right from the definition; type id:

$$\begin{aligned}\langle (R_A^*\theta^a)_e, w\rangle &= \langle (\theta^a)_{eA}, R_{A*}w\rangle = \langle (eA)^a, \pi_* R_{A*} w\rangle = (A^{-1})^a_b \langle e^b, (\pi \circ R_A)_* w\rangle \\ &= (A^{-1})^a_b \langle \theta^b_e, w\rangle\end{aligned}$$

(iii) $\sigma^*(k_a\theta^a_e) = k_a e^a$; (iv) the type of the components: Ω^a_b are of type $\text{Ad} \sim \text{id} \otimes \check{\text{id}}$ according to (20.4.4), so that $R_A^*\Omega^a_b = (A^{-1})^a_c \Omega^c_d A^d_b$; on the other hand, $R_A^*(\frac{1}{2} R^a{}_{bcd}\theta^c \wedge \theta^d) = \cdots$; the pull-back to the base: (21.7.4) and (15.6.3). □

• The 1-form θ is clearly independent of the connection, it enjoys each day of life on LM, whether or not there is any connection.[464] If there is nevertheless a connection available, we get on LM another canonical object, the horizontal 2-form $\Theta \equiv D\theta$.

21.7.2 Let θ be the canonical 1-form on LM and denote $\Theta := D\theta$. Check that

(i) it is a horizontal 2-form of type id, so that it may be decomposed with respect to θ^a as

$$\Theta = \Theta^a E_a = \frac{1}{2} T^a_{bc}\theta^b \wedge \theta^c E_a$$

the components $T^a_{bc}(e)$ constitute a (global) *function* of type $\text{id} \otimes \check{\text{id}} \otimes \check{\text{id}}$ on LM

(ii) the form Θ uniquely encodes the *torsion* of the linear connection by

$$\sigma^*\Theta^a = \hat{T}^a \equiv \frac{1}{2}\hat{T}^a_{bc}(x) e^b \wedge e^c$$

[464] It has lived there since long ago, when the connection was not even on the drawing board of evolution. Several of the world's top natural history museums pride themselves on having a few intact components (mostly θ^1) found in Palaeozoic layers (at those times θ fed on trilobites).

where $\hat{T}^a$ denotes the *torsion 2-forms* with respect to the frame field $e_a \leftrightarrow \sigma$ introduced in (15.6.3) and $\hat{T}^a_{bc}(x)$ are "ordinary" components of the torsion *tensor* with respect to the frame field e_a (they are only defined locally)

$$\hat{T}^a_{bc} := \sigma^* T^a_{bc}$$

Hint: (i) the property of D and θ; (ii) since θ is of type $\rho = \text{id}$, we have $D\theta^a \equiv d\theta^a + (\rho'(\omega)\theta)^a = d\theta^a + \omega^a_b \wedge \theta^b$; therefore $\sigma^*\Theta^a \equiv \sigma^* D\theta^a = \sigma^*(d\theta^a + \omega^a_b \wedge \theta^b) = de^a + \hat{\omega}^a_b \wedge e^b \equiv T^a$ according to (15.6.7). □

21.7.3 Consider the manifold LM and the following vector and covector fields:

vector fields:	$\xi_{E^a_b}$	the generators of the action of $GL(n, \mathbb{R})$	(19.1.4)
	$\hat{\mathcal{E}}_a$	the standard horizontal fields	(19.4.8)
covector fields:	ω^a_b	(component forms of) the connection form	(19.2.1)
	θ^a	(component forms of) the canonical 1-form	(21.7.1)

Check that

(i) $(\xi_{E^a_b}, \hat{\mathcal{E}}_a)$ and (ω^a_b, θ^a) constitute global frame and coframe fields on LM, moreover they are *dual* to each other

$$\langle \omega^a_b, \xi_{E^c_d} \rangle = \delta^a_d \delta^c_b \qquad \langle \omega^a_b, \hat{\mathcal{E}}_c \rangle = 0$$
$$\langle \theta^a, \xi_{E^c_d} \rangle = 0 \qquad \langle \theta^a, \hat{\mathcal{E}}_b \rangle = \delta^a_b$$

and adapted to the vertical and horizontal distribution (subspaces): $(\xi_{E^a_b}, \omega^a_b)$ are vertical and $(\hat{\mathcal{E}}_a, \theta^a)$ are horizontal

(ii) LM is always *parallelizable* and consequently also an *orientable* manifold (even if M itself fails to have these properties)

(iii) the commutators of the frame field turn out to be

$$[\xi_{E^a_b}, \xi_{E^c_d}] = \delta^a_d \xi_{E^c_b} - \delta^c_b \xi_{E^a_d}$$
$$[\xi_{E^a_b}, \hat{\mathcal{E}}_c] = \delta^a_c \hat{\mathcal{E}}_b$$
$$[\hat{\mathcal{E}}_a, \hat{\mathcal{E}}_b] = -T^c_{ab} \hat{\mathcal{E}}_c - R^d{}_{cab} \xi_{E^c_d}$$

Hint: (i) right from the definitions and basic properties of the quantities involved; ω^a_b span *vertical* forms (annihilated by horizontal vectors) and θ^a on the contrary *horizontal* forms (annihilated by vertical vectors); (ii) the definition, (19.4.9); (iii) $[\xi_{E^a_b}, \cdot] = \cdots$ are nothing but the infinitesimal versions of $R^*_A(\cdot) = \cdots$ (set $A = \mathbb{I} + \epsilon E^a_b, \ldots$); the last one: the duality of the frames gives $[\hat{\mathcal{E}}_a, \hat{\mathcal{E}}_b] = \langle \omega^c_d, [\hat{\mathcal{E}}_a, \hat{\mathcal{E}}_b] \rangle \xi_{E^c_d} + \langle \theta^c, [\hat{\mathcal{E}}_a, \hat{\mathcal{E}}_b] \rangle \hat{\mathcal{E}}_c$; the coefficients:

$$
\begin{aligned}
R^c_{dab} &= \Omega^c_d(\hat{\mathcal{E}}_a, \hat{\mathcal{E}}_b) \equiv D\omega^c_d(\hat{\mathcal{E}}_a, \hat{\mathcal{E}}_b) && \text{(21.7.1)} \\
&= d\omega^c_d(\hat{\mathcal{E}}_a, \hat{\mathcal{E}}_b) && \text{since } \hat{\mathcal{E}}_a \text{ are horizontal} \\
&= \hat{\mathcal{E}}_a \langle \omega^c_d, \hat{\mathcal{E}}_b \rangle - \hat{\mathcal{E}}_b \langle \omega^c_d, \hat{\mathcal{E}}_a \rangle - \langle \omega^c_d, [\hat{\mathcal{E}}_a, \hat{\mathcal{E}}_b] \rangle && \text{Cartan formulas (6.2.13)} \\
&= -\langle \omega^c_d, [\hat{\mathcal{E}}_a, \hat{\mathcal{E}}_b] \rangle && \text{duality of frames}
\end{aligned}
$$

and similarly $T^c_{ab} = D\theta^c(\hat{\mathcal{E}}_a, \hat{\mathcal{E}}_b) = \cdots = -\langle \theta^c, [\hat{\mathcal{E}}_a, \hat{\mathcal{E}}_b] \rangle$. □

• It is fairly useful to realize that some expressions and manipulations may considerably simplify if we use the operator $\mathcal{D}$, the exterior covariant derivative *on the base* (21.2.4).

21.7.4 Check that the Cartan structure equations (15.6.7) may be written in terms of $\mathcal{D}$ as

$$\mathcal{D}e^a = \hat{T}^a \qquad \mathcal{D}\hat{\omega}^a_b = \hat{\Omega}^a_b \qquad \text{i.e.} \quad \mathcal{D}e = \hat{T} \qquad \mathcal{D}\hat{\omega} = \hat{\Omega}$$

and the Ricci and Bianchi identities then take the form

$$\mathcal{D}\hat{T} = \hat{\Omega} \wedge e \qquad \mathcal{D}\hat{\Omega} = 0$$

Hint: apply σ^* to the equations $D\theta \equiv d\theta + \omega \wedge \theta = \Theta$ and $D\omega \equiv d\omega + \frac{1}{2}[\omega \wedge \omega] = \Omega$ (see also (21.7.2), (21.2.4) and (21.2.7)). □

• It is convenient to learn a simple mnemonic when working with $\mathcal{D}$ which may be applied to "indexed" forms, which transform "nicely" under the change of frame field (for example, the 1-forms e^a, the 0-forms g_{ab}, the 2-forms $\hat{T}^a$ and so on). The "nice" transformation rule means that they are "scrambled" by means of a particular *representation* of the group $GL(n, \mathbb{R})$ according to the scheme

$$\alpha(eA) = \rho(A^{-1})\alpha(e)$$

($\alpha(e)$ means that the form is expressed *with respect to the frame field* $e \equiv e_a$; compare with (12.4.14)). As a rule ρ is a multiple tensor product of the simplest representations $\mathrm{id} \equiv \rho^1_0$ (due to upper indices on the form) and the contragredient representation $\check{\mathrm{id}} \equiv \rho^0_1$ (due to lower indices on the form); for example, the particular cases mentioned above correspond to

$$e^a \mapsto (A^{-1})^a_b e^b \quad \Rightarrow \quad \rho = \rho^1_0$$
$$g_{ab} \mapsto A^c_a A^d_b g_{cd} \quad \Rightarrow \quad \rho = \rho^0_1 \otimes \rho^0_1$$
$$\hat{T}^a \mapsto (A^{-1})^a_b \hat{T}^b \quad \Rightarrow \quad \rho = \rho^1_0$$

21.7.5 Check that the rules for the computation of $\mathcal{D}$ of such forms may be concisely summarized in the form of a table – a recipe for cooking the house speciality $\mathcal{D}\alpha^{a...b}_{c...d}$:

	for preparation of $\mathcal{D}\alpha^{a...b}_{c...d}$
first put on the bottom of a pan	$d\alpha^{a...b}_{c...d}$
plus for each $\alpha^{...a...}$ add	$+\ \hat{\omega}^a_b \wedge \alpha^{...b...}$
plus for each $\alpha_{...a...}$ add	$-\ \hat{\omega}^b_a \wedge \alpha_{...b...}$

i.e. there is the first term (flat amount), plus there is one term to be added for each index (with a plus sign for an upper index and a minus sign for a lower one); compare with (15.2.7).

Hint: in general we may write for the forms under consideration $\mathcal{D}\alpha = d\alpha + \rho'(\hat{\omega}) \dot{\wedge} \alpha$; according to (12.4.14) the corresponding ρ' reads explicitly as $(\rho'(C)t)^{a...b}_{c...d} = C^a_i t^{i...b}_{c...d} + \cdots + C^b_i t^{a...i}_{c...d} - C^i_c t^{a...b}_{i...d} - \cdots - C^i_d t^{a...b}_{c...i}$. □

21.7.6 Use the table from (21.7.5) to obtain explicit expressions for the exterior covariant derivative of the following p-forms:[465]

$$\begin{array}{lll} g_{ab} & p=0 & \mathcal{D}g_{ab} = dg_{ab} - (\hat{\omega}_{ab} + \hat{\omega}_{ba}) \\ \omega_{a...b} & p=0 & \mathcal{D}\omega_{a...b} = d\omega_{a...b} - \hat{\omega}^c_c \omega_{a...b} \\ e^a & p=1 & \mathcal{D}e^a = de^a + \hat{\omega}^a_b \wedge e^b \\ \hat{T}^a & p=2 & \mathcal{D}\hat{T}^a = d\hat{T}^a + \hat{\omega}^a_b \wedge \hat{T}^b \\ \hat{\Omega}^a_b & p=2 & \mathcal{D}\hat{\Omega}^a_b = d\hat{\Omega}^a_b + \hat{\omega}^a_c \wedge \hat{\Omega}^c_b - \hat{\omega}^c_b \wedge \hat{\Omega}^a_c \end{array}$$

where g_{ab} and $\omega_{a...b}$ denote the components of a metric tensor and a volume form respectively, e^a is a coframe field, $\hat{T}^a$ are the torsion forms and $\hat{\Omega}^a_b$ are the curvature forms.

Hint: *use* the table and realize that $\hat{\omega}^c_a g_{cb} + \hat{\omega}^c_b g_{ac} = \hat{\omega}_{ba} + \hat{\omega}_{ab}$ and $\hat{\omega}^c_a \omega_{c...b} + \cdots + \hat{\omega}^c_b \omega_{a...c} = \hat{\omega}^c_c \omega_{a...b}$. □

21.7.7 Check that the equations

$$\mathcal{D}g_{ab} = 0 \qquad \mathcal{D}\omega_{a...b} = 0 \qquad \mathcal{D}t^{a...}_{...b} = 0$$

express the conditions of the *compatibility* of a connection with a metric, a volume form and in general with a tensor $t^{a...}_{...b} = 0$, or alternatively of the covariant constancy of the tensors.

Hint: according to (21.3.7) there holds $\mathcal{D}g_{ab} = g_{ab;c}e^c$ so that $\mathcal{D}g_{ab} = 0$ actually means $g_{ab;c} = 0$, (15.3.1); $\omega_{a...b}$ and $t^{a...}_{...b}$ in full analogy (15.3.11). □

• We are now in a position to complete the two missing points in calculations concerning the Einstein tensor and the Einstein equations in Chapter 16. The first missing point was the fact that the variation of the Hilbert action (16.5.4) with respect to the metric tensor occurring in the *curvature forms* contributes just an *exact* form in the total variation of the form under the integral sign.

21.7.8 Consider the Hilbert action, (16.5.3) and (16.5.4),

$$S_H[g] = \int g_{ac}\hat{\Omega}^c_b(g) \wedge *_g(e^a \wedge e^b)$$

Check that the variation with respect to g occurring in the curvature form $\hat{\Omega}^c_b(g) \mapsto \hat{\Omega}^c_b(g + \epsilon h)$ is irrelevant since it only adds an exact form to the form under the integral sign. In order to do this check that

[465] One has to distinguish ω as a volume n-form and $\hat{\omega}^a_b$ as connection 1-forms.

(i) the variation of the curvature forms turns out to be

$$g_{ab} \mapsto g_{ab} + \epsilon h_{ab} \quad \Rightarrow \quad \hat{\Omega}^a_b \mapsto \hat{\Omega}^a_b + \epsilon \mathcal{D}\hat{\sigma}^a_b \qquad \mathcal{D}\hat{\sigma}^a_b \equiv d\hat{\sigma}^a_b + \hat{\omega}^a_c \wedge \hat{\sigma}^c_b - \hat{\omega}^c_b \wedge \hat{\sigma}^a_c$$

where $\hat{\sigma}^a_b$ is a certain 1-form of type ρ^1_1 (in the sense of (21.7.5); it might also be calculated explicitly in terms of h, but we will not need it)

(ii) the variation of the form under the integral sign then reads

$$g_{ac}\hat{\Omega}^c_b \wedge *_g(e^a \wedge e^b) \mapsto g_{ac}\hat{\Omega}^c_b \wedge *_g(e^a \wedge e^b) + \epsilon d\alpha \qquad \alpha \equiv g_{ac}\hat{\sigma}^c_b \wedge *_g(e^a \wedge e^b)$$

Hint: (i) imagine performing the variation "upstairs" on LM (see (21.5.3)); then $g_{ab} \mapsto g_{ab} + \epsilon h_{ab}$ gives rise to $\omega^a_b \mapsto \omega^a_b + \epsilon\sigma^a_b$; so as the new ω were again a connection form, σ^a_b is to be a horizontal 1-form of type Ad; the pull-back $\hat{\sigma}^a_b$ ("downstairs" on M) is of type ρ^1_1 in the sense of (21.7.5) so that $\mathcal{D}\hat{\sigma}^a_b = d\hat{\sigma}^a_b + \hat{\omega}^a_c \wedge \hat{\sigma}^c_b - \hat{\omega}^c_b \wedge \hat{\sigma}^a_c$; then $\hat{\Omega} \mapsto d(\hat{\omega} + \epsilon\hat{\sigma}) + (\hat{\omega} + \epsilon\hat{\sigma}) \wedge (\hat{\omega} + \epsilon\hat{\sigma}) = \hat{\Omega} + \epsilon\mathcal{D}\hat{\sigma}$; (ii) the form under the integral sign changes[466] by the term

$$\begin{aligned} g_{ac}\mathcal{D}\hat{\sigma}^c_b \wedge *_g(e^a \wedge e^b) &= g_{af}\mathcal{D}\hat{\sigma}^f_b \wedge \omega^{ab}{}_{cd} e^c \wedge e^d \\ &= \mathcal{D}\{g_{af}\hat{\sigma}^f_b \wedge \omega^{ab}{}_{cd} e^c \wedge e^d\} + \hat{\sigma}^f_b \wedge \mathcal{D}\{g_{af}\omega^{ab}{}_{cd} e^c \wedge e^d\} \\ &\equiv \mathcal{D}\alpha - \hat{\sigma}^a_b \wedge \mathcal{D}\{\omega^b{}_{acd} e^c \wedge e^d\} \end{aligned}$$

Due to $\mathcal{D}g_{ac} = 0 = \mathcal{D}\omega_{abcd}$ (the connection is metric, (21.7.7)) and $\mathcal{D}e^c \equiv T^c = 0$ (the connection is symmetric) the second term *vanishes*:

$$\mathcal{D}\{\omega^b{}_{acd} e^c \wedge e^d\} = \mathcal{D}\omega^b{}_{acd} \wedge e^c \wedge e^d + 2\omega^b{}_{acd}\mathcal{D}e^c \wedge e^d = 0$$

Since α is of type ρ^0_0 (there are no "free indices"), there holds $\mathcal{D}\alpha = d\alpha$ (see the table in (21.7.5)). □

- The second missing point concerned the variation of the Cartan action $S_C[e, \omega]$ with respect to the forms ω_{ab} (16.5.9).

21.7.9 Consider the Cartan action of a gravitational field

$$S_C \equiv S_C[e, \omega] := \int_{\mathcal{U}} \tau[e, \omega] \equiv \int_{\mathcal{U}} \frac{1}{2}\epsilon_{abcd}\Omega^{ab}(\omega) \wedge e^c \wedge e^d$$

Check that the response of the form under the integral sign to the variation of ω is as follows:

$$\tau[e, \omega + \epsilon\sigma] = \tau[e, \omega] + d\{\cdots\} - \epsilon\sigma^{ab} \wedge \epsilon_{abcd} T^c \wedge e^d$$

Hint: $\omega \mapsto \omega + \epsilon\sigma$ results in the change of the curvature 2-form

$$\begin{aligned} \Omega_{ab}(\omega) \mapsto \Omega_{ab}(\omega + \epsilon\sigma) &= d(\omega_{ab} + \epsilon\sigma_{ab}) + (\omega_{ac} + \epsilon\sigma_{ac}) \wedge \left(\omega^c_b + \epsilon\eta^{cd}\sigma_{db}\right) \\ &= \Omega_{ab}(\omega) + \epsilon\left(d\sigma_{ab} - \omega^c_a \wedge \sigma_{cb} - \omega^c_b \wedge \sigma_{ac}\right) \\ &= \Omega_{ab}(\omega) + \epsilon\mathcal{D}\sigma_{ab} \end{aligned}$$

[466] In the *four-dimensional* space-time; in n dimensions one should just add two times three dots, $\omega^{ab}{}_{cd} e^c \wedge e^d \mapsto \omega^{ab}{}_{c\ldots d} e^c \wedge \cdots \wedge e^d$ and everything goes through as before.

Then,

$$\begin{aligned}\tau[e, \omega + \epsilon\sigma] &= \frac{1}{2}\epsilon_{abcd}(\Omega^{ab} + \epsilon\mathcal{D}\sigma^{ab}) \wedge (e^c \wedge e^d) \\ &= \tau[e, \omega] + \epsilon\frac{1}{2}\{\epsilon_{abcd}\mathcal{D}\sigma^{ab} \wedge e^c \wedge e^d\} \\ &= \tau[e, \omega] + \mathcal{D}\left\{\epsilon\frac{1}{2}\epsilon_{abcd}\sigma^{ab} \wedge e^c \wedge e^d\right\} - \epsilon\frac{1}{2}\sigma^{ab} \wedge \mathcal{D}\{\epsilon_{abcd}e^c \wedge e^d\} \\ &= \tau[e, \omega] + \mathcal{D}\{\cdots\} - \frac{\epsilon}{2}\sigma^{ab} \wedge \{\mathcal{D}\epsilon_{abcd} \wedge e^c \wedge e^d + 2\epsilon_{abcd}\mathcal{D}e^c \wedge e^d\} \\ &= \tau[e, \omega] + d\{\cdots\} - \epsilon\sigma^{ab} \wedge \epsilon_{abcd}T^c \wedge e^d\end{aligned}$$

(we used $\mathcal{D}\epsilon_{abcd} = \mathcal{D}\omega_{abcd} = 0$, since $\omega_{g(e)}$ is metric (21.7.7)). □

• Eventually, let us compare the actions for the pure gauge field (21.5.3) and for the pure gravitational field (16.5.3). If we write them side by side (and omit irrelevant coefficients in front of them)

$$\int k_{ij}\mathcal{F}^i \wedge *_g\mathcal{F}^j \qquad \int \hat{\Omega}_{ab} \wedge *_g(e^a \wedge e^b)$$

we do not see, at first glance, too many similarities. A moment later we realize, however, that actually $\hat{\Omega}$ in the second action is a particular case of $\mathcal{F}$ in the first one (for the *linear* connection) and also the summation over the indices in the second one is a particular case of the summation in the first one (for the *orthogonal* Lie algebra, see (12.3.11), since the connection is metric, (15.6.6)). The similarity is thus considerably closer than it seemed to be at first glance: *both* the action integrals have the structure of the scalar product of 2-forms with values in a Lie algebra

$$\int k_{ij}\alpha^i \wedge *_g\beta^j \equiv \langle\alpha, \beta\rangle_k$$

Now we can see clearly the *essential difference* between them: whereas the first action contains the curvature 2-form twice (it is *quadratic* in the curvature, $\alpha = \beta = \mathcal{F}$), the second one contains it only once (being thus *linear* in the curvature, $\alpha = \mathcal{F}$) and the second needed 2-form is realized *differently* there. The role of the form β is played by the 2-form $e^a \wedge e^b$; note that this form satisfies all the necessary transformational properties (namely type Ad) in spite of being completely independent of the connection form $\mathcal{A}$. We see that the action of the gravitational field turns out to be *exceptional* among the actions of gauge fields and that its exceptionality (i.e. linearity in the curvature) may be ultimately traced back to the specific feature of the frame bundle LM, the existence of the above-mentioned "horizontal building material" with just the needed transformation type[467] even without connection.

We close this section with a problem which returns to the concept of *scalar density*.

[467] "Upstairs" on LM (or on the subbundle OM (20.5.5)) we are speaking about the 2-form $\theta^a \wedge \theta^b$, which happens to be a *horizontal* 2-form of just the needed type, being at the same time constructed exclusively from the "canonical building material" (which is available regardless of any *connection* whatsoever).

21.7.10 Contemplate an equivariant function Φ of type ρ on LM, where $\rho(A) = (\det A)^\lambda$, i.e. obeying (for any $A \in GL(n, \mathbb{R})$ and some $\lambda \in \mathbb{R}$)

$$\Phi(eA) = \rho(A^{-1})\Phi(e) \equiv (\det A)^{-\lambda}\Phi(e)$$

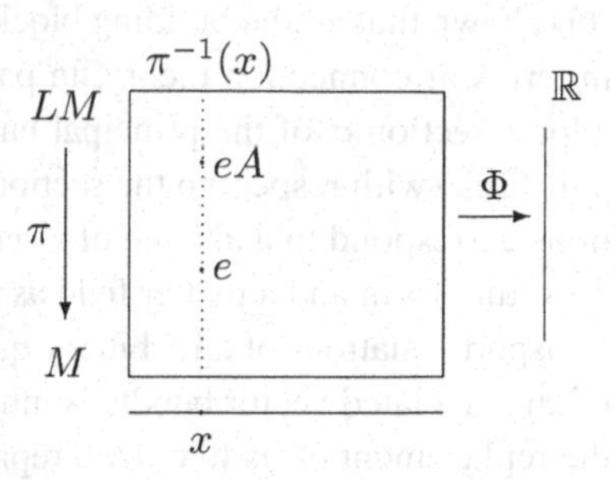

Such a field, as well as its pull-back living on M, is called the *scalar density of weight* λ (see (5.7.1) and the text before (6.3.8)). Consider two local sections σ and $\hat{\sigma}$ associated with two frame fields e_a and $\hat{e}_a$ respectively. Assume that they are related by $\hat{\sigma}(x) = \sigma(x)A(x)$ (i.e. $\hat{e}_a(x) = A^b_a(x)e_b(x) \equiv (R_A e)_a(x)$). Denote the corresponding pull-backs by $\varphi(x) = (\sigma^*\Phi)(x) \equiv \Phi(\sigma(x))$ and $\hat{\varphi}(x) = (\hat{\sigma}^*\Phi)(x) \equiv \Phi(\hat{\sigma}(x))$. Check that

(i) the fields $\varphi(x)$ and $\hat{\varphi}(x)$ are related by

$$\hat{\varphi}(x) = (\det A(x))^{-\lambda}\varphi(x)$$

In particular, if the sections correspond to *coordinate* frame fields ∂_μ and ∂'_μ, the formula reads

$$\varphi'(x') = (J(x))^\lambda \varphi(x) \equiv (\det(\partial x'/\partial x))^\lambda \varphi(x)$$

(φ' denotes the field with respect to the coordinates x'^μ; for $\lambda = 0$ the representation ρ is trivial and we obtain a true function (scalar field) on M)

(ii) the covariant derivative $\mathcal{D}\varphi(x)$ of $\varphi(x)$ is given by

$$\mathcal{D}\varphi = d\varphi + \lambda\hat{\omega}^a_a\varphi \equiv d\varphi + \lambda(\operatorname{Tr}\hat{\omega})\varphi$$

and, in particular (for the "coordinate" section σ),

$$\mathcal{D}\varphi = d\varphi + \lambda\left(\Gamma^\nu_{\nu\mu}\,dx^\mu\right)\varphi \qquad \text{i.e.} \quad \mathcal{D}_\mu\varphi = \partial_\mu\varphi + \lambda\Gamma^\nu_{\nu\mu}\varphi$$

Hint: (i) $\partial'_\mu = (\partial x^\nu/\partial x'^\mu)\partial_\nu \equiv (J^{-1})^\nu_\mu(x)\partial_\nu \overset{!}{=} A^\nu_\mu(x)\partial_\nu \Rightarrow A^\nu_\mu(x) = (J^{-1})^\nu_\mu(x)$; (ii) according to (12.1.7), (20.4.6) and (21.2.4) we have $\sigma^*(d\Phi + \rho'(\omega)\Phi) = d\varphi + \lambda\operatorname{Tr}(\sigma^*\omega)\sigma^*\Phi = d\varphi + \lambda(\hat{\omega}^a_a)\varphi$. □

• The case of *tensor* densities (where $\rho(A) = (\det A)^\lambda \rho^r_s(A)$, see Section 19.6) may be treated by a straightforward modification of the approach discussed above (and it is left as an exercise for the reader).

Summary of Chapter 21

A link between connections on a principal G-bundle and gauge field theory (known from physics) is systematically built here. First, a standard "physical" approach is briefly introduced for the convenience of the reader who is not familiar with these ideas from physics. Namely, a "global" G-symmetry of an action is made "local." This is achieved by adding new fields with quite definite transformation properties and interaction with the initial fields.

It is shown that all the building blocks of the gauge scheme possess a natural interpretation in terms of connection theory. In particular, fixing of the gauge is given by the choice of a local section σ of the principal bundle, gauge potentials (in this gauge) are obtained by pull-back (with respect to the section) of a connection form to the base, gauge transformations correspond to a change of a section, field strength is obtained as the pull-back of the curvature form and a matter field as the pull-back of an equivariant function on P. Parallel transport equations of an arbitrary quantity of type ρ in a gauge σ are derived. The concept of an associated vector bundle is introduced (it arises from a principal bundle as a result of the replacement of its fiber by a representation space of the group G). The structure of the (locally) gauge invariant action is given and the equations of motion are derived (they generalize Maxwell's equations of electrodynamics, which turns out to be a gauge theory with group $U(1)$). Noether's theorem is introduced, providing a link between the symmetries of the action integral and the conserved quantities. The theorem sheds new light on some older results in this direction – the relation between conservation laws and the energy–momentum tensor in the field theory as well as exact Cartan symmetries in Hamiltonian mechanics. In the last section we return to the frame bundle LM and introduce the canonical 1-form θ with values in $\mathbb{R}^n$ which is related to the torsion on M and learn how to use the exterior covariant derivative $\mathcal{D}$ on the base M.

$\phi \mapsto e^{-i\alpha(x)}\phi, \quad A \mapsto A + d\alpha(x)$	$U(1)$-local gauge transformation	(21.1.2)
$\hat{\sigma}(x) = \sigma(x)S(x) \equiv R_{S(x)}\sigma(x)$	Two sections related via $S \in G^{\mathcal{U}}$	(21.2.1)
$\hat{\phi} = B^{-1}\phi$	Local gauge transformation of a matter field	(21.2.5)
$\hat{\mathcal{A}} = B^{-1}\mathcal{A}B + B^{-1}dB$	The same for the gauge potential	(21.2.5)
$\hat{\mathcal{F}} = B^{-1}\mathcal{F}B$	The same for the field strength	(21.2.5)
$\dot{v} + \langle \mathcal{A}, \dot{\gamma}\rangle v = 0$	Equation of parallel transport	(21.3.2)
$S[\phi, \mathcal{A}] = -\frac{1}{2}\langle \mathcal{D}\mathcal{A}, \mathcal{D}\mathcal{A}\rangle_k$ $+\frac{1}{2}\langle \mathcal{D}\phi, \mathcal{D}\phi\rangle_h - (m^2/2)\langle \phi, \phi\rangle_h$	Action of the coupled system $(\phi, \mathcal{A})$	(21.5.6)
$\mathcal{D}^{+}\mathcal{F} = -\mathcal{J}, \quad \mathcal{D}\mathcal{F} = 0,$ $(\mathcal{D}^{+}\mathcal{D} - m^2)\phi = 0$	Corresponding field equations	(21.5.6)
$S[\rho(e^{\epsilon s(x)})\psi] = S[\psi] + \epsilon\langle ds, j\rangle_k$	Computation of Noether currents j	(21.6.1)
$j_i = T(\xi_{E_i}, \cdot)$	Noether currents due to Killing vectors	(21.6.6)
$\Theta := D\theta$	Where torsion sits in the LM formalism	(21.7.2)

22*

Spinor fields and the Dirac operator

• So far we have not mentioned yet another type of geometrical object, which may live on (some Riemannian) manifolds, the *spinor field*. These fields play an important role both in physics and in mathematics. The way in which Paul Dirac arrived at this concept may serve as an amusing (as well as highly instructive) example of how sometimes a discovery of truly the highest importance[468] may originate from assumptions which are actually regarded as erroneous from a present-day perspective.

Dirac tried to find a relativistic formulation of quantum mechanics, i.e. to generalize the non-relativistic theory, based on the Schrödinger equation. This problem was also attacked before by several eminent physicists and their activity resulted in the equation nowadays known as the *Klein–Gordon* equation (16.3.7).

(In addition to Klein and Gordon it was independently discovered by other people, for example Fock. Amazingly, it was even discovered by Schrödinger himself before he found the equation that now bears his name. However, when he computed the energy spectrum of the hydrogen atom from the relativistic equation, he found that although it comes out "roughly correct," it differs slightly from experiment at the finer level of accuracy. That is why he only published the non-relativistic approximation of the equation (leading just to the well-known "roughly correct" formula for the spectrum). *This* equation was then "relativistically generalized" by other authors.)

There were some serious problems with the physical interpretation of the Klein–Gordon equation resulting from the fact that it is a *second-order* equation in time (contrary to the Schrödinger equation, which is only first order in time). In spite of this fact the generally accepted opinion in the community was that the relativistic equation was already known. This was also the belief of Niels Bohr. When he asked Dirac at some conference what he was working on just now he was indeed very much surprised to hear that Dirac was trying to find *the* relativistic wave equation. Dirac's surprise was, however, no less. He was not able to understand why the community accepted a second-order equation suffering from serious interpretational problems. He strongly believed that the true relativistic equation should be a *first-order* differential equation. Eventually he indeed succeeded in

[468] That is, the Dirac equation and the predictions (which were later confirmed) which stem from the equation (just to mention the existence and properties of antiparticles, corrections to the hydrogen energy spectrum and automatic appearance of spin $\frac{1}{2}$).

achieving this objective and since then we have known the Dirac equation and spinor fields.[469]

A possible way of deriving the Dirac equation consists in "splitting" of the Klein–Gordon second-order operator[470] $\Box - m^2$, i.e. in writing down this operator as a *product* of *two first-order* operators $\hat{A}$, $\hat{B}$; the Klein–Gordon equation then would become

$$(\Box - m^2)\psi \equiv \hat{A}\hat{B}\psi = 0$$

and we could postulate a "deeper" equation

$$\hat{B}\psi = 0$$

Clearly the solutions of this equation are at the same time also solutions of the original equation (just apply $\hat{A}$ from the left), however, the converse may not be true. This "deeper" equation then represents the first-order equation we are looking for.

If the product $\hat{A}\hat{B}$ is to give $\Box - m^2$, the individual factors $\hat{A}$ and $\hat{B}$ might (according to the formula $u^2 - v^2 = (u+v)(u-v)$) look like

$$\hat{A} = \hat{a} + m\hat{1} \qquad \hat{B} = \hat{a} - m\hat{1}$$

with $\hat{a}$ being a strictly differential operator, which may be regarded as a "square root" of the d'Alembert operator $\Box$, i.e. it satisfies

$$\hat{a}\hat{a} = \Box \equiv -\partial_\mu \partial^\mu \equiv -\eta^{\mu\nu}\partial_\mu \partial_\nu \equiv \Delta - \partial_t{}^2$$

Now, since $\hat{a}$ is a first-order operator in the *time* variable and in special relativity time and space form a whole in which these two parts are to be regarded in many respects as being equal, it is natural to expect that $\hat{a}$ would be a first-order operator in the *spatial* variables as well. Considerations of this type eventually lead to an ansatz

$$\hat{A} = i\gamma^\mu \partial_\mu + m\hat{1} \qquad \hat{B} = i\gamma^\mu \partial_\mu - m\hat{1}$$

where γ^μ are unknown *numerical* coefficients. The coefficients are to be fixed so that $\hat{a}\hat{a} = \Box$ holds. This gives the following restrictions for the unknown coefficients:

$$\gamma^\mu \gamma^\nu + \gamma^\nu \gamma^\mu = 2\eta^{\mu\nu}$$

They might look fairly innocent at first sight, but appearances are sometimes deceptive. Actually these equations clearly say that γ^a *cannot be numbers*, since one can hardly find numbers which would willingly have a good mind to satisfy (for example)

$$\gamma^2\gamma^3 + \gamma^3\gamma^2 = 2\eta^{23} \equiv 0 \qquad \text{i.e.} \quad \gamma^2\gamma^3 = -\gamma^3\gamma^2$$

[469] The present-day point of view is based on relativistic quantum *field theory* (rather than relativistic quantum *mechanics*). Here *both* equations are accepted to be correct. So Dirac was in fact lucky that he was not aware of the fact that the blame with respect to the Klein–Gordon equation is based on its erroneous interpretation and that the equation itself is perfectly all right – he would not have been motivated enough to replace it by another one and he would have missed the chance to discover an *additional* fundamental physical equation.

[470] We mean the operator in the usual ("flat") *Minkowski* space-time $E^{1,3}$, i.e. for the moment "only" the *special* theory of relativity is considered.

Dirac realized, however, that the conditions *may* still be fulfilled by square *matrices* (the function ψ thus inevitably becomes a *column* vector) and the *Dirac equation*

$$(i\partial\!\!\!/ - m\hat{1})\psi = 0 \qquad \partial\!\!\!/ \equiv \gamma^\mu \partial_\mu \equiv \text{Dirac operator}$$

came to light (the matrices γ^μ form, as we will see, a representation of the *Clifford algebra*; sometimes also the factor of i enters the definition of the Dirac operator). The following detailed analysis of the equation revealed that in order for everything to be relativistic invariant the column vector ψ should transform according to a new "spinor" representation of the group $SO(1,3)$ (its rather peculiar properties reflect the fact that it is actually a representation of the *covering* group and, if treated as a representation of $SO(1,3)$, it becomes *two-valued* (13.3.16)) and the spinors (column vectors $\psi(x)$) came to light as well.

All of this analysis took place in the Minkowski space-time and in Cartesian coordinates. Recall now the situation with the parallel transport of vectors and tensors. In the (pseudo-)Euclidean spaces in the Cartesian coordinates the procedure turned out to be technically completely trivial, but in a more general situation, when the two restrictions were relaxed, this idea brought a fruitful development and enough material for not only the whole of Chapter 15, but even for a far-reaching further generalization in subsequent chapters (Chapters 19–21). A similar situation is repeated here. After a short time works (by Fock and Weyl) appeared which generalized the formalism of the spinor fields from the Minkowski space-time $E^{1,3}$ to any "curved" space-time (M,g) (motivated by applications in the *general* theory of relativity; orthonormal tetrad fields and Ricci coefficients of rotations were used for that). This already becomes slightly more complicated (however, just as was the case with the connection, also more interesting and *richer*) and it needs once more a separate chapter to become familiar with these ideas.

In this chapter we introduce an approach of modern differential geometry based on fiber bundles. Spinor fields will be described in the same way as we have already described tensor fields (and other matter fields in the theory of gauge fields), as "objects of type ρ." So they are related to an appropriate principal bundle and a function of (some particular) type ρ on the total space of the bundle (they may then be pulled back to the space-time by means of a local *section*). The description of this important bundle needs, however, some preparatory material at the level of linear algebra – we learn necessary facts about the covering groups of orthogonal groups (generalizing the covering $SU(2) \to SO(3)$, which we studied in Chapter 13). The whole approach is based on the theory of Clifford algebras. That is why we start with this material right now.

22.1 Clifford algebras $C(p,q)$

• The real *Clifford algebra* $C(L,g)$ is a certain finite-dimensional associative algebra with unit (see Appendix A.2), which may be canonically associated with the pair (L,g), a linear space L endowed with a symmetric bilinear form g. We will restrict ourselves to the most important case, when g is non-degenerate;[471] so g will be a *metric tensor* with signature (p, q) in what follows. In particular, if we take $(\mathbb{R}^n, \eta^{ab})$ (the "canonical" coordinate linear space, $n = p + q$) as (L, g), the corresponding algebra $C(L, g)$ will be denoted by $C(p, q)$. Let us turn to the description of the algebra $C(L, g)$.

Recall that in (5.3.4) we obtained the *exterior* algebra ΛL^* by means of the *factorization* of the "strictly covariant" tensor algebra with respect to an appropriate ideal I. In order to construct the algebra $C(L, g)$ we may proceed in like manner, modifying somewhat, however, the ideal of the same starting algebra. So, consider again the linear space

$$T_{(\cdot)}(L) := \bigoplus_{r=0}^{\infty} T_r^0(L) \equiv T_0^0(L) \oplus T_1^0(L) \oplus T_2^0(L) \oplus \cdots$$
$$\equiv \mathbb{R} \oplus L^* \oplus T_2^0(L) \oplus \cdots \text{ (up to infinity)}$$

endowed with the product induced by the tensor product of the homogeneous terms (just as it was the case in Section 2.4). However, here we will consider the ideal J generated by elements of the form

$$\alpha \otimes \alpha - g(\alpha, \alpha) \qquad \alpha \in L^* \quad g(\alpha, \alpha) \equiv g^{ab}\alpha_a\alpha_b$$

so that the elements of the ideal J are sums of terms of the form

$$t_1 \otimes (\alpha \otimes \alpha - g(\alpha, \alpha)) \otimes t_2 \qquad t_1, t_2 \in T_{(\cdot)}(L) \quad \alpha \in L^*$$

22.1.1 Consider the algebra $T_{(\cdot)}(L)$ and a set J described above. Check that

(i) the set J is indeed a two-sided ideal of the algebra $T_{(\cdot)}(L)$
(ii) the same ideal is also generated by elements of the form

$$\alpha \otimes \beta + \beta \otimes \alpha - 2g(\alpha, \beta) \qquad \alpha, \beta \in L^*$$

(iii) the factorization under consideration leads to a (bilinear and associative) multiplication of classes, which satisfies the relation

$$[e^a][e^b] + [e^b][e^a] = 2g^{ab}[1]$$

Hint: as in (5.3.4). □

- If we already deal with the resulting factor-algebra

$$C(L, g) := T_{(\cdot)}(L)/J \qquad \text{Clifford algebra}$$

then the square brackets denoting equivalence classes are usually *omitted* and what remains is the key relation for the (bilinear and associative) *Clifford product*

$$e^a e^b + e^b e^a = 2g^{ab}$$

So at the level of manipulations of letters the element of the Clifford algebra is an expression

$$u = \hat{u} + \hat{u}_a e^a + \hat{u}_{ab} e^a e^b + \hat{u}_{abc} e^a e^b e^c + \cdots \qquad \hat{u}, \hat{u}_a, \hat{u}_{ab}, \ldots \in \mathbb{R}$$

[471] Also the other extreme case is of considerable interest, when g *vanishes*. Then we get the good old *exterior* algebra ΛL^* from Section 5.3.

and in order to multiply two such elements we multiply each term of the first one with each term of the second one, always being allowed to simplify the result using the basic relation $e^a e^b + e^b e^a = 2g^{ab}$. For example, we get

$$\begin{aligned}(1 + 2e^1 e^2)(e^1 + 3e^2 e^3) &= e^1 + 3e^2 e^3 + 2e^1 e^2 e^1 + 6e^1 e^2 e^2 e^3 \\ &= e^1 + 3e^2 e^3 - 2e^1 e^1 e^2 + 6e^1 g^{22} e^3 \\ &= (e^1 - 2g^{11} e^2) + (3e^2 e^3 + 6g^{22} e^1 e^3) \\ &\in C^1(L, g) \oplus C^2(L, g)\end{aligned}$$

where, at the end of the computation, we have grouped together the terms which contain the product of an *equal number* of e^as, all the products being always expressed in order of *increasing* values of indices (for example, $e^2 e^3$ rather than $-e^3 e^2$, see below). The space of products of just p entries (ordered in this way), is denoted by $C^p(L, g)$.

22.1.2 Consider the Clifford algebra $C(L, g)$. Check that

(i) $C(L, g)$ may be written as the direct sum of the subspaces $C^p(L, g)$, where $p = 0, 1, \dots, n$, in which we may use a basis in the form of

$$\begin{array}{lll} C^0(L, g) = \mathbb{R} & 1 & \\ C^1(L, g) = L^* & e^a & a = 1, \dots, n \\ C^2(L, g) & e^a e^b & a < b \\ C^3(L, g) & e^a e^b e^c & a < b < c \\ \dots & & \dots \\ C^n(L, g) & e^1 \dots e^n & \end{array}$$

so that $C(L, g)$ is isomorphic *as a linear space* to the *exterior* algebra[472] $\wedge L^*$ of the space L^* (and just as in $\wedge L^*$, the subspace $C^0(L, g)$ may be identified with $\mathbb{R}$ and $C^1(L, g)$ with L^*)

(ii) the resulting algebra $C(L, g)$ has dimension 2^n, where $n \equiv p + q$ is the dimension of the space L

(iii) the ideal J is generated by *inhomogeneous* elements of the initial $\mathbb{Z}$-graded algebra $T_{(\cdot)}(L)$ so that $\mathbb{Z}$-*grading* does *not* pass to the factor-algebra $C(L, g)$;[473] however, if regarded from the point of view of the weaker $\mathbb{Z}_2$-grading, the generators *are* homogeneous (even), and so this $\mathbb{Z}_2$-grading *passes* to the quotient $C(L, g)$

$$C(L, g) = C^0(L, g) \oplus C^1(L, g) \oplus C^2(L, g) \oplus \dots \oplus C^n(L, g)$$

(only) as a linear space

$$C(L, g) = C^+(L, g) \oplus C^-(L, g)$$

(also) as an algebra

[472] However, when regarded as an *algebra* it is substantially different, since the *multiplication* is different. Put another way, the elements of the Clifford algebra are (inhomogeneous) exterior forms. They are, however, multiplied using the Clifford rule rather than using the exterior product $\wedge$ (nor is the exterior product used for their presentation, and therefore the fact that they actually "are" exterior forms is somewhat obscure). There is also an algebra in which *both* products are defined in the linear space of forms $\wedge L^*$ (and in addition also the standard *scalar* product of forms (α, β) form (5.8.4)); it is called the *Kähler–Atiyah algebra*.

[473] The algebra $C(L, g)$ *is* $\mathbb{Z}$-graded at the level of the *linear space* (it is the direct sum of $C^p(L, g)$); however, this grading is not inherited from the tensor algebra and (more importantly) it is *not respected* by the *product*.

where

$$C^{+}(L,g) := C^{0}(L,g) \oplus C^{2}(L,g) \oplus C^{4}(L,g) \oplus \cdots$$
$$C^{-}(L,g) := C^{1}(L,g) \oplus C^{3}(L,g) \oplus C^{5}(L,g) \oplus \cdots$$

Hint: (i) the rule $e^a e^b + e^b e^a = 2g^{ab}$ evidently enables one to rewrite any element $C(L, g)$ in the form which corresponds to an inhomogeneous form, for example

$$e^a e^b = \frac{1}{2}\left(e^a e^b + e^b e^a + e^a e^b - e^b e^a\right) = \frac{1}{2}\left(2g^{ab} + e^a \wedge e^b\right)$$
$$= g^{ab} + \frac{1}{2} e^a \wedge e^b \in \Lambda^0 L^* \oplus \Lambda^2 L^*$$

There is some extra work needed for products of more basis covectors, but it is clear that each *symmetric part* may be rewritten by means of $e^a e^b + e^b e^a = 2g^{ab}$ as an expression which contains two *fewer* terms e^a, so that at the very end only a sum of *completely antisymmetric* terms, which correspond ($e^a \dots e^b \leftrightarrow e^a \wedge \dots \wedge e^b$, $a < \dots < b$) to some unique elements of the *exterior* algebra, remains, (5.2.13); (ii) and this algebra *has* the dimension 2^n; (iii) proceed as in (5.3.4). □

• Now, let us look in detail at what the Clifford algebras with the lowest values of (p, q) "indeed" look like. We will begin with the most trivial case, $C(0, 0)$. We have a zero-dimensional L, the whole tensor algebra $T_{(\cdot)}(L)$ reduces to $\mathbb{R}$ and there is no ideal to make a factorization, so that this $\mathbb{R}$ is at the same time the final result of all this backbreaking labor: we get $C(0, 0) = \mathbb{R}$.

Let us try to take a courageous step to $C(0, 1)$. We have a one-dimensional space, a single basis covector e^1 and the general relation $e^a e^b + e^b e^a = 2\eta^{ab}$ reduces to a single restriction $e^1 e^1 = -1$. So the general element of the algebra reads $a + be^1$ and the linear combination and the multiplication of such elements give

$$(a + be^1) + \lambda(\hat{a} + \hat{b}e^1) := (a + \lambda\hat{a}) + (b + \lambda\hat{b})e^1$$
$$(a + be^1)(\hat{a} + \hat{b}e^1) := a\hat{a} + a\hat{b}e^1 + b\hat{a}e^1 + b\hat{b}e^1 e^1$$
$$\equiv (a\hat{a} - b\hat{b}) + (a\hat{b} + b\hat{a})e^1 \qquad a, \hat{a}, b, \hat{b}, \lambda \in \mathbb{R}$$

If we envisage in these formulas the imaginary unit $i \equiv \sqrt{-1}$ instead of e^1, we will obtain just the standard rules for the basic manipulations with *complex numbers*, so that $a + be^1 \leftrightarrow a + bi$ is an isomorphism of the two-dimensional algebra $C(0, 1)$ with the algebra of complex numbers $\mathbb{C}$ (regarded as a two-dimensional algebra over $\mathbb{R}$).

In a similar way we may also sense other Clifford algebras with sufficiently small $n \equiv p + q$. Let us see how this comes out.

22.1.3 Find explicitly how Clifford algebras work for $n \leq 2$ (with every possible signature (p, q) for given $n = p + q$). Check that the following isomorphisms hold:

$n = 0$	$C(0,0) = \mathbb{R}$		
$n = 1$	$C(1,0) = \mathbb{R} \oplus \mathbb{R}$	$C(0,1) = \mathbb{C}$	
$n = 2$	$C(2,0) = \mathbb{R}(2)$	$C(1,1) = \mathbb{R}(2)$	$C(0,2) = \mathbb{H}$

where $\mathbb{R}(n) = M_n(\mathbb{R})$ = the algebra of real $n \times n$ matrices and $\mathbb{H}$ is the *algebra of quaternions* (see Appendix A.2).

Hint: all of them by an appropriate choice of bases; for example, $(1, 0)$: we have $e^1 e^1 = +1 \Rightarrow$ in the basis $\frac{1}{2}(1 \pm e^1)$, i.e. if the elements are written in the form $(a, b) \leftrightarrow \frac{1}{2}a(1 + e^1) + \frac{1}{2}b(1 - e^1)$, the multiplication reads $(a, b)(\hat{a}, \hat{b}) = (a\hat{a}, b\hat{b})$, which corresponds (see Appendix A.2) to $\mathbb{R} \oplus \mathbb{R}$; $(2, 0)$: we have $e^1 e^1 = e^2 e^2 = +1 \Rightarrow$ if $e^1 \leftrightarrow \sigma_1$ and $e^2 \leftrightarrow \sigma_3$, we have an isomorphism with $\mathbb{R}(2)$; for $(1, 1)$ set $e^1 \leftrightarrow \sigma_1$ and $e^2 \leftrightarrow i\sigma_2$, leading again to $\mathbb{R}(2)$ and finally for $(0, 2)$ we have $e^1 \leftrightarrow -i\sigma_1$ and $e^2 \leftrightarrow -i\sigma_2$, from where $e^1 e^2 \leftrightarrow -i\sigma_3$; the general element then corresponds to the matrix $a_0\mathbb{I} + a_l(-i\sigma_l)$, which may be identified with the quaternions, since $-i\sigma_l$, $l = 1, 2, 3$ behave exactly like quaternionic "imaginary units" i, j, k. □

• The isomorphisms described above are non-canonical. Let us look, for example, at the isomorphism $C(1, 1) = \mathbb{R}(2)$.

22.1.4 A straightforward check shows that for each ϕ and an arbitrary choice of $\pm$ the assignment

$$1 \mapsto \mathbb{I} \qquad e^1 \mapsto \sigma_1 \cos\phi + \sigma_3 \sin\phi \qquad e^2 \mapsto \pm i\sigma_2$$
$$(\Rightarrow \quad e^1 e^2 \mapsto \mp\sigma_3 \cos\phi \pm \sigma_1 \sin\phi)$$

describes an isomorphism $C(1, 1) \to \mathbb{R}(2)$ (so that the choice made in the hint to problem (22.1.3) corresponds to $\phi = 0, \pm = +$). □

• In principle we might go on in this way for higher n. However, actually our passion would ooze away very soon. A fairly passable way is offered by recurrent formulas.

22.1.5 Let us see, for example, how the tensor product $C(p, q) \otimes C(2, 0)$ works. Check that the following isomorphism of algebras holds:

$$C(p, q) \otimes C(2, 0) = C(q + 2, p)$$

Hint: let

$$\begin{array}{llll} C(p, q) & \text{have the generators} & E^i E^j + E^j E^i = 2\eta^{ij} & \eta^{ij} \text{ of type } (p, q) \\ C(2, 0) & \text{have the generators} & e^a e^b + e^b e^a = 2\eta^{ab} & \eta^{ab} \text{ of type } (2, 0) \end{array}$$

Then we may generate a complete basis of the tensor product of the algebras from the elements $\mathcal{E}^A = (\mathcal{E}^i, \mathcal{E}^a)$

$$\mathcal{E}^i := E^i \otimes e^1 e^2 \qquad \mathcal{E}^a := 1 \otimes e^a$$

and moreover there holds

$$\mathcal{E}^A \mathcal{E}^B + \mathcal{E}^B \mathcal{E}^A = 2\eta^{AB} 1 \otimes 1 \qquad \eta^{\text{AB}} = \text{diag}\,(-\eta^{ij}, \eta^{ab}) \leftrightarrow (q + 2, p)$$

so that actually they generate the whole Clifford algebra $C(q + 2, p)$. □

• In a similar way we may verify another two useful recurrence formulas; when all three together are displayed they read

$$C(p,q)\otimes C(2,0)=C(q+2,p)$$
$$C(p,q)\otimes C(1,1)=C(p+1,q+1)$$
$$C(p,q)\otimes C(0,2)=C(q,p+2)$$

Now consider a non-negative quadrant in the pq-plane. Interpret these formulas as appropriate shifts in the plane (the first one, say, performs the shift $(p,q)\mapsto(q+2,p)$). One can see easily (as is often the case, a small diagram might be useful) that the three steps are enough (and, moreover, all of them are needed) to get to any point on the integer lattice lying in the quadrant under consideration, starting from the small triangle near the origin which corresponds to the algebras examined in problem (22.1.3). Put differently, these formulas enable one to express whatever Clifford algebra we need as a tensor product of those particular small algebras which we discussed in (22.1.3). For example,

$$C(3,0)=C(0,1)\otimes C(2,0)=\mathbb{C}\otimes\mathbb{R}(2)$$
$$C(1,3)=C(0,2)\otimes C(1,1)=\mathbb{H}\otimes\mathbb{R}(2)$$

The results are already written in the form of a tensor product of *known* algebras. The products may even be simplified further with the help of several isomorphisms which may be easily verified.

22.1.6 Check that the following isomorphisms of algebras (where $\otimes$ is over $\mathbb{R}$) hold:

$$\mathbb{C}\otimes\mathbb{C}=\mathbb{C}\oplus\mathbb{C}\qquad \mathbb{H}\otimes\mathbb{C}=\mathbb{C}(2)\qquad \mathbb{H}\otimes\mathbb{H}=\mathbb{R}(4)$$
$$\mathbb{R}(m)\otimes\mathbb{R}(n)=\mathbb{R}(mn)\qquad \mathbb{R}(m)\otimes\mathbb{C}=\mathbb{C}(m)\qquad \mathbb{R}(m)\otimes\mathbb{H}=\mathbb{H}(m)$$

Hint: all of them by a convenient choice of bases; for example, for $\mathbb{C}\otimes\mathbb{C}=\mathbb{C}\oplus\mathbb{C}$: let $(1,i)$ denote the basis in the first copy of $\mathbb{C}$ and $(1',i')$ in the second one; then in $\mathbb{C}\otimes\mathbb{C}$ we may choose the basis $\hat{1}=1\otimes 1'$, $\hat{\imath}=i\otimes 1'$, $\hat{\jmath}=1\otimes i'$, $\hat{k}=i\otimes i'$; for the new basis

$$A=\frac{1}{2}(\hat{1}-\hat{k})\qquad B=\frac{1}{2}(\hat{\imath}+\hat{\jmath})\qquad a=\frac{1}{2}(\hat{1}+\hat{k})\qquad b=\frac{1}{2}(\hat{\imath}-\hat{\jmath})$$

already holds

$$A^2=A,\ AB=BA=B\qquad a^2=a,\ ab=ba=b\qquad Cc=cC=0$$
$$B^2=-A\qquad\qquad b^2=-a$$

where C denotes anything "big" (A or B) and c is anything "small"; this is, however, just a basis in $\mathbb{C}\oplus\mathbb{C}$; $\mathbb{R}(n)\otimes A=A(n)$: the general element from the left algebra is $x=x_j^{i\alpha}E_i^j\otimes e_\alpha=E_i^j\otimes x_j^{i\alpha}e_\alpha=:E_i^j\otimes x_j^i$ and this defines a bijection $\mathbb{R}(n)\otimes A\to A(n)$, $x\mapsto x_j^i$; check that the multiplication works; $\mathbb{H}\otimes\mathbb{H}=\mathbb{R}(4)$: the question is whether $\mathbb{H}\otimes\mathbb{H}\overset{?}{=}\mathbb{R}(2)\otimes\mathbb{R}(2)$; a basis in $\mathbb{H}$ is $(\mathbb{I}\equiv\sigma_0,-i\sigma_1,-i\sigma_2,-i\sigma_3)\equiv E_\mu$ and in $\mathbb{R}(2)$ in turn $(\mathbb{I}\equiv\sigma_0,-\sigma_1,-i\sigma_2,-\sigma_3)\equiv\mathcal{E}_\mu$, so that in $\mathbb{H}\otimes\mathbb{H}$ it is $E_\mu\otimes E_\nu$ and in $\mathbb{R}(2)\otimes\mathbb{R}(2)$ it is $\mathcal{E}_\mu\otimes\mathcal{E}_\nu$; check that $E_\mu\otimes E_\nu\mapsto\mathcal{E}_\mu\otimes\mathcal{E}_\nu$ is an isomorphism, etc. □

• Now we may express the tensor products mentioned above more explicitly: $\mathbb{C} \otimes \mathbb{R}(2) = \mathbb{C}(2)$ and $\mathbb{H} \otimes \mathbb{R}(2) = \mathbb{H}(2)$, so that we get simple results

$$C(3,0) = \mathbb{C}(2)$$
$$C(1,3) = \mathbb{H}(2)$$

In this way we may express step by step *all* real Clifford algebras $C(p,q)$. Let us look first at the "edge" cases, $C(p,0)$ and $C(0,p)$.

22.1.7 Use this technique to check that we get the following results:

(i) for the values $p = 0, \ldots, 8$ there holds

p	0	1	2	3	4	5	6	7	8
$C(p,0)$	$\mathbb{R}$	$\mathbb{R} \oplus \mathbb{R}$	$\mathbb{R}(2)$	$\mathbb{C}(2)$	$\mathbb{H}(2)$	$\mathbb{H}(2) \oplus \mathbb{H}(2)$	$\mathbb{H}(4)$	$\mathbb{C}(8)$	$\mathbb{R}(16)$
$C(0,p)$	$\mathbb{R}$	$\mathbb{C}$	$\mathbb{H}$	$\mathbb{H} \oplus \mathbb{H}$	$\mathbb{H}(2)$	$\mathbb{C}(4)$	$\mathbb{R}(8)$	$\mathbb{R}(8) \oplus \mathbb{R}(8)$	$\mathbb{R}(16)$

so that all of these algebras turn out to be either $\mathbb{K}(l)$ or $\mathbb{K}(l) \oplus \mathbb{K}(l)$ for $\mathbb{K} = \mathbb{R}, \mathbb{C}$ or $\mathbb{H}$ and for some $l = 0, 1, 2, \ldots$

(ii) one can get beyond eight by means of the recurrence formulas

$$C(p+8,0) = C(p,0) \otimes \mathbb{R}(16) \qquad \text{i.e.} \quad C(p+m.8,0) = C(p,0) \otimes \mathbb{R}(16^m)$$
$$C(0,p+8) = C(0,p) \otimes \mathbb{R}(16) \qquad \text{i.e.} \quad C(0,p+m.8) = C(0,p) \otimes \mathbb{R}(16^m)$$

(iii) there holds

$$\mathbb{K}(l) \otimes \mathbb{R}(16^m) = \mathbb{K}(l \cdot 16^m) \qquad (\mathbb{K}(l) \oplus \mathbb{K}(l)) \otimes \mathbb{R}(16^m) = \mathbb{K}(l \cdot 16^m) \oplus \mathbb{K}(l \cdot 16^m)$$

which enables one to find explicitly the algebras $C(p,0)$ and $C(0,p)$ for an *arbitrary* p (i.e. we already know all the algebras situated at the points *on both axes* of the quadrant pq)

(iv) so that, for example,

$$C(0,14) = \mathbb{R}(8.16)$$
$$C(0,41) = \mathbb{C}(16^5)$$
$$C(1957,0) = \mathbb{H}(4.16^{244}) \oplus \mathbb{H}(4.16^{244})$$

Hint: (ii) $C(p+8,0) = C(0,p+6) \otimes C(2,0) = \cdots = C(p,0) \otimes \mathbb{H} \otimes \mathbb{R}(2) \otimes \mathbb{H} \otimes \mathbb{R}(2) = C(p,0) \otimes (\mathbb{H} \otimes \mathbb{H}) \otimes (\mathbb{R}(2) \otimes \mathbb{R}(2)) = C(p,0) \otimes \mathbb{R}(16)$; (iii) $\mathbb{K}(l) = \mathbb{K} \otimes \mathbb{R}(l)$; (iv) $14 = 1.8 + 6$, $41 = 5.8 + 1$, $1957 = 244.8 + 5$. □

• Finally, the relation $C(p+1, q+1) = C(p,q) \otimes C(1,1) \equiv C(p,q) \otimes \mathbb{R}(2)$ shows that, walking with a dandified gait, we are able to move obliquely upwards (parallel to the symmetry axis of the quadrant) and go from appropriate points on the coordinate axes p and q to an arbitrary point *inside* the quadrant; in particular, in order to come to (p,q), we should start

$$\text{from } C(p-q,0) \qquad \text{if } p \geq q$$
$$\text{from } C(0,q-p) \qquad \text{if } p \leq q$$

Note that *all* the algebras $C(p, q)$ are either of the "type" $\mathbb{K}(l)$ or $\mathbb{K}(l) \oplus \mathbb{K}(l)$, for some possible $\mathbb{K}$ mentioned above and some $l = 0, 1, 2, \ldots$, since

1. this is the case on the coordinate axes for p and q ranging from 0 to 8
2. it turned out that a shift along the coordinate axes by 8 yields the tensor product with $\mathbb{R}(16)$, which *does not change* this "type"[474]
3. moving inwards the quadrant *once more* does not change the "type" of an algebra, since this only amounts to the tensor product (possibly iterated) with the algebra $\mathbb{R}(2)$.

We see from this consideration that the "type" of the algebra depends crucially on the number[475]

$$(p - q) \bmod 8$$

So if we need to know what $C(p, q)$ looks like for a given pair (p, q), we are to slip down to the place where the line along which we move crosses a coordinate axis and we look at the algebra which sits there. We find either $\mathbb{K}(r)$ or $\mathbb{K}(r) \oplus \mathbb{K}(r)$ with some particular $r = 0, 1, 2, \ldots$. Now we start to take steps backward along the same line and we multiply the value r by 2 for each step by $(1, 1)$.

[22.1.8] Verify[476] that all the real Clifford algebras may be summarized in the following concise table, generalizing the table from problem (22.1.7):

0	1	2	3	4	5	6	7
$\mathbb{R}(2^l)$	$\mathbb{R}(2^l) \oplus \mathbb{R}(2^l)$	$\mathbb{R}(2^l)$	$\mathbb{C}(2^l)$	$\mathbb{H}(2^{l-1})$	$\mathbb{H}(2^{l-1}) \oplus \mathbb{H}(2^{l-1})$	$\mathbb{H}(2^{l-1})$	$\mathbb{C}(2^l)$

In the upper line the numbers $(p - q) \bmod 8$ are displayed, the lower line then shows the corresponding Clifford algebras $C(p, q)$. The number l is given by $l \equiv [(p + q)/2]$, where $[k]$ denotes the integer part of k. □

• Now let us see how the situation *simplifies* for *complex* Clifford algebras. The general element may still be expressed in terms of the basis from (22.1.2), but complex coefficients are now allowed. Then one can easily see that the resulting algebra turns out to be "the same" (i.e. isomorphic) as if the initial real algebra were multiplied by $\otimes\, \mathbb{C}$. It is also useful to observe that the detailed signature (p, q) is completely irrelevant in the complex Clifford algebra; what really matters is only the total[477] dimension $n = p + q$, so that one should actually speak about $\bar{C}(n)$. We get the complete list of these algebras by simply tensor multiplying the real versions (with any signature whatsoever within a given n) by the algebra $\mathbb{C}$. The list turns out to be fairly short.

[474] By the "type" we understand the particular $\mathbb{K}$ as well as whether the sum $\oplus$ of two terms is present or not.

[475] The number $(p - q)$ itself labels the particular *slant line* (parallel to the symmetry axis of the quadrant, for which $(p - q) = 0$) on which the point (p, q) lies. We learned that for the every eighth line there is the same "type" of algebra.

[476] This problem is especially recommended for lifelong prisoners and shipwrecked persons living on desert islands to help pass the time.

[477] For example, consider the term $e^3 e^3 = \eta^{33}$. If we choose a new basis element to be $\hat{e}^3 = i e^3$, we get $\hat{e}^3 \hat{e}^3 = -\eta^{33}$; the choice of a new basis thus enables one to adjust the signature at will, $\hat{e}^a \hat{e}^b + \hat{e}^b \hat{e}^a = 2\hat{\eta}^{ab}$.

22.1.9 Check that the situation indeed simplifies greatly and the result may be summarized as follows:

$n \mod 2$	0	1
$\bar{C}(n)$	$\mathbb{C}(2^{[n/2]})$	$\mathbb{C}(2^{[n/2]}) \oplus \mathbb{C}(2^{[n/2]})$

Put another way, for *even* $n = 2m$ it is $\mathbb{C}(2^m)$ and for *odd* $n = 2m+1$ it is a direct sum of two copies of $\mathbb{C}(2^m)$.

Hint: for example, for $n = 3$ we have $C(3,0) = \mathbb{C}(2)$, $C(2,1) = \mathbb{R}(2) \oplus \mathbb{R}(2)$, $C(1,2) = \mathbb{C}(2)$, $C(0,3) = \mathbb{H} \oplus \mathbb{H}$; if *any* of these algebras is multiplied by $\otimes\mathbb{C}$, we get $\mathbb{C}(2) \oplus \mathbb{C}(2)$. □

22.2 Clifford groups Pin (p,q) and Spin (p,q)

• In Clifford algebra elements may be multiplied in an associative way. However, an associative multiplication also occurs in groups. In groups each element should have (by definition) its inverse, but this already need not be true for each element of $C(p,q)$. However, for some elements this *is* true and they then form a group with respect to the multiplication in $C(p,q)$. A special subgroup of the group turns out to be of particular interest. It may be described as follows: consider first the elements of $C^1(L,g) \equiv L^*$, which are normalized to unity, i.e. such $\alpha \in C^1(L,g)$ for which $g(\alpha,\alpha) = \pm 1$. We will check that the finite products of such elements constitute a group and eventually we will find that the group is fairly interesting.

22.2.1 Consider finite products of elements of $C^1(L,g)$, which are normalized to unity

$$u = \alpha_1\alpha_2\ldots\alpha_k \qquad g(\alpha_j,\alpha_j) = \pm 1 \quad j = 1,\ldots,k$$

Check that

(i) such products constitute a group – it is denoted by Pin (p,q)
(ii) the elements which contain an *even* number of factors constitute a subgroup in its own right – this subgroup is called the *spinor group of* $E^{p,q}$ and it is denoted by Spin (p,q)
(iii) the group Spin (p,q) lies in the even subalgebra $C^+(p,q)$
(iv) direct computations right from the definition lead for $p+q \equiv n \leq 2$ to the following results:

$n = 0$	not defined ($C^1(0,0)$ missing)	
$n = 1$	Pin $(1,0) = \mathbb{Z}_2 \times \mathbb{Z}_2$	Spin $(1,0) = \mathbb{Z}_2$
	Pin $(0,1) = C_4$	Spin $(0,1) = C_2 = \mathbb{Z}_2$
$n = 2$	Pin $(2,0) = O(2,0) \equiv O(2)$	Spin $(2,0) = SO(2,0) = SO(2)$
	Pin $(1,1) = O(1,1)$	Spin $(1,1) = SO(1,1)$
	Pin $(0,2) = O(0,2) \equiv O(2)$	Spin $(0,2) = SO(0,2) = SO(2)$

(C_k is the cyclic group of order k).

Hint: (i) closure with respect to the product is evident; since $\alpha_j\alpha_j = g(\alpha_j,\alpha_j) = \pm 1$, we have $\alpha_j^{-1} = \pm\alpha_j$ and so $(\alpha_1\alpha_2\ldots\alpha_k)^{-1} = \pm\alpha_k\ldots\alpha_2\alpha_1 \in$ Pin (p,q); (iii) $\alpha_j \in C^-(p,q)$

and $C(p,q)$ is $\mathbb{Z}_2$-graded; (iv) Pin (1, 0): $\alpha_j = ae^1$, $a \in \mathbb{R}$, $g(\alpha_j, \alpha_j) = 1$ gives $a^2 = 1$, so that $\alpha_j = \pm 1e^1$ and the group contains $(1, -1, e^1, -e^1)$; Pin (2, 0): normalized α_j has the form $\alpha \equiv \alpha(\phi) = e^1 \cos\phi + e^2 \sin\phi$; if we denote $\beta(\phi) = 1\cos\phi + e^1e^2\sin\phi$, the Clifford multiplication gives

$$\alpha(\phi)\alpha(\psi) = \beta(\phi+\psi) \qquad \alpha(\phi)\beta(\psi) = \alpha(\phi+\psi) \qquad \beta(\phi)\beta(\psi) = \beta(\phi+\psi)$$

Thus, there are only two kinds of elements in the group Pin (2, 0), namely $\alpha(\phi) = \alpha(\phi + 2\pi)$ and $\beta(\phi) = \beta(\phi + 2\pi)$ (with vanishing intersection $\Rightarrow$ the group as a manifold consists of two circles), their multiplication just corresponding to $O(2)$ ($\beta(\phi)$, the rotation by ϕ; $\alpha(\phi)$, rotation combined with a reflection (det $= -1$)); the group Spin (2, 0) contains only $\beta(\phi)$ and is isomorphic to $SO(2)$; Pin (1, 1): similarly, start from a normalized α_j of the form $\alpha(\phi) = e^1\cosh\phi + e^2\sinh\phi$ (to $+1$) or $\hat{\alpha}(\phi) = e^1\sinh\phi + e^2\cosh\phi$ (to -1). □

• The discrete groups at the beginning are not very interesting, but we see that for $n = 2$ we already get *Lie* groups, namely *orthogonal* groups $O(2)$ and $SO(2)$. And what does the situation look like for higher n? A deeper insight is provided by means of the following representation of the group Pin (p, q) in $C^1(p,q) \equiv L^*$.

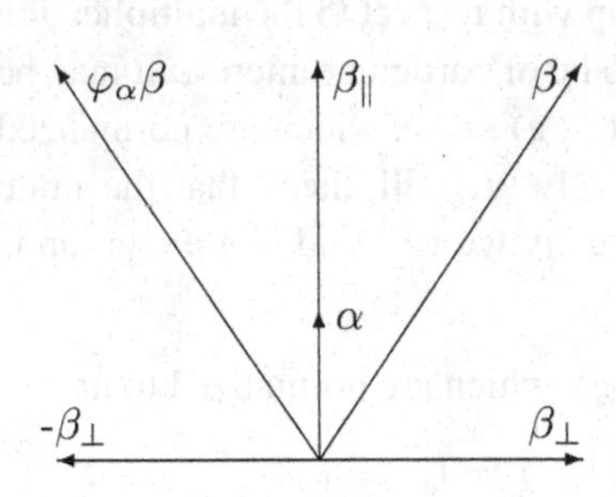

22.2.2 Consider an arbitrary element $u \in$ Pin (p,q) and define a linear map

$$\varphi_u : C^1(p,q) \to C^1(p,q) \quad \beta \mapsto u\beta u^{-1}$$

Check that

(i) if u contains *just one* normalized (co)vector, $u = \alpha = \alpha_1$, then the resulting (co)vector $\varphi_\alpha(\beta) \equiv \alpha\beta\alpha^{-1}$ is obtained from the initial β by the *reflection* with respect to α

(ii) for general $u \in$ Pin (p,q) it is obtained by a composition of several reflections, i.e. together with an *orthogonal transformation* $\hat{A}$ in the space[478] $C^1(p,q) \equiv (L^*, g)$

(iii) if an element $u = \alpha_1\alpha_2\ldots\alpha_k$ contains an even number of factors, i.e. if $u \in$ Spin (p,q), then $\hat{A}$ is a *special* orthogonal transformation (rotation) in $C^1(p,q)$

(iv) the prescription $u \mapsto \varphi_u$ is a *representation*[479] of Pin (p,q) in $C^1(p,q)$.

Hint: (i) decompose β into the part $\beta_\parallel = \lambda\alpha$ parallel to α and $\beta_\perp$ perpendicular to α; then

$$\beta \equiv \beta_\parallel + \beta_\perp \mapsto \alpha(\beta_\parallel + \beta_\perp)\alpha^{-1} = \lambda\alpha\alpha\alpha^{-1} + \alpha\beta_\perp\alpha^{-1} = \lambda\alpha + [-\beta_\perp\alpha + 2g(\alpha,\beta_\perp)]\alpha^{-1}$$
$$= \beta_\parallel - \beta_\perp$$

(ii) $\varphi_u\beta = \varphi_{\alpha_1\alpha_2\ldots\alpha_k}\beta = \cdots = \varphi_{\alpha_1}\varphi_{\alpha_2}\cdots\varphi_{\alpha_k}\beta$; a composition of orthogonal transformations (each reflection *is* an orthogonal transformation) is again an orthogonal transformation; (iii) any reflection changes the orientation $\Rightarrow$ it has determinant of $= -1$, so that the product

[478] We can see from this that the image of φ_u is *indeed* (only) in C^1 (from the definition it is only clear that it is somewhere in (the whole algebra) $C(p,q)$).

[479] It is sometimes called the *twisted adjoint representation* of Pin (p,q) and is denoted by $\widetilde{\text{Ad}}$.

of an even number of reflections has a determinant of $+1$; (iv) if $u_1 = \alpha_1\alpha_2 \dots \alpha_k$ and $u_2 = \alpha_{k+1}\alpha_{k+2} \dots \alpha_{k+r}$ then $u_1 u_2 = \alpha_1\alpha_2 \dots \alpha_{k+r}$ and $\varphi_{u_1 u_2} = \varphi_{\alpha_1\alpha_2 \dots \alpha_{k+r}} = \varphi_{\alpha_1} \dots \varphi_{\alpha_{k+r}} = \varphi_{u_1}\varphi_{u_2}$. □

22.2.3 Consider again the map φ_u from the last problem. Check that

(i) if we assign to the element $u \in$ Pin (p, q) the orthogonal transformation φ_u in $C^1(p, q)$ and if the matrices A^a_b are associated with this transformation in a standard way, i.e. by the prescription[480]

$$ue^a u^{-1} =: (A^{-1})^a_b e^b \qquad A^a_b \in O(p, q)$$

then the assignment $\pi : u \mapsto A^a_b(u)$ is a homomorphism of groups

$$\pi : \text{Pin}\,(p, q) \to O(p, q) \qquad \pi : \text{Spin}\,(p, q) \to SO(p, q)$$

(the second homomorphism is just the restriction of the first one to the subgroup)
(ii) this homomorphism is surjective
(iii) however, *it is not* injective; its kernel contains at least the *two-element* subgroup $(1, -1) \subset$ Spin (p, q).

Hint: (i) $u \mapsto \varphi_u$ is a representation by orthogonal operators or directly $u_1 u_2 e^a u_2^{-1} u_1^{-1} = (A_2^{-1})^a_b u_1 e^b u_1^{-1} = \cdots$; (ii) the geometrical folklore says[481] that *each* rotation (orthogonal transformation with determinant $+1$) may be realized as a sequence of an even number of reflections; a sequence of an odd number of reflections then leads to a composition of a rotation and a reflection, i.e. an element of $O(p, q)$, which is not contained in $SO(p, q)$. □

• For further material it is useful to know the following facts (we mention them without any detailed analysis):

1. starting from dimension $n = 2$ both the groups Pin (p, q) and Spin (p, q) turn out to be *Lie* groups (they constitute a submanifold of the vector space $C(p, q)$)
2. the kernel of their homomorphic maps onto the orthogonal groups is given by *exactly* the two-element subgroup $(1, -1) \subset$ Spin (p, q) mentioned above, so that according to Section 13.3 Spin (p, q) is a *two-sheeted covering* of $SO(p, q)$ and according to (13.3.1) their Lie algebras are *isomorphic*:

$$\text{spin}\,(p, q) = so(p, q)$$

3. for $n \geq 3$ the connected component of the identity of the group Spin (p, q) is also *simply* connected, so that the homomorphism actually provides a *universal covering* of the connected component of the identity in $SO(p, q)$ (the "*proper Lorentz group*")
4. Spin $(3, 0)$ is isomorphic to $SU(2)$ and the connected component of the identity of the group Spin $(1, 3)$ is isomorphic to the group $SL(2, \mathbb{C})$ (which is consistent with our previous knowledge that it is just $SU(2)$ which is the two-sheeted covering of $SO(3)$ and $SL(2, \mathbb{C})$ which covers the

[480] The inverse matrix is needed in order to obtain a *left* action.

[481] In *two* dimensions, draw two straight lines intersecting at an angle ψ and check that the composition of two reflections with respect to the lines results in the rotation by an angle of 2ψ around the point at which the lines intersect. In *three* dimensions one should replace the straight lines by planes. For more dimensions consult any official treatment.

proper Lorentz group $L_+^\uparrow$); next, Spin (0, 6) is isomorphic to $SU(4)$. The spinor groups in higher dimensions are already no longer isomorphic to any "classical" matrix groups (like $SU(\cdot)$, $SO(\cdot)$ etc.; they turn out to be simply different).

22.2.4 Check that also in the case where $(p, q) = (2, 0)$ we have a two-sheeted covering: Spin $(2, 0) = SO(2)$ covers $SO(2)$ (the group covers "itself").

Hint: according to (22.2.1) we have $u \equiv u(\psi) = 1 \cos\psi + e^1 e^2 \sin\psi$; check that $e^a \mapsto u e^a u^{-1} \equiv (A^{-1})^a_b e^b$ yields a matrix A, which describes the rotation by 2ψ, so that the circle = Spin (2, 0) is wound *twice* around the circle $SO(2)$. □

• Now let us see how the isomorphism of the Lie algebras spin $(p, q) = so(p, q)$ looks explicitly (we will need it, among other things, for the Dirac operator). Since this isomorphism is actually a derived homomorphism with respect to the covering π of the groups, we may proceed as follows: we first introduce a convenient basis in $so(p, q)$. Then we find what the one-parameter groups corresponding to the basis look like in $SO(p, q)$ as well as the one-parameter groups in Spin (p, q), which project on them. Eventually, we read off the basis elements of the Lie algebra spin (p, q), which serve as preimages with respect to the basis in $so(p, q)$ mentioned above.

22.2.5 Let E^a_b be a standard basis in $gl(n, \mathbb{R})$, so that $(E^a_b)^c_d = \delta^a_d \delta^c_b$ and let η denote the Minkowski metric tensor matrix of signature (p, q). Check that

(i) the matrices $E^{ab} := \eta^{ac} E^b_c$ also constitute a basis of $gl(n, \mathbb{R})$

(ii) the matrices

$$\mathcal{E}^{ab} := (E^{ab} - E^{ba}) \qquad \text{i.e.} \quad (\mathcal{E}^{ab})^c_d = \eta^{ac}\delta^b_d - \eta^{bc}\delta^a_d$$

constitute a basis of $so(p, q)$ (they are pseudo-antisymmetric)

(iii) the commutation relations in $so(p, q)$ in this basis read as follows:

$$[\mathcal{E}^{ab}, \mathcal{E}^{cd}] = -2\left(\eta^{c[a}\mathcal{E}^{b]d} - (c \leftrightarrow d)\right) \equiv \eta^{ac}\mathcal{E}^{db} + \eta^{cb}\mathcal{E}^{ad} + \eta^{ad}\mathcal{E}^{bc} + \eta^{db}\mathcal{E}^{ca}$$

(iv) in the particular case of the Lie algebra $so(3)$ (i.e. $p = 3, q = 0$, so that $\eta^{ab} = \delta^{ab}$) is the basis $\mathcal{E}^{ab}$ related to the basis l_a introduced earlier (11.7.13), $(l_a)_{bc} = -\epsilon_{abc}$, by

$$l_a = -\frac{1}{2}\epsilon_{abc}\mathcal{E}^{bc} \qquad \mathcal{E}^{ab} = -\epsilon_{abc} l_c$$

Hint: E^a_b has unity at the place ba and elsewhere zeros, E^{ab} has the number $\eta^{aa} \equiv \pm 1$ at the place ab and elsewhere zeros, $\mathcal{E}^{ab}$ has at the place ab the number $\eta^{aa} \equiv \pm 1$ and at the place ba the number $-\eta^{bb}$ and elsewhere zeros; the commutation relations follow straightforwardly. □

22.2.6 Let us look at what the matrices $\mathcal{E}^{ab}$ actually "do." For definiteness, think of $\mathcal{E}^{12}$. Check that the matrix $A(\epsilon) = (\mathbb{I} + \epsilon\mathcal{E}^{12})$ performs on the basis e^a the transformation

$$e^1 \mapsto e^1 + \epsilon\eta^{11} e^2$$
$$e^2 \mapsto e^2 - \epsilon\eta^{22} e^1$$
$$e^3 \mapsto e^3 \text{ etc. up to } e^n \mapsto e^n$$

It thus generates the (pseudo)rotation by an "angle" ϵ in the 12 plane; it is clear that in a similar way the matrix $\mathcal{E}^{ab}$ generates the (pseudo)rotation by an "angle" ϵ in the ab plane. Hint: $(\mathbb{I} + \epsilon\mathcal{E}^{12})^c_d e^d = \cdots$. □

• The rotation by an "angle" ϵ in the 12 plane may be replaced by a composition of two reflections, namely with respect to (arbitrary) two vectors in the 12 plane, which make the *half* "angle," i.e. $\epsilon/2$. For the case of a signature in which there holds $\eta^{11}\eta^{22} = 1$, we may choose, for example, e^1 and $e^1 \cos(\epsilon/2) + e^2 \sin(\epsilon/2)$ (for $\eta^{11}\eta^{22} = -1$ the *hyperbolic* cosine and sine occur). According to (22.2.2), such reflections in $C^1(p, q) \equiv L^*$ are produced by the operators φ_{α_1} and φ_{α_2} with $\alpha_1 = e^1$ and $\alpha_2 = e^1 \cos(\epsilon/2) + e^2 \sin(\epsilon/2)$ and the composition of the two reflections is performed by φ_u with $u = \alpha_1\alpha_2$. However, within first-order accuracy in ϵ we have

$$\alpha_1\alpha_2 = e^1[e^1 \cos(\epsilon/2) + e^2 \sin(\epsilon/2)] = e^1e^1 \cos(\epsilon/2) + e^1e^2 \sin(\epsilon/2) \doteq 1 + \epsilon\frac{1}{2}e^1e^2$$

and from this expression we can eventually see that the element $\hat{\mathcal{E}}^{12}$ from the Lie algebra spin (p, q), which is "over" the basis element $\mathcal{E}^{12}$ from $so(p, q)$ is $\frac{1}{2}e^1e^2 \in C^2(p, q)$. Again, it is clear that in the ab plane we would get $\frac{1}{2}e^ae^b \in C^2(p, q)$, $a < b$ in this way and since just these elements constitute a basis of $C^2(p, q)$, we may identify the Lie algebra spin (p, q) with the subspace $C^2(p, q)$ (with respect to the commutator induced by the associative Clifford product in $C(p, q)$).

22.2.7 Consider the subspace $C^2(p, q) \subset C(p, q)$ and introduce a commutator into $C(p, q)$ in terms of the associative (= here Clifford) product, i.e. by the prescription $[x, y] := xy - yx$ (see Appendix A.3). Check that

(i) the subspace $C^2(p, q)$ turns out to be closed with respect to the commutator[482]
(ii) the map

$$\pi' : \text{spin}\,(p, q) \to so(p, q) \qquad \frac{1}{2}e^ae^b \mapsto \mathcal{E}^{ab} \quad a < b$$

is an isomorphism of Lie algebras (it is the derived homomorphism of the two-sheeted universal covering $\pi : \text{Spin}\,(p, q) \to SO(p, q)$)
(iii) verify once more explicitly that the defining relation for the covering homomorphism

$$ue^au^{-1} = (A^{-1})^a_b e^b \qquad \text{i.e.} \quad A = \pi(u)$$

gives for

$$u = 1 + \epsilon\frac{1}{2}e^ae^b \qquad \text{the matrix} \quad A = \mathbb{I} + \epsilon\mathcal{E}^{ab} \quad a < b$$

[482] So that it is a Lie subalgebra of the Lie algebra $C(p, q)$ (with the commutator $[x, y] := xy - yx$ mentioned above). The (whole) Lie algebra $C(p, q)$ is easily seen to correspond to the group of *all* invertible elements of the Clifford algebra $C(p, q)$. Those of them which happen to be products of normalized elements from $C^1(p, q)$ constitute the subgroup Pin (p, q) with the Lie algebra $C^2(p, q)$.

Hint: (iii)

$$(1 + \epsilon/2\, e^a e^b) e^c (1 - \epsilon/2\, e^a e^b) = \cdots = (\mathbb{I} - \epsilon \mathcal{E}^{ab})^c_d e^d \stackrel{!}{=} (A^{-1})^c_d e^d$$

□

22.3 Spinors: linear algebra

• The (non-canonical) isomorphisms of Clifford algebras to the matrix algebras, which are summarized in tables in (22.1.8) and (22.1.9), provide automatically their (faithful) matrix *representations* ρ. The elements of the spaces in which these representations act as matrices are called *spinors*. So spinors are ordinary real or complex column vectors, taken together with a rule saying how the elements of the Clifford *algebra* act on them (put another way they are the elements of the representation *module* of the Clifford algebra).

Since in the Clifford algebra there are also hidden (as subsets) Clifford *groups* Pin (p, q) and Spin (p, q) as well as their (common) *Lie algebra* spin (p, q), the operations in the latter structures being simply barefacedly "stolen" from the Clifford algebra, each representation of the Clifford algebra also yields free of charge representations of the "substructures" under consideration, of the Clifford groups Pin (p, q) and Spin (p, q) as well as of the Lie algebra spin (p, q); they are called the *spinor representation* (of the groups and the Lie algebra) and we will denote all of them by the same letter ρ.

Recall that the whole Clifford algebra may be generated from a basis e^a of the subspace $C^1(p, q)$ (all the other elements of $C(p, q)$ then being linear combinations of the products of e^a). This means, however, that a representation of the whole algebra $C(p, q)$ is actually fully specified if we represent the basis e^a (for example, $\rho(3e^2e^4) = 3\rho(e^2)\rho(e^4)$); these key objects, the images of an orthonormal basis with respect to the spinor representation ρ

$$\gamma^a := \rho(e^a)$$

are called the γ*-matrices*.

22.3.1 Check that the γ-matrices satisfy the following identities:

$$\gamma^a \gamma^b + \gamma^b \gamma^a = 2\eta^{ab} \qquad \rho(u)\gamma^a \rho(u^{-1}) = (A^{-1})^a_b \gamma^b$$

where $A^a_b \in SO(p, q)$ is the image of $u \in$ Spin (p, q) with respect to the two-sheeted covering.

Hint: the first one is just the ρ-image of the essential relation obeyed by the Clifford multiplication of an orthonormal basis, $e^a e^b + e^b e^a = 2\eta^{ab}$, the second one is the ρ-image of the relation $ue^a u^{-1} = (A^{-1})^a_b e^b$ from (22.2.3). □

• The properties of the γ-matrices as well as of the spinors on which they act depend on what the concrete isomorphism of the corresponding Clifford algebra looks like. For example, if $(p-q)$ mod 8 is 0 or 2, we can see from the table in (22.1.8) that the Clifford algebra is isomorphic to the *real matrix* algebra $\mathbb{R}(2^l)$. This means that in an *appropriate* representation ρ all the elements of the Clifford algebra, in particular also the basis elements e^a, are represented by *real* matrices $2^l \times 2^l$ and the spinors themselves may then be realized by *real* column vectors. This particular representation is called the *Majorana representation* of the corresponding Clifford algebra (and, consequently, also the related group and the Lie algebra). We see that this only occurs for rather specific values of (p, q).

22.3.2 Consider the two-dimensional space with signature $(p, q) = (1, 1)$ and the four-dimensional space with signature $(p, q) = (3, 1)$. Check that

(i) the Majorana representation exists for both of these signatures; write down explicitly the corresponding γ-matrices
(ii) the Clifford algebra $C(1, 1)$ may also be represented "improperly," by means of matrices which *are not* real.

Hint: (i) according to the hint to (22.1.3) for $C(1, 1)$ the following (real) representation works (for example) $e^1 \mapsto \gamma^1 \equiv \rho(e^1) = \sigma_1, e^2 \mapsto \gamma^2 \equiv \rho(e^2) = i\sigma_2$; according to (22.1.5) there holds $C(3, 1) = C(1, 1) \otimes C(2, 0) = \mathbb{R}(2) \otimes \mathbb{R}(2) = \mathbb{R}(4)$; the algorithm from the hint to problem (22.1.5) gives $\mathcal{E}^1 = E^1 \otimes e^1 e^2, \mathcal{E}^2 = E^2 \otimes e^1 e^2, \mathcal{E}^3 = 1 \otimes e^1, \mathcal{E}^4 = 1 \otimes e^2$; since according to (22.1.3) one can assign $E^1 \mapsto \sigma_1, E^2 \mapsto i\sigma_2, e^1 \mapsto \sigma_1, e^2 \mapsto \sigma_3$, we eventually get the matrices $\Gamma^A := \rho(\mathcal{E}^A), A = 1, 2, 3, 4$ in the form $\Gamma^1 = \sigma_1 \otimes \sigma_1\sigma_3, \Gamma^2 = i\sigma_2 \otimes \sigma_1\sigma_3$, $\Gamma^3 = \mathbb{I} \otimes \sigma_1, \Gamma^4 = \mathbb{I} \otimes \sigma_3$, i.e.

$$\hat{\Gamma}^1 = \begin{pmatrix} 0 & -i\sigma_2 \\ -i\sigma_2 & 0 \end{pmatrix} \qquad \hat{\Gamma}^3 = \begin{pmatrix} \sigma_1 & 0 \\ 0 & \sigma_1 \end{pmatrix}$$

$$\hat{\Gamma}^2 = \begin{pmatrix} 0 & -i\sigma_2 \\ i\sigma_2 & 0 \end{pmatrix} \qquad \hat{\Gamma}^4 = \begin{pmatrix} \sigma_3 & 0 \\ 0 & \sigma_3 \end{pmatrix}$$

which satisfy the relations "of type" (3, 1)

$$\Gamma^A\Gamma^B + \Gamma^B\Gamma^A = 2\eta^{AB}\mathbb{I} \qquad \eta^{AB} \equiv \text{diag}\,(-1, 1, 1, 1)$$

(ii) the assignment $e^1 \mapsto \sigma_1, e^2 \mapsto i\sigma_3$ (rather than the *real* matrices $i\sigma_2$) realizes $C(1, 1)$ as a *subalgebra* of the matrix algebra $\mathbb{C}(2)$, where it is not evident at first sight that it is "actually" $\mathbb{R}(2)$. □

• If there exist real γ-matrices for signature (p, q), then it is clear that for signature (q, p) by contrast *purely imaginary* γ-matrices exist (it is enough to take i-multiples of the real matrices). So purely imaginary γ-matrices exist just for such (p, q) for which real matrices exist for (q, p); this, however, occurs just for $(q-p)$ mod $8 = 0, 2$, i.e. for $(p-q)$ mod $8 = 0, 6$.

22.3.3 Check that in the space with ordinary Minkowski signature $(1, 3)$ purely imaginary γ-matrices exist and find their explicit form.

Hint: $(1-3) \mod 8 = (-2) \mod 8 = 6 \mod 8$; take the matrices from (22.3.2) and multiply all of them by i; we get

$$\hat{\Gamma}^1 = \begin{pmatrix} 0 & \sigma_2 \\ \sigma_2 & 0 \end{pmatrix} \qquad \hat{\Gamma}^3 = \begin{pmatrix} i\sigma_1 & 0 \\ 0 & i\sigma_1 \end{pmatrix}$$
$$\hat{\Gamma}^2 = \begin{pmatrix} 0 & \sigma_2 \\ -\sigma_2 & 0 \end{pmatrix} \qquad \hat{\Gamma}^4 = \begin{pmatrix} i\sigma_3 & 0 \\ 0 & i\sigma_3 \end{pmatrix}$$

□

• The isomorphism (22.1.9) of the complex Clifford algebra $\bar{C}(2m)$ for even-dimensional space with the matrix algebra $\mathbb{C}(2^m)$ is called the *Dirac representation* ρ of the Clifford algebra and the elements of the representation space (module) are *Dirac spinors*.

This representation, when regarded as a representation of the Clifford *algebra*, is irreducible.[483] However, "the same" representation, when only regarded as a representation of the Clifford *group* Spin (p, q), already turns out to be reducible – the representation space (the space of Dirac spinors) invariantly decomposes with respect to the action of the group to two halves (in which *inequivalent* representations operate). We say that *chiral spinors* exist here (they are called right and left spinors). Let us see how this arises.

22.3.4 Consider the element $z \equiv e^1 e^2 \dots e^{2m} \in C^{2m}(p, q) \subset C(p, q)$ (the product of *all* basis elements of $C^1(p, q)$; it constitutes a basis of the subspace $C^{2m}(p, q)$). Its image in the Dirac representation thus amounts to the product of all γ-matrices[484]

$$\gamma_5 := \gamma^1 \gamma^2 \dots \gamma^{2m} \equiv \rho(e^1 e^2 \dots e^{2m})$$

Check that

(i) the element z anticommutes with all the basis elements e^a, so that γ_5 anticommutes with all matrices γ^a (note that for $\bar{C}(2m+1)$ it *commutes*)
(ii) consequently this element commutes with all the elements of the group Spin (p, q) (being itself from the group); upon the representation this means that γ_5 commutes with all operators of the representation and therefore, according to Schur's lemma, the representation is reducible
(iii) the square of the element z is the number ± 1 (the concrete value depends on the signature and it may be easily computed in the general case), so that upon the representation we have

$$\gamma_5 \gamma_5 = \pm \mathbb{I}$$

and this means that

$$L := \frac{1}{2}(\mathbb{I} - \gamma_5) \qquad P := \frac{1}{2}(\mathbb{I} + \gamma_5) \qquad \text{if} \quad \gamma_5\gamma_5 = +\mathbb{I}$$
$$L := \frac{1}{2}(\mathbb{I} - i\gamma_5) \qquad P := \frac{1}{2}(\mathbb{I} + i\gamma_5) \qquad \text{if} \quad \gamma_5\gamma_5 = -\mathbb{I}$$

[483] The algebra of *all* square matrices $\mathbb{C}(2^m)$ evidently does not preserve any non-trivial subspace in the space of column vectors of appropriate dimension. To see this consider a non-trivial subspace that stands as a candidate for being invariant and an operator which casts out at least one vector from the subspace. *This* operator certainly belongs to *all* operators and at the same time it does not preserve the subspace. (Note that the odd-dimensional case is reducible (22.1.9).)

[484] The subscript 5 is a residue from the time where all this was only treated in "ordinary" four-dimensional Minkowski space, where the indices ran through the values 1, 2, 3, 4 (rather than 0, 1, 2, 3, as is done as a rule today), so that the "ordinary" γ-matrices were $\gamma^1, \gamma^2, \gamma^3, \gamma^4$. Their product was then denoted by γ^5. If we, however, consider more dimensions, there is also γ^5 which corresponds to the fifth dimension. In order for these two "gamma fives" not to be confused, we use the subscript 5 to denote the "product" one.

act as projection operators onto the invariant subspaces of *right and left spinors*[485]

$$\psi = \psi_L + \psi_P := L\psi + P\psi$$

Such spinors are also said to be *chiral* or *Weyl spinors*.

Hint: (i) e^a equals just one particular element of the product; the interchange of e^a with those elements inside z which differ from e^a brings just a change of sign, the interchange with "itself" does nothing at all; (iii) $zz = e^1e^2 \dots e^{2m}e^1e^2 \dots e^{2m} = \pm e^1e^1e^2e^2 \dots e^{2m}e^{2m}$. □

22.3.5 Let us see how this works for the two-dimensional case with signatures (2, 0) and (1, 1). Write down explicitly the projection operators P, L and specify the structure of right and left spinors for these two cases.

Solution: for (2, 0) we have according to (22.1.3) $\gamma_5 \equiv \gamma^1\gamma^2 = \sigma_1\sigma_3 \equiv -i\sigma_2$, so that $\gamma_5\gamma_5 = -\mathbb{I}$; similarly for (1, 1) we have $\gamma_5 \equiv \gamma^1\gamma^2 = \sigma_1 i\sigma_2 \equiv -\sigma_3$, so that $\gamma_5\gamma_5 = +\mathbb{I}$; consequently,

$$(2,0): \qquad L = \frac{1}{2}(\mathbb{I} - \sigma_2) \equiv \frac{1}{2}\begin{pmatrix} 1 & i \\ -i & 1 \end{pmatrix} \qquad \psi_L = a\begin{pmatrix} 1 \\ -i \end{pmatrix}$$

$$P = \frac{1}{2}(\mathbb{I} + \sigma_2) \equiv \frac{1}{2}\begin{pmatrix} 1 & -i \\ i & 1 \end{pmatrix} \qquad \psi_P = a\begin{pmatrix} 1 \\ i \end{pmatrix}$$

$$(1,1): \qquad L = \frac{1}{2}(\mathbb{I} + \sigma_3) \equiv \begin{pmatrix} 1 & 0 \\ 0 & 0 \end{pmatrix} \qquad \psi_L = a\begin{pmatrix} 1 \\ 0 \end{pmatrix}$$

$$P = \frac{1}{2}(\mathbb{I} - \sigma_3) \equiv \begin{pmatrix} 0 & 0 \\ 0 & 1 \end{pmatrix} \qquad \psi_P = a\begin{pmatrix} 0 \\ 1 \end{pmatrix}$$

where $a \in \mathbb{C}$ is an arbitrary complex number. □

22.3.6 Check that for signatures (1, 3) and (3, 1) we get the matrix γ_5 in the form

$$(1,3): \qquad \Gamma_5 \equiv \Gamma^1\Gamma^2\Gamma^3\Gamma^4 = \sigma_3 \otimes (-i\sigma_2) \qquad \text{so that} \quad \Gamma_5\Gamma_5 = -\mathbb{I}$$

$$(3,1): \qquad \hat{\Gamma}_5 \equiv \hat{\Gamma}^1\hat{\Gamma}^2\hat{\Gamma}^3\hat{\Gamma}^4 = \sigma_3 \otimes (-i\sigma_2) \qquad \text{so that} \quad \hat{\Gamma}_5\hat{\Gamma}_5 = -\mathbb{I}$$

In both cases specify the structure of Weyl spinors ψ_L and ψ_P.

Hint: see (22.3.2), (22.3.3) and (22.3.5); solve $L\psi_L = \psi_L$ and $R\psi_R = \psi_R$. □

• Note that in the case of signature (1, 1) the projectors P, L turn out to be given by real matrices (22.3.5), so that in this particular case spinors exist which are *at the same time* Majorana and Weyl. This happens in general in those cases when in addition to the requirement for the existence of Majorana spinors, $(p - q) \bmod 8 = 0, 2$, also the condition $\gamma_5\gamma_5 = +\mathbb{I}$ is satisfied (so as not to allow i to enter into the projectors). A computation of

[485] This dividing turns out to be of great importance in physics. For example, in the "standard model" of electroweak interactions (which won the Nobel prize for Glashow, Salam and Weinberg) the right and left fermions (represented by right and left spinor fields) behave very differently (they transform in different representations of the gauge symmetry and this results in different interactions; see the text after (21.1.5)).

the relevant square shows (the reader is invited to work out the details as homework) that this excludes the 2 in the Majorana condition. *Majorana–Weyl spinors* thus exist only in spaces with signatures (p, q) such that $(p - q) \bmod 8 = 0$; for example, in the case where $(p, q) = (1, 1)$.

Remark: the concept of chirality may also be regarded as a $\mathbb{Z}_2$-*grading* in the space of spinors; right and left spinors constitute the subspaces labeled by the elements [0] and [1]. The grading in the space of spinors also induces the grading in the space of *operators* acting on them (see Appendix A.5) – operators decompose into even and odd parts (which respect and change the chirality respectively). In particular, the main celebrity of this chapter, the Dirac operator, turns out to be odd in this sense (22.5.12).

22.3.7 A useful piece of wisdom for life consists of realizing that *Pauli matrices* σ_a, $a = 1, 2, 3$, actually represent particular γ-*matrices*. Check that this is indeed the case, namely for the Clifford algebra $C(3, 0)$ (i.e. the algebra associated with the most mundane, but at the same time the most important three-dimensional[486] space E^3 and the usual action of the rotation group $SO(3, 0) \equiv SO(3)$ in this space).

Hint: according to (22.1.8) we have $C(3, 0) = \mathbb{C}(2)$, so that γ-matrices are to be 2×2 complex matrices (and consequently the spinors are two-component complex column vectors; we already mentioned that Spin $(3, 0) \approx SU(2)$, so that the action of the spinor group on spinors actually coincides with the natural action of $SU(2)$ on two-component column vectors); with regard to the identity $\sigma_a \sigma_b + \sigma_b \sigma_a = 2\delta_{ab}$ we may take the matrices to be just *Pauli* matrices. □

22.4 Spin bundle $\pi : SM \to M$ and spinor fields on M

• Recall that a *vector* field on a manifold M may be described in several ways, differing so much at first sight that it is not immediately clear whether we are even speaking about the same object.

In the first approach (2.2) the concept of a vector at the point $x \in M$ was introduced first (this in turn may be done in a highly intuitive way in terms of tangent curves on M) and the vector field was then identified with a rule providing "a vector at each point on M" (plus an appropriate introduction of smoothness). Then, step by step, various different possibilities arose: the vector field as a derivation of the algebra of functions $\mathcal{F}(M)$ on M (2.2.8), as a section of the tangent bundle (17.2.6), as an equivariant function of type ρ_0^1 on LM (19.6), or eventually as a section of an appropriate bundle associated with the principal bundle LM (the last alternative being, however, isomorphic to that with TM). All of these possibilities of introducing a vector field on M are equivalent to each other and although some of them might seem unnatural and even unduly complicated, each of them actually has its virtue when used in the appropriate context.

[486] We saw already before (e.g. for $C(2, 0)$ in (22.1.3)) that in terms of *some* Pauli matrices the γ-matrices may be realized for "smaller" Clifford algebras; here all three of them are understood.

Now we would like to introduce a *spinor* field on M. We might try once again to begin with a "spinor at each point on M." However, contrary to the vector, now we are not able to give any simple and *intuitive* meaning to the concept of a spinor at the point x (like tangent curves, refining good old arrows for a vector), so that the first approach cannot be imitated. Therefore we would appreciate any alternative way of introducing a vector field which *can be* modified so that it might serve for introducing a spinor field as well. A convenient (and conceptually fairly simple) alternative turns out to be the method of equivariant functions on principal bundles.

A spinor in linear algebra has been introduced as an element of the representation space of the Clifford algebra $C(p, q)$, and consequently also the Clifford group Spin (p, q). So in terms of equivariant functions we should then first introduce a spin bundle, i.e. a principal bundle $\pi : SM \to M$ with the group Spin (p, q) and then an equivariant function of type ρ on SM, where ρ denotes the spinor representation of the group Spin (p, q) (a representation in the space of column vectors acted on by the matrices given by the isomorphism of $C(p, q)$ with a particular matrix algebra). However, the group Spin (p, q) is just a two-sheeted covering φ of the special orthogonal group $SO(p, q)$, so that a fiber in the principal bundle with group Spin (p, q) should also be a two-sheeted covering of the fiber in some principal bundle with group $SO(p, q)$.

Such a bundle, which is canonically associated with the *metric tensor* (recall that the metric tensor is essential for the very existence of the algebra $C(p, q)$ as well as the group Spin (p, q)) and the orientation on a manifold M is, however, well known from Section 20.5 – it is the bundle $\pi : OM \to M$ of (right-handed) *orthonormal* frames. So the total space SM (to be found) of the principal bundle needed for spinors is to be a two-sheeted covering of the manifold OM. Recall, however, that the two-sheeted covering of the groups φ is moreover a homomorphism. We eventually realize that all of this exactly matches the general scheme which we treated at the end of Section 20.5 under the name "morphisms of principal bundles"; and we see at last a non-trivial example of *surjective* f (here, in particular, the two-sheeted covering).

$$\begin{array}{ccc} SM \times \mathrm{Spin}\,(p,q) & \xrightarrow{\;f \times \varphi\;} & OM \times SO(p,q) \\ R \downarrow & & \downarrow \tilde{R} \\ SM & \xrightarrow{\;f\;} & OM \\ & {\scriptstyle\pi} \searrow \quad M \quad \swarrow {\scriptstyle\tilde{\pi}} & \end{array}$$

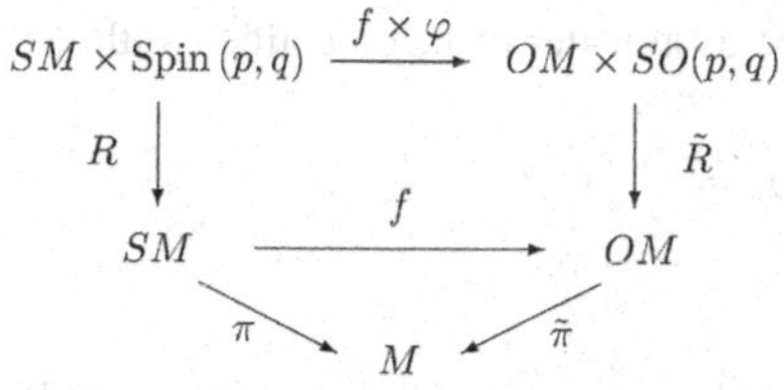

So consider a (pseudo-)Riemannian manifold (M, g) with signature (p, q). Then the *spin bundle*

$$\pi : SM \to M$$

is, in the terminology introduced in Section 20.5, a *prolongation* of the bundle of (right-handed) orthonormal frames $\tilde{\pi} : OM \to M$ such that its structure group is Spin (p, q) and the needed homomorphism φ : Spin $(p, q) \to SO(p, q)$ is the two-sheeted covering mentioned in Section 22.2. This bundle (which is also known as the *spin frame bundle*) is often identified with the *spin structure* on M. In analogy to the case of the restriction the existence of the prolongation of a given bundle is not guaranteed; there may exist various topological "obstructions." In particular, the spin structure may not exist for a given (pseudo-)Riemannian manifold. We will not treat the details of the issue,[487] but instead we will simply assume that *our* manifold (M, g) *admits* a spin structure.

22.4.1 Be sure to understand that for the base manifold $(M, g) = E^{p,q}$ (ordinary pseudo-Euclidean space) both the bundle of right-handed orthonormal frames as well as the spin bundle are *trivial*, i.e. both of them are equivalent to *product* bundles

$$\tilde{\pi} : E^{p,q} \times SO(p, q) \to E^{p,q} \qquad \pi : E^{p,q} \times \text{Spin}\,(p, q) \to E^{p,q}$$

Hint: on $E^{p,q}$ there is the (right-handed orthonormal) global Cartesian coordinate frame field ∂_μ; any (right-handed orthonormal) frame e in the point $x \in E^{p,q}$ uniquely decomposes with respect to ∂_μ at this point, $e_a = e^\mu_a \partial_\mu, e^\mu_a \in SO(p, q)$, so that $e \mapsto (x, e^\mu_a)$ is a global trivialization; $E^{p,q} \times \text{Spin}\,(p, q)$ satisfies the commutative diagram for $OM \equiv OE^{p,q} = E^{p,q} \times SO(p, q)$. □

22.4.2 However, there are also fairly simple manifolds for which both the bundles, $\tilde{\pi} : OM \to M$ as well as $\pi : SM \to M$, already turn out to be non-trivial. For example, let us see how this works for a sphere (S^2, g) with the standard "round" metric (induced from E^3). Check that

(i) the total space of the bundle of the (right-handed) orthonormal frames may be identified as a manifold with the group $SO(3)$
(ii) its two-sheeted covering space, the total space of the spin bundle, may be identified as a manifold with the group $SU(2) \equiv S^3$
(iii) the spin bundle $\pi : SM \to M$ for $M = S^2$ may thus be as a matter of fact identified with the good old *Hopf bundle*

$$\pi : SU(2) \to S^2$$

from (20.1.9).

Hint: (i) consider in E^3 a fixed (fiducial) right-handed orthonormal basis $E_a, a = 1, 2, 3$. If we act on the basis by an element $A \in SO(3)$ according to $E_a \mapsto e_a := A^b_a E_b$, we again get a (right-handed) orthonormal basis and we thus clearly obtain a bijection $A \leftrightarrow e$ between $SO(3)$ and the set of all right-handed orthonormal bases in E^3. However, the set of bases may be easily identified with the total space of the bundle of right-handed orthonormal frames on the *sphere* S^2: regard the end of the vector e_3 as a *point* s on S^2 and the remaining pair (e_1, e_2) (after a shift to the point s on the sphere) as a right-handed orthonormal frame in $s \in S^2$. By reversing this procedure we may in turn associate a unique element $A \in SO(3)$ with each right-handed orthonormal frame (e_1, e_2) in s on S^2. $SO(2)$ acts in the fiber over $e_3 = s$ as the group of such $B \in SO(3)$, which do not change e_3 (the stabilizer of the point e_3 with respect to the action $e_a \mapsto B^b_a e_b$, i.e. $A \mapsto AB$); (ii) let us look at what happens, say, over the north pole; this point corresponds to the matrices $A \in SO(3)$ such that $e_3 = E_3$,

[487] For the convenience of the reader who intended to actively join with nonchalance a debate of experts at an evening party, we mention that it is advisable to remember that the object which acts as an obstruction to introducing the spin structure is "the second Stiefel–Whitney class" $w_2(M)$ of the manifold M. Just after saying this we recommend leaving the group of experts as soon as possible under the guise of, say, tasting "that marvellous cake."

i.e. (13.1.11) $A = \text{diag}(\tilde{A}, 1) = \exp\{\alpha\, l_3\}$, $\tilde{A} \in SO(2)$, $l_3 \in so(3)$ (they form the fiber over the north pole of the sphere S^2 in the bundle of right-handed orthonormal frames). In the fiber the group $SO(2)$ acts by $\tilde{A} \mapsto \tilde{A}\tilde{B}$, $\tilde{B} = \exp\{\beta\, i\sigma_2\} \in SO(2)$. The covering (13.3.6) shows that over the fiber there are the elements in $SU(2)$ of the form $U = \exp\{-i\alpha/2\, \sigma_3\}$; they form in turn the fiber in the spin bundle; the half-angle indicates that this fiber realizes a *two-sheeted covering* of the fiber in the first bundle, $\exp\{-i(\alpha/2)\sigma_3\} \mapsto \exp\{\alpha l_3\}$ (see also (22.2.4)); the group $U(1) = SO(2) = \text{Spin}\,(2,0)$ acts there by $U \mapsto Ue^{-i\beta/2}$. □

• Now, when the matter of a principal bundle needed for spinor fields is already settled, we can introduce the fields themselves. They are, as already mentioned before, the equivariant functions Φ of type ρ on the total space SM, where ρ denotes the spinor representation of the group Spin (p, q) which acts in the fibers of SM. If we want to work with them "downstairs" on M (and this would indeed be desirable if, for example, (M, g) represents the space-time and we want to describe fermions there), we need to choose a gauge, i.e. to choose a (local) *section* $\sigma : \mathcal{U} \to SM$ and pull-back Φ with the help of the section to the base

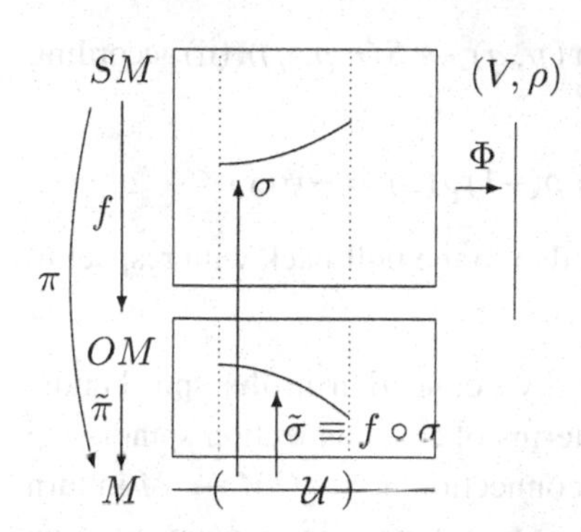

$$\psi := \sigma^* \Phi \equiv \Phi \circ \sigma \qquad \text{spinor field on the base}$$

Notice that the composition of the section σ with the covering map f of the total spaces $SM \to OM$ also gives automatically a section $\tilde{\sigma} := f \circ \sigma$ of the bundle of the orthonormal frames, i.e. an orthonormal *frame field* on $\mathcal{U} \subset M$. The basic property of the two-sheeted covering immediately leads to the conclusion that there is *one more* section of the bundle $\pi : SM \to M$ which would yield the same section $\tilde{\sigma}$, namely the section $\sigma' : \mathcal{U} \to SM$ which we obtain from the initial one σ by means of the action[488] of the element $-1 \in \text{Spin}\,(p, q)$. And what would be the result of a pull-back of the same spinor field Φ with the help of the "primed" section?

22.4.3 Let Φ be a spinor field on SM and $\sigma : \mathcal{U} \to SM$ a local section. Denote by $\sigma' : \mathcal{U} \to SM$ another local section which may be obtained from σ by the "gauge function" $S(x) = -1 \in \text{Spin}\,(p, q)$. Denote further by ψ and ψ' the spinor fields on the base which we get by pull-back of (the same) Φ with respect to σ and σ'

$$\psi := \sigma^* \Phi \qquad \psi' := \sigma'^* \Phi$$

Check that

(i) the composition of these sections with the covering map f indeed gives the *same* section $\tilde{\sigma}$ of the bundle of orthonormal frames, i.e. the same (right-handed) orthonormal frame field $e_a(x)$ in the domain $\mathcal{U} \subset M$

[488] Recall (21.2.1) the relation $\sigma'(x) = \sigma(x)S(x)$, which holds for two arbitrary sections, where $S : \mathcal{U} \to G$; here $S(x) = -1 \in \text{Spin}\,(p, q)$, so that it may be regarded as a "global" gauge transformation ($S(x)$ does not depend on x).

(ii) the fields $\psi(x)$ and $\psi'(x)$ differ just by a *sign*

$$\psi'(x) = -\psi(x)$$

(iii) so, the other way round, if we fix a (right-handed) orthonormal frame field $e_a(x)$ in the domain $\mathcal{U}$, *exactly two* sections of the spin bundle are induced by this choice and moreover for a given Φ we get a spinor field $\psi(x)$ *up to a sign*.

Hint: (i) a consequence of "equivariance" of f, expressed by the commutative diagram at the beginning of the section (see also the general diagram in (20.5.11)):

$$\tilde{\sigma}'(x) = f(\sigma(x)S(x)) = f(\sigma(x))\varphi(S(x)) = f(\sigma(x))\varphi(-1) = f(\sigma(x)) \equiv \tilde{\sigma}(x)$$

(since -1 is in the kernel of the covering group $\varphi : \text{Spin}\,(p,q) \to SO(p,q)$); (ii) according to (21.2.3) we may write in general

$$\psi'(x) := \sigma'^{*}\Phi(x) = \rho((S(x))^{-1})\psi(x) = \rho(-1)\psi(x) = -\psi(x)$$

(iii) there are just two preimages of $\tilde{\sigma}(x)$; uncertain $\pm$ due to the pull-back with respect to the uncertain section (out of two possibilities). □

• Now let us turn to the question of connection. Any connection in the spin bundle $\pi : SM \to M$ will be called the *spin connection* (in terms of this connection spinors are parallel transported and covariantly differentiated). A connection in $\tilde{\pi} : OM \to M$ in turn corresponds (as we know from the end of Section 20.5) to a *metric* linear connection on M (it may have non-vanishing torsion, so that it is not necessarily the RLC connection). It is important to realize that if we fix a particular connection in $\tilde{\pi} : OM \to M$, a unique connection in the spin bundle $\pi : SM \to M$ is automatically induced. And, the other way round, a connection in $\pi : SM \to M$ induces a unique metric linear connection on M (connection in $\tilde{\pi} : OM \to M$).[489]

22.4.4 Be sure to understand exactly how the one-to-one correspondence between connections in the bundles $\tilde{\pi} : OM \to M$ and $\pi : SM \to M$ works in terms of vertical and horizontal subspaces.

Hint: f is a two-sheeted covering and consequently also a local diffeomorphism, so that f_* is an *isomorphism* of tangent spaces; it moreover preserves vertical subspaces (due to the equivariance of f, i.e. the property $\tilde{R}_{\varphi(u)} \circ f = f \circ R_u, u \in \text{Spin}\,(p,q)$). This isomorphism enables one to *define horizontal* subspaces on either of the bundles, if they are available on the other one, as isomorphic images of the known subspaces ($\text{Hor}_{f(p)}OM = f_*\text{Hor}_p SM$; the equivariance of f then guarantees the needed G-invariance of the constructed horizontal distribution, i.e. the validity of the relation $\tilde{R}_{\varphi(u)*}\text{Hor}_{f(p)}OM = \text{Hor}_{f(p)\varphi(u)}OM$). □

22.4.5 Be sure to understand how exactly the one-to-one correspondence between connections in the bundles $\tilde{\pi} : OM \to M$ and $\pi : SM \to M$ works in terms of connection

[489] Clearly this one-to-one relation between the spin and metric connections works only if the prolongation of OM to SM exists, i.e. if (M, g) admits the spin structure.

forms: namely that if ω is a connection form on $\pi : SM \to M$ and $\tilde{\omega}$ is a connection form on $\tilde{\pi} : OM \to M$, the two connections being related in the way described in (22.4.4), then there is a simple relation between them[490]

$$f^*\tilde{\omega} = \varphi'\omega \qquad \text{i.e.} \quad (f^*\tilde{\omega}^i)\tilde{E}_i = \omega^i\varphi'(E_i)$$

(E_i being a basis of spin (p, q) and $\tilde{E}_i$ a basis of $so(p, q)$, such that $\varphi'(E_i) = \tilde{E}_i$; see more details in (22.4.6)).

Hint: if $\tilde{X} = \varphi'(X) \in so(p, q)$, then the corresponding fundamental fields are related by $\xi_{\tilde{X}} = f_*\xi_X$; one checks that $f^*\tilde{\omega} = \varphi'\omega$ is consistent with all the requirements demanded of a connection form

$$R_u^*\omega = \text{Ad}_{u^{-1}}\omega \qquad \langle\omega, \xi_X\rangle = X$$
$$\tilde{R}^*_{\varphi(u)}\tilde{\omega} = \text{Ad}_{(\varphi(u))^{-1}}\tilde{\omega} \qquad \langle\tilde{\omega}, \xi_{\tilde{X}}\rangle = \tilde{X}$$

One has still to check whether there holds $\text{Hor}_{f(p)}OM = f_*\text{Hor}_p SM$, i.e. whether $\text{Ker}\,\tilde{\omega}_{f(p)} = f_*\text{Ker}\,\omega_p$. However, this follows from $\omega_p = (f^*\tilde{\omega})_p \equiv \tilde{\omega}_{f(p)} \circ f_{*p}$ and the fact that f_{*p} is an isomorphism. □

22.4.6 The relation between the connection forms on SM and OM may also be stated more explicitly, since we know from the result of (22.2.7) what the adapted bases (such that $\varphi'(E_i) = \tilde{E}_i$) in the Lie algebras actually look like. Check that there holds (E^a being the generators of the "abstract" Clifford algebra, a basis of $C^1(p, q)$)

(i) $$\omega = \frac{1}{4}\omega_{ab}E^aE^b \equiv \frac{1}{8}\omega_{ab}(E^aE^b - E^bE^a) \qquad \omega_{ab} = -\omega_{ba} \in \Omega^1(SM)$$

(ii) $$\tilde{\omega} = \frac{1}{2}\tilde{\omega}_{ab}\mathcal{E}^{ab} \qquad \tilde{\omega}_{ab} = -\tilde{\omega}_{ba} \in \Omega^1(OM)$$

(iii) $$\omega_{ab} = f^*\tilde{\omega}_{ab}$$

(iv) $$\rho(\omega) = \frac{1}{4}\omega_{ab}\gamma^a\gamma^b \equiv \frac{1}{8}\omega_{ab}(\gamma^a\gamma^b - \gamma^b\gamma^a) \qquad \omega_{ab} = -\omega_{ba} \in \Omega^1(SM)$$

Hint: (i) according to (22.2.7) a basis of spin (p, q) is given by the elements $\frac{1}{2}E^aE^b$, $a < b$; then

$$\omega = \sum_{a<b}\omega_{ab}\frac{1}{2}E^aE^b = \sum_{a,b}\omega_{ab}\frac{1}{4}E^aE^b \equiv \frac{1}{4}\omega_{ab}E^aE^b$$

(ii) in a similar way a basis of $so(p, q)$ is given by the matrices $\mathcal{E}^{ab}$, $a < b$, so that $\tilde{\omega} = \sum_{a<b}\tilde{\omega}_{ab}\mathcal{E}^{ab} = \frac{1}{2}\sum_{a,b}\tilde{\omega}_{ab}\mathcal{E}^{ab} \equiv \frac{1}{2}\tilde{\omega}_{ab}\mathcal{E}^{ab}$; (iii) the isomorphism φ' of the Lie algebras (it was denoted as π' there) reads $\varphi'(\frac{1}{2}E^aE^b) = \mathcal{E}^{ab}$, so that $f^*\tilde{\omega} = \varphi'\omega$ yields $\omega_{ab} = f^*\tilde{\omega}_{ab}$; (iv) after the representation ρ of the Clifford algebra (and as a consequence also of the Clifford group as well as its Lie algebra)[491] in terms of the isomorphic matrix (22.3)

[490] The formal complication with the isomorphism of Lie algebras φ' is needed so that the connection forms have their values in due Lie algebras: ω in spin (p, q) and $\tilde{\omega}$ in $so(p, q)$.

[491] All three representations are denoted by the same letter ρ. However, the representation of spin (p, q) is actually the *derived* representation to the representation of Spin (p, q), so that it should, in fact, be written (in the notation used so far) by ρ'. Here

algebra (introduction of γ-matrices by the relations $\gamma^a := \rho(E^a)$) we get the expression $\rho(\omega) = \frac{1}{4}\omega_{ab}\gamma^a\gamma^b$. □

• The decompositions of the connection forms ω and $\tilde{\omega}$ with respect to the bases in Lie algebras yield the component 1-forms ω_{ab} and $\tilde{\omega}_{ab}$. These forms live in the total spaces of the corresponding bundles and they may be pulled back onto the base M. If the sections $\sigma : \mathcal{U} \to SM$ and $\tilde{\sigma} : \mathcal{U} \to OM$ are not chosen at random, but rather they correspond to one another in the sense of (22.4.3) (i.e. $\tilde{\sigma} = f \circ \sigma$), we will find out that actually both pull-backs lead to *the same result*.

22.4.7 Consider sections σ and $\tilde{\sigma}$ from (22.4.3) and use them to pull-back the forms ω and $\tilde{\omega}$ onto the base (a domain $\mathcal{U}$). Check that

$$\tilde{\sigma}^*\tilde{\omega} = \varphi'(\sigma^*\omega)$$

i.e. at the level of component forms

$$\tilde{\sigma}^*\tilde{\omega}_{ab} = \sigma^*\omega_{ab}$$

Hint: $\tilde{\sigma}^* = (f \circ \sigma)^*$, $f^*\tilde{\omega} = \varphi'\omega$, see (22.4.6). □

• The 1-forms pulled back onto the base

$$\hat{\omega}_{ab} := \tilde{\sigma}^*\tilde{\omega}_{ab}$$

already live in the domain $\mathcal{U} \subset M$ where the section $\tilde{\sigma}$ is defined and, consequently, also where the corresponding orthonormal frame field lives. And if they seem to be somehow familiar to you, you are perfectly right.

22.4.8 Let $\tilde{\sigma} : \mathcal{U} \to OM$ be the section which corresponds to a (right-handed orthonormal) frame field e_a on $\mathcal{U}$ and let $\tilde{\omega}$ be a connection form on OM. Be sure to understand that the component 1-forms $\hat{\omega}_{ab} := \tilde{\sigma}^*\tilde{\omega}_{ab}$, given by the decomposition

$$\hat{\omega} := \tilde{\sigma}^*\tilde{\omega} = \frac{1}{2}(\tilde{\sigma}^*\tilde{\omega}_{ab})\mathcal{E}^{ab} \equiv \frac{1}{2}\hat{\omega}_{ab}\mathcal{E}^{ab}$$

are nothing but the good old 1-forms of the *metric* connection with respect to an *orthonormal* frame field e_a from (15.6.6).

Hint: according to (19.2.3), the connection forms[492] ω^a_b introduced in (15.6.1) and living in the domain $\mathcal{O} \subset M$ may be pulled back from LM by means of a section of the bundle $\pi : LM \to M$; if the connection happens to be metric, it is enough to know it (according to the considerations presented at the end of Section 20.5) on OM, and it may thus be pulled back by means of a section of the subbundle $\pi : OM \to M$; due to the antisymmetry $\omega_{ab} = -\omega_{ba}$ (15.6.6), the basis of antisymmetric matrices is sufficient: according to (19.2.1)

it concerns the term $\rho(\omega)$, which should be written more precisely as $\rho'(\omega)$ – one should keep this in mind when writing the exterior covariant derivative of the spinor field in (22.5.1).

[492] Just to be sure, fix the notation: in (15.6.6) they were denoted by ω_{ab}. Here, as many as *three* versions of similar objects knock about: ω_{ab} on SM, $\tilde{\omega}_{ab}$ on OM and $\hat{\omega}_{ab} := \tilde{\sigma}^*\tilde{\omega}_{ab} = \sigma^*\omega_{ab}$ on $\mathcal{U} \subset M$. Those of them denoted here by $\hat{\omega}_{ab}$ actually coincide with ω_{ab} from (15.6.6).

and (22.2.5) we may write

$$\omega = \omega^a_b E^b_a = \eta^{ac}\omega_{cb}E^b_a = \omega_{ab}E^{ab} = \frac{1}{2}\omega_{ab}(E^{ab} - E^{ba}) \equiv \frac{1}{2}\omega_{ab}\mathcal{E}^{ab}$$

□

• We see that the component forms of the spin connection coincide, when pulled back onto the base, with the component forms of the metric connection. This is a useful observation. It means that if we want to compute covariant derivatives of spinor fields $\psi(x)$ on the base, no new objects are necessary for doing so, we need just the good old forms of the *metric* connection, since they happen to play *at the same time* the role of the forms of the *spin* connection; both of them read

$$\hat{\omega}_{ab} \equiv \tilde{\sigma}^*\tilde{\omega}_{ab} \equiv \sigma^*\omega_{ab} =: \omega_{ab\mu}(x)\,dx^\mu$$

Recall that the coordinate basis components $\omega_{ab\mu}(x)$ were already called (thus manifesting an admirable foresight, indeed) "spin connection" in (15.6.19).

And finally, let us introduce important vector fields on SM, which may be canonically associated with the bundle $\pi : SM \to M$ endowed with a connection.

22.4.9 Consider a point $p \in SM$. Its f-image is $e = f(p) \in OM$, i.e. an orthonormal frame e_a in the tangent space T_xM. Now the individual vectors e_a may in turn be *horizontally* lifted to the point p, making use of the connection; we get the vectors $\mathcal{E}_a \in T_pSM$

$$\mathcal{E}_a(p) := e^h_a(p) \qquad e_a \leftrightarrow f(p)$$

Check that

(i) in this way we get on the manifold SM an n-tuple of global horizontal vector fields $\mathcal{E}_a$
(ii) under the action of the group Spin (p,q) on SM the fields behave as follows:

$$R^*_u\mathcal{E}_a = A^b_a\mathcal{E}_b \qquad u \in \text{Spin}\,(p,q) \quad \varphi(u) \equiv A \in SO(p,q)$$

Hint: (ii) if $f(p) = e$ and $\varphi(u) = A$, then $f(pu) = f(p)\varphi(u) \equiv eA$; then $\mathcal{E}_a(pu) = (eA)^h_a(pu) = A^b_a e^h_b(pu) = A^b_a R_{u*}e^h_b(p) \equiv A^b_a R_{u*}\mathcal{E}_b(p)$, from where $R^*_u\mathcal{E}_a(pu) = A^b_a\mathcal{E}_b(p)$.

□

• If the construction of the fields $\mathcal{E}_a$ on SM and their behavior with respect to the action of the group Spin (p,q) seems to be somehow familiar to you, you are perfectly right once again. And if you even noticed that it reminds you of the construction of the fields $\hat{\mathcal{E}}_a$ on LM from (19.4.8) and their behavior with respect to the action of the group $GL(n,\mathbb{R})$, you are undoubtedly showing a promising observation ability. It is not hard to check that the f_*-image of the field $\mathcal{E}_a$ happens to be a field on OM which is the restriction of the field $\hat{\mathcal{E}}_a$ from LM to the submanifold OM (provided that there is a *metric* connection on LM).

Now everything is already prepared to write down *routinely* the exterior covariant derivative of spinors $D\Phi$ (or, after the pull-back to the base, $\mathcal{D}\psi$). In doing this, *surprisingly enough* an interesting combination of operators D, γ^a and $i_{\mathcal{E}_a}$ arises (or, after the pull-back to the base, $\mathcal{D}$, γ^a and i_{e_a}), which is called the *Dirac* operator and which will be denoted by $\not{D}$ (and its version on the base $\not{\mathcal{D}}$).

22.5 Dirac operator

• A spinor field Φ is, according to its introduction in Section 22.4, a field of type ρ on SM (where ρ is the spinor representation). This field (as well as the field ψ, which arises by pull-back of Φ onto the base) thus has its values in (V, ρ), the representation space of the Clifford algebra $C(p,q)$ (and automatically also of the group Spin (p,q) and the Lie algebra spin (p,q)). The space (V, ρ) coincides with the space in which γ-matrices act as linear operators (since $\gamma^a := \rho(E^a)$). We may thus regard this space as the space of column vectors of appropriate type (either real or complex) and dimension, according to what exactly the isomorphism ρ of a given Clifford algebra $C(p,q)$ with the matrix algebra in (22.1.8) and (22.1.9) looks like. If a basis E_α is introduced in this space, we obtain in a standard way the decomposition of both the field Φ and ψ as well as of the γ-matrices γ^a

$$\Phi = \Phi^\alpha E_\alpha \qquad \psi = \psi^\alpha E_\alpha \qquad \gamma^a E_\alpha = \gamma^{a\beta}_{\ \alpha} E_\beta$$

We will call indices α of this type the *spinor indices*. Of course, all the common rules of index gymnastics also apply for this particular type of index. The components of an object depend (as the world, at least since Section 2.4, goes), on the choice of a basis and they transform under the change of basis in a well-known way. For example, if we regarded γ^a as a linear operator in V, it is a tensor of type (1, 1), so that its matrix elements with respect to E_α are labeled by *one upper and one lower* "spinor" index

$$\gamma^a = \gamma^{a\beta}_{\ \alpha} E^\alpha \otimes E_\beta$$

Now, let us look at the exterior covariant derivative of a spinor field. This is already child's play since for the fields of type ρ we have the universal formula (20.4.6)

$$D\Phi = d\Phi + \rho'(\omega)\Phi$$

which also holds for the case of a spinor field on SM; we just need to write down explicitly the term $\rho'(\omega)$.

22.5.1 Let $\Phi \equiv \Phi^\alpha E_\alpha$ be a spinor field on SM, $\sigma : \mathcal{U} \to SM$ a section of the spin bundle, corresponding to a frame field e_a on $\mathcal{U}$ (as a section $f \circ \sigma : \mathcal{U} \to OM$) and let ψ be a spinor field in the domain $\mathcal{U} \subset M$, resulting from the pull-back of Φ by means of the section σ. Check that the explicit formulas for the exterior covariant derivative read

$$D\Phi = d\Phi + \frac{1}{4}\omega_{ab}\gamma^a\gamma^b\Phi \equiv d\Phi + \frac{1}{8}\omega_{ab}[\gamma^a, \gamma^b]\Phi$$

$$\mathcal{D}\psi = d\psi + \frac{1}{4}\hat{\omega}_{ab}\gamma^a\gamma^b\psi \equiv d\psi + \frac{1}{8}\hat{\omega}_{ab}[\gamma^a, \gamma^b]\psi \qquad \hat{\omega}_{ab} = \sigma^*\omega_{ab}$$

so that

$$\mathcal{D}_\mu \psi^\alpha = \partial_\mu \psi^\alpha + \frac{1}{4}\hat{\omega}_{ab\mu}(\gamma^a\gamma^b)^\alpha_\beta \psi^\beta$$
$$\equiv \partial_\mu \psi^\alpha + \frac{1}{8}\hat{\omega}_{ab\mu}[\gamma^a, \gamma^b]^\alpha_\beta \psi^\beta$$

Hint: see (22.4.6); according to (21.3.7) we have $\mathcal{D}\psi =: (\mathcal{D}_\mu \psi^\alpha)\, dx^\mu\, E_\alpha$. □

• In addition to the spinor field which is a *function* (0-form) of type ρ on SM, it is also convenient to consider more general objects, *horizontal forms of type* ρ on SM (the space of such forms[493] will be denoted by $\bar{\Omega}^p(SM, \rho)$). The representation ρ remains the same *spinor* representation of the group Spin (p, q). When pulled back to the base, such p-forms χ have in the component language a single upper "spinor" index α (due to the values lying in (V, ρ)) and p lower "ordinary tensor" indices (either coordinate μ, or a general frame a; due to being forms). In particular, for $p = 1$ this gives, for example,

$$\chi = \chi^\alpha_\mu(x)\, dx^\mu\, E_\alpha \equiv \chi^\alpha_a(x) e^a(x) E_\alpha$$

Such a field $\chi^\alpha_\mu(x)$ is in physics usually called the[494] *Rarita–Schwinger field.*

Now, there is a remarkable operator $i_{\not{\mathcal{E}}}$ which may be applied to such fields. The operator preserves both the horizontality and type ρ of the forms, with the first property being trivial, whereas the second one is not visible in an instantaneous view but rather it is to be verified by a computation.

22.5.2 Consider the space $\bar{\Omega}^p(SM, \rho)$ of horizontal p-forms of type ρ (where ρ is the spinor representation) on SM and introduce the operator

$$i_{\not{\mathcal{E}}} := \gamma^a i_{\mathcal{E}_a}$$

with $\mathcal{E}_a$ being the canonical horizontal vector fields on SM, studied in problem (22.4.9). Check that

(i) $i_{\not{\mathcal{E}}}$ preserves the horizontality of forms

$$\chi \text{ is horizontal} \quad \Rightarrow \quad i_{\not{\mathcal{E}}}\chi \text{ is horizontal}$$

(ii) $i_{\not{\mathcal{E}}}$ preserves the type ρ of forms

$$\chi \text{ is of type } \rho \quad \Rightarrow \quad i_{\not{\mathcal{E}}}\chi \text{ is of type } \rho$$

This means that $i_{\not{\mathcal{E}}}$ may be regarded as an operator

$$i_{\not{\mathcal{E}}} : \bar{\Omega}^p(SM, \rho) \to \bar{\Omega}^{p-1}(SM, \rho)$$

(iii) if pulled back to the base (by means of a local section σ), the operator takes the form

$$i_{\not{e}} := \gamma^a i_{e_a}$$

[493] In the general context of principal bundles we already recognized their usefulness in Chapter 20.

[494] An "ordinary" spinor field ψ^α describes particles with spin $\frac{1}{2}$, whereas the field χ^α_μ corresponds to particles with spin $\frac{3}{2}$ (*gravitino*).

where $e_a(x)$ is the orthonormal frame field, which corresponds to the section σ according to (22.4.3)

Solution: (ii) if $R_u^*\chi = \rho(u^{-1})\chi$, then

$$\begin{aligned}
R_u^*(i_{\not{\mathcal{E}}}\chi) &= \gamma^a R_u^*(i_{\mathcal{E}_a}\chi) \\
&= \gamma^a i_{R_u^*\mathcal{E}_a} R_u^*\chi && (8.3.6) \\
&= \gamma^a i_{A_a^b\mathcal{E}_b}\rho(u^{-1})\chi && (22.4.9) \text{ and the type } \rho \\
&= A_a^b\rho(u^{-1})\rho(u)\gamma^a\rho(u^{-1})i_{\mathcal{E}_b}\chi && (5.4.1) \text{ and the type } \rho \\
&= A_a^b\rho(u^{-1})(A^{-1})_c^a\gamma^c i_{\mathcal{E}_b}\chi && (22.3.1) \\
&= \rho(u^{-1})\gamma^a i_{\mathcal{E}_a}\chi \equiv \rho(u^{-1})(i_{\not{\mathcal{E}}}\chi)
\end{aligned}$$

(iii)

$$\begin{aligned}
(\sigma^* i_{\not{\mathcal{E}}}\chi)(w,\dots) &= \gamma^a\chi(\mathcal{E}_a,\sigma_* w,\dots) = \gamma^a\chi(\sigma_* e_a,\sigma_* w,\dots) = \gamma^a(\sigma^*\chi)(e_a,w,\dots) \\
&= \gamma^a i_{e_a}(\sigma^*\chi)(w,\dots)
\end{aligned}$$

so that $\sigma^* i_{\not{\mathcal{E}}} = i_{\not{e}}\sigma^*$; we used the fact that we may replace the *horizontal* lift of a vector by the lift obtained by the *section* ($\sigma_* e_a$ instead of $\mathcal{E}_a \equiv e_a^h$) if used as an argument of a horizontal form, since they only differ by a vertical vector (21.3.1). □

22.5.3 Consider the space $\bar{\Omega}^p(SM,\rho)$ of horizontal p-forms of (spinor) type ρ and introduce an operator which results by the composition of the exterior covariant derivative D with the operator $i_{\not{\mathcal{E}}}$

$$\not{D} := i_{\not{\mathcal{E}}} \circ D \equiv \gamma^a i_{\mathcal{E}_a} D \qquad \text{Dirac operator on } SM$$

Check that

(i) it indeed acts in the space mentioned above

$$\not{D} : \bar{\Omega}^p(SM,\rho) \to \bar{\Omega}^p(SM,\rho)$$

(ii) in the most important (and the simplest) case, on spinor fields $\Phi \in \Omega^0(SM,\rho)$ (i.e. for $p = 0$), this gives a strictly differential operator

$$\not{D}\Phi = \gamma^a\mathcal{E}_a\Phi \equiv \not{\mathcal{E}}\Phi$$

(iii) when pulled back onto the base (in terms of a local section σ), the operator takes the form[495]

$$\not{\mathcal{D}} = i_{\not{e}} \circ \mathcal{D} \equiv \gamma^a i_{e_a}\mathcal{D} \qquad \text{Dirac operator on } M$$

where $e_a(x)$ is the orthonormal frame field, which corresponds to the section σ (22.4.3).

Hint: (i) D raises the degree by 1 and $i_{\not{\mathcal{E}}}$ returns it back right away (similarly, as it works, say, in the identities $\mathcal{L}_V = i_V d + d i_V$, $\Delta = -(\delta d + d\delta)$ and in numerous further examples),

[495] Note that this is *no longer* a strictly differential operator even for $p = 0$, see (22.5.4).

both of them preserve the type ρ; (ii)

$$\not{D}\Phi = \gamma^a i_{\mathcal{E}_a} D\Phi = \gamma^a \langle D\Phi, \mathcal{E}_a\rangle = \gamma^a \langle d\Phi, \mathcal{E}_a\rangle = \gamma^a \mathcal{E}_a \Phi$$

(the operator D may be replaced by d, since $\mathcal{E}_a$ is horizontal); (iii) $\sigma^* \not{D} = \sigma^* i_{\not{\mathcal{E}}} D = i_{\not{e}} \sigma^* D = i_{\not{e}} \mathcal{D}\sigma^*$ according to (22.5.2) and (21.2.4). □

• Let us look in detail at various explicit expressions of the Dirac operator which acts on a spinor field $\psi(x) \equiv \sigma^*\Phi$ on the base. In these expressions several objects occur which we have already encountered earlier, namely the *vielbein field* e_a^μ (in four-dimensional space the *tetrad field*), the *Ricci coefficients of rotation* $\gamma_{abc} \equiv \Gamma_{abc}$, the "*spin connection*" $\omega^a_{b\mu}(x)$ and the *coefficients of anholonomy* $c^a_{bc}(x)$. For the convenience of the reader we summarize the definitions in terms of the notation used here (where connection forms *on the base* are denoted by (22.4.8) with the hat); their mutual interrelations were studied in problems (15.6.19) and (15.6.20):

$$\begin{array}{lll}
e_a =: e_a^\mu(x)\partial_\mu & & (4.5.5), (15.6.19) \\
\hat{\omega}_{ab} =: \omega_{ab\mu}(x)dx^\mu & \hat{\omega}_{ab} \equiv \eta_{ac}\hat{\omega}^c_b & (15.6.19) \\
\nabla_a e_b =: \Gamma^c_{ba}(x) e_c & \Gamma_{abc} \equiv \eta_{ad}\Gamma^d_{bc} & (15.6.20) \\
[e_a, e_b] =: c^c_{ab}(x) e_c & c_{abc} \equiv \eta_{ad} c^d_{bc} & (9.2.10)
\end{array}$$

22.5.4 Let x^μ be local coordinates and $e_a \equiv e_a^\mu \partial_\mu$ an orthonormal frame field. Check that the action of the Dirac operator $\not{\mathcal{D}}$ on a spinor field $\psi(x)$ on the base may be explicitly expressed by either of the following formulas:

$$\begin{aligned}
\not{\mathcal{D}}\psi &\overset{1}{=} \gamma^a \left(e_a\psi + \frac{1}{4}\langle\hat{\omega}_{bc}, e_a\rangle \gamma^b\gamma^c\psi \right) \\
&\overset{2}{=} \gamma^a \left(e_a\psi + \frac{1}{4}\Gamma_{bca}\gamma^b\gamma^c\psi \right) \\
&\overset{3}{=} \gamma^a e_a^\mu \left(\partial_\mu\psi + \frac{1}{4}\omega_{bc\mu}\gamma^b\gamma^c\psi \right) \\
&\overset{4}{=} \gamma^a \left(e_a\psi + \frac{1}{8}(c_{abc} + c_{cba} - c_{bca})\gamma^b\gamma^c\psi \right)
\end{aligned}$$

Hint: the expression of $\mathcal{D}\psi$ from (22.5.1); see (15.6.19) and (15.6.20). □

• It may not have escaped your notice that the last formula (which contains the coefficients of anholonomy $c^a_{bc}(x)$) actually does not need any objects *explicitly* characterizing the *connection*; it is just enough to evaluate (all) mutual commutators of the frame field (being sort of a simple homework from (quasi-)quantum mechanics).[496] In this way we can get rid of the computation of, say, connection forms $\hat{\omega}_{ab}$ starting from the Cartan structure equations, or Ricci coefficients of rotations from the formula $\gamma_{abc} := e_a^\mu e_{b\mu;\nu} e_c^\nu$ (15.6.20)

[496] This kind of computation is recommended to be performed, in order to save time, in parallel with watching the evening news (except for breaking news coverage – this often leads to sign errors), a weather forecast or financial reports.

(one has to compute all semi-colons and this in turn needs computation of Christoffel symbols from the formula $2\Gamma^{\mu}_{\nu\sigma} = g^{\mu\rho}(g_{\rho\nu,\sigma} + \cdots)$ etc.).

22.5.5 Consider the ordinary flat space $E^{p,q}$ and the *Cartesian* frame field $e_a = \partial_a$. Check that *then* we get for the Dirac operator the following well-known simple formula:

$$\not{D} = \gamma^a \partial_a \equiv \not{\partial} \qquad \text{in } E^{p,q}, \text{ Cartesian frame field}$$

(i.e. *then* it coincides with the operator mentioned at the beginning of the chapter) and that its *square* yields just the Laplace[497] operator Δ in $E^{p,q}$ (in Minkowski space $\Delta \equiv -\Box$)

$$\not{D}\not{D} = \Delta \equiv -\Box \equiv \partial_\mu \partial^\mu \qquad \text{in } E^{p,q}$$

In this sense the Dirac operator may be regarded as a "square root" of the Laplacian.
Hint: see (22.5.4), the basis ∂_a is orthonormal and at the same time holonomic (i.e. $c_{abc} = 0$); the square gives $\not{\partial}\not{\partial} = \gamma^\mu \gamma^\nu \partial_\mu \partial_\nu = \eta^{\mu\nu} \partial_\mu \partial_\nu = \Delta \equiv -\Box$. □

(Another square root reads $\Delta \equiv -(\delta d + d\delta) = (d - \delta)^2$ and it provides the basis of an alternative formalism to the standard Dirac approach, known as *Kähler fermions*. It was first proposed shortly after Dirac's paper in traditional "tensor" language by Ivanenko and Landau; some quarter of a century later it was then independently rediscovered (and elaborated) in terms of non-homogeneous differential forms by Kähler; in real physics it is, however, practically not used. The role of the "Dirac" operator is played by the *Dirac–Kähler operator* $(d - \delta)$.)

• As the world goes, in *two dimensions* things used to be considerably simpler (we have seen this for example when treating the Cartan structure equations (15.6.10)). This also applies here. There are just two γ-matrices γ^1, γ^2, and all antisymmetric objects become proportional to ϵ_{ab}, etc. – not bad news, indeed.

22.5.6 Consider a *two-dimensional* manifold and an orthonormal frame field e_a. Check that

(i) the exterior covariant derivative of a spinor field turns out to be

$$\mathcal{D}\psi = d\psi + \frac{1}{2}\alpha\gamma_5\psi \qquad \hat{\omega}_{ab} =: \alpha\epsilon_{ab}$$

$$\text{i.e.} \quad \mathcal{D}_\mu \psi^\alpha = \partial_\mu \psi^\alpha + \frac{1}{2}\alpha_\mu (\gamma_5)^\alpha_\beta \psi^\beta$$

(ii) for γ-matrices there holds the identity

$$\epsilon_{ab}\gamma^b = \eta_{ab}\gamma^b\gamma_5 =: \gamma_a\gamma_5$$

(iii) the coefficients of anholonomy may be parametrized by a single one-index (non-tensor; neither does c_{abc} correspond to a tensor) quantity $c_a(x)$

$$c_{abc}(x) =: c_a(x)\epsilon_{bc}$$

[497] Also in the general case the square of the Dirac operator is closely related to the Laplace operator. The computation is, however, more complicated and both the curvature and torsion enter the final formula.

(iv) the relevant term in $\not{D}\psi$ from the option 4 in (22.5.4) may be written in the form

$$\frac{1}{8}\gamma^a(c_{abc}+c_{cba}-c_{bca})\gamma^b\gamma^c = \frac{1}{2}c_a\epsilon^a{}_b\gamma^b \equiv \frac{1}{2}\not{c}\,\gamma_5 \qquad \not{c} := c_a\gamma^a \quad \epsilon^a_{\ b} := \eta^{ac}\epsilon_{cb}$$

(v) so that we get for the Dirac operator the following expression:

$$\not{D}\psi = \gamma^a\left(e_a + \frac{1}{2}c_a\gamma_5\right)\psi \equiv \gamma^a\left(e_a+\frac{1}{2}c_b\epsilon^b{}_a\right)\psi$$

Hint: (iii) $c_{abc} = -c_{acb}$; (iv) e.g. $c_{abc}\gamma^b\gamma^c = c_a\epsilon_{bc}\gamma^b\gamma^c = 2c_a\epsilon_{12}\gamma^1\gamma^2 = 2c_a\gamma_5$. □

22.5.7 Check that the explicit form of the Dirac operator

(i) in the two-dimensional Euclidean plane with respect to the orthonormal frame field $e^1 = dr$, $e^2 = r\,d\varphi$, based on the *polar* coordinates (r,φ) reads

$$\not{D}\psi = \left\{\gamma^1\left(e_1+\frac{1}{2r}\right)+\gamma^2 e_2\right\}\psi \equiv \left\{\gamma^1\left(\partial_r+\frac{1}{2r}\right)+\frac{1}{r}\gamma^2\partial_\varphi\right\}\psi$$
$$\psi = \psi(r,\varphi)$$

(ii) on the ordinary two-dimensional sphere of radius R with respect to the orthonormal frame field $e^1 = R\,d\vartheta$, $e^2 = R\sin\vartheta\,d\varphi$, based on standard *spherical* coordinates (ϑ,φ) is

$$\not{D}\psi = \left\{\gamma^1\left(e_1+\frac{1}{2R}\cot\vartheta\right)+\gamma^2 e_2\right\}\psi \equiv \frac{1}{R}\left\{\gamma^1\left(\partial_\vartheta+\frac{1}{2}\cot\vartheta\right)+\frac{1}{\sin\vartheta}\gamma^2\partial_\varphi\right\}\psi$$
$$\psi = \psi(\vartheta,\varphi)$$

(iii) a possible choice of the γ-matrices is (in both cases)[498]

$$\gamma^1 = \sigma_1 \quad \gamma^2 = \sigma_3 \qquad \text{if we consider *Majorana* spinor fields } \psi$$
$$\gamma^1 = \sigma_1 \quad \gamma^2 = \sigma_2 \qquad \text{if we consider *Dirac* spinor fields } \psi$$

Hint: (i) $[e_r, e_\varphi] = -e_\varphi/r \equiv c_2\epsilon_{12}e_\varphi$, so that $\not{c}\,\gamma_5 = c_2\gamma^2\gamma^1\gamma^2 = \gamma^1/r$; (ii) $[e_\vartheta, e_\varphi] = -e_\varphi/R\cot\vartheta \equiv c_2\epsilon_{12}e_\varphi$, therefore $\not{c}\,\gamma_5 = c_2\gamma^2\gamma^1\gamma^2 = \gamma^1/R\cot\vartheta$; (iii) both the plane and the sphere have signature $++$, so that we are to represent $C(2,0)$, (22.1.3); for Majorana spinors the γ-matrices are to be real and for Dirac spinors complex matrices are also allowed. □

22.5.8 Consider the standard orthonormal (co)frame fields in the ordinary four-dimensional Minkowski space, based on the *spherical* and *cylindrical* coordinates

$$e^0 = dt \quad e^1 = dr \quad e^2 = r\,d\vartheta \quad e^3 = r\sin\vartheta\,d\varphi \qquad \text{spherical}$$
$$e^0 = dt \quad e^1 = dr \quad e^2 = r\,d\varphi \quad e^3 = dz \qquad \text{cylindrical}$$

Check that the explicit form of the Dirac operator with respect to these frame fields reads

$$\not{D} = \gamma^0 e_0 + \gamma^1\left(e_1+\frac{1}{r}\right)+\gamma^2\left(e_2+\frac{\cot\vartheta}{2r}\right)+\gamma^3 e_3 \qquad \text{spherical}$$
$$\not{D} = \gamma^0 e_0 + \gamma^1\left(e_1+\frac{2}{r}\right)+\gamma^2 e_2+\gamma^3 e_3 \qquad \text{cylindrical}$$

[498] The first choice is also suitable for Dirac spinors, however, the second one is *not* suitable for Majorana spinors.

Hint: see (22.5.4) and either use $\hat{\omega}_{ab}$ from (15.6.14), or compute the coefficients of anholonomy for e_a. □

22.5.9 Check that the explicit form of the Dirac operator on a semi-simple Lie group with respect to the *left-invariant* (orthonormal) frame field e_a is

$$\not{D} = \gamma^a \left(e_a - \frac{1}{8} c_{abc} \gamma^b \gamma^c \right)$$

(c_{abc} being the *structure constants* with respect to e_a) and, in particular, on $SU(2)$ it may be expressed in terms of Pauli matrices σ_a and the fields e_a from (11.7.23) in the form

$$\not{D} = \sigma_a e_a - \frac{3i}{4} \mathbb{I}$$

Hint: the coefficients of anholonomy in (22.5.4) then coincide with the structure constants (11.2.1), the latter being *completely* antisymmetric (12.3.9), $g = -\frac{1}{2}\mathcal{K}$; on $SU(2)$ we may take as γ-matrices the σ-matrices (12.3.15), (22.3.7) and we have $\gamma^a c_{abc} \gamma^b \gamma^c = \sigma_a \epsilon_{abc} \sigma_b \sigma_c = 6\sigma_1\sigma_2\sigma_3 = 6i\mathbb{I}$. □

• Note that the expressions of the Dirac operator on the base $\not{D}$ in (22.5.4) are uniquely given once we choose a frame field e_a. Recall that the choice of a frame field defines a section $\hat{\sigma}$ of (only) the bundle of orthonormal frames $\hat{\pi} : OM \to M$, whereas the spinor fields ψ on the base as well as the operators acting on them are given by a section σ of the *spin* bundle $\pi : SM \to M$. Secondly, recall (see (22.4.3)) that as many as *two* sections σ_1 and σ_2 may be associated with a single section $\hat{\sigma}$. So it looks like the Dirac operator $\not{D}$ on the base does not feel the difference between the sections σ_1 and σ_2 (which is fairly convenient from the practical point of view, since everybody would probably agree that it is simpler to fix (only) a frame field than to make still an additional choice of one of the (two possible) sections σ, which correspond to the frame field). We will verify that this is indeed the case. Actually we will derive a more general formula for the transformation of the Dirac operator on the base under an arbitrary change of a section of the spin bundle.

22.5.10 Consider two sections σ and σ' of the spin bundle, which are related (in the sense of (21.2.1)) by an element $S(x)$ of the groups $\{\text{Spin}\,(p,q)\}^{\mathcal{U}}$ of the local gauge transformations ($S : \mathcal{U} \to \text{Spin}\,(p,q)$ is a function on $\mathcal{U}$ with values in the group $\text{Spin}\,(p,q)$). Denote (see (21.2.5)) also the image of $S(x)$ with respect to the spinor representation ρ by $B(x)$

$$B := \rho \circ S : \mathcal{U} \to GL(N \equiv \dim \rho, \mathbb{C}) \qquad \text{i.e.} \qquad B(x) := \rho(S(x))$$

Check that

(i) the Dirac operators $\not{D}$ and $\not{D}'$ on the base, which result from the pull-back by means of the sections σ and σ' respectively, are related by

$$\not{D}' = B^{-1}(x) \not{D} B(x)$$

(ii) this formula indeed enables one to transform, say, the Dirac operator with respect to the "spherical polar" (orthonormal) frame field (22.5.7) to the corresponding operator with respect to the "Cartesian" frame field (22.5.5)

(iii) it follows from the general formula that if the two sections σ_1 and σ_2 correspond *to the same frame* field (section $\hat{\sigma}$), then the Dirac operator turns out to be *the same*

$$\not{D}_1 = \not{D}_2$$

Hint: (i) according to (21.2.5), any spinor field transforms according to the prescription $\psi \mapsto B^{-1}(x)\psi \equiv \psi'$; if any, then also $\not{D}\psi \mapsto (\not{D}\psi)' = B^{-1}(\not{D}\psi) \equiv B^{-1}(\not{D}BB^{-1}\psi) \equiv (B^{-1}\not{D}B)\psi' =: \not{D}'\psi'$; (ii) first realize that for the choice $\gamma^1 = \sigma_1, \gamma^2 = \sigma_3$ we have $\gamma_5 = -i\sigma_2 = -\mathcal{E}^{12}$; the (orthonormal) spherical polar basis $e_a \equiv (e_r, e_\varphi)$ and the Cartesian basis $e_i \equiv (e_x, e_y)$ in the xy-plane are related by $e_a = A^i_a e_i$, where

$$A^i_a(r,\varphi) = A^i_a(\varphi) = \begin{pmatrix} \cos\varphi & -\sin\varphi \\ \sin\varphi & \cos\varphi \end{pmatrix} = e^{-\varphi\mathcal{E}^{12}}$$

This matrix from $SO(2)$ relates the sections of OM; the corresponding sections of SM are then related by such S, which covers this A; since $\mathcal{E}^{12} \leftrightarrow \frac{1}{2}E^1E^2$, a covering S is $S(r,\varphi) = S(\varphi) = \exp\{-\varphi\frac{1}{2}E^1E^2\}$, so that

$$B(r,\varphi) \equiv \rho(S(\varphi)) = e^{-\varphi\frac{1}{2}\gamma^1\gamma^2} \equiv e^{-\varphi\frac{1}{2}\gamma_5} = e^{-\varphi\frac{1}{2}(-\mathcal{E}^{12})} \equiv A(-\varphi/2)$$

Then the "spherical polar" and the "Cartesian" Dirac operators should be related by $\not{D}_{\text{pol}} = B^{-1}\not{D}_{\text{Cart}}B$; indeed,

$$\begin{aligned} B\not{D}_{\text{pol}}B^{-1} &= e^{-\frac{1}{2}\varphi\gamma_5}\left(\gamma^1\left(e_r + \frac{1}{2r}\right) + \gamma^2 e_\varphi\right)e^{\frac{1}{2}\varphi\gamma_5} = \cdots \\ &= \gamma^1 e^{\varphi\gamma_5}e_r + \gamma^2 e^{\varphi\gamma_5}e_\varphi = \cdots = \gamma^1 e_x + \gamma^2 e_y \equiv \not{D}_{\text{Cart}} \end{aligned}$$

(iii) $S(x) = -1 \Rightarrow B(x) = -\mathbb{I} \equiv B^{-1}(x)$. □

• In (12.5.3) we learned that the matrix elements of Pauli matrices may be regarded as the components of an *invariant* $SU(2)$-tensor. But Pauli matrices are nothing but special γ-matrices (22.3.7) and $SU(2) = \text{Spin}\,(3,0)$ (see Section 22.2). We will verify that such an invariance also holds in general: the matrix elements of γ-matrices may be regarded as the components of Spin (p,q)-invariant tensors. This means that they can be used (in the sense of Section 12.5) for "transmutation of types" of objects.

22.5.11 According to the beginning of this section, the matrix elements of the γ-matrices carry the indices $\gamma^{a\alpha}_{\beta}$. Thus we may associate with them a vector of the representation space of $\varphi \otimes \rho \otimes \check{\rho}$, where φ is the covering Spin $(p,q) \to SO(p,q)$, ρ is the spinor representation and $\check{\rho}$ is the contragredient one. Check that

(i) the vector under consideration is an *invariant* element of the space, i.e.

$$\rho(u)^\alpha_\tau \rho(u^{-1})^\sigma_\beta A^a_b \gamma^{b\tau}_{\sigma} = \gamma^{a\alpha}_{\beta} \qquad A \equiv \varphi(u)$$

(ii) if we regard $\gamma^{a\alpha}_{\beta}$ as a (constant) *function of type* $\varphi \otimes \rho \otimes \check{\rho}$ on SM (which *is* possible according to item (i)), then the function has vanishing exterior covariant derivative:

$$D\gamma^a = 0 \qquad \text{i.e.} \quad D\gamma^{a\alpha}_{\beta} = 0$$

(iii) the same is also true for γ_5

(iv) the Dirac operator anticommutes with the matrix γ_5

$$\not{D}\gamma_5 = -\gamma_5\not{D}$$

Hint: (i) this is exactly what the equation $\rho(u)\gamma^a\rho(u^{-1}) = (A^{-1})^a_b\gamma^b$ from (22.3.1) says; (ii) the general formula $D(\cdot) = d(\cdot) + \rho'(\omega)\,\dot{\wedge}\,(\cdot)$ gives $D\gamma^{a\alpha}_{\beta} = d\gamma^{a\alpha}_{\beta} + ((\varphi \otimes \rho \otimes \check{\rho})'(\omega)\gamma)^{a\alpha}_{\beta}$; the first term vanishes due to *constancy* of $\gamma^{a\alpha}_{\beta}$, the second one due to *invariance* of the latter (in general if $\rho(g)C = C$, then $\rho'(X)C = 0$); (iii) this is true for *any* product of γ-matrices; (iv) making use of $\gamma^a\gamma_5 = -\gamma_5\gamma^a$ (22.3.4) we have $\not{D}\,\gamma_5\Phi = i_{\not{e}}D\gamma_5\Phi = i_{\not{e}}\{(D\gamma_5)\Phi + \gamma_5 D\Phi\} = i_{\not{e}}(\gamma_5 D\Phi) = -\gamma_5 i_{\not{e}} D\Phi \equiv -\gamma_5\not{D}\Phi$. □

22.5.12 Check that the Dirac operator is *odd with respect to chirality*, i.e. it behaves with respect to the projectors to the right and left spinors as follows:

$$\not{D}\,P = L\,\not{D} \quad \not{D}\,L = P\,\not{D}$$

This says that $\not{D}$ maps right spinors to left spinors and vice versa.

Hint: according to (22.3.4) the projectors L and P read $L = \frac{1}{2}(\mathbb{I} - \lambda\gamma_5)$, $P = \frac{1}{2}(\mathbb{I} + \lambda\gamma_5)$ (where $\lambda = 1$ or i) and according to (22.5.11) $\not{D}$ anticommutes with γ_5; if $L\Phi = \Phi$, then $R(\not{D}\Phi) = \not{D}L\Phi = \not{D}\Phi$ and vice versa. □

Summary of Chapter 22

The special orthogonal group $SO(p, q)$ admits the two-sheeted universal covering group, which is called the spin group and is denoted by Spin (p, q). An elementary theory of spin groups is systematically developed with the help of Clifford algebras. An isomorphism of these algebras with appropriate matrix algebras (a faithful representation) is constructed. This leads naturally to the concept of a spinor as an element (vector) of the representation space of the Clifford algebra. Since the spin groups are subsets of the Clifford algebras, restriction of the representation of the algebra is automatically a representation of the spin group. Consequently, spinors also carry a representation of spin groups (and also the two-valued representation of the orthogonal groups). This is called the spinor representation. For some particular values of (p, q) special kinds of spinors may exist (Weyl, Majorana, etc.). The term spin structure on M is sometimes used as a synonym for a principal bundle over M (the spin bundle), whose total space is a two-sheeted covering of the total space of the bundle of right-handed orthonormal frames and in the fibers of which the spin group acts. There are also manifolds which do not admit the spin structure. Equivariant functions of type ρ on the total space of the spin bundle (as well as their pull-backs to the base with the help of a section), where ρ is the spinor representation, are called spinor fields on M.

The Rarita–Schwinger field then corresponds to a 1-form of type ρ. There is a specific first-order operator which acts on spinor fields, the Dirac operator. Its historical origin is in physics, in the quantum theory of the relativistic electron and it enters the Dirac equation.

$e^a e^b + e^b e^a = 2g^{ab}$	Fundamental relations in Clifford product	(22.1.1)
$u = \alpha_1 \dots \alpha_k, \;\; g(\alpha_j, \alpha_j) = \pm 1$	Elements of the group Pin (p, q)	(22.2.1)
$ue^a u^{-1} =: (A^{-1})^a_b e^b$	Two-sheeted covering Spin $(p, q) \to SO(p, q)$	(22.2.3)
$\frac{1}{2} e^a e^b \mapsto \mathcal{E}^{ab}$	Derived isomorphism spin $(p, q) \to so(p, q)$	(22.2.7)
$\gamma^a := \rho(e^a)$	γ-matrices	(22.3.1)
$\mathcal{D}\psi = d\psi + \frac{1}{4}\hat{\omega}_{ab}\gamma^a\gamma^b\psi$	Exterior covariant derivative of a spinor field	(22.5.1)
$\chi^\alpha_\mu(x)\, dx^\mu E_\alpha \equiv \chi^\alpha_a(x) e^a(x) E_\alpha$	Rarita–Schwinger field	Sec. 22.5
$\not{D} := i_{\mathcal{E}} \circ D \equiv \gamma^a i_{\mathcal{E}_a} D$	Dirac operator on SM	(22.5.3)
$\not{\mathcal{D}} = i_{\not{e}} \circ \mathcal{D} \equiv \gamma^a i_{e_a} \mathcal{D}$	Dirac operator on M	(22.5.3)
$\not{\mathcal{D}}\psi = \gamma^a e^\mu_a (\partial_\mu \psi + \frac{1}{4}\omega_{bc\mu}\gamma^b\gamma^c\psi)$	Action of the Dirac operator on spinor fields	(22.5.4)
$\not{\mathcal{D}}\psi = \gamma^a e^\mu_a (\partial_\mu \psi + \frac{1}{2}\alpha_\mu \gamma_5 \psi)$	How it simplifies for two dimensional M	(22.5.4)
$\rho(u)^\alpha_\tau \rho(u^{-1})^\sigma_\beta A^a_b \gamma^{b\tau}_\sigma = \gamma^{a\alpha}_\beta$	γ-matrices are Spin (p, q)-invariant tensors	(22.5.11)

Appendix A

Some relevant algebraic structures

In these appendices elementary facts are collected concerning those algebraic structures which are most frequently used in the text. As a rule the proofs of the statements are straightforward and the reader is invited to carry them out her(him)self.

A.1 Linear spaces

In a *linear space* (i.e. vector space) V over the field F, we may perform linear combinations $x + \lambda y \in V$, where $x, y \in V$, $\lambda \in F$. In an overwhelming majority of cases we will encounter the field $F = \mathbb{R}$ in this book, sometimes the field $F = \mathbb{C}$. We will denote by $\mathcal{L}(V, W)$ (or $\mathrm{Hom}(V, W)$) linear maps from V to W. They form in their own right a linear space of dimension nm (if $\dim V = n$, $\dim W = m$).

Given two linear spaces V, W, their *direct sum* is a linear space denoted[499] by $V \oplus W$. Its elements are ordered pairs (v, w) (so that it is a Cartesian product $V \times W$ as a set) and the linear combination is given by components

$$(v, w) + \lambda(v', w') := (v + \lambda v', w + \lambda w')$$

The dimension of the space $V \oplus W$ is the sum of the dimensions of V and W (if they happen to be finite). Each element (v, w) may be uniquely written as the sum $(v, w) = (v, 0) + (0, w)$. This means that there is a subspace $\tilde{V}$ in $V \oplus W$ which is isomorphic to V (vectors $(v, 0)$) and $\tilde{W}$ isomorphic to W (vectors $(0, w)$), their intersection containing the single vector $0 \equiv (0, 0) \in V \oplus W$. If e_i is a basis of V and e_α a basis of W, in $V \oplus W$ we may use the basis E_i, E_α ($n + m$ vectors), where $E_i = (e_i, 0)$, $E_\alpha = (0, e_\alpha)$ (it is moreover adapted to the structure of the direct sum). A general vector $u \in V \oplus W$ then admits the expression $u = u^i E_i + u^\alpha E_\alpha$. If A and B are linear operators in V and W respectively, $Ae_i = A^j_i e_j$ and $Be_\alpha = B^\beta_\alpha e_\beta$, we may introduce their sum $A \oplus B$ by $(A \oplus B)(v, w) := (Av, Bw)$. Then $(A \oplus B)E_i = A^j_i E_j$, $(A \oplus B)E_\alpha = B^\beta_\alpha E_\beta$, so that the matrix of the operator $A \oplus B$ with respect to the basis E_i, E_α has the block diagonal form

$$A \oplus B \quad \leftrightarrow \quad \begin{pmatrix} A^j_i & 0 \\ 0 & B^\beta_\alpha \end{pmatrix}$$

[499] If there is an additional structure in the linear space (see, for example, Section A.3 on Lie algebras), the sum of the spaces themselves is sometimes denoted by $V \dotplus W$ and then $V \oplus W$ means more, namely the sum of the spaces including the compatibility of the additional structure with the sum (for Lie algebras this requires that the commutator of elements from V was again from V and similarly for W).

We can also easily see that the following useful relations hold for operators of the form of a sum of two other operators:

$$(A \oplus B)(C \oplus D) = AC \oplus BD$$
$$\lambda(A \oplus B) = (\lambda A) \oplus (\lambda B)$$
$$[(A \oplus \mathbb{I}), (\mathbb{I} \oplus B)] = 0$$

Given two vector spaces V, W, there is also another way of obtaining a new space, namely their *tensor product* $V \otimes W$ (i.e. a direct product). A possible definition is that this is the (linear) space of bilinear maps $V^* \times W^* \to \mathbb{R}$. So this is reminiscent of second rank tensors from Section 2.4. Note, however, that here the two spaces V, W may not be interrelated at all, whereas there the second one either coincides with the first one or it is its dual. (There holds $T_2^0(L) = L^* \otimes L^*$, $T_0^2(L) = L \otimes L$, $T_1^1(L) = L^* \otimes L \sim L \otimes L^*$.) By repeating the considerations from Section 2.4, special elements of the space $V \otimes W$ have the form $v \otimes w$, $v \in V$, $w \in W$, where

$$(v \otimes w)(a, b) := \langle a, v\rangle\langle b, w\rangle \qquad a \in V^*, b \in W^*$$

or at the level of components

$$(v \otimes w)^{i\alpha} := v^i w^\alpha$$

(so that $\otimes : V \times W \to V \otimes W$) and a general element is in turn a linear combination of such special ones; if e_i, e_α constitute bases in V, W, a general element $u \in V \otimes W$ may be uniquely expressed as

$$u = u^{i\alpha} e_i \otimes e_\alpha \qquad u^{i\alpha} := u(e^i, e^\alpha) \in \mathbb{R}$$

This means that the elements $e_i \otimes e_\alpha$ form a basis in $V \otimes W$ ($\Rightarrow \dim V \otimes W =$ the product of the dimensions of the corresponding spaces) and a general element is specified by the values of the components $u^{i\alpha}$ (they may be displayed as an $n \times m$ matrix) with respect to the basis. Given two operators A and B in V and W respectively we may construct their tensor product $A \otimes B$. This is a linear operator in $V \otimes W$ defined by

$$(A \otimes B)(v \otimes w) := Av \otimes Bw$$

or equivalently

$$\begin{aligned}(A \otimes B)u \equiv (A \otimes B)(u^{i\alpha} e_i \otimes e_\alpha) &:= u^{i\alpha}(Ae_i) \otimes (Be_\alpha) = A_i^j B_\alpha^\beta u^{i\alpha} e_j \otimes e_\beta \\ &\equiv (A^i{}_j u^{j\beta} B^\alpha_\beta) e_i \otimes e_\alpha\end{aligned}$$

For operators in the form of a tensor product of the other two operators there holds

$$(A \otimes B)(C \otimes D) = AC \otimes BD$$
$$\lambda(A \otimes B) = (\lambda A) \otimes B = A \otimes (\lambda B)$$
$$(A + C) \otimes (B + D) = A \otimes B + A \otimes D + C \otimes B + C \otimes D$$
$$[(A \otimes \mathbb{I}), (\mathbb{I} \otimes B)] = 0$$

The matrix elements of the operators in $V \otimes W$ naturally carry *four* indices (since the basis vectors carry *two*)

$$H(e_i \otimes e_\alpha) =: H_{i\alpha}^{j\beta}(e_j \otimes e_\beta)$$

In particular, for $H \equiv A \otimes B$ we have

$$(A \otimes B)_{i\alpha}^{j\beta} = A_i^j B_\alpha^\beta$$

so that

$$u \mapsto (A \otimes B)u \qquad \leftrightarrow \qquad u^{i\alpha} \mapsto ((A \otimes B)u)^{i\alpha} = A^i_{\ j} B^\alpha_\beta u^{j\beta} \equiv A^i_{\ j} u^{j\beta} B^\alpha_\beta$$

and this may be written in terms of matrices as[500]

$$u \mapsto A u B^{\mathrm{T}}$$

If we labeled the basis $e_i \otimes e_\alpha$ in terms of *a single* index $a = 1, \dots, nm$

$$E_a = E_1, \dots, E_{nm} \equiv e_1 \otimes e_1, \dots, e_1 \otimes e_m, e_2 \otimes e_1, \dots, e_n \otimes e_m$$

then also the matrix of the operator $A \otimes B$ would only have two indices $(A \otimes B)^b_a$, which are given by its action on the basis, $(A \otimes B)E_a =: (A \otimes B)^b_a E_b$. This $nm \times nm$ matrix is said to be the *tensor product* of the matrices A^j_i and B^β_α; explicitly it reads

$$A \otimes B = \begin{pmatrix} A^1_1 B & \dots & A^1_n B \\ \vdots & \ddots & \vdots \\ A^n_1 B & \dots & A^n_n B \end{pmatrix} \qquad B = \text{the whole } m \times m \text{ matrix}$$

so that, for example,

$$\begin{pmatrix} 1 & 2 \\ 3 & 4 \end{pmatrix} \otimes \begin{pmatrix} 5 & 6 \\ 7 & 8 \end{pmatrix} = \begin{pmatrix} 1\begin{pmatrix} 5 & 6 \\ 7 & 8 \end{pmatrix} & 2\begin{pmatrix} 5 & 6 \\ 7 & 8 \end{pmatrix} \\ 3\begin{pmatrix} 5 & 6 \\ 7 & 8 \end{pmatrix} & 4\begin{pmatrix} 5 & 6 \\ 7 & 8 \end{pmatrix} \end{pmatrix} = \begin{pmatrix} 5 & 6 & 10 & 12 \\ 7 & 8 & 14 & 16 \\ 15 & 18 & 20 & 24 \\ 21 & 24 & 28 & 32 \end{pmatrix}$$

An element B of the space $V^* \otimes W^*$ is also called a *bilinear pairing* of the spaces V and W. It is thus a bilinear map[501]

$$V \times W \to \mathbb{R} \qquad (v, w) \mapsto B(v, w) \in \mathbb{R}$$

Its expression with respect to bases $e_i \in V$, $e_\alpha \in W$ reads

$$B = B_{i\alpha} e^i \otimes e^\alpha \qquad B_{i\alpha} := B(e_i, e_\alpha)$$

so that the components $B_{i\alpha}$ form a matrix (not necessarily square). A pairing naturally induces the linear maps

$$\begin{aligned} \hat{B} &: V \to W^* \qquad v \mapsto B(v, \cdot\,) \qquad \text{i.e.} \quad v^i \mapsto v^i B_{i\alpha} \\ \check{B} &: W \to V^* \qquad w \mapsto B(\,\cdot\,, w) \qquad \text{i.e.} \quad w^\alpha \mapsto B_{i\alpha} w^\alpha \end{aligned}$$

A pairing is said to be *non-degenerate* (sometimes also called *dual*) if there holds

$$\begin{aligned} B(v, w) = 0 \quad &\text{for all } v \in V \quad \Rightarrow \quad w = 0 \\ B(v, w) = 0 \quad &\text{for all } w \in W \quad \Rightarrow \quad v = 0 \end{aligned}$$

In this case V and W^* are (canonically) isomorphic ($\hat{B}$ is the isomorphism, the matrix $B_{i\alpha}$ is square and non-singular). The existence of a non-degenerate pairing thus automatically

[500] This "matrix" presentation is specific for the product of *two* operators. The *general* rule is that each of the initial matrices acts on "its own index"; for example, $u^{i\alpha a} \mapsto ((A \otimes B \otimes C)u)^{i\alpha a} = A^i_{\ j} B^\alpha_\beta C^a_b u^{j\beta b}$.

[501] A bilinear *form* is a special case where both arguments belong to *the same* space, i.e. where $V = W$.

means[502] that V and W^* are isomorphic (and vice versa). The standard example of a non-degenerate pairing is provided by the *canonical pairing* of the spaces V^* and V, $B(\alpha, v) := \langle \alpha, v \rangle$, introduced in Section 2.4.

Consider a subspace $W \subset V$ of a vector space V. A subspace U is said to be *complementary to* W if each vector $v \in V$ may be uniquely written in the form of a sum $w + u$, where $w \in W, u \in U$ (so that V is a direct sum $W \oplus U$). Given the subspace W, there are in general an infinite number of complementary subspaces; any other subspace U' may be obtained from a fixed one U by $U' = U + hU$, for an appropriate linear map $h : U \to W$ (if $u' \in U'$, there is a decomposition $u' = w + u \equiv hu + u$; the vector $hu \in W$ corrects the vector u so that it belongs to the new complementary subspace).

If $W \subset V$, we may introduce in V an equivalence $\hat{v} \sim v \Leftrightarrow \hat{v} = v + w$ for some $w \in W$. In the factor *set* V/W (i.e. the set of equivalence classes) a linear structure is introduced by means of representatives ($[v_1] + \lambda[v_2] := [v_1 + \lambda v_2]$), so that the factor *space* V/W arises with dimension dim V minus dim W. This space is isomorphic to an arbitrary complementary subspace U (if $v = w + u$, then $[v] = [u]$ and $[v] \leftrightarrow u$ is the isomorphism; for another choice, U', we have $u' = u + hu$ and consequently $[u] = [u']$).

Given $V = W \oplus U$, let P, Q be the projection operators onto W, U. Then an arbitrary operator A in V may be expressed as a sum of four parts,

$$A = (P + Q)A(P + Q) = PAP + PAQ + QAP + QAQ \equiv a + b + c + d$$

where the operators a, b, c, d effectively only act as $a : W \to W, b : U \to W, c : W \to U, d : U \to U$ (their matrices with respect to the adapted basis (e_i, e_α) in V being $A^i_j, A^i_\alpha, A^\alpha_i, A^\alpha_\beta$). If, in contrast, we know these four operators, we can in turn reconstruct A. If W is invariant with respect to A, then $c = 0$ and if U is invariant, then $b = 0$.

More generally, consider *two* linear spaces which happen to be direct sums of their subspaces, $L = \oplus_i L_i$ and $\hat{L} = \oplus_\alpha \hat{L}_\alpha$, and denote by P_i and $\hat{P}_\alpha$ the projection operators onto the corresponding subspaces. Then an arbitrary linear map $A : L \to \hat{L}$ has a decomposition $A = (\sum_i P_i)A(\sum_\alpha \hat{P}_\alpha) = \sum_{i\alpha} P_i A \hat{P}_\alpha = \sum_{i\alpha} \hat{A}_{i\alpha}$, where $\hat{A}_{i\alpha}$ may already be regarded as a map only acting between the relevant subspaces $A_{i\alpha} : L_i \to \hat{L}_\alpha$. If, vice versa, we know all of these partial linear maps, we can reconstruct the whole A.

A.2 Associative algebras

An *algebra* $\mathcal{A}$ is a linear space endowed with an additional binary operation $\mathcal{A} \times \mathcal{A} \to \mathcal{A}$ ("multiplication") $a, b \mapsto ab$, $a, b, ab \in \mathcal{A}$. Compatibility with the linear structure is required in the sense that

$$a(b + \lambda c) = ab + \lambda ac \qquad (b + \lambda c)a = ba + \lambda ca$$

i.e. the multiplication is to be bilinear. If there exists an identity in $\mathcal{A}$ (an element $1 \in \mathcal{A}$, satisfying $a1 = 1a = a, a \in \mathcal{A}$), the algebra is said to be *unital*, if the multiplication happens to be associative ($(a(bc) = (ab)c \equiv abc$), we speak about the *associative algebra.* In what follows what we will have in mind will be as a rule a real associative algebra (except for the *Lie* algebra, which is *not* associative, see Section A.3). Standard examples are provided by the *complete matrix algebra* $\mathbb{R}(n) \equiv M_n(\mathbb{R})$ of all real $n \times n$ matrices (with the common matrix multiplication), the algebra End$V$ of endomorphisms of a linear

[502] Also $\check{B}$ is an isomorphism so that a non-degenerate pairing also says that V^* and W are canonically isomorphic.

space V (= all linear maps $V \to V$; the choice of a basis yields a non-canonical isomorphism with $\mathbb{R}(n)$), the algebra $\mathcal{F}(M)$ of smooth functions on a manifold M, the tensor algebra $T(V)$ of a linear space V, the algebra $\mathcal{T}(M)$ of tensor fields on M, the exterior algebra ΛV^* of forms in V, the Cartan algebra $\Omega(M)$ of differential forms on M or the Clifford algebra $C(V, g)$ associated with a linear space V endowed with a symmetric (as a rule non-degenerate) bilinear form g.

The real numbers $\mathbb{R}$, the complex numbers $\mathbb{C}$ and the *quaternions* $\mathbb{H}$ may be regarded as algebras over $\mathbb{R}$ (of dimensions one, two and four respectively). A basis in $\mathbb{C}$ is $(1, e_1 = i)$, where $e_1 e_1 = -1$. A matrix realization is, for example, $1 = \mathbb{I}, e_1 = i\sigma_2$. A basis in $\mathbb{H}$ is $(1, e_1 \equiv i, e_2 \equiv j, e_3 \equiv k)$ and the multiplication satisfies $e_a e_b = -\delta_{ab} + \epsilon_{abc} e_c$, $a = 1, 2, 3$ (e_a represent sort of three imaginary units). A general element of $\mathbb{H}$ is $w = a + w^a e_a \equiv a + w^1 i + w^2 j + w^3 k, a, w^a \in \mathbb{R}$. A simple matrix realization is $1 = \mathbb{I}$, $e_a = -i\sigma_a$. Further useful algebras are $\mathbb{R}(n)$, $\mathbb{C}(n)$ and $\mathbb{H}(n)$, $n \times n$ matrices with entries from $\mathbb{R}$, $\mathbb{C}$ and $\mathbb{H}$ respectively.

For the finite-dimensional case, if e_a is a basis in $\mathcal{A}$ as a linear space, complete information about the multiplication ($\Rightarrow$ about the algebra) is contained in the *structure constants* c^c_{ab}, given by the decomposition of the products of the basis elements $e_a e_b =: c^c_{ab} e_c$. These constants comprise the components of a tensor of type $\binom{1}{2}$.

A subspace $\mathcal{B} \subset \mathcal{A}$ is a *subalgebra* if it (also) happens to be closed with respect to the multiplication and a subalgebra $\mathcal{I}$ is an *ideal* (left, right, two-sided) if the multiplication of an arbitrary element $a \in \mathcal{A}$ by an element $i \in \mathcal{I}$ (from the left, from the right and both) results in an element in $\mathcal{I}$ (for example, for the left ideal $ia = i' \in \mathcal{I}$ for any $a \in \mathcal{A}$).[503] Given a two-sided ideal $\mathcal{I}$ in the algebra $\mathcal{A}$, we may introduce the multiplication into the factor space $\mathcal{A}/\mathcal{I}$ by means of representatives and obtain the *factor-algebra* ($[a][b] := [ab]$; for another choice of representatives[504] we get $[a + i][b + i'] = [ab + ai' + ib + ii'] = [ab]$, if $\mathcal{I}$ is a two-sided ideal).

A linear bijective map $A : \mathcal{A} \to \mathcal{A}$ is called an *automorphism* of the algebra $\mathcal{A}$ if it (also) respects the product, $A(ab) = A(a)A(b)$. All the automorphisms of $\mathcal{A}$ form a group Aut $\mathcal{A}$ (with respect to composition of maps).

A linear map $D : \mathcal{A} \to \mathcal{A}$ is called a *derivation* of the algebra $\mathcal{A}$, if it behaves on a product according to the Leibniz rule, i.e. if $D(ab) = (Da)b + a(Db)$. All the derivations of $\mathcal{A}$ form a *Lie* algebra Der $\mathcal{A}$ (with the commutator $[D_1, D_2] := D_1 D_2 - D_2 D_1$, see Section A.3), which is just the Lie algebra of the group Aut $\mathcal{A}$.

Given two algebras $\mathcal{A}$ and $\mathcal{B}$, we may form two other algebras, their *direct sum* $\mathcal{A} \oplus \mathcal{B}$ and their *tensor product* $\mathcal{A} \otimes \mathcal{B}$. Regarded as linear spaces they represent the constructions bearing the same name with the linear spaces of the original algebras (see Section A.1). The multiplication in the direct sum reads $(a, b)(a', b') := (aa', bb')$ (in the original algebras as before, e.g. $(a, 0)(a', 0) := (aa', 0)$, and the mutual products vanish, $(a, 0)(0, b) := (0, 0)$).

In the tensor product we define on decomposable elements (i.e. of the form of a product)

$$(a \otimes b)(a' \otimes b') := (aa' \otimes bb')$$

and this is extended then by linearity, so that for general elements we get

$$(K^{i\alpha} e_i \otimes e_\alpha)(k^{j\beta} e_j \otimes e_\beta) := K^{i\alpha} k^{j\beta} e_i e_j \otimes e_\alpha e_\beta = K^{i\alpha} k^{j\beta} c^k_{ij} c^\gamma_{\alpha\beta} e_k \otimes e_\gamma$$

[503] If elements of the ideal $\mathcal{I}$ are regarded as carriers of the gene of *idealism*, then the offspring from mating (multiplication) of an idealist with any other element of $\mathcal{A}$ (including realists, pragmatists and so on) consists again only of idealists.

[504] The square brackets in $[a]$ denote the *class* given by the representative a; it is not related to the commutator from Section A.3.

Making use of adapted bases one easily verifies the isomorphisms

$$(A \oplus B) \otimes C = (A \otimes C) \oplus (B \otimes C)$$
$$C \otimes (A \oplus B) = (C \otimes A) \oplus (C \otimes B)$$

A.3 Lie algebras

An algebra in which the multiplication is antisymmetric and satisfies the Jacobi identity is called the *Lie algebra.* So it is a linear space $\mathcal{G}$ in which we have moreover a bilinear "multiplication" $\mathcal{G} \times \mathcal{G} \to \mathcal{G}$ (here, the product xy is usually called *commutator* and denoted by $[x, y]$) with the properties

$$[x, y] = -[y, x] \qquad \text{antisymmetry}$$
$$[x, [y, z]] + [z, [x, y]] + [y, [z, x]] = 0 \qquad \text{Jacobi identity}$$

If $\mathcal{A}$ is an *associative* algebra, the prescription $[a, b] := ab - ba$ makes from this algebra (also) the *Lie* algebra (however, not every Lie algebra has this origin; for example, the commutator of vector fields does not arise from their associative multiplication – the latter is even not defined).

The concept of an ideal, factor-algebra, automorphism and derivation are introduced in close analogy with the corresponding definitions in associative algebras.

A subspace $\mathcal{B} \subset \mathcal{G}$ is a Lie *subalgebra* if it is (also) closed with respect to the commutator and a subalgebra $\mathcal{I}$ is an *ideal* (only two-sided here) if the commutator with any element yields an element in $\mathcal{I}$ (i.e. $[i, x] = i' \in \mathcal{I}$ for arbitrary $i \in \mathcal{I}, x \in \mathcal{G}$). Given an ideal $\mathcal{I}$ in a Lie algebra $\mathcal{G}$, we may introduce the commutator into the factor space $\mathcal{G}/\mathcal{I}$ by means of representatives and obtain the *Lie factor-algebra.*

A map $D : \mathcal{G} \to \mathcal{G}$ is called a *derivation* of the (Lie) algebra $\mathcal{G}$, if it is linear and if it behaves on the "product" (commutator) according to the Leibniz rule

$$D(x + \lambda y) = D(x) + \lambda D(y) \qquad D([x, y]) = [D(x), y] + [x, D(y)]$$

An important example of the derivation is the *inner derivation* $D \equiv \mathrm{ad}_x$, defined by

$$\mathrm{ad}_x := [x, \cdot\,] \qquad \text{i.e.} \quad \mathrm{ad}_x(y) := [x, y]$$

(the Jacobi identity may be regarded as just expressing the fact that it is indeed a derivation).

A map $A : \mathcal{G} \to \mathcal{G}$ is called an *automorphism* of a (Lie) algebra $\mathcal{G}$, if it is bijective, linear and if it preserves the "product" (commutator)

$$A(x + \lambda y) = A(x) + \lambda A(y) \qquad A([x, y]) = [A(x), A(y)]$$

An important example is provided by the *inner automorphism* $A \equiv \mathrm{Ad}_g$, which is (for $g \in G \equiv$ the Lie *group*) defined by (12.3.1)

$$I_g e^x \equiv g e^x g^{-1} =: e^{\mathrm{Ad}_g(x)}$$

All the automorphisms of $\mathcal{G}$ form a group Aut $\mathcal{G}$ (they are closed with respect to compositions), whose Lie algebra Der $\mathcal{G}$ turns out to be given by just all the derivations of $\mathcal{G}$ (the commutator in Der $\mathcal{G}$ arises from the associative multiplication of linear maps, i.e. $[D_1, D_2] := D_1 D_2 - D_2 D_1$). Infinitesimal automorphisms thus have the form $\hat{1} + \epsilon D$, $D \in \mathrm{Der}\, \mathcal{G}$.

If E_i is a basis of $\mathcal{G}$, the formula $[E_i, E_j] =: c^k_{ij} E_k$ defines *structure constants*. They form the components of a tensor of type $\binom{1}{2}$. Antisymmetry and the Jacobi identity are reflected in the properties

$$c_{ij}^k = -c_{ji}^k \qquad c_{si}^r c_{jk}^s + c_{sk}^r c_{ij}^s + c_{sj}^r c_{ki}^s = 0$$

Given two Lie algebras, $\mathcal{G}_1$ and $\mathcal{G}_2$, we may construct their *direct sum* $\mathcal{G} \equiv \mathcal{G}_1 \oplus \mathcal{G}_2$ as the linear space $\mathcal{G}_1 + \mathcal{G}_2$, where the commutator is introduced by the prescription

$$[(x_1, x_2), (y_1, y_2)] := ([x_1, y_1], [x_2, y_2]) \qquad x_j, y_j \in \mathcal{G}_j \quad j = 1, 2$$

The elements of the form $(x_1, 0)$ constitute a Lie subalgebra (a subspace closed with respect to the commutator) $\tilde{\mathcal{G}}_1$ isomorphic to the algebra $\mathcal{G}_1$, and similarly the elements $(0, x_2)$ form $\tilde{\mathcal{G}}_2 \subset \mathcal{G}$, $\tilde{\mathcal{G}}_2 \approx \mathcal{G}_2$. The subalgebras $\tilde{\mathcal{G}}_j$ are moreover ideals in $\mathcal{G}$ and there holds

$$\mathcal{G}/\tilde{\mathcal{G}}_1 \approx \mathcal{G}_2 \qquad \mathcal{G}/\tilde{\mathcal{G}}_2 \approx \mathcal{G}_1$$

If $\mathcal{G}_1$ and $\mathcal{G}_2$ happen to be matrix Lie algebras, their direct sum may be realized by block-diagonal matrices

$$(x_1, x_2) \quad \leftrightarrow \quad \begin{pmatrix} x_1 & 0 \\ 0 & x_2 \end{pmatrix}$$

A.4 Modules

A module may be regarded as a generalization of the concept of a linear space in which the elements may be linearly combined with coefficients from an *algebra* $\mathcal{A}$ (rather than from $\mathbb{R}$, as is the case in a real linear space). A left *$\mathcal{A}$-module M* is thus a linear space (over $\mathbb{R}$), in which the *algebra $\mathcal{A}$ acts from the left*, i.e. there exists a prescription (map) $\mathcal{A} \times M \to M$, $(a, m) \mapsto am$, such that

$$a(bm) = (ab)m \qquad (a + \lambda b)m = am + \lambda bm \qquad a(m + \lambda n) = am + \lambda an$$
$$a, b \in \mathcal{A} \quad \lambda \in \mathbb{R} \quad m, n \in M$$

In a right $\mathcal{A}$-module the algebra acts from the right, so that we have a prescription $M \times \mathcal{A} \to M$, $(m, a) \mapsto ma$ such that

$$(ma)b = m(ab) \qquad m(a + \lambda b) = ma + \lambda mb \qquad (m + \lambda n)a = ma + \lambda na$$
$$a, b \in \mathcal{A} \quad \lambda \in \mathbb{R} \quad m, n \in M$$

A standard example is given by $\mathbb{R}^k$ as M and $\mathbb{R}(k) \equiv M_k(\mathbb{R})$ ($k \times k$ real matrices) as $\mathcal{A}$ (the left action being $x \mapsto Bx$ and the right action given by $x^T \mapsto x^T B$). Other examples are provided by $\mathcal{T}_s^r(\mathcal{M})$ (tensor fields of type $\binom{r}{s}$ on a manifold $\mathcal{M}$) and, in particular, $\Omega^p(\mathcal{M})$ (differential p-forms on M) as two-sided $\mathcal{F}(\mathcal{M})$-modules. The algebra $\mathcal{A}$ itself is an $\mathcal{A}$-module (left action $b \mapsto ab$, right one $b \mapsto ba$; just as $\mathbb{R}$ is a linear space over $\mathbb{R}$). Given two modules $M, \tilde{M}$ over the same algebra $\mathcal{A}$, we may speak about $\mathcal{A}$-linear maps; they are maps which preserve linear combinations over $\mathcal{A}$, i.e. such $f : M \to \tilde{M}$ that $f(m + an) = f(m) + af(n)$, $a \in \mathcal{A}$, whereas for (only) $\mathbb{R}$-linear maps (only) $f(m + \lambda n) = f(m) + \lambda f(n)$, $\lambda \in \mathbb{R}$ holds. For example, tensor fields on a manifold $\mathcal{M}$ are $\mathcal{A} \equiv \mathcal{F}(\mathcal{M})$-linear with respect to each argument.

A.5 Grading

If G is a commutative group then a G-graded vector space is a space which is a direct sum of subspaces "labeled" by elements of the group G (homogeneous subspaces). Actually we only encounter the grading with respect to three groups in this book, namely $\mathbb{Z}$ (the subspaces labeled by integers), $\mathbb{Z} \times \mathbb{Z}$ (pairs of integers) and $\mathbb{Z}_2$ (the subspaces labeled by zero or one; their elements are then called even and odd vectors respectively)

$$\mathbb{Z} : L = \bigoplus_{p=-\infty}^{\infty} L^p \qquad \mathbb{Z} \times Z : L = \bigoplus_{p,q=-\infty}^{\infty} L^p_q \qquad \mathbb{Z}_2 : L = L^{[0]} \oplus L^{[1]}$$

The group elements are usually called degree in this context. The grading naturally passes to linear maps. A general linear map of G-graded spaces $L \to \mathcal{L}$ (in particular, an endomorphism $L \to L$) may be regarded as a system of maps of the homogeneous subspaces (in the way described at the end of Section A.1). These maps are said to be homogeneous and we may assign a degree g to a map from the subspace with degree $\hat{g}$ to the subspace with degree $\hat{g}g$. For example, in the case of $\mathbb{Z}$-grading a map from the subspace L^3 to $\mathcal{L}^7$ has degree 4, in the $\mathbb{Z}_2$-case the maps which preserve parity are even, and the maps which reverse parity are odd.

If there is moreover the structure of an algebra in a graded linear space and this structure happens to be compatible with grading (the product of homogeneous elements is again homogeneous, its degree being the product of the original degrees in the sense of G), we speak about a *graded algebra*. Examples: $\mathbb{Z} \times \mathbb{Z}$-graded algebra $\mathcal{T}(\mathcal{M})$ of tensor fields on a manifold $\mathcal{M}$, $\mathbb{Z}$-graded algebra of strictly covariant fields, $\mathbb{Z}$-graded Cartan algebra $\Omega(\mathcal{M})$ of differential forms on M (all of them may also be regarded as (only) $\mathbb{Z}_2$-graded) and the algebra of endomorphisms of a graded vector space. A $\mathbb{Z}_2$-graded algebra is also known as a *superalgebra*.

Consider a $\mathbb{Z}$-graded algebra $\mathcal{A}$. A *derivation of degree* k is a linear map $\mathcal{A} \to \mathcal{A}$ of degree k which behaves on products according to the graded Leibniz rule, i.e.

$$D_k : \mathcal{A}^l \to \mathcal{A}^{l+k} \qquad D_k(a_i b) = (D_k a_i)b + (-1)^{ik} a_i (D_k b) \qquad a_i \in \mathcal{A}_i \quad b \in \mathcal{A}$$

For the $\mathbb{Z}_2$-case we get in this way (only) even and odd derivations of a superalgebra $\mathcal{A}$.

In a graded *Lie* algebra (in particular, in a *Lie superalgebra*) the role of "multiplication" is played by the graded commutator (in particular, the *supercommutator*), which is bilinear and on homogeneous elements it satisfies

$$[a_i, a_j] = -(-1)^{ij}[a_j, a_i]$$

$$(-1)^{ik}[a_i, [a_j, a_k]] + (-1)^{kj}[a_k, [a_i, a_j]] + (-1)^{ji}[a_j, [a_k, a_i]] = 0$$

so that it is antisymmetric except for two elements of odd degrees, where it becomes symmetric and it satisfies the graded Jacobi identity (it says that $\mathrm{ad}_{a_i} \equiv [a_i, \cdot\,]$ is a derivation of degree i). Such a *graded Lie algebra* is realized for example by derivations (of all degrees k) of a graded (associative) algebra (see (6.1.6)). In particular, the triple of the operators $i_V, \mathcal{L}_V, d$ acting on the Cartan algebra $\Omega(\mathcal{M})$ (as a_{-1}, a_0, a_1, all the others vanish).

If V is a $\mathbb{Z}$-graded linear space, then its general subspace may not be compatible with the grading.[505] However, if *it is* compatible, i.e. if $W = \oplus W_i$, $W_i \subset V_i$, then the grading

[505] Consider a two-dimensional space V. Let e_1 span the subspace of degree 1 and e_2 span the subspace of degree 2 (all the others being trivial, i.e. zero spaces); if a one-dimensional subspace W is spanned by $e_1 + e_2$, no non-zero vector $w \in W$ may be written as a sum of *homogeneous* terms, all of them being from W. The structures of $\mathbb{Z}$-grading and a subspace are not compatible.

also passes to the factor space V/W, $\deg[v_k] := \deg v_k \equiv k$ and $V/W = \oplus V_k/W_k$. If, moreover, $V \equiv \mathcal{A}$ is an algebra and the (graded) subspace is an ideal, the grading passes to the factor algebra, $\mathcal{A}/\mathcal{I} = \oplus \mathcal{A}_k/\mathcal{I}_k$, $[a_i][a_j] = [a_i a_j] \in \mathcal{A}_{i+j}/\mathcal{I}_{i+j}$.

A.6 Categories and functors

In modern mathematics and mathematical physics the language of category theory is often used to discuss properties of different mathematical structures in a unified way. In this book these concepts are encountered only seldom and they are only used as a *language* (no non-trivial results or statements of the theory are needed). The aim of this appendix is then just to make some simple statements which concern geometry understandable to the beginner (so that the reader is not forced immediately to put aside a potentially useful and simple text about geometry as being "unreadable" just because it uses this (natural and convenient) language).

To define a particular category K amounts to describing two things, the class of its objects ObK and, for each pair A, B of objects, the set Mor(A, B) of morphisms of the object A to the object B ($f \in$ Mor(A, B) is standardly drawn as an arrow $A \xrightarrow{f} B$). For each triple of objects A, B, C a prescription should be given of how to compose $A \xrightarrow{f} B$ with $B \xrightarrow{g} C$; this results in a composed morphism $A \xrightarrow{g \circ f} C$. For each object A the "identity" morphism should exist (such that for each $f \in$ Mor(A, A) there holds $1_A \circ f = f = f \circ 1_A$) and the composition is to be associative, $f \circ (g \circ h) = (f \circ g) \circ h$. If for $f \in$ Mor(A, B) the inverse morphism exists, f is said to be an isomorphism.

All the categories encountered in this book have as objects some structured sets (vector spaces, groups, manifolds, etc.) and as morphisms (arrows) *homo*morphic *maps* of these structured sets. For example, the objects of the category of linear spaces are (all the possible) linear spaces and the morphisms of this category are *linear* maps between vector spaces; the objects of the category of Lie groups are (all the possible) Lie groups and the morphisms are smooth maps, which at the same time happen to be homomorphisms of groups.

Given two categories, K and $\hat{K}$, a (covariant) functor F from K to $\hat{K}$ is a prescription which assigns to objects of the first category objects of the second one ($A \mapsto F(A) \in$ Ob$\hat{K}$), but also to morphisms of the first category morphisms of the second one ($f \in$ Mor $(A, B) \mapsto F(f) \in$ Mor $(F(A), F(B))$); one requires the preserving of the identity morphism as well as of the composition rule for morphisms ($F(f \circ g) = F(f) \circ F(g)$; for a contravariant functor one requires $F(f \circ g) = F(g) \circ F(f)$). Let us mention some examples encountered in the main text of the book.

The assignment "a Lie group $\mapsto$ (its) Lie algebra" is a functor from the category of Lie groups to the category of Lie algebras. We assign to a morphism of the category of Lie groups (a homomorphism $f : G \to H$) a morphism of the category of Lie algebras – the *derived* (homo)morphism $f' : \mathcal{G} \to \mathcal{H}$ (so that using the notation from the main text, $F(G) = \mathcal{G}$, $F(f) = f'$). The point is that f' is indeed a homomorphism of Lie algebras as well as that $(f \circ g)' = f' \circ g'$ (so that it is a *covariant* functor; these are technical facts, which are not evident (although not surprising) and they needed to be proved in the main text).

The assignment "a representation of a Lie group $\mapsto$ the derived representation of its Lie algebra" is a functor from the category of representations of a Lie group G (objects being representations of the Lie group G, morphisms are intertwining operators between them) to the category of representations of the Lie algebra $\mathcal{G}$ (objects are representations of the Lie algebra $\mathcal{G}$, morphisms are intertwining operators between them).

The assignment T "a manifold $M \mapsto$ (its) tangent bundle TM" is a functor (known as the tangent functor) from the category of smooth manifolds (objects are smooth manifolds, morphisms are their *smooth* maps) to the category of vector bundles (objects are vector bundles, morphisms are maps of their total spaces which *preserve fibers* and are *linear* in fibers). We learned in the main text that Tf indeed preserves fibers, it is linear on fibers and that $T(f \circ g) = Tf \circ Tg$.

The assignment "a manifold $M \mapsto$ (its) algebra of functions $\mathcal{F}(M)$" is a (contravariant) functor F from the category of manifolds to the category of associative (and commutative) algebras ($F(M) = \mathcal{F}(M)$, $F(f) = f^*$).

The assignment "a manifold $M \mapsto$ (its) Cartan algebra $\Omega(M)$" is a (contravariant) functor from the category of manifolds to the category of $\mathbb{Z}$-graded (graded) commutative algebras.

The assignment "a manifold $\mapsto$ (its) deRham complex" is a functor from the category of manifolds to the category of complexes (the morphisms being maps of complexes, which preserve degree and commute with the differentials).

A standard trick of algebraic topology is the construction of a functor from the category of topological spaces to the category of some algebraic structure (the category of groups, algebras, etc.). For example, the assignment "a manifold $M \mapsto H^p(M)$" (its pth deRham cohomology group) is the functor from the category of manifolds to the category of groups. This enables one to formulate important statements about a particular topological space (possibly enriched by an additional structure, for example a manifold) in the language and by technical means of the other category (groups, algebras, etc.).

Appendix B
Starring

Abel, Niels Hendrik, 1802 Finnøy–1829 Froland
Ampère, Marie André, 1775 Lyon–1836 Marseilles
Atiyah, Michael Francis, 1929 London
Beltrami, Eugenio, 1835 Cremona–1900 Rome
Betti, Enrico, 1823 Pistoia–1892 Soiana
Bianchi, Luigi, 1856 Parma–1928 Pisa
Carathéodory, Constantin, 1873 Berlin–1950 Munich
Cartan, Elie Joseph, 1869 Dolomieu–1951 Paris
Cauchy, Augustin Louis, 1789 Paris–1857 Sceaux (Paris)
Christoffel, Elwin Bruno, 1829 Montjoie Aachen–1900 Strasbourg
Clebsch, Rudolf Friedrich Alfred, 1833 Königsberg–1872 Göttingen
Clifford, William Kingdon, 1845 Exeter–1879 Madeira
Coriolis, Gaspard Gustave de, 1792 Paris–1843 Paris
Coulomb, Charles Augustin de, 1736 Angoulême–1806 Paris
d'Alembert, Jean Le Rond, 1717 Paris–1783 Paris
Darboux, Jean Gaston, 1842 Nimes–1917 Paris
deRham, Georges, 1903 Roche–1990 Lausanne
Descartes, René, 1596 La Haye–1650 Stockholm
Dirac, Paul Adrien Maurice, 1902 Bristol–1984 Tallahassee
Einstein, Albert, 1879 Ulm–1955 Princeton
Euclid, circa 325 BC (?)–circa 265 BC (?) Alexandria
Euler, Leonhard, 1707 Basel–1783 St Petersburg
Faraday, Michael, 1791 Newington Butts–1867 Hampton Court
Foucault, Jean Bernard Léon, 1819 Paris–1868 Paris
Fourier, Jean Baptiste Joseph, 1768 Auxerre–1830 Paris
Frobenius, Ferdinand Georg, 1849 Berlin-Charlottenburg–1917 Berlin
Gauss, Johann Carl Friedrich, 1777 Brunswick–1855 Göttingen
Gordan, Paul Albert, 1837 Breslau–1912 Erlangen
Gordon, Walter, 1893 Apolda–1939 Stockholm
Grassmann, Herman Günter, 1809 Stettin–1877 Stettin
Green, George, 1793 Sneinton–1841 Sneinton
Hamilton, William Rowan, 1805 Dublin–1865 Dublin
Hausdorff, Felix, 1868 Breslau–1942 Bonn
Heaviside, Oliver, 1850 Camden Town (London)–1925 Torquay
Heisenberg, Werner Karl, 1901 Würzburg–1976 Münich
Helmholtz, Hermann Ludwig Ferdinand von, 1821 Potsdam–1894 Berlin

Hilbert, David, 1862 Königsberg–1943 Göttingen
Hodge, William Vallance Douglas, 1903 Edinburgh–1975 Cambridge
Hopf, Eberhard Frederich Ferdinand, 1902 Salzburg–1983
Jacobi, Carl Gustav Jacob, 1804 Potsdam–1851 Berlin
Kähler, Erich, 1906 Leipzig–2000
Killing, Wilhelm Carl Joseph, 1847 Burbach–1923 Münster
Klein, Felix Christian, 1849 Düsseldorf–1925 Göttingen
Klein, Oscar, 1894 Stockholm–1977 Stockholm
Kronecker, Leopold, 1823 Liegnitz–1891 Berlin
Lagrange, Joseph-Louis, 1736 Turin–1813 Paris
Lamé, Gabriel, 1795 Tours–1870 Paris
Laplace, Pierre-Simon, 1749 Beaumont-en-Auge–1827 Paris
Legendre, Adrien-Marie, 1752 Paris–1833 Paris
Leibniz, Gottfried Wilhelm von, 1646 Leipzig–1716 Hannover
Levi-Civita, Tullio, 1873 Padua–1941 Rome
Lie, Marius Sophus, 1842 Nordfjorderde–1899 Kristiania
Liouville, Joseph, 1809 Saint-Omer–1882 Paris
Lorentz, Hendrik Antoon, 1853 Arnhem–1928 Haarlem
Majorana, Ettore, 1906 Catania–1938 (disappeared)
Maurer, Ludwig, 1859–1927
Maxwell, James Clerk, 1831 Edinburgh–1879 Cambridge
Minkowski, Hermann, 1864 Alexotas (Kaunas)–1909 Göttingen
Moivre, Abraham de, 1667 Vitry–1754 London
Möbius, August Ferdinand, 1790 Schulpforta–1868 Leipzig
Nambu, Yoichiro, 1921 Tokyo
Newton, Isaac, 1643 Woolsthorpe–1727 London
Nijenhuis, Albert, 1926 Eindhoven
Noether, Emmy Amalie, 1882 Erlangen–1935 Bryn Mawr
Pauli, Wolfgang Ernst, 1900 Wien–1958 Zürich
Pfaff, Johann Friedrich, 1765 Stuttgart–1825 Halle
Poincaré, Jules Henri, 1854 Nancy–1912 Paris
Poisson, Simeon Denis, 1781 Pithiviers–1840 Sceaux (Paris)
Proca, Alexandru, 1897–1955
Pythagoras, circa 580 BC Samos–circa 500 BC Metapontum
Rarita, William, 1906–1999
Ricci-Curbastro, Gregorio, 1853 Lugo–1925 Bologna
Riemann, Georg Friedrich Bernhard, 1828 Breselenz–1866 Selasca
Schrödinger, Erwin Rudolf Josef Alexander, 1887 Wien–1961 Wien
Schur, Issai, 1875 Mogilyov–1941 Tel Aviv
Schwinger, Julian, 1918 New York–1994 Los Angeles
Stokes, George Gabriel, 1819 Skreen–1903 Cambridge
Sylvester, James Joseph, 1814 London–1897 London
Taylor, Brook, 1685 Edmonton–1731 Somerset House (London)
Weyl, Hermann Klaus Hugo, 1885 Elmshorn–1955 Zürich
Whitehead, John Henry Constantine, 1904 Madras–1960 Princeton
Whitney, Hassler, 1907 New York–1989 Mount Dent Blanches
Wigner, Eugene Paul, 1902 Budapest–1995 Princeton

Bibliography

Arnold, V. I. (1978). *Mathematical Methods of Classical Mechanics*. New York: Springer-Verlag.

Benn, I. M. and Tucker, R. W. (1989). *An Introduction to Spinors and Geometry with Applications in Physics*. Bristol: Adam Hilger.

Birkhoff, G. and Mac Lane, S. (1965). *A Survey of Modern Algebra*. New York: Macmillan Company, Inc.

Chandrasekhar, S. (1983). *The Mathematical Theory of Black Holes*. New York: Clarendon.

Crampin, M. and Pirani, F. A. E. (1987). *Applicable Differential Geometry*. Cambridge: Cambridge University Press.

Dubrovin, B. A., Novikov, S. P. and Fomenko, A. T. (1984, 1985). *Modern Geometry: Methods and Applications*. New York: Springer-Verlag.

Flanders, H. (1963). *Differential Forms (with Applications to Physical Sciences)*. New York: Academic Press (1989, New York: Dover Publications).

Frankel, T. (2004). *The Geometry of Physics: an Introduction*. Cambridge: Cambridge University Press.

Freund, P. G. O. (1986). *Supersymmetry*. Cambridge: Cambridge University Press.

Fulton, W. and Harris, J. (1991). *Representation Theory (A First Course)*. New York: Springer-Verlag.

Gelfand, I. M. (1961). *Lectures on Linear Algebra*. New York: Interscience Publishers.

Göckeler, M. and Schücker, T. (1987). *Differential Geometry, Gauge Theories and Gravity*. Cambridge: Cambridge University Press.

Goto, M. and Grosshans, F. D. (1978). *Semisimple Lie Algebras*. New York: Marcel Dekker, Inc.

Hawking, S. W. and Ellis, G. F. R. (1973). *The Large Scale Structure of Space-time*. Cambridge: Cambridge University Press.

Isham, Ch. J. (1989, 1999). *Modern Differential Geometry for Physicists*. Singapore: World Scientific.

Lightman, A. P., Press, W. H., Price, R. H. and Teukolsky, S. A. (1975). *Problem Book in Relativity and Gravitation*. Princeton: Princeton University Press.

Marathe, K. B. and Martucci, G. (1992). *The Mathematical Foundations of Gauge Theories*. Amsterdam: Elsevier Science Publishers.

Marder, L. (1970). *Vector Analysis*. London: George Allen and Unwin Ltd.

Marsden, J. E., Ratiu, T. and Abraham, R. (2001). *Manifolds, Tensor Analysis, and Applications*. New York: Springer-Verlag.

Misner, Ch. W., Thorne, K. S. and Wheeler, J. A. (1973). *Gravitation*. San Francisco: Freeman.
Nash, Ch. and Sen, S. (1992). *Topology and Geometry for Physicists*. London: Academic Press.
Rubakov, V. A. (2002). *Classical Theory of Gauge Fields*. Princeton: Princeton University Press.
Schutz, B. F. (1982). *Geometrical Methods of Mathematical Physics*. Cambridge: Cambridge University Press.
Schwarz, A. S. (1993). *Quantum Field Theory and Topology*. Berlin: Springer-Verlag.
Spivak, M. (1965). *Calculus on Manifolds*. Reading, MA: Addison-Wesley Publishing Company.
Sternberg, S. (1995). *Group Theory and Physics*. Cambridge: Cambridge University Press.
Straumann, N. (1984, 1991). *General Relativity and Relativistic Astrophysics*. Berlin: Springer-Verlag.
Thirring, W. (1978). *A Course in Mathematical Physics*, vol. 1. *Classical Dynamical Systems*. New York: Springer-Verlag.
(1986). *A Course in Mathematical Physics*, vol. 2. *Classical Field Theory*. New York: Springer-Verlag.
Trautman, A. (1984). *Differential Geometry for Physicists*. Naples: Bibliopolis.
(1980). Fiber bundles, gauge fields, and gravitation. In *General Relativity and Gravitation (One Hundred Years After the Birth of Albert Einstein)*, ed. A. Held. New York: Plenum Press, pp. 287–308.
Wallace, A. H. (1968). *Differential Topology (First Steps)*. New York and Amsterdam: W. A. Benjamin.
Woodhouse, N. M. J. (1991). *Geometric Quantization*. Oxford: Oxford University Press.

Index of (frequently used) symbols

$T_x M$	Tangent space at the point $x \in M$	(2.2.2)
$T_s^r(L)$	Tensors of type (r, s) in L	(2.4.5)
$T_s^r(M)$	Tensor fields of type (r, s) on M	(2.5.2)
T^a	Torsion forms	(15.6.3)
F	2-form of the electromagnetic field	(16.2.1)
$U(n)$	Unitary group	(10.1.12)

Index